NATIONAL COLORS

WORN BY

UNITED STATES VESSELS.

ENSIGN.

UNION JACK.

THE NORRIS PETERS CO., WASHINGTON, D. C.

No. 87

International Code of Signals

American Edition

PUBLISHED BY THE HYDROGRAPHIC OFFICE
UNDER THE AUTHORITY OF THE SECRETARY OF THE NAVY

Washington : Government Printing Office : 1909

INTRODUCTION.

The International Code of Signals consists of twenty-six flags—one for each letter of the alphabet—and a Code Pennant

Urgent and important signals are two-flag signals

General signals are three-flag signals

Geographical, Alphabetical Spelling Tables, and Vessels' Numbers are four-flag signals

The book is divided into three parts The first part contains urgent and important signals and all the tables of money, weights, barometric heights, etc , together with a geographical list and a table of phrases formed with the auxiliary verbs

The second part is an index It consists of a general vocabulary and a geographical index It is arranged alphabetically

The third part gives lists of the United States storm-warning, life-saving, time-signal, and wireless telegraph stations, and of Lloyd's signal stations of the world It also contains semaphore and distant signal codes and the United States Army and Navy and Morse Wigwag Codes

CONTENTS.

CONTENTS.

PART III

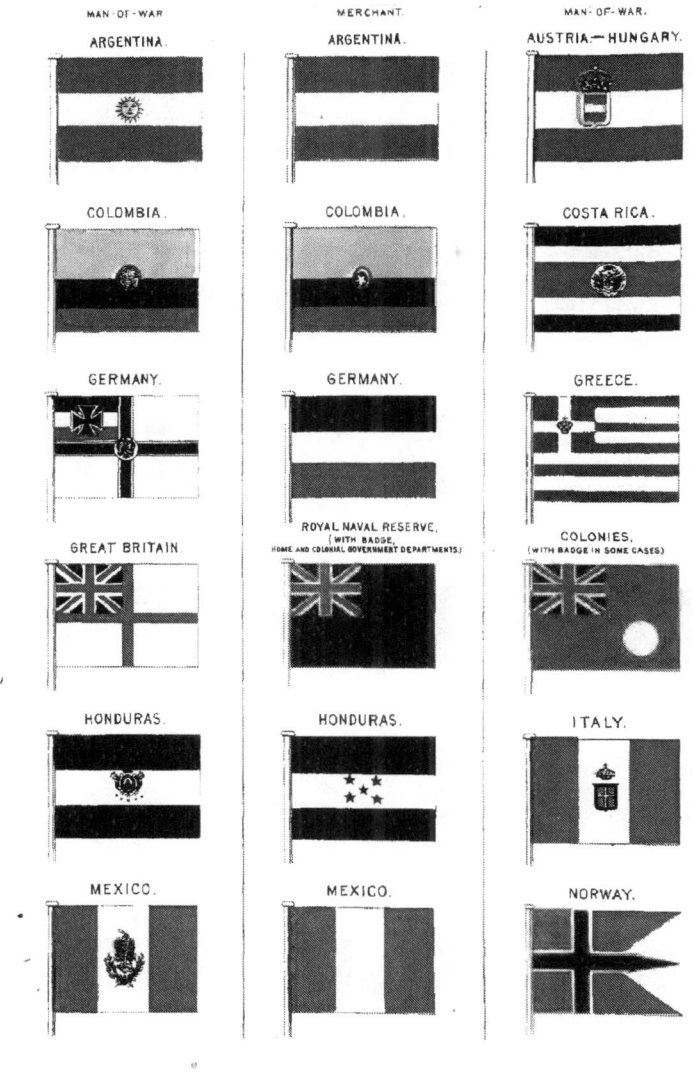

MAN-OF-WAR.	MERCHANT.	MAN-OF-WAR.
ARGENTINA.	ARGENTINA.	AUSTRIA.—HUNGARY.
COLOMBIA.	COLOMBIA.	COSTA RICA.
GERMANY.	GERMANY.	GREECE.
GREAT BRITAIN.	ROYAL NAVAL RESERVE, (WITH BADGE, HOME AND COLONIAL GOVERNMENT DEPARTMENTS.)	COLONIES, (WITH BADGE IN SOME CASES)
HONDURAS.	HONDURAS.	ITALY.
MEXICO.	MEXICO.	NORWAY.

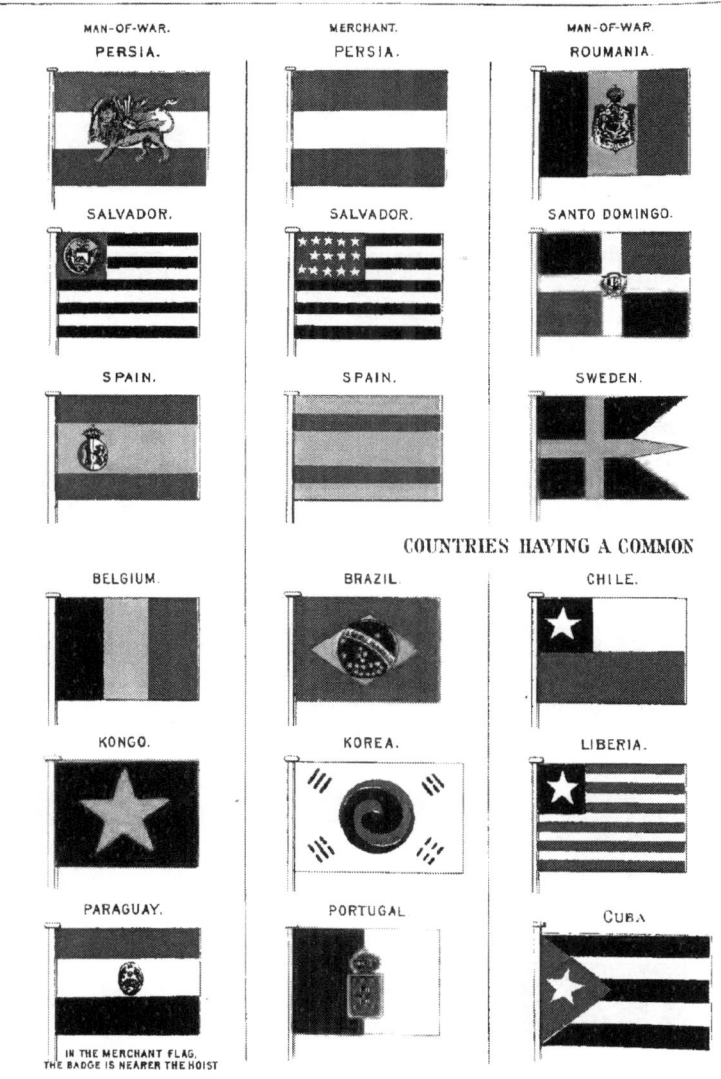

MAN-OF-WAR.
PERSIA.

MERCHANT.
PERSIA.

MAN-OF-WAR.
ROUMANIA.

SALVADOR.

SALVADOR.

SANTO DOMINGO.

SPAIN.

SPAIN.

SWEDEN.

COUNTRIES HAVING A COMMON

BELGIUM.

BRAZIL.

CHILE.

KONGO.

KOREA.

LIBERIA.

PARAGUAY.

PORTUGAL

CUBA

IN THE MERCHANT FLAG,
THE BADGE IS NEARER THE HOIST

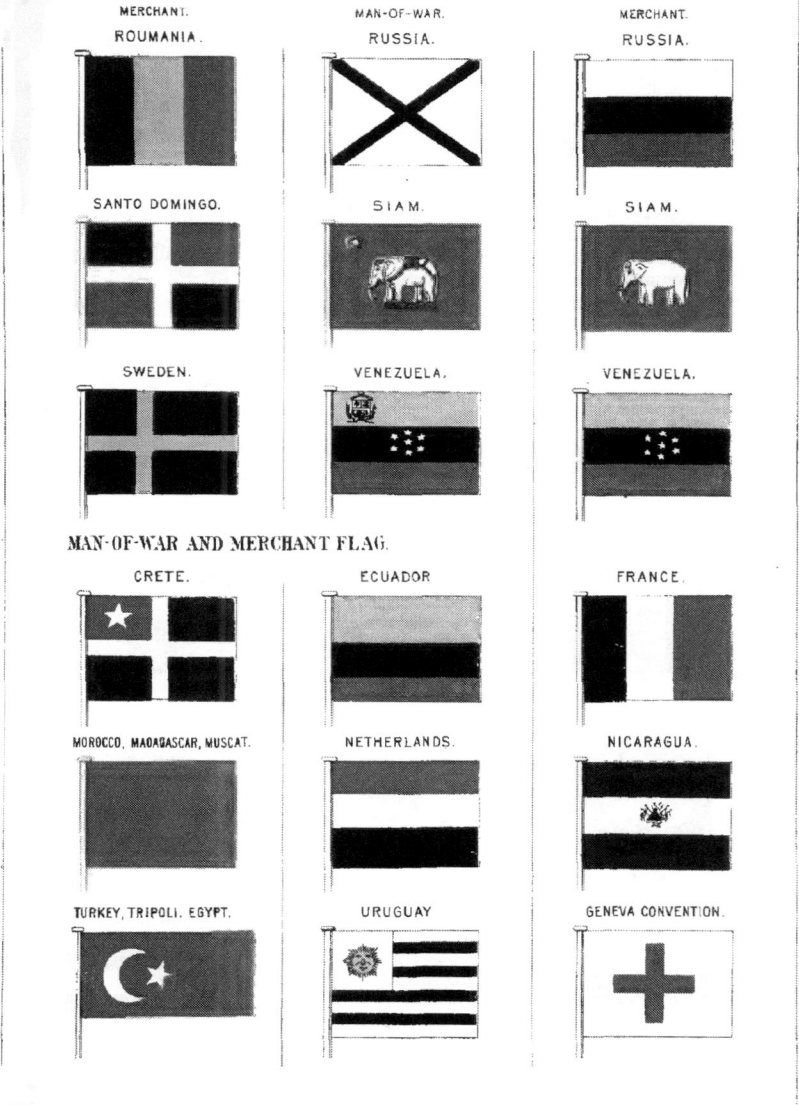

MERCHANT.
ROUMANIA.

MAN-OF-WAR.
RUSSIA.

MERCHANT.
RUSSIA.

SANTO DOMINGO.

SIAM.

SIAM.

SWEDEN.

VENEZUELA.

VENEZUELA.

MAN-OF-WAR AND MERCHANT FLAG.

CRETE.

ECUADOR

FRANCE.

MOROCCO, MADAGASCAR, MUSCAT.

NETHERLANDS.

NICARAGUA.

TURKEY, TRIPOLI. EGYPT.

URUGUAY

GENEVA CONVENTION.

MAN-OF-WAR.

FEDERATED MALAY STATES.

CODE FLAGS AND PENNANTS.

PLATE III

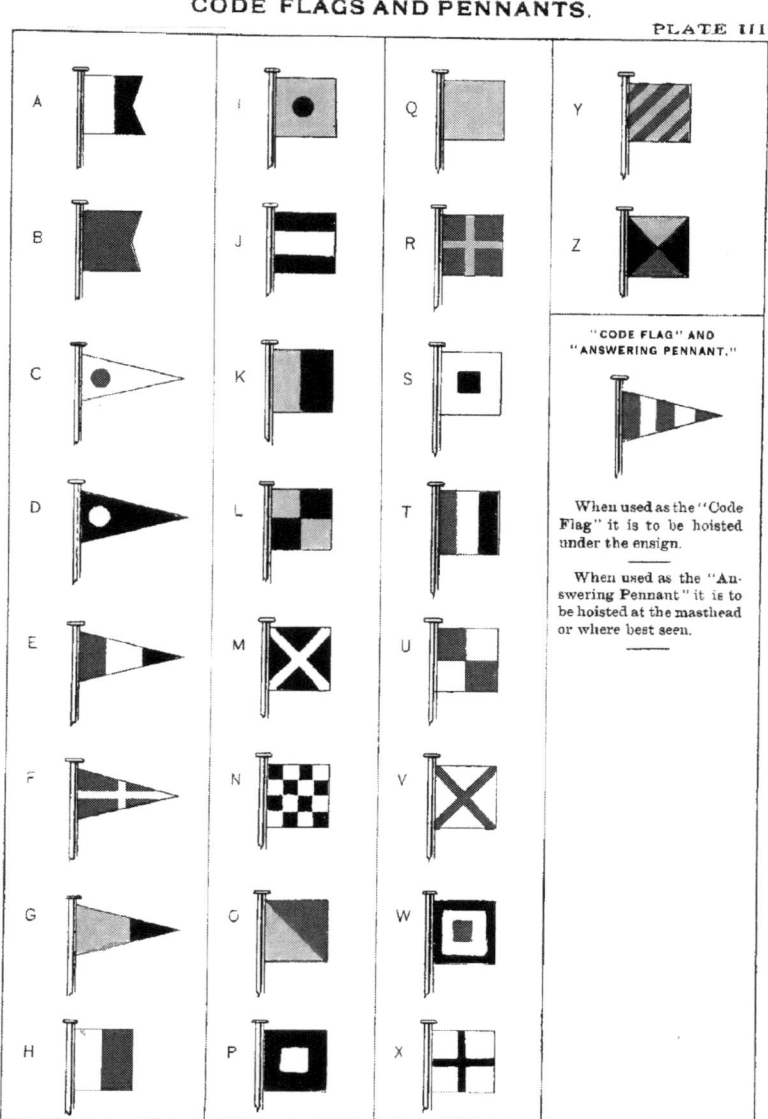

"CODE FLAG" AND "ANSWERING PENNANT."

When used as the "Code Flag" it is to be hoisted under the ensign.

When used as the "Answering Pennant" it is to be hoisted at the masthead or where best seen.

INTERNATIONAL CODE OF SIGNALS.

MEANINGS OF FLAGS AND PENNANTS HOISTED SINGLY

Signal	Meaning
B ---- ------ -----	—I am taking in (or, discharging) explosives
C ---------------	—Yes, or, Affirmative
D -------- --------	—No, or, Negative
L ------------------	—I have (or, have had) some dangerous infectious disease on board
P --------- --------	—I am about to sail, all persons to report on board
Q --------- -- ------	—I have a clean bill of health, but am liable to quarantine
S-------------------	—I want a pilot

MEANINGS OF FLAGS AND PENNANTS HOISTED WITH CODE FLAG

CODE FLAG over A	—I am on full speed trial
" " " B	—I am taking in (or, discharging) explosives
" " " C	—Yes, or, Affirmative
" " " D	—No, or, Negative
" " " E	—Alphabetical Signal No 1 *
" " " F	—Alphabetical Signal No 2 *
" " " G	—Alphabetical Signal No 3 *
" " " H	—Stop, Heave to, or, Come nearer, I have something important to communicate
" " " I	—I have not a clean bill of health
" " " J	—I have headway
" " " K	—I have stern-board.
" " " L	—I have (or, have had) some dangerous infectious disease on board
" " " M	—Numeral Signal No 1 †
" " " N	—Numeral Signal No 2 †
" " " O	—Numeral Signal No 3 †
" " " P	—I am about to sail, all persons to report on board
" " " Q	—I have a clean bill of health, but am liable to quarantine
" " " R	—Do not pass ahead of me
" " " S	—I want a pilot
" " " T	—Do not overtake me
" " " U	—My engines are stopped
" " " V	—My engines are going astern
" " " W	—All boats are to return to the ship
" " " X	—I will pass ahead of you
" " " Y	—All ships of the convoy are to rejoin company (See NXP, p 254)
" " " Z	—I will pass astern of you

* For instructions as to the use of these Signals, see page 18   † For instructions as to the use of these Signals, see page 32

INTERNATIONAL CODE SIGNALS OF DISTRESS

(1) The International Code Signal of Distress indicated by NC,

(2) The distant signal, consisting of a square flag, having either above or below it a ball or anything resembling a ball,

(3) The distant signal, consisting of a cone point upward, having either above it or below it a ball or anything resembling a ball

For other signals of distress see article 31 of the International Rules to Prevent Collisions and article 31 of the Inland Rules to Prevent Collisions

INTERNATIONAL CODE SIGNALS FOR A PILOT.

(1) The International Code Pilot Signal indicated by PT,

(2) The International Code Flag S, with or without the Code Pennant over it,

(3) The distant signal, consisting of a cone point upward, having above it two balls or shapes resembling balls.

INTERNATIONAL CODE OF SIGNALS.

PART I.

SIGNALS MADE BY FLAGS OF THE CODE

PART I.

INSTRUCTIONS HOW TO SIGNAL

In the following instructions the ship making the signal is called A *, the ship signaled to is called* B

HOW TO MAKE A SIGNAL.

1 Ship A, wishing to make a signal, hoists her Ensign with the Code Flag under it

2 If more than one vessel or signal station is in sight, and the signal is intended for a particular vessel or signal station, ship A should indicate which vessel or signal station she is addressing by making the distinguishing signal (*i. e* , the signal letters) of the vessel or station with which she desires to communicate

3. If the distinguishing signal is not known, ship A should make use of one of the signals DI to DQ (*page* 37)

4. When ship A has been answered by the vessel she is addressing (*see paragraph 9*), she proceeds with the signal which she desires to make, first hauling down her Code Flag if it is required for making the signal

5. Signals should always be hoisted where they can best be seen, and not necessarily at the masthead.

6 Each hoist should be kept flying until ship B hoists her Answering Pennant "CLOSE UP" (*see paragraph* 10)   .

7 When ship A has finished signaling she hauls down her Ensign, and her Code Flag, if the latter has not already been hauled down (*see paragraph* 4)

8. When it is desired to make a signal it should be looked out in the General Vocabulary (*pages* 143–451), which is the index to the Signal Book

HOW TO ANSWER A SIGNAL.

9. Ship B (the ship signaled to), on seeing the signal made by ship A, hoists her Answering Pennant at the "DIP"

(*A flag is at the* "DIP" *when it is hoisted about two-thirds of the way up, that is, some little distance below where it would be when hoisted* "CLOSE UP")

The Answering Pennant should always be hoisted where it can best be seen

10. When A's hoist has been taken in, looked out in the Signal Book, and is understood, B hoists her Answering Pennant "CLOSE UP" and keeps it there until A hauls her hoist down

11 B then lowers her Answering Pennant to the "DIP," and waits for the next hoist

12. If the flags in A's hoist can not be made out, or if, when the flags are made out, the purport of the signal is not understood, B keeps her Answering Pennant at the "DIP" and hoists the signal OWL or WCX, or such other signal as may meet the case, and when A has repeated or rectified her signal, and B thoroughly understands it, B hoists her Answering Pennant "CLOSE UP."

NOTES ON SIGNALING.

1. **Plurals.**—To facilitate the translation of the Code into foreign languages, the plurals of words given in the Signal Book have been omitted The words should be regarded as being used in the singular, unless the contrary is indicated by the context

2 **When making Signals for comparing Chronometers (page 58) or showing the Mean Time** vessel A is to hoist the signal denoting the *hour* (*see page* 58), and shortly after the signal has been answered by B dip it sharply to denote the precise instant used for comparison The signals denoting the *minutes* and *seconds shown by the chronometer at the instant of dipping* are then to be hoisted To insure accuracy, a second comparison should be made

3 **In signaling Longitude or Time** vessels should always reckon from the meridian of Greenwich, except French and Spanish vessels, which will use the meridian of Paris or Cadiz, respectively. If any doubt is entertained, the vessel to which the signal is made should hoist NBL = What is your first meridian?

4 **Meridians.**—The British Meridian is that of Greenwich The Meridian of Paris (Observatory) is 2° 20′ 15″ east of Greenwich, or 0 h 9 m 21 secs The Meridian of Cadiz (San Fernando Observatory) is 6° 12′ 24″ west of Greenwich, or 0 h. 24 m. 49 6 secs

5 **Passing Vessels.**—Ships passing one another or Signal Stations will do well to hoist the following signals in the order shown

      (1) National Colors with the Code Signal under them (*The Ensign should be kept flying until all communication is ended, the Code Flag may be hauled down if it is required for making a signal*)
      (2) Ship's name (signal letters)
      (3) Where from
      (4) Where bound.
      (5) Number of days out
      (6) My longitude by chronometer is — ·
      The ensign should be dipped and rehoisted as a farewell.

When vessels are passing each other quickly, time will be saved if, instead of hoisting the Answering Pennant, they exchange signals in the following manner
      On reading A's name (that is, distinguishing signal), B should hoist hers. A should not haul down until she understands B's hoist, when both ships should haul down together and proceed in the order suggested above

6. **Procedure when signaling Names and Addresses.**—The following course is to be followed when making a signal which contains the ship's name and the owner's name and address.

Ship A, wishing to obtain orders from her owner, will make—
      (1) Her distinguishing signal (signal letters)
      (2) The signal SW = I wish to obtain orders from my owner, **Mr** —, at —
      (3) The owner's name, by spelling it letter by letter (*see page* 13) or by using the Alphabetical Spelling Table (*pages* 14–15)
      (4) The owner's address, by spelling it letter by letter (*see page* 13) or by using the Alphabetical Spelling Table (*pages* 14–15) Signals from the Geographical Table (*page* 89) can also be used in some cases Figures can be made by the Numeral Signals on page 32 or by the Numeral Table on page 33.

EXAMPLE

Ship A wishes to get orders from her owner (say), Mr C Thomson, at 25 Broad street, New York

Having hoisted her national colors with the Code Signal under them, she makes the following signals

1st hoist,	*Her distinguishing signal (signal letters)*
2d hoist,	SW = *I wish to obtain orders from my owner, Mr ——, at ——*
3d hoist,	Code flag over E = *The signals which follow are alphabetical* (*Before this signal*
4th hoist,	C = C *is made the Code Flag under the Ensign may be hauled down*)
5th hoist,	Code Flag over F = *Dot between initials*
6th hoist,	THOM ⎫
7th hoist,	SON ⎬ = *Thomson*
8th hoist,	Code Flag over M = *The signal which follows is a numeral signal, and is to be*
9th hoist,	BE = 25 *looked out in the Numeral Table (page 32)*
10th hoist,	Code Flag over E = *The signals which follow are alphabetical*
11th hoist,	BRO ⎫
12th hoist,	AD ⎬ = *Broad*
13th hoist,	Code Flag over G = *The Alphabetical Signals are ended*
14th hoist,	WZN = *Street*
15th hoist,	AZOT = *New York.*

ALPHABETICAL (SPELLING) SIGNALS

Under the arrangement explained below, every flag hoisted after Alphabetical Signal No 1 has been made, and until Alphabetical Signal No 3, or Numeral Signal No 1 (*see page* 32) is made, represents the letter of the alphabet which has been allotted to it in the Code. As each of the 26 letters of the alphabet is now represented by a flag, any word can be spelled by this system

If the word to be spelled consists of more than four letters, two or more hoists must be used, as no hoist is to contain more than four flags, and, if any letter occurs more than once in the word, this letter must on its second occurrence begin or be in a second hoist, and on its third occurrence must begin or be in a third hoist

The following are the signals to be used

Signal	Meaning
CODE FLAG OVER FLAG E____	ALPHABETICAL SIGNAL NO 1, indicating that the flags hoisted after it until Alphabetical Signal No 3 or Numeral Signal No 1 is made do not represent the signals in the Code, but are to be understood as having their alphabetical meanings and express individual letters of the alphabet which are to form words
CODE FLAG OVER FLAG F____	ALPHABETICAL SIGNAL NO 2, indicating the end of a word made by Alphabetical Signals, or dot between initials
CODE FLAG OVER FLAG G____	ALPHABETICAL SIGNAL NO 3, indicating that the Alphabetical Signals are ended, the signals which follow are to be looked out in the Code in the usual manner

To spell "William J Perry "

1st hoist,	Code Flag over E=*The signals which follow are alphabetical.*	
2d hoist,	WIL	} =*William*
3d hoist,	LIAM	
4th hoist,	Code Flag over F=*End of the word (also means dot between initials)*	
5th hoist,	J=*J*	
6th hoist,	Code Flag over F=*Dot between initials (also means end of the word)*	
7th hoist,	PER	} =*Perry*
8th hoist,	RY	
9th hoist,	Code Flag over G=*Alphabetical Signals are ended*	

An alternative method of spelling words is provided by the Alphabetical Spelling Table on page 15

ALPHABETICAL SPELLING TABLE.

This Table can be used in communications between Vessels of all Countries employing the Roman characters, A, B, C, etc See also page 13 for alternative system of making Alphabetical Signals

	A										
		CBHP	Adc	CBMD	Afm	CBQR	Ahw	CBVG	Akf	CBYS	Amp
		CBHQ	Add	CBMF	Afn	CBQS	Ahx	CBVH	Akg	CBYT	Amq
CBDF	Aa	CBHR	Ade	CBMG	Afo	CBQT	Ahy	CBVJ	Akh	CBYV	Amr
CBDG	Aab	CBHS	Adf	CBMH	Afp	CBQV	Ahz	CBVK	Aki	CBYW	Ams
CBDH	Aac	CBHT	Adg	CBMJ	Afq	CBQW	Ai	CBVL	Akj	CBYX	Amt
CBDJ	Aad	CBHV	Adh	CBMK	Afr	CBQX	Aia	CBVM	Akk	CBYZ	Amu
CBDK	Aam	CBHW	Adi	CBML	Afs	CBQY	Aib	CBVN	Akl	CBZD	Amv
CBDL	Ab	CBHX	Adj	CBMN	Aft	CBQZ	Aic	CBVP	Akm	CBZF	Amw
CBDM	Aba	CBHY	Adk	CBMP	Afu	CBRD	Aid	CBVQ	Akn	CBZG	Amx
CBDN	Abb	CBHZ	Adl	CBMQ	Afv	CBRF	Aie	CBVR	Ako	CBZH	Amy
CBDP	Abc	CBJD	Adm	CBMR	Afw	CBRG	Aif	CBVS	Akp	CBZJ	Amz
CBDQ	Abd	CBJF	Adn	CBMS	Afx	CBRH	Aig	CBVT	Akq	CBZK	An
CBDR	Abe	CBJG	Ado	CBMT	Afy	CBRJ	Aih	CBVW	Akr	CBZL	Ana
CBDS	Abf	CBJH	Adp	CBMV	Afz	CBRK	Aii	CBVX	Aks	CBZM	Anb
CBDT	Abg	CBJK	Adq	CBMW	Ag	CBRL	Aij	CBVY	Akt	CBZN	Anc
CBDV	Abh	CBJL	Adr	CBMX	Aga	CBRM	Aik	CBVZ	Aku	CBZP	And
CBDW	Abi	CBJM	Ads	CBMY	Agb	CBRN	Ail	CBWD	Akv	CBZQ	Anf
CBDX	Abj	CBJN	Adt	CBMZ	Agc	CBRP	Aim	CBWF	Akw	CBZR	Ang
CBDY	Abk	CBJP	Adu	CBND	Agd	CBRQ	Ain	CBWG	Akx	CBZS	Anh
CBDZ	Abl	CBJQ	Adv	CBNF	Age	CBRS	Aio	CBWH	Aky	CBZT	Ani
CBFD	Abm	CBJR	Adw	CBNG	Agf	CBRT	Aip	CBWJ	Akz	CBZV	Anj
CBFG	Abn	CBJS	Adx	CBNH	Agg	CBRV	Aiq	CBWK	Al	CBZW	Ank
CBFH	Abo	CBJT	Ady	CBNJ	Agh	CBRW	Air	CBWL	Ala	CBZX	Anl
CBFJ	Abp	CBJV	Adz	CBNK	Agi	CBRX	Ais	CBWM	Alb	CBZY	Anm
CBFK	Abq	CBJW	Ae	CBNL	Agj	CBRY	Ait	CBWN	Alc	CDBF	Ann
CBFL	Abr	CBJX	Aea	CBNM	Agk	CBRZ	Aiu	CBWP	Ald	CDBG	Anp
CBFM	Abs	CBJY	Aeb	CBNP	Agl	CBSD	Aiv	CBWQ	Ale	CDBH	Anq
CBFN	Abt	CBJZ	Aec	CBNQ	Agm	CBSF	Aiw	CBWR	Alf	CDBJ	Anr
CBFP	Abu	CBKD	Aed	CBNR	Agn	CBSG	Aix	CBWS	Alg	CDBK	Ans
CBFQ	Abv	CBKF	Aee	CBNS	Ago	CBSH	Aiy	CBWT	Alh	CDBL	Ant
CBFR	Abw	CBKG	Aef	CBNT	Agp	CBSJ	Aj	CBWV	Ali	CDBM	Anu
CBFS	Abx	CBKH	Aeg	CBNV	Agq	CBSK	Aja	CBWX	Alj	CDBN	Anv
CBFT	Aby	CBKJ	Aeh	CBNW	Agr	CBSL	Ajb	CBWY	Alk	CDBP	Anw
CBFV	Abz	CBKL	Aei	CBNX	Ags	CBSM	Ajc	CBWZ	All	CDBQ	Anx
CBFW	Ac	CBKM	Aej	CBNY	Agt	CBSN	Ajd	CBXD	Alm	CDBR	Any
CBFX	Aca	CBKN	Aek	CBNZ	Agu	CBSP	Aje	CBXF	Aln	CDBS	Anz
CBFY	Acb	CBKP	Ael	CBPD	Agv	CBSQ	Ajf	CBXG	Alo	CDBT	Ao
CBFZ	Acc	CBKQ	Aem	CBPF	Agw	CBSR	Ajg	CBXH	Alp	CDBV	Aoa
CBGD	Acd	CBKR	Aen	CBPG	Agx	CBST	Ajh	CBXJ	Alq	CDBW	Aob
CBGF	Ace	CBKS	Aeo	CBPH	Agy	CBSV	Aji	CBXK	Alr	CDBX	Aoc
CBGH	Acf	CBKT	Aep	CBPJ	Agz	CBSW	Ajj	CBXL	Als	CDBY	Aod
CBGJ	Acg	CBKV	Aeq	CBPK	Ah	CBSX	Ajk	CBXM	Alt	CDBZ	Aoe
CBGK	Ach	CBKW	Aer	CBPL	Aha	CBSY	Ajl	CBXN	Alu	CDFB	Aof
CBGL	Aci	CBKX	Aes	CBPM	Ahb	CBSZ	Ajm	CBXP	Alv	CDFG	Aog
CBGM	Acj	CBKY	Aet	CBPN	Ahc	CBTD	Ajn	CBXQ	Alw	CDFH	Aoh
CBGN	Ack	CBKZ	Aeu	CBPQ	Ahd	CBTF	Ajo	CBXR	Alx	CDFJ	Aoi
CBGP	Acl	CBLD	Aev	CBPR	Ahe	CBTG	Ajp	CBXS	Aly	CDFK	Aoj
CBGQ	Acm	CBLF	Aew	CBPS	Ahf	CBTH	Ajq	CBXT	Alz	CDFL	Aok
CBGR	Acn	CBLG	Aex	CBPT	Ahg	CBTJ	Ajr	CBXV	Am	CDFM	Aol
CBGS	Aco	CBLH	Aey	CBPV	Ahh	CBTK	Ajs	CBXW	Ama	CDFN	Aom
CBGT	Acp	CBLJ	Aez	CBPW	Ahi	CBTL	Ajt	CBXY	Amb	CDFP	Aon
CBGV	Acq	CBLK	Af	CBPX	Ahj	CBTM	Aju	CBXZ	Amc	CDFQ	Aoo
CBGW	Acr	CBLM	Afa	CBPY	Ahk	CBTN	Ajv	CBYD	Amd	CDFR	Aop
CBGX	Acs	CBLN	Afb	CBPZ	Ahl	CBTP	Ajw	CBYF	Ame	CDFS	Aoq
CBGY	Act	CBLP	Afc	CBQD	Ahm	CBTQ	Ajx	CBYG	Amf	CDFT	Aor
CBGZ	Acu	CBLQ	Afd	CBQF	Ahn	CBTR	Ajy	CBYH	Amg	CDFV	Aos
CBHD	Acv	CBLR	Afe	CBQG	Aho	CBTS	Ajz	CBYJ	Amh	CDFW	Aot
CBHF	Acw	CBLS	Aff	CBQH	Ahp	CBTV	Ak	CBYK	Ami	CDFX	Aou
CBHG	Acx	CBLT	Afg	CBQJ	Ahq	CBTW	Aka	CBYL	Amj	CDFY	Aov
CBHJ	Acy	CBLV	Afh	CBQK	Ahr	CBTX	Akb	CBYM	Amk	CDFZ	Aow
CBHK	Acz	CBLW	Afi	CBQL	Ahs	CBTY	Akc	CBYN	Aml	CDGB	Aox
CBHL	Ad	CBLX	Afj	CBQM	Aht	CBTZ	Akd	CBYP	Amm	CDGF	Aoy
CBHM	Ada	CBLY	Afk	CBQN	Ahu	CBVD	Ake	CBYQ	Amn	CDGH	Aoz
CBHN	Adb	CBLZ	Afl	CBQP	Ahv	CBVF		CBYR	Amo	CDGJ	Ap

CDGK	Apa	CDLM	Aru	CDQN	Aun	CDVQ	Axg	CDZS		CFHJ	Bf
CDGL	Apb	CDLN	Arv	CDQP	Auo	CDVR	Axh	CDZT		CFHK	Bg
CDGM	Apc	CDLP	Arw	CDQR	Aup	CDVS	Axi	CDZV		CFHL	Bh
CDGN	Apd	CDLQ	Arx	CDQS	Auq	CDVT	Axj	CDZW		CFHM	Bi
CDGP	Ape	CDLR	Ary	CDQT	Aur	CDVW	Axk	CDZX		CFHN	Bia
CDGQ	Apf	CDLS	Arz	CDQV	Aus	CDVX	Axl	CDZY		CFHP	Bib
CDGR	Apg	CDLT	As	CDQW	Aut	CDVY	Axm			CFHQ	Bic
CDGS	Aph	CDLV	Asa	CDQX	Auu	CDVZ	Axn			CFHR	Bid
CDGT	Api	CDLW	Asb	CDQY	Auv	CDWB	Axo			CFHS	Bie
CDGV	Apj	CDLX	Asc	CDQZ	Auw	CDWF	Axp			CFHT	Bif
CDGW	Apk	CDLY	Asd	CDRB	Aux	CDWG	Axq			CFHV	Big
CDGX	Apl	CDLZ	Ase	CDRF	Auy	CDWH	Axr			CFHW	Bih
CDGY	Apm	CDMB	Asf	CDRG	Auz	CDWJ	Axs			CFHX	Bii
CDGZ	Apn	CDMF	Asg	CDRH	Av	CDWK	Axt			CFHY	Bij
CDHB	Apo	CDMG	Ash	CDRJ	Ava	CDWL	Axu			CFHZ	Bik
CDHF	App	CDMH	Asi	CDRK	Avb	CDWM	Axv			CFJB	Bim
CDHG	Apq	CDMJ	Asj	CDRL	Avc	CDWN	Axw		B	CFJD	Bin
CDHJ	Apr	CDMK	Ask	CDRM	Avd	CDWP	Axx	CFBD	Ba	CFJG	Bio
CDHK	Aps	CDML	Asl	CDRN	Ave	CDWQ	Axy	CFBG	Baa	CFJH	Bip
CDHL	Apt	CDMN	Asm	CDRP	Avf	CDWR	Axz	CFBH	Bab	CFJK	Biq
CDHM	Apu	CDMP	Asn	CDRQ	Avg	CDWS	Ay	CFBJ	Bac	CFJL	Bir
CDHN	Apv	CDMQ	Aso	CDRS	Avh	CDWT	Aya	CFBK	Bad	CFJM	Bis
CDHP	Apw	CDMR	Asp	CDRT	Avi	CDWV	Ayb	CFBL	Bae	CFJN	Bit
CDHQ	Apx	CDMS	Asq	CDRV	Avj	CDWX	Ayc	CFBM	Baf	CFJP	Biu
CDHR	Apy	CDMT	Asr	CDRW	Avk	CDWY	Ayd	CFBN	Bag	CFJQ	Biv
CDHS	Apz	CDMV	Ass	CDRX	Avl	CDWZ	Aye	CFBP	Bah	CFJR	Biw
CDHT	Aq	CDMW	Ast	CDRY	Avm	CDXB	Ayf	CFBQ	Bai	CFJS	Bix
CDHV	Aqa	CDMX	Asu	CDRZ	Avn	CDXF	Ayg	CFBR	Baj	CFJT	Biy
CDHW	Aqb	CDMY	Asv	CDSB	Avo	CDXG	Ayh	CFBS	Bak	CFJV	Biz
CDHX	Aqc	CDMZ	Asw	CDSF	Avp	CDXH	Ayi	CFBT	Bal	CFJW	Bj
CDHY	Aqd	CDNB	Asx	CDSG	Avq	CDXJ	Ayj	CFBV	Bam	CFJX	Bk
CDHZ	Aqe	CDNF	Asy	CDSH	Avr	CDXK	Ayk	CFBW	Ban	CFJY	Bl
CDJB	Aqf	CDNG	Asz	CDSJ	Avs	CDXL	Ayl	CFBX	Bao	CFJZ	Bla
CDJF	Aqg	CDNH	At	CDSK	Avt	CDXM	Aym	CFBY	Bap	CFKB	Ble
CDJG	Aqh	CDNJ	Ata	CDSL	Avu	CDXN	Ayn	CFBZ	Baq	CFKD	Bli
CDJH	Aqi	CDNK	Atb	CDSM	Avv	CDXP	Ayo	CFDB	Bar	CFKG	Blo
CDJK	Aqj	CDNL	Atc	CDSN	Avw	CDXQ	Ayp	CFDG	Bas	CFKH	Blu
CDJL	Aqk	CDNM	Atd	CDSP	Avx	CDXR	Ayq	CFDH	Bat	CFKJ	Bly
CDJM	Aql	CDNP	Ate	CDSQ	Avy	CDXS	Ayr	CFDJ	Bau	CFKL	Bm
CDJN	Aqm	CDNQ	Atf	CDSR	Avz	CDXT	Ays	CFDK	Bav	CFKM	Bn
CDJP	Aqn	CDNR	Atg	CDST	Aw	CDXV	Ayt	CFDL	Baw	CFKN	Bo
CDJQ	Aqo	CDNS	Ath	CDSV	Awa	CDXW	Ayu	CFDM	Bax	CFKP	Boa
CDJR	Aqp	CDNT	Ati	CDSW	Awb	CDXY	Ayv	CFDN	Bay	CFKQ	Bob
CDJS	Aqq	CDNV	Atj	CDSX	Awc	CDXZ	Ayw	CFDP	Baz	CFKR	Boc
CDJT	Aqr	CDNW	Atk	CDSY	Awd	CDYB	Ayx	CFDQ	Bb	CFKS	Bod
CDJV	Aqs	CDNX	Atl	CDSZ	Awe	CDYF	Ayy	CFDR	Bc	CFKT	Boe
CDJW	Aqt	CDNY	Atm	CDTB	Awf	CDYG	Ayz	CFDS	Bd	CFKV	Bof
CDJX	Aqu	CDNZ	Atn	CDTF	Awg	CDYH	Az	CFDT	Be	CFKW	Bog
CDJY	Aqv	CDPB	Ato	CDTG	Awh	CDYJ	Aza	CFDV	Bea	CFKX	Boh
CDJZ	Aqw	CDPF	Atp	CDTH	Awi	CDYK	Azb	CFDW	Beb	CFKY	Boi
CDKB	Aqx	CDPG	Atq	CDTJ	Awj	CDYL	Azc	CFDX	Bec	CFKZ	Boj
CDKF	Aqy	CDPH	Atr	CDTK	Awk	CDYM	Azd	CFDY	Bed	CFLB	Bok
CDKG	Aqz	CDPJ	Ats	CDTL	Awl	CDYN	Aze	CFDZ	Bee	CFLD	Bol
CDKH	Ar	CDPK	Att	CDTM	Awm	CDYP	Azf	CFGB	Bef	CFLG	Bom
CDKJ	Ara	CDPL	Atu	CDTN	Awn	CDYQ	Azg	CFGD	Beg	CFLH	Bon
CDKL	Arb	CDPM	Atv	CDTP	Awo	CDYR	Azh	CFGH	Beh	CFLJ	Bop
CDKM	Arc	CDPN	Atw	CDTQ	Awp	CDYS	Azi	CFGJ	Bei	CFLK	Boq
CDKN	Ard	CDPQ	Atx	CDTR	Awq	CDYT	Azj	CFGK	Bej	CFLM	Bor
CDKP	Are	CDPR	Aty	CDTS	Awr	CDYV	Azk	CFGL	Bek	CFLN	Bos
CDKQ	Arf	CDPS	Atz	CDTV	Aws	CDYW	Azl	CFGM	Bel	CFLP	Bot
CDKR	Arg	CDPT	Au	CDTW	Awt	CDYX	Azm	CFGN	Bem	CFLQ	Bou
CDKS	Arh	CDPV	Aua	CDTX	Awu	CDYZ	Azn	CFGP	Ben	CFLR	Bov
CDKT	Ari	CDPW	Aub	CDTY	Awv	CDZB	Azo	CFGQ	Beo	CFLS	Bow
CDKV	Arj	CDPX	Auc	CDTZ	Aww	CDZF	Azp	CFGR	Bep	CFLT	Box
CDKW	Ark	CDPY	Aud	CDVB	Awx	CDZG	Azq	CFGS	Beq	CFLV	Boy
CDKX	Arl	CDPZ	Aue	CDVF	Awy	CDZH	Azr	CFGT	Ber	CFLW	Boz
CDKY	Arm	CDQB	Auf	CDVG	Awz	CDZJ	Azs	CFGV	Bes	CFLX	Bp
CDKZ	Arn	CDQF	Aug	CDVH	Ax	CDZK	Azt	CFGW	Bet	CFLY	Bq
CDLB	Aro	CDQG	Auh	CDVJ	Axa	CDZL	Azu	CFGX	Beu	CFLZ	Br
CDLF	Arp	CDQH	Aui	CDVK	Axb	CDZM	Azv	CFGY	Bev	CFMB	Bra
CDLG	Arq	CDQJ	Auj	CDVL	Axc	CDZN	Azw	CFGZ	Bew	CFMD	Bre
CDLH	Arr	CDQK	Auk	CDVM	Axd	CDZP	Azx	CFHB	Bex	CFMG	Bri
CDLJ	Ars	CDQL	Aul	CDVN	Axe	CDZQ	Azy	CFHD	Bey	CFMH	Bro
CDLK	Art	CDQM	Aum	CDVP	Axf	CDZR	Azz	CFHG	Bez	CFMJ	Bru

CFMK	Bry	CFQW	Cao	CFVY	Cic	CGBD	Cua	CGJF	Czl	CGMS	Deq
CFML	Bs	CFQX	Cap	CFVZ	Cid	CGBF	Cub	CGJH	Czm	CGMT	Der
CFMN	Bt	CFQY	Caq	CFWB	Cie	CGBH	Cuc	CGJK	Czn	CGMV	Des
CFMP	Bu	CFQZ	Car	CFWD	Cif	CGBJ	Cud	CGJL	Czo	CGMW	Det
CFMQ	Bua	CFRB	Cas	CFWG	Cig	CGBK	Cue	CGJM	Czp	CGMX	Deu
CFMR	Bub	CFRD	Cat	CFWH	Cih	CGBL	Cuf	CGJN	Czq	CGMY	Dev
CFMS	Buc	CFRG	Cau	CFWJ	Cii	CGBM	Cug	CGJP	Czr	CGMZ	Dew
CFMT	Bud	CFRH	Cav	CFWK	Cij	CGBN	Cuh	CGJQ	Czs	CGNB	Dex
CFMV	Bue	CFRJ	Caw	CFWL	Cik	CGBP	Cui	CGJR	Czt	CGND	Dey
CFMW	Buf	CFRK	Cax	CFWM	Cil	CGBQ	Cuj	CGJS	Czu	CGNF	Dez
CFMX	Bug	CFRL	Cay	CFWN	Cim	CGBR	Cuk	CGJT	Czv	CGNH	Df
CFMY	Buh	CFRM	Caz	CFWP	Cin	CGBS	Cul	CGJV	Czw	CGNJ	Dg
CFMZ	Bui	CFRN	Cb	CFWQ	Cio	CGBT	Cum	CGJW	Czx	CGNK	Dh
CFNB	Buj	CFRP	Cc	CFWR	Cip	CGBV	Cun	CGJX	Czy	CGNL	Dha
CFND	Buk	CFRQ	Cd	CFWS	Ciq	CGBW	Cuo	CGJY	Czz	CGNM	Dhe
CFNG	Bul	CFRS	Ce	CFWT	Cir	CGBX	Cup	CGJZ		CGNP	Dhi
CFNH	Bum	CFRT	Cea	CFWV	Cis	CGBY	Cuq			CGNQ	Dho
CFNJ	Bun	CFRV	Ceb	CFWX	Cit	CGBZ	Cur			CGNR	Dhu
CFNK	Buo	CFRW	Cec	CFWY	Ciu	CGDB	Cus			CGNS	Dhy
CFNL	Bup	CFRX	Ced	CFWZ	Civ	CGDF	Cut			CGNT	Di
CFNM	Buq	CFRY	Cee	CFXB	Ciw	CGDH	Cuu			CGNV	Dia
CFNP	Bur	CFRZ	Cef	CFXD	Cix	CGDJ	Cuv			CGNW	Dib
CFNQ	Bus	CFSB	Ceg	CFXG	Ciy	CGDK	Cuw			CGNX	Dic
CFNR	But	CFSD	Ceh	CFXH	Ciz	CGDL	Cux			CGNY	Did
CFNS	Buu	CFSG	Cei	CFXJ	Cj	CGDM	Cuy			CGNZ	Die
CFNT	Buv	CFSH	Cej	CFXK	Ck	CGDN	Cuz			CGPB	Dif
CFNV	Buw	CFSJ	Cek	CFXL	Cl	CGDP	Cv	**D.**		CGPD	Dig
CFNW	Bux	CFSK	Cel	CFXM	Cla	CGDQ	Cva	CGKB	Da	CGPF	Dih
CFNX	Buy	CFSL	Cem	CFXN	Cle	CGDR	Cve	CGKD	Daa	CGPH	Dii
CFNY	Buz	CFSM	Cen	CFXP	Cli	CGDS	Cvi	CGKF	Dab	CGPJ	Dij
CFNZ	Bv	CFSN	Ceo	CFXQ	Clo	CGDT	Cvo	CGKH	Dac	CGPK	Dik
CFPB	Bw	CFSP	Cep	CFXR	Clu	CGDV	Cvu	CGKJ	Dad	CGPL	Dil
CFPD	Bx	CFSQ	Ceq	CFXS	Cly	CGDW	Cvy	CGKL	Dae	CGPM	Dim
CFPG	By	CFSR	Cer	CFXT	Cm	CGDX	Cw	CGKM	Daf	CGPN	Din
CFPH	Bya	CFST	Ces	CFXV	Cn	CGDY	Cx	CGKN	Dag	CGPQ	Dio
CFPJ	Bye	CFSV	Cet	CFXW	Co	CGDZ	Cy	CGKP	Dah	CGPR	Dip
CFPK	Byi	CFSW	Ceu	CFXY	Coa	CGFB	Cya	CGKQ	Dai	CGPS	Diq
CFPL	Byo	CFSX	Cev	CFXZ	Cob	CGFD	Cyb	CGKR	Daj	CGPT	Dir
CFPM	Byu	CFSY	Cew	CFYB	Coc	CGFH	Cyc	CGKS	Dak	CGPV	Dis
CFPN	Bz	CFSZ	Cex	CFYD	Cod	CGFJ	Cyd	CGKT	Dal	CGPW	Dit
CFPQ		CFTB	Cey	CFYG	Coe	CGFK	Cye	CGKV	Dam	CGPX	Diu
CFPR		CFTD	Cez	CFYH	Cof	CGFL	Cyf	CGKW	Dan	CGPY	Div
CFPS		CFTG	Cf	CFYJ	Cog	CGFM	Cyg	CGKX	Dao	CGPZ	Diw
CFPT		CFTH	Cg	CFYK	Coh	CGFN	Cyh	CGKY	Dap	CGQB	Dix
CFPV		CFTJ	Ch	CFYL	Coi	CGFP	Cyi	CGKZ	Daq	CGQD	Diy
CFPW		CFTK	Cha	CFYM	Coj	CGFQ	Cyj	CGLB	Dar	CGQF	Diz
CFPX		CFTL	Chb	CFYN	Cok	CGFR	Cyk	CGLD	Das	CGQH	Dj
CFPY		CFTM	Chc	CFYP	Col	CGFS	Cyl	CGLF	Dat	CGQJ	Dk
CFPZ		CFTN	Chd	CFYQ	Com	CGFT	Cym	CGLH	Dau	CGQK	Dl
		CFTP	Che	CFYR	Con	CGFV	Cyn	CGLJ	Dav	CGQL	Dla
		CFTQ	Chf	CFYS	Coo	CGFW	Cyo	CGLK	Daw	CGQM	Dle
		CFTR	Chg	CFYT	Cop	CGFX	Cyp	CGLM	Dax	CGQN	Dli
		CFTS	Chh	CFYV	Coq	CGFY	Cyq	CGLN	Day	CGQP	Dlo
		CFTV	Chi	CFYW	Cor	CGFZ	Cyr	CGLP	Daz	CGQR	Dlu
		CFTW	Chj	CFYX	Cos	CGHB	Cys	CGLQ	Db	CGQS	Dly
		CFTX	Chk	CFYZ	Cot	CGHD	Cyt	CGLR	Dc	CGQT	Dm
		CFTY	Chl	CFZB	Cou	CGHF	Cyu	CGLS	Dd	CGQV	Dn
		CFTZ	Chm	CFZD	Cov	CGHJ	Cyv	CGLT	De	CGQW	Do
		CFVB	Chn	CFZG	Cow	CGHK	Cyw	CGLV	Dea	CGQX	Doa
C.		CFVD	Cho	CFZH	Cox	CGHL	Cyx	CGLW	Deb	CGQY	Dob
CFQB	Ca	CFVG	Chp	CFZJ	Coy	CGHM	Cyy	CGLX	Dec	CGQZ	Doc
CFQD	Caa	CFVH	Chq	CFZK	Coz	CGHN	Cyz	CGLY	Ded	CGRB	Dod
CFQG	Cab	CFVJ	Chr	CFZL	Cp	CGHP	Cz	CGLZ	Dee	CGRD	Doe
CFQH	Cac	CFVK	Chs	CFZM	Cq	CGHQ	Cza	CGMB	Def	CGRF	Dof
CFQJ	Cad	CFVM	Cht	CFZN	Cr	CGHR	Czb	CGMD	Deg	CGRH	Dog
CFQK	Cae	CFVN	Chu	CFZP	Cra	CGHS	Czc	CGMF	Deh	CGRJ	Doh
CFQL	Caf	CFVP	Chv	CFZQ	Cre	CGHT	Czd	CGMH	Dei	CGRK	Doi
CFQM	Cag	CFVQ	Chw	CFZR	Cri	CGHV	Cze	CGMJ	Dej	CGRL	Doj
CFQN	Cah	CFVR	Chx	CFZS	Cro	CGHW	Czf	CGMK	Dek	CGRM	Dok
CFQP	Cai	CFVT	Ci	CFZT	Cru	CGHX	Czg	CGML	Del	CGRN	Dol
CFQR	Caj	CFVW	Cia	CFZV	Cry	CGHY	Czh	CGMN	Dem	CGRP	Dom
CFQS	Cal	CFVX	Cib	CFZW	Cs	CGHZ	Czi	CGMP	Den	CGRQ	Don
CFQT	Cam			CFZX	Ct	CGJB	Czj	CGMQ	Deo	CGRS	Doo
CFQV	Can			CFZY	Cu	CGJD	Czk	CGMR	Dep	CGRT	Dop

CGRV	Doq	CGWJ	Ead	CHBM	Ecx	CHJP	Egk	CHNR	Ejd	CHST	Elx
CGRW	Dor	CGWK	Eae	CHBN	Ecy	CHJQ	Egl	CHNS	Eje	CHSV	Ely
CGRX	Dos	CGWL	Eaf	CHBP	Ecz	CHJR	Egm	CHNT	Ejf	CHSW	Elz
CGRY	Dot	CGWM	Eag	CHBQ	Ed	CHJS	Egn	CHNV	Ejg	CHSX	Em
CGRZ	Dou	CGWN	Eah	CHBR	Eda	CHJT	Ego	CHNW	Ejh	CHSY	Ema
CGSB	Dov	CGWP	Eai	CHBS	Edb	CHJV	Egp	CHNX	Eji	CHSZ	Emb
CGSD	Dow	CGWQ	Eaj	CHBT	Edc	CHJW	Egq	CHNY	Ejj	CHTB	Emc
CGSF	Dox	CGWR	Eak	CHBV	Edd	CHJX	Egr	CHNZ	Ejk	CHTD	Emd
CGSH	Doy	CGWS	Eal	CHBW	Ede	CHJY	Egs	CHPB	Ejl	CHTF	Eme
CGSJ	Doz	CGWT	Eam	CHBX	Edf	CHJZ	Egt	CHPD	Ejm	CHTG	Emf
CGSK	Dp	CGWV	Ean	CHBY	Edg	CHKB	Egu	CHPF	Ejn	CHTJ	Emg
CGSL	Dq	CGWX	Eao	CHBZ	Edh	CHKD	Egv	CHPG	Ejo	CHTK	Emh
CGSM	Dr	CGWY	Eap	CHDB	Edi	CHKF	Egw	CHPJ	Ejp	CHTL	Emi
CGSN	Dra	CGWZ	Eaq	CHDF	Edj	CHKG	Egx	CHPK	Ejq	CHTM	Emj
CGSP	Dre	CGXB	Ear	CHDG	Edk	CHKJ	Egy	CHPL	Ejr	CHTN	Emk
CGSQ	Dri	CGXD	Eas	CHDJ	Edl	CHKL	Egz	CHPM	Ejs	CHTP	Eml
CGSR	Dro	CGXF	Eat	CHDK	Edm	CHKM	Eh	CHPN	Ejt	CHTQ	Emm
CGST	Dru	CGXH	Eau	CHDL	Edn	CHKN	Eha	CHPQ	Eju	CHTR	Emn
CGSV	Dry	CGXJ	Eav	CHDM	Edo	CHKP	Ehb	CHPR	Ejv	CHTS	Emo
CGSW	Ds	CGXK	Eaw	CHDN	Edp	CHKQ	Ehc	CHPS	Ejw	CHTV	Emp
CGSX	Dt	CGXL	Eax	CHDP	Edq	CHKR	Ehd	CHPT	Ejx	CHTW	Emq
CGSY	Du	CGXM	Eay	CHDQ	Edr	CHKS	Ehe	CHPV	Ejy	CHTX	Emr
CGSZ	Dua	CGXN	Eaz	CHDR	Eds	CHKT	Ehf	CHPW	Ejz	CHTY	Ems
CGTB	Dub	CGXP	Eb	CHDS	Edt	CHKV	Ehg	CHPX	Ek	CHTZ	Emt
CGTD	Duc	CGXQ	Eba	CHDT	Edu	CHKW	Ehh	CHPY	Eka	CHVB	Emu
CGTF	Dud	CGXR	Ebb	CHDV	Edv	CHKX	Ehi	CHPZ	Ekb	CHVD	Emv
CGTH	Due	CGXS	Ebc	CHDW	Edw	CHKY	Ehj	CHQB	Ekc	CHVF	Emw
CGTJ	Duf	CGXT	Ebd	CHDX	Edx	CHKZ	Ehk	CHQD	Ekd	CHVG	Emx
CGTK	Dug	CGXV	Ebe	CHDY	Edy	CHLB	Ehl	CHQF	Eke	CHVK	Emy
CGTL	Duh	CGXW	Ebf	CHDZ	Edz	CHLD	Ehm	CHQG	Ekf	CHVM	Emz
CGTM	Dui	CGXY	Ebg	CHFB	Ee	CHLF	Ehn	CHQJ	Ekg	CHVN	En
CGTN	Duj	CGXZ	Ebh	CHFD	Eea	CHLG	Eho	CHQK	Ekh	CHVP	Ena
CGTP	Duk	CGYB	Ebi	CHFG	Eeb	CHLJ	Ehp	CHQL	Eki	CHVQ	Enb
CGTQ	Dul	CGYD	Ebj	CHFJ	Eec	CHLK	Ehq	CHQM	Ekj	CHVR	Enc
CGTR	Dum	CGYF	Ebk	CHFK	Eed	CHLM	Ehr	CHQN	Ekk	CHVS	End
CGTS	Dun	CGYH	Ebl	CHFL	Eem	CHLN	Ehs	CHQP	Ekl	CHVT	Ene
CGTV	Duo	CGYJ	Ebm	CHFM	Ef	CHLP	Eht	CHQR	Ekm	CHVV	Enf
CGTW	Dup	CGYK	Ebn	CHFN	Efa	CHLQ	Ehu	CHQS	Ekn	CHVW	Eng
CGTX	Duq	CGYL	Ebo	CHFP	Efb	CHLR	Ehv	CHQT	Eko	CHVX	Enh
CGTY	Dur	CGYM	Ebp	CHFQ	Efc	CHLS	Ehw	CHQV	Ekp	CHVY	Eni
CGTZ	Dus	CGYN	Ebq	CHFR	Efd	CHLT	Ehx	CHQW	Ekq	CHVZ	Enj
CGVB	Dut	CGYP	Ebr	CHFS	Efe	CHLV	Ehy	CHQX	Ekr	CHWB	Enk
CGVD	Duu	CGYQ	Ebs	CHFT	Eff	CHLW	Ehz	CHQY	Eks	CHWD	Enl
CGVF	Duv	CGYR	Ebt	CHFV	Efg	CHLX	Ei	CHQZ	Ekt	CHWF	Enm
CGVH	Duw	CGYS	Ebu	CHFW	Efh	CHLY	Eia	CHRB	Eku	CHWG	Enn
CGVJ	Dux	CGYT	Ebv	CHFX	Efi	CHLZ	Eib	CHRD	Ekv	CHWJ	Eno
CGVK	Duy	CGYV	Ebw	CHFY	Efj	CHMB	Eic	CHRF	Ekw	CHWK	Enp
CGVL	Duz	CGYW	Ebx	CHFZ	Efk	CHMD	Eid	CHRG	Ekx	CHWL	Enq
CGVM	Dv	CGYX	Eby	CHGB	Efl	CHMF	Eie	CHRJ	Eky	CHWM	Enr
CGVN	Dw	CGYZ	Ebz	CHGD	Efm	CHMG	Eif	CHRK	Ekz	CHWN	Ens
CGVP	Dx	CGZB	Ec	CHGF	Efn	CHMJ	Eig	CHRL	El	CHWP	Ent
CGVQ	Dy	CGZD	Eca	CHGJ	Efo	CHMK	Eih	CHRM	Ela	CHWQ	Enu
CGVR	Dya	CGZF	Ecb	CHGK	Efp	CHML	Eii	CHRN	Elb	CHWR	Env
CGVS	Dye	CGZH	Ecc	CHGL	Efq	CHMN	Eij	CHRP	Elc	CHWS	Enw
CGVT	Dyi	CGZJ	Ecd	CHGM	Efr	CHMP	Eik	CHRQ	Eld	CHWT	Enx
CGVW	Dyo	CGZK	Ece	CHGN	Efs	CHMQ	Eil	CHRS	Ele	CHWV	Eny
CGVX	Dyu	CGZL	Ecf	CHGP	Eft	CHMR	Eim	CHRT	Elf	CHWW	Enz
CGVY	Dz	CGZM	Ecg	CHGQ	Efu	CHMS	Ein	CHRV	Elg	CHWX	Eo
CGVZ		CGZN	Ech	CHGR	Efv	CHMT	Eio	CHRW	Elh	CHWY	Eoa
		CGZP	Eci	CHGS	Efw	CHMV	Eip	CHRX	Eli	CHWZ	Eob
		CGZQ	Ecj	CHGT	Efx	CHMW	Eiq	CHRY	Elj	CHXB	Eoc
		CGZR	Eck	CHGV	Efy	CHMX	Eir	CHRZ	Elk	CHXD	Eod
		CGZS	Ecl	CHGW	Efz	CHMY	Eis	CHSB	Ell	CHXF	Eoe
		CGZT	Ecm	CHGX	Eg	CHMZ	Eit	CHSD	Elm	CHXG	Eof
		CGZV	Ecn	CHGY	Ega	CHNB	Eiu	CHSF	Eln	CHXJ	Eog
		CGZW	Eco	CHGZ	Egb	CHND	Eiv	CHSG	Elo	CHXK	Eoh
		CGZX	Ecp	CHJB	Egc	CHNF	Eiw	CHSJ	Elp	CHXL	Eoi
		CGZY	Ecq	CHJD	Egd	CHNG	Eix	CHSK	Elq	CHXM	Eoj
		CHBD	Ecr	CHJF	Ege	CHNJ	Eiy	CHSL	Elr	CHXN	Eok
E.		CHBF	Ecs	CHJG	Egf	CHNK	Eiz	CHSM	Els	CHXP	Eol
CGWB	Ea	CHBG	Ect	CHJK	Egg	CHNL	Ej	CHSN	Elt	CHXQ	Eom
CGWD	Eaa	CHBJ	Ecu	CHJL	Egh	CHNM	Eja	CHSP	Elu	CHXR	Eon
CGWF	Eab	CHBK	Ecv	CHJM	Egi	CHNP	Ejb	CHSQ	Elv	CHXS	Eoo
CGWH	Eac	CHBL	Ecw	CHJN	Egj	CHNQ	Ejc	CHSR	Elw	CHXT	Eop

CHXV	Eoq	CJDY	Erj	CJLB	Euc	CJQF	Eww	CJVH	Ezp	CJYT	Fes
CHXW	Eor	CJDZ	Erk	CJLD	Eud	CJQG	Ewx	CJVK	Ezq	CJYV	Fet
CHXY	Eos	CJFB	Erl	CJLF	Eue	CJQH	Ewy	CJVL	Ezr	CJYW	Feu
CHXZ	Eot	CJFD	Erm	CJLG	Euf	CJQK	Ewz	CJVM	Ezs	CJYX	Fev
CHYB	Eou	CJFG	Ern	CJLH	Eug	CJQL	Ex	CJVN	Ezt	CJYZ	Few
CHYD	Eov	CJFH	Ero	CJLK	Euh	CJQM	Exa	CJVP	Ezu	CJZB	Fex
CHYF	Eow	CJFK	Erp	CJLM	Eui	CJQN	Exb	CJVQ	Ezv	CJZD	Fey
CHYG	Eox	CJFL	Erq	CJLN	Euj	CJQP	Exc	CJVR	Ezw	CJZF	Fez
CHYJ	Eoy	CJFM	Err	CJLP	Euk	CJQR	Exd	CJVS	Ezx	CJZG	Ff
CHYK	Eoz	CJFN	Ers	CJLQ	Eul	CJQS	Exe	CJVT	Ezy	CJZH	Fg
CHYL	Ep	CJFP	Ert	CJLR	Eum	CJQT	Exf	CJVW	Ezz	CJZK	Fh
CHYM	Epa	CJFQ	Eru	CJLS	Eun	CJQV	Exg	CJVX		CJZL	Fha
CHYN	Epb	CJFR	Erv	CJLT	Euo	CJQW	Exh	CJVY		CJZM	Fhe
CHYP	Epc	CJFS	Erw	CJLV	Eup	CJQX	Exi	CJVZ		CJZN	Fhi
CHYQ	Epd	CJFT	Erx	CJLW	Euq	CJQY	Exj			CJZP	Fho
CHYR	Epe	CJFV	Ery	CJLX	Eur	CJQZ	Exk			CJZQ	Fhu
CHYS	Epf	CJFW	Erz	CJLY	Eus	CJRB	Exl			CJZR	Fhy
CHYT	Epg	CJFX	Es	CJLZ	Eut	CJRD	Exm			CJZS	Fi
CHYV	Eph	CJFY	Esa	CJMB	Euu	CJRF	Exn			CJZT	Fia
CHYW	Epi	CJFZ	Esb	CJMD	Euv	CJRG	Exo			CJZV	Fib
CHYX	Epj	CJGB	Esc	CJMF	Euw	CJRH	Exp			CJZW	Fic
CHYZ	Epk	CJGD	Esd	CJMG	Eux	CJRK	Exq			CJZX	Fid
CHZB	Epl	CJGF	Ese	CJMH	Euy	CJRL	Exr			CKBD	Fie
CHZD	Epm	CJGH	Esf	CJMK	Euz	CJRM	Exs	F		CKBG	Fif
CHZF	Epn	CJGK	Esg	CJML	Ev	CJRN	Ext	CJWB	Fa	CKBH	Fig
CHZG	Epo	CJGL	Esh	CJMN	Eva	CJRP	Exu	CJWD	Faa	CKBJ	Fih
CHZJ	Epp	CJGM	Esi	CJMP	Evb	CJRQ	Exv	CJWF	Fab	CKBL	Fii
CHZK	Epq	CJGN	Esj	CJMQ	Evc	CJRS	Exw	CJWG	Fac	CKBM	Fij
CHZL	Epr	CJGP	Esk	CJMR	Evd	CJRT	Exx	CJWH	Fad	CKBN	Fik
CHZM	Eps	CJGQ	Esl	CJMS	Eve	CJRV	Exy	CJWK	Fae	CKBP	Fil
CHZN	Ept	CJGR	Esm	CJMT	Evf	CJRW	Exz	CJWL	Faf	CKBQ	Fim
CHZP	Epu	CJGS	Esn	CJMV	Evg	CJRX	Ey	CJWM	Fag	CKBR	Fin
CHZQ	Epv	CJGT	Eso	CJMW	Evh	CJRY	Eya	CJWN	Fah	CKBS	Fio
CHZR	Epw	CJGV	Esp	CJMX	Evi	CJRZ.	Eyb	CJWP	Fai	CKBT	Fip
CHZS	Epx	CJGW	Esq	CJMY	Evj	CJSB	Eyc	CJWQ	Faj	CKBV	Fiq
CHZT	Epy	CJGX	Esr	CJMZ	Evk	CJSD	Eyd	CJWR	Fak	CKBW	Fir
CHZV	Epz	CJGY	Ess	CJNB	Evl	CJSF	Eye	CJWS	Fal	CKBX	Fis
CHZW	Eq	CJGZ	Est	CJND	Evm	CJSH	Eyf	CJWT	Fam	CKBY	Fit
CHZX	Eqa	CJHB	Esu	CJNF	Evn	CJSK	Eyg	CJWV	Fan	CKBZ	Fiu
CHZY	Eqb	CJHD	Esv	CJNG	Evo	CJSL	Eyh	CJWX	Fao	CKDB	Fiv
CJBD	Eqc	CJHF	Esw	CJNH	Evp	CJSN	Eyi	CJWY	Fap	CKDF	Fiw
CJBF	Eqd	CJHG	Esx	CJNK	Evq	CJSP	Eyj	CJWZ	Faq	CKDG	Fix
CJBG	Eqe	CJHK	Esy	CJNL	Evr	CJSQ	Eym	CJXB	Far	CKDH	Fiy
CJBH	Eqf	CJHL	Esz	CJNM	Evs	CJSR	Eyn	CJXD	Fas	CKDJ	Fiz
CJBK	Eqg	CJHM	Et	CJNP	Evt	CJST	Eyo	CJXF	Fat	CKDL	Fj
CJBL	Eqh	CJHN	Eta	CJNQ	Evu	CJSV	Eyp	CJXG	Fau	CKDM	Fk
CJBM	Eqi	CJHP	Etb	CJNR	Evv	CJSW	Eyq	CJXH	Fav	CKDN	Fla
CJBN	Eqj	CJHQ	Etc	CJNS	Evw	CJSX	Eyr	CJXK	Faw	CKDP	Fle
CJBP	Eqk	CJHR	Etd	CJNT	Evx	CJSY	Eys	CJXL	Fax	CKDQ	Fli
CJBQ	Eql	CJHS	Ete	CJNV	Evy	CJSZ	Eyt	CJXM	Fay	CKDR	Flo
CJBR	Eqm	CJHT	Etf	CJNW	Evz	CJTB	Eyu	CJXN	Faz	CKDS	Flu
CJBS	Eqn	CJHV	Etg	CJNX	Ew	CJTD	Eyv	CJXP	Fb	CKDT	Fly
CJBT	Eqo	CJHW	Eth	CJNY	Ewa	CJTF	Eyw	CJXQ	Fc	CKDV	Fm
CJBV	Eqp	CJHX	Eti	CJNZ	Ewb	CJTG	Eyx	CJXR	Fd	CKDW	Fn
CJBW	Eqq	CJHY	Etj	CJPB	Ewc	CJTH	Eyy	CJXS	Fe	CKDX	Fo
CJBX	Eqr	CJHZ	Etk	CJPD	Ewd	CJTK	Eyz	CJXT	Fea	CKDY	Foa
CJBY	Eqs	CJKB	Etl	CJPF	Ewe	CJTL	Ez	CJXV	Feb	CKDZ	Fob
CJBZ	Eqt	CJKD	Etm	CJPG	Ewf	CJTM	Eza	CJXW	Fec	CKFB	Foc
CJDB	Equ	CJKF	Etn	CJPH	Ewg	CJTN	Ezb	CJXY	Fed	CKFD	Fod
CJDF	Eqv	CJKG	Eto	CJPK	Ewh	CJTP	Ezc	CJXZ	Fee	CKFG	Foe
CJDG	Eqw	CJKH	Etp	CJPL	Ewi	CJTQ	Ezd	CJYB	Fef	CKFH	Fof
CJDH	Eqx	CJKL	Etq	CJPM	Ewj	CJTR	Eze	CJYD	Feg	CKFJ	Fog
CJDK	Eqy	CJKM	Etr	CJPN	Ewk	CJTS	Ezf	CJYF	Feh	CKFL	Foh
CJDL	Eqz	CJKN	Ets	CJPQ	Ewl	CJTV	Ezg	CJYG	Fei	CKFM	Foi
CJDM	Er	CJKP	Ett	CJPR	Ewm	CJTW	Ezh	CJYH	Fej	CKFN	Foj
CJDN	Era	CJKQ	Etu	CJPS	Ewn	CJTX	Ezi	CJYK	Fek	CKFP	Fok
CJDP	Erb	CJKR	Etv	CJPT	Ewo	CJTY	Ezj	CJYL	Fel	CKFQ	Fol
CJDQ	Erc	CJKS	Etw	CJPV	Ewp	CJTZ	Ezk	CJYM	Fem	CKFR	Fom
CJDR	Erd	CJKT	Etx	CJPW	Ewq	CJVB	Ezl	CJYN	Fen	CKFS	Fon
CJDS	Ere	CJKV	Ety	CJPX	Ewr	CJVD	Ezm	CJYP	Feo	CKFT	Foo
CJDT	Erf	CJKW	Etz	CJPY	Ews	CJVF	Ezn	CJYQ	Fep	CKFV	Fop
CJDV	Erg	CJKX	Eua	CJPZ	Ewt	CJVG	Ezo	CJYR	Feq	CKFW	Foq
CJDW	Erh	CJKY	Eua	CJQB	Ewu			CJYS	Fei	CKFX	For
CJDX	Eri	CJKZ	Eub	CJQD	Ewv					CKFY	Fos

CKFZ	Fot	CKLP	Gah	CKQR	Gip
CKGB	Fou	CKLQ	Gai	CKQS	Giq
CKGD	Fov	CKLR	Gaj	CKQT	Gir
CKGF	Fow	CKLS	Gak	CKQV	Gis
CKGH	Fox	CKLT	Gal	CKQW	Git
CKGJ	Foy	CKLV	Gam	CKQX	Giu
CKGL	Foz	CKLW	Gan	CKQY	Giv
CKGM	Fp	CKLX	Gao	CKQZ	Giw
CKGN	Fq	CKLY	Gap	CKRB	Gix
CKGP	Fr	CKLZ	Gaq	CKRD	Giy
CKGQ	Fra	CKMB	Gar	CKRF	Giz
CKGR	Fre	CKMD	Gas	CKRG	Gj
CKGS	Fri	CKMF	Gat	CKRH	Gk
CKGT	Fro	CKMG	Gau	CKRJ	Gl
CKGV	Fru	CKMH	Gav	CKRL	Gla
CKGW	Fry	CKMJ	Gaw	CKRM	Gle
CKGX	Fs	CKML	Gax	CKRN	Gli
CKGY	Ft	CKMN	Gay	CKRP	Glo
CKGZ	Fu	CKMP	Gaz	CKRQ	Glu
CKHB	Fua	CKMQ	Gb	CKRS	Gly
CKHD	Fub	CKMR	Gc	CKRT	Gm
CKHF	Fuc	CKMS	Gd	CKRV	Gn
CKHG	Fud	CKMT	Ge	CKRW	Go
CKHJ	Fue	CKMV	Gea	CKRX	Goa
CKHL	Fuf	CKMW	Geb	CKRY	Gob
CKHM	Fug	CKMX	Gec	CKRZ	Goc
CKHN	Fuh	CKMY	Ged	CKSB	God
CKHP	Fui	CKMZ	Gee	CKSD	Goe
CKHQ	Fuj	CKNB	Gef	CKSF	Gof
CKHR	Fuk	CKND	Geg	CKSG	Gog
CKHS	Ful	CKNF	Geh	CKSH	Goh
CKHT	Fum	CKNG	Gei	CKSJ	Goi
CKHV	Fun	CKNH	Gej	CKSL	Goj
CKHW	Fuo	CKNJ	Gek	CKSM	Gok
CKHX	Fup	CKNL	Gel	CKSN	Gol
CKHY	Fuq	CKNM	Gem	CKSP	Gom
CKHZ	Fur	CKNP	Gen	CKSQ	Gon
CKJB	Fus	CKNQ	Geo	CKSR	Goo
CKJD	Fut	CKNR	Gep	CKST	Gop
CKJF	Fuu	CKNS	Geq	CKSV	Goq
CKJG	Fuv	CKNT	Ger	CKSW	Gor
CKJH	Fuw	CKNV	Ges	CKSX	Gos
CKJL	Fux	CKNW	Get	CKSY	Got
CKJM	Fuy	CKNX	Geu	CKSZ	Gou
CKJN	Fuz	CKNY	Gev	CKTB	Gov
CKJP	Fv	CKNZ	Gew	CKTD	Gow
CKJQ	Fw	CKPB	Gex	CKTF	Gox
CKJR	Fx	CKPD	Gey	CKTG	Goy
CKJS	Fy	CKPF	Gez	CKTH	Goz
CKJT	Fya	CKPG	Gf	CKTJ	Gp
CKJV	Fye	CKPH	Gg	CKTL	Gq
CKJW	Fyi	CKPJ	Gh	CKTM	Gr
CKJX	Fyo	CKPL	Gha	CKTN	Gra
CKJY	Fyu	CKPM	Ghe	CKTP	Gre
CKJZ	Fz	CKPN	Ghi	CKTQ	Gri
		CKPQ	Gho	CKTR	Gro
		CKPR	Ghu	CKTS	Gru
		CKPS	Ghy	CKTV	Gry
		CKPT	Gi	CKTW	Gs
		CKPV	Gia	CKTX	Gt
		CKPW	Gib	CKTY	Gu
		CKPX	Gic	CKTZ	Gua
		CKPY	Gid	CKVB	Gub
G		CKPZ	Gie	CKVD	Guc
		CKQB	Gif	CKVF	Gud
CKLB	Ga	CKQD	Gig	CKVG	Gue
CKLD	Gaa	CKQF	Gih	CKVH	Guf
CKLF	Gab	CKQG	Gii	CKVJ	Gug
CKLG	Gac	CKQH	Gij	CKVL	Guh
CKLH	Gad	CKQJ	Gik	CKVM	Gui
CKLJ	Gae	CKQL	Gil	CKVN	Guj
CKLM	Gaf	CKQM	Gim	CKVP	Guk
CKLN	Gag	CKQN	Gin	CKVQ	Gul
		CKQP	Gio	CKVR	Gum

CKVS	Gun	CKZG	Hei	CLGK	Hoo
CKVT	Guo	CKZH	Hej	CLGM	Hop
CKVW	Gup	CKZJ	Hek	CLGN	Hoq
CKVX	Guq	CKZL	Hel	CLGP	Hor
CKVY	Gur	CKZM	Hem	CLGQ	Hos
CKVZ	Gus	CKZN	Hen	CLGR	Hot
CKWB	Gut	CKZP	Heo	CLGS	Hou
CKWD	Guu	CKZQ	Hep	CLGT	Hov
CKWF	Guv	CKZR	Heq	CLGV	How
CKWG	Guw	CKZS	Her	CLGW	Hox
CKWH	Gux	CKZT	Hes	CLGX	Hoy
CKWJ	Guy	CKZV	Het	CLGY	Hoz
CKWL	Guz	CKZW	Heu	CLGZ	Hp
CKWM	Gv	CKZX	Hev	CLHB	Hq
CKWN	Gw	CKZY	Hew	CLHD	Hr
CKWP	Gx	CLBD	Hex	CLHF	Hra
CKWQ	Gy	CLBF	Hey	CLHG	Hre
CKWR	Gya	CLBG	Hez	CLHJ	Hri
CKWS	Gye	CLBH	Hf	CLHK	Hro
CKWT	Gyi	CLBJ	Hg	CLHM	Hru
CKWV	Gyo	CLBK	Hh	CLHN	Hry
CKWX	Gyu	CLBM	Hi	CLHP	Hs
CKWY	Gz	CLBN	Hia	CLHQ	Ht
CKWZ		CLBP	Hib	CLHR	Hu
		CLBQ	Hic	CLHS	Hua
		CLBR	Hid	CLHT	Hub
		CLBS	Hie	CLHV	Huc
		CLBT	Hif	CLHW	Hud
		CLBV	Hig	CLHX	Hue
		CLBW	Hih	CLHY	Huf
		CLBX	Hii	CLHZ	Hug
		CLBY	Hij	CLJB	Huh
		CLBZ	Hik	CLJD	Hui
H.		CLDB	Hil	CLJF	Huj
		CLDF	Him	CLJG	Huk
CKXB	Ha	CLDG	Hin	CLJH	Hul
CKXD	Haa	CLDH	Hio	CLJK	Hum
CKXF	Hab	CLDJ	Hip	CLJM	Hun
CKXG	Hac	CLDK	Hiq	CLJN	Huo
CKXH	Had	CLDM	Hir	CLJP	Hup
CKXJ	Hae	CLDN	His	CLJQ	Huq
CKXL	Haf	CLDP	Hit	CLJR	Hur
CKXM	Hag	CLDQ	Hiu	CLJS	Hus
CKXN	Hah	CLDR	Hiv	CLJT	Hut
CKXP	Hai	CLDS	Hiw	CLJV	Huu
CKXQ	Haj	CLDT	Hix	CLJW	Huv
CKXR	Hak	CLDV	Hiy	CLJX	Huw
CKXS	Hal	CLDW	Hiz	CLJY	Hux
CKXT	Ham	CLDX	Hj	CLJZ	Huy
CKXV	Han	CLDY	Hk	CLKB	Huz
CKXW	Hao	CLDZ	Hl	CLKD	Hv
CKXY	Hap	CLFB	Hla	CLKF	Hw
CKXZ	Haq	CLFD	Hle	CLKG	Hx
CKYB	Har	CLFG	Hli	CLKH	Hy
CKYD	Has	CLFH	Hlo	CLKJ	Hya
CKYF	Hat	CLFJ	Hlu	CLKM	Hye
CKYG	Hau	CLFK	Hly	CLKN	Hyi
CKYH	Hav	CLFM	Hm	CLKP	Hyo
CKYJ	Haw	CLFN	Hn	CLKQ	Hyu
CKYL	Hax	CLFP	Ho	CLKR	Hyy
CKYM	Hay	CLFQ	Hoa	CLKS	Hz
CKYN	Haz	CLFR	Hob	CLKT	
CKYP	Hb	CLFS	Hoc	CLKV	
CKYQ	Hc	CLFT	Hod	CLKW	
CKYR	Hd	CLFV	Hoe	CLKX	
CKYS	He	CLFW	Hof	CLKY	
CKYT	Hea	CLFX	Hog	CLKZ	
CKYV	Heb	CLFY	Hoh		
CKYW	Hec	CLFZ	Hoi		
CKYX	Hed	CLGB	Hoj		
CKYZ	Hee	CLGD	Hok		
CKZB	Hef	CLGF	Hol		
CKZD	Heg	CLGH	Hom		
CKZF	Heh	CLGJ	Hon		

	L.										
CLMB	Ia	CLRB	Iil	CLWF	Ile	CMBJ	Iny	CMHL	Iqr	CMNP	Itk
CLMD	Iaa	CLRD	Iim	CLWG	Ilf	CMBK	Inz	CMHN	Iqs	CMNQ	Itl
CLMF	Iab	CLRF	Iin	CLWH	Ilg	CMBL	Io	CMHP	Iqt	CMNR	Itm
CLMG	Iac	CLRG	Iio	CLWJ	Ilh	CMBN	Ioa	CMHQ	Iqu	CMNS	Itn
CLMH	Iad	CLRH	Iip	CLWK	Ili	CMBP	Iob	CMHR	Iqv	CMNT	Ito
CLMJ	Iae	CLRJ	Iiq	CLWM	Ilj	CMBQ	Ioc	CMHS	Iqw	CMNV	Itp
CLMK	Iaf	CLRK	Iir	CLWN	Ilk	CMBR	Iod	CMHT	Iqx	CMNW	Itq
CLMN	Iag	CLRM	Iis	CLWP	Ill	CMBS	Ioe	CMHV	Iqy	CMNX	Itr
CLMP	Iah	CLRN	Iit	CLWQ	Ilm	CMBT	Iof	CMHW	Iqz	CMNY	Its
CLMQ	Iai	CLRP	Iiu	CLWR	Iln	CMBV	Iog	CMHX	Ir	CMNZ	Itt
CLMR	Iaj	CLRQ	Iiv	CLWS	Ilo	CMBW	Ioh	CMHY	Ira	CMPB	Itu
CLMS	Iak	CLRS	Iiw	CLWT	Ilp	CMBX	Ioi	CMHZ	Irb	CMPD	Itv
CLMT	Ial	CLRT	Iix	CLWV	Ilq	CMBY	Ioj	CMJB	Irc	CMPF	Itw
CLMV	Iam	CLRV	Iiy	CLWX	Ilr	CMBZ	Iok	CMJD	Ird	CMPG	Itx
CLMW	Ian	CLRW	Iiz	CLWY	Ils	CMDB	Iol	CMJF	Ire	CMPH	Ity
CLMX	Iao	CLRX	Ij	CLWZ	Ilt	CMDF	Iom	CMJG	Irf	CMPJ	Itz
CLMY	Iap	CLRY	Ija	CLXB	Ilu	CMDG	Ion	CMJH	Irg	CMPK	Iu
CLMZ	Iaq	CLRZ	Ijb	CLXD	Ilv	CMDH	Ioo	CMJK	Irh	CMPL	Iua
CLNB	Iar	CLSB	Ijc	CLXF	Ilw	CMDJ	Iop	CMJL	Iri	CMPN	Iub
CLND	Ias	CLSD	Ijd	CLXG	Ilx	CMDK	Ioq	CMJN	Irj	CMPQ	Iuc
CLNF	Iat	CLSF	Ije	CLXH	Ily	CMDL	Ior	CMJP	Irk	CMPR	Iud
CLNG	Iau	CLSG	Ijf	CLXJ	Ilz	CMDN	Ios	CMJQ	Irl	CMPS	Iue
CLNH	Iav	CLSH	Ijg	CLXK	Im	CMDP	Iot	CMJR	Irm	CMPT	Iuf
CLNJ	Iaw	CLSJ	Ijh	CLXM	Ima	CMDQ	Iou	CMJS	Irn	CMPV	Iug
CLNK	Iax	CLSK	Iji	CLXN	Imb	CMDR	Iov	CMJT	Iro	CMPW	Iuh
CLNM	Iay	CLSM	Ijj	CLXP	Imc	CMDS	Iow	CMJV	Irp	CMPX	Iui
CLNP	Iaz	CLSN	Ijk	CLXQ	Imd	CMDT	Iox	CMJW	Irq	CMPY	Iuj
CLNQ	Ib	CLSP	Ijl	CLXR	Ime	CMDV	Ioy	CMJX	Irr	CMPZ	Iuk
CLNR	Ic	CLSQ	Ijm	CLXS	Imf	CMDW	Ioz	CMJY	Irs	CMQB	Iul
CLNS	Id	CLSR	Ijn	CLXT	Img	CMDX	Ip	CMJZ	Irt	CMQD	Ium
CLNT	Ie	CLST	Ijo	CLXV	Imh	CMDY	Ipa	CMKB	Iru	CMQF	Iun
CLNV	Iea	CLSV	Ijp	CLXW	Imi	CMDZ	Ipb	CMKD	Irv	CMQG	Iuo
CLNW	Ieb	CLSW	Ijq	CLXY	Imj	CMFB	Ipc	CMKF	Irw	CMQH	Iup
CLNX	Iec	CLSX	Ijr	CLXZ	Imk	CMFD	Ipd	CMKG	Irx	CMQJ	Iuq
CLNY	Ied	CLSY	Ijs	CLYB	Iml	CMFG	Ipe	CMKH	Iry	CMQK	Iur
CLNZ	Iee	CLSZ	Ijt	CLYD	Imm	CMFH	Ipf	CMKJ	Irz	CMQL	Ius
CLPB	Ief	CLTB	Iju	CLYF	Imn	CMFJ	Ipg	CMKL	Is	CMQN	Iut
CLPD	Ieg	CLTD	Ijv	CLYG	Imo	CMFK	Iph	CMKN	Isa	CMQP	Iuu
CLPF	Ieh	CLTF	Ijw	CLYH	Imp	CMFL	Ipi	CMKP	Isb	CMQR	Iuv
CLPG	Iei	CLTG	Ijx	CLYJ	Imq	CMFN	Ipj	CMKQ	Isc	CMQS	Iuw
CLPH	Iej	CLTH	Ijy	CLYK	Imr	CMFP	Ipk	CMKR	Isd	CMQT	Iux
CLPJ	Iek	CLTJ	Ijz	CLYM	Ims	CMFQ	Ipl	CMKS	Ise	CMQV	Iuy
CLPK	Iel	CLTK	Ik	CLYN	Imt	CMFR	Ipm	CMKT	Isf	CMQW	Iuz
CLPM	Iem	CLTM	Ika	CLYP	Imu	CMFS	Ipn	CMKV	Isg	CMQX	Iv
CLPN	Ien	CLTN	Ikb	CLYQ	Imv	CMFT	Ipo	CMKW	Ish	CMQY	Iva
CLPQ	Ieo	CLTP	Ikc	CLYR	Imw	CMFV	Ipp	CMKX	Isi	CMQZ	Ivb
CLPR	Iep	CLTQ	Ikd	CLYS	Imx	CMFW	Ipq	CMKY	Isj	CMRB	Ivc
CLPS	Ieq	CLTR	Ike	CLYT	Imy	CMFX	Ipr	CMKZ	Isk	CMRD	Ivd
CLPT	Ier	CLTS	Ikf	CLYV	Imz	CMFY	Ips	CMLB	Isl	CMRF	Ive
CLPV	Ies	CLTV	Ikg	CLYW	In	CMFZ	Ipt	CMLD	Ism	CMRG	Ivf
CLPW	Iet	CLTW	Ikh	CLYX	Ina	CMGB	Ipu	CMLF	Isn	CMRH	Ivg
CLPX	Ieu	CLTX	Iki	CLYZ	Inb	CMGD	Ipv	CMLG	Iso	CMRJ	Ivh
CLPY	Iev	CLTY	Ikj	CLZB	Inc	CMGF	Ipw	CMLH	Isp	CMRK	Ivi
CLPZ	Iew	CLTZ	Ikk	CLZD	Ind	CMGH	Ipx	CMLJ	Isq	CMRL	Ivj
CLQB	Iex	CLVB	Ikl	CLZF	Ine	CMGJ	Ipy	CMLK	Isr	CMRN	Ivk
CLQD	Iey	CLVD	Ikm	CLZG	Inf	CMGK	Ipz	CMLN	Iss	CMRP	Ivl
CLQF	Iez	CLVF	Ikn	CLZH	Ing	CMGN	Iqa	CMLP	Ist	CMRQ	Ivm
CLQG	If	CLVG	Iko	CLZJ	Inh	CMGP	Iqb	CMLQ	Isu	CMRS	Ivn
CLQH	Ig	CLVH	Ikp	CLZK	Ini	CMGQ	Iqc	CMLR	Isv	CMRT	Ivo
CLQJ	Ih	CLVJ	Ikq	CLZM	Inj	CMGR	Iqd	CMLS	Isw	CMRV	Ivp
CLQK	Ii	CLVK	Ikr	CLZN	Ink	CMGS	Iqe	CMLT	Isx	CMRW	Ivq
CLQM	Iia	CLVM	Iks	CLZP	Inl	CMGT	Iqf	CMLV	Isy	CMRX	Ivr
CLQN	Iib	CLVN	Ikt	CLZQ	Inm	CMGV	Iqg	CMLW	Isz	CMRY	Ivs
CLQP	Iic	CLVP	Iku	CLZR	Inn	CMGW	Iqh	CMLX	It	CMRZ	Ivt
CLQR	Iid	CLVQ	Ikv	CLZS	Ino	CMGX	Iqi	CMLY	Ita	CMSB	Ivu
CLQS	Iie	CLVR	Ikw	CLZT	Inp	CMGY	Iqj	CMLZ	Itb	CMSD	Ivv
CLQT	Iif	CLVS	Ikx	CLZV	Inq	CMGZ	Iqk	CMNB	Itc	CMSF	Ivw
CLQV	Iig	CLVT	Iky	CLZW	Inr	CMHB	Iql	CMND	Itd	CMSG	Ivx
CLQW	Iih	CLVW	Ikz	CLZX	Ins	CMHD	Iqm	CMNF	Ite	CMSH	Ivy
CLQX	Iii	CLVX	Il	CLZY	Int	CMHF	Iqn	CMNG	Itf	CMSJ	Ivz
CLQY	Iij	CLVY	Ila	CMBD	Inu	CMHG	Iqo	CMNH	Itg	CMSK	Iw
CLQZ	Iik	CLVZ	Ilb	CMBF	Inv	CMHJ	Iqp	CMNJ	Ith	CMSL	Iwa
		CLWB	Ilc	CMBG	Inw	CMHK	Iqq	CMNK	Iti	CMSN	Iwb
		CLWD	Ild	CMBH	Inx			CMNL	Itj	CMSP	Iwc

				J.						K.		
CMSQ	Iwd	CMXS	Iyx	CNDH	Jau	CNJK	Jl	CNPL	Jv	CNSX	Keu
CMSR	Iwe	CMXT	Iyy	CNDJ	Jav	CNJL	Jla	CNPM	Jw	CNSY	Kev
CMST	Iwf	CMXV	Iyz	CNDK	Jaw	CNJM	Jle	CNPQ	Jx	CNSZ	Kew
CMSV	Iwg	CMXW	Iz	CNDL	Jax	CNJP	Jli	CNPR	Jy	CNTB	Kex
CMSW	Iwh	CMXY	Iza	CNDM	Jay	CNJQ	Jlo	CNPS	Jya	CNTD	Key
CMSX	Iwi	CMXZ	Izb	CNDP	Jaz	CNJR	Jlu	CNPT	Jye	CNTF	Kez
CMSY	Iwj	CMYB	Izc	CNDQ	Jb	CNJS	Jly	CNPV	Jyi	CNTG	Kf
CMSZ	Iwk	CMYD	Izd	CNDR	Jc	CNJT	Jm	CNPW	Jyo	CNTH	Kg
CMTB	Iwl	CMYF	Ize	CNDS	Jd	CNJV	Jn	CNPX	Jyu	CNTJ	Kh
CMTD	Iwm	CMYG	Izf	CNDT	Je	CNJW	Jo	CNPY	Jyy	CNTK	Kha
CMTF	Iwn	CMYH	Izg	CNDV	Jea	CNJX	Joa	CNPZ	Jz	CNTL	Khe
CMTG	Iwo	CMYJ	Izh	CNDW	Jeb	CNJY	Job			CNTM	Khi
CMTH	Iwp	CMYK	Izi	CNDX	Jec	CNJZ	Joc			CNTP	Kho
CMTJ	Iwq	CMYL	Izj	CNDY	Jed	CNKB	Jod			CNTQ	Khu
CMTK	Iwr	CMYN	Izk	CNDZ	Jee	CNKD	Joe			CNTR	Khy
CMTL	Iws	CMYP	Izl	CNFB	Jef	CNKF	Jof			CNTS	Ki
CMTN	Iwt	CMYQ	Izm	CNFD	Jeg	CNKG	Jog			CNTV	Kia
CMTP	Iwu	CMYR	Izn	CNFG	Jeh	CNKH	Joh			CNTW	Kib
CMTQ	Iwv	CMYS	Izo	CNFH	Jei	CNKJ	Joi			CNTX	Kic
CMTR	Iww	CMYT	Izp	CNFJ	Jej	CNKL	Joj			CNTY	Kid
CMTS	Iwx	CMYV	Izq	CNFK	Jek	CNKM	Jok			CNTZ	Kie
CMTV	Iwy	CMYW	Izr	CNFL	Jel	CNKP	Jol			CNVB	Kif
CMTW	Iwz	CMYX	Izs	CNFM	Jem	CNKQ	Jom			CNVD	Kig
CMTX	Ix	CMYZ	Izt	CNFP	Jen	CNKR	Jon	CNQB	Ka	CNVF	Kih
CMTY	Ixa	CMZB	Izu	CNFQ	Jeo	CNKS	Joo	CNQD	Kaa	CNVG	Kii
CMTZ	Ixb	CMZD	Izv	CNFR	Jep	CNKT	Jop	CNQF	Kab	CNVH	Kij
CMVB	Ixc	CMZF	Izw	CNFS	Jeq	CNKV	Joq	CNQG	Kac	CNVJ	Kik
CMVD	Ixd	CMZG	Izx	CNFT	Jer	CNKW	Jor	CNQH	Kad	CNVK	Kil
CMVF	Ixe	CMZH	Izy	CNFV	Jes	CNKX	Jos	CNQJ	Kae	CNVL	Kim
CMVG	Ixf	CMZJ	Izz	CNFW	Jet	CNKY	Jot	CNQK	Kaf	CNVM	Kin
CMVH	Ixg	CMZK		CNFX	Jeu	CNKZ	Jou	CNQL	Kag	CNVP	Kio
CMVJ	Ixh	CMZL		CNFY	Jev	CNLB	Jov	CNQM	Kah	CNVQ	Kip
CMVK	Ixi	CMZN		CNFZ	Jew	CNLD	Jow	CNQP	Kai	CNVR	Kiq
CMVL	Ixj	CMZP		CNGB	Jex	CNLF	Jox	CNQR	Kaj	CNVS	Kir
CMVN	Ixk	CMZQ		CNGD	Jey	CNLG	Joy	CNQS	Kak	CNVT	Kis
CMVP	Ixl	CMZR		CNGF	Jez	CNLH	Joz	CNQT	Kal	CNVW	Kit
CMVQ	Ixm	CMZS		CNGH	Jf	CNLJ	Jp	CNQV	Kam	CNVX	Kiu
CMVR	Ixn	CMZT		CNGJ	Jg	CNLK	Jq	CNQW	Kan	CNVY	Kiv
CMVS	Ixo	CMZV		CNGK	Jh	CNLM	Jr	CNQX	Kao	CNVZ	Kiw
CMVT	Ixp	CMZW		CNGL	Jha	CNLP	Jra	CNQY	Kap	CNWB	Kix
CMVW	Ixq	CMZX		CNGM	Jhe	CNLQ	Jre	CNQZ	Kaq	CNWD	Kiy
CMVX	Ixr	CMZY		CNGP	Jhi	CNLR	Jri	CNRB	Kar	CNWF	Kiz
CMVY	Ixs			CNGQ	Jho	CNLS	Jro	CNRD	Kas	CNWG	Kj
CMVZ	Ixt			CNGR	Jhu	CNLT	Jru	CNRF	Kat	CNWH	Kk
CMWB	Ixu			CNGS	Jhy	CNLV	Jry	CNRG	Kau	CNWJ	Kl
CMWD	Ixv			CNGT	Ji	CNLW	Js	CNRH	Kav	CNWK	Kla
CMWF	Ixw			CNGV	Jia	CNLX	Jt	CNRJ	Kaw	CNWL	Kle
CMWG	Ixx			CNGW	Jib	CNLY	Ju	CNRK	Kax	CNWM	Kli
CMWH	Ixy			CNGX	Jic	CNLZ	Jua	CNRL	Kay	CNWP	Klo
CMWJ	Ixz			CNGY	Jid	CNMB	Jub	CNRM	Kaz	CNWQ	Klu
CMWK	Iy			CNGZ	Jie	CNMD	Juc	CNRP	Kb	CNWR	Kly
CMWL	Iya	J.		CNHB	Jif	CNMF	Jud	CNRQ	Kc	CNWS	Km
CMWN	Iyb			CNHD	Jig	CNMG	Jue	CNRS	Kd	CNWT	Kn
CMWP	Iyc	CNBD	Ja	CNHF	Jih	CNMH	Juf	CNRT	Ke	CNWV	Ko
CMWQ	Iyd	CNBF	Jaa	CNHG	Jii	CNMJ	Jug	CNRV	Kea	CNWX	Koa
CMWR	Iye	CNBG	Jab	CNHJ	Jij	CNMK	Juh	CNRW	Keb	CNWY	Kob
CMWS	Iyf	CNBH	Jac	CNHK	Jik	CNML	Jui	CNRX	Kec	CNWZ	Koc
CMWT	Iyg	CNBJ	Jad	CNHL	Jil	CNMP	Juj	CNRY	Ked	CNXB	Kod
CMWV	Iyh	CNBK	Jae	CNHM	Jim	CNMQ	Juk	CNRZ	Kee	CNXD	Koe
CMWX	Iyi	CNBL	Jaf	CNHP	Jin	CNMR	Jul	CNSB	Kef	CNXF	Kof
CMWY	Iyj	CNBM	Jag	CNHQ	Jio	CNMS	Jum	CNSD	Keg	CNXG	Kog
CMWZ	Iyk	CNBP	Jah	CNHR	Jip	CNMT	Jun	CNSF	Keh	CNXH	Koh
CMXB	Iyl	CNBQ	Jai	CNHS	Jiq	CNMV	Juo	CNSG	Kei	CNXJ	Koi
CMXD	Iym	CNBR	Jaj	CNHT	Jir	CNMW	Jup	CNSJ	Kej	CNXK	Koj
CMXF	Iyn	CNBS	Jak	CNHV	Jis	CNMX	Juq	CNSK	Kek	CNXL	Kok
CMXG	Iyo	CNBT	Jal	CNHW	Jit	CNMY	Jur	CNSL	Kel	CNXM	Kol
CMXH	Iyp	CNBV	Jam	CNHX	Jiu	CNMZ	Jus	CNSM	Kem	CNXP	Kom
CMXJ	Iyq	CNBW	Jan	CNHY	Jiv	CNPB	Jut	CNSP	Ken	CNXQ	Kon
CMXK	Iyr	CNBX	Jao	CNHZ	Jiw	CNPD	Juu	CNSQ	Keo	CNXR	Koo
CMXL	Iys	CNBY	Jap	CNJB	Jix	CNPF	Juv	CNSR	Kep	CNXS	Kop
CMXN	Iyt	CNBZ	Jaq	CNJD	Jiy	CNPG	Juw	CNST	Keq	CNXT	Koq
CMXP	Iyu	CNDB	Jar	CNJF	Jiz	CNPH	Jux	CNSV	Ker	CNXV	Kor
CMXQ	Iyv	CNDF	Jas	CNJG	Jj	CNPJ	Juy	CNSW	Kes	CNXW	Kos
CMXR	Iyw	CNDG	Jat	CNJH	Jk	CNPK	Juz	CNSW	Ket	CNXY	Kot

CNXQ	Kou	CPDN	Lah	CPJR	Liv	CPNT	Lut	CPTG	Mei	CPYJ	Moo
CNYB	Kov	CPDQ	Lai	CPJS	Liw	CPNV	Luu	CPTH	Mej	CPYK	Mop
CNYD	Kow	CPDR	Laj	CPJT	Lix	CPNW	Luv	CPTJ	Mek	CPYL	Moq
CNYF	Kox	CPDS	Lak	CPJV	Liy	CPNX	Luw	CPTK	Mel	CPYM	Mor
CNYG	Koy	CPDT	Lal	CPJW	Liz	CPNY	Lux	CPTL	Mem	CPYN	Mos
CNYH	Koz	CPDV	Lam	CPJX	Lj	CPNZ	Luy	CPTM	Men	CPYQ	Mot
CNYJ	Kp	CPDW	Lan	CPJY	Lk	CPQB	Luz	CPTN	Meo	CPYR	Mou
CNYK	Kq	CPDX	Lao	CPJZ	Ll	CPQD	Lv	CPTQ	Mep	CPYS	Mov
CNYL	Kr	CPDY	Lap	CPKB	Lla	CPQF	Lw	CPTR	Meq	CPYT	Mow
CNYM	Kra	CPDZ	Laq	CPKD	Lle	CPQG	Lx	CPTS	Mer	CPYV	Moy
CNYP	Kre	CPFB	Lar	CPKF	Lli	CPQH	Ly	CPTV	Mes	CPYX	Moz
CNYQ	Kri	CPFD	Las	CPKG	Llo	CPQJ	Lya	CPTW	Met	CPYZ	Mp
CNYR	Kro	CPFG	Lat	CPKH	Llu	CPQK	Lye	CPTX	Meu	CPZB	Mq
CNYS	Kru	CPFH	Lau	CPKJ	Lly	CPQL	Lyi	CPTY	Mev	CPZD	Mr
CNYT	Kry	CPFJ	Lav	CPKL	Lm	CPQM	Lyo	CPTZ	Mew	CPZF	Mra
CNYV	Ks	CPFK	Law	CPKM	Ln	CPQN	Lyu	CPVB	Mex	CPZG	Mre
CNYW	Kt	CPFL	Lax	CPKN	Lo	CPQR	Lyy	CPVD	Mey	CPZH	Mri
CNYX	Ku	CPFM	Lay	CPKQ	Loa	CPQS	Lz	CPVF	Mez	CPZJ	Mro
CNYZ	Kua	CPFN	Laz	CPKR	Lob	CPQT		CPVG	Mf	CPZK	Mru
CNZB	Kub	CPFQ	Lb	CPKS	Loc	CPQV		CPVH	Mg	CPZL	Mry
CNZD	Kuc	CPFR	Lc	CPKT	Lod	CPQW		CPVJ	Mh	CPZM	Ms
CNZF	Kud	CPFS	Ld	CPKV	Loe	CPQX		CPVK	Mi	CPZN	Mt
CNZG	Kue	CPFT	Le	CPKW	Lof	CPQY		CPVL	Mia	CPZQ	Mu
CNZH	Kuf	CPFV	Lea	CPKX	Log	CPQZ		CPVM	Mib	CPZR	Mua
CNZJ	Kug	CPFW	Leb	CPKY	Loh			CPVN	Mic	CPZS	Mub
CNZK	Kuh	CPFX	Lec	CPKZ	Loi			CPVQ	Mid	CPZT	Muc
CNZL	Kui	CPFY	Led	CPLB	Loj			CPVR	Mie	CPZV	Mud
CNZM	Kuj	CPFZ	Lee	CPLD	Lok			CPVS	Mif	CPZW	Mue
CNZP	Kuk	CPGB	Lef	CPLF	Lol			CPVT	Mig	CPZX	Muf
CNZQ	Kul	CPGD	Leg	CPLG	Lom			CPVW	Mih	CPZY	Mug
CNZR	Kum	CPGF	Leh	CPLH	Lon			CPVX	Mii	CQBD	Muh
CNZS	Kun	CPGH	Lei	CPLJ	Loo			CPVY	Mij	CQBF	Mui
CNZT	Kuo	CPGJ	Lej	CPLK	Lop			CPVZ	Mik	CQBG	Muj
CNZV	Kup	CPGK	Lek	CPLM	Loq			CPWB	Mil	CQBH	Muk
CNZW	Kuq	CPGL	Lel	CPLN	Lor	M		CPWD	Mim	CQBJ	Mul
CNZX	Kur	CPGM	Lem	CPLQ	Los	CPRB	Ma	CPWF	Min	CQBK	Mum
CNZY	Kus	CPGN	Len	CPLR	Lot	CPRD	Maa	CPWG	Mio	CQBL	Mun
CPBD	Kut	CPGQ	Leo	CPLS	Lou	CPRF	Mab	CPWH	Mip	CQBM	Muo
CPBF	Kuu	CPGR	Lep	CPLT	Lov	CPRG	Mac	CPWJ	Miq	CQBN	Mup
CPBG	Kuv	CPGS	Leq	CPLV	Low	CPRH	Mad	CPWK	Mir	CQBP	Muq
CPBH	Kuw	CPGT	Ler	CPLW	Lox	CPRJ	Mae	CPWL	Mis	CQBR	Mur
CPBJ	Kux	CPGV	Les	CPLX	Loy	CPRK	Maf	CPWM	Mit	CQBS	Mus
CPBK	Kuy	CPGW	Let	CPLY	Loz	CPRL	Mag	CPWN	Miu	CQBT	Mut
CPBL	Kuz	CPGX	Leu	CPLZ	Lp	CPRM	Mah	CPWQ	Miv	CQBV	Muu
CPBM	Kv	CPGY	Lev	CPMB	Lq	CPRN	Mai	CPWR	Miw	CQBW	Muv
CPBN	Kw	CPGZ	Lew	CPMD	Lr	CPRQ	Maj	CPWS	Mix	CQBX	Muw
CPBQ	Kx	CPHB	Lex	CPMF	Lra	CPRS	Mak	CPWT	Miy	CQBZ	Mux
CPBR	Ky	CPHD	Ley	CPMG	Lre	CPRT	Mal	CPWV	Miz	CQDB	Muy
CPBS	Kya	CPHF	Lez	CPMH	Lri	CPRV	Mam	CPWX	Mj	CQDF	Muz
CPBT	Kye	CPHG	Lf	CPMJ	Lro	CPRW	Man	CPWY	Mk	CQDG	Mv
CPBV	Kyi	CPHJ	Lg	CPMK	Lru	CPRX	Mao	CPWZ	Ml	CQDH	Mw
CPBW	Kyo	CPHK	Lh	CPML	Lry	CPRY	Map	CPXB	Mla	CQDJ	Mx
CPBX	Kyu	CPHL	Li	CPMN	Ls	CPRZ	Maq	CPXD	Mle	CQDK	My
CPBY	Kyy	CPHM	Lia	CPMQ	Lt	CPSB	Mar	CPXF	Mli	CQDL	Mya
CPBZ	Kz	CPHN	Lib	CPMR	Lu	CPSD	Mas	CPXG	Mlo	CQDM	Mye
		CPHQ	Lic	CPMS	Lua	CPSF	Mat	CPXH	Mlu	CQDN	Myi
		CPHR	Lid	CPMT	Lub	CPSG	Mau	CPXJ	Mly	CQDP	Myo
		CPHS	Lie	CPMV	Luc	CPSH	Mav	CPXK	Mm	CQDR	Myu
		CPHT	Lif	CPMW	Lud	CPSJ	Maw	CPXL	Mn	CQDS	Myy
		CPHV	Lig	CPMX	Lue	CPSK	Max	CPXM	Mo	CQDT	Mz
		CPHW	Lih	CPMY	Luf	CPSL	May	CPXN	Moa	CQDV	
		CPHX	Lii	CPMZ	Lug	CPSM	Maz	CPXQ	Mob	CQDW	
		CPHY	Lij	CPNB	Luh	CPSN	Mb	CPXR	Moc	CQDX	
		CPHZ	Lik	CPND	Lui	CPSQ	Mc	CPXS	Mod	CQDY	
		CPJB	Lil	CPNF	Luj	CPSR	Md	CPXT	Moe	CQDZ	
L.		CPJD	Lim	CPNG	Luk	CPST	Me	CPXV	Mof		
CPDB	La	CPJF	Lin	CPNH	Lul	CPSV	Mea	CPXW	Mog		
CPDF	Laa	CPJG	Lio	CPNJ	Lum	CPSW	Meb	CPXY	Moh		
CPDG	Lab	CPJH	Lip	CPNK	Lun	CPSX	Mec	CPXZ	Moi		
CPDH	Lac	CPJK	Liq	CPNL	Luo	CPSY	Med	CPYB	Moj		
CPDJ	Lad	CPJL	Lir	CPNM	Lup	CPSZ	Mee	CPYD	Mok		
CPDK	Lae	CPJM	Lis	CPNQ	Luq	CPTB	Mef	CPYF	Mol		
CPDL	Laf	CPJN	Lit	CPNR	Lur	CPTD	Meg	CPYG	Mom		
CPDM	Lag	CPJQ	Liu	CPNS	Lus	CPTF	Meh	CPYH	Mon		

N (left section)   **O.** (right section)

	N	CQKB	N1l	CQPF	Nuj	CQTR	Obb	CQYT	Odv	CRFX	Ogo
		CQKD	N1m	CQPG	Nuk	CQTS	Obc	CQYV	Odw	CRFY	Ogp
CQFB	Na	CQKF	N1n	CQPH	Nul	CQTV	Obd	CQYW	Odx	CRFZ	Ogq
CQFD	Naa	CQKG	N1o	CQPJ	Num	CQTW	Obe	CQYX	Ody	CRGB	Ogr
CQFG	Nab	CQKH	N1p	CQPK	Nun	CQTX	Obf	CQYZ	Odz	CRGD	Ogs
CQFH	Nac	CQKJ	N1q	CQPL	Nuo	CQTY	Obg	CQZB	Oe	CRGF	Ogt
CQFJ	Nad	CQKL	N1r	CQPM	Nup	CQTZ	Obh	CQZD	Oea	CRGH	Ogu
CQFK	Nae	CQKM	N1s	CQPN	Nuq	CQVB	Obi	CQZF	Oeb	CRGJ	Ogv
CQFL	Naf	CQKN	N1t	CQPR	Nur	CQVD	Obj	CQZG	Oec	CRGK	Ogw
CQFM	Nag	CQKP	N1u	CQPS	Nus	CQVF	Obk	CQZH	Oed	CRGL	Ogx
CQFN	Nah	CQKR	N1v	CQPT	Nut	CQVG	Obl	CQZJ	Oee	CRGM	Ogy
CQFP	Nai	CQKS	N1w	CQPV	Nuu	CQVH	Obm	CQZK	Oef	CRGN	Ogz
CQFR	Naj	CQKT	N1x	CQPW	Nuv	CQVJ	Obn	CQZL	Oeg	CRGP	Oh
CQFS	Nak	CQKV	N1y	CQPX	Nuw	CQVK	Obo	CQZM	Oeh	CRGQ	Oha
CQFT	Nal	CQKW	Niz	CQPY	Nux	CQVL	Obp	CQZN	Oei	CRGS	Ohb
CQFV	Nam	CQKX	Nj	CQPZ	Nuy	CQVM	Obq	CQZP	Oej	CRGT	Ohc
CQFW	Nan	CQKY	Nk	CQRB	Nuz	CQVN	Obr	CQZR	Oek	CRGV	Ohd
CQFX	Nao	CQKZ	Nl	CQRD	Nv	CQVP	Obs	CQZS	Oel	CRGW	Ohe
CQFY	Nap	CQLB	Nla	CQRF	Nw	CQVR	Obt	CQZT	Oem	CRGX	Ohf
CQFZ	Naq	CQLD	Nle	CQRG	Nx	CQVS	Obu	CQZV	Oen	CRGY	Ohg
CQGB	Nar	CQLF	Nli	CQRH	Ny	CQVT	Obv	CQZW	Oeo	CRGZ	Ohh
CQGD	Nas	CQLG	Nlo	CQRJ	Nya	CQVW	Obw	CQZX	Oep	CRHB	Ohi
CQGF	Nat	CQLH	Nlu	CQRK	Nye	CQVX	Obx	CQZY	Oeq	CRHD	Ohj
CQGH	Nau	CQLJ	Nly	CQRL	Nyi	CQVY	Oby	CRBD	Oer	CRHF	Ohk
CQGJ	Nav	CQLK	Nm	CQRM	Nyo	CQVZ	Obz	CRBF	Oes	CRHG	Ohl
CQGK	Naw	CQLM	Nn	CQRN	Nyu	CQWB	Oc	CRBG	Oet	CRHJ	Ohm
CQGL	Nax	CQLN	No	CQRP	Nyy	CQWD	Oca	CRBH	Oeu	CRHK	Ohn
CQGM	Nay	CQLP	Noa	CQRS	Nz	CQWF	Ocb	CRBJ	Oev	CRHL	Oho
CQGN	Naz	CQLR	Nob	CQRT		CQWG	Occ	CRBK	Oew	CRHM	Ohp
CQGP	Nb	CQLS	Noc	CQRV		CQWH	Ocd	CRBL	Oex	CRHN	Ohq
CQGR	Nc	CQLT	Nod	CQRW		CQWJ	Oce	CRBM	Oey	CRHP	Ohr
CQGS	Nd	CQLV	Noe	CQRX		CQWK	Ocf	CRBN	Oez	CRHQ	Ohs
CQGT	Ne	CQLW	Nof	CQRY		CQWL	Ocg	CRBP	Of	CRHS	Oht
CQGV	Nea	CQLX	Nog	CQRZ		CQWM	Och	CRBQ	Ofa	CRHT	Ohu
CQGW	Neb	CQLY	Noh			CQWN	Oci	CRBS	Ofb	CRHV	Ohv
CQGX	Nec	CQLZ	Noi			CQWP	Ocj	CRBT	Ofc	CRHW	Ohw
CQGY	Ned	CQMB	Noj			CQWR	Ock	CRBV	Ofd	CRHX	Ohx
CQGZ	Nee	CQMD	Nok			CQWS	Ocl	CRBW	Ofe	CRHY	Ohy
CQHB	Nef	CQMF	Nol			CQWT	Ocm	CRBX	Off	CRHZ	Ohz
CQHD	Neg	CQMG	Nom			CQWV	Ocn	CRBY	Ofg	CRJB	Oi
CQHF	Neh	CQMH	Non			CQWX	Oco	CRBZ	Ofh	CRJD	Oia
CQHG	Nei	CQMJ	Noo			CQWZ	Ocq	CRDB	Ofi	CRJF	Oib
CQHJ	Nej	CQMK	Nop	**O.**		CQXB	Ocr	CRDF	Ofj	CRJG	Oic
CQHK	Nek	CQML	Noq			CQXD	Ocs	CRDG	Ofk	CRJH	Oid
CQHL	Nel	CQMN	Nor			CQXF	Oct	CRDH	Ofl	CRJK	Oie
CQHM	Nem	CQMP	Nos	CQSB	Oa	CQXG	Ocu	CRDJ	Ofm	CRJL	Oif
CQHN	Nen	CQMR	Not	CQSD	Oaa	CQXH	Ocv	CRDK	Ofn	CRJM	Oig
CQHP	Neo	CQMS	Nou	CQSF	Oab	CQXJ	Ocw	CRDL	Ofo	CRJN	Oih
CQHR	Nep	CQMT	Nov	CQSG	Oac	CQXK	Ocx	CRDM	Ofp	CRJP	Oii
CQHS	Neq	CQMV	Now	CQSH	Oad	CQXL	Ocy	CRDN	Ofq	CRJQ	Oij
CQHT	Ner	CQMW	Nox	CQSJ	Oae	CQXM	Ocz	CRDP	Ofr	CRJS	Oik
CQHV	Nes	CQMX	Noy	CQSK	Oaf	CQXN	Od	CRDQ	Ofs	CRJT	Oil
CQHW	Net	CQMY	Noz	CQSL	Oag	CQXP	Oda	CRDS	Oft	CRJV	Oim
CQHX	Neu	CQMZ	Np	CQSM	Oah	CQXR	Odb	CRDT	Ofu	CRJW	Oin
CQHY	Nev	CQNB	Nq	CQSN	Oai	CQXS	Odc	CRDV	Ofv	CRJX	Oio
CQHZ	New	CQND	Nr	CQSP	Oaj	CQXT	Odd	CRDW	Ofw	CRJY	Oip
CQJB	Nex	CQNF	Nra	CQSR	Oak	CQXV	Ode	CRDX	Ofx	CRJZ	Oiq
CQJD	Ney	CQNG	Nre	CQST	Oal	CQXW	Odf	CRDY	Ofy	CRKB	Oir
CQJF	Nez	CQNH	Nri	CQSV	Oam	CQXY	Odg	CRDZ	Ofz	CRKD	Ois
CQJG	Nf	CQNJ	Nro	CQSW	Oan	CQXZ	Odh	CRFB	Og	CRKF	Oit
CQJH	Ng	CQNK	Nru	CQSX	Oao	CQYB	Odi	CRFD	Oga	CRKG	Oiu
CQJK	Nh	CQNL	Nry	CQSY	Oap	CQYD	Odj	CRFG	Ogb	CRKH	Oiv
CQJL	Ni	CQNM	Ns	CQSZ	Oaq	CQYF	Odk	CRFH	Ogc	CRKJ	Oiw
CQJM	Nia	CQNP	Nt	CQTB	Oar	CQYG	Odl	CRFJ	Ogd	CRKL	Oix
CQJN	Nib	CQNR	Nu	CQTD	Oas	CQYH	Odm	CRFK	Oge	CRKM	Oiy
CQJP	Nic	CQNS	Nua	CQTF	Oat	CQYJ	Odn	CRFL	Ogf	CRKN	Oiz
CQJR	Nid	CQNT	Nub	CQTG	Oau	CQYK	Odo	CRFM	Ogg	CRKP	Oja
CQJS	Nie	CQNV	Nuc	CQTH	Oav	CQYL	Odp	CRFN	Ogh	CRKQ	Ojb
CQJT	Nif	CQNW	Nud	CQTJ	Oaw	CQYM	Odq	CRFP	Ogi	CRKS	Ojc
CQJV	Nig	CQNX	Nue	CQTK	Oax	CQYN	Odr	CRFQ	Ogj	CRKT	Ojd
CQJW	Nih	CQNY	Nuf	CQTL	Oay	CQYP	Ods	CRFS	Ogk	CRKV	Oje
CQJX	Nii	CQNZ	Nug	CQTM	Oaz	CQYR	Odt	CRFT	Ogl	CRKW	Ojf
CQJY	Nij	CQPB	Nuh	CQTN	Ob	CQYS	Odu	CRFV	Ogm	CRKX	Ojg
CQJZ	Nik	CQPD	Nui	CQTP	Oba			CRFW	Ogn	CRKY	Ojh

Code	Spell	Code	Spell	Code	Spell	Code	Spell	Code	Spell	Code	Spell
CRKZ	Oji	CRQD	Omb	CRWG	Opm	CSBK	Osg	CSHM	Ov	CSMP	Oxt
CRLB	Ojj	CRQF	Omc	CRWH	Opn	CSBL	Osh	CSHN	Ova	CSMQ	Oxu
CRLD	Ojk	CRQG	Omd	CRWJ	Opo	CSBM	Osi	CSHP	Ovb	CSMR	Oxv
CRLF	Ojl	CRQH	Ome	CRWK	Opp	CSBN	Osj	CSHQ	Ovc	CSMT	Oxw
CRLG	Ojm	CRQJ	Omf	CRWL	Opq	CSBP	Osk	CSHR	Ovd	CSMV	Oxx
CRLH	Ojn	CRQK	Omg	CRWM	Opr	CSBQ	Osl	CSHT	Ove	CSMW	Oxy
CRLJ	Ojo	CRQL	Omh	CRWN	Ops	CSBR	Osm	CSHV	Ovf	CSMX	Oxz
CRLK	Ojp	CRQM	Omi	CRWP	Opt	CSBT	Osn	CSHW	Ovg	CSMY	Oy
CRLM	Ojq	CRQN	Omj	CRWQ	Opu	CSBV	Oso	CSHX	Ovh	CSMZ	Oya
CRLN	Ojr	CRQP	Omk	CRWS	Opv	CSBW	Osp	CSHY	Ovi	CSNB	Oyb
CRLP	Ojs	CRQS	Oml	CRWT	Opw	CSBX	Osq	CSHZ	Ovj	CSND	Oyc
CRLQ	Ojt	CRQT	Omm	CRWV	Opx	CSBY	Osr	CSJB	Ovk	CSNF	Oyd
CRLS	Oju	CRQV	Omn	CRWX	Opy	CSBZ	Oss	CSJD	Ovl	CSNG	Oye
CRLT	Ojv	CRQW	Omo	CRWY	Opz	CSDB	Ost	CSJF	Ovm	CSNH	Oyf
CRLV	Ojw	CRQX	Omp	CRWZ	Oq	CSDF	Osu	CSJG	Ovn	CSNJ	Oyg
CRLW	Ojx	CRQY	Omq	CRXB	Oqa	CSDG	Osv	CSJH	Ovo	CSNK	Oyh
CRLX	Ojy	CRQZ	Omr	CRXD	Oqb	CSDH	Osw	CSJK	Ovp	CSNL	Oyi
CRLY	Ojz	CRSB	Oms	CRXF	Oqc	CSDJ	Osx	CSJL	Ovq	CSNM	Oyj
CRLZ	Ok	CRSD	Omt	CRXG	Oqd	CSDK	Osy	CSJM	Ovr	CSNP	Oyk
CRMB	Oka	CRSF	Omu	CRXH	Oqe	CSDL	Osz	CSJN	Ovs	CSNQ	Oyl
CRMD	Okb	CRSG	Omv	CRXJ	Oqf	CSDM	Ot	CSJP	Ovt	CSNR	Oym
CRMF	Okc	CRSH	Omw	CRXK	Oqg	CSDN	Ota	CSJQ	Ovu	CSNT	Oyn
CRMG	Okd	CRSJ	Omx	CRXL	Oqh	CSDP	Otb	CSJR	Ovv	CSNV	Oyo
CRMH	Oke	CRSK	Omy	CRXM	Oqi	CSDQ	Otc	CSJT	Ovw	CSNW	Oyp
CRMJ	Okf	CRSL	Omz	CRXN	Oqj	CSDR	Otd	CSJV	Ovx	CSNX	Oyq
CRMK	Okg	CRSM	On	CRXP	Oqk	CSDT	Ote	CSJW	Ovy	CSNY	Oyr
CRML	Okh	CRSN	Ona	CRXQ	Oql	CSDV	Otf	CSJX	Ovz	CSNZ	Oys
CRMN	Oki	CRSP	Onb	CRXS	Oqm	CSDW	Otg	CSJY	Ow	CSPB	Oyt
CRMP	Okj	CRSQ	Onc	CRXT	Oqn	CSDX	Oth	CSJZ	Owa	CSPD	Oyu
CRMQ	Okk	CRST	Ond	CRXV	Oqo	CSDY	Oti	CSKB	Owb	CSPF	Oyv
CRMS	Okl	CRSV	One	CRXW	Oqp	CSDZ	Otj	CSKD	Owc	CSPG	Oyw
CRMT	Okm	CRSW	Onf	CRXY	Oqr	CSFB	Otk	CSKF	Owd	CSPH	Oyx
CRMV	Okn	CRSX	Ong	CRXZ	Oqs	CSFD	Otl	CSKG	Owe	CSPK	Oyy
CRMW	Oko	CRSY	Onh	CRYB	Oqt	CSFG	Otm	CSKH	Owf	CSPL	Oyz
CRMX	Okp	CRSZ	Oni	CRYD	Oqu	CSFH	Otn	CSKJ	Owg	CSPM	Oz
CRMY	Okq	CRTB	Oo	CRYF	Oqv	CSFJ	Oto	CSKL	Owh	CSPN	Oza
CRMZ	Okr	CRTD	Ooa	CRYG	Oqw	CSFK	Otp	CSKM	Owi	CSPQ	Ozb
CRNB	Oks	CRTF	Oob	CRYH	Oqx	CSFL	Otq	CSKN	Owj	CSPR	Ozc
CRND	Okt	CRTG	Ooc	CRYJ	Oqy	CSFM	Otr	CSKP	Owk	CSPT	Ozd
CRNF	Oku	CRTH	Ood	CRYK	Oqz	CSFN	Ots	CSKQ	Owl	CSPV	Oze
CRNG	Okv	CRTJ	Ooe	CRYL	Or	CSFP	Ott	CSKR	Owm	CSPW	Ozf
CRNH	Okw	CRTK	Oof	CRYM	Ora	CSFQ	Otu	CSKT	Own	CSPX	Ozg
CRNJ	Okx	CRTL	Oog	CRYN	Orb	CSFR	Otv	CSKV	Owo	CSPZ	Ozh
CRNK	Oky	CRTM	Ooh	CRYP	Orc	CSFT	Otw	CSKW	Owp	CSQB	Ozi
CRNL	Okz	CRTN	Ooi	CRYQ	Ord	CSFV	Otx	CSKX	Owq	CSQD	Ozj
CRNM	Ol	CRTP	Ooj	CRYS	Ore	CSFW	Oty	CSKY	Owr	CSQF	Ozk
CRNP	Ola	CRTQ	Ook	CRYT	Orf	CSFX	Otz	CSKZ	Ows	CSQG	Ozl
CRNQ	Olb	CRTS	Ool	CRYV	Org	CSFY	Ou	CSLB	Owt	CSQH	Ozm
CRNS	Olc	CRTV	Oom	CRYW	Orh	CSFZ	Oua	CSLD	Owu	CSQJ	Ozn
CRNT	Old	CRTW	Oon	CRYX	Ori	CSGB	Oub	CSLF	Owv	CSQK	Ozo
CRNV	Ole	CRTX	Oop	CRYZ	Orj	CSGD	Ouc	CSLG	Oww	CSQL	Ozp
CRNW	Olf	CRTY	Ooq	CRZB	Ork	CSGF	Oud	CSLH	Owx	CSQM	Ozq
CRNX	Olg	CRTZ	Oor	CRZD	Orl	CSGH	Oue	CSLJ	Owy	CSQN	Ozr
CRNY	Olh	CRVB	Oos	CRZF	Orm	CSGJ	Ouf	CSLK	Owz	CSQP	Ozs
CRNZ	Oli	CRVD	Oot	CRZG	Orn	CSGK	Oug	CSLM	Ox	CSQR	Ozt
CRPB	Olj	CRVF	Oou	CRZH	Oro	CSGL	Ouh	CSLN	Oxa	CSQT	Ozu
CRPD	Olk	CRVG	Oov	CRZJ	Orp	CSGM	Oui	CSLP	Oxb	CSQV	Ozv
CRPF	Oll	CRVH	Oow	CRZK	Orq	CSGN	Ouj	CSLQ	Oxc	CSQW	Ozw
CRPG	Olm	CRVJ	Oox	CRZL	Orr	CSGP	Ouk	CSLR	Oxd	CSQX	Ozx
CRPH	Oln	CRVK	Ooy	CRZM	Ors	CSGQ	Oul	CSLT	Oxe	CSQY	Ozy
CRPJ	Olo	CRVL	Ooz	CRZN	Ort	CSGR	Oum	CSLV	Oxf	CSQZ	Ozz
CRPK	Olp	CRVM	Op	CRZP	Oru	CSGT	Oun	CSLW	Oxg		
CRPL	Olq	CRVN	Opa	CRZQ	Orv	CSGV	Ouo	CSLX	Oxh		
CRPM	Olr	CRVP	Opb	CRZS	Orw	CSGW	Oup	CSLY	Oxi		
CRPN	Ols	CRVQ	Opc	CRZT	Orx	CSGX	Ouq	CSMB	Oxj		
CRPQ	Olt	CRVS	Opd	CRZV	Ory	CSGY	Our	CSMD	Oxk		
CRPS	Olu	CRVT	Ope	CRZW	Orz	CSGZ	Ous	CSMF	Oxl		
CRPT	Olv	CRVW	Opf	CRZX	Os	CSHB	Out	CSMG	Oxm		
CRPV	Olw	CRVX	Opg	CRZY	Osa	CSHD	Ouu	CSMH	Oxn		
CRPW	Olx	CRVY	Oph	CSBD	Osb	CSHF	Ouv	CSMJ	Oxo		
CRPX	Oly	CRVZ	Opi	CSBF	Osc	CSHG	Ouw	CSMK	Oxp		
CRPY	Olz	CRWB	Opj	CSBG	Osd	CSHJ	Oux	CSML	Oxq		
CRPZ	Om	CRWD	Opk	CSBH	Ose	CSHK	Ouy	CSMN	Oxs		
CRQB	Oma	CRWF	Opl	CSBJ	Osf	CSHL	Ouz				

	P										
		CSXB	Pic	CTDG	Pse	CTHR	Qal	CTMV	Qj	CTRX	Qyy
		CSXD	Pid	CTDH	Psh	CTHS	Qam	CTMW	Qk	CTRY	Qz
CSRB	Pa	CSXF	Pie	CTDJ	Psi	CTHV	Qan	CTMX	Qla	CTRZ	
CSBD	Paa	CSXG	Pif	CTDK	Psj	CTHW	Qao	CTMY	Qle		
CSRF	Pab	CSXH	Pig	CTDL	Pso	CTHX	Qap	CTMZ	Qli		
CSRG	Pac	CSXJ	Pih	CTDM	Pst	CTHY	Qaq	CTNB	Qlo		
CSRH	Pad	CSXK	Pii	CTDN	Psu	CTHZ	Qar	CTND	Qlu		
CSRJ	Pae	CSXL	Pij	CTDP	Psy	CTJB	Qas	CTNF	Qly		
CSRK	Paf	CSXM	Pik	CTDQ	Psz	CTJD	Qat	CTNG	Qm		
CSRL	Pag	CSXN	Pil	CTDR	Pt	CTJF	Qau	CTNH	Qn		
CSRM	Pah	CSXP	Pim	CTDS	Pta	CTJG	Qav	CTNJ	Qoa		
CSRN	Pai	CSXQ	Pin	CTDV	Pte	CTJH	Qaw	CTNK	Qob		
CSRP	Paj	CSXR	Pio	CTDW	Pti	CTJK	Qax	CTNL	Qoc		**R.**
CSRQ	Pak	CSXT	Pip	CTDX	Pto	CTJL	Qay	CTNM	Qod		
CSRT	Pal	CSXV	Piq	CTDY	Ptu	CTJM	Qaz	CTNP	Qoe	CTSB	Ra
CSRV	Pam	CSXW	Pir	CTDZ	Pty	CTJN	Qb	CTNQ	Qof	CTSD	Raa
CSRW	Pan	CSXY	Pis	CTFB	Pu	CTJP	Qc	CTNR	Qog	CTSF	Rab
CSRX	Pao	CSXZ	Pit	CTFD	Pua	CTJQ	Qd	CTNS	Qoh	CTSG	Rac
CSRY	Pap	CSYB	Piu	CTFG	Pub	CTJR	Qea	CTNV	Qoi	CTSH	Rad
CSRZ	Paq	CSYD	Piv	CTFH	Puc	CTJS	Qeb	CTNW	Qoj	CTSJ	Rae
CSTB	Par	CSYF	Piw	CTFJ	Pud	CTJV	Qec	CTNX	Qok	CTSK	Raf
CSTD	Pas	CSYG	Pix	CTFK	Pue	CTJW	Qed	CTNZ	Qol	CTSL	Rag
CSTF	Pat	CSYH	Piy	CTFL	Puf	CTJX	Qee	CTPB	Qom	CTSM	Rah
CSTG	Pau	CSYJ	Piz	CTFM	Pug	CTJY	Qef	CTPD	Qop	CTSN	Rai
CSTH	Pav	CSYK	Pj	CTFN	Puh	CTJZ	Qeg	CTPF	Qoq	CTSP	Raj
CSTJ	Paw	CSYL	Pk	CTFP	Pui	CTKB	Qeh	CTPG	Qor	CTSQ	Rak
CSTK	Pax	CSYM	Pl	CTFQ	Puj	CTKD	Qei	CTPH	Qos	CTSR	Ral
CSTL	Pay	CSYN	Pla	CTFR	Puk	CTKF	Qej	CTPJ	Qot	CTSV	Ram
CSTM	Paz	CSYP	Ple	CTFS	Pul	CTKG	Qek	CTPK	Qou	CTSW	Ran
CSTN	Pb	CSYQ	Pli	CTFV	Pum	CTKH	Qel	CTPL	Qov	CTSX	Rao
CSTP	Pc	CSYR	Plo	CTFW	Pun	CTKJ	Qem	CTPM	Qow	CTSY	Rap
CSTQ	Pd	CSYT	Plu	CTFX	Puo	CTKL	Qen	CTPN	Qox	CTSZ	Raq
CSTR	Pe	CSYV	Ply	CTFY	Pup	CTKM	Qeo	CTPQ	Qoy	CTVB	Rar
CSTV	Pea	CSYW	Pm	CTFZ	Puq	CTKN	Qep	CTPR	Qoz	CTVD	Ras
CSTW	Peb	CSYX	Pn	CTGB	Pur	CTKP	Qeq	CTPS	Qp	CTVF	Rat
CSTX	Pec	CSYZ	Po	CTGD	Pus	CTKQ	Qer	CTPV	Qr	CTVG	Rau
CSTY	Ped	CSZB	Poa	CTGF	Put	CTKR	Qes	CTPW	Qs	CTVH	Rav
CSTZ	Pee	CSZD	Pob	CTGH	Puu	CTKS	Qet	CTPX	Qt	CTVJ	Raw
CSVB	Pef	CSZF	Poc	CTGJ	Puv	CTKV	Qeu	CTPY	Qua	CTVK	Rax
CSVD	Peg	CSZG	Pod	CTGK	Puw	CTKW	Qev	CTPZ	Qub	CTVL	Ray
CSVF	Peh	CSZJ	Poe	CTGL	Pux	CTKX	Qew	CTQB	Quc	CTVM	Raz
CSVG	Pei	CSZK	Pof	CTGM	Puy	CTKY	Qex	CTQD	Qud	CTVN	Rb
CSVH	Pej	CSZL	Pog	CTGN	Puz	CTKZ	Qey	CTQF	Que	CTVP	Rc
CSVJ	Pek	CSZM	Poh	CTGP	Pv	CTLB	Qez	CTQG	Quf	CTVQ	Rd
CSVK	Pel	CSZN	Poi	CTGQ	Pw	CTLD	Qf	CTQH	Qug	CTVR	Re
CSVL	Pem	CSZP	Poj	CTGR	Px	CTLF	Qg	CTQJ	Quh	CTVS	Rea
CSVM	Pen	CSZQ	Pok	CTGS	Pya	CTLG	Qh	CTQK	Qui	CTVW	Reb
CSVN	Peo	CSZR	Pol	CTGV	Pye	CTLH	Qi	CTQL	Quj	CTVX	Rec
CSVP	Pep	CSZT	Pom	CTGW	Pyi	CTLJ	Qia	CTQM	Quk	CTVY	Red
CSVQ	Peq	CSZV	Pon	CTGX	Pyo	CTLK	Qib	CTQN	Qul	CTVZ	Ree
CSVR	Per	CSZW	Poo	CTGY	Pyu	CTLM	Qic	CTQP	Qum	CTWB	Ref
CSVT	Pes	CSZX	Pop	CTGZ	Pz	CTLN	Qid	CTQR	Qun	CTWD	Reg
CSVW	Pet	CSZY	Poq			CTLP	Qie	CTQS	Quo	CTWF	Reh
CSVX	Peu	CTBD	Por			CTLQ	Qif	CTQV	Qup	CTWG	Rei
CSVY	Pev	CTBF	Pos			CTLR	Qig	CTQW	Quq	CTWH	Rej
CSVZ	Pew	CTBG	Pot			CTLS	Qih	CTQX	Qur	CTWJ	Rek
CSWB	Pex	CTBH	Pou			CTLV	Qii	CTQY	Qus	CTWK	Rel
CSWD	Pez	CTBJ	Pov			CTLW	Qij	CTQZ	Qut	CTWL	Rem
CSWF	Pf	CTBK	Pow			CTLX	Qik	CTRB	Quu	CTWM	Ren
CSWG	Pg	CTBL	Pox			CTLY	Qil	CTRD	Quv	CTWN	Reo
CSWH	Ph	CTBM	Poy			CTLZ	Qim	CTRF	Quw	CTWP	Rep
CSWJ	Pha	CTBN	Poz		**Q**	CTMB	Qin	CTRG	Qux	CTWQ	Req
CSWK	Phe	CTBP	Pp			CTMD	Qio	CTRH	Quy	CTWR	Rer
CSWL	Phf	CTBQ	Pq	CTHB	Qa	CTMF	Qip	CTRJ	Quz	CTWS	Res
CSWM	Phi	CTBR	Pr	CTHD	Qaa	CTMG	Qiq	CTRK	Qv	CTWV	Ret
CSWN	Phl	CTBS	Pra	CTHF	Qac	CTMH	Qir	CTRL	Qw	CTWX	Reu
CSWP	Pho	CTBV	Pre	CTHG	Qad	CTMJ	Qis	CTRM	Qx	CTWY	Rev
CSWQ	Phr	CTBW	Pri	CTHJ	Qae	CTMK	Qit	CTRN	Qy	CTWZ	Rew
CSWR	Pht	CTBX	Pro	CTHK	Qaf	CTML	Qiu	CTRP	Qya	CTXB	Rex
CSWT	Phu	CTBY	Pru	CTHL	Qag	CTMN	Qiv	CTRQ	Qye	CTXD	Rey
CSWV	Phy	CTBZ	Pry	CTHM	Qah	CTMP	Qiw	CTRS	Qyi	CTXF	Rez
CSWX	Pi	CTDB	Ps	CTHN	Qai	CTMQ	Qix	CTRV	Qyo	CTXG	Rf
CSWY	Pia	CTDF	Psc	CTHP	Qaj	CTMR	Qiy	CTRW	Qyu	CTXH	Rg
CSWZ	Pib			CTHQ	Qak	CTMS	Qiz			CTXJ	Rh

CTXK	Ri	CVDN	Rud	CVHZ	Sca	CVND	Ska	CVSG	Suo	CVXR	Tch
CTXL	Ria	CVDP	Rue	CVJB	Sce	CVNF	Ske	CVSH	Sup	CVXS	Tci
CTXM	Rib	CVDQ	Ruf	CVJD	Sch	CVNG	Ski	CVSJ	Suq	CVXT	Tco
CTXN	Ric	CVDR	Rug	CVJF	Sci	CVNH	Sko	CVSK	Sur	CVXW	Tcu
CTXP	Rid	CVDS	Ruh	CVJG	Scl	CVNJ	Sku	CVSL	Sus	CVXY	Tcy
CTXQ	Rie	CVDT	Rui	CVJH	Sco	CVNK	Sky	CVSM	Sut	CVXZ	Tcz
CTXR	Rif	CVDW	Ruj	CVJK	Scr	CVNL	Sla	CVSN	Suu	CVYB	Td
CTXS	Rig	CVDX	Ruk	CVJL	Scu	CVNM	Sle	CVSP	Suv	CVYD	Te
CTXV	Rih	CVDY	Rul	CVJM	Scy	CVNP	Sli	CVSQ	Suw	CVYF	Tea
CTXW	Rii	CVDZ	Rum	CVJN	Sd	CVNQ	Slo	CVSR	Sux	CVYG	Teb
CTXY	Rij	CVFB	Run	CVJP	Se	CVNR	Slu	CVST	Suy	CVYH	Tec
CTXZ	Rik	CVFD	Ruo	CVJQ	Sea	CVNS	Sly	CVSW	Suz	CVYJ	Ted
CTYB	Ril	CVFG	Rup	CVJR	Seb	CVNT	Sm	CVSX	Sv	CVYK	Tee
CTYD	Rim	CVFH	Ruq	CVJS	Sec	CVNW	Sn	CVSY	Sw	CVYL	Tef
CTYF	Rin	CVFJ	Rur	CVJT	Sed	CVNX	So	CVSZ	Swa	CVYM	Teg
CTYG	Rio	CVFK	Rus	CVJW	See	CVNY	Soa	CVTB	Swe	CVYN	Teh
CTYH	Rip	CVFL	Rut	CVJX	Sef	CVNZ	Sob	CVTD	Swi	CVYP	Tei
CTYJ	Riq	CVFM	Ruu	CVJY	Seg	CVPB	Soc	CVTF	Swo	CVYQ	Tej
CTYK	Rir	CVFN	Ruv	CVJZ	Seh	CVPD	Sod	CVTG	Swu	CVYR	Tek
CTYL	Ris	CVFP	Ruw	CVKB	Sei	CVPF	Soe	CVTH	Sx	CVYS	Tel
CTYM	Rit	CVFQ	Rux	CVKD	Sej	CVPG	Sof	CVTJ	Syl	CVYT	Tem
CTYN	Riu	CVFR	Ruy	CVKF	Sek	CVPH	Sog	CVTK	Sym	CVYW	Ten
CTYP	Riv	CVFS	Ruz	CVKG	Sel	CVPJ	Soh	CVTL	Syn	CVYX	Teo
CTYQ	Riw	CVFT	Rv	CVKH	Sem	CVPK	Soi	CVTM	Syr	CVYZ	Tep
CTYR	Rix	CVFW	Rw	CVKJ	Sen	CVPL	Soj	CVTN	Sys	CVZB	Teq
CTYS	Riy	CVFX	Rx	CVKL	Seo	CVPM	Sok	CVTP	Sz	CVZD	Ter
CTYV	Riz	CVFY	Ry	CVKM	Sep	CVPN	Sol	CVTQ		CVZF	Tes
CTYW	Rj	CVFZ	Rya	CVKN	Seq	CVPQ	Som	CVTR		CVZG	Tet
CTYX	Rk	CVGB	Rye	CVKP	Ser	CVPR	Son	CVTS		CVZH	Teu
CTYZ	Rl	CVGD	Ryi	CVKQ	Ses	CVPS	Soo	CVTW		CVZJ	Tev
CTZB	Rla	CVGF	Ryo	CVKR	Set	CVPT	Sop	CVTX		CVZK	Tew
CTZD	Rle	CVGH	Ryu	CVKS	Seu	CVPW	Soq	CVTY		CVZL	Tex
CTZF	Rli	CVGJ	Ryy	CVKT	Sev	CVPX	Sor	CVTZ		CVZM	Tey
CTZG	Rlo	CVGK	Rz	CVKW	Sew	CVPY	Sos			CVZN	Tez
CTZH	Rlu			CVKX	Sex	CVPZ	Sot			CVZP	Tf
CTZJ	Rly			CVKY	Sey	CVQB	Sou			CVZQ	Tg
CTZK	Rm			CVKZ	Sez	CVQD	Sov			CVZR	Th
CTZL	Rn			CVLB	Sf	CVQF	Sow			CVZS	Tha
CTZM	Ro			CVLD	Sg	CVQG	Sox			CVZT	The
CTZN	Roa			CVLF	Sha	CVQH	Soy			CVZW	Thi
CTZP	Rob			CVLG	She	CVQJ	Soz			CVZX	Tho
CTZQ	Roc			CVLH	Shi	CVQK	Sp			CVZY	Thr
CTZR	Rod			CVLJ	Sho	CVQL	Spa			CWBD	Thu
CTZS	Roe	**S**		CVLK	Sht	CVQM	Spe	**T**		CWBF	Thw
CTZV	Rof	CVGL	Sa	CVLM	Shu	CVQN	Spi	CVWB	Ta	CWBG	Thy
CTZW	Rog	CVGM	Saa	CVLN	Shy	CVQP	Spo	CVWD	Taa	CWBH	Ti
CTZX	Roh	CVGN	Sab	CVLP	Si	CVQR	Spy	CVWF	Tab	CWBJ	Tia
CTZY	Roi	CVGP	Sac	CVLQ	Sia	CVQS	Sq	CVWG	Tac	CWBK	Tib
CVBD	Roj	CVGQ	Sad	CVLR	Sib	CVQT	Squ	CVWH	Tad	CWBL	Tic
CVBF	Rok	CVGR	Sae	CVLS	Sic	CVQW	Sr	CVWJ	Tae	CWBM	Tid
CVBG	Rol	CVGS	Saf	CVLT	Sid	CVQX	Ss	CVWK	Taf	CWBN	Tie
CVBH	Rom	CVGT	Sag	CVLW	Sie	CVQY	St	CVWL	Tag	CWBP	Tif
CVBJ	Ron	CVGW	Sah	CVLX	Sif	CVQZ	Sta	CVWM	Tah	CWBQ	Tig
CVBK	Roo	CVGX	Sai	CVLY	Sig	CVRB	Ste	CVWN	Tai	CWBR	Tih
CVBL	Rop	CVGY	Saj	CVLZ	Sih	CVRD	Sti	CVWP	Taj	CWBS	Tii
CVBM	Roq	CVGZ	Sak	CVMB	Sii	CVRF	Sto	CVWQ	Tak	CWBT	Tij
CVBN	Ror	CVHB	Sal	CVMD	Sij	CVRG	Str	CVWR	Tal	CWBV	Tik
CVBP	Ros	CVHD	Sam	CVMF	Sik	CVRH	Stu	CVWS	Tam	CWBX	Til
CVBQ	Rot	CVHF	San	CVMG	Sil	CVRJ	Sty	CVWT	Tan	CWBY	Tim
CVBR	Rou	CVHG	Sao	CVMH	Sim	CVRK	Su	CVWX	Tao	CWBZ	Tin
CVBS	Rov	CVHJ	Sap	CVMJ	Sin	CVRL	Sua	CVWY	Tap	CWDB	Tio
CVBT	Row	CVHK	Saq	CVMK	Sio	CVRM	Sub	CVWZ	Taq	CWDF	Tip
CVBW	Rox	CVHL	Sar	CVML	Sip	CVRN	Suc	CVXB	Tar	CWDG	Tiq
CVBX	Roy	CVHM	Sas	CVMN	Siq	CVRP	Sud	CVXD	Tas	CWDH	Tir
CVBY	Roz	CVHN	Sat	CVMP	Sir	CVRQ	Sue	CVXF	Tat	CWDJ	Tis
CVBZ	Rp	CVHP	Sau	CVMQ	Sis	CVRS	Suf	CVXG	Tau	CWDK	Tit
CVDB	Rq	CVHQ	Sav	CVMR	Sit	CVRT	Sug	CVXH	Tav	CWDL	Tiu
CVDF	Rr	CVHR	Saw	CVMS	Siu	CVRW	Suh	CVXJ	Taw	CWDM	Tiv
CVDG	Rs	CVHS	Sax	CVMT	Siv	CVRX	Sui	CVXK	Tax	CWDN	Tiw
CVDH	Rt	CVHT	Say	CVMW	Siw	CVRY	Suj	CVXL	Tay	CWDP	Tix
CVDJ	Ru	CVHW	Saz	CVMX	Six	CVRZ	Suk	CVXM	Taz	CWDQ	Tiy
CVDK	Rua	CVHX	Sb	CVMY	Siy	CVSB	Sul	CVXN	Tb	CWDR	Tiz
CVDL	Rub	CVHY	Sc	CVMZ	Siz	CVSD	Sum	CVXP	Tc	CWDS	Tj
CVDM	Ruc			CVNB	Sj	CVSF	Sun	CVXQ	Tce	CWDT	Tja

CWDV	Tje	CWJY	Tug	CWNL	Uay	CWSN	Udr	CWYQ	Ugk	CXFT	Ujd
CWDX	Tji	CWJZ	Tuh	CWNM	Uaz	CWSP	Uds	CWYR	Ugl	CXFV	Uje
CWDY	Tjo	CWKB	Tui	CWNP	Ub	CWSQ	Udt	CWYS	Ugm	CXFW	Ujf
CWDZ	Tju	CWKD	Tuj	CWNQ	Uba	CWSR	Udu	CWYT	Ugn	CXFY	Ujg
CWFB		CWKF	Tuk	CWNR	Ubb	CWST	Udv	CWYV	Ugo	CXFZ	Ujh
CWFD	Tjz	CWKG	Tul	CWNS	Ubc	CWSV	Udw	CWYX	Ugp	CXGB	Uji
CWFG	Tk	CWKH	Tum	CWNT	Ubd	CWSX	Udx	CWYZ	Ugq	CXGD	Ujj
CWFH	Tl	CWKJ	Tun	CWNV	Ube	CWSY	Udy	CWZB	Ugr	CXGF	Ujk
CWFJ	Tla	CWKL	Tuo	CWNX	Ubf	CWSZ	Udz	CWZD	Ugs	CXGH	Ujl
CWFK	Tle	CWKM	Tup	CWNY	Ubg	CWTB	Ue	CWZF	Ugt	CXGJ	Ujm
CWFL	Tli	CWKN	Tuq	CWNZ	Ubh	CWTD	Uea	CWZG	Ugu	CXGK	Ujn
CWFM	Tlo	CWKP	Tur	CWPB	Ubi	CWTF	Ueb	CWZH	Ugv	CXGL	Ujo
CWFN	Tlu	CWKQ	Tus	CWPD	Ubj	CWTG	Uec	CWZJ	Ugw	CXGM	Ujp
CWFP	Tly	CWKR	Tut	CWPF	Ubk	CWTH	Ued	CWZK	Ugx	CXGN	Ujq
CWFQ	Tm	CWKS	Tuu	CWPG	Ubl	CWTJ	Uee	CWZL	Ugy	CXGP	Ujr
CWFR	Tn	CWKT	Tuv	CWPH	Ubm	CWTK	Uef	CWZM	Ugz	CXGQ	Ujs
CWFS	To	CWKV	Tuw	CWPJ	Ubn	CWTL	Ueg	CWZN	Uh	CXGR	Ujt
CWFT	Toa	CWKX	Tux	CWPK	Ubo	CWTM	Ueh	CWZP	Uha	CXGS	Uju
CWFV	Tob	CWKY	Tuy	CWPL	Ubp	CWTN	Uei	CWZQ	Uhb	CXGT	Ujv
CWFX	Toc	CWKZ	Tuz	CWPM	Ubq	CWTP	Uej	CWZR	Uhc	CXGV	Ujw
CWFY	Tod	CWLB	Tv	CWPN	Ubr	CWTQ	Uek	CWZS	Uhd	CXGW	Ujx
CWFZ	Toe	CWLD	Tva	CWPQ	Ubs	CWTR	Uel	CWZT	Uhe	CXGY	Ujy
CWGB	Tof	CWLF	Tve	CWPR	Ubt	CWTS	Uem	CWZV	Uhf	CXGZ	Ujz
CWGD	Tog	CWLG	Tvi	CWPS	Ubu	CWTV	Uen	CWZX	Uhg	CXHB	Uk
CWGF	Toh	CWLH	Tvo	CWPT	Ubv	CWTX	Ueo	CWZY	Uhh	CXHD	Uka
CWGH	Toi	CWLJ	Tvu	CWPV	Ubw	CWTY	Uep	CXBD	Uhi	CXHF	Ukb
CWGJ	Toj	CWLK	Tvy	CWPX	Ubx	CWTZ	Ueq	CXBF	Uhj	CXHG	Ukc
CWGK	Tok	CWLM	Tw	CWPY	Uby	CWVB	Uer	CXBG	Uhk	CXHJ	Ukd
CWGL	Tol	CWLN	Tx	CWPZ	Ubz	CWVD	Ues	CXBH	Uhl	CXHK	Uke
CWGM	Tom	CWLP	Ty	CWQB	Uc	CWVF	Uet	CXBJ	Uhm	CXHL	Ukf
CWGN	Ton	CWLQ	Tya	CWQD	Uca	CWVG	Ueu	CXBK	Uhn	CXHM	Ukg
CWGP	Too	CWLR	Tye	CWQF	Ucb	CWVH	Uev	CXBL	Uho	CXHN	Ukh
CWGQ	Top	CWLS	Tyi	CWQG	Ucc	CWVJ	Uew	CXBM	Uhp	CXHP	Uki
CWGR	Toq	CWLT	Tyo	CWQH	Ucd	CWVK	Uex	CXBN	Uhq	CXHQ	Ukj
CWGS	Tor	CWLV	Tyu	CWQJ	Uce	CWVL	Uey	CXBP	Uhr	CXHR	Ukk
CWGT	Tos	CWLX	Tyy	CWQK	Ucf	CWVM	Uez	CXBQ	Uhs	CXHS	Ukl
CWGV	Tot	CWLY	Tyz	CWQL	Ucg	CWVN	Uf	CXBR	Uht	CXHT	Ukm
CWGX	Tou	CWLZ		CWQM	Uch	CWVP	Ufa	CXBS	Uhu	CXHV	Ukn
CWGY	Tov			CWQN	Uci	CWVQ	Ufb	CXBT	Uhv	CXHW	Uko
CWGZ	Tow			CWQP	Ucj	CWVR	Ufc	CXBV	Uhw	CXHY	Ukp
CWHB	Tox			CWQR	Uck	CWVS	Ufd	CXBW	Uhx	CXHZ	Ukq
CWHD	Toy			CWQS	Ucl	CWVT	Ufe	CXBY	Uhy	CXJB	Ukr
CWHF	Toz			CWQT	Ucm	CWVX	Uff	CXBZ	Uhz	CXJD	Uks
CWHG	Tp			CWQV	Ucn	CWVY	Ufg	CXDB	Ui	CXJF	Ukt
CWHJ	Tq			CWQX	Uco	CWVZ	Ufh	CXDF	Uia	CXJG	Uku
CWHK	Tr			CWQY	Ucp	CWXB	Ufi	CXDG	Uib	CXJH	Ukv
CWHL	Tra			CWQZ	Ucq	CWXD	Ufj	CXDH	Uic	CXJK	Ukw
CWHM	Tre			CWRB	Ucr	CWXF	Ufk	CXDJ	Uid	CXJL	Ukx
CWHN	Tri		**U**	CWRD	Ucs	CWXG	Ufl	CXDK	Uie	CXJM	Uky
CWHP	Tro	CWMB	Ua	CWRF	Uct	CWXH	Ufm	CXDL	Uif	CXJN	Ukz
CWHQ	Tru	CWMD	Uaa	CWRG	Ucu	CWXJ	Ufn	CXDM	Uig	CXJP	Ul
CWHR	Try	CWMF	Uab	CWRH	Ucv	CWXL	Ufo	CXDN	Uih	CXJQ	Ula
CWHS	Ts	CWMG	Uac	CWRJ	Ucw	CWXM	Ufp	CXDP	Uii	CXJR	Ulb
CWHT	Tsa	CWMH	Uad	CWRK	Ucx	CWXN	Ufq	CXDQ	Uij	CXJS	Ulc
CWHV	Tsc	CWMJ	Uae	CWRL	Ucy	CWXP	Ufr	CXDR	Uik	CXJT	Uld
CWHX	Tse	CWMK	Uaf	CWRM	Ucz	CWXQ	Ufs	CXDS	Uil	CXJV	Ule
CWHY	Tsh	CWML	Uag	CWRN	Ud	CWXR	Uft	CXDT	Uim	CXJW	Ulf
CWHZ	Tsi	CWMN	Uah	CWRP	Uda	CWXS	Ufu	CXDV	Uin	CXJY	Ulg
CWJB	Tsj	CWMP	Uai	CWRQ	Udb	CWXV	Ufv	CXDW	Uio	CXJZ	Ulh
CWJD	Tsk	CWMQ	Uaj	CWRS	Udc	CWXW	Ufw	CXDY	Uip	CXKB	Uli
CWJF	Tsl	CWMR	Uak	CWRT	Udd	CWXX	Ufx	CXDZ	Uiq	CXKD	Ulj
CWJG	Tso	CWMS	Ual	CWRV	Ude	CWXY	Ufy	CXFB	Uir	CXKF	Ulk
CWJH	Tss	CWMT	Uam	CWRX	Udf	CWXZ	Ufz	CXFD	Uis	CXKG	Ull
CWJK	Tst	CWMV	Uan	CWRY	Udg	CWYB	Ug	CXFG	Uit	CXKH	Ulm
CWJL	Tsu	CWMX	Uao	CWRZ	Udh	CWYD	Uga	CXFH	Uiu	CXKJ	Uln
CWJM	Tsz	CWMY	Uap	CWSB	Udi	CWYF	Ugb	CXFJ	Uiv	CXKL	Ulo
CWJN	Tt	CWMZ	Uaq	CWSD	Udj	CWYG	Ugc	CXFK	Uiw	CXKM	Ulp
CWJP	Tu	CWNB	Uar	CWSF	Udk	CWYH	Ugd	CXFL	Uix	CXKN	Ulq
CWJQ	Tua	CWND	Uas	CWSG	Udl	CWYJ	Uge	CXFM	Uiy	CXKP	Ulr
CWJR	Tub	CWNF	Uat	CWSH	Udm	CWYK	Ugf	CXFN	Uiz	CXKQ	Uls
CWJS	Tuc	CWNG	Uau	CWSJ	Udn	CWYL	Ugg	CXFP	Uj	CXKR	Ult
CWJT	Tud	CWNH	Uav	CWSK	Udo	CWYM	Ugh	XFQ	Uja	CXKS	Ulu
CWJV	Tue	CWNJ	Uaw	CWSL	Udp	CWYN	Ugi	CXFR	Ujb	CXKT	Ulv
CWJX	Tuf	CWNK	Uax	CWSM	Udq	CWYP	Ugj	CXFS	Ujc	CXKV	Ulw

CXKW	Ulx	CXPZ	Uoq	CXVD	Urj	CYBH	Uuc	CYHK	Uww	CYLS	Vea
CXKY	Uly	CXQB	Uor	CXVF	Urk	CYBJ	Uud	CYHL	Uwx	CYLT	Veb
CXKZ	Ulz	CXQD	Uos	CXVG	Url	CYBK	Uue	CYHM	Uwy	CYLV	Vec
CXLB	Um	CXQF	Uot	CXVH	Urm	CYBL	Uuf	CYHN	Uwz	CYLW	Ved
CXLD	Mma	CXQG	Uon	CXVJ	Urn	CYBM	Uug	CYHP	Ux	CYLX	Vee
CXLF	Umb	CXQH	Uov	CXVK	Uro	CYBN	Uuh	CYHQ	Uxa	CYLZ	Vef
CXLG	Umc	CXQJ	Uow	CXVL	Urp	CYBP	Uui	CYHR	Uxb	CYMB	Veg
CXLH	Umd	CXQK	Uox	CXVM	Urq	CYBQ	Uuj	CYHS	Uxc	CYMD	Veh
CXLJ	Ume	CXQL	Uoy	CXVN	Urr	CYBR	Uuk	CYHT	Uxd	CYMF	Vei
CXLK	Umf	CXQM	Uoz	CXVP	Urs	CYBS	Uul	CYHV	Uxe	CYMG	Vej
CXLM	Umg	CXQN	Up	CXVQ	Urt	CYBT	Uum	CYHW	Uxf	CYMH	Vek
CXLN	Umh	CXQP	Upa	CXVR	Uru	CYBV	Uun	CYHX	Uxg	CYMJ	Vel
CXLP	Umi	CXQR	Upb	CXVS	Urv	CYBW	Uuo	CYHZ	Uxh	CYMK	Vem
CXLQ	Umj	CXQS	Upc	CXVT	Urw	CYBX	Uup	CYJB	Uxi	CYML	Ven
CXLR	Umk	CXQT	Upd	CXVW	Urx	CYBZ	Uuq	CYJD	Uxj	CYMP	Veo
CXLS	Uml	CXQV	Upe	CXVY	Ury	CYDB	Uur	CYJF	Uxk	CYMQ	Vep
CXLT	Umm	CXQW	Upf	CXVZ	Urz	CYDF	Uus	CYJG	Uxl	CYMR	Veq
CXLV	Umn	CXQY	Upg	CXWB	Us	CYDG	Uut	CYJH	Uxm	CYMS	Ver
CXLW	Umo	CXQZ	Uph	CXWD	Usa	CYDH	Uuu	CYJK	Uxn	CYMT	Ves
CXLY	Ump	CXRB	Upi	CXWF	Usb	CYDJ	Uuv	CYJL	Uxo	CYMV	Vet
CXLZ	Umq	CXRD	Upj	CXWG	Usc	CYDK	Uuw	CYJM	Uxp	CYMW	Veu
CXMB	Umr	CXRF	Upk	CXWH	Usd	CYDL	Uux	CYJN	Uxq	CYMX	Vev
CXMD	Ums	CXRG	Upl	CXWJ	Use	CYDM	Uuy	CYJP	Uxr	CYMZ	Vew
CXMF	Umt	CXRH	Upm	CXWK	Usf	CYDN	Uuz	CYJQ	Uxs	CYNB	Vex
CXMG	Umu	CXRJ	Upn	CXWL	Usg	CYDP	Uv	CYJR	Uxt	CYND	Vey
CXMH	Umv	CXRK	Upo	CXWM	Ush	CYDQ	Uva	CYJS	Uxu	CYNF	Vez
CXMJ	Umw	CXRL	Upp	CXWN	Usi	CYDR	Uvb	CYJT	Uxv	CYNG	Vf
CXMK	Umx	CXRM	Upq	CXWP	Usj	CYDS	Uvc	CYJV	Uxw	CYNH	Vg
CXML	Umy	CXRN	Upr	CXWQ	Usk	CYDT	Uvd	CYJW	Uxx	CYNJ	Vh
CXMN	Umz	CXRP	Ups	CXWR	Usl	CYDV	Uve	CYJX	Uxy	CYNK	Vi
CXMP	Un	CXRQ	Upt	CXWS	Usm	CYDW	Uvf	CYJZ	Uxz	CYNL	Via
CXMQ	Una	CXRS	Upu	CXWT	Usn	CYDX	Uvg			CYNM	Vib
CXMR	Unb	CXRT	Upv	CXWV	Uso	CYDZ	Uvh			CYNP	Vic
CXMS	Unc	CXRV	Upw	CXWY	Usp	CYFB	Uvi			CYNQ	Vid
CXMT	Und	CXRW	Upx	CXWZ	Usq	CYFD	Uvj			CYNR	Vie
CXMV	Une	CXRY	Upy	CXYB	Usr	CYFG	Uvk			CYNS	Vif
CXMW	Unf	CXRZ	Upz	CXYD	Uss	CYFH	Uvl			CYNT	Vig
CXMY	Ung	CXSB	Uq	CXYF	Ust	CYFJ	Uvm			CYNV	Vih
CXMZ	Unh	CXSD	Uqa	CXYG	Usu	CYFK	Uvn			CYNW	Vii
CXNB	Uni	CXSF	Uqb	CXYH	Usv	CYFL	Uvo			CYNX	Vij
CXND	Unj	CXSG	Uqc	CXYJ	Usw	CYFM	Uvp			CYNZ	Vik
CXNF	Unk	CXSH	Uqd	CXYK	Usx	CYFN	Uvq			CYPB	Vil
CXNG	Unl	CXSJ	Uqe	CXYL	Usy	CYFP	Uvr			CYPD	Vim
CXNH	Unm	CXSK	Uqf	CXYM	Usz	CYFQ	Uvs	**V**		CYPF	Vin
CXNJ	Unn	CXSL	Uqg	CXYN	Ut	CYFR	Uvt	CYKB	Va	CYPG	Vio
CXNK	Uno	CXSM	Uqh	CXYP	Uta	CYFS	Uvu	CYKD	Vaa	CYPH	Vip
CXNL	Unp	CXSN	Uqi	CXYQ	Utb	CYFT	Uvv	CYKF	Vab	CYPJ	Viq
CXNM	Unq	CXSP	Uqj	CXYR	Utc	CYFV	Uvw	CYKG	Vad	CYPK	Vir
CXNP	Unr	CXSQ	Uqk	CXYS	Utd	CYFW	Uvx	CYKH	Vae	CYPL	Vis
CXNQ	Uns	CXSR	Uql	CXYT	Ute	CYFX	Uvy	CYKJ	Vaf	CYPM	Vit
CXNR	Unt	CXST	Uqm	CXYV	Utf	CYFZ	Uvz	CYKL	Vag	CYPN	Viu
CXNS	Unu	CXSV	Uqn	CXYW	Utg	CYGB	Uw	CYKM	Vah	CYPQ	Viv
CXNT	Unv	CXSW	Uqo	CXYZ	Uth	CYGD	Uwa	CYKN	Vai	CYPR	Viw
CXNV	Unw	CXSY	Uqp	CXZB	Uti	CYGF	Uwb	CYKP	Vaj	CYPS	Vix
CXNW	Unx	CXSZ	Uqq	CXZD	Utj	CYGH	Uwc	CYKQ	Vak	CYPT	Viy
CXNY	Uny	CXTB	Uqr	CXZF	Utk	CYGJ	Uwd	CYKR	Val	CYPV	Viz
CXNZ	Unz	CXTD	Uqs	CXZG	Utl	CYGK	Uwe	CYKS	Vam	CYPW	Vj
CXPB	Uo	CXTF	Uqt	CXZH	Utm	CYGL	Uwf	CYKT	Van	CYPX	Vk
CXPD	Uoa	CXTG	Uqu	CXZJ	Utn	CYGM	Uwg	CYKV	Vao	CYPZ	Vla
CXPF	Uob	CXTH	Uqv	CXZK	Uto	CYGN	Uwh	CYKW	Vap	CYQB	Vlb
CXPG	Uoc	CXTJ	Uqw	CXZL	Utp	CYGP	Uwi	CYKY	Vaq	CYQD	Vlc
CXPH	Uod	CXTK	Uqx	CXZM	Utq	CYGQ	Uwj	CYKZ	Var	CYQF	Vld
CXPJ	Uoe	CXTL	Uqy	CXZN	Utr	CYGR	Uwk	CYLB	Vas	CYQG	Vle
CXPK	Uof	CXTM	Uqz	CXZP	Uts	CYGS	Uwl	CYLD	Vat	CYQH	Vm
CXPL	Uog	CXTN	Ur	CXZQ	Utt	CYGT	Uwm	CYLF	Vau	CYQJ	Vn
CXPM	Uoh	CXTP	Ura	CXZR	Utu	CYGV	Uwn	CYLG	Vav	CYQK	Vo
CXPN	Uoi	CXTQ	Urb	CXZS	Utv	CYGW	Uwo	CYLH	Vaw	CYQL	Voa
CXPQ	Uoj	CXTR	Urc	CXZT	Utw	CYGX	Uwp	CYLJ	Vax	CYQM	Vob
CXPR	Uok	CXTS	Urd	CXZV	Utx	CYGZ	Uwq	CYLK	Vay	CYQN	Voc
CXPS	Uol	CXTV	Ure	CXZW	Uty	CYHB	Uwr	CYLM	Vaz	CYQP	Vod
CXPT	Uom	CXTW	Urf	CXZY	Utz	CYHD	Uws	CYLN	Vb	CYQR	Voe
CXPV	Uon	CXTY	Urg	CYBD	Uu	CYHF	Uwt	CYLP	Vc	CYQS	Vof
CXPW	Uoo	CXTZ	Urh	CYBF	Uua	CYHG	Uwu	CYLQ	Vd	CYQT	Vog
CXPY	Uop	CXVB	Uri	CYBG	Uub	CYHJ	Uwv	CYLR	Ve		

CYQV	Voh	CYVX	Vyz	CZBM	Who	CZHP	Wro
CYQW	Voi	CYVZ	Vz	CZBN	Whu	CZHQ	Wru
CYQX	Voj			CZBP	Why	CZHR	Wry
CYQZ	Vok			CZBQ	Wi	CZHS	Ws
CYRB	Vol			CZBR	Wia	CZHT	Wt
CYRD	Vom			CZBS	Wib	CZHV	Wu
CYRF	Von			CZBT	Wic	CZHW	Wua
CYRG	Voo			CZBV	Wid	CZHX	Wub
CYRH	Vop			CZBW	Wie	CZHY	Wuc
CYRJ	Voq			CZBX	Wif	CZJB	Wud
CYRK	Vor			CZBY	Wig	CZJD	Wue
CYRL	Vos		**W**	CZDB	Wih	CZJF	Wuf
CYRM	Vot			CZDF	Wii	CZJG	Wug
CYRN	Vov	CYWB	Wa	CZDG	Wij	CZJH	Wuh
CYRP	Vow	CYWD	Waa	CZDH	Wik	CZJK	Wui
CYRQ	Vox	CYWF	Wab	CZDJ	Wil	CZJL	Wuj
CYRS	Voy	CYWG	Wac	CZDK	Wim	CZJM	Wuk
CYRT	Voz	CYWH	Wad	CZDL	Win	CZJN	Wul
CYRV	Vp	CYWJ	Wae	CZDM	Wio	CZJP	Wum
CYRW	Vq	CYWK	Waf	CZDN	Wip	CZJQ	Wun
CYRX	Vr	CYWL	Wag	CZDP	Wiq	CZJR	Wuo
CYRZ	Vra	CYWM	Wah	CZDQ	Wir	CZJS	Wup
CYSB	Vre	CYWN	Wai	CZDR	Wis	CZJT	Wur
CYSD	Vri	CYWP	Waj	CZDS	Wit	CZJV	Wus
CYSF	Vro	CYWQ	Wak	CZDT	Wiu	CZJW	Wut
CYSG	Vru	CYWR	Wal	CZDV	Wiv	CZJX	Wuv
CYSH	Vry	CYWS	Wam	CZDW	Wiw	CZJY	Wux
CYSJ	Vs	CYWT	Wan	CZDX	Wix	CZKB	Wuy
CYSK	Vt	CYWV	Wao	CZDY	Wiy	CZKD	Wuz
CYSL	Vu	CYWX	Wap	CZFB	Wiz	CZKF	Wv
CYSM	Vua	CYWZ	Waq	CZFD	Wj	CZKG	Wva
CYSN	Vub	CYXB	War	CZFG	Wk	CZKH	Wx
CYSP	Vuc	CYXD	Was	CZFH	Wl	CZKJ	Wxa
CYSQ	Vud	CYXF	Wat	CZFJ	Wla	CZKL	Wxo
CYSR	Vue	CYXG	Wau	CZFK	Wle	CZKM	Wxu
CYST	Vuf	CYXH	Wav	CZFL	Wli	CZKN	Wy
CYSV	Vug	CYXJ	Waw	CZFM	Wlo	CZKP	Wya
CYSW	Vuh	CYXK	Wax	CZFN	Wlu	CZKQ	Wye
CYSX	Vui	CYXL	Way	CZFP	Wly	CZKR	Wyi
CYSZ	Vuj	CYXM	Waz	CZFQ	Wm	CZKS	Wyj
CYTB	Vuk	CYXN	Wb	CZFS	Wn	CZKT	Wyl
CYTD	Vul	CYXP	Wc	CZFT	Wo	CZKV	Wym
CYTF	Vum	CYXQ	Wd	CZFV	Woa	CZKW	Wyn
CYTG	Vun	CYXR	We	CZFW	Wob	CZKX	Wyo
CYTH	Vuo	CYXS	Wea	CZFX	Woc	CZKY	Wyp
CYTJ	Vup	CYXT	Web	CZFY	Wod	CZLB	Wyr
CYTK	Vuq	CYXV	Wec	CZGB	Woe	CZLD	Wys
CYTL	Vur	CYXW	Wed	CZGD	Wof	CZLF	Wyt
CYTM	Vus	CYXZ	Wee	CZGF	Wog	CZLG	Wyv
CYTN	Vut	CYZB	Weh	CZGG	Woh	CZLH	Wyx
CYTP	Vuu	CYZD	Wei	CZGH	Woi	CZLJ	Wyz
CYTQ	Vuv	CYZF	Wej	CZGJ	Woj	CZLK	Wz
CYTR	Vuw	CYZG	Wek	CZGK	Wok	CZLM	Wza
CYTS	Vux	CYZH	Wel	CZGL	Wol	CZLN	Wze
CYTV	Vuy	CYZJ	Wem	CZGM	Wom	CZLP	Wzi
CYTW	Vuz	CYZK	Wen	CZGN	Won	CZLQ	Wzo
CYTX	Vv	CYZL	Weo	CZGP	Woo	CZLR	Wzu
CYTZ	Vw	CYZM	Wep	CZGQ	Wop	CZLS	W,zy
CYVB	Vx	CYZN	Weq	CZGR	Woq	CZLT	
CYVD	Vy	CYZP	Wer	CZGS	Wor	CZLV	
CYVF	Vya	CYZQ	Wes	CZGT	Wos	CZLW	
CYVG	Vyd	CYZR	Wet	CZGV	Wou	CZLX	
CYVH	Vye	CYZS	Weu	CZGW	Wou	CZLY	
CYVJ	Vyi	CYZT	Wev	CZGX	Wov		
CYVK	Vyl	CYZV	Wew	CZGY	Wow		
CYVL	Vym	CYZW	Wex	CZHB	Wox		
CYVM	Vyn	CYZX	Wey	CZHD	Woy		
CYVN	Vyo	CZBD	Wez	CZHF	Woz		
CYVP	Vyr	CZBF	Wf	CZHJ	Wp		
CYVQ	Vys	CZBG	Wg	CZHL	Wq		
CYVR	Vyt	CZBH	Wh	CZHM	Wr		
CYVS	Vyu	CZBJ	Wha	CZHN	Wri		
CYVT	Vyv	CZBK	Whe				
CYVW	Vyx	CZBL	Whi				

	X.	Code			**Y**
CZMB	Xa	CZRB	Xui	CZSB	Ya
CZMD	Xac	CZRD	Xuo	CZSD	Yal
CZMF	Xae	CZRF	Xy	CZSF	Yam
CZMG	Xai	CZRG	Xya	CZSG	Yan
CZMH	Xain	CZRH	Xyb	CZSH	Yas
CZMJ	Xan	CZRJ	Xyc	CZSJ	Yat
CZMK	Xao	CZRK	Xyd	CZSK	Yav
CZML	Xap	CZRL	Xye	CZSL	Yaw
CZMN	Xar	CZRM	Xyo	CZSM	Ye
CZMP	Xas	CZRN	Xyr	CZSP	Yea
CZMQ	Xat	CZRP	Xz	CZSQ	Yed
CZMR	Xau	CZRQ	Xza	CZSR	Yei
CZMS	Xav	CZRS	Xze	CZST	Yeo
CZMT	Xaw	CZRT	Xzi	CZSV	Yi
CZMV	Xay	CZRV	Xzo	CZSW	Yl
CZMW	Xaz	CZRW	Xzu	CZSX	Yla
CZMX	Xb	CZRX		CZSY	Ylo
CZMY	Xba	CZRY		CZTB	Ym
CZNB	Xbe			CZTD	Yma
CZND	Xbi			CZTF	Ymo
CZNF	Xbo			CZTG	Yn
CZNG	Xbn			CZTH	Yna
CZNH	Xc			CZTJ	Yno
CZNJ	Xca			CZTK	Yo
CZNK	Xce			CZTL	Yp
CZNL	Xci			CZTM	Yr
CZNM	Xco			CZTN	Ys
CZNP	Xcu			CZTP	Ysa
CZNQ	Xd			CZTQ	Yse
CZNR	Xe			CZTR	Ysi
CZNS	Xea			CZTS	Yso
CZNT	Xei			CZTV	Ysn
CZNV	Xeo			CZTW	Yt
CZNW	Xer			CZTX	Yta
CZNX	Xes			CZTY	Yte
CZNY	Xet			CZVB	Yti
CZPB	Xeu			CZVF	Yto
CZPD	Xh			CZVG	Ytu
CZPF	Xi			CZVH	Yu
CZPG	Xia			CZVJ	Yua
CZPH	Xie			CZVK	Yus
CZPJ	Xio			CZVL	Yut
CZPK	Xiu			CZVM	Yx
CZPL	Xiz				
CZPM	Xla				
CZPN	Xle				
CZPQ	Xli				
CZPR	Xlo				
CZPS	Xlu				
CZPT	Xm				
CZPV	Xma				
CZPW	Xme				
CZPX	Xmi				
CZPY	Xmo				
CZQB	Xmu				
CZQD	Xn				
CZQF	Xna				
CZQG	Xni				
CZQH	Xno				
CZQJ	Xo				
CZQK	Xoa				
CZQL	Xoe				
CZQM	Xoi				
CZQN	Xou				
CZQP	Xr				
CZQR	Xra				
CZQS	Xre				
CZQT	Xri				
CZQV	Xro				
CZQW	Xu				
CZQX	Xua				
CZQY	Xue				

			z						
CZVN	Yxa			CZXB	Zoa	CZYF	Zyd	CZYT	a
CZVP	Yxo			CZXD	Zoe	CZYG	Zye	CZYV	o
OZVQ	Yz	CZWB	Za	CZXF	Zon	CZYH	Zyf	CZYW	u
OZVR	Yza	CZWD	Zac	CZXG	Zoi	CZYJ	Zyi	CZYX	å
CZVS	Yze	CZWF	Zai	CZXH	Zos	CZYK	Zyl		
OZVT	Yzi	CZWG	Zao	CZXJ	Zot	CZYL	Zym		
OZVW	Yzo	CZWH	Zd	CZXK	Zov	CZYM	Zyn		
CZVX	Yzu	CZWJ	Ze	CZXL	Zow	CZYN	Zyo		
CZVY		CZWK	Zea	CZXM	Zi	CZYP	Zyp		
		CZWL	Zee	CZXN	Zry	CZYQ	Zyt		
		CZWM	Zei	CZXP	Zu	CZYR			
		CZWN	Zeo	CZXQ	Zua	CZYS			
		CZWP	Zi	CZXR	Zue				
		CZWQ	Zia	CZXS	Zui				
		CZWR	Zie	CZXT	Zuo				
		CZWS	Zio	CZXV	Zv				
		CZWT	Zl	CZXW	Zva				
		CZWV	Zla	CZXY	Zvo				
		CZWX	Zlo	CZYB	Zy				
		CZWY	Zo	CZYD	Zya				

NUMERAL SIGNALS.

Under the arrangement shown below, every flag hoisted after Numeral Signal No 1 has been made, and until Numeral Signal No. 3 or Alphabetical Signal No 1 (*see page* 13) has been made, represents one or more figures, as indicated in the Numeral Table below

If the number to be signalled consists of more than four figures, two or more hoists must be used, as no hoist is to contain more than four flags, and if any figure occurs more than once in the same number, the figure must on its second occurrence begin or be in a second hoist, and on its third occurrence it must begin or be in a third hoist, unless use can be made of the signals K to Z in the Numeral Table below

The following are the signals to be used

Signal	Meaning
CODE FLAG OVER FLAG M	NUMERAL SIGNAL No 1, indicating that the flags hoisted after it, until Numeral Signal No 3 or Alphabetical Signal No 1 is made, do not represent the signals in the Code, but express figures, as indicated in the Numeral Table below, and have the special numerical values there given to them
CODE FLAG OVER FLAG N	NUMERAL SIGNAL No 2, indicating the Decimal Point
CODE FLAG OVER FLAG O	NUMERAL SIGNAL No 3, indicating that the Numeral Signals are ended, the signals which follow are to be looked out in the Code in the usual manner

Numeral Table		Examples
A	1	To make 78,865
B	2	1st hoist, Code Flag over M = *The signals which follow are Numeral Signals and are to be looked out in the Numeral Table*
C	3	
D	4	
E	5	2d hoist, GRFE = 78,865
F	6	3d hoist, Code Flag over O = *The Numeral Signals are ended*
G	7	
H	8	
I	9	To make 787
J	10	1st hoist, Code Flag over M = *The signals which follow are Numeral Signals, and are to be looked out in the Numeral Table*
K	11	
L	22	
M	33	2d hoist, GH = 78 } = 787
N	44	3d hoist, G = 7 }
O	55	4th hoist, Code Flag over O = *The Numeral Signals are ended*
P	66	
Q	77	
R	88	
S	99	To make 9 99876
T	100	1st hoist, Code Flag over M = *The signals which follow are Numeral Signals, and are to be looked out in the Numeral Table*
U	0	
V	00	
W	000	2d hoist, I = 9 }
X	0000	3d hoist, Code Flag over N = *Decimal point* } = 9 99876
Y	00000	4th hoist, SHGF = 99876 }
Z	000000	5th hoist, Code Flag over O = *The Numeral Signals are ended.*

Signals for certain Decimals and Fractions will be found on page 34.

(32)

NUMERAL TABLE.

CODE FLAG UNDER TWO FLAGS.

Signal	Number	Signal	Number	Signal	Number
Code Flag under UA	0	Code Flag under WA	50	Code Flag under YA	100
" " UB	1	" " WB	51	" " YB	200
" " UC	2	" " WC	52	" " YC	300
" " UD	3	" " WD	53	" " YD	400
" " UE	4	" " WE	54	" " YE	500
" " UF	5	" " WF	55	" " YF	600
" " UG	6	" " WG	56	" " YG	700
" " UH	7	" " WH	57	" " YH	800
" " UI	8	" " WI	58	" " YI	900
" " UJ	9	" " WJ	59	" " YJ	1,000
" " UK	10	" " WK	60	" " YK	2,000
" " UL	11	" " WL	61	" " YL	3,000
" " UM	12	" " WM	62	" " YM	4,000
" " UN	13	" " WN	63	" " YN	5,000
" " UO	14	" " WO	64	" " YO	6,000
" " UP	15	" " WP	65	" " YP	7,000
" " UQ	16	" " WQ	66	" " YQ	8,000
" " UR	17	" " WR	67	" " YR	9,000
" " US	18	" " WS	68	" " YS	10,000
" " UT	19	" " WT	69	" " YT	11,000
" " UV	20	" " WU	70	" " YU	12,000
" " UW	21	" " WV	71	" " YV	13,000
" " UX	22	" " WX	72	" " YW	14,000
" " UY	23	" " WY	73	" " YX	15,000
" " UZ	24	" " WZ	74	" " YZ	16,000
" " VA	25	" " XA	75	" " ZA	17,000
" " VB	26	" " XB	76	" " ZB	18,000
" " VC	27	" " XC	77	" " ZC	19,000
" " VD	28	" " XD	78	" " ZD	20,000
" " VE	29	" " XE	79	" " ZE	30,000
" " VF	30	" " XF	80	" " ZF	40,000
" " VG	31	" " XG	81	" " ZG	50,000
" " VH	32	" " XH	82	" " ZH	60,000
" " VI	33	" " XI	83	" " ZI	70,000
" " VJ	34	" " XJ	84	" " ZJ	80,000
" " VK	35	" " XK	85	" " ZK	90,000
" " VL	36	" " XL	86	" " ZL	100,000
" " VM	37	" " XM	87	" " ZM	200,000
" " VN	38	" " XN	88	" " ZN	300,000
" " VO	39	" " XO	89	" " ZO	400,000
" " VP	40	" " XP	90	" " ZP	500,000
" " VQ	41	" " XQ	91	" " ZQ	600,000
" " VR	42	" " XR	92	" " ZR	700,000
" " VS	43	" " XS	93	" " ZS	800,000
" " VT	44	" " XT	94	" " ZT	900,000
" " VU	45	" " XU	95	" " ZU	1,000,000
" " VW	46	" " XV	96	" " ZV	2,000,000
" " VX	47	" " XW	97	" " ZW	3,000,000
" " VY	48	" " XY	98	" " ZX	4,000,000
" " VZ	49	" " XZ	99	" " ZY	5,000,000

See also page 32 for alternative method of making Numeral Signals
For Decimals and Fractions, see page 34

EXAMPLES

Numbers which can not be represented in one hoist to be shown as follows

To make 10,011
YS over Code Flag = 10,000
UL over Code Flag = 11
_____
10,011

To make 97½
XW over Code Flag = 97
BDF = ½
_____
97½

To make 165 09
YA over Code Flag = 100
WP over Code Flag = 65
BCX = 09
_____
165 09

(33)

DECIMALS AND FRACTIONS

BCO	DECIMAL			BDK	FRACTION—*Continued*
BCP	— 01, *or*, $\frac{1}{100}$			BDL	— $\frac{1}{16}$
BCQ	— 02, *or*, $\frac{2}{100}$, *or*, $\frac{1}{50}$			BDM	— $\frac{2}{16}$
BCR	— 03, *or*, $\frac{3}{100}$			BDN	— $\frac{3}{16}$
BCS	— 04, *or*, $\frac{4}{100}$, *or*, $\frac{1}{25}$			BDO	— $\frac{4}{16}$
BCT	— 05, *or*, $\frac{5}{100}$, *or*, $\frac{1}{20}$	} 100ths Hundredths		BDP	— $\frac{5}{16}$
BCU	—.06, *or*, $\frac{6}{100}$, *or*, $\frac{3}{50}$			BDQ	— $\frac{6}{16}$ } 16ths Sixteenths.
BCV	— 07, *or*, $\frac{7}{100}$			BDR	— $\frac{7}{16}$
BCW	— 08, *or*, $\frac{8}{100}$, *or*, $\frac{2}{25}$			BDS	— $\frac{8}{16}$
BCX	— 09, *or*, $\frac{9}{100}$			BDT	— $\frac{9}{16}$
				BDU	— $\frac{10}{16}$
BCY	FRACTION			BDV	— $\frac{11}{16}$
BCZ	— $\frac{1}{10}$, *or*, 1			BDW	— $\frac{12}{16}$
BDA	— $\frac{2}{10}$, *or*, 2			BDX	— $\frac{13}{16}$
BDC	— $\frac{3}{10}$, *or*, 3				
BDE	— $\frac{4}{10}$, *or*, 4				
BDF	— $\frac{5}{10}$, *or*, 5	} 10ths Tenths.		BDY	— $\frac{1}{3}$ } 3ds Thirds
BDG	— $\frac{6}{10}$, *or*, 6			BDZ	— $\frac{2}{3}$
BDH	— $\frac{7}{10}$, *or*, 7				
BDI	— $\frac{8}{10}$, *or*, 8				
BDJ	— $\frac{9}{10}$, *or*, 9				

SIGNALS MADE BY ONE FLAG.

***

TO BE USED ONLY BETWEEN VESSELS TOWING AND BEING TOWED.

The flag is to be held in the hand, and only to be shown just above the Gunwale

Signals about—	Flag	Meaning when made—	
		By the ship towed	By the ship towing
CABLES, or HAWSERS	A	—Is the towing cable (or, hawser) fast?	Is the towing cable (or, hawser) fast?
	B	—The towing cable (or, hawser) is fast	The towing cable (or, hawser) is fast
	C	—The towing cable (or, hawser) is not fast	The towing cable (or, hawser) is not fast
	D	—Equalize the strain on the cables (or, hawsers)	Equalize the strain on the cables (or, hawsers)
	E	—Can not slack more cable (or, hawser)	Can not slack more cable (or, hawser)
	F	—Shorten in the port cable (or, hawser)	Shorten in the port cable (or, hawser)
	G	—Shorten in the starboard cable (or, hawser)	Shorten in the starboard cable (or, hawser)
	H	—Shorten in both cables (or, hawsers)	Shorten in both cables (or, hawsers)
	I	—Veer the port cable (or, hawser)	Veer the port cable (or, hawser)
	J	—Veer the starboard cable (or, hawser)	Veer the starboard cable (or, hawser)
	K	—Veer both cables (or, hawsers)	Veer both cables (or, hawsers)
	L	—Cast off the port hawser	Cast off the port hawser
	M	—Cast off the starboard hawser	Cast off the starboard hawser
	N	—Cast off both hawsers	Cast off both hawsers.
SPEED	O	—Go as slow as possible	Am going as slow as possible
	P	—Go easy	Am going easy
	Q	—Go half speed	Am going half speed
	R	—Go full speed	Am going full speed
	S	—Stop the engines	Must stop
STEERING	T	—Steer more to port	Steer more to port
	U	—Steer more to starboard	Steer more to starboard
	V	—Steady as you go	Steady as you go
	W	—Steer six points from the wind	Ship will not steer properly
SAIL	X	—Make sail	Make sail
	Y	—Set fore and aft sails	Set fore and aft sails
SPECIAL	Z	—Man overboard	Man overboard

***

SIGNALS BY NIGHT.

To be used only between ships towing and being towed, and to be made by short flashes of light or blasts of the whistle one second in duration

***

Steer more to port   – –   Steer more to starboard   –

Cast off hawsers   – – – –

(35)

URGENT AND IMPORTANT SIGNALS.

TWO-FLAG SIGNALS.

ABANDON—ASSISTANCE

ABANDON

AB	—Abandon the vessel as fast as possible
AC	—Do not abandon the vessel
AD	—Do not abandon the vessel until the tide has ebbed
AE	—Do you intend to abandon ?
AF	—I do not intend to abandon the vessel
AG	—I must abandon the vessel
AH	—I shall abandon my vessel unless you will keep by us
AI	—I will not abandon you I will remain by you (or vessel indicated)
AJ	—I (or, They) wish to abandon, but have not the means
AK	—Is she (or vessel indicated) abandoned ?

ACCIDENT, or DAMAGE

AL	—Accident, dangerously wounded (or, hurt)
AM	—Accident, want a surgeon (or, doctor)
AN	—Are you (or, Is —, vessel indicated) in a condition to proceed ?
AO	—Are you (or, Is — vessel indicated) mate rially damaged (or, injured)?
AP	—Boat is capsized (bearing, if necessary)
AQ	—Boat is lost
AR	—Boat is stove
AS	—Boat is swamped
AT	—Boiler burst, — men killed
AU	—Boiler burst, — men killed — wounded
AV	—Boiler burst, no one hurt
AW	—Boiler burst, no one seriously hurt
AX	—Boiler burst, number of killed and wounded not yet known
AY	—Boiler burst, several men killed
AZ	—Boiler burst, several men wounded
BA	—Can damage be repaired at sea ?
BC	—Can damage be repaired by yourselves ?
BD	—Damage can be put to rights in — hours
BE	—Damage can be repaired at sea
BF	—Damage can not be repaired at sea (or place indicated)
BG	—Damage or, Defects can not be repaired without assistance
BH	—Damaged, or, Sprung mast, can not carry sail
BI	—Damaged rudder, can not steer
BJ	—Engines broken down, I am disabled
BK	—Has any accident happened ? Is anything the matter ?
BL	—Have you lost all your boats ?
BM	—How long will you (or vessel indicated) be repairing damage ?
BN	—I am very seriously damaged
BO	—I have lost all my boats
BP	—In case of accident (or, necessity)
BQ	—Is your cargo much damaged ?
BR	—Man overboard
BS	—Met with an accident
BT	—Want a boat, man overboard

ACCIDENT—Continued

BU	—When did the accident happen ?
BV	—Where will you go to repair damage ?

AFLOAT AND AGROUND

BW	—Afloat aft
BX	—Afloat forward
BY	—Am afloat
BZ	—Am aground
CA	—Am aground, likely to break up, require immediate assistance
CB	—Are you afloat ?
CD	—Can not be got off (or, afloat) by any means now available
CE	—I am aground, send what immediate as sistance you can
CF	—Lose no time in shoring up
CG	—May be got off (or, afloat) if prompt as sistance be given
CH	—Vessel (or vessel indicated) ashore near — (or, on —)
CI	—Was the tide high when you grounded ?
CJ	—Was the tide low when you grounded ?
CK	—When did you ground ?

ANCHORAGE   (See HARBOR)

ASSISTANCE

	IN DISTRESS, WANT IMMEDIATE ASSIST- ANCE .... NC
	NOTE —See DISTRESS SIGNALS, page 7
CL	—Am disabled, communicate with me
CM	—Am drifting, want assistance
CN	—Can I procure any assistance in the way of — ?
CO	—Can you assist — (vessel indicated) ?
CP	—Can not assist
CQ	—Could not render assistance
CR	—Do not require assistance (or, further assistance)
CS	—Do you require any assistance (or, help) of (or, from)— ?
CT	—Do you require further assistance ?
CU	—Do you require immediate assistance ?
CV	—I require assistance from Lloyd's agent
CW	—Lighthouse, or, Light ship at — wants immediate assistance
CX	—No assistance can be rendered, do the best you can for yourselves
CY	—Render all the assistance possible—to —
CZ	—Require help (or assistance) of (or, from)—
DA	—Rudder disabled, will you assist me (into — port indicated, if necessary)?
DB	—Send immediate assistance
DC	—We are coming to your assistance
DE	—Will you assist me (or vessel indicated) ?
DF	—With some assistance I shall be able to set things to rights

ATTENTION—CAUTION

	ATTENTION or, **DEMAND**
DG	—Attend to signals
DH	—Attention is called to page —, paragraph —, of the INTERNATIONAL CODE
	N B —*The page to be signaled first and then the paragraph This signal should be employed only between vessels using the United States edition, as otherwise the pages referred to will not correspond*
DI	—*Calls* the attention of the FIRST (or, nearest) ship on the bearing (FROM *the person signaling*) pointed out by compass signal
DJ	—*Calls* the attention of the SECOND ship, etc
DK	—*Calls* the attention of the THIRD ship, etc
DL	—*Calls* the attention of the FOURTH ship, etc
DM	—*Calls* the attention of the FIFTH ship, etc
DN	—*Calls* the attention of the SIXTH ship, etc
DO	—*Calls* the attention of the shore signal station on the bearing (FROM *the person signaling*) pointed out by compass signal
DP	—*Calls* the attention of the vessel COMING into the anchorage on the bearing (FROM *the person signaling*) pointed out by compass signal
DQ	—*Calls* the attention of vessel LEAVING the anchorage on the bearing (FROM *the person signaling*) pointed out by compass signal
DR	—*Calls* the attention of the vessel whose distinguishing signal follows
DS	—Look out Pay attention
DT	—Pay strict attention to signals during the night
DU	—Repeat ship s name, your flags were not made out
DV	—Show your distinguishing signal
DW	—Show your ensign
DX	—Vessels just arrived, show your distinguishing signals
DY	—Vessels just arrived, show your ensigns
DZ	—Vessels just weighed (or, leaving), make your distinguishing signals
EA	—Vessels that wish to be reported make your distinguishing signals
EB	—Vessels wishing to be reported on leaving, make your distinguishing signals
EC	—What ship is that (*bearing to be indicated, if necessary*)?
	BEARINGS
	NOTE —*Unless otherwise indicated bearings in points and ¼-points are to be understood as* COMPASS *bearings, i e, bearings as shown by the Compass and uncorrected for either local attraction or variation But bearings when given in degrees are to be considered as* MAGNETIC *bearings, i e, Compass bearings corrected for local attraction due to the vessel (deviation) Both are to be reckoned* FROM *the vessel signaling the bearing*
ED	—At — (*time indicated*), — (*place indicated*) bore —, distant — miles
EF	—Best berth (or, anchorage) bears from me —
EG	—Entrance bears —
EH	—How did the land bear when last seen?
EJ	—How does the entrance bear?
EK	—Place indicated bears —

	BEARINGS—*Continued.*
EL	—There are moorings on — (*bearing indicated*)
EM	—There is a strange sail on — (*bearing indicated*)
EN	—There is a telegraph cable on — (*bearing indicated*)
EO	—What are the bearing and distance of — (*ship or place indicated*)?
	BOATS
EP	—Bank is encumbered by fishing boats
EQ	—Bar is impassable for boats on the ebb tide
ER	—Boat, or Lifeboat, can not come
ES	—Boat in distress (*bearing, if necessary*)
ET	—Boat is coming to your assistance
EU	—Boat is going to you
EV	—Boat is on board (or, alongside)
EW	—Can boats land? Can landing be effected?
EX	—Can not send a boat
EY	—Do not attempt to land in your own boats
EZ	—Do you want a lifeboat?
FA	—Have no lifeboat No lifeboat here
FB	—Have you seen (or, heard) anything of my (or, the) boat?
FC	—I should like to know the nature of the sickness, if any, before I send my boat (or, communicate)
FD	—Let your boats keep to windward until they are picked up
FE	—Lifeboat is going to you
FG	—No boat available
FH	—Send a boat (*number, if necessary*)
FI	—Send boats suitable for landing passengers
FJ	—Send lifeboat to save crew
FK	—Vessel indicated wants boats
FL	—We have sent for a lifeboat
	CAUTION, or, **DANGER**
FM	—Allow no communication Allow no person on board
FN	—Appearances are threatening, be on your guard
FO	—Are you in danger?
FP	—Bad weather is expected
FQ	—Bar, or Entrance, is dangerous
FR	—Bar is impassable
FS	—Be very careful in your intercourse with strangers
FT	—Caution is requisite, take care
FU	—Channel, or Fairway, is dangerous
FV	—Dangerous to (or, Do not) come into less than — feet of water
FW	—Do not attempt to make the harbor (or, anchorage)
FX	—Do not risk an anchorage unless you have very good ground tackle
FY	—Do not trust to the weather ; it has not done yet
FZ	—Heavy weather coming, look sharp
GA	—Is there any danger—in — (of, or, from)?
GB	—It is dangerous to allow too many people on board at once
GC	—It is dangerous to lose the wind
GD	—Prepare for a hurricane
GE	—Reef or Shoal stretches a long way out
GF	—Shoal water, or Danger, in direction indicated
GH	—Stranger (or vessel on bearing indicated) is suspicious
GI	—The attempt is dangerous

CAUTION—DAMAGE

CAUTION—*Continued*

GJ —Thick fog coming on
GK —Wait until the weather moderates
GL —We are in danger of parting
GM —You are all clear of danger
GN —You are in a dangerous (or, unsafe) position   Your position is dangerous
You are standing into danger ....JD
GO —You are within gunshot, or, You are within reach of guns (or, of batteries)

CHRONOMETER

GP —My chronometer gains daily —
GQ —My chronometer has run down
GR —My chronometer is fast on Greenwich (or, first meridian) mean time
GS —My chronometer is slow on Greenwich (or, first meridian) mean time
GT —My chronometer loses daily —
GU —Will you give me a comparison? Wish to get a rate for my chronometer
NOTE —*The vessel showing the mean time will hoist the signal indicating the hour, and dip it sharply shortly after it has been answered The minutes and seconds at the instant of dipping will immediately follow To insure accuracy, a second comparison should be made*
GV —Would like to have another comparison

COAL

GW —Are you (or, Is —, *vessel indicated*) in want of coal?
GX —Can coal be had at — ?
GY —Can you spare me coal ?
GZ —Coal — a ton at —
HA —Coal can be got at — (or, from —)
HB —In want of coal (*amount to follow*)
HC —Indicate nearest place I can get coal
HD —No coal to be got at —
HE —Owners desire me to inform you to coal at —
HF —What quantity of coal have you left ?

COLLISION

HG —Has the vessel with which you have been in collision proceeded on her voyage ?
HI —Have you been in collision ?
HJ —I have been in collision with —
HK —I have sunk (or, run down) a vessel
HL —My ship has lost her headway, you may feel your way past me
HM —Vessel seriously damaged, wish to transfer passengers
HN —With what vessel have you been in collision ?

COMBUSTIBLES

HO —Have you any combustibles on board?
HP —I am loaded with combustibles
HQ —I have no combustibles on board
HR —No combustibles (or, explosives) near the fire
HS —Vessels with combustibles not allowed in

COMMUNICATE COMMUNICATION

HT —Ask stranger (or vessel *indicated*) if he will communicate

COMMUNICATE—*Continued*

HU —Can you forward my communication by telegraph ?
Come nearer, Stop, or, Heave-to I have something important to communicate
Code Flag over H
HV —Do not have any communication with —
HW —Do you (or, Does —, *vessel indicated*) call anywhere (or, at —)?
HX —Following communication is confidential (or, private)
HY —Forward my communication by telegraph, and pay for its transmission
HZ —Have no communication with the shore (or vessel *indicated*).
IA —Have received the following communication (or, instructions) from your owner (or, agents)
IB —Have you any message (telegram, orders, or, communication) for me?
IC —Have you had any communication with infected vessels (or, places)?
ID —Heave to, or I will fire into you
IE —I am going to signal the contents of an important telegram, to be communicated to you
IF —I can not stop to have (or receive) any communication
IG —I have had no communication with — (place or vessel *indicated*)
IH —Important to communicate by (or, with) —
IJ —Unless your communication is very important, I must be excused
IK —Will you forward (or, communicate) the following signal for me ?
IL —Will you leave a message for me ?
IM —Will you telegraph to — (*ship or person named*) the intelligence I am about to communicate ?

COMPANY

IN —Let us keep together (or, in company) for mutual support (or, protection)
IO —Parted company with — on the — (or, at —)
IP —Shall we keep company ?
IQ —What company do you belong to ?
IR —When did you part company from — ?

CONVOY

IS —Convoy to disperse and make for destination
IT —Convoy to disperse and reassemble at rendezvous (*or place indicated*)
IU —Convoy to keep ahead of escort
IV —Convoy to keep as close as possible
IW —Convoy to keep astern of leader (or, escort)
IX —Convoy to keep on port side of leader (or, escort)
IY —Convoy to keep on starboard side of leader (or, escort)
IZ —Convoy to spread as far as possible, keeping within signal distance
JA —Do not go through (or, near) — (*place indicated*) without a convoy
JB —Rendezvous at — (*port indicated or on bearing indicated*) from — (*place indicated*) distant — miles
JC —Rendezvous in latitude —, and longitude —

DAMAGE   (*See* ACCIDENT)

DANGER—DISTRESS

	DANGER (*See also* CAUTION)			**DIRECTIONS TO A VESSEL UNDER WAY**
JD	—You are standing into danger		KY	—Anchor as convenient
			KZ	—Anchor instantly
	DERELICT		LA	—Bar is not dangerous
JE	—Beware of derelict, dangerous to naviga-		LB	—Bear up instantly
	tion		LC	—Buoy, *or* Mark, is not in its proper position
JF	—Derelict reported off — (*or*, in the vicinity		LD	—Channel dangerous without a pilot
	of —)		LE	—Do not anchor on any account
JG	—Have you seen derelict?		LF	—Get her on the other tack (*or*, Get her head
JH	—Saw derelict in latitude — and longitude			round) or you will be on shore
	— on the —		LG	—Go ahead
			LH	—Go ahead easy Easy ahead
			LI	—Go ahead full speed
	DESPATCH DESPATCHES		LJ	—Go astern
JI	—Any letters (*or*, papers) from —?		LK	—Go astern easy Easy astern
JK	—Despatch is necessary		LM	—Go astern full speed
JL	—Go for despatches		LN	—Haul your wind on the port tack
JM	—Had any letters (*or*, despatches) arrived		LO	—Haul your wind on the starboard tack
	at —?		LP	—Heave all aback. Go astern.
JN	—Have (*or* vessel indicated has) despatches		LQ	—Heave to, head off shore
	for —		LR	—Heave to until —
JO	—Have (*or* vessel indicated has) despatches		LS	—It is not safe to go so fast
	from —		LT	—Keep, *or*, Stand closer in, *or*, Keep under
JP	—Have letters (mails, *or* despatches) for			the land
	you (*or vessel or person indicated*)		LU	—Keep in the center of the channel
JQ	—Have you letter (mails, *or*, despatches) for		LV	—Keep on the port side of channel
	me (*or for vessel, person, or place indi-*		LW	—Keep on the starboard side of the channel
	cated)?		LX	—Keep the vessel under command
JR	—I am the bearer of important despatches		LY	—Keep to windward
JS	—I have Government despatches (*or*, —)		LZ	—Leave the buoy (*or*, beacon) to port
JT	—I shall make for the anchorage (*or*, —)		MA	—Leave the buoy (*or*, beacon) to starboard
	with all despatch		MB	—Not room to wear
JU	—Refer to my despatches (*or* letters)		MC	—Open Get farther away
JV	—Refer to your despatches (*or*, letters)		MD	—Put your helm hard-a-port, ship's head to
JW	—Send for despatches (*or*, letters)			go to starboard
JX	—Telegraphic despatches arrived from —		ME	—Put your helm hard-a-starboard, ship's
JY	—Will you take despatches (*or*, letters)?			head to go to port
			MF	—Reduce speed Ease her
	DIRECTIONS FOR SAVING CREW (*See also*		MG	—Stand off Get an offing Put to sea at once
	ABANDON)		MH	—Stand on
JZ	—Boat should endeavor to land where flag		MI	—Steady your helm
	is waved (*or*, light is shown)		MJ	—Steer more to port (*To be kept flying until*
KA	—Endeavor to send a line by boat (cask, kite,			*course is sufficiently altered*)
	raft, etc)		MK	—Steer more to starboard (*To be kept fly-*
KB	—Is the line fast?			*ing until course is sufficiently altered*)
KC	—Keep a light burning		ML	—Stop her
KD	—Landing is impossible		MN	—Stop instantly
KE	—Lights, *or* Fires, will be kept at the best		MO	—Tack instantly
	place for coming on shore		MP	—Wear instantly
KF	—Look out for rocket-line (*or*, line)			
KG	—Look-out will be kept on the beach all			**DISABLED STEAMER** (*See also* ACCIDENT)
	night		MQ	—Engines completely disabled
KH	—Remain by the ship		MR	—Have broken main shaft
			MS	—Have lost screw, *or*, Screw disabled
	DIRECTIONS TO A VESSEL AT ANCHOR		MT	—Have passed steamer with machinery dis-
KI	—Beach your vessel at all risks			abled
KJ	—Cut away your masts		MU	—Have passed steamer with steering gear
KL	—Get steam up as fast as possible			disabled
KM	—Get steam up, report when ready		MV	—Have you seen disabled steamer?
KN	—Get under way as fast as you can		MW	—One screw disabled, can work the other
KO	—Hoist the blue-peter		MX	—Passed disabled steamer at —
KP	—Hold on until high water		MY	—Ship disabled, will you tow me? (Into —
KQ	—Keep steam ready (*or*, up)			*port indicated if necessary*)
KR	—Let go another anchor		MZ	—Steering gear disabled
KS	—Ride it out if possible			
KT	—Shift your berth Your berth is not safe			**DISTRESS**
KU	—Veer more cable			IN DISTRESS, WANT IMMEDIATE ASSIST-
KV	—Weigh and proceed as ordered			ANCE............................NC
KW	—Weigh, Cut, *or*, Ship Wait for nothing			NOTE —*See* DISTRESS SIGNALS, *page 7*
	Get an offing			A vessel (*or vessel indicated*) ashore near —
KX	—You will be aground at low water			(*or*, on —)....................CH

DISTRESS—ICE

	DISTRESS—*Continued*
NA	—Aground, want immediate assistance
	Am aground, likely to break up, require immediate assistance _____ _____ CA
	Am drifting, want assistance _____ CM
	Boat is capsized (*bearing, if necessary*)..AP
NB	Boat in distress (*bearing, if necessary*)..ES
	—Can not save the ship, take people off
NC	Can not stop the leak _____ QH
	—In distress, want immediate assistance
ND	(*See* DISTRESS SIGNALS, *page* 7)
NE	—Did you see a vessel in distress?
	—Do you think we ought to pass a vessel in this distressed condition ?
NF	—Dying for want of water
NG	—Fire gains rapidly, take people off
NH	—Fire, *or,* Leak, want immediate assistance
NI	—Have lost all boats can you take off crew ?
	Have you seen or heard of a vessel wrecked or in distress _____ _____ ZU
NJ	—I am attacked, want assistance Help, I am attacked
NK	—I am dragging, I can veer no more cable, no more anchors to let go
NL	—I am in danger (*or,* shoal water), direct me how to steer
NM	—I am on fire
NO	—I am sinking (*or* on fire), send all available boats to save passengers and crew
NP	—I am unmanageable
NQ	—I have sprung a leak
	I must abandon the vessel _____ ____AG
NR	—I want assistance, please remain by me
	I will not abandon you, I will remain by you (*or vessel indicated*) _____ AI
NS	—In distress, want assistance
	In distress, want immediate assistance_NO
	NOTE—*See* DISTRESS SIGNALS, *page* 7
NT	—Is not vessel (*bearing, if necessary*) in distress?
NU	—Leak is gaining rapidly
	Light-house, *or,* Light ship at — wants immediate assistance____ _____ CW
	Rudder disabled Will you assist me ? (into — *port indicated, if necessary*) _____ DA
	Ship disabled Will you tow me ? (into — *port indicated if necessary*) _____ MY
NV	—Short of provisions Starving
NW	—Vessel indicated in distress, wants immediate assistance
NX	—Vessel indicated appears in distress
NY	—With immediate assistance fire can be extinguished
	DRAFT OF WATER
NZ	—Drawing too much water
OA	—I draw — feet aft and — feet forward
OB	—What draft of water in feet could you lighten to?
OC	—What is your draft of water?
	ENEMY
OD	—Enemy is closing with you, *or,* You are closing the enemy
OE	—Enemy is in sight
OF	—Enemy's cruisers have been seen to the —, steering to the —
OG	—Enemy's fleet has been seen to the — steering to the —
OH	—Enemy's torpedo boats have been seen to the —, steering to the —.

	ENEMY—*Continued*
OI	—Have been chased by the enemy
OJ	—Keep a good lookout, as it is reported that enemy's war vessels are going about disguised as merchantmen
OK	—Saw the enemy s fleet off —
OL	—Vessel in sight is an enemy
	ENGINES
	Ease her Reduce speed _____MF
	Go ahead _____LG
	Go ahead easy Easy ahead_____LH
	Go ahead full speed_____ __ __ _____LI
	Go astern _____LJ
	Go astern easy Easy astern ____ ____LK
	Go astern full speed_____LM
	Stop her _____ML
	Stop instantly_____MN
	FIRE
OM	—Are you on fire ?
ON	—Fire can easily be got at
OP	—Fire difficult to get at
OQ	—Fire gaining
	Fire gains rapidly, take people off_____NG
OR	—Fire in hold amongst the cargo
	Fire, *or* Leak, want immediate assistance _____ _____ NH
	Heave to or I will fire into you _____ID
	I am on fire _____NM
	I am sinking (*or,* on fire), send all available boats to save passengers and crew___NO
OS	—Is fire extinguished ?
	No combustibles (*or,* explosives) near the fire _____HR
OT	—Vessel indicated is on fire
OU	—Where is the fire ?
	With immediate assistance fire can be extinguished _ ____ _____NY
	HARBOR, *or,* ANCHORAGE
	Best anchorage (*or* berth) bears from me — ___ _____EF
OV	—Best berth (*or* anchorage) is in — fathoms
OW	—Channel *or* Fairway, is buoyed
OX	—Harbor, *or* Anchorage, is good enough with winds from —
OY	—Harbor, *or* Anchorage is indifferent
OZ	—Harbor *or* Anchorage, is safe with all winds
PA	—How does the harbor (*or* anchorage) bear?
PB	—Is the anchorage (*or* my anchorage) safe with all winds (*or,* winds from —)?
PC	—Is the channel buoyed?
PD	—Permission is urgently requested to enter harbor
PE	—Unsafe anchorage
	ICE
PF	—Have encountered ice between 30° and 35° of longitude on the —*
PG	—Have encountered ice between 35° and 40° of longitude on the —*
PH	—Have encountered ice between 40 and 45° of longitude on the —*
PI	—Have encountered ice between 45° and 50° of longitude on the —*
PJ	—Have encountered ice between 50° and 55° of longitude on the —*
PK	—Have encountered ice between 55° and 60° of longitude on the —*

*Date to be indicated if necessary

ICE—NEWS AND NEWSPAPERS

	ICE—*Continued*
PL	—Have encountered ice between 60° and 65° of longitude on the —*
PM	—Have encountered ice between 65° and 70° of longitude on the —*
PN	—Have encountered ice between 70° and 75° of longitude on the —*
PO	—Have encountered ice between 75° and 80° of longitude on the —*
PQ	—Have passed ice in latitude — and longitude — on the —*
PR	—Have you fallen in with ice? (*State whether berg or field*)
PS	—What latitude and longitude did you (*or*, they) have ice in?
PT	**WANT A PILOT** NOTE —*See* PILOT SIGNALS, *page 8*
	LAND, LANDMARKS, AND LIGHTS
PU	—All the lights are out along the coast of —
PV	—At — (*time indicated*) light bore —
PW	—Can you give me a leading mark (*or*, direction) for making the land hereabouts (*or at place indicated*)?
PX	—Do not bring the light to the — (*bearing indicated*) of — (*bearing indicated*)
PY	—Do you *or*, Did you get a good look at the land to know exactly where we are?
PZ	—Had we not better run in and make the land?
QA	—I shall keep hold of the land (*or*, light), *or*, I shall keep the land (*or*, light) bearing —
QB	—Indicate the bearing of light (lighthouse, *or*, lightship)
QC	—Keep the light between — (*bearing indicated*) and — (*bearing indicated*)
QD	—Light is not to be depended on
QE	—The lightship at — is not at anchor on her station, *or*, Lightship at — is out of position
QF	—What bearing shall I keep the light (*or*, landmark) on?
	LATITUDE AND LONGITUDE (*See page* 56)
	LEAK
QG	—Can you stop the leak?
QH	—Can not stop the leak
QI	—Do you consider leak dangerous? Fire, *or*, Leak, want immediate assistance ....NH
QJ	—Have you sprung a leak? Are you (*or*, Is —, *vessel indicated*) leaking? I am sinking (*or*, on fire), send all available boats to save the passengers and crew.................... ... NO I have sprung a leak...... ....NQ Leak is gaining rapidly.................NU
QK	—Leak is stopped
QL	—Vessel indicated has sprung a leak
QM	— — feet water in the hold
	LETTERS, *or*, **MAILS** (*See also* DESPATCHES)
QN	—Has the mail arrived—(from — *if necessary*)?
QO	—Has the mail sailed—(for — *if necessary*)?
QP	—I will take mails for you
QR	—Send your letters
QS	—Tell — (*person indicated*) not to forward any more letters for me

	LETTERS—*Continued*
QT	—When does the mail leave?
QU	—Will you forward my letters to —?
	LIFEBOAT
	Boat, *or*, Lifeboat, can not come ......ER
	Do you want a lifeboat . . ..EZ
	Have no lifeboat No lifeboat here....FA
	Lifeboat is going to you ... ...FB
	Send lifeboat to save crew ....FJ
	We have sent for a lifeboat ....FL
	LIGHTS
QV	—Do not show a light on any account
QW	—I will show a light to-night when I alter course
QX	—I will show a light to-night when I tack
QY	—Keep a light at your bowsprit end
	LIGHTS (MASTHEAD AND SIDE) Your lights are out (*or*, want trimming) *Made by* - —— - ——— - *flashes or blasts of a steam whistle* (*See page* 548)
	LIGHTERS
QZ	—Lighter coming off
RA	—Lighter is adrift
RB	—There are no lighters available
RC	—Will you send off lighters as fast as possible?
	MACHINERY BOILERS AND ENGINEERS (*See also* ACCIDENT)
RD	—Anything wrong with the engines?
RE	—Boiler can not be repaired
RF	—Boiler leaking seriously
RG	—Boiler tubes leaking (*or* leaky)
RH	—Can you supply me with anyone to take charge (*or*, act) as engineer?
RI	—Hot bearings
RJ	—Machinery out of order
RK	—I am going to stop, machinery requires adjusting
RL	—The damage to the machinery is not serious, and is such as can be repaired by the vessel's own engineers
	NAVIGATION
RM	—Did you get good observations?
RN	—Have you a Book of Navigation Tables?
RO	—Have you a Book of Sailing Directions?
RP	—Have you a chart of coast (*or place indicated*)?
RQ	—Have you a Nautical Almanac for the current year (*or year indicated*)?
RS	—May I depend on your time?
RT	—No recent observations I did not get an observation
RU	—Reckoning not to be depended on
RV	—What is my position by bearings?
RW	—What is the name of lighthouse (lightship, *or*, point) in sight?
RX	—When did you see the land?
RY	—When were your last observations for time?
RZ	—Where am I? What is my present position?
	NEWS AND NEWSPAPERS
SA	—Are there any men-of-war about?
SB	—Can you lend (*or*, give) me a newspaper?

* Date to be indicated if necessary

NEWS AND NEWSPAPERS—SIGNALS

	NEWS AND NEWSPAPERS—*Continued*
SC	—Newspapers up to — (*date indicated, state where from, if necessary*)
SD	—No news of —
SE	—What is the date of latest news from — ?
SF	—What is the news from — ?
SG	—When did you sail ?
SH	—Where are you bound ?
SI	—Where are you from ?
SJ	—Where is the admiral (or, senior officer)?
SK	—Where is the — fleet (or, squadron)?
SL	—Would you like to see our newspapers ?

	ORDERS
SM	—Have orders (*or*, telegram) for you
SN	—Have orders for you not to touch (*or*, call) at —
SO	—Have orders to telegraph you passing
SP	—Have received orders for you not to proceed without further instructions
SQ	—Have received orders for you to await instructions from owners at —
SR	—Have received orders for you to discharge cargo at —
ST	—Have telegraphed for your orders
SU	—Have you received any orders from — ?
SV	—I will telegraph for your orders if you will await reply
SW	—I wish to obtain orders from my owner, Mr — at —
	N B —*This signal is to be followed by—*
	1 *Vessel's distinguishing signal, if not already made*
	2 *Owner's name by Spelling*
	3 *Owner's address by Spelling, if not in the Geographical Table*
SX	—Keep off and on and await instructions
SY	—Owners desire me to inform you to discharge cargo first at — (*place indicated*)
SZ	—Proceed in execution of previous orders
TA	—Proceed to — (*port indicated*)
TB	—Shall I telegraph for your orders ?
TC	—Telegraph instructions—to —
TD	—There are no orders for you here
TE	—Wait for orders
TF	—You are ordered to proceed to —
TG	—Your orders are at — (*or*, will be at —)
TH	—Your original orders are cancelled , I am directed to inform you to proceed to —

	PILOT
TI	—Dangerous without a pilot
TJ	—Have you seen a pilot vessel (*or*, boat)?
TK	—Pilot boat most likely—(*on bearing indicated*) (*or*, off —)
	Want a pilot ------------------------ PT
	N B —*See* PILOT SIGNALS, page 8
TL	—Where can I get (*or*, look for) a pilot ?

	PORT
TM	—Wish to enter port or dock

	PROVISIONS
TN	—Are you in want of provisions ?
TO	—I have only — days' provisions

	QUARANTINE AND PRATIQUE
TP	—Am, *or*, Is in quarantine (*Number of days to be indicated, if necessary*)
TQ	—Clean bill of health—from —

	QUARANTINE AND PRATIQUE—*Continued.*
TR	—Do you come from any port putting you in quarantine ?
TS	—Have (*or vessel indicated* has) pratique.
TU	—Have you a clean bill of health ?
TV	—Hoist a quarantine flag
TW	—You have pratique

	REPLY
TX	—Wait for answer (*or*, reply)
TY	—What was the answer (*or*, reply)?
TZ	—Will await answer (*or*, reply)
UA	—Will you await answer (*or*, reply)?

	REPORT
UB	—Do you wish to be reported ?
UC	—Report me by post to my owners (*or*, to Mr. —) at —
UD	—Report me by telegraph to Lloyd's
UE	—Report me by telegraph to owners (*or* to Mr —) at —
UF	—Report me by telegraph to "Shipping Gazette"
UG	—Report me to Lloyd's
UH	—Report me (*or*, my communication) to Lloyd's agent at —
UI	—Report me to "New York Herald" Office, London
UJ	—Report me to "New York Herald" Office, New York
UK	—Report me to "New York Herald" Office, Paris
UL	—Report me to the Borsenhalle at Hamburg
UM	—Report me to—
UN	—Report (*the vessel indicated*) all well, in latitude — and longitude — (*indicated*), on the — (*date indicated*)

	SEA
UO	—Calm sea
UP	—Considerable sea
UQ	—Heavy sea Much sea.
UR	—Much swell on
US	—Not much sea
UT	—Rollers Rollers setting in.
UV	—Too much sea.
UW	—Very heavy sea
UX	—What is the sea like ?

	SICK SICKNESS
UY	—Any sickness at — ?
UZ	—Can I land my sick ?
VA	—Is the sickness contagious ?
VB	—Sickness is contagious
VC	—Sickness is not contagious
VD	—What is the sickness ?

	SIGNALS
VE	—Annul the last hoist, I will repeat it
VF	—Annul the whole signal
VG	—What is the name of ship (*or*, signal station) in sight ?
VH	—I am unable to repeat signal from — (*hoist indicated*)
VI	—Repeat your signal, *or*. Repeat your signal from — (*hoist indicated*)
VJ	—I wish to signal, will you come within easy signal distance ?
VK	—Signal is answered
VL	—What is the date of your signal book ?

SIGNALS—TOWING AND TUGS

SIGNALS—Continued

VM	—Can not distinguish your flags, come nearer, or, make distant signals
VN	—Will you repeat the signals being made to me?
VO	—You may work the semaphore

SOUNDINGS

N B —*Depth of water is always to be shown in feet*

VP	—I am in — feet of water
VQ	—What is the depth of water on the bar?
VR	—What is the least depth of water we shall have (or, ought to have)?
VS	—What soundings have you got?
VT	—You will have water enough over the bar (*depth in feet to follow*)
VU	—You will not have less than — feet of water

SPARS

| VW | —Can you rig a jury mast? |
| VX | —Have you any spars that will do for — ? |

STEAM (*See also* KL and KM)

VY	—How long will you be getting up steam?
VZ	—Is steam up? Are you at full speed?
WA	—Shall I get up steam?
WB	—Steam is not ready, will be up in —
WC	—Steam is ready

STEAM WHISTLE, or, SIREN

| WD | —Blow steam whistle (or, siren) at intervals |
| WE | —Will signal with steam whistle (or, siren) during fog (or, darkness) |

STEER. (*See also* DIRECTIONS TO A VESSEL UNDER WAY)

WF	—Alter course to — (*point indicated*)
WG	—My helm is hard-a-port, ship's head going to starboard
WH	—My helm is hard-a-starboard, ship's head going to port
WI	—Port your helm Ship's head to go to starboard
WJ	—Starboard your helm Ship's head to go to port

SURGEON, or, DOCTOR

WK	—Have you a surgeon?
WL	—May I send a sick person to see your surgeon?
WM	—No surgeon available
WN	—Surgeon will come immediately
WO	—Want a surgeon, send me one from the nearest place
WP	—Will your surgeon come on board?

TELEGRAPH AND MESSAGES

WQ	—Answer by telegraph
WR	—Can telegram be forwarded from — ?
WS	—Communication by telegraph is stopped
WT	—Forward reply to my message by telegraph to —
WU	—I have picked up telegraph cable
WV	—In anchoring look out for telegraph cable
WX	—Open telegram for me and signal its contents
WY	—Send following message by post to owners (or, to Mr —) at —

TELEGRAPH AND MESSAGES—Continued

WZ	—Send following message by post to owners (or, to Mr —) at — by signal letters instead of at length in writing
XA	—Send following message by telegraph to owners (or, to Mr —) at —
XB	—Send my message through the telegraph by signal letters
XC	—Shall I open your telegram and signal its contents?
XD	—Telegraph following message to — (*ship or person named*) at —

TIDE

| XE | —Asks precise time of high water and minimum depth at that time |

NOTE —*Reply will be—*
1 *By time signal*
2 *A numeral signal indicating number of feet*

XF	—How is the tide? What tide have we now?
XG	—Keep out of the tide, or current
XH	—What is the rise and fall of the tide?
XI	—When will the tide ebb?
XJ	—When will the tide flow?
XK	—When will the tide turn?

TIME

XL	—My Greenwich (or, first meridian) mean time is —.*
XM	—When will be the best time for crossing the bar?
XN	—Will you show me your Greenwich (or, first meridian) time?*

*In signaling longitude or time, vessels should always reckon from the meridian of Greenwich, except French vessels, which will use the meridian of Paris and Spanish vessels, which will use the meridian of Cadiz If any doubt is entertained, the vessel to which the signal is made should hoist NBL (What is your first meridian?)

NOTE —*The vessel showing the mean time will hoist the signal denoting the hour (see page 58) and dip it sharply shortly after it has been answered, the minutes and seconds at the instant of dipping will immediately follow To assure accuracy, a second comparison should be made*

TORPEDOES AND MINES

XO	—Beware of torpedo boats
XP	—Beware of torpedoes, channel (or, fairway) is mined
XQ	—Channel, or Fairway, is not mined
XR	—Have you seen torpedo boat (or, boats)?
XS	—Is there danger of mines (or, torpedo boats)?
XT	—Saw torpedo boat (or, boats) (*number, if necessary*) at — (or, near —)

TOWING AND TUGS

XU	—Can you take me in tow?
XV	—Can not take you in tow
XW	—I wish to be taken in tow
XY	—Shall I take you in tow?
XZ	—There are no tugs available

TOWING AND TUGS—WRECK

TOWING AND TUGS—*Continued*

YA —Tug is going to you
Want a tug .... .. ----------------YP
NOTE.—*For Signals between ships towing and being towed, see page 35*

WANT
YB —Want a cable (*size to follow*)
YC —Want an anchor
YD —Want an anchor and cable
YE —Want assistance
YF —Want assistance, mutiny
YG —Want a boat immediately (*if more than one, number to follow*)
YH —Want a chart of channel (*or, harbor, etc*)
YI —Want coal immediately
YJ —Want hands
Want immediate assistance. ....NC
(*See* DISTRESS SIGNALS, *page 7*)
YK —Want immediate instructions
YL —Want immediate medical assistance
YM —Want a lighter immediately (*If more than one, number to follow*)
Want a pilot - .. .. --------PT
(*See* PILOT SIGNALS, *page 8*)
YN —Want police
YO —Want provisions immediately
YP —Want a tug (*if more than one, number to follow*)
YQ —Want warp run out
YR —Want water immediately

WAR AND PEACE
YS —Armistice has been arranged
YT - Is peace proclaimed?
YU —Is war declared? *or,* Has war commenced?
YV —Peace is proclaimed
YW —War between — and —
YX —War is declared (*or*, has commenced)

WATER. (*See also* SOUNDINGS)
YZ —Are you in want of water?
ZA —I have only — days' water
Want water immediately.-----------YR

WEATHER
ZB —Bad weather
ZC —Foggy weather
ZD —Gale is expected from —
ZE —I do not like the look of the weather
ZF —Moderate weather
ZG —No weather report (forecast)
Prepare for a hurricane.------------.GD
ZH —Squally weather, blowing hard
ZI —Stormy (boisterous) weather from the—
ZJ —Thick weather
ZK —What is the meteorological forecast for to-day?
ZL —What is the meteorological forecast for to-morrow? ,

WEIGHING
ZM —Are you (*or*, Is — *vessel indicated*) aweigh?
ZN —I am aweigh
ZO —I shall weigh as soon as weather permits (*or*, moderates)
ZP —I shall weigh immediately (*or at time indicated*)
ZQ —Prepare to weigh
ZR —Ready for weighing (*or*, to weigh)
ZS —Weigh immediately (*or, at time indicated*)
ZT —When do you propose sailing?

WRECK
ZU —Have you seen *or* heard of a vessel wrecked *or* in distress?
ZV —Passed a wreck (*date, latitude, and longitude to follow if necessary*), but could not render any assistance, people still on board
ZW —Passed a wreck (*date, latitude, and longitude to follow, if necessary*), but was unable to ascertain whether any people were on board
ZX —Passed a wreck (*date, latitude, and longitude to follow, if necessary*), no one on board
ZY —Will you go to the assistance of wreck? (*Bearing to follow, if necessary*)

COMPASS IN DEGREES.

THREE FLAG SIGNALS.

NOTE —Unless otherwise indicated, bearings when given in degrees are to be understood as being magnetic bearings, i e , compass bearings corrected for deviation

Signal	Degree.	Signal	Degree.	Signal	Degree.	Signal	Degree	Signal	Degree
ABC	—North	ADM	—N 58° E	AFS	—S 68° E	AID	—S 10° E	AKJ	—S 44° W
ABD	—N 1° E	ADN	— 59°	AFT	— 67°	AIE	— 9°	AKL	— 45°
ABE	— 2°	ADO	— 60°	AFU	— 66°	AIF	— 8°	AKM	— 46°
ABF	— 3°	ADP	— 61°	AFV	— 65°	AIG	— 7°	AKN	— 47°
ABG	— 4°	ADQ	— 62°	AFW	— 64°	AIH	— 6°	AKO	— 48°
ABH	— 5°	ADR	— 63°	AFX	— 63°	AIJ	— 5°	AKP	— 49°
ABI	— 6°	ADS	— 64°	AFY	— 62°	AIK	— 4°	AKQ	— 50°
ABJ	— 7°	ADT	— 65°	AFZ	— 61°	AIL	— 3°	AKR	— 51°
ABK	— 8°	ADU	— 66°	AGB	— 60°	AIM	— 2°	AKS	— 52°
ABL	— 9°	ADV	— 67°	AGC	— 59°	AIN	—S 1° E.	AKT	— 53°
ABM	— 10°	ADW	— 68°	AGD	— 58°	AIO	—South	AKU	— 54°
ABN	— 11°	ADX	— 69°	AGE	— 57°			AKV	— 55°
ABO	— 12°	ADY	— 70°	AGF	— 56°			AKW	— 56°
ABP	— 13°	ADZ	— 71°	AGH	— 55°			AKX	— 57°
ABQ	— 14°	AEB	— 72°	AGI	— 54°			AKY	— 58°
ABR	— 15°	AEC	— 73°	AGJ	— 53°	AIP	—S 1° W	AKZ	— 59°
ABS	— 16°	AED	— 74°	AGK	— 52°	AIQ	— 2°	ALB	— 60°
ABT	— 17°	AEF	— 75°	AGL	— 51°	AIR	— 3°	ALC	— 61°
ABU	— 18°	AEG	— 76°	AGM	— 50°	AIS	— 4°	ALD	— 62°
ABV	— 19°	AEH	— 77°	AGN	— 49°	AIT	— 5°	ALE	— 63°
ABW	— 20°	AEI	— 78°	AGO	— 48°	AIU	— 6°	ALF	— 64°
ABX	— 21°	AEJ	— 79°	AGP	— 47°	AIV	— 7°	ALG	— 65°
ABY	— 22°	AEK	— 80°	AGQ	— 46°	AIW	— 8°	ALH	— 66°
ABZ	— 23°	AEL	— 81°	AGR	— 45°	AIX	— 9°	ALI	— 67°
ACB	— 24°	AEM	— 82°	AGS	— 44°	AIY	— 10°	ALJ	— 68°
ACD	— 25°	AEN	— 83°	AGT	— 43°	AIZ	— 11°	ALK	— 69°
ACE	— 26°	AEO	— 84°	AGU	— 42°	AJB	— 12°	ALM	— 70°
ACF	— 27°	AEP	— 85°	AGV	— 41°	AJC	— 13°	ALN	— 71°
ACG	— 28°	AEQ	— 86°	AGW	— 40°	AJD	— 14°	ALO	— 72°
ACH	— 29°	AER	— 87°	AGX	— 39°	AJE	— 15°	ALP	— 73°
ACI	— 30°	AES	— 88°	AGY	— 38°	AJF	— 16°	ALQ	— 74°
ACJ	— 31°	AET	—N 89° E	AGZ	— 37°	AJG	— 17°	ALR	— 75°
ACK	— 32°	AEU	—East	AHB	— 36°	AJH	— 18°	ALS	— 76°
ACL	— 33°			AHC	— 35°	AJI	— 19°	ALT	— 77°
ACM	— 34°			AHD	— 34°	AJK	— 20°	ALU	— 78°
ACN	— 35°			AHE	— 33°	AJL	— 21°	ALV	— 79°
ACO	— 36°			AHF	— 32°	AJM	— 22°	ALW	— 80°
ACP	— 37°	AEV	—S 89° E	AHG	— 31°	AJN	— 23°	ALX	— 81°
ACQ	— 38°	AEW	— 88°	AHI	— 30°	AJO	— 24°	ALY	— 82°
ACR	— 39°	AEX	— 87°	AHJ	— 29°	AJP	— 25°	ALZ	— 83°
ACS	— 40°	AEY	— 86°	AHK	— 28°	AJQ	— 26°	AMB	— 84°
ACT	— 41°	AEZ	— 85°	AHL	— 27°	AJR	— 27°	AMC	— 85°
ACU	— 42°	AFB	— 84°	AHM	— 26°	AJS	— 28°	AMD	— 86°
ACV	— 43°	AFC	— 83°	AHN	— 25°	AJT	— 29°	AME	— 87°
ACW	— 44°	AFD	— 82°	AHO	— 24°	AJU	— 30°	AMF	— 88°
ACX	— 45°	AFE	— 81°	AHP	— 23°	AJV	— 31°	AMG	— 89°
ACY	— 46°	AFG	— 80°	AHQ	— 22°	AJW	— 32°	AMH	—West
ACZ	— 47°	AFH	— 79°	AHR	— 21°	AJX	— 33°		
ADB	— 48°	AFI	— 78°	AHS	— 20°	AJY	— 34°		
ADC	— 49°	AFJ	— 77°	AHT	— 19°	AJZ	— 35°		
ADE	— 50°	AFK	— 76°	AHU	— 18°	AKB	— 36°		
ADF	— 51°	AFL	— 75°	AHV	— 17°	AKC	— 37°	AMI	—N 89° W
ADG	— 52°	AFM	— 74°	AHW	— 16°	AKD	— 38°	AMJ	— 88°
ADH	— 53°	AFN	— 73°	AHX	— 15°	AKE	— 39°	AMK	— 87°
ADI	— 54°	AFO	— 72°	AHY	— 14°	AKF	— 40°	AML	— 86°
ADJ	— 55°	AFP	— 71°	AHZ	— 13°	AKG	— 41°	AMN	— 85°
ADK	— 56°	AFQ	— 70°	AIB	— 12°	AKH	— 42°	AMO	— 84°
ADL	—N 57° E	AFR	—S 69° E	AIC	—S 11° E	AKI	—S 43° W	AMP	—N 83° W

Signal.	Degree	Signal	Degree	Signal	Degree	Signal	Degree	Signal	Degree.
AMQ	—N 82° W	ANJ	—N 65° W	AOD	—N 48° W	AOV	—N 31° W	APN	—N 14° W
AMR	— 81°	ANK	— 64°	AOE	— 47°	AOW	— 30°	APO	— 13°
AMS	— 80°	ANL	— 63°	AOF	— 46°	AOX	— 29°	APQ	— 12°
AMT	— 79°	ANM	— 62°	AOG	— 45°	AOY	— 28°	APR	— 11°
AMU	— 78°	ANP	— 61°	AOH	— 44°	AOZ	— 27°	APS	— 10°
AMV	— 77°	ANQ	— 60°	AOI	— 43°	APB	— 26°	APT	— 9°
AMW	— 76°	ANR	— 59°	AOJ	— 42°	APC	— 25°	APU	— 8°
AMX	— 75°	ANS	— 58°	AOK	— 41°	APD	— 24°	APV	— 7°
AMY	— 74°	ANT	— 57°	AOL	— 40°	APE	— 23°	APW	— 6°
AMZ	— 73°	ANU	— 56°	AOM	— 39°	APF	— 22°	APX	— 5°
ANB	— 72°	ANV	— 55°	AON	— 38°	APG	— 21°	APY	— 4°
ANC	— 71°	ANW	— 54°	AOP	— 37°	APH	— 20°	APZ	— 3°
AND	— 70°	ANX	— 53°	AOQ	— 36°	API	— 19°	AQB	— 2°
ANF	— 69°	ANY	— 52°	AOR	— 35°	APJ	— 18°	AQC	— 1°
ANG	— 68°	ANZ	— 51°	AOS	— 34°	APK	— 17°		North ABC
ANH	— 67°	AOB	— 50°	AOT	— 33°	APL	— 16°		
ANI	—N 66° W	AOC	—N 49° W	AOU	—N 32° W	APM	—N 15° W		

NOTE —True bearings are compass bearings corrected for deviation and variation
Magnetic bearings are compass bearings corrected for deviation only
Compass bearings are bearings as observed and are uncorrected

COMPASS IN POINTS AND HALF POINTS.

(COMPASS BEARINGS)

NOTE —Unless otherwise indicated, bearings when given in points and ½ points are to be understood as being compass bearings, i e, bearings as shown by the compass and uncorrected for either deviation or variation

Signal	Point	Signal	Point.	Signal	Point.	Signal	Point.
AQD	North	AQU	East	ARL	South	ASD	West
AQE	N ½ E	AQV	E ½ S	ARM	S ½ W	ASE	W ½ N
AQF	N by E	AQW	E by S	ARN	S by W	ASF	W by N
AQG	N by E ½ E	AQX	E S E ½ E	ARO	S. by W ½ W	ASG	W N W ½ W
AQH	N N E	AQY	E S. E	ARP	S S W	ASH	W N W
AQI	N N E ½ E	AQZ	S E by E ½ E	ARQ	S S W ½ W	ASI	N W by W ½ W
AQJ	N E by N	ARB	S E by E	ARS	S W by S	ASJ	N W. by W
AQK	N E ½ N	ARC	S. E. ½ E	ART	S W ½ S	ASK	N W ½ W
AQL	N E	ARD	S E	ARU	S W	ASL	N W
AQM	N E ½ E	ARE	S E ½ S	ARV	S W ½ W	ASM	N W ½ N
AQN	N E by E	ARF	S E by S	ARW	S W by W	ASN	N W by N
AQO	N E by E ½ E	ARG	S S E ½ E	ARX	S W by W ½ W	ASO	N N W ½ W
AQP	E N E	ARH	S S E	ARY	W S W	ASP	N N W
AQR	E N E ½ E	ARI	S by E ½ E	ARZ	W S W ½ W	ASQ	N by W ½ W
AQS	E by N	ARJ	S by E	ASB	W by S	ASR	N by W
AQT	E ½ N	ARK	S. ½ E	ASC	W. ½ S	AST	N ½ W
							North ..... AQD

MONEY TABLE.

PRINCIPAL MONEYS IN USE THROUGHOUT THE WORLD.

(See also page 49)

Signal		
ASU	—MONEY	
ASV	—BANK NOTE, or, PAPER MONEY	
ASW	—COIN	

Signal	Coin	Country
ASX	—Anna	{ East Africa / India
ASY	—Argentino	Argentina
ASZ	—Bolivar	Venezuela
ATB	—Candareen, or, Fun	China
ATC	—Cash, Le, or, Sapeque	China
ATD	—Cent	{ Canada / Ceylon / China / Mauritius / Netherlands / Newfoundland / Straits Settlements / United States / Zanzibar
ATE	—Centavo	{ Argentina / Bolivia / Chile / Colombia / Costa Rica / Ecuador / Guatemala / Haiti / Honduras / Mexico / Nicaragua / Paraguay / Peru / Santo Domingo / San Salvador / Uruguay / Venezuela.
ATF	{ Centesimo / Centime / Centimo	{ Italy / Belgium / France / Tunis / Spain }
ATG	—Condor	{ Bolivia / Chile / Colombia
ATH	—Conto	{ Brazil / Portugal

Signal.	Coin	Country
ATI	—Dime	United States
	Dinar (see Leo) AUB	Servia
ATJ	—Dollar	{ Argentina / Bolivia / Canada. / Chile / China / Colombia / Costa Rica / Ecuador / Guatemala / Haiti / Honduras / Mauritius / Mexico / Newfoundland / Nicaragua. / Paraguay / Peru / Santo Domingo / San Salvador / Straits Settlements / United States / Uruguay / Venezuela / Zanzibar
ATK	—Doubloon	{ Chile. / Mexico
ATL	—Drachma	Greece
ATM	—Ducat	{ Austria-Hungary / Netherlands East Indies
ATN	—Eagle	United States
ATO	—Egyptian Pound, or, Lira	Egypt
ATP	—Farthing	Great Britain
ATQ	—Florin, Guilder, Gulden	{ Austria-Hungary / Netherlands
ATR	—Franc	{ Belgium / France / Tunis
	Fun, or, Candareen ATB	China
	Guilder, Gulden, Florin ATQ	{ Austria-Hungary / Netherlands
ATS	—Heller	Austria-Hungary
ATU	—Imperial	Russia
ATV	—Keran, or, Kran	Persia
ATW	—Kopeck	Russia
ATX	—Kreutzer	Austria Hungary
ATY	{ Krona / Krone	{ Sweden / Austria-Hungary / Denmark / Norway }
	Kroner ZNK	Germany

NOTE — The name alone is common to two or more countries, the value may differ

(47)

Signal	Coin	Country	Signal	Coin	Country
ATZ	—Lac --------------- India		AUP	—Pfennig ------------- Germany	
	Le, Cash, or, Sapeque		AUQ	—Piaster ------------- { Egypt, Tripoli, Turkey }	
	Leang, or, Tael -- ATC China		AUR	—Pie, or, Pice ------- { East Africa, India }	
	Leo, or, Leu --------- Roumania		AUS	—Pound, or, Sovereign -- Great Britain	
AUB	{ Dinar ---------- Servia }		AUT	—Real-reis --------- { Brazil, Portugal }	
	{ Lev, or, Lew -------- Bulgaria }		AUV	—Rin ------------- Japan	
AUC	—Lepton ------------ Greece		AUW	—Rouble (paper) ------- Russia	
AUD	—Lira ------------- Italy		AUX	—Rouble (silver) ------ Russia	
	Lira egiziana, or, Egyptian Pound		AUY	—Rupee ------------ { Ceylon, East Africa, India }	
AUE	—Lira turca --------- Turkey			Sapeque, Cash, or, Le	
AUF	—Louis, or, Napoleon -- France			ATC China	
AUG	—Mace, or, Tsin ------ China		AUZ	—Sen ------------- Japan	
AUH	—Mark ------------- Germany		AVB	—Shilling ----------- Great Britain	
AUI	—Milreis ----------- { Brazil, Portugal }			Sol or, Dollar -- ATJ Peru	
AUJ	—Mithkal ----------- Morocco			Sovereign, or, Pound	
	Napoleon, or, Louis			AUS Great Britain	
	AUF France		AVC	—Tael, or, Leang ------ China	
AUK	{ Ochr-el-guerch, or, Ushr-el-ghirsh } ---- Egypt		AVD	—Thaler ------------- { Austria-Hungary }	
			AVE	—Tical ------------- { Burmah, Siam }	
AUL	—Öre -------------- { Denmark, Norway, Sweden }		AVF	—Toman ----------- Persia	
				Tsin, or, Mace -- AUG China	
AUM	—Para ------------- { Egypt, Servia, Tripoli, Turkey }			Ushr-el-ghirsh, or, Ochr el-guerch } AUK Egypt	
				Venezolano, or, Dollar ---------- ATJ Venezuela	
AUN	—Penny-pence --------- Great Britain				
AUO	—Peseta ------------ Spain		AVG	—Yen ----- ----------- Japan	
	Peso, or, Dollar -- ATJ { Chile, Colombia, Paraguay, Venezuela }		AVH		
			AVI		
			AVJ		

(For list of moneys arranged alphabetically, see page 47)

UNITED STATES

10 cents (ATD)	= 1 dime (ATI)
100 cents	= 1 dollar (ATJ)
10 dollars	= 1 eagle (ATN)

ARGENTINE REPUBLIC

100 centavos (ATE)	= 1 dollar (ATJ)
5 dollars	= 1 Argentino (ASY)

AUSTRIA-HUNGARY

100 kreutzer (ATX)	= 1 florin, *or*, gulden (ATQ)
2 1/10 florins	= 1 thaler (AVD)
4 1/2 florins	= 1 ducat (ATM)
	and
100 heller (ATS)	= 1 krone (ATY)

BELGIUM

100 centimes (ATF)	= 1 franc (ATR)

BOLIVIA

100 centavos (ATE)	= 1 dollar (ATJ)
10 dollars	= 1 condor (ATG)

BRAZIL

1,000 reis (AUT)	= 1 milreis (AUI)
1,000 milreis	= 1 conto (ATH)

BULGARIA

Lev, *or*, lew (AUB)

BURMAH

Tical (AVE)
Also currency of British India

CANADA

100 cents (ATD)	= 1 dollar (ATJ)

CEYLON

100 cents (ATD)	= 1 rupee (AUY)

CHILE

100 centavos (ATE)	= 1 dollar, *or*, peso (ATJ)
5 dollars	= 1 doubloon (ATK)
10 dollars	= 1 condor (ATG)

CHINA

10 cash, le, *or*, sapeque	= 1 candareen, *or*, fun (ATC) (ATB)
10 candareens	= 1 mace, *or* tsin (AUG)
10 mace	= 1 tael, *or*, leang (AVC)
	and
100 cents (ATD)	= 1 dollar (ATJ)

COLOMBIA

100 centavos (ATE)	= 1 dollar, *or*, peso (ATJ)
10 pesos	= 1 condor (ATG)

COSTA RICA

100 centavos (ATE)	= 1 dollar (ATJ)

DENMARK

100 ore (AUL)	= 1 krone (ATY)

EAST AFRICA (British and German territories)
The same as British India

ECUADOR

100 centavos (ATE)	= 1 dollar (ATJ)

EGYPT

10 ochr-el-guerch (AUK), *or*,	= 1 piaster (AUQ)
40 pares (AUM)	
100 piasters	= 1 Egyptian pound, *or*, Lira egiziana (ATO)

FRANCE

100 centimes (ATF)	= 1 franc (ATR)
20 francs	= 1 Louis, *or*, Napoleon (AUF)

GERMANY

100 pfennigs (AUP)	= 1 mark (AUH)

GREAT BRITAIN *

4 farthings (ATP)	= 1 penny (AUN)
12 pence	= 1 shilling (AVB)
20 shillings	= 1 pound, *or*, sovereign (AUS)

GREECE

100 lepta (AUC)	= 1 drachma (ATL)

GUATEMALA

100 centavos (ATE)	= 1 dollar (ATJ)

HAITI

100 centavos (ATE)	= 1 dollar (ATJ)

HONDURAS

100 centavos (ATE)	= 1 dollar (ATJ)

HOLLAND (*see* Netherlands)

HUNGARY (*see* Austria-Hungary)

INDIA (BRITISH)

12 pies, *or*, pice (AUR)	= 1 anna (ASX)
16 annas	= 1 rupee (AUY)
100,000 rupees	= 1 lac (ATZ)

ITALY

100 centesimi (ATF)	= 1 lira (AUD)

JAPAN

10 rin (AUV)	= 1 sen (AUZ)
100 sen	= 1 yen (AVG)

MAURITIUS

100 cents (ATD)	= 1 dollar (ATJ)

* This system is used by the British Colonies in Australasia and South Africa.

MEXICO
100 centavos (ATE)	= 1 dollar (ATJ)
16 dollars	= 1 doubloon (ATK)

MOROCCO
Mithkal (AUJ)

NETHERLANDS
100 cents (ATD)	= 1 guilder, or, florin (ATQ)
5¼ gulden	= 1 ducat (ATM) (used in Netherlands East Indies)

NEWFOUNDLAND
100 cents (ATD)	= 1 dollar (ATJ)

NICARAGUA
100 centavos (ATE)	= 1 dollar (ATJ)

NORWAY (see Sweden and Norway).

PARAGUAY
100 centavos (ATE)	= 1 dollar, or, peso (ATJ)

PERSIA
10 kerans (ATV)	= 1 toman (AVF)

PERU
100 centavos (ATE)	= 1 dollar, or, sol (ATJ)

PORTUGAL
1,000 reis (AUT)	= 1 milreis (AUI)
1,000 milreis	= 1 conto (ATH)

ROUMANIA
Leo, or, leu (AUB)

RUSSIA
100 kopecks (ATW)	= 1 rouble (AUX)
10 roubles	= 1 imperial (ATU)

SANTO DOMINGO
100 centavos (ATE)	= 1 dollar (ATJ)

SAN SALVADOR.
100 centavos (ATE)	= 1 dollar (ATJ)

SERVIA
100 paras (AUM)	= 1 dinar (AUB)

SIAM
Tical (AVE)

SPAIN
100 centimos (ATF)	= 1 peseta (AUO)
5 pesetas (AUO)	= 1 peso (ATJ)

STRAITS SETTLEMENTS
100 cents (ATD)	= 1 dollar (ATJ)

SWEDEN AND NORWAY
100 ore (AUL)	= 1 krona, or, krone (ATY)

TRIPOLI
40 paras (AUM)	= 1 piaster (AUQ)

TUNIS
100 centimes (ATF)	= 1 franc (ATR)

TURKEY
40 paras (AUM)	= 1 piaster (AUQ)
100 piasters	= 1 lira turca (AUE)

URUGUAY
100 centavos (ATE)	= 1 dollar (ATJ)

VENEZUELA
100 centavos (ATE) or, 5 Bolivars (ASZ)	= 1 Venezolano, dollar, or, peso (ATJ)

ZANZIBAR
100 cents (ATD)	= 1 dollar (ATJ)

MEASURES AND WEIGHTS OF DIFFERENT COUNTRIES.

MEASURES OF LENGTH.

Signal.	Measure.	Country or System.	Signal	Measure	Country or System.
AVK	—MEASURE OF LENGTH			Knot, *or*, Nautical Mile......AWO	{ United States and Great Britain
AVL	—Alen	{ Denmark Norway.	AWH	—League	{ United States and Great Britain
AVM	—Archine	Russia	AWI	—Li	China
	Archine (*or*, Meter) AWJ	Turkish Metrical	AWJ	{ Meter	Metrical
AVN	—Cable	{ United States and Great Britain		*or* Archine	Turkish Metrical
AVO	{ Centimeter *or* Daktylas	Metrical / Greek Metrical		*or* El	Dutch Metrical
	or Duim	Dutch Metrical.		*or* Pecheus	Greek Metrical.
	or Khat	Turkish Metrical.		Mijle (*or*, Kilometer) AWG	Dutch Metrical
AVP	—Chang, *or*, Cheung...China			Mil (*or*, Kilometer,) AWG	Turkish Metrical
AVQ	—Chih.	China	AWK	—Mil	Denmark
AVR	—Cho, *or*, Tcho	Japan	AWL	—Mil	{ Norway and Sweden
	Daktylas (*or*, Centi meter)......AVO	Greek Metrical	AWM	—Mile (German)	Germany.
AVS	{ Decameter *or* Roede	Metrical / Dutch Metrical.	AWN	—Mile, Statute	{ United States and Great Britain
	Decimeter *or* Palame	Metrical. / Greek Metrical	AWO	{ Mile, Geographical or Nautical, *or*, Knot	{ United States and Great Britain
AVT	*or* Palm	Dutch Metrical		Millimeter *or*	Metrical
	or Parmak	Turkish Metrical	AWP	Gramme *or*	Greek Metrical.
AVU	—Draa, *or*, Pike	{ Egypt. Morocco Tripoli Tunis Turkey		Nokta *or* Streep	Turkish Metrical / Dutch Metrical
	Duim (*or*, Centimeter) AVO	Dutch Metrical	AWQ	Myriameter *or* Skoins	Metrical / Greek Metrical
	El (*or*, Meter) AWJ	Dutch Metrical.		Nokta (*or*, Millimeter)......AWP	TurkishMetrical
AVW	—Fathom	{ United States and Great Britain		Palame (*or*, Decimeter)......AVT	Greek Metrical
AVX	—Fod	{ Denmark Norway		Palm (*or*, Decimeter)......AVT	Dutch Metrical
AVY	—Foot	{ United States and Great Britain		Parmak (*or*, Decimeter)......AVT	Turkish Metrical
AVZ	—Fot	Sweden		Pecheus (*or*, Meter), AWJ	Greek Metrical.
AWB	—Furlong	{ United States and Great Britain.		Pike, *or*, Draa, AVU	{ Egypt. Morocco Tripoli Tunis Turkey
	Gramme (*or*, Millimeter)......AWP	Greek Metrical	AWR	—Ri	Japan
AWC	{ Guz......India / Guz, *or*, Zer...Persia			Roede (*or*, Decameter)......AVS	Dutch Metrical
AWD	—Hectometer	Metrical	AWS	—Saading, *or*, Toung..	Burmah
AWE	—Inch	{ United States and Great Britain.	AWT	—Sagene, *or*, Saschene..	Russia
AWF	—Ken	Japan	AWU	—Shaku	Japan.
	Khat (*or*, Centimeter) AVO	Turkish Metrical		Skoins (*or*, Myriameter......AWQ	Greek Metrical
	Kilometer *or*	Metrical		Stadion (*or*, Kilometer......AWG	Greek Metrical
AWG	{ Mijle *or*	Dutch Metrical	AWV	—Stopa	Russia
	Mil *or*	Turkish Metrical		Streep, *or*, Millimeter......AWP	Dutch Metrical
	Stadion	Greek Metrical			

Signal	Measure	Country or System	Signal	Measure	Country or System
AWX	—Sun	Japan	AXC	—Vershok	Russia
	Tcho, or, Cho...AVR	Japan	AXD	—Verst.........	Russia
AWY	—Tomme...............	Denmark	AXE	—Yard	{ United States and Great Britain
	Toung, or, Saading, AWS	Burmah	AXF	—Yin	China
AWZ	—Tsun	China		Zers, or, Guz...AWC	Persia
AXB	—Vara ----------------	{ Argentina, Brazil, Chile, Mexico, Peru, Portugal, Uruguay		*For list of countries in which the Metrical System is used, and for Table for Converting Metrical Measures and Weights into their American Equivalents, see page 55*	

SQUARE OR SURFACE MEASURES

Signal	Measure	Country or System	Signal	Measure	Country or System
AXH	—SQUARE, OR, SURFACE MEASURE		AXS	—Square Decimeter....	Metrical
AXI	—Acre --------------	{ United States and Great Britain	AXT	—Square Foot	{ United States and Great Britain
AXJ	{ Are, or, Sq Decameter . Metrical or Deunum .... Turkish Metrical or Stremma .... Greek Metrical			Square Hectometer (or, Hectaire) ....AXP	Metrical
	Bunder (or, Hectare) AXP Dutch Metrical		AXU	—Square Inch	{ United States and Great Britain
AXK	—Centiare, or, Sq Meter.	Metrical	AXV	—Square Kilometer....	Metrical
AXL	—Cho ----------------	Japan		Square Meter (or, Centiare).... ...AXK	Metrical
AXM	—Decare	Metrical	AXW	—Square Mile.... ...	{ United States and Great Britain
AXN	—Declare....	Metrical	AXY	—Square Millimeter ...	Metrical
AXO	—Dessatine	Russia	AXZ	—Square Yard.......	{ United States and Great Britain
	Deunum (or, Are) AXJ Turkish Metrical			Stremma (or, Are) AXJ	Greek Metrical
	Djerib (or, Hectare) AXP Turkish Metrical				
AXP	{ Hectare, or, Sq Hectometer .. Metrical or Bunder ------------ Dutch Metrical or Djerib.............. Turkish Metrical			*For list of countries in which the Metrical System is used, and for Table for Converting Metrical Measures and Weights into their American Equivalents, see page 55*	
AXQ	—Mon --------------	China			
AXR	—Square Centimeter..	Metrical			
	Square Decameter (or Are) .... AXJ	Metrical			

MEASURES OF CUBIC CAPACITY OR SOLID MEASURES.

Signal	Measure	Country or System	Signal	Measure	Country or System
AYB	—MEASURE OF CUBIC CAPACITY, OR, SOLID MEASURE Cubic Centimeter ZNM	Metrical	AYL	—Ardeb	Egypt
			AYM	—Arroba	{ Chile, Spain
AYC	—Cubic Foot	{ United States and Great Britain	AYN	—Bale.............	{ United States and Great Britain
AYD	—Cubic Inch	{ United States and Great Britain	AYO	—Barrel.............	{ United States and Great Britain
	Cubic Meter....ZNP Cubic Millimeter ZNO	Metrical Metrical	AYP	—Bushel	{ United States and Great Britain
AYE	—Cubic Yard........	{ United States and Great Britain	AYQ	—Butt............. ...	Great Britain
AYF	—Decastere..........	Metrical	AYR	{ Centiliter.......... Metrical or Mystron............. Greek Metrical or Vingerhoed......... Dutch Metrical or Zarf............... Turkish Metrical	
AYG	—Decistere	Metrical			
AYH	—Stere	Metrical			
AYI	—Ton (ship measurement)	{ United States and Great Britain		*For list of countries in which the Metrical System is used, see page 55*	
AYJ	—MEASURE OF CAPACITY				
AYK	—Almuda	{ Brazil, Portugal			

Signal	Measure	Country or System	Signal.	Measure	Country or System
AYS	—Chtof ....Russia			Maatje (or, Deciliter)	
	Cronchka, or, Kruschka				AYU Dutch Metrical
		AZD Russia	AZG	{ Milliliter ....Metrical	
AYT	{ Decaliter ....Metrical or Kile ....Turkish Metrical or Schepel ....Dutch Metrical			or Kybos ....Greek Metrical	
			AZH	—Minim.......... { United States and Great Britain	
AYU	{ Deciliter......... Metrical or Kotyle .... Greek Metrical or Maatje ....Dutch Metrical			Mud, Vat, or, Zac (or, Hectoliter)...AYZ Dutch Metrical Mystron(or, Centiliter)	
					AYR Greek Metrical
AYV	—Firkin........ { United States and Great Britain		AZI	—Oxehoved.......... Denmark	
AYW	—Gallon .. .. { United States and Great Britain		AZJ	—Peck .... { United States and Great Britain	
AYX	—Gill .... { United States and Great Britain		AZK	—Pint............ { United States and Great Britain	
			AZL	—Puncheon . ...Great Britain	
AYZ	{ Hectoliter ....Metrical or Kileh ....Turkish Metrical or Koilon ....Greek Metrical or Mud, Vat, or, Zac....Dutch Metrical		AZM	{ Quart ....United States and Great Britain or Sheng ....China	
AZB	—Hogshead........... { United States and Great Britain		AZN	—Quarter .... { United States and Great Britain	
	Kan, or, Kop (or, Liter)			Schepel (or, Decaliter)	
		AZF Dutch Metrical			AYT Dutch Metrical
	Kile (or, Decaliter)		AZQ	—Seer . ...India	
		AYT Turkish Metrical		Ser (or, Liter) ..AZF India	
	Kileh (or, Hectoliter)			Sheng(or, Quart)AZM China	
		AYZ Turkish Metrical	AZR	—Shoo ....Japan	
AZC	—Kilohter. ....Metrical			Sultchek (or, Liter)	
	Koilon (or, Hectoliter)				AZF Turkish Metrical
		AYZ Greek Metrical	AZS	—Tierce ....Great Britain.	
	Kop or, Kan (or, Liter)		AZT	—Tonde .... Denmark	
		AZF Dutch Metrical	AZU	—Tou . ...China	
	Kotyle (or, Deciliter)		AZV	—Tschetwerik........... Russia.	
		AYU Greek Metrical		Vat, Mud, or, Zac (or,	
AZD	—Kruschka, or, Cron- chka ... . Russia			Hectoliter) ..AYZ Dutch Metrical	
	Kybos (or, Milliliter)		AZW	—Vedro ....Russia	
		AZG Greek Metrical		Vingerhoed (or, Cen- tiliter) ....AYR Dutch Metrical	
AZE	—Last........ ....Netherlands			Zac, Mud, or, Vat (or, Hectoliter).. AYZ Dutch Metrical	
AZF	{ Liter .... Metrical or Kan, or, Kop ....Dutch Metrical or Ser ....India or Sultchek .... Turkish Metrical			Zarf (or, Centiliter)	
					AYR Turkish Metrical
				For list of countries in which the Metrical System is used and for Table for Con- verting Metrical Measures and Weights into their American Equivalents, see page 55	

MEASURES OF WEIGHT.

Signal	Measure	Country or System.	Signal	Measure	Country or System
AZX	—MEASURES OF WEIGHT		BAD	—Cantar .... { Egypt Turkey	
AZY	—Arroba .... { Argentina Brazil Portugal Spain		BAE	—Carat............... { United States and Great Britain	
			BAF	—Catty, or, Chin........China	
	Batman (or, Milia- gramme).....BAY Turkish Metrical		BAG	{ Centigramme .... Metrical or Habbe . ...Turkish Metrical or Kokkos...... ......Greek Metrical	
BAC	—Berkovitz............Russia				
	Boughdais (or, Deci- gramme) .....BAJ Turkish Metrical				

Signal	Measure	Country or System	Signal.	Measure	Country or System.
BAH	—Centner -----------	{ Denmark / Sweden	BAV	{ Millier --------------	Metrical
	Centner, or, Quintal			or	
	BCI	United States		Tcheki ------------	Turkish Metrical
	Chin, or, Catty__BAF	China		or	
BAI	{ Decagramme --------	Metrical		Ton ----------------	United States and / Great Britain
	or			or	
	Drachma -----------	Turkish Metrical		Tonne --------------	German Metrical
	or		BAW	—Milligramme --------	Metrical
	Lood --------------	Dutch Metrical	BAX	—Mna--------------	Greece
.BAJ	{ Decigramme--------	Metrical	BAY	{ Myriagramme -------	Metrical
	or			or	
	Boughdais-----------	Turkish Metrical		Batman ------------	Turkish Metrical
	or			Obolos (or, Decigramme)	
	Korrel.--------------	Dutch Metrical		BAJ	Greek Metrical
	or			Ock, or, Oke (or, Kilo-	
	Obolos--------------	Greek Metrical		gramme)_____BAQ	Turkish Metrical
	Denk (or, Gramme)			Onze (or, Hectogramme)	
	BAN	Turkish Metrical		BAO	Dutch Metrical
	Doppelzentner__ZNL	Germany	BAZ	—Ounce ------------	{ United States and / Great Britain
BAK	—Drachm ----------	{ United States and / Great Britain		Pecul, or, Tan__BCM	China
	Drachma (or, Deca-		BCA	—Pennyweight(dwt)	{ United States and / Great Britain
	gramme)_____BAI	Turkish Metrical	BCD	—Pfund -------.----	Austria-Hungary
	Drachme(or, Gramme)			Pond (or, Kilogramme)	
BAL	BAN	Greek Metrical		BAQ	Dutch Metrical
	—Funt --------------	Russia.	BCE	—Pound (lb)--------	{ United States and / Great Britain
BAM	—Grain -----------	{ United States and / Great Britain	BCF	—Pud, or, Pood-------	Russia
BAN	{ Gramme-----------	Metrical	BCG	—Pund------------	{ Denmark / Sweden
	or		BCH	—Quarter ----------	{ United States and / Great Britain
	Denk --------------	Turkish Metrical	BCI	—Quintal -----------	{ United States / Argentina / Brazil / Chile / Portugal / Spain.
	or				
	Drachme -----------	Greek Metrical			
	or				
	Wigtje -------------	Dutch Metrical			
	Habbe (or, Centi-				
	gramme)_____BAG	Turkish Metrical	BCJ	{ Quintal Metric.-------	Metrical
BAO	{ Hectogramme ------	Metrical		or	
	or			Kantar -------------	Turkish Metrical
	Onze ------	Dutch Metrical	BCK	—Rottolo-----------	{ Egypt / Turkey
BAP	—Hundredweight(cwt)	Great Britain		Ser (or, Kilogramme)	
	Kantar (or, Quintal		BCL	BAQ	India
	Metric)_____BCJ	Turkish Metrical		—Stone--------------	Great Britain
BAQ	{ Kilogramme.--------	Metrical	BCM	—Tan, or, Pecul -------	China
	or			Tcheki (or, Millier)	
	Ock, or, Oke. -------	Turkish Metrical		BAV	Turkish Metrical
	or			Ton (or, Millier)	{ United States and / Great Britain
	Pond --------------	Dutch Metrical		BAV	
	or			Tonne (or, Millier)	
	Ser ----------------	India		BAV	German Metrical
BAR	—Kin --------------	Japan		Wigtje (or, Gramme)	
	Kokkos (or, Centi-			BAN	Dutch Metrical
	gramme) ---BAG	Greek Metrical	BCN	—Zolotnick---- -------	Russia
	Korrel (or Decigramme)				
	BAJ	Dutch Metrical			
BAS	—Kvint--------------	Denmark			
BAT	—Kwamme----------	Japan			
BAU	—Libra ----- -------	{ Argentina / Chile / Portugal / Spain			
	Lood (or, Decagramme)				
	BAI	Dutch Metrical			

For list of countries in which the Metrical System is used, and for Table for Converting Metrical Measures and Weights into their American Equivalents, see page 55

LIST OF THE PRINCIPAL COUNTRIES IN WHICH THE METRICAL SYSTEM IS USED.

United States (permissive)	France	Paraguay
Argentina	German Empire (with German	Peru
Austria-Hungary	names in some cases)	Portugal
Belgium	Great Britain (permissive) .	Roumania
Bolivia	Greece (with Greek names)	Santo Domingo.
Brazil.	Guatemala	San Salvador.
Bulgaria	Haiti	Servia
Canada (permissive)	Honduras	Spain
Chile	Italy	Sweden
Colombia	Mexico.	Switzerland
Costa Rica	Netherlands (with Dutch	Tripoli
Denmark	names)	Turkey (with Turkish names)
Ecuador	Nicaragua.	Uruguay
Egypt.	Norway	Venezuela

TABLES FOR CONVERTING METRICAL MEASURES AND WEIGHTS INTO THEIR AMERICAN EQUIVALENTS.

Meters		Yards	Kilometers		Stat Miles	Yards	Hectares		Acres	r	p
1	=	1 094	1	=	0	1094	1	=	2	1	35
2	=	2 187	2	=	1	427	2	=	4	3	31
3	=	3 281	3	=	1	1521	3	=	7	1	26
4	=	4.374	4	=	2	855	4	=	9	3	22
5	=	5 468	5	=	3	188	5	=	12	1	17
6	=	6 562	6	=	3	1282	6	=	14	3	12
7	=	7 655	7	=	4	615	7	=	17	1	8
8	=	8 749	8	=	4	1709	8	=	19	3	3
9	=	9 843	9	=	5	1048	9	=	22	0	38
10	=	10 936	10	=	6	376	10	=	24	2	34
20	=	21 873	20	=	12	753	20	=	49	1	28
30	=	32 809	30	=	18	1139	30	=	74	0	21
40	=	43 745	40	=	24	1505	40	=	98	3	15
50	=	54 682	50	=	31	122	50	=	123	2	9
60	=	65 618	60	=	37	498	60	=	148	1	3
70	=	76 554	70	=	43	874	70	=	172	3	37
80	=	87 491	80	=	49	1251	80	=	197	2	32
90	=	98 427	90	=	55	1627	90	=	222	1	24
100	=	109 363	100	=	62	243	100	=	247	0	18
200	=	218 727	200	=	124	487	200	=	494	0	37
300	=	328 090	300	=	186	730	300	=	741	1	15
400	=	437 453	400	=	248	973	400	=	988	1	33
500	=	546 816	500	=	310	1217	500	=	1235	2	11

Liters		Galls	Quarts	Hectoliters		Qrs	Bushels	Kilogrammes		Cwts	Qrs	Lbs	Ozs
1	=	0	0 880	1	=	0	2 751	1	=	0	0	2	3¼
2	=	0	1 761	2	=	0	5 502	2	=	0	0	4	6¼
3	=	0	2 641	3	=	1	0 254	3	=	0	0	6	9¼
4	=	0	3 521	4	=	1	3 005	4	=	0	0	8	13
5	=	1	0 402	5	=	1	5.756	5	=	0	0	11	0¼
6	=	1	1 282	6	=	2	0 507	6	=	0	0	13	3¼
7	=	1	2 163	7	=	2	3 258	7	=	0	0	15	7
8	=	1	3 043	8	=	2	6 010	8	=	0	0	17	10¼
9	=	1	3 923	9	=	3	0 761	9	=	0	0	19	13¼
10	=	2	0 804	10	=	3	3 512	10	=	0	0	22	0¼
20	=	4	1 608	20	=	6	7 024	20	=	0	1	16	1¼
30	=	6	2 412	30	=	10	2 536	30	=	0	2	10	2¼
40	=	8	3 215	40	=	13	6 048	40	=	0	3	4	3
50	=	11	0 019	50	=	17	1 560	50	=	0	3	26	3¼
60	=	13	0 823	60	=	20	5 072	60	=	1	0	20	4¼
70	=	15	1 627	70	=	24	0 585	70	=	1	1	14	5¼
80	=	17	2 431	80	=	27	4 097	80	=	1	2	8	6
90	=	19	3 335	90	=	30	7 609	90	=	1	3	2	6¼
100	=	22	0 089	100	=	34	3 121	100	=	1	3	24	7
200	=	44	0 077	200	=	68	6 242	200	=	3	3	20	15
300	=	66	0 116	300	=	103	1 362	300	=	5	3	17	6
400	=	88	0 155	400	=	137	4 483	400	=	7	3	13	14
500	=	110	0 193	500	=	171	7 604	500	=	9	3	10	5

DEGREES OF LATITUDE AND LONGITUDE

Vessels other than those of Spain or France use the meridian of Greenwich. Spanish and French vessels use the meridians of Cadiz and Paris, respectively. In case of doubt, the vessel to which the signal is made should hoist the signal **NBL** = What is your First Meridian? Which can be answered by one of the following

TBL = My first meridian is Paris, east of Greenwich 2° 20′ 15″ = 0 h 9 m 21 secs

HSI = My first meridian is Cadiz, west of Greenwich 6° 12′ 24″ = 0 h 24 m 49 secs

CODE FLAG OVER TWO FLAGS

LATITUDE AND LONGITUDE.

CODE FLAG OVER—	LATITUDE	CODE FLAG OVER—	LATITUDE—Cont'd	CODE FLAG OVER—	LONGITUDE—Cont'd
AB		BT	—42° latitude	DL	— 2° longitude
	North latitude.. QHX	BU	—43° latitude	DM	— 3° longitude
	South latitude.. QHZ	BV	—44° latitude	DN	— 4° longitude.
AC	— 1° north latitude	BW	—45° latitude	DO	— 5° longitude
AD	— 1° south latitude	BX	—46° latitude	DP	— 6° longitude
AE	— 2° latitude	BY	—47° latitude	DQ	— 7° longitude.
AF	— 3° latitude	BZ	—48° latitude	DR	— 8° longitude
AG	— 4° latitude	CA	—49° latitude	DS	— 9° longitude
AH	— 5° latitude	CB	—50° latitude	DT	—10° longitude
AI	— 6° latitude	CD	—51° latitude	DU	—11° longitude
AJ	— 7° latitude	CE	—52° latitude	DV	—12° longitude
AK	— 8° latitude	CF	—53° latitude	DW	—13° longitude
AL	— 9° latitude	CG	—54° latitude	DX	—14° longitude
AM	—10° latitude	CH	—55° latitude	DY	—15° longitude.
AN	—11° latitude	CI	—56° latitude	DZ	—16° longitude
AO	—12° latitude	CJ	—57° latitude	EA	—17° longitude
AP	—13° latitude	CK	—58° latitude	EB	—18° longitude
AQ	—14° latitude	CL	—59° latitude	EC	—19° longitude.
AR	—15° latitude.	CM	—60° latitude	ED	—20° longitude
AS	—16° latitude	CN	—61° latitude	EF	—21° longitude
AT	—17° latitude	CO	—62° latitude	EG	—22° longitude
AU	—18° latitude	CP	—63° latitude	EH	—23° longitude
AV	—19° latitude	CQ	—64° latitude	EI	—24° longitude
AW	—20° latitude	CR	—65° latitude	EJ	—25° longitude
AX	—21° latitude	CS	—66° latitude	EK	—26° longitude
AY	—22° latitude	CT	—67° latitude	EL	—27° longitude
AZ	—23° latitude	CU	—68° latitude	EM	—28° longitude
BA	—24° latitude	CV	—69° latitude	EN	—29° longitude
BC	—25° latitude	CW	—70° latitude	EO	—30° longitude
BD	—26° latitude	CX	—71° latitude	EP	—31° longitude
BE	—27° latitude	CY	—72° latitude	EQ	—32° longitude
BF	—28° latitude	CZ	—73° latitude	ER	—33° longitude
BG	—29° latitude	DA	—74° latitude	ES	—34° longitude
BH	—30° latitude.	DB	—75° latitude	ET	—35° longitude
BI	—31° latitude	DC	—76° latitude	EU	—36° longitude
BJ	—32° latitude	DE	—77° latitude	EV	—37° longitude
BK	—33° latitude	DF	—78° latitude	EW	—38° longitude
BL	—34° latitude	DG	—79° latitude	EX	—39° longitude
BM	—35° latitude	DH	—80° latitude	EY	—40° longitude
BN	—36° latitude			EZ	—41° longitude
BO	—37° latitude	DI	LONGITUDE	FA	—42° longitude
BP	—38° latitude		East longitude.. QYZ	FB	—43° longitude
BQ	—39° latitude		West longitude.. QZI	FC	—44° longitude
BR	—40° latitude	DJ	— 1° east longitude	FD	—45° longitude
BS	—41° latitude	DK	— 1° west longitude	FE	—46° longitude

NOTE.—If the position to be signaled is within a degree of the equator, or first meridian, only the minutes and seconds need be indicated.

CODE FLAG OVER—	LONGITUDE—Cont'd	CODE FLAG OVER—	LONGITUDE—Cont'd	CODE FLAG OVER—	LONGITUDE—Cont'd
FG	—47° logituude	HA	— 92° longitude	IV	—137° longitude
FH	— 48° longitude	HB	— 93° longitude	IW	—138° longitude
FI	—49° longitude	HC	— 94° longitude	IX	—139° longitude
FJ	—50° longitude	HD	— 95° longitude	IY	—140° longitude
FK	—51° longitude	HE	— 96° longitude	IZ	—141° longitude
FL	—52° longitude	HF	— 97° longitude	JA	—142° longitude
FM	—53° longitude	HG	— 98° longitude	JB	—143° longitude
FN	—54° longitude	HI	— 99° longitude	JC	—144° longitude
FO	—55° longitude	HJ	—100° longitude	JD	—145° longitude
FP	—56° longitude	HK	—101° longitude	JE	—146° longitude
FQ	—57° longitude	HL	—102° longitude	JF	—147° longitude
FR	—58° longitude	HM	—103° longitude	JG	—148° longitude
FS	—59° longitude	HN	—104° longitude	JH	—149° longitude
FT	—60° longitude	HO	—105° longitude	JI	—150° longitude
FU	—61° longitude	HP	—106° longitude	JK	—151° longitude
FV	—62° longitude	HQ	—107° longitude.	JL	—152° longitude
FW	—63° longitude	HR	—108° longitude	JM	—153° longitude
FX	—64° longitude	HS	—109° longitude	JN	—154° longitude
FY	—65° longitude	HT	—110° longitude	JO	—155° longitude
FZ	—66° longitude	HU	—111° longitude	JP	—156° longitude
GA	—67° longitude	HV	—112° longitude	JQ	—157° longitude
GB	—68° longitude	HW	—113° longitude	JR	—158° longitude
GC	—69° longitude	HX	—114° longitude	JS	—159° longitude
GD	—70° longitude	HY	—115° longitude	JT	—160° longitude
GE	—71° longitude	HZ	—116° longitude	JU	—161° longitude.
GF	—72° longitude	IA	—117° longitude	JV	—162° longitude.
GH	—73° longitude	IB	—118° longitude	JW	—163° longitude
GI	—74° longitude	IC	—119° longitude	JX	—164° longitude.
GJ	—75° longitude	ID	—120° longitude	JY	—165° longitude
GK	—76° longitude	IE	—121° longitude	JZ	—166° longitude
GL	—77° longitude	IF	—122° longitude	KA	—167° longitude
GM	—78° longitude	IG	—123° longitude	KB	—168° longitude
GN	—79° longitude	IH	—124° longitude	KC	—169° longitude
GO	—80° longitude	IJ	—125° longitude	KD	—170° longitude
GP	—81° longitude	IK	—126° longitude	KE	—171° longitude
GQ	—82° longitude	IL	—127° longitude	KF	—172° longitude
GR	—83° longitude	IM	—128° longitude	KG	—173° longitude
GS	—84° longitude	IN	—129° longitude	KH	—174° longitude
GT.	—85° longitude	IO	—130° longitude	KI	—175° longitude
GU	—86° longitude	IP	—131° longitude	KJ	—176° longitude
GV	—87° longitude	IQ	—132° longitude	KL	—177° longitude
GW	—88° longitude	IR	—133° longitude	KM	—178° longitude
GX	—89° longitude	IS	—134° longitude	KN	—179° east longitude
GY	—90° longitude	IT	—135° longitude	KO	—179° west longitude
GZ	—91° longitude	IU	—136° longitude	KP	—180° longitude

DIVISIONS OF TIME, ETC.

HOURS OF TIME, AND MINUTES AND SECONDS OF TIME AND ARC.

Civil time is to be understood unless the signal FHW (astronomical) is made

Code Flag over—	Hours	Code Flag over—	Minutes—Cont'd	Code Flag over—	Seconds—Cont'd
KQ	— 0 hours	MQ	—25 minutes	OL	—10 seconds
KR	— 1 hour (or, 1 p m)	MR	—26 minutes	OM	—11 seconds
KS	— 2 hours (or, 2 p m)	MS	—27 minutes	ON	—12 seconds
KT	— 3 hours (or, 3 p m)	MT	—28 minutes	OP	—13 seconds.
KU	— 4 hours (or, 4 p m)	MU	—29 minutes	OQ	—14 seconds
KV	— 5 hours (or, 5 p m)	MV	—30 minutes	OR	—15 seconds.
KW	— 6 hours (or, 6 p m)	MW	—31 minutes	OS	—16 seconds
KX	— 7 hours (or, 7 p m)	MX	—32 minutes	OT	—17 seconds.
KY	— 8 hours (or, 8 p m)	MY	—33 minutes	OU	—18 seconds
KZ	— 9 hours (or, 9 p m)	MZ	—34 minutes.	OV	—19 seconds
LA	—10 hours (or, 10 p m)	NA	—35 minutes	OW	—20 seconds
LB	—11 hours (or, 11 p m)	NB	—36 minutes	OX	—21 seconds
LC	—12 hours (or, midnight)	NC	—37 minutes	OY	—22 seconds
LD	—13 hours (or, 1 a m)	ND	—38 minutes	OZ	—23 seconds
LE	—14 hours (or, 2 a m)	NE	—39 minutes.	PA	—24 seconds.
LF	—15 hours (or, 3 a m)	NF	—40 minutes	PB	—25 seconds
LG	—16 hours (or, 4 a m)	NG	—41 minutes	PC	—26 seconds
LH	—17 hours (or, 5 a m)	NH	—42 minutes	PD	—27 seconds
LI	—18 hours (or, 6 a m)	NI	—43 minutes.	PE	—28 seconds.
LJ	—19 hours (or, 7 a.m)	NJ	—44 minutes.	PF	—29 seconds
LK	—20 hours (or, 8 a. m)	NK	—45 minutes	PG	—30 seconds
LM	—21 hours (or, 9 a m)	NL	—46 minutes	PH	—31 seconds
LN	—22 hours (or, 10 a m)	NM	—47 minutes	PI	—32 seconds.
LO	—23 hours (or, 11 a m)	NO	—48 minutes	PJ	—33 seconds.
LP	—24 hours (or, noon)	NP	—49 minutes	PK	—34 seconds.
		NQ	—50 minutes	PL	—35 seconds
	MINUTES	NR	—51 minutes	PM	—36 seconds
LQ	— 0 minutes	NS	—52 minutes	PN	—37 seconds
LR	— 1 minute	NT	—53 minutes	PO	—38 seconds
LS	— 2 minutes	NU	—54 minutes	PQ	—39 seconds.
LT	— 3 minutes	NV	—55 minutes.	PR	—40 seconds
LU	— 4 minutes	NW	—56 minutes	PS	—41 seconds
LV	— 5 minutes	NX	—57 minutes	PT	—42 seconds.
LW	— 6 minutes	NY	—58 minutes	PU	—43 seconds
LX	— 7 minutes	NZ	—59 minutes	PV	—44 seconds
LY	— 8 minutes		—60 minutes Code flag over KR	PW	—45 seconds.
LZ	— 9 minutes		Degree of latitude KRY	PX	—46 seconds
MA	—10 minutes		Degree of longitude KRZ	PY	—47 seconds
MB	—11 minutes			PZ	—48 seconds
MC	—12 minutes		**SECONDS**	QA	—49 seconds
MD	—13 minutes	OB	—0 seconds	QB	—50 seconds
ME	—14 minutes	OC	—1 second	QC	—51 seconds.
MF	—15 minutes	OD	—2 seconds.	QD	—52 seconds
MG	—16 minutes.	OE	—3 seconds.	QE	—53 seconds
MH	—17 minutes	OF	—4 seconds	QF	—54 seconds
MI	—18 minutes	OG	—5 seconds	QG	—55 seconds
MJ	—19 minutes	OH	—6 seconds	QH	—56 seconds
MK	—20 minutes	OI	—7 seconds	QI	—57 seconds
ML	—21 minutes	OJ	—8 seconds	QJ	—58 seconds
MN	—22 minutes	OK	—9 seconds	QK	—59 seconds
MO	—23 minutes				—60 seconds Code flag over LR
MP	—24 minutes.				

BAROMETER.

[In inches and millimeters]

The readings in inches and millimeters are approximate but are not equivalent to each other

Signal		Inches	Millimeters	Signal		Inches	Millimeters
CODE FLAG OVER	QM	27 8	706	CODE FLAG OVER	SO	29 85	758
"	QN	85	707	"	SP	9	759.
"	QO	9	708	"	SQ	92	760.
"	QP	92	709.	"	SR	96	761
"	QR	.96	710	"	ST	30 0	762
"	QS	28 0	711	"	SU	04	763
'	QT	02	712	"	SV	08	764.
"	QU	06	713	"	SW	1	765
'	QV	1	714	"	SX	15	766
'	QW	15	715	"	SY	2	767.
'.	QX	2	716.	"	SZ	24	768.
"	QY	22	717	"	TA	28	769.
"	QZ	26	718	"	TB	3	770.
"	RA	3	719	"	TC	35	771
"	RB	35	720	"	TD	4	772
"	RC	4	721	"	TE	.42	773
"	RD	42	722.	"	TF	46	774
"	RE	46	723.	"	TG	.5	775
"	RF	5	724.	"	TH	55	776
"	RG	.55	725.	"	TI	6	777
"	RH	6	726	"	TJ	62	778
"	RI	62	727	"	TK	66	779
"	RJ	66	728	"	TL	7	780
"	RK	7	729	"	TM	75	781
"	RL	74	730	"	TN	8	782
"	RM	78	731.	"	TO	82	783.
"	RN	8	732.	"	TP	86	684
"	RO	85	733	"	TQ	9	785.
"	RP	9	734	"	TR	94	786
"	RQ	94	735.	"	TS	98	787
"	RS	98	736.				
"	RT	29 0	737				
"	RU	05	738				
"	RV	1	739.				
"	RW	12	740				
"	RX	16	741				
"	RY	2	742.				
"	RZ	25	743.				
"	SA	3	744				
"	SB	32	745.				
"	SC	36	746				
"	SD	4	747.				
"	SE	44	748				
"	SF	48	749				
"	SG	.5	750				
"	SH	55	751				
"	SI	.6	752				
"	SJ	65	753				
"	SK	7	754.				
"	SL	.72	755				
"	SM	76	756				
"	SN	8	757.				

THERMOMETER

Signal		Fahrenheit.	Centigrade	Réaumur	Signal		Fahrenheit	Centigrade	Réaumur
CODE FLAG OVER	TU	1°	−17° 2	−13° 8	CODE FLAG OVER	WH	44° 6	+ 7°	+ 5° 6
"	TV	1° 4	−17°	−13° 6	"	WI	45°	+ 7° 2	+ 5° 8
"	TW	2°	−16° 6	−13° 3	" ,	WJ	46°	+ 7° 8	+ 6° 2
"	TX	3°	−16° 1	−12° 9	"	WK	46° 4	+ 8°	+ 6° 4
"	TY	3° 2	−16°	−12° 8	"	WL	47°	+ 8° 3	+ 6° 7
"	TZ	4°	−15° 6	−12° 4	"	WM	48°	+ 8° 9	+ 7° 1
"	UA	5°	−15°	−12°	"	WN	48° 25	+ 9°	+ 7° 25
"	UB	6°	−14° 4	−11° 6	"	WO	49°	+ 9° 4	+ 7° 6
"	UC	6° 8	−14°	−11° 2	"	WP	50°	+10°	+ 8°
"	UD	7°	−13° 9	−11° 1	"	WQ	51°	+10° 6	+ 8° 4
"	UE	8°	−13° 3	−10° 7	"	WR	51° 8	+11°	+ 8° 8
"	UF	8° 6	−13°	−10° 4	"	WS	52°	+11° 1	+ 8° 9
"	UG	9°	−12° 8	−10° 2	"	WT	53°	+11° 7	+ 9° 3
"	UH	10°	−12° 2	− 9° 8	"	WU	53° 6	+12°	+ 9° 6
"	UI	10° 4	−12°	− 9° 6	"	WV	54°	+12° 2	+ 9° 8
"	UJ	11°	−11° 7	− 9° 3	"	WX	55°	+12° 8	+10° 2
"	UK	12°	−11° 1	− 8° 9	"	WY	55° 4	+13°	+10° 4
"	UL	12° 2	−11°	− 8° 8	"	WZ	56°	+13° 3	+10° 7
"	UM	13°	−10° 6	− 8° 4	"	XA	57°	+13° 9	+11° 1
"	UN	14°	−10°	− 8°	"	XB	57° 25	+14°	+11° 2
"	UO	15°	− 9° 4	− 7° 6	"	XC	58°	+14° 4	+11° 6
"	UP	15° 8	− 9°	− 7° 2	"	XD	59°	+15°	+12°
"	UQ	16°	− 8° 9	− 7° 1	"	XE	60°	+15° 6	+12° 4
"	UR	17°	− 8° 3	− 6° 7	"	XF	60° 8	+16°	+12° 8
"	US	17° 6	− 8°	− 6° 4	"	XG	61°	+16° 1	+12° 9
"	UT	18°	− 7° 8	− 6° 2	"	XH	62°	+16° 7	+13° 3
"	UV	19°	− 7° 2	− 5° 8	"	XI	62° 6	+17°	+13° 6
"	UW	19° 4	− 7°	− 5° 6	"	XJ	63°	+17° 2	+13° 8
"	UX	20°	− 6° 7	− 5° 3	"	XK	64°	+17° 8	+14° 2
"	UY	21°	− 6° 1	− 4° 9	"	XL	64° 5	+18°	+14° 4
"	UZ	21° 2	− 6°	− 4° 8	"	XM	65°	+18° 3	+14° 7
"	VA	22°	− 5° 6	− 4° 4	"	XN	66°	+18° 9	+15° 1
"	VB	23°	− 5°	− 4°	"	XO	66° 2	+19°	+15° 4
"	VC	24°	− 4° 4	− 3° 6	"	XP	67°	+19° 4	+15° 6
"	VD	24° 8	− 4°	− 3° 2	"	XQ	68°	+20°	+16°
"	VE	25°	− 3° 9	− 3° 1	"	XR	69°	+20° 6	+16° 4
"	VF	26°	− 3° 3	− 2° 7	"	XS	69° 8	+21°	+16° 8
"	VG	26° 6	− 3°	− 2° 4	"	XT	70°	+21° 1	+16° 9
"	VH	27°	− 2° 8	− 2° 2	"	XU	71°	+21° 7	+17° 3
"	VI	28°	− 2° 2	− 1° 8	"	XV	71° 6	+22°	+17° 6
"	VJ	28° 4	− 2°	− 1° 6	"	XW	72°	+22° 2	+17° 8
"	VK	29°	− 1° 7	− 1° 3	"	XY	73°	+22° 8	+18° 2
"	VL	30°	− 1° 1	− 0° 9	"	XZ	73° 4	+23°	+18° 4
"	VM	30° 2	− 1°	− 0° 8	"	YA	74°	+23° 3	+18° 7
"	VN	31°	− 0° 6	− 0° 4	"	YB	75°	+23° 9	+19° 1
"	VO	32°	0° 0	0° 0	"	YC	75° 2	+24°	+19° 2
"	VP	33°	+ 0° 6	+ 0° 4	"	YD	76°	+24° 4	+19° 6
"	VQ	33° 8	+ 1°	+ 0° 8	"	YE	77°	+25°	+20°
"	VR	34°	+ 1° 1	+ 0° 9	"	YF	78°	+25° 6	+20° 4
"	VS	35°	+ 1° 75	+ 1° 3	"	YG	78° 8	+26°	+20° 8
"	VT	35° 6	+ 2°	+ 1° 6	"	YH	79°	+26° 1	+20° 9
"	VU	36°	+ 2° 2	+ 1° 8	"	YI	80°	+26° 7	+21° 3
"	VW	37°	+ 2° 8	+ 2° 2	"	YJ	80° 6	+27°	+21° 6
"	VX	37° 4	+ 3°	+ 2° 4	"	YK	81°	+27° 2	+21° 8
"	VY	38°	+ 3° 3	+ 2° 7	"	YL	82°	+27° 8	+22° 2
"	VZ	39°	+ 3° 9	+ 3° 1	"	YM	82° 4	+28°	+22° 4
"	WA	39° 25	+ 4°	+ 3° 2	"	YN	83°	+28° 3	+22° 7
"	WB	40°	+ 4° 4	+ 3° 6	"	YO	84°	+28° 9	+23° 1
"	WC	41°	+ 5°	+ 4°	"	YP	84° 2	+29°	+23° 2
"	WD	42°	+ 5° 6	+ 4° 4	"	YQ	85°	+29° 4	+23° 6
"	WE	42° 8	+ 6°	+ 4° 9	"	YR	86°	+30°	+24°
"	WF	43°	+ 6° 1	+ 4° 9	"	YS	87°	+30° 6	+24° 8
"	WG	44°	+ 6° 7	+ 5° 3	"	YT	87° 8	+31°	+24° 8

Signal		Fahrenheit.	Centigrade	Réaumur	Signal		Fahrenheit	Centigrade	Réaumur
CODE FLAG OVER					CODE FLAG OVER				
"	YU	88°	+31° 1	+24° 9	"	ZK	98° 6	+37°	+29° 6
"	YV	89°	+31° 7	+25° 3	"	ZL	99°	+37° 2	+29° 8
"	YW	89° 8	+32°	+25° 6	"	ZM	100°	+37° 8	+30° 2
"	YX	90°	+32° 2	+25° 8	"	ZN	100° 4	+88°	+30° 4
"	XZ	91°	+32° 8	+26° 2	"	ZO	101°	+38° 3	+30° 7
"	ZA	91° 4	+33°	+26° 4	"	ZP	102°	+88° 9	+81° 1
"	ZB	92°	+33° 2	+26° 7	"	ZQ	102° 2	+89°	+81° 2
"	ZC	93°	+9d° 9	+27° 1	"	ZR	103°	+39° 4	+81° 6
"	ZD	93° 2	+34°	+27° 2	"	ZS	104°	+40°	+82°
"	ZE	94°	+34° 4	+27° 6	"	ZT	105°	+40° 6	+82° 4
"	ZF	95°	+85°	+28°	"	ZU	105° 8	+41°	+82° 8
"	ZG	96°	+85° 6	+28° 4	"	ZV	106°	+41° 1	+82° 9
"	ZH	96° 8	+86°	+28° 8	"	ZW			
"	ZI	97°	+86° 1	+28° 9	"	ZX			
"	ZJ	98°	+86° 7	+29° 3	"	ZY			

AUXILIARY PHRASES.

(i e , **Phrases containing the Auxiliary Verbs**, or forming sentences or parts of sentences in frequent use)

N B —Many of the following phrases can not be translated into foreign languages They must therefore be used with great discretion when communication is being held with ships or signal stations not using the English language

	AM—ARE	

BEA	Am I am		Are—Continued
BEC	—Am I?		How are they ?......... ....BXH
BED	—Am I not?		How are you ?.....................BXI
BEF	—Am I to?		How is (or, are) he (she, it, or person-s
BEG	—Am, Are, or, Is not.		or thing-s indicated) ?..... ..... BXO
	Am, Are, or, Is not to be......... _ BFD		If he (she, it, or person-s or thing-s indi-
	Am, Are, or, Is not to have........BQD		cated) is (or, are) ....BYL
BEH	—Am, Are, or, Is to		If he (she, it, or person-s or thing-s indi-
	Am, Are, or, Is to be ....BFC		cated) is (or, are) able to.........BYM
	Am, Are, or, Is to have ....BQC		If he (she, it, or person-s or thing-s indi-
	How am I?.......................BXG		cated) is (or, are) not ....BYN
BEI	—I am not		If he (she, it, or person s or thing-s indi-
	If I am.................... ....BYX		cated) is (or, are) not able to.....BYO
	If I am not........ ....BYZ		If they are ....CAF
	What, or, Which am I ? ....CPV		If they are not ....CAG
	When am I to?.....................CQV		If we are.......................CAX
	Where am I (or, are we)? ....CSQ		If we are not.................. ...CAY
	Where I am (or, we are) ....CTG		If you are ....CBQ
	Why am I?......................CUY		If you are not ....CBR
			Is. or, Are, he, (she, it, or person-s or
BEJ	Are		thing-s indicated)? ....BVY
BEK	—Are, or, Is his (her-s, it-s)		There are (or, is)..................CMG
BEL	—Are, or, Is my (mine)		There are (or, is) not..............CMH
BEM	—Are, or, Is our-s, or, Are we?		They are for ....CNM
	Are, Am, or, Is not.................BEG	BET	—They are from
	Are, Am, or, Is not to be ....BFD	BEU	—They are
	Are, Am, or, Is not to have ....BQD	BEV	—They are not
	Are, Am, or, Is to.................BEH	BEW	—We are
	Are, Am, or, Is to be ....BFC	BEX	—We are not
BEN	—Are, or, Is the		What, or, Which are (or, is)?.....CPW
BEO	—Are, or, Is their-s, or, Are they ?		What or, Which are (or, is) you-r-s ?.CPY
BEP	—Are, or, Is there—any ?		What, or Which is (or, are) he (she, it,
BEQ	—Are these (or, those)?		or person-s or thing-s indicated)?.CPX
BER	—Are, or, Is with —		When are (or, is)?.... ....CQW
BES	—Are, or, Is you-r s		When are we ? ....CQX
	He, She, It (or person-s or thing s indi-		When are you?...... ....CQY
	cated) had (has, or, have) done (or, is,		When he (she, it, or person s or thing s
	or, are doing) ....BUL		indicated) is (or, are) ....CRJ
	He, She It (or person-s or thing-s indi-		When is (or, are) he (she, it, or person s
	cated) had (has, or, have) not done		or thing-s indicated)?...........CRN
	(or, is, or, are not doing) ....BUP		When they are....................CSI
	He, She, It (or person s or thing s indi-		When we are....................CSL
	cated) is (or, are) for.......... _ BUR		When you are ....CSN
	He, She, It (or person s or thing s indi-		Where am I (or are we)?..........CSQ
	cated) is (or are) from ....BUS		Where are (or, is)?...............CSR
	He, She, It (or person s or thing s indi-		Where are they ?.................CST
	cated) is (or, are) ....BUT		Where are you from ?.............SI
	He, She, It (or person-s or thing-s indi-		Where he (she, it, or person-s or thing s
	cated) is (or, are) able to.........BUV		indicated) is (or, are).............CTD
	He, She, It (or person-s or thing-s indi-		Where I am (or, we are) ....CTG
	cated) is (or, are) not............BUW		Where is (or, are) he (she, it, or person s
	He, She, It (or person-s or thing-s indi-		or thing-s indicated)?...........CTH
	cated) is (or, are) not able to....BUX		Where they are...................CUA
			Where we are (or, I am)...........CTG

	ARE—BE		
	ARE—*Continued*		**BE**—*Continued*
	Where you are ------------------CUF		He, She, It (*or person-s or thing-s indi-*
	Who are (*or,* is)? ------------------CUI		*cated*) had (has, *or,* have) not been. BUO
	Why are (*or,* is)? ------------------CUZ		He, She, It (*or person-s or thing-s indi-*
	Why are you? ------------------CVA		*cated*) must be ------------------BUZ
	Why are you not? ------------------CVB		He, She, It (*or person-s or thing-s indi-*
BEY	—You are		*cated*) must not be ------------------BVC
BEZ	—You are not		He, She, It (*or person-s or thing-s indi-*
			cated) ought to be------------------BVF
BFA	BE BEEN BEING		He, She, It (*or person-s or thing-s indi-*
BFC	—Am Are, *or,* Is to be		*cated*) ought not to be------------------BVG
BFD	—Am Are, *or,* Is not to be		He, She, It (*or person-s or thing-s indi-*
BFE	—Be the		*cated*) shall (*or,* will) be------------------BVI
BFG	—Can, *or* May be		He, She, It (*or person-s or thing-s indi-*
	Can, *or* May do (*or,* be done)------BLT		*cated*) shall (*or,* will) not be------BVK
BFH	—Can, *or* May have been		He, She, It (*or person-s or thing-s indi-*
	Can, *or* May have done (*or,* been done) BLV		*cated*) should (*or,* would) be------BVM
	Can, *or,* May he (she, it, *or person-s or*		He, She, It (*or person s or thing-s indi-*
	thing-s indicated) be?------------BTK		*cated*) should (*or,* would) do (*or,* be
	Can, *or,* May he (she, it, *or person-s or*		done)------------------BVN
	thing-s indicated) have been?------BTL		He, She, It (*or person s or thing-s indi-*
	Can, *or,* May I be?------------------BKH		*cated*) should (*or,* would) not be. BVQ
	Can, *or,* May it be done?------------BLW		He, She, It (*or person-s or thing-s indi-*
BFI	—Can *or,* May there be?		*cated*) should (*or,* would) not do (*or,*
BFJ	—Can not, *or,* May not be		be done)------------------BVR
	Can not, *or,* May not do (*or,* be done) . BLX		He, She, It (*or person-s or thing-s indi-*
BFK	—Can not *or* May not have been		*cated*) was (*or,* were) to be------BVW
	Can not, *or* May not have done (*or,*		He, She, It (*or person-s or thing-s indi-*
	done)------------------BLZ		*cated*) was (*or,* were) not to be... BVX
	Can, *or,* May not he (she it, *or per-*		
	son-s or thing s indicated) be?---- BTM	BFX	—I can (*or,* may) be
	Can not, *or,* May not he (she, it, *or per-*	BFY	—I can not (*or,* may not) be
	son-s or thing-s indicated) have been?	BFZ	—I could (*or,* might) be
	BTN	BGA	—I could (*or,* might) not he
BFL	—Could, *or,* Might be	BGC	—I had (*or,* have) been
	Could *or* Might do (*or,* be done) ---BMA	BGD	—I had (*or,* have) not been
BFM	—Could, *or,* Might have been	BGE	—I must be
	Could, *or,* Might have done (*or,* been	BGF	—I must not be
	done)------------------BMC	BGH	—I shall (*or,* will) be
	Could, *or,* Might he (she, it, *or person s*	BGI	—I shall (*or,* will) not be
	or thing s indicated) be?----------BTO	BGJ	—I should (*or,* would) be
	Could *or* Might he (she, it, *or person-s*	BGK	—I should (*or,* would) have been
	or thing-s indicated) not be?------BTP	BGL	—I should (*or,* would) not be
BFN	—Could *or,* Might not be	BGM	—I should (*or,* would) not have been
	Could, *or,* Might not do (*or,* be done) BMD		If it can (*or,* may) be done---------BZR
BFO	—Could *or,* Might not have been		If it can not (*or,* may not) be done . BZS
	Could *or* Might not have done (*or,* been		Is, etc (*See* Am)
	done)------------------BME		Is it not to be done?------------------ . BNI
BFP	—Could, *or,* Might there be?		Is it to be done?------------------BNJ
BFQ	—Had, Has, *or* Have been		It, etc '(*See* He)
	Had, Has, *or,* Have done (*or,* been	BGN	—It can be
	done)------------------BMR		It can (*or,* may) be done------- --BNK
	Had Has, *or* Have, he (she, it, *or per-*	BGO	—It can not be
	son s or thing s indicated)--BTU		It can not (*or,* may not) be done...BNL
	Had, Has, *or,* Have, he (she, it, *or per-*		It could (*or* might) be done------BNM
	son-s or thing-s indicated) not been?		It could (*or,* might) not be done...BNO
	BTX	BGP	—It had (*or,* has) been
BFR	—Had, *or,* Have I been?		It had (*or,* has) been done ------BNP
BFS	—Had, Has, *or,* Have not been	BGQ	—It had (*or,* has) not been
	Had, Has, *or,* Have not done (*or,* been		It had (*or,* has) not been done----BNQ
	done)------------------BMS	BGR	—It ought to be
BFT	—Had, *or,* Have they been?		It ought to be done------------------BNR
BFU	—Had, *or,* Have we been?	BGS	—It ought not to be
BFV	—Had, *or,* Have you been?		It ought not to be done ----------- BNS
BFW	—Having been	BGT	—It shall (*or,* will) be
	He, She, It (*or person-s or thing-s indi-*		It shall (*or,* will) be done --- ------BNT
	cated) can (*or,* may) be----------BUC	BGU	—It shall (*or,* will) not be
	He She, It (*or person-s or thing-s indi-*		It shall (*or,* will) not be done ----:.BNU
	cated) can not (*or,* may not) be. BUE	BGV	—Let her be
	He, She, It (*or person s or thing s indi-*		Let him be------------------BXE
	cated) had (has, *or,* have) been----BUK	BGW	—Let them be

BE

	BE—*Continued*		BE—*Continued*
BGX	—Let us be	BHO	—Should, *or*, Would there be ?
	May, etc (*See* Can)		That, *or*, This can (*or*, may) be ___CKR
	Might, etc (*Sec* Could)		That, *or*, This can not (*or*, may not) be
	Must be _____CEM		CKT
	Must do (*or*, be done)_____CEN		That, *or*, This could (*or*, might) be.CKV
	Must have been _____CEP		That, *or*, This could (*or*, might) not be
	Must not be___ _____CEV		CKW
	Must not do (*or*, be done) _____CEW		That, *or*, This had (has, *or*, have) be
BGY	—Need be		CKY
BGZ	—Need not be		That, *or*, This had (has, *or*, have) not
	Not be (been, being)_____CFN		been _____ ___ CKZ
	Not to be _____CGH		That, *or*, This must be_____CLJ
	Not to be done_____CGI		That, *or*, This must not be ___ __ CLM
	Not to be had _____ _____CGJ		That, *or*, This shall (*or*, will) be ___CLN
	Ought not to be _____ ___CHS		That, *or*, This shall (*or*, will) not be.CLO
	Ought to be_____ _____ _____ CHY		That, *or*, This should (*or*, would) be.CLP
	Ought to do (*or*, be done)_____CHZ		That, *or*, This should (*or*, would) not be,
			CLQ
BHA	—Shall, *or*, Will be		There can (*or*, may) be_____CMJ
	Shall, *or*, Will do (*or*, be done)____BNW		There can not (*or*, may not) be ____CML
	Shall, *or*, Will have done (*or*, been done)		There could (*or* might) be_____CMO
	BNX		There could (*or*, might) not be ____CMP
	Shall, *or*, Will he (she, it, *or*, *person-s*		There had (has, *or*, have) been ____CMR
	or thing-s indicated) be ?_____BWE		There had (has, *or*, have) not been.CMS
	Shall *or*, Will he (she, it, *or person-s*		There shall (*or*, will) be _____ ___CMV
	or thing-s indicated) do (*or*, be done)?		There shall (*or*, will) not be_____CMX
	BWF		There should (*or*, would) be_____CMZ
	Shall, *or* Will he (she, it, *or person-s*		There should (*or* would) have been.CNA
	or thing-s indicated) not be ?_ ___BWI		There should (*or*, would) not be ____CND
	Shall, *or*, Will he (she, it, *or person s*		There should (*or*,would) not have been,
	or thing s indicated) not do (*or*, be		CNE
	done) ?____ _____ _____BWJ	BHP	—They can (*or*, may) be
		BHQ	—They can not (*or*, may not) be
BHC	—Shall, *or*, Will I be?	BHR	—They could (*or*, might) be
BHD	—Shall, *or*, Will not be?	BHS	—They could (*or*, might) not be
	Shall, *or*, Will not do (*or*, be done).BNY	BHT	—They had (*or*, have) been
	Shall *or*, Will not have done (*or*, been	BHU	—They had (*or*, have) not been
	done) _____ _____ BNZ	BHV	—They must be
BHE	—Shall, *or*, Will there be ?	BHW	—They must not be
BHF	—Shall, *or*, Will they be ?	BHX	—They ought to be
BHG	—Shall, *or*, Will we be ?	BHY	—They ought not to be
BHI	—Shall, *or*, Will you be ?	BHZ	—They shall (*or*, will) be
BHJ	—Shall, *or*, Will you not be ?	BIA	—They shall (*or*, will) not be
BHK	—Should, *or*, Would be	BIC	—They should (*or*, would) be
	Should, *or*, Would do (*or*, be done).BOE	BID	—They should (*or*, would) not be
		BIE	—They were to be
BHL	—Should, *or*, Would have been	BIF	—They were not to be
	Should, *or*, Would have done (*or*, been		To be _____CNP
	done) __ __ _____BOG		To be at _____CNQ
	Should, *or* Would he (she, it, *or per-*		To be done_____CNR
	son-s or thing s indicated) be____BWN		To be had __ _____CNS
	Should, *or*, Would he (she, it, *or person-s*	BIG	—Was, *or*, Were to be
	or thing s indicated) do (*or*, be done)?	BIH	—We can (*or* may) be
	BWO	BIJ	—We can not (*or*, may not) be
	Should, *or*, Would he (she, it, *or person s*	BIK	—We could (*or*, might) be
	or thing-s indicated) have been ?.BWQ	BIM	—We could (*or*, might) not be
	Should, *or*, Would he (she, it, *or person s*	BIN	—We had (*or*, have) been
	or thing-s indicated) not be ? _ __BWS	BIO	—We had (*or*, have) not been
	Should, *or*, Would he (she, it, *or person s*	BIP	—We must be
	or thing-s indicated) not do (*or*, be	BIQ	—We must not be
	done) ?_____BWT	BIR	—We shall (*or*, will) be
	Should, *or*, Would he (she, it, *or person s*	BIS	—We shall (*or*, will) not be
	or thing s indicated) not have been ?	BIT	—We should (*or*, would) be
	BWV	BIU	—We should (*or*, would) not be
BHM	—Should, *or*, Would not be	BIV	—We were to be
	Should, *or*, Would not do(*or*, be done),	BIW	—We were not to be
	BOH		When shall (*or*, will) be ?_____CRS
BHN	—Should, *or*, Would not have been		When shall (*or*, will) it be done ?__CRX
	Should *or* Would not have done (*or*,		When shall (*or*, will) you be ?_____CSD
	been done) _____BOI		Where shall (*or*, will) be ?_____CTL

	BE—CAN		

BE—Continued

	Where shall (or, will) you be ?____CTU		Can not, or, May not have been____BFK
BIX	Will, etc (See Shall)		Can not, or, May not have done (or, been done) _____ _____BLZ
	—Will it be ?		Can not, or, May not have had ... BQI
BIY	Would, etc (See Should)		Can not, or, May not he (she, it, or person-s or thing-s indicated) be ?____BTM
BIZ	—You can (or, may) be		Can not, or, May not he (she, it, or person-s or thing-s indicated) have been ? BTN
BJA	—You can not (or, may not) be		
BJB	—You could (or, might) be		He, She, It (or person-s or thing-s indicated) can (or, may)_____BUA
BJC	—You could (or, might) not be		He, She, It (or person s or thing s indicated) can (or, may) be_____BUC
BJD	—You had (or, have) been		He, She It (or person-s or thing-s indicated) can not (or, may not)____BUD
BJE	—You had (or, have) not been		He, She, It (or person-s or thing-s indicated) can not (or, may not) be__BUE
BJF	—You must be		How can (or, may)? ___ _____BXJ
BJG	—You must not be		
BJH	—You ought to be	BKQ	—I can (or, may)
BJI	—You ought not to be		I can (or, may) be _____BFX
BJK	—You shall (or, will) be		I can (or, may) have_ _____BRI
BJL	—You shall (or, will) not be	BKR	—I can not (or, may not)
BJM	—You should (or, would) be		I can not (or, may not) be___ _____BFY
BJN	—You should (or, would) not be.		I can not (or, may not) have_____BRJ
BJO	—You were to be	BKS	—I shall (or, will) if I can
BJP	—You were not to be		I will do it if I can__:____ _ ____LIP
			If he (she, it, or person-s or thing-s indicated) can (or, may)____BYD
BJQ	**BY**		If he (she, it, or person-s or thing-s indicated) can not (or, may not)__BYE
BJR	—By him (his, her-s, it-s)		If I can (or, may) _____BZA
BJS	—By me (my, mine)		If I can not (or, may not)_____BZC
BJT	—By that (or, this)		If it can (or, may) be done _____BZR
BJU	—By the		If it can not (or, may not) be done__BZS
BJV	—By them (then-s)		If they can (or may) _____CAH
BJW	—By there		If they can not (or, may not) ___CAI
BJX	—By these (or, those)		If we can (or, may) _____CAZ
BJY	—By us (our-s)		If we can not (or, may not)_____CBA
BJZ	—By what (or, which)		If you can (or, may) ___ _____CBS
BKA	—By whom (or, whose)		If you can not (or, may not) _____CBT
BKC	—By you-r-s		It, etc (See He)
BKD	—Me by		It can be_____ _____ ____BGN
	Not by_____CFO		It can (or, may) be done _____BNK
			It can not be ... _____ _____BGO
BKE	**CAN** (or, MAY)_____ BKE		It can not (or, may not) be done___BNL
	—Can		No I can not _____HUZ
	Can, or, May be_____BFG		Nothing can (or, could) _____SKB
	Can, or, May do (or, be done) _____BLT		That, or, This can (or, may)_____CKQ
	Can, or, May do it_____BLU		That, or, This can (or, may) be_____CKR
	Can, or, May have_____BQR		That, or, This can not (or may not)_CKS
	Can, or, May have been_____BFH		That, or, This can not (or, may not) be .. _____ _____CKT
	Can, or, May have done (or, been done) BLV		There can be_____BGN
	Can, or, May have had _____BQF		There can (or, may)_____CMI
	Can, or, May he (she, it, or person s or thing-s indicated)?_____ _____BTJ		There can (or, may) be ____._____CMJ
	Can, or, May he (she, it, or person-s or thing s indicated) be ?_____ BTK		There can not (or, may not)_____CMK
	Can, or, May he (she, it, or person-s or thing s indicated) have been ? ____BTL		There can not (or, may not) be__CML
BKF	—Can, or, May his (her-s, it-s)?	BKT	—They can (or, may)
BKG	—Can, or, May I (my, mine)?		They can (or, may) be _____BHP
BKH	—Can, or, May I be?		They can (or, may) be_____BSF
	Can, or, May I have — some — ?___BQG	BKU	—They can not (or, may not)
	Can, or, May it be done ?_____BLW		They can not (or, may not) be _____BHQ
BKI	—Can, or, May that (or, this)?		They can not (or, may not) have ___BSG
	Can, or, May the ? _____HUE	BKV	—We can (or, may)
	Can, or May there be ? _____BFI		We can (or, may) be_____ BIH
BKJ	—Can, or, May these (or, those)?		We can (or, may) have _____BSN
BKL	—Can, or, May they (their s)?	BKW	—We can not (or may not)
BKM	—Can, or, May they not?		We can not (or, may not) be_____BIJ
BKN	—Can, or, May we (our-s)?		We can not (or, may not) have ____BSO
BKO	—Can, or, May you-r-s?		What, or, Which can (or, may)?___CPZ
BKP	—Can not, or, May not be?		When can (or, may)?_____CQZ
	Can not, or, May not be _____BFJ		
	Can not, or, May not do (or, be done) BLX		
	Can not, or, May not do it_____BLY		
	Can not, or, May not have_____BQH		

CAN—DO

	Can—*Continued*			Could—*Continued*	
	When I can (*or,* may)	CRM		She, etc. (*See* He)	
	Where can (*or,* may)?	CSV		That, *or* This could (*or,* might)	CKU
	Who can (*or,* may)?	CUJ		That, *or* This could (*or,* might) be	CKV
	Why can (*or,* may)?	HUF		That, *or* This could (*or,* might) not be,	
	Why can not (*or,* may not)	CVD			CKW
	Yes, I can	HUG		There could (*or,* might)	CMN
BKX	—You can, (*or,* may)			There could (*or,* might) be	CMO
	You can (*or,* may) be	BIY		There could (*or,* might) not be	CMP
	You can (*or,* may) have	BSY	BLN	—They could (*or,* might)	
BKY	—You can not (*or,* may not)			They could (*or,* might) be	BHR
	You can not (*or,* may not) be	BIZ	BLO	—They could (*or,* might) not.	
	You can not (*or,* may not) have	BSZ		They could (*or,* might) not be	BHS
			BLP	—We could (*or,* might)	
	Could (*or,* Might)	BKZ		We could (*or,* might) be	BIK
BKZ	—Could		BLQ	—We could (*or,* might) not	
	Could, *or,* Might be	BFL		We could (*or,* might) not be	BIM
	Could, *or,* Might do (*or,* be done)	BMA		What, *or,* Which could (*or,* might)?	CQA
	Could, *or,* Might have	BQJ		When could (*or,* might)?	CRA
	Could, *or,* Might have been	BFM		Where could (*or,* might)?	CSW
	Could, *or,* Might have done (*or,* been			Who could (*or,* might)?	CUM
	done)	BMC		Why could (*or,* might)?	CVE
	Could, *or,* Might have had	BQK		Why could (*or,* might) not?	CVF
	Could, *or,* Might he (she, it, *or person s*		BLR	—You could (*or,* might)	
	or thing-s indicated) be ?	BTO		You could (*or,* might) be	BJA
	Could *or,* Might he (she, it, *or person-s*		BLS	—You could (*or,* might) not	
	or thing s indicated) not be ?	BTP		You could (*or,* might) not be	BJC
BLA	—Could, *or,* Might his (her-s, it-s) ?				
BLC	—Could, *or,* Might I (my, mine)?			Do Does Did Doing Done	BMF
BLD	—Could, *or,* Might not			—Can, *or,* May do (*or,* be done)	
	Could, *or,* Might not be	BFN	BLT	—Can, *or,* May do it	
	Could, *or,* Might not do (*or,* be done)	BMD	BLU	—Can, *or,* May have done (*or,* been done)	
	Could, *or,* Might not have	BQL	BLV	—Can, *or,* May it be done?	
	Could, *or,* Might not have been	BFO	BLW	—Can not, *or,* May not do it	
	Could, *or,* Might not have done (*or,* been		BLX	—Can not, *or,* May not do (*or,* be done)	
	done)	BME	BLY	—Can not, *or,* May not have done (*or,*	
	Could, *or,* Might not have had	BQM	BLZ	been done)	
BLE	—Could, *or,* Might that (*or,* this)?		BMA	—Could, *or,* Might do (*or,* be done)	
BLF	—Could, *or,* Might the?		BMC	—Could, *or,* Might have done (*or,* been	
	Could, *or,* Might there be ?	BFP		done)	
BLG	—Could, *or,* Might these (*or,* those)?		BMD	—Could, *or,* Might not do (*or,* be done)	
BLH	—Could, *or,* Might they (their-s)?		BME	—Could, *or,* Might not have done (*or,* been	
BLI	—Could, *or,* Might we (our-s)?			done)	
BLJ	—Could, *or,* Might you-r s?		BMF	—Do, Does, Did, Doing, Done	
	He, She, It (*or person-s or thing-s indi-*			Do, Does, *or,* Did he (she, it, *or person-s*	
	cated) could (*or,* might)	BUF		*or thing s indicated*)?	BTQ
	He, She, It (*or person s or thing-s indi-*		BMG	—Do, Does, *or,* Did his (her-s, it-s)?	
	cated) could (*or,* might) not	BUG	BMH	—Do, Does, *or,* Did I (my, mine)?	
	How could (*or,* might)?	BXX	BMI	—Do, Does, *or,* Did not	
BLK	—I could (*or,* might)			Do not let him (her, it, *or person-s indi-*	
	I could (*or,* might) be	BFZ		*cated*) have	BTR
	I could (*or,* might) have	BRK	BMJ	—Does, *or,* Did that (*or* this)?	
BLM	—I could (*or,* might) not		BMK	—Do, Does, *or,* Did the?	
	I could (*or,* might) not be	BGA	BML	—Do, Does, *or,* Did there?	
	I could (*or,* might) not have	BRL	BMN	—Do, *or,* Did these (*or,* those)?	
	If he (she, it, *or person-s or thing-s indi-*		BMO	—Do, Does, *or,* Did they (their-s)?	
	cated) could (*or,* might)	BYF	BMP	—Do, Does, *or,* Did we (our-s)?	
	If he (she, it, *or person-s or thing-s indi-*		BMQ	—Do, Does, *or,* Did you-r s?	
	cated) could (*or,* might) not	BYG	BMR	—Had, Has, *or,* Have done (*or,* been	
	If I could (*or,* might)	BZD		done)	
	If I could (*or,* might) not	BZE	BMS	—Had, has, *or,* Have not done (*or,* been	
	If they could (*or,* might)	CAJ		done)	
	If they could (*or,* might) not	CAL	BMT	—Had, *or,* Have I done?	
	If we could (*or,* might)	CBD	BMU	—Had, *or,* Have they done?	
	If we could (*or,* might) not	CBE	BMV	—Had, *or,* Have we done?	
	If you could (*or,* might)	CBU	BMW	—Had, *or,* Have you done?	
	If you could (*or,* might) not	CBV	BMX	—Having done	
	It, etc. (*See* He)			He, She, It (*or person-s or thing s indi-*	
	It could (*or,* might) be done	BNM		*cated*) does (do, *or,* did)	BUH
	It could (*or,* might) not be done	BNO		He, She, It (*or person s or thing s indi-*	
	Nothing could	SKB		*cated*) does (do, *or,* did) not	BUI

DO

	Do—*Continued*		**Do**—*Continued*
	He, She, It (*or person-s or thing s indi cated*) had (has, *or*, have) done (*or*, is, *or*, are doing) ---- ----------- BUL	BNY	—Shall, *or*, Will not do (*or*, be done)
		BNZ	—Shall, *or*, Will not have done (*or*, been done)
	He, She, It (*or person-s or thing-s indi cated*) had (has, *or*, have) not done (*or*, is, *or*, are not doing)---- -- BUP	BOA	—Shall, *or*, Will they do it ?
		BOC	—Shall, *or*, Will they not do it ?
	He, She, It (*or person s or thing-s indi cated*) should (*or*, would) do (*or*, be done) ------------------- ------BVN	BOD	—Shall, *or*, Will you do it ?
		BOE	—Shall, *or*, Will you not do it ?
	He, She, It (*or person-s or thing s indi cated*) should (*or*, would) not do (*or*, be done) --- ---------------------BVR	BOF	—Should, *or*, Would do (*or*, be done)
		BOG	—Should, *or*, Would have done (*or*, been done)
	How do (does *or* did)? ----------BXL		Should, *or*, Would he (she, it, *or person-s or thing s indicated*) do (*or*, be done), BWO
	How do (*or*, did) you?. ----------BXM		
BMY	—I do (*or*, did)		Should, *or*, Would he (she, it, *or person-s or thing s indicated*) not do (*or*, be done)----------------------BWT
BMZ	—I do (*or*, did) not		
BNA	—I had (*or*, have) done	BOH	—Should, *or*, Would not do (*or*, be done)
BNC	—I had (*or*, have) not done	BOI	—Should, *or* Would not have done (*or*, been done)
BND	—I shall (*or* will) do		
BNE	—I shall (*or*, will) not do	BOJ	—They do (*or*, did)
BNF	—I should (*or*, would) do	BOK	—They do (*or*, did) not
BNG	—I should (*or*, would) not do	BOL	—They had (*or*, have) done
	I will do it if I can----------------LIP	BOM	—They had (*or*, have) not done
	If he (she, it, *or person s or thing-s indi cated*) does (do, *or*, did)----------BYH		To be done---- ------------------CNR
			To do -------------------------------CNT
	If he (she it, *or person s or thing-s indi cated*) does (do, *or*, did) not--- ..BYI	BON	—Was, *or*, Were done (*or*, doing)
	If I do (*or*, did)---- . ---- -------BZF	BOP	—Was, *or*, Were not done (*or*, doing)
	If I do (*or*, did) not--- -----------BZG	BOQ	—We do (*or*, did)
	If I were you I should not do it ----LIQ	BOR	—We do (*or*, did) not
	If it can (*or*, may) be done -------BZR	BOS	—We had (*or*, have) done
	If it can not (*or*, may not) be done--BZS	BOT	—We had (*or*, have) not done
	If they do (*or*, did)---- ----------CAM		What, *or*, Which do (does, *or*, did)?--OQB
	If they do (*or* did) not------------CAN		What, *or*, Which does (do, *or*, did) he (she it, *or person-s or thing s indicated*)?--OQD
	If we do (*or*, did) ---- -------------CBF		
	If we do (*or* did) not -------------CBG		What, *or*, Which do (*or*, did) they (we, *or*, you)? ------ ---- ------------CQE
	If you do (*or*, did)-----------------CBW		
	If you do (*or*, did) not .. --------CBX		What, *or* Which shall (*or*, will) I (*or*, we) do ?------------ ------------CQL
BNH	—Is it done?		What *or* Which shall (*or*, will) they (*or*, you) do ?-- -------------------CQN
BNI	—Is it not to be done?		
BNJ	—Is it to be done?		What, *or* Which should (*or*, would) I (*or*, we) do ?---- ---- ------------CQR
	It, etc (*See* He)		
BNK	—It can (*or*, may) be done		What, *or*, Which should (*or*, would) they (*or*, you) do ? ----------------------CQS
BNL	—It can not (*or*, may not) be done		When do (does, *or*, did) ? -------- CRB
BNM	—It could (*or*, might) be done		When does (do, *or*, did) he (she, it, *or*, *person-s or thing-s indicated*)?---CRD
BNO	—It could (*or*, might) not be done		
BNP	—It has (*or* had) been done		When do (*or*, did) I (*or*, we) ? --------CRE
BNQ	—It has (*or* had) not been done		When do (*or*, did) they (*or*, you) ?---CRF
BNR	—It ought to be done		When done with --- ----- ----------CRG
BNS	—It ought not to be done		When shall (*or*, will) I (they, we, *or*, you) do ?----- ----------- -------CRV
BNT	—It shall (*or*, will) be done		
BNU	—It shall (*or*, will) not be done		When shall (*or*, will) it be done ?---CRX
BNV	—It will do		Where do (does, *or*, did)? ----------CSX
	May, etc (*See* Can)		Where do (*or*, did) I (*or*, we) ? ---*-- CSY
	Might, etc (*See* Could)		Where do (*or*, did) they (*or* you) ?--CSZ
	Must do (*or*, be done)----------------CEN		Who do (does, *or*, did) ? ----------CUO
	Must not do (*or*, be done)------------CEW		Why do (does, *or*, did) ?-----------CVG
	Not doing----------- - - ----------CFP		Why do (does, *or*, did) not ? --------CVH
	Not done ---- ------ ---------CFQ		Why do (*or*, did) you ?-----------CVI
	Not to be done-------------------CGI		Why do (*or*, did) you not ?----------CVJ
	Ought I to—do?---- ----------CHQ		Will, etc (*See* Shall)
	Ought to do (*or*, be done) --------CHZ	BOU	—Will it do ?
BNW	—Shall, *or*, Will do, (*or*, be done)	BOV	—Will not do
BNX	—Shall *or* Will have done (*or*, been done)		Will you do it ?--------------------CWE
	Shall, *or*, Will he (she, it, *or person-s or thing s indicated*) do (*or*, be done) ? BWF		Would (*See* Should)
		BOW	—You do (*or*, did)
	Shall, *or*, Will he (she, it, *or*, *person s or thing-s indicated*), not do (*or*, be done)? BWJ	BOX	—You do (*or*, did) not
		BOY	—You had (*or*, have) done
		BOZ	—You had (*or*, have) not done

FOR—HAD, HAS, HAVE

	FOR---- ------------------------BPA
BPA	—For
BPC	—For all
BPD	—For him (his, her-s, it-s, or person-s or thing-s indicated)
BPE	—For me (my, mine)
BPF	—For that (or, this)
BPG	—For the
BPH	—For them (their-s)
BPI	—For these (or, those)
BPJ	—For us (our-s)
BPK	—For what (or, which)
BPL	—For whom (or, whose)
BPM	—For you r-s
	He, She, It (or person s or thing-s indicated) is (or, are) for-----------BUR
	Not for ------------------------CFR
	They are for ----------------------CNM
	Was, or, Were for----- ------------OOP
	Where for ? ----------------- ----- OTA
	FROM ------- ----------------------BPN
BPN	—From
BPO	—From him (his, her s, it-s, or person-s or thing-s indicated)
BPQ	—From me (my, mine)
BPR	—From that (or, this)
BPS	—From the
BPT	—From them (their-s)
BPU	—From there
BPV	—From th·se (or, those)
BPW	—From us (our s)
BPX	—From what (or, which)
BPY	—From whence (or, where)
BPZ	—From whom (or, whose)
BQA	—From you-r-s
	He, She, It (or person-s or thing-s indicated) is (or, are) from----- -- ----BUS
	Not from ... - ----------- ----CFS
	They are from----- --- ----BET
	Was, or, Were from.. --- ------ COQ
	Where are you from ? ---------------SI
	HAD, HAS, or, HAVE---- --------------BQN
BQC	—Am, Are, or, Is to have
BQD	—Am, Are, or, Is not to have
BQE	—Can, or, May have
	Can, or, May have been .. -------BFH
	Can, or, May have done (or, been done) ---- ----- --------- ---- BLV
BQF	—Can, or, May have had
	Can, or, May he (she, it, or person-s or thing-s indicated) have been ? .. BTL
BQG	—Can, or, May I have—some — ?
BQH	—Can not, or, May not have
	Can not, or, May not have been----BFK
	Can not, or, May not have done (or, been done) --- ----------------- BLZ
BQI	—Can not or, May not have had
	Can not or, May not he (she, it, or person-s or thing-s indicated) have been ?... ----------------- BTN
BQJ	—Could, or, Might have
	Could, or, Might have been----BFM
	Could, or, Might have done (or, been done) ---- ----------- ------BMO
BQK	—Could, or, Might have had
BQL	—Could or, Might not have
	Could, or, Might not have been----BFO
	Could, or, Might not have done (or, been done) -------------------------BME

	HAD, HAS, or, HAVE—Continued
	—Could, or, Might not have had
BQM	Do not let him (her, it, or person-s indicated) have----- -----------BTR
	—Had, Has, or, Have
BQN	—Had, Has, or, Have any
BQO	Had, Has, or, Have been --------BFQ
	Had, Has, or Have done (or, been done)---- -- --------------- -- BMR
BQP	—Had Has, or, Have had
	Had, Has, or, Have, he (she, it, or person s or thing s indicated)---------BTS
	Had, Has, or, Have, he (she, it, or person s or thing s indicated) been .. BTU
	Had, Has, or, Have, he (she, it, or person-s or thing s indicated) had . BTV
	Had, Has, or, Have he (she, it, or person-s or thing s indicated) not ...BTW
	Had Has, or, Have, he (she it or person-s or thing-s indicated) not been, BTX
	Had Has, or, Have, he (she, it, or person s or thing-s indicated) not had.BTY
BQR	Had, Has, or, Have, his (her-s, it-s)
BQS	Had, Has, or, Have, I (my mine) ?
	Had, or, Have I been ?--- --------BFR
BQT	Had, or, Have I done ?------------BMT
BQU	—Had, or, Have I had ?
	—Had, Has, or, Have not
	Had, Has, or Have not been-------BFS
	Had Has, or, Have not done (or, been done) ---- ------ ------- ------BMS
BQV	—Had, Has, or, Have not had
BQW	—Had, Has, or, Have that (or this)
BQX	—Had, Has, or, Have the
BQY	—Had, Has, or, Have there
BQZ	—Had, or, Have these (or, those) ?
BRA	—Had, Has or, Have they (them their s) ?
	Had, or Have they been ?---- ----BFV
	Had, or, Have they done ?----- ---BMU
BRC	—Had or Have they had ?
BRD	—Had Has or, Have we (our-s)?
	Had, or, Have we been -----------BFU
	Had, or, Have we done-----------BMV
BRE	—Had, or, Have we had ?
BRF	—Had, Has or, Have you-r-s
	Had, or, Have you been ?---------BFV
BRG	Had or, Have you done ? ... ------BMW
	—Had or Have you had ?
	Having-----------------------------OPI
	Having been-------- --------------BFW
	Having done------- ---------------BMX
BRH	—Having had
	He, She, It (or person-s or thing-s indicated) had (has or, have) - ----BUJ
	He She It (or person s or thing-s indicated) had (has, or, have been)---BUK
	He, She, It (or person s or thing-s indicated) had (has, or, have) done (or, is, or, are doing) -------------- BUL
	He, She, It (or person s or thing s indicated) had (has, or, have) had . BUM
	He, She It (or person s or thing s indicated) had (has, or, have) not ----BUN
	He, She It (or person-s or thing-s indicated) had (has, or, have) not been.BUO
	He, She, It (or person-s or thing s indicated) had (has, or, have) not done (or, is, or, are not doing)----------- BUP
	He, She, It (or person s or thing s indicated) had (has, or, have) not had.BUQ

HAD, HAS, HAVE

HAD, HAS, or, HAVE—*Continued*	
He, She, It (*or person s or thing-s indicated*) should (*or, would*) have___BVO	
He, She, It (*or person-s or thing-s indicated*) should (*or, would*) not have_BVS	
How had (has, *or,* have)?_____BXN	
BRI —I can (*or* may) have	
BRJ —I can not (*or* may not) have	
BRK —I could (*or,* might) have	
BRL —I could (*or,* might) not have	
BRM —I had (*or,* have)	
I had (*or,* have) been _____BGC	
I had (*or,* have) done _____BNA	
BRN —I had (*or,* have) had	
BRO —I had (*or,* have) not	
I had (*or,* have) not been _____BGD	
I had (*or,* have) not done _____BNC	
BRP —I had (*or,* have) not had	
BRQ —I shall (*or,* will) have	
BRS —I shall (*or,* will) not have	
BRT —I should (*or,* would) have	
BRU —I should (*or,* would) not have	
I should (*or,* would) have been _____BGK	
I should (*or,* would) not have been _BGM	
If he (she, it, *or person s or thing-s indicated*) had (has, *or,* have) _____BYJ	
If he (she, it, *or person-s or thing-s indicated*) had (has, *or,* have) not____BYK	
If I had (*or,* have)_____BZH	
If I had (*or,* have) not_____BZI	
If they had (*or,* have) _____OAO	
If they had (*or,* have) not_____CAP	
If we had (*or,* have)_____OBH	
If we had (*or,* have) not_____CBI	
If you had (*or,* have) _____OBY	
If you had (*or,* have) not_____CBZ	
Is, etc (*See* Are)	
It, etc (*See* He)	
It has (*or,* had) been_____BGP	
It has (*or,* had) been done_____BNP	
It has (*or,* had) not been_____BGQ	
It has (*or,* had) not been done_____BNQ	
Let him (her, it, *or person s or thing s indicated*) have _____BWA	
BRV —Let them have	
May, etc (*See* Can)	
Might, etc (*See* Could)	
Must have _____CEO	
Must have been_____CEP	
Must have had_____CEQ	
Not to be had _____CGJ	
Ought not to have_____CHT	
Ought to have_____CIA	
BRW —Shall, *or,* Will have	
Shall, *or,* Will have done (*or,* been done)_____BNX	
Shall, *or,* Will he (she, it, *or person s or thing-s indicated*) have?_____BWG	
Shall, *or,* Will he (she, it, *or person s or thing s indicated*) not have?____BWK	
BRX —Shall, *or,* Will I have?	
BRY —Shall, *or,* Will I not have?	
BRZ —Shall, *or,* Will not have	
Shall, *or,* Will not have done (*or,* been done) _____BNZ	
She, etc (*See* He)	
BSA —Should, *or,* Would have	
Should, *or,* Would have been_____BHL	
Should, *or,* Would have done (*or,* been done) _____BOG	

HAD, HAS, or, HAVE—*Continued*	
BSC —Should, *or,* Would have had	
Should, *or,* Would he (she, it, *or person-s or thing-s indicated*) have?_____BWP	
Should, *or,* Would he (she, it, *or person-s or thing-s indicated*) have been_BWQ	
Should, *or,* Would he (she, it, *or person-s or thing-s indicated*) not have___BWU	
Should, *or,* Would he (she, it, *person-s or thing s indicated*) not have been?_BWV	
BSD —Should, *or,* Would not have	
Should, *or,* Would not have been?_BHN	
Should, *or,* Would not have done (*or,* been done) _____BOI	
BSE —Should, *or,* Would not have had	
That, *or,* This had (has, *or,* have)_CKX	
That, *or,* This had (has, *or,* have) been, _____CKY	
That, *or,* This had (has, *or,* have) not been _____CKZ	
There had (has, *or,* have) _____CMQ	
There had (has, *or,* have) been _____CMR	
There had (has, *or,* have) not been_CMS	
There should (*or,* would) have been_CNA	
There should (*or,* would) not have been, _____CNE	
BSF —They can (*or,* may) have	
BSG —They can not (*or,* may not) have.	
BSH —They had (*or,* have)	
They had (*or,* have) been_____BHT	
They had (*or,* have) done_____BOL	
BSI —They had (*or,* have) had	
BSJ —They had (*or,* have) not	
They had (*or,* have) not been _____BHU	
They had (*or,* have) not done_____BOM	
BSK —They had (*or,* have) not had	
BSL —They should (*or,* would) have	
BSM —They should (*or,* would) not have	
This, etc (*See* That)	
To be had _____CNS	
To have_____CNU	
BSN —We can (*or,* may) have	
BSO —We can not (*or,* may not) have	
BSP —We had (*or,* have)	
We had (*or,* have) been_____BIN	
We had (*or,* have) done_____BOS	
BSQ —We had (*or,* have) had	
BSR —We had (*or,* have) not	
We had (*or,* have) not been_____BIO	
We had (*or,* have) not done_____BOT	
BST —We had (*or,* have) not had	
BSU —We shall (*or,* will) have	
BSV —We shall (*or,* will) not have	
BSW —We should (*or,* would) have	
BSX —We should (*or,* would) not have	
What, *or,* Which had (has, *or,* have)?_CQF	
What, *or,* Which shall (*or,* will) I (*or,* we) have?_____CQM	
What, *or,* Which shall (*or,* will) they (*or,* you) have?_____CQO	
When had (has, *or,* have)?_____CRH	
When had (*or,* have) you?_____CRI	
When shall (*or,* will) I have?_____CRW	
When shall (*or,* will) they (*or,* you) have?_____CRZ	
When shall (*or,* will) we have?_____CSA	
Where had (has, *or,* have)?_____CTB	
Where shall (*or,* will) I have?_____CTO	
Where shall (*or,* will) they (*or,* you) have?_____CTQ	

HAD, HAS, HAVE—HE, SHE, IT.

	HAD, HAS, or, HAVE—Continued			HE, SHE, IT—Continued.
	, Where shall (or, will) we have ?----CTR	BUE		—He, She, It (or person-s or thing-s indicated) can not (or, may not) be
	Which, etc (See What)	BUF		—He, She, It (or person-s or thing-s indicated) could (or, might)
	Who had (has, or, have)? ---------CUP			—He, She, It (or person-s or thing-s indicated) could (or, might) not
	Why had (has, or, have)?--- -----CVK	BUG		—He, She, It (or person-s or thing-s indicated) does (do, or, did)
	Why had (has, or, have) not?------CVL	BUH		—He, She, It (or person-s or thing-s indicated) does (do, or, did) not
	Why had (or, have) you ?---------CVM	BUI		—He, She, It (or person-s or thing-s indicated) had (has, or, have)
	Why had (or, have) you not?-.----CVN	BUJ		—He, She, It (or person-s or thing-s indicated) had (has, or, have) been
	Will it have ?---------------------OWD	BUK		—He, She, It (or person s or thing-s indicated) had (has, or, have) done (or, is,
	Would etc (See Should)			or, are doing)
BSY	—You can (or, may) have	BUL		—He, She, It (or person-s or thing-s indicated) had (has, or have) had
BSZ	—You can not (or, may not) have	BUM		—He, She, It (or person-s or thing s indicated) had (has, or have) had
BTA	—You had (or, have)	BUN		—He, She, It (or person s or thing s indicated) had (has or have) not
	You had (or, have) been ----------BJD	BUO		—He, She, It (or person s or thing s indicated) had (has or have) not been
	You had (or, have) done----------BOY	BUP		—He She It (or person s or thing s indicated) had (has or, have) not done
BTC	—You had (or, have) had			(or is, or are not doing)
BTD	—You had (or, have) not	BUQ		—He She It (or person s or thing-s indicated) had (has, or have) not had
	You had (or, have) not been ----- BJE	BUR		—He, She, It (or person-s or thing-s indicated) is (or, are) for
	You had (or, have) not done------BOZ	BUS		—He, She, It (or person-s or thing-s indicated) is (or, are) from
BTE	—You had (or, have) not had	BUT		—He, She, It (or person-s or thing-s indicated) is (or are)
BTF	—You shall (or will) have	BUV		—He, She, It (or person-s or thing s indicated) is (or, are) able to
BTG	—You shall (or, will) not have	BUW		—He, She, It (or person-s or thing-s indicated) is (or, are) not
BTH	—You should (or, would) have	BUX		—He, She, It (or person s or thing-s indicated) is (or, are) not able to
BTI	—You should (or, would) not have			He, She, It (or person s or thing-s indicated) may (See Can)
				He, She, It (or person s or thing-s indicated) might (See Could)
	HE---------------------------------BTZ	BUY		—He She It (or person s or thing s indicated) must
		BUZ		—He, She, It (or person s or thing s indicated) must be
	HE, SHE, IT, or, PERSONS, or, THINGS	BVA		—He, She, It (or person s or thing s indicated) must not
	Are, or, Is, etc (See Is)	BVC		—He, She, It (or person-s or thing-s indicated) must not be
BTJ	—Can, or, May he (she, it, or person-s or thing s indicated)?	BVD		—He, She, It (or person-s or thing-s indicated) ought to
BTK	—Can, or, May he (she, it, or person-s or thing s indicated) be ?	BVE		—He, She, It (or person-s or thing-s indicated) ought not to
BTL	—Can, or, May he (she, it or person-s or thing s indicated) have been ?	BVF		—He, She, It (or person-s or thing-s indicated) ought to be
BTM	—Can not, or, May not he (she, it or person-s or thing-s indicated) be ?	BVG		—He, She, It (or person-s or thing-s indicated) ought not to be
BTN	—Can not, or, May not he (she, it, or person-s or thing-s indicated) have been?	BVH		—He, She, It (or person-s or thing-s indicated) shall (or, will)
BTO	—Could, or, Might he (she, it, or person-s or thing s indicated) be ?	BVI		—He, She It (or person s or thing s indicated) shall (or, will) be
BTP	—Could, or, Might he (she, it or person-s or thing s indicated) not be?	BVJ		—He, She It (or person s or thing s indicated) shall (or, will) not
BTQ	—Do, Does, or, Did he (she, it, or person-s or thing s indicated)?	BVK		—He, She, It (or person-s or thing-s indicated) shall (or, will) not be
BTR	—Do not let him (her, it, or person-s indicated) have	BVL		—He, She, It (or person s or thing s indicated) should (or, would).
	For him (his, her-s, it-s, or person-s or thing-s indicated)------------BPD			
	From him (his, her-s, it-s, or person-s or thing-s indicated)-- - - - BPO			
BTS	—Had, Has, or, Have, he (she, it, or person-s or thing-s indicated)"			
BTU	—Had, Has, or, Have, he (she, it or person-s or thing-s indicated) been ?			
BTV	—Had, Has, or, Have, he (she it, or person-s or thing s indicated) had ?			
BTW	—Had, Has, or, Have, he (she, it or person s or thing s indicated) not?			
BTX	—Had, has, or, Have he (she, it, or person s or thing s indicated) not been			
BTY	—Had, Has, or, Have, he (she, it, or person-s or thing s indicated) not had ?			
BTZ	—He			
BUA	—He, She, It (or person-s or thing-s indicated) can (or, may)			
BUC	—He, She, It (or person-s or things indicated) can (or, may) be			
BUD	—He, She, It (or person-s or thing-s indicated) can not (or may not)			

	HE, SHE, IT		

	HE, SHE, IT—Continued		
BVM	—He, She It (or person s or thing-s indicated) should (or, would) be		
BVN	—He, She, It (or person s or thing-s indicated) should (or, would) do (or, be done)		
BVO	—He, She, It (or person s or thing-s indicated) should (or, would) have		
BVP	—He, She, It (or person s or thing-s indicated) should (or, would) not		
BVQ	—He, She, It (or person s or thing-s indicated) should (or, would) not be		
BVR	—He, She, It (or person-s or thing-s indicated) should (or, would) not do (or, be done)		
BVS	—He, She, It (or person-s or thing s indicated) should (or, would) not have		
BVT	—He, She, It (or person s or thing-s indicated) was (or, were)		
BVU	—He, She, It (or person s or thing-s indicated) was (or, were) not		
BVW	—He, She. It (or person-s or thing s indicated) was (or, were) to be		
BVX	—He, She. It (or person s or thing-s indicated) was (or, were) not to be		
	He, She, It (or person-s or thing-s indicated) will, etc (See Shall)		
	He, She It (or person-s or thing-s indicated) would, etc (See Should)		
	How is (or are) he (she, it, or person-s or thing-s indicated)? ....BXO		
	How shall (or, will) he (she, it, or person-s or thing-s indicated)? ......BXR		
	If he (she, it, or person s or thing s indicated) can (or, may) ....BYD		
	If he (she it, or person-s or thing-s indicated) can not (or, may not) .....BYE		
	If he (she, it, or person s or thing s indi cated) could (or, might) ....BYF		
	If he (she it or person-s or thing s indicated) could (or, might) not......BYG		
	If he (she it, or person s or thing s indicated) does (do, or, did) ....BYH		
	If he (she, it, or person-s or thing-s indicated) does (do, or, did) not.......BYI		
	If he (she, it or person s or thing s indicated) had (has, or, have) ....BYJ		
	If he (she, it, or person-s or thing-s indicated) had (has, or, have) not.. - BYK		
	If he (she, it, or person-s or thing-s indicated) is (or, are) ....BYL		
	If he (she it, or person-s or thing s indicated) is (or, are) able to...... - - BYM		
	If he (she it, or person-s or thing s indicated) is (or are) not ... .BYN		
	If he (she, it, or person-s or thing-s indicated) is (or, are) not able to.....BYO		
	If he (she, it, or person-s or thing-s indicated) shall (or, will) ....BYP		
	If he (she, it, or person-s or thing s indicated) shall (or, will) not ....BYQ		
	If he (she, it, or person-s or thing-s indicated) should (or, would) ..... BYR		
	If he (she, it, or person-s or thing-s indicated) should (or, would) not.....BYS		
	If he (she, it, or person s or thing-s indicated) was (or, were) ....BYT		
	If he (she, it, or person s or thing-s indicated) was (or, were) not .... ---BYU		
	If he (she, it, or person-s or thing-s indicated) will, etc (See Shall)		

	HE, SHE, IT—Continued		
	If he (she, it, or person-s or thing-s indicated) would, etc (See Should)		
BVY	—Is, or, Are he (she, it, or person-s or thing s indicated)?		
BVZ	—It		
BWA	—Let him (her, it, or person-s or thing-s indicated) have		
	May he (she, it, etc)? (See Can)		
	Might he (she, it, etc)? (See Could)		
	Must he (she, it, or person-s or thing s indicated)? ....CER		
	Ought he (she, it, or person-s or thing-s indicated) to?..............---------CHN		
BWC	—Person-s, or, Thing-s indicated		
BWD	—Shall or, Will he (she, it, or person-s or thing-s indicated)?		
BWE	—Shall, or, Will he, (she it, or person-s or thing-s indicated) be?		
BWF	—Shall, or, Will he (she, it, or person-s or thing s indicated) do (or, be done)?		
BWG	—Shall or, Will he (she, it, or person-s or thing s indicated) have?		
BWH	—Shall, or, Will he (she, it, or person-s or thing-s indicated) not?		
BWI	—Shall, or, Will he (she, it, or person-s or thing s indicated) not be?		
BWJ	—Shall or, Will he (she, it, or person-s or thing-s indicated) not do (or, be done)?		
BWK	—Shall, or, Will he (she, it or person s or thing s indicated) not have?		
BWL	—She		
BWM	—Should, or, Would he (she, it, or person-s or thing s indicated)?		
BWN	—Should, or, Would he (she, it, or person-s or thing s indicated) be?		
BWO	—Should, or, Would he (she, it, or person s or thing-s indicated) do (or, be done)?		
BWP	—Should, or, Would he (she, it, or person-s or thing s indicated) have?		
BWQ	—Should, or, Would he (she, it, or person s or thing s indicated) have been?		
BWR	—Should, or, Would he (she it, or person s or thing-s indicated) not?		
BWS	—Should, or, Would he (she, it, or person s or thing-s indicated) not be?		
BWT	—Should, or, Would he (she, it, or person-s or thing-s indicated) not do (or, be done)?		
BWU	—Should, or, Would he (she, it, or person-s or thing s indicated) not have?		
BWV	—Should or Would he (she, it, or person-s or thing s indicated) not have been?		
	That he (she, it, or person s or thing-s indicated) .. .. ------------------CLA		
BWX	—Was, or Were he (she, it, or person-s or thing-s indicated)?		
BWY	—Was, or Were he (she, it, or person-s or thing-s indicated) not?		
	What, or Which does (do, or did) he (she, it, or person-s or thing-s indicated)? ....CQD		
	What, or Which is (or are) he (she, it, or person s or thing s indicated)?..CPX		
	When does (did, or do) he (she, it, or person-s or thing s indicated)?....CRD		
	When he (she, it, or person-s or thing-s indicated) is (or are) ------ ----CRJ		

HE, SHE, IT—I

HE, SHE, IT—*Continued*
When is (*or* are) he (she, it, *or person-s or thing-s indicated*)?..........CRN
When shall (*or*, will) he (she, it, *or person-s or thing-s indicated*)?........CRT
Where he (she, it, *or person-s or thing-s indicated*) is (*or*, are) ....CTD
Where is (*or* are) he (she, it *or person-s or thing-s indicated*)?..... ....CTH
Where shall (*or*, will) he (she, it, *or person-s or thing-s indicated*)? ......CTM
Will he (she it *or person-s or thing-s indicated*) etc ? (*See* Shall)
Would he (she, it, *or person-s or thing-s indicated*) etc ? (*See* Should)

HER S-SELF   (*See* HIM) ....BXD

HIM   HIS   HIMSELF.............BXC

HIM   HIS   HER S   IT-S
BWZ —After him (his her-s, it)
          Are, *or*, Is his (her-s, it s) . .....BEK
BXA —Before him (his, her s, it-s)
          By him (his, her s, it-s) .... BJR
          Can, *or*, May his (her-s, it-s)?......BKF
          Could, *or*, Might his (her-s, it s)?..BLA
          Do, Does, *or*, Did him (her-s, it-s)?..BMG
          Do not let him (her, it, *or person indicated*) have ....BTR
          For him (his, her-s, it-s, *or person-s or thing-s indicated*) .....BPD
          From him (his, her-s. it s, *or person-s or thing-s indicated*) ....BPO
          Had, Has, *or*, Have his (her-s, it s).BQR
BXC —Him, His, Himself
BXD —Her-s-self
          If he (his, her s it s) .... BYV
          In him (his, her-s, it-s)...........CDJ
          Is, etc   (*See* Are)
          It .... BVZ
          Its-self .... - ....CDY
          Let her be ....BGV
BXE —Let him be
          Let him (her it, *or person-s or thing-s indicated*)—have.. .... BWA
          May, etc   (*See* Can)
          Might, etc   (*See* Could)
          Must his (her-s it-s)?...........CES
          Not his (her-s, it-s) ....CFT
          Of him (his, her-s, it-s)........CGN
          On him (his, her-s, it-s).........CGY
          Ought his (her s, it-s)?.........CHO
          Shall, *or*, Will his (hers, its)?......CIS
          Should, *or*, Would his (her-s, it s)?..CJP
          That his (her-s, it s) ....CLB
          To him (his her-s it s) ....CNV
          Was *or* Were his (her-s, it-s) ....COR
          When his (her-s, it-s) .... .CRK
          Where his (her-s, it-s) ....CTE
          Will etc   (*See* Shall)
          With him (his, her-s, it-s)..........CWG
          Would, etc   (*See* Should)

          How.............BXF
BXF —How
BXG —How am I?
BXH —How are they?
BXI —How are you?
BXJ —How can (*or* may)?
BXK —How could (*or* might)?

How—*Continued*
BXL —How do (does, *or*, did)?
BXM —How do (*or*, did) you?
BXN —How had (has, *or*, have)?
BXO —How is (*or*, are) he (she, it, *or person s or thing s indicated*)?
BXP —How must?
BXQ —How ought?
BXR —How shall (*or*, will) he (she, it, *or person-s or thing-s indicated*)?
BXS —How shall (*or*, will) I ?
BXT —How shall (*or*, will) we ?
BXU —How shall (*or*, will) you ?
BXV —How should (*or*, would)?
BXW —How was (*or*, were)?
BXY —How will they?

I......... .... BYA
BXZ —After I
          Am I? ....BEC
          Am I not? ....BED
          Am I to? ....BEF
          Can, *or*, May I (my, mine)?....... BKG
          Can, *or*, May I be?..... ...... BKH
          Can, *or*, May I have—some?.....BQG
          Could, *or*, Might I (my, mine)? . .BLC
          Do, Does, *or*, Did I (my, mine)?... BMH
          Had, Has, *or* Have I (my mine)?....BRJ
          Had, *or*, Have I been ?......BFR
          Had, *or*, Have I done ?......BMT
          Had, *or*, Have I had ....BQT
          How am I? ....BXG
          How shall (*or*, will) I?........BXS

BYA —I
          I am .... BEA
          I am not ....BEI
          I can (*or*, may) ....BKQ
          I can (*or*, may) be ....BFX
          I can (*or*, may) have...........BRI
          I can not (*or*, may not)........BKR
          I can not (*or*, may not) be ....BFY
          I can not (*or* may not) have ... .BRJ
          I could (*or* might) ....BLK
          I could (*or* might) be.........BFZ
          I could (*or* might) have...........BRK
          I could (*or* might) not ....BLM
          I could (*or*, might) not be ......BGA
          I could (*or*, might) not have ......BRL
          I do (*or*, did) ....BMY
          I do (*or* did) not.............. ...BMZ
          I had (*or* have)...........BRM
          I had (*or*, have) been .... BGC
          I had (*or*, have) done............BNA
          I had (*or*, have) had...... ......BRN
          I had (*or*, have) not ....BRO
          I had (*or*, have) not been ... ...BGD
          I had (*or*, have) not done.........BNC
          I had (*or*, have) not had..........BRP
          I may, etc   (*See* Can)
          I might, etc   (*See* Could)
          I must ....CEK
          I must be ....BGE
          I must not ....CEL
          I must not be .. ..BGF
          I ought to.... .... CHL
          I ought not to.... ....CHM
          I shall (*or*, will) .... CIO
          I shall (*or*, will) be.................BGB
          I shall (*or*, will) do.............BND
          I shall (*or*, will) have.............BRQ

I—IF

I—*Continued*

I shall (*or*, will) if I can	BKS
I shall (*or*, will) not	CIP
I shall (*or*, will) not be	BGI
I shall (*or*, will) not do	BNE
I shall (*or*, will) not have	BRS
I should (*or*, would)	CJM
I should (*or*, would) be	BGJ
I should (*or*, would) do	BNF
I should (*or*, would) have	BRT
I should (*or*, would) have been	BGK
I should (*or*, would) not	CJN
I should (*or*, would) not be	BGL
I should (*or*, would) not do	BNG
I should (*or*, would) not have	BRU
I should (*or*, would) not have been	BGM
I was	COH
I was not	COI
I will, etc (*See* Shall)	
I will do it if I can	LIP
I would, etc (*See* Should)	
If I (my, mine)	BYW
If I am	BYX
If I am not	BYZ
If I can (*or*, may)	BZA
If I can not (*or*, may not)	BZC
If I could (*or*, might)	BZD
If I could (*or*, might) not	BZE
If I do (*or*, did)	BZF
If I do (*or*, did) not	BZG
If I had (*or*, have)	BZH
If I had (*or*, have) not	BZI
If I must	BZJ
If I must not	BZK
If I ought to	BZL
If I ought not to	BZM
If I shall (*or*, will)	BZN
If I shall (*or*, will) not	BZO
If I should (*or*, would)	BZP
If I should (*or*, would) not	BZQ
If I was	BZT
If I was not	BZU
If I were you I should not do it	LIQ
May etc (*See* Can)	
Might, etc (*See* Could)	
Must I (my mine)?	CET
Need I (my, mine)?	CEH
No, I can not	HUZ
Not I (my, mine)	CFU
Ought I (my, mine)?	CHP
Ought I to—do ?	CHQ
Shall, *or*, Will I (my, mine)?	CIT
Shall, *or*, Will I be ?	BHC
Shall *or*, Will I have ?	BRX
Shall *or*, Will I not have	BRY
Should, *or*, Would I (my, mine)?	CJQ
That I (my, mine)	CLD
Was, *or*, Were I (my, mine)?	COS
Was I ?	COT
Was I not?	COU
What, *or*, Which am I ?	CPV
What *or*, Which shall (*or*, will) I ?	CQK
What, *or*, Which shall (*or*, will) I (*or*, we) do?	CQL
What, *or*, Which shall (*or*, will) I (*or*, we) have ?	CQM
What, *or*, Which should (*or*, would) I (*or*, we) do ?	CQR
When am I to ?	CQV
When do (*or*, did) I (*or*, we)?	CRE
When I (my, mine)	CRL

I—*Continued*

When I can (*or*, may)	CRM
When shall (*or*, will) I (*or*, we) ?	CRU
When shall (*or*, will) I have ?	CRW
When shall (*or*, will) I (they, we, *or*, you) do ?	CRV
Where am I (*or*, are we)?	CSQ
Where do (*or*, did) I (*or*, we)?	CSY
Where I (my, mine)	CTF
Where I am (*or*, we are)	CTG
Where shall (*or* will) I (*or*, we)?	CTN
Where shall (*or*, will) I have ?	CTO
Why am I ?	CUY

IF

I shall, *or*, I will if I can		BKS
If he (she, it, *or* person-s *or* thing-s indicated) is (*or*, are) (*See* Is)		
—If he (she, it, *or* person-s *or* thing-s indicated) can (*or*, may)	BYC	
—if he (she, it, *or* person-s *or* thing-s indicated) can not (*or*, may not)	BYD	
—If he (she, it, *or* person s *or* thing-s indicated) could (*or* might)	BYE	
—If he (she, it, *or* person-s *or* thing s indicated) could (*or*, might) not	BYF	
—If he (she, it, *or* person-s *or* thing-s indicated) does (do, *or*, did)	BYG	
—If he (she, it, *or* person-s *or* thing-s indicated) does (do, *or*, did) not	BYH	
—If he (she, it, *or* person-s *or* thing-s indicated) had (has, *or*, have)	BYI	
—If he (she, it, *or* person-s *or* thing s indicated) had (has, *or*, have) not	BYJ	
—If he (she, it, *or* person-s *or* thing-s indicated) is (*or*, are)	BYK	
—If he (she, it, *or* person s *or* thing-s indicated) is (*or*, are) able to	BYL	
—If he (she, it, *or* person-s *or* thing-s indicated) is (*or*, are) not	BYM	
—If he (she, it, *or* person-s *or* thing-s indicated) is (*or*, are) not able to	BYN	
—If he (she it, *or* person s *or* thing-s indicated) shall (*or*, will)	BYO	
—If he (she, it, *or* person-s *or* thing-s indicated) shall (*or*, will) not	BYP	
—If he (she, it, *or* person-s *or* thing s indicated) should (*or*, would)	BYQ	
—If he (she, it, *or* person s *or* thing s indicated) should (*or* would) not	BYR	
—If he (she, it, *or* person-s *or* thing-s indicated) was (*or*, were)	BYS	
—If he (she, it, *or* person-s *or* thing-s indicated) was (*or*, were) not	BYT	
—If he (his, she, her-s, it-s)	BYU	
—If I (my, mine)	BYV	
—If I am	BYW	
—If I am not	BYX	
—If I can (*or*, may)	BYZ	
—If I can not (*or*, may not)	BZA	
—If I could (*or*, might)	BZC	
—If I could (*or*, might) not	BZD	
—If I do (*or*, did)	BZE	
—If I do (*or*, did) not	BZF	
—If I had (*or*, have)	BZG	
—If I had (*or*, have) not	BZH	
—If I must	BZI	
—If I must not	BZJ	
—If I ought to	BZK	
—If I ought not to	BZL	
—If I shall (*or*, will)	BZM	
	BZN	

IF—IS

	IF—Continued
BZO	—If I shall (or, will) not
BZP	—If I should (or, would)
BZQ	—If I should (or, would) not.
BZR	—If it can (or may) be done
BZS	—If it can not (or, may not) be done.
BZT	—If I was
BZU	—If I was not
	If I were you I should not do it....LIQ
BZV	—If not
	If our-s (we)......................CAW
BZW	—If so
BZX	—If that (or, this)
BZY	—If the
CAB	—If there
CAD	—If these (or, those)
CAE	—If they (their-s)
CAF	—If they are
CAG	—If they are not
CAH	—If they can (or, may)
CAI	—If they can not (or, may not).
CAJ	—If they could (or, might)
CAL	—If they could (or, might) not
CAM	—If they do (or, did)
CAN	—If they do (or, did) not.
CAO	—If they had (or, have)
CAP	—If they had (or, have) not.
CAQ	—If they shall (or, will)
CAR	—If they shall (or, will) not
CAS	—If they should (or, would)
CAT	—If they should (or, would) not
CAU	—If they were
CAV	—If they were not
CAW	—If we (our-s)
CAX	—If we are
CAY	—If we are not
CAZ	—If we can (or, may)
CBA	—If we can not (or, may not)
CBD	—If we could (or, might)
CBE	—If we could (or, might) not
CBF	—If we do (or, did)
CBG	—If we do (or, did) not
CBH	—If we had (or, have)
CBJ	—If we had (or, have) not
CBJ	—If we shall (or, will)
CBK	—If we shall (or, will) not
CBL	—If we should (or, would)
CBM	—If we should (or, would) not.
CBN	—If we were
CBO	—If we were not.
CBP	—If you r-s
CBQ	—If you are
CBR	—If you are not
CBS	—If you can (or, may).
CBT	—If you can not (or, may not)
CBU	—If you could (or might)
CBV	—If you could (or, might) not.
CBW	—If you do (or, did)
CBX	—If you do (or, did) not
CBY	—If you had (or have)
CBZ	—If you had (or, have) not.
CDA	—If you shall (or, will)
CDB	—If you shall (or, will) not
CDE	—If you should (or, would)
CDF	—If you should (or, would) not
CDG	—If you were
CDH	—If you were not.
CDI	**IN**
CDJ	—In him (his, her-s, it-s)
CDK	—In me (my, mine)

	IN—Continued
CDL	—In our-s (us)
CDM	—In that (or, this)
CDN	—In the
CDO	—In them (their-s)
CDP	—In these (or, those)
CDQ	—In what (or, which)
CDR	—In you-r-s
	Not in .... CFV
CDS	**IS**
	He, She, It (or person-s or thing-s indicated) had (has, or, have) done (or, is, or, are doing) ....BUL
	He, She, It (or person-s or thing-s indicated) had (has, or, have) not done (or, is, or, are not doing)............BUP
	He, She, It (or person-s or thing-s indicated) is (or, are) .... BUT
	He, She, It (or person-s or thing-s indicated) is (or, are) able to.......BUV
	He, She, It (or person-s or thing-s indicated) is (or, are) for... .... BUR
	He, She, It (or person-s or thing-s indicated) is (or, are) from ....BUS
	He, She, It (or person-s or thing-s indicated) is (or, are) not.. ....BUW
	He, She, It (or person-s or thing-s indicated) is (or, are) not able to ....BUX
	How is (or, are) he (she, it, or person-s or thing-s indicated)?............BXO
	If he (she, it, or person-s or thing-s indicated) is (or, are)........ :.....BYL
	If he (she, it, or person-s or thing-s indicated) is (or, are) able to ....BYM
	If he (she, it, or person-s or thing-s indicated) is (or, are) not...........BYN
	If he (she, it or person-s or thing-s indicated) is (or, are) not able to ....BYO
	Is, Am, or, Are to.. .. ..BEH
	Is, Am, or, Are to be ....BFC
	Is, Am, or, Are to have...........BQO
	Is, Am, or, Are not ....BEG
	Is, Am, or, Are not to be ....BFD
	Is, Am, or, Are not to have........BQD
	Is, or, Are, he (she, it or person-s or thing-s indicated)—to ?.....BVY
	Is, or, Are his (her-s, it-s)...........BEK
CDT	—Is it?
	Is it done? .... BNH
	Is it not to be done? ....BNI
CDU	—Is it so?
	Is it to be done? ....BNJ
	Is, or, Are my (mine)?.. ....BEL
	Is, or, Are our-s or, Are we? ..BEM
CDV	—Is that (or, this)
	Is, or, Are the. ....BEN
	Is, or, Are their-s, or, Are they?...BEO
	Is, or, Are there—any? ....BEP
	Is, or, Are with — ....BER
	Is, or, Are you r-s?...............BES
	It, etc (See He, She, It)
	That or, This is.. ....CLE
	That or, This is not ....CLF
	That it is .... CLG
	That it is not ....CLH
	There is (or, are).................CMG
	There is (or, are) not.............CMH
	There is nothing ....CMT
	What, or, Which are (or, is)?.... CPW

IS—MIGHT

	Is—*Continued*		It—*Continued*	
	What, *or*, Which is (*or*, are) he (she, it, *or person-s or thing-s indicated*)?_CPX		Must it-s (his, her-s)?_____CES	
	What, *or*, Which, is (*or*, are) you-r s?		Not it-s (his, her-s)?_____OFT	
	CPY		Of it s (him, his, her-s)_____CGN	
	What, *or*, Which is it?_____CQG		On it-s (him, his, her-s)_____CGY	
	When are (*or*, is)?_____CQW		Ought it s (his, her-s)?_____CHO	
	When he (she, it, *or, person-s or thing-s indicated*) is (*or*, are)_____CRJ		Shall, *or*, Will it (he, she)?_____BWD	
			Shall, *or*, Will its (his, hers)?_____CIS	
	When is (*or*, are) he (she, it, *or person-s or thing s indicated*)?_____CRN		Shall *or*, Will they do it?_____BOA	
			Shall, *or*, Will they not do it?_____BOC	
	Where are (*or*, is)?_____OSR		Shall, *or*, Will you do it?_____BOD	
	Where he (she, it, *or person-s or thing-s indicated*) is (*or*, are)_____CTD		Shall, *or*, Will you not do it?_____BOE	
			Should, *or*, Would it s (his, her-s)__CJP	
	Where is (*or* are) he (she, it, *or person-s or thing-s indicated*)?_____CTH		That it-s (his, her-s)_____CLB	
			That it is_____CLG	
	Who are (*or*, is)?_____CUI		That it is not_____CLH	
	Why are (*or*, is)?_____CUZ		To it s (him, his, her-s)_____CNV	
			Was, *or*, Were it-s (his, her-s)?_____COR	
	It_____BVZ		What, *or*, Which is it?_____CQG	
			When it s (his, her-s)_____CRK	
	(*See also* HE, SHE, IT, *or*, PERSON-s, *or*, THING-s)		When shall (*or* will) it be done?__CRX	
			Where it-s (his, her s)_____CTE	
CDW	—After it		Will, etc (*See* Shall)	
CDX	—Before it—is.		Will it be?_____BIX	
	By it-s (him, his, her-s)_____BJR		Will it do?_____BOU	
	Can, *or*, May do it_____BLU		Will you do it?_____CWE	
	Can, *or*, May it-s (his, her-s)?_____BKF		With it-s (him, his, her-s)_____CWG	
	Can, *or*, May it be done?_____BLW		Would, etc. (*See* Should)	
	Can not, *or*, May not do it_____BLY	CDZ	MAY (*See* CAN)	
	Could, *or*, Might it-s (his, her-s)?_____BLA			
	Do, Does, *or*, Did it-s (his, her-s)?_____BMG		ME, MY, MINE, MYSELF_____CED	
	For it-s (him, his, her-s, *or person-s or thing-s indicated*)_____BPD	CEA	—After me (my, mine)	
			Are, *or*, Is my (mine)_____BEL	
	From it-s (him, his, her-s, *or person-s or thing s indicated*)_____BPO	CEB	—Before me (my, mine)	
			By me (my, mine)_____BJS	
	Had, Has, *or*, Have, it-s (his, her-s)?_BQR		Can, *or*, May I (my, mine)?_____BKG	
	If it s (his, her-s he, she)_____BZR		Could, *or*, Might I (my, mine)?_____BLC	
	If it can (*or*, may) be done_____BZB		Do, Does, *or*, Did I (my, mine)?___BMH	
	If it can not (*or*, may not) be done_____BZS		For me (my, mine)_____BPE	
	In it-s (him, his, her-s)_____CDJ		From me (my, mine)_____BPQ	
	Is, *or*, Are it-s (his, her-s)_____BEK		Had, Has, *or*, Have I (my, mine)?__BQS	
	Is it?_____CDT		If I (my, mine)_____BYW	
	Is it done?_____BNH		In me (my, mine)_____CDK	
	Is it not to be done_____BNI		Is, etc (*See* Are)	
	Is it so?_____CDU		May, etc (*See* Can)	
CDY	—Its-self	CED	—Me, My, Mine, Myself	
	Is it to be done?_____BNJ		Me by_____BKD	
	It can be_____BGN	CEF	—Me to	
	It can (*or*, may) be done_____BNK	CEG	—Me with	
	It can not be_____BGO		Might, etc (*See* Could)	
	It can not (*or* may not) be done___BNL		Must I (my, mine)?_____CET	
	It could (*or*, might) be done_____BNM	CEH	—Need I (my, mine)?	
	It could (*or*, might) not be done___BNO		Not I (my, mine)_____CFU	
	It had (*or*, has) been done_____BNP		Of me (my, mine)_____CGO	
	It had (*or*, has) not been done_____BNQ		On me (my, mine)_____COZ	
	It ought to be_____BGR		Ought I (my, mine)?_____CIP	
	It ought to be done_____BNR		Shall, *or*, Will I (my, mine)?_____CIT	
	It ought not to be_____BGS		Should, *or*, Would I (my, mine)?___CJQ	
	It ought not to be done_____BNS		That I (my, mine)_____CLD	
	It shall (*or*, will)_____CIQ		To me (my, mine)_____CNW	
	It shall (*or*, will) be_____BGT		Was, *or*, Were I (my, mine)?_____COS	
	It shall (*or*, will) be done_____BNT		When I (my, mine)_____CRL	
	It shall (*or*, will) not be_____BGU		Where I (my, mine)_____CTF	
	It shall (*or*, will) not be done_____BNU		Will, etc (*See* Shall)	
	It will do_____BNV		With me (my, mine)_____CWH	
	Let him (her, it, *or person-s or thing-s indicated*)—have_____BWA		Would, etc (*See* Should)	
	May, etc (*See* Can)	CEI	MIGHT (*See* COULD)	
	Might, etc (*See* Could)			

MUST—NOT

	MUST
CEJ	He, She, It (*or person-s or thing-s indicated*) must............................BUY
	He, She, It (*or person s or thing-s indicated*) must be... ....BUZ
	He, She, It (*or person-s or thing-s indicated*) must not . ...BVA
	He, She, It (*or person-s or thing-s indicated*) must not be ....BVC
	How must?............................BXP
CEK	—I must
	I must be ....BGE
CEL	—I must not
	I must not be ....BGF
	If I must....... ....BŽJ
	If I must not........................BZK
CEM	—Must be
CEN	—Must do (*or*, be done)
CEO	—Must have
CEP	—Must have been
CEQ	—Must have had
CER	—Must he (she, it, *or person-s or thing s indicated*)?
CES	—Must his (her-s, it-s)?
CET	—Must I (my, mine)?
CEU	—Must not
CEV	—Must not be
CEW	—Must not do (*or*, be done)
CEX	—Must that (*or*, this)?
CEY	—Must the?
CEZ	—Must there be?
CFA	—Must these (*or*, those)?
CFB	—Must they (their-s)?
CFD	—Must we (our-s)?
CFE	—Must you-r-s?
	That, *or*, This must..............CLI
	That, *or*, This must be..........CLJ
	That, *or*, This must not ... .CLK
	That *or*, This must not be ......CLM
CFG	—They must
	They must be..........................BHV
CFH	—They must not
	They must not be......................BHW
CFI	—We must
	We must be.............................BIP
CFJ	—We must not
	We must not be.......................BIQ
	What, *or*, Which must?............CQH
	When must?...........................CRO
	Where must?..........................CTI
	Who must?.............................CUQ
	Why must?.............................CVO
CFK	—You must
CFL	You must be............................BJF
	—You must not
CFM	You must not be.......................BJG
	NOT
	Am I not?............................BED
	Am, Is, *or*, Are not............BEG
	Am, Is, *or*, Are not to be ......BFD
	Am, Is, *or*, Are not to have.........BQD
	Can, *or*, May they not?............BKM
	Can not, *or*, May not ....BKP
	Can not, *or*, May not be.. ... .BFJ
	Can not, *or*, May not do (*or*, be done)BLX
	Can not, *or*, May not do it.........BLY
	Can not, *or*, May not have.........BQH
	Can not, *or*, May not have been....BFK
	Can not, *or*, May not have done (*or*, been done) ....BLZ

NOT—*Continued*

Can not, *or*, May not have had .....BQI
Can not, *or*, May not he (she, it, *or person-s or thing-s indicated*) be? ...BTM
Can not, *or*, May not he (she, it, *or person-s or thing-s indicated*) have been? ....BTN
Could, *or* Might be (she, it, *or person-s or thing-s indicated*) not be? .....BTP
Could, *or*, Might not ....BLD
Could, *or*, Might not be..............BFN
Could, *or*, Might not do (*or*, be done).BMD
Could *or*, Might not have ....BQL
Could, *or*, Might not have been ...BFO
Could, *or*, Might not have done (*or*, been done) ....BMC
Could, *or*, Might not have had.....BQM
Do, Does, *or*, Did not ....BMI
Do not let him (her, it, *or person-s or thing-s indicated*) have............BTR
Had, Has, *or*, Have, he (she, it, *or person s or thing-s indicated*) not been? ...BTW
Had, Has, *or*, Have, he(she, it, *or person-s or thing-s indicated*) not been? ...BTX
Had, Has, *or*, Have, he(she, it,*or person-s or thing s indicated*) not had?....BTY
Had, Has, *or*, Have not ....BQU
Had Has, *or*, Have not been.. ......BFS
Had, Has, *or*, Have not done (*or*, been done) .... BMS
Had, Has, *or*, Have not had ......BQV
He, She, It (*or person-s or thing s indicated*) can not (*or*, may not) ....BUD
He, She, It (*or person-s or thing s indicated*) can not (*or*, may not) be ..BUB
He She It (*or person s or thing-s indicated*) could (*or*, might) not .....BUG
He, She, It (*or person-s or thing-s indicated*) does (do, *or*, did) not ..... BUI
He, She, It (*or person-s or thing-s indicated*) had (has, *or*, have) not ....BUN
He, She, It (*or person-s or thing-s indicated*) had (has, *or*, have) not been..BUO
He, She, It (*or person-s or thing s indicated*) had (has, *or*, have) not done (*or*, is, *or*, are not doing) .....BUP
He, She, It (*or person-s or thing-s indicated*) had (has, *or* have) not had ..BUQ
He, She, It (*or person-s or thing-s indicated*) is (*or*, are) not............BUW
He, She, It (*or person-s or thing-s indicated*) is (*or* are) not able to.....BUX
He, She, It, etc , may   (*See* Can)
He, She, It, etc , might   (*See* Could)
He, She, It (*or person s or thing s indicated*) must not ... .BVA
He, She, It (*or person-s or thing-s indicated*) must not be..............BVC
He, She, It (*or person-s or thing-s indicated*) ought not to..............BVE
He, She, It (*or person-s or thing-s indicated*) ought not to be. ....BVG
He, She, It (*or person-s or thing-s indicated*) shall (*or*, will) not ....BVJ
He, She, It (*or person-s or thing-s indicated*) shall (*or* will) not be .....BVK
He, She, It (*or person s or thing s indicated*) should (*or*, would) not ....BVP
He, She, It (*or person-s or thing-s indicated*) should (*or*, would) not be..BVQ

NOT

Not—Continued

He, She, It (or person-s or things indicated) should (or, would) not do (or, be done) ---------- ---------- BVR
He, She, It (or person s or thing-s indicated) should (or, would) not have-BVS
He, She, It (or person-s or thing-s indicated) was (or, were) not---------BVU
He, She, It (or person s or thing-s indicated) was (or, were) not to be ---BVX
I am not----------------------------BEI
I can not (or, may not) ------------BKR
I can not (or, may not) be ---------BFY
I can not (or, may not) have ------BRJ
I could (or, might) not ----------BLM
I could (or, might) not be----------BGA
I could (or, might) not have -------BRL
I do (or, did) not ----------------BMZ
I had (or, have) not---------------BRC
I had (or, have) not been ---------BGD
I had (or, have) not done --------BNC
I had (or, have) not had ----------BRP
I must not ------------------------CBI
I must not be ---------------------BGF
I ought not to --------------------CHM
I shall (or, will) not --------------CIP
I shall (or, will) not be ---------- BGI
I shall (or, will) not do -----------BNE
I shall (or, will) not have ---------BRS
I should (or, would) not-----------CJN
I should (or, would) not be --------BGL
I should (or, would) not do --------BNG
I should (or, would) not have -----BRU
I should (or, would) not have been--BGM
I was not-------------------------COI
If he (she, it, or person-s or thing-s indicated) can not (or, may not) -- -BYE   CFN
If he (she, it, or person-s or thing-s indicated) could (or, might) not -----BYG   CFO
If he (she, it, or person-s or thing s indicated) does (do, or, did) not------BYI   CFP
If he (she, it, or person-s or thing-s indicated) had (has, or, have) not------BYK   CFQ
If he (she, it, or person-s or thing-s indicated) is (or, are) not------------BYN   CFR
If he (she, it, or person-s or thing-s indicated) is (or, are) not able to-----BYO   CFS
If he (she, it, or person-s or thing s indicated) shall (or will) not---------BYQ   CFT
If he (she, it, or person-s or thing-s indicated) should (or, would) not----BYS   CFU
If he (she, it, or person-s or thing-s indicated) was (or, were) not --------BYU   CFV
If I am not ------------- - ------BYZ   CFW
If I can not (or, may not) ---------BZC   CFX
If I could (or, might) not ---------BZE   CFY
If I do (or, did) not --------------BZG   CFZ
If I had (or, have) not-------------BZI   CGA
If I must not--------- -- - -------BZK   CGB
If I ought not to ------------------BZM   CGD
If I shall (or, will) not -----------BZO   CGE
If I should (or, would) not --------BZQ   CGF
If I was not ----------------------BZU   CGH
If it can not (or, may not) be done--BZS   CGI
If not ---- -------------- ---------BZV   CGJ
If they are not ----- --- ----------CAG   CGK
If they can not (or, may not)-------CAI   CGL
If they could (or, might) not ------CAL
If they did (or, do) not ---- ------CAN
If they had (or, have) not----------CAP
If they shall (or, will) not --------CAR

Not—Continued

If they should (or, would) not -----CAT
If they were not ----------- ----CAV
If we are not----------------------CAY
If we can not (or, may not) -------CBA
If we could (or, might) not --------CBE
If we do (or, did) not--------------CBG
If we had (or, have) not-----------CBI
If we shall (or, will) not-----------CBK
If we should (or, would) not --- --CBM
If we were not ----- -- -----------CBO
If you are not ----- --- -----------CBR
If you can not (or, may not) -------CBT
If you could (or, might) not-------CBV
If you do (or, did) not ------------CBX
If you had (or, have) not-----------CBZ
If you shall (or, will) not----------CDB
If you should (or, would) not------CDF
If you were not ----- ------------ CDH
Is, etc   (See Are)
It can not be ---------------------BGO
It can not (or, may not) be done ---BNL
It could (or, might) not be done... BNO
It had (or, has) not been done ------BNQ
It ought not to be -----------------BGS
It ought not to be done -----------BNS
It shall (or, will) not be-----------BGU
It shall (or, will) not be done------BNU
May, etc   (See Can)
Might, etc   (See Could)
Must not----------------------------CEU
Must not be -------------- ----CEV
Must not do (or, be done) ----------CEW
Need not be----- ------------------BGZ
Not----------------------------------CFM
—Not be (been, being)
—Not by
—Not doing
—Not done
—Not for
—Not from
—Not his (her-s, it-s)
—Not I (my, mine)
—Not in
—Not of
—Not on
—Not our-s (us)
—Not that (or, this)
—Not the
—Not there
—Not these (or, those)
—Not they (their-s)
—Not to
—Not to be
—Not to be done
—Not to be had
—Not with
—Not your-s
Ought not to----------------------CHR
Ought not to be -------------------CHS
Ought not to have------------------CHT
Person-s, or, Thing-s, etc   (See He)
Shall, or, Will he (she, it, or person s or thing s indicated) not ?------ . BWH
Shall, or, Will he (she, it, or person-s or thing-s indicated) not be ? ------BWI
Shall, or, Will he (she, it, or person-s or thing-s indicated) not do (or, be done)?BWJ
Shall, or, Will he (she, it, or person-s or thing-s indicated) not have ?-----BWK

NOT.

NOT—*Continued.*

Shall, *or*, Will I not have ? ....BRY
Shall, *or*, Will not.... .... CIU
Shall, *or*, Will not be.............BHD
Shall, *or*, Will not do (*or*, be done)_BNY
Shall, *or*, Will not have...........BRZ
Shall, *or*, Will not have done (*or*, been done) ... . BNZ
Shall, *or*, Will they not ? ....CIZ
Shall, *or*, Will they not do it ? ...BOO
Shall, *or*, Will we not ? ....CJB
Shall, *or*, Will you not ? ....CJE
Shall, *or*, Will you not be ? ....BHJ
Shall, *or*, Will you not do it ?..... BOE
Should, *or*, Would he (she, it, *or person-s or thing-s indicated*) not ? ......BWR
Should, *or*, Would he(she, it, *or person-s or thing s indicated*) not be ? ....BWS
Should, *or*, Would he (she, it, *or person s or thing-s indicated*) not do (*or*, be done*)? ....BWT
Should, *or*, Would he (she, it, *or person-s or thing-s indicated*) not have ? ....BWU
Should, *or*, Would he (she, it, *or person-s or thing-s indicated*) not have been ? ....BWV
Should, *or*, Would not..............CJR
Should, *or*, Would not be ....BHM
Should, *or* Would not do (*or*, be done) ....BOH
Should, *or*, Would not have - .....BSD
Should, *or*, Would not have been.-BHN
Should, *or*, Would not have done (*or*, been done).... ....BOI
Should, *or*, Would not have had.....BSE
That, *or*, This can not (*or*, may not)_CKS
That, *or*, This can not (*or*, may not) be ....CKT
That, *or*, This could (*or*, might) not be ....CKW
That, *or*, This had (has, *or*, have) not been ....CKZ
That, *or*, This is not ....CLF
That it is not.................... CLH
That, *or*, This must not ....CLK
That, *or*, This must not be ....CLM
That, *or*, This shall (*or*, will) not be _CLO
That, *or*, This should (*or*, would) not be...... .. .. CLQ
That, *or*, This was not ....CLS
There are (*or*, is) not ....CMH
There can not (*or*, may not) ......CMK
There can not (*or*, may not) be....CML
There could (*or*, might) not be ....CMP
There had (has, *or*, have) not been.CMS
There is nothing .... CMT
There shall (*or*, will) not ....CMW
There shall (*or*, will) not be........CMX
There should (*or*, would) not .....CNB
There should (*or*, would) not be ...CND
There should (*or*, would) not have been ....CNE
There was (*or*, were) not ....CNG
They are not . ...BEV
They can not (*or*, may not).........BKU
They can not (*or*, may not) be ...-.BHQ
They can not (*or*, may not) have ...BSG
They could (*or*, might) not .... .. BLO
They could (*or*, might) not be . .. BHS
They do (*or*, did) not.... .... ---BOK

NOT—*Continued*

They had (*or*, have) not............ BSJ
They had (*or*, have) not been......BHU
They had (*or*, have) not done .....BOM
They had (*or*, have) not had ....BSK
They must not .... CFH
They must not be ....BHW
They ought not to ....CIF
They ought not to be...............BHY
They shall (*or*, will) not ....CJG
They shall (*or*, will) not be ....BIA
They should (*or*, would) not .....-.CJZ
They should (*or*, would) not be.....BID
They should (*or*, would) not have -BSM
They were not ....COK
They were not to be...................BIF
Was, *or*, Were he (she, it, *or person s or thing s indicated*) not—to ? ......BWY
Was I not—to ?......................COU
Was, *or* Were not ....COV
Was, *or*, Were not done (*or*, doing)_BOP
We are not.......................BEX
We can not (*or*, may not) ....BKW
We can not (*or*, may not) be ...... BIJ
We can not (*or*, may not) have ....BSO
We could (*or*, might) not............BLQ
We could (*or*, might) not be.........BIM
We do (*or*, did) not................BOR
We had (*or*, have) not..............BSR
We had (*or*, have) not been.........RIO
We had (*or*, have) not done ....BOT
We had (*or*, have) not had.........BST
We must not ....CFJ
We must not be ....BIQ
We ought not to ....CIH
We shall (*or*, will) not ... .CJI
We shall (*or*, will) not be .... BIS
We shall (*or*, will) not have.........BSV
We should (*or*, would) not ....CKB
We should (*or*, would) not be.......BIU
We should (*or*, would) not have ....BSX
We were not ....CPF
We were not to be ....BIW
Were they not ?....................CPI
Were we not ? ....CPK
Were you not ?.....................CPM
Why are you not ?..................CVB
Why can not (*or*, may not) ?........CVD
Why could (*or*, might) not ?.........CVF
Why do (does, *or*, did) not ?.......CVH
Why do (*or*, did) you not ? ....CVJ
Why had (has, *or*, have) not ?.......CVL
Why had (*or*, have) you not ? . ....CVN
Why not ?..........................CVP
Why shall (*or*, will) not ? ....CVS
Why should (*or*, would) not ?.......CVU
You are not ....BEZ
You can not (*or*, may not) ... . BKY
You can not (*or*, may not) be ....BIZ
You can not (*or*, may not) have....BSZ
You could (*or*, might) not .. ..BLS
You could (*or*, might) not be ......BJC
You do (*or*, did) not ....BOX
You had (*or* have) not ....BTD
You had (*or*, have) not been.......BJE
You had (*or*, have) not done ......BOZ
You had (*or*, have) not had ....BTE
You must not ....CFL
You must not be ....BJG
You ought not to ....CIK
You ought not to be................BJI

NOT—SHALL

NOT—Continued

You shall (or, will) not .... CJL
You shall (or, will) not be .... BJL
You shall (or, will) not have ...... BTG
You should (or, would) not........ CKE
You should (or, would) not be...... BJN
You should (or, would) not have ...BTI
You were not .... CPO
You were not to be .... BJP

OF .... CGM
   Not of .... CFW
CGM —Of
CGN —Of him (his, her s, it-s)
CGO —Of me (my, mine)
CGP —Of our-s (us)
CGQ —Of that (or, this)
CGR —Of the
CGS —Of them (their s)
CGT —Of these (or, those)
CGU —Of what (or, which)
CGV —Of whom (or, whose)
CGW —Of you r-s
   What, or, Which of the — ?........ CQI

ON .... CGX
   Not on....... .... CFX
CGX —On
CGY —On him (his, her-s, it-s)
CGZ —On me (my, mine)
CHA —On our-s (us)
CHB —On that (or, this)
CHD —On the
CHE —On them (their-s)
CHF —On these (or, those)
CHG —On what (or, which)
CHI —On whom (or, whose)
CHJ —On you-r s

CHK OUGHT TO
He, She, It (or person-s or thing-s indicated) ought to .... BVD
He She, It (or person-s or thing s indicated) ought not to........ BVE
He, She, It (or person-s or thing-s indicated) ought not to be........... BVG
He, She, It (or person-s or thing-s indicated) ought to be........ BVF
How ought ? .... BXQ
CHL —I ought to
CHM —I ought not to
If I ought to .... BZL
If I ought not to .... BZM
It ought not to be .... BGS
It ought not to be done .... BNS
It ought to be .... BGR
It ought to be done......... ...BNR
CHN —Ought he (she, it or person-s or thing-s indicated) to?
CHO —Ought his (her-s, it-s)?
CHP —Ought I (my, mine)?
CHQ —Ought I to—do?
CHR —Ought not to
CHS —Ought not to be
CHT —Ought not to have
CHU —Ought that (or, this)?
CHV —Ought the?
CHW —Ought these (or, those)?
CHX —Ought they to?
CHY —Ought to be
CHZ —Ought to do (or, be done)

OUGHT TO—Continued

CIA —Ought to have
CIB —Ought we (our-s) to?
CID —Ought you-r-s to to?
CIE —They ought to
CIF —They ought not to
   They ought not to be .... BHY
   They ought to be.................. BHX
CIG —We ought to
CIH —We ought not to
   What, or, Which ought to?........ CQJ
   When ought? .... CRP
   Where ought? .... CTJ
   Who ought? .... CUR
   Why ought? .... CVQ
CLJ —You ought to
CIK —You ought not to
   You ought not to be .... BJI
   You ought to be................. BJH

OUR-S-SELVES .... CIN
   —After our-s (us, we)
   Are, or, Is our-s, or, Are we?...... BEM
CIM —Before our-s (us)
   By our-s (us) .... BJY
   Can, or, May we (our-s)? .... BKN
   Could, or, Might we (our-s)? .... BLI
   Do, Does, or, Did we (our s)? ...... BMP
   For our-s (us)........ .... BPJ
   From our-s (us) .... BPW
   Had, Has, or, Have we (our-s)? .... BRD
   If we (our-s)..... .... CAW
   In our-s (us)................... CDL
   Is, etc (See Are)
   May etc (See Can)
   Might, etc (See Could)
   Must we (our-s)? .... CFD
   Not our-s (us) .... CFY
   Of our s (us) .... CGP
   On our-s (us)................... CHA
   Ought we (our s) to?............. CIB
CIN —Our-s-selves
   Shall, or, Will we (our-s)?......... CJA
   Should, or, Would we (our-s)?...... CJW
   Than we (our s) .... CKJ
   That we (our-s) .... CLW
   To our-s (us) .... COB
   Was or, Were our-s (we)?.......... CPA
   When we (our-s)................. CSK
   Where we (our-s) .... CUD
   Why we (our-s).................. CVY
   Will etc (See Shall)
   With our-s (us) .... CWM
   Would, etc (See Should)

SHALL (or WILL) .... CIR
He She, It (or person-s or thing-s indicated) shall (or, will).......... BVH
He, She, It (or person-s or thing s indicated) shall (or, will) be.......... BVI
He, She, It (or person s or thing-s indicated) shall (or, will) not....... BVJ
He, She, It (or person-s or thing-s indicated) shall (or, will) not be .. BVK
How shall (or, will) I?.............. BXS
How shall (or, will) he (she, it, or person s or thing-s indicated)?........... BXR
How shall (or, will) we?............ BXT
How shall (or, will) you?............ BXU
CIO —I shall (or, will)
   I shall (or, will) be .... BGH

SHALL

	SHALL—*Continued*		
	I shall (*or*, will) do ....BND	CJA	
	I shall (*or*, will) have ....BRQ		
	I shall (*or*, will) if I can ....BKS	CJB	
CIP	—I shall (*or*, will) not	CJD	
	I shall (*or*, will) not be ... .BGI		
	I shall (*or*, will) not do ....BNE		
	I shall (*or*, will) not have . ...BRS	CJE	
	If he (she, it, *or person-s or thing-s indi-*		
	cated) shall (*or*, will) ....BYP		
	If he (she, it, *or person-s or thing s indi-*		
	cated) shall (*or*, will) not ....BYQ		
	If I shall (*or*, will) ....BZN		
	If I shall (*or*, will) not..............BZO		
	If they shall (*or*, will).............CAQ		
	If they shall (*or*, will) not.........CAR		
	If we shall (*or*, will) . ...CBJ	CJF	
	If we shall (*or*, will) not... .. ....CBK		
	If you shall (*or*, will) ....CDA	CJG	
	If you shall (*or*, will) not..........CDB		
CIQ	—It shall (*or*, will)	CJH	
	It shall (*or*, will) be..............BGT		
	It shall (*or*, will) be done ... .BNT		
	It shall (*or*, will) not be ....BGU		
	It shall (*or*, will) not be done . ...BNU		
CIR	—Shall	CJI	
	Shall, *or*, Will be. ....BHA		
	Shall, *or*, Will do (*or*, be done) ... BNW		
	Shall, *or*, Will have................BRW		
	Shall, *or*, Will have done (*or*, been done)		
	BNX		
	Shall, *or*, Will he (she, it, *or person-s or*		
	thing-s indicated)? .... BWD		
	Shall, *or*, Will he (she, it, *or person s or*		
	thing-s indicated) be?.............BWE		
	Shall, *or*, Will he (she, it, *or person-s or*		
	thing-s indicated) do (*or*, be done)?		
	BWF		
	Shall, *or*, Will he (she, it, *or person-s or*		
	thing-s indicated) have? .. ..BWG		
	Shall, *or*, Will he (she, it, *or person-s or*		
	thing-s indicated) not?.... .. ...BWH		
	Shall, *or*, Will he (she, it, *or person-s or*		
	thing-s indicated) not be?........BWI		
	Shall, *or*, Will he (she, it, *or person-s or*		
	thing-s indicated) not do (*or*, be done)?		
	BWJ		
	Shall, *or*, Will he (she, it, *or person s or*		
	thing-s indicated) not have? ....BWK		
CIS	—Shall, *or*, Will his (heis, its)?		
CIT	—Shall, *or*, Will I (my, mine)?		
	Shall, *or*, Will I be? . ...BHC		
	Shall, *or*, Will I have? ....BRX		
	Shall, *or*, Will I not have...........BRY		
CIU	Shall, *or*, Will it? (*See* HE.)		
	—Shall, *or*, Will not		
	Shall, *or*, Will not be... .. ..BHD		
	Shall, *or*, Will not do (*or*, be done).BNY		
	Shall, *or*, Will not have?.. ...... BBZ		
	Shall, *or*, Will not have done (*or*, been		
	done) ....BNZ		
CIV	—Shall, *or*, Will that (*or*, this)?		
CIW	—Shall, *or*, Will the?		
	Shall, *or*, Will there be?...........BHE		
CIX	—Shall, *or*, Will these (*or*, those)?		
CIY	—Shall, *or*, Will they (their-s)?	CJK	
	Shall, *or*, Will they be? .... BHF		
	Shall, *or*, Will they do it? ... .BOA		
CIZ	—Shall, *or*, Will they not?	CJL	
	Shall, *or*, Will they not do it?.....BOC		

SHALL—*Continued.*	
—Shall, *or*, Will we (our-s)?	
Shall, *or*, Will we be?..............BHG	
—Shall, *or*, Will we not?	
—Shall, *or*, Will you-r-s?	
Shall, *or*, Will you be?............ BHI	
Shall, *or*, Will you do it?.. ....BOD	
—Shall, *or*, Will you not?	
Shall, *or*, Will you not be?........BHJ	
Shall, *or*, Will you not do it? ......BOE	
That, *or*, This shall (*or*, will) be....CLN	
That, *or*, This shall (*or*, will) not be.CLO	
There shall (*or*, will)... ....CMU	
There shall (*or*, will) be ....CMV	
There shall (*or*, will) not ....CMW	
There shall (*or*, will) not be ._......CMX	
—They shall (*or*, will)	
They shall (*or*, will) be ....BHZ	
—They shall (*or*, will) not	
They shall (*or*, will) not be ....BIA	
—We shall (*or*, will)	
We shall (*or*, will) be...............BIR	
We shall (*or*, will) have......... ...BSU	
—We shall (*or*, will) not	
We shall (*or*, will) not be...........BIS	
We shall (*or*, will) not have ....BSV	
What, *or*, Which shall (*or*, will) I?..CQK	
What, *or*, Which shall (*or*, will) I (*or*,	
we) do? ....CQL	
What, *or*, Which shall (*or*, will) I (*or*,	
we) have.......................... CQM	
What, *or*, Which shall (*or*, will) you (*or*,	
they) do? ....CQN	
What, *or* Which shall (*or*, will) you (*or*,	
they) have?......................CQO	
When shall (*or*, will)? ....CBQ	
When shall (*or*, will) be? ....CRS	
When shall (*or*, will) he (she, it, *or per-*	
son-s or thing-s indicated)? ......CRT	
When shall (*or*, will) I (*or*, we)?.. CRU	
When shall (*or*, will) I have? ......CRW	
When shall (*or*, will) I (they, we, *or*, you)	
do? ....CRV	
When shall (*or*, will) it be done?...CRX	
When shall (*or*, will) they? ....CRY	
When shall (*or*, will) they (*or*, you)	
have? ....CRZ	
When shall (*or*, will) we have? ....CSA	
When shall (*or*, will) you?........CSB	
When shall (*or*, will) you be?.......CSD	
Where shall (*or*, will)? ....CTK	
Where shall (*or*, will) be? ....CTL	
Where shall (*or*, will) he (she, it, *or*	
person-s or thing-s indicated)?....CTM	
Where shall (*or* will) I (*or*, we)?...CTN	
Where shall (*or*, will) I have? .... CTO	
Where shall (*or*, will) they? ....CTP	
Where shall (*or*, will) they (*or*, you)	
have? ....CTQ	
Where shall (*or*, will) you? ....CTR	
Where shall (*or*, will) you? ....CTS	
Where shall (*or*, will) you be?......CTU	
Who shall (*or*, will)? ....ZBF	
Why shall (*or*, will)?................CVB	
Why shall (*or*, will) not? ....CVS	
—You shall (*or*, will)	
You shall (*or*, will) be.............BJK	
You shall (*or*, will) have ....BTF	
—You shall (*or*, will) not	
You shall (*or*, will) not be...... ...BJL	
You shall (*or*, will) not have . .....BTG	

	SHE—SHOULD.		

	SHE (See HE, SHE, IT) .... BWL		SHOULD—Continued.
			Should, or, Would he (she, it, or person-s or thing-s indicated) not be?. BWS
	SHOULD (or, WOULD) .... CJO		Should, or, Would he (she, it, or person-s or thing-s indicated) not do (or, be done)? _____ _____ BWT
	He, She, It (or person-s or thing-s indicated) should (or, would) .... BVL		Should, or, Would he (she, it, or person-s or thing-s indicated) not have? BWU
	He, She, It (or person-s or thing-s indicated) should (or, would) be .... BVM		
	He, She, It. (or person-s or thing-s indicated) should (or, would) do (or, be done) _____ _____ _____ BVN		Should, or, Would he (she, it, or person-s or thing-s indicated) not have been? ____ _____ __ _____ BVV
	He, She, It (or person-s or thing-s indicated) should (or, would) have. _BVO	CJP	—Should, or, Would his (her-s, it-s)?
	He, She, It (or person-s or thing-s indicated) should (or, would) not ....BVP	CJQ	—Should, or, Would I (my, mine)?
	He, She, It (or person-s or thing-s indicated) should (or, would) not be .BVQ	CJR	—Should or, Would not
	He, She, It (or person-s or thing-s indicated) should (or, would) not do (or, be done) _____ . _____ BVR		Should, or, Would not be _____ BHK
			Should, or, Would not do (or, be done), BOH
	He, She, It (or person-s or thing-s indicated) not have. BVS		Should, or, Would not have _____ BSD
	How should (or, would)? _____ RXV		Should or, Would not have been __BHN
CJM	—I should (or, would)		Should, or, Would not have done (or, been done) _____ BOI
	I should (or, would) be _____ BGJ		Should, or, Would not have had ... BSE
	I should (or, would) do _____ BNF	CJS	—Should, or, Would that (or, this)?
	I should (or, would) have _____ BRT	CJT	—Should, or, Would the?
	I should (or, would) have been ... BGK		Should. or, Would there be? ...... BHO
CJN	—I should (or, would) not	CJU	—Should or Would these (or, those)?
	I should (or, would) not be_____ BGL	CJV	—Should, or, Would they (their-s)?
	I should (or, would) not do_____ BNO	CJW	—Should, or, Would we (our-s)?
	I should (or, would) not have .... BRU	CJX	—Should, or, Would you-r-s?
	I should (or, would) not have been. BGM		That, or, This should (or, would) be. CLP
	If he (she, it, or person-s or thing-s indicated) should (or, would)_____ BVR		That, or, This should (or, would) not be CLQ
	If he (she, it, or person s or thing-s indicated) should (or, would) not ....BYS		There should (or, would) .... CMY
	If I should (or, would)_____ _____ BZP		There should (or, would) be ...... CMZ
	If I should (or, would) not_____ BZQ		There should (or, would) have been. CNA
	If I were you I should not do it ... LIQ		There should (or, would) not ...... CNB
	If they should (or, would) ....CAS		There should (or, would) not be.... CND
	If they should (or, would) not .....CAT		There should (or, would) not have been CNE
	If we should (or, would)_____ ____CBL	CJY	—They should (or, would)
	If we should (or, would) not_. _____CBM		They should (or, would) be.......... BIC
	If you should (or, would) _____CDE		They should (or, would) have........ BSL
	If you should (or, would) not ......CDF	CJZ	—They should (or, would) not
	It etc (See He, She, It)		They should (or, would) not be...... BID
	She, etc (See He, She, It)		They should (or, would) not have.... BSM
		CKA	—We should (or, would)
CJO	—Should (or, Would)		We should (or, would) be .... BIT
	Should, or, Would be .... BHK		We should (or, would) have.......... BSW
	Should, or, Would do (or, be done).BOF	CKB	—We should (or, would) not
	Should, or, Would have ....BSA		We should (or, would) not be....... BIU
	Should, or, Would have been ......BHL		We should (or, would) not have.....RSX
	Should, or, Would have done (or, been done) _____ ___ BOG		What, or, Which should (or, would)? CQP
			What, or, Which should (or, would) I (or we) do? _____ _____ CQR
	Should, or, Would have had ...... BSC		What, or, Which should (or, would) you (or, they) do? _____ CQS
	Should, or, Would he (she, it, or person-s or thing-s indicated)?...... BWM		When should (or, would)? _____ CSE
	Should, or, Would he (she, it, or person-s or thing-s indicated) be?_ .BWN		Where should (or, would)? ...... CTV
	Should, or, Would he (she, it, or person-s or thing-s indicated) do (or, be done)? _____ BWO		Where should (or, would) the— ?_CTW
			Where should (or, would) they (their-s)? CTX
			Who should (or, would)? .... CUV
	Should, or, Would he (she, it, or person-s or thing-s indicated) have.. BWP		Why should (or, would)? _____ CVT
	Should, or, Would he (she, it, or person-s or thing-s indicated) have been? BWQ		Why should (or, would) not? _____ CVU
		CKD	—You should (or, would)
			You should (or, would) be .... BJM
			You should (or, would) have........ BTH
	Should, or, Would he (she, it or person-s or thing-s indicated) not? .BWR	CKE	—You should (or, would) not
			You should (or, would) not be...... BJN
			You should (or, would) not have ... BTI

THAN—THERE

	THAN ....CKF			THE ....CME
CKF	—Than		CMB	—After the —
CKG	—Than the		CMD	—And the —
CKH	—Than these (or, those).			Are, or, Is the — ....BEN
CKI	—Than they (their-s)			Be the — ....BFE
CKJ	—Than we (our-s)			By the — ....BJU
CKL	—Than which (or, what).			Can, or May the —? ....HUE
CKM	—Than you r-s			Could, or, Might the —?.. ....BLF
				Do, Does, or, Did the —?............BMK
	THAT... ----CKP (or, THIS........CMA)			For the — ....BPG
CKN	—After that (or, this)			From the — ....BPS
CKO	—Before that (or, this)			Had, Has, or, Have the — ....BQX
	By that (or, this). ....BJT			If the — ....BZY
	Can, or, May that (or, this)? ....BKI			In the — ....CDN
	Could, or, Might that (or, this)?....BLE			Is, etc (See Are)
	Does, or, Did that (or, this)? ....BMJ			May, etc (See Can)
	For that (or, this) ....BFF			Might, etc (See Could)
	From that (or, this) ....BPR			Must the —? ....CEY
	Had, Has or, Have that (or, this)...BQW			Not the — ....CGA
	If that (or, this) ....BZX			Of the — ....CGB
	In that (or, this) ....CDM			On the — ....CHD
	Is that (or, this) ....CDV			Ought the —? ....CHV
	May, etc (See Can)			Shall, or, Will the —?............CIW
	Might, etc (See Could)			Should, or, Would the —?............CJT
	Must that (or, this)?............CEX			Then the — ....CKG
	Not that (or, this) ....CFZ			That the — ....CLT
	Of that (or, this) ....CGQ		CME	—The
	On that (or, this)............CHB			To the — ....CNX
	Ought that (or, this)? ....CHU			Was, or, Were the — ....COW
	Shall, or, Will that (or, this)?............CIV			What, or, Which of the —? ....CQI
	Should, or, Would that (or, this)?... CJS			When the — ....CSF
CKP	—That.			Where should (or, would) the — ?_CTW
CKQ	—That, or, This can (or, may)			Where the — ....CTY
CKR	—That, or, This can (or, may) be			Will, etc (See Shall)
CKS	—That, or, This can not (or, may not)			With the — ....CWI
CKT	—That, or, This can not (or, may not) be			Would, etc (See Should)
CKU	—That, or, This could (or, might)			
CKV	—That, or, This could (or, might) be			THEM THEIR-S (See THEY) ......CNL
CKW	—That, or, This could (or, might) not be			
CKX	—That, or, This had (has, or, have)			THERE ....CMF
CKY	—That, or, This had (has, or, have) been			Are, or, Is there—any? ....BEP
CKZ	—That, or, This had (has, or, have) not been			By there............BJW
CLA	—That he (she, it, or, person-s or things indicated)			Can, or, May there be............BFI
CLB	—That his (her-s, it-s)			Could, or, Might there be?............BFP
CLD	—That I (my, mine).			Do, Does, or, Did there? ....BML
CLE	—That, or, This is			From there ....BPU
CLF	—That, or, This is not			Had, Has, or, Have there ....BQY
CLG	—That it is			If there ....CAB
CLH	—That it is not			Must there be? ....CEZ
CLI	—That, or, This must			Not there ....CGB
CLJ	—That, or, This must be			Shall, or, Will there be ? ....BHE
CLK	—That, or, This must not.			Should, or, Would there be ? ......BHO
CLM	—That, or, This must not be		CMF	—There
CLN	—That, or, This shall (or, will) be,		CMG	—There are (or, is)
CLO	—That, or, This shall (or, will) not be		CMH	—There are (or, is) not.
CLP	—That, or, This should (or, would) be		CMI	—There can (or, may)
CLQ	—That, or, This should (or, would) not be		CMJ	—There can (or, may) be
CLR	—That, or, This was		CMK	—There can not (or, may not)
CLS	—That, or, This was not		CML	—There can not (or, may not) be
CLT	—That the		CMN	—There could (or, might)
CLU	—That these (or, those)		CMO	—There could (or, might) be
CLV	—That they (their-s)		CMP	—There could (or, might) not be
CLW	—That we (our-s)		CMQ	—There had (has, or, have)
CLX	—That which (or, what)		CMR	—There had (has, or, have) been
CLY	—That with		CMS	—There had (has, or, have) not been
CLZ	—That you-r-s		CMT	—There is nothing
CMA	—This		CMU	—There shall (or, will)
	To that (or, this)............COA		CMV	—There shall (or, will) be
	Was that (or, this)............COZ		CMW	—There shall (or, will) not
	With that (or, this)............CWJ		CMX	—There shall (or, will) not be
			CMY	—There should (or, would)

	THERE—THEY

THERE—THEY

CMZ	—There should (or, would) be		If they should (or, would) _____ ___CAS
CNA	—There should (or, would) have been		If they should (or, would) not_____CAT
CNB	—There should (or, would) not		If they were _____CAU
CND	—There should (or, would) not be		If they were not_____CAV
CNE	—There should (or, would) not have been		In them (their-s) _____CDO
CNF	—There was (or, were)		Is, etc (See Are)
CNG	—There was (or, were) not		Let them be_____BGW
	Was, or, Were there? _____ COX		Let them have _____BRV
			Most of them _____SAX
	THESE THOSE _____ ___CN1		Must they (their-s) ? _____CFB
CNH	—After these (or, those),		Not they (their-s)_____CGE
	Are these (or, those)____ _____BEQ		Of them (their-s)_____ _____CGS
	By these (or, those)_ _____BJX		On them (their-s)_____CHE
	Can, or, May these (or, those) ? ___BKJ		Ought they to?_____ _ ____CHX
	Could, or, Might these (or, those) ?_ELG		Shall, or, Will they (their-s) ?____CIY
	Do, or, Did these (or, those)? _____BMN		Shall, or, Will they be ?_____BHF
	For these (or, those) _____BPI		Shall, or, Will they do it ?_____BOA
	From these (or, those)_____BPV		Shall, or, Will they not ?_____CIZ
	Had, or, Have these (or, those) ? ___BQZ		Shall, or, Will they not do it ?_____BOC
	If these (or, those) _____CAD		Should, or, Would they (their-s) ? ___CJV
	In these (or, those) __ _____CDP		Then they (their-s)_____CKI
	May, etc (See Can)		That they (their-s)_____CLV
	Might, etc (See Could)	CNL	—They Them Their-s.
	Must these (or, those)? _____CFA	CNM	—They are for
	Not these (or, those)_____CGD		They are from__ _____BET
	Of these (or those)_____CGT		They are_____BEU
	On these (or, those)_____CHF		They are not _____ _____BEV
	Ought these (or, those)? _____CHW		They can (or, may)_____BKT
	Shall, or, Will these (or, those)? ___CIX		They can (or, may) be. _____BHP
	Should, or, Would these (or, those)?_CJU		They can (or, may) have _____BSF
	Than these (or, those)_____CKH		They can not (or, may not)_____BKU
	That these (or, those)_____CLU		They can not (or, may not) be_____BHQ
CNI	—These. Those		They can not (or, may not) have___BSG
	To these (or, those) _____CNZ		They could (or, might)_____BLN
	Were these (or, those)_____CPG		They could (or, might) be_____BHR
	When these (or, those)_____CSG		They could (or, might) not_____BLO
	Will, etc (See Shall)		They could (or, might) not be_____BHS
	With these (or, those)___ __ ____CWL		They do (or, did)_____BOJ
	Would, etc (See Should)		They do (or, did) not_____ _____BOK
			They had (or, have)_____BSH
	THEY THEM THEIR-S _____CNL		They had (or, have) been_____BHT
CNJ	—After them (them, theirs)		They had (or, have) done_____BOL
	Are, or, Is their-s, or, Are they? ___BEO		They had (or, have) had_____BSI
CNK	—Before them (their-s, they)		They had (or, have) not_____ ____BSJ
	By them (their-s) _____BJV		They had (or, have) not been_____BHU
	Can, or, May they (their-s)? _____BKL		They had (or, have) not done_____BOM
	Can, or, May they not ?_____BKM		They had (or, have) not had_____BSK
	Could, or, Might they (their-s)?____BLH		They must _____CFG
	Do, Does, or, Did they (their-s)? __BMO		They must be _ _____BEV
	For them (their-s)_____BPH		They must not _____CFH
	From them (their-s)____ __ _____BPT		They must not be____ _____BHW
	Had, Has, or, Have they (them, their-s)BRA		They ought to _____CIE
	Had, or, Have they been ?_____BFT		They ought not to_____CIF
	Had, or, Have they done ?_____BMU		They ought to be_____BHX
	Had, or, Have they had ? _____ _BRC		They ought not to be _____/_____BHY
	How are they ? _ _____BXH		They shall (or, will)_____CJF
	How will they ?_____BXY		They shall (or, will) be _____BHZ
	If they (their-s)_____CAE		They shall (or, will) not_____CJG
	If they are _____CAF		They shall (or, will) not be_____BIA
	If they are not_____CAG		They should (or, would)_____CJY
	If they can (or, may)_____CAH		They should (or, would) be _____BIC
	If they can not (or, may not)_____CAI		They should (or, would) have_____BEL
	If they could (or, might)_____CAJ		They should (or, would) not_____ CJZ
	If they could (or, might) not_____CAL		They should (or, would) not be____ BID
	If they do (or, did)_____ ___ CAM		They should (or, would) not have___BSM
	If they do (or, did) not_____CAN		They were_____COJ
	If they had (or, have)_____CAO		They were not_____ _____COK
	If they had (or, have) not_____CAP		They were to be___ ___ ___ _ ____BIE
	If they shall (or, will) _____CAQ		They were not to be_____BIF
	If they shall (or, will) not_____CAR		To them (their-s) _____CNY

THEY—WAS

THEY—Continued

Was, or, Were they (their s)		COY
Were they ?		CPH
Were they not ?		CPI
What, or, Which do (or, did) they (we, or, you) ?		CQE
What, or, Which shall (or, will) they (or, you) do ?		CQN
What, or, Which shall (or, will) they (or, you) have		CQO
What, or, Which should (or, would) they (or, you) do?		CQS
When do (or, did) they (or, you) ?		CRF
When shall (or, will) I (they, we, or, you) do?		CRV
When shall (or, will) they ?		CRY
When shall (or, will) they (or, you) have ?		CRZ
When they (their-s)		CSH
When they are		CSI
Where are they ?		CST
Where do (or, did) they (or, you) ?		CSZ
Where shall (or, will) they ?		CTP
Where shall (or, will) they (or, you) have ?		CTQ
Where should (or, would) they (their-s)?		CTX
Where they (their-s)		CTZ
Where they are		CUA
Why they (their s) ?		CVW
With them (their-s)		CWK
THIS (See THAT)		CMA
THOSE (See THESE)		CNI
To		CNO
Am I to ?		BEF
Am, Is, or, Are to		BEH
CNO	—To	
CNP	—To be	
CNQ	—To be at	
CNR	—To be done	
CNS	—To be had	
CNT	—To do	
CNU	—To have	
CNV	—To him (his, her-s, it-s)	
CNW	—To me (my, mine)	
CNX	—To the	
CNY	—To them (their-s)	
CNZ	—To these (or, those)	
COA	—To this (or, that)	
COB	—To us (our-s)	
COD	—To which (or, what)	
COE	—To whom (or, whose)	
COF	—To you r-s	
	Me to	CEF
	Not to	CGF
	Not to be	CGH
	Not to be done	CGI
	Not to be had	CGJ
COG	Us (See OURS)	
	WAS (or, WERE)	COM
	He, She It (or person s or things s indicated) was (or, were)	BVT
	He, She It (or person s or thing-s indicated) was (or, were) to be	BVW
	He, She, It (or person s or thing s indicated) was (or, were) not	BVU

WAS—Continued

He, She, It (or person s or thing-s indicated) was (or, were) not to be		BVX
How was (or, were) ?		BXW
COH	—I was	
COI	—I was not	
	If he (she, it, or person-s or thing s indicated) was (or, were)	BYT
	If he (she, it, or person s or thing-s indicated) was (or, were) not	BYU
	If I was	BZT
	If I was not	BZU
	If they were	CAU
	If they were not	CAV
	If we were	CBN
	If we were not	CBO
	If you were	CDG
	If you were not	CDH
	That, or, This was	CLR
	That, or, This was not	CLS
	There was (or, were)	CNF
	There was (or, were) not	CNG
COJ	—They were	
COK	—They were not	
	They were not to be	BIF
	They were to be	BIE
COM	—Was, or, Were	
	Was, or, Were done (or, doing)	BON
COP	—Was, or, Were for	
COQ	—Was, or, Were from	
	Was, or, Were he (she, it, or person-s or thing-s indicated)?	BWX
	Was, or, Were he (she, it, or persons or thing-s indicated) not ?	BWY
COR	—Was, or, Were his (her-s, it-s)	
COS	—Was, or, Were I (my, mine)?	
COT	—Was I ?	
COU	—Was I not ?	
COV	—Was, or, Were not	
	Was, or, Were not done (or, doing)	BOP
COW	—Was, or, Were the	
COX	—Was, or, Were there	
COY	—Was, or, Were they (their s)	
COZ	—Was this (or, that)	
	Was, or, Were to be	BIG
CPA	—Was, or, Were our-s (we)?	
CPB	—Was, or, Were with	
CPD	—Was, or, Were you-r s	
CPE	—We were	
CPF	—We were not	
	We were to be	BIV
	We were not to be	BIW
CPG	—Were these (or, those)	
CPH	—Were they ?	
CPI	—Were they not ?	
CPJ	—Were we ?	
CPK	—Were we not ?	
CPL	—Were you?	
CPM	—Were you not ?	
	What, or, Which was (or, were)?	CQT
	When was (or, were) ?	CSJ
	Where was (or, were)?	CUB
	Who was (or, were)?	CUW
	Why was (or, were)?	CVX
	Why were you ?	CVZ
CPN	—You were	
CPO	—You were not	
	You were to be	BJO
	You were not to be	BJP

WE—WHAT

WE (See also OURS) ---------------- CPQ	**WE**—Continued
After we (our-s, us) --- -- ----------CIL	We shall (or, will) be -------------BIR
Are, or, Is our-s, or, Are we ? ------BEM	We shall (or, will) have ----------BSU
Can, or, May we (our-s)? --- ------BKN	We shall (or, will) not--------------CJI
Could, or, Might we (our-s)? --------BLI	We shall (or, will) not be----------BIS
Do, Does, or, Did we (our s)? ------BMP	We shall (or, will) not have--------BSV
Had, Has, or, Have we (our-s)?----BRD	We should (or, would) -----------CKA
Had, or, Have we been ? ----------BFU	We should (or, would) be ----------BIT
Had, or, Have we done ? ------- ---BMV	We should (or, would) have------BSW
Had, or, Have we had ?----- ------BRE	We should (or, would) not --------CKB
How shall (or, will) we ?----------BXT	We should (or, would) not be ------BIU
If we (our-s)---------- -----------CAW	We should (or, would) not have---BSX
If we are----- ---------- --------CAX	We were ----------------------CPE
If we are not---------- ----------CAY	We were not ------------------CPF
If we can (or, may)----- --------CAZ	We were not to be -------------BIW
If we can not (or may not) --------CBA	We were to be-----------------BIV
If we could (or, might)-----------CBD	—We who
If we could (or, might) not --- ----CBE	Were we? ---------------------CPJ
If we do (or, did) ---- --------- - CBF	Were we not? ------------------CPK
If we do (or, did) not-------------CBG	What, or, Which do (or, did) they (we,
If we had (or, have) -------------CBH	or, you)? --------------------CQE
If we had (or, have) not----- -----CBI	What, or, Which shall (or, will) I (or,
If we shall (or, will) ----- - --- --CBJ	we) do?----------- -----------CQL
If we shall (or, will) not----------CBK	What, or, Which shall (or, will) I (or,
If we should (or, would)-----------CBL	we) have?--------- -----------CQM
If we should (or, would) not----- --CBM	What, or, Which should (or, would) I
If we were --------------------CBN	(or, we) do?---------- ---------CQR
If we were not ----- ------- ------CBO	When are we? ------- ----------CQX
May, etc (See Can.)	When do (or, did) I (or, we)? ------CRE
Might, etc (See Could)	When shall (or, will) I (or, we)? ----CRU
Must we (our-s)? -----------------CFD	When shall (or, will) I (they, we, or,
Ought we (our-s) to? --------------CIB	you) do? --------------------CRV
Shall, or, Will we (our-s)? ---------CJA	When shall (or, will) we have?-----CSA
Shall, or, Will we be? ------------BHG	When we (our-s) ----------------CSK
Shall, or, Will we not? -------------CJB	When we are-----------------CSL
Should, or, Would we (our-s)?-----CJW	Where am I (or, are we)? --- - ---CSQ
Than we (our-s)----------------CKJ	Where do (or, did) I (or, we)? --- --CSY
That we (our-s)---- -- ----------CLW	Where I am (or, we are) ---------CTG
Was, or, Were our-s (we)?------ ---CPA	Where shall (or, will) I (or, we)?----CTN
—We	Where shall (or, will) we have?----CTR
We are----------- ---- ---------BEW	Where we (our-s) ---------------CUD
We are not ---------- ----------BEX	Why we (our-s)----------------CVY
We can (or, may) ---------------BKV	Will, etc (See Shall)
We can (or, may) be--------------BIH	Would, etc (See Should)
We can (or, may) have-----------BSN	
We can not (or, may not)---- ----BKW	**WERE** (See Was) ----------------COM
We can not (or, may not) be ------BIJ	
We can not (or, may not) have----BSO	**WHAT**----- CPU (or, **WHICH**------CUG)
We could (or, might) ------------BLP	—After what (or, which)
We could (or, might) be ---------BIK	—Before what (or, which)
We could (or, might) not --------BLQ	By what (or, which)--------------BJZ
We could (or, might) not be -----BIM	For what (or, which)-------------BPK
We do (or, did) --------- --------BOQ	From what (or, which)-----------BPX
We do (or did) not--------------BOR	Of what (or, which)--------------CGU
We had (or, have) --------------BSP	On what (or, which) ------------CHG
We had (or, have) been ---------BIN	Than what (or, which)-----------CKL
We had (or, have) done----------BOS	That what (or, which) ---------CLX
We had (or, have) had -------- ---BSQ	To what (or, which) ------ ------COD
We had (or, have) not -----------BSR	—What
We had (or, have) not been -------BIO	—What, or, Which am I?
We had (or, have) not done------BOT	—What, or, Which are (or, is)?
We had (or have) not had --------BST	—What, or, Which are (or, is) he (she, it
We may etc (See Can)	or person-s or thing-s indicated)?
We might, etc (See Could)	—What, or, Which are (or, is) you-r-s?
We must --------------------CFI	—What, or, Which can (or, may)?
We must be ---------- ----------BIP	—What, or, Which could (or, might)?
We must not ------------------CFJ	—What, or, Which do (does, or, did)?
We must not be ----- ----------BIQ	—What, or, Which does (do, or, did) he
We ought to-------------------CIG	(she, it, or person-s or thing s indi-
We ought not to----------------CIH	cated)?
We shall (or, will) ----------------CJH	

Left margin labels: CPQ

Right column margin labels: CPR, CPS, CPT, CPU, CPV, CPW, CPX, CPY, CPZ, CQA, CQB, CQD

	WHAT—WHICH

	WHAT—*Continued*		WHEN—*Continued*
CQE	—What, (*or*, Which do (*or*, did) they (we, *or*, you)?	CSA	—When shall (*or*, will) we have?
CQF	—What, *or*, Which had (has, *or*, have)?	CSB	—When shall (*or*, will) you?
	What, *or*, Which is? (*See* Are)	CSD	—When shall (*or*, will) you be?
CQG	—What, *or*, Which is it?	CSE	—When should (*or*, would)?
	What, *or*, Which may? (*See* Can)	CSF	—When the —
	What, *or*, Which might? (*See* Could)	CSG	—When these (*or*, those)
CQH	—What, *or*, Which must?	CSH	—When they (their-s)
CQI	—What, *or*, Which of the — ?	CSI	—When they are
CQJ	—What, *or*, Which ought to?	CFJ	—When was (*or*, were)?
CQK	—What, *or*, Which shall (*or*, will)—I?	CSK	—When we (our-s)
CQL	—What, *or*, Which shall (*or*, will) I (*or*, we) do?	CSL	—When we are
			When will, etc ? (*See* Shall)
			When would, etc ? (*See* Should)
CQM	—What, *or*, Which shall (*or*, will) I (*or*, we) have?	CSM	—When you-r-s
CQN	—What, *or*, Which shall (*or*, will) they (*or*, you) do?	CSN	—When you are
CQO	—What, *or*, Which shall (*or*, will) they (*or*, you) have?		
CQP	—What, *or*, Which should (*or*, would)?	WHERE ....CSO	
CQR	—What, *or*, Which should (*or*, would) I (*or*, we) do?	From whence (*or*, where)............BPY	
		CSO	—Where
CQS	—What, *or*, Which should (*or*, would) they (*or*, you) do?	CSP	—Whereabouts
		CSQ	—Where am I (*or*, are we)?
CQT	—What, *or*, Which was (*or* were)?	CSR	—Where are (*or*, is)?
	What *or*, Which will? (*See* Shall)	CST	—Where are they?
	What, *or*, Which would? (*See* Should)		Where are you from?............SI
	Which ....CUG	CSU	—Where are you going?
	With what (*or*, which)? ....OWN	CSV	—Where can (*or*, may)?
		CSW	—Where could (*or*, might)?
		CSX	—Where do (does, *or*, did)
	WHEN ....CQU	CSY	—Where do (*or*, did) I (*or*, we)?
CQU	—When	CSZ	—Where do (*or*, did) they (*or*, you)?
CQV	—When am I to?	CTA	—Where for?
CQW	—When are (*or*, is)?	CTB	—Where had (has *or* have)?
CQX	—When are we?	CTD	—Where he (she, it, *or person-s or thing s indicated*) is (*or*, are)
CQY	—When are you?		
CQZ	—When can (*or*, may)?	CTE	—Where his (her-s, it-s)
CRA	—When could (*or*, might)?	CTF	—Where I (my, mine)
CRB	—When do (does, *or*, did)	CTG	—Where I am (*or*, we are)
CRD	—When does (do, *or*, did) he (she, it, *or person-s or thing s indicated*)?	CTH	—Where is (*or*, are) he (she, it, *or person s or thing-s indicated*)?
CRE	—When do (*or*, did) I (*or*, we)?		Where is, etc ? (*See* Are)
CRF	—When do (*or*, did) they (*or*, you)?		Where may, etc ? (*See* Can)
CRG	—When done—with		Where might etc ? (*See* Could)
CRH	—When had (has, *or*, have)?	CTI	—Where must?
CRI	—When had (*or*, have) y. u?	CTJ	—Where ought?
CRJ	—When he (she, it, *or person s or thing-s indicated*) is (*or*, are)	CTK	—Where shall (*or*, will)?
		CTL	—Where shall (*or*, will) be?
CRK	—When his (her-s, it s)	CTM	—Where shall (*or*, will) he (she, it, *or person s or thing-s indicated*)?
CRL	—When I (my, mine)		
CRM	—When I can (*or*, may)	CTN	—Where shall (*or*, will) I (*or*, we)?
	When is? (*See* Are)	CTO	—Where shall (*or*, will) I have?
CRN	—When is (*or*, are) he (she, it, *or person-s or thing-s indicated*)?	CTP	—Where shall (*or*, will) they?
		CTQ	—Where shall (*or*, will) they (*or*, you) have?
	When may, etc ? (*See* Can)		
	When might, etc ? (*See* Could)	CTR	—Where shall (*or*, will) we have?
CRO	—When must?	CTS	—Where shall (*or*, will) you?
CRP	—When ought?	CTU	—Where shall (*or*, will) you be?
CRQ	—When shall (*or*, will)?	CTV	—Where should (*or*, would)?
CRS	—When shall (*or*, will) be?	CTW	—Where should (*or*, would) the —?
CRT	—When shall (*or*, will) he (she, it, *or per son s or thing-s indicated*)?	CTX	—Where should (*or*, would) they (their s)?
		CTY	—Where the
CRU	—When shall (*or*, will) I (*or*, we)?	CTZ	—Where they (their-s)
CRV	—When shall (*or*, will) I (they, we, *or*, you) do?	CUA	—Where they are
		CUB	—Where was (*or*, were)?
CRW	—When shall (*or*, will) I have?	CUD	—Where we (our-s)
CRX	—When shall (*or*, will) it be done?		Where we are ....CTG
CRY	—When shall (*or*, will) they?	CUE	—Where you-r-s
CRZ	—When shall (*or*, will) they (*or*, you) have?	CUF	—Where you are
		CUG	WHICH (*See* WHAT)

WHO—YOU.

	WHO WHOM .... CUH
	By whom (or, whose).............. BKA
	For whom (or, whose) .... BPL
	From whom (or, whose) .... BPZ
	Of whom (or, whose) .... CGV
	On whom (or, whose) .... CHI
	To whom (or whose) .... OOE
	We who .. .. OPR
CUH	—Who　Whom
CUI	—Who are (or, is)?
CUJ	—Who can (or, may)?
CUM	—Who could (or, might)?
CUO	—Who do (does, or, did)?
CUP	—Who had (has, or, have)?
	Who is, etc? (See Are)
	Who may, etc? (See Can)
	Who might, etc? (See Could)
CUQ	—Who must?
CUR	—Who ought?
	Who shall (or, will)? .... ZEF
CUV	—Who should (or, would)?
CUW	—Who was (or, were)?
	Who will, etc? (See Shall)
	Who would, etc? (See Should.)
	Whose? .... ZDU
	With whom (or, whose)? .... cwo

	WHY .... CUX
CUX	—Why?
CUY	—Why am I?
CUZ	—Why are (or, is)?
CVA	—Why are you?
CVB	—Why are you not?
	Why can (or, may)? .... HUF
CVD	—Why can not (or, may not)?
CVE	—Why could (or, might)?
CVF	—Why could (or, might) not?
CVG	—Why do (does, or, did)?
CVH	—Why do (does, or, did) not?
CVI	—Why do (or, did) you?
CVJ	—Why do (or, did) you not?
CVK	—Why had (has, or, have)?
CVL	—Why had (has, or, have) not?
CVM	—Why had (or, have) you?
CVN	—Why had (or, have) you not?
	Why is, etc? (See Are)
	Why may, etc? (See Can)
	Why might, etc? (See Could.)
CVO	—Why must?
CVP	—Why not?
CVQ	—Why ought?
CVR	—Why shall (or, will)?
CVS	—Why shall (or, will) not?
CVT	—Why should (or, would)?
CVU	—Why should (or, would) not?
CVW	—Why they (their-s)?
CVX	—Why was (or, were)?
CVY	—Why we (our-s)?
CVZ	—Why were you?
	Why will, etc? (See Shall)
	Why would, etc? (See Should)
CWA	—Why you-r-s.

	WILL (See SHALL)
CWB	How will they? .... BXY
	It will do .... BNV
	Will .... CWB
	Will it be? .... BIX
	Will it do? .... BOU

	WILL——Continued.
CWD	—Will it have?
	Will not .... CIU
	Will not do .... BOV
CWE	—Will you do it?

	WITH .... CWF
	Are, or, Is with— .... BER
	Me with .... CEG
	Not with .... CGK
	That with .... CLY
	Was, or, Were with— .... CPB
	When done—with .... CRG
CWF	—With
CWG	—With him (his, her-s, it-s)
CWH	—With me (my, mine)
CWI	—With the.
CWJ	—With that (or, this).
CWK	—With them (their-s)
CWL	—With these (or, those)
CWM	—With us (our-s)
CWN	—With what (or, which).
CWO	—With whom (or, whose).
CWP	—With you-r-s.

	WOULD　(See SHOULD) .... CJO

	YOU-R S .... CWT
CWR	—After you-r-s
	Are, or, Is, you-r-s .... BES
CWS	—Before you-r-s
	By you-r-s .... BKC
	Can, or, May you-r-s? .... BKO
	Could, or, Might you-r-s? ...... BLJ
	Do, Does, or, Did you-r-s? ...... BMQ
	For you-r-s .... BPM
	From you-r-s .... BQA
	Had, Has, or, Have you-r-s? ..... BRF
	Had, or, Have you been? .... BFV
	Had, or, Have you done? .... BMW
	Had, or, Have you had? .... BRG
	How are you? .... BXI
	How do (or, did) you? .... BXM
	How shall (or, will) you? .... BXU
	If you-r-s .... CBP
	If you are .... CBQ
	If you are not .... CRR
	If you can (or, may) .... CBS
	If you can not (or, may not) ..... CBT
	If you could (or, might) .... CBU
	If you could (or, might not) ..... CBV
	If you do (or, did) .... CBW
	If you do (or, did) not .... CBX
	If you had (or, have) .... CBY
	If you had (or, have) not .... CBZ
	If you shall (or, will) .... CDA
	If you shall (or, will) not .... CDB
	If you should (or, would). ...... CDE
	If you should (or, would) not ..... CDF
	If you were .... CDG
	If you were not .... CDH
	In your s .... CDR
	Is, etc　(See Are)
	May, etc　(See Can)
	Might, etc　(See Could)
	Must you-r-s? .... CFE
	Not you-r-s .... CGL
	Of you-r-s .... CGW
	On you-r-s .... CHJ
	Ought you-r-s to? .... CID
	Shall, or, Will you r-s? .... CJD

YOU

You—*Continued.*
Shall, *or,* Will you be?_ ----------BHI
Shall, *or,* Will you do it?----------BOD
Shall, *or,* Will you not?------------CJE
Shall, *or,* Will you not be?--------BHJ
Shall, *or,* Will you not do it?------BOE
Should, *or,* Would you-r-s?--------CJX
Than you-r-s----------------------CKM
That you r s----------------------CLZ
To you-r-s-----------------------COF
Was, *or,* Were you-r s------------CPD
Were you?------------ -----------CPL
Were you not?---------------------CPM
What, *or,* Which are(*or,* is) you-r-s? CPY
What, *or,* Which do (*or,* did) you (they, *or,* we)?--------------------- CQE
What, *or,* Which shall (*or,* will) you (*or,* they) do?------------------- CQN
What, *or,* Which shall (*or,* will) you (*or,* they) have?----------------CQO
What, *or,* Which should (*or,* would) you (*or,* they) do?----- ---------CQS
When are you?--------- -----------CQY
When do (*or,* did) you (*or,* they)?__CRF
When had (*or,* have) you?----------CRI
When shall (*or,* will) I (you, they, *or,* we) do?----------------------CRV
When shall (*or,* will) you?--------- CSB
When shall (*or,* will) you be?----- CSD
When shall (*or,* will) you (*or,* they) have?--------------------------CRZ
When you-r-s --------------------- CSM
When you are--------------------CSN
Where are you from?-------- -------SI
Where are you going?-------------CSU
Where do (*or,* did) you (*or,* they)?-CSZ
Where shall (*or,* will) you?---------CTS
Where shall (*or,* will) you be?-----CTU
Where shall (*or,* will) you (*or,* they) have? - ----------------------CTQ
Where you-r-s-------------------CUE
Where you are-------------------CUF
Why are you?--------------------OVA
Why are you not?----------------CVB
Why do (*or,* did) you?-----------CVI
Why do (*or,* did) you not?---------CVJ
Why had (*or,* have) you?----------CVM
Why had (*or,* have) you not?----- CVN
Why were you?------------------CVZ
Why you-r-s?------ -------------CWA
Will you do it?------- ----------CWE
With you-r-s -------------------CWP

CWT

You—*Continued.*
—You-r-s
You are----------------------------BEY
You are not------------------------BEZ
You can (*or,* may) -----------------BKX
You can (*or,* may) be--------------BIY
You can (*or,* may) have------------BSY
You can not (*or,* may not)----------BKY
You can not (*or,* may not) have----BSZ
You could (*or,* might)--------------BLE
You could (*or,* might) be----------BJA
You could (*or,* might) not----------BLS
You could (*or,* might) not be------BJC
You do (*or,* did) --------------------BOW
You do (*or,* did) not ---------------BOX
You had (*or,* have) ----------------BTA
You had (*or,* have) been-------- ---BJD
You had (*or,* have) done----------BOY
You had (*or,* have) had -----------BTC
You had (*or,* have) not-----------BTD
You had (*or,* have) not been------BJE
You had (*or,* have) not done-------BOZ
You had (*or,* have) not had--------BTE
You may, etc (*See* Can)
You might, etc (*See* Could)
You must_ ------------------------CFK
You must be------------------------BJF
You must not ---------------------CFL
You must not be-------------------BJG
You ought to ---------------------CIJ
You ought not to------------------CIK
You ought to be-------------------BJH
You ought not to be---------------BJI
You shall (*or,* will)----------------CJK
You shall (*or,* will) be -----------BJK
You shall (*or,* will) have----------BTF
You shall (*or,* will) not-----------CJL
You shall (*or,* will) not be--------BJL
You shall (*or,* will) not have------BTG
You should (*or,* would)------------CKD
You should (*or,* would) be--------BJM
You should (*or,* would) have-------BTH
You should (*or,* would) not--------CKE
You should (*or,* would) not be----BJN
You should (*or,* would) not have----BTI
You were ------------ --------- CFN
You were not--------------------CFO
You were to be-------------------BJO
You were not to be.._ ---------- BJP
You will, etc (*See* Shall)
You would, etc (*See* Should)

GEOGRAPHICAL SIGNALS

NAMES OF PLACES ARRANGED ACCORDING TO THEIR GEOGRAPHICAL POSITIONS

Notations

B Bay	Hr Harbor	Pt Point, Pointe
C, Cape, Capo, Cap	I Island	R River
Ch Channel	Lt Light	Sh Shoal
G Gulf	Lt. V Light vessel	Str Strait
F Fort	Mt Mount, Mont	
Hd Head	P. Port, Porto Puerte	

The System of Orthography is that adopted by the Board of Geographic Names

ARCTIC OCEAN		NORTH COAST RUSSIA		WEST COAST NORWAY	
ABCD	Arctic Ocean	ABEJ	Zimnegorski	ABGM	Gjesvær
ABCE	C Cheluskin	ABEK	Modyugski I *d*	ABGN	Fruholm
ABCF	Taimur G	ABEM	Dwina (N) Lt V *a*	ABGP	Alten Fiord
ABCH	Pjasina R	ABEN	Dwina R	ABGQ	Fuglenæs
ABCI	Kuskino	ABEO	Arkhangel.	ABGR	Hammerfest
ABCJ	Yenisei	ABEQ	Zhizhginsk I	ABGT	Alten
ABCL	Tolstonow.	ABER	Solovetski	ABGU	Hasvig.
ABCM	Ob	ABET	Bol Shuzhmin	ABGV	Skibotten
ABCO	Obdorsk	ABEU	Onega	ABGW	Loppen
ABCP	Beresow	ABEV	Kush	ABGX	Lyngen
ABCQ	Samarowsk	ABEW	Nyukcha	ABHC	Tromsö
ABCS	Yalmal Promontory	ABEX	Suma	ABHD	Malangen
ABCT	Krusenstern B	ABEY	Soroka.	ABHE	Hekkingen
ABCU	Bahydarat B	ABEZ	Shuya	ABHG	Stöneshotten
ABCW	Franz Josef Land.	ABFC	Kem	ABHI	Berg
ABCY	Spitzbergen	ABFD	Kalgalaksha	ABHJ	Reisen
ABDC	Bel Sound	ABFG	Kandalak G	ABHL	Andenæs Lt
ABDE	Magdalena B	ABFH	Knivaia	ABHM	Haarbjerget
ABDG	Hakluyts Hd	ABFI	Kandalaksha	ABHN	Nordland
ABDH	Hinlopen Str	ABFK	Lapland	ABHP	Sandtorv
ABDI	Seven Is , The	ABFL	Sosnovetz I	ABHQ	Vesteraalen
ABDK	Parry I	ABFM	Orlov C	ABHR	Hassel
ABDL	Ross I	ABFO	Orlovka	ABHT	Gisund
ABDM	Bear I	ABFP	Svyatoi Nos *d*	ABHU	Harstad
ABDO	Kara Sea	ABFQ	Alexandrovsk (Kola)	ABHV	Stangholm
ABDP	Nova Zembla			ABHW	Tranö
ABDQ	Matochkin Shar (Str)	ABFS	Norway	ABHX	Hammerö
ABDR	Moller B *d*	ABFT	Finmarken	ABHY	Grytö
ABDS	Kara Str	ABFU	Varanger Fiord	ABHZ	Flado
		ABFV	Jarfjord	ABIC	Lofoten Is
ABDT	Europe.	ABFW	Vadsö	ABID	Lodingen
		ABFX	Vardo	ABIE	Vaagen.
ABDU	Russia	ABFY	Tana Fiord	ABIF	Skraaven
ABDV	Waigatch I	ABGC	Berlevaag	ABIJ	Flagstad.
ABDW	Yugorski Str	ABGD	Gamvig	ABIL	Kabelvaag
ABDY	Chabarova	ABGE	Kollefjord	ABIM	Svolvær
ABDZ	Petchora R	ABGF	Lebesby	ABIP	Örsvaag
ABEC	Oussa	ABGH	Repvaag	ABIQ	Hennings Vær
ABED	White Sea	ABGI	Kistrand	ABIR	Gimsö
ABEF	Rusânovka	ABGJ	Magerosund	ABIT	Valberg
ABEG	Mezen	ABGK	Havosund	ABIU	Stamsund
ABEI	Morzhavetz I *d*	ABGL	North C	ABIV	Balstad

a Signifies an International Signal Station, *b* a Time Signal Station, *c* a Weather, Tide, or Ice Signal Station, *d* a Life-Saving Station, *e* a Lloyd's Signal Station, *f* a Wireless Telegraph Station

WEST COAST NORWAY

ABIX	Reine	ABMS	Ringholm	ABQS	Fresvik
ABIY	Moskenæs	ABMU	Grip	ABQT	Sogndal
ABIZ	Glopen	ABMV	Stavenes	ABQV	Kaupanger
ABJC	Sörvaag	ABMW	Kristiansund	ABQW	Lerdalsören
ABJE	Ofoten Fiord	ABMY	Kvitnes	ABQX	Aardals
ABJF	Victoria Haven (Narvick)	ABMZ	Hestskjær	ABQZ	Solvorn
		ABNC	Little Sando.	ABRC	Marifjæren
ABJG	Bodo	ABNE	Kvitholm	ABRD	Holmengraa
ABJH	Biörnö	ABNF	Hustad	ABRE	Gudvangen
ABJI	Salten Fiord	ABNG	Bud	ABRF	Helliso
ABJK	Nyholm	ABNI	Björnsund	ABRG	Ostrejm
ABJL	Hellig Vær	ABNJ	Ona	ABRH	Lindaas
ABJM	Sulitjelma Mines	ABNK	Rödsbugt	ABRJ	Sjelanger
ABJO	Blixvær	ABNM	Molde	ABRL	Bergen
ABJP	Skomvæi	ABNO	Molde Fiord	ABRM	Lerö
ABJR	Rost	ABNP	Veblüngsnæs	ABRP	Marsten
ABJS	Femris	ABNS	Mooen	ABRQ	Pirholm
ABJT	Stotsund	ABNT	Röd	ABRS	Slottero
ABJV	Gilleskaal	ABNV	Visdal	ABRT	Bekkervig
ABJX	Grinö	ABNW	Rodven	ABRU	Trano
ABJY	Rödö	ABNX	Fiksdal	ABRV	Folgero
ABJZ	Selsovik	ABNZ	Kværnholm	ABRW	Lervig
ABKC	Træn I	ABOC	Hellevik	ABRY	Langenuen
ABKD	Soholm	ABOD	Lepsorev	ABRZ	Hardanger Fiord
ABKE	Aas Vær	ABOF	Rosholm	ABSC	Kvindherred.
ABKG	Donnæso	ABOG	Erkno	ABSE	Strandebarm.
ABKH	Kobberdal	ABOH	Synæs	ABSF	Jondal
ABKI	Mosjoen	ABOJ	Aalesund	ABSG	Yikor
ABKJ	Buholm	ABOK	Hogsten, or Hogstenen	ABSI	Östenso
ABKL	Bronnösund	ABOL	Alnæs	ABSJ	Kinservik
ABKN	Bronno	ABON	Græsholmen	ABSK	Ullensvang
ABKO	Melstenen	ABOP	Rundo	ABSM	Odde
ABKP	Leka	ABOQ	Ulfsten	ABSN	Holmedal
ABKQ	Gutvik	ABOS	Flaavær	ABSO	Olen
ABKR	Fjeldvik	ABOT	Sando	ABSQ	Mosterhavn
ABKS	Dolma	ABOU	Örsten	ABSR	Bommelo
ABKT	Vigten Is	ABOW	Volden	ABST	Nordoerne
ABKU	Nordoerne	ABOX	Sövde	ABSV	Lille Blegen
ABKW	Rörvig	ABOY	Hogsholm	ABSW	Espevær
ABKX	Gjoeslingerne	ABOZ	Lekanger	ABSX	Ryvarden
ABKY	Otterö	ABPO	Stat	ABSZ	Haugesund
ABLC	Namsos	ABPD	Skongsnæs	ABTC	Rovær
ABLD	Statland	ABPE	Ulvesund	ABTD	Sorhaugo
ABLF	Villa	ABPG	Moldo	ABTF	Sandeid
ABLH	Reko	ABPH	Nord Fiord	ABTG	Neerstrand
ABLI	Björo	ABPI	Rugsund	ABTH	Stolmen
ABLK	Ramso	ABPK	Daviken	ABTK	Fæö
ABLM	Halten Is	ABPL	Starheim	ABTL	Udsire
ABLN	Bessaker	ABPM	Smörhavn	ABTN	Hoievarde
ABLP	Syd Krogo	ABPN	Olden	ABTO	Vigsnæs Mines
ABLQ	Lundö	ABPO	Ytteröerns	ABTP	Skudesnæs
ABLR	Stokköen	ABPQ	Stabben	ABTR	Hvidingso
ABLS	Valdersund	ABPR	Floro	ABTS	Fjeldo
ABLT	Kjeungskjær	ABPS	Kinn	ABTU	Tungenæs
ABLU	Orland B	ABPU	Eke Fiord	ABTW	Stavanger
ABLV	Beian	ABPV	Svanö	ABTX	Sandnæs
ABLX	Agdenes	ABPW	Vofring	ABTY	Fladholm
ABLY	Lensviken	ABPY	Naustdal	ABUC	Osa
ABLZ	Rödberg	ABPZ	Askevold	ABUD	Lille Feisten
ABMD	Trondhjem b c	ABQC	Dale	ABUE	Jæderens Pt d
ABME	Munkholm	ABQE	Tansö	ABUF	Vig d
ABMF	Stenkjær	ABQF	Hindo	ABUG	Obrestad a d
ABMH	Hommelviken	ABQG	Stensündene.	ABUH	Husvegg d
ABMI	Levanger	ABQI	Sogne Fiord	ABUI	Kvalbein d
ABMJ	Boroholm	ABQJ	Bratholm	ABUJ	Rauna d
ABML	Tranholm	ABQK	Bö	ABUK	Ekero
ABMN	Terningen	ABQM	Ladvik	ABUL	Ekersund
ABMO	Væröerne	ABQN	Viksören	ABUM	Vibberodden
ABMQ	Lyen	ABQO	Vangsnæs	ABUN	Soggendalstrand
ABMR	Edö	ABQR	Lekanger	ABUO	Reke Fiord

a Signifies an International Signal Station, b a Time Signal Station, c a Weather, Tide, or Ice Signal Station, d a Life-Saving Station, e a Lloyd's Signal Station, f a Wireless Telegraph Station

SKAGERRACK		SWEDEN AND DENMARK			
ABUP	Flekkefiord	ABYL	Dröbak	ACEG	Barsebäck
ABUQ	Hittero	ABYM	Hvidsteen	ACEH	Lomma.
ABUR	Lister	ABYN	Soon	ACEJ	Malmö b
ABUS	Stave d	ABYO	Moss	ACEK	Kalkgrundet Lt V
ABUT	Osthasselstranden d	ABYQ	Laurkollen	ACEM	Oscargrundet Lt V
ABUW	Kvilo d	ABYR	Ellinggaard Kilen	ACEN	Limhamn
ABUX	Farsund	ABYS	Hankö	ACEO	Klagstorp
ABUY	Katland	ABYU	Slevik	ACEP	Skanor d
ABUZ	Lindesnæs (The Naze)	ABYV	Torgauten	ACEQ	Falsterbo i
ABVC	Svinör	ABYW	Thorbjörns kjær	ACER	Trelleborg
ABVD	Remesvigen	ABYX	Frederikstad c	ACES	Falsterboref Lt V
ABVE	Hatholm	ABYZ	Sandesund	ACET	Smyge Pt
ABVF	Mandal	ABZC	Sarpsborg		
ABVG	Risö Bank	ABZE	Fredrikshald	ACEU	BALTIC SEA
ABVH	Kleven	ABZF	Hvaler Is	ACEV	Ystad d
ABVI	Mannefjord	ABZG	Lauersvælgen	ACEX	Sandhammar Pt d
ABVJ	Ryvingen			ACEY	Bornholm
ABVL	Songvaar Lt	ABZH	SWEDEN	ACEZ	Rönne b d
ABVM	Höllen	ABZJ	Bohus B	ACFD	Hasle d
ABVP	Oxö a c e	ABZL	Koster I.	ACFE	Hammershus a c e
ABVQ	Grönningen	ABZM	Stromstad	ACFG	Hammer Pt
ABVR	Odderö	ABZN	Ursholm.	ACFI	Christianso
		ABZP	Svangen.	ACFJ	Svaneke d
ABVT	CHRISTIANSAND	ABZQ	Vaderobod	ACFK	Nexo
ABVW	Vestre Havn	ABZR	Hällö c d	ACFM	Due Oddt
ABVX	Topdals Fiord	ABZT	Måseskar	ACFO	Simrishamn
ABVZ	Lillesand	ABZU	Uddevalla	ACFP	Hanobugten
ABWC	Homborgsund	ABZV	Mollo	ACFQ	Ahus
ABWE	Grimstad	ABZX	Vuderoarne	ACFR	Solvesborg
ABWF	Arendal	ABZY	Grönskaren.	ACFS	Christianstad (Kristianstad)
ABWG	Torungen	ACBD	Kladesholm		
ABWI	Naresto	ACBE	Hamnskar	ACFT	Brantevik d
ABWJ	Ydre Mokkelas	ACBF	Lysekil I c	ACFU	Hano I
ABWK	Tvedestrand	ACBH	Pater Noster	ACFV	Karlshamn
ABWM	Lyngör	ACBI	Marstrand	ACFW	KARLSKRONA b f
ABWN	Risör	ACBJ	KATTEGAT	ACFZ	Utklippan
ABWO	Kran Fiord	ACBL	East Ch (Östra rannan)	ACGD	Utlangan
ABWQ	Kragero	ACBM	West Ch (Vestra rannan)	ACGE	Ut Grunden Lt V
ABWR	Helle	ACBN	Winga (Vinga) a c e	ACGF	Kalmar Sound
ABWS	Iomfruland	ACBP	Kanso	ACGH	Garpen
ABWU	Langesund	ACBQ	Buskar I	ACGI	Grimskär
ABWV	Langötangen	ACBR	GOTTENBURG b	ACGJ	Kalmar
ABWX	Brevik	ACBT	Waro	ACGK	Skagganas
ABWY	Stathelle	ACBU	Tistlarne	ACGL	Dämman
ABWZ	Porsgrund	ACBV	Nidingen	ACGM	Oscarshamn
ABXC	Skien	ACBW	Fladen Lt V	ACGO	Furön
ABXD	Helgeraaen	ACBX	Warberg c	ACGP	Figeholm
ABXE	Nevlunghavn	ACBY	Morup Tånge c	ACGQ	Öland
ABXF	SKAGERRACK.	ACBZ	Falkenberg	ACGS	Morbylånga
ABXG	Fredriksværn.	ACDE	Tylo	ACGT	Ispe Pt
ABXH	Staværnsodden	ACDF	Halmstad d	ACGU	Ekerums
ABXI	Laurvik	ACDG	Hallands Waderö	ACGW	Borgholm
ABXJ	Svenoer	ACDI	Vingaskar	ACGX	Biornabben Rock
ABXL	Sandefiord	ACDJ	Skelder B	ACGY	Kappel Pt (Olands
ABXN	Tönsberg	ACDK	Engelholm d		Östra)
ABXO	Færder a e	ACDL	Arildslage d	ACGZ	Grasgard d
ABXP	Vrængen			ACHB	Segerstad
ABXQ	Vallo	ACDM	THE SOUND (SUNDET)	ACHD	GOTLAND
ABXS	Fulehuk	ACDN	Kullen	ACHF	Hoborg
ABXT	Aasgaardstrand	ACDO	Molle	ACHG	Carlso
ABXV	Basto	ACDQ	Lerhamn	ACHI	Klintehamn
ABXW	Horten	ACDR	Hoganäs d	ACHK	Vstholmen
ABXY	Carljohansværn	ACDS	Svinbåden Lt V	ACHL	Visby d
ABYC	Holmestrand	ACDU	Wikon	ACHM	Stenkyrkehuk
ABYD	Drammen	ACDV	Helsingborg a c d e	ACHN	Skarsande d
ABYE	Svelvik	ACDW	Råå	ACHO	Fåro f
ABYG	Rödtangen	ACDX	Haken	ACHP	Holmudden
ABYH	Fildtvedt	ACDY	Hven I	ACHQ	Gotska Sandön
ABYI	CHRISTIANIA b c	ACEB	Landskrona	ACHS	Kopparstenarne Lt V
ABYK	Sandviken	ACEF	Flint Ch (Flint rannan)		

a Signifies an International Signal Station, b a Time Signal Station, c a Weather, Tide, or Ice Signal Station, d a Life-Saving Station, e a Lloyd's Signal Station f a Wireless Telegraph Station

SWEDEN		FINLAND			
ACHT	Slitehamn	ACLV	Gran I	ACPO	Biörneborg
AOHU	Ostergarn	ACLW	Bremo a	ACPR	Kallo I
ACHW	Ljugarn	ACLX	Sundsvall	ACPS	Raumo
ACHX	Ronehamn	ACLZ	Balsö	ACPT	Relandersgrund
ACHY	Falluden d	ACMB	Alnö	ACPV	Nystad
ACHZ	Spåro	ACMD	Juniskaren	ACPW	Enskar
ACIB	Vesterwik	ACME	Vifsta (Wifsta) Wharf	ACPX	Åland Isles
ACID	Ido	ACMF	Hernosand	ACPZ	Skalskär
ACIE	Stedsholm	ACMG	Lungo	ACQB	Market Rock
ACIG	Syrsan	ACMH	Angermannaelf	ACQD	Hellman
ACIH	Valdemarsvik	ACMJ	Ullånger	ACQE	Nodendal
ACIJ	Haradsekar	ACMK	Bredskar	ACQG	Åbo b
ACIK	Storklappen	ACML	Skagens Hamn	ACQH	Uto
ACIL	Söderköping	ACMO	Ornskoldsvik	ACQI	Korsö
ACIM	Götha (Gota) Canal	ACMP	Nordmaling	ACQK*	Bogskär.
ACIN	Norrkoping	ACMQ	Sydostbrotten Lt V	ACQL	Lågskär
ACIP	Arko.	ACMS	Umeå	ACQM	Led Sund.
ACIQ	Oxelö	ACMT	Holmögadd	ACQO	Bomarsund.
ACIR	Hafringe.	ACMU	Norrbyskär	ACQP	Hangö a c
ACIT	Nykoping	ACMW	Stora Fjaderägg	ACQR	Rusaro a d
ACIU	Landsort	AOMX	Bergudden Pt	ACQT	Eknas
ACIV	Almagrundet Lt V	ACMZ	Holmon	ACQU	Renskär
ACIY	Södertelje	ACNB	Snipan Lt V.	ACQV	Gröhara
ACIZ	Måsknuf	ACND	Ratan	ACQX	Helsingfors a b c d
ACJD	Hufvudskar	ACNF	Biurö (Bjuro) Klubb		
ACJE	Ålandskär.	ACNG	Gåsoren		
ACJF	Dalarö	ACNH	Furugrund	ACQY	GULF OF FINLAND
ACJH	Gronskar	ACNJ	Ursvik.	ACQZ	Sveaborg
ACJI	Sandhamn.	ACNK	Skellefteå	ACRD	Soder Skars
ACJK	Fredriksborg	ACNL	Kåge	ACRE	Kalbåden Lt V d
ACJM	Vaxholm	ACNO	Piteå	ACRF	Glosholm
ACJQ	STOCKHOLM b f	ACNP	Leskar	ACRH	Borgå
ACJT	Svenska Högarne	ACNQ	Rodkallen	ACRI	Lovisa
ACJU	Svenska Bjorn Lt V	ACNR	Norströmsgrund Lt V	ACRJ	Kotka I
ACJV	Soderarm	ACNS	Germandö	ACRL	Fredrikshamn
ACJW	Fejan	ACNT	Luleå	ACRM	Villa
ACJX	Norretelje	ACNU	Salmis	ACRN	Viborg
ACJY	Vato	ACNV	Haparanda	ACRP	Biorkö
ACKB	Waddo (Vaddo)	ACNW	Råneå	ACRQ	Verko Matala Lt V
ACKD	Arholma	ACNX	Malören	ACRS	Stirs Pt
ACKE	Simpnasklubb	ACNY	Töre	ACRU	Sestroretzk
ACKG	Grisselhamn	ACNZ	Kalix	ACRV	Kronstadt a b c f
ACKH	Svartklubben	ACOB	Seskaro	ACRW	Tolboukin
ACKI	Allsta, or, Hallsta			ACRY	ST PETERSBURG b d f
ACKI	Understen	ACOD	FINLAND	ACSB	Moscow
		ACOE	Kemi	ACSE	Neva d
ACKO	GULF OF BOTHNIA	ACOG	Plevna.	ACSF	Strelna
ACKQ	Harg	ACOH	Simå	ACSH	Oranienbaum d
ACKS	Öregrund	ACOI	Knivaniemi	ACSI	Neva or Korableny Lt V d
ACKU	Djursten	ACOK	Uleåborg b		
ACKV	Grundkallen Lt V	ACOL	Karlö	ACSK	Kronslot
ACKW	Grepen Lt V	ACOM	Brahestad.	ACSL	London Shoals d
ACKY	Orskär	ACOP	Ulko Kalla	ACSM	Krasnaia Gorka a c e
ACKZ	Björn	ACOG	Kalajoki	ACSO	Seskar d
ACLB	Lofsta	ACOR	Gamla (Old) Karleby	ACSP	Nerva Rock
ACLE	Skutskar	AOOT	Tankar	ACSQ	Sommars
ACLF	Gefle	ACOU	Jacobstadt	ACST	Narva a c d
ACLG	Upsala	ACOV	Nya (New) Karleby	ACSU	Hogland d
ACLI	Eggegrundet	ACOX	Helsingkallan	ACSV	Rodekär
ACLJ	Axmar	ACOY	Storskar	ACSX	Stenskar
ACLK	Finngrundet Lt V	ACOZ	Norrskar	ACSY	Exholm
ACLN	West Finngrund (Vestra Banken) Lt V	ACPD	North Kallan I	ACSZ	Kokskar
		ACPE	Nikolaistad d	ACTD	Revel Stone Lt V.
ACLO	Norrsund	ACPF	Wasa	ACTE	Revel a c d
ACLP	Storjungfrun	ACPH	Storkallagrund	ACTF	Nargen I
ACLQ	Liusne	ACPI	Skalgrund	ACTH	Cape Sourop
ACLR	Söderhamn	ACPJ	Kaskö	ACTI	Paker Ort a c
ACLS	Agön	ACPL	Khristinestad	ACTJ	Port Baltic c d
ACLT	Hudiksvall	ACPM	Sidby	ACTL	Odens-holm d
ACLU	Iggesund	ACPN	Sastmola	ACTM	Paternoster I

a Signifies an International Signal Station. b a Time Signal Station. c a Weather, Tide, or Ice Signal Station. d a Life-Saving Station; c a Lloyd's Signal Station, f a Wireless Telegraph Station

EAST AND SOUTH COASTS BALTIC SEA

ACTN	Wormsö d	ACXE	Balga a c	ADBM	C. Arkona. a c d f
ACTP	Hapsal	ACXG	Passarge R	ADBN	Adler Grund Lt V
ACTQ	Saxbi Ness	ACXH	Frauenburg	ADBO	Wittower a c
ACTR	Takhkona Pt	ACXI	Tolkemit	ADBP	Dornbusch Pt c
ACTS	Dago d	ACXJ	Probbernau d	ADBQ	Hiddensee I d
ACTV	Kassar.	ACXK	Elbing	ADBR	Proser Bay
ACTW	C Dagerort a c	ACXL	Pasewark d	ADBS	Dranske d
ACTX	Tiefen Hafen d	ACXM	Vistula R	ADBT	Stralsund c d
ACTY	Osel.	ACXN	Schewenhorst c	ADBU	Barth
ACUB	Filsand d	ACXO	Dirschau	ADBV	Prerow d
ACUD	Arensburg	ACXP	Danzig a f	ADBW	Zingst d
ACUE	Swalfer Ort d	ACXQ	Neufahrwasser a b c d	ADBX	Dars Pt a c d
ACUF	Möön	ACXR	Westerplatte d	ADBY	MECKLENBURG
ACUH	Kinö d	ACXS	Oxhöft Pt c	ADCB	Wustrow d
ACUI	Pernau a c d	ACXT	Putziger	ADCE	Ribnitz
ACUJ	Sallis (Alt Salis) d	ACXU	Putzig	ADCF	Warnemünde a c d
ACUL	Dwina R d	ACXV	Neufahr c d	ADCG	Arendsee d
ACUM	RIGA a b c d	ACXW	Hela Pt c d	ADCH	Rostok c
ACUO	Dunamund I	ACXY	Heisternest a c d	ADCI	Buk Pt a
ACUQ	Runo I d	ACXZ	Rixhoft a c	ADCJ	Poel d
ACUR	Messaragotsem Pt d	ACYB	Kopalin d	ADCK	Timmendorf c
ACUS	Domesness a d	ACYD	Stilow Beacon	ADCL	Kirchdorf
ACUV	Kourland	ACYE	POMERANIA	ADCM	Wismar c
ACUW	Michael (Pissen)	ACYF	Leba c d	ADCN	Klütz Hd
ACUX	Lyser Ort a c	ACYH	Scholpin d	ADCP	Dassow a
ACUZ	Windau a c d	ACYI	Stolpmunde a c d	ADCQ	Travemunde a c d
ACVB	Backofen Lt	ACYJ	Jershoft d f	ADCR	Lubeck
ACVE	Steinort C d	ACYL	Rugenwalde a c d	ADCT	Trave R d
ACVF	Libau c d f	ACYM	Funkenhagen d	ADCU	SCHLESWIG-HOLSTEIN
ACVG	Dunamund Fortress	ACYN	Colberg a c d	ADCV	Neustadt
ACVI	Kowno	ACYO	Gross Ziegenort c	ADCX	Pelzer Pt a
ACVJ	Wilna	ACYP	Horst a c	ADCY	Dame Hd a
		ACYQ	Dievenow c d	ADCZ	Staaken
ACVL	POLAND	ACYR	Cammin	ADEC	Femern I d
ACVM	Warsaw	ACYS	Neuendorf. d	ADEF	Marien c d f
		ACYT	Wollin c	ADEG	Markelsdorf Pt a
ACVP	GERMANY	ACYU	Oder R.	ADEI	Flugge
		ACYV	Grosses Haff	ADEJ	Strukamp Pt
ACVQ	PRUSSIA	ACYW	Kleines Haff	ADEK	Femern Sound
ACVR	Bohemia	ACYX	Stettin	ADEL	Fehmarnbelt Lt V c
ACVS	Bavaria	ACYZ	Starkenhorst	ADEM	Heiligenhafen c d
ACVT	Saxony	ACZB	Swinemunde a b c d	ADEN	Labo c d
ACVW	Wurtemberg	ACZD	Swine R	ADEO	KIEL b c
ACVX	Berlin	ACZE	Lebbin	ADEQ	Rendsburg
ACVY	Munich	ACZF	Kaiserfahrt	ADES	Wik b
ACVZ	Strasburg	ACZG	Woitzig Lt V	ADET	Holtenau a c e
ACWB	Leipzig	ACZH	Swanterwitz Lt V	ADEU	Friedrichsort c
ACWD	Dresden	ACZI	Altwarp	ADEW	Bulk Pt a f
ACWE	Nimmersatt c d	ACZJ	Ahlbeck c	ADEX	Stollergrund
ACWF	Memel a c d f	ACZK	Ruden I a d	ADEY	Eckernförde
ACWG	Melneraggen d	ACZL	Zinnowitz d	ADFB	Shmunde a c
ACWH	Konig Wilhelm Canal.	ACZM	Peene R	ADFO	Kappeln
ACWJ	Windenburger c	ACZN	Peenemunde	ADFE	Arnis
ACWK	Russ	ACZO	Wolgast c	ADFH	Schleswig
ACWL	Memel R	ACZP	Greifswalder Oie c d	ADFI	Fredrichsbarg
ACWN	Tilsit	ACZQ	Usedom	ADFJ	Sli (Schlei) Fiord
ACWO	Labiau	ACZR	Lassan	ADFL	Maesholm
ACWP	Schwarzort c d	ACZS	Anclam	ADFP	Kalk Ground Lt V a
ACWQ	Rositten d	ACZT	Streckelsberg c	ADFQ	Haber Ness
ACWR	Nidden a c d	ACZU	Uckermunde c	ADFR	Flensborg c
ACWS	Cranz c d	ACZV	Greifswald	ADFS	Hol Ness
ACWT	Brüster Ort c f	ACZW	RUGEN I	ADFT	Eken Sound
ACWU	Kraxtepellen d	ACZY	Thiessow a c d	ADFU	Nubel
ACWV	Palmnicken c	ADBC	Palmer Ort	ADFV	Gt Borns Hd
ACWX	Pillau a c d	ADBE	Peerd Pt	ADFX	Sonderburg c
ACWY	Fischausen a c	ADBG	Göhren c d	ADFZ	Keke Ness
ACWZ	Tenkitten. d	ADBH	Lanterbach	ADGB	Hörup Haff
ACXB	Grossbruch d	ADBI	Bergen	ADGC	Pol Pt
ACXD	Königsberg	ADBK	Sasnitz c d	ADGE	Als I
		ADBL	Jasmund	ADGF	Augustenburg

a Signifies an International Signal Station b a Time Signal Station. c a Weather Tide, or Ice Signal Station d a Life-Saving Station e a Lloyd's Signal Station, f a Wireless Telegraph Station

DENMARK

Code	Place	Code	Place	Code	Place
ADGH	Als Sound	ADKF	Skiods Hd.	ADNU	Faxo
ADGI	Apenrade c	ADKH	Sletterhage	ADNW	Rödvig.
ADGJ	Barso	ADKI	Lyngsbek	ADNX	C Stevns, or, Stevns
ADGK	Gienner Fiord	ADKJ	Ebeltofte		Klint
ADGM	Aarosund c	ADKL	Hasenore Hd	ADNY	Kiöge
ADGN	Aaro Is	ADKN	Hielm (Hjelm)	ADOB	Sorte Ch
ADGO	Hadersleben	ADKO	Grenaa d	ADOC	Aflandshage
ADGQ	Brandso	ADKP	Fornæs a c d e	ADOE	Drogden Ch
		ADKR	Schultz Ground Lt V	ADOG	Drogör, or, Drogden Lt.
ADGS	DENMARK	ADKS	Hesselo I		V
ADGT	Kolding	ADKT	Lyse Ground	ADOH	Nordre Röse
ADGU	Fredericia b			ADOI	Kastrup
ADGW	Kesserodde	ADKV	SIÆLLAND (ZEALAND)	ADOK	Saltholm
ADGY	Little Belt	ADKW	Gilbierg Hd	ADOL	Middlegrund.
ADGZ	FYEN	ADKX	Spotsbierg	ADOM	COPENHAGEN b c
ADHB	Æbelo	ADKZ	Sætteriet Road.	ADOP	Frederiksberg
ADHE	Bogenæe	ADLB	Ise Fiord	ADOQ	Trekroner
ADHF	Strib Pt	ADLC	Frederikssund	ADOR	Konge Deep
ADHG	Middlefart	ADLF	Roskilde	ADOS	Kalveboderne (Kalle-
ADHJ	Fænö	ADLG	Holbek		boer) Lt V
ADHK	Fonskov Pt	ADLH	Nykiöbing b	ADOT	Holländer Deep
ADHM	Aale Hd	ADLJ	Korshagen	ADOU	Vedbæk
ADHN	Vedelsborg Hd	ADLK	Rönnen	ADOV	Nivaa
ADHO	Baagö	ADLN	Great Belt	ADOX	Sletten
ADHQ	Assens	ADLO	Seiro c	ADOY	Helsingor (Elsinore) a
ADHS	Lydo	ADLP	Siælland Reef Lt		b c e f
ADHU	Faaborg	ADLQ	Revsnæs	ADOZ	Kronborg.
ADHV	Avernako	ADLR	Kallundborg b	ADPC	Lappe Grund Lt V
ADHW	Svendborg b	ADLS	As Ness	ADPE	Hornbæk
ADHY	Thorö	ADLU	Jammerland B.	ADPF	Nakke Hd
ADHZ	Troense	ADLV	Reersö		
ADIB	Skiold Ness	ADLW	Halskov	ADPH	Anholt I d
ADIE	Æroskiobing	ADLX	Korsor Road Lt V	ADPI	Anholt Knob Lt V
ADIF	Veis Ness	ADLY	Korsör b	ADPJ	Stauns Hd
ADIG	Marstal	ADLZ	Skielskör	ADPK	Uddyhöj Lt
ADIH	Langeland	ADMB	Agersö	ADPL	Randers b
ADIJ	Taasinge I	ADME	Helleholm	ADPM	Mariager
ADIK	Rudkiobing b	ADMF	Omö	ADPN	Hobro
ADIL	Bagnkop	ADMG	Stig Ness	ADPO	Gjerrild Lt
ADIM	Kjeldsnor c	ADMI	Karrebæk	ADPQ	Als
ADIO	Fakkebierg	ADMJ	Væiro	ADPR	Hurup
ADIP	Tranekicer			ADPS	Egense
ADIR	Hov Sand	ADMK	LAALAND	ADPU	Liim Fiord.
ADIT	Vresen	ADMN	Saxkiobing	ADPV	Aalborg c
ADIU	Knuds Hd	ADMO	Banholm	ADPW	Hals
ADIV	Nyborg b	ADMP	Taars	ADPY	Rimmen
ADIW	Sprogö c	ADMR	Nakskov	ADPZ	Læsö Rende.
ADIY	Kerteminde	ADMS	Albue Pt	ADQB	Læsö d
ADIZ	Romsö	ADMT	Femern Belt	ADQE	North Rouner
ADJB	Korshavn	ADMV	Rödby	ADQF	Dvale Ground, or, Læsö
ADJC	Skoven	ADMW	Nysted		Ch Lt V
ADJF	Enebærodde.	ADMX	Guldborg Sound	ADQG	Kobber Grund Lt V
ADJG	Odense b c			ADQI	Trindelen Lt V
ADJH	Gersö	ADMZ	FALSTER	ADQJ	Frederikshavn c
ADJI	Agernæss.	ADNB	Nykiobing b	ADQK	Hirtsholm
ADJL	Klinte	ADNC	Giedser Reefs Lt V f	ADQM	Aalbæk d
		ADNE	Giedser Pt Lt V f	ADQN	THE SKAW, or, SKAGEN
ADJM	JUTLAND	ADNF	Varsko		a c d e
ADJN	Veile b	ADNG	Heste Hd	ADQO	Hoien d
ADJP	Horsens b	ADNH	Stubbekiobing	ADQR	Hirtshals a c d e
ADJQ	Endelave I	ADNJ	Grön Sound	ADQS	Jammer B d
ADJR	Hov Huk			ADQT	Lildstrand d
ADJT	Samso	ADNK	MÖEN d	ADQV	Hanstholm a c d e
ADJU	Isse Hd			ADQW	Klitmoller d
ADJV	Langore	ADNL	Hellehaven Pt	ADQX	North Vorupor d
ADJX	Vestborg Pt	ADNO	Ulvshale	ADQZ	Lodbjerg
ADJY	Kyholm	ADNP	Vordingborg	ADRB	Agger d
ADJZ	Thunö	ADNQ	Stege	ADRC	Thybo Ron. c d
ADKC	Aarhus b c	ADNS	Præsto	ADRF	Thisted
ADKE	Kalö B	ADNT	Hylleholt	ADRG	Lögstör

a Signifies an International Signal Station, b a Time Signal Station, c a Weather, Tide, or Ice Signal Station, d a Life-Saving Station, e a Lloyd's Signal Station, f a Wireless Telegraph Station

NORTH SEA   HOLLAND

ADRH	Mors	ADUQ	Schulau Lt V	ADYJ	NETHERLANDS
ADRJ	Nyklöbing	ADUR	Blankenese c	ADYK	Delfzyl c
ADRK	Lemvig	ADUS	Neuenfelde	ADYM	Pilsum
ADRL	Bovbjerg d	ADUV	HAMBURG bc	ADYN	Campen
ADRN	Sondervig d	ADUW	Altona c	ADYO	Watum
ADRO	Nyminde Gab d	ADUX	Wandsbek	ADYP	Borkum Flat (Borkum
ADRP	Blaavand Pt cdf	ADUZ	Finkenwerder		Riff) Lt V acf
ADRS	Horn Reefs Lt V cf	ADVB	Harburg c	ADYQ	Schiermonnikoog acd
ADRT	Vyl Lt V c	ADVC	Hanover	ADYR	Lauwerzee
ADRU	Ssedenstrand	ADVF	Luneburg	ADYS	Zoutkamp
ADRV	Graa Deep	ADVH	Neuwerk d	ADYU	Gröningen
ADRX	Fanö			ADYV	Ameland acd
ADRY	Esbjerg c	ADVJ	HELGOLAND acdef	ADYW	Hallam
				ADYX	Terschelling Bank Lt V
ADSB	NORTH SEA	ADVK	Weser R	ADYZ	Terschelling acd
ADSC	Rom I cd	ADVL	Weser Lt \ f	ADZB	Brandaris
ADSE	Lister Deep	ADVN	Rothersand ace	ADZC	Harlingen c
ADSG	Kirkeby d	ADVO	Salzhorn	ADZF	Vlieland ac
ADSH	Sylt I acd	ADVP	Wremen Tief d	ADZG	Eyerland ac
ADSK	Ellenbogen cd	ADVR	Bremerhaven abcdf	ADZH	Texel I d
ADSL	Hoyer	ADVS	Blexen	ADZI	Schilbols Nol
ADSM	Rothe Kliff af	ADVT	Nordenham c	ADZJ	The Texel a
ADSO	Kampen d	ADVW	Geestemunde c	ADZK	Helder d
ADSP	Amrum I acd	ADVX	Brake c	ADZL	Nieuwe Diep bcd
ADSQ	Nebel	ADVY	Elsfleth	ADZN	Great North Holland
ADSR	Pelworm I e	ADWB	Vegesack c		Canal
ADST	Föhr I	ADWC	Bremen abccf	ADZO	Alkmaar.
ADSU	Wyk c	ADWE	Kaiser Haib, or Kaiser-	ADZP	Wullemsoord
ADSV	Nordstrand I		hafen	ADZR	Wieringen f
ADSW	Ording Pt d			ADZS	Medemblik
ADSX	Husum c	ADWG	Meyers Legde a	ADZT	Enkhuizen f
ADSY	Hever R	ADWH	Eversand a	ADZV	Hoorn
ADSZ	Eider R	ADWI	Hohe Weg Flat ace	ADZW	Edam
ADTB	Suderhoft cd	ADWJ	Fedderwarder Siel d	ADZX	Zaandam
ADTC	Outer Eider (Eider) Lt	ADWK	Jade R	AEBC	Marken I
	V acf	ADWL	Mellum Flat d	AEBD	AMSTERDAM bc
ADTE	Eider Galliot Lt V ac	ADWM	Aussen Jade Lt V adf	AEBF	Hoek e
ADTF	Tonning c	ADWO	Minsener Sand Lt V d	AEBH	Muiden
ADTG	Eider Canal	ADWP	Wilhelmshaven abcdf	AEBI	Naarden
ADTI	Fredrichstadt	ADWQ	Heppens	AEBJ	Utrecht
ADTK	Norder Piep.			AEBL	Harderwijk
ADTL	Busum cd	ADWS	OLDENBURG	AEBM	Elburg
ADTM	Neufeld d	ADWT	Gellen	AEBN	Kampen
ADTN	Suder Piep	ADWU	Varel	AEBP	Zuider Zee
		ADWX	Ems Jade Canal	AEBQ	Schokland
ADTO	ELBE R	ADWY	Vareler Siel	AEBR	Urk
ADTP	Outer Elbe or Elbe No	ADWZ	Schilighorn acd	AEBT	Kragchenburg
	1 Lt V df	ADXC	Genius Bank Lt V d	AEBU	De Lemmer c
		ADXE	Minsener Old Oog (Res-	AEBV	Stavoren cf
ADTQ	Elbe Pilot Galliot d		cue Bn)	AEBX	Kykduin a
ADTS	Elbe No 2 Lt V d	ADXF	Wangeroog I acdf	AEBY	Dirkoomsduin
ADTU	Elbe No 3 Lt V	ADXI	Spiekeroog I d	AEBZ	Zanddijk c
ADTV	Elbe No 4 Lt V d	ADXJ	Langeoog I d	AECD	Haaks Lt V
ADTW	Duhnen d	ADXL	Friedrichsschleuse d	AECF	Egmond aan Zee acd
ADTX	Vogelsand	ADXM	Carolinensiel c	AECG	Ymuiden acd
ADTY	Cuxhaven abcdef	ADXN	Neuharrlingersiel cd	AECI	North Sea Ship Canal
ADTZ	Oste Reef (Oste Riff)			AECJ	Zandvoort cd
	Lt V c	ADXP	NORTH FRIESLAND	AECK	Haarlem
ADUB	Neuhaus	ADXQ	Baltrum I d	AECM	Leyden
ADUC	Brunsbuttelkoog acef	ADXR	Norderney I cd	AECN	Scheveningen acdf
ADUE	Brunsbuttel	ADXS	Norddeich cdf	AECO	The Hague
ADUF	Kaiser Wilhelm Canal	ADXT	Juist I d	AECP	Maas Lt V
ADUG	Bösch	ADXU	Aurich	AECQ	Maas R
ADUH	Stor R	ADXV	Papenburg	AECR	New Rotterdam Canal
ADUI	Freiburg	ADXW	Norderney Lt V	AECS	Noorder Hoofd
ADUJ	Gluckstadt c	ADXZ	Weener.	AECU	Rhine R
ADUK	Kraut Sand Lt V	ADYB	Ems R	AECV	Hook of Holland acdf
ADUL	Pagen Sand	ADYE	Borkum I acdf	AECW	ROTTERDAM b
ADUM	Brunshausen c	ADYF	Norden	AECY	Schiedam
ADUN	Stade	ADYG	Emden cd	AECZ	Delfshaven
ADUO	Grünendeich	ADYI	Leer		

a Signifies an International Signal Station, b a Time Signal Station, c a Weather, Tide, or Ice Signal Station, d a Life-Saving Station, e a Lloyd's Signal Station, f a Wireless Telegraph Station

NORTH SEA		ENGLAND			
AEDB	Vlaardingen	AEHQ	GREAT BRITAIN	AELF	Beachy Hd c e
AEDF	Maasluis c			AELG	Newhaven c d
AEDG	Nieuwe Maas	AEHR	ENGLAND	AELI	Brighton c d
AEDH	Brielle d	AEHS	English Ch	AELJ	Shoreham d
AEDJ	Goeree	AEHU	Thames R	AELK	Owers Lt V
AEDK	Goedereede a c	AEHV	LONDON	AELM	Worthing d
AEDL	Molenpolder	AEHW	Westminster	AELN	Selsea Bill d
AEDN	Yzeren Baak	AEHX	The City	AELO	Chichester
AEDO	Hellevoetsluis c d	AEHZ	Windsor	AELP	Langston Hr
AEDP	Voorne Canal	AEIB	Maplin Sands	AELR	Spithead e
AEDR	Middelharnis	AEIC	Tongue Lt V	AELS	Portsmouth b c
AEDS	Ooltgensplaat	AEID	Mouse Lt V.	AELT	Nab Lt V
AEDT	Willemstad	AEIF	Edinburgh Ch	AELV	Warner Lt V
AEDU	Stryensas	AEIG	Prince's Ch	AELW	SOUTHAMPTON b c
AEDW	Moerdyk	AEIH	Nore Lt V	AELX	Solent, The
AEDX	Dordrecht c	AEIJ	Girdler Lt V	AELY	Calshot Lt V
AEDY	Schouwen a c	AEIK	Garrison Pt b	AELZ	Ryde c d
AEDZ	Renesse	AEIL	Shoeburyness	AEMB	Osborne
AEFC	Brouwershaven c d	AEIM	Southend d e	AEMC	Cowes c
AEFD	Hook of Schouwen	AEIO	Tilbury Docks	AEMF	Lymington c d
AEFG	Scheldt	AEIP	Gravesend	AEMG	Hurst Castle c
AEFH	Schouwen Bank Lt V	AEIQ	Northfleet	AEMH	Needles, The e
AEFI	Zierikzee c	AEIR	Black Deep Lt V	AEMJ	Isle of Wight.
AEFJ	Tholen	AEIS	Greenhithe	AEMK	Bembridge d
AEFK	Bergen-op-Zoom	AEIT	Purfleet	AEML	St Helens
AEFM	Goes	AEIU	Erith	AEMO	St Catherine Pt c d e f
AEFN	Domburg	AEIW	Woolwich	AEMP	Christchurch d
AEFO	Walcheren	AEIX	Blackwall	AEMQ	Poole c d f
AEFQ	Veere	AEIY	GREENWICH	AEMR	Swanage d
AEFR	Westkapelle a c	AEJB	Greenwich Observatory b	AEMS	Anvil Pt c
AEFS	Zoutelande			AEMT	St Alban's Hd d
AEFU	Flushing (Vlissingen) b c d e	AEJC	Deptford	AEMU	Weymouth c d
		AEJD	Royal Albert Docks	AEMW	PORTLAND b d f
AEFV	Middelburg	AEJG	Royal Victoria Docks	AEMX	Bill of Portland c e
AEFW	Arnemuiden	AEJH	East India Docks	AEMY	Shambles Lt V
AEFY	Ellewoutsdijk	AEJI	West India Docks	AENB	Bridport
AEFZ	Breskens	AEJL	Millwall Docks	AENC	Lyme Regis d
AEGB	Hoedekenskerke	AEJM	Commercial Docks	AEND	Exmouth c d
AEGD	Hanswest (Hansweerd)	AEJN	London Docks	AENG	Teignmouth c d
AEGF	Terneuse (Neuzen)	AEJP	St Katharine Docks	AENH	Torquay d
AEGH	Bath	AEJQ	Grand Surrey Canal	AENI	Tor Bay
AEGJ	Doel	AEJR	Medway, The	AENK	Brixham d
AEGK	Zeeland	AEJT	Sheerness b c	AENL	Dartmouth d
AEGL	Nieuwe Sluis c	AEJU	Queenborough	AENM	Start Pt
AEGN	Sluis	AEJV	Port Victoria	AENP	Prawl Pt c d e
		AEJX	Stangate Creek	AENQ	Salcombe d
		AEJY	Chatham	AENR	Bolt Hd d
		AEJZ	Strood	AENT	Eddystone d
AEGO	BELGIUM	AEKC	Rochester	AENU	Plymouth c d
AEGP	Knocke c d	AEKD	Whitstable c	AENV	Stonehouse
AEGQ	Lillo	AEKF	Margate c d	AENX	Cattewater
AEGS	ANTWERP (ANVERS) b	AEKH	North Foreland c f	AENY	Devonport c
AEGT	Tête de Flandre	AEKI	Broadstairs d	AENZ	Mt Wise (Devonport) b
AEGU	Burght	AEKJ	Ramsgate c d	AEOC	Keyham
AEGV	Boom	AEKL	North Goodwin Lt V	AEOD	CORNWALL
AEGX	Tamise	AEKM	Goodwin Sands	AEOF	Whitsand B
AEGY	Ghent	AEKN	Deal b c d	AEOH	Fowey
AEGZ	Brussels	AEKO	Downs, The	AEOI	Dodman Pt
AEHD	Wielingen Lt V	AEKQ	Gull Lt V	AEOJ	Falmouth b c d
AEHF	Wandelaar Lt V	AEKR	East Goodwin Lt V	AEOK	Manacle Rks
AEHG	North (Noord) Hinder Lt V	AEKS	South Goodwin Lt V	AEOL	Lizard, The c d e f
		AEKU	South Foreland	AEOM	Penzance c d e
AEHI	West Hinder Lt V	AEKV	Dover b c d e	AEON	Mount's B
AEHJ	Heyst a d e	AEKW	Dover Harbor Lt V	AEOQ	Longships Lt d
AEHK	Blankenberghe c d	AEKX	Varne Lt V	AEOR	Land's End d
AEHL	Bruges v	AEKY	Folkestone c	AEOS	Wolf Rock
AEHM	Ostende c d	AEKZ	Sandgate Road c d e		
AEHN	Nieuport c d f	AELB	Dungeness d e	AEOT	SCILLY Is e f
AEHP	La Panne, or Adin Kerke c d	AELC	Hastings c d	AEOU	St Mary's I c d
		AELD	Royal Sovereign Lt V	AEOV	St Agnes d

a Signifies an International Signal Station, b a Time Signal Station, c a Weather, Tide, or Ice Signal Station d a Life-Saving Station, e a Lloyd's Signal Station, f a Wireless Telegraph Station

WEST COAST ENGLAND				WEST COAST SCOTLAND	
AEOW	Bishop Rock d	AESP	Newport d	AEWH	Cumberland
AEOY	St Mary's Road	AESQ	Cardigan d	AEWI	Duddon R
AEOZ	Round I	AESR	Cardigan B Lt V	AEWK	St Bees Hd
AEPB	Seven Stones Lt. V.	AEST	Aberaeron	AEWL	Whitehaven c d
		AESU	Aberystwith d	AEWM	Harrington
AEPD	St Ives c d	'AESV	Aberdovey d	AEWN	Selker Lt V
AEPF	Hayle c d	AESX	Barmouth d	AEWO	Workington c
AEPG	Godrevy I c d	AESY	P Madoc	AEWP	Maryport c d
AEPI	Padstow d	AESZ	Pwilheli d	AEWQ	Solway Firth d
AEPJ	Trevose Hd d	AETC	St Tudwall Rds d	AEWS	Silloth c
AEPK	Hartland Pt c	AETD	Bardsey I	AEWT	Carlisle
AEPM	Lundy I c d e	AETF	Carnarvon		
AEPN	Bideford	AETG	Carnarvon B Lt V	AEWU	ISLE OF MAN
AEPO	Appledore e d	AETH	Porth Dinlleyn d	AEWX	Ayre Pt c
AEPR	Barnstaple c	AETI	South Stack c d	AEWY	Calf of Man
AEPS	Bull Pt c	AETJ	Anglesea	AEWZ	Chicken Rock
AEPT	Ilfracombe c d	AETL	Holyhead c d f	AEXC	Castletown c d
AEPV	The Foreland	AETN	Skerries d	AEXD	Lang Ness
AEPW	Watchet d	AETP	Menai Str	AEXF	Douglas c d
AEPX	Burnham e d	AETQ	Bangor	AEXH	Bahama Bank
AEPZ	Bridgewater c	AETR	Conway		
AEQB	Flatholm d	AETU	Beaumaris	AEXI	SCOTLAND
AEQC	BRISTOL	AETV	Lynus Pt c	AEXJ	Dumfries
AEQF	Portishead	AETW	Great Orme Hd	AEXL	Southerness
AEQG	Avon R	AETY	Abergele	AEXM	Little Ross
AEQI	Avonmouth	AETZ	Flint	AEXN	Kirkcudbright
AEQJ	King Road	AEUB	Dee R	AEXP	Wigton
AEQK	Severn R	AEUC	Dee Lt V	AEXQ	Burrow Hd
AEQM	Sharpness Docks	AEUD	Air Pt d	AEXR	Mull of Galloway c
AEQN	Gloucester	AEUF	Connah's Quay c	AEXT	P Patrick d
AEQO	Monmouth	AEUG	Saltney	AEXU	Corsewall Pt c
AEQR	Chepstow			AEXV	Loch Ryan c
AEQS	Newport	AEUI	CHESTER	AEXY	Stranraer
AEQT	Usk	AEUJ	North West Lt V	AEXZ	Ailsa Craig
AEQU	English and Welsh grounds Lt V	AEUK	Hoylake d	AEYB	Girvan d
		AEUL	Bar Lt V	AEYD	Turnberry Pt d
		AEUM	Leasowe	AEYF	CLYDE
AEQV	WALES	AEUN	Bidston	AEYG	Ayr d
AEQW	CARDIFF c	AEUO	Crosby Lt V	AEYI	Troon d
AEQX	Bute Docks	AEUQ	North Wall (Mersey)	AEYJ	Irvine d
AEQZ	Penarth c d	AEUR	Formby Lt V	AEYK	Ardrossan c d
AERB	Barry Docks e			AEYM	Cloch Pt
AERC	Breaksea Pt	AEUS	MERSEY	AEYN	Gourock
AERD	Breaksea Lt V	AEUV	Liverpool d	AEYO	Toward Pt
AERF	Nash Pt c	AEUW	Garston	AEYQ	Rothesay c
AERH	Porthcawl d	AEUX	Manchester Ship Canal	AEYR	Cumbraes
AERJ	Scarweather Lt V	AEUZ	Widnes	AEYV	Greenock c
AERK	P Talbot	AEVB	Runcorn	AEYW	P. Glasgow
AERL	Neath	AEVC	Ellesmere Pt	AEYZ	Glasgow c
AERN	Briton Ferry	AEVD	Prince's Landing Stage	AEZB	Dumbarton
AERO	Swansea b	AEVF	Prince's Dock	AEZC	Gare Loch
AERP	Mumbles c d e	AEVG	Queen's Dock	AEZF	Helensburg
AERS	Helwick Lt V	AEVH	Brunswick Dock	AEZG	Loch Fyne
AERT	Worm's Hd.	AEVJ	North Dock	AEZH	Inverary
AERU	Burry Port d	AEVK	Birkenhead	AEZJ	Arran
AERW	Llanelly e	AEVL	Morpeth Dock b	AEZK	Lamlash c
AERX	Caermarthen	AEVN	Southport e d	AEZL	Pladda c
AERY	Caldy I c d	4EVO	Ribble R	AEZM	Kildonan Pt d e
AESB	Tenby d	AEVP	Lytham e d	AEZN	Campbelton c d
AESC	Old Castle Hd	AEVR	Preston	AEZO	Sanda
AESD	St Ann's Hd d e	AEVS	MANCHESTER	AEZP	MULL OF CANTYRE c
AESF	Milford Haven	AEVT	Birmingham	AEZR	Islay c
AESG	Neyland	AEVU	Blackpool e d	AEZS	P Ellen
AESH	Pembroke Dock	AEVW	Fleetwood d	AEZT	P Charlotte
		AEVX	Wyre R	AEZV	Jura Sound
AESI	St George's Ch	AEVY	LANCASTER	AEZW	Loch In Dail
AESK	Smalls d	AEWB	Morecambe B c	AEZX	Oversay I
AESL	St Bride's B	AEWC	Piel Hr d	AFBC	McArthur Hd
AESM	Bishops & Clerks	AEWD	Walney I c	AFBD	Rudha Mhail c
AESO	Fishguard d	AEWG	Barrow c	AFBE	Islay Sound

a Signifies an International Signal Station, b a Time Signal Station, c a Weather, Tide, or Ice Signal Station, d a Life-Saving Station, e a Lloyd's Signal Station, f a Wireless Telegraph Station

	SCOTLAND		EAST COAST SCOTLAND		
AFBH	Sgeir Maiole (Sgeir Vuile)	AFEW	Cantick Hd c	AFJS	Queensferry d
		AFEX	Auskerry	AFJT	St Margaret's Hope
AFBI	Crinan Canal	AFGB	Start Pt	AFJU	Alloa
AFBJ	Fladda Hr	AFGC	Pierowall Road	AFJW	EDINBURGH b
AFBL	Colonsay	AFGD	Noup Hd	AFJX	Stirling
AFBM	OBAN	AFGH	Mull Hd	AFJY	Forth and Clyde Canal
AFBN	Loch Linnhe	AFGI	North Ronaldsay	AFKB	Grangemouth c
AFBP	F William	AFGJ	Fair I d	AFKC	Borrowstounness (Bo'ness) c
AFBQ	Lismore I				
AFBR	Caledonian Canal	AFGL	SHETLAND Is	AFKD	Leith b c
AFBT	Iona	AFGM	Sumburgh Hd c	AFKG	Granton Hr c
AFBU	Dubh Artach	AFGN	Bressay I	AFKH	Newhaven e
AFBV	Skerryvore	AFGP	Lerwick c d	AFKI	Musselburgh
AFBX	Coll	AFGQ	Dury Voe	AFKL	North Berwick d
AFBY	Mull	AFGR	Whalsey	AFKM	Fidra
AFBZ	Ardnamurchan	AFGT	The Out Skerries	AFKN	Dunbar c d
AFCD	Rum I	AFGU	Balta I	AFKP	St Abb's Hd c e
AFCE	Maleg Hr	AFGV	Hillswick	AFKQ	Eyemouth c d
AFCG	Loch Nevis	AFGX	Scalloway c		
AFCI	Ornsay I	AFGY	Foula	AFKR	Berwick c d
AFCJ	Kyle Akin	AFGZ	Ve Skerries	AFKT	Tweed R
AFCK	Strome Ferry	AFHC	Papa Stour I	AFKU	Holy I d
AFCM	Isle of Skye	AFHD	Unst	AFKV	Longstone
AFCN	Portree			AFKX	Farn I
AFCO	The Minch	AFHE	FAROE Is	AFKY	Alnmouth d
		AFHI	Thorshavn	AFKZ	Coquet I
AFCQ	HEBRIDES	AFHJ	Vestmanhavn	AFLC	Warkworth c
AFCR	Stornoway c d e	AFHK	Sudero	AFLD	Newbiggin d
AFCS	Lewis	AFHM	Strömö	AFLE	BLYTH c d
AFCU	Carloway			AFLH	Tyne R
AFCV	Butt of Lewis e f	AFHN	Noss Hd	AFLI	Tynemouth d c
AFCW	Glas I c	AFHO	Wick c d	AFLJ	North Shields b d
AFCY	Flannan Is f	AFHQ	Dornoch	AFLM	South Shields d
AFCZ	North Uist	AFHR	Tarbet Ness c	AFLN	Jarrow
AFDB	Loch Maddy	AFHS	Moray Firth	AFLO	NEWCASTLE
AFDE	Monach Is	AFHU	Cromarty	AFLP	Hepburn
AFDG	South Uist	AFHV	Invergordon	AFLQ	Souter Pt c
AFDH	Ushinish	AFHW	Dingwall	AFLR	SUNDERLAND c d
AFDJ	Loch Boisdale	AFHY	Chanonry Pt	AFLS	Seaham c d
AFDK	Barra	AFHZ	Inverness c	AFLU	HARTLEPOOL c d
AFDL	Castle B	AFIB	Nairn c d	AFLV	Heugh
AFDN	Vatersay I	AFID	Covesea	AFLW	Tees R
AFDO	St Kilda	AFIE	Banff c d	AFLY	Stockton
AFDP	Rockall	AFIG	Kinnard Hd	AFLZ	MIDDLESBROUGH
AFDQ	North Rona	AFIJ	Fraserburgh c d f	AFMB	Whitby c d
AFDR	Sulisker	AFIK	Rattray Hd d	AFMD	Scarborough c d
		AFIL	Peterhead c d	AFME	Filey d
AFDS	SUTHERLAND	AFIN	Buchan Ness	AFMG	Flamborough Hd d e
AFDT	Ru Stoer	AFIO	Newburg d	AFMI	Bridlington a c d
AFDU	C Wrath c	AFIP	Aberdeen c d	AFMJ	York
AFDV	Loch Eriboll	AFIR	Girdle Ness c	AFMK	Withernsea d f
AFDW	Sule Skerries	AFIS	Stonehaven c d	AFMN	Humber R
AFDX	Strathie Pt	AFIT	Montrose c d	AFMO	Spurn Pt d e
AFDY	Scrabster d	AFIV	Arbroath c d	AFMP	Grimsby c d
AFDZ	Thurso d	AFIW	Bell Rock	AFMQ	Spurn Lt. V
AFEB	Dunnet Hd c e	AFIX	Buddon Ness d	AFMR	HULL c
AFEC	Stroma	AFIZ	Broughty Ferry c d	AFMS	Goole
AFEG	Duncansby Hd	AFJB	Tay R	AFMT	Bradford
AFEH	Pentland Firth	AFJC	Dundee b c	AFMV	Sheffield
AFEI	Pentland Skerries	AFJD	Perth	AFMX	Leeds
		AFJE	St Andrews c d	AFMY	Gainsborough
AFEK	ORKNEY Is	AFJG	Fife Ness	AFNB	The Wash
AFEL	South Ronaldsay	AFJH	Firth of Forth	AFNC	Outer Dowsing Lt V
AFEM	Stronsay	AFJI	Methil	AFND	Inner Dowsing Lt V
AFEO	Pomona	AFJK	May I	AFNE	Lynn Well Lt V
AFEP	Helliar Holm	AFJL	Fife	AFNG	Dudgeon Lt V
AFEQ	Kirkwall c	AFJM	Inch Keith	AFNH	Wainfleet
AFES	Hoy.	AFJO	Kirkcaldy	AFNI	Boston c
AFET	Stromness c d	AFJP	Burntisland c d	AFNK	King's Lynn c
AFEU	Long Hope d	AFJQ	Forth Bridge	AFNL	Wisbeach

a Signifies an International Signal Station, b a Time Signal Station, c a Weather, Tide, or Ice Signal Station, d a Life-Saving Station, e a Lloyd's Signal Station, f a Wireless Telegraph Station

IRELAND		NORTH COAST FRANCE	
AFNM	Hunstanton d	AFRI	Eeragh I
AFNP	Wells.d	AFRK	Galway c
AFNQ	Cromer d	AFRL	Arran Is
AFNR	Haisborough Lt V.	AFRM	Slyne Hd
AFNS	Bacton d	AFRO	Killary B
AFNT	Would Lt V	AFRP	Clare I
AFNU	Winterton d	AFRQ	Inishgort
AFNV	Leman and Ower Lt V	AFRT	Clew B
AFNX	Smiths Knoll Lt V	AFRU	Westport
AFNY	Yarmouth c d	AFRV	Achill Hd
AFNZ	Cockle Lt V	AFRX	Blacksod B
AFOC	St Nicholas Lt V	AFRY	Mayo
AFOD	Corton Lt V .	AFRZ	Eagle I
AFOE	Newarp Lt V	AFSC	Broadhaven. : .
AFOH	Cross Sand Lt V	AFSD	Killala
AFOI	Lowestoft c d	AFSE	Ballina
AFOJ	Pakefield	AFSH	Sligo
AFOL	Southwold d	AFSI	Ballyshannon
AFOM	Aldborough d e	AFSJ	Donegal
AFON	Orford Ness c d	AFSL	Killybegs c
AFOQ	Shipwash Lt V	AFSM	Rathlin O'Birne I
AFOR	Hollesley B	AFSN	Bloody Foreland
AFOS	Harwich d	AFSP	Arranmore (Aran I) d
AFOU	Orwell R	AFSQ	Tory 1 c e
AFOV	Ipswich	AFSR	Lough Swilly
AFOW	Manningtree	AFSU	Fanad Pt c
AFOY	Dovercourt	AFSV	Buncrana
AFOZ	Outer Gabbard Lt V	AFSW	Rathmullan c
AFPB	Felixstowe	AFSY	Malin Hd c e f
AFPD	Cork Lt V	AFSZ	Inishtrahull e f
AFPE	Long Sand Lt V	AFTB	Lough Foyle c d
AFPG	Sunk Lt V	AFTD	Inishowen Hd
AFPH	Gunfleet c	AFTE	Londonderry
AFPI	Wivenhoe.	AFTH	Moville
AFPJ	Colchester	AFTI	Coleraine
AFPK	Kentish Knock Lt V	AFTI	Rathlin I
AFPL	Swin Middle Lt V	AFTL	Altacarry Hd
AFPM	Maldon	AFTN	Ballycastle c
AFPN	Galloper Lt V	AFTO	Tor Pt e
		AFTP	Maidens
AFPO	IRELAND	AFTQ	Larne
AFPR	O Clear	AFTR	BELFAST
AFPS	Fastnet c e f	AFTS	Black Hd.
AFPT	Crookhaven d f	AFTU	Carrickfergus d
AFPV	Brow Hd c e f	AFTV	Bangor
AFPW	Bantry B	AFTW	Mew I
AFPX	Mizen Hd	AFTY	Donaghadee c
AFPZ	Berehaven b	AFTZ	Skulmartin Lt V d
AFQB	Ardnakinna Pt	AFUB	South Rock Lt V d
AFQC	Roancarrig I	AFUD	Strangford
AFQE	Glengariff Hr	AFUE	Ardglass Hr
AFQG	Bull Rock	AFUG	St John's Pt
AFQH	Kenmare	AFUI	Dundrum B d
AFQJ	Ballinskellig d	AFUJ	Carlingford d
AFQK	Skelligs	AFUK	Greenore. d
AFQL	Valentia d	AFUM	Newry
AFQN	Dingle B d	AFUO	Dundalk
AFQO	Castlemaine	AFUQ	Boyne R d
AFQP	Tearaght I.	AFUR	Drogheda
AFQS	Tralee c	AFUS	Balbriggan d
AFQT	Shannon R	AFUV	Rockabill
AFQU	Loop Hd. c	AFUX	Howth Bailey c d
AFQW	Kilcradan Hd	AFUZ	Kish Lt V
AFQX	Scattery I	AFVB	DUBLIN b c d
AFQY	Foynes Hr.	AFVD	Poolbeg
AFRB	Tarbert	AFVE	Liffey R
AFRC	Clare.	AFVG	Kingstown c d
AFRD	LIMERICK c	AFVI	Wicklow d
AFRG	Liscanor	AFVJ	Codling Bank Lt V
AFRH	Inisheer	AFVK	Arklow d

	NORTH COAST FRANCE continued
AFVM	Blackwater Bank Lt V d
AFVN	Lucifer Lt V d
AFVO	Wexford d
AFVP	North Arklow Bk Lt V
AFVQ	Rosslare d f
AFVR	South Arklow Bk Lt V
AFVS	Tuskar Rk. c
AFVU	Carnsore Pt d
AFVW	Barrels Rock Lt V
AFVX	Coningbeg Lt V
AFVZ	Saltees
AFWB	Bannow B
AFWC	New Ross c
AFWE	Waterford
AFWG	Hook Pt
AFWH	Duncannon
AFWJ	Dunmore c d
AFWK	Dungarvan c
AFWL	Mine Hd c
AFWN	Ardmore. d
AFWO	Youghal c d
AFWP	Ballycottin I
AFWR	CORK b c
AFWS	Queenstown b c d
AFWT	Haulbowline I
AFWV	Roche Pt a e f
AFWX	Daunt Rock Lt V
AFWY	Kinsale c
AFXB	Old Head of Kinsale c d e
AFXC	Clonakilty
AFXD	Galley Hd c
AFXG	Glandore d
AFXH	Baltimore d
AFXJ	FRANCE
AFXL	DUNKERQUE a b c d
AFXM	Lisle
AFXN	Bergues
AFXP	Zuydcoote a c
AFXQ	Ruytingen Lt V d
AFXR	Snouw Lt V d
AFXT	Dyck Lt V d
AFXU	Gravelines a c d
AFXV	Walde d
AFXY	Calais a c d
AFXZ	C Blanc Nez
AFYB	C Gris Nez a c d e
AFYG	BOULOGNE c d
AFYI	C Alprech a
AFYJ	Etaples
AFYK	Pt du Touquet a
AFYM	Berck d
AFYN	P d'Abbeville
AFYO	Somme R
AFYP	St Valery d
AFYR	Cayeux a d
AFYS	Le Tréport a d
AFYU	Dieppe a d f
AFYV	Pt d'Ailly a
AFYW	St Valery-en-Caux a d
AFYZ	Fécamp a d
AFZB	Fagnet c
AFZC	Etretat a d
AFZE	C d'Antifer a
AFZG	C la Hève a
AFZH	HAVRE (LE HAVRE) c d
AFZJ	Harfleur
AFZK	Seine R
AFZM	Fatouville

a Signifies an International Signal Station, b a Time Signal Station, c a Weather, Tide, or Ice Signal Station, d a Life-Saving Station, e a Lloyd's Signal Station, f a Wireless Telegraph Station

NORTH AND WEST COASTS FRANCE

AFZN	Berville.	AGEO	Granville d	AGJO	USHANT, or ILE D'OUESSANT d f
AFZO	Pt de la Roque	AGEP	Roc Pt a	AGJQ	Stiff Pt a d
AFZQ	Quilleboeuf	AGER	Mt St Michel	AGJR	Creac'h Pt a d e
AFZR	Caudebec	AGES	Avranches		
AFZS	Guerbaville	AGET	Bretagne	AGJS	BISCAY, BAY OF, or GULF OF GASCOGNE
AFZU	Rouen	AGEV	Cancale		
AFZW	PARIS	AGEW	Pierre de Herpin d	AGJU	Corsen Pt a
AFZY	Versailles	AGEX	Grouin Pt a	AGJV	P Conquet d
AGBC	The Elysée	AGEZ	Bénard Pt a	AGJW	St Mathieu Pt a
AGBE	Honfleur c d	AGFB	Roches Bonnes	AGJY	Les Pierres Noires
AGBF	Trouville d	AGFC	ST MALO d	AGJZ	Bertheaume
AGBH	Pt Beuzeval a	AGFE	St Servan	AGKB	Créac'hmeur Pt. a
AGBJ	Dives	AGFH	Grand Jardin Islet d	AGKD	Minou Pt
AGBK	Oyestreham a c d	AGFI	La Rance R	AGKE	Portzic Pt a
AGBL	Orne R	AGFK	Dinan	AGKH	BREST b d
AGBN	Caen	AGFL	Dinard d	AGKJ	Lanninon
AGBO	St Aubin a	AGFM	St Briac	AGKL	Penfeld R
AGBP	Courseulles c d	AGFO	Décollé Pt a	AGKM	Rade de Brest
AGBR	Pt de Ver	AGFP	St Cast Pt a d	AGKO	Kerhuon
AGBS	Bernieres a	AGFQ	C Fréhel a	AGKP	Landerneau
AGBT	Pt en Bessin a d	AGFS	Erqui a d	AGKQ	Chateaulin
AGBV	Percée Pt a	AGFT	Le Légué	AGKS	Roscanvel
AGBW	Isigny.	AGFU	St Brieuc	AGKT	Capucins Pt
AGBX	Carentan.	AGFV	Grand Léjon	AGKU	Camaret a d
AGBZ	St Marcouf	AGFW	Roselier Pt a d	AGKW	Toulinguet Pt
AGCB	La Hougue a	AGFX	Portrieux d	AGKX	Pt des Pois a
AGCD	St Vaast d	AGFY	St Quay a	AGKY	Chèvre C a
AGCF	Réville	AGHB	Plouzec Pt a	AGLB	Douarnenez. d
AGCH	Barfleur a d	AGHC	Paimpol	AGLC	Millier Pt
AGCI	C Barfleur	AGHD	Ost-Pic Rock	AGLD	Ar-Men Rock d
AGCK	Gatteville d	AGHF	Trieux R	AGLF	Ile de Sein d
AGCL	C Levi a d	AGHI	Croix Rocks	AGLH	Tévennec d
AGCM	CHERBOURG b d	AGHJ	Bréhat I a	AGLI	La Vieille d
AGCO	Pelée I a	AGHL	Roches d'Ouvres d	AGLK	Bec du Raz de Sein a
AGCP	Pt de Querqueville a d	AGHM	Héaux de Bréhat d	AGLM	Pt de l'Ervily a
AGCQ	C de la Hague a f	AGHN	Créach-ar-Maout a	AGLN	'Audierne d
AGCS	Jardeheu Pt a	AGHP	Tréguier	AGLP	Poulgoazec
AGCT	Nez de Jobourg a	AGHQ	P Blanc a d	AGLQ	Penmarc'h a d
		AGHR	Anse de Perros	AGLR	Gulfinec d
AGCU	CHANNEL Is	AGHT	Perros-Guirrec d	AGLT	L'Escoril a d
AGCW	ALDERNEY	AGHU	Ploumanac'h a d	AGLU	Pont l'Abbé.
AGCX	St Anne	AGHV	Les Sept-Iles	AGLV	Combrit
AGCY	Casquets	AGHX	Triagoz d	AGLX	Bénodet d
AGDB	GUERNSEY	AGHY	Bégleguer Pt	AGLY	Quimper
AGDC	St Peter Port c d	AGHZ	Lannion	AGLZ	Beg Meil Pt a
AGDE	Hanois	AGIC	Bihit Pt a	AGMC	La Forest
AGDH	St. Sampson's	AGID	Primel Pt a d	AGMD	Concarneau d
AGDI	Les Ecréhos	AGIE	Ile Noire d	AGME	Médée Rock
AGDJ	Herm	AGIH	Tour de la Lande	AGMH	Ile aux Moutons d
AGDL	Serk	AGIJ	Morlaix	AGMI	Penfret I. a d
AGDM	Creux Hr	AGIK	Penpoull	AGMJ	Trévignon Pt d
AGDN	JERSEY	AGIM	St Pol de Léon	AGML	Beg Morg Pt a
AGDP	St Helier a d	AGIN	Bloscon Pt. a	AGMN	Douélan d
AGDQ	St Aubin	AGIO	Roscoff d	AGMO	Anse du Pouldu a
AGDR	La Corbière.	AGIQ	Ile de Bas a d	AGMQ	Quimperlé R
AGDT	Gorey a	AGIR	St Jacques	AGMR	Groix I d
AGDU	Rozel	AGIS	P Nevez	AGMS	Pen Men Pt
AGDV	St Catherine B	AGIU	Pouléazec Pt	AGMU	Croix a
AGDX	Minquiers	AGIV	Kernic	AGMV	Bec Melen Pt a
		AGIX	Goulven	AGMY	Talut Pt (Lorient)
AGDY	P DE DIELETTE d	AGIY	Plounéour	AGMZ	P Louis
AGDZ	C de Flamanville a	AGIZ	Pontusval d	AGNB	Grave Pt a
AGEB	C de Carteret a d	AGJC	P Tresseny.	AGND	Lorient b
AGED	P Bail	AGJD	Kerisoc Pt a	AGNE	Kéroman
AGEF	St Germain a	AGJE	Vierge I	AGNF	Blavet
AGEH	Le Sénéquet d	AGJH	Abervrac'h R a d	AGNI	La Peyrière
AGEJ	Pt d'Agon a	AGJI	Lanvaon Height	AGNJ	La Teignouse d
AGEK	Régnéville d	AGJK	L'Aberbenoit	AGNK	Belle-Ile
AGEL	Tourville (Tourneville)	AGJM	Landunevés Pt a	AGNM	Loc Maria d
AGEN	Chausey Is a d	AGJN	Le Four d		

a Signifies an International Signal Station, b a Time Signal Station, c a Weather, Tide, or Ice Signal Station, d a Life-Saving Station, e a Lloyd's Signal Station, f a Wireless Telegraph Station

WEST COAST FRANCE		SPAIN		PORTUGAL	
AGNO	Arzic Pt	AGRM	Maumusson Ch.	AGVQ	St Martin de la Arena a
AGNP	Taillefer Pt a	AGRN	Seudre R	AGVR	Suances
AGNQ	Hastellic Pt a	AGRP	Gironde R	AGVT	San Vicente de la Barquera
AGNR	Goulfar B (Talut Pt) a	AGRQ	Coubre Pt a d f		
AGNS	Kerdonis Pt	AGRS	Grand Banc Lt V d	AGVU	Tina Mayor
AGNT	Le Palais d	AGRU	Corduan d	AGVW	Llanes. d
AGNV	Sauzon	AGRV	Palmyre Pt	AGVY	C Prieto
AGNW	Quiberon (Loc-Maria) a d	AGRW	La Falaise	AGVZ	Rivadesella d
AGNX	St Pierre	AGRY	Terre-Negre.	AGWB	Villaviciosa.
AGNZ	Kerpenhir Pt a	AGRZ	Chay	AGWD	Gijon a d
AGOB	Auray d	AGSB	St Pierre de Royan	AGWE	Candas
AGOC	P Navallo d	AGSD	Royan d	AGWF	Luanco
AGOE	Vannes	AGSE	St Nicolas d	AGWI	Asturias
AGOF	Morbihan	AGSF	Grave Pt a d	AGWJ	C Peñas
AGOH	Grand Mont Pt a	AGSI	Talais Bank Lt. V. d	AGWK	Avilés. a
AGOJ	Penerf	AGSJ	Vallière Pt	AGWM	Pravia R
AGOK	Penlan Pt	AGSK	P St Georges	AGWN	Cudillero
AGOL	Haedik I a	AGSM	Richard	AGWO	C Busto
AGON	La Vilaine R	AGSN	By Lt. V	AGWQ	Luarca
AGOP	La Roche Bernard	AGSQ	Calonge	AGWR	Navia
AGOQ	Redon	AGSR	Mousset	AGWS	Orrio de Tapia I
AGOS	Rennes.	AGST	La Gaet	AGWU	Castropol
AGOT	Le Four. d	AGSV	Pauillac d	AGWV	Rivadeo. d
AGOU	Grands Cardinaux d	AGSW	Patiras I	AGWX	San Ciprian
AGOX	Pt du Castelli a	AGSX	Blaye	AGWZ	Vivero
AGOY	Le Croisic c d	AGSZ	Garonne R	AGXB	Barquero
AGPB	La Banche d	AGTB	Dordogne R	AGXC	C Vares a
AGPC	Aiguillon	AGTD	Bordeaux d	AGXE	Estaca Pt
AGPD	Ville-es-Martin	AGTF	Hourtin	AGXF	C Ortegal
AGPF	La Loire R	AGTH	C Ferret d	AGXH	C Prior.
AGPH	St Nazaire c	AGTI	Arcachon a d	AGXJ	Mt. Ventoso a
AGPI	Chemoulin Pt a d	AGTK	Leyre R	AGXK	C Prioriño
AGPK	Paimbœuf	AGTL	La Teste de Buch	AGXL	Ferrol f
AGPL	Nantes	AGTM	Contis d	AGXN	Jubia B
AGPM	Angers	AGTO	C Breton a d	AGXO	Serantes R
AGPO	St. Gildas Pt a	AGTP	Adour R c	AGXP	Mugardos
AGPQ	Pornic. c d	AGTQ	Bayonne a d	AGXR	Eume R
AGPR	Pilier I a d	AGTS	Biarritz a d	AGXS	Ares
AGPT	Noirmoutier I d	AGTU	St Jean de Luz d	AGXT	Redes
AGPU	Pt des Dames	AGTV	St Barbe Pt	AGXV	Betanzos
AGPV	P Breton c d	AGTX	Socoa a d	AGXW	Sada
AGPW	St Sauveur a	AGTY	Pyrénées	AGXY	Coruña d f
AGPX	Ile d'Yeu a			AGYB	Sisargas I
AGPY	Croix de Vic	AGUB	Spain	AGYC	Corme
AGPZ	St. Gilles sur Vie c d	AGUD	Madrid	AGYD	C Villano.
AGQC	Sables d Olonne a c d	AGUF	Bidassoa	AGYF	Camariñas d
AGQD	Luçon	AGUH	C Higuera	AGYH	C Toriñana
AGQE	Ile de Ré	AGUI	Fuenterrabia	AGYI	Nemiña
AGQH	Baleines Pt a d	AGUK	P Pasages	AGYK	C Finisterre a e f
AGQI	Haut Banc du Nord a	AGUL	San Sebastian a d	AGYL	Corcubion d
AGQJ	Chauveau Pt.	AGUM	Guetaria	AGYM	Sadineiro (Lobera)
AGQL	Roche Bonne Lt V	AGUO	Sumaya, or, Zumaya	AGYO	Ezaro
AGQM	La Pallice P d	AGUP	Motrico	AGYP	Queijal Pt
AGQN	La Rochelle. d	AGUQ	Lequeitio	AGYQ	Santiago de Compostella
AGQP	Marans	AGUS	Mondaca R		
AGQR	Niort	AGUT	Bermeo	AGYS	Muros d
AGQS	Antioche	AGUV	C Machichaco	AGYT	C Corrobedo
AGQU	Basques Roads, or, P des Barques	AGUX	Galea a	AGYU	Caramiñal
		AGUY	Portugalete d	AGYW	Arosa
AGQV	Ile d'Aix a	AGUZ	Bilbao a c	AGYX	Carril.
AGQW	Fouras b	AGVC	Castro Urdiales d	AGYZ	Padron
AGQY	Rochefort b d	AGVD	Viscaya	AGZC	Villagarcia
AGQZ	La Charente R	AGVE	Laredo d	AGZD	Salvora I
AGRB	Saintes	AGVH	Santoña	AGZE	Rua I
AGRD	Marennes	AGVI	Caballo Pt	AGZH	Ons I
AGRE	Ile d'Oleron. d	AGVJ	Pescador Pt	AGZI	Pontevedra
AGRH	St Pierre	AGVL	C Quexo	AGZJ	Marin.
AGRI	Le Château	AGVM	Santander a d	AGZL	Aldan B
AGRJ	Chassiron a d	AGVN	Mouro I	AGZM	Cies I
AGRL	Boursefrank	AGVP	C Mayor a	AGZN	Bayona

a Signifies an International Signal Station, b a Time Signal Station, c a Weather, Tide, or Ice Signal Station, d a Life-Saving Station, e a Lloyd's Signal Station, f a Wireless Telegraph Station

	PORTUGAL		SOUTH AND EAST COASTS SPAIN.		
AGZP	Vigo c	AHEN	Rota	AHJI	Estacio
AGZQ	C Silleiro	AHEO	Santa Maria	AHJK	Torrevieja d
AGZR	La Guardia	AHEP	CADIZ d	AHJM	Segura R
		AHER	Xeres	AHJN	Tabarca, or, Plana I
AGZT	PORTUGAL	AHES	Haste Afuera Sh	AHJO	Santa Pola
AGZU	Caminha d	AHET	P Real	AHJQ	Alicante
AGZV	Miño R	AHEV	Carraca	AHJR	C Huertas
AGZX	Valenca	AHEW	Trocadero	AHJS	Villajoyosa
AGZY	Lima R	AHEX	Puntales Castle	AHJU	Benidorme
AHBC	Vianna a c d	AHEZ	San Fernando b	AHJV	Albir Pt
AHBE	Villa de Conde	AHFB	Las Puercas	AHJW	Altea
AHBF	P Leixões a	AHFC	San Sebastian d	AHJY	Calpe
AHBG	Douro R	AHFE	Sancti Petri	AHJZ	Morayra
AHBJ	OPORTO a c	AHFG	Conil	AHKB	Javea d
AHBK	Aveiro d	AHFI	C Trafalgar	AHKD	C San Antonio a f
AHBL	C Mondego	AHFK	Aceitera Sh	AHKE	Denia d
AHBN	Figueira	AHFL	Tarifa a d e	AHKF	Almadabra
AHBO	San Martinho	AHFM	Palmones R	AHKI	Gandia
AHBP	Burling I	AHFO	Carnero	AHKJ	Jucar R.
AHBR	C Carvoeiro a c	AHFP	Mayorga	AHKL	Cullera
AHBS	Peniche d e	AHFQ	Algeciras d	AHKN	Albufera
AHBT	Ericeira	AHFS	Verde I	AHKO	VALENCIA d
AHBV	C Roca	AHFT	San Roque	AHKP	Grao de Valencio.
AHBW	Oitavos a c e	AHFU	GIBRALTAR a b e f	AHKR	Cabañal.
AHBX	Cascaes B a c d f	AHFW	Rosia B	AHKS	Burriana
AHBZ	Tagus R	AHFX	Europa Pt	AHKT	Castellon de la Plana
AHCB	Belem.			AHKV	C Oropesa
AHCE	LISBON b c			AHKW	Columbretes Is.
AHCG	Guia	AHFZ	MEDITERRANEAN	AHKZ	Peñiscola
AHCI	San Julian a c			AHLB	Benicarló
AHCJ	Caxias (Observatory)	AHGB	Blackstmp B	AHLC	Vinaroz d
AHCL	Bugio	AHGC	Tuiiara d	AHLE	P Alfaques
AHCM	Ajuda	AHGE	Sabinilla	AHLF	St Carlos
AHCN	Cacilhas Pt	AHGF	Doncella Pt	AHLG	Baña Pt
AHCP	C Espichel a c	AHGI	Estepona	AHLJ	Ebro R
AHCQ	Setubal (St Ubes) d	AHGK	Marbella	AHLK	Amposta.
AHCR	C Sines	AHGL	Calaburras Pt	AHLM	C Tortosa
AHCT	Odemira	AHGM	Fuengirola	AHLO	Escudel, or, Fango Pt
AHCU	C Sardao	AHGO	Malaga d	AHLP	P Fangar.
AHCV	C St Vincent	AHGP	Velez Malaga	AHLQ	Cambrils d
AHCX	Sagres a c e	AHGQ	Tortox	AHLS	Salou.
AHCY	Lagos d	AHGS	Nerja	AHLT	Tarragona d
AHCZ	Portimao d	AHGT	Almuñecar	AHLU	C Gros
AHDC	C Carvoeiro	AHGU	Salobreña	AHLW	Palomera
AHDE	Balieira Pt	AHGW	Motril	AHLX	Villanueva. d
AHDF	Faro d	AHGX	C Sacratif	AHLY	Sitges
AHDI	C St Mary	AHGY	GRANADA	AHMB	Llobregat a
AHDJ	Tavira	AHIB	Alboran I	AHMC	BARCELONA d
AHDK	Villa Reale	AHIC	Ferro Castel	AHMD	Besós R
AHDM	Guadiana R	AHID	Adra	AHMF	Badalona
AHDN	Pomarão	AHIF	Sabinal Pt	AHMG	Mongat Pt
		AHIG	Roquetas.	AHMI	Masnou
AHDO	AYAMONTE	AHIJ	Almeria f	AHMK	Mataró d
AHDQ	Higuerita Bar	AHIL	C DE GATA	AHML	Arenys de Mar
AHDR	Rompido Pt	AHIM	Genoves	AHMN	Calella
AHDS	Cartaya	AHIN	San Pedro	AHMP	Tordera
AHDU	Huelva	AHIP	Mesa de Roldan.	AHMQ	Blanes
AHDV	Odiel R	AHIQ	Carboneras	AHMR	Lloret
AHDW	Tinto R	AHIR	Mojacar	AHMS	C Bagur
AHDY	Palos	AHIS	Garrucha d	AHMT	San Feliu de Guixols d
AHDZ	Picacho	AHIT	Villaricos	AHMU	Palamos d
AHEB	Morro (Cano) Pt	AHIU	Agulas	AHMV	Molino Pt
AHED	Padre Santo Pt	AHIV	Cope	AHMX	C San Sebastian
AHEF	Oro R	AHIX	Mazarron	AHMY	Torruella
AHEG	Guadalquivir R	AHIY	C Tiñoso	AHMZ	Meda I
AHEI	San Lucar de Barra-meda d	AHIZ	CARTAGENA a d	AHNC	La Escala d
		AHJC	Escombrera	AHND	Ampurias
AHEJ	San Lucar	AHJD	Portman B	AHNE	Rosas
AHEK	Seville d	AHJE	C Palos d	AHNG	Cadaqués d
AHEL	Chipiona	AHJG	Las Hormigas	AHNI	C 'Creus

a Signifies an International Signal Station, b a Time Signal Station, c a Weather, Tide, or Ice Signal Station d a Life-Saving Station, e a Lloyd's Signal Station, f a Wireless Telegraph Station

	SOUTH COAST FRANCE				GULF OF GENOA	
AHNJ	St Cruz de la Selva (P Selva) d		AHQY	Saintes Maries d f	AHVE	Antibes d
AHNL	Llansá		AHQZ	La Gacholle	AHVF	Nice d
AHNM	Portbou		AHRC	Faraman (Camargue) a d	AHVG	Villefranche d
			AHRD	Rhône R d	AHVJ	C Ferrat a
AHNO	BALEARIC Is		AHRE	Arles	AHVK	St Hospice Pt d
AHNQ	Formentera I		AHRG	Tarascon	AHVL	Monaco
AHNR	Codolar Pt		AHRJ	Avignon	AHVN	C Martin a
AHNS	IVIZA		AHRL	Lyon	AHVO	Mentone (Menton) d
AHNU	San Antonio P		AHRM	Foz		
AHNV	Ahorcados I		AHRN	St Louis Canal	AHVQ	SWITZERLAND
AHNW	Puercos I		AHRP	Bouc d	AHVR	Berne
AHNY	Botafoch I		AHRQ	Martigues	AHVS	Geneva
AHNZ	Grosa Pt		AHRS	C Couronne a		
AHOB	Conejera I		AHRU	C Janet	AHVW	ITALY
AHOD	MAJORCA		AHRW	MARSEILLE d	AHVY	Ventimiglia d
AHOE	Palma d		AHRY	National Basin	AHVZ	San Remo d
AHOF	C Blanco		AHRZ	Maritime Basin	AHWB	Maurizio d
AHOI	P Pi		AHSC	Lazaret Basin	AHWD	Oneglia d
AHOJ	C Cala Figuera		AHSD	Joliette Basin	AHWE	C Mele a f
AHOK	Dragonera I		AHSE	Old Basin	AHWF	Alassio d
AHOM	P Soller d		AHSG	Château d'If I	AHWI	Albenga
AHON	Grosa, or, Llarga Pt		AHSI	Ratonneau I	AHWJ	Noli a d
AHOP	C Formenton		AHSJ	P du Frioul	AHWK	Spotorno d
AHOR	Pollensa		AHSL	Pomègues I a e	AHWM	Vado d
AHOS	Alcudia d		AHSM	Planier I d	AHWN	Savona d
AHOT	Alcanada, or, Aucanda Islet		AHSN	C Croisette a	AHWO	Varazze
AHOV	C Pera		AHSP	Riou I	AHWQ	Voltri
AHOW	P Colon		AHSQ	Cassis	AHWR	GENOA b d
AHOX	C Salinas		AHSR	Bec de l'Aigle a	AHWS	Piedmont
AHOZ	Cabrera		AHSU	Ciotat d	AHWU	Turin
AHPB	Ansiola Pt		AHSV	Bandol	AHWX	C San Benigno (Ferro, or, Faro) a
AHPC	MINORCA		AHSW	Sanary		
AHPD	C Bajoli		AHSY	Six-Fours a	AHWZ	St Giacomo Pt
AHPE	P Mahon		AHSZ	Brusc d	AHXB	P Fino a
AHPF	San Carlos Pt		AHTB	Embiez I d	AHXC	Santa Margherita d
AHPG	Ayre I		AHTD	C Sicié a	AHXE	Rapallo
AHPJ	C Dartuch		AHTE	C Sepet a	AHXF	Chiavari a
AHPK	Ciudadela		AHTF	TOULON b	AHXG	Sestri Levante
AHPL	C Caballeria, or, Cavalleria		AHTI	Lazaretto Road	AHXJ	P Venere d
			AHTK	La Seyne B	AHXK	Levanto
AHPN	P Fornells		AHTM	Inner Road	AHXL	Mesco Pt
			AHTN	Mourillon	AHXN	Palmaria I a f
			AHTP	Missiessy	AHXO	Tino I
AHPO	BANYULS d		AHTQ	L'Ane Bank	AHXP	Spezia b d
AHPQ	C Béar (Béarn) a		AHTR	Giens a	AHXR	Lerici d
AHPS	P Vendres d		AHTU	Great Ribaud I	AHXS	Avenza
AHPT	Collioure d		AHTV	Hyères	AHXT	Viareggio a
AHPU	Perpignan		AHTW	C Bénat a	AHXV	TUSCANY
AHPW	St Nazaire		AHTY	Fourches Mt	AHXW	Arno R
AHPX	Leucate. a		AHTZ	Porquerolles a d f	AHXY	Pisa
AHPY	La Nouvelle a d		AHUB	P Cros	AHYB	Florence
AHQB	Narbonne		AHUD	Levant Is a	AHYC	LEGHORN d f
AHQC	Béziers		AHUE	Bormes Road	AHYE	Meloria Sh
AHQD	Toulouse		AHUF	Lavandou a	AHYF	Gorgona I a
AHQF	Hérault R		AHUI	Cavalaire d	AHYG	Vada
AHQG	Agde a d		AHUJ	C Camarat a d	AHYJ	Barratti P
			AHUK	St Tropez	AHYK	Piombino Pt a
AHQI	GULF OF LYON		AHUM	St Maxime	AHYL	Vecchio P
			AHUN	Sardimières Pt a	AHYN	Troja I
AHQK	Brescou I		AHUO	Fréjus	AHYO	Castiglione
AHQL	Bessan		AHUQ	St Raphael d	AHYP	Talamone B
AHQM	Marseillan		AHUR	C Drummond a	AHYQ	Campo alle Serre a
AHQO	Mèze		AHUS	Agay Road d	AHYR	Elba I a f
AHQP	Cette a c d		AHUV	Napoule G	AHYS	Ferrajo P
AHQR	Etang de Thau		AHUW	Cannes d	AHYT	P Longone
AHQT	Frontignan		AHUX	St Marguerite Ile a	AHYU	Monte Grosso a
AHQU	Montpellier		AHUZ	Jouan G d	AHYV	Focardo F
AHQV	Aigues-Mortes d		AHVB	Ilette Pt	AHYW	Palmajola I
AHQX	Espiguette Pt a		AHVC	Garoupe a	AHYX	Pianosa I

a Signifies an International Signal Station, b a Time Signal Station, c a Weather, Tide, or Ice Signal Station, d a Life-Saving Station
e a Lloyd's Signal Station, f a Wireless Telegraph Station

CORSICA		SARDINIA	ITALY	SICILY	
AHZB	Africa Rock	AIDW	Quartu	AIHQ	P Torre dell' Annunziata
AHZC	Monte Christo	AIDX	C St Elia *a*		
AHZD	Giglio *a*	AIDY	CAGLIARI	AIHS	Castellamare
AHZF	Fenaio Pt	AIEB	C Spartivento *a*	AIHT	Sorrento
AHZG	Caporosso Pt	AIEC	P Malfatano	AIHU	Massa Lubrense
AHZI	Giannutri	AIED	Rossa B	AIHW	Campanella Pt
AHZK	Capel rosso Pt	AIEG	C Teulada	AIHX	Capri *a*
AHZL	Capraia *a*	AIEH	Palmas B	AIHY	Carena Pt
AHZM	Castello Mt	AIEJ	San Antioco I	AIJB	Amalfi
AHZO	C Ferraione	AIEL	C Sperone *a f*	AIJC	C Orso
		AIEM	P Scuso	AIJD	SALERNO *d*
		AIEN	St Pietro I	AIJF	Agropoli
AHZP	CORSICA (CORSE)	AIEP	C Sandalo	AIJG	Licosa Pt
AHZQ	C Corso (Corse) *a e*	AIEQ	Carlo Forte	AIJH	C Palinuro *a*
AHZS	Giraglia I *d*	AIER	Iglesias	AIJL	Infreschi P
AHZT	C Sagro *a*	AIET	Oristano	AIJM	Scario
AHZU	Bastia *d*	AIEU	C San Marco	AIJN	Policastro
AHZW	P St Nicolas	AIEV	Bosa	AIJP	Sapri
AHZX	Alistro *a*	AIEX	Rossa I	AIJQ	C Bonifati
AHZY	Favone	AIEY	Alghero	AIJR	C Suvero
AIBD	Pinarello B	AIEZ	P. Conte	AIJT	Pizzo.
AIBE	Chiappa Pt *a d*	AIFC	C Caccia *a*	AIJU	Santa Venere P
AIBF	P Vecchio *d*	AIFD	C Argentiera	AIJV	Monteleone
AIBH	Cavallo I	AIFE	Asinara I *a f*	AIJX	C Vaticano
AIBJ	Lavezzi I *d*	AIFH	C Caprara *a*	AIJY	Gioja *d*
AIBK	C Pertusato *a e*	AIFJ	Reale Road	AIJZ	Bagnara
AIBM	Bonifacio *d*	AIFK	Fornelli Road	AIKC	Scilla *d*
AIBN	C Feno *d*	AIFM	P. Torres	AIKD	Pezzo
AIBO	Sénétose Pt	AIFN	Sassari	AIKE	Reggio *f*
AIBQ	Propriano *d*	AIFO	Castel Sardo	AIKG	C dell' Armi *a e*
AIBR	Ajaccio *d f*			AIKH	C Spartivento *u*
AIBS	Sanguinaire I *a*				
AIBU	Galéria *d*	AIFQ	CALA GRANDE I		
AIBV	C Cavalo *a*	AIFR	Argentaro Monte *a*	AIKJ	SICILY
AIBW	Revellata Pt *d*	AIFS	Orbetello	AIKM	C Pelero, *or*, Faro
AIBY	Calvi *d*	AIFU	Ercole Pt	AIKN	Mazzone Pt
AIBZ	Mt Grosso	AIFV	Civita Vecchia *d*	AIKO	Str of Messina
AICB	Algajola	AIFW	C Linaro	AIKQ	Spurla *a e f*
AICE	Rousse I *a d*	AIFY	Chiaruccia Tower *a*	AIKR	Canzirri
AICF	Mortella Pt *a d*	AIFZ	Tiber R *c*	AIKS	MESSINA *b*
AICG	St Florent *d*	AIGB	Fiumicino	AIKU	Scaletta
		AIGD	ROME *f*	AIKV	Taormina *u*
		AIGE	Fiumara Grande	AIKW	Riposto
AICJ	SARDINIA	AIGF	Ostia	AIKY	Aci Reale
AICK	C Testa *a e*	AIGJ	Anzio *d*	AIKZ	Mt Ætna
AICL	Santa Teresa Gallura	AIGK	Mt Circeo Circello *a*	AILB	Catania *b*
AICN	Falcone Pt	AIGL	San Felice	AILD	C Santa Croce
AICO	Razzoli I	AIGM	Terracina	AILE	P Augusta
AICP	Vacche I	AIGN	Gaeta *a*	AILF	Avolos I (Torre Avolos)
AICR	Liscia P	AIGO	Ponza *a f*	AILH	Magnisi
AICS	Arsachena G	AIGP	Guardia Pt	AILJ	Syracuse
AICT	Maddalena I *a*	AIGQ	Torre Orlando *a*	AILK	Belvedere Tower (Syracuse) *a*
AICV	Guardia Vecchia Hill *a*	AIGR	Zannone		
AICW	Caprera I *f*	AIGS	Ventotene I *a*	AILN	C Murro di Porco
AICX	St Stefano	AIGT	C Miseno	AILO	Avola
AICZ	C Ferro *a*	AIGV	Procida *a*	AILP	Noto
AIDB	Cognena G	AIGW	Mt Solaro	AILR	Cozzo Spadaro *a f*
AIDC	C Figari *a*	AIGX	Propetto Pt	AILS	C Passaro
AIDF	Aranci B	AIGZ	Ischia	AILT	Correnti I
AIDG	Terranova	AIHB	Forio.	AILV	Pozzalo
AIDH	Della Bocca I	AIHC	Imperatore Pt *a*	AILW	C Scalambri.
AIDK	Bianca I	AIHE	P Bagno	AILX	Terranova
AIDL	Tavolara I	AIHF	Baia	AILZ	Licata *u*
AIDM	P Brandinchi	AIHG	Pozzuoli	AIMB	Palma
AIDO	Siniscola	AIHK	Nisita	AIMC	Girgenti
AIDP	C Comino	AIHL	NAPLES *b*	AIME	P Empedocle
AIDQ	Tortoli	AIHM	Portici	AIMF	C Rossello
AIDS	C Bellavista *a*	AIHN	Granatello	AIMG	Sciacca
AIDT	C Carbonara *a*	AIHO	Mt Vesuvius	AIMJ	C St Marco
AIDU	Cavoli I	AIHP	Torre del Greco	AIMK	C Granitola

a Signifies an International Signal Station, *b* a Time Signal Station, *c* a Weather, Tide, or Ice Signal Station, *d* a Life-Saving Station, *e* a Lloyd's Signal Station, *f* a Wireless Telegraph Station

MALTA		ADRIATIC.			
AIML	Mazzara	AIQH	Palascia Tower	AIUF	Vicenza
AIMO	MARSALA	AIQJ	Missipezza Rk	AIUG	Piave Vecchia
AIMP	Tramontana Pt	AIQK	Brindisi a	AIUJ	Cortellazzo
AIMQ	Favignana a			AIUK	Cavetta Canal
AIMS	Sottile Pt	AIQL	ADRIATIC	AIUL	Piave R
AIMT	Maritimo			AIUN	Falconera
AIMU	Libeccio Pt	AIQN	C Gallo	AIUO	Baseleghe
AIMW	Levanzo	AIQO	Monopoli	AIUP	Tagliamento R
AIMX	Grosso Pt	AIQP	Mola	AIUR	Lignano
AIMY	Formica	AIQS	San Cataldo Pt	AIUS	Marano
AINB	Trapani a f	AIQT	Bari a f	AIUT	Alps
AINC	Palumbo I	AIQU	Giovinazzo	AIUW	Lombardy
AIND	Columbaia I	AIQW	Molfetta	AIUX	Milan
AINF	Ronciglio Pt	AIQX	Bisceglie	AIUY	P Buso
AING	C St Vito	AIQY	Trani		
AINH	Castellamare	AIRB	Barletta	AIVB	AUSTRIA
AINL	C Gallo a	AIRC	Manfredonia	AIVC	Vienna
AINM	Mt Pellegrino a	AIRD	Mattinata	AIVD	Grado
AINP	Palermo	AIRF	Rossa Pt	AIVF	P Primero
AINQ	C Zaffrano	AIRG	Gargano Hd	AIVG	Isonzo R
AINR	Termini Imerese	AIRH	Vieste a f	AIVH	Panzano B
AINT	Cefalu	AIRK	Peschici	AIVK	Gorizia or. Goritz
AINU	C Orlando	AIRL	TREMITI Is a	AIVL	Duino
AINV	Patti	AIRM	San Nicola I	AIVM	Grignano B
AINX	Oliveri	AIRO	Pianosa I	AIVO	TRIESTE a b d
AINY	Milazzo	AIRP	Caprara I	AIVP	Muggia
AINZ	Ustica I a	AIRQ	Pelagosa I	AIVQ	Servola
AIOC	Uomo-Morto Pt	AIRT	Mileto Pt	AIVS	Zaole
AIOD	Gavazzi Pt	AIRU	Termoli	AIVT	Sottile Pt
		AIRV	Vasto	AIVU	Istria
AIOE	LIPARI I	AIRX	Ortona	AIVX	Capo d'Istria B
AIOG	Vulcano I	AIRY	Francavilla	AIVY	Isola a
AIOH	Saraceno Mt	AIRZ	Pescara	AIVZ	Pirano
AIOJ	Stromboli I a	AISC	Fermo	AIWC	Madonna Pt
AIOL	Salina I a	AISD	Mt Conero	AIWD	Salvore Pt a
AIOM	Alicudi	AISE	Ancona a d f	AIWE	Pegolotta Pt
		AISG	Mt Cappuccini a	AIWG	Umago
AIOQ	MALTA e	AISH	Senigallia d	AIWH	Daila
AIOR	Valetta b f	AISJ	Fano	AIWJ	Quieto
AIOS	Floriana.	AISL	Pesaro	AIWL	Cittanuova.
AIOU	St Elmo	AISM	Cattolica	AIWM	Torre
AIOV	Sliema	AISN	Rimini	AIWN	Cervera
AIOW	Marsa Scirocco	AISP	Cesenatico	AIWP	Parenzo
AIOY	Comino I	AISQ	Cervia	AIWQ	Fontane
AIOZ	Gozo	AISR	Bologna	AIWR	Orsera
AIPB	LAMPEDUSA	AISU	Ravenna	AIWT	Leme Ch
AIPD	Cavallo Bianco Pt	AISV	P Corsini a	AIWU	Rovigno
AIPE	C Grecale	AISW	Primaro	AIWY	San Giovanni di Pelago
AIPF	Linosa	AISY	Comacchio	AIWZ	Brioni Is
AIPH	PANTELLARIA e	AISZ	Po R	AIXB	Pedena Pt
AIPJ	Currina (Spadillo) Pt	AITB	Volano	AIXD	Fasana
		AITD	Goro Pt	AIXE	POLA b d
AIPK	CALABRIA	AITE	Gorino.	AIXF	Olivi Islet
AIPM	Bovalino	AITF	Tolle	AIXH	C Compare
AIPN	Gioiosa Jonica	AITH	Maestra Pt	AIXJ	P Veruda
AIPO	Squillace	AITJ	Contarina	AIXK	Porer Rock a
AIPR	Catanzaro	AITK	Ferrara	AIXM	C Promontore
AIPS	C Rizzuto	AITM	Adige R	AIXN	Medolino G
AIPT	C Colonne a	AITN	Chioggia d	AIXO	Ronze
AIPV	Cotrone	AITO	Malamocco f	AIXQ	Merlera Pt
AIPW	Taranto b	AITQ	Rocchetta Ch a	AIXR	Quarnero G
AIPX	San Paolo I	AITR	Alberoni	AIXS	Bado
AIPZ	C St'Vito. a	AITS	VENICE b	AIXU	Arsa Canal
AIQB	Gallipoli	AITV	San Nicolo Del Lido a	AIXV	Gradaz
AIQC	San Andrea I	AITW	Tre Porti	AIXW	Ubas Pi
		AITX	Murano	AIXZ	Nera Pt
AIQE	C SANTA MARIA DI LEUCA. a f	AITZ	Burano	AIYB	Lungo (Istria)
		AIUB	Giudecca Canal	AIYC	Rabaz
AIQF	Montelungo	AIUC	Mestre	AIYE	Fianona.
AIQG	Otranto a	AIUE	Padua	AIYF	Farasina

a Signifies an International Signal Station, b a Time Signal Station, c a Weather, Tide or Ice Signal Station, d a Life-Saving Station, e a Lloyd's Signal Station, f a Wireless Telegraph Station

ADRIATIC				GREECE	
AIYG	Moscenice	AJDG	Lucietta I	AJHB	P Palermo
AIYJ	Volosca	AJDH	Capocesto	AJHC	Santa Quaranta
AIYK	Castua	AJDI	Rogosnizza	AJHD	Pagania P
		AJDL	Mulo I	AJHF	Gomenizza P.
AIYL	HUNGARY	AJDM	Zirona		
AIYN	Budapest	AJDN	Bossiljina	AJHG	IONIAN Is
AIYO	Prague	AJDP	Trau	AJHK	Fano I
AIYP	CROATIA	AJDQ	Castel nuovo	AJHL	Kastri Pt
AIYR	FIUME b d	AJDR	Spalato d	AJHM	Samothraki I.
AIYS	Maria Teresa Mole	AJDT	Almissa	AJHO	Corfu
AIYT	Buccari B	AJDU	Makarska	AJHP	Tignoso I.
AIYU	P Re	AJDV	Brazza I	AJHQ	Paxo I
AIYV	Cherso	AJDX	Milna	AJHS	Laka P
AIYW	Zaglava Rock	AJDY	St Pietro	AJHT	Madonna I
AIYX	Glavina Pt	AJDZ	Juja Pt.	AJHU	P Gayo
AIZB	Prestenizze Pt	AJEC	Spalmadori Ch	AJHW	Anti-Paxo
AIZC	Galiola Is	AJED	Lesina	AJHX	Sivota I
AIZD	Unie I	AJEF	Citta Vecchia	AJHY	Parga P
AIZF	Sansego	AJEH	Verboska	AJIB	P St Giovanni
AIZG	Ossero	AJEI	Gelsa	AJIC	Phanari P
AIZH	Martinscica or, Martincica	AJEK	LISSA	AJID	Prevesa.
		AJEL	St Georgio P	AJIE	GREECE
AIZK	Lussin Piccolo b c d	AJEM	Comisa	AJIF	Kervasara
AIZL	Kolorat Hr	AJEN	Stoncica Pt a d	AJIG	Arta G
AIZM	Plaunick I	AJEO	Ourzola a	AJIH	Vonitza
AIZO	Mortar I	AJEQ	Valle Grande	AJIL	Demata B
AIZP	Sansego I a f	AJER	Tre Pozzi P	AJIM	P S Nikolo
AIZQ	St Pietro di Nembo	AJES	Badia P	AJIN	Santa Maura
AIZR	Gruizza, or, Gruica I	AJEU	Lagosta a	AJIP	Vliko P
AIZT	Terstenik I	AJEV	P Rosso	AJIQ	Meganisi Ch
AIZU	Veglia	AJEW	C Gomena	AJIR	C Dukato
AIZV	Negritto Pt	AJEY	Narenta Ch c	AJIT	Drepano
AIZX	Malinska	AJEZ	Tolero	AJIU	Zaverda B
AIZY	Verbenico	AJFB	Sabbioncello	AJIV	Vurko B
AJBC	New Besca	AJFD	Meleda I	AJIX	Kalomo.
AJBF	Novi.	AJFE	P Palazzo	AJIY	Cephalonia
AJBG	Morlacca Ch	AJFG	Palma P	AJIZ	Asso
AJBI,	Segna, or, Zengg	AJFI	Stagno	AJKC	Samos
AJBL	Pervicchio I	AJFK	Lagostini I	AJKD	San Theodoro Pt.
AJBN	Arbe P	AJFL	Olipa I	AJKE	Anastasia B
AJBO	Dolin I	AJFN	P Gravosa	AJKG	Argostoli
AJBP	Gabovacatrida	AJFO	Pettini I	AJKH	Lixuri
AJBR	P Jablanaz	AJFP	RAGUSA d	AJKI	Guardiana I
AJBS	Pago I	AJFR	Old Ragusa. (Ragusa Vecchia)	AJKM	Ithaca I.
AJBT	Novalja			AJKN	P Vathi
AJBV	Saska B	AJFS	Pt D'Ostro a	AJKO	Frikes P
AJBW	Skerda I	AJFT	Topla B	AJKQ	Zante.
AJBX	Poklib I	AJFV	Castelnuovo	AJKR	Krionero Pt
AJBZ	Ulbo	AJFW	Teodo B	AJKS	Strovathi (Stamphani Is)
AJCB	Selve	AJFX	Risano		
AJCD	San Antonio Pt	AJFZ	Cattaro	AJKU	Astoko
AJCF	Lutostrak I	AJGB	Traste	AJKV	Pandelemona P
AJCG	Puntadura	AJGC	Budua	AJKW	Petala P
AJCH	Brevilacqua			AJKY	Missolonghi
AJCK	Berguglie B	AJGE	BOSNIA	AJKZ	Sosti I
		AJGF	HERZEGOVINA	AJLB	Patras
AJCL	DALMATIA	AJGH	MONTENEGRO	AJLD	CORINTH
AJCM	Grossa I	AJGK	Antivari f	AJLE	Lepanto
AJCO	P Lungo	AJGL	SERVIA	AJLF	Galaxidi
AJCP	Bianche Pt			AJLH	Salona G
AJCQ	Manzo.	AJGM	ALBANIA.	AJLI	Aspra Spitia, or, Isidoro B
AJCS	Premuda	AJGO	Dulcigno		
AJCT	Amica Pt	AJGP	Bojana R	AJLK	Vostitza
AJCU	Zara d	AJGQ	Skutari	AJLN	C Papas
AJCW	P St Cassano	AJGS	C Rodoni	AJLO	Glarenza
AJCX	Babac I	AJGT	Durazzo	AJLP	Katakolo
AJCY	Pasman	AJGU	Samana Pt (Semeni R)	AJLR	Pyrgos
AJDB	Morter I	AJGW	Saseno I	AJLV	Arcadia G
AJDC	Sebenico d	AJGX	Valona B	AJLW	Kyparissia
AJDE	P Tajer	AJGY	Dukati P	AJLX	Proti I

a Signifies an International Signal Station, b a Time Signal Station c a Weather, Tide, or Ice Signal Station, d a Life-Saving Station, e a Lloyd's Signal Station, f a Wireless Telegraph Station

ARCHIPELAGO			SEA OF MARMORA	
AJLZ	Navarin, or, Neo Kastro	AJPR	C Sitia	AJTN	Pasha I
	tro	AJPS	C Sidero	AJTO	Kastro
AJMB	Sapienza I	AJPT	Hierapetra	AJTQ	Kolokithia
AJMC	Mothoni	AJPV	Sphakia	AJTR	Paspargo I
AJME	Morea	AJPW	Gavdo I	AJTS	Boghazi P
AJMF	Koroni			AJTV	Mitylene
AJMG	Petalidi			AJTW	P Kalloni
AJMI	Kalamata	AJPX	Greek Archipelago	AJTX	Sigri P
AJMK	C Kitries	AJPY	Milo	AJTZ	Elecs I
AJML	Limeni	AJPZ	St Georgio Islet	AJUB	Iero P
AJMO	C Matapan	AJQB	Siphano	AJUC	Strati
AJMP	Kolokithia B	AJQC	Serpho	AJUE	Lemnos
AJMQ	Skutari	AJQE	P Livadhi	AJUF	Mudros B
AJMS	Gythium	AJQF	Thermia	AJUG	Kondia P
AJMT	Crance I.	AJQG	Zea e	AJUI	Kastro
AJMU	Iris, or, Vasili R	AJQI	P. St Nikolo	AJUK	Purnea B
AJMW	Sparta	AJQK	Tinos	AJUL	Tenedos
		AJQL	St Nikolas	AJUN	Ponente Pt
AJMX	Cerigo I	AJQN	Planumi I	AJUO	Gadaro I
AJMY	C Spathi	AJQO	Steno Pass	AJUP	Imbros
AJNB	Kapsali B	AJQP	Andros	AJUR	C Kephalo
AJNC	Cerigotto	AJQS	Gavrion P	AJUS	Samothraki
AJND	C Malea	AJQT	Kordion B	AJUT	Thaso I
AJNF	Belo Pulo I	AJQU	Doro Ch.		
AJNG	Ieraka P	AJQW	Syra		
AJNH	Pulithra P	AJQX	Gaidaro Islet	AJUW	Euboea I, or, Negro-
		AJQY	Hermopolis		Font
AJNK	Nauplia	AJRB	Mykoni	AJUX	Strongilo I
AJNL	Argos	AJRC	C Armenisti	AJUY	Zeitun, or, Stylida G
AJNM	Tolon B	AJRD	Rhenea	AJVB	C Kiliomeli
AJNP	Khaidari P	AJRF	Delos	AJVC	C Vasilina
AJNQ	Kiladia P	AJRG	Paros	AJVD	Mijella, or, Amaliopolis
AJNR	Spezzia I	AJRH	Naussa P	AJVF	Volo
AJNU	Hydra	AJRK	Parekhia P	AJVG	Skyro
AJNV	C Zurva	AJRL	Antiparos	AJVH	Skiatho
AJNX	Mandraki	AJRM	Naxos	AJVK	Skopelo
AJNY	Poros	AJRO	Naxia	AJVL	Guruni Hd
AJNZ	Dana Pt	AJRP	Amorgo	AJVM	Khelidromi I
AJOC	Pogon P	AJRQ	P Vathy	AJVO	Pelago I
AJOD	Ægina	AJRT	Nio		
AJOE	Sophiko P	AJRU	Santorin	AJVP	Turkey
AJOG	Kalamaki	AJRV	C Akroteri	AJVQ	Macedonia
AJOH	Salamis B	AJRX	Anaphi	AJVS	Roumelia
AJOI	Lipso I	AJRY	Rhodes f	AJVT	Saloniki
AJOL	Eleusis	AJRZ	Lindos	AJVU	Vardar Bank Lt V
AJOM	Piraeus	AJSC	C Prasonisi	AJVX	C Kara
AJON	Phalerum B	AJSD	Scarpanto	AJVY	Panomi Pt
AJOP	Athens	AJSE	Tristoma	AJVZ	Kassandra G
AJOQ	C Themistocles	AJSG	Livadia B	AJWC	Problaka B
AJOR	Phleva I	AJSH	Stampalia	AJWD	M Athos
AJOS	C Colonna	AJSI	Maltezana	AJWE	Plati
AJOU	Agastira B	AJSL	Kandeliusa	AJWG	Denthero Cove
AJOV	P Mandri	AJSM	Niseros	AJWH	Kavala
AJOW	P Raphti	AJSN	Symi	AJWI	Kalamuti Hr
AJOX	Petali G	AJSP	Kalolimno	AJWL	Fenar Pt
AJOY	Euripo Ch	AJSQ	Kalimno	AJWM	Dédéagatch
AJOZ	Talanta B	AJSR	Levitha	AJWN	Enos
AJPB	Berdugi I	AJSU	Lero	AJWP	G of Xeros
		AJSV	Lipso	AJWQ	Baklar P
AJPD	Candia (Crete)	AJSW	Patmos		
AJPE	Khania	AJSY	Skala	AJWR	The Dardanelles e
AJPF	C Drepano	AJSZ	Stavros P	AJWT	C. Helles
AJPH	Suda B	AJTB	Gialo P	AJWU	C Yeni-Shehr
AJPI	Azizieh	AJTD	Nikaria	AJWV	Kum Kalessi
AJPK	Rithymno, or, Retimo	AJTE	Furni	AJWY	Kephez Pt
AJPM	Megalo Kastron, or,	AJTF	Samos	AJWZ	Sari Siglar
	Candia Town	AJTH	P. Tigani	AJXB	Chanak
AJPN	C St John, or, Spina-	AJTI	Vathi P	AJXD	Nagara Liman
	longa	AJTK	Psara	AJXE	Nagara Kalessi Abydos
AJPO	P Nikolo	AJTM	Khios (Scio)		(Castle)

a Signifies an International Signal Station   b a Time Signal Station   c a Weather, Tide, or Ice Signal Station,   d a Life-Saving Station,
e a Lloyd's Signal Station   f a Wireless Telegraph Station

	SEA OF MARMORA			BLACK SEA	
AJXH	GALLIPOLI	AKCF	Therapia	AKFN	Moldava
AJXI	Dohan Aslan Bank Lt V	AKCG	Buyukbéré	AKFO	Dobrogea.
AJXK	Sar Kioi	AKCI	Buyuk B		
AJXM	Myriophyto	AKCJ	Fanaraki	AKFQ	BESSARABIA
AJXN	Hora	AKCL	Kandilli	AKFR	Dniester B
AJXO	Tekfur Dagh, or, Ro-	AKCN	Anadolu Hissári I	AKFS	Tsarigrad Mouth d
	dosto	AKCO	Khanlijeh	AKFU	Akerman d
AJXQ	Erekli	AKCP	Injir B	AKFV	C Fontana
AJXR	Silivri	AKCR	Beikos		
AJXS	Bayuk Chekmejeh B	AKCS	Selvi Burnu	AKFW	ODESSA b c d
AJXU	Stefano Pt	AKCT	Umur Banks I t V	AKFY	Dnieper R c
AJXV	Moda Liman	AKCV	Majar B	AKFZ	Berezan I
AJXW	Fanar Burnu	AKCW	Rumili Kavak	AKGB	Victorovski
AJXZ	Princes I	AKCX	Rumili	AKGD	Suvorovski
AJYB	Halki	AKCZ	C Anatoli	AKGE	Ochákov a c d
AJYC	Prinkipo			AKGF	Nikolaev b d
AJYE	Kartal	AKDB	BLACK SEA	AKGI	Kinburn
AJYF	Pendik			AKGJ	Ajighiol Spit Lt V
AJYG	Paulo Liman	AKDC	Kara Burnu d	AKGL	Bug R c
AJYI	Tuzla (Touzla)	AKDF	Inada	AKGN	Sviatotroitski
AJYK	Tutun Liman	AKDG	C Kuri	AKGO	Voloshski or, Volosh-
AJYL	Ismid	AKDH	Sizepoli (Pribachi ( )		skaia
AJYN	Karamusal	AKDJ	Megalo Nisi I	AKGP	Aogoyavlensk
AJYO	Dil Burnu	AKDL	Poros Pt	AKGR	Siversov
AJYP	Topche B	AKDM	Burghaz B	AKGS	Boggonosky
AJYR	Boz Burnu	AKDO	Anastatia, or, Papas I	AKGT	F Nikolaev b
AJYS	Gemlik, or, Kios	AKDP	Messemvria	AKGV	Konstantinovski
AJYT	Mudania	AKDQ	C Emineh.	AKGW	Alexandrovsk
AJYV	Kalolimno	AKDS	Kamchy R	AKGX	Stanislav
AJYW	Brusa	AKDT	BULGARIA	AKGZ	Kherson
AJYX	Panderma.	AKDU	C Galata	AKHB	Tendra B a
AJZB	Peramo B	AKDV	Varna	AKHC	Perekop G
AJZC	Marmara I	AKDW	Baljik		
AJZD	Palatia B	AKDX	Kavárna	AKHE	KRIMEA, or CRIMEA
AJZF	Mermerjik B	AKDY	C Kaliakra	AKHF	C Tarkhan a c d
AJZG	Galimi	AKEB	C Shableh	AKHG	Eupatoria d
AJZH	Fanar Adasi Islet	AKEC	Kustenjeh (Kustendji)	AKHJ	SEVASTOPOL b d f
AJZK	Kalazak, or, Klazati	AKED	ROUMANIA	AKHL	Inkerman
AJZL	Alonyi	AKEF	Mangalia (Mangolia)	AKHM	C Khersonese a c f
AJZM	Pasha Liman	AKEG	C Tuzla	AKHO	Balaklava d
AJZO	Kutali Road	AKEH	Constantza f	AKHP	C Aitodor a d
AJZP	Arablar	AKEI	Cara-Irman (Kara-Her-	AKHQ	Yalta c d
AJZQ	Palios Pt		man)	AKHS	Urzuf
AJZS	Rhoda	AKEJ	Portiza (Portici)	AKHT	Sudak
AJZT	Artaki B	AKEL	St George Mouth.	AKHW	Theodosia (Kaffa) d
AJZU	Karabuga	AKEM	Sulina Mouth d	AKHX	C Chauda a
AJZW	Kamir	AKEN	Serpents I (Fidonisi)	AKHY	Kyz Aul Pt a c
		AKEO	Kilia Mouth	AKIB	C Takly
AJZX	BOSPHORUS a	AKEP	Tulcea	AKIC	Tuslinski Lt V
AJZY	Seraglio Pt	AKEQ	Isaccea	AKID	C St Paul d
AKBD	Kadi Kioi	AKER	Galatz	AKIF	Kamish
AKBE	Haidar Pasha	AKES	Braila	AKIG	Churubash
AKBF	Kavak Burnu	AKET	Macin		
AKBH	Leander Tower	AKEU	Borcea	AKIH	KERTCH c
AKBI	Skutari	AKEV	Hirsova (Hirssova)	AKIL	Yenikale C d
AKBJ	CONSTANTINOPLE, or,	AKEW	Cernavoda	AKIM	Arabat B
	STAMBUL	AKEX	Calarasi	AKIN	Ghenitchesk Str
AKBN	Adrianople	AKEY	Olteniza	AKIP	Beriuch Spit
AKBO	Golden Horn	AKEZ	Giurgiu	AKIQ	Obitotchna Spit
AKBQ	Pera	AKFB	Turnu-Magurele	AKIR	Berdiansk d
AKBR	Galata	AKFC	Corabia	AKIT	Bielosarai Spit d
AKBS	Topkhana	AKFD	Calafat	AKIU	Mariupol c d
AKBU	Fundukli	AKFE	Turnu-Severin	AKIV	Beglitzkaia Lt V c
AKBV	Orta Kioi	AKFG	DANUBE e	AKIX	Taganrog c d
AKBW	Kuru Chesmeh	AKFH	Bucharest	AKIY	Rostov d
AKBY	Arnaut Kioi	AKFI	Jassy	AKIZ	Azov.
AKBZ	Bebek	AKFJ	Craiova	AKJC	Don R c d
AKCB	Rumili Hissar	AKFL	Oltenia	AKJD	Pechany Lt V
AKCE	Yeni-Kioi	AKFM	Mutenia	AKJE	Eisk Spit d

a Signifies an International Signal Station   b a Time Signal Station,   c a Weather, Tide, or Ice Signal Station   d a Life Saving Station,   e a Lloyd's Signal Station, f a Wireless Telegraph Station

ASIA MINOR				NORTH AFRICA	
AKJG	Dolga Spit	AKOB	Ali-Agha P	AKRV	Latakiyah.
AKJH	Temriuk B	AKOD	Foggia Nova	AKRW	Markhab
AKJI	Tamau Lake	AKOE	Foujes	AKRX	Tartus
		AKOF	Oglak I	AKRZ	Ruad I
AKJM	ASIA MINOR	AKOH	C Merminji	AKSB	Tripoli
AKJN	Caucasus	AKOI	Pelican Spit Lt V	AKSC	Ramkine I
AKJO	Circassia	AKOJ	Sanjak Spit Lt V	AKSE	Jebeil
AKJQ	Tuzla Bank	AKOM	SMYRNA	AKSF	BEIRUT
AKJR	Anápa	AKON	Vourlah	AKSG	Damascus
AKJS	Doob Pt	AKOP	Sahib P	AKSI	Saida
AKJU	Penai Pt	AKOR	Eritra	AKSJ	Jezireh
AKJW	Novorossisk, or, Sujak Bay	AKOS	Kezil Pt	AKSL	Sûr, or, Tyre
AKJY	Ghelenjik B	AKOT	Chesme	AKSN	Akka, or, Acre
AKJZ	Kadosh, or, Chardak Pt	AKOV	Egrilar	AKSO	Haifa
AKLD	F St Duka	AKOW	Mersin P	AKSP	C Carmel
AKLE	Sochinski (Socha Bitke Pt)	AKOX	Sikia P	AKSR	Kaisariyeh (Cæsarea).
AKLF	Pitsunda Pt	AKOZ	Sighajik P	AKST	Arsuf (Appolonia)
AKLH	Bombori F	AKPB	Ephesus	AKSU	Yafa
AKLI	Guduat	AKPC	Skala Nuova G	AKSV	Jericho
AKLJ	Sukhúm B	AKPE	Hussein Pt	AKSW	Palestine
AKLN	Redoute Kalessi	AKPF	Budrum	AKSX	Jerusalem
AKLO	Poti F c d	AKPG	Giova P	AKSY	Yebnah
AKLP	BATUM c d	AKPI	Gallipoli P	AKSZ	Dead Sea
AKLR	Gumeh	AKPJ	Kos		
		AKPL	Kum Pt	AKTB	AFRICA
AKLS	CASPIAN SEA.	AKPN	Doris	AKTC	EGYPT
AKLT	Volga R	AKPO	Ipsera Islet	AKTD	Cairo
AKLV	Baku			AKTF	El Arish
AKLW	Tiflis	AKPQ	LEVANT	AKTG	P SAID a b e f
AKLX	Georgia.	AKPS	Karamania	AKTI	SUEZ CANAL
AKLZ	Armenia.	AKPT	Marmarice	AKTL	Lake Menzaleh
		AKPU	Karaghatch Hr	AKTM	Ismaila
AKMB	Rizo	AKPW	Makri Hr f	AKTO	Lake Timsah
AKMC	Trebizond	AKPX	Kastelorizo I	AKTQ	Great Bitter Lake
AKMD	Erzerúm	AKPY	Tristomos	AKTR	Little Bitter Lake
AKME	Platana	AKQB	Yali B	AKTU	Damietta
AKMF	C Ierós.	AKQC	Adália	AKTW	Nile R
AKMG	Tereboli	AKQD	Kara Burnu	AKTX	Brulos
AKMI	Kerassond, or, Kerasunda	AKQF	C Kiloarda	AKTZ	Rosetta
AKML	Vona B	AKQG	Aláya	AKUB	Abukir B
AKMO	Unieh	AKQH	Anamur C	AKUC	ALEXANDRIA a b d e
AKMP	Samsun	AKQJ	Chelindreh P	AKUE	El Amaid
AKMQ	Halys Pt , or, C Bafra	AKQL	Cavalière C	AKUF	Kanais G
AKMS	Kizil Irmak	AKQM	Aghaliman	AKUG	Marsa Omrakum.
AKMT	Sinub, or, Sinope	AKQO	Bagasse Pt	AKUI	Marsa Tebruk
AKMW	C Injeh	AKQP	Mersina	AKUJ	Bombah G
AKMX	C Ineboli	AKQR	Karadash Burnu	AKUL	Menelaus Hr
AKMY	C Kerempeh				
AKNB	Kara-Agatch	AKQT	CYPRUS	AKUN	TRIPOLI
AKNC	Amastra	AKQU	Kerynia	AKUO	Dernah f
AKND	Koslu B	AKQV	Nicosia	AKUP	Marsa Sousah
AKNF	Bender Erekli (Heraklea)	AKQW	C Arnauti	AKUR	Gureina (Cyrene)
		AKQX	Paphos	AKUS	Tolmetta
AKNH	C Kefken	AKQY	C Gata	AKUT	Ben Ghazi (Berenice)
AKNJ	Kirpen I	AKQZ	Limasol	AKUW	Marsa Zafran, or, Chebek
AKNL	Kilia Pt	AKRC	C Kiti	AKUX	Misratah
		AKRD	Larnaka	AKUY	Ras Zorug
AKNM	ANATOLIA	AKRE	C Greco	AKVB	Lebidah
AKNP	Bashika B	AKRF	Famagusta		
AKNQ	Yukyeri Pt	AKRH	C Andreas	AKVD	TUNIS
AKNR	C Baba			AKVE	Zarzis
AKNT	Sivriji Pt	AKRI	SYRIA	AKVF	Jerba I (Djerba)
AKNU	Adramyti	AKRJ	Ayas	AKVG	Bahiret el bu Grara
AKNV	Dikili	AKRM	Bayas	AKVH	Humt Adjim
AKNX	P Ajano	AKRN	Iskanderún G	AKVI	Gabes, or, Kabes d
AKNY	Sandarli G	AKRO	Alexandretta (Iskanderún)	AKVJ	Skira
AKNZ	Rema B	AKRQ	Antioch	AKVM	Sur-Kenis B
		AKRS	Bazit Pt	AKVN	Sfax d
		AKRT	Ibn Hâni Pt	AKVO	Kerkenah Is

NORTH AFRICA		ATLANTIC OCEAN		WEST AFRICA	
AKVQ	Sberki I	AKZV	Velez de la Gomera	ALEM	Arrecife Pt
AKVR	Mahedia	AKZX	Alcala B	ALEN	Lobos I
AKVS	Kuriat Is	AKZY	Tetuan	ALEP	Fuerteventura
AKVU	Monastir	ALBC	Almadabra, or, Madraga	ALEQ	Jandia Pt
AKVW	Susa (Sousse) d		B	ALER	Gran Canaria
AKVY	Herkla (Hergla)	ALBE	Ceuta d	ALET	Isleta Pt e
AKVZ	Hammamet	ALBF	Almina Pt	ALEU	P de la Luz
AKWB	Kalibia	ALBG	Malabata Pt	ALEV	Palmas
AKWD	C Bon a e	ALBI	Tangier a	ALEX	Maspalomas Pt
AKWE	Goletta (La Goulette) d	ALBJ	C Spartel a e	ALEY	Sardina Pt
AKWG	Carthage C a	ALBK	Jeremias Anchorage	ALEZ	Tenerife
AKWH	Piana I (El Kamela)	ALBN	Arsila, or, Arzila	ALFC	Anaga Pt e
AKWI	Porto Farina a	ALBO	El Araish	ALFD	Santa Cruz a c
AKWJ	Cani Is	ALBP	Mehediya	ALFE	Orotava P
AKWM	Benzert (Bizerta) a d	ALBS	Rabat and Sali	ALFH	Gomera
AKWN	Fratelli Rocks	ALBT	Dar el Beida	ALFI	San Sebastian
AKWO	Ras Engela	ALBV	Azimur	ALFJ	Hierro, or, Ferro
AKWP	C Serrat	ALBW	Mazigban	ALFM	Palma
AKWR	Galita (La Galite) d	ALBX	C Cantin, or, Ras iil	ALFN	Cumplida Pt
AKWS	Sorelle Rocks		Hadik	ALFO	Salvage Is
AKWU	C Negro	ALBZ	Safi (Asafi)		
AKWV	Tabarka d	ALCB	Mogador	ALFQ	C Bojador
		ALCD	Fez	ALFR	Ouro R
AKWY	Algeria	ALCF	Agadir, or, Santa Cruz	ALFS	C Barbas
AKWZ	La Calle a d	ALCG	Tiznitz	ALFU	C Blanco
AKXC	C Rosa (Rose).	ALCH	Wadi Draa (Nun R)	ALFV	Arguin B
AKXD	Bona (Bône) d	ALCJ	C Juby	ALFW	C Mirik
AKXF	F Génois			ALFY	Portendick
AKXG	C de Garde a	ALCK	Atlantic Ocean		
AKXI	Takush B			ALFZ	Barbary Coast
AKXJ	C de Fer a	ALCM	Azores		
AKXM	Philippeville d	ALCN	Corvo	ALGB	Sénégal R
AKXN	Stora d	ALCO	Flores	ALGD	St Louis
AKXP	Constantine	ALCP	Santa Cruz	ALGE	Almadi, or, Almadies
AKXQ	Srigina I	ALCR	Fayal		Pt
AKXS	Collo d	ALCS	Horta d e	ALGF	Senegambia
AKXT	C el Jerba	ALCT	Espalamaca Pt	ALGI	C Verde a
AKXV	C Bougaroni a	ALCU	Pico	ALGJ	Gorée
AKXW	Jidjelli (Djidjelli) d	ALCW	San Jorge	ALGK	Dakar a b
AKXZ	Ras Afia	ALCX	Graciosa	ALGN	Rufisque
AKYB	Bougie d	ALCY	Terceira		
AKYD	C Carbon a	ALDB	Praya	ALGO	Cape Verde Is
AKYE	Dellys d	ALDC	Angra c d	ALGP	St Antonio.
AKYG	C de l'Aiguille a	ALDE	San Miguel (St Michaels)	ALGR	Bull Pt
AKYH	C Bengut			ALGS	Tarrafal B
AKYJ	C Matifu a d	ALDG	Ponta Delgada a b d	ALGT	Santa Cruz
AKYL	Bouzaréah a	ALDH	Villa Franca	ALGV	St Vincent
AKYN	Algier (Alger) b d f	ALDI	Arnel Pt a c e	ALGW	P Grande
AKYO	C Caxine	ALDK	Ferraria Pt e	ALGX	Bird I a
AKYQ	Tipaza P d	ALDM	Santa Maria I	ALGZ	Santa Luzia
AKYR	Shershel (Cherchell) d	ALDN	Villa do Porto	ALHB	St Nicholas
AKYT	Tènèz, or, Tènès a d	ALDP	Formigas	ALHC	Preguiza
AKYU	C Ivi			ALHE	Sal (Salt) I
AKYW	Mostaganem d	ALDQ	Madeira	ALHF	Bonavista
AKYX	Arzeu, or, Arzew d	ALDR	Fora I a d	ALHG	English Road
AKZB	Oran d	ALDS	Loo Rock a c	ALHJ	Mayo I
AKZC	Marsa el Kébir d	ALDT	Funchal d	ALHK	St Jago
AKZE	C Falcon a	ALDU	Pargo Pt a c	ALHM	P Praya
AKZF	Habibas Is	ALDV	Cruz Pt	ALHO	Tarrafal Pt
AKZH	Beni Saf Hr d	ALDX	P Santo	ALHP	Leton Rock
AKZI	Raschgun I	ALDY	Bezerta Grande	ALHQ	Fogo I
AKZL	Nemours d	ALDZ	Bugio	ALHS	Brava I
AKZM	Morocco	ALEB	Canary Is	ALHT	Gambia
AKZN	Milonia R	ALEC	Allegranza	ALHU	Bathurst
AKZP	Mulaya R	ALED	Delgada Pt	ALHW	Albreda
AKZQ	Zafarin I	ALEG	Graciosa	ALHX	Yarbutenda
AKZR	Melilla d f	ALEH	Lanzarote	ALHY	Kasamanze R
AKZT	C Tre Forcas	ALEI	Pechiquera Pt	ALIB	Karabane I
AKZU	Alhucemas.	ALEK	Naos P	ALIC	Zighinkor

a Signifies an International Signal Station, b a Time Signal Station, c a Weather, Tide, or Ice Signal Station, d a Life-Saving Station, e a Lloyd's Signal Station, f a Wireless Telegraph Station

WEST AFRICA   SOUTH AFRICA

Code	Place	Code	Place	Code	Place
ALID	Brin Factory	ALND	Prah R	ALQT	Bata.
ALIF	Sedhiu	ALNE	C Coast Castle a	ALQU	San Bento R
ALIG	Dianna.	ALNF	Anamaboe	ALQV	Medondo
ALIH	C Roxo	ALNG	ASHANTI.	ALQW	Corisco B
ALIK	Cacheo R	ALNH	Akkra a b	ALQX	Elobey I
ALIM	Salsang	ALNI	Kumassi	ALQY	Muni
ALIN	Farin	ALNJ	SOUDAN	ALQZ	Gaboon R.
ALIP	Batur	ALNK	Volta R	ALRC	Gombé Pt.
ALIQ	Bissao I	ALNM	Adda	ALRD	Libreville
ALIR	Jeba R	ALNO	C St Paul	ALRE	Komo R
ALIU	Bijouga Is	ALNP	Jella Koffi	ALRG	Ogowé R
ALIV	Bulama	ALNQ	Kitta	ALRH	Nazareth B
ALIX	Alcatraz I	ALNR	P Seguro	ALRI	Lopez C
ALIY	Nuñez R	ALNS	Agweh	ALRK	Fernand Vaz
ALIZ	C Verga	ALNT	Togo.	ALRM	Settée R a
ALJC	Ponga R	ALNU	Little Popo	ALRN	Mayumba B
ALJD	Tabuia, or, Taboria.	ALNV	Whyda	ALRP	Kulin R
ALJE	Bangalong	ALNX	Kotonu	ALRQ	Loango B
ALJG	Brameya R	ALNY	P Novo	ALRS	Massabé
ALJH	Isles de Los			ALRU	Chiloango R
ALJI	Bulbineh	ALNZ	DAHOMEY	ALRV	Malemba B
ALJM	Konakri.	ALOB	Abomeh (Abomey)	ALRW	Kabinda
ALJN	Matakong I	ALOC	Badagri	ALRY	Kongo R
ALJO	Mellakori R.	ALOD	Lekki	ALRZ	French Pt a
ALJQ	Benti Pt	ALOE	Lagos	ALSB	Shark Pt (Padrão)
ALJR	Great Skarcies R	ALOF	Palm Oil Rs		
ALJS	Mambolo	ALOH	Benin R	ALSD	KONGO FREE STATE
ALJU	Yellaboi Sound	ALOI	Wari.	ALSE	Banana Creek
		ALOJ	Forcados R	ALSF	P da Lenha
ALJV	SIERRA LEONE	ALOM	Gana Gana.	ALSI	Embomma (Bomma)
ALJW	Free Town	ALON	Middleton R	ALSJ	Brazzaville
ALJY	Banana Is.			ALSM	Leopoldville
ALJZ	Sherbro R	ALOP	NIGER	ALSN	Stanley Pool
ALKB	O St Ann	ALOR	Akedo	ALSO	Padron Pt
ALKD	Manna R	ALOS	Nun Entrance		
ALKF	C Mount	ALOT	Akassa (Baracoon Pt)	ALSQ	ANGOLA.
ALKH	LIBERIA	ALOV	Abo	ALSR	Ambrizette
ALKI	Caldwell	ALOW	Ndoni	ALST	Foreland Bluff
ALKJ	C Mesurado	ALOX	Onitsha	ALSV	Ambriz B a
ALKN	Monrovia	ALOZ	Asaba	ALSW	Dandé Pt
ALKO	Grand Bassa	ALPB	Ida	ALSX	Funta B
ALKP	Trade Town	ALPC	Binue R.	ALSZ	Bengo B
ALKR	Sestos R	ALPE	Yola	ALTB	LOANDA (P OF) b,
ALKS	Kru	ALPG	C Formosa.	ALTC	San Miguel F a
ALKT	Nifu	ALPI	Brass R	ALTE	Loanda I
ALKV	Grand Sesters.	ALPJ	Bonny R	ALTF	Palmeirinhas Pt.
ALKW	Garnaway	ALPK	New Calabar R	ALTG	Coanza R
ALKX	Mano R	ALPN	Calabar	ALTI	Calumbo
ALKZ	C PALMAS	ALPO	Opobo R	ALTJ	Massangano
ALMB	Grain Coast	ALPQ	Old Calabar R	ALTK	Dondo
ALMC	Tabu	ALPS	Duke Town	ALTN	Cassanha
ALME	Drewin	ALPT	Cross R	ALTO	Nova Redonda a
ALMF	Sassandra, or, St Andrew R	ALPU	Rio del Rey	ALTP	Kwi Kombo
				ALTR	Lobito B
ALMG	Amokwa	ALPW	BIGHT OF BIAFRA	ALTS	Benguela
ALMI	Akba	ALPX	Fernando Po	ALTU	Salinas Pt
ALMJ	Grand Bassam	ALPY	Fernanda Pt	ALTW	Elephant B
ALMK	Assini R a	ALQB	Santa Isabel	ALTX	C St Mary
ALMO	Newtown.	ALQC	Princes I	ALTY	C Martha
ALMP	Albani R	ALQD	Santo Antonio B	ALUB	Little Fish B
		ALQF	St Thomé (Thomas I)	ALUC	Mossamedes
ALMQ	IVORY COAST	ALQG	Anna de Chaves B	ALUD	P Alexander
ALMS	Ankobra R	ALQH	Annobon I	ALUE	Albino Pt
ALMT	Axim			ALUF	Great Fish B
ALMU	C Three Points.	ALQJ	KAMERUN R	ALUG	Nourse, or, Cunene R
ALMW	Gold Coast	ALQK	Malimba R		
ALMX	Guinea G	ALQM	Batanga	ALUH	DAMARALAND
ALMY	Dix Cove	ALQO	Beundo R	ALUI	C Frio
ALNB	Sekondi	ALQP	Batonga Roadstead	ALUJ	Swakopmund Road
ALNC	Elmina	ALQR	Campo R	ALUK	Walfisch B

a Signifies an International Signal Station, b a Time Signal Station, c a Weather, Tide, or Ice Signal Station, d a Life Saving Station, e a Lloyd's Signal Station, f a Wireless Telegraph Station

SOUTH AFRICA				EAST AFRICA	
ALUM	Hollam's Bird I	ALYK	EAST LONDON a b c d	AMCS	Mazimbwa B
ALUO	Spencer B	ALYM	Buffalo R	AMCU	Marongo
ALUP	Ichabo	ALYN	Kei R	AMCV	Tunghi B
ALUQ	Angra Pequeña	ALYP	Mazeppa Pt	AMCY	COMORO I (COMORES)
ALUS	Robert Hr	ALYQ	Bashee R	AMCZ	Maroni
ALUT	Possession I	ALYR	St John R	AMDB	Mohilla I (Mohéli)
ALUV	Great Namaqualand	ALYS	Grovenor P	AMDC	Duéni Cove (Fumboni
ALUW	Orange R	ALYT	P Shepstone		B)
		ALYU	Umpambinyoni R	AMDE	Johanna I (Aujouan)
ALUX	ASCENSION I e	ALYV	Amahlongwana R	AMDF	Pomoni Hr
ALUY	Georgetown	ALYW	Ahwal Shoal	AMDG	Mayotta
ALUZ	Clarence B			AMDJ	Pamanzi, or, Zaudzi
ALVC	St HELENA e	ALYX	NATAL		Road
ALVD	James Town b	ALYZ	The Bluff a b c e	AMDK	Mamutzu
ALVE	Prosperous B a	ALZB	DURBAN d	AMDN	Glorieuse Isle
ALVG	Tristan da Cunha	ALZC	Pietermaritzburg		
ALVH	Inaccessible I.	ALZD	Tugela R	AMDO	MADAGASCAR
ALVI	Nightingale I	ALZE	Durnford B	AMDP	Tananarivo.
ALVK	Gough I	ALZF	St Lucia B	AMDR	C Amber (Ambre) a
		ALZH	Sordwana Road	AMDS	Diego Suarez B a
ALVM	CAPE COLONY	ALZI	Kosi R	AMDT	P Nievre
		ALZJ	RHODESIA	AMDU	Oronjia a
ALVN	ORANGE FREE STATE			AMDV	Antsirana c
ALVP	TRANSVAAL	ALZK	DELAGOA B	AMDW	Ambavarane
ALVQ	Kaffaria	ALZM	Reuben Pt a	AMDX	P Looké (Loqué)
ALVR	Zululand	ALZN	English R	AMDZ	P Leven
ALVT	P Nolloth	ALZO	Lourenço Marquez b	AMEB	Andrava B
ALVU	Hondeklip B	ALZP	Maputa R	AMEC	Vohimao, or, Vohemar
ALVW	Olifant R	ALZR	P Melville		B
ALVZ	St Helena B	ALZS	Limpopo R	AMEF	Antongil B
ALWB	Saldanha B	ALZT	C Corrientes	AMEG	P Choiseul
ALWC	Dassen I	ALZV	Burra	AMEH	Tangtang, or, Tintingue
ALWE	Robben I	ALZW	Innambán R		
ALWG	Table B b d	ALZX	Bezaruto C	AMEJ	St Mary I
ALWI	CAPE TOWN c	AMBC	Marsha, or, S Carolina I	AMEK	P St Mary
ALWK	Lions Rump a			AMEL	Blève Reef
ALWN	Mouillé Pt	AMBD	Sabi R	AMEN	Fénérive
ALWP	Green Pt	AMBE	Chiluán I	AMEO	Mahambo Pt
ALWQ	C of Good Hope e	AMBG	Sofala R	AMEP	Foulepointe
ALWS	Roman Rocks	AMBH	Pungue R	AMEQ	Tamatave, or, Toamasina
ALWT	False B	AMBI	Beira		
ALWU	Simon's B b	AMBK	Inhamboio	AMER	Vatomandri a
ALWV	C Hangklip	AMBL	Thornton R	AMES	Mahanoro a
ALWX	Danger Pt a	AMBN	Zambezi R	AMET	Maninzari (Mananjara
ALWY	C Agulhas e	AMBO	MASHONALAND		R)
ALWZ	Struys B	AMBP	Kangoni	AMEU	Matitanana R
ALXC	Bredasdorp	AMBQ	Conceicao R	AMEW	Farafangana
ALXD	P Beaufort	AMBR	Chinde R	AMEX	St Lucia (Luce)
ALXE	Breede R			AMEY	Ytapere, or, Itapérina
ALXG	St Sebastian B	AMBS	BRITISH CENTRAL AFRICA		B
ALXH	Kaffir Kuyl B			AMFB	F Dauphin
ALXJ	C St Blaize a	AMBT	Sena	AMFC	C St Mary, or, St Marie
ALXK	Ahwal a c	AMBU	Tete		
ALXM	Mossel B	AMBV	LAKE NYASSA	AMFD	St Augustine B
ALXO	George Town	AMBW	Shiré R	AMFG	Saolara
ALXP	Knysna Hr	AMBX	Linde R	AMFH	Tullear, or, Ankatsaoka
ALXQ	Plettenberg B	AMBY	Kilimán (Quilimane) R		
ALXS	C St Francis e	AMBZ	Angoche R , or, Mlui	AMFI	Ranobé P
ALXT	Krom B	AMCD	P Mokambo	AMFK	Fanemotra B
ALXU	Humansdorp			AMFL	Morondava.
ALXW	C Recife e	AMCE	MOZAMBIQUE a c e	AMFN	Ambondro R
ALXY	Algoa B	AMCF	St George I	AMFP	Maintirano, or, Myanterano R
ALXZ	P Elizabeth a b c	AMCH	Conducia B		
ALYC	Bird I	AMCI	Nakala Pt	AMFQ	C St Andrew
ALYD	P Alfred a b c d	AMCJ	Belmore Hr	AMFR	Boyanna B
ALYE	Kowie R	AMCL	Memba B	AMFT	Makambi R
ALYF	Bathurst	AMON	Almeida B	AMFU	Bombetoke B
ALYG	Grahams Town	AMCO	Pomba B (Mwambi)	AMFV	Mojanga c
ALYH	Great Fish R	AMCQ	Montepes B	AMFX	Betsiboka R
ALYI	Keiskamma R	AMCR	Ibo	AMFY	Ikopa R

a Signifies an International Signal Station. b a Time Signal Station. c a Weather, Tide, or Ice Signal Station, d a Life-Saving Station? e a Lloyd's Signal Station, f a Wireless Telegraph Station

SOUTH INDIAN OCEAN		RED SEA		EAST AFRICA	
AMFZ	Mevatanana	AMJL	CHAGOS ARCHIPELAGO	AMNU	Tula
AMGC	Suberhieville	AMJN	Diego Garcia	AMNV	Kisimayu B
AMGD	Katsépé			AMNW	Juba
AMGE	Majamba B	AMJO	Prince Edward I		
AMGH	Miramba P	AMJQ	Crozet I d	AMNX	ITALIAN SOMALILAND
AMGI	Narendri	AMJR	Kerguelen I d	AMNY	Brawa
AMGJ	Raminitok B	AMJS	Heard I	AMNZ	Merka
AMGL	Radama P	AMJT	St Paul I d	AMOC	Mogdishu (Magadoxa)
AMGN	Ambavatobi B	AMJU	Amsterdam I d	AMOD	Oblat
AMGO	Pasindava B			AMOE	Ras Hafún
AMGQ	Nosi-Bé	AMJV	GERMAN EAST AFRICA	AMOG	C Guardafui, or, Ras
AMGR	Hellville	AMJW	C Delgado		Asir
AMGS	Nosi Vauru, or, Vorona	AMJX	Keonga B	AMOH	Sokótra I
	Islet	AMJY	Rovuma R	AMOI	Tamrída
AMGU	Tani-Keli Islet	AMJZ	Msimbati Hr	AMOJ	Ras Radressa
AMGV	C St Sébastien	AMKB	Mikindani Hr	AMOK	Abd-al-kuri
AMGW	William Pitt B (An-	AMKD	Mgau Mwania		
	dramatuba)	AMKE	Lindi R	AMOL	G OF ADEN
AMGY	P Liverpool	AMKG	Mchinga B	AMON	Somali
AMGZ	Europa I	AMKH	Kiswere Hr	AMOQ	Ras Alula
AMHB	Bassas da India	AMKI	Kilwa Kisiwani	AMOR	Bander Marayeh
AMHC	Juan de Nova I	AMKJ	P Beaver	AMOS	Burnt I, or, Ar-Eabbah
		AMKN	Rukyira B	AMOU	Berbera
AMHD	RÉUNION I	AMKO	Kilwa Kivinje	AMOV	Bulhar
AMHE	St Denis a c	AMKP	Songa Songa I	AMOW	Zeila
AMHF	St Marie a	AMKQ	Mohoro B	AMOY	Jibuti
AMHI	Bel-Air Pt	AMKR	Rufiji R	AMOZ	Tajura
AMHJ	St Suzanne a	AMKS	Mafia I	AMPB	Obokh
AMHK	Bois Rouge Pt a	AMKT	Ras Mkumbi	AMPD	Ras al Bir
AMHN	Champborne Pt a	AMKU	Latham I	AMPE	Str of Bab el Mandeb
AMHO	St Benoit a	AMKV	Ras Kanzi		
AMHP	St Rose a	AMKW	Dar es Salaam	AMPF	RED SEA
AMHR	St Pierre a c	AMKX	Bagamoyo	AMPH	Perim d e
AMHS	St Louis (St Leu) a	AMKY	Ruva, or, Kingani R	AMPI	Obstruction Pt
AMHT	St Paul a c			AMPJ	Asab
AMHX	Pt des Galets, or, Shin-	AMLB	ZANZIBAR	AMPL	Shummá I
	gle Pt a e	AMLC	Kiungani	AMPN	Sheik ul Abu
		AMLE	Mwana Mwana a	AMPO	Annesley B
AMHZ	MAURITIUS	AMLF	Kokotoni Hr	AMPR	Difnein I
AMIB	P Louis a b c d	AMLG	Mungopani a	AMPS	ABYSSINIA
AMIC	P Louis Lt V	AMLI	Ras Nungwe	AMPT	Massaua Hr
AMID	Signal Mt e	AMLJ	Pemba	AMPV	Trinkitat Hr
AMIE	C Malheureux	AMLK	Chaki Chaki	AMPW	Suakin
AMIF	Canonnier Pt e	AMLN	Ras Kegomacha	AMPX	Raweiya Ras
AMIG	Grand Bay	AMLO	Ras Upemba	AMPZ	Berenice P
AMIJ	C Brabant	AMLP	Saadani	AMQB	St Johns, or, Zeberjed I
AMIK	Grand P	AMLQ	Pangani	AMQC	Dædalus Sh
AMIL	Mahébourg	AMLS	Tanga B	AMQE	Koseir
AMIN	Roche Isle Beacon	AMLT	Ulenge	AMQF	The Brothers
AMIO	Fouquets Isle			AMQG	Shadwan I
AMIP	Flat I e	AMLU	BRITISH EAST AFRICA	AMQI	Ashrafi Reef
AMIQ	Black River B	AMLW	Vanga	AMQJ	Str of Jubal
AMIS	Rodriguez I.	AMLX	Wasin	AMQK	Suez G
AMIT	Mathurin B	AMLY	Mombasa	AMQL	Ras Gharib
AMIU	P South East	AMLZ	Mvita	AMQN	Ras Zafarana
AMIV	Albatross I (Cargados	AMNB	P Kilindini	AMQO	Newport Rock Lt V
	Carajos)	AMNC	P Reitz	AMQP	P Ibrahim
AMIW	Agalega I	AMND	P Tudor	AMQS	SUEZ a e f
AMIX	Tromelin I	AMNE	Lake Tanganyika	AMQT	Suez B
		AMNF	Lake Victoria Nyanza		
AMIY	SEYCHELLES	AMNG	Matapwa R	AMQU	ASIA
AMIZ	Dennis I	AMNH	Kilifi R		
AMJB	Mahé I	AMNI	Owyombo R	AMQW	Tor
AMJC	P Victoria	AMNJ	Malindi	AMQX	Mt Sinai
AMJD	Amirante Is	AMNK	Tana R	AMQY	Hedjaz Province
AMJE	Farquhar Is	AMNL	Kipini	AMRB	Akabah G
AMJG	Cosmoledo Is	AMNP	Lamu	AMRC	Mowila
AMJH	Assumption I	AMNQ	Manda Roads	AMRD	Sherm Wej
AMJI	Aldabra I	AMNR	Patta	AMRF	C Baridi
AMJK	Providence and St	AMNT	P Durnford	AMRG	Yenbo f
	Pierre Is				

a Signifies an International Signal Station, b a Time Signal Station, c a Weather, Tide, or Ice Signal Station, d a Life-Saving Station, e a Lloyd's Signal Station, f a Wireless Telegraph Station

ARABIAN SEA		PERSIA		INDIA	
AMRH	Medina f	AMVB	Bushire e	AMYV	Daman P
AMRJ	Jidda, or, Jeddah f	AMVC	Dayyir	AMYW	Dábanu
AMRK	Mekka	AMVD	Kangún, or, Kangán	AMYX	Tarapur Pt
AMRL	Lith	AMVF	Charák	AMZB	Bassein
AMRO	Kuntida	AMVG	Mughú	AMZC	Tanna
AMRP	Loherya	AMVH	Linja	AMZD	BOMBAY b c f
AMRQ	Kamaran Hr	AMVJ	Básidu	AMZE	Malabar Pt
AMRT	Jebel Teir	AMVK	Khamír	AMZF	Outer Lt V
AMRU	Zebayir Is	AMVL	Jezírat Nábiyu Tanb	AMZG	Sunk Rock Lt
AMRV	Hodeida	AMVN	Clarence Str	AMZH	Southwest Prong Lt
AMRW	Abu Ail Is	AMVO	Kishm, or, Jezírat at	AMZI	Alibagh c d
AMRX	Hanish Is		Tawíla e	AMZJ	Mazagon
AMRY	Jebel Zukur	AMVP	Bander Abbas	AMZK	Victoria Dock
AMRZ	Mokha, or, Mocha	AMVQ	Ras al Kuh	AMZL	Princes Dock
AMSB	C Bab el Mandeb	AMVT	Maksa, or, C Jáshak e	AMZO	Kolaba Pt f
AMSC	Yemen Province	AMVU	Makrán Coast	AMZP	Kolaba Oby
		AMVX	Khor Rabij	AMZQ	Kundari I a c d
AMSD	ARABIA	AMVY	Chahbár B	AMZR	P Chaul
		AMVZ	Gwatar B ·	AMZS	Janjira Hr
AMSE	ADEN e			AMZT	Marad
AMSG	Bal-haf	AMWC	BALUCHISTAN	AMZU	Rajpuri
AMSH	Makalla	AMWD	Gwádar	AMZV	Bankot P , or, Savitri R
AMSI	Ash-Shehr	AMWE	Pasni	AMZW	Savaradrug
AMSK	Kishin	AMWG	Ormára	AMZX	Dabhcl or, Washishti R
AMSL	Rás Farták	AMWH	Sunmiyani B	AMZY	Tolkeshwar Pt
AMSP	Damkut	AMWI	Baila	ANBC	Shastri R
AMSQ	Khorya Morya Is	AMWK	Ras Muwári (C Monze)	ANBD	Ratnagiri c
AMSR	Ras Madraka			ANBE	Rajapur
AMSU	Ghubbet Hashish	AMWL	INDIA	ANBF	Viziadrug
AMSV	Masíra G	AMWN	Afghanistan	ANBG	Malwan
AMSW	Ras al-Hadd	AMWP	KARACHI b	ANBH	Vengurla Rocks, or,
AMSY	Sur	AMWQ	Manora Pt a b c		Burnt Is
AMSZ	G OF OMAN	AMWR	Indus R	ANBI	Vengurla,or,Vingorla a c
AMTB	Maskat	AMWT	Keti	ANBJ	Aguada B
AMTD	Barka	AMWU	Haiderabad	ANBK	Panjim
AMTE	Sohár	AMWV	Sind Coast	ANBL	Mandavi R
AMTF	Shinás	AMWY	Khori, or, Lakhpat R		
AMTH	Khor Fakán.	AMWZ	Kutch G	ANBM	GOA
AMTI	Dibba	AMXB	Mandvi	ANBP	Murmagao
		AMXD	Navinar Pt	ANBQ	Oyster Rock
AMTJ	PERSIAN G	AMXE	Karumbhar I	ANBR	Sadashivgad c
AMTL	Quoin Is	AMXF	Khambhalia	ANBT	Karwar Hd
AMTN	Ras al Khaima	AMXH	Nawanagar	ANBU	Tadri R
AMTO	Shárja	AMXI	Beit Hr	ANBV	Kumpta
AMTQ	Dabai	AMXJ	Guzerat	ANBX	Honawar
AMTR	Abu Thabi	AMXL	Dwarka Pt	ANBY	Bhatkul Cove
AMTS	Al Wakra	AMXN	Porbandar c	ANBZ	Kundapur
AMTW	Doha	AMXO	Navibandar	ANCD	Mulpy
AMTW	Ras Rakkin	AMXQ	Mangrol	ANCE	Mulki (Primeira) Rocks
AMTX	Bahrain Hr	AMXR	Verawal c	ANCF	Mangalore
AMTZ	Al Manáma	AMXS	Diu	ANCH	MALABAR COAST
AMUB	Dammám	AMXU	Jáfarabad	ANCI	Kannanur
AMUC	Al Katif	AMXV	Mahuwa	ANCJ	Tellicherri c
AMUE	Kuweit, or, Al Kuweit	AMXW	Kutpur I	ANCL	Mahé
	Hr	AMXZ	Goapnáth Pt	ANCM	Calicut c
AMUF	Shatt-al-Arab	AMYB	Piram I	ANCO	Beipur
AMUG	Fao e	AMYC	Gogha	ANCQ	Narakal
AMUI	Basra	AMYE	Bhaunagar Creek c	ANCR	Cochin c
AMUJ	Euphrates R	AMYF	Kunhbandar	ANCS	Alleppi
AMUK	Makil	AMYG	Cambay G	ANCU	Travancore Coast.
AMUN	Baghdád	AMYI	Baroda	ANCV	Quilon
AMUO	Karún R	AMYJ	Mahi R	ANCW	Trivandrum
AMUP	Muhammera	AMYK	Tankari	ANCY	Kolachel
AMUR	Ahwaz	AMYN	Nerbudda, or, Narbada	ANCZ	Muttum Pt a
AMUS	Tigris R		R	ANDB	C Comorin
		AMYO	Broach		
AMUT	PERSIA	AMYP	Bhagwadandi	ANDF	LACCADIVE Is
AMUW	Teheran	AMYR	Tapti R	ANDG	Androth I
AMUX	Dilam	AMYS	Surat	ANDI	Aucutta I
AMUY	Kharij, or, Kharag I	AMYT	Bulsar	ANDJ	Minikoi I

a Signifies an International Signal Station b a Time Signal Station, c a Weather, Tide, or Ice Signal Station d a Life-Saving Station,
e a Lloyd's Signal Station, f a Wireless Telegraph Station

BAY OF BENGAL				MALAY PENINSULA	
ANDK	Maldive Is	ANHD	Cocanada c	ANKV	Mayu R
ANDM	Mále, or, Sultan's I	ANHF	Vakalpudi a c	ANKX	Oyster I
ANDO	Nine Degree Ch	ANHG	Vizagapatam a c		
ANDP	Eight Degree Ch	ANHI	Bimlipatam a c	ANKY	BURMA.
ANDR	One and a half Degree Ch	ANHJ	Santapilli	ANKZ	Akyab c
		ANHK	Kalingapatam a	ANLC	Arakan R
ANDS	Equatorial Ch	ANHM	Púndi a	ANLD	Fakir Pt a
		ANHO	Baruva (Barwa). a	ANLE	Combermere B
ANDT	G OF MANAR.	ANHP	Sonapur a	ANLG	Kyauk Pyu Hr
ANDV	Manapad Pt	ANHR	Gopalpur a c	ANLH	Southampton Road
ANDW	Coilsinipatam	ANHS	Berhampur	ANLI	The Terribles
ANDX	Coilnapatam	ANHT	Ganjám	ANLK	Ramree Hr
ANDZ	Tuticorin a c	ANHV	Rambha	ANLM	Chedúba Str
ANEB	Hare I	ANHW	Chilka Lake	ANLO	Sandoway
ANED	Pamban a	ANHX	Puri a c	ANLQ	Khwa, or, Gwa
ANEF	Manar	ANHZ	Devi R	ANLR	C Negrais
		ANIB	Mahanadi R	ANLS	Pegu
ANEG	CEYLON	ANIC	Cuttack	ANLU	Bassein R
ANEI	Putlam	ANID	False Pt c e	ANLV	Diamond I c e
ANEJ	Chilaw	ANIF	False B	ANLW	Alguada Reef
ANEK	Negombo	ANIG	Palmyras Pt	ANLX	Irrawady R
ANEM	COLOMBO b	ANIH	Dhamra R	ANLY	Barragua Lt V
ANEO	Kandi	ANIJ	Shorti I	ANLZ	Krishna Lt V
ANEP	Kaltura	ANIK	Chandbali a c	ANMB	China Bakir
ANER	Barberyn I	ANIM	Balasor a c	ANMD	RANGOON b c f
ANES	POINT DE GALLE d e			ANME	Spit Lt V
ANET	Neptune Battery			ANMF	Elephant Pt e
ANEV	Matura	ANIO	BENGAL B	ANMH	Eastern Grove Lt
ANEW	Dondra Hd			ANMI	Henzada
ANEX	Tangalle	ANIP	Pilots Ridge Lt V	ANMJ	Prome
ANEZ	Hambantotti	ANIR	Húgli, or, Hoogly R	ANML	Thayetmyo
ANFB	Great Basses	ANIS	Eastern Ch Lt V	ANMO	Ava
ANFC	Little Basses	ANIT	Intermediate Lt V	ANMP	Mandalay
ANFD	Batticaloa Roads a	ANIU	Lower Gaspar Lt V	ANMR	Bhamo
ANFE	Trincomali	ANIV	Upper Gaspar Lt V	ANMS	Yunan
ANFG	Pt Pedro	ANIW	Long Sand Lt V		
ANFI	Palmyra Pt.	ANIX	Saugor c e	ANMT	ANDAMAN IS
		ANIZ	Cowcolli, or, Kaukhali	ANMV	Table I
ANFJ	PALK STR	ANJB	Mud Pt c e	ANMW	P Blair f
ANFK	Jafnapatam	ANJC	Kalpi Anchorage	ANMX	P Campbell
ANFM	Tondi	ANJE	Diamond Hi c c		
ANFO	Adrampatam	ANJF	Húgli Pt e	ANMY	NICOBAR IS
ANFP	Pt Calimere	ANJG	Garden Reach	ANOB	Kar Nicobar
ANFR	Coromandel Coast	ANJH	Kidderpur Wet Docks	ANOC	Nankauri Hr
ANFS	Negapatam	ANJI	CALCUTTA b	ANOD	Kamorta
ANFT	Tanjore	ANJK	Sibpur c		
ANFV	Trichinopoly	ANJL	Howrah	ANOF	MARTABAN G
ANFW	Nagar	ANJO	Kidderpur b c	ANOG	Sittang
ANFX	Kárikal a	ANJP	Chandernagore	ANOH	Moulmein
ANFZ	Tranquebar.	ANJQ	The Sundarbans	ANOJ	Amherst b c e
ANGB	Colerun	ANJS	Mutlah Lt V	ANOK	Pagata
ANGC	P Novo	ANJT	P Canning	ANOL	Salween R
ANGD	Cuddalore	ANJU	Haringhata R	ANOP	Tong Aing
ANGE	Pondicherri a	ANJW	Morelganj	ANOQ	Double I
ANGH	Mahábalipúr Pagodas	ANJX	Barisál	ANOR	Yé
ANGI	Covelong	ANJY	Ganges R	ANOT	Tavoy
ANGJ	MADRAS a b c	ANKB	Tetulia R	ANOU	Reef I (Gogaligyun)
ANGK	Pulicat	ANKC	Brahmaputra R	ANOV	Nwa la bo (The Cone)
ANGM	Armeghon	ANKD	Meghna R	ANOX	Mergui
ANGO	Nellore	ANKF	Narainganj	ANOY	Tenasserim
ANGP	Kottapatam	ANKG	Dacca	ANOZ	Morrison B
ANGR	Nizampatam	ANKH	CHITTAGONG	ANPC	Hastings Hr
ANGS	Kistna R	ANKJ	Karnafuli R a	ANPD	Renong
ANGT	Divi Pt	ANKL	Júldia Hill c	ANPE	Kra
ANGV	Masulipatam a c	ANKM	Sudder-ghat c	ANPG	Pakchan R
ANGW	Narsapur	ANKP	Kutabdia I		
ANGX	Yanaon	ANKQ	Maskhal I	ANPH	MALAY PENINSULA
ANGZ	Godavari R.	ANKR	Cox Bazár	ANPI	Kopah Inlet
ANHB	Coringa B	ANKT	Elephant Pt.	ANPK	Taukopah
ANHC	Hope I	ANKU	Naaf R	ANPL	Klong Bagatae

a Signifies an International Signal Station, b a Time Signal Station, c a Weather, Tide, or Ice Signal Station, d a Life-Saving Station, e a Lloyd's Signal Station, f a Wireless Telegraph Station

MALAY PENINSULA		SIAM		CHINA	
ANPM	Salang I, or, Junkseylon	ANTS	Redang Is	ANXJ	CHINA
ANPR	Khelong B	ANTU	Kalantan	ANXK	Pakhoi
ANPS	Tongka	ANTV	Singora	ANXL	Hainan
ANPU	Puket Hr	ANTW	Lakon	ANXO	Kienchu fu
ANPV	Tharún Hr	ANTY	Bandon	ANXP	Hoihau B
ANPW	Kasom R	ANTZ	Meuangpran	ANXQ	Hausui B
ANPY	Panga R	ANUB	Faichabun	ANXS	Yulinkau B
ANPZ	Malakka Str	ANUD	Maikulaung	ANXT	C Bastion
ANQB	Tarang	ANUE	Ratburi	ANXU	Leichau
ANQD	Butang Group.	ANUF	Kanburi	ANXV	Kwangchauwan
ANQE	Langkawi I	ANUG	Tachin R	ANXW	Tihenpien, or, Tienpak Hr
ANQF	Pulo Lada				
ANQH	Bass Hr	ANUI	SIAM	ANXZ	Hnilingsan Hr.
ANQI	Kedah R	ANUJ	Bangkok. f	ANYC	Namo Hr
ANQJ	Muda R	ANUK	Bangkok R Lt V	ANYD	St John I
ANQL	Penang e	ANUL	Bang pa Kong R	ANYE	Ladrone Is
ANQM	Government Hill a	ANUM	C Liant	ANYG	Macao
ANQO	Georgetown	ANUO	Chuen I	ANYH	F Guia
ANQR	Muka Hd a	ANUP	Chentabun R	ANYI	CANTON c
ANQS	Fort Pt	ANUQ	Chang I , or, Koh Chang	ANYJ	Canton R (Chukiang).
ANQT	Province Wellesley	ANUR	Tung Yai R	ANYK	Whampoa
ANQV	P Weld or, Sapetang	ANUT	Soton	ANYL	Lantau, or, Taiho I
ANQX	Lárut R.	ANUV	Sala	ANYM	HONGKONG b
ANQZ	Thaipeng	ANUW	Kampong Ton R	ANYP	Victoria c
ANRB	Pulo Pangkor	ANUY	Hatien	ANYQ	Kaulng b
ANRC	Dinding R	ANUZ	Kankao R	ANYR	Aberdeen Docks
ANRE	Pulo Katak	ANVD	Cambodia, or, Kamao Pt	ANYT	Lema Is
ANRF	Perak R.	ANVE	Mitho	ANYU	Gap Rock a
ANRG	Telok Anson	ANVF	Angiang	ANYV	Green I
ANRH	Durian Sabatang	ANVH	Oudon	ANYX	C D'Aguilar f
ANRJ	Kwala Kangsa	ANVI	Pulo Condore	ANYZ	C Collinson
ANRK	Bernam R			ANZB	St Kiang, or, West R
ANRL	Selángor R	ANVJ	COCHIN CHINA	ANZD	Samshui
ANRO	Pulo Angsa	ANVL	SAIGON b c	ANZE	Wuchu fu
ANRP	One Fathom Bank Lt V	ANVM	Haonbaikan I	ANZF	Waglan Islet
ANRQ	Klang	ANVO	Bienhoa	ANZH	Mirs B
ANRS	P Swettenham	ANVQ	C St James a c	ANZI	Bias B
ANRT	Kwala Lumpor	ANVR	Kangio Bank	ANZJ	Honghai B
ANRU	Langat R	ANVS	Fuokbinkiang Lt	ANZL	Hiechechin B
ANRV	Jugru R	ANVT	Pulo Sapatu	ANZM	Kiahtsz.
ANRX	Arang Arang Anchorage	ANVU	Pulo Ceicer de Mer	ANZO	Tungao.
ANRY	C Rachado			ANZQ	Breaker Pt
ANRZ	Lingi R	ANVW	ANAM	ANZR	C Good Hope
ANSC	Malakka e	ANVX	Kega Pt	ANZS	Sugarloaf I
ANSD	St Paul Hill	ANVY	C Padaran a c	ANZU	SWATAU b
ANSE	Muar, or, Moar R	ANVZ	Fanrang B	ANZV	Chauchu fu
ANSG	Tanjong Panchur	ANWB	Kamranh B	ANZW	Han R
ANSH	Pulo Undan	ANWC	Hon Kohe P	ANZY	Namoa I
ANSI	Formosa Mt	ANWE	C Varella	AOBC	Lamock Is
ANSK	Pulo Pisang	ANWF	Fuyen Hr	AOBD	Tongsang Hr
ANSL	Tanjong Bulus	ANWG	Vungchao	AOBE	Brothers
ANSM	SINGAPORE	ANWI	Kinhon, or, Thinai Hr	AOBF	Chapel I
ANSP	Pulo Brani b	ANWJ	Kikik B (Vungkuit)	AOBG	Tsingseu I
ANSQ	Sultan Shoal Lt	ANWK	Tourane c	AOBH	Taitan I
ANSR	Raffles Lt	ANWM	HUÉ.	AOBI	Amoy b c
ANSU	F Canning b e	ANWO	TONKIN G	AOBJ	Chinchu
ANSV	Singapore New Hr	ANWP	Hatinh	AOBK	Dodd I
ANSW	Mt Faber e	ANWR	Songhoi, or, Kuahoi R	AOBM	Ockseu I
ANSY	Johore R	ANWS	Songka R	AOBN	Hunghwa Ch
ANSZ	Horsburgh Lt	ANWT	Nam Dinh	AOBP	Turnabout I
ANTB	Pedra Branca Lt	ANWV	Hanoi	AOBR	Haitan Str
ANTD	Old Strait of Singapore	ANWX	Bacht Ninh		
ANTE	New Johór	ANWY	Kua Kam R	AOBS	FORMOSA
ANTF	Sibu Is	ANXB	Hon Dau I	AOBT	Kelung Hr f
ANTH	Blair Hr	ANXC	Haifong a b c	AOBV	Syauki Pt
ANTI	Endau R	ANXD	Norway Is	AOBW	Tamsui Hr a c
ANTJ	Pahang R	ANXF	Ha Long B	AOBX	Mengka
ANTL	Pakan	ANXG	Paracel Is	AOBZ	Teukcham
ANTQ	Kuantan R	ANXH	Pratas I	AOCB	Heongsan P
ANTR	Tringano	ANXI	Macclesfield Bank	AOCD	Tyka

a Signifies an International Signal Station, b a Time Signal Station, c a Weather, Tide, or Ice Signal Station, d a Life-Saving Station, e a Lloyd's Signal Station, f a Wireless Telegraph Station

CHINA		PE CHILE GULF		KOREA	
AOCF	Goché	AOGB	Kwachau	AOJR	Shaliutien I
AOCG	Lokiang	AOGC	Sienniumiao	AOJS	Laumuho
AOCH	P Koksi	AOGE	Golden I	AOJU	Ninghai
AOCJ	Taiwan fu	AOGF	Pisinchau	AOJV	Shanhainuan
AOCK	Amping	AOGH	Bethune Pt	AOJW	Great Wall of China
AOCL	Taksu	AOGI	Tsauhia I	AOJX	Liautung Province
AOCN	Tangkiang	AOGJ	NANKIN	AOJY	Kingchu fu
AOCP	Saracen Hd	AOGL	Wade I	AOKB	Liaoho c
AOCQ	Nansha, or, South C	AOGM	Taiping fu	AOKC	Newchwang c
AOCS	Sauo B	AOGN	Wuhu	AOKD	Newchwang Lt V
AOCT	Kaliwan R	AOGP	Hains Pt	AOKE	Mukden
AOCU	PESCADORES Is	AOGR	Pihmatsui	AOKG	Yinkoa
AOCV	Makung Hr	AOGS	Kieuhien	AOKH	P Adams
AOCW	Ponghau Hr	AOGT	Buckminster I	AOKI	Kinchau
AOCX	Meiaco sima Is	AOGV	Wild Boar Reach	AOKL	Liautishan Promontory
		AOGW	Fitzroy I	AOKM	P ARTHUR f
AOCY	MIN R a	AOGX	Taitzuchi Beacon	AOKN	Taihenwan B
AODB	Wufu	AOGZ	Langkiangki, or HenPt	AOKQ	Victoria B
AODC	Tungkuen	AOHB	Nganking Reach	AOKR	Yentoa B
AODE	Tungsha	AOHC	Eagle I Beacon	AOKS	Kwanglotau I
AODG	Mingan Pass	AOHD	Christmas I	AOKU	Blonde Is
AODH	Pagoda Anchorage c	AOHE	Tungliu	AOKV	Haiyuntau I
AODI	Fuehau	AGHG	Spencer Rock	AOKW	Thornton Haven
AODK	Nantai I	AOHI	Dove Pt Beacon	AOKY	Beurchier Group
AODL	Samsa Inlet	AOHJ	Northeast Crossing Bn Lt V	AOKZ	Takushan
AODM	Tongshu I			AOLB	Tayangho
AODP	Fuyan I	AOHK	Otter Pt	AOLD	Yalu R
AODQ	Namkwan Hr	AOHM	Hukau	AOLE	Aichiu
AODR	Wenchau fu	AOHN	Poyang Lake		
AODT	Smpe I	AOHP	Wuching	AOLF	KOREA
AODU	Ngaukiang	AOHR	Kiukiang	AOLG	Chenampo (Chel tan)
AODV	Ou Kiang	AOHS	Wusueh	AOLH	Wiju
AODX	Wenchau B.	AOHT	Liyushan	AOLI	Pingyang Inlet
AODY	Sanmun B	AOHU	Low Pt	AOLJ	Sir James Hall Group.
AODZ	Shipu Hr	AOHV	Kichau	AOLM	Seoul
AOEC	Nimrod Sound	AOHW	Hwangchau.	AOLN	Jinchuen, or, In chön
AOED	Chusan Archipelago.	AOHX	HANKOW	AOLP	Salée R
AOEF	Barren Is	AOHY	Chungking	AOLR	CHEMULPO c
AOEG	Loka I	AOIB	Wuchang fu	AOLS	Wonsan
AOEH	Sinkeamun Hr.	AOIC	Singti Reach	AOLT	Prince Jérôme G
AOEI	Tinghai Hr	AOID	Yochau fu	AOLV	Asan Anchorage
AOEJ	Yung R	AOIE	Tungting Lake	AOLW	Marjoribanks Hr
AOEL	Chinhai	AOHF	Ichang fu	AOLX	Keum gang
AOEM	Ning po	AOIG	Sha Sze	AOLZ	Conference Group
AOEN	Chapu	AOIH	Suchau	AOMB	Guérin I
AOEQ	Wantaokwan	AOIJ	YELLOW SEA, or, HWANG HAI	AOMC	Kokuntan Is
AOER	Hangchau B			AOMD	Kunsan
AOES	Tsientangkiang	AOIK	Kyauchau Hr b c	AOME	Naju Group
AOEU	Yenchu	AOIL	Haiyang	AOMF	Mokpo f
AOEV	Steep I	AOIN	Aylen B.	AOMG	Yong San Gang R
AOEW	West Volcano I	AOIP	Yungching	AOMI	Modeste I.
AOEY	Bonham I	AOIQ	Shantung Promontory	AOMJ	Mackau I
AOEZ	North Saddle I	AOIS	Weihaiwei	AOMK	Washington G
AOFB	Gutzlaff I c	AOIT	Leukungtau	AOMN	Chindo
AOFD	Yangtse R	AOIU	Lungmun Hr	AOMP	Chepn
AOFE	Tungsha Lt V	AOJW	Chefoo b c	AOMQ	Kuper Hr
AOFG	Kiutoan Lt V	AOIX	Tengchau fu	AOMS	Quelpart I
AOFI	Wusung c	AOIY	Miautau Is	AOMT	Beaufort I
AOFJ	SHANGHAI b c f			AOMU	P Hamilton f
AOFK	Paoushan Pt	AOJB	G OF PE CHILI	AOMW	Cargodo I (Kojedo)
AOFM	Langshan Crossing	AOJC	Tatsingho, or, Yellow R	AOMX	Broughton Hr
AOFN	North Tree Beacon	AOJD	Tiemunkwau	AOMY	Douglas Inlet
AOFP	Kiushan Beacon	AOJF	Takoho, or, Tasanho	AONB	Masanpo f
AOFR	Tunglotu Beacon	AOJG	Peiho R	AONC	Tsu sima
AOFS	Kiangyin	AOJH	Taku a c f	AOND	Fusan
AOFT	Fishbourne I	AOJK	Taku Bar Lt V	AONF	Fusan Kai
AOFV	Pottinger I	AOJL	Tientsin f	AONG	Sorio
AOFW	Tantu	AOJM	Tungchau	AONH	Unkofsky B
AOFX	Silver I	AOJP	PEKING f	AONJ	C Duroch
AOFZ	Chinkiang fu	AOJQ	Petang	AONK	Chaguchiendogu

a Signifies an International Signal Station   b a Time Signal Station, c a Weather, Tide, or Ice Signal Station, d a Life-Saving Station, e a Lloyd's Signal Station   f a Wireless Telegraph Station

SEA OF OKHOTSK		JAPAN			
AONL	Gensan (Yuensan)	AOSH	Tausk B	AOWK	Yobuko
AONP	P Lazaref	AOSJ	Arnam R	AOWL	Kariya
AONQ	Dungan R	AOSK	Babushkin G	AOWM	Hibi Ch
AONR	P Chestakof	AOSL	Ghijinsk B	AOWP	Imari
AONS	Song Chiu	AOSN	Gighiga	AOWQ	Hirado no Seto (Spex Str)
AONT	Tizenko B	AOSP	Penjinsk G		
AONU	Sivutch B	AOSQ	Tigil R	AOWR	Hirado
AONV	Anna B			AOWT	Goto Is
AONX	Kornilof G	AOSZ	JAPAN	AOWU	Koshiki-sima
AONY	Nadjini			AOWV	Nama ura
AONZ	Goshkevitch B	AOTB	HOUSHU	AOWX	Kisakasima (Kuga sima)
		AOTC	Tsugaru Str	AOWY	Ose saki, or, C Goto
AOPC	MANCHURIA	AOTE	Omasaki sima	AOWZ	Fukai
		AOTF	Rikuoku G	AOXB	Omura
AOPD	RUSSIAN TARTARY	AOTH	Ominato	AOXD	Otawa
AOPE	Tumen R (Tumen Ula)	AOTI	Yokohama (Rikuoku G)	AOXE	NAGASAKI c f
AOPG	Hunchun	AOTJ	Nohechi	AOXF	Kageno sima
AOPH	Posiette B	AOTL	Shiranai B	AOXH	Iwo sima
AOPI	Expedition B	AOTM	Moura B	AOXI	Fukabori
AOPK	Novgorod	AOTN	Azamushi	AOXJ	Kuchinotsu
AOPL	Pallada Road	AOTQ	Aomori c	AOXL	Shimabara
AOPM	Trinity B	AOTR	Tappi Saki (C Tateupi)	AOXM	Miike c
AOPQ	Gamova B	AOTS	Yonesiro Gawa c	AOXN	Haya saki Ch
AOPR	Slavianski B	AOTV	Sakata c	AOXQ	Misumi Hr
AOPS	Amur B	AOTW	Kamo	AOXR	Yatsushiro
AOPU	Sedimi B	AOTX	Sado I	AOXS	Sakitsu ura
AOPV	Pestchannui Pt	AOTZ	Niigata Roadstead c	AOXU	Ushibuka Hr
AOPW	Retchnoi I d	AOUB	Fushiki c	AOXV	Naga sima
AOPY	C Tokarofski	AOUC	Noto Peninsula	AOXW	Koshiki Is
AOPZ	VLADIVOSTOK a b c f	AOUE	Nanao c	AOXZ	Kata ura
AOQB	Golden Horn Hr	AOUF	Nakai	AOYB	Tomari ura
AOQD	C Galdobin d	AOUG	Sudzu misaki	AOYC	Kagosima c
AOQE	Kazakavitch	AOUI	Kanazawa c	AOYE	Sata no misaki, or, C Chichakoff
AOQF	Novik B	AOUJ	Mikuni, or, Sakai		
AOQH	Skrypleff I a	AOUK	Tate ishi zaki	AOYF	Van Diemen Str (Osumi Str)
AOQI	Eugénie Archipelago	AOUM	Wakasa B		
AOQJ	Peter the Great B	AOUN	Tsuruga B c	AOYG	Linschoten Is
AOQL	Onsuri B	AOUP	Ohama	AOYI	Riu Kiu
AOQM	Askold I a	AOUR	Miyadsu	AOYJ	Okinawa sima
AOQN	Strelok B	AOUS	Oki sima	AOYK	Kerama Group
AOQR	Puttatin I	AOUT	Saigo Hr	AOYL	Me Sima Group
AOQS	America B	AOUW	Tsuno sima	AOYM	Aburatsu Hr
AOQT	Povorotny Pt	AOUX	Miho B	AOYN	Osima d
AOQV	Olga B	AOUY	Hamada	AOYP	P Inokushi
AOQW	St Vladimir B	AOVB	P Susa	AOYR	P Yonodzu
AOQX	G OF TARTARY	AOVC	Hagi	AOYS	Saiki
AOQZ	Pestchanii Pt	AOVD	Sen Zaki uchi	AOYT	Bungo Ch
AORB	Barracouta Hr (Imperial Hr)	AOVF	Ouro Anchorage	AOYV	Tsukumi
		AOVG	Aburatani B	AOYW	Usuki
AORC	Menschikoff Pt	AOVH	Siro simas	AOYX	Sagano seki
AORE	Castries B d	AOVJ	Rokuren I	AOZB	Oita
AORF	Alexandroveki	AOVK	Mutsure c	AOZC	Beppu
AORG	Amur R	AOVL	Shirasu	AOZD	Hiji
AORI	North Ch	AOVN	Kosedo Str		
AORJ	Nikolaevek	AOVP	Shimonoseki c	AOZF	SETO UCHI, or, INLAND SEA
AORK	Sakhalin I (Karafuto)	AOVQ	Hiku sima		
AORM	Jonquière C a			AOZG	Hime sima
AORN	Dui	AOVS	KIUSHU	AOZH	Nakatsu
AORP	Kosounai	AOVT	Moji	AOZJ	Isaki
AORS	Mauka Cove	AOVU	SHIMONOSEKI STR	AOZK	Suwo nada
AORT	C Notoro a	AOVX	Hayatomo Str	AOZL	Mitazuri
AORU	La Pérouse Str	AOVY	Dairi	AOZN	Tokuyama.
AORW	Koreakovek	AOVZ	Kokura	AOZP	Kasado sima
AORX	C Elizabeth	AOWC	Wakamatsu c	AOZQ	Iyo nada
AORY	Shantarski Is	AOWD	Fukuoka c	AOZS	Osima (Seto Uchi)
AOSB	Tougourski G	AOWE	Hakata	AOZT	Misima nada
AOSC	Udeki B	AOWG	Iki sima	AOZU	Hirosima c
AOSD	P Aian	AOWH	Yebosi I	AOZW	O Sima
AOSF	SIBERIA	AOWI	Karatsu	AOZX	Mitzura
AOSG	Okhotsk	AOWJ	Taka sima	AOZY	Miwara

a Signifies an International Signal Station, b a Time Signal Station, c a Weather, Tide, or Ice Signal Station, d a Life-Saving Station, e a Lloyd's Signal Station, f a Wireless Telegraph Station

JAPAN		BERING SEA	
APBD	Onomichi no seto c	APEY	Owasi	APIZ	Shiogama
APBE	Mekari seto	APEZ	Mura Hr	APJD	Tona Canal
APBF	Tomo	APFC	Goza Hr	APJE	Ishi no maki c d
APBH	Fukuyama (Honshu)	APFD	Matoya	APJG	Kinkuwasan, or, Yama-dori Ch
APBI	Okayama	APFE	Toba		
APBJ	Katakami	APFH	Suga sima		
APBL	Sakoshi B	APFI	OWARA B (ISE SEA)	APJH	Onagawa
APBM	Oo-Ura B	APFJ	Owari	APJI	Okatsu
APBN	Akashi	APFL	Yamada	APJL	Oppa B
APBQ	Wada no misaki	APFM	Okuchi B	APJM	Kita kami R
APBR	KOBE b c	APFN	Tsu	APJN	Shidzukawa
APBS	Hiogo	APFQ	Yokka ichi c	APJQ	Koidzumi
APBU	Kawatsi	APFR	Kuwana	APJR	Kesennuma
APBV	Osaka c	APFS	Nagoya	APJS	Hirota B
APBW	Kizu gawa	APFU	Atsuta c	APJU	Imaidzumi gawa
APBY	Sakai c	APFV	Chita	APJV	Ofunato B
APBZ	Tomago sima	APFW	Taketoyo c	APJW	Kamaishi
APCB	Isumi Str	APFY	Handa	APJY	Riyotshi
APCE	Hachken R	APFZ	Mikawa	APJZ	Odzuchi
APCF	Wakayana c d	APGB	Toyohashi	APKB	Funakoshi
		APGD	Oyama	APKD	Kuji B
APCG	SHIKOKU	APGE	Hamamatsu	APKE	Yamada
APCI	P Ikata	APGF	Kakesuka minato.c	APKF	Miyako
APCJ	P Amai	APGI	Omai saki	APKH	Shiriya saki
APCK	Yawatahama Hr	APGJ	Suruga G	APKI	Ohata
APCM	P Shitama	APGK	Shizuoka		
APCN	Okuchi B	APGM	Shimidzu Hr c	APKJ	HOKUSHU
APCO	P Yoshida	APGN	Mumadsu	APKM	Fukuyama, or, Matsu-mai (Hokushu)
APCR	Uwajima	APGO	Yeno ura		
APCS	Kure b	APGR	Heda Hr	APKN	Sirakami saki
APCT	Susaki c	APGS	Arari	APKO	Fuku shima c
APCV	Riudsu saki	APGT	Tago B	APKR	Kattoji saki
APCW	Urado Hr c	APGV	Iro o saki	APKS	HAKODATE b c
APCX	Kochi	APGW	Mikomoto, or, Rock I	APKT	Anama saki spit Lt V
APCZ	Moroto Saki	APGX	Simoda	APKU	Shiokubi mi saki
APDB	Kamata saki	APGZ	Sagami ura	APKV	Yesan Saki c
APDC	I shima	APHB	Atami.	APKW	Volcano B
APDF	Tachibana ura	APHC	Odawara	APKX	Mori c
APDG	Naka shima gawa	APHE	Kotawa B	APKZ	Simmororan, or, Ender-mo Hr c
APDH	Komatsushima An-chorage.	APHF	Joka sima.		
		APHG	Tsurugi saki	APLB	Saru
APDJ	Toku shima	APHJ	Kaneda B	APLC	Sutsini
APDK	Naruto Passage	APHK	Uraga Ch	APLE	Urakawa
APDL	Awaji sima	APHL	Kanonsaki	APLF	Yerimo saki
APDN	Fuk ura	APHN	Yokosuka c	APLG	Sirunoki
APDO	Yura Hr	APHO	Susaki	APLH	Kushiro
APDQ	Tsuda B	APHQ	Ngisi	APLI	Keramot Saki
APDS	Takamatsu c	APHS	Honmoku Lt V	APLJ	Akishi
APDT	Nabe sima	APHT	YOKOHAMA b c	APLK	Hamanaka
		APHU	Takasimacho	APLN	Otchisi saki, or, C Usu
APDW	Sakaide	APHW	Kanagawa	APLO	Noshapp Saki.
APDX	Marugame	APHX	Kawa saki	APLQ	Nemoro Anchorage c
APDY	BINGO NADA.	APHY	Haneda	APLS	Natske B
APEB	Imaban	APIB	Shinagawa c	APLT	Kinashiri
APEC	Matsuyama	APIC	TOKYO	APLU	Kuril Is
APED	Mitsuga hama c	APID	Tsukiji	APLW	Shakotan Hr
APEG	Tsuru sima	APIF	Tone gawa	APLX	Matsugahama B
APEH	Nagahama	APIG	Funabashi	APLY	Anama Hr
		APIH	Chiba	APMB	Yetorup I
APEI	KII CH	APIK	Iwaga saki	APMC	Hitokatpu B
APEK	Hino misaki	APIL	Kisaradzu	APMD	Naibo Hr
APEL	Tanabe	APIM	Futsu saki	APMF	Urup I
APEM	Siwo misaki	APIO	Tate yama	APMG	Soya saki
APEO	Kusimoto d	APIQ	No sima saki	APMH	C Nosseyab
APEQ	Oo sima.	APIR	Dzunan Is	APMJ	Refunsiri I
APER	Kashi no saki c	APIT	Fusi yama	APMK	Risiri I
APET	Ura Kami Hr	APIU	Inuboye saki c	APML	Mashike, or, Maski c
APEU	Taize Ura	APIV	Naka Hr	APMQ	Ishikari R
APEV	Singo c	APIX	Sendai B	APMR	Oterranai c
APEX	Kada B	APIY	Matsu shima	APMT	Maloyama

a Signifies an International Signal Station   b a Time Signal Station,   c a Weather, Tide, or Ice Signal Station,   d a Life-Saving Station,
e a Lloyd's Signal Station,   f a Wireless Telegraph Station

SUMATRA		JAVA		BORNEO	
APMU	Kamoi saki, or, C Novosilzov	APRG	Siak	APUZ	Babi I
		APRH	Durian Str	APVB	Payung I
APMV	Siru unku (I wani B) c	APRJ	Pulo Batam.	APVC	Edam I
APMX	Bebkei misaki (Sitt B) c	APRK	Rhio Str.	APVE	Batavia c f
		APRL	Linga	APVF	Tanjong Priok b
APMY	Okusiri	APRN	Little Karas I	APVG	Buitenzorg.
APMZ	Yesashi c	APRO	Terkolei I	APVJ	Boompjes I , or, Menjawak
APND	C Sirakami	APRQ	Pulo Sau		
		APRT	Indragirie R	APVK	Inderamayu Pt
APNE	KAMCHATKA	APRU	Jambie B	APVM	Cheribon f
APNG	C Lopatka (Kapury).	APRV	Berhala Str	APVN	Tegal
APNH	Avatcha B	APRX	Sungi Banju Asing	APVO	Pekalongan
APNI	Dalni Pt	APRY	Sungi Sungsang, or, Palembang R	APVR	Samarang
APNK	Petropavlovsk			APVS	Krimon Java Is
APNL	Ukinsk B	APRZ	BANKA	APVT	Katang Islet
APNM	Komandorski Is	APSB	Muntok	APVW	Japara Islet
APNQ	Bering I	APSC	Marawang Road	APVX	Mandalike I
APNR	Medni, or, Copper I	APSD	Pulo Dupur	APVY	Juana
APNS	Karaga Hr	APSE	Gaspar Str.	APWB	Rembang.
APNU	Karaginski I	APSG	Pulo Mendanao	APWC	Tuban
APNV	Baron Korfa G , or, Kultuznay B	APSH	Langwas I	APWD	C Piring
		APSI	Belitung, or, Billiton I	APWF	Madura I
APNW	Bering Sea	APSJ	Pulo Leat	APWG	Sapudi
APNY	Lavrova Hr	APSK	North Watcher Lt	APWH	Geresik.
APNZ	Oliutorski C	APSL	Masui R	APWJ	SURABAYA b e
APOB	Natalie B	APSM	Tulang Bawang R	APWK	Zwaantjes Drooghte, or, Koko Reef Lt
APOD	C Navarin	APSO	Saputi R.		
APOE	Anadyr R	APSQ	SUNDA STRAIT	APWL	Pasuruan.
APOF	Klinkowstroem B	APSR	Telok Betung	APWN	Probolingo
APOH	Kresta G	APSU	Ratai	APWO	Besuki
APOI	Providence B.	APSV	Pundu	APWQ	Panarukan
APOJ	Emma Hr	APSW	Lagundi Str	APWS	Karang Mas, or, Meinderts Reef
APOL	St Lawrence I	APSY	Semangka B		
APOM	Bering Str	APSZ	Flat C	APWT	Duiven I , or, Tabuan
APON	Mechigme B	APTB	Belimbing	APWU	Bansering
APOR	Lutke Hr	APTD	Pulo Pisang Hr	APWX	Banjuwangi a
APOS	East C	APTE	Kroé	APWY	Segoro Wedi B
APOT	C Serdze Kamen	APTF	Kawur	APWZ	Chilachap Inlet
APOV	Wrangell I	APTG	Pulo Tikus (Rat I)	APXC	Chitando Inlet
APOW	Kolyma R	APTH	Benkulen	APXD	Zand, or, Chiletu B
APOX	Indigirka R	APTI	MENTAWI Is.		
APOZ	Iana R	APTJ	Vekens B	APXF	BORNEO
APQB	Liakhov, or, New Siberia Is	APTL	Hurlock B	APXH	Sambar Pt
		APTM	Katorei B	APXI	Pawan R.
APQC	Kotelnoi I	APTN	Trussan B	APXJ	CARIMATA STR.
APQE	Lena R	APTQ	Bunga B	APXL	Panjang I (Borneo)
APQF	Oleneka R	APTR	Pulo Katang Katang	APXM	Great Kapuas R
APQG	Hatanga R	APTS	Padang a	APXN	Pontianak
		APTV	Priaman	APXQ	Sambas R
		APTW	Tiko	APXR	Palo R
APQI	SUMATRA	APTX	Coros, or, Keeling Is	APXS	Tambelan Is
APQJ	Acheh Hd , or, Acheen Hd	APTZ	Ayer Bangies	APXU	Anamba Is
		APUC	Natal	APXV	Natuna Is
APQK	Pulo Bras	APUD	Batu Is	APXW	Tanjong Datu
APQM	Kota Raja	APUF	Pulo Nias	APXZ	Saráwak
APQN	Olehleh B	APUG	Baros	APYB	Moratabas Entrance
APQO	Pulo Buru	APUH	Tapanuli B	APYC	Lupar R
APQR	Sabang B e	APUI	Siboga	APYE	Rajang R
APQS	Jambu Ayer, or, Diamond Pt	APUJ	Singkel	APYF	C Sirik
		APUK	Melabu B	APYG	Bruit R
APQT	Edie Rajut R	APUL	Rigas	APYI	Ballang R
APQU	Langksa R	APUN	Christmas I	APYJ	Kidorong Pt
APQW	Langkat R			APYK	Barram R
APQX	Medan	APUQ	JAVA	APYM	Muara I
APQY	Deli River Lt V	APUR	First Pt	APYN	Brunei
APRB	Batu Bara	APUS	Fourth Pt (Anjer) e	APYO	LABÚAN
APRC	Lidung B	APUT	Old Anjer	APYR	Victoria Hr
APRD	Sungi Rokan	APUV	New Anjer	APYS	Kimánis B.
APRE	Pulo Jarak	APUW	St Nicolas Pt	APYT	Gaya B.
APRF	Brewer Str	APUX	Bantam	APYV	Sapangar B

a Signifies an International Signal Station, b a Time Signal Station c a Weather, Tide, or Ice Signal Station, d a Life-Saving Station e a Lloyd's Signal Station, f a Wireless Telegraph Station

PHILIPPINES		CELEBES	
APYW	Ambong	AQDW	P Mariveles
APYX	Kudat B	AQDX	P Olongapó
APZB	Marudu B	AQDY	Subig P
APZC	Kina Balu	AQEB	Capones Pt
APZD	Banguey I	AQEC	Pampanga
APZE	Mitford Hr	AQED	Iba
APZF	Mallawallé I	AQEG	Zambales Province
APZG	Balambángan I.	AQEH	Masinglok P
		AQEI	Santa Cruz
APZH	BALABAC I	AQEK	Dazol B
APZJ	Marchesa B	AQEL	Bolinao
APZK	Sugut R	AQEM	Piedra Pt
APZL	Labuk R	AQEO	Sual
APZN	Kalagaan R	AQEP	Lingayen
APZO	Sandakan Hr	AQER	Ilocos (St Thomas Peak)
APZQ	Elopura		
APZS	Kinabatangan R	AQET	San Fernando Pt
APZT	Dent Haven	AQEU	Vigan
APZU	Darvel B	AQEV	Salomague P
APZW	Silam Hr	AQEX	C Bojeador c
APZX	Sibutu I	AQEY	Rio Grande de Cagayan.
APZY	Sibuko B	AQEZ	Aparri
AQBD	Sesajab R	AQFC	Nueva Ecija
AQBE	Mahakan B	AQFD	C Engaño
AQBF	Pasir	AQFE	P St Vincente
AQBH	Kelumpang B	AQFG	Babuyan Is
AQBI	Pulo Laut	AQFH	Camiguin I
AQBJ	Kota Baharu	AQFI	BATAN Is
AQBL	C Selatan	AQFK	Bashi Is
AQBM	Baritto R	AQFL	P Lampon
AQBN	Banjermasin	AQFM	Gumaka
AQBP	Sampit	AQFO	Calagua
AQBR	Pambuang	AQFP	Capalonga
AQBS	Kumai R	AQFR	Mambulao P
AQBU	Kotawaringin R	AQFT	San Miguel B
AQBV	Jelai R	AQFU	Nueva Carceres
		AQFV	Sisiran P
		AQFX	Catanduanes I
ABQX	PHILIPPINE Is	AQFY	Tobako B
AQBY	C Melville	AQFZ	Albay
AQBZ	Calandorang B	AQGC	Polki B
AQCD	PALAWAN I	AQGD	San Bernardino Str
AQCE	P Barton	AQGE	Bulusan
AQCF	Malampaya Sound.	AQGF	Magnok B
AQCH	Bakit B	AQGI	Tikao
AQCI	Talindak	AQGK	Sorsogon
AQCJ	Taitai B	AQGL	Donsol
AQCL	Dumaran I	AQGM	Pasakao Anchorage
AQCM	P Royalist	AQGO	BURIAS I
AQCN	Calamianes Group	AQGP	Ragai G
AQCP	Culion I	AQGR	Marinduque I
AQCR	Busuanga	AQGT	Buac
AQCS	MINDORO	AQGU	P Laguimmanoc
AQCU	Ilin I	AQGV	Pagbilao B
AQCV	Mangarin	AQGX	Tayabas R
AQCW	Sablayan	AQGY	Batangas
AQCY	Paluan Bay	AQGZ	Taal
AQCZ	Calapan	AQHC	Tablas I
AQDB	Polak	AQHD	Loog P
AQDE	Lubang I	AQHE	Romblon P
AQDF	P Tilig	AQHG	Sibuyan I
AQDG	Cabra I	AQHI	Magallanes
AQDJ	LUZON I.	AQHJ	MASBATE I
AQDK	C Santiago	AQHK	Arroroi
AQDL	Restinga Pt	AQHL	Mandao
AQDN	Corregidor I	AQHM	Palanog P
AQDO	Caballo I	AQHO	Uson B
AQDS	Cavite b f	AQHP	Catangan P
AQDT	MANILA b c	AQHR	PANAY I
AQDU	Pasig R	AQHT	Tilat, or, Buena Vista.

AQHU	Pandan
AQHV	Colasi
AQHX	Batan P
AQHY	Capis
AQHZ	Calabazas Isles
AQIC	Siete Pecados
AQID	Iloilo
AQIE	Guimarás
AQIG	Bondulan Pt
AQIH	Lusaran Pt
AQIJ	NEGROS I
AQIL	Bantayan I.
AQIM	SEBU (ZEBU)
AQIN	Barili
AQIO	Maktan I
AQIP	Tanón Str
AQIR	Dumaguete
AQIS	Argao Pt
AQIU	Sibonga
AQIV	Carcar
AQIW	Naga
AQIY	Bagacai Pt
AQIZ	Bohul I
AQJB	Tagbilaran.
AQJD	Laon
AQJE	Sikijor I
AQJF	Leite
AQJH	P Bello
AQJI	Tacloban
AQJK	SAMAR
AQJM	Calbayoc
AQJN	Catbalogan
AQJO	Guiuan
AQJR	Palapa P
AQJS	Laguan B
AQJT	Candolu Islet
AQJV	Dinagat I
AQJW	P Gabó
AQJX	Siargao I
AQJZ	Bucas Is
AQKC	MINDANAO I
AQKD	Dapitan B
AQKE	Tubud Pt
AQKG	P Misamis
AQKH	Butuan
AQKI	Surigao
AQKL	Caraga B
AQKM	Pujada B
AQKN	Mati
AQKP	P Balete
AQKR	Hijo
AQKS	Davao
AQKU	Malalag P
AQKV	P Lebak
AQKW	Linao B
AQKY	Kota-batú
AQKZ	Palak
AQLB	Pagadian B
AQLD	P Sambulauan
AQLE	Dumankilas B
AQLF	Zamboanga f
AQLI	JOLO f
AQLJ	Basilan I
AQLK	P Isabela
AQLN	Dalrymple Hr
AQLO	Buol
AQLP	Pangasinan I
AQLS	Taptil Group
AQLT	Siasen
AQLU	Tawi Tawi

OCEANIA

AQLV	Bongao I	AQPI	Arru Is.	AQTY	Geelvink B
AQLW	Talautse Is	AQPJ	Wokam I	AQUB	Dorei P
AQLX	Talauer Is	AQPL	Dobbo		
		AQPM	Ké Dulan I	AQUC	BISMARCK ARCHIPELAGO.
AQLY	CELEBES	AQPN	Banda		
AQLZ	Manado	AQPR	Amboina	AQUD	NEW POMERANIA
AQMB	Banka I	AQPS	Ceram I'	AQUF	Stephansort
AQMD	Kema	AQPT	Sawai Hr	AQUG	Herbertshöh.
AQME	Pondang Islets	AQPV	Waru B	AQUH	Blanche B
AQMF	Gorontalo	AQPW	Hoya (Teluti) B	AQUJ	Matupi I.
AQMH	Peling I	AQPX	Elaputi B		
AQMI	Bangaai Archipelago	AQPZ	Amahoi	AQUK	NEW MECKLENBURG
AQMJ	Kendari B	AQRB	Kaibobo Road.	AQUL	Meoko Hr
AQML	Buton	AQRC	Piru B	AQUM	Nusa Hr
AQMN	Palopo	AQRE	Buru I	AQUO	Kahu.
AQMO	Palima	AQRF	Kayeh B		
AQMR	Boni	AQRG	Pitt Passage	AQUR	PACIFIC OCEAN (OCEA-
AQMS	Salengketa Pt	AQRI	GILLOLO I		NIA)
AQMT	Balang Nipa (Sinjai)	AQRJ	Maba Road	AQUS	Admiralty I
AQMU	Makassar Str	AQRK	Weda	AQUT	Nares Hr.
AQMV	Salayar	AQRM	Gebi I	AQUW	Hermit Is
AQMW	Malsoro B	AQRN	Molucca Passage	AQUX	Caroline Is
AQMX	MAKASSAR f	AQRO	Bachian	AQUY	Kusaie I
AQMY	Taka Ramata (Brill	AQRP	Ternate	AQVB	Chabrol Hr
	Rk)	AQRS	Waigiu I	AQVC	Ponapi I
AQMZ	Mamuju B	AQRT	Gillolo Passage	AQVD	Jamestown Hr (Je-
AQNB	Palos	AQRU	Dampier Str		koits)
AQNC	Wani B			AQVF	Truk Is
AQNF	Dondo B	AQRV	NEW GUINEA (PAPUA)	AQVG	Yap I
AQNG	BALI			AQVH	Tomil B
AQNH	Beliling (Buleleng)	AQRW	BRITISH NEW GUINEA	AQVJ	PELEW (PALOA) Is.
	Roadstead	AQRX	Salwatti	AQVK	Marianas
AQNI	Amok (Lahuan Amok)	AQRY	Segaar B.	AQVL	Guam I f
AQNK	Panti Barat (Badong B)	AQSB	McCluer Inlet) Berau	AQVN	Agaña
AQNL	Lombok		G)	AQVO	P San Luis D'Apra
AQNM	Ampenan B	AQSD	Sekru	AQVP	Bonin Is (Arzobispo)
AQNP	Labuan Tering B	AQSE	Kapau.	AQVS	Hillsborough I (Bailly
AQNR	Piju B	AQSF	Frederick Henry I		Group)
AQNS	Alas Str	AQSH	C Valsche, or, False C	AQVT	Peel I (Chicijima)
AQNU	Sumbawa	AQSI	Dourga Str		(Beechy Group)
AQNV	Bima	AQSJ	Pisang B	AQVU	P Lloyd
AQNW	Sapeh B	AQSL	Wassi Kussa	AQVX	MARSHALL Is
AQNX	Pulo Tenga (Paternos-	AQSM	Mai Kussa	AQVY	Jaluit
	ter) Is	AQSN	Boigu, or, Talbot I	AQVZ	GILBERT Is
AQNY	Sumba I	AQSP	Daudai	AQWC	Apaiang
AQNZ	Nangamessi B	AQSR	Tauan		
AQOB	FLORES	AQST	Saibai I	AQWE	SOLOMON Is
AQOD	Solor I	AQSV	Pahoturi, or, Kaua R	AQWF	Buaro (San Cristoval) I
AQOE	Terang B	AQSW	Brinatauri, or, Katau R	AQWG	Ugi I
AQOF	Bari B	AQSX	G of Papua	AQWI	Guadalcanar I
AQOG	Sabalanga (Postillion) Is	AQSZ	Fly R	AQWJ	Malaita I
AQOH	Potta Road	AQTB	Aird R	AQWK	Ysabel I
AQOI	Karakaiki B	AQTC	Lolo I	AQWM	Choiseul I
AQOJ	Dondo B	AQTE	Hall Sound	AQWN	Bougainville I
AQOL	Ombai, or, Allor.	AQTF	P Moresby	AQWO	Santa Cruz Is (Nitendi
AQOM	Kebula B	AQTG	Granville		Is)
AQON	Timor.	AQTI	P Glasgow	AQWR	Vanikoro I
AQOP	Koepang	AQTJ	South C		
AQOS	Dilhi	AQTK	Samarai I	AQWS	NEW HEBRIDES
AQOT	Semao (Kurong)	AQTM	Louisiade Archipelago	AQWT	Banks Is
AQOU	Rotti			AQWV	Ureparapara I
AQOV	Serwatti Is	AQTN	KAISER WILHELMSLAND	AQWX	Avreas B
AQOW	Wetta I	AQTO	Finsch Hr	AQWY	P Patteson.
AQOX	Kissa (Makisar).	AQTR	Astrolabe B	AQXB	Mota I
AQOZ	Letti I	AQTS	Frederich Wilhelm Hr	AQXC	Lakona B
AQPB	Damma I	AQTU	Hatzfeldt Hr	AQXD	Marina, or, Espiritu
AQPC	Tepa	AQTW	Empress Augusta R		Santo I
AQPE	Baba I	AQTX	Amberno (Mamberon-	AQXF	Bougainville Str.
AQPF	Timor Laut		mo) R	AQXG	Tangoa
AQPH	ARAFURA SEA			AQXH	Mallicolo (Malekula)

a Signifies an International Signal Station, b a Time Signal Station, c a Weather, Tide, or Ice Signal Station d a Life-Saving Station, e a Lloyd's Signal Station, f a Wireless Telegraph Station

OCEANIA				AUSTRALIA	
AQXJ	P Stanley	ARCG	Viti Levu I t. V.	ARGC	MARQUESAS Is
AQXK	P Sandwich.	ARCH	Suva a	ARGE	Hiva-Oa.
AQXL	Epi I , or, Taaiko I	ARCJ	Rewa	ARGF	Périgot B
AQXN	Efáte, or, Sandwich I	ARCK	Naailai	ARGH	Taa hu ku B
AQXO	Fila Hr	ARCL	Mbau Waters.	ARGJ	Nukuhiva
AQXP	Meli Anchorage	ARCN	Lovuka	ARGK	Comptroller B
AQXS	Havannah Hr	ARCO	Nandi Waters.	ARGL	Tai-ó-haé, or, Anna Ma-
AQXT	Eromanga	ARCP	Likuri Hr.		ria B
AQXU	Dillon B	ARCS	Singatoka R	ARGN	P Tai oa.
AQXW	Polenia B	ARCT	Rovondrau B.	ARGO	Ua Huku I
AQXY	Cook B	ARCU	Vanua Levu	ARGP	Ua Pu I,
AQXZ	Tanna	ARCW	Savu Savu B		
AQYO	Kwamera	ARCX	Nandi B	ARGQ	TUAMOTU
AQYD	Wea Sisi	ARCY	Mbua B.	ARGS	Pitcairn I
AQYE	Erronan, or, Futuna	ARDB	Ruku Ruku B.	ARGT	Mangareva Is
AQYG	Aneityum (Annatàm)	ARDC	No Kinasi	ARGV	Anaa atoll
AQYH	P Patrick	ARDE	Vicuna Hr	ARGW	Fakarava atoll
		ARDG	Nateva B	ARGX	Rotoava Anchorage.
AQYJ	NEW CALEDONIA	ARDH	Somo Somo Str	ARGZ	Society Is
AQYL	Goro P	ARDI	Nanuku Passage.	ARHB	Tahiti
AQYM	Prony B	ARDJ	Wailangilala I	ARHC	Venus Pt
AQYN	Sebert Cove.	ARDK	Taviuni	ARHE	Papieté Hr
AQYP	Grand P	ARDL	Vanua Mbalavu	ARHF	P Phaeton
AQYR	Amédée I	ARDM	Lomaloma	ARHG	Huaheine I
AQYS	Bulari B	ARDO	Rotumah	ARHJ	Faré Hr (Owharre).
AQYU	Noumea.c			ARHK	Raiatea I
AQYV	P Laguerre	ARDQ	TONGA Is	ARHL	Teavarua Hr
AQYW	Muendu B.	ARDS	Tongatàbu I	ARHN	Suvarov Is
AQYZ	P Burai	ARDT	Nukualofa	ARHO	Cook Is
AQZB	P Muéo.	ARDV	Niue I	ARHP	Rarotonga I.
AQZC	Kataviti B	ARDW	Uea, or Wallis I	ARHS	Tubuai, or, Austral Is
AQZD	Chasseloup B			ARHT	Rapa (Oparo) I
AQZF	Gomen B	ARDX	SAMOA Is	ARHU	Easter I
AQZG	Néhoué B	ARDZ	Tutuila I		
AQZH	Tanlé B	AREB	Fungasar Hr		
AQZJ	Banaré B	AREC	Pago Pago Hr.	ARHW	AUSTRALIA
AQZK	Art I	AREF	Vailoa		
AQZL	Aue	AREG	Upolu I	ARHX	NORTH AUSTRALIA
AQZN	Harcourt B.	AREH	Saiatu Hr	ARHY	Carpentaria G
AQZO	Arama.	AREJ	Apia.	ARIB	Batavia R
AQZP	Bondé R.	AREK	Saluafata Hr.	ARIC	Mitchell R
AQZS	Balade.	AREL	Falifa Hr.	ARID	Normanton
AQZT	Pouebo	AREN	Fangaloa B.	ARIE	Norman R Lt V.
AQZU	Ubati	AREO	Savaii I.	ARIF	Kimberley.
AQZW	Yengen.	AREP	Union (Tokelau) Is	ARIG	Burketown
AQZX	Bâ B	ARES	Atáfu atoll	ARIJ	Wellesley Is
AQZY	Kuaua B	ARET	Phoenix Is	ARIL	McArthur R.
ARBC	P Duperré	AREU	Ellice Is	ARIN	Roper
ARBD	Kanala B.	AREW	Christmas I	ARIP	Arnhem B
ARBE	Lavaissière B.			ARIS	C Wessel
ARBG	Nékété B	AREY	HAWAIIAN Is.	ARIU	Goyder R
ARBH	Tchio	AREZ	Hawaii	ARIW	Liverpool R
ARBI	P. Bouquet	ARFB	Hilo	ARIY	Van Diemen G
ARBK	Kuakue B.	ARFD	Mahukona	ARJB	Melville I
ARBL	P. Yate	ARFE	Kawaihae f	ARJD	Alligator R
ARBM	Isle of Pines.	ARFG	Kailua	ARJF	Adelaide R
ARBO	Vao	ARFI	Kealakekua	ARJH	P Daly.
ARBP	Gadji	ARFJ	Maui I	ARJK	P Darwin
		ARFK	Kahului Hr	ARJO	Palmerston.
ARBQ	LOYALTY Is	ARFM	Lahaina f	ARJQ	Southport
ARBT	Bishop Sound.	ARFN	Kanahena Pt	ARJT	Bynoe Hr
ARBU	Tandine B.	ARFO	Lanai	ARJV	P Patterson
ARBV	Norfolk I e	ARFQ	Molokai f	ARJX	P Keats
ARBW	Lord Howe I.	ARFS	Lae o ka Laau	ARJZ	Fitzmaurice R.
ARBX	Kermadec Is	ARFT	Kaunakakai	ARKC	Victoria R
		ARFV	Oahu		
ARBZ	Fiji Is	ARFW	Waialua	ARKG	WEST AUSTRALIA.
ARCB	Kandavu	ARFX	Honolulu a b e f	ARKI	Cambridge G.
ARCD	Ngaloa Hr	ARFZ	Puuloa	ARKL	Ord R
ARCF	Solo Rock	ARGB	Kauai	ARKN	Wyndham a

AUSTRALIA				TASMANIA	
ARKP	Admiralty G	ARQG	SOUTH AUSTRALIA.	ARWZ	Glenelg R
ARKS	P Warrender	ARQI	P Eyre		
ARKU	York Sound	ARQK	Fowler B	ARXC	VICTORIA
ARKW	Prince Frederick Hr	ARQM	Nuyts Archipelago	ARXE	C Nelson e
ARKY	P Nelson	ARQO	Streaky B d	ARXG	Portland a d
ARLB	St George Basin	ARQS	Blanche P	ARXI	P Fairy a
ARLD	Buccaneer Archipelago	ARQU	Flinders	ARXK	Belfast d
ARLF	King Sound	ARQW	Venus Hr	ARXM	Moyne R
ARLH	Kimberley	ARQY	P. Douglas.	ARXO	Warrnambool a c d
ARLJ	Derby a	ARSB	C. Catastrophe	ARXQ	P Campbell
ARLK	Roebuck B	ARSD	Spencer G	ARXS	C Otway e
ARLM	Broome a	ARSF	P. Lincoln d	ARXU	Appollo B
ARLO	Bedout I	ARSG	Neptune Is	ARXW	Louttit B
ARLQ	P Walcott	ARSH	Franklin Hr	ARXZ	Eagle Nest Pt
ARLT	Jarman I	ARSI	Middle Bank Lt V	ARYC	P PHILLIP
ARLV	Rowley Shs.	ARSJ	Lowly Pt a c	ARYE	Lonsdale Pt c d e
ARLX	Cossack a	ARSL	Augusta	ARYG	Shortland Bluff
ARLZ	Roebourne a	ARSN	Germein B	ARYI	Queenscliff b d e
ARMC	Dampier Archipelago	ARSP	P Pirie c	ARYK	Swan I
ARME	C Preston	ARST	P Broughton	ARYM	West Sand
ARMG	Shark B	ARSV	Wallaroo c	ARYO	Portarlington
ARMI	Gascoyne Road	ARSX	Tipara B a	ARYQ	Geelong b
ARMK	Carnarvon a	ARSZ	Moonta	ARYT	Hobson B
ARMN	Dirk Hartog I.	ARTC	P. Victoria	ARYV	Gellibrand Pt a b
ARMP	Murchison R	ARTE	Wauraltee	ARYW	Gellibrand Pt Lt V
ARMS	Geraldine	ARTG	Hardwicke B	ARYX	Williamstown
ARMU	P Gregory	ARTI	Corny Pt a	ARZB	Yarra R
ARMW	Champion B	ARTK	P. Turton	ARZD	MELBOURNE
ARMY	Houtman Rocks.	ARTM	Althorpe I a	ARZF	Williamstown Docks
ARNB	Geraldton a	ARTO	Investigator Str	ARZG	St Kilda Pier
ARND	Northampton	ARTQ	P. Moorowie	ARZH	P Melbourne (Sand-
ARNF	P Grey	ARTS	Troubridge Shs a c		ridge)
ARNH	Leander Pt	ARTU	Kangaroo I	ARZJ	Mornington
ARNJ	Dongara a	ARTW	C Borda c d e	ARZL	Rye
ARNL	Rottnest I e	ARTY	Kingscote d	ARZN	Sorrento
ARNO	Garden I	ARUB	C Willoughby c e	ARZP	Nepean Pt
ARNQ	Swan R	ARUD	G of St Vincent	ARZS	C Schank e
ARNT	Gage Road	ARUF	Edithburgh	ARZU	P Western
ARNV	Warnbro Sound.	ARUH	Macdonnell Sound	ARZW	Orib Pt a
		ARUJ	P Vincent	ARZY	Liptrap C
ARNX	FREMANTLE a b	ARUL	P Alfred	ASBC	Wilson Promontory. e
ARNZ	Perth	ARUN	Ardrossan	ASBE	Cliffy I
AROC	Guildford.	ARUP	Clinton	ASBG	Corner Inlet
AROE	Cockburn Sound.	ARUS	Wakefield	ASBI	Welshpool
AROG	Rockingham	ARUV	P. Gawler	ASBK	Clonmel I a
AROI	P Kennedy	ARUX	Adelaide	ASBM	P Albert a c d
AROK	Mandurah	ARUY	P ADELAIDE	ASBO	Alberton
AROM	Koombanah B	ARUZ	Semaphore a b d	ASBQ	Tarraville
AROP	Bunbury a	ARVC	Glenelg c	ASBU	Lake Reeve
AROS	Lockeville.	ARVE	Holdfast B,	ASBW	Sale
AROU	Géographe B.	ARVF	P. Noarlunga	ASBY	Mitchell R
AROW	Busselton a	ARVG	Willunga	ASCB	Bairnsdale
AROY	Quindalup	ARVI	Yankalilla	ASCD	C Everard
ARPB	C Naturaliste e	ARVK	C Jervis c c	ASCF	Mallagoota Inlet
ARPD	Hamelin B	ARVM	Backstairs Passage	ASCH	Gabo I e
ARPF	C Leeuwin e	ARVO	P Victor c d	ASCJ	C Howe
ARPH	Flinders B	ARVQ	P Elliot		
ARPI	West C Howe	ARVT	Murray R		
ARPJ	King George Sound	ARVW	Lake Alexandrina		
ARPL	Breaksea I e	ARVY	Goolwa	ASCN	BASS STR
ARPN	King Pt	ARWB	Lacepede B	ASCP	King I
		ARWD	Kingston	ASCR	C Wickham e
		ARWF	C Jaffa a	ASCU	Curtis I
ARPQ	ALBANY	ARWJ	Robe c d	ASCW	Hogan Group
ARPT	Princess Royal Hr	ARWL	Rivoli B	ASCX	Kent Group e
ARPV	Bremer B	ARWN	Penguin I a	ASCZ	Deal I
ARPX	Esperance B	ARWP	Beachport.d	ASDC	Flinders I a
ARPZ	Malcolm Pt	ARWS	C Banks a	ASDE	Franklin Sound
ARQC	Israelite B	ARWU	C Northumberland e	ASDF	Goose I a
ARQE	Eucla	ARWX	P Macdonnell c d	ASDG	Banks Str

a Signifies an International Signal Station, b a Time Signal Station, c a Weather, Tide, or Ice Signal Station, d a Life-Saving Station, e a Lloyd's Signal Station, f a Wireless Telegraph Station

QUEENSLAND

ASDK	TASMANIA	ASIT	Bateman B	ASOP	Ballina
ASDM	Swan I a	ASIV	Bundoo, or, Clyde R	ASOR	Wardell
ASDO	Bridport	ASIX	Nelligan	ASOT	C Byron
ASDP	Low Hd e	ASIZ	Ulladulla a	ASOV	Brunswick R
ASDQ	P Dalrymple	ASJC	St Georges Hd	ASOX	Cavvanba
ASDU	Tamar R	ASJE	Jervis B	ASOZ	Fingal Pt
ASDW	Georgetown	ASJG	North Shoalhaven R a	ASEC	Tweed R.
ASDY	Ilfracombe	ASJI	Goulburn	ASPE	Danger Pt
ASEB	Sidmouth.	ASJK	Kiama		
ASED	Dorchester	ASJM	Wollongong a	ASPG	QUEENSLAND
ASEF	Exeter	ASJO	P Hacking	ASPI	Burleigh
ASEH	Swan B	ASJQ	Botany a	ASPK	Nerang R
ASEJ	Launceston	ASJU	George R	ASPM	Southport
ASEL	P Sorell e	ASJW	P Jackson	ASPO	Sandy Pt a c
ASEN	Burges	ASJY	Inner South Hd a c	ASPQ	Moreton B
ASEP	P Frederick	ASKB	SYDNEY b	ASPU	C Moreton e
ASER	Mersey Bluff e	ASKD	Macquarie Pt	ASPW	Comboyuro Pt
ASET	Table C e	ASKF	Sow and Pigs Lt V	ASPY	Bulwer Pilot Station
ASEV	Stanley	ASKH	Bradley Hd.	ASQB	Cowan Cowan Pt
ASEX	Robbins I	ASKL	Watson B d	ASQD	Tangaluma Roads.
ASEZ	Hunter I	ASKN	Garden I	ASQF	Howe Ch
ASFB	Three Hummock I	ASKP	Wooloomooloo B.	ASQH	Freeman Ch
ASFC	Arthur R	ASKR	Farm Cove	ASQJ	Yule Roads
ASFE	Remine a	ASKT	Darling Hr	ASQL	Brisbane Road Pile Lt
ASFG	Macquarie Hr c	ASKV	Pyrmont		Ho a c
ASFI	Strahan a	ASKX	Balmain	ASQN	Lytton.
ASFK	Gordon R	ASKZ	Cockatoo I	ASQP	BRISBANE b
ASFM	P Davey	ASLC	Fitzroy Dock	ASQR	Ipswich
ASFO	Maatsuyker I	ASLE	Southerland Dock	ASQT	Cleveland
ASFQ	D'Entrecasteaux Ch	ASLG	Goat I	ASQV	Redland
ASFU	Recherche B a	ASLI	Waterview B	ASQX	Logan R
ASFW	Bruny I.e	ASLK	Parramatta	ASQZ	Alberton
ASFY	Hythe.	ASLM	St Leonards	ASRC	Waterford
ASGB	Hastings	ASLO	Middle Hr	ASRE	Mooloolah R.
ASGD	Esperance P	ASLQ	Manly Beach	ASRG	Laguna B
ASGF	Huon R	ASLU	Baranju Hd	ASRI	Tewantin
ASGH	P Cygnet.	ASLW	Hawkesbury R	ASRK	Double Island Pt a
ASGJ	Tasman Hd	ASLY	Broken B	ASRM	Wide Bay Hr
ASGL	Adventure B	ASMB	Hales Bluff	ASRO	Great Sandy Str.
ASGN	Derwent R	ASMD	Catherine Hill B	ASRQ	Inskip Pt
ASGP	HOBART b	ASMF	Moon Islet	ASRU	Tyroom Road
ASGR	Storm B	ASMH	Lake Macquarie	ASRW	Mary R
ASGT	Burnett Hr	ASMJ	Nobby Hd	ASRY	Maryborough
ASGV	Frederick Henry B.	ASML	NEWCASTLE a b c d	ASTB	Woody I
ASGW	Eagle Hawk B a	ASMN	Hunter R	ASTD	Sandy C a
ASGX	Pitt Water	ASMP	Stockton	ASTF	Hervey B
ASGZ	P Arthur	ASMR	Hexham	ASTH	Burrum R
ASHC	Prosser B	ASMT	Maitland	ASTJ	Burnett R a c
ASHD	Swansea a	ASMV	Stephens Pt	ASTL	Bundaberg
ASHE	Oyster B	ASMX	P Stephens	ASTN	Lady Elliot I a
ASHG	Great Swan P.	ASMZ	Sugarloaf Pt	ASTP	Baffle Creek
ASHI	Long Pt	ASNC	C Hawke	ASTR	Bustard Hd a c
ASHK	Falmouth a	ASNE	Harrington Inlet	ASTV	Rodd Hr
ASHM	George B	ASNF	Manning R	ASTX	P Curtis
ASHO	Eddystone Pt e	ASNG	Crowdy Hd	ASTZ	Boyne R
		ASNI	Camden Haven	ASUC	Gatcombe Hd a
ASHQ	NEW SOUTH WALES	ASNK	Tacking Pt	ASUE	Gladstone
ASHU	C Green a	ASNM	P Macquarie.	ASUG	Calliope R
ASHW	Twofold B a c	ASNO	Hastings R	ASUI	Hamilton
ASHY	Eden	ASNQ	Marias R	ASUK	C Capricorn a e
ASIB	Kiyerr Inlet	ASNU	Smoky C	ASUM	Keppel C a
ASID	Panbula	ASNW	Trial B	ASUO	Little Sea Hill
ASIF	Merimbula	ASNY	MacLeay R	ASUQ	Fitzroy R Lt V
ASIH	Bega	ASOB	Nambuckra R.	ASUW	Mackenzie R
ASIJ	Barunguba, or, Monta-gu I a	ASOD	Bellinger R	ASUY	Broad Mount Hr
ASIL	Turos Inlet	ASOF	South Solitary I.	ASVB	Herbert
ASIN	Bodalla	ASOH	Clarence R	ASVD	Rockhampton
ASIP	Toragy Pt	ASOJ	MacLean	ASVF	Hewittville
ASIR	Moruya a	ASOL	Grafton	ASVH	North Reef Lt a
		ASON	Richmond R	ASVJ	P Bowen

a Signifies an International Signal Station, b a Time Signal Station, c a Weather, Tide, or Ice Signal Station, d a Life-Saving Station
e a Lloyd's Signal Station, f a Wireless Telegraph Station

NEW ZEALAND			WEST COAST NORTH AMERICA	
ASVL	Shoalwater B	ATCB	Whaingaroa Hr	ATHU	Molyneux B
ASVN	Broad Sound	ATCD	Raglan	ATHW	Balclutha
ASVP	St Lawrence	ATCF	Aotea Hi	ATHY	Kaitan
ASVR	Percy I	ATCH	Kawhia	ATIB	C Saunders a
ASVT	Pine I a	ATCJ	Mokau R a	ATID	OTAGO HR
ASVX	Flat-Top I a	ATCL	Waitara R a	ATIF	Taiaroa Hd a c
ASVZ	Pioneer R.	ATCN	Raleigh	ATIG	Otago Lt V
ASWC	East Pt a	ATCR	New Plymouth (Taranaki) a c	ATIH	P Chalmers b
ASWE	Mackay			ATIJ	Dunedin
ASWG	Repulse B	ATCV	C Egmont a	ATIL	Blueskin
ASWI	Dent I a	ATCX	Opunake a c	ATIN	Waikouaiti
ASWK	P Denison	ATCZ	Patea R c	ATIP	Hawksbury
ASWM	Dalrymple Pt a	ATDC	Wanganui a c	ATIR	Moeraki
ASWO	Bowen	ATDE	Wangaehu	ATIV	Oamaru a c
ASWQ	C Bowling Green. a	ATDG	Rangitikei R	ATIX	Timaru a c
ASWR	C Cleveland a	ATDI	Manawatu R a c	ATIZ	Ashburton
ASWT	Townsville a c	ATDK	Foxton	ATJC	Akaroa Hr a
ASWV	Bay Rock	ATDM	Kapiti I	ATJE	Banks' Peninsula
ASWX	Halifax B	ATDO	Porirua Hr	ATJG	Lyttelton b
ASWZ	Hinchinbrook c	ATDQ	COOK STR	ATJI	CHRISTCHURCH
ASXC	Herbert R.	ATDS	Queen Charlotte Sound	ATJK	Godley Hd a c
ASXE	Dungeness a	ATDU	Brothers Lt	ATJM	Gore R
ASXG	Rockingham B	ATDW	Picton	ATJO	Waiauous R
ASXI	Cardwell	ATDY	Havelock	ATJQ	Kaikoura
ASXK	Moresby R	ATEB	D'Urville I	ATJS	Gooch B
ASXM	Mourilyan Hr	ATED	P Hardy.	ATJU	C Campbell a c
ASXO	Johnstone R	ATEF	French Pass	ATJW	Awatere R
ASXQ	Geraldton	ATEH	C Stephens	ATJY	Wairau R. a
ASXU	Mulgrave R	ATEJ	NELSON a	ATKB	Blenheim
ASXW	C Grafton	ATEL	Motueka R	ATKD	Tuamarina
ASXY	CAIRNS a	ATEN	Astrolabe Road	ATKF	Cloudy B
ASYB	P Douglas	ATEP	Waitapu	ATKH	P Underwood
ASYD	Low I	ATER	Clifton	ATKJ	Tory Ch
ASYF	Daintree R	ATEV	Collingwood	ATKL	P Nicholson
ASYH	Archer Pt	ATEX	C Farewell Spit a c e	ATKN	Pencarrow Hd
ASYJ	Rocky I	ATEZ	Wanganui Inlet c	ATKP	Somes I
ASYL	Endeavour R a	ATFC	Karamea R	ATKR	Mt Victoria u
ASYN	Cooktown	ATFE	Mokihinui	ATKV	WELLINGTON b
ASYP	Lizard I	ATFG	Hector	ATKX	Lambton Hr
ASYQ	Channel R (C Melville) Lt V	ATFI	Buller R	ATKZ	Pitini
		ATFK	Westport u	ATLC	C Palliser
ASYR	Kennedy R	ATFM	C Foulwind a c	ATLE	Ahuriri Bluff a
ASYT	Claremont I Lt V	ATFO	Grey R a	ATLG	Napier
ASYV	Piper Is Lt V	ATFQ	Cobden	ATLI	Hawkes B
ASYX	Raine I	ATFS	Taylorville.	ATLK	Mohaka R
ASYZ	Somerset	ATFU	Greymouth	ATLM	Portland I a
ASZC	C York	ATFW	Hokitika a	ATLO	Waipaoa R
ASZE	Torres Str	ATFY	Puysegur Pt	ATLQ	Turanganui R
ASZG	Bligh Boat Entrance	ATGB	FOVEAUX STR	ATLS	Gisborne
ASZI	Prince of Wales Ch	ATGD	Centre I	ATLU	Opotiki R
ASZK	Goode I (Pálñug) c	ATGF	Riverton	ATLW	Tauranga
ASZM	Thursday I	ATGH	Oreti R	ATLY	Mercury B
ASZO	P Kennedy	ATGJ	Invercargill	ATMB	Whitianga
ASZQ	Endeavour Str	ATGL	Campbelltown	ATMD	Cuvier I
ASZU	Booby I	ATGN	Awarua (Bluff) Hr a c e	ATMF	Great Barrier I (Aotea)
ASZV	Proudfoot Shoal Lt V	ATGO	Awarua Lt V	ATMH	Waihau (Coromandel) Hr
		ATGP	Mataura R		
ASZW	NEW ZEALAND	ATGR	Dog I	ATMJ	Thames R
ASZY	C Maria Van Diemen c e	ATGV	Ruapuke I	ATML	Grahamstown
ATBD	Hokianga R a	ATGX	Stewart I a	ATMN	Pauhenehene Spit
ATBE	Kaipara Hr a c	ATGZ	P Pegasus	ATMP	Waiheke Ch
ATBH	Helensville	ATHC	P Adventure	ATMR	Tehmaki Str
ATBJ	Dargaville	ATHE	Paterson Inlet	ATMS	AUCKLAND b
ATBL	P Albert	ATHG	P William	ATMV	Mt Victoria
ATBN	Manukau a c	ATHI	THE SNARES d	ATMX	Rangitoto I
ATBP	Onehunga	ATHK	Waipapapa Pt	ATMZ	Tiri tiri Matangi I a c
ATBR	Waikato R	ATHM	Waikawa	ATNC	Kawau
ATBU	Mercer	ATHO	Catlin R	ATNE	Burgess Islet
ATBW	Huntly	ATHQ	Newhaven	ATNG	Whangarei Hr
ATBY	Taupiri	ATHS	Nugget Pt c e	ATNI	Tutukaka Hr

a Signifies an International Signal Station   b a Time Signal Station   c a Weather, Tide, or Ice Signal Station,   d a Life-Saving Station, e a Lloyd's Signal Station, f a Wireless Telegraph Station

WEST COAST NORTH AMERICA

ATNK	Whangaruru Hr	ATSR	P Rothsay	ATZG	Admiralty Inlet
ATNM	B of Islands	ATSV	Tongass P	ATZI	Wilson Pt
ATNO	Opua			ATZK	P Discovery
ATNQ	Paihia	ATSX	BRITISH COLUMBIA	ATZM	Admiralty Hd
ATNS	Kororarika B (P Russell) a	ATSZ	P Simpson	ATZO	P Townsend c
		ATUC	Naas B	ATZQ	P Hadlock
ATNU	Whangoroa	ATUD	Dixon Entrance	ATZS	P Ludlow c
ATNW	Mongunui Hr	ATUE	Queen Charlotte Is	ATZU	Hood Canal
ATNY	Rangaunu (Awanui) R	ATUG	Skeena R	ATZW	Skokomish
ATOB	Ohora R	ATUI	P Essington	ATZY	Union City
ATOD	Parengarenga Hr.	ATUK	Fitzhugh Sound	AUBC	P Susan
ATOF	North C	ATUM	P Blakeney	AUBD	Tulalip
ATOJ	Auckland Is d	ATUO	Namu Hr	AUBF	Marysville
ATOL	Campbell I d	ATUQ	Owikino	AUBH	Snohomish
ATON	Antipodes Is d	ATUS	Queen Charlotte Sound	AUBJ	P No Point
ATOP	Bounty Is d	ATUV	VANCOUVER I	AUBL	Puget Sound
ATOR	Chatham Is	ATUW	Str of Georgia	AUBN	P Madison
ATOV	Whangaroa (P Hutt)	ATUY	Atkinson Pt	AUBP	SEATTLE c
ATOX	Macquarie I	ATVB	Burrard Inlet	AUBQ	Bremerton f
		ATVD	Brocton Pt b	AUBR	TACOMA c
ATOZ	NORTH AMERICA	ATVF	English B	AUBT	Steilacoom
		ATVH	VANCOUVER	AUBV	Nisqually
ATPC	C Bathurst	ATVJ	P Moody	AUBX	Olympia
ATPE	C Dalhousie	ATVL	Fraser R	AUBZ	Budd Inlet
ATPG	Mackenzie R	ATVN	New Westminster	AUCD	New Dungeness.
ATPI	Alaska	ATVP	Pitt R	AUCF	P Angeles c
ATPK	Pt Barrow	ATVR	Derby, or, New Lang-	AUCH	Neah B c
ATPM	C Lisburne		ley	AUCJ	C Flattery
ATPO	Hope Pt	ATVW	P Augusta	AUCL	Tatoosh I a c f
ATPQ	Kotzebue Sound	ATVY	Comox	AUCN	Destruction I
ATPS	C Prince of Wales	ATWB	Yellow I (Baynes	AUCP	Gray Hr a d
ATPU	Diomede Is		Sound)	AUCR	Willapa R a d
ATPW	P Clarence	ATWD	Nanoose Hr	AUCT	Shoalwater
ATPY	Imuruk Lake	ATWF	Departure B	AUCV	Bruceport
ATQB	Norton B.	ATWH	Nanaimo Hr		
ATQD	F Unalakhk	ATWJ	Grave Pt	AUCX	OREGON
ATQF	Ketchumville	ATWL	Entrance I	AUCZ	Columbia R a c d
ATQH	Pastolik	ATWN	Oyster Hr	AUDC	C Hancock
ATQJ	Kusilvak	ATWP	Georgina Pt	AUDE	Adams Pt a c d
ATQK	Juneau	ATWR	East Pt (Saturna I)	AUDG	ASTORIA c
ATQL	Yukon R	ATWV	Cowitchin	AUDI	Cathlamet
ATQN	Andreievsky	ATWX	Saminos	AUDK	Kalama
ATQO	KLONDIKE R	ATWZ	Haro Str	AUDM	PORTLAND b c
ATQP	Nulato	ATXC	Discovery I	AUDO	Tillamook Rock
ATQR	Nowikakat	ATXE	VICTORIA f	AUDQ	C Meares
ATQS	Dawson	ATXG	ESQUIMALT HR.	AUDS	C Foulweather
ATQU	Circle City	ATXI	Race Is	AUDW	Yaquina a d
ATQV	C Romanzof	ATXK	Juan de Fuca Str	AUDY	Heceta Hd
ATQW	St Michael P f	ATXM	Sooke Inlet	AUEB	Umpqua R a d
ATQX	St Matthews I	ATXO	P San Juan	AUED	Koos R (Coos)
ATRC	Nunivak I	ATXQ	Carmanah d	AUEF	Empire City
ATRE	Kuskoquim R	ATXS	Barclay Sound	AUEH	C Gregory, or, Arago
ATRG	Kolmakoff F	ATXU	C Beale d	AUEJ	Coquille R a d
ATRI	Khramtshenko			AUEL	C Orford, or, Blanco f
ATRK	F Alexander	ATXW	WASHINGTON STATE	AUEN	P Orford
ATRM	P Möller	ATXY	Roberts Pt	AUEP	Ellensburg
ATRO	ALEUTIAN Is	ATYB	Boundary B	AUER	Hunter Cove
ATRQ	Pribiloff Is	ATYD	Drayton Hr	AUET	Mack Reef
ATRS	Attu I	ATYF	Bellingham B	AUEV	Chetko R
ATRU	Shelikoff Str	ATYH	New Whatcom c		
ATRW	Kadiak I	ATYJ	Fairhaven c	AUEX	CALIFORNIA
ATRY	St Paul Hr	ATYL	Padilla	AUEZ	St Georges Reef
ATSB	St Orlovsk	ATYN	Rosario Str	AUFC	Crescent City
ATSD	Alsentia	ATYP	Patos I	AUFE	Trinidad Hr
ATSF	P Graham	ATYR	Orcas I	AUFG	Humboldt (Arcata) c d
ATSH	P Mulgrave	ATYV	Stuart I	AUFI	Eureka c
ATSJ	Chichagoff I	ATYX	San Juan I	AUFK	Table Bluff f
ATSL	Baranof I	ATYZ	Lopez I	AUFM	Noyo
ATSN	Sitka f	ATZC	Whidbey I	AUFO	Mendocino C c
ATSP	Stikine R (P Wrangel).	ATZE	Smith, or, Blunt I	AUFQ	Arena Pt

a Signifies an International Signal Station, b a Time Signal Station, c a Weather, Tide, or Ice Signal Station d a Life-Saving Station, e a Lloyd's Signal Station, f a Wireless Telegraph Station

WEST COAST NORTH AMERICA

AUFS	F Ross Cove
AUFW	Bodega B
AUFY	Tomales B
AUGB	Reyes Pt a c d
AUGD	Bonita Pt d
AUGE	San Francisco Lt V
AUGF	SAN FRANCISCO b c
AUGH	Fort Pt a d
AUGJ	Alcatraz I
AUGK	Hunter Pt Dock
AUGL	Goat I f
AUGM	Union Iron Works Docks
AUGN	Oakland
AUGP	San Pablo B
AUGR	Petaluma
AUGT	Mare I b f
AUGV	Vallejo
AUGX	Napa
AUGZ	Karquines
AUHC	Roe I
AUHE	Rio Vista
AUHG	Sacramento
AUHI	Farallone Islets a f
AUHK	Pigeon Pt
AUHM	Año Nuevo Pt
AUHO	Santa Cruz
AUHQ	Pinos Pt
AUHS	Monterey
AUHW	Carmel
AUHY	Pt Sur
AUIB	Piedras Blancas
AUID	San Simeon
AUIF	Esteros
AUIH	San Luis Pt
AUIJ	San Luis Obispo
AUIL	Arguello Pt f
AUIN	Conception Pt
AUIP	Santa Barbara Ch c
AUIR	San Buenaventura
AUIT	Hueneme Pt
AUIV	Santa Monica c
AUIX	Ballona P
AUIY	Los Angeles
AUIZ	San Pedro.c
AUJC	Wilmington
AUJE	Newport
AUJG	San Juan Pt
AUJI	Oceanside
AUJK	Loma Pt f
AUJM	Ballast Pt
AUJO	San Diego c
AUJQ	San Miguel I
AUJS	Santa Rosa I
AUJW	Santa Cruz I
AUJY	Anacapa I
AUKB	Santa Catalina
AUKD	San Clemente I
AUKF	San Nicolas I
AUKH	MEXICO
AUKJ	Lower California
AUKL	Los Coronados
AUKN	Ensenada
AUKP	Todos Santos B
AUKR	San Quentin B
AUKT	Sebastian Viscaino B
AUKV	Cerros I
AUKX	P San Bartolomé
AUKZ	Magdalena B

WEST COAST CENTRAL AND SOUTH AMERICA

AULC	C San Lucas
AULD	G OF CALIFORNIA
AULE	San José del Cabo B
AULG	La Paz B
AULI	Concepcion B
AULK	Santa Rosalia f
AULM	San Felipe B
AULO	Colorado R
AULQ	ARIZONA
AULS	Yuma
AULW	Eureka
AULY	Ehrenberg
AUMB	C Haro f
AUMD	Guaymas
AUMF	Santa Barbara B
AUMH	Agiabampo
AUMJ	Topolobampo B
AUML	Culiacan R
AUMN	Altata
AUMP	Mazatlan
AUMR	Creston I
AUMT	Chamatla R
AUMV	Palmito
AUMX	San Blas.
AUMY	Las Tres Marias Is
AUMZ	C Corrientes
AUNC	REVILLA GIGEDO Is
AUNE	Chamela
AUNG	Tenacatita B
AUNI	Navidad B
AUNK	Manzanilla B
AUNM	Cayutlan Lagoon
AUNO	Colima
AUNQ	Sacatula R
AUNS	Orilla
AUNW	Sihuatanejo
AUNY	Tequepa B
AUOB	ACAPULCO
AUOD	Tecoanapa.
AUOF	Maldonado
AUOH	Chacahua B
AUOJ	Escondido B.
AUOL	P Angeles
AUON	Sacrificios
AUOP	P Guatulco
AUOR	Tangola tangola B
AUOT	Morro Ayuca.
AUOV	SALINE CRUZ
AUOX	Ventosa B
AUOZ	Tehuantepec
AUPC	San Juan
AUPE	Tonala
AUPG	San Benito
AUPI	GUATEMALA
AUPK	Ocos R
AUPM	Champerico
AUPO	San Luis
AUPQ	Sesecapa
AUPS	Tecojate
AUPW	San Jeronimo
AUPY	San José de Guatemala
AUQB	Istapa
AUQD	SALVADOR
AUQF	CENTRAL AMERICA
AUQH	Acajutla
AUQJ	Sonsonate
AUQL	Libertad

AUQN	La Concordia
AUQP	Jaltepeque Lagoon
AUQR	G of Fonseca
AUQT	P La Union
AUQV	San Lorenzo
AUQX	NICARAGUA
AURC	Corinto P
AURE	Managua
AURG	San Juan del Sur
AURK	Salinas B
AURM	P Elena
AURO	P Culebra
AURQ	C Blanco
AURS	Nicoya G
AURW	Colorado G
AURY	Arenas Pt f
AUSB	P Herredura
AUSD	Uvita Hi
AUSF	G of Dulce
AUSH	San Domingo
AUSJ	El Rincon
AUSL	Golfito
AUSN	PANAMA
AUSP	Ciudad de David
AUSR	Bahia Honda
AUST	Estéro
AUSV	Chamé B.
AUSX	Chorrera
AUSZ	Perico I
AUTC	Naos I
AUTE	Chepo R
AUTG	San Miguel B
AUTI	Darien Hr
AUTK	Perlas Is
AUTM	NUEVA GRANADA
AUTO	San Juan R
AUTQ	Buenaventura
AUTS	P Tumaco
AUTW	ECUADOR
AUTY	Esmeralda
AUVB	Atacamas
AUVD	Caracas
AUVF	Manta B
AUVH	GALAPAGOS Is
AUVJ	Albemarle I
AUVL	Chatham I
AUVN	Stephens B (P Grande)
AUVP	Wreck-B (P Chico)
AUVR	Monte Christo
AUVT	La Plata I
AUVX	Santa Elena
AUVZ	Guayaquil
AUWC	Quito
AUWE	Puna
AUWG	Mandinga Pt
AUWI	Española Pt
AUWK	Arena Pt
AUWM	Santa Rosa R
AUWO	SOUTH AMERICA
AUWQ	PERU
AUWS	Tumbez
AUWV	Talara
AUWX	Negrito B.
AUWZ	Paita
AUXC	Piura R
AUXE	Eten Pt

a Signifies an International Signal Station, b a Time Signal Station c a Weather, Tide, or Ice Signal Station, d a Life-Saving Station, e a Lloyd's Signal Station f a Wireless Telegraph Station

WEST AND EAST COASTS SOUTH AMERICA

AUXF	Lobos de Afuera	AVDM	Chañaral	AVIH	Ancud
AUXG	Lambayeque	AVDO	P Flamenco	AVIJ	Calbuco Ch
AUXI	Pacasmayo	AVDQ	P Caldera a	AVIL	P Montt
AUXK	Malabrigo Road	AVDS	Copiapó	AVIM	Huafo
AUXM	Huanchaco	AVDU	Carrizal Bajo B	AVIN	C Tres Montes
AUXO	Salaverry a	AVDW	P Huasco	AVIO	Melinka P
AUXQ	Truxillo	AVDX	St Ambrose I	AVIP	G of Peñas
AUXS	Guañape Is	AVDY	C Bascuñan	AVIR	Fallos Ch
AUXW	Santa	AVEB	Pajaros Islets	AVIT	Stosch Ch
AUXY	P Chimbote	AVED	Tortoralillo	AVIW	P Otway
AUYB	Ferrol B	AVEF	Coquimbo. a b	AVIX	Messier Ch
AUYD	Samanco B.	AVEH	P Herradura	AVIZ	G of Trinidad
AUYF	Casma B.	AVEJ	P Tongoi	AVJC	Smyth Ch
AUYH	Huarmey.	AVEL	Maitencillo	AVJE	Evangelistas
AUYJ	Supé	AVEN	Conchali B	AVJG	MAGELLAN STR
AUYL	Huacho.	AVEO	Los Vilos	AVJH	Borja B
AUYN	Salinas B	AVEP	Huevos I	AVJI	C Pillar
AUYP	Chancay	AVER	Pichidanque B.	AVJK	C Upright
AUYR	Ancon	AVET	Lagua B	AVJL	P Tamar
AUYT	CALLAO	AVEX	P Papudo	AVJM	Sandy Pt. (Punta Arenas)
AUYV	Lima	AVEZ	Quintero B		
AUYX	C San Lorenzo	AVFC	VALPARAISO b c d	AVJN	C Gregory
AUYZ	Chorillos B	AVFE	Angeles Pt	AVJO	C San Isidro
AUZC	Chilca	AVFG	Curaumilla Pt e	AVJP	Delgada Pt a
AUZE	Cerro Azul	AVFI	Casablanca	AVJQ	Dungeness e
AUZG	Tambo de Mora P	AVFK	Santiago	AVJS	C Virgins d e
AUZI	Chincha Is	AVFM	JUAN FERNANDEZ I	AVJU	Tierra del Fuego
AUZK	Pisco	AVFO	Cumberland B	AVJW	Ushuwaia
AUZM	Ica	AVFQ	Algarrobo	AVJY	C Horn
AUZO	Paracas	AVFR	San Antonio Nuevo	AVKB	Staten I
AUZQ	San Gallan I	AVFS	San Antonio Pt	AVKD	St John Hr
AUZS	Independencia B	AVFT	Matanzas Caleta	AVKF	Laserre Pt
AUZW	P San Nicolas	AVFU	Tuman B	AVKG	Diego Ramirez
AUZY	P San Juan	AVFW	Pichilemu		
AVBD	C Lomas Roadstead	AVFX	San Antonio del Petrel	AVKH	South Shetland
AVBF	Chala	AVFY	Lloco	AVKJ	South Orkneys
AVBH	Atico	AVGB	Constitucion	AVKL	FALKLAND IS
AVBJ	Camana	AVGD	Talca	AVKN	C Pembroke
AVBL	Quilca R	AVGF	C Carranza a	AVKP	Stanley Hr
AVBN	Islay (Ilay)	AVGH	Puchepo Pt	AVKR	Cranmar
AVBP	Arequipa	AVGI	Buchupero. a	AVKT	P Stephens
AVBR	P Mollendo	AVGJ	Tomé		
AVBT	Tambo R	AVGK	Guarampe	AVKX	ARGENTINA
AVBU	Ylo Road	AVGL	Penco	AVKZ	Patagonia
AVBX	Puno	AVGN	TALCAHUANO e	AVLC	P Gallegos
		AVGP	Concepcion	AVLE	Coy Inlet
AVBZ	CHILE	AVGR	Quiriquina I	AVLG	P Santa Cruz
AVCD	Arica	AVGT	Bio Bio R	AVLI	Chico R
AVCE	Huanillos.	AVGX	Santa Maria I	AVLK	P San Julian
AVCF	Tacna	AVGZ	Puchoco.	AVLM	P Desire
AVCH	Huaina Pisagua B	AVHC	CORONEL	AVLO	C Three Points (Tres Puntas)
AVCJ	Junin	AVHE	Lota		
AVCL	Mejillones Cove	AVHF	Puerto Cuarenta Dias	AVLQ	P Malaspina
AVCN	Caleta Buena	AVHG	Laraquete	AVLS	Tova I
AVCO	Fabellion de Pica	AVHI	Arauco	AVLU	P Melo
AVCP	IQUIQUE	AVHK	Lavapié Pt a	AVLW	Egg Hr (P San Antonio)
AVCR	Tocopilla	AVHL	Yanez B.		
AVCT	Algodonales B	AVHM	Lebu I	AVLY	Gill B
AVCU	Cobija B	AVHN	P Quidico, or, Nena	AVMB	Leones Isle
AVCX	Mejillones del Sur B	AVHO	Mocha I	AVMD	P Santa Elena
AVCZ	Constitucion Hr	AVHP	Imperial R	AVMF	Chupat R
AVDC	Moreno B	AVHQ	Tolten	AVMH	P Madryn
AVDE	ANTOFAGASTA c	AVHS	Corral	AVMJ	Delgado Pt
AVDF	Oliva Road.	AVHU	Valdivia	AVML	P San Antonio
AVDG	Taltal	AVHW	Galera Pt	AVMN	Parana
AVDH	Pan de Azucar	AVHX	Bueno R	AVMP	San Blas Hr
AVDI	Lavata B	AVHY	Maullin	AVMR	Union B
AVDJ	Sarco	AVIB	Chiloc I	AVMT	Rio Colorado
AVDK	Paposo Mines	AVID	C Quilan	AVMX	Mt Hermoso
AVDL	Peña Blanca Cove	AVIF	Corona Ild	AVMY	Bahia Blanca Lt V.

a Signifies an International Signal Station, b a Time Signal Station, c a Weather, Tide, or Ice Signal Station, d a Life-Saving Station, e a Lloyd's Signal Station, f a Wireless Telegraph Station

EAST AND NORTHEAST COASTS SOUTH AMERICA

Code	Place	Code	Place	Code	Place
AVMZ	P Belgrano (Bahia Blanca)	AVSL	Paranagua	AVYZ	Alagoas Province
AVNC	Mogotes Pt e	AVSN	Larangeiras	AVZC	P de Pedras
AVNE	C Corrientes	AVSP	Mel I	AVZE	Barra Grande
AVNG	Medano Pt	AVSR	Bom Abrigo I	AVZG	Una R
AVNI	C San Antonio e	AVST	Cananea	AVZI	Tamandare
AVNK	PLATA R	AVSX	SANTOS	AVZK	San Aleixo I
AVNM	English Bank Lt V	AVSZ	Moella I	AVZM	C St Agostinho
AVNO	Cuirassier Bank, or, Indio Pt Lt V	AVTC	Ubatuba	AVZO	PERNAMBUCO a
AVNQ	Chico Bank Lt V	AVTE	Parati	AVZS	Olinda
AVNS	Ensenada de Barragan	AVTG	Ilha Grande a f	AVZU	Itamaraca I
AVNU	Traiventos	AVTI	Albrahoa B	AVZW	Catuáma
AVNW	La Plata b c	AVTK	Angra dos Reis	AVZY	Goiana R
AVNY	P Rio Santiago f	AVTM	Sapetiba	AWBC	Cabedello
AVOB	Lara Pt '	AVTO	Raza I	AWBE	Parahiba R
AVOD	BUENOS AIRES b f	AVTQ	RIO DE JANEIRO b	AWBG	Mamanguape
AVOF	Riachuelo R c	AVTR	Villegagnon F	AWBI	Formosa
AVOH	San Nicolas	AVTS	F Santa Cruz	AWBK	Cunhahu R
AVOL	Campana	AVTU	Nitheroy	AWBM	Pirangi R
AVON	Rosario	AVTW	Saude Dock	AWBO	Rio Grande do Norte
AVOP	Colastine	AVTX	Mucangue Pequena (Coal I)	AWBQ	Natal
AVOR	Parana			AWBS	C St Roque
AVOT	La Paz	AVTY	C Frio a	AWBU	FERNANDO NORONHA I a e
AVOX	Corrientes.	AVUB	Busios C		
AVOZ	Paraguay	AVUD	Imbetiba	AWBY	As Rocas
AVPC	Villa Pilar	AVUF	Macahé	AWCB	St Paul Rocks.
AVPE	Asuncion	AVUH	C St Thomé	AWCD	Trinidad Is
AVPG	Colonia	AVUJ	Campos	AWCE	Martin Vas I
AVPI	Farallon I	AVUL	St Joao da Barra	AWCF	Amargoso R
AVPK	Martin Garcia I	AVUN	Itabapuana	AWCG	Macáu
		AVUP	Itapemerim	AWCH	Retiro B
AVPM	URUGUAY	AVUR	Francesa I	AWCJ	Aracati
AVPO	Fray Bentos	AVUT	Benevente	AWCL	Jaguarybe R
AVPQ	Concepcion	AVUX	Guarapari	AWCN	Ceará
AVPS	Paysandu	AVUZ	Espirito Santo B	AWCP	Macoripe Pt
AVPU	Concordia	AVWC	Villa Velha B	AWCR	Paranahyba
AVPW	Pañela Rock Lt V	AVWE	Victoria	AWCT	Amaraçao
AVPY	MONTEVIDEO c f	AVWG	Doce R	AWCV	Tutoia
AVQB	Cibil Dock	AVWI	Portalegre	AWCX	Santa Anna I
AVQD	Mana Dock	AVWK	Caravellas	AWCZ	Alcantara
AVQF	Flores I a	AVWM	Abrolhos Rocks	AWDC	Maranhão
AVQH	Maldonado a	AVWO	Prado	AWDG	San Marcos B
AVQJ	Lobos I	AVWQ	P Seguro.	AWDI	San Joao I
AVQL	San José Ignacio Pt	AVWS	Cabral B	AWDK	Turyassu
AVQN	C Santa Maria	AVWU	Santa Cruz B	AWDM	Gurupi R
AVQP	C Polonio	AVWY	Belmonte a	AWDO	Vizeu
AVQR	Castillo B	AVXB	Canaveiras	AWDQ	Atalaia Pt
		AVXC	Commandatubu	AWDS	PARA R
AVQT	BRAZIL	AVXE	Una	AWDU	Braganza Shoal Lt V
AVQX	Rio Grande do Sul a d	AVXG	Olivenca	AWDV	Gaviotas (Taipu Pt) Lt V
AVQZ	San Pedro do Sul	AVXI	St Jorge dos Ilheos		
AVRC	São José do Norte	AVXM	Contas	AWDX	Chapeo Virado Lt
AVRE	Pelotas	AVXM	Camamú	AWDY	Pará f
AVRG	Lagoa Dos Patos	AVXO	Morro San Paulo	AWEB	Guama R
AVRI	Guahyba R	AVXQ	Valença	AWED	Guajara R
AVRK	P Alegre	AVXS	BAHIA	AWEF	Tocantins R
AVRM	Triumpho	AVXU	San Antonio Pt	AWEH	Breves
AVRO	Mostardas	AVXW	Itaparica I	AWEJ	AMAZON R
AVRQ	C Santa Marta Grande	AVXY	Tapagipe B	AWEL	Macapa
AVRS	Imbituba	AVYB	San Francisco R	AWEN	Gurupá R
AVRU	Santa Catharina I	AVYD	Para Guassu	AWEP	P do Moz
AVRW	Nossa Senhora do Desterro	AVYF	Itapuan Pt	AWER	Xingu R
		AVYH	Garcia de Avila	AWET	Mt. Alégre
AVRY	P dos Naufragados	AVYJ	Real R	AWEV	Santarem f
AVSB	Anhatomirim Islet	AVYL	Sergipe	AWEX	Tapajos R
AVSD	F Santa Cruz a	AVYN	Cotinguiba R	AWEZ	Obidos
AVSF	Arvoredo I	AVYP	P Aracaju	AWFC	Manáos f
AVSH	Itapacaroya B	AVYR	San Francisco do Norte R	AWFE	Negro R
AVSJ	Saõ Francisco	AVYT	Peñedo	AWFG	Balique I
		AVYX	Maceió	AWFI	Madeira R
				AWFK	Oyapok R

a Signifies an International Signal Station,   b a Time Signal Station,   c a Weather, Tide, or Ice Signal Station,   d a Life-Saving Station,   e a Lloyd's Signal Station,   f a Wireless Telegraph Station

	GULF OF MEXICO					

AWFM	GUIANA	AWKV	Santa Ana Hr b	AWQJ	Izabel
		AWKX	Little Curaçao I	AWQL	BRITISH HONDURAS
AWFO	FRENCH GUIANA	AWKZ	Oruba I	AWQM	Sarstoon R
AWFQ	Apruague R	AWLC	Sierra Colorado Pt.	AWQN	Half-Moon cay
AWFS	Guisanburg	AWLE	Orangestadt	AWQP	Northern Two cays
AWFU	Cayenne			AWQR	Turneffe Is
AWFY	-Enfant Perdu.	AWLG	C San Roman	AWQT	Mauger cay
AWGB	Cépérou F. a	AWLI	Estanques B	AWQV	Bokel cay
AWGD	Salut Is	AWLK	Maracaibo	AWQX	Icacos Road
AWGF	Kuru R	AWLM	Altagracia	AWQZ	Monkey R
AWGH	Sinnamari			AWRC	Belize
AWGJ	Maroni R	AWLO	COLOMBIA	AWRE	Corosal
AWGL	Les Hattes	AWLQ	Bahia Honda		
		AWLS	La Hacha	AWRG	YUCATAN.
AWGN	DUTCH GUIANA	AWLU	Santa Marta f	AWRI	Cozumel I
AWGP	Surinam R Lt V.	AWLY	Magdalena R	AWRK	Mugeres Hr
AWGR	Galibi Pt	AWMB	BOGOTA	AWRM	C Catoche
AWGT	Paramaribo b	AWMD	Barranquilla	AWRO	Yalahau
AWGV	Coppename R	AWMF	Honda	AWRQ	Progreso
AWGX	Saramacca R	AWMH	Caracoli	AWRS	Sisal
AWGZ	Coronie	AWMJ	Belillo	AWRU	Celestun
AWHC	Corentyn R	AWML	Savanilla	AWRY	Campeche
AWHE	New Rotterdam	AWMN	Salgar		
AWHG	Nickerie R	AWMP	Galera de Zamba B	AWRZ	G OF MEXICO
		AWMR	Cartagena	AWSB	Mérida
AWHI	BRITISH GUIANA	AWMT	Boca Chica	AWSD	,Champoton
AWHK	Berbice	AWMV	Barbacoas B	AWSF	Carmen I
AWHM	New Amsterdam	AWMX	Morrosquillo G	AWSH	Terminos Lagoon
AWHN	Demerara b	AWMZ	P Cispata	AWSJ	Tabasco
AWHO	Demerara Lt V	AWNC	G of Darien	AWSL	Frontera
AWHQ	Georgetown b			AWSN	San Juan de Bautista
AWHS	Christianburg	AWNE	P Escocés	AWSP	Coatzacoalcos R
AWHU	Essequibo R	AWNG	Caledonia Hr	AWSR	Minatitlan
AWHY	Cuyuni R	AWNI	Sasardi B	AWST	Roca Partida
AWIB	Bartika	AWNK	Mandinga Hr	AWSV	Alvarado
AWID	Mazaroni R	AWNM	Farallon Sucio	AWSX	VERA CRUZ c
		AWNO	P Bello (P Velo)	AWSZ	Tuxpan, or, Tuspan
AWIF	VENEZUELA	AWNQ	COLON f	AWTC	Tampico
AWIH	Orinoco R	AWNS	Manzanilla I	AWTE	Rio Grande
AWIJ	Barrancas	AWNU	Toro Pt	AWTG	Bagdad
AWIL	Las Tablas	AWNY	Chagres	AWTI	Matamoros
AWIN	Bolivar	AWOB	Bonaventura Cove		
AWIP	Camaguan.	AWOD	Chiriqui Lagoon	AWTK	TEXAS
AWIR	Macareo R	AWOF	Almirante B	AWTM	Brownsville c
AWIT	Paria G	AWOH	Boca del Toro f	AWTP	Brazos Santiago a d
AWIV	Maturin	AWOJ	Shepherd Hr	AWTS	Corpus Christi c
AWIX	P Santo.	AWOL	COSTA RICA	AWTV	Mustang I
AWIZ	Testigos Is.	AWON	P Limon f	AWTY	Aransas c d
AWJC	Carupano	AWOP	San José	AWUB	Guadalupe R
AWJE	Margarita I	AWOR	SAN JUAN DEL NORTE	AWUE	Cavallo Pass
AWJG	Cariaco	AWOT	San Juan de Nicaragua R	AWUH	Matagorda
AWJI	Cumaná			AWUK	Colorado R
AWJK	Mochima P	AWOV	Bluefields f	AWUN	Indianola
AWJM	Guanta B	AWOX	Mosquito Coast	AWUQ	Lavaca c
AWJO	Barcelona	AWOZ	C Gracias a Dios Hr	AWUT	Brazos R
AWJQ	Tortuga I	AWPC	Old Providence Is	AWUV	Velasco a c d
AWJS	C Codera	AWPE	HONDURAS.	AWUZ	San Luis Pass a d
AWJU	LA GUAIRA	AWPG	Truxillo	AWVC	GALVESTON a b c d f
AWJY	CARACAS	AWPI	Bonacca I	AWVD	Galveston Lt V
AWKB	Carabobo Province	AWPK	Roatan I	AWVF	Houston
AWKD	P El Roque	AWPM	P Royal	AWVI	Bolivar Pt
AWKF	Turiamo P	AWPO	Coxen Hole	AWVL	Half-Moon Shoal Lt
AWKH	P CABELLO	AWPQ	Utilla I	AWVO	Red Fish Bar Lt
AWKJ	Cumarebo	AWPS	Ulua R	AWVR	Lynchburg
AWKL	Coro Province.	AWPU	Caballos Pt	AWVU	Sabine a c d
AWKN	C La Vela	AWPY	P Cortez.		
		AWQB	Omoa	AWVX	LOUISIANA
AWKP	BONAIRE I	AWQD	St Thomas Bight	AWXD	Calcasieu R
AWKR	Lacre Pt	AWQF	Dulce R	AWXJ	Vermilion B
AWKT	CURAÇAO I	AWQH	Livingstone	AWXM	Avery

a Signifies an International Signal Station   b a Time Signal Station, c a Weather Tide, or Ice Signal Station, d a Life-Saving Station, e a Lloyd's Signal Station, f a Wireless Telegraph Station

GULF OF MEXICO		WEST INDIES			
AWXP	Cypres Mort	AXGU	KEY WEST b c f	AXMP	Maternillos Pt
AWXS	Atchafalaya R	AXHD	Sand cay c	AXMR	Mucaras Reef
AWXV	Morgan City c	AXHG	American Shs	AXMT	Lobos cay (Bahamas)
AWXY	Ship Island Shoal	AXHJ	Sombrero cay	AXMV	Guinchos cay
AWYB	Timbalier I	AXHM	Alligator Reef	AXMW	Paredon Grande cay
AWYE	Barataria B	AXHP	Carysfort Reef	AXMZ	Frances cay
AWYH	NEW ORLEANS b c f	AXHS	Fowey Rocks Beacon	AXNC	P Caybarien c
AWYK	Algiers	AXHV	C Florida	AXNE	P Sagua la Grande
AWYN	Deer I	AXHY	FLORIDA STR	AXNG	Bahia de Cadiz Cay
AWYQ	South Pass.	AXIB	Jupiter Inlet a c f	AXNI	Santa Clara B
AWYT	P Eads c	AXIE	Indian R d	AXNK	Cruz del Padre cay
AWYZ	Pass a Loutre	AXIH	C Canaveral	AXNM	Diana cay
AWZC	Chandeleur Is			AXNO	Piedras cay
		AXIJ	WEST INDIES	AXNQ	Cárdenas
AWZF	MISSISSIPPI			AXNS	Siguapa
AWZI	Lake Borgne	AXIL	BAHAMA Is	AXNU	Matanzas
AWZL	Proctorville	AXIN	Great Abaco I	AXNW	HABANA a b c e f
AWZO	Lake Pontchartrain	AXIP	Elbow cay	AXNY	P Mariel
AWZR	Pass Manchac	AXIR	Abaco	AXOB	P Cabañas
AWZU	Jackson	AXIT	Providence Ch	AXOD	Bahia Honda
AWZX	Vicksburg	AXIV	Gun cay	AXOF	Mulata
AXBD	Chefuncte R	AXIW	Great Isaac	AXOH	Santa Rosa B
AXBE	Pt aux Herbes	AXIZ	Stirrup cays	AXOJ	Guadiana B
AXBH	The Rigolets	AXJC	Andros I	AXOL	C San Antonio
AXBK	Lower Pt Clear	AXJE	Santaren Ch	AXON	C Corrientes
AXBN	Shieldsboro	AXJG	Cay Sal Bank	AXOP	C Pepe
AXBQ	Merrill Shell Bank	AXJI	Nicolas Ch	AXOR	ISLA DE PINOS
AXBT	Cat I	AXJK	New Providence I	AXOT	Santa Fé R
AXBW	Mississippi City	AXJM	NASSAU a	AXOV	Calonna R
AXBZ	Biloxi c	AXJO	Harbour I	AXOZ	Batabano
AXCF	Ship I	AXJQ	Egg I	AXPC	Rosario cay
AXCI	East Pascagoula. c	AXJS	Eleuthera I	AXPE	Gorda Pt
AXCL	Scranton c	AXJU	San Salvador, or, Watling I	AXPG	Piedras cay
AXCO	Round I			AXPI	Cochinos B
AXCR	Horn I	AXJW	Cat I	AXPK	P Xagua
		AXJY	Exuma I	AXPM	Cienfuegos
AXCU	ALABAMA	AXKB	Long I	AXPO	P Casilda
AXCV	Alabama City	AXKD	CROOKED ISLAND PASSAGE	AXPQ	P Mosio
AXCW	MOBILE c			AXPS	Trinidad de Cuba
AXDG	Montgomery	AXKF	Bird Rock	AXPU	Tunas
AXDJ	Coosa R	AXKH	Acklin I	AXPW	Zarza R
AXDM	Sand I	AXKJ	Castle I	AXPY	Júcaro
AXDP	Siguenza Pt.	AXKL	Mariguana I	AXPZ	Santa Cruz del Sur
AXDS	Warrington	AXKN	Caicos Is	AXQB	Jobabo
		AXKP	Cockburn Hr	AXQD	Virama
AXDV	FLORIDA	AXKR	Turk Is	AXQF	Cauto R
AXDY	Pensacola c f	AXKT	Grand Turk I e	AXQH	Manzanillo
AXEB	P Barancas	AXKV	Great Inagua I	AXQJ	Limones R
AXEH	St Andrew City e	AXKZ	Mathew Town	AXQL	C Cruz
AXEK	St Joseph			AXQN	El Portillo
AXEN	C San Blas	AXLC	CUBA	AXQP	SANTIAGO DE CUBA
AXFQ	St Vincent I	AXLE	C Maysi	AXQR	Guantanamo f
AXET	Apalachicola c	AXLG	Yumuri B	AXQT	Caymanera
AXEW	C St George	AXLI	P Mata	AXQV	Grand Cayman
AXEZ	St Marks	AXLK	P Boma	AXQW	Little Cayman
AXFB	Cedar cays e	AXLM	P Baracoa		
AXFE	Anclote cays	AXLO	Pt Plata	AXQZ	JAMAICA
AXFH	Egmont cay c	AXLQ	P Navas	AXRC	P Antonio
AXFK	Tampa c	AXLS	P Taco	AXRE	Annatto B
AXFN	Palmasola B	AXLU	P Tanamo	AXRG	P Maria
AXFQ	Sarasota B	AXLW	Cabonico	AXRI	Ocho Rios B
AXFT	Gasparilla I	AXLY	Livisa	AXRK	St Anns
AXFW	Charlotte Hr	AXMB	P Nipe	AXRM	Rio Bueno
AXFZ	Pease Creek	AXMD	P Banes	AXRO	Falmouth.
AXGC	Sanibel I	AXMF	Lucrecia Pt	AXRQ	Montego
AXGF	San Carlos Hr	AXMH	Naranjo P	AXRS	St Luces Hr
AXGI	C Sable	AXMJ	Gibara P	AXRU	Negril Hr
AXGL	Tortugae Is c	AXML	P Padre	AXRW	Savannah-la-Mar
AXGO	Loggerhead cay	AXMN	P Nuevitas del Principe	AXRY	Bluefields B
AXGR	Rebecca Sh			AXSB	Black R

a Signifies an International Signal Station, b a Time Signal Station, c a Weather, Tide, or Ice Signal Station, d a Life-Saving Station, e a Lloyd's Signal Station, f a Wireless Telegraph Station

WEST INDIES				EAST COAST UNITED STATES	
AXSD	Pedro Bluff.	AXYS	Culebrita	AYFE	WINDWARD Is
AXSF	Milk R	AXYU	Vieques I	AYFG	St LUCIA
AXSH	Carlisle B	AXYW	P Mula	AYFI	P Castries b
AXSJ	Portland Bight	AXZB	St THOMAS	AYFK	Vigie Summit
AXSL	Old Harbour B	AXZD	Charlotte Amalia	AYFM	Soufrière B
AXSN	P Royal	AXZF	Savana I	AYFO	Vieux F
AXSP	Plum Pt.	AXZI	Myhlenfels Pt	AYFQ	St VINCENT
AXSR	Kingston	AXZK	St John I	AYFS	Kingstown
AXST	Morant a	AXZM	Tortola	AYFU	THE GRENADINES.
AXSV	Blue Mountain	AXZO	Virgin Gorda	AYFW	Carriacou
AXSZ	MORANT CAYS	AXZQ	Anegada	AYGB	GRENADA
		AXZS	SANTA CRUZ (ST. CROIX)	AYGD	St George
AXTC	HAITI		I	AYGF	BARBADOS.
AXTD	Navassa I	AXZU	Frederichsted	AYGH	Carlisle B
AXTE	Port au Prince c	AXZW	Christiansted	AYGJ	Bridgetown
AXTG	Lemantin Pt	AXZY	Sombrero I	AYGL	Needham Pt a
AXTI	Arcadins	AYBC	Anguilla	AYGN	South Pt
AXTK	Gonaives	AYBD	Crocus B	AYGP	Ragged Pt
AXTM	Tortuga I	AYBE	Seal Rocks	AYGR	TOBAGO f
AXTO	P Paix	AYBF	St MARTIN	AYGT	Scarborough
AXTQ	Acul	AYBH	Marigot	AYGV	Rockly B
AXTS	C Haiti c	AYBJ	Grande B	AYGX	Plymouth
AXTU	Caracol	AYBL	Philipsburg.	AYGZ	TRINIDAD
AXTW	F Dauphin P	AYBN	St BARTHOLOMEW	AYHC	P of Spain b c f
AXTY	Manzanillo B	AYBP	Gustaf Hr	AYHE	North Post Signal Sta-
AXUB	Monte Christi	AYBR	Saba I		tion
AXUD	P Plata	AYBT	St Eustatius	AYHG	San Fernando
AXUF	Samaná B	AYBV	Orange Town	AYHI	Icacos Pt
AXUH	P Santa Bárbara	AYBX	St KITTS (ST CHRISTO-	AYHK	Galera Pt
AXUJ	C Engaño		PHER) I	AYHM	Dragons Mouth
AXUL	Saona I	AYBZ	Basseterre	AYHO	Serpent Mouth
AXUN	Macoris	AYCB	Nevis		
AXUP	Santo Domingo Hr c	AYCD	Charlestown		
AXUR	Escondido P	AYCF	Cades Pt	AYHQ	BERMUDA
AXUT	Neyba B	AYCH	Barbuda	AYHS	Gibbs Hill e
AXUV	Barahona	AYCJ	Codrington	AYHU	Mt Hill
AXUZ	Jacmel c	AYCL	Palmetto Pt	AYHW	The Narrows
AXVC	Aquin	AYCN	ANTIGUA e	AYIB	St George Hr
AXVE	St Louis B	AYCP	St John Hr	AYID	Grassy B
AXVG	Cayes	AYCR	Sandy I	AYIF	Ireland I
AXVI	Châteaudin Road	AYCT	English Hr	AYIH	Hamilton
AXVK	Principe Pt	AYCV	Redonda		
AXVM	C Dame Marie	AYCX	MONTSERRAT	AYIJ	GULF STREAM
AXVO	Cayemites	AYCZ	Plymouth		
AXVQ	Mona	AYDC	GUADELOUPE	AYIL	EAST COAST OF FLORIDA.
AXVS	Desecheo I	AYDE	P Moule a	AYIP	Mosquito Inlet
		AYDG	P Louis	AYIS	New Smyrna
AXVU	PORTO RICO	AYD1	Lamentin	AYJV	Matanzas
AXVW	C Rojo	AYDK	Barque Cove	AYJB	St Augustine c f
AXVY	Mayaguez	AYDM	Basse Terre	AYJE	Anastasia I
AXWB	Aguadilla	AYDO	Pt á Pitre	AYJH	St John R c
AXWD	C Borinquen	AYDQ	Gozier I	AYJK	Jacksonville c
AXWF	Arecibo d	AYDS	Isles des Saintes	AYJN	Dame Pt
AXWH	Manati R	AYDU	Désirade	AYJT	Nassau R
AXWJ	P San Juan a c d f	AYDW	Petite Terre	AYJW	Cumberland Sound
AXWL	Cabeza de San Juan	AYEB	Marie Galante	AYJZ	Amelia I
AXWN	Great Hr	AYED	Grand Bourg	AYKC	Fernandina c
AXWP	P Naguabo	AYEF	Dominica		
AXWR	Humacao	AYEH	Roseau	AYKF	GEORGIA
AXWV	Tuna Pt	AYEJ	MARTINIQUE	AYKI	St Mary
AXWZ	Arroyo	AYEL	Fort de France B	AYKL	Little Cumberland I
AXYC	P Jobos	AYEN	Fort de France	AYKO	St Andrew Sound
AXYE	Muertos I	AYEP	F St Louis a	AYKR	St Simon Sound c
AXYG	P Ponce	AYER	St Pierre	AYKU	Brunswick c
AXYI	P Guayanilla	AYET	La Trinité B	AYKX	Altamaha R
AXYK	P Guanica	AYEV	Havre du Robert.	AYLD	Doboy Sound
		AYEX	François	AYLG	Wolf I
AXYM	LEEWARD Is	AYEZ	Marin	AYLJ	Sapelo
		AYFC	Le Diamant, or, Dia-	AYLM	St Catherine Sound
AXYO	VIRGIN Is		mond Rock		
AXYQ	Culebra f				

a Signifies an International Signal Station, b a Time Signal Station, c a Weather, Tide, or Ice Signal Station, d a Life-Saving Station, e a Lloyd's Signal Station, f a Wireless Telegraph Station

EAST COAST UNITED STATES

AYLP	Medway R	AYTX	Chowan R	AZDX	Havre de Grace.
AYLS	Ossabaw Sound	AYUD	Elizabeth City c	AZED	Susquehanna R
AYLV	Ogeechee R	AYUG	Dismal Swamp Canal	AZEG	North East R
AYMB	Hardwick	AYUJ	Wade Pt	AZEJ	Elk R
AYME	Vernon R	AYUM	North R	AZEM	French Town
AYMH	Warsaw Sound			AZEP	Sassafras R
AYMK	Wilmington	AYUP	VIRGINIA	AZES	Chester R
AYMN	SAVANNAH b c f	AYUS	False C a d	AZEV	Kent I
AYMQ	Tybee c	AYUV	Chesapeake B	AZEY	Tred Haven Creek.
AYMT	Oyster Beds	AYVB	C Henry a c d	AZFB	Sharps I
AYMW	Cockspur I	AYVE	Lynn Haven	AZFE	Choptank R
AYNC	F Jackson	AYVH	F Monroe e	AZFH	Castle Haven
		AYVK	Old Pt Comfort	AZFK	Honga R
AYNF	SOUTH CAROLINA	AYVN	Hampton Roads	AZFN	Hooper Str
AYNI	P Royal c	AYVQ	Craney I	AZFQ	Nanticoke R
AYNL	Martins Industry Lt V	AYVT	Elizabeth R	AZFT	Monie B
AYNO	Paris I	AYVW	NORFOLK b c f	AZFW	Manokin R
AYNR	Hilton Head I	AYVZ	PORTSMOUTH	AZGC	Somers Cove
AYNU	Beaufort f	AYWC	Norfolk Navy Yard	AZGF	Pocomoke R
AYNX	St Helena Sound	AYWF	Chesapeake and Albe-	AZGI	Tangier I
AYOD	Combahee R		marle Canal	AZGL	Cherrystone Inlet
AYOG	Hunting I	AYWI	Nansemond R	AZGO	C Charles a d
AYOJ	Pon-pon R	AYWL	James R	AZGP	C Charles Lt V
AYOM	Edisto R	AYWO	Petersburg	AZGR	Machipongo Inlets
AYOP	Stono Inlet	AYWR	NEWPORT NEWS b c	AZGU	Chincoteague Inlet
AYOQ	Charleston Lt V	AYWU	Appomattox R	AZGX	Assateague a d
AYOS	CHARLESTON c f	AYWX	City Pt	AZHD	Winter Quarter Sh Lt V
AYOV	F Sumter	AYXD	Richmond	AZHG	Ocean City a d
AYPB	Moultrieville a c d	AYXG	Back R	AZHJ	Fenwick I Sh Lt V
AYPE	Ashley R	AYXJ	York R		
AYPH	Cooper R	AYXL	Yorktown	AZHM	DELAWARE
AYPK	Wando R	AYXM	West Pt (York R)	AZHP	Five Fathom Bank Lt V
AYPN	Rattlesnake Sh	AYXP	Delaware		
AYPQ	Bull B	AYXS	New Pt Comfort	AZHQ	North-East End Lt V
AYPT	C Romain	AYXV	Wolf Trap Shs	AZHS	C Henlopen a d f
AYPW	Racoon cay	AYZB	Piankatank R	AZHV	Breakwater Hr
AYPZ	Peedee R	AYZE	Rappahannock R	AZHY	Delaware Breakwater a c
AYQC	Georgetown c	AYZH	Windmill Pt	AZIB	Lewes a d
AYQF	Columbia (Congaree R)	AYZK	Corrotoman R	AZIE	Brandywine Sh
		AYZN	Fredericksburg	AZIH	Fourteen-feet Bank
AYQI	NORTH CAROLINA	AYZQ	POTOMAC R	AZIK	Mispillion Creek
AYQL	C Fear a d	AYZT	Piney Pt	AZIN	Cross Ledge Sh
AYQO	Bald Head Ch	AYZW	Blakistone I	AZIQ	Ship John Sh
AYQR	Frying-pan Shs Lt V	AYZX	St Clement B	AZIT	Bombay Hook
AYQU	Federal Pt	AZBD	Chaptico	AZIW	Reedy I c
AYQX	Oak I c d	AZBG	Leonardstown B	AZJC	Delaware City
AYRD	Wilmington c	AZBJ	Wicomico R	AZJF	New Castle.
AYRG	Beaufort c f	AZBM	Mathias Pt	AZJI	Christiana Creek
AYRJ	Morehead City c	AZBP	P. Tobacco R	AZJL	Wilmington
AYRM	C Lookout a c d	AZBS	Upper Cedar Pt	AZJO	Marcus Hook
AYRP	Core Sound c d	AZBV	Maryland Pt		
AYRS	Ocracoke Inlet a d	AZBY	Mt Vernon	AZJR	PENNSYLVANIA c
AYRV	Neuse R	AZCB	F Washington	AZJU	Chester
AYSB	New Berne c	AZCE	Alexandria	AZJX	Schuylkill
AYSE	Goldsboro			AZKD	League I
AYSH	Raleigh	AZCH	WASHINGTON, D C b f	AZKG	PHILADELPHIA b f
AYSK	Hayti				
AYSN	Bay R	AZCK	DISTRICT OF COLUMBIA	AZKJ	NEW JERSEY
AYSQ	Pamplico			AZKM	Trenton
AYST	Bath	AZCN	UNITED STATES	AZKP	Bordentown
AYSW	Washington c			AZKS	Burlington
AYSZ	Hatteras Inlet a d	AZCQ	MARYLAND	AZKV	Camden
AYTB	Diamond Shoal Lt V c f	AZCT	Patuxent R	AZKY	Gloucester
AYTC	C Hatteras a c d f	AZDC	Jones Point	AZLB	Billingsport
AYTF	Oregon Inlet a d	AZDF	BALTIMORE b c	AZLE	Penn Grove
AYTI	Bodie I a d	AZDI	Thomas Pt	AZLH	Salem
AYTL	Albemarle Sound	AZDL	Annapolis c f	AZLK	Cohansey
AYTO	Plymouth	AZDO	Severn R	AZLN	Maurice R
AYTR	Roanoke R	AZDR	Magothy R	AZLQ	C MAY a c d f
AYTU	Edenton c	AZDU	Patapsco R	AZLT	Hereford Inlet a c d

a Signifies an International Signal Station   b a Time Signal Station,   c a Weather, Tide, or Ice Signal Station,   d a Life-Saving Station, e a Lloyd's Signal Station, f a Wireless Telegraph Station

EAST COAST UNITED STATES

Code	Place	Code	Place	Code	Place
AZLW	Ludlam Beach	AZTJ	New Haven c	BAEM	Handkerchief Lt V.
AZMC	Atlantic City a d f	AZTM	Falkner I.	BAEP	Shovelful Sh Lt V.
AZMF	Absecon a d	AZTO	Cornfield Pt Lt V	BAES	Cross Rip Lt V
AZMI	Tucker Beach.	AZTP	Bartlett Reef Lt V.	BAEV	Succonnesset Sh. Lt V
AZML	Little Egg Hr a c d	AZTS	NEW LONDON c	BAEY	Nantucket c f
AZMO	Barnegat d	AZTV	Thames R	BAFE	Nantucket Sh Lt V c f
AZMR	Long Branch a c d	AZTY	Race Rock	BAFH	Sankaty Hd
AZMS	Fire I. Lt V	AZUB	Ram I Reef Lt. V.	BAFK	Great Round Sh Lt V
AZMU	Navesink	AZUE	Latimer Reef.	BAFN	Pollock Rip Lt V.
AZMX	Sandy Hook a c d f	AZUH	Fisher I a d	BAFQ	Nauset Beach a d
AZMY	Sandy Hook Lt V	AZUK	Stonington c	BAFW	C Cod (Highlands) a d f
AZNB	Scotland Lt V	AZUN	Watch Hill Pt a d	BAFZ	Race Pt a c d
AZND	Gedney Ch			BAGC	Wood End a d
AZNG	Raritan	AZUQ	RHODE ISLAND	BAGF	Provincetown Hr c
AZNJ	Staten I	AZUT	Block I a c d f	BAGI	Long Pt
AZNM	Narrows	AZUW	Judith Pt a c d f	BAGL	Billingsgate I
AZNP	Stapleton	AZVC	Whale Rock	BAGO	Barnstable
AZNS	Tompkinsville	AZVF	Narragansett B a e d	BAGR	Plymouth
AZNV	New Brighton	AZVI	Dutch I.	BAGU	Duxbury Pier
AZNY	Elizabethtown	AZVL	Conanicut I	BAGX	Minots Ledge
AZOB	Newark	AZVO	Wickford	BAGY	Boston Lt V
AZOE	Bergen	AZVR	Connicut Pt	BAHD	Deer I
AZOF	Robbins Reef Lt	AZVU	Greenwich.	BAHG	Long Island Hd.
AZOH	Bedloe I	AZVX	Sabine Pt	BAHJ	Hull c
AZOK	Jersey City f	AZWD	Pawtuxet	BAHM	Hingham
AZON	Hudson City	AZWG	Fuller Rock	BAHP	Neponset
AZOQ	Hoboken	AZWJ	PROVIDENCE c f	BAHS	BOSTON a b c d f
AZOR	Harlem	AZWM	Warren	BAHV	Charlestown
		AZWP	Hog I Lt V	BAHY	Medford
AZOS	NEW YORK STATE	AZWS	Bristol	BAIE	Winnissimmet
AZOT	New York City a b c f	AZWV	Prudence I	BAIH	Egg Rock.
AZOW	Hudson R	AZWY	Taunton R	BAIK	Lynn
AZPC	Governors I c	AZXB	Fall River, Mass c	BAIN	Nahant
		AZXE	Newport b c f	BAIQ	Marblehead c
AZPF	BROOKLYN	AZXH	Goat I	BAIT	Salem
AZPI	East R	AZXK	Brenton Reef Lt V	BAIW	Baker I
AZPL	Long I	AZXN	Sakonnet Ch	BAIZ	Gloucester c
AZPO	Astoria			BAJC	Eastern Pt
AZPR	Hell Gate	AZXQ	MASSACHUSETTS	BAJF	C Ann
AZPS	Harlem Ship Canal	AZXT	Hen and Chickens Reef Lt V	BAJI	Straitsmouth I a d
AZPU	P Morris			BAJL	Annisquam
AZPX	Coney I a c d	AZXW	Buzzard B	BAJO	Ipswich
AZQD	Brighton	AZYC	Dumpling Rock	BAJR	Newburyport a c d
AZQG	Manhattan Beach c	AZYF	Clark Pt		
AZQJ	Fire I a d f	AZYI	NEW BEDFORD.c	BAJU	NEW HAMPSHIRE
AZQK	LONG ISLAND SOUND	AZYL	Palmer I	BAJX	Hampton
AZQM	Montauk Pt a d	AZYO	Fairhaven	BAKD	Isle of Shoal, Me c
AZQP	Sands Pt	AZYR	Mattapoiset.	BAKG	PORTSMOUTH c f
AZQS	Gardiner I	AZYU	Bird I	BAKJ	Concord
AZQV	Plum I	AZYX	Sippican	BAKM	Piscataqua R
AZQY	Little Gull I	BACD	Wings Neck		
AZRB	Sag Hr	BACF	Cuttyhunk a e d	BAKP	MAINE
AZRE	Peconic B	BACI	Vineyard Sound Lt V	BAKS	Boon I.
AZRH	Greenport		(Sow and Pigs)	BAKV	York R
AZRK	Southold	BACL	Tarpaulin Cove c	BAKY	C Neddick (Knubble)
AZRN	James P	BACO	Nobska Pt c	BALE	Kennebunk
AZRQ	Riverhead	BACR	Woods Holl c	BALH	Augusta.
AZRT	Huntington	BACU	MARTHAS VINEYARD	BALK	Goat I'
AZRW	Oyster B	BACX	Gay Hd a d	BALN	Porpoise Hr
AZSC	Hempstead.	BADC	West Chop Pt c	BALQ	Wood I
AZSF	Throgs Neck	BADF	Vineyard Haven	BALT	Saco
AZSI	Execution Rocks	BADI	East Chop Pt c	BALW	C Elizabeth a d f
AZSL	Great Captain I	BADL	Edgar Town	BALZ	Casco B
		BADO	Cottage City	BAMC	PORTLAND c
AZSO	CONNECTICUT	BADR	C Poge	BAMF	Halfway Rock
AZSR	Stamford Hr	BADU	Bishop and Clerks	BAMI	Seguin I
AZSU	Norwalk	BADX	Hyannis c	BAML	Kennebec R
AZSX	Black Rock Hr	BAED	Bass R	BAMO	Pond I
AZTD	Bridgeport c f	BAEG	Chatham a e d	BAMR	Bath
AZTG	Stratford Shs	BAEJ	Monomoy Pt a c d	BAMU	Wiscasset

a Signifies an International Signal Station   b a Time Signal Station,   c a Weather, Tide, or Ice Signal Station,   d a Life-Saving Station   e a Lloyd's Signal Station,   f a Wireless Telegraph Station.

CANADA.

Code	Place	Code	Place	Code	Place
BAMX	Hendrick Hd	BAVI	Avery Rock	BCFJ	Prim Pt
BAND	Damariscotta	BAVL	Machiasport c	BCFM	Petit Passage
BANG	Medomak R	BAVO	Machias	BCFP	Boars Hd
BANJ	St George R.	BAVR	Quoddy Head a c d	BCFS	Brier I
BANM	Thomaston.	BAVU	Welchpool	BCFV	St Mary B
BANP	Burnt I a d	BAVX	Lubeck	BCFY	Weymouth
BANS	Townsend	BAWC	Eastport c	BCGE	Sissibou R
BANV	Pemaquid Pt	BAWF	St Croix R	BCGH	Church Pt
BANY	Franklin I	BAWI	St Stephen	BCGK	Yarmouth c
BAOB	Monhegan I			BCGN	C Fourchu
BAOE	Marshall Pt c	BAWL	CANADA	BCGQ	Pease I
BAOH	Tennant Hr.			BCGT	Seal I d
BAOK	Whitehead I a c d	BAWO	NEW BRUNSWICK	BCGW	Tusket R
BAON	Matinicus Rock	BAWR	Passamaquoddy B	BCGZ	Argyle
BAOQ	Owls Hd	BAWU	St Andrew c	BCHF	Pubnico Hr
BAOT	Rockland	BAWX	Digdequash R	BCHI	Bon Portage I
BAOW	Rockport	BAXD	St George	BCHL	C Sable d f
BAOZ	Indian I	BAXG	L'Etang Hr	BCHO	Barrington B
BAPC	Negro I	BAXJ	Campobello I	BCHR	Baccaro Pt
BAPF	Camden	BAXM	Machias Seal I.	BCHU	P Latour
BAPI	Northaven	BAXP	Gannet Rock	BCHX	Negro I d
BAPL	Grindel Pt	BAXS	Grand Manan I	BCIA	SHELBURNE
BAPO	Belfast	BAXV	Wolves Is	BCID	C Roseway
BAPR	Searsport	BAXY	Beaver Hr	BCIG	Jordon R
BAPU	Stockton	BAYE	B of Fundy	BCIJ	Green Hr
BAPX	Winterport	BAYH	Lepreau Pt c	BCIM	Ragged Island Hr
BAQD	Bangor c	BAYK	Musquash Hr	BCIP	Gull Rock
BAQG	Brewer	BAYN	St John b c f	BCIS	P Hebert
BAQJ	Bucksport	BAYQ	Partridge I	BCIV	Little Hope Islet.
BAQM	Orland	BAYT	C Spencer	BCIY	Mouton P d
BAQP	Dice Hd	BAYW	Quaco	BCJE	Liverpool c
BAQS	Castine	BAYZ	C Enrage	BCJH	Coffins I
BAQV	Penobscot	BAZC	Chignecto B	BCJK	Medway
BAQY	Pumpkin I	BAZF	Harvey Corner	BCJN	La Have R
BARB	Eggemoggin Reach	BAZI	Grindstone I	BCJQ	Bridgewater
BARE	Eagle I	BAZL	Petitcodiac R	BCJT	West Ironbound I
BARH	Deer I	BAZO	Folly Pt	BCJW	LUNENBURG
BARK	Heron Neck (Green I)	BAZR	Hillsborough	BCJZ	Cross I
BARN	Saddleback Ledge	BAZU	Moncton	BCKA	Green I (Nova Scotia)
BARQ	Blue Hill B	BAZX	Salisbury	BCKF	Mahone
BART	Mt Desert Rk			BCKI	Quaker I
BARW	Surry	BCAD	NOVA SCOTIA	BCKL	Chester
BARZ	Ellsworth	BCAE	Charters Corner	BCKO	East Ironbound I
BASC	Union River B.	BCAH	Dorchester	BCKR	St Margaret B
BASF	Great Duck I	BCAK	Sackville	BCKU	Hubbard Cove
BASI	Baker I	BCAN	Cumberland Basin	BCKX	Crouchers I.
BASL	Somes Hr	BCAQ	Marine Transport Railway	BCLA	Dover
BASO	Bar Hr			BCLD	Betty I
BASR	Frenchman B	BCAT	Amherst	BCLG	Turner B
BASU	Skilling R	BCAW	South Joggins	BCLJ	Sambro I
BASX	Sullivan	BCAZ	Apple Hd	BCLM	Chebucto Hd.
BATD	Stave I	BCDA	Haute I		
BATG	Winter Hr	BCDE	C d'Or	BCLN	SABLE I d f
BATH	Egg Rock	BCDH	Black Rock		
BATJ	Prospect Hr	BCDK	C Split	BCLP	HALIFAX b c d f
BATM	Gouldsborough	BCDN	BASIN OF MINES	BCLS	Bedford Basin
BATP	Dyer B	BCDQ	Parrsboro	BCLV	Dartmouth
BATS	PETIT 'MANAN	BCDT	Cobequid B	BCLY	George I
BATV	Narraguagus R.	BCDW	Truro.	BCME	Mauger Beach Lt
BATY	Millbridge	BCDZ	Shubenacadie R	BCMH	Devil I d
BAUE	Cherryfield	BCEF	Burntcoat Hd	BCMK	Jeddore Rock
BAUH	Harrington	BCEI	Walton	BCMN	Egg I
BAUK	Pleasant R	BCEL	Avon R	BCMQ	Ship Hr
BAUN	Addison Pt	BCEO	Kingsport	BCMT	Pope Hr
BAUQ	Columbia Falls	BCER	Canning	BCMW	Sheet Hr
BAUT	Nash I	BCEU	Horton.	BCMX	Sheet Rock
BAUW	Moose Peak (Mistake I)	BCEX	WINDSOR.	BCMZ	Beaver I
BAUZ	Moose-a-bec Reach	BCFA	P Lorne	BCNF	Newtonquaddy
BAVC	Chandler B	BCFD	Annapolis	BCNI	Liscomb Hr c
BAVF	Libby I	BCFG	Digby c	BCNL	Wedge I

a Signifies an International Signal Station, b a Time Signal Station, c a Weather, Tide, or Ice Signal Station, d a Life-Saving Station, e a Lloyd's Signal Station, f a Wireless Telegraph Station

GULF AND RIVER OF ST LAWRENCE

Code	Place	Code	Place	Code	Place
BCNO	St Mary R	BCWF	Merigomish	BDGF	MIRAMICHI B
BCNR	Sherbrooke	BCWI	Pictou	BDGI	Escuminac Pt a
BCNU	Country Hr.	BCWL	East Arm		
BCNX	Isaac Hi	BCWO	New Glasgow	BDGL	Preston beach Lts
BCQA	Green I			BDGO	Fox I
BCOD	Tor B	BCWR	NORTHUMBERLAND STR	BDGR	Vin I
BCOE	Berry Hd	BCWU	Caribou	BBGU	Portage I
BCOG	Whitehead I e	BCWX	Amet Isle	BDGV	Miramichi B Lt V
BCOJ	Canso c	BCXA	Tatamagouche	BDGX	Sheldrake I
BCOM	Cranberry I	BCXD	John R	BDHA	Grant beach Lts
BCOP	Crow Hi	BCXG	Wallace	BDHG	Chatham
BCOS	Chedabucto B	BCXJ	Pugwash	BDHJ	Newcastle
BCOV	Guysborough	BCXM	Cold Spring Hd	BDHM	Nelson Town
BCOY	GUT OF CANSO	BCXP	B Verte	BDHP	Neguac gully
BCPE	Eddy Pt	BCXS	C Tormentine d	BDHS	Tracadie
BCPH	P Mulgrave	BCXV	Jourimain I	BDHV	Pokemouche
BCPK	Havre Bouche	BCXY	Shemogue	BDHY	Birch Pt a
BCPN	CAPE BRETON ISLAND	BCYE	Shediac	BDIE	Miscou Hi a
BCPQ	P Hastings	BCYH	Cassie Pt	BDIH	Pokesudi I Lt
BCPT	P Hawkesbury	BCYJ	Cocagne	BDIK	Caraquette I
BCPW	Habitants Hr	BCYK	Buctouche R	BDIN	Shippigan
BCPZ	Arichat	BCYN	Richibucto Hd	BDIQ	CHALEUR B
BCQF	Creighton Hd	BCYQ	Liverpool	BDIT	Clifton
BCQI	Green I	BCYT	St Louis	BDIW	Nipisighit B
BCQL	C Round			BDIZ	Bathurst
BCQO	Big Arrow I	BCYW	PRINCE EDWARD I	BDJC	Petit Rocher Lt
BCQR	St Peter B	BCYZ	West Pt	BDJF	Little Belledune Pt
BCQU	Haulover Isthmus	BCZF	C Egmont	BDJI	Heron I
BCQX	M'Kenzie Pt	BCZI	Bedeque B	BDJL	Dalhousie
BCRA	Ouetique I	BCZL	Summerside	BDJO	Restigouche R
BCRD	St Esprit	BCZO	C Traverse	BDJR	Campbell town
BCRG	Guyon I	BCZR	Tryon	BDJU	New Carlisle
BCRJ	LOUISBURG HR	BCZU	Crapaud	BDJX	Richmond
BCRM	Scatari I d	BCZX	Hillsborough B	BDKA	Carleton
BCRP	Menadon Passage	BDAC	CHARLOTTE TOWN	BDKE	Pasbebiac
BCRS	Mira	BDAF	Haszard Pt	BDKG	Maquereau Pt a
BCRV	Flint I	BDAI	Prim Pt	BDKJ	Pabou
BCRY	C Percy	BDAL	C Bear f	BDKM	C d'Espoir a
BCSE	Schooner Cove	BDAO	Murray Hr e	BDKP	Percé
BCSH	Lingan Hd	BDAR	Cardigan B	BDKS	Pt Peter (Flat I)
BCSK	Bridgeport a	BDAU	GEORGE TOWN	BDKT	C Gaspé
BCSN	Flat Pt a	BDAX	St Andrew Pt	BDKU	Gaspé B Lt V
BCSQ	North Sidney c	BDCA	Panmure Hd	BDKV	Gaspé
BCST	Victoria Pier	BDCG	Boughton R	BDKY	C Rozier e
BCSW	SYDNEY	BDCJ	Colville R	BDLE	Fame Pt e f
BCSZ	Battery Pt	BDCM	Souris		
BCTF	Bird Rock	BDCP	East Pt	BDLH	ANTICOSTI I
BCTI	P Aconia	BDCS	St Peter Hr	BDLK	Heath Pt c d e f
BCTL	Bras d Or	BDCV	Tracadie	BDLN	Bagot Pt d
BCTO	St Andrew Ch	BDCY	Grand Rustico	BDLQ	Salt Lake B
BCTR	Blackrock Pt	BDEA	Grenville Hr	BDLT	South Pt e
BCTU	Barra Str	BDEG	Stanley R	BDLW	South West Pt c e
BCTX	Mac Kinnon Hr	BDEJ	Bullhook I	BDLZ	Ellis B
BCUA	Bird I Lt (Ciboux I)	BDEM	Richmond Hi	BDMC	West Pt c
BCUD	St Anne Hr	BDEP	Malpeque		
BCUG	Ingonish	BDES	Cascumpeque Hr	BDMF	St LAWRENCE R
BCUJ	C North	BDEV	Alberton	BDMI	Natashquan R
BCUM	Meat Cove a	BDEY	Tignish	BDML	Mingan Hr
BCUP	C St Lawrence e	BDFE	North Pt	BDMO	Perroquets Lt
BCUS	St Paul I d e	BDFH	MAGDALEN IS	BDMR	Manitou R
BCUV	CABOT STR	BDFK	Entry I	BDMU	Seven I B f
BCUY	Chetican Pt	BDFN	Amherst Hr	BDMX	Egg I
BCVE	Sea Wolf I	BDFO	South Pt (Amherst I) e	BDNA	C Magdalen e
BCVH	Mabou	BDFQ	Grindstone I	BDNG	Martin R a
BCVK	P Hood	BDFT	Grosse I a	BDNJ	C Chatte a
BCVN	George B	BDFW	Étang du Nord a	BDNM	Pt de Monts a
BCVQ	Tracadie	BDFZ	Great Bird Rock	BDNP	Matane a
BCVT	Pomquet			BDNS	Metis Pt a
BCVW	Antigonish.			BDNV	Manicouagan a
BCVZ	C George	BDGC	G OF ST LAWRENCE	BDNY	Bersimis R

a Signifies an International Signal Station,   b a Time Signal Station,   c a Weather, Tide, or Ice Signal Station,   d a Life-Saving Station, e a Lloyd's Signal Station,   f a Wireless Telegraph Station

GREAT LAKES · NEWFOUNDLAND

BDOE	Father Pt a f	BDWA	Clayton	BEGA	Bay City.
BDOH	P Neuf a	BDWG	C Vincent c	BEGF	Saginaw B
BDOK	Pilot Station			BEGI	Thunder B c d
BDON	Rimouski a	BDWJ	LAKE ONTARIO	BEGL	Alpena c
BDOQ	Bicquette I			BEGO	Duncan City
BDOT	Bic I	BDWM	KINGSTON	BEGR	Détour Passage
BDOW	Red Islet Bank Lt V	BDWP	Napanee R		
BDOX	Red Islet Lt	BDWS	Belleville.	BEGU	Mississauga I
BDOZ	Saguenay R	BDWV	Pigeon I	BEGX	Manitoulin I.
BDPC	Tadousac	BDWY	False Duck I	BEHC	Southampton
BDPF	Chicoutimi	BDXE	Peter Pt	BEHG	Kincardine
BDPI	Green I	BDXH	Egg I	BEHJ	Goderich
BDPL	Rivière du Loup a	BDXK	Newcastle Hr	BEHM	Georgian B.
BDPM	White I Reef Lt V	BDXN	Coburg.	BEHP	Collingwood
BDPO	Brandy Pots a	BDXQ	P Hope	BEHS	Matchedash B
BDPR	Long Pilgrim I	BDXT	TORONTO	BEHV	Parry
BDPU	Grand I	BDXW	Hamilton		
BDPX	Kamouraska	BDXZ	P Dalhousie	BEHY	LAKE MICHIGAN
BDQA	Orignaux Pt	BDYC	Welland Canal	BEID	Green Bay c
BDQG	Goose C	BDYF	NIAGARA d	BEIH	Kewaunee d
BDQJ	Coudres I	BDYI	Lewiston.	BEIK	Manitowoc
BDQM	Ouelle R	BDYL	Genesee R	BEIN	Sheboygan d
BDQP	St Paul B	BDYO	Soduspoint c	BEIQ	P Washington
BDQS	LOWER TRAVERSE Lt V	BDYR	Glasgow	BEIT	Milwaukee c d
BDQV	Upper Traverse Lt V	BDYU	Fair Haven c	BEIW	Racine d
BDQY	Stone Pillar	BDYX	Oswego c d	BEIZ	Kenosha d
BDRE	Beaujeu Ch	BDZA	Stony Pt	BEJA	Waukegan
BDRH	St Ignace	BDZG	Galloo I	BEJF	Evanston d
BDRK	L'Islet a	BDZJ	Sacketts Hr c	BEJI	CHICAGO d
BDRN	Crane I	BDZM	Tibbett Pt	BEJL	Illinois
BDRQ	St Thomas			BEJO	Calumet
BDRT	Grosse Isle a	BDZP	LAKE ERIE	BEJR	Michigan City c d
BDRW	C Brulé	BDZS	Buffalo b c d f	BEJU	Indiana
BDRZ	Orleans I	BDZV	P Colborne	BEJX	St Joseph d
BDSC	St François	BDZY	Mohawk I	BEKC	South Haven d
BDSF	Bellechasse I	BEAC	P Maitland	BEKG	Saugatuck
BDSI	St Jean	BEAF	Long Pt	BEKJ	Holland
BDSL	St Laurent	BEAI	P Burwell	BEKM	Grand Haven c d
BDSO	St Famille	BEAL	P Stanley	BEKP	Muskegon d
BDSR	St Pierre	BEAO	Pt Pelée	BEKS	Whitehall
BDSU	L'Ange Gardien	BEAR	Gibraltar	BEKV	Pentwater d
BDSX	Beauport	BEAU	Trenton	BEKY	Ludington d
BDTA	QUEBEC b	BEAX	Wyandotte.	BELD	Manistee c d
BDTG	Levis	BECD	Windsor	BELH	Traverse City
BDTJ	New Liverpool	BECH	DETROIT f	BELK	St Mary R
BDTM	Carouge R	BECK	Michigan State	BELN	Neebish Rapids
BDTN	Trembles Sh	BECN	Monroe.		
BDTP	Portneuf	BECO	Toledo	BELQ	LAKE SUPERIOR
BDTS	Champlain	BECT	Ohio	BELT	Sault Sainte Marie b
BDTV	Three Rivers	BECW	West Sister I	BELW	Iroquois Pt
BDTY	P St Francis	BECZ	P Clinton	BELZ	Corbay Pt
BDUE	Sorel	BEDC	Strontian I	BEMA	Michipicoten I
BDUH	Richelieu I	BEDG	Marblehead d	BEMF	Battle I
BDUK	Longueuil	BEDJ	Sandusky	BEMI	Lamb I
BDUN	Maisonneuve	BEDM	Vermilion	BEML	Porphyry Pt
		BEDP	Lorain c	BEMO	THUNDER C
BDUO	OTTAWA	BEDS	CLEVELAND c d f	BEMR	Victoria I
		BEDV	Ashtabula c d	BEMU	Minnesota
BDUQ	MONTREAL b	BEDY	Erie City d	BEMV	Isle Royale
BDUT	Lachine Canal			BEMW	Passage I
BDUW	Cornwall	BEFD	Dunkirk	BEMX	DULUTH b d
BDUZ	Morrisburg	BEFH	Lake St Clair	BENC	Superior City
BDVC	Waddington	BEFK	Marine City	BENG	Wisconsin
BDVF	Matilda	BEFN	St Clair	BENJ	Sand I
BDVI	Edwardsburg			BENM	Outer Apostle I
BDVL	Ogdensburg c	BEFQ	LAKE HURON	BENP	Michigan I
BDVO	Prescott	BEFR	P Huron c f	BENS	Chaquamegon Pt.
BDVR	Morristown	BEFT	Sarnia	BENV	Ontonagon
BDVU	Brockville	BEFW	Gratiot F	BENY	Portage Lake Canal d
BDVX	The Thousand Is	BEFZ	Ottawa	BEOD	Eagle Hr

a Signifies an International Signal Station b a Time Signal Station c a Weather Tide, or Ice Signal Station, d a Life-Saving Station, e a Lloyd's Signal Station, f a Wireless Telegraph Station

LABRADOR		ICELAND	

BEOH	Agate Hr	BETM	Brunet I a	BEYA	Hamilton Inlet
BEOK	Copper Hr	BETP	Fortune B.	BEYC	Rigoulette
BEON	Manitou I d	BETS	Long Hr.	BEYF	Davis Inlet
BEOQ	Keweenaw	BETV	MIQUELON. c	BEYG	Hopedale.
BEOT	L'Anse	BETY	St Pierre c	BEYH	Nain
BEOW	Huron I	BEUD	Galantry Hd a	BEYJ	Okkak
BEOZ	Granite I	BEUH	Burin I a	BEYK	C. Chidley.
BEPA	Marquette d	BEUK	Placentia Hr		
BEPF	Grand I	BEUN	St Mary a		
BEPI	P au Sable	BEUQ	C Pine a	BEYL	HUDSON BAY.
BEPL	White Fish Pt.	BEUT	C Race e f	BEYN	F Chimo
BEPO	BELLE ISLE STR	BEUW	Ferryland Hd a	BEYP	Baffin Land
BEPR	Wash-sheecootai B.	BEUZ	C Spear a	BEYR	James B
BEPU	Coacoacho B	BEVA	ST JOHNS a b	BEYT	East Main F
BEPX	Watagheistic Sound	BEVF	C St Francis a	BEYV	Ruperts House.
BEQC	Mecattina C.	BEVI	Brigus B	BEYX	Albany F
BEQG	Greenly I	BEVL	Harbour Grace	BEYZ	Moose F
BEQJ	Armour Pt a f	BEVO	Carbonear I	BEZA	Severn F
BEQM	Red B	BEVR	Baccalieu I a	BEZC	York Factory
BEQP	Chateau B f	BEVU	Trinity Hr	BEZF	F Churchill.
BEQS	BELLE ISLE e f	BEVX	Catalina Hr a	BEZI	Chesterfield Inlet
BEQV	La Talus	BEWC	C Bonavista a	BEZK	Davis Str.
		BEWG	Little Denier I	BEZM	Baffin B.
BEQY	NEWFOUNDLAND	BEWJ	Greenspond		
BERD	C Bauld	BEWL	Stinking I		
BERH	Pistolet B.	BEWM	Cabot I a	BEZO	GREENLAND
BERK	C Norman	BEWN	Funk I	BEZQ	Godhavn (P. Lievety)
BERN	St John B	BEWP	Offer Wadham I a	BEZS	Disko B
BERQ	Rich Pt f	BEWS	Fogo Hr	BEZU	Egedesminde
BERT	P Saunders	BEWV	Toulinguet I a	BEZW	Holsteinborg (Sisimiut)
BERW	Ingornachoix B	BEWY	Gull I a	BEZY	C Farewell
BERZ	Bonne B a	BEXD	Croc Hr		
BESF	B of Islands	BEXH	Betts Cove		
BESI	P au Port	BEXK	Tilt Cove	BFAC	ICELAND.
BESL	St George Hr.	BEXN	Hare B	BFAE	Reykjavik
BESO	C RAY e f			BFAH	Svendseyre
BESU	P Basque a	BEXQ	LABRADOR	BFAK	Talkna Fiord
BESX	Rose Blanche B	BEXT	St Lewis Sound	BFAM	Dyre Fiord
BETC	La Poile B	BEXV	Cartright Hr	BFAO	Oe Fiord
BETG	Boar I (Burgeo) a	BEXY	Indian Tickle.	BFAQ	Holar
BETJ	Pass I a	BEXZ	Sandwich B.	BFAS	Seidis Fiord.
				BFAU	JAN MAYEN I

a Signifies an International Signal Station. b a Time Signal Station, c a Weather, Tide, or Ice Signal Station, d a Life-Saving Station, e a Lloyd's Signal Station, f a Wireless Telegraph Station

INTERNATIONAL CODE OF SIGNALS.

PART II.

INDEX.

GENERAL VOCABULARY AND GEOGRAPHICAL LIST.

(141)

PART II.

GENERAL VOCABULARY

	A—ABOUND		
CXA	A (Letter) (For new method of spelling, see page 13)	CXZ	ABIDE ING ABODE
CXB	A, or, AN	CYA CYB	ABILITY —I doubt your (or, my) ability to —
CXD	A1	CYD	—Your ability—to
CXE	A. B ABLE SEAMAN	CYE	ABLE-Y ABLE TO TO BE ABLE Able seaman A B_____CXE
	A M BEFORE NOON _____ EDU	CYF	—Am, Is, or, Are able to.
		CYG	—Am Is, or, Are not able to
CXF	ABACK	CYH	—Are you able to?
	Heave all aback Go astern _____LP		Can I cross the bar? Shall I be able to
CXG	—Taken aback		get over the bar?_____ _____FVY
			I have not been able to make out the
CXH	ABAFT		buoy or beacon_____GAJ
CXI	—Abaft the beam	CYI	—I was able to
	Afloat abaft _____BW	CYJ	—I was not able to.
			If he (she, it, or person-s or thing-s indi-
CXJ	ABANDON-ING-MENT		cated) is (or, are) able to_____BYM
	Abandon the vessel as fast as possible_AB		If he (she, it, or person-s or thing-s indi-
CXK	—Abandoned		cated) is (or, are) not able to ___BYO
CXL	—Do not abandon me	CYK	—If I am able to
	Do not abandon the vessel _____AC	CYL	—If I am not able to
	Do not abandon the vessel until the tide	CYM	—If you are able to.
	has ebbed _____ ____ _____AD	CYN	—If you are not able to
	Do you intend to abandon?_____AE	CYO	—Is able to.
CXM	—Enemy has abandoned		Not able to make such short tacks__XIL
CXN	—Has she (or vessel indicated) been long	CYP	—Ought not to be able to
	abandoned?	CYQ	—Ought to be able to
CXO	—I did not abandon her until I was forced	CYR	—Shall, or, Will be able to.
	I do not intend to abandon the vessel AF	CYS	—Shall, or, Will not be able to
	I must abandon the vessel _____AG		Vessel (or person indicated) is not able
	I shall abandon my vessel unless you		to comply _____JLG
	will keep by us_____AH	CYT	—When will you be able to?
	I will not abandon you (or vessel indi	CYU	—Will you be able to?
	cated) I will remain by you _____AI		With some assistance I shall be able to
	I (or, They) wish to abandon but have		set things right_____DF
	not the means _____AJ	CYV	—You are able to
CXP	—Intend to abandon ship to Lloyd's agent.	CYW	—You are not able to
	Is she (or vessel indicated) abandoned?		
	AK	CYX	ABOARD (See also BOARD),
CXQ	—They have abandoned	CYZ	—Above board
CXR	—Vessel indicated is not abandoned		Admiral, or, Senior officer is on board
			(or, at —)_____DKV
CXS	ABATE-D ING-MENT.	CZA	—All on board
	If the wind abates_____ZEQ	CZB	—I am coming (or, sending) aboard
CXT	—No abatement	CZD	—I will come alongside (or, aboard) if you
	Wind has abated outside_____ZFJ		will allow me
	Wind has not abated outside _____ZFK	CZE	—Sending Sent a responsible person on
CXU	—You may make some abatement		board
	ABBREVIATE-D ING-ION (See ABRIDGE)		ABODE _____CXZ
	DAQ		
CXV	ABEAM, or, ABREAST OF	CZF	ABOLISH-ED-ING-TION
CXW	—At — (time indicated) we were abeam		
	of —	CZG	ABOUND-ED-ING
CXY	—I was abeam.		

	ABOUT—ACCOMMODATE		
CZH	ABOUT		ABSENT—*Continued*
CZI	—About a mile (*or number of miles indicated*)	DBA	—In the absence of —
		DBC	—Is he (*or person indicated*) absent ?
CZJ	—About April the —		Leave of absence .... QLX
CZK	—About August the —	DBE	—My, *or*, Our absence
CZL	—About December the —	DBF	—Shall, *or*, Will be absent
CZM	—About February the —	DBG	—Your absence.
CZN	—About January the —		
CZO	—About July the —	DBH	ABSOLUTE-LY
CZP	—About June the —	DBI	—Absolutely necessary
CZQ	—About March the —		
CZR	—About May the —	DBJ	ABSTAIN-ED-ING-ER
CZS	—About mid-channel		
CZT	—About November the —.	DBK	ABSURD-ITY-LY
CZU	—About October the —		
CZV	—About September the —.	DBL	ABUNDANT-LY-ANCE
CZW	—About the —		Supply abundant .... XEV
CZX	—About the average	DBM	—There is abundance of
CZY	—About the size		
DAB	—About the time	DBN	ABUSE-D-ING-IVE-LY
DAC	—About to —		
DAE	—About to assemble	DBO	ACCEDE-D-ING-SSION
	About to veer cable .... HQM	DBP	—Can accede to
	Admiral, *or*, Senior officer is generally about .... DKT	DBQ	—Can not accede to
	Am about to sail .P, *or*, Code Flag over P	DBR	ACCEPT ED-ING
	Are there any men-of-war about ? ...SA	DBS	—Acceptable-y-ance
	Could you understand what they were	DBT	—Accepted by —.
	signaling about ? .... WCI	DBU	—Can accept (*or* be accepted)
	Go about.... .... OBQ	DBV	—Decline accepting any —
DAF	—I am about to —	DBW	—Offer not accepted
	I had a glimpse of the land about—..QDU	DBX	—Very acceptable
DAG	—In about —	DBY	—Will accept your offer
DAH	—In about a fortnight	DBZ	—Will you accept ?
	Keep about mid-channel............LU	DCA	—With pleasure, I will accept
	Need be under no anxiety about—..EOJ		
DAI	—Round about	DCB	ACCESS-IBLE
	Under no anxiety to (for, *or*, about) —		
	EOK	DCE	ACCIDENT
DAJ	—Very anxious—about	DCF	—Accident happened, crushed, fall, jammed
DAK	—What are you about ?	DCG	—Accident happened, cut, stab
	Whereabout................. CSP		Accident happened, dangerously
DAL	ABOVE		wounded (*or*, hurt) ....AL
	Above board .... CYZ	DCH	—Accident happened, drowned, drowning
DAM	—Above the —		
DAN	—Above the bow	DCI	—Accident happened, gunshot wound
DAO	—Above the horizon On the horizon	DCJ	—Accident to boat
			Accidents to boilers (*see page* 178)
	ABREAST, *or*, ABRAM OF — CXV		Accidents to engines or machinery (*see page* 233)
	At — (*time indicated*) we were abreast (*or*, abeam) of — ....CXW		Accidents to screw propeller (*see page* 352)
	I was abreast ....CXY		Accidents to steering gear (*see page* 399)
DAP	—When do you expect to be off (*or*, abreast of) ?		Accident, want a surgeon (*or*, doctor)...................AM
DAQ	ABRIDGE D ING-MENT		Has any accident happened? Is anything the matter ....BK
DAR	—You have abridged the message so much that it is unintelligible		In case of accident (*or*, necessity) ...BP
DAS	ABROAD		Met with an accident ....BS
			When did the accident happen ? ....BU
DAT	ABSCOND ED ING		
		DCK	ACCIDENTAL-LY
DAU	ABSENT ABSENCE		
DAV	—Absent from home	DCL	ACCOMMODATE-D ING ION
DAW	—Absent without leave	DCM	—Accommodation ladder
	Consul is absent ....JRN	DCN	—Bad accommodation
DAX	—During my absence (*or*, the absence of)	DCO	—Can you accommodate?
DAY	—Have, *or*, Has been absent	DCP	—Can you accommodate a passenger?
DAZ	—His, Her, *or*, Their absence.		(*Number to be shown if more than one*)

	ACCOMMODATE—ACQUAINT		

	ACCOMMODATE—*Continued*	DFP	ON ACCOUNT OF
DCQ	—Can not accommodate No accommodation for —		Do not anchor on any account ------LB
			Do not show a light on any account.QV
DCR	—Good accommodation		Do not show a light on my account.QRT
DCS	—I can accommodate a person (*or persons indicated*)		Light-ship has been withdrawn on account of ice ---- --------- ------PCB
DCT	—Indifferent accommodation	DFQ	—On account of the fog (*or,* weather)
DCU	—No accommodation for passengers.	DFR	—On account
DCV	ACCOMPANY-IED-ING IMENT	DFS	ACCOUNTABLE.
DCW	—Accompanied by (*or,* with) —	DFT	—I hold you accountable
DCX	—Can accompany	DFU	—Will be held accountable
DCY	—I can not accompany	DFV	—Will not be accountable
DCZ	—I will accompany	DFW	—Will not be accountable for boat hire
DEA	—Will you accompany?		
		DFX	ACCOUNTANT
		DFY	—Accountant General
DEB	ACCOMPLICE		
		DFZ	ACCOUTERMENTS
DEC	ACCOMPLISHED-ED-ING	DGA	ACCUMULATOR.
DEF	—Can you (*or person indicated*) accomplish (*or,* do) it?	DGB	ACCUMULATE-D-ING-ION-IVE
	Can not be accomplished (*or,* done)..BLX		
DEG	—Easily done (*or,* accomplished)	DGC	ACCURATE-LY-ACY-NESS
DEH	—Has it been done (*or,* accomplished)?	DGE	—Is it accurate?
	If it can be done (*or,* accomplished).BZR	DGF	—It is not accurate
	If it can not be done (*or,* accomplished) ----------------------BZS	DGH	ACCUSE-D-ING-ATION (*See also* CHARGE)
	May be done (*or,* accomplished)----BLT	DGI	—Accused of having killed some one
DEI	—Not easily done (*or,* accomplished)	DGJ	—Is accused of —
		DGK	—Wh.. is he (*or person indicated*) accused of?
DEJ	ACCORD-ED-ING		
DEK	—According to	DGL	ACCUSTOM-ED-ING
DEL	—Accordingly		
DEM	—According to directions (*or,* orders)	DGM	ACID-ITY
	According to regulations ----------UOF		Carbolic acid --------------------HYD
DEN	—According to seniority		Muriatic acid----------- --- ----SCK
DEO	—According to the usual practice		Nitric acid----------------------SIQ
DEP	—In accordance with		Picric acid-----------------------THV
DEQ	—Not according to		Sulphuric acid-------------------XDI
DER	—Not according to orders, *or,* Against instructions		Tartaric acid --------------------XKV
		DGN	ACKNOWLEDGE-ING-MENT
		DGO	—Acknowledge, *or,* Answer my signal (*or,* message).
	ACCORDINGLY ----------------------DEL	DGP	—Acknowledge the receipt of —
DES	ACCOUNT-ED-ING TO ACCOUNT (*See also* ACCOUNT *below*)	DGQ	—Has the receipt been acknowledged?
	Account (*Bill*)-------------- --GKA	DGR	—I beg to acknowledge
	Account (*A Report*)------------DEW	DGS	—Make an acknowledgement
DET	—Account sent is correct	DGT	—Must be acknowledged
DEU	—Account sent is incorrect		
DEV	—Alarming account of —	DGU	ACORN
DEW	—An account (*See also* Report)		
DEX	—Bad account	DGV	ACQUAINT-ED-ING-ANCE (*See also* INFORM.)
DEY	—Can account for —	DGW	—Acquainted with
DEZ	—Can you account for —?	DGX	—Am acquainted with the anchorage
DFA	—Good account of —	DGY	—Am acquainted with the coast
DFB	—Has brought the account of —	DGZ	—Am desired to acquaint you (*or person indicated*)
DFC	—Have, *or,* Has received account of —		
DFE	—I have private account of —	DHA	—Am not acquainted with the anchorage
DFG	—Indifferent account of —	DHB	—Am not acquainted with the coast.
DFH	—Last account	DHC	—Are you acquainted with the anchorage (place, *or,* channel)?
DFI	—Next account	DHE	—Are you acquainted with the coast?
DFJ	—No account of —	DHF	—Desire you will acquaint
DFK	—No account received	DHG	—Did you acquaint?
DFL	—Nothing important by last account	DHI	—I am directed to acquaint you
DFM	—Quarterly account	DHJ	—Not acquainted—with
DFN	—Send an (*or,* your) account	DHK	—Shall, *or,* Will acquaint
DFO	—Send account	DHL	—Will you acquaint?

76564—09——10

ACQUIRE—ADMIRAL

DHM	ACQUIRE-D-ING-SITION *(See also* OBTAIN *)*
DHN	—No acquisition
DHO	ACQUIT-TED-TING-TAL
DHP	—He (*or person indicated*) is acquitted
DHQ	—He (*or person indicated*) is not acquitted.
DHR	—He (*or person indicated*) will be acquitted
DHS	—Is he (*or person indicated*) likely to be acquitted?
DHT	—Police (*or other*) authorities have acquitted
DHU	ACQUITTANCE
	ACRE (*Square Measure*) ---------- AXI
DHV	ACROSS
DHW	—Across channel
DHX	—Swayed across
DHY	—There is (*or*, are) a boom (*or*, booms) across
DHZ	ACT-ED ING To ACT
DIA	—Act (*An Act*)
DIB	—Act as your judgment directs
DIC	—Act of Parliament (*or*, Congress) Be certain before you act ---------- IHD
DIE	—By an act of Parliament (*or*, Congress) Can you supply me with anyone to take charge (*or*, act) as engineer ------- RH
DIF	—Foreign Enlistment Act
DIG	—Have you the Navigation Laws or, Merchant Shipping Act of the country of vessel signaling?
DIH	—How would you act?
DIJ	—In the act of —
DIK	—Navigation Laws, *or*, Merchant Shipping Act of the country of vessel signaling
DIL	—Recent Act of Congress Parliament, Assembly, or other legislative body
DIM	—Self-acting
DIN	—Shall we act in concert?
DIO	—Will they act?
DIP	—Will you act for me (*or*, —)?
DIQ	ACTION
DIR	—Can you renew the action?
DIS	—Commence action Direct-action engine ------------ MAO Double action ing ------------ LKW
DIT	—I am not in a condition for action
DIU	—Opportunities for action
DIV	—Prepare for action
DIW	—Radius of action
DIX	—Ready for action
DIY	ACTIVE-LY NESS-ITY
DIZ	—Active service
DJA	ACTUAL-LY-ITY.
DJB	ADAPT-ED-ING-ATION.
DJC	ADD-ED-ITION-ITIONAL LY. In addition—to — ------------ GHO
DJE	—Nothing additional.
DJF	—To be added.
DJG	ADDRESS-ED-ING ADDRESSED TO.
DJH	—Address (*a direction*)
DJI	—Address my letters to me at —
DJK	—Could not find the address
DJL	—Have you the address? *or*, What is the address of —?
DJM	—Insufficient address
DJN	—Leave your address at —.
DJO	—Send your address to — Spell the address ------------ WNC
DJP	—Telegraph address
DJQ	—The address is incorrect
DJR	—What is your address?
DJS	—What is your owner's address?
DJT	ADEQUATE-LY
DJU	ADHERE-D-ING-ENCE
DJV	ADHESIVE-ION.
DJW	—Adhesive plaster, *or*, Diachylon.
DJX	ADIEU FAREWELL
DJY	ADJACENT ADJOINING
DJZ	ADJOURN-ED-ING MENT
DKA	—Have, *or* Has adjourned Parliament, Congress, Assembly (*or other legislative body*) adjourns (*or*, adjourned) on the ------------ FEI
DKB	—When do they (*or*, does —) adjourn?
DKC	ADJUDGE D-ING-ICATE-ICATION
DKE	ADJUST ED-ING MENT Adjust ed-ing compasses ---------- JHD Are your compasses adjusted -- --- JHE Compasses not adjusted lately ------ JHL I am going to stop, as machinery requires adjusting --- ------------ RK
DKF	—Machinery requires adjusting Must be swung to adjust compasses ------------ JBV
DKG	—Small adjustment required to machinery
DKH	—Stop engines to adjust towing cables Stopping only for small adjustment to machinery ------------ RDY Was swung for adjustment of compasses at ------------ JIM When, *or*, Where were your compasses last adjusted? ------------ JIQ
DKI	ADJUTANT-CY
DKJ	ADMINISTER-ED-ING-RATION TOR-RIX
DKL	ADMIRAL
DKM	—Admiral's flag
DKN	—Admiral's house Admiral's office ------------ SPF
DKO	—Admiral's secretary.
DKP	—Admiral, *or*, Senior officer has commands for you
DKQ	—Admiral, *or*, Senior officer has no commands for you
DKR	—Admiral, *or*, Senior officer has requested
DKS	—Admiral, *or*, Senior officer intends

ADMIRAL—ADVISE

	ADMIRAL—*Continued*
DKT	—Admiral, or, Senior officer is generally about—
DKU	—Admiral, or, Senior officer is off (or at) —
DKV	—Admiral, or Senior officer is on board (or, at —)
DKW	—Admiral, or, Senior officer is on shore
DKX	—Admiral, or, Senior officer requests the pleasure of your company to dinner.
DKY	—Admiral, or, Senior officer wishes
DKZ	—Admiral, or, Senior officer wishes to know
DLA	—Admiral, or, Senior officer wishes to see you (or *person indicated*)
DLB	—Admiral, or, Senior officer wishes to see you (or, —) as soon as convenient.
DLC	—Admiral's, or, Senior officer's memo
DLE	—Admiral's, or, Senior officer's orders
DLF	—Go to the admiral's office
DLG	—Have not seen the admiral (or, senior officer)
DLH	—Have you seen the admiral (or, senior officer)?
DLI	—Port admiral
DLJ	—Rear admiral
DLK	—Send to the admiral's office
DLM	—There are orders for you at the admiral's office
DLN	—Vice admiral.
DLO	—What is the name of the admiral (or, senior officer)? Where is the admiral (or, senior officer)? ....SJ
DLP	**ADMIRALTY** Navy Department ....SEO
DLQ	—Admiralty, or, H O chart
DLR	—Admiralty court
DLS	—Admiralty, or, Secretary of the Navy's flag
DLT	—Admiralty instructions
DLU	—Admiralty memorandum
DLV	—Admiralty, or, Navy Department, telegraph-ed-ing Board of Admiralty ....GOW
DLW	—Droits of Admiralty
DLX	—Secretary of the Navy, or, of the Admiralty. Telegraph Navy Department, or, Admiralty, to — ....XLW
DLY	**ADMIRE-ABLE-Y.**
DLZ	**ADMISSION-IBLE**
DMA	**ADMIT-TED-TING** (*See also* ALLOW *and* PERMIT) Admission-ible ....DLZ
DLB	—Admitted on Dangerous to admit too many on board at a time ....GB
DMC	—If time will admit
DME	—It admits of.
DMF	—Ticket of admission
DMG	—Will it be admitted?
DMH	—Will not admit of
DMI	—Will not be admitted.
DMJ	**ADOPT-ED-ING ION.**
DMK	—Must be adopted.
DML	—Will you adopt?
DMN	**ADRIFT** Beacon, or, Buoy has broken adrift (or, is gone) ....GAE
	Boat is adrift ....GRN
	Boiler loose, (or, adrift) ....GWD
	Boom, or, Obstruction broke adrift..GZM
DMO	—Broke adrift
DMP	—Go-ing, Gone adrift
DMQ	—Is adrift—near — Lighter is adrift ....RA
DMR	—Spars are all adrift
DMS	**ADULT**
DMT	—Statute adult
DMU	**ADULTERATE-D-ING-ION.**
DMV	**ADVANCE-D-ING-MENT**
DMW	—Advance note
DMX	—Advance post
DMY	—Advanced guard
DMZ	—An advance in price
DNA	—Army is advancing
DNB	—Can you advance?
DNC	—Do not advance
DNE	—Enemy is advancing
DNF	—Have, or, Has advanced
DNG	—Month's advance
DNH	—No advance
DNI	—Pilot boat is advancing towards you
DNJ	—Prevent advance—of
DNK	—To advance-d-ing (*money*)
DNL	**ADVANTAGE** or, **BENEFIT**
DNM	—A great advantage
DNO	—Advantageous-ly
DNP	—Can not see the advantage
DNQ	—Gained advantage
DNR	—Has been attended with no advantage
DNS	—I took advantage of.
DNT	—Is there any advantage to be gained?
DNU	—No advantage
DNV	—Take advantage—of
DNW	**ADVENTURE-R**
DNX	**ADVERSARY.**
DNY	**ADVERSE**
DNZ	—Adverse circumstances
DOA	**ADVERTISE-D-ING**
DOB	—Advertise cargo and day of sale.
DOC	—Advertise for
DOE	—Advertisement.
DOF	—Have they advertised for any vessel to carry the mail
DOG	**ADVICE**
DOH	—His advice.
DOI	—I wait for advice
DOJ	—What advice have you received?
DOK	**ADVISE-D-ING**
DOL	—Advised to
DOM	—Do you advise me to—?
DON	—I would advise you—to—
DOP	—I would not advise—you I strongly recommend you not to— Lloyds's agent advises you to—...QWT
DOQ	—What would you advise?

ADVISE—AGAINST.

	ADVISE—*Continued*
DOR	—Would advise Would recommend
DOS	—Would be advisable
DOT	—Would not be advisable.
DOU	ADVOCATE-D-ING-CY
DOV	ADZE
DOW	—Can you lend me hatchets (*or*, adzes)?
DOX	AFAR
DOY	AFFAIR
DOZ	—Affairs appear more settled (look more peaceful)
DPA	—Affairs are very unsettled
DPB	—Affairs look more unsettled
DPC	—Foreign affairs
DPE	—Partnership affairs
DPF	—Present state of affairs
DPG	—Private affairs
DPH	—Privy to the whole affair
DPI	—State of affairs at —
DPJ	—The authorities will put an end to the affair
DPK	—You had better put your affairs in the hands of —
DPL	AFFECT-ED-ING
	Does disconnecting affect your steering? LDC
DPM	AFFIDAVIT
DPN	—Be prepared with affidavits
DPO	AFFIRM-ATIVE-LY
DPQ	—Answered affirmatively
	Yes, *or*, Affirmative.
	C, *or*, Code Flag over C
DPR	AFFLICT-ED-ING-ION
DPS	AFFORD-ED-ING.
DPT	AFFIX-ED ING
DPU	AFFRAY.
DPV	AFLOAT
	Afloat aft (astern)..................BW
	Afloat forward ....BX
	Am afloat ....BY
	Are you afloat ....CB
DPW	—As soon as she is off (*or*, afloat).
	Can not be got off (*or*, afloat) by any means now available..................CD
DPX	—Expect (*or vessel indicated* expects) to be got off (*or* afloat)
	Floated ....NFA
DPY	—I shall get off (*or*, afloat) with assistance.
	May be got off (*or*, afloat) if prompt assistance be given ....CG
DPZ	—May be got off (*or*, afloat) without assistance
DQA	—Nearly afloat
DQB	—Not afloat
DQC	—Not worth getting off (*or*, afloat)
	The tide will take her off (float her) NFE
DQE	—When will you be afloat?
	When you float ....NFG
DQF	—Will be afloat
	Will not float ....NFH
DQG	—You will not be afloat all the tide

DQH	AFRAID
DQI	—Do not be afraid—of
DQJ	—I am afraid—of
DQK	AFRICA-N.
DQL	AFT
	Afloat aft..................BW
	Fore and aft..................NIT
	Fore and aft sails. ....NIU
	I draw—feet aft and—feet forward_OA
	My draught of water is — feet — inches aft..................LNC
DQM	AFTER AFTERWARDS.
DQN	—After a consultation
DQO	—After a passage—of—
DQP	—After breakfast
DQR	—After Christmas.
DQS	—After coaling
DQT	—After dark
DQU	—After dinner.
DQV	—After Easter
	After him (his, her-s, it-s) ....RWZ
DQW	—After-hold
DQX	—After-hours
	After I — ....BXZ
DQY	—After it was done.
DQZ	—After magazine
	After me (my, mine) ....CEA
DRA	—After midnight
DRB	—After much persuasion
DRC	—After, *or* Jigger mast
DRE	—After *or*, On arrival of — (*or*, at).
DRF	—After order
DRG	—After-part
DRH	—After running — (*miles indicated*)
DRI	—After sunrise
DRJ	—After sunset
DRK	—After supper
	After the ....CMB
DRL	—After the battle
DRM	—After the departure of.
DRN	—After the gale.
	After these (*or*, those) ....CNH
	After they (their them) ....CNJ
	After this (*or*, that) ....CKN
	After we (our s, us) CIL
	After which (*or*, what) ....CPS
	After you-r-s ....OWR
	Any day—after.. ....EOS
DRO	—Day after Next day
DRP	—How long after?
DRQ	—Long after
DRS	—Morning after
DRT	—Shortly after
DRU	—Steer after me, I have a pilot
DRV	—The day after to morrow
DRW	AFTERNOON IN THE AFTERNOON P M.
DRX	—Afternoon watch.
DRY	—This afternoon
DRZ	—To morrow afternoon
DSA	—Yesterday afternoon
DSB	AGAIN
	Try again..................YCS
DSC	AGAINST
	Against instructions Not according to orders ....DER

AGAINST—AGUE

	AGAINST—*Continued*
DSE	—Against the current (tide, or, stream)
DSF	—Against the monsoon
	Can you beat up against?----------GCD
	Can not beat up against ----------GCE
DSG	—Run against, or, Foul of
DSH	AGE AGED
DSI	—What age?
DSJ	AGENCY
DSK	AGENT.
DSL	—Agent is bankrupt
DSM	—Agent's instructions
DSN	—Agent's request
DSO	—Colonial agent.
	Consular agent----------------JRQ
DSP	—Diplomatic agent
	Have received the following communication (or, instructions) from your agent (or owner)----------------IA
DSQ	—I have no business agent (or, I do not know of any agent) at —
DSR	—Inform agent—of —.
DST	—My agents are —
DSU	—Prize agent-cy
DSV	—Telegraph to my agents
DSW	—The best agents are —
DSX	—Transport agent
DSY	—Who are the agents for —?
DSZ	—Who are the best agents (or, people) to do business with?
DTA	—Who are your agents?
DTB	—Who generally does business for — ?
DTC	—Who do you do business with?
DTE	—Your agents desired me to inform you
DTF	—Your agent (or agent specified) has discontinued business.
DTG	—Your agent (or agent specified) has suspended business
	You had better put your affairs in the hands of —----------------DPK
	LLOYD'S (AGENT)
	Condemned by Lloyd's agent------QWF
	Consult Lloyd's agent ----------QWG
	Do not interfere with (or, take the matter out of the hands of) Lloyd's agent.QWH
	French Lloyd's Veritas----------NQP
	Have you been visited by Lloyd's agent? QWI
	Have you seen Lloyd's agent?------QWJ
	I require assistance from Lloyd's agent, CV
	Inform Lloyd's agent ------------QWP
	Intend to abandon the ship to Lloyd's agent ----------------CXP
	Leave the matter in the hands of Lloyd's agent ----------------QWR
	Lloyd's agent advises you to —----QWT
	Lloyd's agent is—----------------QWS
	Report me (or, my communication) to Lloyd's agent at— ----------UH
	Surveyed by Lloyd's agent--------QXI
	Who is Lloyd's agent?------------QXJ
	Will Lloyd's agent come on board?--QXK
DTH	AGGRIEVE-D
DTI	AGITATE-D-ING-ION

DTJ	AGO
DTK	—How long ago (or, since)?
DTL	—Long time ago (or, since)
DTM	—Months ago
DTN	—Not long ago
DTO	—Some time ago
DTP	AGONY
DTQ	AGREE-D-ING (See also CONSENT and PERMIT)
DTR	—Agree-d-ing with him (or, them)
DTS	—Agree-d-ing with me (or, mine)
DTU	—Agree-d-ing with you-r-s
DTV	—Am I to understand you agree?
DTW	—Can not agree
DTX	—Dead reckoning agrees with observation.
DTY	—Do you agree?
DTZ	—Does not agree
DUA	—I agree
DUB	—Vessel (or person indicated) agrees
DUC	—Will you (or person indicated) agree?
DUE	AGREEABLE-Y.
DUF	AGREEMENT
DUG	—Articles of agreement
	Lloyd's Salvage Agreement --------QWY
DUH	—Make, or, Made an agreement with
DUI	—The agreement
DUJ	—Will you make an agreement?
DUK	AGRICULTURE-IST-AL
DUL	—Agricultural district—of
DUM	—Agricultural implement
DUN	—Agricultural produce
	Department or Board of Agriculture.GOX
DUO	AGROUND
	Aground, want immediate assistance.NA
	Am aground ----------------------BZ
	Am aground, likely to break up, require immediate assistance----------CA
DUP	—Are you aground?
	Floated off ----------------------NFA
	Has she sewed?
DUQ	—Have you been aground?
DUR	—I am aground Is aground.
	I am aground, send what immediate assistance you can ------------CE
DUS	—I have (or vessel indicated has) been aground
DUT	—I have (or vessel indicated has) not touched the ground
DUV	—Is — (vessel indicated) aground?
	Lose no time in shoring up--------CF
DUW	—Steamboat is aground
	Vessel (or vessel indicated) ashore near (or, on) —----------------CH
	Was the tide high when you grounded?.CI
	Was the tide low when you grounded?.CJ
	When did you ground? What time and date did you (or vessel indicated) ground?----------------OK
DUX	—You will be aground
	You will be aground at low water.....KX
DUY	AGUE-ISH
DUZ	—Fever and ague.

AHEAD—ALL

DVA	Ahead		DWU	Alert-ness Alacrity
DVB	—Ahead of the —		DWV	—Be on the alert
DVC	—Am, Is, or, Are ahead—of		DWX	—Enemy on the alert
	Boats can not pull ahead ....GRK			
	Breakers ahead Reef, Rock, or, Shoal		DWY	Alias
	ahead of you ....HFB			
DVE	—Can you steam ahead—of?		DWZ	Alien
	Convoy is ahead....................JUK			
	Convoy to keep ahead of escort......IU		DXA	Alike
	Do not pass ahead of me		DXB	—Both alike
	Code Flag over R			
DVF	—Farther ahead of.		DXC	Alive
	Go ahead...........................LG			
	Go ahead easy Easy ahead ....LH		DXE	Alkaline Alkali
	Go ahead full speed................LI			
DVG	—Go ahead of me		DXF	All
DVH	—Go ahead of—		DXG	—All captains
DVI	—Keep ahead—of		DXH	—All classes
	I will pass ahead of you		DXI	—All clear
	Code Flag over X		DXJ	—All fast.
DVJ	—Is she (or, it) ahead ?		DXK	—All gone
DVK	—Keep ahead and carry a light		DXL	—All hands
DVL	—Keep going ahead		DXM	—All hands at the pumps.
DVM	—Keep going ahead slowly		DXN	—All included
DVN	—Make sail, or, Go ahead, and drop a boat		DXO	—All lost Total loss
	on board		DXP	—All nets lost
DVO	—Next ahead.			All on board ....CZA
DVP	—Pass ahead of me		DXQ	—All on board perished All hands lost
DVQ	—She (vessel indicated) is ahead		DXR	—All on board saved All hands saved
			DXS	—All possible information—of (or, from)
DVR	Aid ed ing (See also Help and Assist)		DXT	—All quiet at (or, in) — (or, on shore)
			DXU	—All quiet with —
DVS	Aid-de-camp		DXV	—All right
			DXW	—All safe Saved
DVT	Aim-ed-ing To aim		DXY	—All snug
DVU	—Aimed at		DXZ	—All sorts
			DYA	—All the crew not on board
DVW	Air-ed ing-y		DYB	—All the passengers not on board
DVX	—Air bedding		DYC	—All to pieces
DVY	—Air-compressing pumps		DYE	—All well All well at —
DVZ	—Air-pipe			Are all your bills of lading complete and
DWA	—Air pump.			signed ?.......................GEQ
DWB	—Air pump gear.			Are all your crew on board ?.......GFT
DWC	—Air-pump levers out of order			Are all your passengers on board ?..GPU
DWE	—Air-pump out of order		DYF	—Are you all well?
DWF	—Air-pump rod broken		DYG	—Are your passengers all well?
DWG	—Air reservoir		DYH	—At all events
DWH	—Air space (or, cells)		DYI	—At all (or, every)
DWI	—Air surface		DYJ	—At all times
DWJ	—Air tight.			Bring all your papers ....LJZ
DWK	—Air valve			Buoys have just been examined, they
DWL	—Cold air chamber			are all right..................HMJ
DWM	—Compressed air		DYK	—By all means
				Crew are all on board..............GQB
DWN	Alabaster			Do all ....LIK
				First of all In the first place ......NBH
DWO	Alarm-ed-ing.			Have you all your papers ?.........TAA
	Alarm of fire.....................MZN		DYL	—Is all included?
	Alarming account of ....DEV		DYM	—Is she all right?
DWP	—Do not be alarmed		DYN	—Make all snug
DWQ	—False alarm		DYO	—Most of all
			DYP	—Not at all
DWR	Alcohol ic		DYQ	—Over alls
				Passengers all on board............GQR
DWS	Alderman Civic authorities			Passengers are not all on board ... DYB
				Report (the vessel indicated) all well, in
DWT	Ale Beer			latitude — and longitude — (indi-
				cated) on the — (date indicated)...UN
	Alen (Measure of length) ....AVL		DYR	—She has righted, she is all safe
			DYS	—That is all

ALL—ALSO

DYT	ALL—*Continued* —Vessels wishing to be reported all well, show your distinguishing signals Was all well when you left ? ....QLN You will not lie afloat all the tide..DQG	EAG EAH	ALOFT—*Continued.* —Keep a good lookout aloft —Man aloft
		EAI	ALONE
DYU	ALLEGIANCE ALLIANCE (*See* ALLY) ....DZS	EAJ	ALONG—THE All the lights are out along the coast — of — . ...PU
DYV	ALLIGATOR　CROCODILE	EAK EAL	—Along the bank—of —Along the lines
DYW DYX	ALLOT-TED-TING　To allot —Allotment	EAM	—Along the road
DYZ	ALLOW-ED ING-ABLE　To allow for　(*See also* ADMIT, LET, *and* PERMIT) Allowance ....DZO	EAN EAO	ALONGSHORE —Alongshore to the eastward of —(*distance in miles may follow*)
DZA	—Allow me to congratulate you Allow no communication　Allow no person on board ....FM	EAP	—Alongshore to the northward of —(*distance in miles may follow*)
DZB	—Allow no lights to be seen　Extinguish all lights	EAQ	—Alongshore to the southward of —(*distance in miles may follow*)
DZC	—Allowance Armed men not allowed to land ...EUV	EAR	—Alongshore to the westward of —(*distance in miles may follow*)
DZE DZF DZG DZH DZI	—Can allow　Can permit —Can not allow it —Communication is allowed with —Half allowance —Have been on short allowance for some time	EAS EAT EAU EAV EAW	ALONGSIDE —Alongside the jetty (wharf, or, pier) —Alongside the mole —As soon as alongside —Assemble alongside Boat is alongside (is on board)......EV
	I will come alongside (*or*, on board) if you will allow me...............CZD It is dangerous to allow too many people on board at once..................GE	EAX EAY EAZ EBA EBC	—Call alongside —Can I come alongside? —Can not come alongside —Come alongside —Do not come alongside
DZJ	—It is expected the vessel (*or vessels indicated*) will be allowed to leave	EBD EBF	—Endeavor to come alongside —Go alongside
DZK DZL DZM DZN	—Make allowance—to (*or*, for) —Not allowed　Disallowed —Put your people on an allowance. —Short allowance Vessels with combustibles not allowed in ....HS		I shall fire into the boats if they persist in coming alongside......... .....NAK I will come alongside (*or*, on board) if you will allow me ....CZD
DZO	—What allowance are you on ? What variation do you allow ? .... JIP	EBG EBH	—Lashed alongside —Let no boat come alongside Lighter alongside ....QTC
DZP DZQ	—Will an allowance be made ? —Will you allow ?　Will you permit ?	EBJ EBK	—Not alongside —To be alongside — at— What is the depth of water alongside the — ?...,..............KVI
DZR	ALLUDE-D-ING-SION	EBL	ALOOF　KEEP ALOOF
DZS DZT DZU	ALLY-IED-ING-IANCE —Allied with —Our ally	EBM	ALOUD
DZV	ALMANAC Have you a Nautical Almanac for the current year (*or for the year indicated*)? ....RQ Lloyd's Seaman's Almanac ....QWZ	EBN	ALPHABET-ICAL-LY Alphabetical signal, No 1 　　　　　Code Flag over E Alphabetical signal, No 2 　　　　　Code Flag over F Alphabetical signal, No 3 　　　　　Code Flag over G NOTE —*For method of using Alphabetical Signals, see page 18*
DZW	—Nautical Almanac		
DZX	ALMOND	EBO EBP	—Alphabetical table (*page* 15) —Morse alphabet Use alphabetical (spelling) table...WNF
DZY EAB	ALMOST —Almost anything may be got	EBQ	—Your name is not on my list, spell it alphabetically (*or*, by alphabetical table)
	ALMUDA　(*Measure of capacity*)......AYK		
EAC EAD EAF	ALOFT —Fell from aloft —Gear aloft.	EBR	ALREADY
		EBS	ALSO

ALTER—AMMUNITION

EBT	ALTER-ED-ING TO ALTER
EBU	—Alter course
	Alter course one point (or *number indicated*) to port .... JZD
	Alter course one point (or *number indicated*) to starboard. .... JZE
	Alter course to—(*point indicated*) WF
EBV	—Alteration
EBW	—Alters the circumstances
EBX	—An alteration has taken place
EBY	—Any alteration in the position of the armies?
EBZ	—Any alteration in the position of the squadron (or, fleet)?
ECA	—Any alteration (or, change)?
ECB	—Are there any alterations in the lights (or, marks)?
ECD	—Channel has altered, do not try it
ECF	—Do not alter until —
ECG	—Do not make any alteration—until
ECH	—Duties are altered
ECI	—I do not intend to alter course at present (or until)
ECJ	—I shall alter course
ECK	—I shall not alter course
ECL	—I think we had better alter course to —(*point indicated*)
	I will show a light to-night when I alter course .... QW
ECM	—Is there any alteration of marks?
ECN	—Is there any alteration in the market?
ECO	—It alters the case
ECP	—Make an alteration
ECQ	—Much altered
ECR	—Must be altered
ECS	—No alteration has taken place
ECT	—Shall, or, Will alter (or, be altered)
ECU	—Vessel indicated has altered course
ECV	—We had better alter course to — (*point indicated*)
ECW	—What alteration?
ECX	—When do you intend to alter course?
	When I alter course to-night I will show a light................. QW
ECY	—When will you alter course?
	When you alter course (or, tack) to-night show a light .... JZS
ECZ	—Will be altered
EDA	ALTERNATE-LY.
EDB	ALTERNATIVE
EDC	—Alternatively
	ALTERNATING LIGHT............. QRP
EDF	ALTHOUGH
EDG	ALTITUDE-INAL ALLY
EDH	—Did you get a meridian altitude?
EDI	—Double altitudes (Sumner's line)
EDJ	—Equal altitudes
EDK	—Ex meridian altitude
EDL	—Meridian altitude.
EDM	—Sun's altitude.
EDN	ALTOGETHER
EDO	—We have lost altogether
EDP	ALUM.

EDQ	ALUMINIUM
EDR	ALWAYS
EDS	—Is it always?
EDT	—Not always
EDU	A M BEFORE NOON IN THE FORENOON
	Am I AM ....BEA
	Am, Are, or, Is able to ....CYF
	Am I? .. ..BEC
	Am I not?..................... BED
	Am I to? ....BEF
	Am, Are, or, Is not...............BEG
	Am, Are, or, Is not able to ....CYG
	Am, Are, or, Is not to be ....BFD
	Am, Are, or, Is not to have........BQD
	Am, Are, or, Is to................BEH
	Am, Are, or, Is to be ....BFC
	Am, Are, or, Is to have.............BQC
	How am I? ....BXG
	I am about to — ....DAF
	I am not......................BEI
	If I am ....BYX
	If I am able to ....CYK
	If I am not ....BYZ
	If I am not able to ....CYL
	What, or, Which am I?............CPV
	When am I to?................. ..CQV
	Where am I (or, are we)?............CSQ
	Where I am (or, we are)?............CTG
	Why am I? ....CUY
EDV	AMBASSADOR, or, MINISTER
EDW	—Our ambassador (or, minister)—at —
EDX	AMBULANCE
EDY	—Ambulance required to convey a patient to hospital
EDZ	—Prepare ambulance—for —
EFA	AMBUSH-CADE
EFB	AMEND-ED ING-MENT
EFC	AMERICA N AMERICAN COLORS
EFD	AMICABLE-Y-ILITY
EFG	—Amicably settled (or, arranged)
EFH	AMID-ST
EFI	AMIDSHIPS
EFJ	—Helm amidships
EFK	AMISS
EFL	—Not amiss.
EFM	—Something amiss.
EFN	—What is amiss?
EFO	AMMONIA IACAL
	Sal ammoniac ....VFI
EFP	AMMUNITION
EFQ	—Ammunition is damaged.
EFR	—Ammunition is nearly expended
EFS	—Can spare ammunition.
EFT	—Can you spare ammunition?
EFU	—I have (or *vessel indicated* has) no ammunition

AMMUNITION—ANCHOR

	AMMUNITION—*Continued*
EFV	—I will supply you with ammunition for —
EFW	—In want of ammunition
EFX	—Quick-firing ammunition
	Revolver ammunition ICX
EFY	—Rifle ammunition
EFZ	—Rounds of ammunition for —
EGA	—Rounds of ammunition for heavy guns
EGB	—What quantity of ammunition have you?
EGC	**AMNESTY**
EGD	**AMONG ST**
EGF	—Among the breakers
	Fire in hold amongst the cargo .....OR
	I have been amongst the ice ....PBM
EGH	**AMOUNT-ING**
EGI	—Amount to —
EGJ	—Does not amount to an average of —
EGK	—Not amounting to —
EGL	—To a great amount
EGM	—What amount of —?
EGN	—What is the amount of —?
EGO	—What is the total amount?
EGP	**AMPÉRE** (*Measure of electrical current*)
EGQ	**AMPLITUDE**
EGR	**AMPUTATE-D-ING²ION**
EGS	—Amputation necessary
AGT	**AMUSE-D-ING-MENT**
	AN, *or*, A .. ..CXB
EGU	**ANALYSE-D-ING-TIC-AL-IS.**
EGV	**ANARCHY**
EGW	**ANCHOR-ED-ING TO ANCHOR.**
EGX	—Anchor (*An Anchor*)
	Anchor as convenient ....KY
EGY	—Anchor buoy
EGZ	—Anchor crane
EHA	—Anchor drag-ged ging
EHB	—Anchor fished. *or*, Fish the anchor
EHC	—Anchor fluke
	Anchor instantly .. ..KZ
EHD	—Anchor light (lantern, *or*, lamp)
EHF	—Anchor on bearing (*indicated*)
EHG	—Anchor shank
EHI	—Anchor stock
EHJ	—Anchor stowed, *or*, Stow anchor
EHK	—Anchor watch.
EHL	—Anchor where you are
EHM	—Are you going to anchor?
EHN	—At anchor
EHO	—Back the anchor
EHP	—Best bower anchor
EHQ	—Bower anchor (*weight, in cwts, if necessary*)
EHR	—Buoy rope (*of anchor*)
EHS	—Can we anchor with safety?
EHT	—Cat the anchor
EHU	—Clear anchor
EHV	—Directly you get bottom, let go your anchor.
EHW	—Do not anchor Keep under weigh

	ANCHOR—*Continued*
	Do not anchor on any account ......LE
EHX	—Do not anchor yet
EHY	—Do not hold on to your anchor, but let her beat up on the beach
EHZ	—Do you intend anchoring?
	Driving I can veer no more cable
	Have no more anchors to let go....NK
EIA	—Driving, parted from anchor
EIB	—Drove from her anchor
	Fish the anchor, *or*, Anchor fished.EHB
EIC	—Floating anchor Sea anchor
EID	—Foul anchor Fouled her (*or*, my) anchor
EIF	—Foundered at her anchors
EIG	—Have all your anchors and cables ready for use
EIH	—Have you anchored?
EIJ	—Have you parted from your anchor?
EIK	—Help me to lay out an anchor,
	I am at anchor .... - - EIM
EIL	—I have hooked your anchor
EIM	—I have let go my anchor, *or*, I am at anchor
EIN	—I have lost an anchor
EIO	—I have no means of laying out an anchor
EIP	—I have no more anchors to let go
EIQ	—I have slipped my anchor, but it is buoyed
EIR	—I must come to an anchor
EIS	—I must recover my anchor
EIT	—I shall anchor if circumstances permit (*place or bearing to follow, if necessary*)
EIU	—I shall not anchor
EIV	—If you will lay out an anchor for me, I can get off
EIW	—In anchoring look out for moorings
	In anchoring look out for telegraph cable, WV
EIX	—It is not very safe where you are anchored
EIY	—Kedge anchors (*weight, in cwts, if necessary*)
EIZ	—Lay out an anchor
	Let go another anchor...... .....KR
EJA	—Let go your anchor
	Light-ship at — is not at anchor on her station *or*, The light-ship at — is out of position............. ......QE
EJB	—Lost an anchor
EJC	—Mushroom anchor
EJD	—Must anchor
EJF	—No anchor left
EJG	—Only one anchor left
EJH	—Patent anchor
EJI	—Parted from her anchor
EJK	—Parted in the night
EJL	—Port anchor
	Sea anchor, *or*, Floating anchor ....EIC
EJM	—Send an anchor and cable off immediately (*indicate if more than one*)
EJN	—Sending, Sent an anchor and cable
EJO	—Shall anchor in the bay
EJP	—Shall I anchor?
EJQ	—Sheet anchor
EJR	—Sight your anchor Make sure your anchor is clear
EJS	—Small bower anchor
EJT	—Spare anchor
EJU	—Starboard anchor
	Stow anchor, *or*, Anchor is stowed.EHJ

ANCHOR—ANNUL

	ANCHOR—*Continued*
EJV	—Stream, *or*, Small anchor (*weight in cwts , if necessary*)
EJW	—Tell them to recover my anchor
	The light ship at — is not at anchor on her station, *or*, Light-ship at — is out of position _ _ _ _ _ _ _ _ _ _ _ _ _ _ _ QE
EJX	—Tripped her anchor
EJY	—Unable to weigh—anchor
EJZ	—Vessel indicated is at anchor (is anchoring)
EKA	—Vessel indicated parted from her anchor
	Want an anchor_ _ _ _ _ _ _ _ _ _ _ _ _ _ _ YC
	Want an anchor and cable_ _ _ _ _ _ _ _ _ YD
EKB	—Water is too deep to anchor
EKC	—We had better anchor
EKD	—When anchored
EKF	—When do you intend to anchor?
EKG	—When shall we anchor?
EKH	—Where do you intend to anchor?
EKI	—Where shall we anchor?
EKJ	—Why do you anchor?
EKL	—Will you assist me (*or vessel indicated*) in laying-out an anchor?
EKM	—Will you pick up my anchor for me?
EKN	—You may come to an anchor
EKO	—You will overlay my anchor
EKP	**ANCHORAGE** (*See also* BERTH *and* HARBOR)
	Am acquainted with the anchorage_DGX
	Am not acquainted with anchorage_DHA
	Anchorage, *or*, Harbor is good enough with wind from —_ _ _ _ _ _ _ _ _ _ _OX
	Anchorage, *or*, Harbor is indifferent_OY
	Anchorage, *or*, Harbor is safe with all winds _ _ _ _ _ _ _ _ _ _ _ _ _ _ _ _ OZ
EKQ	—Any anchorage in the bay?
	Are you acquainted with the anchorage (place, *or*, channel)?_ _ _ _ _ _ _ _ _DHC
EKR	—Bad anchorage
	Best berth, (*or*, anchorage) bears from me_ _ _ _ _ _ _ _ _ _ _ _ _ _ _ _ _ _ EF
	Best anchorage is in — fathoms_ _ _ _ _OV
	Calls the attention of vessels COMING into the anchorage on the bearing (FROM *the person signaling*) pointed out by compass signal _ _ _ _ _ _ _ _ _ DP
	Calls the attention of vessels LEAVING the anchorage on the bearing (FROM *the person signaling*) pointed out by compass signal _ _ _ _ _ _ _ _ _DQ
EKS	—Can, *or*, will you point out (*or*, take me to) a good anchorage?
EKT	—Change your anchorage
	Do not attempt to make the anchorage (*or*, harbor)_ _ _ _ _ _ _ _ _ _ _ _ _FW
EKU	—Do not know the anchorage (*or*, harbor)
	Do not risk an anchorage unless you have very good ground tackle_ _ _ _ _FX
EKV	—Do you know of any anchorage (*or*, harbor)?
EKW	—Do you know the anchorage (*or*, harbor)?
	Good anchorage_ _ _ _ _ _ _ _ _ _ _ _ _ _ODU
EKX	—Good bottom for anchorage
	How does the anchorage (*or*, harbor) bear?_ _ _ _ _ _ _ _ _ _ _ _ _ _ _ _PA
EKY	—I do not know the anchorage (*or*, harbor)
EKZ	—I have (*or vessel indicated* has) knowledge of the anchorage

	ANCHORAGE—*Continued*
	I shall make for the anchorage (*or*, —) with all despatch _ _ _ _ _ _ _ _ _ _ _ _JT
ELA	—I shall try the anchorage
ELB	—Is it a difficult anchorage to get away from?
	Is the anchorage (*or*, my anchorage) safe with all winds (*specify which wind, if necessary*)?_ _ _ _ _ _ _ _ _ _ _PB
	It is not very safe where you are anchored _ _ _ _ _ _ _ _ _ _ _ _ _ _ _ _ _ _EIX
ELC	—Lose no time in getting to the anchorage
ELD	—Safe anchorage
ELF	—Shall we get to the anchorage to-night?
	Unsafe anchorage _ _ _ _ _ _ _ _ _ _ _ _ _ _ _PE
ELG	—Unsafe to remain at anchorage—until—
ELH	—We are to leeward of the anchorage
ELI	—We are to windward of the anchorage
ELJ	**ANCHOVY**
ELK	**AND**
	And the _ _ _ _ _ _ _ _ _ _ _ _ _ _ _ _ _ _CMD
ELM	—And they
	In and out_ _ _ _ _ _ _ _ _ _ _ _ _ _ _ _ _ SVW
ELN	—You and I (*or*, me)
ELO	**ANEMOMETER WIND GAUGE**
ELP	**ANEROID BAROMETER**
ELQ	**ANEURISM**
ELR	**ANGER-BY-ILY**
ELS	**ANGLE ANGLEWAYS**
ELT	—Angle iron
ELU	—Beveled to an angle of—.
ELV	—Helm angle
ELW	**ANGULAR-LY**
ELX	**ANIMAL**
	ANNA (*Coin*) _ _ _ _ _ _ _ _ _ _ _ _ _ _ _ASX
ELY	**ANNEX-ED-ING-ATION**
ELZ	**ANNIHILATE-D-ING-ION**
EMA	**ANNIVERSARY**
EMB	**ANNOUNCE-D MENT**
EMC	**ANNOY-ED-ING-ANCE**
EMD	**ANNUAL-LY**
EMF	**ANNUITY**
EMG	**ANNUL-LED-LING**
EMH	—Annul present formation
EMI	—Annul the hoist indicated by numeral signal
	Annul the last hoist, I will repeat it_VE
EMJ	—Annul the last signal The last signal is annulled
	Annul the whole signal_ _ _ _ _ _ _ _ _ _VF

ANOTHER—ANY

EMK	ANOTHER	EOB	ANTI-PYRINE
EML	—Another time	EOC	ANTI SCORBUTIC
	Let go another anchor.... .... KR	EOD	ANTISEPTIC
	Should like to have another comparison ....GV	EOF	ANVIL
	Want another tug (*indicate if more than one*).. ... .YTB	EOG	ANXIOUS LY-IETY—TO (FOR, *or*, ABOUT).
EMN	ANSWER-ED ING TO ANSWER (*See also* REPLY)	EOH	—Am anxious for you to —
	Acknowledge, *or*, Answer my signal (*or*, message)....................DGO	EOI	—Anxious to hear of (*or*, from) —.
EMO	—Answer (*A reply*)	EOJ	—Need be under no anxiety about —
	Answered affirmatively..........DPQ	EOK	—Under no anxiety to (for, *or*, about) —.
	Answered by telegraph.............WQ		Very anxious—about —...... .....DAJ
EMP	—Answerable	EOL	ANY
EMQ	—Answered, *or*, Replied by telegraph		Any alteration (*or*, change)?..... ..ECA
EMR	—Answer the purpose		Any alteration in the position of the armies?.......... .. ..EBY
EMS	—Answered negatively		Any alteration in the position of the fleet (*or*, squadron)?............. BBZ
EMT	—Answering pennant (*when spoken of*)		Any anchorage in the bay?........EKQ
EMU	—Be very careful in your answer	EOM	—Any arrival from?
EMV	—Can you await answer (*or*, reply)?		Any bones broken?................GYS
EMW	—Forward answer by telegraph to signal station at —	EON	—Any cases of fever on board (*or*, at —)?
EMX	—He (*or vessel indicated*) has not had an answer	EOP	—Any casualties?
EMY	—How long ought I to be getting an answer from — ?	EOQ	—Any chance of war?
EMZ	—I am waiting for your answer	EOR	—Any commands?
ENA	—I must decline answering	EOS	—Any day—after —
ENB	—I think something might be got to answer the purpose		Any deception ....KPG
ENC	—I will answer presently	EOT	—Any doubt
	In answer—to..... - ....URP	EOU	—Any extras?
END	—Is answer suitable (*or*, sufficient)?	EOV	—Any farther.
ENF	—It will not answer It is not suitable	EOW	—Any fees (*or*, charges)?
ENG	—Must give, *or*, Gives an evasive answer	EOX	—Any fighting at?
ENH	—Not answered. No reply	EOY	—Any forts?
ENI	—Perhaps it might be made to answer	EOZ	—Any freight offering?
ENJ	—Received an answer		Any letters (*or*, despatches) on board? KWX
ENK	—Send an answer (*or*, reply)	EPA	—Any letters (*or*, papers) for me (*or person indicated*)?
ENL	—Shall, *or*, Will answer (*or*, be answered)		Any letters (*or*, papers) from — ?.....JI
	Signal is answered................ VK	EPB	—Any longer
ENM	—Signal is not answered	EPC	—Any missing?
ENO	—Tell my owner ship answers remarkably well	EPD	—Any mistake?
	Wait for answer (*or*, reply)..........TX	EPF	—Any more?
ENP	—What answer (*or*, reply)?	EPG	—Any necessity for —?
	What was the answer (*or*, reply)?....TY	EPH	—Any news—of —?
ENQ	—When do you expect an answer?	EPI	—Any news of my family?
	Will await answer (*or*, reply)..........TZ	EPJ	—Any news of the — mail?
ENR	—Will it answer the purpose?	EPK	—Any opposition—to (*or*, from —)?
ENS	—Will not answer her helm	EPL	—Any other
	Will you await answer (*or*, reply)?....UA		Any person Anybody............EPQ
ENT	—You can communicate with — and get an answer in —	EPM	—Any person (anyone) on board
ENU	—You can get an answer in —	EPN	—Any rate
ENV	ANT	EPO	—Any sick on board?
ENW	ANTARCTIC		Any sickness at — ?...... ....UY
	ANTE-MERIDIAN, *or*, A.M... ....EDU	EPQ	—Anybody Any person.
ENX	ANTHRACITE COAL		Anyone .... (*see* EPZ)
ENY	ANTICIPATE-D ING-ION		Anything ... . (*see* EQD)
ENZ	ANTI FOULING COMPOSITION		Anywhere ....(*see* EQR)
EOA	ANTIMONY		Are, *or*, Is there—any?.............BEP
			Are there any alterations in the lights (*or*, marks)?...................ECB
			Are there any lights (*or*, landmarks)? ....QFA
			Are there any men of war about? ...SA
			Are there any orders?..... ....STL
			At any time ....XTI
			Can I get any —?.... ....NYI
			Can, *or*, Will you take any —?.....XJE

	ANY—APPEAR	

	ANY—*Continued*	
	Can not be got off (*or*, afloat) by any means now available................CD	
	Did you do any business?..........HNX	
	Do not have any communications with—........... ....HV	
	Do not make any alterations—until — ECG	
	Do not show a light on any account..QV	
	Had any letters (or, despatches) arrived—at?........... ....JM	
EPR	—Have you (*or person indicated*) any— ?	
	Have you any combustibles on board?.HO	
	Have you any message (telegram, orders or, communication) for me? ....IB	
	Have you any rockets?..............UZE	
	Have you had any births?..........GLH	
	Have you had any sickness on board? WBA	
	Have you received any orders—from —?........ ... .SU	
	I can not stop to have any communication ....IF	
EPS	—If you have any doubt	
	Is, or, Are there any?.... ....BEP	
	Is there any advantage to be gained?.DNT	
	Is there any alteration in the market? ECN	
	Is there any alteration in the marks?.ECM	
	Is there any danger—in — (of, or, from)?........... ... .GA	
	Is there any naval news?...........SEM	
	Is there any opportunity?.........SRX	
	Is there any risk?...............UYB	
	Is there any shelter?..............VWL	
EPT	—Not any None	
EPU	—Of any use	
EPV	—Scarcely any	
EPW	—Were there any?	
EPX	—Will it be of any use?	
EPY	—Will there be any?	
	ANYBODY ANY PERSON.............EPQ	
EPZ	ANYONE	
EQA	—Anyone else	
EQB	—Anyone hurt (or, wounded)	
	Anyone on board?..................BPM	
	Can you supply me with anyone to take charge (or, act) as engineer?......RH	
	Have you anyone that can work at a forge?........................NLE	
EQC	—Is, or, Was there anyone on board?	
	Is anyone wounded (or, hurt)? ... .EQB	
	Not anyone..............., ...SIU	
EQD	ANYTHING	
	Almost anything may be got ......EAB	
EQF	—Anything belonging to—	
EQG	—Anything else?	
EQH	—Anything extra?	
EQI	—Anything for me (*or person indicated*)?	
EQJ	—Anything further?	
EQK	—Anything in sight?	
EQL	—Anything of (or, from) —?	
EQM	—Anything to (or, for) —?	
EQN	—Anything to send?	
	Anything wrong with the engines...RD	
	Do not leave anything unsettled...QXR	
	Have you anything for sale?.......VFL	
	Have you anything to report? ....URS	

	ANYTHING—*Continued*	
	Have you seen (or, heard) anything of my (or, the) boat?..............FB	
EQO	—I decline having anything further to do (or, say) in the matter	
	Is anything the matter? Has any accident happened? ....BK	
EQP	—Is there anything to bring me up?	
EQR, EQS	ANYWHERE —Anywhere else	
	Do you, or, Does—(*vessel indicated*) call anywhere (or, at —)?............HW	
EQT	—Have you called anywhere (or, at —)?	
EQU	—I shall not touch (or, call) anywhere (or, at —)	
EQV	APART	
EQW	APARTMENTS	
EQX	APEAK	
EQY	APERIENT	
EQZ	APERTURE	
ERA	—Screw aperture Screw well.	
ERB	APOLOGIZE-D-ING APOLOGY	
ERC	APOPLEXY-ECTIC	
ERD	APPARATUS	
ERF	—Apparatus for casting oil on the sea to quiet it	
ERG	—Diving apparatus	
ERH	—Life-saving (*Rocket*) apparatus	
	Life-saving apparatus (*Rocket*) Station, WRP	
ERI	—Photographic apparatus	
ERJ	—Rocket (*Life saving*) apparatus coming	
ERK	—Sounding apparatus	
	Thomson's (Kelvin) patent sounding apparatus................ ....TEG	
ERL	APPARENT-LY	
ERM	—Apparent time	
ERN	—Apparently favorable-y	
ERO	—Apparently shoal water to the —.	
ERP	APPEAL ED-ING TO APPEAL	
	Crew have appealed to the authorities, FNE	
ERQ	—Have, or, Has appealed	
ERS	—No use appealing	
ERT	—Shall, or, Will appeal	
ERU	APPEAR ED-ING ANCE	
	Affairs appear more settled.........DOZ	
ERV	—Appear ed-ing-ance favorable	
	Appearances are threatening Be on your guard........ ....FN	
	Buoy, or, Mark does not appear to be in its proper position ....HMG	
ERW	—Do, or, Does not like the appearance of	
ERX	—Does not appear	
ERY	—Does, or Did it appear?	
ERZ	—Every appearance of foul weather	
ESA	—Has, or, Have the appearance of	
ESB	—It appears	
ESC	—No appearance of	

	APPEAR—ARE		
	APPEAR—*Continued*	ETZ	APRIL
ESD	—Should it appear		About April the—------------------CZJ
ESF	—The appearance is not satisfactory	EUA	—Beginning of April
ESG	—The appearance is satisfactory	EUB	—End of April.
	Vessel indicated appears in distress.NX	EUC	—Next April
ESH	—Vessel indicated (*bearing, if necessary*) appears to be—		
	What state does the battery appear to be in? -------------------- FYV	EUD	ARAB-IAN-IC
	Wind appears to be steady--------ZFH	EUF	APRON
ESI	APPETITE. APPETIZE-ING	EUG	ARBITRATE-D-ING-OR-ION
			Court of Arbitration -------------KAB
ESJ	APPLE		
		EUH	ARC
ESK	APPLIANCE	EUI	—Arc light
ESL	—Life-saving appliance		
		EUJ	ARCHBISHOP, *or*, BISHOP
ESM	APPLICANT		
			ARCHINE (*Russian measure of length*).AVM
ESN	APPLY-IED-ING—TO (*or*, FOR)		
ESO	—Application		ARCHINE (*Turkish measure of length*).AWJ
ESP	—Forward-ed-ing application to (*or*, for)		
ESQ	—Have applied (*or*, spoken) to—?	EUK	ARCHIPELAGO
ESR	—Have you applied (*or*, spoken) to—?		
EST	—Shall apply (*or*, speak) to	EUL	ARCTIC
ESU	—Will you apply (*or*, speak) to?		Antarctic ------------------------ENW
ESV	APPOINT-ED-ING-MENT—TO (*nomination*)		ARDEB (*measure of capacity*)--------AYL
ESW	—Appointment is made		
ESX	—Have, *or*, has an appointment		ARE (*Square or surface measure*)..AXJ
ESY	—Is the appointment settled?		
ESZ	—Make an appointment for—		**ARE** --------------------------BEJ
ETA	—Shall appoint—		Are, Am, *or*, Is able to -----------CYF
ETB	—The appointment has not taken place		Are, Am, *or*, Is not--------------BEG
ETC	—Time appointed		Are, Am, *or*, Is not able to -------CYG
ETD	—Will appoint		Are, Am, *or*, Is not to be ---------BFD
ETF	—Will you appoint?		Are, Am, *or*, Is not to have ------BQD
			Are, Am, *or*, Is to----------- ----BEH
ETG	APPREHEND-ED-ING		Are, Am, *or*, Is to be -------,-----BFC
	Apprehended (*arrested*) by the police, TNW		Are, Am, *or*, Is to blame ----- ----GMF
			Are, Am, *or*, Is to have-----------BQC
ETH	—Apprehension-ive		Are, *or*, Is his (her s, its)----------BEK
ETI	—Do, *or*, Does not apprehend		Are, *or*, Is my (mine)-------------BEL
ETJ	—Do you (he, *or*, they) apprehend?		Are, *or*, Is our-s, *or*, Are we?-- ---BEM
			Are, *or*, Is the? ----------------BRN
ETK	APPRENTICE-D-ING-SHIP		Are, *or*, Is their-s, *or*, Are they?...BEO
			Are, *or*, Is there—any? -----------BEP
ETL	APPROACH-ED-ING-ABLE (*See also* CLOSE)		Are, *or*, Is with — ---------------BER
	Approaching land--------------QDM		Are, *or*, Is you-r-s ----------------BES
ETM	—Can not approach		Are there? ---------- ------------BEP
ETN	—Do not approach the coast, because it is mined		Are there any orders?----------STL
			Are these (*or*, those)? ------------BEQ
ETO	—Do not approach too near		Are they? ------ -----------------BEO
ETP	—Meteorological Office (*or*, Weather Bureau) reports depression approaching from the —		Are to ---------------------- ------BEH
			Are to be --------------------------BFC
ETQ	—Meteorological Office (*or*, Weather Bureau) reports gale approaching from the —		Are we (our-s)? ----- -----------BEM
			Are you-r-s? ---------------------BES
	Pilot boat is approaching you ------DNI		Are you able to?------------------CYH
			He, She, It (*or person-s or thing s indicated*) had (*or*, have) done (*or*, is, *or*, are doing*) ----------------BUL
ETR	APPROPRIATE-D-ING-LY		He, She, It (*or person s or thing-s indicated*) had (has, *or*, have) not done (*or*, is, *or*, are not doing*) ----------BUP
ETS	APPROVE-BATION—OF		
ETU	—Approved		He, She, It (*or person-s or thing-s indicated*) is (*or*, are) for.. ---------BUR
ETV	—Do you approve—of?		
ETW	—Not approved of		He, She, It (*or person-s or thing s indicated*) is (*or*, are) from ----------BUS
ETX	APPROXIMATE-LY		He, She, It (*or person-s or thing-s indicated*) is (*or*, are)----------------BUT
ETY	—Approximation		

ARE—ARMY.

ARE—*Continued*		EUM	AREA. *or*, SURFACE
He, She, It (*or person-s or thing-s indi-cated*) is (*or*, are) able to _____BUV		EUN	ARGENTINA ARGENTINA COLORS
He, She, It (*or person-s or thing-s indi-cated*) is (*or*, are) not_____BUW			ARGENTINO (*Coin*)_____ASY
He, She, It (*or person-s or thing s indi-cated*) is (*or*, are) not able to ____BUX		EUO	ARGUE-D ING-MENT.
How are they? _____BXH		EUP	ARISE N-ING AROSE
How are you? _____ BXI			
How is (*or*, are) he (she, it, *or person-s or thing-s indicated*)? _____BXO		EUQ	ARM-ED-ING TO ARM.
If he (she, it, *or person-s or thing-s in-dicated*) is (*or*, are) _____ ___ ___ BYL		EUR	—Arms (*weapons*)
		EUS	—Arm boats
If he (she, it, *or person s or thing s in-dicated*) is (*or*, are) able to _____BYM		EUT	—Arm chest
			Arm the lead_____QJG
If he (she, it, *or person s or thing-s in dicated*) is (*or*, are) not_____BYN		EUV	—Armed men not allowed to land
		EUW	—Armed merchant ship
If he (she, it, *or person-s or thing-s in-dicated*) is (*or*, are) not able to___BYO		EUX	—Armed with —
			Fire-arms _____MZY
If they are _____CAF		EUY	—How are you armed ?
If they are not _____CAG		EUZ	—How is she armed?
If we are _____CAX		EVA	—I have no arms
If we are not _____CAY			Not armed Unarmed _____YGJ
If you are _____ _____CBQ			Man and arm boats_____GTV
If you are not _____ _____CBR			Send an armed boat_____ ___ ___GUK
If, *or*, Are, he (she, it, *or person s or thing-s indicated*)? ____ _____BVY		EVB	—She is armed They are armed
Is, *or* Are his (her-s, it s) _____BEK		EVC	—Small-arm men
		EVD	—Small arms (muskets, etc)
There are (*or*, is)_____CMG		EVF	—Small arm magazine
There are (*or*, is) not___ _____ CMH		EVG	—With arms
They are for _ _____CNM			Without arms Unarmed_____YGJ
They are from _____BET			
They are _____ BEU		EVH	ARM (*Limb*)
They are not_____ BEV		EVI	—Arm broken
We are _ ___ _____BEW		EVJ	—Arm dislocated
We are not _____ _____BEX		EVK	—Left arm
What, *or*, Which are (*or*, is)? ____CPW		EVL	—Right arm
What, *or*, Which are (*or*, is) you-r-a?.CPY			Yard arm _____ZKP
What, *or*, Which is (*or*, are) he (she, it, *or person s or thing s indicated*)? CPX		EVM	ARMAMENT
When are (*or*, is)? _____CQW		EVN	ARMATURE
When are we? _____CQX			
When are you? _____ __ CQY		EVO	ARMISTICE
When he (she, it, *or person s or thing-s indicated*) is (*or* are) _____CRJ			Armistice has been arranged_____YS
		EVP	—Is an armistice probable ?
When is (*or* are) he (she, it, *or person-s or thing s indicated*)? _____CRN		EVQ	ARMOR-ED.
When they are_____ __ ___CSI		EVR	—Armor belt
When we are_____CSL		EVS	—Armor-clad Iron-clad.
When you are_____CSN		EVT	—Armor-plate-d ing
Where am I (*or*, are we)? _____CSQ		EVU	—Armored cable
Where are (*or*, is)? _____CSR		EVW	—Armored cruiser
Where are they? _ ___ _____CST		EVX	—Armored deck
Where are you bound? _____SH		EVY	—Armored ship
Where are you from? _____SI		EVZ	—Armored train
Where are you going? _____CSU			
Where he (she, it, *or person-s or thing-s indicated*) is (*or*, are)__ _____CTD		EWA	ARMORER
Where I am (*or*, we are) _____CTG		EWB	ARMY
			Any alteration in the position of the armies ? _____EBY
Where is (*or*, are) he (she, it, *or person-s or thing-s indicated*)? _____ __ CTH			Army is advancing_____DNA
Where they are _ _____CUA		EWC	—Army is retreating.
Where we are (*or*, I am) _____ __ CTG		EWD	—Army List
Where you are_____ ___CUF		EWF	—Army Service Corps
Who are (*or*, is)? _____CUI		EWG	—By, *or*, With the army
Why are (*or*, is)? ____,_____ CUZ		EWH	—Captain in the army
Why are you? _____CVA		EWI	—Colonel
Why are you not? _____CVB		EWJ	—Commander-in Chief
You are ___ _____BEY		EWK	—General
You are not _____BEZ			

ARMY—ARTICLES OF AGREEMENT

	ARMY—*Continued*
EWL	—Lieutenant in the army
EWM	—Major
EWN	—Of, or, From the army
EWO	—To, or, For the army.
EWP	—Where is the — army ?
	AROSE ARISE-N-ING ....BUP
EWQ	AROUND.
EWR	ARRACK.
EWS	ARRANGE-D-ING To ARRANGE— TO (or, FOR) (*See* also SETTLE)
	Amicably arranged (or, settled)....EFG
	Armistice has been arranged........YS
EWT	—Arrange d-ing mails
EWU	—Arrange-d-ing to receive (or, for reception of)
EWV	—Arrangement
EWX	—Arrangements completed
EWY	—Can it be arranged ?
EWZ	—Can the difference be arranged ?
EXA	—Has arrangement been made ?
EXB	—I can arrange it
EXC	—I leave it to your arrangement.
EXD	—Is all settled (or arranged)?
EXF	—Is everything arranged—for ?
EXG	—It was arranged that I should
EXH	—It was arranged that I should not
EXI	—Make-ing Made arrangement—to—(or, for —)
EXJ	—No arrangement — to — (or, for)
EXK	—Unable to arrange
EXL	—What are the time arrangements for—?
EXM	—What arrangements to (or, for —) ?
EXN	ARREAR
EXO	ARREST-ED-ING
	Arrested (apprehended) by the police, TNW
EXP	—Has been arrested
EXQ	—Has not been arrested
EXR	—Under arrest
EXS	ARRIVE D-ING—AT To ARRIVE
	After, or, On arrival of—(or, at—) .DRE
	An express has arrived from — --- MPH
	Any arrivals from — ?............BOM
EXT	—Arrival
EXU	—Arrived at destination
EXV	—Arrived during the day
EXW	—Arrived during the night
EXY	—Arrived safely
	Await my arrival at — ....FPA
	Await the arrival of — ....FPC
EXZ	—Before arrival — of (or, at)
EYA	—Can you await arrival of — ?
EYB	—Despatches arrived.
EYC	—Do you wish your arrival reported ?
EYD	—Due to arrive — at
EYF	—Expect to arrive
	Had any letters (or, despatches) arrived — at ?.......JM
EYG	—Had the mail arrived when you left ?
EYH	—Has arrived off
EYI	—Has he (she, or, it) arrived — at ?
	Has the mail arrived —(from— *if necessary*) ?............ QN

	ARRIVE—*Continued*
EYJ	—Have you heard of the arrival of — ?
EYK	—I (or *vessel in question or now indicated*) arrived
EYL	—I shall await their arrival (or, the arrival of —)
EYM	—Important intelligence has arrived
EYN	—Large arrivals of —
	Mail arrived—on the—(*date indicated*), RFC
	Mails have (or, had) arrived from — (*dates up to* —)..........RFG
	Mail has or, had) not arrived from — (*place indicated*) ....RFI
	Mail steamer arrives-d at —.........RFL
EYO	—Not arrived
	On, or, After arrival—of — (or, at—) DRE
	Orders have not arrived yet — for—.SUK
	Reinforcements constantly arriving.UOM
	Report my arrival (or, arrival of—).USC
EYP	—Shall I await your arrival?
EYQ	—Shall you await her arrival (or, the arrival of —)?
EYR	—She (or *vessel indicated*) had (or, has) arrived
EYS	—She (or *vessel indicated*) had (or, has) not arrived.
	Telegraphic dispatches arrived from — JX
EYT	—Very few arrivals
	Vessels just arrived, show your ensigns, DY
	Vessels just arrived show your numbers (or, distinguishing signals).......DX
	Wait for arrival—of.............FPC
EYU	—When did he (she, it, or *vessel indicated*) arrive?
EYV	—When did you arrive?
	When does the — mail arrive?.....RFZ
	When ought the next mail to arrive? RGA
EYW	—Will arrive—at
	Will you await her arrival (or, the arrival or —)? ....EYQ
	Your orders have not arrived yet ..SVA
	ARROBA (*Measure of capacity*) ....AYM
	ARROBA (*Measure of weight*) ......AZY
EYX	ARROW
EYZ	ARROWROOT.
EZA	ARSENAL
EZB	ARSENIC AL
EZC	ARTERY-ERIAL
EZD	ARTICLE (*Object*)
EZF	—Article indicated can be supplied, but it will require fitting
EZG	—Article indicated may be had at —(*place indicated*)
EZH	—Send the following articles
EZI	—The following articles
	ARTICLES OF AGREEMENT ....DUG
EZJ	—Send ship's articles

ARTICLES OF AGREEMENT—ASK

	ARTICLES OF AGREEMENT—*Continued.*	FBZ	ASCEND-ED-ING
EZK	—Ship's articles (*or*, papers)		
EZL	—Signed articles	FCA	ASCERTAIN ED-ING
EZM	—Termination of articles	FCB	—Ascertain what
		FCD	—Ascertain when
		FCE	—Can you ascertain?
EZN	ARTIFICE-IAL-LY	FCG	—Can not ascertain
EZO	—Artificial horizon.	FCH	—Could it not be ascertained?
			Damage not ascertained ---------- KIY
EZP	ARTIFICER	FCI	—Did you ascertain?
EZQ	—Artificer diver.	FCJ	—Endeavor-ed-ing to ascertain
EZR	—Engine-room artificer	FCK	—Have you ascertained?
EZS	—Skilled artificer	FCL	—It has been ascertained
EZT	—What artificers have you?		Passed a wreck(*date, latitude, and longi-*
			tude, if necessary) but was unable to
EZU	ARTILLERY-IST-MAN.		ascertain whether any people were on
EZV	—Artillery has been landed.		board ---------------------------- ZW
EZW	—Artillery, *or*, Ordnance stores	FCM	—Shall, *or*, Will ascertain
EZX	—In want of artillery	FCN	—When it is ascertained
EZY	—No want of artillery		
		FCO	ASCRIBE-D-ING TO ASCRIBE
FAB	As	FCP	—What do you ascribe it to?
FAC	—As a matter of course		
FAD	—As close as.	FCQ	ASH (*Wood*)
FAE	—As convenient		
FAG	—As far as	FCR	ASH CINDER
FAH	—As fast as	FCS	—Ash pit
FAI	—As fast as possible	FCT	—Ash shoot
FAJ	—As follows		
FAK	—As he (she, it, his, her-s, it-s)	FCU	ASHAMED
FAL	—As he is	FCV	—Ought to be ashamed
FAM	—As I (my, mine)		
FAN	—As it happens.	FCW	ASHORE (*See also* SHORE *and* AGROUND)
FAO	—As it is		Am on shore, likely to break up, require
FAP	—As it is not		immediate assistance ----------- CA
FAQ	—As it may	FCX	—Are you going ashore?
FAR	—As it was		Do not send boats ashore after dark.GST
FAS	—As it was not		I am a stranger here, will you let me
FAU	—As it will be		go ashore with you ?---------- OCH
	As little as -------------------- QVU	FCY	—Let him bring ashore (on shore)
FAV	—As long as	FCZ	—Send ashore—to (*or*, for)
FAW	—As many as		Send the letter bag on shore (*or to vessel*
FAX	—As many as convenient		*indicated*) ---- ----- ------- FSN
FAY	—As many (*or*, much) as possible		The ship's bags are sent ashore----- FSO
FAZ	—As much as		Vessel (*or vessel indicated*) ashore near
FBA	—As much as can be taken		(*or*, on)— -------------------- CH
FBC	—As near as		
FBD	—As often as.	FDA	ASIA-TIC
FBE	—As quick as		
FBG	—As quick as possible	FDB	ASIDE
FBH	—As requested		
FBI	—As requisite	FDC	ASK—FOR— TO ASK (*See also* REQUEST)
FBJ	—As slow as possible		Asked-ing --- ------------------ FDM
FBK	—As soon as	FDE	—Ask permission
	As soon as alongside -------------- EAV		Ask stranger (*or vessel indicated*) if he
FBL	—As soon as convenient		will communicate ------ ------- HT
FBM	—As soon as it is dark	FDG	—Ask-(*person or vessel indicated*) not
FBN	—As soon as it is dusk		to —
FBO	—As soon as possible	FDH	—Ask-(*person or vessel indicated*) to —
	As soon as she is off (*or*, afloat)----DPW	FDI	—Ask him if
FBP	—As soon as the flood tide makes	FDJ	—Ask the
FBQ	—As soon as there is a breeze	FDK	—Ask you not to
FBR	—As they (their-s)	FDL	—Ask you to
FBS	—As to	FDM	—Asked-ing
FBT	—As usual		*Asks* name of ship (*or*, signal station) in
FBU	—As we (our-s)		sight. ---- --------------- VG
FBV	—As well as		*Asks* precise time of high water, and
FBW	—As you-r-s		minimum depth at that time------ XE
FBX	—As you please		NOTE.—*Reply will be*
			1 *By an hour signal* (*see page* 58)
FBY	ASBESTOS		2 *By a numeral signal denoting feet*

ASK—ASSIST.

	ASK—*Continued*	**FEX**	**ASSIST-ANCE**
FDN	—Better not ask the reason		Aground, want immediate assistance,
FDO	—Hail the strange vessel, and ask—if —		NA
FDP	—Have, *or*, Has not asked		Am drifting want assistance........CM
FDQ	—Shall, *or*, Will ask		Am on shore, likely to break up, re-
FDR	—Unable to ask		quire immediate assistance......CA
FDS	—Will you ask?		
		FEY	—Assist—*(vessel indicated)*
		FEZ	—Assistance was refused
FDT	**ASLEEP** (*See also* SLEEP)		Assistant ....FGY
			Boat is coming to your assistance ...ET
FDU	**ASSASSIN-ATE-ION**	FGA	—Can assist
FDV	—An attempt to assassinate	FGB	—Can be assisted (*or*, helped)
	Assassinated (*or murdered*)..........SCJ	FGC	—Can I assist?
			Can I procure any assistance in the way
FDW	**ASSAULT ED-ING TO ASSAULT** (*See also*		of — ?..................................CN
	ATTACK)	FGD	—Can you assist?
FDX	—Assault did not succeed		Can you assist—(*vessel indicated*)?...CO
FDY	—Commit-ted-ting an assault		Can not assist... ....CP
			Could not render assistance...........CQ
FDZ	**ASSEMBLE-D-ING**		Damage, *or*, Defects can not be repaired
	About to assemble ....DAE		without assistance..... ....BG
	Assemble alongside ....EAW		Do not require assistance (*or*, further
FEA	—Assembled on the		assistance)... .. ..CR
	Boats to assemble........ ... .GSE		Do the best you can for yourselves, no
	Convoy will assemble at port (*or place*		assistance can be given CX
	indicated) ....JUO		Do you require any assistance (*or*, help)
FEB	—Have, *or*, Has assembled		of — (*or*, from —)? ....OS
FEC	—Shall *or*, Will assemble		Do you require further assistance ?..CT
	Should you part, endeavor to beach your		Do you require immediate assistance?
	vessel where people are assembled (*or*		CU
	as pointed out by Compass Signal	FGE	—Every assistance will be given
	FROM *you*)...............FZU		Fire, *or* Leak, want immediate assist-
FED	—Was, *or*, Were assembled		ance ....NH
FEG	—When they assemble •	FGH	—Give every assistance to —
			I am aground, send what immediate
FEH	**ASSEMBLY**		assistance you can ....CE
FEI	—Assembly Congress, Parliament (*or other*		I am attacked want assistance Help,
	legislative body) adjourns (*or*, ad-		I am attacked ....NJ
	journed) on the —		I can not assist (help) you..........OTX
FEJ	—Assembly, Congress, Parliament (*or other*		I can not take you in tow, but will
	legislative body) is dissolved		report you at — and send immediate
	Assembly, Congress, Parliament (*or other*		assistance.........................XYQ
	legislative body) met (*or*, meets) on		I require assistance from Lloyd's agent,
	the — ....JPB		CV
FEK	—House of Assembly		I shall get off (*or*, afloat) with assist-
			ance ....DPY
FEL	**ASSENT-ED-ING**		I want assistance, please remain by
	Yes, *or*, Affirmative		me...............................NR
	c, *or*, Code Flag over c		
		FGI	—I will assist you (*or vessel in distress*)
FEM	**ASSERT-ED-ING ASSERTION**	FGJ	—If I have not assistance
FEN	—Have, *or*, Has asserted		In distress; want assistance..........NS
FEO	—It is asserted		In distress, want immediate assistance *a*
			NO
FEP	**ASSESS-ED-ING-MENT**	FGK	—Is — (*vessel indicated*) likely to assist
FEQ	—Damages are assessed at —		(*or*, supply) me?
			Lighthouse, *or*, Light-ship at — wants
FER	**ASSESSOR**		immediate assistance........ .....CW
			Light-ship, *or*, Lighthouse at — wants
FET	**ASSETS**		immediate assistance...............CW
			May be got off (*or*, afloat) if prompt
FEU	**ASSIGN-ING-MENT**		assistance be given.................CG
FEV	Assigned		May be got off (*or*, afloat) without
	Cause, *or*, Reason must be assigned.IFO		assistance ....DPZ
	The cause (*or*, reason) assigned, is ..IFT	FGL	—Medical assistance wanted, want a sur-
	The cause (*or*, reason) assigned is suf		geon
	ficient . ...IFU		Mutiny, want assistance............YF
			No assistance can be rendered, do the
FEW	**ASSIGNEE**		best you can for yourselves CX

a See Distress Signals, page 7

	ASSIST—AT		
	ASSIST—*Continued.*	FHG	**ASTERN**—*Continued*
FGM	—Offered assistance, but they declined		—Astern of the
	Passed a wreck (*date, latitude and longitude to follow, if necessary*) but could not render any assistance, people still on board........................ ZV		Convoy is astern ....JUL
			Convoy to keep astern of leader (*or, escort*) ....IW
	Render all the assistance possible to . CY	FHI	—Drop astern—of
	Require help (*or, assistance*) of (*or, from*) .... CZ	FHJ	—Farther astern of
			Go astern .. . .LJ
	Rudder disabled, will you assist me ? (*into port indicated, if necessary*) .DA		Go astern Heave all aback ....LP
	Send immediate assistance.... ....DB		Go astern easy Easy astern ...... LK
FGN	—Send boat to assist		Go astern full speed........ ... .LM
	Ship disabled, will you tow me ? (*into port indicated, if necessary*) ......MY	FHK	—Go astern of—
			Go astern of me... ....FHQ
FGO	—Should you require any assistance (*or help*)?	FHL	—Has fallen astern.
			Heave all aBack Go astern ....LP
FGP	—Surgeon, *or*, Doctor wants assistance		I will pass astern of you
	The ice is so solid I can not break through, send help (*or, assistance*). PCF		Code Flag over Z
		FHM	—Keep farther astern
	Vessel indicated (*or*, in distress) wants immediate assistance............. NW	FHN	—Keep right astern
FGQ	—Vessel indicated does not require further assistance	FHO	—Keep going astern
			My engines are going astern
			Code Flag over V
FGR	—Vessel indicated not likely to supply (*or*, assist) you	FHP	—Next astern
		FHQ	—Pass astern of me
FGS	—Vessel indicated wants assistance	FHR	—Stern way Going astern.
	Want a surgeon (*or*, doctor) Medical assistance ....FGL	FHS	—Tow astern.
			Veer a boat astern ....GVE
	Want assistance..................YE	•	Veer a breaker astern ....FXZ
	Want assistance, mutiny..... .....YF		Veer a rope astern ....VAC
	Want immediate assistance In distress *a*........................ NO	FHT	**ASTONISH-ED-ING MENT**
		FHU	—Am, Is, *or*, Are astonished
	Want immediate medical assistance.YL	FHV	—It is astonishing.
	We are coming to your assistance...DC	FHW	**ASTRONOMY ICAL-LY**
FGT	—What assistance do you require ? What do you want ?	FHX	**ASUNDER**
FGU	—Will assist	FHY	**ASYLUM**
	Will you assist me (*or vessel indicated*)?	FHZ	—Lunatic asylum
	DE	FIA	**AT AT THE**
	Will you assist me (*or vessel indicated*) in laying out an anchor ? .... EKL		At all (*or*, every)..................DYI
			At all events....................DYH
	Will you go to the assistance of wreck (*bearing to follow, if necessary*)...ZY		At all times..... ....DYJ
FGV	—With assistance fire can be subdued		At anchor ....EHN
	With immediate assistance fire can be extinguished ....NY		At any time........ .. ..XTI
		FIB	—At dawn Daybreak Dawn
	With some assistance I shall be able to set things right ....DF		At daylight Daylight............KNB
FGW	—With the assistance of	FID	—At first.
FGX	—Without assistance	FIE	—At home.
		FIG	—At last
FGY	**ASSISTANT**	FIH	—At least
		FIJ	—At midnight
FGZ	**ASSOCIATE-D-ING-ION.**	FIK	—At night
		FIL	—At no time.
		FIM	—At noon Noon
		FIN	—At once Now
FHA	**ASSORT-ED ING-MENT**	FIO	—At one.
	Assorted cargo ....HYV	FIP	—At, *or*, In station.
FHB	**ASSUME-D-ING-PTION**	FIQ	—At place specified
			At present For the present ......NHJ
FHC	**ASSURANCE.**	FIR	—At quarters
		FIT	—At random
FHD	**ASSURE D ING**	FIU	—At sea
		FIV	—At sunrise Sunrise
FHE	**ASTERN—OF**	FIW	—At sunset Sunset
	Afloat astern (aft)..................BW	FIX	—At the extremity (*or*, end) of
	Astern easy ....LK	FIY	—At the landing place
		FIZ	—At the latter end of— The latter end—of.

a See Distress Signals, page 7

	AT—ATTEST		
	AT—*Continued*	FKR	**ATTEMPT**—*Continued*
	At the nearest landing place........QDO		—Shall I try (*or*, attempt)—to ?
	At the railway station ....UHE		The attempt is dangerous .... GI
FJA	—At the (*or*, that) time specified.	FKS	—Useless to attempt (*or*, try)—to
	At this port......... .... TOQ		When may the bar (*or*, entrance) be
FJB	—At top of		attempted ? .. ..FWD
FJC	—At what time ? At what hour ?	FKT	—Will attempt—to (*or*, try)
FJD	—Is at—the	FKU	—Will you attempt to—(*or*, try)?
	Not at all....................DYP	FKV	—You may make an attempt (*or*, try)
	To be at ....ONQ		
		FKW	**ATTEND-ED-ING-ANCE** (*To be present at*)
FJE	**ATHWART**	FKX	—Attend on board.
FJG	—Athwart hawse	FKY	—Attend survey
FJH	—Athwart ship	FKZ	—Attendant
FJI	—Athwart the tide (current, *or*, stream)		Has been attended with no advantage,
			DNR
FJK	**ATLANTIC**		
FJL	—North Atlantic.	FLA	**ATTEND-ED ING** (*To pay attention*) (*See*
FJM	—South Atlantic		*also* ATTENTION)
	Trans-Atlantic ....XZW		Attend to signals ....DG
		FLB	—I will attend to your signal
FJN	**ATLAS**		
		FLC	**ATTENTION-IVE**
FJO	**ATMOSPHERE**		Attention is called to page —, paragraph
FJP	—Atmospheric disturbance		—, of the INTERNATIONAL CODE...DH
FJQ	—Atmospheric pressure		N. B —*The page is to be shown first, and*
			then the paragraph This signal
FJR	**ATROCIOUS-LY-NESS-OCITY.**		*should be employed only between ves-*
			sels using the United States edition,
FJS	**ATTACH-ED-ING-MENT**		*as otherwise the pages referred to will*
			not correspond
FJT	**ATTACHÉ**		*Calls* the attention of the FIRST (*or*,
			nearest) ship on the bearing (FROM *the*
FJU	**ATTACK-ED-ING** TO ATTACK (*See also*		*person signaling*) pointed out by com-
	ASSAULT)		pass signal .... ----DI
FJV	—(An) Attack		*Calls* the attention of SECOND ship etc.DJ
	Coup de-main . ... JYX		*Calls* the attention of THIRD ship, etc.DK
	False attack..................MSD		*Calls* the attention of FOURTH ship, etc.DL
FJW	—Have, *or*, Has been attacked		*Calls* the attention of FIFTH ship, etc.DM
FJX	—Have you (*or vessel indicated*) been at-		*Calls* the attention of SIXTH ship, etc.DN
	tacked ?		*Calls* the attention of the shore signal
	I am attacked , want assistance Help,		station on the bearing (FROM *the per-*
	I am attacked.................. NJ		*son signaling*) pointed out by compass
FJY	—I attacked		signal ....DO
FJZ	—I (*or vessel indicated*) have been at-		*Calls* the attention of vessel COMING into
	tacked—by		the anchorage on the bearing (FROM
	Night attack.....................SID		*the person signaling*) pointed out by
FKA	—Serious attack		compass signal . ...DP
FKB	—Shall I attack ?		*Calls* the attention of vessel LEAVING the
FKC	—Slight attack		anchorage on the bearing (FROM *the*
	Support my attack.............XFA		*person signaling*) pointed out by com-
FKD	—To be attacked		pass signal . ...DQ
	Torpedo attack ....XWO		*Calls* the attention of vessel whose dis-
FKE	—Wait for the attack		tinguishing signal will immediately
FKG	—Will you attack ?		follow ....DR
FKH	— — made an unsuccessful attack on —	FLD	—For your attention
		FLE	—Have, or, Has paid particular atten-
FKI	**ATTEMPT-ED ING TO** (*See also* TRY)		tion—to
	An attempt to assassinate ....FDV	FLG	—I am paying every attention
FKJ	—Attempted to put to sea		Look out Pay attention ....DS
	Do not attempt landing in your own	FLH	—Pay attention to orders
	boats...... .. ..EY		Pay attention to signal . ...DG
	Do not attempt to make the anchorage	FLI	—Pay attention to the helm
	(*or*, harbor)........ ....FW	FLJ	—Pay no attention to
FKL	—Do not attempt without a pilot		Pay strict attention to signals during
FKM	—Have, *or*, Has attempted		the night.................. ....DT
FKN	—I (*or vessel indicated*) made the attempt	FLK	—Pay the greatest attention
	but did not succeed	FLM	—You do (*or*, did) not pay attention—to
FKO	—Make an (*or*, the) attempt		
FKP	—Shall attempt to get away	FLN	**ATTEST-ED ING-ATION**
FKQ	—Shall I make another attempt ?		

ATTORNEY—AVOID

FLO	ATTORNEY (*Solicitor*)		AUTHORITIES—*Continued*
FLP	—Power of attorney	FNH	—Dock-yard authorities
		FNI	—Harbor authorities
FLQ	ATTRACT-ED-ION	FNJ	—Inform harbor authorities.
	Have you any local attraction (deviation)----------------------------JHR	FNK	—Legal authorities
		FNL	—Local authorities
		FNM	—Medical authorities
FLR	ATTRIBUTE-D-ING-ABLE	FNO	—Police authorities
			Police (*or*, other) authorities have acquitted ---------- ----- ------DHT
FLS	AUCTION		Police (*or*, other) authorities have decided against the crew ---------KDG
FLT	—Sales by auction		
FLU	—Sold by auction	FNP	—Police (*or*, other) authorities have fined
FLV	AUDIENCE AUDIBLE-Y.		Police (*or*, other) authorities have interfered --------------------- -----FNB
FLW	AUDIT-ED ING		Police (*or*, other) authorities have taken some of the crew out of the ship-KDH
FLX	AUDITOR	FNQ	—Police (*or*, other) authorities will not give any redress.
FLY	AUGER	FNR	—Port authorities
FLZ	AUGMENT-ED-ING-ATION		
		FNS	AUTHORIZE D-ING (*See also* AUTHORITY)
FMA	AUGUST	FNT	—Have you authorized?
	About August the — --------------CZK		
FMB	—Beginning of August	FNU	AUTOMATIC-ALLY.
FMC	—End of August		
FMD	—Next August	FNV	AUTUMN—OF
		FNW	—Autumnal
FME	AUSTRALIA-N	FNX	—Last autumn
FMG	—Australian meat	FNY	—Next autumn
FMH	—Australian wine		
		FNZ	AUXILIARY
FMI	AUSTRIA-N AUSTRIAN AND HUNGARIAN COLORS	FOA	—Auxiliary engine
		FOB	—Auxiliary screw.
	Bank of Austria ---------------- -----FUM	FOC	AVAIL-ED
FMJ	AUSTRO-HUNGARIAN LLOYD'S	FOD	—Avail yourself of
		FOE	—Available
FMK	AUTHOR		Can not be got off (*or*, afloat) by any means now available -------------CD
FML	AUTHORITY (*Delegated power*)		I am sinking (or, on fire), send all available boats to save passengers and crew -------------------------NO
FMN	—Authority insufficient		
FMO	—Authority is (or, will be) questioned		
FMP	—Authority sufficient for the occasion	FOG	—I shall (or, will) avail myself of
	By what authority? What is your authority? ---------------------FMY	FOH	—If available
		FOI	—It is of no avail.
FMQ	—Exceed-ed his (or, their) authority	FOJ	—It will not avail much
FMR	—Exercise d his (or, their) authority		No boat available ----------------- FG
	Give authority (*authorize*) ---------FNS		No doctor (or, surgeon) available --WM
FMS	—Have (or, has) no authority to —		No interpreters available - --------PRF
FMT	—Have you authority to —?		No lighters available ---------------RB
FMU	—I have authority from, *or*, Am authorized by —	FOK	—Send all available spars for — to
			There are no tugs available -------¨-XZ
FMV	—I have no authority		
FMW	—No authority—to—	FOL	AVAST
FMX	—On the authority of —	FOM	—Avast heaving
FMY	—What is your authority? By what authority?	FON	AVENUE
FMZ	—Without authority Unauthorized	FOP	AVERAGE AN AVERAGE
			About the average ----------------CZX
FNA	AUTHORITIES. (*Persons in authority*)	FOQ	—An average number (or, quantity) of —
FNB	—Authorities (*or*, Police authorities) have interfered		Does not amount to an average of—-EGJ
		FOR	—General average
FNC	—Authorities must interfere		
FND	—Authorities will not interfere	FOS	AVERSE-ION
	Authorities will put an end to the affair ------- ---------------DPJ		AVISO, *or* DESPATCH, VESSEL_____KWY
	Civic authorities Alderman -----DWS		
FNE	—Crew have appealed to the authorities	FOT	AVOID-ED ING ABLE
FNG	—Dock authorities	FOU	—Avoid, if possible

AVOID—BACK

	AVOID—*Continued*		AWAY—*Continued.*
FOV	—Avoid-ed-ing confusion		My sails are all blown away (or, split),
FOW	—Can avoid (or, be avoided)		....VEC
FOX	—Can not avoid (or, be avoided).		Open Get farther away ....MC
	Unavoidable-ly ....YGK		Ran, or, Run away................UHO
			Shall attempt to get away ....FKP
FOY	AVOIRDUPOIS (*Weight*)		Slack away ....WFQ
			The breeze will die away .... HGF
FOZ	AWAIT-ED-ING (*See also* WAIT) TO AWAIT		You run away from me.............VCL
FPA	—Await my arrival at —		
FPB	—Await outside the inspection of the offi-	FPR	AWEIGH (*See also* WEIGH)
	cials		Are you, or, Is — (*vessel indicated*)
FPC	—Await the arrival of —		aweigh?...... . ...ZM
FPD	—Awaiting coroner's inquest		I am aweigh...........................ZN
	Can you await answer (or, reply)?...EMV	FPS	—Vessel indicated is aweigh
	Can you await arrival of —?.......EYA		
	I have received orders for you to await	FPT	AWHILE WHILE-ST.
	instructions from owners at — ... SQ		
	I shall await their arrival (or, the arrival	FPU	AWKWARD-LY-NESS
	of —) ....EYL		
	I will telegraph for your orders if you	FPV	AWNING
	will await reply ....SV	FPW	—Deck awnings
	Keep off and on, and await instruc-	FPX	—Furl awnings.
	tions ....SX	FPY	—Slope awnings.
	Shall I await your arrival ? ....EYP	FPZ	—Spread awnings
	Will await answer (or, reply) ....TZ		
	Will you await answer (or, reply)?..UA		AWOKE AWAKE-N-ING.... ....PPE
	Will you await her (or, the arrival		
	of —)?.............................EYQ	FQA	AXE
	Your orders will be awaiting you off —		Pickaxe ....THR
	(or, at —) ....SVB		
		FQB	AXIS-IAL
FPE	AWAKE-N-ING AWOKE		
		FQC	AXLE AXLETREE
FPG	AWARD-ED-ING		
		FQD	AZIMUTH-AL
FPH	AWARE	FQE	—Azimuth compass.
FPI	—Am, Is, or, Are aware—of.	FQG	—Azimuth table.
FPJ	—Am, Is, or, Are not aware—of.	FQH	—Have you an azimuth compass?
FPK	—Are you aware?	FQI	—I have an azimuth compass
FPL	—I was not aware of it	FQJ	—Variation by azimuth
FPM	—Is he aware?		
FPN	—Not aware.		
FPO	AWASH.		
FPQ	AWAY		
	Bring away....................HIP		
	Carried away ....ICO		
	Cast away ....IEB		
	Castaways....................IEC		
	Clear away ....ITB		
	Could any vessel have got away after	FQK	B (*Letter*) (*For new method of spelling*
	you ?................. ....OFC		*see page* 13)
	Cut away your masts ....KJ		
	Cut away................ -- ....KHG	FQL	BACK-ED-ING
	Do you think we could get away		Back main topsail ....XWC
	from — ?........................XPZ	FQM	—Back the
	Doctor, or, Surgeon is away ....LJV		Back the anchor ....EHO
	Funnel carried away ....NTO		Ordered back.....................SUF
	Go away.........................OBR		Put back ....UCW
	Have, or, Has carried away ....ICQ		Put back damaged ....UCX
	Heave away.....................ORZ		Send back ....VPQ
	Is it a difficult anchorage to get away	FQN	—Was obliged to put back
	from ?.................. ....ELB		What is the cause of your putting back?
	Is keeping away............. .....PYP		UDE
	Is my berth a good one for getting away		
	from ?...... ....GFW		
	Keep away.......................PYM		
	Keep more away........ ....PYW		
	Lower away....................RBZ		

BACKSTAY—BANK

FQO	BACKSTAY	FSW	BAKE-D-ING
FQP	—Topmast backstay	FSX	—Bakery Bake house
FQR	BACKWARD-LY-NESS	FSY	BAKER
FQS	BACON	FSZ	BALANCE-D-ING
FQT	BAD-LY	FTA	—Balanced rudder
FQU	—A bad method		BALE (*Measure of Capacity*)_____AYN
	Bad accommodation_____DCN	FTB	—Bale of—
	Bad account _____DEX	FTC	—Bales of cotton
	Bad anchorage_____ _____EKR		
FQV	—Bad berth.	FTD	BALE-D-ING-R TO BALE OUT.
FQW	—Bad case		
FQX	—Bad chance	FTE	BALL
FQY	—Bad coal		Ball (*dance*) _____KJW
FQZ	—Bad condition	FTG	—Ball cartridge
FRA	—Bad conduct	FTH	—Is there a time ball at —?
FRB	—Bad example	FTI	—Time ball
FRC	—Bad fuel.	FTJ	—Time ball drops at —
FRD	—Bad habit	FTK	—When does time ball drop?
FRE	—Bad landing		
FRG	—Bad lookout	FTL	BALLAST-ED-ING
FRH	—Bad news—of	FTM	—Ballast can be got at
FRI	—Bad order In bad order.	FTN	—Ballast, *or,* Cargo has shifted
	Bad weather _____ZB	FTO	—Ballasted with —
	Bad weather is expected _____FP	FTP	—Can I obtain ballast at —?
FRJ	—Badly done	FTQ	—I am (*or vessel indicated* is) in ballast.
FRK	—Badly stowed.	FTR	—Light in ballast
FRL	—Badly, *or,* Seriously wounded	FTS	—Must take in more ballast
FRM	—Engines working badly	FTU	—Pig of ballast
FRN	—He, She, *or,* It is very bad	FTV	—Water ballast
FRO	—In a bad way	FTW	—Water ballast tank.
FRP	—In bad health		
FRQ	—Is he (she, *or,* it) bad?	FTX	BALLOON
FRS	—Markets are very bad (very low).	FTY	—Captive balloon
FRT	—Not bad-ly.		
FRU	—Very bad-ly		BALSAM OF COPAIBA_____JVO
FRV	—Very bad draught		Friars Balsam _____NRF
FRW	—Very bad weather.		
FRX	—Work badly	FTZ	BAMBOO
FRY	—Worked-ing very badly.		
		FUA	BAND (*Music*)
FRZ	BADGE		
FSA	—Badge of office.	FUB	BAND-ED-ING
FSB	BAFFLE-D-ING	FUC	BANDAGE-D-ING
FSC	—Baffling wind		
		FUD	BANDIT-TI
FSD	BAG		
FSE	—Bread bag Biscuit bag.	FUE	BANISH-ED-ING-MENT
FSG	—Carpet bag		
FSH	—Coal bag.	FUG	BANJO, *or,* SCREW FRAME
FSI	—Hand bag		
FSJ	—Mail bag Letter bag.	FUH	BANK ED-ING TO BANK
FSK	—Oil bag	FUI	—Bank, *or,* Pile of —
FSL	—Sand bag	FUJ	—Bank up
FSM	—Send bag for	FUK	—Bank up fires
FSN	—Send the letter bag on shore (*or to vessel*		Clouds are banking up _____IWA
	indicated)		Keep your fires banked up_____PZV
FSO	—The ship's bags are sent ashore		
FSP	—What is the number of mail bags you	FUL	BANK (*Office*)
	have on board?	FUM	—Bank of Austria
		FUN	—Bank of England
FSQ	BAGGAGE	FUO	—Bank of France
FSR	—Baggage port.	FUP	—Bank of Germany
		FUQ	—Bank of Italy.
FST	BAIL-ED-ING-ABLE	FUR	—Bank of Russia
		FUS	—Bank of Spain
FSU	BAILIFF.	FUV	—Bank of Turkey
		FUW	—Bank of —
FSV	BAIT		

	BANK—BARRICADE		
	BANK—*Continued.*	FWG	**BAR　BARRED-ING**
FUX	—Bank (*name indicated*) has suspended	FWH	—Capstan bar
	payment	FWI	—Crowbar
	Bank note, *or,* paper money......ASV	FWJ	—Eccentric bar broken
FUY	—Bank rate	FWK	—Fire, *or,* Furnace bar
FUZ	—Banker	FWL	—Radius bars
	Savings bank.............VIG		
		FWM	**BARBER**
FVA	**BANK　(Shoal)**		
	Along the bank of ....BAK	FWN	**BARBETTE**
FVB	—Deepest water is nearest the bank (*or,*		
	reef)	FWO	**BARE-NESS**
FVC	—I have had thick fog on the bank	FWP	—Barely
FVD	—On a sand bank	FWQ	—Bare poles
FVE	—On the left bank of the river *	FWR	—Barely possible
FVG	—On the right bank of the river *		Barely room to wear ....UZO
	NOTE—The left or the right hand bank	FWS	—Barely sufficient
	in descending the stream		Barely time to save the mail ......RRY
FVH	—Sand bank		Running under bare poles ....VCJ
	Stand nearer the bank (reef, shoal, *or,*	FWT	—You have barely room
	bar).................FWA		
	The bank is encumbered by fishing	FWU	**BARGAIN-ED-ING　TO BARGAIN.**
	boats ....EP	FWV	—Better bargain beforehand
FVI	—The bank was clear of fog		
		FWX	**BARGE**
FVJ	**BANKRUPT-CY**	FWY	—Bargeman
	Agent is bankrupt.............DSL		
FVK	—Bankrupt, *or,* Failed	FWZ	**BARILLA**
FVL	—Bankruptcies, *or,* Failures		
FVM	—Have any bankruptcies taken place?		**BARK　(See BARQUE)** ....FXO
FVN	**BANQUET-ED-ING**	FXA	**BARK (*of a tree*)**
FVO	**BAPTIZE-D ING-ISM**	FXB	**BARLEY**
FVP	**BAR (*of a harbor, etc*)**	FXC	**BARNACLE**
FVQ	—At—(*time specified*) there will be—feet		
	water over the bar	FXD	**BAROMETER**
FVR	—Bar can not be crossed until —		NOTE—*For Barometer Table, see page* 59
FVS	—Bar harbor		Aneroid barometer.............ELP
	Bar, *or,* Entrance is dangerous......FQ	FXE	—How is the barometer? What is the
	Bar is impassable.............FR		barometer doing?
	Bar is impassable for boats on the ebb	FXG	—I have not a barometer
	tide.............BQ	FXH	—The barometer has fallen very rapidly
	Bar is not dangerous ....LA	FXI	—The barometer is falling.
FVT	—Bar *or,* Entrance not safe except just	FXJ	—The barometer is rising.
	at slack water (*or,* at —, *time indi-*	FXK	—The barometer is steady
	cated)	FXL	—The height of the barometer is —
FVU	—Bar, *or,* Sands shifted	FXM	—Watch the barometer
FVW	—Bar passable		What is the barometer doing?......FXE
FVX	—Can I clear the bar?	FXN	—What is the height of barometer?
FVY	—Can I cross the bar?　Shall I be able to		
	get over the bar?	FXO	**BARQUE.**
FVZ	—On the bar	FXP	—Barquentine
FWA	—Stand nearer the shoal (bank, reef, *or,*	FXQ	—Four-masted barque
	bar)		
FWB	—Vessel has struck on the bar	FXR	**BARRACK**
FWC	—What are the leading marks for cross-		
	ing the bar?		**BARREL　(*Measure of capacity*)** ....AYO
	What is the depth of water on the bar	FXS	—Barrel of —
	(*in feet*)?.............VQ	FXT	—Barrel of flour
FWD	—When may the bar (*or,* entrance) be	FXU	—Barrel of oil.
	attempted?		Empty barrel ....LXW
	When will be the best time for crossing		Powder barrel......　.........TRJ
	the bar? ....　.............XM		Tar barrel ....XKQ
	You will have water enough over the		Water barrel ....YVM
	bar (*depth in feet to follow*) V1		
FWE	—You will not have water enough over	FXV	**BARREN-NESS**
	the bar (*or,* into the harbor) (*depth in*		
	feet to follow)	FXW	**BARRICADE-D-ING.　BARRIER**

BARRICOE—BE

FXY	BARRICOE (*Breaker or small cask*)
FXZ	—Veer a barricoe astern
FYA	BARROW.
FYB	BARTER-ED-ING
FYC	BASE
FYD	—Base of operations
FYE	BASIN (*See also* DOCK)
FYG	BASIS
FYH	BASKET
FYI	BATHE-D-ING-R BATH
	BATMAN. (*Measure of Weight*) ..... BAY
FYJ	BATTEN-ED-ING
FYK	—Batten-ed-ing down
	Rolling batten .... UZK
FYL	BATTER-ED ING TO BATTER
FYM	BATTERY
FYN	—Are there any batteries?
	Daniell's batteries ....KLM
	Electric batteries ....LUQ
	Firing batteries ....NAE
	Le Clanché's batteries ....QLZ
FYO	—Masked batteries
FYP	—Mortar batteries
FYQ	—Range of batteries.
FYR	—Test batteries
FYS	—The enemy is throwing up batteries.
FYT	—There are no batteries
FYU	—Under the batteries
FYV	—What state does the battery appear to be in?
	You are within range of the batteries (*or*, of the guns) ....GO
FYW	BATTLE
	After the battle............DRL
FYX	—Battle ship Line of battle ship
FYZ	—Before the battle
FZA	—Indecisive battle between the — and — (*names indicated*)
FZB	—When was the battle fought?
FZC	—Where was the battle fought?
FZD	— has won a great battle at —, with a loss of killed and wounded reported at —
FZE	BAULK ED-ING BALK
FZG	—Baulk of timber
FZH	BAY BAY OF
	Any anchorage in the bay? ....EKQ
FZI	—In the bay of —
	Shall anchor in the bay............EJO
FZJ	BAYONET
FZK	BAZAAR
	BE BEEN BEING ....BFA
	Am, Are, *or*, Is not to be .... BFD

BE—*Continued*
- Am, Are, *or*, Is to be ....BFC
- Be on the alert ....DWV
- Be the ....BFE
- Can, *or* May be ....BFG
- Can, *or*, May do (*or*, be done) .....BLT
- Can, *or*, May have been. ....BFH
- Can, *or*, May have done (*or*, been done) BLV
- Can, *or*, May he (she, it, *or person s or thing s indicated*) be? ....BTK
- Can, *or*, May he (she, it, *or person-s or thing-s indicated*) have been? .....BTL
- Can, *or*, May I be? ....BKH
- Can, *or*, May it be done? ....BLW
- Can, *or*, May there be? ....BFI
- Can not *or*, May not be ....BFJ
- Can not, *or*, May not do(*or*, be done).BLX
- Can not, *or*, May not have been ...BFK
- Can not, *or*, May not have done (*or*, been done) ....BLZ
- Can not *or* May not he (she it *or person s or thing-s indicated*) be?....BTM
- Can not *or* May not he (she it *or person s or thing-s indicated*) have been? BTN
- Could, *or*, Might be .. ..BFL
- Could, *or*, Might do (*or*, be done)..BMA
- Could, *or*, Might have been........BFM
- Could, *or*, Might have done (*or* been done) ....BMC
- Could, *or*, Might he (she, it, *or person s or thing-s indicated*) be? ....BTO
- Could, *or*, Might he (she, it *or person-s or thing-s indicated*) not be?.....BTP
- Could, *or*, Might not be... ....BFN
- Could, *or*, Might not do (*or* be done).BMD
- Could, *or*, Might not have been... BFO
- Could *or*, Might not have done (*or*, been done) ....BMR
- Could *or*, Might there be?............BFP
- Had, Has, *or*, Have been............BFQ
- Had, Has, *or*, Have done (*or*, been done) BMR
- Had, Has, *or*, Have, he (she, it *or person s or thing-s indicated*) been?..BTU
- Had Has *or*, Have, he (she, it *or person s or thing s indicated*) not been? BTX
- Had, *or*, Have I been ....BFR
- Had, Has, *or*, Have not been............BFS
- Had, Has, *or* Have not done (*or*, been done) ....BMS
- Had, *or*, Have they been?...... . BFT
- Had, *or*, Have we been? ....BFU
- Had, *or*, Have you been? ....BFV
- Having been.... ....BFW
- He, She, It (*or person-s or things-s indicated*) can (*or*, may) be............BUC
- He, She, It (*or person-s or things-s indicated*) can not (*or*, may not) be...BUP
- He, She, It (*or person.s or things-s indicated*) had (has, or — been............BUR
- He, She, It (*or person-s or thing s indicated*) had (has, *or*, have) not been.BUO
- He, She, It (*or person-s or things-s indicated*) must be ... .BUZ
- He, She, It (*or person-s or things-s indicated*) must not be............BVC
- He, She, It (*or person-s or thing-s indicated*) ought to be ....BVF

BE

BE—Continued

He, She. It (or *person-s or thing-s indicated*) ought not to be .... ---BVG
He, She, It (or *person-s or thing-s indicated*) shall (*or*, will) be............BVI
He, She, It (or *person-s or thing-s indicated*) shall (*or*, will) not be ....BVK
He, She, It (or *person-s or thing-s indicated*) should (*or*, would) be.....BVM
He, She, It (or *person s or thing-s indicated*) should (*or*, would) do (*or*, be done) .... ---------------------BVN
He, She It (or *person-s or thing-s indicated*) should (*or*, would) not be...BVQ
He, She, It (or *person-s or thing-s indicated*) should (*or*, would) not do (*or*, be done) -------------------- --- BVR
He, She. It (or *person-s or thing-s indicated*) was (*or*, were) to be ......BVW
He, She It (or *person-s or thing-s indicated*) was (*or*, were) not to be...BVX
I can (*or*, may) be .... BFX
I can not (*or*, may not) be...........BFY
I could (*or*, might) be.............BFZ
I could (*or*, might) not be`............BGA
I had (*or*, have) been ....BGC
I had (*or*, have) not been ....BGD
I must be ....BGE
I must not be......... ....BGF
I shall (*or*, will) be.................BGH
I shall (*or*, will) not be............BGI
I should (*or*, would) be......... ....BGJ
I should (*or*, would) have been ......BGK
I should (*or*, would) not be . ... BGL
I should (*or*, would) not have been .BGM
If it can (*or*, may) be done..........BZR
If it can not (*or* may not) be done.. .BZS
Is, etc (*See Am, above*)
Is it not to be done?.................BNI
Is it to be done? .. ..BNJ
It, etc (*See He, above*)
It can be.... .... BGN
It can (*or*, may) be done............ BNK
It can not be ....BGO
It can not (*or*, may not) be done ...BNL
It could (*or* might) be done........BNM
It could (*or*, might) not be done.. BNO
It had (*or*, has) been ....BGP
It had (*or*, has) been done...........BNP
It had (*or*, has) not been...... ....BGQ
It had (*or*, has) not been done......BNQ
It ought to be .... BGR
It ought to be done..................BNR
It ought not to be ....BGS
It ought not to be done ....BNS
It shall (*or*, will) be..............BGT
It shall (*or*, will) be done ....BNT
It shall (*or*, will) not be ....BGU
It shall (*or*, will) not be done.......BNU
Let her be ....BGV
Let him be................ ....BXE
Let them be ....BGW
Let us be ....BGX
May, etc (*See Can, above*.)
Might, etc (*See Could, above*)
Must be ....CEM
Must do (*or*, be done).... ....CEN
Must have been ....CEP
Must not be......... ....CEV
Must not do (*or* be done).........CEW
Need be ....BGY

BE—Continued.

Need not be......................BGZ
Not be (been, being) ....CFN
Not to be ....CGH
Not to be done......................CGI
Not to be had .. ..CGJ
Ought not to be ....CHS
Ought to be...... ....CHY
Ought to do (*or*, be done)........ --CHZ
Shall, *or*, Will be...................BHA
Shall, *or*, Will do (*or*, be done).....BNW
Shall, *or*, Will have done (*or*, been done), BNX
Shall, *or*, Will he (she, it, or *person-s or thing-s indicated*) be?...........BWE
Shall, *or*, Will he, (she it, or *person s or thing-s indicated*) do (*or*, be done)? BWF
Shall, *or*, Will he (she, it, or *person-s or thing-s indicated*) not be? ......BWI
Shall, *or*, Will he (she, it, or *person-s or thing-s indicated*) not do (*or*, be done)? BWJ
Shall, *or*, Will I be?..................BHC
Shall, *or*, Will not be ....BHD
Shall, *or*, Will not do (*or*, be done)..BNY
Shall, *or*, Will not have done (*or*, been done) ....BNZ
Shall, *or*, Will there be?...............BHE
Shall, *or*, Will they be? ....BHF
Shall, *or*, Will we be? ....BFG
Shall, *or*, Will you be? ....BHI
Shall, *or* Will you not be? ....BHJ
Should, *or*, Would be.................BHK
Should, *or*, Would do (*or*, be done)..BOF
Should, *or*, Would have been ......BHL
Should, *or*, Would have done (*or*, been done) ....BOG
Should, *or*, Would he (she, it, *or person-s or thing-s indicated*) be? ....BWN
Should, *or* Would he (she, it, *or person-s or thing-s indicated*) do (*or*, be done)? BWO
Should, *or*, Would he (she, it, *or person-s or thing-s indicated*) have been?..BWQ
Should, *or*, Would he (she, it, *or person-s or thing-s indicated*) not be?....BWS
Should *or* Would he (she, it, *or person-s or things indicated*) not do (*or*, be done)?..................BWT
Should *or*, Would he (she, it, *or person-s or thing-s indicated*) not have been? BWV
Should, *or*, Would not be ....BHM
Should, *or*, Would not do (*or*, be done), BOH
Should, *or*, Would not have been...BHN
Should, *or* Would not have done (*or* been done) ....BOI
Should, *or* Would there be?..........BHO
That, *or*, This can (*or*, may) be.....CKR
That, *or*, This can not (*or*, may not) be, CKT
That, *or*, This could (*or*, might) be.CKV
That, *or*, This could (*or*, might) not be, CKW
That, *or*, This had has, (*or*, have) been, CKY
That, *or*, This had (has *or*, have) not been ....CKZ
That, *or*, This must be .... ---CLJ

	BE—BEAM.

BE—*Continued*
That, or, This must not be_ _ _ _ _ _ _ _CLM
That, or, This shall (or, will) be_ _ _ _CLN
That, or, This shall (or, will) not be_CLO
That, or, This should (or, would) be_CLP
That, or, This should (or, would) not be,
　　　　　　　　　　　　　　　CLQ
There can (or, may) be _ _ _ _ _ _ _ _ _ _CMJ
There can not (or, may not) be_ _ _ _CML
There could (or, might) be _ _ _ _ _ _ _ _CMO
There could (or, might) not be_ _ _ _CMP
There had (has, or, have) been. _ _CMR
There had (has, or, have) not been_CMS
There shall (or, will) be_ _ _ _ _ _ _ _ _ _CMV
There shall (or, will) not be _ _ _ _ _CMX
There should (or, would) be_ _ _ _ _ _CMZ
There should (or, would) have been_CNA
There should (or, would) not be_ _ _CND
There should (or, would) not have been,
　　　　　　　　　　　　　　　CNE
They can (or, may) be_ _ _ _ _ _ _ _ _ _ _BHP
They can not (or, may not) be_ _ _ _ _ BHQ
They could (or, might) be_ _ _ _ _ _ _ _ _BHR
They could (or, might) not be_ _ _ _ _ _BHS
They had (or, have) been_ _ _ _ _ _ _ _ _ _BHT
They had (or, have) not been_ _ _ _ _ _BHU
They must be_ _ _ _ _ _ _ _ _ _ _ _ _ _ _ _ _BHV
They must not be. _ _ _ _ _ _ _ _ _ _ _ _ _BHW
They ought to be _ _ _ _ _ _ _ _ _ _ _ _ _ _BHX
They ought not to be_ _ _ _ _ _ _ _ _ _ _BHY
They shall (or, will) be_ _ _ _ _ _ _ _ _ _BHZ
They shall (or, will) not be_ _ _ _ _ _ _BIA
They should (or, would) be_ _ _ _ _ _ _BIC
They should (or, would) not be_ _ _ _ BID
They were to be_ _ _ _ _ _ _ _ _ _ _ _ _ _BIE
They were not to be_ _ _ _ _ _ _ _ _ _ _BIF
To be _CNP
To be at_ _ _ _ _ _ _ _ _ _ _ _ _ _ _ _ _ _ _CNQ
To be done _ _ _ _ _ _ _ _ _ _ _ _ _ _ _ _ _CNE
To be had _ _ _ _ _ _ _ _ _ _ _ _ _ _ _ _ _ _CNS
Was, or, Were to be _ _ _ _ _ _ _ _ _ _ _BIG
We can (or, may) be_ _ _ _ _ _ _ _ _ _ _BIH
We can not (or, may not) be_ _ _ _ _ _BIJ
We could (or, might) be_ _ _ _ _ _ _ _ _BIK
We could (or, might) not be_ _ _ _ _ _BIM
We had (or, have) been _ _ _ _ _ _ _ _ _BIN
We had (or, have) not been_ _ _ _ _ _ _BIO
We must be _ _ _ _ _ _ _ _ _ _ _ _ _ _ _ _ _BIP
We must not be _ _ _ _ _ _ _ _ _ _ _ _ _ _BIQ
We shall (or, will) be_ _ _ _ _ _ _ _ _ _ _BIR
We shall (or, will) not be _ _ _ _ _ _ _ _BIS
We should (or, would) be _ _ _ _ _ _ _ _BIT
We should (or, would) not be_ _ _ _ _ _BIU
We were to be_ _ _ _ _ _ _ _ _ _ _ _ _ _ _BIV
We were not to be _ _ _ _ _ _ _ _ _ _ _ BIW
When shall (or, will) be?_ _ _ _ _ _ _ _ _CRS
When shall (or, will) it be done?_ _ _CRX
When shall (or, will) you be? _ _ _ _ _ _CSD
Where shall (or, will) be?_ _ _ _ _ _ _ _CTL
Where shall (or, will) you be? _ _ _ _ _CTU
Will, etc (See Shall, above)
Will it be?_ _ _ _ _ _ _ _ _ _ _ _ _ _ _ _ _ _BIX
Would, etc (See Should, above)
You can (or, may) be_ _ _ _ _ _ _ _ _ _ _BIY
You can not (or, may not) be_ _ _ _ _ _BIZ
You could (or, might) be _ _ _ _ _ _ _ _BJA
You could (or, might) not be_ _ _ _ _ _BJC
You had (or, have) been. _ _ _ _ _ _ _ _BJD
You had (or, have) not been_ _ _ _ _ _BJE
You must be _ _ _ _ _ _ _ _ _ _ _ _ _ _ _ _ BJF
You must not be _ _ _ _ _ _ _ _ _ _ _ _ _BJG

BE—*Continued*
You ought to be_ _ _ _ _ _ _ _ _ _ _ _ _ _ BJH
You ought not to be_ _ _ _ _ _ _ _ _ _ _ _BJI
You shall (or, will) be _ _ _ _ _ _ _ _ _ _ _BJK
You shall (or, will) not be_ _ _ _ _ _ _ _BJL
You should (or, would) be_ _ _ _ _ _ _ _BJM
You should (or, would) not be_ _ _ _ _BJN
You were to be_ _ _ _ _ _ _ _ _ _ _ _ _ _ _BJO
You were not to be_ _ _ _ _ _ _ _ _ _ _ _BJP

FZL　　**BEACH-ED-ING**
FZM　　—A rocky beach
FZN　　—A sandy beach
FZO　　—Beach (*indicated*) is good for watering
FZP　　—Beach (*indicated*) is not good for water-
　　　　　ing
FZQ　　—Beach master
FZR　　—Beach party
FZS　　—Beach the vessel　　Run on the beach
FZT　　—Beach the vessel where flag is waved (or,
　　　　　light is shown)
　　　　Beach your vessel at all risks_ _ _ _ _ _ _KI
　　　　Do not hold on to your anchor, but let
　　　　　her beat up on the beach _ _ _ _ _ _ _ _EHY
FZU　　—If you part, endeavour to beach your ves-
　　　　　sel where people are assembled (or,
　　　　　as pointed out by compass signal
　　　　　FROM you)
FZV　　—Is there much surf on the beach?
FZW　　—Keep lights (or, fires) on the beach all
　　　　　night
　　　　Lookout will be kept on the beach all
　　　　　night _ _ _ _ _ _ _ _ _ _ _ _ _ _ _ _ _ _ _KG
FZX　　—Not easy landing on the beach, scarcely
　　　　　prudent to land
FZY　　—On the beach
　　　　Run on the beach　　Beach the vessel_FZS

GAB　　**BEACON** (See also BUOY)
GAC　　—Beacon bears
GAD　　—Beacon buoy
GAE　　—Beacon, or, Buoy has broken adrift (or,
　　　　　is gone)
　　　　Can see the beacon (or, mark)_ _ _ _ _ _BLI
GAF　　—Can you see the beacon (or, mark)?
GAH　　—How does the beacon (buoy, or, mark)
　　　　　bear?
GAI　　—How must I bring the beacon (buoy, or,
　　　　　mark) to bear?
GAJ　　—I have not been able to make out the
　　　　　beacon (or, buoy)
　　　　I think I must have passed the buoy (or,
　　　　　beacon)_ _ _ _ _ _ _ _ _ _ _ _ _ _ _ _ _ _HML
GAK　　—Keep the beacon (or, buoy) on your port
　　　　　hand
GAL　　—Keep the beacon (or, buoy) on your star-
　　　　　board hand
　　　　Leave the beacon (or, buoy) to port_ _LZ
　　　　Leave the beacon (or, buoy) to starboard,
　　　　　　　　　　　　　　　　　　　MA
GAM　　—Steer directly for the beacon (or, buoy)

GAN　　**BEAM**
　　　　Abaft the beam_ _ _ _ _ _ _ _ _ _ _ _ _ _CXI
　　　　Abeam, or, Abreast of_ _ _ _ _ _ _ _ _ _CXV
GAO　　—Beam ends
GAP　　—Before the beam
GAQ　　—Keep on lee beam
GAR　　—Keep on weather beam
GAS　　—Lee beam
GAU　　—On port beam

BEAM—BEAUTY

	BEAM—*Continued*
GAV	—On starboard beam
GAW	—On the beam
	Sway beam ....XGZ
GAX	—Was thrown on her (*or*, my) beam ends
GAY	—Weather beam
GAZ	BEAN
GBA	BEAR-ING TO BEAR (*or*, CARRY) BEARER
	Borne ....HAI
	I am the bearer of important despatches, JR
GBC	BEARING
	NOTE —*Unless otherwise indicated, bearings in points and half points are to be understood as* COMPASS *bearings, i e , bearings as shown by the compass and uncorrected for either deviation or variation But bearings when given in degrees are to be considered as* MAGNETIC *bearings, i e , compass bearings corrected for deviation Both are to be reckoned* FROM *the vessel signaling the bearing*
	Anchor on bearing (*indicated*) ....EHF
GBD	—Are your bearings true or magnetic bearings?
	At — (*time indicated*) light bore — .PV
	At — (*time indicated*) — (*place indicated*) bore — distant — miles......ED
	Beacon bears — ....GAC
GBE	—Bear-ing Bore To bear
GBF	—Bearing and distance of
GBH	—Bearings by compass are —
GBI	—Bears Bears to the —
	Best berth (*or*, anchorage) bears from me — ....BF
GBJ	—Buoy bears
	Calls the attention of the FIRST (*or*, nearest) ship on the bearing (FROM *the person signaling*) pointed out by compass signal ... .DI
	Calls the attention of the SECOND ship, etc ....DJ
	Calls the attention of the THIRD ship, etc............................DK
	Calls the attention of the FOURTH ship, etc..........................DL
	Calls the attention of the FIFTH ship, etc.............................DM
	Calls the attention of the SIXTH ship, etc.............................DN
	Calls the attention of the shore signal station on the bearing (FROM *the person signaling*) pointed out by compass signal.............................DO
	Calls the attention of the vessel COMING into the anchorage on the bearing (FROM *the person signaling*) pointed out by compass signal......... ...DP
	Calls the attention of the vessel LEAVING the anchorage on the bearing, etc..DQ
	Center of hurricane bears —........IGX
	Compass bearing (*i e the bearing shown by the compass and uncorrected for either deviation or variation*) ....JHF
	Cross bearings ....KEA
	Do not bring the light to the — of — (*bearing indicated*) ....PX

	BEARING—*Continued.*
	Entrance bears — ....EG
	How did the land bear when last seen? EH
	How does the beacon (buoy, *or*, mark) bear?.............................GAE
	How does the entrance bear?.........EJ
	How does the harbor (*or*, anchorage) bear? ....PA
	How does the land bear?. ....QDS
	How must I bring the beacon (*or*, buoy) to bear? ....GAI
	I saw the land bearing — (*add time, if necessary*)...........................VNS
	I shall keep the land (*or*, light) bearing — ,or, I shall keep hold of the land (*or*, light) ... .QA
	Indicate bearing of light (lighthouse, or, light-ship)...................QB
	Indicate the bearing of — What is the bearing and distance of —?..........EO
	Keep the light bearing between — and — (*bearings indicated*) ....QC
	Keep the — bearing — ....PZM
	Look out for a boat bearing — .....GTS
	Magnetic bearing (*i e , compass bearing corrected for deviation*)...........JHX
GBK	—On the bearing of —
	Place indicated bears... ....EK
	There is a strange sail bearing —....EM
	True bearing (*i e , compass bearing corrected for deviation and variation*), IIK
	Vessel bearing — appears to be......ESH
GBL	—Vessel indicated bears
	What bearing shall I keep the light (*or*, landmark) on?.................QF
	What is my position by bearings?...RV
	What are the bearing and distance of (*place or vessel indicated*)?.........EO
GBM	—What is the bearing by compass?
GBN	BEARINGS OF AN ENGINE
	Hot bearings ....RI
GBO	—Main bearing brasses broken (*or*, out of order)
GBP	—Shaft bearing
GBQ	—Thrust bearing
GBR	BEAR UP—FOR
	Bear up instantly...................LB
GBS	—Bore up—for.
GBT	—I shall bear up.
GBU	—I shall not bear up
GBV	—May, or, Can I bear up?
GBW	—Shall we bear up?
	Sprung my foremast and must bear up, NKW
GBX	—Vessel indicated was compelled to bear up
GBY	BEAT-EN ING
GBZ	—Beat off a pirate
GCA	—Beating off
GCB	—Beating up
GCD	—Can you beat up against?
GCE	—Can not beat up against.
GCF	BEAUTY-IFUL

BECALMED—BELAY

GCH	BECALMED		BEFORE—*Continued*
GCI	—Becalmed off — for — days		Before Thursday .... XRS
GCJ	—Several vessels becalmed off —		Before to-morrow .... SAH
GCK	—There is a likelihood of being becalmed.		Before Tuesday............ .... YDI
			Before us (our-s) .... CIM
GCL	BECAUSE		Before Wednesday .... ZAG
			Before you-r-s .... CWS
GCM	BECKET		Before you decide .... KPQ
			Before you sail............... VDR
GCN	BECOME-ING BECAME		Breakfast before you start... ... HFJ
GCO	—What became of the proceeds—of —?		Day before................... KMZ
GCP	—What has become of the crew?		In the same state as before _ .... VGK
GCQ	—What has become of wreck?		Intend sounding (*place indicated*) be-
			fore I try it..................... POT
GCR	BED-DING		Just before .... PXK
	Air bedding ....---.....DVX		Keep before the wind .... ZEU
GCS	—Bed clothes		Morning before................ SAB
GCT	—Bed linen	GDJ	—Not before
GCU	—Oyster bed		Running before the wind........ VCH
			The day before yesterday..... ... KNU
GCV	BEE	GDK	—Was it before?
GCW	—Beeswax		
		GDL	BEG-GED-GING TO BEG
GCX	BEEF		I beg to acknowledge ....DGR
GCY	—Beef is in good order	GDM	—I (*or person-s indicated*) beg to be ex-
GCZ	—Can fresh beef be procured?		cused
GDA	—Can you supply me with salt beef?		
	(*Number of casks indicated*)	GDN	BEGIN-NING BEGAN. BEGUN
GDB	—Fresh beef	GDO	—Can begin at once(*or on date specified*)
GDC	—Fresh beef and vegetables	GDP	—Beginning of — the
	Salt beef ....VFS		Beginning of April............. EUA
			Beginning of August........ .... FMB
	BEEN (*See* BE) ... .BFA		Beginning of December .... KPL
			Beginning of February............ MUQ
	BEER ALE ....DWT		Beginning of January ....PUS
			Beginning of July ....PWN
GDE	BEFALL-EN-ING BEFELL		Beginning of June. ....PWV
			Beginning of March ....RKW
GDF	BEFORE		Beginning of May........... .RNF
	Before arrival—of (*or*, at) ....EXZ		Beginning of November ....SKV
	Before breakfast ....HFI		Beginning of October............SOH
	Before Christmas ....IQB		Beginning of September ....VRL
	Before dark.... ... .KLP	GDQ	—Do not begin yet
	Before departure—of —.. .. ..KTS	GDR	—Have, *or*, Has begun.
	Before dinner ....LAS	GDS	—May I begin—to?
	Before Easter ....LSH	GDT	—Not begun yet
	Before Friday ... .NRI		When do you begin (*or*, commence)?
GDH	—Beforehand		JEW
	Before him (his, her-s, it-s) ....BXA	GDU	—When does it begin (*or*, commence)?
	Before it—is ....CDX		
	Before it is dusk....... ....LPZ	GDV	BEHALF. ON BEHALF OF
	Before it is too late .. ..QHC		
	Before leaving ....QKL	GDW	BEHAVE-D-ING-IOR
	Before me (my, mine)............CEB	GDX	—Have, *or* Has behaved very gallantly
	Before midnight.... ....RTI		
	Before Monday ....RXZ	GDY	BEHIND
	Before night................... SIC		Behind the land...... .... QDP
	Before noon...................EDU		Leave behind .. ..QKZ
	Before sailing....................VDQ		We have left men behind (*number to*
	Before Saturday ....VHT		*be shown*) ....QLO
	Before sunrise ....XDS	GDZ	—You have left a person behind
	Before sunset ....XDT		
	Before supper...... ....XER	GEA	BEHOLD-ING-ER BEHELD
	Before that (*or*, this)..............CKO		
GDI	—Before the — In front of		BEING (*See* BE) ....BFA
	Before the battle ....FYZ		
	Before the beam .... .GAP	GEB	BELAY-ED
	Before the end—of ... .LYI	GEC	—Belayed to vessel
	Before the wind ....ZEM	GED	—Belaying pin
	Before them (their-s, they). ......CNK		Is the cable (*or*, rope) belayed (fast)?
	Before this (*or*, that)..............CKO		HRK

BELGIUM—BEST

GEF	BELGIUM BELGIAN BELGIAN COLORS
GEH	BELIEF
GEI	BELIEVE-D-ING-ER
GEJ	—Do, *or*, Does not believe.
GEK	—Do you (*or does person indicated*) believe?
GEL	BELL
GEM	—Bell buoy
GEN	—Bell buoy damaged, not working
GEO	—Diving bell
GEP	—Electric bell
	Submarine signal bell ....ZNR
GEQ	BELLIGERENT
GER	BELONG ED-ING—TO
	Anything belonging to —? ....EQF
GES	—Belong-ing to Government
GET	—Belong-ing to me
GEU	—Belonging to this port
GEV	—Do you belong to the convoy?
GEW	—Does it belong?
	Does not belong to................XUF
	I belong to the convoy of—(*ships indicated*) ....JUS
	I have goods belonging to you (*or*,—).ORX
GEX	—Nothing belonging to you (*or*, —)
	Stranger belongs to — ....WYX
	What company do you belong to?...IQ
	What nation do you (*or vessel indicated*) belong to? . ...SEC
GEY	—Where do you (*or vessel indicated*) belong?
GEZ	—Whom does it belong to?
GFA	—Whom does she (*or vessel indicated*) belong to?
GFB	BELOW BENEATH.
	Watch below -----------------YVA
GFC	BELT-ED
	Armor belt...... .. ..RVR
GFD	—Belted cruiser
	I have no life belts. ....QRB
	Life belt-----------------------QRH
GFE	BENCH
GFH	—King's Bench, *or*, Queen's Bench
GFI	BEND-ING BENT (*To fix*)
	Bend a new sail -----------------VDS
GFJ	—Bend cables
	Bend sails------------------------VDT
	BENEATH .... ---------------------GFB
	BENEFICIAL-LY (*Advantageous*) ..DNO
	BENEFIT, *or*, ADVANTAGE..... ....DNL
GFK	—For the benefit of —
GFL	—For the benefit of the insurers (underwriters)
	BENT Crooked ----------------KDR
	Eccentric rod bent... - ....LSW
GFM	BERG
	Have you fallen in with ice? (*State whether berg or field*).............PR
	Iceberg ----------------------PBY

	BERKOVITZ (*Measure of weight*)...BAC
GFN	BERTH (*See also* ANCHORAGE)
GFO	—Am I in a good berth?
	Bad berth... ----------- ....FQV
	Best berth (*or*, anchorage) bears from me — .... EF
	Best berth (*or*, anchorage) is in — fathoms........ ....OV
GFP	—Clear berth
GFQ	—Foul berth.
GFR	—Good berth
GFS	—I shall shift my berth
GFT	—I would shift my berth as soon as possible
GFU	—Inside berth
GFV	—Is it worth while shifting berth?
GFW	—Is my berth a good one for getting away from?
GFX	—Old berth
GFY	—Outside berth
GFZ	—Shall I in my present berth have room for weighing if the wind shifts?
GHA	—Shift your berth farther in
GHB	—Shift your berth farther out
GHC	—Shift your berth farther to the —
	Shift your berth, your berth is not safe . -----------------------KT
GHD	—Sick berth
GHE	—The berth you are now in is not safe
GHF	—Wide berth
GHI	—Will you lead into (*or*, point out) a good berth?
GHJ	—You are in a very fair berth.
GHK	—You are not in a good berth (are in a foul berth)
GHL	—You have given me a foul berth
GHM	BERTHON BOAT, *or*, COLLAPSIBLE BOAT
GHN	BESEECH-ING BESOUGHT
	BESIDE ALONGSIDE...............EAS
GHO	BESIDES IN ADDITION TO
GHP	BESIEGE-ING-R
GHQ	BESPEAK-ING BESPOKE
GHR	BEST.
	Best berth (*or*, anchorage) bears from me — ....EF
	Best berth (*or*, anchorage) is in — fathoms ----------- -----------OV
	Best bower ----------------------EHP
GHS	—Best chance
GHT	—Best watering place is — ·
	Do the best you can for yourselves, no assistance can be given (rendered)..CX
	Hoist your flags where best seen..NDE
GHU	—It will be best
	Light, *or*, Fires will be kept at the best place for coming on shore........KE
GHV	—My best respects
	The best agents are —....-------DSW
GHW	—The best place
GHX	—The best plan will be ·
GHY	—The best time
GHZ	—The best way.
	The course steered is the best for the present (*or until time indicated*)..JZM

	BEST—BIRD		

GIA	BEST—*Continued.*		BICARB OF SODA .... WIZ
GIB	—This is my best point.		BICKFORD'S FUZE .... NUV
	—This is not my best point.		
	Try your best ....YCZ	GJK	BICYCLE.
	What is your best point of sailing?..VEX		
	When will be the best time for crossing	GJL	BID-DING-DER BADE
	the bar? . _ . . ..XM		
GIC	—Where is the best place to get —?	GJM	BIG-GER-GEST-NESS
GID	—Where is the best place to land?	GJN	—Too big (*or*, large).
GIE	—Which do you think the best?		
GIF	—Which, *or*, What is the best?	GJO	BIGHT (*of a rope*)
	Who are the best agents (*or*, people) to		
	do business with?.................DSZ		BIGHT (*a cove*).............. .... KAF
GIH	BET-TED-TING		
		GJP	BILGE-D-ING
GIJ	BETRAY-ED-ING-ER-AL	GJQ	—Are you (*or vessel indicated*) bilged?
		GJR	—Bilge keel
		GJS	—Bilge pump
GIK	BETTER-ED-ING TO BETTER	GJT	—Bilge pump (*donkey*)
GIL	—Better	GJU	—Bilge pump (*main*)
	Better bargain beforehand........FWV	GJV	—Bilge pump out of order
GIM	—Better not	GJW	—Bilge pump valves out of order
	Better not ask the reason..........FDN	GJX	—Bilge water
GIN	—Better not send.	GJY	—I am, *or*, She is bilged
GIO	—Better separate	GJZ	—I am, *or*, She is not bilged
	Freights looking better...........NQF		
	Had we not better run in and make the	GKA	BILL
	land? ... .PZ	GKB	—Bill of — from —
GIP	—Had you not better?	GKC	—Bill of health.
GIQ	—Is there any better?		Clean bill of health—from —.......TQ
GIR	—Little better		Have you a clean bill of health?....TU
GIS	—Much better		I have a clean bill of health, but am lia-
GIT	—The sooner the better		ble to quarantine.. Code Flag over Q
	We had better anchor ....EKC		I have not a clean bill of health
	We shall do better on the other tack..XIN		Code Flag over I
	You had better get a sick nurse ..WBK		Not clean bill of health flag (*when spoken*
	You had better put your affairs in the		*of*)............. .... ISW
	hands of —.................. DPK	GKD	BILL-BOOK
	You had better take a pilot at once.TJX		
GIU	BETWEEN	GKE	BILL OF EXCHANGE
GIV	—Between decks	GKF	—Bill at — days
GIW	—Between the two	GKH	—Bill at sight
GIX	—Between whiles.	GKI	—Can I get a bill cashed here?
GIY	—Between wind and water	GKJ	—Can not get a bill cashed here
	Hostilities commenced between —	GKL	—I will cash (*or*, take) your bill
	and —........ ....OZC	GKM	—Will you cash a bill on — for me?
GIZ	—Just between	GKN	—Will you take a bill on —?
	War between — and — ....YW	GKO	—Will you take my bill?
GJA	BEVEL-LED-LING	GKP	BILL OF LADING
	Bevelled to an angle of —..... ...ELU	GKQ	—Are all your bills of lading complete
			and signed?
GJB	BEWARE—OF	GKR	—Bills of lading not signed
GJC	—Beware of boom (*or*, obstruction).	GKS	—False bill of lading.
GJD	—Beware of brigands	GKT	—Have you bill of lading for —?
	Beware of derelict, dangerous to navi-	GKU	—I have no bill of lading for —
	gation .. ..JB	GKV	—I will not give up the bill of lading
	Beware of torpedo boats. ....XO	GKW	—I will not sign the bill of lading unless—
	Beware of torpedoes, channel is		
	mined. ....XP	GKX	BILLET-ED-ING
GJE	BEYOND.	GKY	BILLY BOAT
GJF	—Just beyond		
	Nothing to be depended on beyond	GKZ	BIND-ING (*See also* BOUND.)
	your own resources ....KUM	GLA	BINNACLE
GJH	BIAS-ED-ING.	GLB	BINOCULAR FIELD GLASS,
GJI	BIBLE	GLC	BIRD

	BIRTH—BOARD		
GLD	BIRTH BORN		BLOCK—Continued
GLE	—Birthday	GMX	—Purchase block
GLF	—Certificate of birth	GMY	—Snatch block.
GLH	—Have you had any births?	GMZ	—Thrust block broken.
GLI	—Registrar of births, deaths, and mar-	GNA	—Thrust block damaged
	riages	GNB	—Thrust block rings defective.
GLJ	BISCUIT	GNC	BLOCKADE-D
	Biscuit bag Bread bag _____FSE	GND	—Blockade is not taken off
GLK	—Biscuit is not good Bread is not good.	GNE	—Blockade is taken off
GLM	—Can you spare any biscuit?	GNF	—Blockade runner
GLN	—In want of biscuit?	GNH	—Blockading
		GNI	—Blockading squadron
	BISHOP, or, ARCHBISHOP_____EUJ	GNJ	—Entrance of — blockaded.
		GNK	—Entrance of — is not blockaded
GLO	BITE-ING-TEN BIT	GNL	—Has the blockade been raised (or, taken
	Frost bitten _____NSC		off)?
		GNM	—I have (or vessel indicated has) run the
GLP	BITT-ED-ING		blockade
	Damaged the bitts_____KJD	GNO	—Is blockaded
	Damaged the windlass_____KJE	GNP	—Under blockade
		GNQ	—When does the blockade commence?
GLQ	BITTER-LY-NESS	GNR	—You will be stopped by the blockading
			ships
GLR	BLACK-ED-ING. TO BLACK		
GLS	—Black	GNS	BLOOD-Y. BLOOD VESSEL.
GLT	—Black buoy		
	Black draught _____LNB	GNT	BLOW-N-ING BLEW
GLU	—Black paint	GNU	—Blow out
	Black wash _____YUQ	GNV	—Blow out — boilers
		GNW	—Blow-off cocks (or, gear)
GLV	BLACKSMITH	GNX	—Blow-off cock out of order
		GNY	—Blow-off pipe burst.
GLW	BLADE	GNZ	—Blow-off valves.
GLX	—How many blades have you to your	GOA	—Blow off your steam
	screw?		Blow siren (or, steam whistle) at in-
GLY	—Lost all propeller blades		tervals _____WD
GLZ	—Lost number of blades (indicated) of	GOB	—Blow through pipe
	screw propeller.		Blowing hard Squally weather____ZH
GMA	—One blade of propeller broken	GOC	—Blowing too hard
GMB	—Three blades of propeller broken	GOE	—I think it will blow
GMC	—Two blades of propeller broken	GOF	—If it blows
		GOH	—If it continues to blow
GMD	BLAME-ABLE	GOI	—It blew too hard
GME	—Blamed for		My sails are all blown away (or, split),
GMF	—Is, or, Are to blame		VEC
GMH	—Much to blame	GOJ	—Shall I blow off my steam?
GMI	—Not much to blame	GOK	—Should it come on to blow—from (quar-
			ter to be indicated, if necessary)
GMJ	BLANK		
GMK	—Blank cartridge	GOL	BLUBBER
GML	—Point blank.		
		GOM	BLUE
GMN	BLANKET	GON	—Blue light
		GOP	—Blue serge
GMO	BLAST	GOQ	—Burn a blue light, or, Flash a little
GMP	—Blasting powder		powder.
			Did boat take a blue light (lantern, or,
	BLAZE-ING-D (Flame)_____NDU		any means of making a signal)?__GSP
		GOR	—Have you any blue lights?
GMQ	BLEED-ING BLED		
		GOS	BLUE PETER (When spoken of)
GMR	BLIND-ED-ING BLINDNESS		Hoist the blue peter_____KO
GMS	—Color blind-ness		Blue peter (the flag)_____P
GMT	BLOCK-ED ING TO BLOCK	GOT	BLUFF
	Canal is blocked _____HUJ		
GMU	—Place (indicated) is blocked up by ice	GOU	BLUNDER-ED-ING
GMV	BLOCK (A block for purchase)	GOV	BOARD (A body of officials)
GMW	—Plomer block.	GOW	—Board of Admiralty

BOARD

	BOARD—*Continued*		BOARD—*Continued*
GOX	—Board or Department of Agriculture	GQD	—Go on board
GOY	—Board or Collector of Customs	GQE	—Got on board
GOZ	—Board of Directors		Has she any people on board? ......TFO
GPA	—Board of Health		Have infectious cases on board.....PJH
GPB	—Board of Survey.		Have military stores on board .....RTV
GPC	—Board of Trade	GQF	—Have, *or*, Has on board
GPD	—Board of Trade Officer (*or*, Inspector)		Have shipwrecked crew on board, will
GPE	—Board of Underwriters		you let me transfer them to you?
GPF	—Board of Works		(*Number to follow*) ....KDA
GPH	—Harbor Board		Have you a Custom House official on
GPI	—Local Board		board? ....KGX
GPJ	—Local Government Board		Have you a pilot on board? ....TJA
GPK	—Local Medical Board.		Have you any combustibles on board?
GPL	—Marine Board		HO
GPM	—Medical Board	GQH	—Have you any women on board?
GPN	—Trinity House (*or*, Light-House Board)	GQI	—Have you boarded (*or*, visited)?
			Have you had any sickness on board?
GPO	BOARD (*Timber*) (*See also* BOARD below)		WBA
	Freeboard.............. ....NPW	GQJ	—Have you on board?
	Gangboard....................NWQ		He has not many hands on board? ..OMD
	BOARD-ED ING TO BOARD (*See also*		I am coming (*or*, sending) on board,
	Board below)..................GPW		CZB
			I have had cases of yellow fever on
	BOARD (ON) ABOARD ....CYX		board ....MWK
GPQ	—A contagious disease on board (*or at place indicated*)		I have had no cases of yellow fever on board ....MWL
GPR	—A case of fever has broken out on board (*or on ship indicated*)		I have illness on board ....PCU
			I have no combustibles on board....HQ
	Above board ....CYZ	GQK	—I have no women on board
	Admiral, *or*, Senior officer is on board (*or*, at —) ....DEV	GQL	—I have wrong papers on board .....TAK
		GQM	—I prefer stopping on board
	All on board ....CZA	GQN	—I (*or vessel indicated*) was boarded
	All on board perished All hands lost, DXQ		—I will board
	All on board saved. All hands saved, DXR		I will come on board (*or*, alongside) if you will allow me ....CZD
		GQO	—Is— (*person indicated*) on board?
	Allow no communication Allow no person on board ....FM		Is the captain on board? ....HXE
GPS	—Am I, *or*, Are we to board?		Is, *or*, Was there anyone on board?..EQC
	Any cases of fever on board (*or*, at —)? EON		It is dangerous to allow too many people on board at once ....GB
	Any letters (*or*, despatches) on board? KWX		Leave on board ....QLC
		GQP	—Log board
	Any person (anyone) on board? ...EPM		Make sail, *or*, Go ahead and drop a boat on board....................DVN
	Any sick on board? ....EPO		No boat on board ....GTX
GPT	—Are all your crew on board?		No intelligence obtained from vessel boarded ....POG
GPU	—Are all your passengers on board?	GQR	—Passengers all on board
	Attend on board ....FKX		Passengers not all on board........DYB
GPV	—Board (*or* visit) steamer (*or*, vessel)	GQS	—Put on board by —
GPW	—Board ed-ing To board	GQT	—Repair on board
GPX	—Boarder	GQU	—Run on board by —
	Boarding officer... ....RPI	GQV	—Send a rope on board
GPY	—Boarding pike		Send a responsible person on board..CZE
	Boat is on board (*or* alongside)......EV		Send carpenters on board—(*ship indicated*) ....IBY
GPZ	—Bring on board	GQW	—Send on board—to (*or*, for)
	Captain is not on board ....HWK	GQX	—Several (*or number indicated*) sick on board
	Captain is on board..............HWL	GQY	—Shipped on board
	Captain is requested to come on board, HWN	GQZ	—Stay on board
GQA	—Come on board		The consul is desired to come (*or*, send) on board ....JRP
GQB	—Crew are all on board		There is a spare propeller on board..TZD
	Crew not all on board ....DYA		To whom shall I telegraph that all are well on board? ....XMH
	Crew refuse to go on board........KCU		Troops on board. ....YBP
	Dangerous to allow too many people on board at a time ....GB	GRA	—Was not boarded
	Females on board ....MVQ	GRB	—Was on board Went on board
GQC	—Fire has broken out on board—of—		Went on board Was on board....GRB
	Freeboard...... ....NPW		
	Gangboard...................NWQ		

	BOARD—BOAT		
	BOARD—*Continued*	GSI	—Can you send a boat?
GRC	—Were you boarded?	GSJ	—Can not get my boat out
	What is the number of mail bags you		Can not send a boat ------- -----EX
	have on board? --------------FSP	GSK	—Cargo boat
GRD	—What vessels have you boarded?	GSL	—Coal boat
GRE	—When will you board?		Collapsible boat, or, Berthon boat__GHM
	Will board (or, visit) --- ----------YRD		Custom house boat Revenue boat..KGU
	Will Lloyd's agent come on board?_QXX	GSM	—Deal boat
GRF	—Will you (or person indicated) come on	GSN	—Despatch boat
	board?	GSO	—Did boat take?
	Will your doctor (or, surgeon) come on	GSP	—Did boat take blue light (lantern, or, any
	board?.__ ----------- ----------WP		means of making a signal)?
	Your friend is on board............NRS	GSQ	—Did boat take water?
GRH	**BOAT**	GSR	—Divisional boat.
	Accident to boat. ----------------DCJ		Do not attempt to land in your own
	All boats to return to the ship		boats -----------------------EY
	Code Flag over w	GST	—Do not send boats ashore after dark
	Am a mail boat ------ ------------REX		Do you want a lifeboat?. ----------EZ
	Arm boat ----------- ----------EUS	GSU	—Drop boat
	Bank is encumbered by fishing boats,	GSV	—Endeavor to pick up boat.
	EP		Endeavor to send a line by boat (cask,
	Bar is impassable for boats on the ebb		kite, or, raft, etc) --------------KA
	tide ------ ----------------------EQ		Enemy's boat._____LZF
	Berthon boat --------------------GHM		Enemy's gunboat ---------------LZI
	Beware of torpedo boats.............XO		Enemy's torpedo boat . ----------LZM
	Billy boat ----------------------GKY		Enemy's torpedo boats have been seen
	Boat is coming to your assistance ---ET		to the —, steering to the ————OH
GRI	—Boat bottom up (bearing if necessary)	GSW	—Every boat
GRJ	—Boat can not be used		Ferry boat.--------------------MVY
	Boat, or, Lifeboat can not come..... ER	GSX	—First-class torpedo boat
GRK	—Boat can not pull ahead		Fishing boat (craft, or, vessel).....KAZ
GRL	—Boat come-ing (came)	GSY	—Flat bottomed boat
GRM	—Boat hook		Government steamboat ----------OFX
	Boat in distress (bearing, if necessary)	GSZ	—Great risk in sending a boat
	ES	GTA	—Guard boat
GRN	—Boat is adrift	GTB	—Gunboat
	Boat is alongside (or, on board)......EV		Have lost all boats, can you take people
	Boat is capsized (bearing, if necessary) AP		off?----------------------------NI
	Boat is going to you ---------------EU		Have no lifeboat No lifeboat here..FA
	Boat is lost ------------------ --AQ	GTC	—Have only one boat (indicate other num-
	Boat is on board (or, alongside) -----EV		ber, if necessary)
GRO	—Boat is on shore	GTD	—Have you any lifeboats?
GRP	—Boat is safe		Have you lost all your boats? -------BL
	Boat is stove ------------------AR		Have you seen a pilot boat (or, vessel)?
GRQ	—Boat is sunk		TJ
	Boat is swamped..................AS		Have you seen any torpedo boats?....XR
GRS	—Boat oars.		Have you seen (or, heard) anything of
GRT	—Boat race		my (or, the) boat?--------------FB
GRU	—Boat rope	GTE	—Heave-to, I will send a boat.
	Boat should endeavor to land where flag	GTF	—His boat
	is waved (or, light shown)..........JZ	GTH	—Hoist up your boat.
GRV	—Boat's compass	GTI	—Horse boat
GRW	—Boat's crew	GTJ	—How many boats?
GRX	—Boat's crew will not come	GTK	—How many torpedo boats passed in sight?
GRY	—Boat's davit.		I am sinking (or, on fire), send all avail-
GRZ	—Boat's gear		able boats to save passengers and crew
GSA	—Boat's gun		NO
	Boats' recall All boats to return to the		I have lost all my boats. ----------BO
	ship..................Code Flag over w	GTL	—I have the boats ready
GSB	—Boat's sail.		I shall fire into the boats if they persist
GSC	—Boat's slings		in coming alongside ----- ----NAK
GSD	—Boats are furnished with — (articles		I should like to know the nature of the
	indicated)		sickness, if any, before I send my boat
GSE	—Boats to assemble		(or, communicate) ----------------FC
	Bridge of boats ....HGO	GTM	—I will send a boat
	Bring-to, I will send a boat.........HIX	GTN	—In the boat
GSF	—Bumboat		Is there any danger of mines or torpedo
	Can boats land? Can landing be effected?		boats?---------- ----------------XS
	EW	GTO	—Jolly boat.
GSH	—Can send boat	GTP	—Keep the boat

	BOAT—BOILER		

	BOAT—*Continued*		BOAT—*Continued*
GTQ	—Lend me a boat		The bank is encumbered by fishing
	Let no boat come alongside ....EBH		boats ....EP
	Let your boats keep to windward until	GVA	—They are all in the boat
	they are picked up.................FD	GVB	They (*boats*) are making for the shore.RIV
	Let your boats pull to — (*bearing indi-*		—Top-up boat.
	cated) ....QOA		Torpedo boat ....XWP
	Lifeboat ....QRC		Torpedo gunboat (*or*, vessel) ......XWU
	Life-Saving Service.................QBI	GVC	—Troop boat
	Lifeboat is going to you ....FE		Tugboat ....YDR
	Lifeboat sent for ....QRJ	GVD	—Use-d-ing steamboats—for
	Lifeboat unable to come...........ER	GVE	—Veer a boat astern
		GVF	—Want a boat
GTR	—Long boat		Want a boat, man overboard ....BT
GTS	—Look out for a boat bearing —		Want a boat immediately (*if more than*
GTU	—Make a signal when you want a boat		*one, number to follow*) ....YG
	Make sail, *or*, Go ahead, and drop a boat		We have sent for lifeboat...........FL
	on board............................DVN	GVH	—Whaleboat Whaler
GTV	—Man and arm boats		Will not be accountable for boat
	No boat available................ ......FG		hire ....DFW
GTW	—No boat could live	GVI	—Will send boat
	No boat fit for this work ... . ZHV		Will you go on shore in my boat?..VZN
GTX	—No boat on board	GVJ	—You can not have a steamboat.
	—(*number*) torpedo boats (*give nation-*	GVK	—Your boat
	ality) passed in sight......... .XYA		
GTY	—Pick up boat	GVL	BOATMAN
GTZ	—Picket (*or*, piquet) boat		
GUA	—Pilot boat (*or*, vessel)	GVM	BOATSWAIN
GUB	—Pilot boat (*steam*).	GVN	—Boatswain's mate
	Pilot boat's flag (*when spoken of*)..NDK	GVO	—Boatswain's stores
	Pilot boat is advancing towards you.DNI		
	Pilot boat is most likely in direction in-	GVP	BOBSTAY.
	dicated (*or*, off —)................TK		
GUC	—Prepare boat—to (*or*. for) —	GVQ	BODY-ILY
	Put a compass in the boat...........JID		Anybody Any person ....EPQ
GUD	—Quarter boat		Anybody on board ... .EPM
GUE	—Saw boat (*indicate place and time, if*	GVR	—Body post, *or*, Propeller post
	necessary)		Everybody *or*, Every one...........MIU
	Saw torpedo boat (*number, if necessary*)		Nobody, *or*, No one ....SIU
	at — (*or*, near —)................XT		Somebody, *or*, Some one...........WJP
GUF	—Seen nothing of your boat		
	Send a boat (*number, if necessary*)..FH	GVS	BOIL-ED-ING
	Send life boat to save crew...........FJ		
GUH	—Send me a boat from the shore (a shore	GVT	BOILER.
	boat)		Blow out boiler ....GNV
GUI	—Send a boat within hail	GVU	—Boiler burst
GUJ	—Send a water boat		Boiler burst, — men killed..........AT
	Sending a water boat................YVJ		Boiler burst, — men killed, — others
GUK	—Send an armed boat		wounded............................AU
GUL	—Send boat at —		Boiler burst, no one hurt...........AV
	Send boat suitable for landing passen-		Boiler burst, no one seriously hurt.AW
	gers ....FI		Boiler burst, number of killed and
GUM	—Send boat for telegram		wounded not yet known ....AX
GUN	—Send boat to (*or, for*)		Boiler burst, several men killed ... AY
	Send boat to assist..................FGN		Boiler burst, several men wounded..AZ
GUO	—Send boat to point out reef (rock, *or*,	GVW	Boiler can not be repaired ....RE
	shoal)	GVX	—Boiler damaged
GUP	—Send boat to take off the crew	GVY	—Boiler has been repaired
GUQ	—Send boat to tow	GVZ	—Boiler has not been repaired
GUR	—Send boat with hawser	GWA	—Boiler leaking
	Ship indicated wants a boat..........FK		Boiler leaking seriously.............RF
GUS	—Shore boat	GWB	—Boiler leaking slightly
GUT	—Small-er boat	GWC	—Boiler leaky, must be blown off
GUV	—Steam boat (*large*), launch	GWD	—Boiler loose (*or*, adrift)
GUW	—Steam boat (*small*), cutter, *or* gig	GWE	—Boiler maker
	Steam boat is aground.............DUW	GWF	—Boiler must be repaired
GUX	—Steam guard boat	GWH	—Boiler plate
GUY	—Steam lifeboat	GWI	—Boiler plates damaged
	Steam pilot boat ....GUB	GWJ	—Boiler primes
GUZ	—Steam pinnace	GWK	—Boiler requires cleaning
	Submarine boat...................XBI		

BOILER—BORN

	BOILER—*Continued.*		BOND—*Continued*
GWL	—Boiler requires repairs.	GYO	—In bond
GWM	—Boiler room	GYP	—Out of bond
GWN	—Boiler tube	GYQ	—Under bond
GWO	—Boiler tube ferrules.		
GWP	—Boiler tubes'burst.	GYR	BONE-Y.
	Boiler tubes leaking-y ....RG	GYS	—Any bones broken?
GWQ	—Boiler tubes salted	GYT	—Bones broken
GWR	—Boiler working on seating		
GWS	—Clean the boiler.	GYU	BONFIRE
GWT	—Condensing boiler burst		
GWU	—Condensing boiler damaged.	GYV	BOOK-ED-ING To BOOK.
	Condensing boiler tubes burst .....YDE	GYW	—Book (*a book*)
GWV	—Donkey boiler.	GYX	—Book of Sailing Directions (*indicate place, if necessary*)
GWX	—Donkey boiler burst		Have you a Book of Navigation Tables?..... ....RN
GWY	—Donkey boiler damaged		Have you a Book of Sailing Directions?............ ....RO
GWZ	—Flue boiler	GYZ	—Log book.
	Furnace crown of boiler collapsed..NUC	GZA	—Meteorological log book
GXA	—Had only — boiler-s serviceable	GZB	—Muster book(*or*, roll) Roster.
GXB	—Have only one (*or number shown*) boiler-s.	GZC	—New book
GXC	—High-pressure boiler		Not on Lloyd's Registry Book.....QXF
	Laid up to have a new boiler put in..QIU	GZD	—Official log book
GXD	—Locomotive boiler	GZE	—Order book
GXE	—Main boiler	GZF	—Signal book
GXF	—Multitubular boiler	GZH	—Steam log book
GXH	—Must have a new boiler		What is the date of your signal book? VL
GXI	—New boiler has been put in.		
GXJ	—Port boiler.	GZI	BOOK-SELLER.
GXK	—Port boiler defective		
GXL	—Starboard boiler	GZJ	BOOM
GXM	—Starboard boiler defective		Beware of boom(*or*, obstruction)...GJO
GXN	—Tubular boiler	GZK	—Boom broken
		GZL	—Boom torpedo.
GXO	BOISTEROUS-LY.	GZM	—Boom, *or*, Obstruction broke adrift.
	Stormy (boisterous) weather from the — ....ZI	GZN	—Both booms broken.
		GZO	—Both booms with nets lost.
GXP	BOLD-LY-NESS-ER EST	GZP	—Flying jib boom.
	Is the coast bold?....................1YC	GZQ	—Jib boom
		GZR	—Jib boom sprung
	BOLIVAR (*coin*).. ....ASZ	GZS	—Lower boom
			Main boom ....RGH
GXQ	BOLIVIA-N BOLIVIAN COLORS	GZT	—One boom broken
		GZU	—One boom with net lost
GXR	BOLLARD TOWING BOLLARD	GZV	—Spanker, *or*, Driver boom
		GZW	—Spanker boom sprung
GXS	BOLT-ED ING To BOLT	GZX	—Studding-sail boom.
GXT	—Bolt (*a bolt*)		There is a boom across ....DHY
GXU	—Bolt rope.	GZY	—Torpedo boom, *or*, Spars for nets
GXV	—Connecting-rod bolts broken.		
GXW	—Coupling bolts broken	HAB	BOOT, *or*, SHOE.
GXY	—Engine bolts	HAC	—Sea boots
GXZ	—Eyebolt		
GYA	—Holding-down bolts loose	HAD	BORAX
GYB	—Piston-rod bolts broken		
GYC	—Ringbolt.		BORDER (*Boundary*) ....HBZ
GYD	—Screw bolt		
GYE	—Shackle bolt.	HAE	BORE-D-ING (*To pierce*)
GYF	—Shaft coupling bolts broken		
		HAF	BORE (*Tidal wave*)
GYH	BOMB BOMB SHELL		
GYI	—Bomb vessel	HAG	BORE (*To bear*)
			At — (*hour indicated*), — (*place indicated*) bore —, distant — miles....ED
GYJ	BOMBARD-ED-ING-MENT		At — (*time indicated*) light bore —..PV
GYK	— — has been bombarded.		Bear up for ....GBR
			Bore up for............. ....GBS
GYL	BONA FIDE		
			BORN (See BIRTH) ....GLD
GYM	BOND-ED-ING		
GYN	—Have you given bond?		

BORNE—BOX

HAI	BORNE	HBQ	BOUND-TO (or, TOWARDS). Homeward bound................OXL
HAJ HAK	BORROW ED-ING —Can I borrow ?	HBR	—I am (or, vessel indicated is) bound to — (place indicated)
HAL	BORROWER	HBS	—I am (or, vessel indicated is) from —, and bound to — (places indicated). I have reported you to the vessel I spoke, she is bound to — ....URT
HAM	BORSENHALLE Report me to the Borsenhalle at Ham- burg ....UL	HBT	Outward bound................SWY —Strange sail (or, vessel indicated) is bound to —
HAN HAO	BOSS —Boss of screw	HBU	Where are you bound ? ....SH —Where was she bound ?
HAP	BOTH Both alike ....DXB Both Houses of Congress or Parlia- ment.....................OZP	HBV HBW HBX	BOUND TO (See also OBLIGE and COMPEL) —Bound to be given up —Bound to give it up Will be bound (compelled) to — .....JIW
HAQ HAR	—Both sides Shorten in both hawsers.............OPX Veer both hawsers ....OPZ	HBY	BOUND (Bind.) Weather bound — at ....YZF Wind bound — at — ....ZFQ
HAS HAT HAU	BOTTLE —Empty bottles. —Water bottle	HBZ	BOUNDARY
HAV	BOTTOM Boat bottom up (bearing if necessary), GRI	HCA	BOUNTIFUL-LY NESS
		HCB	BOUNTY
HAW HAX	—Bottom (ship's) foul —Coarse bottom Coarse sand and stones Directly you get bottom, let go your anchor..................... EHV	HCD	BOW (of a ship) Above the bow...............DAN
		HCE HCF HCG	—Bow compartment —Bow injured —Bow light (or lantern, or, lamp)
HAY	—Double bottom Flat-bottom boat ....GSY		Breakers, Reefs, Rocks, or, Shoal on port bow ....HFD
HAZ	—Foul ground Rocky bottom Good bottom for anchorage .. ..EKX		Breakers, Reefs, Rocks, or, Shoal on star board bow ....HFE
HBA HBC	—Got bottom with — feet —Hard bottom		Cross the bow ....KEG
	I can not make out the bottom flag.NDF	HCI	Is your bow much damaged?......KJP —Keep her bows on
HBD HBE	—Muddy bottom —No bottom with — feet Require diver to examine bottom...LIE	HCJ HCK HCL	—Look out on lee bow —Look out on weather bow. —Name on the bow
HBF	Rocky bottom Foul Ground .....HAZ —Sandy bottom Sheathing for bottom (number of sheets	HCM HCN HCO	—No name on the bow —On lee bow Lee bow —On port bow
HBG	to follow) ....VUG —Ship's bottom must be looked at.	HCP HCQ	—On starboard bow —On weather bow
HBI HBJ HBK	—Soft bottom —Weedy bottom —What is the nature of the bottom ? or	HCR HCS	—Port bow —Port-bow light Ram bow ....UHM
	What kind of bottom have (or, had) you ?	HCT HCU	—Starboard bow —Starboard-bow light
HBL	BOTTOMRY	HCV	BOWER Best bower................EHP Bower anchor (weight in cwts , if nec
	BOUGHDAIS (Measure of weight) ....BAJ	HCW	essary)............... ......EHQ —Bower cable
HBM	BOUGHT BUY-ING (See BUY) Can I buy ? ....HOV		Small bower................EJS
	I can buy ....HOW	HCX HCY HCZ	BOWSPRIT —Bowsprit cap —Bowsprit sprung
HBN	—I have bought I intend to buy ....HOX I wish to buy ....HOY	HDA	Keep a light on your bowsprit end..QY —With loss of bowsprit
HBO	—Was not to be bought Will you (or person indicated) buy?..UCF	HDB HDC HDE	BOX —Despatch box. —Junction box
HBP	BOULEVARD, or, TERRACE		

BOX—BRIBE

	Box—*Continued*		**Break**—*Continued*
HDF	—Paddle box	HEW	—Break in cable
HDG	—Stuffing box		Break-ing Broke the quarantine laws
HDI	—Stuffing box broken		UEB
HDJ	—Stuffing-box glands out of order.	HEX	—Breaking up
			Breakwater................HFM
HDK	Boxhaul-ed-ing	HEY	—Can one break through the ice ?
HDL	Boy		Enemy is breaking up his encampment,
			LZC
HDM	Brace d-ing To brace		Ice breaker...............PBV
HDN	—Brace round		The ice is so solid, I can not break
HDO	—Brace round your head yards.		through, send help.............PCF
HDP	—Brace up	HEZ	—When did sickness break out ?
HDQ	Bracket		Breakage (*Fracture*)............NPJ
HDR	,Brackish		Breaker Barricoe (*a small cask*).FXY
			Veer a breaker astern.............FXZ
HDS	Brail-ed-ing—up	HFA	Breakers
HDT	—Brail up the spanker		Among the breakers.............EGF
		HFB	—Breakers ahead Reef, Rock, *or*, Shoal
HDU	Brake		ahead of you
HDV	—Pump brakes	HFC	—Breakers on the —
		HFD	—Breakers, Reef, Rock, *or*, Shoal on your
HDW	Branch-ed-ing		port bow.
		HFE	—Breakers Reef, Rock, *or*, Shoal on your
HDX	Brand ed ing		starboard bow
HDY	—What brand (*or*, trade-mark)?		Ice breakers.............PBV
HDZ	Brandy	HFG	Breakfast To breakfast
			After breakfast ....DQP
HEA	Brass-y	HFI	—Before breakfast
HEB	—Connecting-rod brasses broken (*or*, out	HFJ	—Breakfast before you start
	of order)	HFK	—Have you breakfasted ?
HEC	—Crank brasses broken (*or* out of order)	HFL	—Will you breakfast with me ?
	Crank-pin brasses broken ....KBN		
	Main-bearing brasses broken (*or*, out of	HFM	Breakwater
	order) ....GBO		
HED	—Piston-rod brasses broken (*or*, out of	HFN	Breast
	order)		Abreast, *or* Abeam of CXV
		HFO	—Breast high
HEF	Brave-ly-ry-r-st	HFP	—Breast hook
		HFQ	—Breast rope
HEG	Brazil-ian Brazilian Colors	HFR	Breath-f-ed-ing.
HEI	Breach-ed-ing	HFS	Breech-ing
HEJ	—Made a breach	HFT	Breeches buoy
HEK	Bread	HFU	Breech-loader-ing.
HEL	—Bread and water		
	Bread bag Biscuit bag ....FSE	HFV	Breed-ing Bred
	Bread is not good Biscuit is not		
	good ....GLK	HFW	Breeze (*See also* Wind)
HEM	—Bread room	HFX	—A stiff breeze
HEN	—Breadstuffs		As soon as there is a breeze ....FBQ
HEO	—Fresh bread Soft bread	HFY	—Breeze will freshen
HEP	—No bread.	HFZ	—Fresh breeze
	Soft bread Fresh bread ....HEO	HGA	—I shall weigh the moment the breeze
	Want bread In want of biscuit ..GLN		springs up
		HGB	—Land breeze
HEQ	Breadth Broad (*See also* Broad)		Moderate breeze ....RXK
HER	—Hair's breadth	HGC	—Sea breeze
HES	—The breadth of —	HGD	—Slight breeze
HET	—What width (*or*, breadth)?	HGE	—Strong breeze.
		HGF	—The breeze will die away.
HEU	Break-ing To break (*See also* Broken)	HGI	—When the breeze goes down
HEV	—A regular break up An end to the fine	HGJ	—When the wind (*or*, breeze)
	weather		
	Am aground, likely to break up, require	HGK	Bribe-d-ing-ry
	immediate assistance ....CA		

BRICK—BROKE

HGL	BRICK	HJF	BRING—*Continued*
HGM	BRIDGE (*of a ship*)	HJG	—Shall I bring my ship's papers with me?
			—She (*or vessel indicated*) has brought
HGN	BRIDGE (*over a river, etc*)		The captain must bring the ship's papers
HGO	—Bridge of boats		with him ----------------------HXJ
HGP	—Drawbridge		When do you intend to bring up
HGQ	—Floating bridge.		(anchor)?----------------------EKF
HGR	—Foot bridge.	•	Where do you intend to bring up
			(anchor)?----------------------EKH
HGS	BRIDLE	HJI	BRITAIN-ANNIA-IC-ON
HGT	BRIEF-LY　BREVITY	HJK	BRITISH　BRITISH COLORS　(*See also*
HGU	—Be as brief as possible		ENGLAND)
			British Code List of Ships --------- IZD
HGV	BRIG		Lloyd's Register of British and Foreign
HGW	—Did you see a brig?		Shipping ----------------------QWV
HGX	—Spoke a brig	HJL	BROACH-ED ING—TO
HGY	BRIGADE		BROAD　BREADTH ----------------HEQ
HGZ	—Fire brigade	HJM	—Broader est
HIA	—Naval brigade　Brigade of seamen	HJN	—How broad is —?
HIB	—Rocket brigade	HJO	BROADSIDE
HIC	BRIGAND.	HJP	—Broadside on (*or*, to)
	Beware of brigands------- --------GJD	HJQ	—Port broadside
HID	BRIGANTINE	HJR	—Starboard broadside
HIE	BRIGHT EN-ER-EST-LY-NESS	HJS	BROKE-N　(*See also* BREAK)
			A case of fever has broken out on
HIF	BRINE-Y.		board — (*or on ship indicated*)---GPR
HIG	—Brine cock		Air-pump rod broken ------------DWF
HIJ	—Brine pump		Any bones broken?--------------GYS
			Arm broken --------------------FVI
HIK	BRING-ING　TO BRING		Beacon, or, Buoy has broken adrift (*or*,
HIL	—Bring all the letters and papers		is gone)----------------------GAE
	Bring all your papers-------------LJZ		Bone broken--------------------GYT
HIM	—Bring it to the windlass		Boom broken------- ------------GZK
HIN	—Bring the papers to me		Boom, *or*, Obstruction broken adrift,
HIO	—Bring the ship in as close as possible		GZM
	Bring your papers on board--------TAF		Both booms broken---------------GZN
HIP	—Bring away		Break-ing, Broke the quarantine laws,
HIQ	—Bring, *or*, Brought despatches.		UEB
HIR	—Bring, *or*, Brought intelligence		Broke adrift --------------------DMO
HIS	—Bring nothing.		Broke from ------- ------------NRV
HIT	—Bring off	HJT	—Broke from her moorings
	Bring on board------------------GPZ	HJU	—Broke loose
HIU	—Bring on shore.	HJV	—Broke out.
HIV	—Bring out　•	HJW	—Broke up
HIW	—Bring up	HJX	—Broken down
	Bring-to　Heave-to—until----------LR		Cable is cut (*or*, broken) ---- ---HQP
HIX	—Bring-to, I will send a boat		Condenser broken (damaged, *or*, out of
	Bring-to with a kedge　Drop a kedge,		order) ------------------------JMR
	PXV		Connecting-rod bolts broken ----- GXV
	Bring your papers on board--------TAF		Connecting-rod brasses broken (*or*, out
HIY	—Brought		of order) ----------------------HEB
HIZ	—Can, *or*, Shall I bring?		Coupling bolts broken------------GXW
	Come nearer, Stop, *or*, Bring to, I have		Crank brasses broken (*or*, out of order),
	something important to communi-		HEC
	cate ----------------Code Flag over H		Crank broken------- ------------KBI
	Do not bring the light to the — (*bearing		Crank pin broken ---------------KBO
	indicated*) of — (*bearing indicated*)-PX		Crank shaft broken (*or*, defective) -KBQ
	Has brought the account of — ---- DFB		Cylinder cover broken ---- ------KIC
HJA	—Have you, *or*, Has he brought?		Eccentric bar broken------ -------FWJ
HJB	—I shall bring to at —		Eccentric rod broken ------------LSX
HJC	—I shall not go far before I bring up		Eccentric strap broken ----------LSZ
	Is there anything to bring me up? -EQP		Engines broken down ------------MOE
	Let him bring on shore (ashore)----FCY		Engines broken down, I am disabled-BJ
HJD	—Let them bring their tools with them		Engines broken down, obliged to pro-
HJE	—Shall I bring?	•	ceed under sail ----------------MCI

BROKE—BUMP

	BROKE—*Continued*	HKD	BROOM
	Engines broken down, towed in by —, MCJ	HKE	BROTHER
	Feed pipe broken (*or*, out of order) . MUZ		BROUGHT (*See* BRING) ---------------HIY
	Fire has broken out on board of — ..GQC		
	Has the ice broken up? Has the winter broken up—at — ? ....PBJ	HKF	BROW
	Have broken main shaft............MR		
	Hawse pipe is broken ....OPN	HKG	BROWN-NESS
	High-pressure cylinder cover broken, KAI	HKI	BRUISE-D-ING
	High-pressure piston broken........OUX	HKJ	—Severely bruised.
HJY	—I have broken the (*or*, my)		
	Intermediate pressure cylinder cover broken................KAJ	HKL	BRUNT
	Low-pressure slide jacket cover broken, KAM	HKM	BRUSH-ED ING
		HKN	—Electric brush
	Machinery broken down............MCH	HKO	—Hair brush.
	Machinery broken down, irreparable at sea ....RDV	HKP	—Paint brush (*or*, tool)
		HKQ	—Scrubbing brush
	Machinery supposed to have broken down................MCA	HKR	—Tar brush
	Main-bearing brasses broken (*or*, out of order) ....GBO	HKS	BRUSHWOOD
	Main shaft broken ....MR	HKT	BUBBLE.
	Main shaft broken in stern tube.....RGT		
	One blade of propeller (*or*, screw) broken, GMA	HKU	BUCKET
		HKV	—Fire bucket
	One boom broken..................GZT		
	Packing ring of cylinder cover broken, KAN	HKW	BUFFALO
	Piston broken....................TKX	HKX	BUGLE-D-ING
	Piston rod bolts broken............GYB		
	Piston rod brasses broken (*or*, out of order) ....HED	HKY	BUGLER
	Piston rod broken (*or*, cracked)....KAX		
	Piston rod crossheads damaged (*or*, broken)....................KEJ	HKZ	BUILD-ING BUILT To BUILD
		HLA	—Builder
	Propeller broken ....TYW	HLB	—Clinker built
	Propeller frame broken............TYZ	HLC	—Clipper built
	Propeller shaft broken............MR	HLD	—Contract built
	Pump lever links broken ....UBL	HLE	—Manufactured, Made, *or*, Built by
	Sea inlet valve (*or*, cock) broken (*or*, damaged)................VLA		Ship building yards -------------VXY
			Steel built vessel.. ---------------WTL
	Shaft broken ....VTJ		
	Shaft-coupling bolt broken ....GYF	HLF	BULGARIA-N BULGARIAN COLORS.
	Slide valve rod broken............WGO		
	Slide valves broken ....WGJ	HLG	BULGE-D-ING
	Stuffing box broken... ... .HDI	HLI	BULK-Y-NESS
	Supposed with engines broken down..MCA	HLJ	—In bulk
	Tail end of main shaft broken.....LYO		
	The plague has broken out at —... TLS	HLK	BULKHEAD
	Three blades of propeller (*or*, screw) broken ....GMB	HLM	—Collision bulkhead
		HLN	—Engine room bulkhead
	Thrust block broken ....GMZ		
	Thrust shaft broken ....XRN	HLO	BULLET
	Tunnel shaft broken . ...YDX		
	Two blades of propeller (*or*, screw) broken ....GMC	HLP	BULLION
		HLQ	BULLOCK OX CATTLE
HJZ	BROKER		
	Stockbroker (*or*, jobber)..........PVY	HLR	BULL'S EYE
		HLS	BULWARK
HKA	BROKERAGE	HLT	—Loss of bulwarks
	BROMIDE OF POTASSIUM............ TQY		BUMBOAT..............................GBF
HKB	BRONCHITIS		
		HLU	BUMP-ED-ING
HKC	BRONZE		

BUMPKIN—BURST

HLV	BUMPKIN	HMK	BUOY—*Continued*
	BUNDER (*Square Measure*)_____AXP		—I do not think buoys (or, marks) can be in their proper positions
HLW	BUNDLE-D-ING		I have not been able to make out the buoy (or, beacon)_____ _____GAJ
HLX	BUNG		I have slipped my anchor, but it is buoyed _____ __ _____EIQ
		HML	—I think I must have passed the buoy (or, beacon).
HLY	BUNKER COAL BUNKER.	HMN	—Is it, or, Are they buoyed?
HLZ	—Bunker coal		Is the channel buoyed? _____PO
		HMO	—Is there a buoy (or, mark) on shoal (or, on —)?
HMA	BUNT		Keep buoy (or, beacon) on your port hand __ _____ ____GAK
HMB	BUNTING		Keep buoy (or, beacon) on your starboard hand _____GAL
HMC	BUOY-ING TO BUOY.	HMP	—Leave, or, Put a buoy on —
HMD	—Buoy (a buoy)	HMQ	—Leave buoy—at —
HME	—Buoyed		Leave the buoy (or, beacon) to port._LZ
	Anchor buoy_____ EGY		Leave the buoy (or, beacon) to starboard _____ _____ ____MA
	Beacon buoy__ _ _____GAD	HMR	—Let go the life buoy
	Bell buoy_____ ____ _____GEM	HMS	—Make fast to a buoy.
	Black buoy_____ _____GLT		Marks are all gone There are no marks (or, buoys)_____RLM
	Breeches buoy_____HFT	HMT	—Missed the buoy
	Buoy (a buoy)_____HMD	HMU	—Near the buoy
	Buoyed_____HME	HMV	—Off the buoy
	Can buoy_____HUD		Slip your cable and buoy the end__HRQ
	Channel buoy_____ILA		Steer directly for the buoy (or, beacon) _____GAM
	Chequered buoy____ _____IOK		The shoal is buoyed_____ __VYX
	Conical buoy _____JPD	HMW	—They have a buoy (or, mark) on
	Cork buoy_____ _____JWM	HMX	—To the buoy
	Fairway buoy_____ _____MRG		
	Fog buoy_____ _____NGH	HMY	BUOYANT-CY
	Gas, or, Light buoy_____NXA	HMZ	—Buoyant mine
	Green buoy_____ ____OHS		
	Harbor, or, Entrance buoy_____ __MFR	HNA	BURDEN, or, BURTHEN ED ING-SOME
	Life buoy _____QRD	HNB	—Tonnage burden
	Luminous buoy _____ RDE		
	Mine, or, Torpedo buoy_____XWN	HNC	BUREAU DE CHANGE EXCHANGE OFFICE
	Nun buoy_____SLQ		BUREAU VERITAS_____NQF
	Pillar buoy _____TIR		
	Red buoy_____ ____ _____UMD	HND	BURGEE
	Sounding buoy _____ ___WLD		
	Spar buoy_____WLZ		BURGUNDY WINE _____ZGB
	Spherical buoy _____WNK		
	Striped buoy_____ _____XAN		BURIAL GROUND CEMETERY_____IGS
	Submerged torpedo buoy_____ XBL		
	Telegraph buoy_____XLY	HNE	BURN-ING BURNT TO BURN
	Torpedo buoy_____XWN		Burn a blue light, or, Flash powder, GGQ
	Whistling buoy _ _____ZDE	HNF	—Fire has been burning these — hours
	White buoy_____ZDG		Fire or, Lights will be kept burning at the best place for landing_____KE
	Wreck buoy_____ZIY		Keep a light burning _____ _____KO
	Bell buoy damaged, not working__GEN	HNG	—Vessel indicated was most likely burnt
	Buoy bears_____GBJ	HNI	—Will not burn.
HMF	—Buoy end of cable		You must keep a light burning _____KO
	Buoy, or, Beacon has broken adrift (or, is gone)_____ _____ __ GAE	HNJ	BURST-ING
	Buoy rope (of anchor)_____EHR		Blow-off pipe burst _____GNY
	Buoy, or, Mark is not in its proper place_____LO		Boiler burst_____GVU
HMG	—Buoy, or, Mark does not appear to be in its proper position		Boiler burst, — men killed_____AT
HMI	—Buoys have all disappeared		Boiler burst, — men killed, — others wounded_____AU
HMJ	—Buoys have just been examined, they are all right		Boiler burst, no one hurt_____AV
	Buoyed__ _____ ____HME		Boiler burst, no one seriously hurt_AW
	Channel is buoyed _____ ___ OW		
	How does the beacon (buoy, or, mark) bear?_____GAH		
	How must I bring the buoy (mark, or, beacon) to bear?_ _____ _____GAI		

	BURST—BY		
	BURST—*Continued*	HOP	**BUTTER**
	Boiler burst, number of killed and wounded not yet known _____AX	HOQ	—Irish butter
		HOS	—Fresh butter.
	Boiler burst, several men killed ___AY	HOT	—Salt butter.
	Boiler burst, several men wounded _AZ		
	Boiler burst, very slight damage _GVW		
	Boiler tubes burst _____GWP	HOU	BUTTON-ED-ING.
	Condensing boiler burst_____GWT		
	Condensing boiler tubes burst _____YDB		
	Condenser tubes burst _____JMU		BUY ING. BOUGHT To BUY (*See also*
	Donkey boiler burst _____GWX		PURCHASE)_____ _____HBM
	Steam pipe burst _____TKO	HOV	—Can I buy?
HNK	—Will burst	HOW	—I can buy
			I have bought_____HBN
		HOX	—I intend to buy
	BURTHEN. (*See* BURDEN)_____HNA	HOY	—I wish to buy
			Was not to be bought ____ _____HBO
HNL	BURTON		Will you (*or person indicated*) buy (*or,* purchase)? _____UCF
HNM	BURY-IED-ING-IAL		
	Burial ground Cemetery _____IGS	HOZ	BUYER
HNO	—Buried under —		
	When does burial (funeral) take place? NTL	HPA	BY _____BJQ
			—By a shot
HNP	BUSH-ED		By all means_____DYK
HNQ	—In the bush		By an Act of Congress (*or*, Parliament), DIE
		HPB	By and bye Not yet _____HPX
	BUSHEL. (*Measure of capacity*) _____AYP	HPC	—By day
			—By degrees
HNR	BUSINESS.		By him (his, her-s, it-s) _____BJR
HNS	—Busy-ily		By me (my, mine)_____BJS
HNT	—Are you, *or*, Will you be busy	HPD	By messenger (*or*, bearer) _____RQU
HNU	—Business dull	HPE	—By night
HNV	—Business good.		—By no means
HNW	—Business transactions	HPF	By our-s (us) _____BJY
HNX	—Did you do any business?	HPG	—By post
HNY	—How is business going?		—By rail
HNZ	—I do business with —	HPI	By steamer _____WTF
	I have no business agent (*or*, I do not know of any agent) at — _____DSQ		—By telegraph
			By that (*or*, this)_____BJT
HOA	—If you are busy		By the _____BJU
HOB	—Interferes with business		By, *or*, With the army_____EWG
HOC	—Person indicated has not suspended business.	HPJ	—By, *or*, With the convoy.
			By the crew _____ __KCL
HOD	—Person indicated has suspended business	HPK	—By the head
HOE	—Suspended business	HPL	—By the packet (*or*, mail)
HOF	—They do business with	HPM	—By the side of
HOG	—Very busy-ily	HPN	—By the stern. Inches by the stern
	Who are the best agents (*or*, people) to do business with? _____DS7	HPO	—By the sun
		HPQ	—By, *or*, To the wind
	Whom do you do business with?___DTC		By them (their-s)_____BJV
	Who generally does business for—?_DTB		By there_____BJW
HOI	—Will you be, *or*, Are you busy?		By these (*or*, those) _____BJX
HOJ	—You had better do business with —		By this (*or*, that) _____BJT
	Your agent, (*or agent specified*) has (*or*, have) discontinued business_____DTP	HPR	—By this means
		HPS	By us (our-s) _____BJY
	Your agent (*or agent specified*) has (*or*, have) suspended business_____DTG		—By way of
			By what authority ?_____FMY
		HPT	By what means ?_____RNV
HOK	BUT		—By what time?
HOM	—But not		By which (*or*, what) _____BJZ
			By whom (*or*, whose)_____BKA
HON	BUTCHER		By you-r-s _____BKC
	Someone to look after stock _____RAD	HPU	—Byword
		HPV	—Gone by Passed by
	BUTT (*Measure of capacity*)_____AYQ		Me by _____BKD
			Not by_____CFG
	BUTT-END _____LYJ		Passed by Gone by_____HPV
			Towed by, *or*, In tow of_____XYU

BYE—CAKE

	BYE
HPX	—By and bye Not yet
HPY	—By-law
HPZ	—Good-bye
	I shall come off by and bye ....JDG
HQA	C (*Letter*) (*For new method of spelling, see page 13*)
HQB	CAB
HQC	CABBAGE
HQD	CABIN
HQE	—Cabin passenger—for
HQF	—Cabin table requisites.
HQG	—Cabin window
HQI	—Deck cabin
HQJ	—State room
HQK	CABINET
	CABLE (*Measure of length*)........AVN
HQL	CABLE
HQM	—About to veer cable
	Are towing cables fast ? Are you (*or, your hawsers*) fast ?...........XYJ
HQN	—Are your cables clinched ?
	Armor ed cable ....EVU
HQO	—Be ready to slip your cable
	Bend cable ....GFJ
	Bower cable...................HCW
	Break in cable....................HEW
	Buoy end of cable ....HMF
	Cable (*to telegraph*).............XLO
HQP	—Cable is cut (*or,* broken)
	Cable is repaired........UQY
HQR	—Cable laid Cable-laid rope
HQT	—Cable parted
HQU	—Cable tank
HQV	—Cablegram
HQW	—Cable (*chain*) (*Circumference in ¼ of an inch, and length in fathoms, to be indicated, if necessary*)
HQX	—Cable, *or,* Hawser (*hemp*) (*Size and length to be indicated, if necessary*)
HQY	—Can not veer any more cable
	Cast off towing cables..........XYN
HQZ	—Coir, *or,* Grass cable
HRA	—Cut, *or,* Slip your cable
	Driving, I can veer no more cable
	No more anchors to let go ....NK
HRB	—Electric cable

	CABLE—*Continued*
HRC	—Fasten your chain to towing cable
	Have all your anchors and cables ready for use ....EIG
HRD	—Heave-ing, Hove short
	Hemp cable, *or,* Hawser ....HQX
HRE	—How much cable have you out ?
	I am laying (repairing, *or,* picking up) a telegraph cable Keep out of my way ....XLQ
HRF	—I am obliged to slip my cable, pick it up for me
	I have picked up telegraph cable ...WU
HRG	—I mean to slip cable
HRI	—I must cut my cable
HRJ	—Is there a telegraph cable near me ?
	In anchoring, look out for telegraph cable.......................WV
	Insulate-d-ing wire (*or,* cable).....PNT
HRK	—Is the cable fast (*or,* rope belayed) ?
HRL	—Main cable
HRM	—Not cable enough, give more scope
	Pay out cable ....KU
HRN	—Range your cable
	Send an anchor and cable off immediately (*indicate if more than one*)..EJM
	Sending, Sent an anchor and cable..EJN
HRO	—Sheet cable
HRP	—Shorten in cable to the number of shackles indicated
HRQ	—Slip your cable and buoy the end
HRS	—Spring on the cable
	Steel wire cable.....................WTO
	Stop engines to adjust towing cables, DKH
HRT	—Stream cable (chain).
HRU	—Submarine cable.
HRV	—Survey chain cable
HRW	—Telegraph cable
	There is a telegraph cable on — (*bearing indicated*)...................EN
HRX	—Towing cable is damaged (*or,* stranded)
	Towing cable is fast ....XYV
HRY	—Unbend cables
HRZ	—Unshackle cables
HSA	—Veer cable to — (*number of shackles indicated*)
	Veer more cable.....................KU
	Veered-ing my cable
HSB	—Veered-ing my cable (*indicate which, and number of shackles, if necessary*)
	Want a cable (*size to follow*).........YB
	Want a cable and anchor ....YD
HSC	—Want a hemp cable
HSD	—What length are your cables (*or cable indicated*) ?
HSE	—Wire cable (*or,* hawsers)
HSF	CADET
HSG	CADIZ
HSI	—My first meridian is Cadiz, west of Greenwich 6° 12′ 24′ = 0h. 24m. 49 6 secs
HSJ	CAFÉ RESTAURANT
HSK	CAISSON
HSL	CAKE.
	Oil cake ....SQE

CALAMITY—CAN

	CALAMITY-OUS (*Disaster*) ....LCK	HTJ	—Where do you intend calling for orders?
HSM	CALAVANCES	HTK	—Will you call?
HSN	CALCULATE-ING ION	HTL	CALM-ED-ING-LY-NESS
HSO	—Have, *or*; Has made no calculation	HTM	—Calm and light winds
HSP	—Have you, *or*, Has he calculated?		Calm sea.........................UO
HSQ	—I have, *or*, He has calculated	HTN	—Calm the sea by pouring oil on it
HSR	—What is your calculation?	HTO	—Calms prevail
		HTP	—Fall-ing, calm
HST	CALF	HTQ	—If it should be calm
			Meteorological report for to-day gives
HSU	CALIBER.		calm ....RSG
HSV	—Guns of the caliber of —.		Meteorological report for to-morrow
			gives calm ....RSK
HSW	CALICO		
		HTR	CALORIC
HSX	CALL-ED ING To call	HTS	—Caloric engine.
	Attention is called to page —, paragraph		
	—, of the INTERNATIONAL CODE .. DH	HTU	CAM
	N. B —The page is to be shown first, and		
	then the paragraph This signal	HTV	CAMBER
	should be employed only between vessels		
	using the United States edition, as	HTW	CAMBRIC
	otherwise the pages referred to will not		
	correspond		CAME (*See* COMB) ....JCF
	Call alongside.....................EAX		
HSY	—Call at the post office for letters	HTX	CAMEL
HSZ	—Call for my letters.		
HTA	—Call off — for orders	HTY	CAMP
HTB	—Call upon (*or*, see)	HTZ	—Camp equipage
HTC	—Called for —		
	Calls the attention of the FIRST (*or*,	HUA	CAMPAIGN
	nearest) ship on the bearing (FROM *the*		
	person signaling) pointed out by com-	HUB	CAMPHOR-ATE-D
	pass signal. ....DI		
	Calls the attention of SECOND ship, etc..DJ	HUC	CAN-NED A CAN (*Tin*)
	Calls the attention of THIRD ship, etc..DK	HUD	—Can buoy
	Calls the attention of FOURTH ship, etc..DL		
	Calls the attention of FIFTH ship, etc..DM		CAN (*or*, MAY) ....BKE
	Calls the attention of SIXTH ship, etc..DN		Can, *or*, May be. ....BFG
	Calls the attention of the shore signal		Can, *or*, May do (*or*, be done)......BLT
	station on the bearing (FROM *the per-*		Can, *or*, May do it..............BLU
	son signaling) pointed out by compass		Can, *or*, May have...............BQE
	signal ....DO		Can, *or*, May have been............BFH
	Calls the attention of vessel COMING into		Can, *or*, May have done (*or*, been done),
	the anchorage on the bearing (FROM		BLV
	the person signaling) pointed out by		Can, *or*, May have had.............BQF
	compass signal ....DP		Can, *or*, May he (she, it, *or person-s or*
	Calls the attention of vessel LEAVING		*thing-s indicated*)?.... ....BTJ
	the anchorage on the bearing (FROM		Can, *or*, May he (she, it, *or person-s or*
	the person signaling) pointed out by		*thing-s indicated*)?...............BTK
	compass signal ....DQ		Can, *or*, May he (she, it, *or person-s or*
	Calls the attention of vessel whose dis-		*thing-s indicated*) have been? ....BTL
	tinguishing signal will be immedi-		Can, *or*, May his (her-s, it-s)? .. ..BKF
	ately shown....... - ....DR		Can, *or*, May I (my, mine)?.........BKG
HTD	—Do, *or*, Does not touch (*or*, call) any-		Can, *or* May I be?.........BKH
	where (*or*, at —)		Can, *or*, May I have some — ?. ....BQG
	Do you, *or* Does — (*vessel indicated*)		Can, *or*, May it be done? ....BLW
	call anywhere (*or*, at —)?.........HW		Can, *or*, May that (*or*, this)?.......BKI
HTE	—Do you call at any port before destina-	HUE	—Can, *or*, May the?
	tion?		Can, *or*, May there be?.........BFI
	Have you called anywhere (*or*, at —)?		Can, *or* May these (*or*, those)?.....BKJ
	EQT		Can, *or*, May they (their s) ...... BKL
	I have orders for you not to touch (*or*,		Can, *or* May they not? .. ..BKM
	call at) — 8N		Can, *or*, May we (ou s)?.........BKN
HTF	—I shall call at —		Can, *or*, May you (your-s)?........BKO
	I shall not touch (*or*, call) anywhere (*or*,		Can not, *or*, May not..........BKP
	at —).......... . ...EQU		Can not, *or*, May not be...........BFJ
HTG	—Port of call		Can not, *or*, May not do (*or*, be done),
HTI	—Shall, *or*, Will call (*or*, be called)		BLX

CAN—CAPABLE

CAN—*Continued*
Can not, *or*, May not do it_____BLY
Can not, *or*, May not have_____BQH
Can not, *or*, May not have been___ BFK
Can not, *or*, May not have done (*or*,
   been done) __ _____BLZ
Can not, *or*, May not have had__ ..BQI   **HUF**
Can not, *or*, May not he (she, it, *or per-*
   son-s or thing s indicated) be ? _ .BTM   **HUG**
Can not, *or*, May not he (she, it, *or per-*
   sons-s or thing s indicated) have been ?
                    BTN
He, She, It (*or person-s or thing-s indi-*
   cated) can (*or may*) _____ ____ BUA
He, She, It (*or person-s or thing-s indi-*
   cated) can (*or*, may) be _____BUC
He, She It (*or person s or thing-s indi-*
   cated) can not (*or*, may not)_____BUD   **HUI**
He, She, It (*or person-s or thing-s indi-*   **HUJ**
   cated) can not (*or*, may not) be ..BUE   **HUK**
How can (*or*, may)?_____BXJ   **HUL**
I can (*or*, may)_____BKQ
I can (*or*, may) be_____BFX   **HUM**
I can (*or*, may) have_____BRI   **HUN**
I can not (*or*, may not)_____ BKR   **HUO**
I can not (*or*, may not) be _____BFY   **HUP**
I can not (*or*, may not) have _____BRJ   **HUQ**
I shall (*or*, will) if I can_____BKS
I will do it if I can. _____LIP
If he (she, it, *or person-s or thing-s indi-*
   cated) can (*or*, may)_____BYD
If he (she, it, *or person s or thing s indi-*
   cated) can not (*or*, may not)____ BYE   **HUR**
If I can (*or*, may) _____BZA   **HUS**
If I can not (*or*, may not)__ _____BZC
If it can (*or*, may) be done_____BZR
If it can not (*or*, may not) be done_BZS   **HUT**
If they can (*or*, may)_____CAH
If they can not (*or*, may not) _____CAI   **HUV**
If we can (*or*, may)_____CAZ
If we can not (*or*, may not)_____CBA   **HUW**
If you can (*or*, may) _____CBS
If you can not (*or*, may not) _____CBT   **HUX**
It, etc   (*See* HE, *above*)
It can be_____BGN   **HUY**
It can (*or*, may) be done_____BNK
It can not be _____BGO
It can not (*or*, may not) be done ___BNL   **HUZ**
No, I can not _____HUZ
Nothing can (*or*, could)_____SKB
That, *or*, This can (*or*, may)_____CKQ
That, *or*, This can (*or*, may) be ____CKR   **HVB**
That, *or*, This can not (*or*, may not)_CKS
That, *or*, This can not (*or*, may not) be,
              CKT
There can (*or*, may)_____CMI
There can (*or* may) be_____CMJ
There can not (*or* may not)_____CMK   **HVD**
There can not (*or*, may not) be_____CML   **HVE**
They can (*or*, may)_____BKT   **HVF**
They can (*or*, may) be _____BHP   **HVG**
They can (*or*, may) have_____BSF   **HVI**
They can not (*or*, may not) _____BKU   **HVJ**
They can not (*or*, may not) be.. _BHQ
They can not (*or*, may not) have___BSG   **HVK**
We can (*or*, may)_____BKV
We can (*or*, may) be _____BIH   **HVL**
We can (*or*, may) have_____BSN   **HVM**
We can not (*or*, may not)_____BKW
We can not (*or*, may not) be _____BIJ   **HVN**
We can not (*or*, may not) have ____BSO   **HVO**

CAN—*Continued*
What, *or*, Which can (*or*, may)?___CPZ
When can (*or*, may)?... _____CQZ
When I can (*or*, may)_____CRM
Where can (*or*, may)? _____CSV
Who can (*or*, may)? _____CUJ
—Why can (*or*, may)?
Why can not (*or*, may not)? _____CVD
—Yes, I can
You can (*or*, may) _____BKX
You can (*or*, may) be_____BIY
You can (*or*, may) have _____BSY
You can not (*or*, may not) _____BKY
You can not (*or*, may not) be_____BIZ
You can not (*or*, may not) have ____BSZ

CANAL.
—Canal is blocked
—Canal is clear
—Is canal clear?

CANCEL ED-ING
—Cancel-ed-ing the indentures
—Certificate canceled
—Order canceled
—Your orders are canceled
   Your original orders are canceled, I
   am directed to inform you to proceed
   to — .. _____TH

CANDAREEN, *or*, **FUN**   (*Coin*) _____ATB

CANDIDATE FOR
—Examination of candidate for — (*or*,
   at —)

CANDLE.

CANE-D-ING.

CANNIBAL-ISM

CANNON   (*See also* GUN)

CANNONADE-D-ING

CAN NOT   (*See also* CAN)_____BKP
—No, I can not

CANOE

CANT-ED-ING

CANTAR   (*Measure of weight*) _____BAD

CANTEEN

CANVAS
—Canvas to repair sails
—Is, *or*, Are not stiff under canvas
—Is, *or*, Are very stiff under canvas
—Old canvas
—Want No — canvas (*length indicated*)

CAP.
   Bowsprit cap _____HCY
—Lower cap
—Topmast cap

CAPABLE-ILITY
—Is he (*or*, it) capable?

	CAPACIOUS—CARGO

HVP	Capacious Capacity Measure of cubic capacity (*see page* 52), AYB Measure of capacity (*see page* 52)..AYJ	HXP	Captive-ivity Captive balloon ------------------FTY
HVQ HVR HVS	Cape (*Point*) —Off the cape —Round-ing the cape	HXQ HXR HXS HXT HXU HXV HXW HXY HXZ	Capture-d-ing Captor —Captured by —Has been captured —Have any vessels been captured? —Vessel and cargo have been captured —Was, *or*, Were captured —Will certainly be captured —Will perhaps be captured —You run the risk of capture (*or*, seizure)
HVT	Capital		
HVU	Capitulate-d-ing-ion	HYA HYB HYC	Car —Car, *or*, Omnibus leaves at — —When does car (*or*, omnibus) leave for — ?
HVW	Capsize-d-ing Boat capsized (*bearing, if necessary*) _AP		
HVX HVY	—Did you see her capsize? —Vessel indicated (*by bearing*) is capsized		Carat (*Measure of weight*).........BAR
HVZ	Capstan Capstan bar ------- ------------FWH	HYD	Carbolic acid
HWA HWB	—Capstan disabled —Steam capstan	HYE HYF HYG	Carbon ated-ic-ize-d —Carbon plate —Carbon point
HWC	Captain All captains -------------------DXG	HYI	Card Compass card ------------------JHG
HWD HWE	—By the captain —Captain desires me to inform you Captain in the Army ------------EWH	HYJ	Cardinal

HWF
HWG
HWH
HWI
HWJ
HWK
HWL
HWM
HWN
HWO
HWP
HWQ
	—Captain in the Navy
	—Captain is
	—Captain is dead (*or*, died on —)
	—Captain is not
	—Captain is not on board
	—Captain is on board
	—Captain is on shore
	—Captain is requested to come on board
	—Captain is sick
	—Captain is well
	—Captain, *or*, Master in the Mercantile
	  Marine
		HYK
		HYL
			Care-ful-ly-ness
			—Be careful not to give offense
			  Be very careful in your answer....EMU
			  Be very careful in your intercourse with
			    strangers . .. ---------- ----------FS
			  Take care, *or*, Be careful, caution is
			    requisite --------------- --------FT

HWR
HWS
HWT
HWU
HWV
HWX
	—Captain of —.
	—Captain of the port, *or*, Harbor master
	—Captain requests.
	—Captain will not return until —
	—Consult captain
	—Flag captain
	  For, *or*, to the captain ------------HXK
		HYM
		HYN
		HYO
		HYP
		HYQ
			—Take care range is clear
			—The greatest care (*or* caution)
			—Under the care (*or*, charge)—of
			—Under your care (*or*, charge)
			—Utmost care must be taken
			  You may go by your soundings, if you
			    use care ------------ ----------WLA

HWY	—From the captain
	  Harbor master, *or*, Captain of the
	    port--------- ---- - --------HWS
		HYR	Careen-ed-ing

HWZ
HXA
HXB
HXC
HXD
HXE
HXF
	—Have not seen the captain
	—Have seen the captain
	—Have you seen the captain?
	—Inform the captain of
	—Is the captain coming on shore?
	—Is the captain on board?
	—Request captain to come here (*or to
	  place indicated*)
		HYS
		HYT
		HYU
		HYV
		HYW
		HYX
			Career
			Careless-ly ness.
			Cargo
			—Assorted cargo
			—Coal cargo
			—Cotton cargo
			  Dead weight cargo ---------------KOJ

HXG	—Request the captain of — (*vessel indi- cated*)	HYZ HZA HZB HZC HZD	—Deck cargo —East India cargo —Frozen meat cargo —General cargo —Grain cargo
HXI HXJ	—Should the captain —The captain must bring the ship's papers with him	HZE HZF HZG	—Guano cargo —Inflammable cargo —Live cargo
HXK HXL HXM HXN	—To, *or*, for the captain —When the captain —Where is the captain? —Who is captain of ship signaled (*or indi- cated*)? *or*, Who is your captain?	HZI HZJ HZK HZL HZM	—Miscellaneous cargo —Not inflammable cargo —Perishable cargo —Petroleum cargo —Return cargo
HXO	—With the captain	HZN	—Timber cargo

CARGO—CARRIED AWAY

	CARGO—*Continued*
HZO	—West India cargo
HZP	—Wheat cargo
	Advertise cargo and day of sale....DOB
HZQ	—An opportunity for disposing of cargo
	Cargo boat ------------------------GSK
HZR	—Cargo consists of—
HZS	—Cargo damaged
HZT	—Cargo damaged, extent not yet known
HZU	—Cargo expected to be saved
HZV	—Cargo expected to be sold
HZW	—Cargo has been reshipped
HZX	—Cargo in very good order
HZY	—Cargo lost
IAB	—Cargo must be reshipped
IAC	—Cargo must be shipped
	Cargo not inflammable...............HZJ
IAD	—Cargo not much damaged (slightly damaged)
IAE	—Cargo not salable
	Cargo, or, Ballast has shifted ------FTN
IAF	—Cargo saved
IAG	—Cargo so badly stowed that I am not seaworthy
IAH	—Cargo sold well
IAJ	—Cargo under weight
IAK	—Condition of cargo
IAL	—Discharge d-ing cargo
IAM	—Discharge cargo as soon as possible
IAN	—Dispose of cargo as soon as possible
	Fire in hold amongst cargo ---------OR
IAO	—Have lost deck cargo
IAP	—Have room for no more cargo
IAQ	—Have you cargo for more places than one (*or than one specified*)?
IAR	—Have you room for more cargo?
IAS	—How many tons dead weight of cargo must you have?
	How many tons of measurement goods can you take? -----------------OEV
	I have no cargo to disturb my compasses...................JHU
	I have received orders for you to discharge cargo at-----------------SR
IAT	—Is reshipping her cargo
IAU	—Is the cargo sold?
	Is your cargo much damaged?........BQ
IAV	—Is your cargo very inflammable?
	Manifest of cargo. --- ------------RJO
	Must discharge cargo to repair damage, KJR
	My cargo is — -------------------HZR
	Owners desire me to inform you to discharge cargo first at — (*place indicated*) -----------------SY
IAW	—Part of cargo saved
IAX	—Proceeds of the cargo
IAY	—Sell cargo as soon as possible
IAZ	—Send flats to discharge cargo
	Sold her cargo -----------------WJF
	Supercargo -------------------XEG
IBA	—The cargo is not yet sold.
IBC	—Valuable cargo
	Vessel and cargo have been seized (*or*, captured)----------------HXU
IBD	—Waiting for cargo
	Want lighters to take my cargo.....QTE
IBE	—Want — tons dead weight of cargo
IBF	—What cargo do you want?
IBG	—What cargo have you (*or*, has — *vessel indicated*)?

	CARGO—*Continued*
IBH	—What condition is the cargo in?
IBJ	—When do you commence discharging cargo?
IRK	—When will you have the cargo (*or*, the —)?
IBL	—Where did you discharge cargo?
IBM	—Where would my cargo sell?
IBN	—Whom is your cargo consigned to?
IBO	—Without discharging cargo
IBP	—Would my cargo sell here (*or at place indicated*)?
IBQ	—You will discharge cargo — at —
IBR	CARLING
IBS	CARPENTER
IBT	—Can I get carpenters? Want carpenters
IBU	—Carpenters and caulkers may be obtained at —
IBV	—Carpenter's stores
IBW	—Carpenter's tools
IBX	—I have no carpenter's tools Want carpenter's tools
IBY	—Send carpenters on board — (*ship indicated*)
IBZ	—Vessel (*indicated*) can lend you carpenter's tools
ICA	CARPET-ING.
	Carpet bag ---------------------FSG
ICB	CARRIAGE.
ICD	—Gun carriage
ICE	—Railway carriage
ICF	—Torpedo carriage.
ICG	CARRIER.
ICH	—Carrier pigeon
ICJ	CARRONADE
ICK	CARROT
ICL	CARRY-IED-ING
ICM	—Can carry
ICN	—Can you carry?
	Can not carry more sail ----------VDU
	Carried out ------- ----------------SVT
	Damaged, *or*, Sprung mast, can not carry sail ---- -----------------BH
	Have they advertised for any vessel to carry the mail?----------------DOF
	How do you carry your helm? ......OTS
	How many guns does she carry?...RKH
	I mean to carry sail all night ..... VEA
	I will carry a light I will show a light, QRY
	If you do not carry more sail we shall part company -----------------JGN
	Keep ahead and carry a light......DVK
	No use carrying so much sail......VED
	She carries guns ----------------OKM
	Will you carry a light?............QSR
ICO	CARRIED AWAY
ICP	—Carried away stem
	Funnel carried away -------------NTO
ICQ	—Have, or, Has carried away.
	I have carried away fore yard .....NKH
	I have carried away main yard ...RGH

	CARRIED AWAY—CATTLE	

	CARRIED AWAY—*Continued*.			CASH—*Continued*
	I have carried away topsail yard...ZKN			I will cash (*or*, take) your bill .....GKL
	Stokehole, *or*, Fire-room ventilator			Will you cash a bill on—for me?..GKM
	carried away ....WVY	IDV		
	Tackle, *or*, Purchase carried away..XIT	IDW		CASHIER-ED.
ICR	CART	IDX		CASK
				—Can you supply me with casks of (*indicate number of casks, if necessary*)?
ICS	CARTE BLANCHE	IDY		—Cask of water
ICT	CARTRIDGE	IDZ		—Empty casks
	Ball cartridge ....PTG			Endeavor to send a line by boat (cask,
	Blank cartridge.............. ..GMK			kite, *or*, raft, etc).................KA
ICU	—Empty cartridge cases	IEA		CAST ING
ICV	—Service rifle cartridge.			Apparatus for casting oil on the sea to
ICW	—In want of cartridges			quiet it ....ERP
ICX	—Pistol cartridges (revolver)	IEB		—Cast away
	Quick firing cartridges(or, amm'n't or),	IEC		—Castaways
	EFX			Cast iron ....IEK
ICY	—Rounds of rifle ball cartridge	IED		—Cast off
	Want pistol (revolver) cartridges ..TKV			Cast off towing cables............XYN
		IEF		—Cast to starboard
ICZ	CASH-D ING			Cast steel.................IEL
	Case (*a chest or box*) ....HDB	IEG		—Cast to port
IDA	—Case shot			Downcast ....LMG
IDB	—Cot case	IEH		—I must cast off
	Empty cartridge cases ....ICU	IEJ		—Shall I cast off ?
	Empty cases ....LXY			
IDC	—Have you a case for — ?	IEK		CAST IRON
	Mine case.... . ...RUN			
	Powder case ....TRK	IEL		CAST STEEL
IDE	—Send a — dozen cases — of —	IEM		CASTE
IDF	—Small case			
	Torpedo case ....XWQ	IEN		CASTLE
		IEO		—Forecastle
IDG	CASE (*Instance, etc*)	IEP		CASTOR OIL
	A case of fever has broken out on board			
	(*or on ship indicated*)...........GPR	IEQ		CASUALTY
IDH	—A good many slight cases			Any casualties ? ....EOP
	Any cases of fever on board (*or*, at —)?	IER		CAT (*Animal*)
	EON			
	Bad case......................FQW	IES		CAT-TED-TING TO CAT.
IDJ	—Case is for hospital	IET		—Cat fall
IDK	—Case is infectious.	IEU		—Cat head
	Case of cholera ....IPR			Cat the anchor.................KHT
IDL	—Case of dysentery	IEV		CATAMARAN
IDM	—Case of fever	IEW		CATASTROPHE
IDN	—Case of smallpox			
	Have infectious case on board .....PJH	IEX		CATCH-ING. CAUGHT
	Hospital case.....................OYR			Caught fire ... .IFG
	I have had cases of yellow fever on			There are plenty of fish to be caught
	board ....MWK			here ....NCD
	I have had no cases of yellow fever on			Torpedo catcher (*or*, chaser, *or*, destroyer)
	board ....MWL			XWR
	It alters the case ....ECO	IEY		CATHEDRAL
IDO	—Other case	IEZ		CATHOLIC ROMAN CATHOLIC
IDP	—Slight case			A visit from a Roman Catholic priest
				would be much valued ....TVY
IDQ	CASE IN CASE	IFA		CATOPTRIC LIGHT.
IDR	—In case of In the event of	IFB		CATSPAW
	In case of accident (*or*, necessity)...RP			
	In case of emergency..............LXF			CATTLE BULLOCK OX ....HLQ
	In case we part company .. ..JGO	IFC		—Cattle all well
IDS	—In that case	IFD		—Cattle men
		IFE		—Have lost — head of cattle
	CASH, SAFEQUE, *or*, LE (*Coin*).. ......ATO			
IDT	CASH			
	Can I get a bill cashed here ?........GKI			
	Can not get a bill cashed here .....GKJ			
IDU	—Cashed-ing To cash			

	CATTY—CERTIFICATE		
	CATTY (*Measure of weight*) ....BAF	IGP	CEMENT-ED-ING.
		IGQ	—Portland cement
IFG	CAUGHT CATCH-ING ....IEX	IGR	—Roman cement
	—Caught fire.	IGS	CEMETERY BURIAL GROUND
	There are plenty of fish to be caught here .. ..NCD	IGT	CENSURE-D-ING
		IGU	—Much censure for
IFH	CAULK-ED TO CAULK		CENT (*Coin*). ....ATD
IFJ	—Caulker		Per cent Percentage............TFQ
IFK	—Caulker's tools		
IFL	—Caulking.		CENTAVO (*Coin*)........ .... ATE
IFM	—Thorough caulking necessary		CENTESIMO (*Coin*) ....ATF
	You can procure caulkers and carpenters at— . ...IBU		
			CENTIARE (*Square or surface measure*), AXK
IFN	CAUSE-D-ING (*See also* REASON)		
	Better not ask the cause (reason)...FDN	IGV	CENTIGRADE (*For* THERMOMETER TABLE
IFO	—Cause, or, Reason must be assigned.		*see page* 80)
IFP	—Cause of complaint.		
IFQ	—Cause of delay (*or*, detention)		CENTIGRAMME (*Measure of weight*)..BAG
IFR	—Cause unknown		
	Causeway...................IFW		CENTILITRE (*Measure of capacity*)..AYR
	No cause of detention. ....KYD		
IFS	—Other cause		CENTIME (*Coin*) .. ..ATF
IFT	—The cause (*or*, reason) assigned is.		
IFU	—The cause (*or*, reason) assigned is sufficient		CENTIMETRE (*Measure of length*) ...AVO
	There has been so much fog as to cause considerable detention. ....NGL		Cubic centimetre (*cubic measure*)..ZNM
	What is the cause of your putting back? UDE		Square centimetre (*square measure*)AXR
IFV	—What is the cause (reason) —of —?		CENTIMO (*Coin*) ....ATF
			CENTNER (*Measure of weight*).......BAH
IFW	CAUSEWAY.		CENTNER, or, QUINTAL.
IFX	CAUSTIC		(*Measure of weight*) .... BCI
IFY	CAUTION-OUS-LY (*See also* CARE)	IGW	CENTRE-D-ING-AL ICAL
IFZ	—Cannot be too cautious	IGX	—Centre of hurricane bears —
	Caution is requisite, take care ......FT		Keep in the centre of the channel...LU
IGA	—Greatest caution is necessary.	IGY	—Storm centre in—(*direction indicated*),
IGB	—Not sufficiently cautious.		probable course towards — (*direction indicated*)
IGC	—Sufficient caution		You are steering right into the centre
	Take care, or, Be careful, caution is requisite ....FT		of cyclone (*or*, hurricane) ....PAN
	The greatest caution (*or*, care) ....HYN	IGZ	CENTRIFUGAL
IGD	—They are not cautious enough.		
IGE	—You must be cautious	IHA	CENTURY CENTENARY
IGF	CAVALRY	IHB	CEREMONY IOUS
IGH	CAVE CAVERN	IHC	CERTAIN-LY-TY (*See also* SURE)
IGJ	CEASE-D-ING		Am, Is, or, Are certain (*sure*).......XFK
IGK	—Cease firing		Are you certain (*sure*) — of — ?....XFL
	Cessation....................LJG	IHD	—Be certain before you act
	Cessation of hostilities............OZB	IHE	—Certainly not.
	Has the plague ceased at — (*place indicated*)?..................TLR	IHF	—I am certain.
		IHG	—I am not certain
		IHJ	—If you are certain
IGL	CEDAR	IHK	—Is he certain
IGM	CEDE-D-ING CESSION	IHL	—Most certain-ly
		IHM	—Not certain
IGN	CELEBRATE-D-ING-TION CELEBRITY		Will be certainly seized (*or*, captured), HXW
IGO	CELL-ULAR	IHN	CERTIFICATE
	Air cells (*or*, spaces) ....DWH		Certificate cancelled ....HUO
		IHO	—Certificate irregular
			Certificate of birth ....GLF

CERTIFICATE—CHANNEL

	CERTIFICATE—*Continued*		CHANCE—*Continued*
IHP	—Certificate of character.		If we separate, we may have a chance
IHQ	—Certificate of competency		of escape _____ ____MGY
IHR	—Certificate of death	IKE	—No chance (*or*, opportunity)
IHS	—Certificate of marriage	IKF	—No chance of peace
IHT	—Certificate of pilotage	IKG	—Only chance of safety
IHU	—Certificate of service		
IHV	—Certificate returned		
IHW	—Certificate suspended	IKH	CHANCELLOR
IHX	—Certificates are sufficient		
IHY	—Has no certificate	IKJ	CHANCERY
IHZ	—Have you, *or*, Has —(*person indicated*)		
	the necessary certificate (*or*, creden-	IKL	CHANDLER
	tials)?	IKM	—Ship chandler
IJA	—Have you, *or*, Has —(*person indicated*)		
	the proper certificate of competency?		
IJB	—Lost his (*or*, the) certificate.		CHANG, *or*, CHEUNG (*Measure of length*),
IJC	—Passing certificate—of (*or*, for)		AVP
IJD	—Proper certificate	IKN	CHANGE-D-ING-ABLE
			Any change (*or*, alteration)_____ECA
		IKO	—Change (*a change*)
IJE	CERTIFY-IED ING	IKP	—Change of mind
IJF	—Officially confirmed (*or*, certified)	IKQ	—Change of Ministry
		IKR	—Change the key of cipher to —
IJG	CESSATION		Change your anchorage_____ ___EKT
	Cessation of hostilities_____OZB	IKS	—Change your course
			Changeable weather _____YXZ
	CESSION CEDE-D-ING_____IGM		Exchange office Bureau de change.HNC
			Has there been any change in the Gov-
IJH	CHAFE D-ING		ernment in —?_____OGA
IJK	—Chafing gear		Likely to change _____QTR
IJL	—Chafing mats	IKT	—May I change (*or*, shift)?
			The wind is going to change_____ZEY
IJM	CHAIN-ED-ING.	IKU	—There will be a change
	Chain cable(*circumference in ½ inch, and*	IKV	—There will be no change until
	length in fathoms to be indicated if	IKW	—To change
	necessary) _____ _____HQW	IKX	—Unless a change takes place
IJN	—Chain locker		When wind changes_____ZFE
IJO	—Chain obstruction	IKY	—Will you change?
IJP	—Chains Channels Chain plates		
	Fasten your chain to towing cable__HRC	IKZ	CHANNEL, *or*, FAIRWAY.
IJQ	—Fore chains		About mid-channel ____ _____CZS
IJR	—Fourth-mast chains		Across channel _____DHW
	Length of chain_____QNE		Are you acquainted with the anchorage
IJS	—Main chains		(place, *or*, channel)?_____DHC
IJT	—Mizen chains		Beware of torpedoes, channel is mined,
IJU	—Mooring chain		XP
IJV	—Rudder chain	ILA	—Channel buoy
	Stream chain (cable) _____HRT	ILB	—Channel Fleet (*or*, Squadron)
	Survey chain cable_____HRV		Channel has altered, do not try it__ECD
	Want chain cable _____YB		Channel, *or*, Fairway is buoyed____OW
		ILC	—Channel, *or*, Fairway is clear
	CHAIR-MAN, *or*, PRESIDENT_____TUQ		Channel, *or*, Fairway is dangerous__FU
			Channel is dangerous without a pilot.LD
IJW	CHALK-ED-ING CHALKY.	ILD	—Channel, *or*, Fairway is likely to be
			mined
IJX	CHALLENGE-D-ING	ILE	—Channel, *or*, Fairway is narrow
			Channel, *or*, Fairway is not mined__XQ
IJY	CHAMBER.	ILF	—Channel, *or*, Fairway is obstructed
IJZ	—Chamber of Commerce	ILG	—Channel, *or*, Strait of
IKA	—Chamber of Deputies		Channel pilot____ _____TIY
IKB	—Chamber of Shipping		Clear channel—by (*or*, with) _____ITC
	Refrigerator, *or*, Cold-air chamber_DWL	ILH	—Do not know the channel (*or*, passage).
		ILJ	—Do you know the channel (*or*, passage)?
IKC	CHAMPAGNE	ILK	—Down channel
			I am in want of a channel pilot—for—TJB
IKD	CHANCE-D-ING		I know the channel (*or*, passage)___QCD
	Any chance of war? _____EOQ	ILM	—In the channel (*or*, strait) of —.
	Bad chance _____FWX	ILN	—Inner entrance (channel, *or*, passage)
	Best chance_____GHS		Is the channel buoyed?_____PC
	Good chance _____ODV	ILO	—Is the channel difficult?

CHANNEL—CHIEF

ILP	CHANNEL—*Continued*
	—Is the channel (*or*, fairway) mined?
	Keep in the centre of the channel...LU
	Keep on port shore (*or*, side of channel,
	                   LV
	Keep on starboard shore (*or*, side of
	  channel ---------------------LW
		IMX
		IMY
		IMZ
			CHART—*Continued*
			—Indifferent chart
			—Track chart—of
			—Very good chart
			Want chart of channel (*or*, harbor, etc)
			                     YH

ILQ	—Keep on the — (*compass signal to follow*) side of channel Mid channel ... ---- -- ----------RTA	INA INB INC IND INE	CHARTER-ED—BY — —Are you chartered? —Can I procure charter? —Charter is to be procured —Chartered, *or*, Employed by Govern
ILR ILS ILT ILU	—Most water is in the — channel —Outer entrance (passage, *or*, channel) —The channel is obstructed by a ship. —There is a channel (*or*, passage)—(*bear- ing indicated*)	INF ING INH INJ	ment —Charterer —Charter-party —I am chartered for — —I am not chartered
ILV	—Through, *or*, Into the channel Up channel ---------------------VLU Want chart of channel (*or*, harbor, etc) YH	INK INL INM	CHASE-D-ING —Chase-r proves-d-ing to be —Have been chased by a privateer Have been chased by the enemy-----OI
ILW	—Which channel shall I take?	INO INP	—I have been chased by — —I have been chased by a disguised war vessel
	CHANNELS CHAIN-PLATES CHAINS...IJP	INQ INR	—In pursuit (*or*, chase) of —I shall give chase
	CHAPEL ---------------------------ZNF		Keep to leeward of vessel chasing..PZN Keep to windward of vessel chasing.PZO
	CHAPLAIN, *or*, CLERGYMAN----------ITX	INS INT	—Should the chase —Strangers are chasing.
ILX	CHARACTER Certificate of character ----- -----IHP	INU INV	—Vessel chasing is a friend —Vessel chasing is an enemy
ILY	—Objectionable character	INW INX	—We hold our own with vessel chasing us —What do you think of vessel chasing?
ILZ	CHARCOAL	INY	CHASSE-MARÉE
IMA IMB	CHARGE-D-ING (*See also* ACCUSE)..DGH —Charged with desertion —Charged with drunkenness.	INZ IOA	CHEAP-LY-NESS-ER EST. —Which is cheapest?
IMC	CHARGE-D-ING (*Entrust*) Can you supply me with anyone to take charge as engineer?-------------RH Clerk in charge------------------- IUB	IOB IOC IOD IOE	CHEAT-ED-ING CHECK-ED-ING —Feed-check valves out of order —Main check valves out of order
IMD IME IMF	—In charge of — —Person to take charge —Take charge—of — Under charge (*or*, care)—of — ----HYO Under your charge (*or*, care) ------HYP	IOF IOG	CHEER-ED-ING-FUL-LY NESS CHEESE
IMG IMH IMJ IMK	CHARGE (*Of gunpowder, etc*) —Full charge —Priming charge —Reduced charge	IOH	CHEMIST (CHYMIST) CHEMICALS CHEM- ISTRY (*See also* MEDICINE)
IML	CHARGE-D-ING (*Ask money for*) Any fees (*or*, charges)? -----------EOW	IOJ IOK	CHEQUR. CHEQUER-ED —Chequered buoy
IMN IMO IMP IMQ	—Are the charges moderate? —Charge (*amount demanded*) —Charges (*or*, Expenses) are moderate —What are the charges (*or*, expenses)?	IOL	CHEST (*Part of body*)
IMR	CHARGÉ D'AFFAIRES		CHEST (*Box*) ---------------------HDB Arm chest ---------------- ------EUT
IMS	CHARITY-ABLE-Y	IOM ION	—Medicine chest. —Steam chest
IMT	CHART (*Indicate place, if necessary*) Admiralty chart-------------------DLQ		CHEUNG, *or*, CHANG (*Measure of length*), AVP
IMU IMV	—Can you spare me a chart of —? —Have, *or*, Has chart of — Have you a chart of coast (*or place indicated*) ------- ---- -----------RP	IOP	CHICKEN FOWL POULTRY .
IMW	—I have no chart of coast (*or place indi- cated*)	IOQ IOR	CHIEF LY —Chief of the State (President, Emperor, King, Queen, etc) of — has (*or*, is) —. —Chief officer
		IOS	

CHIEF—CIRCLE

	CHIEF—*Continued*
IOT	—Chief rate ing
IOU	—Commander in chief is —
	Commander in chief's office_____JEQ
IOV	—The commander in chief is at —.
IOW	—Where is commander in chief?
	CHIH. (*Measure of length*) _____AVQ
IOX	CHILD-ISH-LY-NESS
IOY	CHILI-AN　CHILIAN COLORS
IOZ	CHILL-ED-ING
IPA	CHIMNEY (*See also* FUNNEL)
IPB	—Chimney smith
	CHIN (*Measure of weight*)_____BAF
IPC	CHINA (*Earthenware*)
IPD	CHINA-ESE　CHINESE COLORS
IPE	CHISEL
IPF	CHLORIDE OF —.
IPG	—Chloride of potassium
IPH	—Chloride of zinc.
IPJ	CHLORODYNE
IPK	CHLOROFORM
	CHO, or, TCHO (*Measure of length*) _AVR
	CHO (*Square or surface measure*) __AXL
IPL	CHOCK-ED-ING
IPM	—Chock-up, or, Chock-a-block
	Rudder chock_____VBK
IPN	CHOCOLATE
	CHOICE (*See* CHOOSE) _____IPV
IPO	CHOKE-D-ING
	Condenser tubes choked (or salted)_JMV
	Pump choked _____ _____UBF
IPQ	—Sea inlet choked
IPR	CHOLERA　CASE OF CHOLERA.
IPS	—Cholera raging at —
IPT	—Ports suspected of cholera, etc
IPU	CHOOSE-ING　CHOSE-N
IPV	—Choice.
IPW	—No choice
IPX	—Take your choice
	CHRISTEN-ED-ING (*Baptize*) _____FVO
IPY	CHRISTIAN-ITY
IPZ	—Christian name
IQA	CHRISTMAS.
	After Christmas_____ _____DQR
IQB	—Before Christmas
	Christmas day _____KMW

IQC	CHRONOMETER
IQD	—Can you spare me a chronometer?
IQE	—Have great faith in my chronometer
IQF	—How many chronometers have you?
IQG	—I have a chronometer (*number to follow,*
	if necessary).
IQH	—I have no chronometer
	My chronometer gains daily — _____GP
	My chronometer has run down ____GQ
	My chronometer is fast of Greenwich
	(or, first meridian) mean time____GR
	My chronometer is slow of Greenwich
	(or, first meridian) mean time ___GS
	N B—*See note under* GREENWICH,
	page 258
	My chronometer loses daily — _____GT
IQJ	—My chronometers were rated — days
	ago at — (*indicate place, if necessary*)
	My longitude by chronometer is —.QZF
	What is your longitude by chronometer?
	QZL
IQK	—When were your chronometers last
	rated?
	Will you give me a comparison? Wish
	to get a rate for my chronometer *_GU
IQL	—Your chronometer must be fast
IQM	—Your chronometer must be slow
	NOTE —The vessel showing the mean
	time will hoist the signal denoting the
	hour, and dip it sharply shortly after it
	has been answered, the minutes and
	seconds, at the instant of dipping, will
	immediately follow To insure accu-
	racy, a second comparison should be
	made.
	CHTOF (*Measure of capacity*)_____AYS
IQN	CHURCH
	CHYMIST (*See* CHEMIST)_____ IOH
IQO	CIGAR
IQP	CIGARETTE
	CINDER　ASH_____FCR
IQR	CIPHER
	Change the key of cipher—to —_____IKR
IQS	—Cipher is ended
IQT	—Cipher message
IQU	—Communicate in cipher.
IQV	—Following communication (signal, or,
	message) is secret and in cipher
	NOTE.—*The cipher will continue until*
	the person signaling hoists IQS=Ci-
	pher is ended
IQW	—Give me the key to cipher
IQX	—Have not got the key to cipher
IQY	CIRCLE-ULAR
IQZ	—Circulate-d-ing-ation
IRA	—Circulating engine
IRB	—Circulating pump
	Great circle sailing _____OHA
	Turning circle _____YEM

CIRCUIT—CLEAR

IRC	CIRCUIT-OUS-LY	
IRD	—Circuit closer	
	Earth circuit ....DNZ	LRM
	Electric circuit ....LUR	
	Short circuit.....................VZU	
IRE	CIRCUMFERENCE	
IRF	CIRCUMSTANCE	
	Adverse circumstances ....DNZ	
	Alters the circumstances...........EBW	
IRG	—Depends upon circumstances	
IRH	—Existing circumstances	
	I shall anchor if circumstances permit *(place or bearing to follow, if necessary)* ....EIT	
IRJ	—Under no circumstances	
IRK	—Under the circumstances.	
IRL	CISTERN	
IRM	CITADEL	
IRN	CITY *(Town)*	
IRO	—Citizen	
	CIVIC AUTHORITIES ALDERMAN.....DWS	
IRP	CIVIL LY-ITY	
IRQ	—Civil service	
IRS	—Civil war	
IRT	CIVILIAN	
IRU	CLACK VALVE	
IRV	CLAD *(See also* CLOTHE *)*	
	Iron-clad Armor-clad............EVS	
IRW	CLAIM-ED-ING-ABLE-ANT	
IRX	—Has, *or,* Have been claimed by	
IRY	—I have a claim on	
	Prior claim....................TWG	
	What is the demand (*or,* claim)?...KTH	
IRZ	CLAMP-ED-ING	
ISA	—Clamping screw	
ISB	CLAP-PER	
ISC	—Clap on Clapping on.	
ISD	CLARET	
ISE	CLASS-ED-ING TO CLASS	
	All classes....................DXH	
ISF	—First class	
	I am not classed at Lloyd's ....QWM	
ISG	—Not classed	
ISH	—Second class	
ISJ	—Third class	
ISK	CLASSIFY-IED-ING-ICATION	
ISL	CLAUSE	
ISM	CLAW-ED-ING OFF	
ISN	CLAY	
ISO	—Stiff clay	

ISP	CLEAN-ED-ING-LY-NESS	
	Boilers require cleaning ....GWK	
	Clean bill of health—from —........TQ	
	Clean boiler ....GWS	
ISQ	—Clean engine	
ISR	—Clean fires	
IST	—Clean ship	
ISU	—Clean shirts Clean linen (*or,* clothes).	
ISV	—Cleaning-gear for decks	
	Have you a clean bill of health?.....TU	
	How long do you require to clean your engine?......................QYM	
	I have a clean bill of health, but am liable to quarantine ...Code Flag over Q	
	I have not a clean bill of health. Code Flag over I	
	My engine requires cleaning.......MBQ	
ISW	—Not clean bill of health flag (*when spoken of)*	
ISX	—Requires cleaning	
ISY	CLEAR-ED-ING-ANCE	
	All clear ....DXI	
ISZ	—Are you clear?	
	Bar can not be cleared until — ....FVR	
	Can I clear the bar?.............FVX	
	Canal is clear...................HUK	
ITA	—Can not clear to-night (*or before time indicated*)	
	Channel, *or,* Fairway is clear ......ILO	
	Clear anchor............. ....EHU	
ITB	—Clear away	
	Clear berth....................GFP	
ITC	—Clear channel (by, *or,* with)	
	Clear hawse...... ....OPL	
ITD	—Clear sky	
ITE	—Cleared at the custom-house	
ITF	—Cleared for	
ITG	—Clearing house	
ITH	—Clearing papers	
	Clearing the custom-house.........KGS	
ITJ	—Clearing up	
ITK	—Clearly make out	
ITL	—Clearly seen	
	Have you cleared the custom-house?.KGY	
	I have received information — (*place next indicated*) is clear of ice.....FBO	
	I only got clear of the ice (*or,* icebergs) on — ....PBS	
ITM	—I shall clear (*or,* go clear)	
	Is canal clear?..................HUL	
ITN	—Is, *or,* Are clear of	
ITO	—Nearly clear — of —	
ITP	—Not able to clear to-day (*or until date specified*)	
ITQ	—Not clear	
	Not cleared at custom-house ......KHA	
	Prepare to be towed clear ....XYS	
	Ran clear......................UHP	
	Require diver to clear screw ......LID	
	Shall clear at custom-house........EHB	
	Sight your anchor Make sure your anchor is clear..................EJR	
	Take care range is clear..........HYM	
	The bank was clear of fog ....FVI	
ITR	—Want my clearance (*or,* papers)	
	When will be the best time for clearing the bar? ....XM	
ITS	—When will you be clear?	
ITU	—Will not clear us	
	You are all clear of danger ....GM	

	CLEAR—COAL			

ITV	CLEAR—*Continued* —You will not clear Your flags are not clear ....NDT		CLOSE—*Continued* Keep close during the fog ....PYO Keep close order ....PYQ Keep close to...................PYR	
ITW	CLEAT		Keep him (*or*, them) in close confine- ment.........................PYU	
ITX	CLERGYMAN, PARSON, *or*, MINISTER		Keep under the land Keep, *or*, Stand	
ITY	—A visit from a Protestant clergyman would be much valued		closer in ....LT Mail closes—at.................RFD	
		IVH	—Office is closed	
ITZ	CLERK		Stand closer in................. WQF	
IUA	—A clerk from	IVJ	—Too close (*or*, near).	
IUB	—Clerk in charge		Will close (*or*, be closed) ....ZEG	
	Telegraph clerk (*or*, official) ....XLZ	IVK	—Will you keep close to me during the night?	
IUC	CLEVER-LY-NESS	IVL	—You are too close in , keep farther off	
IUD	CLEW-ED-ING		Your port of destination is closed , your owners desire you to proceed to..KXJ	
IUE	—Clew-lines, *or*, Clew-garnets.			
IUF	—Clew up	IVM	CLOTH	
IUG	—Fore clew-garnets	IVN	—Emery cloth	
IUH	—Main clew-garnets	IVO	—Sail cloth	
IUJ	CLIENT CLIENTÈLE	IVP	CLOTHE-D-ING Bed clothes..................GCS	
IUK	CLIFF	IVQ	—Can I get clothes washed?	
IUL	CLIMATE-IC		Clad ....IRV	
IUM	CLINCH-ED-ING	IVR	Clean clothes (linen, *or*, shirts).....ISU —Clothing for female passengers	
	Are your cables clinched? ....HQN		Women's clothing	
		IVS	—Clothing for men Slops	
	CLINKER BUILT...............HLB	IVT	Dirty clothes (linen, *or*, shirts) ....VYO	
		IVU	—Night clothing.	
IUN	CLINOMETER	IVW	—Plain clothes —Summer clothing	
IUO	CLIPPER		Want clothes.................YTS	
	Clipper built..................HLC	IVX	—Warm, *or*, Winter clothing—for	
		IVY	—Wet clothes	
IUP	CLOCK	IVZ	CLOUD-ED-Y	
IUQ	—At — o'clock I shall steer	IWA	—Clouds are banking up.	
IUR	—Clockwork	IWB	—Cloudy sky	
	I shall tack at — o'clock ....XIK			
	O'clock ....ZNG	IWC	CLUB-BED-BING	
IUS	—What o'clock is it?	IWD	—Club Clubhouse.	
		IWE	—Yacht club	
IUT	CLOSE-D-ING TO CLOSE (IN, *or*, TO)			
	As close as....................FAD	IWF	CLUBHAUL-ED-ING	
	Bring the ship in as close as possible .HIO	IWG	—Try to clubhaul.	
	Circuit closer..................IRD			
IUV	—Close confinement	IWH	CLUE	
IUW	—Close hauled			
IUX	—Close , I wish to communicate	IWJ	COACH-ED-ING	
IUY	—Close luff			
IUZ	—Close on the	IWK	COAL ED-ING TO COAL	
IVA	—Close quarters		After coaling.................DQS	
IVB	—Close reef-ed ing		Anthracite coal ....ENX	
IVC	—Close to (near)		Are you, *or*, Is — (*vessel indicated*) in want of coal? ....GW	
IVD	—Closer-est		Bad coal ....FQY	
	Convoy to keep as close as possible ..IV		Bunker coal..................HLZ	
IVE	—Do not pass too close to me		Can coal be had at —?...........GX	
	Enemy is closing with you, *or*, You are closing with the enemy .... OD		Can you spare me coal?...........GY	
	Gate is closed at ....NXF	IWL	—Coal	
	I shall close reef ....UMO		Coal bag ....FSH	
	I shall keep close hauled...........OPH		Coal boat ....GSL	
IVF	—I will keep close to you during the night		Coal bunker ....HLY	
	I wish to communicate, close......IUX		Coal can be got at—(*or*, from—)....HA	
IVG	—Is — (*port indicated*) closed?	IWM	—Coal can not be got in any quantity— at—	
	Keep as close as you can to pick up my people ....PYL		Coal cargo ....HYW	

COAL—CODE

	COAL—*Continued*
IWN	—Coal depot
IWO	—Coal hulk
IWP	—Coal lighter
IWQ	—Coal mines
IWR	—Coal requires much draft
IWS	—Coal tar
IWT	—Coal tip
IWU	—Coal trimmer, *or*, Coal passer
	—Coal — shillings a ton at — _____GZ
IWV	—Coaling party
IWX	—Coaling station.
IWY	—Coaling suit
IWZ	—Collier
IXA	—Colliery
	Complete with coal_____ JKN
	Distressed for want of coal _____LHD
	Do you want coal?_____OW
	Good coal _____ _____ODW
IXB	—Good coal may be got at —
IXC	—How many hours' coal have you (*state at what speed*)?
IXD	—I have — hours' coal, for full speed (*or for speed indicated*)
	In want of coal (*state amount*)_____HB
	Indicate nearest place where I can procure coal ._____ . HC
IXE	—Indifferent coal
	No coal to be got at —_____HD
IXF	—North country coal
	Owners desire me to inform you to coal at — _____HE
IXG	—Plenty of coal to be got—at —
IXH	—Prepare to coal
IXJ	—Steam coal
IXK	—Tons of coal
	Want coal immediately _____YI
IXL	—Welsh coal
IXM	—What is your consumption of coal per hour, at full speed?
IXN	—What price are coals at?
	What quantity of coal have you left? HF
IXO	—What quantity of coal remains in the collier (coal depot, *or*, lighter)?
IXP	—When finished coaling
IXQ	—When will you have finished coaling?
IXR	COALESCE-D-ING COALITION
IXS	COAMING COMB-ING
IXT	COARSE-LY-NESS-R
	Coarse bottom Coarse sand and stones, HAX
IXU	COAST COAST OF
	All the lights are out along the coast— of — _____PU
	Am acquainted with the coast_____DGY
	Am not acquainted with the coast__DHB
	Are you acquainted with the coast?.DHE
IXV	—Coast-defense ship
IXW	—Coast is dangerous
IXY	—Coast is not dangerous
IXZ	—Coasted-ing To coast
IYA	—Coasting trade
IYB	—Coasting vessel Coaster
	Do not approach the coast because it is mined ____ _____ _____ _____ETN
	Have you a chart ot coast (*or place indicated*)?_____ ___ _____RP

	COAST—*Continued*
	I have no chart of coast (*or place indicated*) _____ _____IMW
	I know (*or*, am acquainted with) the coast _____ _ _____ ____DGY
IYC	—Is the coast bold?
IYD	—Off the coast Off shore
IYE	—On the coast
IYF	—On, *or*, At the North coast of —
IYG	—On, *or*, At the NE coast of —
IYH	—On, *or*, At the East coast of —
IYJ	—On, *or*, At the SE coast of —
IYK	—On, *or*, At the South coast of —
IYL	—On, *or*, At the SW coast of —
IYM	—On, *or*, At the West coast of —
IYN	—On, *or*, At the NW coast of —
	Rocky coast_____ _____ UYZ
IYO	—Wild coast
	Uninhabited island (*or*, coast)_____PTS
	Wrecked on the coast_____ ZJD
	COASTER (*Coasting vessel*) _____IYB
IYP	COASTGUARD-MAN.
IYQ	—Coastguard station
IYR	COAT-ED ING
	Rudder coat__ _____VBL
IYS	COBBLE-ER
IYT	COCHINEAL.
IYU	COCK
	Blow off cock (*or*, gear) _____GNW
	Blow-off cock out of order _____GNX
	Brine cock _____ _____HIG
	Flooding cock_____NFP
	Gauge cock _____ _____NXJ
	Sea cock_____ _____VKY
	Sea inlet valve (*or*, cock) broken (*or* damaged)_____VLA
IYV	COCKSWAIN COXSWAIN
IYW	COCOA.
IYX	COCOANUT
IYZ	—Copra
IZA	COD
IZB	—Cod fishery
IZC	CODE
	Attention is called to page —, paragraph —, of the INTERNATIONAL CODE ___DH
	N B.—*The page to be shown first, and then the paragraph This signal should be employed only between vessels using the United States edition, as otherwise the pages referred to will not correspond*
IZD	—Code List of ships
IZE	—Can you furnish (*or*, lend me) a Code of Signals?
IZF	—Code Flag (*when spoken of*)
	Do not understand Morse Code ____ SAI
	Do you understand Morse Code?____SAJ
	International Code of Signals_____PRB
	Morse Code _____SAK

	CODE—COLZA		
	CODE—*Continued*	JAS	COLLIERY ....IXA
IZG	—Show your Code Signal		COLLISION
	Telegraph code ....XMA		Collision bulkhead ....HLM
	Use Morse Code ....SAL	JAT	—Collision mat
IZH	—What is the date of your Code List?		Has the vessel with which you have been
	Will use Morse Code ....SAM		in collision proceeded on her voy-
IZJ	COFFEE		age?..........................HG
		JAU	—Have been in collision
IZK	COFFERDAM	JAV	—Have, or, Has been in collision with —
			(*vessel indicated*)
IZL	COFFIN		Have you been in collision? ....HI
		JAW	—Have you been in collision with — (*vessel*
IZM	COG COGWHEEL		*indicated*)?
			I have been in collision with — ....HJ
IZN	COIL-ED-ING	JAX	—No collision
IZO	—Coil of rope	JAY	—There has been a collision between —
	Resistance coil ....UTV		(*vessels indicated*)
	Test coil.......... ....XOK	JAZ	—Vessel indicated has been in collision.
		JBA	—When did the collision take place?
	COIN (*Money*)..............ASW	JBC	—Where did the collision take place?
		JBD	—Where is the vessel that collided with
IZP	COIN-ED-ING ER AGE TO COIN		you?
IZQ	—What coins are used?		With what vessel have you been in col-
			lision? ....HN
IZR	COIR	JBE	COLLODION
	Coir, or, Grass cable ....HQZ		
IZS	—Coir, or, Grass hawser	JBF	COLOMBIA COLORS
IZT	—Coir, or, Grass matting		
IZU	—Coir, or, Grass rope		COLONEL ....EWI
		JBG	—Colonel commandant
IZV	COKE	JBH	—Lieutenant colonel-cy
IZW	COLD-LY-NESS-ISH-ER-EST	JBI	COLONY-IAL
	Cold-air chamber ....DWL		Colonial Agent ....DSO
IZX	—Cold weather	JBK	—Colonial Government
		JBL	—Colonial Office
IZY	COLIC	JBM	—Colonial Secretary
JAB	COLLAPSE-D-ING-IBLE	JBN	COLONIZE-D-ING-ATION
	Collapsible, or, Berthon boat .... GHM		
	Furnace crown of boiler collapsed..NUC	JBO	COLOR
	Furnaces collapsed..............NUB		Color-ed ing To color ....JBR
JAC	COLLAR		Color blind ness..................GMS
JAD	—Thrust collar damaged		Color vision....................YRB
		JBP	—The same color (*or*, colors)
JAE	COLLECT-ED ING	JBQ	—What color?
	Collection-or ....JAK		
JAF	—Collect all the —	JBR	COLOR-ED-ING TO COLOR
JAG	—Collect what you can		
JAH	—Could not be collected	JBS	COLORS (ENSIGN)
	Enemy is collecting in force at —..LZD		Can not distinguish her (*or*, your)
JAI	—Will you collect?		colors........................LGX
			Dip your colors ....LBC
JAK	COLLECTION-OR		Foreign colors ....MEK
	Collector, or, Board of customs....GOY		He has hoisted — colors..........JBT
			Hoist your colors Show your ensign.DW
JAL	COLLEGE-IAN		Neutral colors ....MER
			Strike your colors ....XAJ
JAM	COLLIDE-D-ING (*See also* COLLISION)	JBT	—They have, or He has hoisted — colors
		JBU	—What colors has she hoisted?
	COLLIER ....IWZ	JBV	—What colors shall we hoist?
JAN	—Collier to proceed to —		
JAO	—Have you seen a (*or*, the) collier?	JBW	COLUMN-AL
JAP	—Sailing collier	JBX	—Number of columns
JAQ	—Steam collier.		Rear column.......... ......UKL
	What quantity of coal remains in the		
	collier (coal depot, or, lighter)?..IXO	JBY	COLZA
JAR	—Where is the collier?	JBZ	—Colza oil

COMB—COMMAND

JCA	COMB-ING. COAMING _____ IXS —Honeycomb ed
JCB JCD	COM⁻INE-ING-ATION—WITH —Combined

JCE	COMBUSTIBLE-ILITY TIÓN
	Have you any combustibles on board?
	    HO
	I am loaded with combustibles_____HP
	I have no combustibles on board ____HQ
	No combustibles (or, explosives) near
	    the fire_____HR
	Spontaneous combustion _____WOJ
	Vessels with combustibles not allowed
	    in _____HS

JCF JCG	COME CAME To COME —Are you coming? Boat, or, Lifeboat can not come_____HR Boat come-ing (came) _____GRL Boat is coming to your assistance ___HT Boat's crew will not come _____GRX
JCH	—Came, Coming by rail
JCI	—Came, Coming by steamer
JCK	—Came, Coming from
JCL	—Can come
JCM	—Can I come? Can I come alongside?_____HAY

JCN	—Can not come
	Can not come alongside_____HAZ
	Can not distinguish your flags, come
	    nearer (or, make distant signals)__VM
	Captain is requested to come on board,
	    HWN
	Come nearer, Stop, or, Heave-to, I have
	    something important to communicate,
	    Code Flag over H

JCO	—Come under the stern
JCP	—Come up Come alongside _____HBA Come directly (immediately)
JCQ	—Come home
JCR	—Come in
JCS	—Come nearer
JCT	Come on board_____GQA
JCU	—Come on shore
JCV	—Come out
JCW	—Come out of port
JCX	—Come round
JCY	—Come to grief
JCZ	—Come to hand
JDA	—Come within hail
JDB	—Coming.
JDC	—Coming at once Dangerous to, or, Do not come into less than — feet of water _____FV Do not come alongside_____HBC
JDE	—Do not come so near. Do you come from any port putting you in quarantine? _____TR Endeavor to come alongside _____ ___HBD
JDF	—Have, or, Has come Heavy weather coming, look sharp__FZ I am coming (or, sending) aboard __CZH I must come to an anchor _____HIR

JDG	—I shall come off by and bye (or at time
	    indicated)
	I shall fire into the boats if they persist
	    in coming alongside _____ __NAK

JDH	COME—Continued
	—I will come
	I will come on board (or, alongside), if
	    you will allow me _____CZD
	I wish to signal, will you come within
	    easy signal distance? _ _____VJ
	Is the captain coming on shore? ___HXD
	Let no boat come alongside_____HBH
	Lifeboat unable to come_____HR
	Lights, or, Fires will be kept at the best
	    place for coming on shore_____KE
	Lighter coming off _ _____ ___QZ

JDI JDK	—Must come —Not come Pilot has not come off_____TJK Request captain to come here (or to place indicated) __ _____ ____HXF

JDL
JDM
	—Shall, or, Will come
	—Shall I come ?
	Sheer off a little when you come to the
	    point ---- ----------- ---------VUP
	Should it come on to blow—from —
	    (quarter indicated, if necessary)__GOK
	Stop, Heave-to, or, Come nearer, I have
	    something important to communicate,
	    Code Flag over H
	Surgeon, or, Doctor will come imme-
	    diately _____WN
	The consul is desired to come (or, send)
	    on board _____JRP
	The orders are coming off_____SUP
	The wind will come from the —. ..ZFA
	Thick fog coming on_____GJ
	Want to come into harbor _____ONC
	We are coming to your assistance___DC
	What water shall I come into ?____YWO

JDN	—When did you come off?
JDO	—When do you come off?
JDP	—When you come Where, or, What source did the news come from? _ _____SEN Where do you come from?. _____SI Will Lloyd's agent come on board ?_QXK
JDQ	—Will you (or person indicated) come ? Will you (or person indicated) come on board ?_____GRF

JDR	—Will you (or person indicated) come on
	    shore to-day (specify hour, if neces
	    sary)?

JDS	—Will you come (or go) on shore?
	Will your pilot come on to me? ____TJW
	Will your surgeon (or, doctor) come on
	    board? _____WP
	You are ordered not to come into
	    port _____SUY

JDT	—You may come into port. You may, or, We may come to an anchor _____ _____HKN
JDU	COMET
JDV JDW	COMFORT-ED-ING —Comfortable-y. Medical comforts . _____ROH Medical comforts running short____ROI

JDX	COMMAND-ED-ING To COMMAND
	Admiral, or, Senior officer has com-
	    mands for you_____DKP
	Admiral, or, Senior officer has no com-
	    mands for you _____ __ _____DKQ
	Any commands _____HOR

COMMAND—COMMUNICATE

	COMMAND—*Continued*
JDY	—Commanding officer
JDZ	—Commanding officer desires
JEA	—Have you any commands?
JEB	—I have no commands
	Keep the vessel under command LX
JEC	—Not under command lantern (lamp, *or*, light)
JED	—Not under command
	Not under command shapes ....\ IU
	N B —"*Not under control or command*" signal is two black balls or shapes vertical
JEF	—She is commanded by —
JEG	—Take command—*of*
JEH	—Under command of —
JEI	—Where is the commanding officer?
JEK	—Whom is the — (*vessel indicated*) commanded by?
JEL	COMMANDANT
	Colonel commandant ....JBG
JEM	COMMANDER.
JEN	—Inform the commander
JEO	—Senior commander
JEP	—What is commander's name?
	COMMANDER IN CHIEF ....EWJ
JEQ	—Commander in chief's office
	Commander in chief is — ....ICU
	The commander in chief is at —....ICV
	Where is the commander in chief? .IOW
JER	COMMEMORATE-D ING
JES	COMMENCE-D-ING-MENT
JET	—At the commencement of —
	Commence action ....DIS
JEU	—Commencement of hostilities.
	Demurrage commenced on the — ..KTN
	Has war commenced? *or*, Is war declared? ....YU
	Hostilities commenced—between — and —........................OZC
	Light indicated will commence on (*date to follow*) ....QSF
JEV	—Not commenced.
	Open fire, *or*, Commence fire.......SRD
	Operations have commenced ....SRG
	War has commenced *or*, War is declared ....YX
	When does it commence? ....GDU
	When do you commence discharging cargo? ....IBJ
JEW	—When do you begin (*or*, commence)?
	When does the blockade commence?.GNQ
JEX	COMMERCE-IAL-LY
	Chamber of Commerce ....LJZ
JEY	COMMISSAIRE OF MARITIME INSCRIPTION
JEZ	COMMISSARY-IAT
JFA	—Commissariat department
JFB	—Commissariat stores
JFC	COMMISSION-ED-ING
JFD	—Commissioned officer
JFE	—Commissioner

	COMMISSION—*Continued*
	Commissioner of Navigation, Treasury Department, or Registrar General of Shipping and Seamen ....UNX
JFG	COMMIT-TED TING
	Commit-ted-ting an assault.........FDY
JFH	COMMITTEE
	Committee of Lloyd's. ....QWE
JFI	COMMODITY
JFK	COMMODORE
JFL	COMMON-LY-NESS-ALTY
JFM	—House of Commons
JFN	COMMUNICATE-D-ING TO COMMUNICATE.
	Allow no communication Allow no one on board ....FM
	Am disabled, communicate with me,CL
	Ask stranger (*or*, *vessel indicated*) if he will communicate ....HT
JFO	—Be more concise in your communication
	Be very careful in your intercourse with strangers ....PS
	Can you forward my communication by telegraph? .... HU
	Close, I wish to communicate ......IUX
	Come nearer, Stop, *or*, Heave-to, I have something important to communicate ....Code Flag over H
	Communicate by distant signals ...LGC
JFP	—Communicate by telegraph
JFQ	Communicate in cipher ....IQU
JFR	—Communicate the following
JFS	—Communication
JFT	—Communication by telegraph is
	—Communication by telegraph is restored
	Communication by telegraph is stopped, WS
	Communication is allowed with....DZG
JFU	—Communication is established—by (*or*, with) —
JFV	—Communication is interrupted—by (*or*, with) —
JFW	—Comprehend the communication perfectly
JFX	—Confidential communication (*or*, report)
	Do not communicate with —.......HV
JFY	—Does telegraphic communication go all the way?
	Electrical communication LVC
	Following communication is confidential (*or*, private) ....HX
	Following communication (signal, *or*, message) is secret, and in cipher..IQV
	NOTE —*The cipher will continue until ended by the person signaling making the hoist* IQS=Cipher is ended
	Forward my communication by telegraph, and pay for its transmission, HY
JFZ	—Had communication with
	Have no communication with — ....HV
	Have no communication with the shore (*or with vessel indicated*)... ... ..HZ
	Have received the following communication (*or*, instructions) from your agent (*or*, owner) ....IA

COMMUNICATE—COMPASS

	COMMUNICATE—Continued
	Have you any message (telegram, order, or, communication) for me? ....IB
	Have you had any communication with infected vessel (or, place)? IC
JGA	—Have you had any communication with the shore?
	I am going to communicate by distant signals................. ....LGF
	I am going to signal the contents of an important telegram, to be communicated to you .... .IE
	I can not stop to receive (or, have) any communication ....IF
	I have had no communication with — (place or vessel).....................IG
JGB	—I shall communicate with —
	I should wish to know the nature of the sickness, if any, before I send my boat (or, communicate) ....FC
	I wish to communicate, close........IUX
JGC	—I wish to communicate personally
	Important to communicate — by (or, with)—........... ....IH
JGD	—Is the communication with — regular?
	Line of communication ....QUN
	No telegraphic communication.....XLS
	Report me (or, my communication) to Lloyd's agent at — ....UH
	Stop, Heave-to, or, Come nearer, I have something important to communicate, Code Flag over H
	Stranger, or, Vessel bearing — wishes to communicate....................WZD
	Telegraphic communication ..... .XLN
	Telephonic communication ... .XMN
	The following signal (or, communication) is private...................HX
JGE	—This communication is private
	Unless your communication is very important, I must be excused.........LJ
	Will you forward (or, communicate) the following signal for me? ....IK
	Will you telegraph to — (ship or person named) the intelligence I am about to communicate? ....IM
JGF	—Wish to communicate by (or, with)
	You can communicate with —, and get an answer in —.................ENT
JGH	**COMMUTATOR**
JGI	**COMPANY-ION**
	All ships of the convoy are to rejoin company...........Code Flag over Y
JGK	—And Co (company) Messrs & Co
JGL	—Better part company
	Company (and Co)...............JGK
JGM	—I shall part company
JGN	—If you do not carry more sail we shall part company
JGO	—In case we part company
JGP	—Keep company
	Let us keep together (or, in company) for mutual support (or, protection) .IN
JGQ	—Part company
	Parted company with — on the — (or, at —) ....IO
JGR	—Permission to part company
	Shall we keep company? ....IP

	COMPANY—Continued
	The admiral (or, senior officer) requests the pleasure of your company to dinner ....DKX
JGS	—The pleasure of your company
JGT	—Very glad of your company
JGU	—Vessel, or, Ship may part company.
	What company do you belong to? ...IQ
	When did you part company—from?.IR
JGV	—Will you give me the pleasure of your company at —?
JGW	**COMPARE-D-ISON**
JGX	—Comparative-ly
	Should like to have another comparison, GV
	Wish to get a rate for my chronometer, will you give me a comparison?...GU
JGY	**COMPARTMENT**
	Bow compartment ....HCE
JGZ	—Engine-room compartment
JHA	—Stoke-hold compartment
JHB	—Water-tight compartment
JHC	**COMPASS**
	(For **COMPASS SIGNALS,** see pages 45 and 46)
	NOTE—*Unless otherwise indicated, bearings if in points and half points are to be understood as* COMPASS BEARINGS, *i e , bearings as shown by the compass and uncorrected for local attraction or variation. But when shown in degrees they are to be understood as* MAGNETIC BEARINGS
JHD	—Adjust-ed-ing compasses
	Are your bearings true or magnetic bearings? ....GBD
JHE	—Are your compasses adjusted?
	Azimuth compass ....FQE
	Bearings by compass are —........GBH
	Boat's compass ....GRV
JHF	—Compass bearing (i e the bearing shown by the compass and uncorrected for either deviation or variation)
JHG	—Compass card
JHI	—Compass signals in degrees
JHK	—Compass signals in points and half points
JHL	—Compasses not adjusted lately
JHM	—Compensation of compass
JHN	—Deviation of compass
JHO	—Deviation table
JHP	—Floating, or, Liquid compass
JHQ	—Gimbal of a compass
	Have you an azimuth compass?....FQH
JHR	—Have you any local attraction (or, deviation)?
JHS	—Have you taken observations for variation?
JHT	—I am using variation of—degrees
	I have an azimuth compass.........FQI
JHU	—I have no cargo to disturb my compasses
JHV	—I have had no observations for variation lately
JHW	—Is there any variation?
JHX	—Magnetic bearing
	NOTE—*Magnetic bearings are compass bearings corrected for deviation*

	COMPASS—*Continued*
JHY	—Must be swung to adjust compasses
JHZ	—My compasses were adjusted at —
JIA	—No variation
	Point (*Of compass*)........ ....TNL
JIB	—Pole, *or*, Standard compass
JIC	—Prismatic compass
JID	—Put a compass in the boat
JIE	—Ship's head by compass—is —
JIF	—Steer course indicated by compass signal
JIG	—Thomson's (Kelvin) patent compass
JIH	—True bearing (*i e compass bearing corrected for deviation and variation* *All bearings are to be reckoned* FROM *the vessel making the signal*)
JIK	—Variation of the compass
JIL	—Vessel indicated bears
JIM	—Was swung for adjustment of compasses at—
	What is the bearing by compass?...GBM
JIN	—What is the variation?
JIO	—What is your deviation?
JIP	—What variation do you allow?
JIQ	—When, *or*, Where were your compasses last adjusted?
JIR	COMPEL LED - LING - PULSION (*See also* BOUND *and* OBLIGE)
JIS	—Am I compelled by law to have a pilot?
JIT	—Can not be compelled
JIU	—Compelled them to
JIV	—Not without compulsion
	Vessel indicated was compelled to bear up ------------------------------GBX
JIW	—Will be compelled
JIX	COMPENSATE-D ING-ATION
	Compensation of compass ---------JHM
JIY	COMPETE-ITION
JIZ	COMPETENT-CE-CY
	Certificate of competency----------IHQ
	Have you, *or*, Has —(*person indicated*) the proper certificate of competency? IJA
JKA	COMPLAIN-ED ING
	Cause of complaint ----------------IFP
JKB	—Have, *or*, Has made a complaint
JKC	—No complaint—of
	Passengers, *or*, Crew sick with infectious complaint----------------------PCX
JKD	COMPLEMENT
JKE	—Short of complement
JKF	—What is your complement?
JKG	COMLETE-D-ING TO COMPLETE—WITH
JKH	—Completely-ion
	A complete wreck-----------------ZIS
	Are all your bills of lading complete and signed? ----------------GKQ
JKI	—Are you complete?
	Arrangement completed --------- EWX
JKL	—Can you complete?
JKM	—Complete refit necessary
JKN	—Complete with coal
JKO	—Complete with water
	Engines completely disabled--------MQ
JKP	—I am (*or vessel indicated* is) a complete wreck

	COMPLETE—*Continued*
JKQ	—I shall be complete in
JKR	—Must be complete d—with
JKS	—Report when finished (*or*, complete-d)—with
JKT	—When completed
JKU	—When will you (*or vessel indicated*) be complete?
JKV	—When you are complete—with
JKW	—Will take time—to complete
JKX	—With complete success
JKY	COMPLICATE-D-ING-ION
JKZ	COMPLIMENT-ED-ING-ARY
	Make my compliments—to---------RIM
JLA	COMPLY-IED ING. TO COMPLY
JLB	—Can comply—with
JLC	—Can not comply—with
JLD	—Comply with request (*or*, signal)
JLE	—In compliance with
	Request can be complied with------USP
JLF	—Shall be complied with
	Sorry I am unable to comply with your request -------- ----------------USX
JLG	—Vessel (*or person indicated*) is not able to comply
JLH	COMPOSE-D-ING-ITE-ITION
	A squadron (*give nationality*) composed of —ships passed in sight-------WPF
	Antifouling composition---- ---- ENZ
	Of how many ships was the squadron seen composed? -----------------VXM
JLI	—To be composed of
JLK	COMPOUND ED-ING
JLM	—Compound engine
JLN	COMPREHEND-ED-ING TO COMPREHEND
	Comprehend the communication perfectly ---------------------JFW
JLO	—Do you, *or*, Does he (*or—person indicated*) comprehend (*or*, understand)?
JLP	—Not sufficiently comprehensible It is impossible to comprehend
JLQ	COMPRESS-ED-ING TO COMPRESS
	Air-compressing pump-----------DVY
	Compressed air---- ----------- ---DWM
JLR	COMPRESSOR
JLS	COMPROMISE-D-ING
JLT	—Do not compromise yourself
JLU	COMPULSORY-ION
	Not without compulsion -----------JIV
	Pilotage is compulsory-------------TJO
JLV	CONCEAL-ED-ING-MENT
JLW	CONCEDE-D-ING
JLX	CONCEIVE ABLE-Y
JLY	CONCENTRATE D ING-ION
JLZ	CONCERN-ED ING

CONCERT—CONGRESS

JMA	CONCERT-ED-ING		JNK	CONDY'S FLUID CRIMSON FLUID
JMB	—In concert with		JNL	CONE
	Shall we act in concert?_____ DIN		JNM	—North cone hoisted
JMC	CONCESSION		JNO	—South cone hoisted
			JNP	—Steam cone
JMD	CONCILIATE-D-TION			
			JNQ	CONFEDERATE-D-ING-ION-CY
JME	CONCISE-LY-NESS			
	Be more concise in your communica-		JNR	CONFER-RED-RING-ENCE
	tion _____JFO		JNS	CONFESS-ED-ING-ION-EDLY
JMF	CONCLUDE-D-ING TO CONCLUDE			
JMG	—Conclusion-ive		JNT	CONFIDENCE-T
JMH	—Not yet concluded		JNU	—Am, Is, or, Are confident
JMI	—Shall conclude		JNV	—Confidencia.ly
				Confidential report (or, communicaticn,
JMK	CONDEMN-ED ING			JFK
JML	—Condemned as unseaworthy			Following signal (or, communication)
	Condemned by Lloyd's agent_____QWF			is confidential (or, private)_____HX
JMN	—Condemned stores		JNW	—Have you any confidence in — ?
JMO	—Have condemned a part of my provisions		JNX	—No confidence in (or, dependence upon)
JMP	CONDENSE D ING-ATION TO CONDENSE		JNY	CONFINE-D-ING
JMQ	—Condenser			Close confinement _____IUV
JMR	—Condenser broken (damaged, or, out of		JNZ	—Confinement
	order)			Keep him (or, them) in close confine-
JMS	—Condenser pipes			ment _____ _____PYU
JMT	—Condenser tubes		JOA	CONFIRM-ED-ING-ATION
JMU	—Condenser tubes burst		JOB	—Confirmation, or, Ratification takes
JMV	—Condenser tubes choked (or, salted).			place at — (or, on —)
JMW	—Condenser tubes leaking		JOC	—Is it confirmed that (vessel indicated)
JMX	—Condenser tubes out of order			has foundered ?
	Condensing boiler burst_____GWT		JOD	—Is there any confirmation ?
	Condensing boiler damaged_____GWU		JOE	—News, or, Report, or, Information is
	Condensing boiler tubes burst ____ YDE			confirmed
	Jet condenser _____PVH			Officially confirmed (or, certified)___IJF
	Surface condenser _ _____XFW		JOF	—Report requires confirmation.
	Surface condenser tubes_____XFY		JOG	—Want-ed-ing confirmation
			JOH	—Want official confirmation
JMY	CONDITION (See also STATE)			
	Are you, or, Is — (vessel indicated) in a		JOI	CONFISCATE-D-ING-ION
	condition to proceed?_____AN		JOK	CONFLAGRATION
	Bad condition _____FQZ			
	Condition of cargo_____IAK		JOL	CONFLICT-ED ING
	Do you think we ought to pass a vessel		JOM	—Conflicting reports
	in this distressed condition ?_____NE			
	Good condition_____OEB		JON	CONFORM-ED-ING ABLE-Y-ITY
	I am not in a condition for action___DIT		JOP	—In conformity with
JMZ	—In what condition—is —?			
	Passed a derelict vessel, apparently in		JOQ	CONFUSE-D-ING-ION
	good condition, hull well out of			Avoid-ed-ing confusion ___ _____ .FOV
	water, dangerous to navigation__KVS		JOR	—In great confusion
JNA	—What condition?		JOS	—Without the least confusion
	What condition is the cargo in?____IBH			
JNB	—What is the condition (state) of —?		JOT	CONGEST-ED-ING-ION
JNC	CONDITION-AL-ALLY (Agreement)		JOU	CONGO (KONGO) STATE COLORS
JND	—On, or, upon condition			
	Unconditional-ly_____YGU		JOV	CONGRATULATE D ING ION
				Allow me to congratulate you ___ _DZA
	CONDOR (Coin) _____ _____ ATG		JOW	—Congratulate-d-ing you
JNE	CONDUCT-ED-ING TO CONDUCT		JOX	—To congratulate you
	Bad conduct_____ __ _____FRA		JOY	CONGRESS (See also PARLIAMENT)
JNF	—Conductor			Act of Congress (or, Parliament)___DIC
JNG	—Gallant Conduct			By an act of Congress (or, Parliament),
JNH	—Good conduct			DIE
	Lightning conductor_____QTJ		JOZ	—Has Congress (or, Parliament) met ?
	Nonconductor-ing _____ SIZ		JPA	—Member of Congress
JNI	—The conductor (or, guard)			

CONGRESS—CONSUME

	CONGRESS—*Continued*
	Congress, Parliament, Assembly (*or other legislative body*) adjourns (*or, adjourned*) on the — .... FEI
	Congress, Parliament, Assembly (*or other legislative body*) is dissolved. FEJ
JPB	—Congress, Parliament, Assembly (*or other legislative body*) met (*or, meets*) on the —
	Recent act of Congress, Parliament, Assembly (*or other legislative body*), DIL
JPC	CONICAL
JPD	—Conical buoy
JPE	CONN-ED-ING
JPF	—Conning tower
JPG	CONNECT-ED-ING ION
JPH	—Connect your screw propeller, *or*, paddles
JPI	—Connected with
JPK	—Connecting rod
	Connecting-rod bolts broken ...... GXV
	Connecting-rod brasses broken (*or, out of order*) .... HEB
JPL	CONNIVE-ANCE
JPM	CONQUER-ED-ING
JPN	CONSCIOUS
JPO	CONSCRIPT-ION
JPQ	CONSECRATE-ED-ING CONSECRATION
JPR	CONSECUTIVE-LY
JPS	CONSENT-ED-ING—TO (*See also* PERMIT *and* AGREE)
JPT	—Can not, *or*, Will not consent (permit it)
JPU	—Have, *or*, Has consented
JPV	—I will consent
JPW	—Vessel (*or person indicated*) consents
	Will you (*or person-s indicated*, consent (*or*, agree) ? ....DUC
JPX	—Would not consent
JPY	CONSEQUENCE-T-LY
JPZ	—Happened in consequence of Occasioned by
	Heave-to, or take the consequences. OSG
JQA	—In consequence of
JQB	—Of not much consequence
JQC	—Of the utmost consequence (*or*, importance)
JQD	—The consequence will be
JQE	—Took place in consequence of
JQF	—What was the consequence?
JQG	CONSIDER-ED-ING TO CONSIDER
JQH	—Consider it (*or*, them) necessary
JQI	—Considerable fault found—with
JQK	—Considerable-y
	Considerable sea .... UP
JQL	—Considerate-ly ion
JQM	—Do you consider, *or*, Have you considered?
	Do you consider leak dangerous? ....QI

	CONSIDER—*Continued.*
	—For consideration—of —
JQN	Furnace considered dangerous..... KLB
JQO	—Has been considered
	Hurricane did considerable damage. PAK
JQP	—Is, *or*, Are not considered
JQR	—Shall, *or*, Will consider
	There has been so much fog as to cause considerable detention . ... NGL
	Under consideration ... . YHD
JQS	CONSIGN-ED-ING-TO TO CONSIGN
	To whom is your cargo consigned?. IBN
JQT	—Whom are you (*or vessel indicated*) consigned to?
JQU	—Consignee
JQV	—Consignment
JQW	CONSIST-ED-ING—OF
	Cargo consists of — .... HZR
	Fleet consists of — .... NEL
JQX	CONSOLS
JQY	CONSORT
JQZ	CONSPICUOUS-LY-NESS-ITY
	I can not make out your flags, hoist them in a more conspicuous position, OWL
JRA	CONSPIRE-D-ING-ACY-ATOR
JRB	CONSTABLE-ULARY
JRC	CONSTANT-LY-CY
	Reinforcements constantly arriving, UOM
JRD	CONSTITUTE-D-ING CONSTITUENT
JRE	CONSTITUTION-AL-ALLY
JRF	CONSTRAIN-ED-ING
JRG	CONSTRUCT-ED-ING-ION
JRH	—Constructor
JRI	—Have, *or*, Has been constructed
JRK	—Shall, *or*, Will be constructed
JRL	CONSUL-AR
JRM	—Consul General
JRN	—Consul is absent
JRO	—Consular Court
JRP	—The consul is desired to come (*or*, send) on board
JRQ	—Vice-consul, *or*, Consular Agent
JRS	CONSULATE
JRT	CONSULT-ATION—WITH
	After a consultation. .... DQN
	Consult captain .... HWV
	Consult Lloyd's agent .... QWG
JRU	—Consult owner
JRV	—Consult underwriters
JRW	—Have, *or*, Has been consulted
JRX	—Shall, *or*, Will be consulted.
JRY	CONSUME-D-ING-PTION
	What is your consumption of coal per hour full speed? .... IXM

CONSUMPTIVE—CONVOY

JRZ	CONSUMPTIVE	JTD	CONTRIVE-D-ING-ANCE
JSA	CONTACT	JTE	—Can you contrive?
	Electro contact .. ..LVE	JTF	—How do (or, did) you contrive (or, manage)—to?
	Electro contact mine .... LVF	JTG	—Must contrive
JSB	CONTAGION-IOUS	JTH	CONTROL-LED-LING LER.
	A contagious disease on board (or at place indicated) ....GPQ		Not under control ....JED
JSC	—Contagious disease		Not under control light (lamp, or, lantern) ... .JEC
	Disease, or, Sickness is contagious .. VB		NOTE —Not under control signal is two black balls or shapes vertical
	Disease, or, Sickness is not contagious, VC		
	Is the sickness contagious?..........VA		CONTROVERSY (Discussion) ....LDY
JSD	CONTAIN-ED-ING	JTI	CONTUSE-D-ING-ION
JSE	CONTEND-ED-ING-TION	JTK	CONVALESCENT-CE
JSF	CONTENT-ED-ING-EDLY-MENT	JTL	CONVENIENT-LY-CE
JSG	CONTENTS		Admiral, or, Senior officer wishes to see you (or, —)as soon as convenient..DLB
	I am going to signal the contents of an important telegram, to be communicated to you ....IE		Anchor as convenient ....KY
			As convenient... ....FAE
			As many as convenient.... ....FAX
	Open telegram for me and signal its contents.................. ....WX		As soon as convenient ....FBL
		JTM	—If convenient.
	Shall I open your telegram and signal its contents?.................XC	JTN	—Is there a convenient watering place?
		JTO	—It will be a great convenience
		JTP	—When most convenient
		JTQ	—Will it be convenient?
JSH	CONTEST-ED-ING	JTR	CONVENT
JSI	CONTINENT-AL-LY	JTS	CONVENTION-AL-LY
JSK	CONTINGENT-CY		Geneva Convention........... ... .NXS
			Geneva Convention flag (Plate II)..NXT
JSL	CONTINUE-D-ING-ATION-AL OUS	JTU	CONVERSE-D-ING-ATION—WITH
JSM	—Continue your present course Stand on	JTV	—A conversation with
	If it continues to blow ....GOH	JTW	CONVERT-ED-ING-IBLE
	If the wind continues as at present..ZER	JTX	CONVEY-ED ING-ANCE—TO
	CONTO (Money) ....ATH		Ambulance required to convey a patient to hospital ....EDY
JSN	CONTRABAND	JTY	—Are there any public conveyances— to — ?
JSO	—Contraband goods		
JSP	CONTRACT-ED-ING-ION	JTZ	—By the first conveyance
	Contract built ....HLD	JUA	—For conveyance—to —
JSQ	—Contractor	JUB	—No conveyances—to —
JSR	CONTRADICT-ED-ING CONTRADICTION	JUC	CONVICT-ED-ING-ION TO CONVICT
JST	—Contradict the report	JUD	—Convict (a convict)
JSU	—Contradictory	JUE	—Will be convicted
JSV	—It has not been contradicted	JUF	—Will not be convicted
JSW	CONTRARY	JUG	CONVINCE D-ING
JSX	—Contrary to order.	JUH	CONVOY-ED ING TO CONVOY
	Contrary to regulations ....UOG		All ships of the convoy are to rejoin company ....Code Flag over V
	Delayed by contrary winds Detained by contrary winds ....KSD		By, or, With the convoy...........HPJ
JSY	—On the contrary	JUI	—Convoy (a convoy)
JSZ	CONTRAST-ED-ING	JUK	—Convoy is ahead
		JUL	—Convoy is astern
JTA	CONTRIBUTE D ING ION—TO	JUM	—Convoy is from
JTB	—Liberal contribution	JUN	—Convoy is going to
JTC	—Will you (or person indicated) contribute?		Convoy to disperse and make for destination ....IS

	CONVOY—CORVETTE		
	Convoy—*Continued*	JVW	Copperas
	Convoy to disperse and reassemble at rendezvous (*or place indicated*)....IT		Copra............................IYZ
	Convoy to keep ahead of escort......IU		
	Convoy to keep as close as possible ..IV	JVX	Copy-ied-ing To copy
	Convoy to keep astern of leader (*or, escort*) ....IW	JVY	—Copy (*A copy*)
		JVZ	—Copy will be wanted
	Convoy to keep on port side of leader (*or, escort*)....................IX	JWA	—Have, or, Has a copy—of
		JWB	—Have you a copy—of?
	Convoy to keep on starboard side of leader (*or, escort*)............IY	JWC	—I have no copy
		JWD	—Take, or, Make a copy (*or, duplicate*)
	Convoy to spread as far as possible, keeping within signal distance.........IZ		
JUO	—Convoy will assemble at port (*or place indicated*)	JWE	Coral line
		JWF	—Coral reef
	Convoy will leave at — (*time to follow*), QKM		
	Convoy will leave on — (*day to follow*), QKN	JWG	Cord-ed-ing-age.
JUP	—Convoy will rendezvous at —, on the —	JWH	—Cordage in lengths
	Convoy will rendezvous in — (*latitude and longitude indicated*)..........UQN	JWI	Cordite
	Do not go through (*or, near*) — (*place indicated*) without a convoy?.....JA	JWK	Corea-n (Korea-n) Korean Colors
	Do you belong to the convoy ....GEV	JWL	Cork
JUQ	—Have you, or, Has — (*vessel indicated*) seen convoy?	JWM	—Cork buoy
		JWN	—Cork jacket
JUR	—How was convoy steering?		
JUS	—I belong to the convoy of — (*ships indicated*)	JWO	Corn . Indian corn ....PHX
JUT	—Is, or, Has the convoy?		
JUV	—It is a convoy of —	JWP	Corned (*Salted*)
JUW	—Of, or, From the convoy		
	Strange vessel following convoy...WYV	JWQ	Corner
JUX	—When did the convoy?		
JUY	—When will the convoy?	JWR	Coroner
JUZ	—Where is the convoy from?		Awaiting coroner's inquest ....FPD
JVA	—Where is the convoy going?	JWS	—Coroner's inquest
	With the convoy By the convoy..HPJ		
		JWT	Corporal
JVB	Cook	JWU	Corporate-ion
JVC	—Cook house	JWV	—Corporation of —
JVD	—Cook-ed provisions		
JVE	—Cooking range, or, Copper	JWX	Corps
			Army Service Corps ....EWF
JVF	Cool-ed-ing-ly ness		
		JWY	Corpse
JVG	Coolie		
		JWZ	Correct-ed ing
JVH	Coop-ed-ing		Account sent is correct ....DBT
			Account sent is incorrect.........DEU
JVI	Cooper-age	JXA	—Correct-ly
JVK	—Cooper's tools	JXB	—Is it correct?
			Is my present course a correct one? (*Course to follow, if necessary*)....JZF
JVL	Cooperate-d-ing-ation-ive		
JVM	—Cooperative society.		Is the latitude correct?..... ...... QHS
JVN	—Shall we cooperate?		Is the longitude correct?.. ... ...QZB
			Is your list of ships correct? ......QVI
JVO	Copaiba (*Drug*)	JXC	—News, or, Report, or, Information is not correct
	Copeck (*Coin*)................ATW	JXD	—Not correct
JVP	Copper-ed-ing-y	JXE	Correspond ed-ing
	Copper, or, Cooking range ....JVE	JXF	—Correspondent-ence
JVQ	—Copperplate		
JVR	—Copper punt	JXG	Corrode d-ing sion-sive
JVS	—Coppersmith		
JVT	—Rubbed off the copper	JXH	Corrupt-ed-ing-ion-ness
	Sulphate of copper ....XDF		
JVU	—When was your ship last coppered?	JXI	Corvette

COST—COUNTERMAND

JXK	COST-LY—OF		
JXL	—What will it cost?		
JXM	COSTA RICA-N COSTA RICAN COLORS		
JXN	COT		
	Cot case .... IDB		
JXO	—Send cot		
JXP	COTTAGE		
JXQ	COTTAGER		
JXR	COTTON		
	Bales of cotton ....FTC		
	Cotton cargo........ ....HYX		
JXS	—Cotton crop		
JXT	—Cotton seed		
JXU	—Cotton twist		
JXV	—Cotton waste (or, tow)		
	Gun cotton......................OJR		
JXW	—Is cotton freight to be obtained?		
	Lamp cotton............. ....QDF		
JXY	—Manufactured cottons Cotton goods		
JXZ	COUGH-ED-ING		
	Whooping (hooping) cough ....OXT		
JYA	COULOMB (Measure of electrical quantity)		
	COULD (or, MIGHT)...................RKZ		
	Could, or, Might be ....BFL		
	Could, or, Might do (or, be done)..BMA		
	Could, or, Might have .... BQJ		
	Could, or, Might have been ....BFM		
	Could or, Might have done (or, been		
	done) ....BMC		
	Could, or, Might have had.........BQK		
	Could or, Might he (she, it or person s		
	or thing s indicated) be? ...... BTO		
	Could, or, Might he (she, it or person-s		
	or thing s indicated) not be? .....BTP		
	Could, or, Might his (her-s, it-s)?....BLA		
	Could, or, Might I (my, mine)? ....BLC		
	Could, or, Might not ....BLD		
	Could, or, Might not be ....BFN	JYB	
	Could, or, Might not do (or, be done),	JYC	
	BMD		
	Could, or, Might not have .. ...BQL		
	Could, or, Might not have been....BFO		
	Could, or, Might not have done (or, been		
	done) ....BME		
	Could, or, Might not have had.....BQM	JYE	
	Could, or, Might that (or, this)?....BLE	JYF	
	Could, or, Might the?.. .... BLF	JYG	
	Could, or, Might there be?.........BFP	JYH	
	Could, or, Might these (or, those)? .BLG		
	Could, or, Might they (their s)?....BLH		
	Could, or, Might we (ours)? ....BLI	JYI	
	Could, or, Might you (your-s)? .....BLJ		
	He, She, It (or person-s or thing-s indi-	JYK	
	cated) could (or, might) ....BUF		
	He, She, It (or person s or thing s indi		
	cated) could (or, might) not......BUG		
	How could (or, might)? ....BXK		
	I am sure I could ....XFM	JYM	
	I could (or, might) ....BLK		
	I could (or, might) be ....BFZ	JYN	
	I could (or, might) have...........BRK		
	I could (or, might) not............BLM	JYO	
	I could (or, might) not be. ...... .BGA	JYP	

COULD—Continued
I could (or, might) not have........BRL
I think I could ....XQF
If he (she, it, or person-s or thing-s indi-
cated) could (or, might)..........BYF
If he (she, it, or person-s or thing-s indi-
rated) could (or, might) not.....BYG
If I could (or, might).............BZD
If I could (or, might) not ....BZE
If they could (or, might) ....CAJ
If they could (or, might) not ......CAL
If we could (or, might)............CBD
If we could (or, might) not........CBE
If you could (or, might)...........CBU
If you could (or, might) not.......CBV
It, etc   (See He, above)
It could (or, might) be done.......BNM
It could (or, might) not be done ...BNO
Nothing could............... .. ..BKB
She, etc   (See He, above)
That, or, This could (or, might)...CKU
That, or, This could (or, might) be.CKV
That, or, This could (or, might) not be,
CKW
There could (or, might)............CMN
There could (or, might) be ....CMO
There could (or, might) not be ....CMP
They could (or, might)............BLN
They could (or, might) be..........BHR
They could (or, might) not ....BLO
They could (or, might) not be .....BHS
We could (or, might) ....BLP
We could (or, might) be............BIK
We could (or, might) not ....BLQ
We could (or, might) not be........BIM
What, or, Which could (or, might)?.CQA
When could (or, might)? ....CRA
Where could (or, might)?...........CSW
Who could (or, might)? ....CUM
Why could (or, might) not?........CVF
You could (or, might) ....BLR
You could (or, might) be ....BJA
You could (or, might) not ....BLS
You could (or, might) not be ......BJC

JYB	COUNCIL
JYC	—Council of war
	Privy Council ....TWQ
JYD	COUNSEL LED LING-LOR
JYE	COUNT ED-ING TO COUNT
JYF	—Can not count
JYG	—Did you count?
JYH	—Have, or, Has counted
JYI	COUNTENANCE-D ING
JYK	COUNTER
	Counter orders ....STO
	Engine counter ....MAZ
JYL	COUNTERACT ED-ING ION
JYM	COUNTERBALANCE-D ING
JYN	COUNTERFEIT-ED-ING
JYO	COUNTERMAND-ED-ING
JYP	—Previous orders are all countermanded.

COUNTERMINE—CRACK

JYQ	COUNTERMINE D-ING		COURSE—*Continued*
JYR	—Countermining launch	JZP	—What course shall you (*or*, will they) steer?
JYS	COUNTERSIGN-ED-ING	JZQ	—What course were you (*or*, they) steering?
	COUNTERWEIGHT --------------ZBD	JZR	—What is the course to —?
SYT	COUNTESS		When do you intend to alter course?-ECX
			When I alter course to-night I will show a light. --------------------QW
JYU	COUNTRY		When will you alter course?-------ECY
	North country coal --------------IXF	JZS	—When you alter course (*or*, tack) to-night show a light
JYV	COUNTY		
JYW	—County court	JZT	COURSES (*Sails*)
			Fore course-----------------------NIV
JYX	COUP-DE-MAIN		Main course.----------------------RGQ
			Mizzen course --------------------RWP
JYZ	COUPLE-D-ING	JZU	—Set the courses
JZA	—Couple of—	JZV	—Shift a course
	Coupling bolts broken -----------GXW		
	Shaft coupling bolts broken -------GYF	JZW	COURSE OF COURSE
			As a matter of course-------------FAC
	COURAGE-OUS-LY (*Bravery*) -------HEF		
JZB	COURIER	JZX	COURT
			Admiralty court -----------------DLR
JZC	COURSE (*See also* STEER)		Consular court -------------------JRO
	Alter course-------- -----------EBU		County court. --------------------JYW
JZD	—Alter course one point (*or number indi-cated*) to port	JZY	—Court-martial
		KAB	—Court of arbitration
JZE	—Alter course one point (*or number indi-cated*) to starboard	KAC	—Court of inquiry •
		KAD	—Court of law
	Alter course to — (*point indicated*) -WF		Naval Court.----------------------SEN
JZF	—Am I steering a proper course? Is my present course a correct one (*indicate the course steered, if necessary*)?		Police court (*or*, station) ----------TNX
		KAE	COURTESY-OUS-LY
	Change your course---------------IKS	KAF	COVE, *or*, BIGHT
	—Continue your present course Stand on ----------------------------JSM	KAG	COVER-ED-ING
JZG	—Course and distance	KAH	—Are you, *or*, Is — (*person indicated*) covered?
JZH	—Course to be followed		Cylinder cover------ -------------KIB
JZI	—Do not stand too far on your present course		Cylinder cover broken ---- --------KIC
			High-pressure cylinder cover ------OUV
JZK	—How long may we stand on our present course?	KAI	—High-pressure cylinder cover broken
			Intermediate-pressure cylinder cover, PQT
	I do not intend altering course for the present (*or*, until —)----------- ---ECI	KAJ	—Intermediate-pressure cylinder cover broken
	I shall alter course-----------------ECJ	KAL	—Low-pressure cylinder cover
	I shall not alter course--- --------ECK	KAM	—Low pressure slide jacket cover broken
	I think we had better alter course to — (*point indicated*)---------- - ---ECL	KAN	—Packing ring of cylinder cover broken
	I will show a light to-night when I alter course- ---------------------QW	KAO	—Under cover—of
	Is my present course a correct one? Am I steering a proper course (*indicate the course steered, if necessary*)?--JZF	KAP	COW
JZL	—Out of the course of —	KAQ	COWARD-LY-ICE
	Race course----------------------UGE		COXSWAIN COCKSWAIN -----------IYV
	Steer course indicated by compass signal ----------------------------JIF	KAR	CRAB
	Storm center in — (*direction indicated*), probable course toward — (*direction indicated*).--------------------IGY	KAS	CRACK-ED ING
		KAT	—Cylinder cracked
JZM	—The course steered is the best for the present (*or until time indicated*)	KAU	—High-pressure cylinder cracked (*or*, out of order)
JZN	—The same course	KAV	—Intermediate-pressure cylinder cracked (*or*, out of order)
	Vessel indicated has altered course-ECU		
	Water course --------------------YVO	KAW	—Low-pressure cylinder cracked (*or*, out of order)
	We had better alter course to — (*point indicated*) --------------------ECV		
JZO	—What course are you (*or*, they) steering?	KAX	—Piston rod cracked (*or*, broken)

KAY	CRAFT-Y-ILY-INESS		CREW—*Continued*
			Crew have appealed to the authorities, FNE
KAZ	CRAFT (*Vessel*) ----------- YPW	KCP	—Crew (*number to be shown*) have left the
KBA	—Fishing craft (boat *or*, vessel),		ship
KBB	—Native craft	KCQ	—Crew have mutinied
KBC	—Small craft	KCR	—Crew healthy
KBD	—Suspicious craft	KCS	—Crew imprisoned
	Vessel seen (*or*, indicated) is native ship		Crew not all on board ----------- DYA
	(*or*, craft) ----------- SEG	KCT	—Crew not heard of
		KCU	—Crew refuse to go on board
KBE	CRAMP-ED	KCV	—Crew saved
		KCW	—Crew sick
KBF	CRANE		Crew, *or*, Passengers sick with infec-
	Anchor crane ----------- EGZ		tious complaint ----------- PCX
KBG	—Steam crane	KCX	—Crew will not leave the vessel
		KCY	—Crew will not pass
KBH	CRANK (*Machinery*)		Foreign crew ----------- NKP
	Crank brasses broken (*or*, out of order), HEC	KCZ	—Full crew
			Gun's crew ----------- OJY
KBI	—Crank broken		Gunner's crew ----------- OKQ
KBJ	—Crank frame		Have lost all boats, can you take crew
KBL	—Crank head		off ? ----------- NI
KBM	—Crank pin	KDA	—Have shipwrecked crew on board, will
KBN	—Crank pin brasses broken		you let me transfer them to you?
KBO	—Crank pin broken		(*Number to follow*)
KBP	—Crank shaft	KDB	—Have you any tidings of the crew?
KBQ	—Crank shaft broken (*or*, defective).		Have you fallen in with crew of dere-
KBR	—Crank webs loose		lict? ----------- MRT
			I am sinking (*or*, on fire), send all
KBS	CRANK-NESS (*Unstable*)		available boats to save passengers
KBT	—Ship (*indicated*) is very crank		and crew ----------- NO
		KDC	—Native crew
KBU	CRATER	KDE	—Not safe to go on with the crew as at
			present
KBV	CREAM	KDF	—Part of the crew (*indicate the number*)
	Cream of tartar ----------- XKW	KDG	—Police (*or other*) authorities have decided
			against the crew
KBW	CREATE-D-ING-OR	KDH	—Police (*or other*) authorities have taken
			some of the crew out of the ship
KBX	CREDENTIAL	KDI	—Relieve-d-ing, Relief crew
KBY	—Credentials are satisfactory		Send lifeboat to save crew ----------- FJ
KBZ	—Credentials not satisfactory		Send boats to take off the crew ----------- GUP
	Have you, *or*, Has —(*person indicated*)	KDJ	—Short manned A weak crew (*number*
	the necessary credentials (*or*, certifi-		*may be indicated*)
	cates)? ----------- IHZ		Some squabble (*or*, fight) on shore with
			crew ----------- MXU
KCA	CREDIT ED-ING		Want hands (*specify number and par*
KCB	—Can you get credit?		*ticulars*) ----------- YJ
KCD	—Have you credit on —?		What has become of the crew? ----------- GCP
	Letter of credit ----------- QOI		— men of the crew have been ill of yel
			low fever (*number to follow*) ----------- MWY
KCE	CREDITOR		
		KDL	CRIME-INAL-LY
KCF	CREEK		
		KDM	CRIMP
KCG	CREEP-ING-ER CREPT		
		KDN	CRIMSON
KCH	CREOSOTE-D		Crimson fluid. Condy's fluid ----------- JNK
KCI	CRESCENT.	KDO	CRIPPLE-D
KCJ	CREW (*See also* HANDS)	KDP	CRISIS CRITICAL
	Are all your crew on board ----------- GPT		
	Boat's crew ----------- GRW	KDQ	CROCKERY
	Boat's crew will not come ----------- GRX		
KCL	—By the crew		CROCODILE ALLIGATOR ----------- DYV
	Crew are all on board ----------- GQB		
KCM	—Crew deserted.		CRONCHKA (*Measure of capacity*) ----------- AZD
KCN	—Crew disaffected, will not work		
KCO	—Crew discontented, will not work	KDR	CROOKED BENT

CROP—CUSTODY.

KDS	Crop
KDT	—A good crop of — Crops look well
KDU	—A short crop of —
	Cotton crop ---------------------JXS
KDV	—Crops destroyed
KDW	—Crops have suffered severely (are much injured)
	Crops look well A good crop of —.KDT
KDX	—How is the — crop?
KDY	—Sugar crop
KDZ	Cross-ed-ing
	Bar can not be crossed until — ....FVR
	Can I cross the bar? Shall I be able to get over the bar?----------------FVY
KEA	—Cross bearings
KEB	—Cross-cut saw
KEC	—Cross-examine-d-ing-ation—of.
KED	—Crosshead
KEF	—Cross-jack yard
KEG	—Cross the bow
KEH	—Crosstree
KEI	—Crossway
KEJ	—Piston-rod crossheads damaged (or, broken)
	Pump crosshead------ ----------UBG
	Pump crosshead out of order------UBH
	Topmast crosstrees ---------------XVY
	What are the leading marks for crossing the bar?---------------------FWC
KEL	—When did you cross the equator?
	When will be the best time for crossing the bar?---------- ----- -------XM
KEM	—Where did you cross the equator?
KEN	—Where will you cross the equator?
KEO	Croton
KEP	—Croton oil
KEQ	Crow
	Crowbar---------------------FWI
KER	—Crowsfoot-feet.
KES	—Crowsnest
KET	Crowd-ed-ing
KEU	Crown-ed-ing
	Furnace crown of boiler collapsed .NUC
KEV	Cruel ly ty
KEW	Cruise d-ing—off
KEX	—Are there many men-of-war now cruising?
	During the cruise ----------------KFA
KEY	Cruiser
	Armored cruiser----------------- EVW
	Belted cruiser----------------GFD
KEZ	—Cruisers are very vigilant
KFA	—During the cruise
	Enemy's cruisers have been seen to the — steering to the —------------OF
	Enemy's ship (or, cruiser)--------LZK
KFB	—Hired cruiser
KFC	—Revenue cruiser (or, vessel)
KFD	—Saw enemy's ship (or, cruiser) off — (or in latitude and longitude indicated)
KFE	—Saw enemy's ship (or, cruiser) to the —
	Torpedo cruiser----- ----------XWS

KFG	Crush-ed ing
	Accident happened Crushed Fall Jammed ----------------------DCF
KFH	Crotch
KFI	Crystal-ise-d line
KFJ	Cube ic-al
	Cubic centimeter ----------------ZNM
	Cubic foot ----------------------AYC
	Cubic inch----------------------AYD
	Cubic meter---------------------ZNP
	Cubic millimeter-----------------ZNO
	Cubic yard -- -------------------AYE
	Measure of cubic capacity (See page 52), AYB
KFL	Cuddy
KFM	—Cuddy passenger
KFN	Cuff ed-ing
KFO	—Handcuff-ed-ing
KFP	Culpable-ility
KFQ	Cultivate-d-ing-ion Culture
KFR	Cunning-ly-ness
KFS	Cup
KFT	Cure-d-ing-able.
KFU	Curious-ly-sity
KFV	Currant
KFW	Current
	Against the current(tide,or,stream)DSE
	Athwart the current (tide, or, stream), FJI
KFX	—Can not stem the current (or, tide)
KFY	—Current runs — miles per hour
KFZ	—Current sets off shore
KGA	—Current sets on shore
KGB	—Current, or, Tide will run very strong (Indicate miles per hour, if necessary)
	Earth current------------------LRN
	Electric current -----------------LUS
	Induced current -----------------PIT
KGC	—Is there much current?
	Keep out of the tide (or, current)----XG
	Price current---------- --------TVQ
KGD	—There is less current inshore (or in direction indicated)
KGE	—Tide, or, Current sets to the —
KGF	—Try the current (or, tide)
	Undercurrent------ ------------YHE
KGH	—What current (rate and direction) do you expect (or, reckon on)?
KGI	—What is the current? Do you, or, Do we feel any current?
KGJ	—What is the set and rate of current (or, tide)?
KGL	Curry
KGM	Curve-d ing-ature
KGN	Custody

CUSTOM—DAM

KGO	CUSTOM
	Customary (*Usual*) ....YMX
KGP	—Customer,
KGQ	—Is it the custom — to —?
KGR	CUSTOMS (*See also* DUTIES)
	Board of Customs.......... ....GOY
	Cleared at the Custom-house ....ITE
KGS	—Clearing the Custom-house
KGT	—Custom house
KGU	—Custom-house boat Revenue boat
KGV	—Custom house duties
KGW	—Customs officer
KGX	—Have you a Custom-house officer on board?
KGY	—Have you cleared the Custom-house?
KGZ	—Is your ship entered at Custom house?
KHA	—Not cleared at Custom-house
KHB	—Shall clear at Custom-house
KHC	CUT TING To CUT
	Accident happened Cut Stab...DCG
KHD	—After cutting away her masts
	Cable is cut (*or*, broken)..........HQP
KHE	—Can be cut off
KHF	—Can not cut
	Cross cut saw ....KEB
KHG	—Cut away
	Cut away the mizzen mast........RWN
	Cut away your masts.............KJ
KHI	—Cut down
KHJ	—Cut off
	Cut out ....SVU
	Cut, weigh, *or*, Ship Get an offing
	Wait for nothing ....KW
	Cut, *or*, slip your cable...........HRA
KHL	—Do not cut
KHM	—Do not cut away your masts
KHN	—Do you think you can cut her (*or*, them) out?
	I must cut my cable..............HRI
KHO	—May, *or*, Can be cut out
KHP	—Much cut up
KHQ	—Obliged to cut away
KHR	—Shall I cut?
	Weigh, cut, *or*, Ship Get an offing
	Wait for nothing ....KW
KHS	CUTLASS
KHT	CUTLERY
KHU	CUTTER
KHV	—Is a Revenue cutter
	Steam cutter Small steamboat ..GUW
	Steam cutter's engine ....MBX
KHW	CUTTING (*An artificial passage*)
KHX	CUTWATER
	CWT HUND'REDWEIGHT.............BAP
KHY	CYCLE
KHZ	CYCLONE
	You are steering right into the center of the hurricane (*or*, cyclone)...... PAN
KIA	CYLINDER
KIB	—Cylinder cover
KIC	—Cylinder cover broken

	CYLINDER—*Continued*
	Cylinder cracked................KAT
KID	—Cylinder damaged
KIE	—Cylinder out of order
KIF	—Cylinder side rod
KIG	—High-pressure cylinder
	High-pressure cylinder cover.. ...OUV
	High-pressure cylinder cover broken, KAI
	High pressure cylinder cracked (*or*, out or order)............KAU
KIH	—Intermediate-pressure cylinder
	Intermediate-pressure cylinder cover, PQT
	Intermediate pressure cylinder cover broken ....KAJ
	Intermediate-pressure cylinder cracked (*or*, out of order)............KAV
KIJ	—Low pressure cylinder
	Low-pressure cylinder cover......KAL
	Low-pressure cylinder cracked (*or*, out of order) ....KAW
	Packing ring of cylinder cover broken, KAN
KIL	CZAR
KIM	—Czarevitch
KIN	—Czarevna
KIO	D (*Letter*) (*For new method of spelling, see page* 18)
KIP	D SLIDE
KIQ	DAILY EVERY DAY
	Daily expected.................KMX
KIR	—Daily paper
KIS	—Daily report
	Daily returns ....KMY
	In daily expectation of an embargo..LWH
	My chronometer gains daily ....GP
	My chronometer loses daily.........GT
KIT	DAIRY
	DAKTYLAS (*Measure of length*).....AVO
KIU	DALE
KIV	DAM-MED-MING
	Cofferdam ....IZK

DAMAGE—DANGER

KIW	DAMAGE-D (*See also* INJURE)
	Ammunition is damaged ....EFQ
	Are you, or, Is — (*vessel indicated*) materi-
	ally damaged (*or, injured*)? ....AO
	Bell buoy damaged, not working.. GEN
	Boiler burst, very slight damage..GVW
	Boiler damaged........................GVX
	Boiler plates damaged...............GWI
	Can damage be repaired at sea? .....BA
	Can the damage be repaired by your-
	selves ....BC
	Can not exactly estimate the damage,
	MHR
	Cargo damaged ....HZS
	Cargo damaged, but extent not known,
	HZT
	Cargo not much damaged (slightly
	damaged) ....IAD
	Condenser broken (damaged *or*, out of
	order)...................... ....JMR
	Condensing boiler damaged. .....GWU
	Cylinder damaged ....KID
	Damage can be put to rights in — hours,
	BD
KIX	Damage can be repaired at sea ..... BE
	Damage can not be repaired at sea (*or*,
	place indicated) ....BF
	Damage, *or*, Defects can not be repaired
	without assistance.................BG
KIX	—Damage is repaired
KIY	—Damage not ascertained
KIZ	—Damage serious (*particulars to follow*,
	if necessary)
KJA	—Damage to machinery (*part to be indi-*
	cated)
KJB	—Damage trifling.
KJC	—Damaged keel
	Damaged, or, Sprung mast, can not carry
	sail ....BH
	Damaged rudder . ...VBI
	Damaged rudder, can not steer......BI
KJD	—Damaged the bitts
KJE	—Damaged the windlass
	Damages (*indemnity*).... ......PHN
	Damages are assessed at — ....FEQ
KJF	—Did you (*or vessel indicated*) receive
	any damage?
	Donkey boiler damaged...........GWY
	Have, *or*, Has received damage (*or*,
	injury)—to ....PLA
	Have you damaged your machinery? MBO
	Have you received any damage (*or*,
	injury)? ....PLB
	How long will you (*or vessel indicated*)
	be repairing damages?............BM
	Hurricane did considerable damage. PAK
	Hurricane did not do much damage. PAL
	I am very seriously damaged........ BN
KJG	—I have damaged (*or*, sprung) —
KJH	—I have damaged (*or*, sprung) fore yard
KJI	—I have damaged (*or*, sprung) lower top
	sail yard
KJL	—I have damaged (*or*, sprung) main yard
KJM	—I have damaged (*or*, sprung) topsail yard
KJN	—I have damaged (*or*, sprung) upper top-
	sail yard
KJO	—I shall repair damage in —
	Is — (*vessel indicated*) *or*, Are you ma-
	terially damaged (*or, injured*)?....AO
KJP	—Is your bow much damaged?
	Is your cargo much damaged?...♦....BQ

	DAMAGE—*Continued*
	Keel damaged ....KJC
	Levers damaged (*or*, out of order)..QOX
KJQ	—Much damage has been done — to (*or*,
	by)
KJR	—Must discharge cargo to repair damages
	No damage (*or*, harm) done........ONJ
	Piston-rod crossheads damaged (*or*,
	broken) ....KEJ
	Powder is damaged ....TRL
	Propeller damaged ....TYX
	Put back damaged ....UCX
	Screw propeller damaged.........TYX
	Sea-inlet valve (*or*, cock) broken (*or*,
	damaged).........................VLA
	Slide-valve gear damaged ....WGL
	Slight damage.....................KJB
	Sprung, *or*, Seriously damaged a lower
	mast ....RCJ
	Sprung, *or*, Seriously damaged lower
	yard (*or spar indicated*) ....RCK
	Steam pipe damaged ... .WSN
	Stem seriously damaged ....WUK
	Stern tube damaged ....WVA
	The damage to machinery is not serious,
	and is such as can be repaired by the
	vessel's own engineers ....RL
	Thrust block damaged ....GNA
	Thrust collar damaged ....JAD
	Towing cable is damaged (*or*, stranded),
	HRX
	Trifling damage.......... ....KJB
	Vessel seriously damaged, wish to
	transfer passengers...............HM
KJS	—What damage have you sustained?
	Where will you go to repair damages?
	BV
KJT	—Without damage
	DAMAGES (*Indemnity*).............PHN
KJU	DAMP
KJV	DAMPER (*For fires*)
KJW	DANCE-D-ING TO DANCE
KJX	DANE DANISH DANISH COLORS. DEN-
	MARK
KJY	DANGER OUS-LY
	Accident, dangerously wounded (*or*,
	hurt)...............................AL
	Appearances are dangerous be on your
	guard ....FN
	Are you in danger? ....FO
	Bar, *or*, Entrance is dangerous. ....FQ
	Bar is not dangerous ... .LA
	Beware of derelict, dangerous to navi-
	gation ....JE
	Channel, *or*, Fairway is dangerous.. FU
	Channel is dangerous without a pilot,
	LD
	Coast is dangerous.... ....IXW
	Coast is not dangerous..............IXY
KJZ	—Danger of parting (*or*, driving)
	Dangerous, *or*, Great risk in sending a
	boat ....GSZ
KLA	—Dangerous position
	Dangerous to admit too many on board
	at a time.........................GB

DANGER—DAY

	DANGER—*Continued*
	Dangerous to, *or*, Do not come into less than — feet of water ....FV
	Dangerous without a pilot............TI
	Dangerously hurt (*or*, wounded), accident ... ------------------------AL
	Derelict reported passed on —, latitude —, longitude —, dangerous to navigation------------- .... KVR
	Do you consider leak dangerous? ....QI
KLB	—Furnace considered dangerous
KLC	—Have you got out of danger?
KLD	—How many have you dangerously ill?
	I am in danger (*or*, shoal water), direct me how to steer .... -----------NL
KLE	—In great danger
	Is there any danger — in — (of, or, from —)? .... ---------------GA
	Is there any danger of mines (*or*, torpedo boats)? --- --------------------XS
	It is dangerous to allow too many people on board at once. ... ----------GB
	It is dangerous to lose the wind- ...GC
	It is dangerous without a pilot.......TI
KLF	—Much danger
KLG	—No danger
	Passed a derelict vessel apparently in good condition, hull well out of the water, dangerous to navigation..KVS
	Shoal water, *or*, Danger in—(*direction indicated*)----------------------GF
KLH	—Shoal water *or*, Danger to leeward
KLI	—Shoal water, *or*, Danger to windward
	The attempt is dangerous. ... .. GI
KLJ	—Vessel indicated is standing into danger
	We are in danger of parting..........GL
	You are all clear of danger... ......GM
	You are in a dangerous (*or*, unsafe) position Your position is dangerous, GN
	You are standing into danger.........JD
KLM	DANIEL'S BATTERY
KLN	DARE-D ING-LY .
KLO	DARK DARKNESS
	After dark .... -----------DQT
	As soon as it is dark----- ----------FBM
KLP	—Before dark
	Do not send boats on shore after dark GST
KLQ	—If the night is dark
	Shall signal with siren (*or*, steam whistle) during fog (*or*, darkness)---------WE
KLR	—When it is dark
KLS	DARKEN-ED-ING TO DARKEN
KLT	DASH-ED-ING
KLU	—Dashed to pieces
KLV	DATE—OF
KLW	—Date not known Without date
KLX	—Date of entry
KLY	—Date of leaving
KLZ	—Dated ing ' To date
KMA	—Dated at — (*or*, from)
KMB	—Dates up to —
KMC	—Has paper of following date

	DATE—*Continued*
KMD	—Have you any news of recent date from (*or*, about) —?
KME	—Is, *or*, Are dated the —
	Newspapers up to date of — (from —, *if necessary*) ------------------------SC
KMF	—On the — (*date indicated*)
KMG	—On what date did you speak the —(*vessel indicated*)?
KMH	—What are the latest dates by the last mail?
KMI	—What is the date of?
	What is the date of latest news from —? SE
	What is the date of your Code List?*.IZR
KMJ	—What is the date of your latest newspaper (*or*, letter)?
	What is the date of your signal book? VL
KML	—What was the date of your departure?
KMN	—What were the latest — dates at — when you last heard?
	Without date Date not known...KLW
KMO	DATE (*Fruit*)
KMP	DAUGHTER
KMQ	DAUNT-ED-ING
KMR	DAVIT
	Boat davit ... ----- - -----------GRY
KMS	—Fish davit
	DAWN DAYBREAK AT DAWN......FIB
KMT	DAY
KMU	—A day's demurrage (*or*, *number indicated*).
KMV	—A few days
	Any day—after ------------------EOS
	Arrived during the day............EXV
	Bill at — days- ....GKF
	By day.. -- .... -HPB
KMW	—Christmas Day
	Daily Every day*.. ....KIQ
KMX	—Daily expected
	Daily report -------------------KIS
KMY	—Daily return
	Day after Next day.............DRO
KMZ	—Day before
KNA	—Day of week
	Daybreak Dawn At dawn ......FIB
KNB	—Daylight. At daylight
KNC	—Day's demurrage—is —
KND	—Day s work.
KNE	—Days from
KNF	—During the day
	Every day Daily............KIQ
	Expected daily.............. ....KMX
	Fine day---------------- ....MYX
KNG	—How many days?
KNH	—How many days since you passed—(*vessel indicated*)?
	I have only — days' provisions ..... TO
	I have only —days' water....... .. .ZA
KNI	—I shall go about — miles an hour till daylight
KNJ	—In about — days
KNL	—Lay-day

DAY—DECK

	DAY——*Continued*	KPA	DEBENTURE
KNM	—Look out at daybreak		
	Made the passage in — days (*number to follow*) ---------------- TDF	KPB	DEBIT-ED-ING
KNO	—Mail day	KPC	DEBT
	Midday ---------------- RTC		
	Next day Day after ---------- DRO	KPD	DEBTOR
KNP	—Not to-day		
KNQ	—On the — day of the month		DECAGRAMME (*Measure of weight*) -- BAI
KNR	—On the same day		
KNS	—Out — days		DECALITRE (*Measure of capacity*) -- AYT
	Passed the mail — day ago -------- RFU		
KNT	—Shall we have daylight enough?		DECAMETRE (*Measure of length*) ---- AVS
	The day after to morrow ---------- DRV		Square decametre (*Square or surface measure*) ---------------- AXJ
KNU	—The day before yesterday		
KNV	—This day		
KNW	—Through the day		DECARE (*Square measure*) -------- AXM
KNX	—To day		
KNY	—To day's post		DECASTERE (*Cubic measure*) -------- AYF
KNZ	—What day?		
KOA	—What day do you leave?	KPE	DECEASE-D
KOB	—What day would be best to —?		
	What is the meteorological forecast for to-day? ---------- ---------- ZK	KPF	DECEIVE-D-ING TO DECEIVE DECEPTION-VE
KOC	Will you be on shore to-day (*specify hour, if necessary*)?	KPG	—Any deception?
		KPH	—Have, or, Has deceived me
KOD	—Will you (*or person indicated*) meet me on shore to-day (*specify time*)?	KPI	—No deception
	You are not likely to receive your orders for some days ---------- SUX	KPJ	DECEMBER
			About December the — ---------- CZL
KOE	DEAD DIED (*See also* DEATH *and* DIE)	KPL	—Beginning of December
	Captain is dead (*or*, died on —) --- HWI	KPM	—End of December
KOF	—Dead-eye	KPN	—Next December
KOG	—Deadlight		
KOH	—Dead reckoning		DECIARE (*Square measure*) -------- AXN
	Dead reckoning agrees with observation ---------- DTX	KPO	DECIDE-D ING (*See also* SETTLE)
KOI	—Dead weight	KPQ	—Before you decide
KOJ	—Dead-weight cargo	KPR	—Decided upon
KOL	—Dead works of a ship	KPS	Decision-ive
KOM	—How many dead?	KPT	—Have you decided—to?
	How many tons dead weight of cargo must you have? -------- --------- IAS	KPU	—Is it decided?
	Latitude by dead reckoning is — -- QHT	KPV	—Nothing decisive yet
	Longitude by dead reckoning is — -QZC		Police (*or other*) authorities have decided against the crew ---------- KDG
	Want — tons dead weight of cargo -- IBE		
KON	—Who is dead?	KPW	—The decision is final
		KPX	—The final decision is
KOP	DEAF-NESS		
			DECIGRAMME (*Measure of weight*) --- BAJ
KOQ	DEAL-ING DEALT		
	A good deal of floating ice—in----- PBG		DECILITRE (*Measure of capacity*) -- AYU
KOR	DEALER		DECIMAL (*See page 34*) ---------- BCO
KOS	DEAL (*Wood*)		DECIMETRE (*Measure of length*) ---- AVT
			Square decimetre (*square or surface measure*) ---------------- AXS
KOT	DEAR-LY-NESS (*Expensive.*)		
KOU	—Dearer-est		
KOV	—Too dear		DECISTERE (*Cubic measure*) -------- AYG
KOW	DEATH (*See also* DEAD)		DECK
	Certificate of death ---------- IHR	KPY	
	How many deaths? ---------- KOM		Armored deck ---------- EVX
KOX	—Last death on the — (*date to follow*)		Between decks ---------- GIV
	Registrar of births, deaths, and marriages ---------- GLI		Deck awning ---------- FPW
			Deck cabin ---------- HQI
			Deck cargo ---------- -------- HYZ
KOY	—When was the last death?	KPZ	—Deck house
KOZ	DEBATE-D-ING	KQA	—Deck passenger

	DECK—DELIVER			

KQB	DECK—*Continued*	KRH	DEFENCE-SIVE-SIBLE	
	—Deck quarters		Coast-defence ship......................IXV	
	Gear for cleaning decks............ ISV	KRI	—Defence net	
	Have lost deck cargo ....IAO		Lines of defence ....QUO	
	Lower deck ....RCB		Local defence...................... .....QXV	
	Main deck...............................RGI	KRJ	—No defence can be made	
KQC	—On deck			
KQD	—Quarter deck	KRL	DEFEND-ED-ING TO DEFEND	
KQE	—Upper deck	KRM	—Defend ship	
			If we keep together, we can defend our-	
KQF	DECLARE-D-ING-ATION		selves ....PYE	
	Declaration of Paris.................TBK			
KQG	—Have, *or*, Has declared	KRN	DEFENDANT	
KQH	—Have, *or*, Has not declared			
	Is war declared? *or*, Has war com-	KRO	DEFER-RED-RING-MENT (*See also* POST-	
	menced?..... ....YU		PONE)	
	War is declared (*or*, has commenced).YX			
		KRP	DEFICIENT-LY-CY	
KQI	—DECLINE D-ING-ATION	KRQ	—Deficit	
	Decline accepting any — ....DBV			
	I decline to have anything further to do	KRS	DEFINE D-ING-ITION	
	(*or*, say) in the matter ....EQO			
	I must decline answering............ENA	KRT	DEFINITE-LY-IVE-LY	
	Offered assistance, but they declined.FGM			
KQJ	—Vessel (*or person indicated*) declines	KRU	DEFRAUD ED ING	
KQL	DECOY-ED-ING TO DECOY	KRV	DEFRAY-ED-ING	
KQM	DECREASE-D-ING	KRW	DEFY-IED-ING-IANCE	
KQN	DECREE-D ING	KRX	DEGREE—OF	
			By degrees...... ....HPC	
KQO	DEDUCT-ED-ING-ION	KRY	Compass signals in degrees(*page* 45).JHI	
		KRZ	—Degree of latitude	
KQP	DEED		—Degree of longitude	
			I am using variation of — degrees.....JHT	
KQR	DEEP-EN ED ING (*See also* DEPTH)		In an ordinary degree...............SVC	
KQS	—Deep sea lead		—degrees below zero..................ZMI	
KQT	—Deep water			
KQU	Deeper-est	KSA	DELAY-ED-ING TO DELAY (*See also*	
	Deepest water is in the — channel...ILR		DETAIN)	
KQV	—Deepest water is nearer the shore		Cause of delay....................IFQ	
	Deepest water is nearest the bank (*or*,	KSB	—Delayed at	
	reef)......... ....FVB	KSC	—Delayed by	
KQW	—Deeply laden	KSD	—Delayed by contrary winds	
KQX	—How deep is —?	KSE	—Delayed by thick fog	
KQY	—Is she deep?	KSF	—Delayed through the recent gale	
KQZ	—Not deep	KSG	—Do not delay (detain)	
KRA	—The deepest water	KSH	—How long will delay be?	
KRB	—Too deep	KSI	—I must not delay (detain)	
	Water is too deep .... YWL	KSJ	—I will not delay (*or*, detain) you long	
	Water is too deep to anchor ......EKB	KSL	—If delayed (*or*, detained)	
		KSM	—What delays-ed (*or*, detains-ed) you?	
	DEFALCATION (*Embezzlement*)......LXD			
		KSN	DELEGATE-D-ING TO DELEGATE	
KRC	DEFEAT-ED-ING TO DEFEAT			
			DELIGHT (*Pleasure*) ....TMS	
KRD	DEFECT IVE (*See also* DAMAGE)		Delighted (*Pleased*)........ .....TMR	
	Can defects be repaired at sea?......BA		Delightful-ly (*Pleasant*) ....TMN	
	Can defects be repaired by yourselves?			
	....................................BC	KSO	DELIRIOUS-UM	
	Defects can be repaired at sea.......BE	KSP	—Delirium tremens	
	Defects can not be repaired at sea (*or*			
	place indicated) ....BF	KSQ	DELIVER-Y	
	Defects, *or*, Damage can not be repaired	KSR	—Delivered	
	without assistance.................BG	KST	—Have, *or* Has been delivered	
	Defects may be set right in — hours.BD	KSU	—Have the mails been delivered?	
KRE	—Found defective	KSV	—Have you delivered?	
KRF	—Send list of defects	KSW	—On delivery	
KRG	—What are your defects?	KSX	—Shall, *or*, Will be delivered	
		KSY	—You are to deliver	

	DEMAND—DERANGE		
KSZ	DEMAND-ED-ING TO DEMAND		DEPEND—*Continued*
KTA	—Demand has been sent	KUJ	—May be trusted (*or*, depended on)
	Have, *or*, Has been demanded (*or*,		May I depend upon your time?......RS
	claimed) by — ....IRW		No confidence in No dependence
KTB	—Make a demand		on ....JNX
KTC	—No demand—for		Not to be depended upon (*not relia-*
KTD	—On demand—of		*ble*) ....UOX
	Salvage demanded ....VGD	KUM	—Nothing to be depended upon beyond
KTE	—Send your demands—to (*or* for)		your own resources
KTF	—Some demand springing up		Reckoning not to be depended upon_RU
KTG	—They will demand explanation		To be depended upon (*reliable*)....UOY
KTH	—What is the demand (*or*, claim)?		
		KUN	DEPLORE-D-ING-ABLE-Y
KTI	DEMOCRAT-ACY-IC	KUO	DEPOSE-D-ING-ITION
KTJ	DEMOLISH-ED-ING	KUP	DEPOSIT-ED-ING
KTL	DEMUR-RED-RING	KUQ	DEPOT
KTM	DEMURRAGE		Coal depot ....IWN
	A day's demurrage (*or number indi-*	KUR	—Naval depot
	cated) ....KMU		What quantity of coal remains in col-
	Day's demurrage—is —.........KNC		lier (lighter, *or*, coal depot)?......IXO
KTN	—Demurrage commences-d on the —		
		KUS	DEPRECIATE-D-ING-ION
	DENK (*Measure of weight*).........BAN		
		KUV	DEPRESS ED ING-ION
	DENMARK (*See* DANE) ....KJX		Meteorological Office, *or*, Weather Bu-
KTO	DENOTE-D-ING		reau, reports depression approaching
	Report when ship denoted is sighted_USD		from the —.......... ....ETP
KTP	DENSE-LY-ITY	KUW	DEPRIVE-D ING-ATION
KTQ	DENY-IED-ING TO DENY	KUX	DEPTH (*See also* DEEP *and* SOUNDINGS)
KTR	DEPART-ED ING-URE		*Asks precise time of high water, and*
	After the departure of —..........DRM		*minimum depth at that time*......XE
KTS	—Before departure—of —		NOTE—*Reply will be*
	Hasten the departure of —.........ONR		1 *By an hour signal* (*See page* 58g)
KTU	—Have you seen any vessel since your		2 *By a numeral signal denoting*
	departure?		*feet*
KTV	—I can not hasten my departure	KUY	—At high water you will have — feet
KTW	—I must hasten my departure	KUZ	—At low water you will have — feet
KTX	—My departure	KVA	—Depth is
	What was the date of your depart-	KVB	—Indicate the depth—of — What depth?
	ure? ....KML	KVC	—Is there sufficient depth of water?
KTY	—Where did you take your last departure	KVD	—The depth here (*or at place indicated*)
	from?		is — feet at low water
KTZ	—Your departure		What depth? Indicate the depth.._KVB
		KVE	—What is the depth at high water?
KUA	DEPARTMENT	KVF	—What is the depth at low water?
	Commissariat department..........JFA	KVG	—What is the depth of water in feet at —?
	Goods department................OBU		What is the depth of water on the bar?
	Head of department......... .....OQH		VQ
	Intelligence department ....POE	KVH	—What is the depth of water over the sill?
	Naval department ....SEO	KVI	—What is the depth of water alongside
	Passenger department ....TDF		the —?
			What is the least depth of water we
KUB	DEPEND-ED ING-ENCE—ON		shall have (*or*, ought to have)?....VR
KUC	—Can depend on	KVJ	—You will have less than — feet of water
KUD	—Can I depend on?		
	Can I depend on your time?.........RS	KVL	DEPUTE-D-ING TO DEPUTE
KUE	—Can you depend on your time?		Chamber of Deputies ....IKA
KUF	—Can not depend on	KVM	—Deputation
	Depends on circumstances ....IRG		Deputy, *or*, Member of Congress or Par-
	Dependence (*reliance*) ....UOW		liament ....TBO
KUH	—I can not depend on my time	KVN	—Deputy, *or*, lieutenant governor
KUI	—I shall depend on		
	Light is not to be depended on......QD	KVO	DERANGE-D-ING-MENT
			Machinery slightly deranged ......MCQ

DERELICT—DESTROY

KVP	DERELICT Beware of derelict, dangerous to navigation_____JE
KVQ	—Derelict has been fired. Derelict reported off —(or, in the vicinity of —) _____JF
KVR	—Derelict reported passed on —, latitude —, longitude —, dangerous to navigation Have you fallen in with crew of derelict?_____MRT Have you seen derelict?_____JG
KVS	—Passed a derelict vessel apparently in good condition, hull well out of water, dangerous to navigation
KVT	—Passed the derelict vessel — on the — Saw derelict in latitude —, longitude —, on the — _____JH
KVU	DERIVE-D-ING-ATION
KVW	DERRICK
KVX	DESCEND-ED-ING DESCENT.
KVY KVZ KWA KWB KWC	DESCRIBE-D-ING —By the description given —Can you describe? —Can not describe —Describe exactly what you want (size, weight, etc)
KWD KWE KWF KWG	—Description —Have you sent description? —Send description —The same description
KWH	DESERT-ED-ING-ION Charged with desertion_____IMA Crew deserted _____KCM
KWI	—Deserter
KWJ	DESERVE-D-ING-LY
KWL	DESIGN-ED-ING-EDLY-ATION
KWM	DESIRE-D ING-OUS-ABLE—TO Am desired to acquaint you (or person indicated) _____DGZ Captain desires me to inform you_HWE Commanding officer desires_____JDZ
KWN	—Desire to see (or, speak)—to Desire you will acquaint_____DHF
KWO KWP KWQ KWR KWS	—Desires me to tell you —Do you desire? —Have, Had, or, Has the desired effect —It is desirable —It is not desirable Owners desire me to inform you to coal at — _____HE Owners desire me to inform you to discharge cargo first at —) place next indicated)_____SY Owners desired me to inform you__SYK
KWT	—That is my desire The consul is desired to come (or, send) on board _____ _____JRP
KWU	—The desired effect Vessel desires to be reported by telegraph to owner (Mr —) at —____USF Your agents desire me to inform you, DTE

KWV	DESPATCH-ED-ING TO DESPATCH (See • also LETTER) Any letters (or, papers) from —?____JI
KWX	—Any letters (or, despatches) on board? Bring (or, Brought) despatches____HIQ Despatch boat__ _____GSN Despatch box _____HDC Despatch is necessary___ _____ _____JK
KWY	—Despatch vessel, or, AVISO
KWZ	—Despatches Despatches arrived_____EYB Forward my despatches_____NMZ Go for despatches_____JL Government despatch_____OFS Had any letters (or, despatches) arrived—at?_____JM Have (or vessel indicated has) despatches for —?_____JN Have(or vessel indicated has)despatches from —_____JO Have letters (mails, or, despatches) for you (or vessel or person indicated)_JP Have you letters (mails, or, despatches) for me (or vessel, person, or place indicated) _____JQ I am the bearer of important despatches, JR I have Government despatches (or, —), JS I shall make for the anchorage (or, —) with all despatch _____JT
KXA	—Important despatches Refer to my despatches (or, letters)__JU Refer to your despatches(or, letters)_JV Send for despatches (or, letters) ____JW
KXB KXC	—Send your despatches —Take despatches Telegraphic despatches ____ _____XLN Telegraphic despatches arrived from —, JX
KXD	—Wait for despatches Will you forward my letters (or, despatches)—to —? _____ _____QU Will you take despatches? _____ __JY
KXE	—Your despatches are ready
KXF	DESPISE-D-ING TO DESPISE
	DESSATINE (square measure)_____AXO
KXG	DESSERT
KXH	DESTINE-D-ING-ATION Arrived at destination_____EXU Convoy to disperse and make for destination _____IS Do you call at any port before destination?_____HTE
KXI KXJ	—What is his (her, or, their) destination? —Your port of destination is closed, your owners desire you to proceed to —
KXL	DESTITUTE-UTION
KXM KXN	DESTROY-ED-ING —Can you destroy? Crops destroyed_____KDV
KXO KXP KXQ KXR	—Destroyed by fire —Have, or, Has destroyed. —Have you destroyed? —Shall I destroy?

DESTROY—DILEMMA

	DESTROY—*Continued*	**KYT**	**DIAMETER-RICAL-LY**
	Torpedo catcher (*or*, chaser, *or*, destroyer) ...... XWR	**KYU**	**DIAMOND**
KXS	—Was, *or*, Were destroyed	**KYV**	**DIARRHŒA** ETIC
KXT	**DESTRUCTIVE-LY-ION**	**KYW**	**DIARY**
KXU	—Destructive radius	**KYX**	**DICTATE-D-ING-ION-TOR**
KXV	**DETACH-ED-ING-MENT**	**KYZ**	**DICTIONARY**
KXW	—Detached squadron		**DID** (*See* DO) ...BMF
KXY	**DETAIL-ED-ING**	**KZA**	**DIE TO DIE** (*See also* DEAD AND DYING)
KXZ	**DETAIN-ED-ING-ENTION** (*See also* DELAY)		Captain died on — ...HWI
	Cause of detention ...IFQ		Died, *or*, Dead ...KOE
	Detained at ...KSB		Dying ...LQM
	Detained by ...KSC		How many have died? ...KOM
	Detained by contrary winds ...KSD		The breeze will die away ...HGF
KYA	—Detained by ice	**KZB**	**DIE**
KYB	—Detained by the emigration officer		—Taps stocks, and dies
	Detained by thick fog ...KSE	**KZC**	**DIET-ED-ING ARY**
KYC	—Detained in harbor	**KZD**	**DIFFER-ED-ING** To differ
	Do not detain (delay) ...KSG		Can the difference be arranged? ...EWZ
	I have been (*or vessel indicated was*) detained — days windbound at —(*place indicated*) ...KYE	**KZE**	—Difference
		KZF	—Difference of latitude
		KZG	—Difference of longitude
	I must not detain (*or*, delay) ...KSI	**KZH**	—Different
	I will not detain you long ...KSJ	**KZI**	—Different size
KYD	If detained ...KSL	**KZJ**	—Different speed
	—No cause of detention	**KZL**	—Differently
	There has been so much fog as to cause considerable detention ...NGL	**KZM**	—No difference
KYE	—Vessel indicated was (*or*, I have been) detained windbound at —(*place indicated*)	**KZN**	—Not much difference—between —and —
		KZO	—Totally *or*, Quite different
	What detains-ed you? ...KSM	**KZP**	—Very different
		KZQ	—We (*or parties indicated*) differ
	DETECT-ED-ING-ION (*Discover*) ...LDM	**KZR**	—What difference?
KYF	**DETECTIVE**	**KZS**	**DIFFICULT-Y**
KYG	**DETERIORATE-D-ING-ATION**		Entrance is difficult ...MFO
KYH	**DETERMINE-D-ING-ATION-ABLE**		Entrance is not difficult ...MFP
KYI	**DETONATE-D-ING-ION**		Fire difficult to get at ...OP
	Detonating fuze ...NUW	**KZT**	—Great difficulty, *or*, Very difficult
KYJ	—Detonator	**KZU**	—I am in difficulties, direct me how to steer
	DEUNUM (*Square measure*) ...AXJ	**KZV**	—I find great difficulty in weighing
KYL	**DEVIATE-D ING-ION**		Is it a difficult anchorage to get away from? ...ELB
	Deviation of compass ...JHN		Is the channel difficult? ...ILO
	Deviation table ...JHO	**KZW**	—Is the river navigation difficult?
	Have you any local attraction (deviation)? ...JHR	**KZX**	—Is there any difficulty?
	What is your deviation? ...JIO	**KZY**	—It is difficult to extricate
KYM	**DEW**	**LAB**	—It is very difficult
			Landing is very difficult ...QEM
KYN	**DHOLL**	**LAC**	—No difficulty
		LAD	—Should there be any difficulty
KYO	**DHOW**	**LAE**	—There is not much difficulty
			You will find great difficulty in getting through the ice at — ...PCI
	DIACHYLON, *or*, ADHESIVE PLASTER...DJW	**LAF**	—You will have great difficulty in obtaining water
KYP	**DIAGONAL-LY**	**LAG**	**DIG-GING-GER** DUG
KYQ	—Diagonal, *or*, Triatic stay	**LAH**	**DIGEST-ION-IVE-IBLE**
KYR	**DIAGRAM—OF**	**LAI**	**DILEMMA**
KYS	—Indicator diagram		

DIM—DISAPPEAR

	DIM-NESS (*Faint*) ------------- --------MRB		DIRECT—*Continued*
	DIME (*Com*) ----------------------ATI		Your original orders are cancelled I am directed to inform you to proceed to — ----------------------------TH
LAJ	DIMENSION	LBK	DIRECT (*Straight*)
LAK	—Dimensions of — yard are — feet in length, — inches in the slings		Directly (See IMMEDIATELY) -----PDG
			Come directly ---------- ---------JCQ
LAM	—External dimensions From out to out		Direct-action engine -------------MAO
LAN	—Internal dimensions (In the clear Inside)	LBM	—Direct to
			Directly (*Immediately*) ----------PDG
LAO	—What are the dimensions of main yard?		Directly you get bottom let go your anchor------- -----------------EHV
LAP	—What are the dimensions of your fore yard?	LBN	—Go direct
			I will heave-to directly I have room.OSN
LAQ	DIMINISH-ED-ING-UTIVE-UTION		Steer direct for the buoy (*or*, beacon) GAM
	Has yellow fever diminished at —?.MWI		
	Yellow fever has diminished at --.MWU	LBO	DIRECTION (*Position*)
		LBP	—In the direction—of
	DINAR (*Com*) ----------------------AUB	LBQ	—Indicate direction (*or*, position)—of —
LAR	DINE D-ING-NER	LBR	DIRT-Y-INESS
	Admiral, or Senior officer requests the pleasure of your company to dinner, DKX		Dirty shirts or linen -------------VYO
	After dinner --------------------DQU	LBS	DISABLE-ING-ILITY
LAS	—Before dinner		Am disabled . communicate with me.CL
LAT	—Keep dinner for	LBT	—Am, Is, *or*, Are disabled
LAU	—People going on service to take their dinners with them	LBU	—Am, Is, *or*, Are not disabled
			Capstan disabled -----------------HWA
LAV	—Send dinner—for	LBV	—Disabled
LAW	—When will dinner be ready? What time is dinner?		Donkey engine disabled ----------LKO
LAX	—Will you dine with me?		Donkey pump disabled -----------LKP
			Engines broken down I am disabled, BJ
LAY	DINGHY		Engines broken down (*or*, disabled)-MCH
			Engines completely disabled--------MQ
LAZ	DIOPTRIC DIOPTRIC LIGHT		Engines disabled , can repair them in — hours----------------------------MCL
			Have lost screw, *or*, Screw disabled.MS
LBA	DIP-PED-PING		Have passed steamer with machinery disabled -----------------------MT
LBC	—Dip your colors Dip your ensign		Have passed steamer with steering gear disabled ---------------------MU
LBD	DIPLOMA-CY-TIC-IST		Have you seen disabled steamer? ---MV
	Diplomatic agent----------------DSP		Machinery disabled ---------------MCN
			One engine disabled---------------MCR
LBE	DIPHTHERIA-IC		One screw disabled , can work the other, MW
LBF	DIRECT ED ING TO DIRECT (*See also* INSTRUCT AND ORDER)	LBW	—Passed disabled ship at —
	According to directions (*or*, orders)-DEM		Passed disabled steamer at —-------MX
	Act as your judgment directs ------DIB		Rudder disabled , will you assist me? (*into port indicated, if necessary*) .DA
LBG	—Am, Is, *or*, Are directed (*or*, ordered) to — (*or*, by—)		Ship disabled , will you tow me (*into port indicated*)?-----------------MY
	Board of directors----- ----------GOZ		Steering gear disabled ------------MZ
	Book of Sailing Directions (*indicate for what place if necessary*) --------GYX	LBX	—Too much disabled
	Direct me how to steer -----------WTV	LBY	—Tug is disabled
	Directible torpedo-- --- ---------XWM		
	Direction (*address*) ------------DJH	LBZ	DISADVANTAGE-OUS-LY
	Directions (*instructions*)---------PNH		
LBH	—Director	LCA	DISAFFECT-ED-ION
LBI	—Directory for —		Crew disaffected, will not work----KON
LBJ	—Give directions—that (for, *or*, to) —	LCB	—Serious disaffection in —
	Have you a Book of Sailing Directions? RO	LCD	DISAGREE-D-ING-MENT-ABLE
	I am directed to acquaint you------DHI	LCE	—Landing safe, but very disagreeable
	I am in danger (*or*, shoal water), direct me how to steer -----------------NL		DISALLOW-ED NOT ALLOWED-------DZL
	I am in difficulties, direct me how to steer------------------------KZU	LCF	DISAPPEAR ED-ING
	Sailing Directions (*indicate for what place, if necessary*) -------------VEK		Buoys have all disappeared -------HMI
		LCG	—Disappearance

DISAPPEAR—DISK

	DISAPPEAR—*Continued*	LDK	DISCOURAGE-D ING
	Has yellow fever disappeared at —?_MWJ		
	Yellow fever has disappeared at —_MWV		DISCOURTESY (*Incivility*)------------PGI
LCH	DISAPPOINT-ED-ING-MENT	LDM	DISCOVER-ED-ING-ABLE-Y
		LDN	—Can you discover?
LCI	DISAPPROVE-D-ING	LDO	—Have, *or*, Has been discovered
		LDP	—Have, *or*, Has not been discovered
LCJ	DISARM-ED-ING-AMENT	LDQ	—Have you been discovered?
		LDR	—Have you discovered?
LCK	DISASTER ROUS-LY	LDS	—I made the discovery
		LDT	—What have you discovered?
LCM	DISBELIEVE-D ING DISBELIEF	LDU	—When did you discover?
LCN	DISC DISK	LDV	DISCREDIT-ED-ING ABLE Y
LCO	DISCHARGE-D-ING	LDW	DISCRETION-AL DISCREETLY.
LCP	—Can I discharge at pier?	LDX	—Use you own discretion
	Discharge-d ing cargo_____IAL		
	Discharge cargo as soon as possible_IAM		
LCQ	—Discharge pipes out of order	LDY	DISCUSS-ED-ING-ION (*Controversy*)
LCR	—Discharge valves out of order		
LCS	—Discharged for	LDZ	DISEASE-D (*See also* SICK)
LCT	—Discharged from		A contagious disease on board (*or at*
LCU	—Discharged in		*place indicated*) ----------------GPQ
LCV	—Discharged on the —		Contagious disease ....JSC
	I am discharging (*or* taking in) powder		Disease, *or*, Sickness is contagious ._VB
	or explosives		Disease, *or*, Sickness is not contagious,
	Flag B, *or*, Code Flag over B		VC
	I have received orders for you to dis-		I have (*or*, have had) a dangerous infec-
	charge cargo at — -----------SR		tious disease on board
LCW	—I shall discharge		Code Flag over L
	Must discharge cargo to repair damages,	LEA	—Infectious disease
	KJR		Is disease (*or*, sickness) contagious?_VA
LCX	—Obliged to discharge	LEB	—Venereal disease
	Owners desire me to inform you to dis-		
	charge cargo first at (*place indicated*),	LEC	DISEMBARK-ATION
	SY	LED	—Disembarked-ing
	Send flats to discharge cargo _____IAZ	LEF	—Have, *or*, Has been disembarked
	When do you commence discharging	LEG	—Passengers will have to be disembarked
	cargo? _____ ... _ .. _ .... IBJ	LEH	—Prepare to disembark—at—
	Where did you discharge cargo?____IBL	LEI	—Shall, *or*, Will be disembarked
	Without discharging cargo_____IBO	LEJ	—To be disembarked
	You will discharge cargo—at — ____IBQ		Troops are ready to disembark_____YBO
			Where will passengers have to be dis-
			embarked? ------------------ --TDV
LCY	DISCIPLINE-D-ING	LEK	DISENGAGE-D-ING-MENT
	DISCLOSE-D-ING (*Divulge*) ----------LIJ	LEM	DISGUISE-D-ING TO DISGUISE
LCZ	DISCOLOR-ED-ING-ATION		I have been chased by a disguised war
			vessel .... -------------------INP
LDA	DISCONNECT	LEN	—I saw war vessel (*or*, vessels) disguised
LDB	—Disconnected-ing		off —
LDC	—Does disconnecting affect your steering?		Keep a good lookout, as it is reported
	Let your wheels (*or*, screw) revolve		enemy's war vessels are going about
	without disconnecting -----------QOB		in disguise of merchantmen _____OJ
		LEO	—Passed a war vessel disguised, funnel
LDE	DISCONTENT-ED LY-MENT		(*or*, funnels)— *Here give the color*
	Crew discontented, will not work _KCO		*or the design of the line of steamers*
			that it may be an imitation of
LDF	DISCONTINUE-D-ING-ATION	LEP	DISGRACE-D-ING-FULLY
LDG	—Discontinue repeating signals		
	Light, *or*, Floating light is discontinued	LEQ	DISHONEST-Y-LY
	(*or*, gone) -------------------QSO		
	Your agent (*or agent specified*) has dis-	LER	DISHONOR-ED-ABLE-ABLY
	continued business -------------DTF		
		LES	DISINFECT-ED-ING-ANT
LDH	DISCOUNT-ED ING		
LDI	—At a discount		DISK DISC _____LCN
LDJ	—What is the discount?		

DISLIKE—DISTINGUISH

LET	DISLIKE-D-ING	LFZ	DISSOLVE-D-ING-UTION
LEU	DISLOCATE D-ING-ION	LGA	—Dissolution of partnership
	Arm dislocated_____EVJ		Congress, Parliament, Assembly (_or other legislative body_) is dissolved_FEJ
LEV	DISMANTLE-D-ING.		
LEW	DISMAST ED ING	LGB	DISTANCE
	Lost the (_or, her_) masts_____RAZ		At — (_time indicated_) — (_place indicated_) bore —, distant — miles____ED
	Vessel seen (_or, indicated_) lost her masts, RBF		Bearings and distance of_____GBF
	When did you lose your masts?____RMJ		Can not make out your flags, come nearer (_or, make Distant Signals_)_VM
LEX	DISMISS-ED-ING-AL	LGC	—Communicate by Distant Signals
LEY	DISOBEY-ED-ING-DIENCE		Convoy to spread as far as possible, keeping within signal distance_____IZ
LEZ	—Disobey-ed-ing-dience of orders	LGD	Course and distance _____ ___JZG
LFA	DISORDER ED-LY	LGE	—Distant
LFB	—Is, or, Are in great disorder		—Distant Signals (_when spoken of_) (See page 523)
	DISPATCH-ED-ING TO DISPATCH (_See_ DESPATCH) _____KWV		Extend-ed-ing your distance_____MPN
LFC	DISPENSE-D-ING—WITH	LGF	How far off? What distance?_____MTA
LFD	—Can spare (_or, dispense with_)		—I am going to communicate by Distant Signals
LFE	—Can you spare (_or, dispense with_)?		I wish to signal, will you come within easy signal distance_____VJ
LFG	DISPERSE-D-ING	LGH	Keep within signal distance_____PZU
	Convoy to disperse and make for destination _____Is	LGI	—Make Distant Signals
	Convoy to disperse and reassemble at rendezvous (_or place indicated_)___ IT	LGJ	—Miles distant (_or, from_)
	Convoy to spread as far as possible, keeping within signal distance _____ __ IZ	LGK	—Not sufficient distance
LFH	DISPLACE-D-ING-MENT	LGM	—Sufficient distance
LFI	DISPLAY ED-ING	LGN	What distance? How far off? _____MTA
LFJ	DISPLEASE-D-ING-URE		—What distance off is?
LFK	DISPOSE D ING-AL		—What distance were you—from?
	An opportunity for selling (_or, disposing of_) _____SRV		What are the bearing and distance of — (_place or vessel indicated_)? _____EO
	An opportunity of disposing of cargo, HZQ	LGO	—What is the distance—to?
LFM	—At my own disposal	LGP	—When I have run my distance I shall put her head off until —
LFN	—At your disposal		
LFO	—Can dispose of	LGQ	DISTIL LED-LING-LER-Y-LATION
LFP	—Can not dispose of	LGR	DISTINCT-LY-ION IVE-NESS
	Dispose of cargo as soon as possible, IAN	LGS	—Distinct enough
LFQ	—For disposal	LGT	—Your signal is not distinct Interpretation doubtful
LFR	—Have you disposed of?	LGU	DISTINGUISH-ED-ING-ABLE (_See also_ MAKE OUT)
LFS	—Not disposed of		Calls the attention of the vessel whose distinguishing signal follows_____DR
LFT	DISPROVE-D	LGV	—Can you, or, Could you distinguish?
LFU	DISPUTE-D-ING-ABLE	LGW	—Can you distinguish what — (_vessel indicated_) has hoisted?
LFV	DISQUALIFY-IED ICATION	LGX	—Can not distinguish her (_or, your_) colors
LFW	DISREGARD ED ING		Can not distinguish your signal flags, come nearer, (_or, make Distant Signals_) __ _____VM
LFX	—Disregard leader's movements Disregard motions	LGY	—Distinguishing vane
	DISSATISFIED DISSATISFACTION (_Discontent_) _____LDE	LGZ	—I can not distinguish (_or, see_)
		LHA	—Plainly distinguishable I saw it very plainly
LFY	DISSENT-ED ING SION		Repeat your distinguishing signal, the flags were not made out _____DU
			Show your distinguishing signal_____DV
			Signal not understood, although the flags are distinguished _____WCX
			Vessel has not shown her number (_or, distinguishing signal_)_____SLH
			Vessels just arrived, show your distinguishing signals____ _____ __DX

DISTINGUISH—DO

	DISTINGUISH—Continued	**LHV**	**DIVE-D ING-R**
	Vessels just weighed (or, leaving), show your distinguishing signals........DZ		Artificer diver.. ....EZQ
	Vessels that wish to be reported all well, show your distinguishing signals.................DYT	**LHW**	—Can you lend me a diver ?
		LHX	—Diver's gear
		LHY	—Diver's helmet
		LHZ	—Diver's pump
	Vessels that wish to be reported on leaving, show your distinguishing signals ....EB	**LIA**	—Diver's telephone
			Diving apparatus ....ERG
			Diving bell..... ....GEO
	Vessels that wish to be reported, show your distinguishing signals........EA	**LIB**	—Diving dress
		LIC	—Have you (or vessel indicated) a diver ?
LHB	—Your signal flags can not be distinguished	**LID**	—Require diver to clear screw
		LIE	—Require diver to examine bottom
LHC	**DISTRESS-ED**		
	Boat in distress (bearing, if necessary), ES	**LIF**	**DIVIDE-D-ING-SION DIVISIBLE**
			Divisional boat...............GSR
LHD	Did you see a vessel in distress ?.....ND		
LHE	—Distressed for want of coal		
LHF	—Distressed for want of food.	**LIG**	**DIVIDEND**
LHG	—Distressed for want of water		
	—Distressed seaman	**LIH**	**DIVINE SERVICE**
	Do you think we ought to pass a vessel in this distressed condition ?.. ....NE		
	Have you seen or heard of a vessel wrecked or in distress ?... ......ZU		**DJERIB** (Square measure)........ AXP
LHI	—How long have you (or vessel indicated) been in such distress ?	**LIJ**	**DIVULGE-D-ING**
	I will assist you (or vessel in distress), FGI		**Do DID, DOING, DONE**........... ... BMF
			After it was done................DQY
			Badly done.......... ... .FRJ
	I will remain by you (or vessel in distress) ....AI		Can, or, May do (or, be done)......BLT
			Can, or, May do it............ . .BLU
	In distress, want assistance........ .NS		Can, or, May have done (or, been done), BLV
	IN DISTRESS, WANT IMMEDIATE ASSISTANCE*................ ....NC		Can, or, May it be done ?..........BLW
			Can you (or person indicated) do it? DEF
	*See Distress Signals, page 7		Can not, or, May not do (or, be done) BLX
	Is not vessel (bearing if necessary) in distress? ....NT		Can not, or, May not do it.........BLY
			Can not, or, May not have done (or, been done) ....BLZ
LHJ	—Put in, in distress		Could, or, Might do (or, be done)...BMA
LHK	—Relieve vessel in distress		Could, or, Might have done (or, been done) .-.......... ....BMC
	Signals of distress ....WCY		Could, or, Might not do (or, be done), BMD
	Signals of distress in—(direction indicated) ....WCZ		Could, or, Might not have done (or, been done) ... ...BME
LHM	—Strange sail (or vessel indicated) is (or, was) in distress	**LIK**	—Do all
	Vessel indicated in distress, wants immediate assistance ....NW		Do, Does, Did, Doing, Done........BMF
			Do Does, or, Did he (she it, or person s or thing s indicated)?......BTQ
	Vessel indicated appears in distress..NX		Do, Does, or, Did his (her-s, it-s)?...BMG
	Vessel on—(bearing indicated) is making signals of distress........ ....WDP		Do, Does, or, Did I (my, mine)? ...BMH
			Do, Does or, Did not........ ......BMI
LHN	**DISTRIBUTE-D-ING-ION.**		Do, Does, or, Did that (or, this) ?..BMJ
			Do, Does, or, Did the ? ....BMK
LHO	**DISTRICT—OF—**		Do, Does, or, Did there ?..........BML
	Agricultural district—of —.........DUL		Do, or, Did these (or, those)? .. .BMN
	Mining district—of —.............RUQ		Do, Does or, Did they (their-s)?...BMO
			Do, Does, or, Did we (our-s)?......BMP
LHP	**DISTRUST-ED-ING-FUL-LY**		Do, Does, or, Did you r-s ?.........BMQ
			Do not let him (her, it, or person indicated)—have ....BTR
LHQ	**DISTURB-ED-ING-ANCE**		
	Atmospheric disturbance ....FJP	**LIM**	—Do what is to be done
LHR	—Disturbance is quelled	**LIN**	Do you think it will be?...........XPY
LHS	—Disturbance on shore (or, at —)		—Doing well
	I have no cargo to disturb my compasses .... JHU		Easily done .... DEG
			Had, Has, or, Have done (or, been done), BMR
LHT	**DITCH**		
LHU	**DITTO**		

DO

Do—*Continued*

Had Has, *or*, Have not done (*or*, been
    done) ---- ---------------------EMS
Had, *or*, Have I done? ------------BMT
Had *or*, Have they done? ---------BMU
Had, *or*, Have we done? -----------BMV
Had, *or*, Have you done? ---------BMW
Has it been done (accomplished)?...DEH
Having done --- --------- ------BMX
He, She, It (*or person-s or thing-s indi-
    cated*) does (do, *or*, did)---------BUH
He, She, It (*or person s or thing-s indi-
    cated*) does (do, *or*, did) not-----BUI
He, She, It, (*or person-s or thing-s indi-
    cated*) had (has, *or*, have) done (*or*,
    is, *or*, are doing) -------------BUL
He, She, It (*or person-s or thing-s indi-
    cated*) had (has, *or*, have) not done
    (*or*, is, *or*, are not doing) --------BUP
He, She, It (*or person-s or thing-s indi-
    cated*) should (*or*, would) do (*or*, be
    done--------------------------BVN
He, She, It (*or person-s or thing-s indi-
    cated*) should (*or*, would) not do (*or*,
    be done) ---- --------------------BVR
How do (does, *or*, did)? ----------BXL
How do (*or*, did) you' -----------BXM
I do (*or*, did) --------------------BMY
I do (*or*, did) not-----------------BMZ

LIO
    —I do not think it will be
I do not think it will do ----------XQB
I had (*or*, have) done----------- ---BNA
I had (*or*, have) not done------ ---BNC
I shall (*or*, will) do... ----- -- --BND
I shall (*or*, will) not do-----------BNE
I should (*or*, would) do------------BNF
I should (*or*, would) not do-- ----BNG

LIP
    —I will do it if I can
If he (she, it, *or person-s or thing-s in-
    dicated*) does (do, *or*, did)- -----BYH
If he (she, it, *or person-s or thing-s in-
    dicated*) does (do, *or*, did) not----BYI
If I do (*or*, did) ------------------BZF
If I do (*or*, did) not -------------BZG

LIQ
    —If I were you, I should not do it.
If it can (*or* may) be done----------BZR
If it can not (*or* may not) be done.BZS
If they do (*or*, did) ---------------CAM
If they do (*or*, did) not -----------CAN
If we do (*or*, did) ----------------CBF
If we do (*or*, did) not-------------CBG
If you do (*or*, did) ---------------CBW
If you do (*or*, did) not------------CBX
Is it done? -----------------------BNH
Is it not to be done? ---------------BNI
Is it to be done? -------------------BNJ
It, etc   (*See* He, *above*)
It can (*or*, may) be done----------BNK
It can not (*or*, may not) be done---BNL
It could (*or*, might) be done ------BNM
It could (*or*, might) not be done---BNO
It has (*or*, had) been done--------ENP
It has (*or*, had) not been done-----BNQ
It ought not to be done ------------BNS
It ought to be done------ ------BNR
It shall (*or*, will) be done----------BNT
It shall (*or*, will) not be done------BNU
It will do ----- --------- -------BNV

LIR
    —It will do very well
May, etc   (*See* Can, *above*)
Might, etc   (*See* Could, *above*)

Do—*Continued.*

Must do (*or*, be done) -----------CEN
Must not do (*or*, be done)---------CEW
Not doing---- ---------------------CFP
Not doing well --------------------ZBL
Not done--------------------------CFQ
Not easily done--------------------DEI
Not to be done --------------------CGI
—Now doing
Ought I to—do?-------------------CHQ
Ought to do (*or*, be done) ---------CHZ
Shall, *or*, Will do (*or*, be done) ---BNW
Shall, *or*, Will have done (*or*, be done),
    BNX
Shall, *or*, 'Will he (she it, *or person-s
    or thing-s indicated*) do (*or*, be done)?
    BWF
Shall, *or*, Will he (she, it, *or person-s
    or thing-s indicated*) not do (*or*, be
    done)?--------- - ----------- BWJ
Shall, *or*, Will not do (*or*, be done).BNY
Shall, *or*, Will not have done (*or*, been
    done) ----------------------------BNZ
Shall, *or*, Will they do it?----------BOA
Shall, *or*, Will they not do it?------BOC
Shall, *or*, Will you do it? ---------BOD
Shall, *or*, Will you not do it?------BOE
Should *or*, Would do (*or*, be done).BOF
Should, *or*, Would have done (*or*, been
    done) ----------------------------BOG
Should, *or*, Would he (she, it, *or per-
    son s or thing-s indicated*) do (*or*, be
    done) --------------------------BWO
Should, *or*, Would he (she, it, *or per-
    son-s or thing s indicated*) not do (*or*,
    be done)?---- ------------------BWT
Should, *or*, Would not do (*or*, be done),
    BOH
Should, *or*, Would not have done (*or*,
    been done) ---------------------BOI
They do (*or*, did) -----------------BOJ
They do (*or*, did) not -------------BOK
They had (*or*, have) done----------BOL
They had (*or*, have) not done -----BOM
To be done-------------------------CNR
To do ---- ------------------------CNT
—Very well done
Was, *or*, Were done (*or*, doing)----BON
Was *or*, Were not done (*or*, doing)-BOP
We do (*or*, did) -------------------BOQ
We do (*or*, did) not ---------------BOR
We had (*or*, have) done -----------BOS
We had (*or*, have) not done--------BOT
—Well done
What, *or*, Which do (does, *or*, did)?.CQB
What, *or*, Which does (do, *or*, did) he
    (she, it, *or person-s or thing-s indi-
    cated*)? ---------------------- -- CQD
What, *or*, Which do (*or*, did) they (we,
    or, you)? ----------------------CQE
What, *or*, Which shall (*or*, will) I (or,
    we) do? ------ -- ------------CQL
What, *or*, Which shall (*or*, will) they
    (*or*, you) do?-------------------CQN
What, *or*, Which should (*or*, would) I
    (*or*, we) do?-------------------CQR
What, *or*, Which should(*or*, would) they
    (*or*, you) do?-------------------CQS
When do (does, *or*, did)?-----------CRB
When does (do, *or*, did) he (she, it, *or*
    person-s or thing-s indicated)? ...CRD

LIS

LIT

LIU

	DO—DOUBLE	

	Do—*Continued*	
	When do (*or*, did) I (*or*, we)?____ .CRE	
	When do (*or*, did) they (*or*, you)?__CBF	
	When done with ... _____CRG	LJX
	When shall (*or*, will) I (they, we, *or*, you) do?_____CRV	
	When shall (*or* will) it be done?____CRX	
	Where do (does, *or*, did)? __ _____CSX	
	Where do (*or*, did) I (*or*, we)? _____CSY	
	Where do (*or*, did) they (*or*, you)?__CSZ	
LIV	—Where is it to take place (*or*, to be done)?	
	Who do (does, *or*, did)?_____CUO	
	Why do (does, *or*, did)? ____ ____ OVG	
	Why do (does *or*, did) not? _____CVH	
	Why do (*or*, did) you?_____CVI	
	Why do (*or*, did) you not?_____CVJ	LJY
	Will, etc (*See* Shall, *above*.)	LJZ
	Will it do? _____ _____BOU	
	Will not do ____ _____ _____BOV	LKA
	Will you do it?_____CWB	LKB
	Would, etc (*See* Should, *above*)	
	You do (*or*, did)_____BOW	
	You do (*or*, did) not?_____BOX	
	You had (*or*, have) done _____BOY	
	You had (*or*, have) not done_____BOZ	LKC
		LKD
LIW	Dock-ed-ing	
LIX	—Do you go into dry dock? Will you have to go into dry dock?	LKE
	Dock authorities _____FNG	LKF
LIY	—Dock dues	
LIZ	—Dock laborer	LKG
LJA	—Dockers' Union	LKH
LJB	—Docks — at —	
LJC	—Dry dock	LKI
LJD	—Floating dock	
LJE	—Graving dock	
LJF	—In dock	LKJ
LJG	—Is there any dry dock here?	
LJH	—Must go into dry dock	
LJI	—No dry dock nearer than —.	
LJK	—Out of dock	LKM
LJM	—Tidal dock	
LJN	—Undock-ed-ing	
LJO	—Wet dock	
	What is the depth over sill? _____KVH	
LJP	—What is the length of dock at —?	LKN
LJQ	—When do you come out of dry dock?	LKO
LJR	—When were you last docked?	LKP
	Will you have to go into dry dock or, Do you go into dry dock? _____LIX	
	Wish to enter port or dock _____TM	LKQ
		LKR
LJS	Docket	LKS
LJT	Dockyard	
	Dockyard authorities_____FNH	LKT
LJU	Doctor	LKU
	Accident, want a surgeon (*or*, doctor), AM	
LJV	—Doctor, *or*, Surgeon is away	LKV
LJW	—Doctor, *or*, Surgeon shall be sent for	LKW
	Doctor, *or*, Surgeon wants assistance FGP	
	Doctor, *or*, Surgeon will come immediately _____WN	
	Have you a doctor (*or*, surgeon)?___WK	LKX
	I am in want of a doctor (*or*, medical assistance) _____FGL	LKY
		LKZ

	Doctor—*Continued*	
	May I send a sick person to see your doctor (*or*, surgeon)?_____WL	
LJX	—No doctor (*or*, surgeon)	
	No doctor (*or*, surgeon) available__WM	
	Surgeon, *or*, Doctor shall be sent for, LJW	
	Want a doctor (*or*, surgeon), send me one from the nearest place . _____WO	
	Want a surgeon (*or*, doctor) Medical assistance wanted_____ ____FGL	
	Will your doctor (*or*, surgeon) come on board?_____ _____WP	
LJY	Document (*See also* Paper)	
LJZ	—Bring all your documents (*or*, papers)	
	Papers and documents_____TAL	
LKA	Dog	
LKB	—Dog watch	
	Doing (*See* Do) ____ _____BMF	
	Dollar. (*Coin*) _____ATJ	
LKC	Dolphin	
LKD	—Dolphin striker.	
LKE	Dome	
LKF	Domestic-ate-d-ing-ity	
LKG	Dominion	
LKH	—His, *or*, Her Majesty's dominions	
LKI	Domingo, Santo Santo Domingo Colors	
LKJ	Don	
	Done (*See* Do) _____BMF	
LKM	Donkey (*Animal*)	
	Donkey Boiler_____GWV	
	Donkey boiler burst ____ _____GWX	
	Donkey boiler damaged_____GWY	
LKN	—Donkey engine	
LKO	—Donkey engine disabled	
LKP	—Donkey pump disabled	
LKQ	Door	
LKR	—Manhole door	
LKS	—Water-tight door	
	Doppelzentner (*Measure of weight*), ZNL	
LKT	Dose	
LKU	Dot-ted-ing	
	Dot between initials__Code Flag over F	
LKV	Double d-y-ing Dual	
LKW	—Double acting-ion	
	Double altitude_____EDI	
	Double bottom__ _____HAY	
LKX	—Double expansion engine	
LKY	—Double reef-ed ·	
LKZ	—Double topsail	
	Double topsail yard_____ ____XWD	

DOUBLOON—DRIVE

	DOUBLOON (*Coin*) _____ATK		DRACHME (*Measure of weight*)____BAN
LMA	DOUBT-ED-FUL-LY	LMW	DRAFT-ED ING
	Any doubt _____EOT	LMX	DRAG-GED-GING TO DRAG
LMB	—Doubtless No doubt		Anchor drag-ged-ging_____EHA
	I doubt if it is possible_____TQB	LMY	—Drag rope
	I doubt your (*or,* my) ability to — _CYB		Dragging (*driving*) _____LNV
	If you have any doubt _____EPS	LMZ	DRAIN-ED-ING AGE
	No doubt Doubtless _ _____LMB		Main drain_____RGJ
	Your signal is not distinct, interpreta-tion doubtful____ _____LGT	LNA	DRAUGHT (*See also* DRAW)
LMC	DOVER'S POWDERS	LNB	—Black draught
LMD	DOWN		Coal requires much draught_____IWR
	Batten-ed-ing down_____ _____FYK		Draught of water _____YVC
	Broken down _____BJX	LNC	—My draught of water is — feet — inches aft
	Chronometer has run down_____GQ	LND	—My draught of water is — feet — inches forward
	Cut down _____ ____ _____KHI		What draught of water in feet could you lighten to?____ _____OB
	Down channel_____ _____TLK		What is your draught of water?_____OC
LME	—Down the river		Very bad draught _____FRV
LMF	—Down to—	LNE	—Very fair draught
LMG	—Downcast	LNF	—Very good draught
LMH	—Downfall		
LMI	—Downhearted	LNG	DRAWING DREW TO DRAW (*Sketch.*)
LMJ	—Downwards		
	Ensign union downwards_____MBJ	LNH	DRAW-ING DREW (*See also Draught.*)
	Fell down_____ ____ _____MVI		Draught _____ _____ ____LNA
	Haul down your ensign _____MEN	LNI	—Draw fires
	Haul the jib down ____ _____GPE		Drawbridge_____ ____ _____HGP
	Haul down _____ _ONY		Drawing too much water _____NZ
	Heave her down_ _____OSA		I draw — feet aft and — feet forward, OA
	Helm is down _ _____OTP		What is your draught of water?_____OC
	Holding-down bolts loose _____ _GYA		
	Hull down _____OZY	LNJ	DRAWBACK
	I have sunk (*or,* run down) a vessel_HK		
	Let your fires down_____NAM	LNK	DREDGE-D-ING
LMK	—Not far down	LNM	—Dredger
	Price has gone down (fallen) _____MRW	LNO	—Keep, *or,* Pass the dredger on your port side
	Sea going, gone down _____VKZ	LNP	—Keep *or,* Pass the dredger on your star-board side
LMN	—Up and down—the —		
LMO	—Upside down	LNQ	DRESS-ED-ING
	Vessel indicated (*or in direction indi-cated*) is supposed to have gone down, NOY		Diving dress _____LIB
	When the breeze goes down_____HGI		Dress-ed-ing ship _____VXI
	When the moon goes down _____ _RZC		Full dress _____NSO
LMP	—When the sea has gone down		Properly dressed_____TZF
LMQ	—When will you drop down?		
			DREW (*See* DRAW)___ _____LNH
	DOWNCAST_____LMG		
LMR	DOWNTON PUMP	LNR	DRIFT-ED-ING
		LNS	I am drifting, want assistance _____CM
	DOWNWARDS . _____LMJ		—You are (*or vessel indicated* is) drifting
LMS	DOZEN	LNT	DRINK ING DRANK
LMT	—Half dozen		Drunk-en ness_____ _____LOJ
LMU	—How many dozen can you spare?		
	Send a — dozen cases of — _____IDE	LNU	DRIVE DROVE DRIVEN (*Carried by wind or sea*)
	DRAA, *or,* PIKE (*Measure of length*)_AVU	LNV	—Driving Dragging
LMV	DRAB (*Color*)		Danger of driving (*or,* parting)_____KJZ
			Driven by stress of weather _____YZA
	DRACHM (*Measure of weight*) _____BAK		Driven on shore near — _ _____VZH
	DRACHMA (*Coin*)_____ _____ATL		Driving, I can veer no more cable I have no more anchors to let go____NK
	DRACHMA (*Measure of weight*)_____BAI		

	DRIVE—DURING	

	DRIVE—*Continued*	LOW	DUE—*Continued*
	Driving, parted from anchors......EIA		—Due on
	Drove on the rocks ....LOE		Due to arrive—at................EYD
LNW	—We are driving		Mail is due on — ....RFE
LNX	DRIVER, *or*, SPANKER	LOX	DUES (*Fees*)
	Spanker, *or*, Driver boom..........GZV		Dock dues..................... ...LIY
	Spanker, *or*, Driver boom sprung ..GZW		Harbor dues .... OMX
LNY	DRIZZLE D-ING		Light dues ....QSB
LNZ	DROGUE		DUG ....LAG
LOA	DROIT		DUIM (*Measure of length*) ....AVO
	Droits of admiralty ....DLW	LOY	DUKE DUCHESS
LOB	DROP-PED-PING	LOZ	DULL-NESS
	Drop a kedge Bring to with a kedge,		Business dull ....HNU
	PXV	LPA	DULY
	Drop astern—of..................FHI	LPB	DUMB
	Drop boat....................... ....GSU	LPC	DUNNAGE
LOC	—Drop down with the tide	LPD	DUPLICATE
	Make sail, *or*, Go ahead and drop a boat		Have you a duplicate (*or*, copy)—of?
	on board.................... ...DVN		JWB
	Time ball drops at — ....FTJ	LPE	—I can supply you with a duplicate of
	When does time ball drop?.........FTK		engine (*part mentioned*)
	When will you drop down?.........LMQ		I have no duplicate ....JWC
LOD	DROUGHT	LPF	—In duplicate
		LPG	—Send a duplicate
	DROVE (*See also* DRIVE)............LNU	LPH	—Send the duplicate
LOE	—Drove on the rocks		Take, *or*, Make a duplicate (*or*, copy),
LOF	DROWN		JWD
	Accident happened, drowned, drown-	LPI	DURABLE-NESS-ILITY
	ing ....DCH	LPJ	DURATION
LOG	—Drowned-ing	LPK	DURING
LOH	DRUG-GED-GING-GIST		Arrived during the day.... ....EXV
LOI	DRUM		Arrived during the night..........EXW
LOJ	DRUNK EN-NESS	LPM	—During a gale
	Charged with drunkenness..........IMB	LPN	—During a thick fog
LOK	—Drunkard	LPO	—During meal hours
LOM	DRY IED-ING TO DRY		During my absence (*or*, the absence of
	Do you go into dry dock? Will you		—) ....DAX
	have to go into dry dock? LIX	LPQ	—During the — (*or*, a —)
LON	—Dry Dryness		During the cruise......... ....KPA
LOP	—Dry at low water		During the day ....KNF
	Dry dock. ....LJC	LPR	—During the fog
LOQ	—Dry rot	LPS	—During the gale Whilst the gale lasts
LOR	—High and dry		During the monsoon (*or*, trades) ..RYJ
	Is there any dry dock here? ....LJG		—During the night
	Must go into dry dock . ...LJH	LPT	—During the passage (*or*, voyage)
	No dry dock nearer than —.... ...LJI	LPU	—During the time in harbor
	When do you come out of dry dock?.LJQ	LPV	—During the war
		LPW	—I shall he-to during the night
	DUAL...... ....LKV	LPX	I shall steer — during the night ...WTZ
			I will keep close to you during the
	DUCAT (*Coin*) . ...ATM		night.... ....IVF
			Keep close during the fog ....PYO
	DUCHESS DUKE................. ...LOY		Keep me in sight during the night
LOS	DUCK		(*bearing to follow, if necessary*) ..PYV
LOT	DUE		Passed during the night ....TCQ
	Duly................ .... LPA		Pay strict attention to signals during
LOU	—Due at		the night.................. ....DT
LOV	—Due from		Shall signal with steam whistle (*or*,
			siren) during fog (*or*, darkness)..WE

	DURING—EAST		
	DURING—*Continued*	LRA	EAGER-LY-NESS
	Spoke several vessels during the passage --- -------------------VSU	LRB	EAGLE (*Bird*)
	Will you keep close to me during the night?-------------------IVK		EAGLE (*Coin*) --------------------ATN
LPY	DUSK-Y-INESS	LRC	EAR
	As soon as it is dusk-------------FBN	LRD	EARING
LPZ	—Before it is dusk	LRE	—Reef earing
LQA	DUST-Y	LRF	EARL
	Gold dust-------------------ODK		
LQB	DUTCH-MAN DUTCH COLORS	LRG	EARLY EARLY—IN.
LQC	DUTY	LRH	—Earlier Earliest
LQD	—A great neglect of duty	LRI	EARN-ED-ING TO EARN
LQE	—Neglect of duty		
LQF	—Will do his (*or*, its) duty very well	LRJ	EARNEST-LY-NESS
LQG	DUTY (*Customs*)	LRK	EARTH
	Custom-house duties -------------EGV	LRM	—Earth circuit
	Duties are altered-----------------ECB	LRN	—Earth current
	Export duties ------------------MPC	LRO	—Earth plate
	Import duties -------------------PEO		
LQH	—Is there any duty on?		EARTHENWARE (*China*) ------------IPC
LQI	—Without duty		
LQJ	DYE-D ING ER	LRP	EARTHQUAKE
LQK	—Dyewoods	LRQ	EARTHWORK
LQM	DYING	LRS	EASE-D-ING TO EASE
LQN	—Dying at the rate of — a day		Ahead easy Go ahead easy--------LH
	Dying for want of water----------NF		Astern easy Go astern easy -------LK
			Ease her Reduce speed ------ ----MF
LQO	DYNAMITE-R	LRT	—Ease off
		LRU	—Easier Easiest
LQP	DYNAMO		Easily accomplished (done)-------DEG
	DYSENTERY CASE OF DYSENTERY----IDL	LRV	—Easy Easily
			Easy ahead -------------------LH
LQR	DYSPEPSIA-TIC		Easy astern ----------------------LK
		LRW	—Easy sail Keep under easy sail
			Fire can easily be got at-----------ON
		LRX	—Go easy
		LRY	—I shall keep under easy sail
			Not easily accomplished (done) ----DEI
			Not easy landing on the beach Scarcely prudent to land ---------------PZX
		LRZ	—Under easy sail
		LSA	—Which is the easiest?
		LSB	—With the greatest ease
			EAST (*Compass in degrees*) --------AEU
			EAST (*Compass in points*)--------- AQU
		LSC	—Easterly-ern-ing-ward
			Alongshore to the eastward of — (*distance in miles may follow*)-------EAO
			East India cargo ---- -------------HZA
			East India Government ------------OFB
			East Indiaman ------------------PHW
			East longitude ----- -------------QYZ
LQS	E (*Letter*) (*For new method of spelling, see page 13*)	LSD	—From the eastward
LQT	EACH		North-east erly-ern ing-ward-er----SJM
LQU	—Each of the (*or*, them)		On *or*, At the east coast of —-------IYH
LQV	—Each of us (*or*, our)		On *or*, At the N. E. coast of —-----IYG
LQW	—Each of you (*or*, your)		On *or*, At the S E. coast of — -----IYJ
LQX	—Each other		South-east erly-ern ing-ward-er----WLI
LQY	—Each ship	LSE	—To, *or*, For the eastward
LQZ	—Each way	LSF	—Too far to the eastward

EASTER—ELIGIBLE

LSG	EASTER After Easter _____DQV	LTX LTY	EFFORT —Use utmost efforts (exertions, or, endeavors)—to
LSH	—Before Easter		
	EASY-ILY (*See Ease*) _____LRV	LTZ	EGG
LSI	EAT-ING-EN-ABLE	LUA	EGYPT-IAN EGYPTIAN COLORS.
LSJ	—Worm eaten		
			EGYPTIAN POUND, or, LIRA _____ATO

LSK	EBB-ED-ING
	    Bar is impassable for boats on the ebb
	    tide ___ . . _____BQ
		LUB	EIGHT-H LY

	Do not abandon the vessel till the tide has ebbed _____AD	LUC	EIGHTEEN-TH
LSM	—Ebb sets —	LUD	EITHER EITHER ONE
LSN	—Ebb tide	LUE	—Either of the (or, them).
LSO	—How will the ebb set?	LUF	—Either of us, our-s
LSP	—When the ebb makes When will the tide ebb? _____XI	LUG	—Either of you-r s
		LUH	EJECT-ED ION-MENT
LSQ	EBONITE	LUI	—Ejector
LSR	EBONY		EL (*Measure of length*)_____AWJ
LST	ECCENTRIC-ITY-AL Eccentric bar broken _____FWJ	LUJ	ELAPSE-D
LSU	—Eccentric gear	LUK	ELASTIC-ITY
LSV	—Eccentric pulleys loose		
LSW	—Eccentric rod bent	LUM	ELBOW
LSX	—Eccentric rod broken		
LSY	—Eccentric sheaves out of order		ELDER (*Older*) _____SQI
LSZ	—Eccentric strap broken		
			ELDEST (*Oldest*) _____SQJ
LTA	ECLIPSE-D-ING		
		LUN	ELECT-ED-ING-ION
LTB	ECONOMY-IC-ICAL-IZE	LUO	—Is, or, are elected — for —
LTC	—Economize-d-ing fuel		
LTD	—Most economical speed	LUP	ELECTRIC-AL-LY ITY
		LUQ	—Electric battery
LTE	ECUADOR ECUADOR COLORS		Electric bell___ . . _____GEP
			Electric brush _____HKN
LTF	EDDY		Electric cable _____HRB
LTG	—Eddy tide	LUR	—Electric circuit
		LUS	—Electric current
LTH	EDGE-D-ING EDGEWISE Water's edge_____YWN	LUT	—Electric engine
		LUV	—Electric fuze
		LUW	—Electric launch
LTI	EDITION EDITOR	LUX	—Electric light
		LUY	—Electric light installation
LTJ	EDUCATE-D-ING-ION	LUZ	—Electric machine
		LVA	—Electric mine
LTK	EDUCE-TION	LVB	—Electric motor
LTM	—Eduction gear	LVC	—Electrical communication
LTN	—Eduction pipe out of order	LVD	—Electrician
LTO	—Eduction port	LVE	—Electro contact
		LVF	—Electro contact mine

LTP	EFFECT-ED-ING
	    Can a landing be effected? Can boats
	    land? _____EW
		LVG	—Electro magnet

		LVH	—Keep electric light on —
	Have, or, Has the desired effect___KWQ	LVI	—Search light, or, Electric projector
	The desired effect _____KWU	LVJ	—Turn off electric light
	Try to effect _____YCV	LVK	—Turn on electric light
LTQ	EFFECTIVE-LY		ELECTRICIAN _____LVD
LTR	EFFECTUAL-LY	LVM	ELEPHANT
LTS	EFFERVESCE-D-ING-NCE	LVN	ELEVATE-D-ING ATION-OR
LTU	—Effervescing drink		
		LVO	ELEVEN-TH
LTV	EFFICIENT-LY-CY		
LTW	—May be made efficient	LVP	ELIGIBLE-ILITY

	ELM—END		
LVQ	ELM	LXO	EMPEROR
LVR	ELSE		Emperor, King Queen, President, Chief
	Anyone else........................EQA		of the State, etc , of—has (or, is) ..IOR
	Anything else..................EQG	LXP	EMPIRE
	Anywhere else ....:.............EQS		
LVS	—Elsewhere	LXQ	EMPLOY ED-ING-MENT
LVT	—Nothing else		Chartered, or, Employed by Govern-
LVU	—Or else		ment............................INE
LVW	—Someone else	LXR	—Employee.
	Something else ....WJR	LXS	—Employer
LVX	—What else?		Government employee (or, officer)..OFT
LVY	—Who, or, Whom else?	LXT	—To be employed (or, used)
LVZ	EMBANK-MENT	LXU	EMPRESS
LWA	EMBARGO	LXV	EMPTY-IED ING
LWB	—An embargo has been laid on at —	LXW	—Empty barrel
LWC	—Does embargo extend to vessels under		Empty bottles ....HAT
	— flag?		Empty cartridge case ....ICU
LWD	—Embargo is not taken off	LXY	—Empty cases
LWE	—Embargo is taken off.		Empty casks......................IDZ
LWF	—Embargo not expected	LXZ	—Water tank is empty
LWG	—Has embargo been taken off at — ?		When empty ied ....ZCK
LWH	—In daily expectation of an embargo		
LWI	—Is embargo likely to be taken off?	LYA	EN ROUTE
LWJ	—It is expected the embargo will be taken	LYB	ENABLE D ING
	off		
LWK	—The (nation specified) have laid an em	LYC	ENCAMP-ED ING-MENT
	bargo on all (nations specified) vessels		Enemy is breaking up his encampment,
LWM	—When is the embargo to be laid on?		LZC
LWN	—When is the embargo to be taken off ?	LYD	ENCLOSE-D ING-URE.
LWO	EMBARK-ED ING	LYE	ENCOUNTER-ED-ING
LWP	—Embarkation		Have encountered ice — (date, latitude,
LWQ	—Enemy's troops embark-ed-ing		and longitude to follow, if necessary)
LWR	—I shall embark at		(See also page 274—ICE) ... .PBL
LWS	—Passengers are embarked		Where did you encounter ice? .....PCG
LWT	—To embark troops		
LWU	—Troops are all embarked	LYF	ENCOURAGE-D-ING-MENT
LWV	—Troops are embarking	LYG	ENCROACH-ED-ING-MENT
LWX	—When do the troops embark?		
LWY	—When do you embark?		ENCUMBER
LWZ	—When will passengers embark?		The bank is encumbered with fishing
			boats.............................EP
LXA	EMBARRASS ED-ING-MENT		
LXB	EMBASSY	LYH	END-ED-ING—OF
			A regular break up　An end to the fine
LXC	EMBAYED		weather ....HEV
LXD	EMBEZZLE-D-ING MENT		At the end (or, extremity)—of .....FIX
			At the latter end—of — The latter
LXE	EMERGENCY		end—of ....FIZ
LXF	—In case of emergency		Authorities will put an end to the affair,
			DPJ
LXG	EMERY		Beam ends........................GAO
	Emery cloth ....IVN	LYI	—Before the end—of
			Buoy end of cable................HMF
LXH	EMETIC	LYJ	—Butt end
	Tartar emetic....................XKY		Cipher is ended ....IQS
		LYK	—End for end
LXI	EMIGRATE D-ING ION-ANT		End of April ....EUB
	Detained by the Emigration Officer.KYB		End of August....................FMC
LXJ	—Emigration, or, Immigration Office		End of December ....KFM
LXK	—Have you settled with the Emigration		End of February ....MUR
	Office?		End of January...................PUV
			End of July ....PWO
LXM	EMINENCE-ENT-LY		End of June......................PWX
			End of March....................RKX
LXN	EMOLUMENT		End of May.......................RNG

END—ENGINE

	END—*Continued*		**ENEMY**—*Continued*
	End of November............SKW		Enemy's torpedo boats have been seen
	End of October ....SOI		to the —, steering to the — ....OH
	End of September...........VRM		Enemy's troops embark-ed-ing ....LWQ
LYM	—End on	LZN	—Enemy's troops land-ed-ing
LYN	—Endless	LZO	—Has the enemy?
	Gaff end........................NVE		Have been chased by the enemy.....OI
LYO	—Tail end of main shaft broken	LZP	—If the enemy
	The authorities will put an end to the	LZQ	—Is, *or*, Was threatened by the enemy
	affair ....DFJ		Keep a good lookout, as it is reported
	The latter end—of ....FIZ		enemy's war vessels are going about
	Was thrown on her (*or*, my) beam		disguised as merchantmen ....OJ
	ends ....GAX	LZR	—Of, or, From the enemy
LYP	—When will the end (*or*, last) be?	•	Saw enemy's ship (*or*, cruiser) off —
			(*or in latitude and longitude indi-*
LYQ	**ENDANGER-ED-ING**		*cated*) ....KFD
			Saw enemy's ship or, (cruiser) to the —,
LYR	**ENDEAVOR-ED-ING TO ENDEAVOR—TO**		KFE
	(*See also* ATTEMPT AND TRY)		Saw the enemy's fleet (*or*, squadron)
	Boats should endeavor to land where	LZS	off — ....OK
	flag is waved (*or*, light is shown)..JZ		—Saw the enemy's fleet (*or*, squadron)
LYS	—Endeavor (*Effort*)	LZT	steering —
	Endeavor-ed-ing to ascertain........FCJ	LZU	—Should the enemy
	Endeavor to come alongside ...... EBD		—To the enemy
	Endeavor to send a line by boat (cask,		Vessel chasing is an enemy ....INV
	kite, *or* raft, etc) . ...KA		Vessel in sight is an enemy ....OL
	Endeavor to pick up boat...........GSV	LZV	—Was the enemy?
	I shall not endeavor to make the land,		With, *or*, By the enemy ....LZA
	QDX	LZW	**ENERGY-ETIC**
LYT	—Shall, *or*, Will endeavor—to		
LYU	—Shall I endeavor—to?	LZX	**SERVICE RIFLE**
	Should you part, endeavor to beach your		Service-rifle cartridges...............ICV
	vessel where people are assembled (*or*		
	as pointed out by compass signal FROM	LZY	**ENFORCE-D-ING**
	you) ....FZU	MAB	—Can not enforce it
	Use utmost endeavors (exertions, or,	MAC	—Try to enforce it
	efforts)—to .... LTY		
LYV	—Utmost endeavor	MAD	**ENGAGE-D-ING-MENT**
	We will endeavor to send a line.. QUW	MAE	—Am, Is, *or*, Are engaged
		MAF	—Are you engaged?
	ENDLESS ....LYN	MAG	—Engage-d-ing with
			Is tug indicated engaged?....... ....YDO
LYW	**ENDORSE-D-ING-MENT**	MAH	—Several engagements
		MAI	—Severe engagement
LYX	**ENDURE-ANCE**	MAJ	—Slight engagement
		MAK	—Sorry am (is, *or*, are) engaged
LYZ	**ENEMY**	MAL	—There has been an engagement between
LZA	—By, *or*, With the enemy		— and —
	Enemy has abandoned ....CXM		
	Enemy is advancing.............DNE	MAN	**ENGINE**
LZB	—Enemy is at sea		KIND OF ENGINE
LZC	—Enemy is breaking up his encampment		
	Enemy is closing with you, *or*, You are		Auxiliary engine ....FOA
	closing the enemy ....OD		Caloric engine.......... . ...HTS
LZD	—Enemy is collecting in force		Circulating engine..... ....IBA
	Enemy is in sight ....OE		Compound engine....................JLM
	Enemy is on the alert............DWX	MAO	—Direct-action engine
LZE	—Enemy is retreating		Donkey engine......................LKN
	Enemy is throwing up batteries ...FYS		Double-expansion engine..........LKX
LZF	—Enemy's boat		Electric engine.....................LUT
	Enemy's cruisers have been seen to the	MAP	—Fire engine
	—, steering to the — ....OF	MAQ	—Gas engine
LZG	—Enemy's fleet (*or*, squadron)	MAR	—Hydraulic engine
	Enemy's fleet has been seen to the —,	MAS	—Locomotive engine
	steering to the — ....OG	MAT	—Motor engine
LZH	—Enemy's force	MAU	—Quadruple-expansion engine
LZI	—Enemy's gunboat	MAV	—Rotary engine
LZJ	—Enemy's infantry	MAW	—Side-lever engine
LZK	—Enemy's ship (*or*, cruiser).	MAX	—Traction engine
LZM	—Enemy's torpedo boat	MAY	—Triple-expansion engine

ENGINE

ENGINE—*Continued*

PARTS OF ENGINES

Air pump	DWA
Air-pump gear	DWB
Air valve	DWK
Bearings of an engine	GBN
Bilge pump	GJS
Bilge pump (*donkey*)	GJT
Bilge pump (*main*)	GJU
Blow-off cocks (*or, gear*)	GNW
Blow-off valve	GNZ
Blow through pipe	GOB
Brine cock	HIG
Circulating pump	IRB
Clack valve	IRU
Cock	IYU
Condenser	JMQ
Condenser pipe	JMS
Condenser tube	JMT
Connecting rod	JPK
Crank	KBH
Crank frame	KBJ
Crank head	KBL
Crank pin	KBM
Crosshead	KED
Cylinder	KIA
Cylinder cover	KIB
Cylinder side rod	KIF
D slide	KIP
Eccentric gear	LSU
Eduction gear	LTM
Eduction port	LTO
Ejector	LUI
Engine bolt	GXY

MAZ	—Engine counter	MBC
MBA	—Engine shaft	MBD
	Escape valve	MGW
	Expansion gear	MLZ
	Flooding cock	NFP
	Gauge cock	NXJ
	High-pressure cylinder	KIG
	High-pressure cylinder cover	OUV
	High-pressure piston	OUW
	Induction gear	PIV
	Injection gear	PKU
	Intermediate-pressure cylinder	KIH
	Intermediate-pressure cylinder cover	PQT
	Intermediate-pressure piston	PQU
	Intermediate shaft	PQV
	Jet condenser	PVH
	Kingston valve	QBD
	Low-pressure cylinder	KIJ
	Low-pressure cylinder cover	KAL
	Low-pressure piston	RBS
	Lubricator	RCO
	Paddle box	HDF
	Paddle wheel	SZG
	Parallel-motion gear	TAX
	Pipe (*steam*)	WSM
	Piston	TKW
	Piston rod	TKY
	Port engine	TOY
	Pump gear	UBI
	Safety valve	VDI
	Sea cock	VKY
	Shaft bearing	GBP
	Slide valve	WGI
	Slide-valve rod	WGN
	Sluice valve	WHJ

ENGINE—*Continued*

Standards or, Supports	WQO
Starboard engine	WQV
Starting gear	WRF
Starting shaft	WRH
Steam port	WSP
Steam pressure gauge	RJW
Steam steering gear	WST
Stop valve	WXK
Stuffing box	HDG
Suction pipe	XCJ
Surface condenser	XFW
Throttle valve	XRG
Thrust bearing	GPQ
Turning gear	YEN
Vacuum gauge	YNM
Valve	YNT
Waste pipe	YUV
Water gauge	YVQ

GENERAL SIGNALS ABOUT ENGINES

	Anything wrong with engine?	RD
	Atmospheric pressure	FJQ
	Can not get a vacuum	YNJ
	Clean engine	ISQ
	Damage to machinery (*part to be indicated*)	KJA
	Defects can not be repaired at sea (*or at place indicated*)	BF
	Defects, or, Damage can not be repaired without assistance	BG
	Defects may be set right in — hours	BD
	Does disconnecting affect your steering?	LDC
MBC	—Engine can be set right in a short time	
MBD	—Engine does not work well	
MBE	—Engine room	
	Engine-room artificer or machinist	EZR
	Engine room bulkhead	HLN
	Engine room compartment	JGZ
MBF	—Engine-room fire	
MBG	—Engine room full of water	
MBH	—Engine room has — feet of water in it.	
MBI	—Engine-room log	
MBJ	—Engine-room telegraph	
MBK	—Engine under temporary repair	
MBL	—Engine wants repair	
	—Engine working badly	FRM
MBN	—Engine works very well	
	Examine-ation (of) machinery	MKA
	Good vacuum	YNE
	Have you a good fire engine?	NAG
MBO	—Have you damaged your machinery?	
	Horsepower	OYE
	Hot bearings	RI
	How long do you require to clean your engines?	QYM
	I am going to stop, machinery requires adjusting	RK
	I can supply you with duplicate of engine (*part mentioned*)	LPE
MBP	—In the engine room	
	Indicated horsepower	OYI
	Indicator diagram	KYS
	Indifferent vacuum	YNL
	Let your wheels (*or, screw*) revolve without disconnecting	QOB
	Making (*number to follow*) revolutions per minute	UWJ
MBQ	—My engine requires cleaning	

ENGINE

ENGINE—*Continued*

My engines are going astern
　Code Flag over V
My engines are stopped
　Code Flag over U
My working pressure is —..........TUS
MBR —Near the engine room
　Packing.........................SZA
　Proceed after repairing machinery.RDW
MBS —Proceed under one engine
MBT —Proceeds under her own steam.
MBU —Repairs of engine can not be done here
MBV —Repairs of engine may be done in —
　hours
　Small adjustment required to machinery,
　　DKG
MBW —Start-ed-ing engine
MBX —Steam cutter's engine
MBY —Steam launch's engine
MBZ —Steam pinnace's engine
　Stoke-hold, or, Fire-room, compartment,
　　JHA
MCA —Supposed with engines broken down
　The damage done to machinery is not
　　serious, and is such as can be repaired
　　by the vessel's own engineers. .. .RL
MCB —The engines must be thoroughly over-
　hauled
MCD —The engines will be taken out.
MCE —Under the engine room.
　Vacuum.......................... YNI
　Water gauge out of order....YVR
　Water pressure..................YVZ
　What is your horsepower (*or that of
　　vessel indicated*)?..............OYL
MCF —Working her own engine
MCG —Working one engine
　What pressure have you on?.......TUW

ACCIDENTS TO ENGINES OR *MACHINERY*

　Air-pump levers out of order .....DWC
　Air pump out of order ....DWE
　Air-pump rod broken . . ..DWF
　Bilge pump out of order...........GJV
　Bilge-pump valves out of order....GJW
　Blow-off cock out of order........GNX
　Blow-off pipe burst.... ....GNY
　Condenser broken (damaged, *or*, out of
　　order) ... .JMR
　Condenser (*surface*) out of order...XFY
　Condenser tubes burst.............JMU
　Condenser tubes choked (*or*, salted)_JMV
　Condenser tubes leaking...... ....JMW
　Condenser tubes out of order. .....JMX
　Connecting-rod bolts broken........GXV
　Connecting-rod brasses broken (*or*, out
　　of order)...... ... ...HEB
　Coupling bolts broken ....GXW
　Crank brasses broken (*or*, out of order),
　　HEC
　Crank broken.....................KBI
　Crank-pin brasses broken....KBN
　Crank pin broken..................KBO
　Crank shaft broken (*or*, defective)..KBQ
　Crank webs loose..................KBR
　Cylinder cover broken........ ....KIC
　Cylinder cracked...............KAT
　Cylinder damaged..................KID
　Cylinder out of order......... ..KIE
　Discharge pipe out of order .... .. LCQ

ENGINE—*Continued*

　Discharge valve out of order.......LCB
　Donkey engine disabled...........LKO
　Donkey pump disabled............LKP
　Eccentric bar broken ....FWJ
　Eccentric pulleys loose............LSV
　Eccentric rod bent ....LSW
　Eccentric rod broken ....LSX
　Eccentric sheaves out of order......LSY
　Eccentric strap broken ....LSZ
　Eduction pipe out of order.........LTN
MCH —Engines broken down
　Engines broken down, I am disabled.BJ
MCI —Engines broken down, obliged to pro-
　ceed under sail
MCJ —Engines broken down, towed in by —
MCK —Engines broken down, under temporary
　repair
　Engines completely disabled.........MQ
MCL —Engines disabled, can repair in — hours
MCN —Engines, *or*, Machinery disabled.
MCO —Engine seating loose
MCP —Engines working on seating
　Expansion valve out of order.......MNA
　Feed check valves out of order......IOD
　Feed pipe broken (*or*, out of order)._MUZ
　Feed pump out of order..... MVB
　Feed-pump valves out of order.... MVC
　Flaw in shaft.....................NEH
　Have broken main shaft............ MR
　High-pressure cylinder cover broken,
　　KAI
　High pressure cylinder cracked (*or*, out
　　of order) ....KAU
　High-pressure piston broken ......OUX
　High-pressure valve out of order ..OUY
　Holding-down bolts loose...........GYA
　Hot bearings.....................RI
　Intermediate pressure cylinder cover
　　broken ....KAJ
　Intermediate-pressure cylinder cracked
　　(*or*, out of order) ....KAV

　Levers damaged (*or*, out of order)..QOX
　Low-pressure cylinder cracked (*or*, out
　　of order) ....KAW
　Low pressure slide jacket cover broken,
　　KAM
　Machinery broken down .. ..MCH
　Machinery broken down, irreparable at
　　sea ....RDV
　Machinery, *or*, Engines disabled...MCN
　Machinery out of order ....RJ
　Machinery requires adjusting ......DKF
MCQ —Machinery slightly deranged
　Machinery supposed to have broken
　　down ....MCA
　Machinery wants repair ....MBL
　Main-bearing brasses broken (*or*, out of
　　order) ....GBO
　Main check valve out of order ....IOE
　Main shaft broken..................MR
　Main shaft broken in stern tube.....RGT
　Must have a new shaft..............VTI
MCR —One engine disabled
　Packing ring of cylinder cover broken.
　　KAN
　Parallel-motion rod out of order.. TAY
　Piston broken.............. ......TKX

ENGINE—ENOUGH

ENGINE—Continued

Piston rod bolts broken ... .GYB
Piston-rod brasses broken (or, out of order) ....HED
Piston rod cracked (or, broken) .. KAX
Piston rod crosshead damaged (or, broken) ... .KEJ
Propeller shaft broken.............MB
Pump choked ....UBF
Pump crosshead out of order ......UBH
Pump gear worn out(or, defective).UBJ
Pump lever links broken . ...UBL
Pump lever out of order...........UBM
Pump wants repairs................UBR
Refrigerating machinery out of order, RDX
Repairs of engine may be done in — hours.... ....MBV
Safety valve out of order ....VDJ
Sea inlet (valve, or, cock) broken (or, damaged). ....VLA
Sea inlet choked ....IPQ
Shaft-coupling bolts broken ....GYF
Slide valve broken ....WGJ
Slide-valve gear damaged (or, out of order)...........................WGL
Slide valve out of order..........WGM
Slide valve rod broken ....WGO
Sluice valve out of order.........WHK
Started a rivet ....UYN
Starting gear out of order...... ...WRG
Steam pipe burst.................TKO
Steam pipe damaged ....WSN
Steam pipe leaky ....WSO
Steering gear disabled.... ....MZ
Stern gland defective.............WUV
Stern tube damaged ....WVA
Stuffing box broken..............HDI
Stuffing-box glands out of order ...HDJ
Surface condenser out of order ...XFY
Tail end of main shaft broken .....LYO
Thrust block broken ....GMZ
Thrust block damaged ....GNA
Thrust block rings defective.......GNB
Thrust collar damaged ....JAD
Thrust shaft broken..............XRN
Tubes out of order ....YDG
Tunnel shaft broken ....YDX
Water gauge out of order ... .YVR

SPEED, OR, MOVEMENTS OF ENGINES

Ahead easy   Go ahead easy ....LH
Astern easy   Go astern easy ......LK
Ease her   Reduce speed ....MF
Full speed ....NSV
Go ahead ....LG
Go ahead easy   Easy ahead.........LH
Go ahead full speed................LI
Go astern ....LJ
Go astern   Heave all aback ....LP
Go astern easy   Easy astern.........LK
Go astern full speed...............LM
Go full speed ....OBS
Go half speed ....OBT
Half speed ....OLM
I am going full speed ....OCI
I am going half speed ....OCJ
Keep-ing, Kept going ahead ....DVL
Keep-ing, Kept going astern.......FHO

ENGINE—Continued

My engines are going astern   Code Flag over V
My engines are stopped   Code Flag over U
Obliged to stop engines... ....WXH
Quarter speed.. ....WMV
Slow speed ....WHE
Stop her ....ML
Stop instantly (Urgent)...........MN
Stop engines to adjust towing cables, DKH
Stopping only for small adjustment of machinery ....RDY

ENGINE ROOM ....MBE
Engine-room artificers, or, machinists, EZR

MCS ENGINEER.
Can you supply me with anyone to take charge (or, act) as engineer?......RH
MCT —Engineer surveyor
MCU —Engineer's stores
MCV —Engineer's tools
The damage to the machinery is not serious, and is such as can be repaired by the vessel's own engineers .....RL

MCW ENGLAND-ISH ENGLISH COLORS (See also BRITISH)
Bank of England..................FUN

MCX ENGLISHMAN

MCY ENJOY-ED-ING-MENT

MCZ ENLARGE-D-ING-MENT

MDA ENLIST-ED-ING-MENT
Foreign Enlistment Act.............DIF

MDB ENOUGH—OF (See also SUFFICIENT)
Distinct enough ....LGS
MDC —Enough, or, Sufficient room
MDE —Enough scope
Enough time............ ....MDZ
MDF —Far enough
MDG —Far enough to windward
Fortunate enough to ....NMO
MDH —Good enough
MDI —Hands enough
MDJ —Have you enough (or sufficient)?
MDK —Have you men enough?
MDL —Have you room enough—for (or, to)—?
MDN —I have enough
MDO —Is it enough?
MDP —Near-ly enough
MDQ —Not enough
Not enough cable, give more scope.HRM
MDR —Not enough wind
MDS —Not far enough
MDT —Not short enough
MDU —Not square enough
Not strong enough ....XAQ
Not time enough.................XTO
MDV —Not water enough
MDW —Not— enough
MDX —Quite enough (or, sufficient)
Room enough...... .. ..MDC
Shall we have daylight enough?...KNT

ENOUGH—EQUIVALENT.

MDY	ENOUGH—*Continued* —Take room enough They are not cautious enough......IGD	MFJ MFK	ENTITLE-D-ING—TO —Is, *or*, Are not entitled
MDZ	—Time enough Water enough -------------------XCU	MFL MFN	ENTRANCE —At the entrance of — the —
MEB	—Will be enough You will have water enough over the bar (*or*, into the harbor) *Depth in feet to follow* ------------------VT You will not have water enough over the bar (*or*, into the harbor) *Depth in feet to follow*------------------FWE	MFO	Entrance bears — ----------- -----EG Entrance, *or*, Bar is dangerous......FQ —Entrance is difficult Entrance, *or*, Bar is not safe except at slack water (*or*, at —, *time indicated*), FVT
MEC	ENQUIRE-Y—INTO (*or*, ABOUT) (*See also* ASK) Enquiry has been made-----------PLV	MFP MFQ 	—Entrance not difficult —Entrance of the port Entrance of — blockaded ---------GNJ Entrance of — is not blockaded....GNK
MED MEF	—Make enquiry —No enquiries—for	MFR	—Harbor, *or*, Entrance buoy How does entrance bear?------------EJ Inner entrance (passage, *or*, channel), ILN
MEG	ENROL-LED-LING-MENT	MFS MFT	—Is entrance —? —Off the entrance
MEH	ENSIGN Can not distinguish her (*or*, your) en sign ----------------------------LGX Dip your ensign ------ ------------LBC	MFU	Outer entrance (passage, *or*, channel), ILS —The entrance of the gulf (*or*, harbor) is mined
MEI MEJ MEK MEL MEN	—Do not show your ensign —Ensign union downward —Foreign ensign —Half-mast your ensign —Haul down your ensign Have you a — ensign?-----------MEV Hoist your ensign (colors) ----------DW	MFV MFW	—We have missed the entrance (*or*, passed the port, *or*, harbor) —What sort of an entrance is it (*or*, is there to —)? What wind leads through the passage (*or*, entrance)?----------------ZFD
MEP MEQ MER MES MET	—Man of-war ensign —Naval Reserve ensign —Neutral colors (*or*, ensign) —Red ensign —She (*or vessel indicated*) has not shown her ensign	MFX	When may the bar (*or*, entrance) be attempted?-------------------FWD ENTRAP-PED-ING
MEU	—She (*or vessel indicated*) has shown her ensign Show your ensign Hoist your colors, DW Vessel just arrived, show your ensign, DY What ensign has she hoisted? ------JBU	MFY MFZ MGA	ENTREAT-ED-ING-Y ENTRY Date of entry --------------------KLX ENVELOPE D-ING
MEV	—Have you a — ensign?	MGB	Envoy (*Delegate*) ---------------KSN EPIDEMIC-AL
MEW MEX	ENSURE-D-ING —Do the utmost you can to ensure holding on, everything will be wanted	MGC MGD	EPILEPSY-TIC EPSOM SALTS
MEY	ENTANGLE-D-ING	MGE MGF	EQUAL-LED-LING-LY-ITY —An equal number (*or*, quantity)—of Equal altitudes-------------------EDJ
MEZ MFA MFB MFC	ENTER-ED-ING —Are you entered? —Can not enter —Have, *or*, Has entered into. Is your ship entered at Custom-house? KGZ Permission is urgently requested to enter harbor---------------------PD	MGH MGI MGJ MGK	EQUALIZE D ING-ATION EQUATOR-IAL —On the equator —The equator The line When did you cross the equator? --KEL Where did you cross the equator?..KEM
MFD	—Try to enter Wish to enter port or dock --------TM	MGL MGN	Where will you cross the equator?..KEN EQUINOX-OCTIAL —Equinoctial gale
MFE	ENTERIC FEVER	MGO	EQUIP-PED-PING-MENT Camp equipage-------------------HTZ
MFG	ENTERPRISE-ING		
MFH MFI	ENTERTAIN-ED-ING —Entertainment Fears are entertained for the safety of ---------------------------MUI	MGP	EQUIVALENT

ERASE—EXACTLY.

MGQ	ERASE-D-ING-URE	MIE	EVENING.
		MIF	—Evening gun
MGR	ERR-ED-ING ERRONEOUS. ERROR	MIG	—Every evening.
MGS	—An error in the —.	MIH	—In the evening.
		MIJ	—The last evening.
MGT	ERYSIPELAS	MIK	—This evening
		MIL	—To-morrow evening
MGU	ESCAPE	MIN	—Yesterday evening.
MGV	—Can escape		
MGW	—Escape valve.		
MGX	—Escaped ing	MIO	EVENT
MGY	—If we separate we may have a chance of escape.		At all events. _ _ _ _ _ _ _ _ _ _ _ _ _DYH
			In the event of (or, case of) _ _ _ _ _ _ _IDR
MGZ	—Narrow-ly escape-d		
MHA	—Try to escape	MIP	EVER
		MIQ	—Did you ever?
MHB	ESCORT-ED ING	MIR	—Have you ever?
	Convoy to keep ahead of escort _ _ _ _ _ _IU	MIS	—Were you ever?
	Convoy to keep astern of leader (or, escort) _ _ _ _ _ _ _ _ _ _ _ _ _ _ _ _ _ _IW		
	Convoy to keep on port side of leader (or, escort) _ _ _ _ _ _ _ _ _ _ _ _ _ IX	MIT	EVERY
			At every (or, all) _ _ _ _ _ _ _ _ _ _ _ _ _ _DYI
			Do the utmost you can to insure holding on, everything will be wanted _ _MEX
	Convoy to keep on starboard side of leader (or, escort) _ _ _ _ _ _ _ _ _ _IY		Every appearance of foul weather _ _ERZ
MHC	—Mounted escort		Every assistance will be given _ _ _ _ FGE
MHD	—Send an escort—to (or, for)		Every boat . _ _ _ _ _ _ _ _ _ _ _ _ _ _ _ _GSW
		MIU	—Everybody, or, Everyone
MHE	ESPLANADE		Every day Daily _ _ _ _ _ _ _ _ _ _ _ _KIQ
			Every evening. _ _ _ _ _ _ _ _ _ _ _ _ _ _MIG
MHF	ESQUIRE	MIV	—Every exertion has been made
		MIW	—Every hour
MHG	ESSENCE-TIAL-LY	MIX	—Every means
MHI	—Is, or, Are not essential	MIY	—Every morning
		MIZ	—Every night
MHJ	ESTABLISH-ED-ING-MENT		Everyone, or, Everybody _ _ _ _ _ _ _ _ _MIU
MHK	—A new establishment	MJA	—Every opportunity
MHL	—A small establishment	MJB	—Every particular
	Communication is established—by (or, with) _ _ _ _ _ _ _ _ _ _ _ _ _ _ _ _ _ _JFU	MJC	—Every place, or, Everywhere
		MJD	—Every precaution
		MJE	—Every precaution has been taken.
MHN	ESTATE	MJF	—Every time
		MJG	—Everything
MHO	ESTEEM-ED-ING	MJH	—Everything is to be sold
MHP	—I should esteem it a favor	MJI	—Everything paid
			Everywhere, or, Every place _ _ _ _ _ _ _MJC
MHQ	ESTIMATE-D-ING-ION	MJK	—Exclusive of everything
MHR	—Cannot exactly estimate the damage		Give every assistance to — _ _ _ _ _ _ _ _FGH
MHS	—Do you estimate?		I am paying every attention _ _ _ _ _ _ _FLG
MHT	—In my estimation, or, In the estimation of—	MJL	—I expect every moment
		MJN	—In every particular
		MJO	—In every respect
MHU	ESTUARY		Is everything arranged—for _ _ _ _ _ _ _EXF
			Keep everything prepared _ _ _ _ _ _ _ _PYS
MHV	EUCALYPTUS	MJP	—Settle everything
			Use every precaution _ _ _ _ _ _ _ _ _ _ _TSD
MHW	EUROPE-AN		
MHX	—European rope	MJQ	EVICT-ED-ING-ION
	Opportunity for sending letters to Europe _QOD	MJR	EVIDENCE
MHY	EVACUATE D ING-ION	MJS	EVIDENT-LY
MHZ	EVADE-D-ING-SION	MJT	EVIL-LY.
MIA	EVAPORATE-D-ING-ION	MJU	EXACT-ED-ING
MIB	EVASIVE-LY	MJV	EXACTLY
	Must Give, or, Gives an evasive answer, ENG		Can not exactly estimate the damage, MHR
MIC	EVEN-LY-NESS		Describe exactly what you want (size, weight, etc.) _ _ _ _ _ _ _ _ _ _ _ _ _ _ _KWC
MID	—Even keel		

EXACTLY—EXPECT

	EXACTLY—*Continued* Did you get (or, have) a good look at the land to know exactly where we are?............................ PY	MLE	EXCITE-D-ING-MENT
MJW	—Not exactly	MLF MLG	EXCUSE —Excused-ing able I (*or person indicated*) beg to be excused, GDM
MJX	EXAGGERATE-D-ING-ION	MLH	—Send an excuse Unless your communication is very important, I must be excused ..... _ LJ
MJY	EXAMINE-D-ING TO EXAMINE Buoys have just been examined, they are all right................. HMJ Cross-examine-d ing-ation—of .....KEC	MLI	• Very excusable ....YPF —You must not (or, You can not) make any excuse
MJZ	—Examination Examination of candidate for — at —, HUS	MLJ	EXECUTE-D-ING Can not be done................. BLX
MKA	—Examine-ation of machinery	MLK	—Execute repairs
MKB	—Have you examined?	MLN	—Execution-ive
MKC	—Is the vessel (object, sail, etc) worth examination?	MLO	Great execution................. OHB —In the execution—of
MKD	—Last examined-ation—of (or, at)		Proceed in execution of previous orders, SZ
MKE	—Must examine (or, be examined)—for Post-mortem examination .. ... TQU Require diver to examine bottom...LIB	MLP	EXECUTOR-SHIP
MKF	—Ship indicated to examine strange sail, and report	MLQ	EXEMPT-ED-ING-ION
MKG	—Still under examination	MLR	EXERCISE-D-ING
MKH	—To be examined—for (or, at)		Exercise d his (or, their) authority.FMR
MKI	—Vessel indicated was minutely examined, no person to be seen, nothing visible	MLS	EXERT-ION Every exertion has been made .....MIV Use utmost exertions (efforts, or, endeavors)—to ....LTY
MKJ	—Your papers will be examined		
MKL	—Your permit will be examined (or, looked at)	MLT	EXHAUST-ED-ING-ION
MKN	EXAMPLE Bad example ....PRB Good example ....ODX Make an example.. ....RIE	MLU	EXHIBIT ED-ING EXHIBITION
		MLV	EXILE-D-ING
MKO	EXCEED-ED-ING-LY Exceed-ed his (or, their) authority.FMQ	MLW	EXIST-ED-ING-ENCE Existing circumstances.............IRH
MKP	—Not to exceed Supply of water is small, not exceeding — (*quantity indicated*) ......YWR		EX-MERIDIAN ALTITUDE ....EDK
		MLX	EXONERATE-D-ING-TION
MKQ	EXCELLENT-LY CE-CY		
MKR	—His, or, Her Excellency.		EXORBITANT-LY-CE (*Excessive*).....MKU
MKS	—Their Excellencies	MLY	EXPAND SION-SIVE
MKT	EXCEPT-ED-ING-ION	MLZ	Double-expansion engine...........LKX —Expansion gear
MKU	EXCESS-IVE-LY	MNA	—Expansion valves out of order Quadruple expansion engine ......MAU Triple-expansion engine...........MAY
MKV	EXCHANGE-D-ING TO EXCHANGE Bill of exchange ....GKE Exchange Office Bureau de Change, HNC	MNB	—Work expansively
MKW	—I, or, They exchanged numbers with — Royal Exchange ....VAX	MNC MND	EXPECT-ED-ING. —At what time do you expect? Bad weather is expected ... . FP Cargo expected to be saved.........HZU
MKX	—The exchange		Cargo expected to be sold ....HZV
MKY	—The rate of exchange is —.		Daily expected . ...KMX
MKZ	—What is the exchange?	MNE	—Do not expect
MLA	EXCHEQUER	MNF	—Do you expect? Embargo not expected ....LWF Expect daily ....KMX
MLB	EXCISE (DUTY)	MNG	—Expect I have to receive
MLC	—Excise officer		Expect to arrive ....EYF Expect (*or vessel indicated expects*) to
MLD	EXCLUDE-D-ING (*See* OMIT).........SQM —Exclusive-ly-ion Exclusive of everything.............MJK	MNH	be got off (or, afloat) ....DPX —Expectation

EXPECT—EXTRA

MNI	EXPECT—*Continued*
MNJ	—Expected next mail
	—Expected next week
	Few expected....................MXB
	Gale is expected—from —...........ZD
	Hourly expected ....OZL
MNK	—I expect
	I expect every moment ....MJL
MNL	—I expect your orders every moment
	In daily expectation of an embargo..LWH
MNO	It is expected the embargo will be taken off....................LWJ
MNP	It is expected the vessels (or *vessels indicated*) will be allowed to leave..DZJ
MNQ	—Many expected
	—May expect (*or,* be expected)
	—May I expect?
	News expected on — ....SHL
MNR	—No expectation of
MNS	—Not expected yet (*or,* until —)
	Reinforcements are expected ..:.. DOL
	War expected...................YUD
MNT	—Was fully expected
MNU	—Were there any vessels, *or,* Was — (*vessel indicated*) expected at —?
	What current (*rate and direction*) do you expect (*or,* reckon on)? .....KGH
MNV	—What do you expect?
MNW	—When do (*or,* did) you expect?
	When do you expect an answer?...ENQ
	When do you expect to be off (*or,* abreast of) — ? ....DAP
MNX	EXPEDIENT ENCY
MNY	EXPEDITE-ING EXPEDITIOUS
MNZ	EXPEDITION
MOA	EXPEND-ED ING ITURE
	Ammunition is nearly expended ...EPR
	Report expenditure—of ....URY
MOB	EXPENSE (*Charges*)
	Expenses, *or,* Charges are moderate, IMP
MOC	—Expensive-ly
	What are the charges (*or,* expenses)? IMQ
MOD	EXPERIENCE D ING
MOE	—Have you experienced?
MOF	—No experience
MOG	EXPERIMENT-AL-LY
MOH	—Experiment has been tried.
MOI	—Experiment will be tried
MOJ	—For, *or,* By experiment
	Under experiment ....YHF
MOK	EXPERT
MOL	EXPIRE-D-ING ATION
MON	—At the expiration of
MOP	EXPLAIN-ED ING TO EXPLAIN
MOQ	—Can explain
MOR	—Can, *or,* Will you explain?
MOS	—Explanation-ory
MOT	—Explanation has been given
MOU	—I can not explain
	Require explanation..............UTB

MOV	EXPLAIN—*Continued*
	—Shall, *or,* Will explain (*or,* be explained)
	They will demand explanations....KTG
	Will fully explain ....NSZ
	Will, *or,* Can you explain? ....MOR
	EXPLICIT-LY (*Definite*) ....KRT
MOW	EXPLODE D-ING-SION-SIVE
MOX	—Explosive gas
MOY	—Explosive grapnel
MOZ	—Explosive rockets War rockets.
	I am taking in (*or,* discharging) explosives.. .Flag B, or. Code Flag over B
	No combustibles (*or,* explosives) near the fire.................... HR
	Vessels with combustibles (*or,* explosives) not allowed in............HS
MPA	EXPLORE-D ING-ATION TO EXPLORE
MPB	EXPORT ED-ING TO EXPORT
MPC	—Export duties
MPD	—Exportation
MPE	EXPOSE-D ING-ITION-URE.
MPF	—Do not expose yourself (*or,* people) unnecessarily.
MPG	EXPRESS
MPH	—An express has arrived from —
MPI	—An express is going to —
	Send an express — with the following intelligence ....POH
MPJ	—Should be express
MPK	EXPRESS-ED-ING-ION-IVE
MPL	EXTEND-ED-ING TO EXTEND
	Cargo damaged, but extent not known, EZT
	Does embargo extend to vessels under — flag? .... LWC
MPN	—Extend-ed-ing your distance
MPO	—Extent-sive-ly
	Reef, Rock, *or,* Shoal extends from — to—.......................... UMS
MPQ	EXTERNAL-LY
	External dimensions From out to out, LAM
MPR	EXTINGUISH-ED-ING
MPS	—Can extinguish fire
	Extinguish all lights Allow no lights to be seen ....DZB
	Fire is extinguished ....MZS
	Is fire extinguished? .. ..OS
	With help I may be able to extinguish fire.... ... .FGV
	With immediate assistance fire can be extinguished ....NY
	Your lights are extinguished, *or,* want trimming (*made by* — —— —— *flashes, or blasts of a steam whistle*) (See page 548)
MPT	EXTORT-ED-ING ION
MPU	EXTRA
	Any extras? ....EOU

	EXTRACT—FALL	

MPV	EXTRACT-ED-ING-ION		
	Goulard's Extract............................OFN	MQY	FAIL—*Continued*
		MQZ	—If it fails
MPW	EXTRADITION	MRA	—Likely to fail
			—Unless it fails
MPX	EXTRAORDINARY-ILY		Without fail...... ....ZHG
MPY	EXTRAVAGANT-LY-ANCE	MRB	FAINT-LY-NESS (*Dim*)
MPZ	EXTREME-LY-ITY		FAINT-ED-ING (*Swoon*)...............ZNI
	At the extremity (*or*, end)—of ......FIX		
MQA	—In the extreme	MRC	FAIR-LY-NESS
MQB	—To, *or*, At the last extremity	MRD	—Fair-lead Fair-leader
		MRE	—Fair wind
MQC	EXTRICATE-D-ING	MRF	—Prices are fair
MQD	—Can it be extricated?		Very fair draught ....LNE
MQE	—Can not extricate		You are in a very fair berth ....GHJ
	It is difficult to extricate ....KZY		
			FAIRWAY, *or*, CHANNEL ....IKZ
MQF	EYE	MRG	—Fairway buoy
	Bull's-eye............................HLR		Fairway, *or*, Channel is buoyed ....OW
	Dead-eye ....KOF		Fairway, *or*, Channel is clear ILC
	Eyebolt..............................GXZ		Fairway, *or*, Channel is dangerous...FU
			Fairway, *or*, Channel is likely to be
MQG	EYELET HOLE		mined ....ILD
			Fairway, *or*, Channel is mined Be-
MQH	EYESORE		ware of torpedoes ....XP
			Fairway, *or*, Channel is narrow ....ILE
MQI	EYEWITNESS		Fairway, *or*, Channel is not mined..XQ
			Fairway, *or*, Channel is obstructed..ILF
			Is the fairway (*or*, channel) mined?.ILP
		MRH	—Warp out to fairway.
		MRI	FAITH-FUL-LY
			Have great faith in my chronometer.IQE
		MRJ	FAITHLESS-LY
		MRK	FALL-EN-ING (*See also* FELL)
			Accident happened, fall...........DCP
			Barometer has fallen very rapidly..FXH
			Barometer is falling ....FXI
			Catfall ....IET
		MRL	—Did you fall in with (*or*, meet) —?
		MRN	—Did you fall in with any men-of-war?
		MRO	Downfall ....LMH
MQJ	F (*Letter*) (*For new method of spell-*		—Falling-ing, Fell in with —
	ing, see page 13)	MRP	—Fall-ing, Fell off
		MRQ	—Fallen overboard
MQK	FABRIC		Fall-ing calm ....HTP
		MRS	—Falls of the river —
MQL	FABRICATE-D-ION		Fell ....MVH
			Fell down...........................MVI
MQN	FACE		Fell from ....MVJ
	Sliding faces ....WGP		Fell from aloft... ....MVJ
			Has fallen astern ....PHL
MQO	FACT	MRT	—Have you fallen in with crew of derelict?
			Have you fallen in with ice? (*State*
MQP	FACTION		*whether berg or field*) ....PE
		MRU	—I shall not go on the other tack unless I
MQR	FACTORY.		fall off—to —
			Landfall ....QEK
MQS	FAHRENHEIT	MRV	—Let fall
		MRW	—Prices have fallen (*or*, gone down)
MQT	FAIL-ED TO FAIL	MRX	—Prices have not fallen
MQU	—Do not fail—to —		Purchase falls Purchases........UBY
	Failed, *or*, Bankrupt ....FVK		Rise and fall is about — feet ......UXT
MQV	—Failing	MRY	—Should I fall in with
MQW	—Failure		Should you fall in with. ....ROW
	Failures, *or*, Bankruptcies ....FVL		The barometer has fallen very rapidly,
MQX	—I have failed to		FXH
			Tide has fallen — feet...............XSI

FALL—FASTEN

FALL—Continued
  Tide is falling ---------------------- XSK
  Water-fall ------------------------- YVP
  What is the rise and fall of the tide? XH
MRZ —When did you fall in with?
MSA —When do you think I am likely to fall in with —(*vessel indicated*)?
  Where did you encounter (fall in with, or, see) ice? --------------------- PCG
  Where did you fall in with the trade (or, monsoon)? ----------------- RYM
MSB —Where do you think I am likely to fall in with —?
  You will fall in with ice if you go beyond — --------------------------- PCH
MSC FALSE-LY-HOOD
  False alarm ---------------------- DWQ
MSD —False attack
  False bill of lading ---------------- GKS
MSE —False keel
MSF —False paper
  The news (or, report) is not correct, JXC
MSG FAME-OUS
MSH FAMILY
  Did you hear of my family? Any news of my family? --------------------- EPI
MSI —I heard that some of your family were seriously ill
MSJ —Inform my family
MSK —Is my family well?
  Telegraph to my family at — the following message ----------------- XMF
MSL —Your family are all well
MSN FAMINE
MSO —Famine at —
MSP —Famine prices
MSQ FAMISH-ED-ING
MSR FANCY-IFUL-LY
MST FAR
MSU —A little farther
  Any farther ---------------------- EOV
  Anything further? ----------------- EQJ
  As far as ------------------------- FAG
  Convoy to spread as far as possible, keeping within signal distance --------- IZ
  Do not stand too far on your present course ------------------------- JZI
  Do not stand too far to the — ----- WQB
  Far enough ---------------------- MDF
  Far enough to windward---------- MDG
  Farther ahead—of --------------- DVF
  Farther astern—of -------------- FHJ
MSV —Farther off (or, from)
  Farther-est --------------------- MTI
MSW —Farther in
MSX —Farther out
MSY —Farther to the
MSZ —How far can light be seen?
MTA —How far off? What distance?
  I shall not go far before I bring up-HJC
MTB —I shall not go much farther
  Ice as far as I can see ----------- PBU
  Keep farther astern ------------- FHM

FAR—Continued
MTC —Keep farther off (or, from)—
  Not far down ------------------- LMK
  Not far enough ----------------- MDS
  Open, get farther away ----------- MC
  Severe squall not far off, look sharp. FZ
  Shift your berth farther in ------- GHA
  Shift your berth farther out ------ GHB
  Shift your berth farther to the — - GHC
MTD —Too far off
  Too far to the eastward ---------- LSF
  Too far to the northward --------- SJQ
  Too far to the southward -------- WLM
  Too far to the westward --------- ZBT
MTE —We are too far to leeward
  You are too close, keep farther off .IVL
  You can telegraph as far as —----- XMI
MTF FARAD (*Measure of electrical capacity*)
  FAREWELL   ADIEU --------------- DJX
MTG FARM
MTH —Farmer
MTI FARTHER-EST
  FARTHING (*Coin*). -------------- ATP
MTJ FASHION-ED-ING-ABLE
MTK FAST (*Quickly*)
  Abandon the vessel as fast as possible, AB
  As fast as ----------------------- FAH
  As fast as possible -------------- FAI
MTL —Faster-est
  Get her head round as fast as possible, or, Get her on the other tack as fast as possible ----------------------- NYU
  Get sail on your ship as fast as possible, NYZ
  Get steam up as fast as possible----- KL
  Get under way as fast as you can--- KN
MTN —Go as fast as you can
  It is not safe to go so fast ---------- LS
  My chronometer is fast on Greenwich (or, first meridian) mean time---- GR
  Reduce speed, you are towing me too fast------------------------------ XYT
  Shorten sail, it is not safe to go so fast, VDK
  Will you send off lighters as fast as possible ------------------------- RC
  Your chronometer must be fast---- IQL
MTO FAST (*Firmly fixed*)
  All fast ------------------------- DXJ
  Are towing cables fast? Are you (your cables) fast? ------------------- XYJ
MTP Fast ice
  Is the cable fast (or, rope belayed)?. HRK
  Is the line fast ------------------- KB
  Make fast—to — ----------------- RIF
  Make fast to a buoy-------------- HMS
  Make fast to the pier ------------ RIG
  Towing cable is fast ------------- XYV
MTQ FASTEN-ED-ING
MTR —Fastening (*A fastening*)
  Fasten your chain to towing cable.HRC

FATHER—FEND

MTS	FATHER-LY		FEET FOOT (*Measure of length*) (*See*	
	FATHOM (*Measure of length*) ......AVW		*also* FOOT)------------------------AVY	
MTU	FATHOM-ED-ING-ABLE		At high water you will have — feet_KUY	
	Best anchorage (*or*, berth) is in —		At low water you will have — feet__KUZ	
	fathoms--------------------------OV		Cubic foot ------------------------AYC	
MTV	—How many fathoms?		Dangerous to,'or, Do not come into less	
			than — feet water----------------FV	
MTW	FATIGUE-D-ING		Footbridge ------------------------HGR	
MTX	—Much fatigued (*or*, tired)		Foot ice .. -----------------------PBI	
MTY	—Too fatigued (*or*, tired).		Footpath --------------------------NHE	
		MVE	Foot sore . ... -------------------NHF	
MTZ	FAULT-Y-ILY		Got bottom—with — feet ---------HBA	
	Considerable fault found—with —__JQI		—How many feet?	
			How many feet at low water? -----KVF	
MUA	FAVOR-ED-ING-ABLE-Y		I am in — feet of water------------VP	
	Apparently favorable-y------------ERN		I draw — feet aft and — feet forward,	
	Appear-ed-ing-ance favorable------ERV		OA	
	I shall not weigh until I get a favorable		Marked in feet--------------------RLK	
	start ------ --------------------ZAW		My draft of water is — feet — inches	
	I should esteem it a favor--------MHP		aft.------------------------------UXT	
MUB	—If favorable		My draught of water is — feet — inches	
MUC	—In favor of		forward.--------------------------LND	
MUD	—Is he (*or person indicated*) in favor of?		No bottom—with — feet----------HBE	
MUE	—Not in favor of		Rise and fall is about — feet ------UXT	
MUF	—Should it be favorable		Square foot ----------------------AXT	
MUG	—Should it not be favorable		The depth of water here (*or at place indi-*	
	Without favor-----------------ZHI		*cated*) is — feet at low water ----KVD	
			Tide has fallen — feet------------XSI	
MUH	FEAR-ED-ING-FUL-LY		Tide rises — feet ----------------XSM	
MUI	—Fears are entertained for the safety of —		What draught of water in feet could	
	Need not fear---------------------SGF		you lighten to? -----------------OB	
MUJ	—No fear		You will have less than — feet of	
			water-----------------------------KVJ	
MUK	FEARNOUGHT, *or*, FELT		You will not have less than — feet of	
			water .. -------------------- -----VU	
MUL	FEAST-ED-ING		— feet water in the hold-----------QM	
		MVF	FEIGN-ED-ING FEINT.	
MUN	FEATHER-ED-ING	MVG	—Only intended for a feint	
MUO	—Feather strengthening			
		MVH	FELL (*See also* FALL)	
MUP	FEBRUARY	MVI	—Fell down	
	About February the —-----------CZM	MVJ	—Fell from —	
MUQ	—Beginning of February		Fell from aloft-------------------EAD	
MUR	—End of February		Fell in with —-------------------MRO	
MUS	—Next February	MVK	—Fell in with a privateer	
			Fell off .... --------------------MRP	
MUT	FEDERAL-IST-ATION			
MUV	—Shipping Federation	MVL	FELLOW.	
MUW	FEE	MVN	FELON-Y-IOUS-LY.	
	Any fee (*or*, charge)?------------EOW			
	What are the pilot fees — for — ? ___TJU		FELT (*See* FEEL)----------------MVD	
	FEEBLE-Y-NESS (*Weak*)-----------YKK		FELT, *or*, FEARNOUGHT------------MUK	
MUX	FEED-ING-ER	MVO	FELUCCA	
	Feed-check valves out of order-----IOD			
MUY	—Feed pipe	MVP	FEMALE	
MUZ	—Feed pipe broken (or, out of order)		Clothing for female passengers Wom-	
MVA	—Feed pump		en's clothing------- --------------IVR	
MVB	—Feed pump out of order	MVQ	—Female on board	
MVC	—Feed-pump valves out of order.	MVR	—Female passenger	
			Have you any females on board?___GQH	
MVD	FEEL-ING FELT TO FEEL		I have no females on board --------GQK	
	The way is off my ship, you may feel			
	your way past me .. ... ----HL	MVS	FEND-ED-ING-ER.	
	What is the current? Do you, or, Do	MVT	—Fend off	
	we feel any current?.. ----- --- KGI	MVU	—Fenders ready	

FERRULE—FINE

MVW	FERRULE Boiler tube ferrules................GWO	MXL	FIFTEEN-TH-LY
MVX	FERRY-IED-ING.	MXN MXO MXP	FIFTH-LY FIVE —Fifth mast —Five-masted ship
MVY	—Ferry boat.		
		MXQ	FIG
MVZ	FETCH ED-ING (*To reach*).		
MWA	—Can fetch	MXR	FIGHT-ING. FOUGHT TO FIGHT
MWB	—Can you fetch?	MXS	—A running fight
MWC	—Fetch on the other tack		Any fighting—at —?____ _____BOX
MWD	—Fetch on this tack.	MXT	—I have had (*or vessel indicated* has had)
MWE	—How high must I lie to fetch?		a fight with a privateer
MWF	—You will not fetch	MXU	—Some squabble (*or*, fight) on shore with
			crew ˮ
	FETCH-ED-ING TO FETCH (*See* BRING)..HIK	MXV	—There has been fighting at —?
MWG	FEVER ISH	MXW	FIGURE
	A case of fever has broken out on board	MXY	—Figurehead.
	this ship (*or ship indicated*)......GPR		
	Any cases of fever on board (*or*,	MXZ	FILE
	at —)?............................BON		
	Case of fever ....IDM	MYA	FILL-ED-ING. TO FILL.
	Enteric fever ....MFE	MYB	—Fill main topsail.
	Fever and ague .... DUZ	MYC	—Fill up water
MWH	—Fever very prevalent at —		
MWI	—Has yellow fever diminished at —?	MYD	FILTER-ED-ING
MWJ	—Has yellow fever disappeared at —?	MYE	—Filter tank.
MWK	—I have had cases of yellow fever on board		
MWL	—I have had no cases of yellow fever on	MYF	FINAL-LY
	board		The decision is final................KPW
MWN	—Intermittent fever		The final decision is...........KPX
MWO	—Is there yellow fever at —?		
MWP	—Jungle fever	MYG	FINANCE-IAL-IER.
MWQ	—Malarious fever		
MWR	—Scarlet fever	MYH	FIND-ING TO FIND (*See also* FOUND)
MWS	—Typhoid fever	MYI	—Can, *or*, May find.
	Typhus fever.....................YFS	MYJ	—Can you find —?
MWT	—Yellow fever	MYK	—Can not find.
MWU	—Yellow fever has diminished at —.	MYL	—Could find
MWV	—Yellow fever has disappeared at —		Could not find the address.........DJK
MWX	—Yellow fever is raging at —	MXN	—Did you find?
MWY	— men of the crew have been ill of	MYO	—Find out.
	yellow fever (*Number to follow*)	MYP	—Find out the mistake
MWZ	— passengers have been ill of yellow		Found ....NOQ
	fever (*Number to follow*)	MYQ	—He (*or person indicated*) will find
			I find great difficulty in weighing. KZV
MXA	FEW. A FEW		Not a piece to be found........ ..TIA
	A few days ....KMV		Range finder...................UHV
MXB	—Few expected	MYR	—Should you (*or person indicated*) find
MXC	—Few hurt.	MYS	—Should you (*or person indicated*) not
	Few opportunities ....SRW		find
MXD	—Fewer		Try to find out ....YCW
MXE	—Fewest The fewest	MYT	—Where shall I find —?
MXF	—Have, *or*, Has a few	MYU	—Where shall I find — (*vessel indicated*)?
	In a few minutes Few minutes...RVC		You will find great difficulty in getting
	Too few ....XVD		through the ice at —..... .... PCI
	Very few ....YPG	MYV	—You will find me at — at — (*time indi-*
	Very few arrivals ....EYT		*cated*)
			You will find the pilot in — (*direction*
MXG	FID-DED-DING		*indicated*) ....TJZ
MXH	FIDDLE D-ING	MYW	FINE (*Good*)
MXI	—Fiddler	MYX	—Fine day
		MYZ	—Fine weather.
MXJ	FIELD-ED-ING	MZA	—If fine
	Field glass Binocular... ....QLB		Very fine ....YPH
MXK	—Field gun		
	Ice field, *or*, Floe.......... ....PBW	MZB	FINE-D-ING (*To mulct*)
	Illuminate-d-ing mine field........PDB		Police (*or other*) authorities have fined—,
	Mine field......................RUO		FNP

FINGER—FIRM

MZC	FINGER
MZD	FINISH
MZE	—Finished-ing To finish
MZF	—Have you finished?
MZG	—Is it finished?
	Report when finished (or, complete-d)— with ....JKS
MZH	—Shall be finished
MZI	—When finished
	When finished coaling.............IXP
MZJ	—When will you (or *vessel indicated*) be finished?
	When will you have finished coaling IXQ
MZK	FIR
MZL	FIRE
MZN	—Alarm of fire
	Are you on fire?....................OM
	Bank up fires......................FUK
	Can extinguish fire...............MPS
MZO	—Can fire be got under?
	Caught fire........................IFG
	Cease firing.......................IGK
	Clean fires........................ISR
	Derelict has been fired..........KVQ
	Destroyed by fire................KXO
	Do not light your fires..........QRS
	Draw fires.........................LN1
	Engine-room fire.................MBF
MZP	—Fire a signal gun
MZQ	—Fire and lights out
	Fire bar Furnace bar.........FWK
	Fire brigade......................HGZ
	Fire bucket.......................HKV
	Fire can easily be got at.........ON
	Fire difficult to get at............OP
	Fire engine.......................MAP
MZR	—Fire-ing To fire
MZS	—Fire is extinguished
	Fire gaining........................OQ
	Fire gains rapidly, take people off..NG
MZT	—Fire grapnel
	Fire has been burning these—hours HNF
	Fire has broken out on board of—..GQC
MZU	—Fire in —
	Fire in hold amongst cargo........OR
	Fire, or, Leak, want immediate assistance........................NH
	Fire, or, Lights will be kept burning at the best place for landing.......KE
MZV	—Fireproof
MZW	—Fire ship (or, raft)
MZX	—Fire upon
MZY	—Firearms
NAB	—Fireman
NAC	—Firewood
NAD	—Fireworks
NAE	—Firing battery
NAF	—Firing gear
	Forced fires......................NHX
	Guns firing in —(*quarter indicated*) OJZ
	Guns firing last night in (*quarter indicated*)...........................OKA
	Guns firing to the —.............OKB
NAG	—Have you a good fire engine?
NAH	—Have you fire hose enough for —?
NAI	—Heard firing to the —
	Heave to, or I will fire into you......JD

	FIRE—*Continued.*
NAJ	—How long have you (or *vessel indicated*) been on fire?
	I am on fire........................NM
	I am sinking (or, on fire), send all available boats to save passengers and crew..............................NO
NAK	—I shall fire into boats if they persist in coming alongside
	Is fire extinguished?................OS
NAL	—Is — (*vessel indicated*) on fire?
	Keep lights (or, fires) on the beach (or, shore all night)...................FZW
	Keep your fires banked up.........PZV
	Lay fires...........................QLV
NAM	—Let your fires down
	Light your fires....................QSJ
	Lighting fires......................QSO
	Lights, or Fires will be kept at the best place for coming on shore.........KE
NAO	—May I fire?
	Minute guns firing in — (*direction indicated*).........................OKH
	Minute guns firing last night in — (*quarter indicated*)................OKI
	Missed fire.........................RVY
	No combustibles (or, explosives) near the fire............................HR
NAP	—On fire
	Open fire, or, Commence fire......SRD
	Port fire..........................TPD
NAQ	—Put out fires
	Quick-firing ammunition..........EFX
	Quick firing gun..................OKL
NAR	—Seamen and Firemen's Union
NAS	—Under fire—of
	Vessel indicated is on fire.........OT
	Where is the fire?..................OU
	With assistance fire can be subdued
	With help I may be able to extinguish fire............................FGV
	With immediate assistance, fire can be extinguished...................NY
NAT	—You can get any quantity of firewood
NAU	—You must not fire
	FIRE BAR FURNACE BAR...........FWK
	FIRE BUCKET.....................HKV
	FIRE ENGINE.....................MAP
	Have you a good fire engine?......NAG
	FIRE SHIP, (or, RAFT).............MZW
	FIREARMS.........................MZY
	FIREMAN..........................NAB
	Firemen and Seamen's Union.....NAR
	FIREWOOD.........................NAC
	You can get any quantity of firewood, NAT
	FIREWORKS.......................NAD
	FIRKIN (*Measure of capacity*).....AYV
NAV	FIRM-LY NESS
NAW	FIRM OF —

FIRST—FLAG

NAX	**FIRST LY**
	At first_____ _____FID
	By the first conveyance _____JTZ
	First class _____ _____ISF
	First class torpedo boat_____GSX
NAY	—First mast
NAZ	—First offer
NBA	—First of all　　In the first place.
NBC	—First officer
NBD	—First opportunity
NBE	—First quarter
NBF	—First quarter of the moon
NBG	—First ship
NBH	—First time
NBI	—First voyage
NBJ	—First watch
	First week in —_____ZAM
	In the first place　First of all _____NBA
	My first meridian (or, Greenwich) mean
	time is _____ _____ __XI.
	My first meridian is Cadiz, west of
	Greenwich 6° 12′ 24″ = 0h 24m 49 6s
	HSI
NBK	—My first meridian is Greenwich
	My first meridian is Paris, east of Green-
	wich 2° 20′ 15′ = 0h 9m 21s_____TBL
	The first lull, or, When it lulls ____RDA
NBL	—What is your first meridian?
NBM	—When first seen
	Will you show me your Greenwich (or,
	first meridian) mean time?____ ．XN
NBO	**FISH**
	Codfish-ery _____ _____IZB
NBP	—Fish-ed-ing　　To fish
	Fish davit _____ _____KMS
NBQ	—Fishhook
NBR	—Fish very scarce
NBS	—Fisherman
NBT	—Fishery
	Fishery steamer _____WTG
	Fishing craft (or, boat) _____KAZ
NBU	—Fishing ground
NBV	—Fishing line
NBW	—Fishing net
NBX	—Fishing season
NBY	—Fresh fish
NBZ	—Good season for fish
	Herring fishery_____OUJ
NCA	—Have you any fish?
NCB	—I have taken—(quantity indicated)—of
	fish
	Pilchard fishery _____TIM
	Preserved fish _____ _____TUI
	Salt fish_____ _____VFT
	Shellfish ＿ _____VWH
	The bank is encumbered with fishing
	boats_____EP
NCD	—There are plenty of fish to be caught
	here
	Whale fishery _____ZBX
NCE	**FISH-ED-ING　TO FISH**　(Repair a spar)
NCF	—A fish for —
NCG	—Can be fished at sea
NCH	—Can you supply a fish for my — (spar
	indicated)?
	Fish davit _____KMS
	Fish the anchor, or, Anchor fished_EHB
	Sprung foremast, but can fish it at
	sea _____NKX

NCI	**FISH**—Continued
	—Will you lend me a fish for my mast?
NCJ	**FIT-TED-TING NESS (TO,'or, FOR)**
	Article indicated can be supplied, but it
	will require fitting__ ．．_____．EZF
NCK	—Fit for the purpose—of
NCL	—Fit-ted ting out.
NCM	—Fit-ted-ting very well
NCO	—Fitting new propeller
	No boats fit for this work _____ZHV
	Send boats fit for landing passengers_FI
NCP	—Will not fit
	FIVE　FIFTH-LY_____MXN
	Fifth mast__ _____MXO
	Five-masted ship_____ _____MXP
NCQ	**FIX-ED-ING**
	Can you fix any time for?_____XTJ
	Can not fix any time—for —_____XTK
NCR	—Fixed and flashing light
NCS	—Fixed light
NCT	**FIXTURE**
NCU	**FLAG**　(See also COLORS)
	Admiral's flag _____DKM
	Admiralty flag _____DLS
	Beach the vessel where flag is waved
	(or, light is shown)_____ _____FZT
	Boat should endeavor to land where flag
	is waved (or, light is shown)_____JZ
NCV	—Can make out flags (or, colors)
NCW	—Can you make out her flag (or, signal)?
	Can not distinguish your flags, come
	nearer, or make Distant Signals___VM
NCX	—Can not make out the flags (or, signals)
	Code flag (when spoken of)_____IZF
	Does embargo extend to vessels under—
	flag?_____ _____LWO
	Flag captain _____HWX
NCY	—Flag of truce
NCZ	—Flagship
NDA	—Flag signal-ling
NDB	—Flagstaff
	Flag union downward _____MEJ
	Flags seen but signal not understood,
	WCX
	Foreign flags (colors, ensigns)_____MEK
	Geneva Convention flag (when spoken
	of)_____NXT
NDC	—Haul down your flags
	Hoist mast-head flags_____OWG
	Hoist quarantine flag_____TV
NDE	—Hoist your flags where best seen
	House flag _____OZQ
	I can not make out the flags (or, signals),
	NCX
	I can not make out the flags, hoist the
	signals in a better position _____OWL
NDF	—I can not make out the bottom flag
NDG	—I can not make out the second flag
NDH	—I can not make out the third flag
NDI	—I can not make out the top flag
	Not clean bill of health flag (when
	spoken of) _____ISW
NDJ	—Numeral flag
NDK	—Pilot-boat flag (when spoken of)
NDL	—Pilot jack (or, flag) (when spoken of)
NDM	—Plague flag (when spoken of)
NDO	—Powder flag (when spoken of)

FLAG—FLUKE

	FLAG—*Continued*			FLEET—*Continued*
NDP	—Private flag			Where is our fleet (*or*, squadron)?..WFI
NDQ	—Quarantine flag (*when spoken of*).			Where is the— fleet(*or*, squadron)?. SK
	Repeat ship's name, your flags were not			Where was the fleet (*or*, squadron) on
	made out ....DU			the —?............................WPJ
	Set of flags ....VSI			FLEW (*See* FLY).....................NGC
	Signal not understood although the		NEW	FLEXIBLE-Y-ILITY
	flags are distinguished ....WCX		NEX	FLIGHT
NDR	—The flag with —			FLING-ING FLUNG (*Throw*) ....XBJ
	Union-jack ....YJE		NEY	FLINT-Y
NDS	—What flag is she (*or*, are they) under?		NEZ	FLOAT-ED-ING
NDT	—Your flags are not clear			A good deal of floating ice in —....PBG
	Your flags seem to be incorrectly hoisted			Am, Is, *or*, Are afloat. ....BY
•	(*or*, hoisted upside down) .....OWM		NFA	—Floated off
	Your signal flags can not be distin-			Floating anchor Sea anchor ......EIC
	guished ....LHB			Floating bridge ....HGQ
NDU	FLAME-D-ING			Floating, *or*, Liquid compass ... ..JHP
NDV	FLANGE			Floating dock ....LJD
NDW	FLANNEL		NFB	—Floating light
NDX	FLARE—UP FLARE LIGHT		NFC	—Floating mine
NDY	—Holme's light (*large flare*)		NFD	—Floating obstruction in (*or*, at) —
NDZ	—Holme's light (*small flare*)			Floating stage....................WPV
	Red flare............................UME			Light, *or*, floating light is discontinued
	White flare..........................ZDH			(*or*, gone).......................QSG
NEA	FLASH-ED-ING			Nearly afloat......................DQA
	Burn blue light, *or*, Flash powder....GOQ			Not afloat....... ....DQB
	Fixed and flashing light ....NCR		NFE	Paddle float ....SZF
NEB	—Flashing light		NFG	—The tide will float her off
NEC	—Group flashing light			—When you float
	Use flashing signals................WDN		NFH	Will be afloat ... .DQF
NED	FLAT-TEN-LY			—Will not float
	Flat bottom boat....................GSY		NFI	FLOCK (*Herd*)
NEF	—Flatter-est		NFJ	FLOE
	Send flats to discharge cargo ....IAZ			Ice field, *or*, Floe................PBW
NEG	FLAW		NFK	FLOG-GED-GING
NEH	—Flaw in shaft		NFL	FLOOD-ED-ING
NEI	FLAX			As soon as the flood tide makes ....FBP
NEJ	FLEET		NFM	—Flood sets
	Any alteration in the position of the		NFO	—Flood tide
	squadron (*or*, fleet)?..... ....EBZ		NFP	—Flooding cock
	Channel fleet (*or*, squadron)........ILB		NFQ	—How will the flood set?
NEK	—Did the fleet (*or*, squadron)?			Next flood ....SHV
	Enemy's fleet (*or*, squadron) ....LZG		NFR	—When the flood makes
	Enemy's fleet has been seen to the —			When will the flood tide make? .....XJ
	steering to the —........ ....OG		NFS	FLOOR-ING
NEL	—Fleet consists of —			FLORIN, GUILDER, *or*, GULDEN (*Coin*).ATQ
NEM	—Fleet, *or*, Squadron is (*or*, was)		NFT	FLOTILLA
NEO	—Fleet, *or*, Squadron when last seen was		NFU	FLOTSAM
	off —		NFV	FLOUR
NEP	—Fleet organization			Barrels of flour....................FXT
NEQ	—For the fleet (*or*, squadron)		NFW	FLOW-ED-ING To FLOW
NER	—From the fleet (*or*, squadron)			Tide flows at — ....XSH
NES	—In the fleet (*or*, squadron)			When will the tide flow?...........XJ
	Letters of fleet (*or*, squadron).....QOM			FLOWN (*See* FLY) ....NGC
	Saw the enemy's fleet (*or*, squadron)		NFX	FLUCTUATE-D-ING-ION
	off — ....OK		NFY	FLUE
	Saw the enemy's fleet (*or*, squadron)			Flue boiler ....GWZ
	steering —........................LZS		NFZ	—Sweeping the flues (*or*, funnels, *or*, tubes)
NET	—Saw the fleet (*or*, squadron) off — steer-		NGA	FLUID
	ing —			Condy's fluid Crimson fluid........JNK
NEU	—The fleet (*or*, squadron) is at —			FLUKE OF ANCHOR ....EHC
NEV	—To the fleet (*or*, squadron)			
	What is the state of health of the fleet			
	(*or*, squadron) at —? ....ORC			
	When is the fleet (*or*, squadron)?..WPH			

	FLUSH—FOR	
NGB	**FLUSH—WITH**	
NGC	**FLY-ING FLEW FLOWN TO FLY**	
NGD	—Flying jib	
	Flying jib-boom _____ GZP	
	Pilot signal flying _____ TJM	
NGE	FOD (*Measure of length*) _____ AVX	
NGE	**FODDER.**	
NGF	**FOG-GY**	
	Delayed by thick fog, *or*, Detained by	
	thick fog _____ KSE	
	During a thick fog _____ LPN	
	During the fog _____ LPR	
NGH	—Fog buoy	
NGI	—Fog horn	
	Fog signal _____ WCJ	
NGJ	—Fog too thick	
	Foggy weather _____ ZC	
	Gun fog-signal _____ OJS	
NGK	—How long have you had this fog?	
	I have had thick fog on the bank ___ FVO	
	Keep close during the fog _____ PYO	
	On account of the fog (*or*, weather) _ DFQ	
	Shall signal with siren (*or*, steam whistle)	
	during fog (*or*, darkness) _____ WE	
	The bank was clear of fog _____ FVI	
NGL	—There has been so much fog as to cause	
	considerable detention	
NGM	—Thick fog	
	Thick fog coming on _____ GJ	
NGO	—Through the fog	
NGP	—We have had a lot of fog	
NGQ	**FOLLOW ED ING-ER**	
	As follows _____ FAJ	
	Communicate the following _____ JFQ	
	Course to be followed _____ JZH	
NGR	—Follow her (it, *or*, them)	
NGS	—Follow me (*or vessel indicated*)	
NGT	—Follow motions	
	Following communication (*or*, signal)	
	is confidential (*or*, private) _____ HX	
	Following communication (signal, *or*,	
	message) is secret and in cipher __ IQV	
	NOTE —*The cipher will continue until*	
	the person signaling hoists IQS	
	Forward following message by tele-	
	graph to— _____ NMW	
	Forward following telegraphic message	
	by signal letters instead of writing it	
	at length _____ XB	
NGU	—Forward the following information to—	
NGV	—Has paper of the following date ___ KMC	
	—Have obtained the following informa-	
	tion from —	
	Have received the following communi-	
	cation (*or*, instructions) from your	
	agent (*or*, owner) _____ IA	
	I have the following information—	
	from— _____ NGV	
NGW	—I will follow	
	Send an express with the following	
	intelligence _____ POH	
	Send the following articles _____ EZH	
	Send the following message by post to	
	owner (*or*, to Mr —) at — _____ WY	
	Send the following message by telegraph	
	to owner (*or*, to Mr —) at — _____ XA	

	FOLLOW—Continued	
	FOLLOW—*Continued*	
	Send the following message through	
	the post to owner (*or*, to Mr —) at —,	
	by signal letters instead of writing it	
	at length _____ _____ WZ	
NGX	—Shall I follow?	
	Strange vessel following convoy _ WYV	
	Telegraph following message to — (*ship*	
	or person indicated) at — _____ XD	
	Telegraph to my family at — the fol	
	lowing message _____ XMF	
	The following articles _____ EZI	
	The following signal (communication,	
	or, message) is in cipher (*secret*) __ IQV	
	NOTE —*The cipher will continue until*	
	the person signaling hoists IQS	
NGY	—Will you follow?	
	Will you forward (*or*, communicate) the	
	following signal for me _____ IK	
NGZ	—With the following information	
NHA	**FOOD**	
	Distressed for want of food _____ LHE	
	Want food (provisions) _____ YTU	
	Want food Starving _____ NV	
	Want food immediately _____ YO	
NHB	**FOOL-ISH LY NESS**	
NHC	**FOOLSCAP PAPER**	
	FOOT (*Measure of length*) (*See also* FEET),	
	_____ AVY	
	Cubic foot (*solid measure*) ___ ___ AYC	
	Square foot (*square or surface measure*),	
	_____ AXT	
NHD	**FOOT** (*See also* FEET)	
	Crowsfoot _____ KER	
	Cubic foot _____ AYC	
	Footbridge _____ HGR	
	Foot ice _____ PBI	
NHE	—Footpath	
NHF	—Foot-sore	
	FOR.	
	For _____ BPA	
	For all _____ BPC	
	For consideration—of — _____ JQN	
	For him (his, her, it-s, *or person-s or*	
	thing-s indicated) _____ BPD	
NHG	—For instance	
	For me (my, mine) _____ BPE	
NHI	—For passage—to (*or*, in)	
	For — (*person-s indicated*) _____ BPD	
	For that (*or*, this) _____ BPF	
	For the _____ BPG	
	For the future _____ NUR	
NHJ	—For the present At present	
NHK	—For the purpose—of	
NHL	—For the quarter	
NHM	—For the safety of	
NHO	—For the sake of.	
NHP	—For the shore	
NHQ	—For the sick.	
	For them (their-s) _____ BPH	
	For these (*or*, those) _____ _ _____ BPI	
	For this (*or*, that) _____ BPF	
NHR	—For this port	
	For us (our-s) _____ BPJ	
	For what (*or*, which)? _____ BPK	
	For, *or*, To whom (*or*, whose)? ____ BPL	
	For you (your-s) _____ BPM	
	He, She, It (*or person s or thing-s indi-*	
	cated) is (*or*, are) for _____ BUR	
	Not for _____ OFB	
	Ready for — (*or*, to —) _____ UJS	

| | FOR—FORCE | |

	FOR—*Continued.*	NHV	FORCE-D-ING TO FORCE.
	Run for Make for........ ....RIH		Enemy is collecting in force—at ...LZD
	Sailed for —....... ....VEI		Enemy's force ....LZH
	Send for ....VPR	NHW	—Forced by
	They are for ....CNM	NHX	—Forced fires
	Used—for —.........................YMV	NHY	—Forcible-y
	Was, *or,* Were for ....COP		I did not abandon her until I was forced,
	Where for?......................OTA		...............................CXO
NHS	—Whom is it for? Whom for		Inferior force.....................PJL
NHT	FORAGE-D-ING TO FORAGE		Superior force ....XEM
NHU	FORBID-DEN-DING FORBADE		United force ....YJH
		NHZ	—What force?

	⌇ FORCE OF WIND

NIA	FORCE OF WIND		
	NOTE—*The signals only represent the numbers Thus·* NIK *is force of wind* 8		
	FORCE		
NIB	— 0	Denotes calm	
NIC	— 1	Light air, just sufficient to give steerage way	
NID	— 2	Light breeze	{ With which a well-conditioned ship under all } 1 to 2 knots.
			sail and clean full would go in smooth water.
NIE	— 3	Gentle breeze	" " " " 3 to 4 knots.
NIF	— 4	Moderate breeze	" " " " 4 to 5 knots.
NIG	— 5	Fresh breeze	{ In which the same ship could just } Royals, etc
			carry closehauled
NIH	— 6	Strong breeze	" " " Single reefs and top-gallant sails
NIJ	— 7	Moderate gale	" " " Double reefs, jib, etc Topgallant sails
NIK	— 8	Fresh gale	" " " Triple reefs, courses, etc Topsails, jib, etc
NIL	— 9	Strong gale	" " " { Close reefs and courses Reefed upper topsails and courses Lower topsails and courses
NIM	—10	Whole gale	With which she could only bear closereefed Lower main topsail maintopsail and reefed foresail and reefed foresail
NIO	—11	Storm	With which she would be reduced to storm staysails
NIP	—12	Hurricane	To which she could show no canvas

| | | | | | | For ships rigged with double topsails |

| | NOTE | | | | | | |

RATE		PRESSURE		RATE		PRESSURE	
Miles per hour	Feet per minute	Force in pounds per square foot.	Description of wind.	Miles per hour	Feet per minute	Force in pounds per square foot	Description of wind
1	88	.005	Hardly perceptible	30	2640	4 429	} Strong gale
2	176	.020	} Light air	35	3080	6 027	
3	264	.044		40	3520	7 870	} Whole gale
4	352	.079	} Gentle breeze	45	3960	9 200	
5	440	.123		50	4400	12.304	Storm
10	880	.492	} Fresh to strong breeze	60	5280	17 733	} Great storm
15	1320	1 107		70	6160	24 153	
20	1760	1 970	} Moderate gale	80	7040	31 490	} Hurricane
25	2200	3 067		100	8800	49 200	

	FORD—FORESIGHT			
NIQ	FORD-ED-ING ABLE		FORE TOPGALLANT SAIL ....NJH	
NIR	FORE		FORE TOPGALLANT YARD............NJI	
NIS	—Can I get a spar for a fore yard at —?			
NIT	—Fore and aft		FORE TOPMAST . ...NJK	
NIU	—Fore and aft sails		Fore topmast sprung ....NJL	
	Fore chains.......... ....IJQ			
	Fore clew garnets ....IUG		FORE TOPSAIL................ ....NJO	
NIV	—Fore course Fore sail.			
	Fore hold ....NJU		FORE TOPSAIL YARD ....NJP	
	Fore magazine ....REL		Lower fore topsail yard..........NKI	
NIW	—Fore part		Upper fore topsail yard........NKJ	
NIX	—Fore rigging			
NIY	—Fore royal		FORE YARD.........................NJQ	
NIZ	—Fore royal yard		I have carried away fore yard .....NKH	
NJA	—Fore runner		I have sprung (or, damaged) fore yard,	
	Fore sail ....NIV		KJH	
NJB	—Fore sheet		What are the dimensions of your fore	
NJC	—Fore stay		yard ....LAP	
NJD	—Fore staysail			
NJE	—Fore tack	NKL	FORECAST (See also WEATHER)	
NJF	—Fore top		What is the meteorological forecast for	
NJG	—Fore topgallant mast		to-day?...... .... ZK	
NJH	—Fore topgallant sail		What is the meteorological forecast for	
NJI	—Fore topgallant yard		to morrow?...................... ZL	
NJK	—Fore topmast	NKM	—What is the weather forecast?	
NJL	—Fore topmast sprung			
NJM	—Fore topmast staysail		FORECASTLE ....IEO	
NJO	—Fore topsail			
NJP	—Fore topsail yard		FOREFOOT ....NJT	
NJQ	—Fore yard			
NJR	—Fore yard gone in the slings		FOREHOLD .... NJU	
NJS	—Fore yard sprung			
	Forecast...................... NKL	NKO	FOREIGN	
	Forecastle....................IEO		Foreign affairs ....DPC	
NJT	—Forefoot		Foreign colors (flags, ensigns).....MEK	
NJU	—Forehold	NKP	—Foreign crew	
NJV	—Foreland		Foreign Enlistment Act............DIF	
NJW	—Forelock	NKQ	—Foreign-going ship	
NJX	—Foreman	NKR	—Foreign Office, State Department	
NJY	—Foremast	NKS	—Foreign officer	
NJZ	—Foremost	NKT	—Foreign telegram	
	Forenoon A M Before noon In the		Foreign trade . ...XZL	
	forenoon ....EDU	NKU	—Foreigner	
NKA	—Forereach-ed-ing		Lloyd's Register of British and Foreign	
NKB	—Foresee-n ing		Shipping ....QWV	
NKC	—Foresight	NKV	—Vessel is a foreigner	
NKD	—Forestall-ed ing			
NKE	—Foretell-ing Foretold		FORELAND ....NJV	
NKF	—Forewarn-ed-ing			
NKG	—Gaff fore sail		FORELOCK ....NJW	
NKH	—I have carried away fore yard			
	I have sprung (or, damaged) fore yard,		FOREMAN ... .NJX	
	KJH			
NKI	—Lower fore topsail yard		FOREMAST.......................NJY	
NKJ	—Upper fore topsail yard	NKW	—Sprung my fore mast and must bear up	
	What are the dimensions of your fore	NKX	—Sprung my fore mast, but can fish it at	
	yard? ....LAP		sea	
		NKY	—With loss of fore mast	
	FORE PART ....NIW		FOREMOST..... ... .NJZ	
	FORE RIGGING ....NIX		FORENOON A M BEFORE NOON IN	
			THE FORENOON ....EDU	
	FORE RUNNER ....NJA	NKZ	—Forenoon watch	
	FORE STAY ....NJC		FOREREACH ED-ING ....NKA	
	FORE TOP ....NJF		FORESEE-N ING .... NKB	
		NLA	—Could not foresee Unforeseen	
	FORE TOPGALLANT MAST.............NJG		FORESIGHT.......................NKC	

	FOREST—FOUL		
NLB	FOREST-ER		FORTUNATE—*Continued*
		NMP	—Fortune
	FORESTALL-ED-ING .... NKD	NMQ	—Have you been fortunate?
		NMR	—I have been fortunate
	FORETELL-ING FORETOLD .... NKE	NMS	—I have not been fortunate
		NMT	—No fortune
	FORETOP .... NJF		Very fortunate .... YPI
	FOREWARN ED-ING .... NKF	NMU	FORWARD-ED ING—TO —. TO FORWARD.
NLC	FORFEIT-FD-ING-URE		Afloat forward .... BX
		NMV	—Can I forward (*or*, take) letters for you?
NLD	FORGE		Can telegraph message be forwarded
NLE	—Have you anyone that can work at a		from —? .... WR
	forge?		Can you forward my message by tele
			graph? .... HU
NLF	FORGE-D-ING-R TO FORGE		Forward answer by telegraph to signal
			station at — .... EMW
NLG	FORGERY		Forward-ed-ing application to (*or*, for),
			ESP
NLH	FORGET-TING FORGOT-TEN		Forward by telegraph and pay for trans-
NLI	—Do not forget.		mission .... HY
NLJ	—I have forgotten	NMW	—Forward following message by tele-
NLK	—I will not forget		graph to —
NLM	—Will not forget		Forward following telegraphic message
NLO	—You have (*or person indicated* has) for-		by signal letters instead of writing it
	gotten		at length .... XB
			Forward immediately .... PDI
NLP	FORGIVE-N-ING-NESS FORGAVE	NMX	—Forward letters—to (*or*, for)
		NMY	—Forward message immediately without
NLQ	FORK-ED-ING IN FORK		loss of time to —
NLR	—Forkhead		Forward my communication by tele-
	In fork .... NLQ		graph, and pay for its transmission,
NLS	—Knives and forks		HY
		NMZ	—Forward my despatches
NLT	FORM-ED-ING TO FORM		Forward my message through the tele-
	Annul present formation .... EMH		graph by signal letters .... XB
NLU	—Form a junction—with		Forward reply to my message by tele-
NLV	—Form columns of divisions in line ahead		graph to — .... WT
NLW	—Form single column in line abreast		Forward the following information to —,
NLX	—Form single column in line ahead		NGU
NLY	—Formation	NOA	—Have my letters been forwarded?
NLZ	—In proper form	NOB	—Have they been forwarded?
	Present form-ation .... TUB		I draw — feet aft — feet forward ... OA
	Preserve formation .... TUG	NOC	—I have forwarded the
		NOD	—I will forward
NMA	FORMAL-LY-ITY	NOE	—I will forward your letter on
			My draught of water is — feet — inches
NMB	FORMER-LY		forward .... LND
		NOF	—My letters are all forwarded
NMC	FORMIDABLE	NOG	—Shall I forward?
		NOH	—Shall, *or*, Will forward (*or*, be for
NMD	FORSAKE-N-ING FORSOOK		warded)
			Tell — (*person indicated*) not to forward
NME	FORT-RESS		any more letters for me .... QS
	Any forts? .... EOY	NOI	—They have been forwarded
	Strong fort .... XAR	NOJ	—Were any letters properly forwarded?
			Will you forward my letters (*or*, des-
NMF	FORTHWITH		patches) to — ? .... QU
			Will you forward (*or*, communicate) the
NMG	FORTIFY-IED-ING		following signal for me? .... IK
NMH	—Fortification		
NMI	—Fortifications have been increased at —		FOT. (*Measure of length*) .... AVZ
	Strongly fortified .... XAU		
			FOUGHT (*See also* FIGHT) ...... MXR
NMJ	FORTNIGHT-LY		When was the battle fought? ..... FZB
NMK	—A fortnight since		Where was the battle fought? ..... FZC
	In about a fortnight .... DAH	NOK	FOUL-ED-ING-NESS.
			Anti fouling composition .... ENZ
NML	FORTUNATE-LY		Bottom (*ship's*) foul .... HAW
NMO	—Fortunate enough to		Every appearance of foul weather ... ERZ

FOUL—FRIAR'S BALSAM

	FOUL—*Continued*		FRANC (*Coin*) _____ ATR
	Foul anchor Fouled her (*or*, my)		
	anchor _____ EID	NPQ	FRANCE (*See also* FRENCH)
	Foul berth _____ GFQ		Bank of France _____ FUO
	Foul ground Rocky bottom_____ HAZ		
NOL	—Foul hawse	NPR	FRAUD-ULENT ,
NOM	—Foul wind		
NOP	—Have *or*, Has been fouled by — (*vessel*	NPS	FREE-D-ING-LY
	indicated)	NPT	—Can not keep free with pumps.
	Hawser fouled screw_____ OPU	NPU	—Free port
	I have hooked your anchor_____ EIL	NPV	—Free trade-r
	Run, *or*, Ran against (*or*, foul of)_ DSG	NPW	—Freeboard
	You are in a foul berth (not in a good	NPX	—Go-ing free
	berth) _____ GHK	NPY	—I can keep free with pumps
	You have given me a foul berth ___GHL		Is the harbor at — free from ice?___FBZ
			The harbor at — is free from ice ___PCE
NOQ	FOUND (*See also* FIND)		
	Considerable fault found—with — _ JQI	NPZ	FREEZE ING FROZEN.
	Found defective _____ KRE	NQA	—Frozen meat
NOR	—Found guilty	NQB	—Is — frozen over?
	Gold has been found—at _____ ODL	NQC	— — is frozen over
	Goods are found _____ OER		Frozen meat cargo _____ HZB
NOS	—Have, *or*, Has found		
NOT	—Have, *or*, Has not found		
	Not a piece has been found_____ TIA	NQD	FREIGHT
	Pieces of the wreck have been found_ TIB		Any freight offering?_____ EOZ
NOU	—Was, *or*, Were found	NQE	—Freight is to be procured.
NOV	—Was, *or*, Were not found	NQF	—Freight looking better.
		NQG	—Freight may be had for — (*place indi-*
NOW	FOUNDATION.		*cated*) at — shillings per ton
		NQH	—Have you any freight for —?
NOX	FOUNDER-ED-ING		Is cotton freight to be procured?___JXW
	Foundered at her anchors_____ EIF	NQI	—Is freight paid?
	Is it confirmed that — (*vessel indicated*)	NQJ	—No freight to be had
	has foundered? _____ JOO	NQK	—There is good freight to be had—at
NOY	—Vessel indicated is supposed to have	NQL	—What are the freights for —?
	foundered.	NQM	—Where can I get any freight?
NOZ	FOUNDRY.	NQO	FRENCH FRENCH COLORS.
NPA	FOUNTAIN.	NQP	FRENCH LLOYD'S VERITAS.
NPB	FOUR-TH-LY	NQR	FREQUENT-LY
	Four-masted barque _____ FXQ	NQS	—I have frequently
NPC	—Four-masted schooner	NQT	—It frequently occurs
NPD	—Four-masted ship.		
NPE	—Fourth mast	NQU	FRESH-LY
	Fourth-mast chain_____ LJR		Can fresh beef be procured?_____ GCZ
NPF	—Fourth week in —.		Fresh beef _____ GDB
			Fresh beef and vegetables _____ GDC
NPG	FOURTEEN-TH.		Fresh bread Soft bread _____ HEO
			Fresh breeze _____ HFZ
	FOWL _____ IOP		Fresh butter _____ HOS
NPH	FOX.		Fresh fish_____ NBY
		NQV	—Fresh gale
	FRACTION (*See page* 84) _____ BCY	NQW	—Fresh meat
NPI	—Fractional.	NQX	—Fresh milk
		NQY	—Fresh mutton
NPJ	FRACTURE-D-ING	NQZ	—Fresh pork
NPK	—A severe fracture.	NRA	—Fresh provisions
		NRB	—Fresh vegetables to be obtained.
NPL	FRAGILE FRAIL	NRC	—Fresh water
			Wind fresh_____ ZFI
NPM	FRAGMENT.		
		NRD	FRESHEN-ED-ING
NPO	FRAME-D-ING		Breeze will freshen_____ HFY
	Crank frame _____ KBJ	NRE	—Freshen the nip
	Propeller frame broken (*or*, gone)_ TYZ		It the wind should freshen_____ ZET
	Screw frame, *or*, Banjo_____ FUG		Wind will freshen_____ ZFN
	Stern frame _____ WUT	NRF	FRIAR'S BALSAM.

FRICTION—FUNERAL

NRG	FRICTION	NSD	FRUIT
	Friction tube _ _ _ _ _ _ _ _ _ _ _ _ _ _ _ _ _ _ YDF	NSE	—Fruit is to be obtained on shore
		NSF	—No fruit is to be obtained on shore
NRH	FRIDAY		Preserved fruit _ _ _ _ _ _ _ _ _ _ _ _ _ _ _ _ _ _ TUJ
NRI	—Before Friday		
NRJ	—Friday morning.	NSG	FRUSTRATE-D-ING-ION
NRK	—Friday night	NSH	—Have, or, Has been frustrated.
NRL	—Good Friday.	NSI	—Was, or, Were frustrated
	Last Friday _ _ _ _ _ _ _ _ _ _ _ _ _ _ QGB		
	Next Friday _ _ _ _ _ _ _ _ _ _ _ _ _ _ _ _ _ _ SHW	NSJ	FUEL (See also COAL)
			Bad fuel _ FRC
NRM	FRIEND-SHIP		Economize-d-ing fuel _ _ _ _ _ _ _ _ _ _ _ _ LTC
NRO	—Are your friends safe?		Patent fuel _ _ _ _ _ _ _ _ _ _ _ _ _ _ _ _ _ _ TEC
NRP	—Friendly with		
NRQ	—Have you any friends—in —?	NSK	FUGITIVE
	Vessel chasing is a friend _ _ _ _ _ _ _ _ _ _ INU		
NRS	—Your friend is on board	NSL	FULFILL ED-ING
NRT	FRIGATE	NSM	FULL
			Are you at full speed? Is steam up?
	FROM _ BPN		VZ
	Broke adrift _ DMO		Engine room full of water _ _ _ _ _ _ _ _ MBG
NRV	—Broke from		Full crew _ KCZ
	Days from _ KNE		Full charge _ _ _ _ _ _ _ _ _ _ _ _ _ _ _ _ _ _ IMH
	From him (his, her-s, or person-s or	NSO	—Full dress
	thing-s indicated) _ _ _ _ _ _ _ _ _ _ _ BPO	NSP	—Full moon
	From me (my, mine) _ _ _ _ _ _ _ _ _ _ _ BPQ	NSQ	—Full of passengers.
	From that (or, this) _ _ _ _ _ _ _ _ _ _ _ BPR	NSR	—Full of water
	From the _ BPS	NST	—Full power
	From the captain _ _ _ _ _ _ _ _ _ _ _ _ HWY	NSU	—Full-rigged ship
	From the eastward _ _ _ _ _ _ _ _ _ _ _ _ LSD	NSV	—Full speed
NRW	—From the island		Full stop _ WXE
	From the northward _ _ _ _ _ _ _ _ _ _ SJL	NSW	—Fully
NRX	—From the rear		Fully insured _ _ _ _ _ _ _ _ _ _ _ _ _ _ _ _ _ PNY
	From the southward _ _ _ _ _ _ _ _ _ _ WLH		Fully tried _ _ _ _ _ _ _ _ _ _ _ _ _ _ _ _ _ _ _ YCM
	From the westward _ _ _ _ _ _ _ _ _ _ _ ZBR		Go ahead full speed _ _ _ _ _ _ _ _ _ _ _ _ LI
NRY	—From, or, Off the wind		Go astern full speed _ _ _ _ _ _ _ _ _ _ _ _ LM
	From them (their-s) _ _ _ _ _ _ _ _ _ _ _ BPT		Go full speed _ _ _ _ _ _ _ _ _ _ _ _ _ _ _ OBS
	From there _ _ _ _ _ _ _ _ _ _ _ _ _ _ _ _ _ _ _ BFU	NSX	—He, or, She is full of men
	From these (or, those) _ _ _ _ _ _ _ _ _ _ BPV		I am going full speed _ _ _ _ _ _ _ _ _ _ _ OCI
	From time to time _ _ _ _ _ _ _ _ _ _ _ XTL		I am on full-speed trial,
	From us (our-s) _ _ _ _ _ _ _ _ _ _ _ _ _ _ BPW		Code Flag over A
	From what (or, which) _ _ _ _ _ _ _ _ _ _ BFX		I have — hours' coal for full speed (or
	From whence (or, where) _ _ _ _ _ _ _ _ BPY		for speed indicated) _ _ _ _ _ _ _ _ _ _ _ IXD
	From whom (or, whose) _ _ _ _ _ _ _ _ BPZ		Is steam up, or, Are you at full speed?
	From you-r-s _ _ _ _ _ _ _ _ _ _ _ _ _ _ _ _ _ BQA		VZ
	He, She, It (or person-s or thing-s indi-	NSY	—Quite full
	cated) is (or, are) from _ _ _ _ _ _ _ _ _ _ BUS		Stokehold full of water _ _ _ _ _ _ _ _ _ _ WVX
	Miles from (or, distant) _ _ _ _ _ _ _ LGI		Was fully expected _ _ _ _ _ _ _ _ _ _ _ _ _ MNT
	Not from _ CFS		What is your consumption of coal per
	Parted from _ _ _ _ _ _ _ _ _ _ _ _ _ _ _ _ _ _ TBU		hour at full speed _ _ _ _ _ _ _ _ _ _ _ _ IXM
	Sailed from — _ _ _ _ _ _ _ _ _ _ _ _ _ _ _ _ VEJ	NSZ	—Will fully explain
	Sails from — _ _ _ _ _ _ _ _ _ _ _ _ _ _ _ _ _ _ VEM		
	They are from _ _ _ _ _ _ _ _ _ _ _ _ _ _ _ _ BET	NTA	FULMINATE-D ING
	Was or, Were from _ _ _ _ _ _ _ _ _ _ _ _ COQ	NTB	—Fulminate of mercury
	Where are you from? _ _ _ _ _ _ _ _ _ _ _ _ SI		
	Where from? From where? _ _ _ _ _ _ _ BPY	NTC	FUMIGATE-D-ING-ION
NRZ	FRONT-ING		FUN, or, CANDAREEN (Com) _ _ _ _ _ _ _ ATB
	In front of — Before the — _ _ _ _ _ _ GDI	NTD	FUND-ED-ING
NSA	FRONTIER	NTE	FUNDS
		NTF	—How are the funds—at —?
NSB	FROST	NTG	—Want of funds
NSC	—Frost bitten	NTH	FUNERAL
		NTI	—Funeral of —
	FROZEN (See FREEZE) _ _ _ _ _ _ _ _ _ _ _ _ NPZ	NTJ	—Funeral party
	Frozen meat _ _ _ _ _ _ _ _ _ _ _ _ _ _ _ _ _ _ NQA	NTK	—Funeral takes place at — (place) and
	Frozen meat cargo _ _ _ _ _ _ _ _ _ _ _ _ HZB		at —(time and date)
	Is — frozen over? _ _ _ _ _ _ _ _ _ _ _ _ NQB	NTL	—When will funeral take place?
	— is frozen over _ _ _ _ _ _ _ _ _ _ _ _ _ NQC		

FUNNEL—GALE

NTM	FUNNEL
NTO	—Funnel carried away
NTP	—Funnel of —(*vessel indicated*) is —
NTQ	—Her funnels are placed —
NTR	—How are her funnels placed?
NTS	—How many funnels has she?
	Passed a war vessel disguised, funnels — (*Here give the color or design of the line of steamers that it may be an imitation of*) ----- ---------LEO
	Sweeping the funnels (*or*, flues, *or*, tubes) ---------------------NFZ
NTU	—Vessel indicated has — funnels
	FUNT (*Measure of weight*) ----------BAL
NTV	FUR-RED-RIER
NTW	FURL-ED-ING
	Do not furl sails----------------VDW
	Furl awnings --------- -----------FPX
NTX	—Furl sails (*or sail indicated*)
NTY	—I shall not furl sails
NTZ	—Shall I furl sails?
	FURLONG (*Measure of length*)-----AWB
NUA	FURNACE
	Furnace bar, *or*, Fire bar ----------FWK
NUB	—Furnace collapsed
	Furnace considered dangerous-----KLB
NUC	—Furnace crown of boiler collapsed
NUD	—Furnace leaky
NUE	—Furnace overheated
NUF	FURNISH-ED-ING—WITH (*See also* SUPPLY)
	Boats are furnished with — (*articles indicated*)--------------------:-GSD
NUG	—Can you furnish me with —?
	Can you furnish me with (*or*, lend me) a Code of Signals?---------------IZE
NUH	—I can furnish you with(*or*, lend you)—
NUI	—I can not furnish you with (*or*, lend you) —
NUJ	—Was, *or*, Were furnished with —
NUK	FURNITURE
NUL	FURTHER-ED-ING.
	Any further ---------------EOV
	Anything further?---------- -----EQJ
	Do not require further assistance ---CR
	Do you require further assistance? .CT
	Furthest----------------------NUO
	Have received orders for you not to proceed without further instructions .SP
	I decline to have anything further to do (*or* say) in the matter --------EQO
	I must remain till further orders---STX
	Repeat my signals till further orders, ---------------------------WCP
	Until further orders ----- -------YLG
	Vessel indicated does not require further assistance --------•-----FGQ
NUM	—You must remain until further orders
NUO	FURTHEST
	FUSE (*See* FUZE) ---------------NUT

NUP	FUTTOCK SHROUD
NUQ	FUTURE
NUR	—In, *or*, For the future
NUS	—Some future opportunity
NUT	FUZE
NUV	—Bickford's fuze.
NUW	—Detonating fuze
	Electric fuze-----------------LUV
NUX	—Instantaneous fuze
NUY	—Patent fuze
NUZ	—Percussion fuze
NVA	—Time fuze.
NVB	G (*Letter*) (*For new method of spelling, see page* 13)
NVC	GAFF
NVD	—At the gaff
NVE	—Gaff end
	Gaff foresail ----------------NKG
NVF	—Gaff mainsail.
NVG	—Gaff topsail.
	Jigger gaff----------------PVR
	Spritsail gaff (*or*, yard) ---------ZKO
NVH	GAGE-D-ING
NVI	—Weather gage
NVJ	GAIN-ED-ING
NVK	—Are you gaining? Have you gained?
NVL	—Can not gain much
NVM	—Can not gain much profit
	Fire gaining --------------OQ
	Fire gains rapidly, take people off--NG
	Gained advantage-------------DNQ
NVO	—Gained a victory
NVP	—Have, *or*, Has gained
NVQ	—Have, *or*, Has not gained
	Have you gained? Are you gaining? ---------------------------XVK
	Is there any advantage to be gained? ---------------------------DNT
	Leak is gaining rapidly-----------NU
	My chronometer gains daily--------GP
	Stranger is gaining -------------WZB
	Stranger is not gaining ----------WZC
	Time will be gained—by----------XTS
NVR	—Who gained?
NVS	GALE
NVT	—A gale is coming on.
	After the gale----------------DRN

GALE—GEAR

	GALE—*Continued*	NWY	**GARRISON**-ED-ING
NVU	—Can — (*vessel indicated*) weather the gale?	NWZ	**GAS**-EOUS
NVW	—Can you ride out the gale?		Explosive gas ....MOX
	Delayed through the recent gale ...KSF	NXA	—Gas, or, Light buoy
NVX	—Do you think the gale is over?		Gas engine ....MAQ
	Do you think we can ride it (*the gale*) out?..............................UWX	NXB	—Gas light
	During a gale ....LPM	NXC	**GASKET**
	During the gale Whilst the gale lasts, LPS	NXD	**GASTRIC**
	Equinoctial gale ....MGN	NXE	**GATE**
	Fresh gale ....NQV	NXF	—Gate is closed at —
	Gale is expected—from —..... .....ZD		
NVY	—Have, or, Has had a heavy gale from —		
NVZ	—Heavy, or, Severe gale	NXG	**GATHER** ED-ING
NWA	—If gale increases	NXH	—Gather ed-ing way
NWB	—In a gale of wind		
	Meteorological Office, or, Weather Bureau, reports gale approaching from the — ....ETQ	NXI	**GAUGE**-D-ING-R
	Moderate gale...................RXL	NXJ	—Gauge cock
NWC	—Nothing but gales of wind		Manometer Steam pressure gauge, RJW
	Severe, or, Heavy gale ....NVZ		Vacuum gauge ....YSM
NWD	—Shall you ride out the gale?		Water gauge ....YVQ
	Strong gale ....XAS		Water gauge out of order..........YVR
	Whole gale ....ZDR		Wind gauge Anemometer.........ELO
NWE	**GALLANT**-LY-RY		**GAVE** (*See* **GIVE**) ....NZV
	Gallant conduct.......... ....JNG		
	Have, or, Has behaved very gallantly, GDX	NXK	**GAZETTE** IN THE **GAZETTE**
NWF	**GALLERY**		Report me by telegraph to "Shipping Gazette"..... ... .UF
NWG	—Quarter gallery		"Shipping Gazette & Lloyd's List," QWU
NWH	**GALLEY**		
NWI	**GALLIOT**	NXL	**GAZETTEER**
NWJ	**GALLIPOLI** OIL	NXM	**GEAR**-ING
			Air-pump gear.....................DWB
	GALLON (*Measure of capacity*) ...AYW		Blow-off cocks (or, gear) ....GNW
NWK	**GALVANIZE** D-ING-ISM-IC		Boat's gear ....GRZ
NWL	—Galvanized iron		Chafing gear......................LJK
			Diver's gear............... ....LHX
NWM	**GALVANOMETER**		Eccentric gear.....................LSU
			Eduction gear.....................LTM
NWO	**GAME**		Expansion gear ....MLZ
			Firing gear ....NAF
NWP	**GANG**		Gear aloft........................EAF
			Gear for cleaning decks ....ISV
NWQ	**GANG** BOARD		Hand steering gear................OMA
			Have passed steamer with steering gear disabled ....MU
NWR	**GANGWAY**		Head gear ....OQG
			Induction gear....................PIV
NWS	**GANGRENE**		Injection gear.....................PKU
			Parallel-motion gear ....TAX
NWT	**GANTLINE**		Pump gear ....UBI
			Pump gear worn out (defective)...UBJ
	GAOL JAIL PRISON...... ....PUK		Slide-valve gear ....WGK
			Slide valve gear damaged (or, out of order)......................WGL
	GAOLER ....PUM		Starting gear ....WRF
NWU	**GAP**		Starting gear out of order.........WRG
			Steam steering gear ....WST
NWV	**GARDEN**-ER		Steam steering gear disabled........MZ
			Steering gear ....WUF
NWX	**GARNET**		Steering gear disabled .. ..MZ
	Fore clew garnets ....IUG		Turning gear ....YEN
	Main clew garnets................IUH		Want gear for ....YTV

GENERAL—GET

NXO	GENERAL-LY		GET—*Continued.*
	Accountant General_____DFY		Do you think we could get away
	Admiral, *or*, Senior officer is generally		from — ?_____XPZ
	about_____DKT		Endeavor to get a line ashore by a boat
	Consul General_____JRM		(cask, kite, raft, *or*, spar, etc)___KA
	General (*Officer*)_____ _____EWK		Expect-ed ing (*or vessel indicated* ex-
	General average_____FOR		pects-ed-ing) to be got off (*or*, afloat),
	General cargo_____HZC		DPX
NXP	—General order		Fire difficult to get at _____ ___OP
	General recall '' signal		Get a preventer on_____TVI
	Code Flag over Y		Get an offing Cut, Weigh, *or*, Ship
	Governor General_____OOD		Wait for nothing_____KW
	Hoist the general recall_____OWH	NYU	—Get her head round as fast as possible,
	Inspector General__ _ _____PMV		*or*, Get her on the other tack as fast
	Lieutenant General _ _____QPV		as possible.
	Registrar General of Shipping and Sea-		Get her on the other tack, *or*, Get her
	men, *or*, Commissioner of Navigation,		head round or you will be on shore,
	Treasury Department _____UNX		LF
NXQ	—What is the general opinion?	NYV	—Getting
	Who generally does business for — ?_DTB	NYW	—Get, *or*, Got pratique
		NYX	—Get ready for sea Prepare for sailing,
NXR	GENEROUS LY	NYZ	—Get sail on your ship as fast as possible
			Get under way as fast as you can ___KN
NXS	GENEVA CONVENTION		Get up steam , report when ready __KM
NXT	—Geneva Convention flag (*when spoken*		Get up steam as fast as possible_____KL
	of)	NZA	—Get what you can
			Got.____ _____ _____OFB
NXU	GENEVA (*Gin*)		How long will you be getting up steam?
			VY
NXV	GENTLE Y NESS		I am getting short of water and must
	Gentle breeze Slight breeze _____HQD		endeavor to get some_____YVG
		NZB	—I can get off
NXW	GENTLEMAN		I can not get at goods required ____OEW
			I shall get off (*or*, afloat) with assist-
NXY	GENUINE NESS LY		ance_____ _____DPY
			I think you might get something to
NXZ	GEOGRAPHY-ICAL LY		answer the purpose_____ENB
	Geographical Index (*See page* 452)	NZC	—If you can get off
	Geographical mile_____AWO		If you will lay out an anchor for me I
NYA	—Refer to the Geographical Signals		can get off_____ _____ _____EIV
			Indicate nearest place I can get coal,
NYB	GERM		BC
			Is it a difficult anchorage to get away
NYC	GERMAN Y GERMAN COLORS		from ?____ _____ELB
	Bank of Germany_____ FUP		Is my berth a good one for getting away
NYD	—German Lloyd's		from ?____ _____ GFW
			Lose no time in getting to the anchor
NYE	GET To GET (*See also* GOT *and*		age_____ _____ELC
	OBTAIN)		Mind you get paid _____RUK
NYF	—Can, *or*, May get	NZD	—No getting out of harbor
	Can I cross the bar ? Shall I be able to		Not worth getting._____OFL
	get over the bar?_____ _ _____FVY		Not worth getting off (*or*, afloat)_DQC
NYG	—Can I get a main yard at — ?		Open Get farther away_____MC
NYH	—Can I get a spar for — mast at — ?		Pilot can not get off ____ _____TJP
NYI	—Can I get any — ?		Put to sea at once Get an offing _ MG
NYJ	—Can I get any sailmakers?		Shall I attempt to get away ? ___FKP
NYK	—Can I get permission?		Shall I get steam up?___ ___:____WA
NYL	—Can you get?		Shall we get to the anchorage to-night?
NYM	—Can you get an introduction to — ?		ELF
NYO	—Can not get		Stand off Get an offing Put to sea
	Can not get my boat out_____GSJ		at once _____ _____MG
NYP	—Could get	NZE	—Vessel (*indicated, if necessary*) has got
	Did not get a mail_____ z		off
NYQ	—Did, *or*, Do you get?	NZF	—Was getting ready for sea Preparing
	Did you get a meridian altitude? _FDH		for sailing
NYR	—Did you get an observation?		What quarantine shall I get?_____ UEJ
	Did you get good observations?____RM		Where can I get (*or*, look for) a pilot ?
NYS	—Did you get pratique — ?		TL
	Do you think I can get through the ice ?		Where can I get any freight?_ ____NQM
	PBH		Where is the best place to get —?___GIC
NYT	—Do you think we could get any — ?	NZG	—Will you get?

	GET—GO		

	GET—*Continued.*		GIVE—*Continued*
	You can get an answer in —........ENU	OAL	—Will give
	You had better get a sick nurse...WBK	OAM	—Will you give?
	You will find great difficulty in getting	OAN	—Will you give me (*or*, my)?
	through the ice at —.............PCI		Will you give me a passage (*on shore*)?
	You will get into shoal water, *or*, You		TDK
	will shoal your water.. ....VZA	OAP	—Will you give me the pleasure?
	You will get quarantine.......... UEL		
	You will have great difficulty in getting	OAQ	GLAD-LY
	water ....LAF	OAR	—Glad to hear—it (or, that)
NZH	GIG	OAS	—Glad to see—it (or, that)
	Steam gig ....GUW	OAT	—Glad to see you Glad to see (*person named*)
	GILL (*Measure of capacity*) ....AYX	OAU	—I shall be very glad
			Very glad................ ....YPJ
	GIMBAL OF A COMPASS ....JHQ		Very glad of your company ....JGT
NZI	GIMLET	OAV	GLAND
			Stern gland defective.............WUV
	GIN (*Spirit*) GENEVA ....NXU		Stuffing-box glands out of order ...HDJ
NZJ	GIN (*For derrick, etc*)	OAW	GLARE-D-ING LY
NZK	GINGER	OAX	GLASS
NZL	GIRD-ER		Binocular glass Field glass ......GLB
NZM	GIRL		Field glass Binocular............GLB
			Hour glass ....OZK
NZO	GIVE TO GIVE		Lamp glass .... QDG
	Are giving ground................OIF		Log glass .... Q)H
	Be careful not to give offense... ..HYL		Panes of glass................ ...SZY
	Bound to be given up.............HBW		Spy glass Telescope ....XMQ
	Bound to give it up ....HBX		Wine glass................ ....ZGD
	By the description given ....KVZ	OAY	GLIMPSE
NZP	—Can, *or*, May give		I had a glimpse of the land about..QDU
NZQ	—Can give no reason	OAZ	GLOBE ULAR
NZR	—Can you give?	OBA	GLOOM-Y-ILY
NZS	—Can you give a passage—to?	OBC	GLOVE
NZT	—Can not give	OBD	GLUE-D ING-TINOUS.
NZU	—Do, *or*, Did you give?	OBE	—Marine glue
NZV	—Gave	OBF	GLUT TED
	Give authority ....FNS	OBG	GLYCERINE
	Give directions—that (for, or, to)..LBJ		Nitro-glycerine ....SJR
	Give every assistance—to..........FGH	OBH	GO. TO GO.
NZW	—Give him (her-s, it-s)		All gone........ ....DXK
NZX	—Give leave—to	OBI	—Am, Is, *or*, Are going (*or*, gone)—to.
NZY	—Give me (my, mine)	OBJ	—Am, Is, *or*, Are to go
OAB	—Give notice		An express is going to—............MPI
OAC	—Give receipt	OBK	—Are you going into port (harbor)?
OAD	—Give them (their-s).	OBL	—Are you going—to?
OAE	—Give us (our-s)		Are you going on shore?..........FCX
	Give way ....YXC		Are you going to anchor?..........EHM
OAF	—Given		Are you going to land all mails at — ?
	Gives, *or*, Must give an evasive answer,		QDN
	BNG		Boat is going to you ....EU
OAG	—Giving.	OBM	—Can you go?
	Have orders been given—to (or, for)?		Can you go up the river?..........UYG
	PNE		Convoy is going to.................JCN
	Have you given bond?.............GYN	OBN	—Do not go
	I shall give chase ....INR		Do not go (*or*, pass) ahead of me
OAH	—I will give you		Code Flag over R
	Notice (*warning*) has been given ..YUL	OBP	—Do not let him (*or*, them) go
	Obliged to be given up ....SMV	OBQ	—Go about
	Obliged to give it up........ ......SMW		Go ahead.. ....LG
	Warning (*notice*) has been given ..YUL		
OAI	—What have you given for?		
OAJ	—What is the highest price you will give?		
OAK	—What will you (*or person indicated*) give?		

GO—GOOD

	Go—*Continued*		
	Go ahead of—		DVH
	Go ahead and drop a boat on board		DVN
	Go ahead easy Easy ahead		LH
	Go ahead full speed		LI
	Go ahead of me		DVG
	Go alongside		EBF
	Go as fast as you can		MTN
	Go astern		LJ
	Go astern Heave all aback		LP
	Go astern of—		FHK
	Go astern easy. Easy astern		LK
	Go astern full speed		LM
	Go astern of me Pass astern of me		FHQ
OBR	—Go away		
	Go direct		LBN
	Go easy		LRX
	Go ing free		NPX
	Go for despatches		JL
OBS	—Go full speed		
OBT	—Go half speed.		
OBU	—Go immediately		
OBV	—Go in (*or*, to)		
OBW	—Go on Proceed		
	Go on board		GQD
OBX	—Go on shore.		
OBY	—Go out (*or*, from)		
OBZ	—Go to—the		
	Go to the admiral's office		DLF
OCA	—Going		
	Going astern Stern way		FHR
OCB	—Going into port (*or*, harbor)		
OCD	—Going round.		
	Going to join		PWA
OCE	—Gone		
	Gone by Passed by		HPV
OCF	—Gone to pieces		
OCG	—Gone to sea		
	How is business going?		HNY
OCH	—I am a stranger here, will you let me go on shore with you?		
OCI	—I am going full speed		
OCJ	—I am going half speed		
OCK	—I am going on shore		
OCL	—I am going—to —		
OCM	—I am not going on shore		
	I shall not go far before I bring up		HJC
	I shall not go much farther		MTB
OCN	—I wish to see you before you go on shore		
OCP	—If, *or*, When you go		
OCQ	—Is, *or*, Are not gone		
OCR	—Is — (*person or ship indicated*) gone?		
	Keep going ahead of		DVL
	Keep going ahead slowly		DVM
	Keep going astern		FHO
	Keep the lead going Keep sounding		QJI
OCS	—Let go		
	Let go another anchor		KR
	Let go the life-buoy		HMR
	Let go your anchor		EJA
OCT	—Let him (*or*, them) go		
	Lifeboat is going to you		FK
OCU	—Make haste, or he will be gone		
OCV	—Must go		
OCW	—Must I go?		
	My engines are going astern		
		Code Flag over v	
OCX	—Not going		
	Sea going (gone) down		VEZ
OCY	—Shall I go?		
	Stand by to let go		WQE

	Go—*Continued*		
	Stock all gone		WVM
	The next mail goes in—(*vessel indicated*),		RFX
	The pilot is going to you		TJR
	—They were gone		
	Tug is going to you		YA
OCZ	—Was, *or*, Were going—to		
	Went—to—		ZBQ
ODA	—When do you go—to?		
ODB	—When do you go on shore?		
ODC	—When going—to?		
ODE	Where are you going?		CSU
ODF	—Where is he (she, *or*, it) gone (*or*, going)?		
ODG	—Will you go		
	Will you go (*or*, come) on shore?		JDS
	Will you go on shore in my boat?		VZN
	Will you have to go into dry dock?		LIX
ODH	GOAT		
ODI	GOD-LY		
ODJ	GOLD EN		
ODK	—Gold dust.		
ODL	—Gold has been found at —		
ODM	—Gold mine		
ODN	—Goldsmith		
ODP	—In gold		
	GONE (*See also* Go)		OCE
ODQ	GOOD (*For* GOODS, *see next page*)		
	A good crop *or*, Crops look well		KDT
	A good many slight cases		IDH
ODR	—A good offing		
ODS	—A good plan (*or*, method)		
ODT	—A good slant		
	Am I in a good berth?		GFO
	Beef in good order		GCY
	Biscuit is not good Bread is not good		GLK
	Business good		HNV
	Can you point out (*or*, take me to) a good anchorage?		EKS
	Cargo in very good order		HZX
	Did you get good observations?		RM
	Good accommodation		DCR
	Good accounts—of —		DFA
ODU	—Good anchorage		
	Good berth		GFR
	Good bottom for anchorage		EKX
	Good-bye		HPZ
ODV	—Good chance		
ODW	—Good coal		
	Good coal may be got at —		IXB
	Good condition		OEB
	Good conduct		JNH
	Good enough		MDH
ODX	—Good example		
	Good Friday		NRL
ODY	—Good landing		
ODZ	—Good lookout		
OEA	—Good news—of		
OEB	—Good order In good order		
OEC	—Good observations		
	Good season for fish		NBZ
	Good shelter from		VWJ
	Good vacuum.		YNK
	Good water Water is good		YVS
	Good water can be got at —		YVD

GOOD—GRANITE

	GOOD—*Continued*		GOT—*Continued*
OED	—Goodness		Expect (*or vessel indicated* expects) to
	Ground tackle not good enough ----OIJ		be got off (*or*, afloat)------------DPX
	Ground tackle very good------------OIK		Get, *or*, Got pratique------------NYW
OEF	—Have the goodness to		Got bottom with — feet------------HBA
	Have you a good fire engine? ----- NAG		Got on board ------------------------GQE
OEG	—I do not think we shall get any good by	OFE	—Have, *or*, Has got
	moving	OFG	—Have, *or*, Has got off
	In good order Good order--------OEB	OFH	—Have, *or*, Has not got
	Is it good holding ground? ---------OIM	OFI	—Have you, *or*, Has — (*vessel indicated*)
OEH	—Is, *or*, Are not good		got pratique?
OEI	—Is, *or*, Are very good	OFJ	—Is it likely you (*or vessel indicated*) will
	Keep a good lookout aloft------ ---EAG		be got off?
	Keep a good lookout, as it is reported	OFK	—May be got off
	enemy's war vessels are going about		May be got off (*or*, afloat) if prompt as-
	disguised as merchantmen --------OJ		sistance be given ------------------CG
	Keep a good lookout for land (*or*, lights),		May be got off (*or*, afloat) without as-
	or, Lookout for land (*or*, lights)	OFL	sistance ----------------------------DPZ
	(*Bearing to follow, if necessary*).PYH		—Not worth getting
	Leave a good margin -- ---------QKX	OFM	Not worth getting off--------------DQC
	Make-ing, Made good—the(*or*, your).RIJ		—Only to be got in small quantities
OEJ	Not a good plan -------------------TLY		Vessel (*indicated, if necessary*) has got
OEK	—Not so good		off ----------------------------------NZE
	—Of good quality		
	Pumps in good order -------------UBP	OFN	GOULARD'S EXTRACT.
	There is good freight to be had —		
	at ---------------------------------NQK	OFP	GOUT-Y
	There is good shelter at (*or*, in)---VWN		
OEL	—Very good	OFQ	GOVERN-ED-ING-MENT
	Very good chart --- ---------------IMZ		Belong-ing to Government------ --GES
	Very good draught ----------------LNF		Chartered, *or*, Employed by Govern-
OEM	—Very good remittances		ment------------------------------INE
OEN	—Was, *or*, Were very good		Colonial Government----------------JBK
	Water is good Good water-------YVS	OFR	—East Indian Government
	Water is not good ----------------YVT	OFS	—Government despatch
OEP	—Will be very good	OFT	—Government employee (*or*, officer)
		OFU	—Government hired vessel
	GOOD-BYE -------------------- ----HPZ	OFV	—Government in minority of —
			Government majority of ---- ----RHV
	GOOD FRIDAY ------- ------------NBL	OFW	—Government ship
		OFX	—Government steamboat, *or*, transport
OEQ	GOODS	OFY	—Government stores
	Contraband goods-----------------JSO	OFZ	—Government tug
OER	—Goods are found	OGA	—Has there been any change in the Gov-
OES	—Goods are landed		ernment in —?
OET	—Goods are shipped	OGB	—The United States, *or*, His, *or*, Her
OEU	—Goods department		Majesty's Government
OEV	—How many tons measurement of goods		I have Government despatches (*or*, —),
	can you take?		----------------------------------JS
OEW	—I can not get at goods required		Local Government Board----------GPJ
OEX	—I have goods belonging to you (*or*, —)		Officer, *or*, Employee of Government,
	Manufactured cottons Cotton goods,		----------------------------------OFT
	JXY	OGC	GOVERNOR
	Stock of goods-------------------WVN		Deputy, *or*, Lieutenant Governor--KVN
OEY	—Tons measurement of goods	OGD	—Governor General
OEZ	GOOSE	OGE	GRADUAL LY
OFA	GOOSENECK		GRAIN (*Measure of weight*)-------BAM
OFB	GOT (*See also* GET)		GRAIN GRAIN CARGO ------------HZD
	Almost anything may be got-------EAB		GRAMME (*Measure of length*)-----AWP
	Articles indicated may be got at —		GRAMME (*Measure of weight*)-----BAN
	(*place*) or, from — (*vessel indi-*		
	cated)----------------------------EZG	OGF	GRANARY
	Can not be got off (*or*, afloat) by any		
	means now available --------- ----CD	OGG	GRANGE
OFC	—Could any vessel have got away after	OGH	GRANGE
	you?	OGI	GRANITE.
OFD	Could she have got to any place of shel-		
	ter?		

GRANT—GROVE

OGJ	GRANT ED-ING Permission is granted. Permitted__TGH
OGK	GRAPE
OGL	GRAPHITE
OGM	GRAPPLE D-ING
OGN	GRAPNEL Explosive grapnel_____MOY Fire grapnel _____MZT
OGP	GRASS-Y Grass, or, Coir cable __ ____ _____ HQZ Grass, or, Coir hawser _____IZS Grass, or, Coir matting _____IZT Grass, or, Coir rope _____IZU
OGQ	GRATEFUL-LY-ITUDE
OGR	GRATIFY-IED-ING-ICATION
OGS	GRATING
QGT	GRATUITY OUS-LY
OGU	GRAVE LY ITY
OGV	GRAVEL-LY GRAVING DOCK, or, DRY DOCK_____LJE
OGW	GRAZE-D-ING
OGX	GREASE-D-ING-Y-INESS
OGY	GREAT-LY-NESS
OGZ	—A great (large) number (or, quantity—of —
OHA	—Great circle sailing
OHB	—Great execution
OHC	—Great hesitation
OHD	—Great hopes—of —
OHE	—Great neglect
OHF	—Great opposition—to (or, from)
OHG	—Great relief
OHI	—Great risk
OHJ	—Great strain (or, stress) on —
OHK	—Greater-est
OHL	—It will be of great service (or, use)
OHM	—Not a great many
OHN	—The greater est number (or, quantity)—of
OHP	—The greater part With the greatest ease_____LSB
OHQ	GREECE-IAN GREEK GREEK COLORS
OHR	GREEN
OHS	—Green buoy
OHT	—Green light
OHU	GREENWICH, My chronometer is fast on Greenwich (or, first meridian) mean time *___GR My chronometer is slow on Greenwich (or, first meridian) mean time *___GS My first meridian is Cadiz, west of Greenwich 6° 12′ 24″ = 0h 24m 49 6s , BSI

	GREENWICH—Continued. My first meridian is Greenwich____NBK My first meridian is Paris east of Greenwich 2° 20′ 15″ = 0h 9m 21s_____THL My Greenwich (or, first meridian) mean time is — *_____XL *NOTE —In signaling longitude or time, vessels should always reckon from the meridian of Greenwich, except French vessels, which will use the meridian of Paris, and Spanish vessels, which will use the meridian of Cadiz If any doubt is entertained, the vessel to whom the signal is made should hoist NBL = What is your first meridian? Will you show me your Greenwich (or, first meridian) mean time? Would like to have another comparison _____GV NOTE —The vessel showing the mean time will hoist the hour signal and dip it sharply shortly after it has been answered The minutes and seconds at the instant of dipping will immediately follow To insure accuracy, a second comparison should be made
OHV	GREY
OHW	GRIDIRON (Grid for hauling up small ships)
OHX	GRIEVE-D-ING GRIEF Come to grief _____JCY
OHY	GRIND-ING GROUND
OHZ	GRINDSTONE
OIA	GRIPE
OIB	GROCER-Y
OIC	GROG
OID	GROSS GROSS TONNAGE
OIE	GROUND-ED ING (See also AGROUND) —Are giving ground
OIF	Burial ground Cemetery_____IGS Fishing ground _____ _____NBU Foul ground Rocky bottom__ ___HAZ
OIG	—Ground mine
OIH	—Ground swell Ground tackle_____ _____XIQ
OIJ	—Ground tackle not good enough
OIK	—Ground tackle very good
OIL	—Ground tier I have (or vessel indicated has) not touched the ground _____DUT
OIM	—Is it good holding ground? Underground _____YHG Was tide high when you grounded?__CI Was tide low when you grounded___CJ When did you ground?_____CK
OIN	GROUNDLESS
OIP	GROUP-ED-ING Group flashing light _____NEC
OIQ	GROVE

GROW—GUZ

OIR	GROW-ING GROWN GREW TO GROW		
OIS	GUANO		
	Guano cargo ------------------- HZE		
OIT	GUARANTEE-D ING		
OIU	GUARD-ED ING		
	Advance guard ------------------DMY		
OIV	—Be on your guard Look out		
OIW	—Be on your guard against long tows		
	Be on your guard, appearances are		
	threatening ----------- ----------FN		
	Be very careful (or, guarded) in your		
	intercourse with strangers --------FS		
OIX	—From the guard ship		
	Guard boat ---------------------GTA		
OIY	—Guard of honor		
OIZ	—Guard ship		
	Horseguards ---------------------OYF		
	Officer of the guard---------------SPL		
	Rear guard -------------- ---------UKM		
OJA	—Ship having the guard		
	Steam guard boat---------- ------GUX		
	The conductor (or, guard) ----------JNI		
	To row guard ------- -------------VAT		
OJB	GUARDA COSTA		
OJC	GUARDIAN-SHIP		
OJD	GUATEMALA GUATEMALIAN COLORS.		
OJE	GUDGEON		
OJF	GUESS-ED-ING		
OJG	—Guess warp		
OJH	GUIDE-D-ING-ANCE		
	Will the soundings be a safe guide?_WKZ		
	GUILDER, GULDEN, or, FLORIN (Coin) ATQ		
OJI	GUILT-Y-INESS		
	Found guilty ---- ----------------NOR		
OJK	--Is, or, Are not guilty.		
OJL	GUINEA (Money)		
OJM	GULF		
	Gulf stream -- ----------------- ------WZL		
	The entrance of the gulf (or, harbor) is		
	mined ------- ------------------MFU		
OJN	GULLY		
OJP	GUM		
OJQ	GUN		
	Boat's gun ---------------------GSA		
	Enemy's gunboat -----------------LZI		
	Evening gun ------------------- MIF		
	Field gun ------- ----------------MXK		
	Fire a signal gun------------------MZP		
	Gunboat----- ------------------GTB		
	Gun carriage ---- ----------------ICD		
OJR	—Gun cotton		
	Gun-cotton rocket----------------UZD		
OJS	—Gun fog signal		
OJT	—Gunroom		

	GUN—Continued
OJU	—Gunshot
	Gunshot wound, accident ---------DCI
OJV	—Gun slide
OJW	—Gun wharf
OJX	—Guns will not reach
OJY	—Gun's crew
OJZ	—Guns firing in — (quarter indicated)
OKA	—Guns firing last night in — (quarter indicated)
OKB	—Guns firing to the —
	Guns of the calibre of —.----------HSV
OKC	—Guns of the weight of —
OKD	—Guns overheated
	Gunwale.. ----- ----------------OKT
	Heard firing to the —.-------------NAI
	Heavy gun (Weight in tons if necessary) OSW
	How many guns?-----------------REF
	How many guns do you mount? ...RKG
	How many guns does she carry?...RKH
	I have a heavy gun (indicate if more)OTC
	I have no heavy guns--- ----------OTD
	Krupp gun ----------------------QCN
OKE	—Light gun
OKF	—Machine gun
	Maxim gun ----------------------RNC
OKG	—Minute gun
OKH	—Minute guns firing in —(direction indicated)
OKI	—Minute guns firing last night in — (quarter indicated)
	Nordenfeldt gun-----------------SJH
OKJ	—Out of gunshot Out of range
OKL	—Quick-firing gun
	Rifled gun ----------------- -----UXE
	Rounds of ammunition for heavy gun, RGA
OKM	—She carries guns.
	She has, or, They have heavy guns_OTF
	She has, or, They have no heavy guns_OTG
	Signal gun----------------------WCT
	Starboard gun -------------------WQX
	Torpedo gun.--------------------XWT
	Torpedo gunboat (or, vessel)-----XWU
	Turret gun ----- : --------------XFT
OKN	—Vessel is within gunshot
	You are within gunshot You are within range of the guns (or, batteries)---GO
	Gunboat----------------------------GTB
	Enemy's gunboat -----------------LZI
OKP	GUNNER-Y
OKQ	—Gunner's crew
OKR	—Gunner's stores
OKS	GUNPOWDER (Weight in lbs , if necessary)
	I am taking in (or, discharging) gunpowder or explosives_ ----Flag B, or, Code Flag over B
OKT	GUNWALE
	GUST-Y (Squall) ----------------WPK
OKU	GUTTA PERCHA, or INDIA RUBBER
OKV	GUY-ED-ING
	GUZ, or, ZER (Measure of length)..AWC

GYBE—HAD

OKW	GYBE-D-ING, or, JIBE-D-ING
OKX	GYMNAST-ASTIC-NASIUM.
OKY	GYPSUM

OKZ	H (Letter) (For new method of spelling, see page 13)
	HABBE, (Measure of weight) _____ BAG
OLA	HABIT
	Bad habit. _____ FRD
OLB	HABITATION
	HAD, HAS, or, HAVE _____ BQN
	Am, Are, or, Is not to have _____ BQD
	Am, Are, or, Is to have _____ BQC
	Can, or, May have _____ BQE
	Can, or, May have been _____ BFH
	Can, or, May have done (or, been done), BLV
	Can, or, May have had ... _____ BQF
	Can, or, May he (she, it, or person s or thing-s indicated) have been? ____ BTL
	Can, or, May I have—some? _____ BQG
	Can not, or, May not have _____ BQH
	Can not, or, May not have been ... BFK
	Can not, or, May not have done (or, been done) _____ BLZ
	Can not, or, May not have had _____ BQI
	Can not, or, May not he (she, it, or person-s or thing-s indicated) have been? BTN
	Could, or, Might have _____ BQJ
	Could, or, Might have been _____ BFM
	Could, or, Might have done (or, been done) _____ BMC
	Could, or, Might have had _____ BQK
	Could, or, Might not have _____ BQL
	Could, or, Might not have been ___ BFO
	Could, or, Might not have done (or, been done) _____ BME
	Could, or, Might not have had _____ BQM
	Do not let him (her, it, or person indicated) have. _____ BTR
	Had, Has, or, Have. _____ BQN
	Had, Has, or, Have any _____ BQO
	Had, Has, or, Have done _____ BFQ
	Had, Has, or, Have done (or, been done), BMR
	Had, Has, or, Have had _____ BQP

HAD—Continued

Had, Has, or, Have, he (she, it, or person s or thing s indicated)? _____ BTS
Had, Has, or, Have, he (she, it, or person-s or thing-s indicated) been? __ BTU
Had, Has, or, Have he (she, it, or person s or thing s indicated) had? _ BTV
Had, Has, or, Have, he (she, it, or person s or thing s indicated) not? _ BTW
Had, Has, or, Have, he (she, it, or person-s or thing s indicated) not been, BTX
Had, Has, or Have, he (she, it, or person s or thing-s indicated) not had? BTY
Had, Has, or, Have his (her-s, it-s)? _ BQR
Had, Has, or, Have I (my, mine)? _ BQS
Had, Has, or, Have I been? _____ BFR
Had, Has, or, Have I done? _____ BMI
Had, Has, or, Have I had? _____ BQT
Had, Has, or, Have not? _____ BQU
Had, Has, or, Have not been? _____ BFS
Had, Has, or, Have not done (or, been done) - _____ BMS
Had, Has, or, Have not had _____ BQV
Had, or, Have that (or, this) _____ BQW
Had, Has, or, Have the _____ BQX
Had, Has, or, Have there? _____ BQY
Had or, Have these (or, those) ____ BQZ
Had, Has, or, Have they (them, their s)? BRA
Had, or Have they been? _____ BFT
Had, or, Have they done? _____ BMU
Had, or, Have they had? _____ BRC
Had, Has, or, Have we (our s)? ___ BRD
Had, or, Have we been? _____ BFU
Had, or, Have we done? _____ BMV
Had, or, Have we had? _____ BRE
Had, Has, or, Have you-r-s? _____ BRF
Had, or, Have you been? _____ BFV
Had, or, Have you done? _____ BMW
Had, or, Have you had? _____ BRG
Have you (or person indicated) any—? EPR
Having _____ . _____ OPI
Having been _____ BFW
Having done _____ BMX
Having had _____ BRH
He, She, It (or person-s or thing s indicated) had (has, or, have) _____ BUJ
He, She, It (or person-s or thing-s indicated) had (has, or, have) been __ BUK
He, She, It (or person-s or thing-s indicated) had (has, or, have) done (or, is, or, are doing) _____ BUL
He, She, It (or person-s or thing-s indicated) had (has, or, have) had ___ BUM
He, She, It (or person-s or thing-s indicated) had (has, or, have) not _ BUN
He, She, It (or person-s or thing-s indicated) had (has, or, have) not been, BUO
He, She, It (or person-s or thing-s indicated) had (has, or, have) not doing (or, is, or, are not doing) _____ BUP
He, She, It (or person s or thing-s indicated) had (has, or, have) not had BUQ
He, She, It (or person s or thing s indicated) should (or, would) have BVO
He, She, It (or person-s or thing s indicated) (should or, would) not have BVS

HAD

HAD—Continued
How had (has, or, have)?..........BXN
I can (or, may) have..................BRI
I can not (or may not) have ....BRJ
I could (or, might) have ....BRK
I could (or, might) not have.......BRL
I had (or, have) .... BRM
I had (or, have) been.... ....BGC
I had (or, have) done..............BNA
I had (or, have) had .... RRN
I had (or, have) not ....BRO
I had (or, have) not been...........BGD
I had (or, have) not done..........BNC
I had (or, have) not had...........PRP
I shall (or, will) have ....BRQ
I shall (or, will) not have....... ..BES
I should (or, would) have..........BRT
I should (or, would) have been....BGK
I should (or, would) not have. ...BRU
I should (or, would) not have been BGM
If he (she, it, or person s or thing s in-
   dicated) had (has, or, have) .....BYJ
If he (she, it, or person s or thing s in
   dicated) had (has, or, have) not .BYK
If I had (or, have) ....BZH
If I had (or, have) not ....BZI
If they had (or, have) ....CAO
If they had (or, have) not ....CAP
If we had (or, have) ....CBH
If we had (or, have) not ....CBI
If you had (or, have) ....CBY
If you had (or, have) not ....CBZ
Is, etc   (See Are, above)
It, etc   (See He, above)
It has (or, had) been ....BGP
It has (or, had) been done ....BNP
It has (or, had) not been ....BGQ
It has (or, had) not been done.....BNQ
Let him (her, it, or person-s indicated)—
   have ....BWA
Let them have ....BRV
May, etc   (See Can, above)
Might, etc   (See Could, above)
Must have ....CEO
Must have been ....CEP
Must have had ....CEQ
Not to be had ....COJ
Ought to have ....CIA
Ought not to have ....CHT
Shall, or, Will have ... .BRW
Shall, or, Will have done (or, been
   done)...............................BNX
Shall, or, Will he (she, it, or person s
   or thing-s indicated) have?........BWG
Shall, or, Will he (she, it, or person s
   or thing s indicated) not have?...BWK
Shall, or, Will I have? ... .BRX
Shall, or, Will I not have? ....BRY
Shall, or, Will not have?...........BRZ
Shall, or, Will not have done (or, been
   done) ....BNZ
She, etc   (See He, above)
Should, or, Would have ....BSA
Should, or, Would have been ......BHL
Should, or, Would have done (or, been
   done) ....BOG
Should, or, Would have had ....BSC
Should, or, Would he (she, it or per-
   son-s or thing-s indicated) have?.BWP
Should, or, Would he (she, it, or person-s
   or thing-s indicated) have been?.BWQ

HAD—Continued
Should, or, Would he (she, it, or per-
   son-s or thing s indicated) not have?
                            BWU
Should, or, Would he, (she, it, or per-
   son-s or thing s indicated) not have
   been? ...... - .... BWV
Should, or, Would not have ....BSD
Should, or, Would not have been ..BHN
Should, or, Would not have done (or,
   been done) ....BOI
Should, or, Would not have had ...BSE
That, or, This had (has, or, have) .CKX
That, or, This had (has, or, have) been,
                            CKY
That, or, This had (has, or, have) not
   been ....CKZ
There had (has, or, have) ....CMQ
There had (has, or, have) been ....CMR
There had (has, or, have) not been.CMS
There should (or, would) have been.CNA
There should (or, would) not have been,
                          CNE
They can (or, may) have ... ...BSF
They can not (or, may not) have...BSG
They had (or, have) ....BSH
They had (or, have been) ....BHT
They had (or, have) done .... BOL
They had (or, have) had ....BSI
They had (or, have) not ....BSJ
They had (or, have) not been ......BHU
They had (or, have) not done .....BOM
They had (or, have) not had.......BSK
They should (or, would) have .....BSL
They should (or, would) not have .BSM
This, etc   (See That, above)
To be had ....CNS
To have ....CNU
We can (or, may) have ....BSN
We can not (or, may not) have ....BSO
We had (or, have) ....BSP
We had (or, have) been............BIN
We have (or, have) done ....BOS
We had (or, have) had ....BSQ
We had (or, have) not.............BSR
We had (or, have) not been....... .BIO
We had (or, have) not done ....BOT
We had (or, have) not had.........BST
We shall (or, will) have ....BSU
We shall (or, will) not have ......BSV
We should (or, would) have ......BSW
We should (or, would) not have ...BSX
What, or, Which had (has, or, have)?
                            CQF
What, or, Which shall (or, will) I (or,
   we) have? ... .CQM
What, or, Which shall (or, will) they
   (or, you) have?...................CQO
When had (has, or, have)? ....CRH
When had (or, have) you? ....CRI
When shall (or, will) I have? .....CRW
When shall (or, will) they (or, you)
   have?. ....CRZ
When shall (or, will) we have?....CSA
Where had (has, or, have)? ....CTB
Where shall (or, will) I have?.....CTO
Where shall (or, will) they (or, you)
   have? ....CTQ
Where shall (or, will) we have?....CTR
Which, etc   (See What, above)
Who had (has, or, have)?..........CUP

HAD—HAPPEN

	HAD—*Continued*
	Why had (has, *or*, have)? ----------CVK
	Why had (has, *or*, have) not? ------CVL
	Why had (*or*, have) you? ----------CVM
	Why had (*or*, have) you not? ------CVN
	Will it have? ------ - ---- ----CWD
	Would, etc (*See* Should, *above*)
	You can (*or*, may) have ----------BSY
	You can not (*or*, may not) have----BSZ
	You had (*or*, have) ----- ----------BTA
	You had (*or*, have) been------ ----BJD
	You had (*or*, have) done ----------BOY
	You had (*or*, have) had------------BTC
	You had (*or*, have) not------------BTD
	You had (*or*, have) not been------BJE
	You had (*or*, have) not done------BOZ
	You had (*or*, have) not had--- ----BTE
	You shall (*or*, will) have --------BTF
	You shall (*or*, will) not have -----BTG
	You should (*or*, would) have ----- BTH
	You should (*or*, would) not have---BTI
OLC	HÆMORRHAGE
OLD	HAIL-ED-ING
	Come within hail----------------JDA
	Hail the strange vessel, and ask if —,
	FDO
OLE	—I shall pass within hail
OLF	—Keep within hail
OLG	—Pass within hail
	Send a boat within hail -----------GUI
OLH	HAILSTORM HAILSTONE
OLI	HAIR-Y
	Hair brushes-------------------HKO
	Hair's breadth -------------------HER
	HAITI-AN HAITIAN COLORS--------OQB
OLJ	HALF-VE
	Go half-speed -------------------OBT
	Going half-speed -------------- -----OCJ
	Half a dozen ----------- ----------LMT
	Half allowance—of ----- ----------DZH
OLK	—Half an hour In half an hour
	Half-mast your ensign -----------MEL
OLM	—Half speed, *or*, Half power
OLN	—Half tide
	Half-year ly -------------------ZKV
	In half an hour ----------------- OLK
OLP	HALL
	Town hall ----------------------XZE
OLQ	HALLIARDS
	HALVE----------------------------OLJ
.	HALYARDS, *or*, HAULYARDS ----------OLQ
OLR	HAM
	HAMBURG
	Report me to the Borsenhalle at Hamburg ------ ------------ ---------UL
OLS	HAMMER
OLT	HAMMOCK
OLU	—Hammock netting

OLV	HAMPER
OLW	HAND-ED-ING-Y TO HAND
	Beforehand------------------------GDH
	Come to hand ------- ------------JCZ
OLX	—Hand (*part of the body*)
	Hand bag-------- ----------------FSI
OLY	—Hand lead
OLZ	—Hand pump
	Hand pump out of order ----------UBD
OMA	—Hand steering gear Handwheel
	Keep the beacon (*or*, buoy) on your port hand -----------------------GAK
	Keep the beacon (*or*, buoy) on your starboard hand -------------------GAL
	Leave the matter in the hands of Lloyd's agent ------------- -------------QWR
	You had better put your affairs into the hands of --------------------DPK
OMB	HANDS (*Crew*)
	All hands-------------- ------DXL
	All hands at the pumps ------ ----DXM
	All hands lost All on board perished, DXQ
	All hands saved All on board saved, DXR
OMC	—Can you spare me a hand or two? (*Give particulars, if necessary*)
	Hands enough---------------------MDI
	Have you hands enough?----------MDK
OMD	—He has not many hands on board
	Short handed A weak crew (*number to be indicated, if necessary*)-----KDJ
OME	—More hands are required at —
	Want hands (*Give particulars, if necessary*) -------------------- ----YJ
	HANDCUFF-ED-ING -------------------KFO
OMF	HANDLE-D-ING
OMG	HANDSPIKE
OMH	HANG ED-ING HUNG
OMI	HANK
OMJ	HAPPEN-ED-ING—TO (*or*, ON)
	Accident happened, crushed, fall, jammed------------------------ DCF
	Accident happened, cut, stab ---- DCG
	Accident happened, dangerously wounded (*or*, hurt)----------------AL
	Accident happened, drowned-ing--DCH
	Accident happened, gunshot wound DCI
	As it happens --------------------FAN
	Happened in consequence of Occasioned by-- ----------- ---- -----JPZ
	Has any accident happened? Is anything the matter?-----------------BK
OMK	—How did it (*or*, that) happen (*or*, occur)?
	Seldom happens--------------------VOM
OML	—Should it (*or*, that) happen— to (*or*, on),
OMN	—What happened (*or*, resulted)?
OMP	—What has happened?
OMQ	—When did it (*or*, that) happen?
	When did the accident happen?-----BU

HAPPY—HAUL

OMR	HAPPY-ILY-INESS (*See also* GLAD)
	  Happy to hear it (*or*, that) ... ----OAR
	  Happy to see you (*or*, —) ----------OAT
	  I am happy to inform you _ -------PKC
		OND	HARBOR—*Continued*.
			—What harbor do you intend to make
			  (*or*, run) for?

OMS	—Shall, *or*, Will be very happy — to	ONE	—Whilst in harbor You will not have water enough over
OMT	—Very happy-ily		the bar (*or* into the harbor) *Depth* *in feet may follow*---------- ------FWE
OMU	HARASS-ED-ING.	ONF	HARD-NESS Blowing hard Squally weather----ZH

OMV	HARBOR.
	  Are you going into port (harbor)?--OBK
	  Bar harbor . -----------------------FVS
	  Detained in harbor-----------------KYC
	  Do not attempt to make the harbor (or
	  anchorage) -----------------------FW
	  Do not know the anchorage (*or*, harbor),
	                           EKU
		ONG	Blowing too hard—to — ----- -----GOC
			  Hard-a-port! (*Urgent*) Head to go to
			    starboard --------------------------MD
			  Hard-a-starboard! (*Urgent*) Head to
			    go to port-----------------------ME
			  Hard bottom -----------------------HBC
			—Harder-est
			  Helm hard up-- ------------------OTN
			  It blew too hard ------------------GOI
			  My helm is hard-a-port, ship's head is

OMW	—Do you intend to go into harbor?
	  Do you know of any harbor (*or* anchor-
	  age)? --------- ---------------EKV
	  Do you know the harbor (*or*, anchorage)?
	                       EKW
	  During the time in harbor. -------LPV
	  Going into harbor (*or*, port) -------OCB
	  Harbor authorities-----------------FNI
	  Harbor board ----------------- GPH
	  Harbor, *or*, Entrance buoy --------MFR
			going to starboard ----------------WG
			  My helm is hard-a-starboard, ship's
			    head is going to port------------WH

		ONH	HARDLY

OMX	—Harbor dues
	  Harbor, *or*, Anchorage is good enough
	  with winds from —----- -----------OX
	  Harbor, *or*, Anchorage is indifferent-OY
	  Harbor, *or*, Anchorage is safe with all
	  winds --- ------------- ------- -- OZ
	  Harbor master Captain of the port,
	                        HWS
			HARDWARE------------------------ZNB

		ONI ONJ	HARM-ED-ING (*See also* DAMAGE) —No harm (*or*, damage) done
		ONK	HARMLESS
		ONL	HARPOON-ED-ING
OMY OMZ	—Harbor master's office —Harbor of refuge	ONM	HARVEST-ED-ING
	Harbor police----------------------YVX		HAS (*See* HAD, HAS, *or*, HAVE) ----BQN

ONA	—Harbor Trust
	  How does the harbor (*or*, anchorage)
	    bear? .. -------------------- -----PA
	  I do not know the anchorage (*or*, har-
	    bor) ------------------------------EKY
	  I know of no safe harbor ---- -------VDA
		ONP
		ONQ
		ONR
			HASTE Y ILY-INESS
			—Hasten-ed-ing
			—Hasten the departure of —
			  I can not hasten my departure-----KTV
			  I must hasten my departure------ KTW

ONB	—In, *or*, Into harbor
	  Inform harbor authorities ---------FNJ
		ONS	—Make haste
			  Make haste, or he will be gone-----OCU

	Inner harbor ----------------------PLM Inside the harbor -----------------PMK	ONT	HAT
	Is the harbor at — free from ice?---PBZ Just out of harbor (*or*, port) ------PXM	ONU	HATCH

	No getting out of harbor -----------NZD
	  Out of harbor ---------------------SVZ
		ONV	HATCHET
			  Can you lend hatchets (*or*, adzes)?-DOW

	Outer harbor ------ --------------SWF Outside the harbor----------------SWV	ONW	HATCHWAY

	Permission urgently requested to enter
	    harbor . ------ ----------- ------PD
	  Proceed into harbor---------------TXA
		ONX	HAUL-ED-ING—OUT
			  Closehauled ----------------- -----IUW
			—Haul down

	Proceed out of harbor ------------TXC
	  The entrance of the gulf (*or*, harbor) is
	    mined -----------------------------MFU
		ONY	Haul down your ensign -----------MEN
			  Haul down your flags --------------NDC

	The harbor at — is free from ice ---PCE Want plan of harbor (channel, etc).-YH	ONZ	—Haul in
ONC	—Want to come into harbor	OPA	—Haul in the slack
	Was in harbor (*or*, port) when I left._QKU	OPB	—Hauling line
	We are to leeward of the harbor (*or*, anchorage) ---------------------ELH	OPC	—Haul off
	We are to windward of the harbor (*or*, anchorage) ---------------------ELI	OPD OPE	—Haul taut Tauten hawsers. —Haul the jib down

	We have missed the entrance (*or*,
	    passed the port, *or*, harbor)------MFV
		OPF
		OPG
			—Haul up
			—Haul your wind Haul to the wind
			  Haul your wind on port tack -------LN
			  Haul your wind on starboard tack -LO

	HAUL—HE, SHE, IT	

HAUL—Continued

OPH　—I shall keep closehauled
　　　Try to clubhaul........... .... IWG

　　Have　(See HAD) .... BQN

　　Haven　(Harbor) .... OMV

OPI　HAVING
　　　Having been................ BFW
　　　Having done................ BMX
　　　Having had................. BRH

OPJ　HAVOC

OPK　HAWSE
OPL　Athwart hawse.............. FJG
　　　—Clear hawse
　　　Foul hawse .... NOL
OPM　—Hawse hole (or, pipe)
OPN　—Hawse pipe is broken
OPQ　—Hawse plug

OPR　HAWSER　(See also TOW and WARP)
　　　For signals between ships towing and
　　　being towed, see page 35
　　　Are you (your towing hawsers) fast?
　　　　　　　　　　　　　　　XYJ
OPS　—Can you spare a hawser?
　　　Cast off hawsers .... XYN
　　　Grass, or, Coir hawser .... IZS
OPT　—Have no other hawser
OPU　—Hawser fouled screw
OPV　—Hawser-laid rope
　　　Hemp cable, or, hawser　(Size and
　　　length, if necessary)........... HQX
OPW　—Large hawser
　　　Prepare another hawser......... TSW
　　　Send boat with hawser.... ...... GUR
OPX　—Shorten in both hawsers
OPY　—Small hawser
　　　Steel wire hawser............. WTO
　　　Tauten hawser　Haul taut...... OPD
　　　Towing hawser................ XYW
OPZ　—Veer both hawsers
　　　Wire cable (or, hawser) .... HSE

OQA　HAY

OQB　HAYTI-AN　(HAITI) HAYTIAN COLORS

OQC　HAZE Y INESS-ILY

　　HE ....BTZ

　　HE, SHE, IT (or PERSON S or THING-S indi-
　　cated)
　　　Are or, Is, etc　(See Is, below)
　　　As he (she it, his, her s, it-s) .... FAK
　　　Can or, May he (she, it, or person-s or
　　　thing s indicated)?.............BTJ
　　　Can, or, May he (she, it, or person-s or
　　　thing-s indicated) be?...........BTK
　　　Can, or May he (she, it or person s or
　　　thing-s indicated) have been?......BTL
　　　Can not, or, May not he (she it, or per-
　　　son s or thing s indicated) be?....BTM
　　　Can not, or, May not he (she, it, or per-
　　　son s or thing-s indicated) have been?
　　　　　　　　　　　　　　　BTN

HE, SHE, IT—Continued

　　Could, or, Might he (she, it, or person-s
　　or thing s indicated) be?..........BTO
　　Could, or, Might he (she, it, or person-s
　　or thing-s indicated) not be?..... BTP
　　Do, Does, or, Did he (she, it, or person-s
　　or thing s indicated)?.... ....BTQ
　　Do not let him (her, it, or person indi-
　　cated) have .. .. BTR
　　For him (his, her-s, it s, or person-s or
　　thing-s indicated)............ ....BPD
　　From him (his, her s, it-s, or person-s or
　　thing s indicated)..............BPO
　　Had, Has, or, Have, he (she, it, or per-
　　son s or thing s indicated).. .. ..BTS
　　Had, Has, or, Have, he (she, it, or per-
　　son s or thing-s indicated) been?..BTU
　　Had, Has, or, Have, he (she, it, or per-
　　son s or thing s indicated) had'..BTV
　　Had, Has, or, Have, he (she, it, or per-
　　son-s or thing-s indicated) not?...BTW
　　Had, Has or Have, he (she, it, or per-
　　son s or thing s indicated) not been?
　　　　　　　　　　　　　　　BTX
　　Had, Has, or, Have, he (she, it, or per-
　　son s or thing-s indicated) not had?
　　　　　　　　　　　　　　　BTY
　　He .. ..BTZ
　　He, She, It (or person-s or thing s indi-
　　cated) can (or, may)............BUA
　　He, She It (or person-s or thing s indi-
　　cated) can (or, may) be..........BLC
　　He, She, It (or person-s or thing-s indi-
　　cated) can not (or, may not) . ...BUD
　　He, She, It (or person-s or thing-s indi-
　　cated) can not (or, may not) be....BUE
　　He, She, It (or person-s or thing s indi-
　　cated) could (or, might) .... BUF
　　He, She, It (or person-s or thing-s indi-
　　cated) could (or, might) not ... BUG
　　He, She, It (or person-s or thing s indi-
　　cated) does (do, or, did) ....BUH
　　He, She, It (or person s or thing s indi-
　　cated) does (do, or, did) not ..... BUI
　　He, She, It (or person-s or thing s indi-
　　cated) had (has, or, have) ....BUJ
　　He, She, It (or person-s or thing-s indi-
　　cated) had (has or, have) been...BUK
　　He, She, It (or person-s or thing s indi-
　　cated) had (has, or, have) done (or, is,
　　or, are doing) ....BUL
　　He, She, It (or person-s or thing-s indi-
　　cated) had (has, or, have) had ...BUM
　　He, She, It (or person-s or thing s indi-
　　cated) had (has, or, have) not ....BUN
　　He, She, It (or person-s or thing-s indi-
　　cated) had (has, or, have) not been,
　　　　　　　　　　　　　　　BUO
　　He, She, It (or person-s or thing-s indi-
　　cated) had (has, or, have) not done
　　(or, is, or, are not doing) ....BUP
　　He, She, It (or person-s or thing s indi-
　　cated) had (has, or, have) not had.BUQ
　　He, She, It (or person-s or thing-s indi-
　　cated) is (or, are)...............BUT
　　He, She, It (or person s or thing s indi-
　　cated) is (or, are) able to........BUV
　　He, She, It (or person-s or thing s indi-
　　cated) is (or, are) for ....BUR
　　He, She, It (or person-s or thing s indi-
　　cated) is (or, are) from... ......BUS

HE, SHE, IT

HE, SHE, IT—Continued

He, She, It (or person-s or thing-s indicated) is (or, are) not........ - BUW
He, She, It (or person-s or thing-s indicated) is (or are) not able to. _ BUX
He, She, It (or person-s or thing-s indicated) may, etc (See CAN, above)
He, She, It (or person-s or thing-s indicated) might, etc (See COULD, above)
He, She, It (or person-s or thing-s indicated) must---------------------BUY
He, She, It (or person-s or thing-s indicated) must be --------------------BUZ
He, She, It (or person-s or thing-s indicated) must not -----------------BVA
He, She, It (or person-s or thing-s indicated) must not be ---------------BVC
He, She, It (or person-s or thing-s indicated) ought to------------------ BVD
He, She, It (or person-s or thing-s indicated) ought not to----- --------BVE
He, She, It (or person-s or thing-s indicated) ought not to be-----------BVG
He, She, It (or person-s or thing-s indicated) ought to be-- -- ---------BVF
He, She, It (or person-s or thing-s indicated) shall (or, will) -----------BVH
He, She, It (or person-s or thing-s indicated) shall (or, will) be---------.BVI
He, She, It (or person-s or thing-s indicated) shall (or, will) not --------BVJ
He, She, It (or person-s or thing-s indicated) shall (or, will) not be ---- .BVK
He, She, It (or person-s or thing-s indicated) should (or, would) ---- ---BVL
He, She, It (or person-s or thing-s indicated) should (or, would) be -----BVM
He, She, It (or person-s or thing-s indicated) should (or, would) do (or, be done) ------------------ ----BVN
He, She, It (or person-s or thing-s indicated) should (or, would) have---BVO
He, She, It (or person-s or thing-s indicated) should (or, would) not ----BVP
He, She, It (or person-s or thing-s indicated) should (or, would) not be--BVQ
He, She, It (or person-s or thing-s indicated) should (or, would) not do (or, be done) ------------ ----------BVR
He, She, It (or person-s or thing-s indicated) should (or, would) not have-BVS
He, She, It (or person-s or thing-s indicated) was (or, were). _ -------BVT
He, She, It (or person-s or thing-s indicated) was (or, were) not----- _BVU
He, She, It (or person-s or thing-s indicated) was (or, were) not to be --BVX
He, She, It (or person-s or thing-s indicated) was (or, were) to be----- BVW
He, She, It (or person-s or thing-s indicated) will, etc (See Shall, above)
He, She, It (or person-s or thing-s indicated) would, etc (See Should, above)
How is (or, are) he (she, it, or person-s or thing-s indicated)?------------BXO
How shall (or, will) he (she, it, or person-s or thing-s indicated)? ------BXR
If he (she, it, or person-s or thing-s indicated) can (or, may) ------------BYD
If he (she, it, or person-s or thing-s indicated) can not (or, may not)-----BYE

HE, SHE, IT—Continued

If he (she, it, or person-s or thing-s indicated) could (or, might)---------BYF
If he (she, it, or person-s or thing-s indicated) could (or, might) not-------BYG
If he (she, it, or person-s or thing-s indicated) does (do, or, did)------ --BYH
If he (she, it, or person-s or thing-s indicated) does (do, or, did) not ---- BYI
If he (she, it, or person-s or thing-s indicated) had (has, or, have) ----- BYJ
If he (she, it, or person-s or thing-s indicated) had (has, or, have) not_ .BYK
If he (she, it, or person-s or thing-s indicated) is (or, are)------ ---- ---BYL
If he (she, it, or person-s or thing-s indicated) is (or, are) able to ------BYM
If he (she, it, or person-s or thing-s indicated) is (or, are) not----------BYN
If he (she, it, or person-s or thing-s indicated) is (or, are) not able to----BYO
If he (she, it, or person-s or thing-s indicated) shall (or, will) ---------BYP
If he (she, it, or person-s or thing-s indicated) shall (or, will) not -------BYQ
If he (she, it, or person-s or thing-s indicated) should (or, would) ------ BYR
If he (she, it or person-s or thing-s indicated) should (or, would) not----BYS
If he (she, it, or person-s or thing-s indicated) was (or, were)----------BYT
If he (she, it, or person-s or thing-s indicated) was (or, were) not -------BYU
If he (she, it, or person-s or thing-s indicated) will, etc (See Shall, above)
If he (she, it, or person-s or thing-s indicated) would, etc (See Should, above)
Is, or, Are, he (she, it, or person-s or thing-s indicated)? ---- ---------BVY
It ------------ --.--------- ----BVZ
Let him (her, it, or person-s or thing-s indicated) have----------------BWA
May he (she, it, etc)? (See Can, above)
Might he (she, it, etc)? (See Could, above)
Must he (she, it, or person-s or thing-s indicated)?------------------ ---CER
Ought he (she, it, or person-s or thing-s indicated) to?------ ----- CHN
Person-s, or, Thing-s indicated . _BWC
Shall, or, Will he (she, it, or person-s or thing-s indicated)? _-----------BWD
Shall, or, Will he (she, it, or person-s or thing-s indicated) be?------ --BWE
Shall, or, Will he (she, it, or person-s or thing-s indicated) do (or, be done)? BWF
Shall, or, Will he (she, it, or person-s or thing-s indicated) have?-----BWG
Shall, or, Will he (she, it, or person-s or thing-s indicated) not?------ BWH
Shall, or, Will he (she, it, or person-s or thing-s indicated) not be---- .BWI
Shall, or, Will he (she, it, or person-s or thing-s indicated) not do (or, be done)? ? ------------------------BWJ
Shall, or, Will he (she, it, or person-s or thing-s indicated) not have? DWK
She ------------------------BWL
Should, or, Would he (she, it, or person-s or thing-s indicated)? -----BWM

HE, SHE, IT—HEAP

HE, SHE, IT—Continued.

Should, or, Would he (she, it, or person-s or thing-s indicated) be?___BWN OQI

Should, or, Would he (she, it, or person s or thing-s indicated) do (or, be done)? ___ ___ _____ _____BWO OQJ

Should, or, Would he (she, it, or person-s or thing-s indicated) have?_BWP OQK

Should, or, Would he (she, it, or person-s or thing-s indicated) have been? OQL

BWQ OQM

Should, or, Would he (she it, or per son-s or thing-s indicated) not?__BWR OQN

Should, or, Would he (she, it, or person-s or thing s indicated) not be? OQP

BWS OQR

Should, or, Would he (she, it, or person s or thing s indicated) not do (or, be done)?_____ __ BWT OQS

Should, or, Would he (she, it, or person-s or thing-s indicated) not have? OQT

BWU

Should, or, Would he (she, it, or person-s or thing s indicated) not have been?_____ _____BWV

That he (she, it, or person-s or thing-s indicated) _____CLA

Was, or, Were, he (she, it, or person s or thing s indicated)?_____BWX

Was, or, Were, he (she, it, or person s or thing-s indicated) not?_____ BWY

What, or, Which does (do, or, did) he (she, it, or person s or thing-s indicated)?_____CQD

What, or, Which is (or, are) he (she, it, or person s or thing-s indicated)_CPX

When does (did, or, do) he (she, it, or person-s or thing s indicated)?_____CRD

When he (she, it, or person-s or thing s indicated) is (or, are)_____CRJ

When is (or, are) he (she, it, or person-s or thing s indicated)?_____ CRN

When shall (or, will) he (she, it, or person-s or thing-s indicated)? _____ CRT

Where he (she, it, or person s or thing s indicated) is (or, are)_____ ____CTD

Where is (or, are) he (she, it, or person-s or thing-s indicated)?_____CTH

Where shall (or, will) he (she, it, or person-s or thing-s indicated)? _____CTM

OQD —While he (she, or, it)

Will he (she, it, etc)? (See Shall, above)

Would he (she, it, etc) (See Should, above)

OQE **HEAD-ED-ING. TO HEAD**

Brace round your head yards_____HDO

By the head _____HPK

Cathead _____ _____1EU

Crank head ___ _____KBL

Crosshead _____ __ ____KED

Figurehead_____ ___ _____MXY

Fork head _____ ___ _____NLR

Get her head round as fast as possible, NYU

Get her head round or you will be on shore _____LF

Have lost — head of cattle _____IFF

OQF —Head (part of body)

OQG —Head gear

OQH —Head of department

HEAD—Continued

—Headquarters

Head rail _____ __ _____UGW

OQJ —Headsails

OQK —Head sea

OQL —Head sheet

OQM —Head to wind

OQN —Head wind

OQP —Head yards

OQR —Headland

Headman_____OQU

OQS —Headmost

OQT —Headway Steerageway.

Heave-to, head offshore _____LQ

I have headway . ___ Code Flag over J

Jetty head_____ _____PVK

Keep head to wind _____PYT

Masthead At masthead_____RMF

Moor head and stern_____ _____RZJ

Pier head_____TID

Piston-rod crossheads damaged (or, broken) _____KEJ

Pump crosshead _____ _ _____UBG

Pump crosshead out of order _____UBH

Rudderhead_____VBM

Ship's head by compass is — _____JIE

Trimmed by the head , — inches by the head _____HPK

Visible from the masthead _____YQZ

When I have run my distance I shall put her head off until —_____LGP

HEADLAND _____ _____ ____OQR

OQU **HEADMAN**

HEADMOST_____OQS

HEADWAY STEERAGEWAY_____OQT

I have headway _____Code Flag over J

OQV **HEAL-ED-ING TO HEAL**

OQW **HEALTH-Y-INESS**

OQX —A healthy season

OQY —Are you healthy?

Bill of health _____GKC

Board of Health _____GPA

Clean bill of health—from — _____TQ

Crew healthy _____KCR

Have you a clean bill of health?____TU

I have a clean bill of health, but am liable to quarantine_Code Flag over Q

I have not a clean bill of health, Code Flag over I

In bad health? _____ _____FRP

Not clean bill of health flag (when spoken of) _ ____ _____ISW

OQZ —Tolerably healthy Tolerably well Troops are in good health_____YBN

ORA —Very healthy

ORB —Was the port you left healthy?

ORC —What is the state of health of the fleet at — ?

What is the state of health of the troops at — ? __ _____ _____YBR

ORD —Where is the health officer?

ORE **HEAP ED-ING**

HEAR—HELM

ORF	**HEAR-ING HEARD**
	Anxious to hear of (*or*, from) —....EOI
	Crew not heard of........ . .. .KCT
ORG	—Did you hear?
	Did you hear if there were any letters for me at — ? ....QOE
	Did you hear of my family?........EPI
	Glad to hear—it (*or*, that) Happy to hear—it (*or*, that) .... OAB
ORH	—Has been heard of
ORI	—Has not been heard of since (*date to follow*)
ORJ	—Have heard—of —
ORK	—Have, *or*, Has not heard of —
ORL	—Have you (*or person-s indicated*) heard —of —?
	Have you heard of the arrival of —? EYJ
	Have you seen (*or*, heard of) a vessel wrecked (*or*, in distress)?... .... ZU
	Have you seen (*or*, heard) anything of my (*or*, the) boat?...... ....FB
	Heard firing to the —.............NAI
	I heard that some of your family were seriously ill..... ....MSI
ORM	—Neither seen nor heard
ORN	—Should you hear (*or*, see)
ORP	—Sorry to hear—it (*or*, that)
	Unless I hear................ ....YJQ
ORQ	—Vessel indicated has not been heard of
ORS	—Vessel indicated has not been heard of since — (*date indicated*), she is supposed to be —.
ORT	—We shall not hear again
	What were the latest dates at — when you last heard?........ ....KMN
ORU	—When, *or*, What did you last hear—of (*or*, from)?
ORV	**HEART**
	Down hearted ....LMI
ORW	**HEAT-ED-ING**
	Superheat-ed-ing.... ....XEH
	The heat of the sun ....XDV
ORX	**HEAVE-ING HOVE**
	Avast heaving ....FOM
ORY	—Can I heave-to?
	Heave all aback Go astern ... .LP
ORZ	—Heave away.
OSA	—Heave her down
	Heave short...................HRD
OSB	—Heave taut
OSC	—Heave to
OSD	—Heaving, Hove-to
	Heave-to, head offshore ... .LQ
	Heave-to, I will send a boat......GTE
OSE	—Heave-to on port tack
OSF	—Heave-to on starboard tack
	Heave-to or I will fire into you ......ID
OSG	—Heave to or take the consequences
	Heave to until — ....LR
OSH	—Heave up
OSI	—I can not heave-to
OSJ	—I have not room to heave to
OSK	—I shall heave-to off — (*or*, at —)
OSL	—I shall heave-to to take soundings (*time indicated, if necessary*)
OSM	—I shall not heave to
OSN	—I will heave-to directly I have room

	HEAVE—*Continued*
OSP	—Shall you heave to?
	Stop Heave-to, *or*, Come nearer, I have something important to communicate . . ..Code Flag over H
	Strange vessel hove in sight........OZS
OSQ	—Vessel indicated has hove to
OSR	**HEAVEN**
OST	**HEAVY**
	Have, *or*, Has had a heavy gale from —, NVY
OSU	—Heavier est
OSV	—Heavily-iness
	Heavy, *or*, Severe gale.............NVZ
OSW	—Heavy gun (*weight in tons, if necessary*)
OSX	—Heavy loss
OSY	—Heavy purchase
	Heavy sea....................UQ
OSZ	—Heavy surf Much surf
OTA	—Heavy swell
	Heavy weather coming, look sharp..FZ
	Heavy weigh!................ZBE
OTB	—How heavy—is — ?
OTC	—I have a heavy gun (*or*, guns) (*number indicated*)
OTD	—I have no heavy guns
OTE	—In a heavy squall
	Rounds of ammunition for heavy gun, EGA
OTF	—She has, *or*, They have heavy gun
OTG	—She has, *or*, They have no heavy gun
	Striking heavily ....WZY
	Struck by a heavy sea.............XAC
	Very heavy sea .. ..UW
OTH	—Which is the heaviest?
	HECTARE (*Square measure*)........AXP
	HECTOGRAMME (*Measure of weight*).BAO
	HECTOLITRE (*Measure of capacity*)..AYZ
	HECTOMETRE (*Measure of length*)..AWD Square hectometre (*Square, or, surface measure*)....................AXP
OTI	**HEEL-ED-ING**
OTJ	**HEIGHT-EN-ED-ING** The height of the barometer is......FXL
OTK	—What is the height—of? What is the height of the barometer? FXN What is the height of the thermometer? What is the temperature?.......XNK
	HELD (*See* HOLD)........ ....OWN
OTL	**HELIOGRAPH-Y** IC
	HELLER (*Coin*)..............ATS
OTM	**HELM** Hard-a-port! (*Urgent*) Head to go to starboard ....MD Hard-a-starboard! (*Urgent*) Head to go to port....................ME Helm amidships ....EFJ

HELM—HIGH

HELM—*Continued*
Helm angle ---------------------------- ELV
OTN —Helm hard up
OTP —Helm is down
OTQ —Helm signal
OTR —Helmsman
OTS —How do you carry your helm?
  Lee helm ------------------------ QMC
  My helm is hard-a-port, ship's head going to starboard ---------- WG
  My helm is hard a starboard, ship's head going to port ---------- WH
  Pay attention to the helm ---------- FLI
  Port Head to go to starboard ----- WI
  Port a little Head to go to starboard, TPG
  Port helm ------------------------- TPE
  Starboard Head to go to port ----- WJ
  Starboard a little Head to go to port, WQU
  Starboard helm ------------------- WRB
  Steady your helm ----------------- MI
  Steer more to port --------------- MJ
  (*To be kept flying until course is sufficiently altered*)
  Steer more to starboard ---------- MK
  (*To be kept flying until course is sufficiently altered*)
OTU —Under steady helm
  Weather helm -------------------- YZO
  Will not answer her helm --------- ENS

OTV HELMET
  Diver's helmet ------------------- LBY

HELMSMAN ----------------------------- OTR

OTW HELP ED ING (*See also* ASSIST)
  Can be helped (assisted) --------- FGB
  Can help (assist) ---------------- FGA
  Can I help? ---------------------- FGC
  Can I procure any help (*or,* assistance) in the way of —? ---------- CN
  Can you help (*or,* assist)? ------ FGD
  Can you help — (*vessel indicated*)? .co
  Can not help (*or,* assist) ------- CP
  Do you want (require) any assistance (*or,* help)—of — (*or,* from —)? ---- CS
  Give every assistance—to (help — to), FGH
  Help, I am attacked, want assistance .NJ
  Help me to lay out an anchor ------ EIK
  Helplessly ----------------------- OTY
OTX —I can not help you
  I will assist you (*or vessel indicated*) in distress ----------------- FGI
  Require help (*or,* assistance)—of — (*or,* from) ------------------- CZ
  Should you require any help (*or,* assistance) --------------------- FGO
  The ice is so solid I can not break through, send help ------------- PCF
  Will help (*assist*) -------------- FGU
  With the help (*or,* assistance) of FGW
  Without help (*or,* assistance) --- FGX

OTY HELPLESS-LY

OTZ HEMP-EN
  Hemp cable (*or,* hawser) *Size and length, if necessary*) --------- HQX
  Want a hemp cable ---------------- HSC

HENBANE, TINCTURE OF ---------------- XUA
OUA HENCOOP

HER S-SELF (*See also* HIM, HIS, HER-S, IT S) ---- BXD
  Her, *or,* His Excellency --------- MKR
  Her, *or,* His Highness ----------- OVL
  Her, *or,* His Majesty ------------ RHO
  Her, *or,* His Majesty's dominions LKB
  Her Royal Highness the Princess of Wales ------------------------ OVK
  Let her be ----------------------- BGV
  On Her, *or,* His Majesty's service RHS
  The United States, *or,* Her, *or,* His Majesty's Government --------- OGB
  The United States, *or,* Her, *or,* His Majesty's service ----------- RHP
  The United States, *or,* Her, *or,* His Majesty's ship ------------- RHQ

OUB HERALD
  Report me to "New York Herald" Office, London ------------------ UI
  Report me to "New York Herald" Office, New York ---------------- UJ
  Report me to "New York Herald" Office, Paris -------------------- UK

HERD (*Flock*) ----------------------- NFI

OUC HERE HEREABOUTS
  Can I get a bill cashed here? ----- GKI
  Can not get a bill cashed here ---- GKJ
OUD —Herewith
  I do not think there is any light hereabouts ----------------------- QRW
OUE —Is, *or,* Are here
  Is there any dry dock here? ------- LJG
OUF —Not here
  There are no orders here for you -- TD
OUG —Was, *or,* Were here
  Will my cargo sell here (*or at place indicated*)? ------------------- IBP
OUH I own owner is here.

OUI HERRING
OUJ —Herring fishery

HERS-SELF (*See also* HIM, HER, IT) BXD

OUK HESITATE-D-ING-ION
OUL —Do not hesitate
  Great hesitation ----------------- OHC
OUM —Without hesitation Unhesitatingly

OUN HIDE-ING HID DEN To HIDE
OUP —In hiding

OUQ HIDE (*Skin*)

OUR HIGH (*See also* HEIGHT)
  Asks precise time of high water and minimum depth at that time ----- XE
  At high water you will have —feet KUY
  Breast high ---------------------- HFO
  Height --------------------------- OTJ
  High and dry --------------------- LOR
OUS —High land
OUT —High pressure
  High-pressure boiler ------------- GXC

HIGH—HOIST

	HIGH—*Continued*
	High pressure cylinder ....KIG
OUV	—High pressure cylinder cover
	High-pressure cylinder cover broken,
	KAI
	High-pressure cylinder cracked (*or*, out
	of order) ... . KAU
OUW	—High pressure piston
OUX	—High-pressure piston broken
OUY	—High-pressure valve out of order
	High tide .... XSD
	High water..................... YVE
	High water mark .... YVF
OUZ	—Higher
OVA	—Highest
OVB	—Highly.
	Hold on till high water............ KP
	How high must I he to fetch?..... MWB
OVC	—Not of high rank
OVD	—Of high rank
OVE	—Prices are high
OVF	—Proceed at high water
OVG	—The highest price is —
	Wages are very high............... YSI
	Wait for high water.............. YSX
	Was tide high when you grounded?..CI
	What is the depth at high water?..KVE
	What is the highest price you will give?
	OAJ
OVH	—When is the highest tide?
OVI	—When will it be high water?
OVJ	**HIGHNESS**
OVK	—Her Royal Highness the Princess of
	Wales
OVL	—His, *or*, Her Highness
OVM	—His Royal Highness the Prince of Wales
OVN	**HILL**
	HIM, HIS, HIMSELF .... BXC
	HIM, HIS, HER-S, IT-S
	After him (his, her-s, it-s) .... BWZ
	Are, *or*, Is his (her-s, it-s)? .... BEK
	As his (her-s, it-s)................. FAK
	Before him (his, her-s, it-s)........ BXA
	By him (his, her-s, it-s)......... . BJR
	Can, *or*, May his (her-s, *it* s)...... BKF
	Could, *or*, Might his (her-s, it-s)?..BLA
	Do, Does, *or*, Did his (her-s, it-s)?..BMG
	Do not let him (her, it, *or person indi-*
	cated) have . ... BIR
	For him (his, her s, it s, *or person s or*
	thing-s indicated) ....BFD
	From him (his, her-s, it s, *or person s or*
	thing-s indicated) .... BFO
	Give him (his, her-s, it-s) .. .. NZW
	Had, Has, *or* Have his (her s, it-s)?
	BQR
	Her-s self .... - .... BXD
	Him self, His .... BXC
	If his (her-s, it-s) . . .. BYV
	In him (his, her-s, it s) ... . CDJ
	Is, etc (*See* Are, *above*)
	It-s self .... CDY
	Let her be .... BGV
	Let him (his, her s, it-s)... ....QNT
	Let him be .... BXE
	Let him (her, it, *or person-s or thing-s*
	indicated)—have .... BWA

	HIM—*Continued*
	May, etc (*See* Can, *above*)
	Might, etc (*See* Could, *above*)
	Must his (her-s, it s) - ....CES
	Not his (her-s, it s).............. CFT
	Of him (his, her-s, it-s).... . .....CGN
	On him (his, her-s, it-s)........... CGY
	Ought his (her-s, it-s)? .... CHO
	Shall, *or*, Will his (her-s, it-s)......CIS
	Should, *or*, Would his (her-s, it-s)..CJP
	That his (her-s, it-s) .. .. CLB
	To him (his, her-s, it-s) ....CNV
	Was, *or*, Were his (her-s, it-s).. .. COR
	When his (her-s, it s.)... .... CRK
	Where his (her-s it-s) ....CTE
	Will, etc (*See* Should *above*)
	With him (his, her-s, it-s) ....CWG
	Would, etc (*See* Should, *above*)
	HINDER-ED-ING-ANCE (*Impede*).....PDY
OVP	**HINGE**
OVQ	**HINT-ED-ING**
OVR	**HIRE-D-ING**
	Government hired vessel ....OFU
	Hired cruiser..................... KFB
	Hired steamer..................... WTH
	Will not be accountable for boat hire,
	DFW
	HIS, HIM-SELF (*See also* HIM, HIS, HER,
	IT-S) .... BXC
	His, *or*, Her Excellency . ... MKR
	His, *or*, Her Majesty .... RHO
	His, *or*, Her Majesty's dominions . LKH
	His, *or*, Her Highness . ... OVM
	His Royal Highness the Prince of Wales,
	OVM
	On His (*or*, Her) Majesty's Service .RHS
	The United States, *or*, His, *or*, Her
	Majesty's Government .... OGB
	The United States, *or*, His, *or*, Her
	Majesty's service ....RHP
	The United States, *or*, His, *or*, Her
	Majesty's ship .. ..RHQ
OVS	**HISTORY-ICAL**
OVT	**HIT-TING**
OVU	**HITCH**
OVW	**HITHERTO**
OVX	**HOARSE-LY-NESS**
OVY	**HOG GED-GING** TO HOG
	HOG (*Pig*) - ....TIF
	HOGSHEAD (*Measure of capacity*).....AZB
OVZ	—Hogsheads of sugar
OWA	**HOIST-ED-ING**
	Annul the hoist indicated by Numeral
	Signal...... ....EMI
	Annul the last hoist, I will repeat it..VE
	Can you distinguish what — (*vessel*
	indicated) has hoisted?... .....LGW
	First flag in hoist is indistinct.......NDI
	He has hoisted — colors..........JBT

HOIST—HORIZON

	HOIST—*Continued*	OXC	HOLD (*Of ship*)
OWB	—Hoist a jack		After hold ....DQW
OWC	—Hoist a light		Fire in hold amongst cargo ....OR
OWD	—Hoist a waft (*or, wheft*)		Fore hold ....NJU
OWE	—Hoist a waft (*or, wheft*) when ready	OXD	—Main hold
OWF	—Hoist in	OXE	—Stoke hold
OWG	—Hoist masthead flags		Stoke-hold compartment ....JHA
	Hoist quarantine flag ....TV		Stoke hold full of water ....WVX
	Hoist the blue peter ....KO		— feet water in hold ....QM
OWH	—Hoist the general recall.		
OWI	—Hoist the jib	OXF	HOLE
OWJ	—Hoist up		Eyelet hole ....MQG
	Hoist up your boats ....GTH		Hawse hole ....OPM
OWK	—Hoist wind sails		Loophole ....RAG
	Hoist your colors. Show your ensign_DW		Porthole...*......................TOW
	Hoist your distinguishing signal....DV		Sheave hole ....VUI
	(*See* CODE LIST OF SHIPS)		Stokehole ....OXE
	Hoist your flags where best seen....NDE		
OWL	—I can not make out the flags, hoist the	OXG	HOLIDAY
	signal in a better position	OXH	HOLLOW-ED-NESS
	Lowest flag of hoist indistinct .... NDF		
	North cone hoisted ....JNM		HOLME'S LIGHT (*large flare*)...........NDY
	Repeat signal, *or*, repeat signal from —		Holme's light (*small flare*) ....NDZ
	(*hoist indicated*) VI		
	Second flag in hoist is indistinct ...NDG	OXI	HOLYSTONE
	South cone hoisted ....JNO	OXJ	HOME
	The last hoist is not understood, repeat		Absent from home ....DAV
	it..............................URK		At home........................FIE
	They have, *or*, She has hoisted — colors,		Come home ....JCR
	JBT	OXK	—Home trade
	Third flag in hoist is indistinct ....NDH	OXL	—Homeward bound
	Top flag in hoist indistinct ....NDI		Sailor's Home..................VFD
	What colors (ensign) has she hoisted?		Sheet home ....VUW
	JBU		
	What colors shall we hoist?........JBV	OXM	HONDURAS HONDURAS COLORS
OWM	—Your flags seem to be incorrectly hoisted	OXN	HONEST-Y-LY
	(*or*, hoisted upside down)		HONEY-COMB ED ....JCA
OWN	HOLD-ING HELD TO HOLD		
OWP	—Better hold on	OXP	HONOR-ED-ING-ABLE.
	Do not hold on to your anchor, but let		Guard of honor....../............OIY
	her beat up on the beach ....EHY		Legion of honor..................QMV
	Do the utmost you can to insure holding		
	on, everything will be wanted ..MEX	OXQ	HOOK-ED-ING TO HOOK
OWQ	—Holder		Billhook .... GKD
OWR	—Hold on		Boat hook ....GBM
	Hold on till high water.............KP		Breasthook ....HFP
OWS	—Hold out		Fishhook ....NBQ
	Holding-down bolts loose..... ....GYA	OXR	—Hook (*a hook*)
OWT	—I fear I can not hold on much longer		I have hooked your anchor.........EIL
	I hold you accountable ....DFT		
	I shall keep hold of the land (*or*, light),	OXS	HOOP
	or, I shall keep the land (*or*, light)	OXT	—Hooping (Whooping) cough.
	bearing —........................QA		
	Inquest will be held—at —........PLS	OXU	HOPE-ING-FUL-LY-NESS—TO (*or*, THAT)
	Is it good holding ground?.........OIM		Great hopes—of — ....OHD
OWU	—Is, *or*, Are to be held ready—to (*or*,	OXV	—Hope you are
	for)	OXW	—Hope you can
	Shall we keep hold of the land?.....QER	OXY	—Hope you will
	Shareholder ....VTZ	OXZ	—Is there any hope—of ?
OWV	—To be held	OYA	—No hope.
	We hold our own with vessel chasing		
	us ....INW	OYB	HOPS
	Will be held accountable ... .DFU		
OWX	—Will hold.	OYC	HORIZON-TAL-LY
OWY	—Will hold together till —		Above the horizon On the horizon,
OWZ	—Will not hold		DAO
OXA	—You will have as much as you can do to		Artificial horizon.................EZO
	hold on		
OXB	—You will hold yourself in readiness—		
	to (*or*, for)		

HORN—HOW.

OYD	HORN-Y		
	Fog horn _____ _____ NGI		
OYE	HORSE		
	Horse boat _____ _____ GTI	OZK	
OYF	—Horse guards	OZL	
OYG	HORSE (*Traveler*)		
OYH	HORSEPOWER		
OYI	—Indicated horsepower		
OYJ	—My horsepower (*or that of vessel indicated*) is —.	OZM	
OYK	—Nominal horsepower		
OYL	—What is your horsepower (*or that of vessel indicated*) ?		
OYM	HOSE (*A flexible tube*)		
	Have you fire hose enough—for — ? NAH		
	Suction hose _____ XCI		
OYN	HOSPITABLE-Y-ALITY	OZN	
OYP	HOSPITAL		
	Ambulance required to convey a patient to hospital _____ EDY	OZP	
OYQ	—Can I send my sick to the hospital (*or, sick quarters*) ?		
	Case is for hospital _____ IDJ		
OYR	—Hospital case		
OYS	—Hospital ship (or, vessel)		
OYT	—Hospital treatment		
OYU	—Require hospital treatment		
OYV	—Send to the hospital		
OYW	—To, or, At the hospital.		
OYX	—Where is the hospital?		
OYZ	HOSTAGE		
OZA	HOSTILE-ILITY (*See also* WAR)	OZQ	
OZB	—Cessation of hostilities		
	Commencement of hostilities _ ____JEU		
OZC	—Hostilities commenced — between — and —.	OZR	
OZD	—In a hostile manner		
OZE	HOT-LY		
	Hot bearings _____ RJ		
OZF	—Hot weather		
OZG	—Hotter-est		
	Red hot _____ _____ UMF		
OZH	HOTEL		
OZI	HOUR-LY		
	A M * Before noon In forenoon__EDU		
	*NOTE —For hours A M and P M see TIME, page 58		
	After hours_____ DQX	OZS	
OZJ	—An hour In an hour		
	At — (*hour indicated*) — (*place indicated*) bore—, distant—miles_____ED		
	At — (*hour indicated*) we were abreast of — _____ CXW		
	At which hour? _ _ _____ FJC		
	Damage can be put to rights in — hours, BD		
	During meal hour_____ LPO		
	Every hour_____ MIW		

HOUR—*Continued*	
Fire has been burning these — hours, HNF	
Half an hour In half an hour____OLK	
—Hourglass	
—Hourly expected	
How many hours' coal have you ? (*state at what speed*)_____ IXC	
I have — hours' coal for full speed (*or for speed indicated*) _____ IXD	
—In a quarter of an hour Quarter of an hour	
In an hour An hour_____ OZJ	
In half an hour Half an hour ___OLK	
P M * Afternoon In the afternoon, DEW	
*NOTE —For hours A M and P M see TIME, page 58	
Quarter of an hour In a quarter of an hour _____ OZM	
What is your consumption of coal per hour, at full speed?_____ IXM	
HOUSE	
Admiral's house _____ DKN	
Bake house Bakery_____ FSX	
—Both Houses of Congress, or, Parliament	
Cleared at the Custom-house _____FUE	
Clearing house_____ ITG	
Clearing the Custom-house _____ KGS	
Club Clubhouse _____ IWD	
Cookhouse _____ JVC	
Custom-house_____ EGT	
Custom-house boat Revenue boat_KGU	
Custom-house duty___ _____ KGV	
Deck house_____ KPZ	
Have you a Custom-house official on board ?_____ KGX	
Have you cleared the Custom-house? KGY	
—House flag	
House of Assembly_____ FEK	
House of Commons (*See also* PARLIAMENT)_____ JFM	
—House of Lords Senate	
House of Representatives _____ IKA	
Houses of Parliament _____ OZP	
Ice house _____ PBX	
Is your ship entered at Custom-house? KOZ	
Not cleared at Custom-house_____ KHA	
Public house_____ UAV	
Shall clear at Custom-house _____ KHB	
Storehouse _____ WXQ	
Trinity House (*or*, Light-house Board), GPN	
Warehouse_____ YUI	
HOVE · (*See also* HEAVE)_____ ORX	
Heaving, Hove-to_____ OSD	
—Strange vessel hove in sight	
Vessel indicated has hove-to_____ OSQ	
HOW_____ BXF	
How am I?_____ BXG	
How are her funnels placed?_____ NTR	
How are the funds at — ? _ _____ NTF	
How are the markets?_____ RLP	
How are the wounded?_____ ZIM	
How are they? ___ _____ BXH	
How are you?_____ BXI	

HOW—HUMAN

How—*Continued*
How are you armed? ....EUY
How are you steering? ....WTX
How broad—is — ? ....HJN
How can (*or*, may)? ....BXJ
How could (*or*, might)? ....BXK
How deep—is — ? - ....KQX
How did it (*or*, that) happen ?......OMK
How did the land bear when last seen?
....EH
How do (does, *or*, did)? ....BXL
How do (*or*, did) you ? ....BXM
How do you carry your helm? ......OTS
How do (*or*, did) you manage (*or*, contrive)—to — ? ....JTF
How does the beacon (buoy, *or*, mark) bear? ....GAH
How does the entrance bear? ....EJ
How does the harbor (*or*, anchorage) bear? ....PA
How does the land bear? ....QDS
How does —(*vessel indicated*) sail? .VDX
How far—off ? What distance?...MTA
How far can light be seen? ....MSZ
How had (has, *or*, have)?..........BXN
How have you had the wind? ....ZEN
How heavy is — ? ....OTB
How high must I lie to fetch?......MWE
How is business going?..............HNY
How is(*or*, are) he (she, it, *or* person-s *or* thing-s indicated)? ....BXO
How is she armed? ....EUZ
How is she rigged? ....UXH
How is(*or*, was)stranger (*or vessel indicated*)steering? ....— ......WTY
How is the barometer? ....FXE
How is the patient? ....TEI
How is the surf?...................XFS
How is the tide? What tide have we now? ....KF
How is your latitude obtained? ....QHN
How large—is (*or*, are)?.... ......QFO
How long (a time)? ....QYL
How long—is(*or*, are)— ? ....QYO
How long after? ....DRP
How long ago? ....DTK
How long do you require to clean your engines ?......................QYM
How long have you(*or vessel indicated*) been in such distress? ....LHI
How long have you (*or vessel indicated*) been on fire ?.................NAI
How long have you been on the voyage? ....QYN
How long have you had this fog? ..NGK
How long may we stand on our present course?.............. ....JZK
How long ought I to be getting an answer from—? ....EMY
How long will delay be? ....KSH
How long will you be getting up steam? ....VY
How long will you (*or vessel indicated*) be repairing damages?..... ......BM
How many?.... ....RKE
How many blades have you to your screw?........ ....GLX
How many boats? ....GTJ
How many days? ... .KNG
How many days since you passed — (*vessel indicated*)? ....KNH

How—*Continued.*
How many deaths?................KOM
How many dozen can you spare? ..LMU
How many feet? ... .MVE
How many funnels has she?........NTS
How many guns?....................RKF
How many guns do you mount? ...RKG
How many guns does she carry?...RKH
How many have you seriously ill?..KLD
How many hours' coal have you? (*State at what speed*)... LXC
How many lighters do you want?..QTA
How many men?.................. ..RKI
How many miles—from —?........RTQ
How many miles an hour are you going? ....RKJ
How many miles an hour can you (*or vessel indicated*) go? ....RTP
How many minutes? .... RVD
How many (*or*, much) more?......RKL
How many passengers? What passengers? ....TDO
How many ships? ....: .....VXK
How many sick? ....WBC
How many times?..................XTM
How many tons—of? ....XUT
How many tons do you take?......XUV
How many torpedo boats passed in sight?. ....GTK
How many were killed?.............QAM
How many wounded?................ZIN
How much?.........................SBR
How much cable have you out?.....HRE
How much longer? .... — ....QYR
How must? ....BXP
How must I bring the beacon (buoy, *or*, mark) to bear?................GAI
How often?...... ....SPX
How old? What age?.... ....DSI
How ought?...... ....BXQ
How rigged?........... ....UXH
How shall (*or*, will)he (she, it, *or* person-s *or* thing-s indicated)? ....BXR
How shall (*or*, will) I? ....BXS
How shall (*or*, will) we? ....BXT
How shall we have the wind? .....ZEP
How shall (*or*, will) you?........BXU
How should (*or*, would)? ....BXV
How soon? ....WJZ
How was (*or*, were)?...............BXW
How will the ebb set?..............LSO
How will the flood set?............NFQ
How will they? ....BXY
How would you act?...............DIH

OZT | HOWEVER
OZU | HOY —Powder hoy (*or*, lighter)
OZV | HUG-GED-GING
OZW | HULK ED-ING
    Coal hulk ....IWO
    Sheer hulk ....VUR
OZX | HULL
OZY | —Hull down
PAB | —Hulled-ing
PAC | HUMAN-E-LY-ITY

	HUMMOCK—I	

PAD	HUMMOCK-Y	PBE	I (*Letter*) (*For new method of spelling, see page 13*)
PAE	HUNDRED		
PAF	—Hundredth		I. (*Personal pronoun*) _____BYA
			After I — _____BXZ
	HUNDREDWEIGHT CWT. (*Measure of weight*) _____BAP		Am I? _____BFC
			Am I not? _____BED
	HUNG _____OMH		Am I to? _____BEF
			Can, *or*, May I (my, mine)? _____BKG
PAG	HUNGER-Y		Can, *or*, May I be? ____ _____BKH
			Can, *or*, May I have—some —?____BQG
PAH	HUNT-ED-ING		Could, *or*, Might I (my, mine)?____BLC
			Do, Does, *or*, Did I (my, mine)?____BMH
PAI	HURRICANE TYPHOON		Had, Has, *or*, Have I (my, mine)? _BQS
	Center of hurricane bears — _____IGX		Had, *or*, Have I been? _____ ___BFR
PAJ	—Has been a hurricane at (*or*, in)—		Had, *or*, Have I done? _____BMT
PAK	—Hurricane did considerable injury		Had, *or*, Have I had? _____BQT
PAL	—Hurricane did not do much damage		How am I? _____BXG
PAM	—Hurricanes have not set in at —(*date to follow, if necessary*)		I ____ _____BYA
	Prepare for a hurricane_____GD		I am _____BEA
PAN	—You are steering right into the centre of the hurricane (*or*, cyclone)		I am not_____BEI
			I can (*or*, may) _____BKQ
	HURRY-IED-ING (*Hasten*) _____ONG		I can (*or*, may) be_____BFX
			I can (*or*, may) have_____BRI
PAO	HURT-ING-FUL-LY (*See also* INJURE)		I can not (*or*, may not) _____BKR
	Anyone hurt (*or*, wounded)?_____EQB		I can not (*or*, may not) be_____BFY
	Boilers burst, no one hurt _____AV		I can not (*or*, may not) have_____BRJ
	Boilers burst, no one seriously hurt.AW		I could (*or*, might) _____BLK
	Dangerously hurt (*or*, wounded), accident _____AL		I could (*or*, might) be _____BFZ
	Few hurt _____MXC		I could (*or*, might) have _____BRK
PAQ	—Is he (she, *or*, it) much hurt?		I could (*or*, might) not_____BLM
PAR	—Is, *or*, Are much hurt		I could (*or*, might) not be _____BGA
PAS	—Is, *or*, Are not much hurt		I could (*or*, might) not have_____BRL
	Much hurt _____SBT		I do (*or*, did) ____ _____BMY
PAT	—No one hurt		I do (*or*, did) not_____ ____ _____BMZ
			I had (*or*, have)_____BRM
PAU	HUSBAND		I had (*or*, have) been ____ __ ____BGC
PAV	—Ship's husband		I had (*or*, have) done _____BNA
			I had (*or*, have) had _____BRN
PAW	HUT-TED-TING		I had (*or*, have) not _____BRO
			I had (*or*, have) not been_____BGD
PAX	HYDRAULIC		I had (*or*, have) not done_____BNC
	Hydraulic engine _____MAR		I had (*or*, have) not had_____BRP
PAY	—Hydraulic jack		I may, etc (*See* Can, *above*)
PAZ	—Hydraulic pump		I might, etc (*See* Could, *above*)
			I must ____ _____CEK
PBA	HYDROGRAPHY-ER-ICAL		I must be __ _____BGE
			I must not _____CEL
PBC	HYDROMETER		I must not be _____BGF
			I ought to_____CHL
PBD	HYDROPHOBIA		I ought not to _____CHM
			I shall (*or*, will) _____CIO
			I shall (*or*, will) be _____BGH
			I shall (*or*, will) do___ _____BND
			I shall (*or*, will) have _____BRQ
			I shall (*or*, will) if I can _____BKS
			I shall (*or*, will) not _____CIP
			I shall (*or*, will) not be _____BGI
			I shall (*or*, will) not do _____BNE
			I shall (*or*, will) not have_____BRS
			I should (*or*, would) ____ _____CJM
			I should (*or*, would) be_____BGJ
			I should (*or*, would) do_____BNF
			I should (*or*, would) have_____BRT
			I should (*or*, would) have been ___BGK
			I should (*or*, would) not ____ ___CJN
			I should (*or*, would) not be. _____BGL
			I should (*or*, would) not do_____BNG
			I should (*or*, would) not have _____BRU
			I should (*or*, would) not have been.BGM
			I was _____COH

I—ICE.

I—_Continued_	**PBF**
I was not ------------------------------COI	**PBG**
I will, etc　(_See Shall, above._)	
I will do it if I can -------------------LIP	
I would, etc　(_See Should, above_)	
If I (my, mine) --------------------BYW	**PBH**
If I am------------------------------BYX	
If I am not ---------------------------BYZ	**PBI**
If I can (_or_, may) -------------------BZA	**PBJ**
If I can not (_or_, may not)------------BZC	
If I could (_or_, might) -------------BZD	**PBK**
If I could (_or_, might) not-----------BZE	**PBL**
If I do (_or_, did) -------------------BZF	
If I do (_or_, did) not ----------------BZG	
If I had (_or_, have) ----------------BZH	
If I had (_or_, have) not -------------BZI	
If I must----------------------------BZJ	
If I must not -------------------------BZK	
If I ought to -----------------------BZL	
If I ought not to ---------------------BZM	
If I shall (_or_, will)----------------BZN	
If I shall (_or_, will) not -------------BZO	
If I should (_or_, would) ------------BZP	
If I should (_or_, would) not --------BZQ	
If I was -----------------------------BZT	
If I was not --------------------------BZU	
If I were you I should not do it----LIQ	
May, etc　(_See Can, above_)	
Might, etc　(_See Could, above_)	
Must I (my, mine)?----------------CET	
Need I (my, mine)?-----------------CEH	
No, I can not-----------------------HUZ	
Not I (my, mine)--------------------CFU	
Ought I (my, mine)?---------------CHP	
Ought I to—do?--------------------CHQ	
Shall, _or_, Will I (my, mine)?-------CIT	
Shall, _or_, Will I be?------------ --BHC	
Shall, _or_, Will I have?------------BRX	
Shall, _or_ will I not have?---------BRY	
Should, _or_, would I (my, mine)?---CJQ	**PBM**
That I (my, mine)------------------CLD	**PBN**
Was, _or_, Were I (my, mine)-------COS	
Was I?-----------------------------COT	
Was I not?--------------------------COU	
What, _or_, Which am I--------------OPV	**PBO**
What, _or_, Which shall (_or_, will)—I?-CQK	**PBQ**
What, _or_, Which shall (_or_ will) I (_or_ we) do?-------------------------CQL	
What, _or_, Which shall (_or_, will) I (_or_ we) have?-----------------------CQM	**PBR**
What, _or_, Which should (_or_, would) I (_or_, we) do?---------------------CQR	**PBS**
When am I to?----------------------CQV	**PBT**
When do (_or_, did) I (_or_, we)?----CRE	
When I (my, mine) ----------------CRL	**PBU**
When I can (_or_, may)-------------CRM	**PBV**
When shall (_or_, will) I (_or_, we)?--CRU	**PBW**
When shall (_or_, will) I (they, we, _or_, you) do?-------------------------CRV	**PBX**
	PBY
When shall (_or_, will) I have?__ --CRW	**PBZ**
Where am I (_or_, are we)?---------CSQ	**PCA**
Where do (_or_, did) I (_or_, we)?----CSY	**PCB**
Where I (my, mine)-----------------CTF	
Where I am (_or_, we are) ----------CTG	**PCD**
Where shall (_or_, will) I (_or_, we)?--CTN	
Where shall (_or_, will) I have?-----CTO	
While I----------------------------ZCU	**PCE**
Why am I?--------------------------CUY	**PCF**
Yes, I can -------------------------HUG	
Yes, I have -----------------------ZLG	
Yes, I will -----------------------ZLH	

ICE

—A good deal of floating ice in —

Can one break through the ice?----HBY

Detained by ice -------------------KYA

—Do you think I could get through the ice?

Fast ice---------------------------MTP

—Foot ice

—Has the ice broken up? Has the winter broken up—at —?

—Has there been much ice at —?

—Have encountered ice (_date, latitude, and longitude to follow, if necessary_)

Have encountered ice between 30° and 35° of longitude on the —*-------PF

Have encountered ice between 35° and 40° of longitude on the —*-------PG

Have encountered ice between 40° and 45° of longitude on the —*-------PH

Have encountered ice between 45° and 50° of longitude on the —*-----PI

Have encountered ice between 50° and 55° of longitude on the —*-------PJ

Have encountered ice between 55° and 60° of longitude on the —*-----PK

Have encountered ice between 60° and 65° of longitude on the —*-----PL

Have encountered ice between 65° and 70° of longitude on the —*-------PM

Have encountered ice between 70° and 75° of longitude on the —*-------PN

Have encountered ice between 75° and 80° of longitude on the —*-------PO

Have passed ice in latitude — and longitude — on the —*----'------PQ

* Date to be indicated if necessary

Have you fallen in with ice (_state whether berg or field_)?------------PR

—I have been amongst the ice

—I have been injured by the ice (_in latitude—, longitude—, to follow, if necessary_)

—I have received information —(_place next indicated_) is clear of ice

—I have received information —(_place next indicated_) is icebound

—I have received orders for you not to proceed to—(_place next indicated_) on account of ice

—I only got clear of the ice (_or_, icebergs)— on

—I saw ice in latitude —(_date to follow, if necessary_)

—Ice as far as I can see

—Ice breaker

—Ice field, _or_, Floe

—Ice house

—Iceberg

—Is the harbor at — free from ice?

—Is there much ice at (_or_, in)—?

—Light-ship has been withdrawn on account of ice

—Not much ice at —(_place next indicated_)

Place next indicated is blocked up by ice -------------------------GMU

—The harbor at — is free from ice

—The ice is so solid that I can not break through, send help

What latitude and longitude did you (_or_, they) have ice in?----------PS

	ICE—IF.	

ICE—*Continued*

PCG —Where did you encounter ice?
PCH —You will fall in with ice if you go be-
       yond —
PCI —You will find great difficulty in getting
       through the ice at —

PCJ IDEA-L

PCK IDENTICAL-LY   IDENTITY

PCL IDENTIFY-IED-ING-ICATION

PCM IDIOT

PCN IDLE-Y-NESS

IF _____BYC
I shall (*or*, will) if I can _____BKS
If he (his, she, her s, it-s) ____ _____BYV
If he (she, it, *or person-s or thing-s indi-*
   cated) is (*or*, are, etc   (*See* Is, below)
If he (she, it, *or person-s or thing-s indi-*
   cated) can (*or*, may) _____BYD
If he (she, it, *or person-s or thing-s indi-*
   cated) can not (or, may not)_____BYE
If he (she, it, *or person-s or thing-s indi-*
   cated) could (*or*, might)_____BYF
If he (she, it, *or person-s or thing-s indi-*
   cated) could (or, might) not_____BYG
If he (she, it, *or person-s or thing-s indi-*
   cated) does (do, or, did)_____BYH
If he (she, it, *or person-s or thing-s indi-*
   cated) does (do, or, did) not_____BYI
If he (she, it, *or person-s or thing-s indi-*
   cated) had (has, or, have) _____BYJ
If he (she, it, *or person-s or thing-s indi-*
   cated) had (has, *or*, have) not __BYK
If he (she it, *or person-s or thing-s indi-*
   cated) is (*or*, are)_____BYL
If he (she, it, *or person-s or thing-s indi-*
   cated) is (*or*, are) able to_____BYM
If he (she, it, *or person-s or thing-s indi-*
   cated) is (*or*, are) not_____BYN
If he (she, it, *or person-s or thing-s indi-*
   cated) is (*or*, are) not able to ____BYO
If he (she, it, *or person-s or thing-s indi-*
   cated) shall (*or*, will) _____BYP
If he (she, it, *or person-s or thing-s indi-*
   cated) shall (*or*, will) not_____BYQ
If he (she, it, *or person-s or thing-s indi-*
   cated) should (*or*, would)_____BYR
If he (she, it, *or person-s or thing-s indi-*
   cated) should (*or*, would) not ___BYS
If he (she, it, *or person-s or thing-s indi-*
   cated) was (*or*, were)_____BYT
If he (she, it, *or person s or thing-s indi-*
   cated) was (*or*, were) not_____BYU
If I (my, mine)_____BYW
If I am_____BYX
If I am not ____ _____ _____BYZ
If I can (*or*, may) _____BZA
If I can not (*or*, may not)_____BZC
If I could (*or*, might)_____BZD
If I could (*or*, might) not _____BZE
If I do (*or*, did)_____BZF
If I do (*or*, did) not_____BZG
If I had (*or*, have)_____BZH
If I had (*or*, have) not_____BZI
If I must _____BZJ

IF—*Continued*

If I must not _____BZK
If I ought to_____BZL
If I ought not to_____BZM
If I shall (*or*, will) _____BZN
If I shall (*or*, will) not_____BZO
If I should (*or*, would)_____BZP
If I should (*or*, would) not_____BZQ
If I was _____BZT
If I was not _____BZU
If I were you I should not do it_____LIQ
If it can (*or*, may) be done _____BZR
If it can not (*or*, may not) be done _BZS
If not _____BZV
If our-s (*or*, we) _____CAW
If so _____BZW
If that (*or*, this)_____BZX
If the _____BZY
If there _____CAB
If these (*or*, those) _____CAD
If they (their-s) _____CAE
If they are _____CAF
If they are not _____CAG
If they can (*or*, may)_____CAH
If they can not (*or*, may not) _____CAI
If they could (*or*, might) _____CAJ
If they could (*or*, might) not _____CAL
If they do (*or*, did)_____CAM
If they do (*or*, did) not_____CAN
If they had (*or*, have)_____CAO
If they had (*or*, have) not _____CAP
If they shall (*or*, will)_____CAQ
If they shall (*or*, will) not_____CAR
If they should (*or*, would)_____CAS
If they should (*or*, would) not ____CAT
If they were___ _____CAU
If they were not_____CAV
If we (our-s)_____CAW
If we are _____CAX
If we are not _____CAY
If we can (*or*, may)_____CAZ
If we can not (*or*, may not)_____CBA
If we could (*or*, might) _____CBD
If we could (*or*, might) not _____CBE
If we do (*or*, did)_____CBF
If we do (*or*, did) not_____CBG
If we had (*or*, have)_____CBH
If we had (*or*, have) not _____CBI
If we shall (*or*, will)_____CBJ
If we shall (*or*, will) not_____CBK
If we should (*or*, would)_____CBL
If we should (*or*, would) not _____CBM
If we were _____CBN
If we were not _____CBO
If you-r-s_____CBP
If you are _____CBQ
If you are not _____CBR
If you can (*or*, may)_____CBS
If you can not (*or*, may not) _____CBT
If you could (*or*, might)_____CBU
If you could (*or*, might) not _____CBV
If you do (*or*, did)_____CBW
If you do (*or*, did) not_____CBX
If you had (*or*, have)_____CBY
If you had (*or*, have) not _____CBZ
If you shall (*or*, will) _____CDA
If you shall (*or*, will) not_____CDB
If you should (*or*, would) _____CDE
If you should (*or*, would) not _____CDF
If you were _____CDG
If you were not _____CDH

	IGNITE—IMPERTINENT		
PCO	IGNITE D-ING-ION	PDK	IMMEDIATE—*Continued.* —Not immediately Send an anchor and cable off immediately .... EJM
PCQ	IGNORANT-LY-CE		
PCR	IGNORE-D ING	PDL	Send immediate assistance... ...... DB —Send immediately—to (or, for) Surgeon, *or,* Doctor will come immediately .... WN Vessel (*indicated*) in distress, wants immediate assistance .... NW
PCS	ILL-NESS　(*See also* SICK) Captain is sick .... HWO		
PCT	—Have, *or,* Has been ill How many have you seriously (*or,* dangerously) ill?... .... KLD		
PCU	—I have illness on board I heard that some of your family were seriously ill .... MSI		Want a boat immediately .... YG Want coal immediately .... YI Want food immediately .... YO Want immediate assistance, in distress,* 　　　　　　　　　　　　　NC
PCV	—Is, *or,* Are ill		
PCW	—Is illness likely to last?		Want immediate instructions........ YK
PCX	—Passenger, *or,* Crew ill (sick) with infectious complaint. Seriously ill .... VRU — men of the crew have been ill with yellow fever　(*Number to follow*) 　　　　　　　　　　　　　MWY — passengers have been ill with yellow fever　(*Number to follow*) ..... MWZ		Want immediate medical assistance. YL
		PDM	—Want immediately Want lighter immediately (*if more than one, number to follow*) .... YM Want provisions immediately .... YO Want water immediately............ YR
		PDN	—Was to sail immediately Weigh immediately (*or at time indicated*) .... ZS With immediate assistance fire can be extinguished .... NY Write immediately .... ZJR
PCY	ILLEGAL-Y-ITY		
PCZ	ILLEGIBLE-Y-ILITY		
		PDO	IMMENSE LY-ITY
PDA	ILLUMINATE-D-ING-ION	PDQ	IMMERSE-D-ING-ION
PDB	—Illuminate-d-ing mine field		
		PDR	IMMIGRATE-D ING-ION-GRANT Immigration, *or,* Emigration Office. LXJ
PDC	ILLUSTRATE-D-ING ION		
PDE	IMAGINE-D-ING-ATION-ARY.	PDS	IMMINENT
PDF	IMITATE-D-ING-ION	PDT	IMMODERATE-LY
PDG	IMMEDIATE-LY Aground , want immediate assistance, 　　　　　　　　　　　　　NA Am aground, likely to break up, require immediate assistance .... OA	PDU	IMMOVABLE-Y-ILITY
		PDV	IMPARTIAL-LY-ITY
	Come immediately (directly) ...... JCQ	PDW	IMPASSABLE-Y Bar is impassable.................... FR Bar is impassable for boats on the ebb tide .... BQ
PDH	—Do it immediately Do you require immediate assistance? 　　　　　　　　　　　　　OU Fire, *or,* Leak, want immediate assistance .... NH		
PDI	—Forward immediately Go immediately..... .... OBU I am aground, send what immediate assistance you can .... OE I can not take you in tow, but will report you at — and send immediate assistance .... XYQ I shall weigh immediately (*or at time indicated*) .... ZP I will telegraph message immediately, 　　　　　　　　　　　　　RQO	PDX	IMPATIENT-LY-OE
		PDY	IMPEDE-D-ING IMENT
		PDZ	IMPEL-LED LING
		PEA	IMPENETRABLE-ILITY
		PEB	IMPERATIVE-LY
		PEC	IMPERCEPTIBLE-LY-ILITY
		PED	IMPERFECT-LY-ION
PDJ	—Immediately on my return (or, on the return of —, *person indicated*) In distress, want immediate assistance,* 　　　　　　　　　　　　　NC Leave immediately—to (or, for) ... QLA Light house, *or,* Light-ship at — wants immediate assistance .... CW Light-ship, *or,* Light-house at — wants immediate assistance .... CW		IMPERIAL　(*Coin*) .... ATU
		PEF	IMPERIAL-IST
			IMPERISHABLE-LY-ILITY　(*Indestructible*), 　　　　　　　　　　　　　PHT
		PEG	IMPERTINENT-LY-OE

* See DISTRESS SIGNALS, page 7

IMPETUOUS—IN

PEH	IMPETUOUS-LY-OSITY	PFN	IMPROBABLE-ILITY
PEI	IMPLEMENT	PFO	IMPROPER-LY-RIETY
	Agricultural implement............DUM	PFQ	IMPROVE-D-ING-MENT
PEJ	IMPLICATE-D-ING-ION	PFR	—Does it (or person indicated) improve?
PEK	IMPLICIT-LY	PFS	—Have, or, Has improved
		PFT	—No improvement.
	IMPLORE-D-ING　(Beg) ....GDL		Not improved ....SJV
PEL	IMPLY-IED-ING	PFU	IMPRUDENT-LY-CE
PEM	IMPOLITIC	PFV	IMPURE-ITY
PEN	IMPORT ED-ING　TO IMPORT		IN ....CDI
PEO	—Import duty		In a bad way.............FRO
PEQ	—Importation		In a quarter of an hour........OZM
PER	—Importer		In an hour ....OZJ
			In and out................SVW
PEU	IMPORTANT-CE		In bad health ....FRP
	I am going to signal the contents of an		In bad order...............FRI
	important telegram, to be communi-		In bond..................GYO
	cated to you ....IE		In bulk ....HLJ
	I am the bearer of important despatches,		In case ....IDQ
	JR		In case of................IDR
	Important despatch ....KXA		In case of accident (or, necessity) ...BP
	Important intelligence—has arrived,		In chase of ....INQ
	EYM		In compliance with ....JLE
PEV	—Important mission		In dock ....LJF
	Important to communicate—by (or,		In every particular...........MJN
	with) ....IH		In every respect.............MJO
PEW	—It is important that it should be so		In favor ot ....MUC
PEX	—No news of importance		In fork ....NLQ
	Nothing important by last accounts..DFL		In front—of ....OLI
	—Of little importance　Unimportant-ce		In future ....NUR
PEY	—Of no importance		In gold..................ODP
PEZ	Of the utmost importance (or, conse-		In good order ....OKB
	quence) ....JQC		In great danger ....KLE
	Some important news has arrived...EYM		In half an hour.............OLK
	Stop, Heave-to, or, Come nearer; I have		In or, Into harbor ....ONB
	something important to communicate,		In him (his, her-s, it-s).........CDJ
	Code Flag over H		In how many minutes?.........RVD
	Unless your communication is very im-		In line ....QUL
	portant I must be excused .....IJ		In making sail..............RID
			In me (my, min) ....CDK
PFA	IMPOSE-D-ING-ITION　IMPOST		In my opinion ....SRL
			In my wake................YTF
PFB	IMPOSSIBLE		In need (or, want) of ....SGE
PFC	—Impossibility.		In order—to...............SUA
PFD	—It is impossible		In our s (us) ....CDL
	It is impossible to comprehend　Not		In place of (Instead of).........PNA
	sufficiently comprehensible ......JLP		In, or, Into port............ONB
PFE	—It is impossible to say		In pursuance of ....UCQ
	Landing is impossible ....KD		In readiness—to (or, for) ....UJN
	Not impossible ....SJU		In shore ....PMG
			In sight—ot ....WBZ
PFG	IMPOSTOR-URE		In, or, At station ....FIP
			In stays ....WSA
PFH	IMPRACTICABLE-Y-ILITY		In that (or, this) ....CDM
			In the..................CDN
PFI	IMPREGNABLE		In the afternoon.............DRW
			In the bay—of ....FZI
PFJ	IMPRESS-ED ING-ION		In the boat................GTN
PFK	—Made no impression		In the channel..............ILM
			In the direction—of..........LBP
PFL	IMPRISON-MENT		In the engine room...........MBP
	Crew imprisoned ....KCS		In the evening ....MIH
PFM	—Imprisoned—for.		In the event (or, case) ot ....IDR
			In the middle of the — ....RTE
			In the month of — ....RYS
			In the morning..............RZY

IN—INDESTRUCTIBLE	

	In—Continued	PGN	INCLUSIVE-LY
	In the offing ....SPV	PGO	INCOGNITO
	In the rear ....UKJ	PGQ	INCOMBUSTIBLE-Y-ILITY
	In the river ....UYI		
	In the same ....VGJ	PGR	INCOME
	In the spring ....WOU	PGS	—Income tax
	In the straits of ....ILM	PGT	INCOMMODE D ING
	In the vocabulary ....YRH		
	In the wake—of ....YTG	PGU	INCOMPETENT-LY-CE-CY
	In the way ....YXD		
	In the winter ....ZGH	PGV	INCOMPLETE-LY-NESS
	In the zenith ....ZMF		
	In them (then s) ....ODO	PGW	INCONSIDERATE-LY-ABLE-CY.
	In these (or, those) ....CDP		
	In this (or, that) ....CDM	PGX	INCONSISTENT-LY-CE-CY
	In this place................TLG		
	In time ....XTN	PGY	INCONVENIENT LY-IENCE
	In tow................XYR		
	In trying to ....YCR	PGZ	INCONVERTIBLE-ILITY
	In turn ....YEG		
	In use ....YMN	PHA	INCORRECT-LY-NESS
	In want of ....SGE		Account sent is incorrect ....DEU
	In wearing........................YXR		Number, or, Quantity is incorrect. SLG
	In what (or, which) ....CDQ		Report incorrect ... .JXC
	In what manner? ... .RJS		The address is incorrect ....DJQ
	In what quarter? ....UEP		Your flags seem to be incorrectly
	In writing.......... ....ZJN		hoisted.....................OWM
	In your-s................ODR		
	In your opinion ....SRM	PHB	INCREASE-D ING
	In your wake ....YTH		Fortifications have been increased at —,
	Not in ....CFV		NMI
PFW	INABILITY		If gale increases ....NWA
		PHC	—Increase speed
PFX	INACCESSIBLE-Y ILITY		Is the sickness on the increase?....WBD
	INACCURATE-LY-CY (Incorrect) ....PHA		Leak increases ....QJY
PFY	INADMISSIBLE	PHD	INCREDIBLE-Y-ILITY
		PHE	INCUR-RED-RING
PFZ	INATTENTION-IVE-LY	PHF	INCURABLE
PGA	INCANDESCENT-CE INCANDESCENT LIGHT	PHG	INDEBTED-NESS.
PGB	INCAPABLE Y-ILITY	PHI	INDECENT-LY-CY
PGC	INCAPACITATE-D-ING	PHJ	INDECISIVE-LY-SION
PGD	INCAUTIOUS-LY-NESS		Indecisive battle between the — and —
PGE	INCENDIARY ISM		(names to follow)................FZA
PGF	INCESSANT-LY	PHK	INDEED
	INCH (Measure of length) ....AWE	PHL	INDEFATIGABLE-Y
	Cubic inch.... ....AYD	PHM	INDEFINITE-LY
	Square inch........................AXU	PHN	INDEMNIFICATION INDEMNITY. (Damages)
PGH	INCITE-D-ING		
PGI	INCIVILITY	PHO	INDEMNIFY-IED-ING
PGJ	INCLINE-D-ING-ATION—TO	PHQ	INDENT-ED-ING
PGK	—Are you inclined to do it?		Cancel-led ing the indentures.....HUM
PGL	INCLOSE-D-ING	PHR	—Indentures
PGM	INCLUDE D ING	PHS	INDEPENDENT-LY-CE.
	All included ....DXN		
	Is all included?..................DYL	PHT	INDESTRUCTIBLE-Y-ILITY
	Not included-ing ....SJW		

	INDEX—INFORM		
PHU	INDEX ED-ING Geographical Index (*See page* 452) "Lloyd's Weekly Shipping Index"--QXD	PJD	INFAMOUS-LY INFAMY
PHV	INDIA-N East India cargo ----------------HZA East India Government-----------OFR	PJE	INFANT-CY
		PJF	INFANTRY Enemy's infantry------- ------ .....LZJ
PHW	—East Indiaman India rubber Gutta-percha ------OKU	PJG	INFECT-ED-ING-IOUS Case is infectious - . ---------------IDK
PHX	—Indian corn	PJH	—Have infectious cases on board
PHY	—Indian meal West India cargo---------------HZO		Have you had any communication with infected vessel (*or, place*) ----------IC I have (*or, have had*) a dangerous in-
PHZ	INDICATE-D-ING-ION		fectious disease on board
PIA	—Indicate number (*or, quantity*)—of.		Code Flag over L
PIB	—Indicate size—of		Infectious disease -----------------LEA
PIC	—Indicate weight—of		Is disease infectious (*or, contagious*)? VA
PID	INDICATOR Indicator diagram---- ------------KYS		Passengers, or, Crew ill with infectious complaint----------------------PCX
PIE	INDICT-ED-ING-MENT	PJI	INFER-RED-RING-ENCE
PIF	INDIFFERENT-LY-CE Anchorage, *or*, Harbor is indifferent-OY Indifferent accommodation --------DCT Indifferent account—of ------------DFG Indifferent chart ----------------IMX Indifferent coal------------------IXE Indifferent vacuum ------------YNL	PJK PJL	INFERIOR-ITY —Inferior force
		PJM	INFEST-ED-ING
		PJN	INFINITE-LY-ITY.
		PJO	INFIRM ITY
PIG	INDIGESTION-IBLE	PJQ	INFLAME D ING-MATION-MABLE Cargo not inflammable -----------HZJ Inflammable cargo ----------------HZF
PIH	IN DIGNANT-LY-ATION	PJR	—Inflammation of the lungs
PIJ	INDIGO		Is your cargo very inflammable? ---IAV Not inflammable cargo -----------HZJ
PIL	INDIRECT-LY-NESS	PJS	INFLEXIBLE-Y-ILITY
PIM	INDISPENSABLE Y	PJT	INFLICT-ED-ING ION.
PIN	INDISPOSE-D ITION	PJU	INFLUENCE-D-ING-TIAL
PIO	INDISTINCT-LY-NESS Bottom flag in hoist indistinct------NDF Second flag in hoist indistinct ------NDG Third flag in hoist indistinct ------NDH Top flag in hoist indistinct----- --NDI	PJV	INFLUENZA
		PJW	INFORM-ED ING-ATION (*See also* AC- QUAINT) All possible information—of (*or, from*), DXS
PIQ	INDOLENT-LY-CE	PJX	—Am, Is, *or*, Are informed—that
PIR	INDUCE-D-ING-MENT	PJY	—Can you inform me?
PIT	—Induced current	PJZ	—Can not inform you
PIU	—Induction		Captain desires me to inform you _HWE
PIV	—Induction gear	PKA	—For the information—of
PIW	—What was (*or, is*) the inducement?	PKB	—For your information
PIX	INDULGE-D-ING-ENCE		Forward the following information to — .... ------------------NGU Have obtained the following informa
PIY	INDUSTRY-IOUS-LY		tion from — -------------------- NGV
PIZ	INEFFICIENT-CY		I am desired to inform you (*or person* *indicated*).... -----------------DGZ
	INEVITABLE Y (*Unavoidable*)-------YGK	PKC	—I am happy to inform you I have received information — (*place* *next indicated*) is clear of ice ----FBO
PJA	INEXCUSABLE-Y		I have received information — (*place* *next indicated*) is icebound ------PBQ
PJB	INEXHAUSTIBLE-Y		I have the following information from, NGV
PJC	INEXPERIENCE-D	PKD	—If I am rightly informed

INFORM—INSERT

PKE	INFORM—*Continued.* Inform agent—of — ....DSR Inform harbor authorities.........FNJ Inform Lloyd's agent ....QWP Inform my family ....MSJ —Inform owner Inform the captain of ....HXC Inform the commander ....JEN Information, News, or, Report is confirmed ....JOE Information, News, or, Report is not correct ....JXC	PLC	INJURE—*Continued.* Hurricane did considerable injury—to, PAK Hurricane did not do much injury.PAL I have been injured by the ice in (*latitude and longitude to follow, if necessary*) ....PBN —Injury Is — (*vessel indicated*), or, Are you materially damaged (*or, injured*)?.AO
PKF	—Informer Owners desire me to inform you to coal at — ....HE Owners desire me to inform you to discharge cargo first at — (*place indicated*) ....SY Owners wished me to inform you..SYK	PLD	—Much injury has been done to (*or, by*)
		PLE	INJUSTICE
		PLF	INK-Y.
		PLG	INLAND
PKG	—Private information Sorry to inform you ....WKG With the following information ...NGZ Your agents desired me to inform you, DTE Your original orders are cancelled, I am directed to inform you to proceed to — ....TH	PLH	INLET Sea inlet choked ....IPQ Sea-inlet valve (*or, cock*) broken (*or, damaged*) ....VLA
		PLI	INMATE
		PLJ	INN
PKH	INFORMAL-ITY-LY.	PLK	INNER-MOST Inner entrance (passage, *or*, channel), ILN
	INFORMER.... ....PKF	PLM PLN	—Inner harbor. —Inner line
PKI	INGENIOUS-LY-UITY	PLO	INNOCENT-LY-CE
PKJ	INGOT		INNUMERABLE (*Numberless*) ....SLI
PKL	INGRATITUDE	PLQ	INOCULATE-D-ING-ION
PKM	INGREDIENT	PLR	INQUEST Awaiting coroner's inquest ....FPD Coroner's inquest ....JWS
PKN	INHABIT-ED-ING-ANCE		
PKO	INHABITANT	PLS	—Inquest will be held—at —
PKQ	INHOSPITABLE-Y	PLT	INQUIRE-D-ING-Y—INTO (*or, ABOUT*) Court of Inquiry (Investigation)...KAC
PKR	INITIAL. Dot between initials..Code Flag over F	PLU PLV	—Did not inquire —Inquiry has been made Make inquiry ....MBD
PKS	INITIATE D ING-ION	PLW	INQUISITIVE-LY-NESS
PKT PKU PKV PKW	INJECT-ED-ING ION —Injection gear —Injection pipe. —Injector	PLX	INROAD
			INSANE-ITY-LY. (*Mad*) ....REA
PKX	INJUDICIOUS-LY	PLY	INSCRIPTION MARITIME Commissaire d'Inscription maritime, JEY
PKY	INJUNCTION	PLZ	—Inscrit maritime
PKZ	INJURE-D-ING IOUS-LY (*See also* DAMAGE *and* HURT) Are you, or, Is — (*vessel indicated*) materially damaged (*or, injured*)?.OA Bow injured ....HCF Crops much injured ....KDW	PMA	INSECT.
		PMB	INSECURE-LY-ITY.
		PMC	INSENSIBLE-Y-ILITY.
PLA	—Have, or, Has received injury (*or, damage*)—to	PMD PME PMF	INSERT-ED-ING ION —Not to be inserted. —To be inserted
PLB	—Have you received any injury (*or, damage*)?		

	INSHORE—INTEMPERATE		
PMG	INSHORE		INSTRUCT—*Continued*
PMH	—Inshore squadron.		I have telegraphed for further instructions (*or*, orders) ------STW
PMI	INSIDE.	PNH	—Instructions
	Inside berth ------GFU		Keep off and on, and wait instructions, SX
PMJ	—Inside of limits	PNI	—No instructions (*or*, orders)
PMK	—Inside the harbor.	PNJ	—Previous instructions (*or*, orders)
PML	—Inside the mole		Private instructions I have private instructions ------PNG
PMN	INSIGHT	PNK	—Send instructions
PMO	INSINUATE-D-ING-ION.		Telegraph instructions (*or*, orders) to —, TC
PMQ	INSIST-ED-ING	PNL	—Urgent instructions (*or*, orders)
			Want immediate instructions------YK
PMR	INSOLUBLE-ILITY	PNM	—Your instructions
PMS	INSOLVENT CY	PNO	INSTRUMENT-AL-LY
			Telegraph instrument------XMB
PMT	INSPECT-ED-ING-ION	PNQ	INSBORDINATE ION
	Await outside the inspection of the officials ------FPB	PNR	INSUFFICIENT-LY-CY
	Board of Trade inspector (*or*, officer), GPD		Authority insufficient ------FMN
PMU	—Inspector		Insufficient address------DJM
PMV	—Inspector general	PNS	INSULATE-D-ING-ION
	Passed the official inspection ------TCU	PNT	—Insulate-d-ing cable (*or*, wire)
	Sanitary inspector ------VHG	PNU	—Insulator
PMW	INSTABILITY UNSTABLE	PNV	INSULT-ED-ING-LY
PMX	INSTAL-LED-LING-MENT-LATION	PNW	INSURE-D-ING
	Electric-light installation ------LUY	PNX	—Are you insured?
PMY	INSTANCE		For the benefit of the insurers (underwriters) ------GFL
	For instance ------NBG	PNY	—Fully insured
PMZ	INSTANT-LY-ANEOUS-LY		I am insured at Lloyd's ------QWL
	Anchor instantly ------KZ		I am not insured at Lloyd's ------QWN
	Bear up instantly ------LB	PNZ	—Insurance
	Instantaneous fuse ------NUX	POA	—Insurer
	Stop instantly------MN	POB	—Not insured
	Tack instantly ------MO		Partially insured ------TBY
	Wear instantly ------MP		Vessel is insured ------YPX
PNA	INSTEAD—OF		INSURGENT (*Rebel*) ------UKS
PNB	INSTINCT-IVE-LY	POC	INSURRECTION-ARY
PNC	INSTITUTE D-ING-ION	POD	INTELLIGENCE-ENT (*See also* NEWS)
	Lifeboat Institution ------QRI		Bring (*or*, Brought) intelligence - HIR
PND	INSTRUCT-ED-ING (*See also* DIRECT *and* ORDER)		Important intelligence — has arrived, KYM
	Admiralty instructions ------DLT	POE	—Intelligence department
	Against instructions Not according to orders ------DER	POF	—No intelligence—of
	Agent's instructions ------DSM	POG	—No intelligence obtained from vessel boarded
	Have received the following communication (*or*, instructions) from your owners (*or*, agents) ------IA		Obtain all the intelligence possible-SNW
PNE	—Have you instructions (Have orders been given)—to (*or*, for)?	POH	—Send an express with the following intelligence
PNF	—I have instructions		Will you telegraph to — (*ship or person indicated*) the intelligence I am about to communicate?------IM
PNG	—I have private instructions		
	I have received orders for you not to proceed without further instructions, SP	POI	INTELLIGIBLE Y ILITY
		POJ	—Not intelligible (unintelligible)
		POK	—Quite intelligible
	I have received orders for you to await instructions from owners at — ------SQ	POL	INTEMPERATE-LY-ANCE

INTEND—INTIMIDATE

POM	INTEND ED-ING-T-TION		INTERFERE—*Continued*
	Admiral, *or*, Senior officer intends__DKS	PQL	—Need not interfere
PON	—Do not intend to weigh		Police (*or other*) authorities will not
POQ	—Do you intend—to?		interfere_____FND
	Do you intend anchoring? _____EHZ	PQM	—Will interfere—with
	Do you intend to abandon?_____AE	PQN	—Will not interfere
	Do you intend to go into harbor?. OMW		
POR	—Do you intend to see?	PQO	INTERIOR
POS	—Do you intend to write from —?		
	I do not intend to —_____PQB	PQR	INTERMEDIATE-LY.
	I do not intend to abandon the vessel,	PQS	—Intermediate pressure
	AF		Intermediate-pressure cylinder_____KIH
	I do not intend to alter course at pres-	PQT	—Intermediate pressure cylinder cover
	ent (*or*, until —)_____ECI		Intermediate-pressure cylinder cover
POT	—I intend sounding — (*place indicated*)		broken___?___ _____KAJ
	before I try it		Intermediate pressure cylinder cracked
POU	—I intend to		(*or*, out of order) ____ _____KAV
	I intend to purchase (*or*, buy) ____HOX	PQU	—Intermediate-pressure piston
POV	—I intend to ship	PQV	—Intermediate shaft
POW	—I intend writing		
POX	—Intend leaving Purpose leaving	PQW	INTERMIT-TED TING
	Intend to abandon the ship to Lloyd's		Intermittent fever_____MWN
	agent _____CXP	PQX	—Intermittent light
POY	—Intend to refit at	PQY	—Intermission
POZ	—Intend to sail		
PQA	—Intentional-ly	PQZ	INTERNAL-LY
	It is my intention—to _____POU		Internal dimensions In the clear In-
PQB	—It is not my intention—to		side._____LAN
	Only intend to — _____ _____SQZ		
	Only intended as _____ ____ SQY	PRA	INTERNATIONAL
	Only intended for a feint_____MVG		Attention is called to page — paragraph
	The signals were not intended for you,		— of the International Code _____DH
	WDM		
PQC	—What are your intentions?		N B —*The page to be signaled first and*
	What harbor do you intend to make (*or*,		*then the paragraph (See note to this*
	run) for ?_____OND		*signal on page 87)*
	What latitude do you intend going into?	PRB	—International Code of Signals
	QID		
	When do you intend to alter course?.ECX		INTERPOSE-D-ING-ITION (*Interfere*)__PQK
	When do you intend to bring up? --EKF		
	Where do you intend calling for orders?	PRC	INTERPRET-ED-ING-ATION
	HTJ	PRD	—Have you an interpreter?
	Where do you intend to bring up? _EKH	PRE	—Interpreter
		PRF	—No interpreter available
PQD	INTENSE-LY-ITY		Require an interpreter _____UTA
			Your signal is not distinct Interpreta-
	INTENTIONAL LY _____PQA		tion doubtful_____ ___ .LGT
PQE	INTERCEDE-D-ING-SSION		
		PRG	INTERROGATE-D-ING-IVE-LY
PQF	INTERCEPT-ED-ING-ION		Note of interrogation Query_____SJZ
PQG	INTERCOURSE	PRH	INTERRUPT ED-ING-ION
	Be very careful (*or*, guarded) in your		Communication by telegraph is stopped,
	intercourse with strangers _____FS		WS
			Communication is interrupted—by (*or*,
PQH	INTEREST-ED-ING		with) ___ _____JFV
PQI	—For your (*or party's named*) interest		Traffic interrupted by snow _____XZP
	Interest in—_____PQH		
PQJ	—No interest in —	PRI	INTERVAL
	Party interested _____ _____TCB		Blow siren (*or*, steam whistle) at inter-
			vals _____WD
PQK	INTERFERE-D-ING-NCE		
	Authorities (*or*, Police authorities) have	PRJ	INTERVENE-D-ING-TION
	interfered_____?____ _____ _____FNB		
	Authorities must interfere _____FNC	PRK	INTERVIEW
	Authorities will not interfere_____FND		
	Do not interfere with (*or*, take the mat-	PRL	INTESTINE-AL
	ter out of the hands of) Lloyd's agent.		
	QWH	PRM	INTIMATE-D ING-LY-ION
	Interferes with business_____HOB	PRN	INTIMIDATE-D-ING-ION

INTO—IS	

PRO	INTO—THE — Into, *or*, In harbor Into, *or*, In port, ONB	PSX	IRISH-MAN Irish butter............HOQ
PRQ	—Into position Into the............PRO Look in (*or*, into)............QZR Ran into............UHQ Run into............VCD	PSY	IRON Angle iron............ELT Cast iron............IEK Galvanized iron............NWL
		PSZ	—Iron mast
		PTA	—Iron plate d
PRS	INTOXICATE-D-ING-ION	PTB	—Iron yard Ironclad Armor-clad ship............EVS Oxide of iron............SYO
PRT	INTRENCH ED-ING-MENT		Pig iron............TIH Railway iron............UGZ
PRU	INTRICATE		Rudder irons............VBN Scrap iron............VJY
PRV	INTRIGUE-D-ING-R		Sulphate of iron............XDG Wrought iron............ZKF
PRW	INTRODUCE-D-ING TION Can you get any introduction to — ? NYM	PTC	IRRECOVERABLE-Y
PRX	INTRUDE-D-ING-SION	PTD	IRREGULAR-LY-ITY Certificate irregular............IHO
PRY	INTRUST-ED-ING	PTE	—Irregular shape
PRZ	INUNDATE-D ING ION	PTF	—Irregular soundings
PSA	INVADE-D-ING-R-SION	PTG	IRREPARABLE Y Machinery broken down, irreparable at sea............RDV
PSB	INVALID (*See also* SICK) Can I land my invalid passengers? QDR	PTH	IRRESISTIBLE-Y-ILITY
PSC	—Invalided Military invalid............RTX Naval invalid............SEP	PTI	IRRESOLUTE-LY-ION
		PTJ	IRRESPECTIVE-LY
PSD	INVALIDATE-D-ING.	PTK	IRRIGATE-D-ING-ION
PSE	INVALUABLE-Y	PTL	IRRITATE-D-ING-ION-ABLE
PSF	INVARIABLE-Y.	PTM	IRRUPTION
PSG	INVENT-ED-ING-ION		Is............CDS
PSH	—Inventor		As it is............FAO As it is not............FAP
PSI	INVENTORY		Before it—is............CDX
PSJ	—Make out an inventory		He, She, It (*or person-s or thing s indicated*) had (has, *or*, have) done (*or*, is, *or*, are doing)............BUL
PSK	INVERT-ED-ING		He, She, It (*or person-s or thing s indicated*) had (has, *or*, have) not done (*or* is, *or*, are not doing)............BUP
PSL	INVEST-ED-ING-MENT		He, She, It (*or person-s or thing-s indicated*) is (*or*, are)............BUT
PSM	INVESTIGATE-D-ING-ION-OR Court of Inquiry (*or*, Investigation), KAC		He, She, It (*or person s or thing-s indicated*) is (*or*, are) able to............BUV
PSN	INVISIBLE-Y-ILITY.		He, She, It (*or person s or thing s indicated*) is (*or*, are) for............BUR
PSO	INVITE-D-ING		He, She, It (*or person-s or thing-s indicated*) is (*or*, are) from............BUS
PSQ	—Invitation		He, She, It (*or person s or thing s indicated*) is (*or*, are) not............BUW
PSR	INVOICE		He, She, It (*or person-s or thing s indicated*) is (*or*, are) not able to............BUX
PST	INVOLUNTARY-ILY.		How is (*or*, are) he (she, it, *or person s or thing s indicated*)?............BXO
PSU	INWARD-S-LY		If he (she, it, *or person s or thing s indicated*) is (*or*, are)............BYL
	IODIDE OF POTASSIUM............TQZ		If he (she, it, *or person-s or thing-s indicated*) is (*or*, are) able to............BYM
PSV	IODINE IODIZE-ING		
PSW	IPECACUANHA		

IS—IT

	Is—*Continued*	**It**	BVZ
	If he (she, it, *or person-s or thing-s indi-cated*) is (*or*, are) not ... ...BYN	Its-self (*See also* HE, SHE, IT, *or* PER-SON S *or* THING-S INDICATED)	CDY
	If he (she, it, *or person-s or thing-s indi-cated*) is (*or*, are) not able to ....BYO	After it	CDW
	Is, Am, *or*, Are to	BEH	As it is	FAO
	Is at—the	FJD	As it is not	FAP
	Is it?	CDT	As it may	FAQ
	Is it done?	BNH	As it was	FAR
	Is it not to be done?	BNI	As it was not	FAS
	Is it so?	CDU	As it will be	FAU
	Is it to be done?	BNJ	Before it—is	CDX
	Is, *or*, Are, he (she, it, *or thing-s indicated*)?	BVY	By it-s (him, his, her-s)	BJR
	Is, *or*, Are, his (her-s, it-s)?	BEK	Can, *or*, May do it	BLU
	Is, *or*, Are my (mine)	BEL	Can, *or*, May it-s (his, her-s)?	BKF
	Is, Am, *or*, Are not	BEG	Can, *or*, May it be done?	BLW
	Is, Am, *or*, Are not to be	BFD	Can not, *or*, May not do it	BLY
	Is, Am, *or*, Are not to have	BQD	Could (*or*, Might it-s (his, her-s)?	BLA
	Is, *or*, Are our s, *or*, Are we?	BEM	Do, Does, *or*, Did it-s (his, her-s)?	BMG
	Is that (*or*, this)	CDV	For it s (him, his, her-s, *or person s or thing-s indicated*)	BPD
	Is, *or*, Are the	BEN	From it-s (him, his, her-s, *or person-s or thing-s indicated*)	BPO
	Is, *or*, Are their-s, *or*, Are they?	BEO	Had, Has, *or*, Have, it-s (his, her-s)?	BQR
	Is, *or*, Are there—any	BEP	
	Is, Am, *or*, Are to be	BFC	If it-s (his, her-s, he, she)	BYV
	Is, Am, *or*, Are to have	BQC	If it can (*or*, may) be done	BZR
	Is, *or*, Are with —	BER	If it can not (*or*, may not) be done.	BZS
	Is, *or*, Are you-r-s	BES	In it-s (him, his, her-s)	CDJ
	It, etc (*See* He, She, It, *above*)		Is, *or*, Are it-s (his, her-s)?	BEK
PTN	—It is as		Is it?	CDT
	That, *or*, This is	CLE	Is it done?	BNH
	That, *or*, This is not	CLF	Is it not to be done?	BNI
	That it is	CLG	Is it so?	CDU
	That it is not	CLH	Is it to done?	BNJ
	There is, (*or*, are)	CMG	It	BVZ
	There is (*or*, are) not	CMH	Its-self	CDY
	There is nothing	CMT	It can be	BGN
	What, *or*, Which are (*or*, is)?	CFW	It can (*or*, may) be done	BNK
	What, *or*, Which is (*or*, are) he (she, it, *or person s or thing-s indicated*)?	CPX	It can not be	BGO
	What, *or*, Which is it?	CQG	It can not (*or*, may not) be done	BNL
	What, *or*, Which is (*or*, are) you-r-s?	CPY	It could (*or*, might) be done	BNM
	When are (*or*, is)?	CQW	It could (*or*, might) not be done	BNO
	When he (she, it, *or person-s or thing s indicated*) is (*or*, are)	CRJ	It had (*or*, has) been done..	BNP
	When is (*or*, are) he (she, it, *or per-son-s or thing-s indicated*)?	CRN	It had (*or*, has) not been done	BNQ
	Where are (*or*, is)?	CSR	It is as	PTN
	Where he (she, it, *or person s or thing-s indicated*) is (*or*, are)	CTD	It ought not to be	BGS
	Where is (*or*, are) he (she, it, *or per-son-s or thing-s indicated*)?	CTH	It ought not to be done	BNS
	Who are (*or*, is)?	CUI	It ought to be	BGE
	Why are (*or*, is)?	CUZ	It ought to be done	BNR
			It shall (*or*, will)	CIQ
PTO	ISINGLASS		It shall (*or*, will) be	BGT
			It shall (*or*, will) be done	BNT
PTQ	ISLAND		It shall (*or*, will) not be	BGU
	From the island	NEW	It shall (*or*, will) not be done	BNU
	Off the island	SOQ	It will do	BNV
PTR	—To the island		Let him (her, it, *or person-s or thing-s indicated*)— have	BWA
PTS	—Uninhabited island (*or*, coast).		May, etc (*See* Can *above*)	
			Might. etc (*See* Could, *above*)	
PTU	ISLANDER		Must it-s (his, her-s)	CES
			Not it-s (his, her-s)	CFT
PTV	ISOLATE-D-ING-ION		Of it-s (him, his, her-s)	CGN
			On it-s (him, his, her s)	CGY
PTW	ISSUE-D-ING		Ought it-s (his, her-s)?	CHO
			Shall, *or*, Will it-s (his, her-s)?	CIS
PTX	ISTHMUS		Shall, *or*, Will they do it?	BOA
			Shall, *or*, Will they not do it?	BOC
			Shall, *or*, Will you do it?	BOD
			Shall, *or*, Will you not do it?	BOE
			Should, *or*, Would it-s (his, her-s)	CJP
			That it-s (his, her-s)	CLB

IT—JIGGER

	I⊤—*Continued*	PUN	JALAP
	That it is ---------------------- CLG		
	That it is not ---------------------- CLH	PUO	JAM (*Preserve*)
	To it-s (him, his, her s) ----------CNV		
	Was, or, Were it-s (his, her-s)? ----COR	PUQ	JAM-MED-MING
	What, or, Which is it? ----------CQG		Accident happened, jammed ------DCF
	When it-s (his, her-s) ----------- CRK		Propeller jammed ----------------TZA
	When shall (or, will) it be done? --CRX		
	Where it-s (his, her-s) ------------CTE	PUR	JANUARY
	Will, etc (*See* Shall, *above*)		About January the—--------------CZN
	Will it be?------------------------BIX	PUS	—Beginning of January
	Will it do? ------ ------------------BOU	PUV	—End of January
	Will you do it?--------------------CWE	PUW	—Next January
	With it-s (him, his, her-s) ------- CWG		
	Would, etc (*See* Should, *above*)		
PTY	ITALIC	PUX	JAPAN-NED NING-NER (*To varnish*)
PTZ	ITALY-IAN ITALIAN COLORS	PUY	JAPAN-ESE JAPANESE COLORS
	Bank of Italy ------------ --------FUQ		
PUA	—Italian Register (*of ships*)	PUZ	JAR RED-RING
PUB	ITCH-ED-ING .	PVA	JAW
PUC	ITEM	PVB	JEALOUS-Y-LY
		PVC	JEER-ED-ING
	ITSELF ITS ----------------------CDY	PVD	JELLY
PUD	IVORY	PVE	JERK-ED-ING
		PVF	JERSEY (*Garment*)
			JEST-ED ING LY JOKE-------------PWE
		PVG	JET
		PVH	JET CONDENSER
		PVI	JETSAM-ED-ING
		PVJ	JETTY
			Alongside the jetty (pier, *or* wharf),
			EAT
		PVK	—Jetty head
		PVL	JEW
PUE	J (*Letter*) (*For new method of spelling,*	PVM	JEWEL-RY-LER
	see page 18)	PVN	JIB
PUF	JACK (*For lifting*)		Flying jib----------------------NGD
	Cross-jack yard --------------- ---KEF		Flying jib boom-----------------GZF
	Hydraulic jack--------------------PAY		Haul the jib down-- -------------OFE
	Jackstay --- --------------- ------- PUI		Hoist the jib--------------------OWI
			Jib boom---- -------------------GZQ
	JACK UNION JACK ------- --------YJE		Jib boom sprung ----------------GZR
	Hoist a jack ----------------------OWB	PVO	—Jib net
PUG	—Jackstaff		Jib sheet ------- ---------------VCT
	Pilot jack (*when spoken of*) -------NDL		
			JIBE-D ING, *or*, GYBE D-ING----------OKW
PUH	JACKET		
	Cork jacket----------------------JWN	PVQ	JIGGER
	Low-pressure slide jacket cover broken,		After, *or*, Jigger mast------------DRC
	KAM	PVR	—Jigger gaff
PUI	JACKSTAY		Jigger mast After mast----------DRC
		PVS	—Jigger royal yard and sail
PUJ	JACOB'S, *or*, ROPE, LADDER	PVT	—Jigger topgallant yard and sail
		PVU	—Jigger topsail yard and sail
PUK	JAIL GAOL PRISON	PVW	—Jigger yard and sail
PUM	—Jailer		

JOB—KEEP.

PVX	JOB BED-BING	PXJ	JUST (*Recently*)	
PVY	—Stock jobber (*or*, broker)	PXK	—Just before	
			Just between _____ GIZ	
PVZ	JOIN-ED-ING TO JOIN		Just beyond _____ GJF	
PWA	—Going to join	PXL	—Just now	
PWB	—Have, *or*, Has not joined	PXM	—Just out of port (*or*, harbor).	
PWC	JOINER	PXN	JUST-LY-ICE.	
PWD	JOINT-ED-LY	PXO	JUSTIFY-IED-ING-ICATION.	
	Smoke joint _____ WHZ	PXQ	—Can not justify	
PWE	JOKE-D-ING JEST-ED-ING	PXR	JUTE	
	JOLLY BOAT _____ GTO			
PWF	JOURNAL-IST			
PWG	JOY-FUL-LY.			
PWH	JUDGE-D ING TO JUDGE			
	Act as your judgment directs _____ DIB			
PWI	—Judgment			
	Want of judgment _____ YTX			
PWJ	JUDICIOUS-LY			
PWK	JUG			
PWL	JUICE-Y			
	Lime, *or*, Lemon juice. _____ QUE			
PWM	JULY			
	About July the — _____ CZO	PXS	K (*Letter*) (*For new method of spelling, see page 13*)	
PWN	—Beginning of July			
PWO	—End of July			
PWQ	—Next July	PXT	KAFFIR KAFFRARIA	
PWR	JUMP-ED-ING		KAN, *or* KOP (*Measure of capacity*) AZF	
PWS	JUNCTION		KANTAR (*Measure of weight*) _____ BCJ	
	Form a junction—with _____ NLU	PXU	KEDGE-D-ING	
	Junction box _____ HDE	PXV	—Drop a kedge Bring to with a kedge	
PWT	JUNCTURE	PXW	—I shall let go a kedge	
PWU	JUNE		Kedge anchor (*weight, if necessary*) EIY	
	About June the — _____ CZP	PXY	KEEL	
PWV	—Beginning of June		Bilge keel _____ GJR	
PWX	—End of June		Damaged keel _____ _____ KJC	
PWY	—Next June		Even keel ____ _____ __ MID	
PWZ	JUNGLE		False keel _____ MSE	
	Jungle fever _____ MWP		KEELSON KELSON _____ QAC	
PXA	JUNIOR JUNIOR OFFICER.			
PXB	—Mr — junior	PXZ	KEEP-ING	
PXC	JUNK (*Chinese vessel*)	PYA	—Can keep (*or*, be kept)	
PXD	JUNK (*Old rope*)	PYB	—Can you keep?	
PXE	JUPITER		Can you keep sight—of? _____ WBU	
PXF	JURISDICTION		Can not keep free with pumps _____ NPT	
PXG	JURY-OR		Can not keep sight—of _____ WBV	
			Convoy to keep ahead of escort _____ IU	
	JURY MAST JURY MASTED _____ RMD		Convoy to keep as close as possible___ IV	
	Can you rig a jury mast? _____ VW		Convoy to keep astern of leader (*or*, escort) _____ IW	
PXH	—Jury rigged		Convoy to keep on port side of leader (*or*, escort)__ _____ IX	
PXI	—Jury rudder		Convoy to keep on starboard side of leader (*or*, escort) _____ IY	

KEEP

KEEP—Continued

Convoy to spread as far as possible, keeping within signal distance.....IZ

Do not anchor Keep under weigh.EHW

PYC —Do not keep so near

Easy sail Keep under easy sail ..LRW

Fire, or, Lights will be kept burning at the best place for landing or coming on shore.................. ....KE

PYD —Have you kept?

I am laying (repairing, or, picking up) a telegraph cable Keep out of my way ....XLQ

I can keep free with pumps.. .....NFY

I do not like to run in, I prefer keeping the sea................................VCA

I shall abandon my vessel unless you will keep by us.....................AH

I shall keep closehauled ....OPH

I shall keep hold of the land (or, lights), or, I shall keep the land (or, lights) bearing —.......................—QA

I shall keep sight of —...............WBY

I shall keep the lead going regularly.QJH

I shall keep under easy sail........LRY

I will keep near you during the night, IVF

PYE —If we keep together, we can defend ourselves

PYF —Is keeping away

PYG —Is, or, Are to keep (or, be kept)

Keep a good lookout aloft ....EAG

Keep a good lookout, as it is reported enemy's war vessels are going about disguised as merchantmen ....OJ

PYH —Keep a good lookout for land (or, lights), or, Look out for land (or, lights) (bearing, if necessary, to follow)

Keep a light at your bowsprit end..QY

Keep a light burning ....KC

PYI —Keep a sharp lookout

PYJ —Keep a strict watch all night

Keep ahead and carry a light.......DVK

Keep ahead of ....DVI

Keep aloof ....EBL

PYK —Keep any letters or papers for me till I return

PYL —Keep as close as you can to pick up my people

PYM —Keep away

Keep before the wind ....ZEU

Keep buoy (or, beacon) on your port hand ....GAK

Keep buoy (or, beacon) on your starboard hand ....GAL

PYN —Keep by the ship

PYO —Keep close during fog

PYQ —Keep close order

PYR —Keep close to

Keep closer in Keep, or, Stand under the land ....LT

Keep company ....JGP

Keep dinner—for..................LAT

Keep electric light on ....LVH

PYS —Keep everything prepared

Keep farther astern................FHM

Keep farther off (or, from)........MTC

Keep going ahead ....DVL

Keep going ahead slowly ....DVM

Keep going astern—of.............FHO

PYT —Keep head to wind.

Keep her bows on..................HCI

PYU —Keep him (or, them) in close confinement

Keep out of the tide (or, current)...XG

Keep in the center of the channel ...LU

Keep lights (or, fires) on the beach (or, shore) all night..............FZW

PYV —Keep me in sight during the night (bearing to follow, if necessary)

PYW —Keep more away

PYX —Keep more to leeward

PYZ —Keep more to port

PZA —Keep more to starboard

PZB —Keep more to windward

PZC —Keep more toward the shore.

PZD —Keep near to me

PZE —Keep nearer—to

Keep off and on, and wait instructions.SX

Keep on lee beam ....GAQ

Keep on port side of channel........LV

PZF —Keep on port tack

Keep on starboard side of the channel, LW

PZG —Keep on starboard tack

Keep on the (compass signal to follow) side of the channel ....ILQ

Keep on weather beam.............GAR

PZH —Keep open order

Keep, or, Pass the dredger on your port side..............................LNO

Keep, or, Pass the dredger on your starboard side.......................LNP

Keep right astern ....FHN

Keep silence........................RIN

PZI —Keep station

Keep steam up (or, ready) ....KQ

Keep the boat.....................GTP

Keep the lead going Keep sounding, QJI

Keep the light between — and — (bearings indicated) ....QC

PZJ —Keep the main topsail shivering

PZK —Keep the mizzen topsail shivering

PZL —Keep the rocks on —

PZM —Keep the — bearing —

Keep the two objects named in one line, QUM

Keep to leeward ....QMB

PZN —Keep to leeward of vessel chasing

Keep to windward ....LY

PZO —Keep to windward of vessel chasing

PZQ —Keep to windward until you are picked up

Keep under the land Keep, or, Stand closer in.........................LT

Keep under weigh Do not anchor.EHW

PZR —Keep vessel going about—miles an hour

Keep vessel under command LX

PZS —Keep what you have

PZT —Keep where you are

Keep within hail ....OLF

Keep within sight—of ....WCB

PZU —Keep within signal distance (bearing to follow, if necessary)

PZV —Keep your fires banked up

PZW —Keeper

PZX —Kept

Let us keep together for mutual support (or, protection) .. ..IN

	KEEP—KNOT		

	KEEP—*Continued*	QAP	KILL—*Continued*
	Let your boats keep to windward until they are picked up_____FD		—Number of killed and wounded not yet known
	Lights, *or*, Fires will be kept at the best place for coming on shore_____KE		— has won a great battle at —, with a loss of killed and wounded reported
	Lookout will be kept for any rafts (*or*, spars) _____QZY	QAR	at — _____ _ _____ _ ____FZD
	Lookout will be kept on the beach all night _____KG		—'s loss in killed and wounded reported at —
	Shall we keep company?. _____IP	QAS	KILN
PZY	Shall we keep hold of the land? ____QER		
	—To be kept secret		KILOGRAMME (*Measure of weight*) __BAQ
	We can defend ourselves if we keep together _____PYE		KILOLITRE (*Measure of capacity*)___AZC
	What bearing shall I keep the light (*or*, landmark) on?_____QF		KILOMETRE (*Measure of length*) ___AWG
	Will you keep close to me during the night?_____ __ _____IVK		Square kilometre (*Square measure*), AXV
	You are too close in Keep farther off, IVL		KIN (*Measure of weight*)_____ BAR
	You must keep a light burning ____KC	QAT	KIN-SHIP
QAB	KEG	QAU	—Next of kin
QAC	KELSON KEELSON	QAV	KIND-LY-NESS.
		QAW	—It is very kind of you
	KELVIN'S PATENT COMPASS, ETC (*See under* THOMSON)	QAX	—Very kind-ly
	KEN (*Measure of length*)_____AWF	QAY	KIND (*See also* SORT) _____WKJ
		QAZ	—The same kind—as —
	KEPT (*See* KEEP)_____PZX		—What kind—of —?
	KERAN, *or*, KRAN (*Coin*)_____ATV		What kind of bottom have(*or*, had)you? *or*, What is the nature of the bottom? HBK
QAD	KEROSINE	QBA	KING
QAE	KETCH		Chief of the State, President, Emperor, King, Queen, etc , of — has (*or*, is) —, IOR
QAF	KETTLE		King's Bench, *or*, Queen's Bench ___GFH
QAG	KEY—OF	QBC	KINGDOM
	Change the key (*of cipher*)— to —__IKR		United Kingdom_____YJI
	Give me the key to cipher_____IQW	QBD	KINGSTON VALVE
	Have not got the key to cipher_____IQX	QBE	KINK
QAH	—Key of propeller lost	QBF	KIT
QAI	—Lock and key.	QBG	KITE
	Send my keys(*or keys indicated*)___VQA		Endeavor to send a line by boat (cask, kite, *or*, raft, etc)_____KA
	KHAT (*Measure of length*) _____AVO	QBH	KNAPSACK
QAJ	KICK-ED ING	QBI	KNEE
QAK	KID	QBJ	KNIFE
			Knives and forks _____NLS
	KILE (*Measure of capacity*)_____AYT	QBK	KNIGHT ED
	KILEH (*Measure of capacity*) _____AYZ	QBL	KNIT-TED-TING
QAL	KILL-ING To KILL	QBM	KNOB
	Accused of killing some person_____DGI	QBN	KNOCK-ED-ING.
	Boiler burst, — men killed_____AT	QBO	KNOLL
	Boiler burst, — men killed, — others wounded_____AU	QBP	KNOT-TED-ING-Y
	Boiler burst, number of killed and wounded not yet known _____AX		
QAM	Boiler burst, several men killed_____AY		
QAN	—How many were killed?		
	—Killed		
QAO.	Killed and wounded __ _____ ZIO		
	—None killed No one killed		

KNOT—LADING

	KNOT SEA MILE (*See also* MILE) ˍAWO	KORREL (*Measure of weight*)ˍˍˍˍˍBAJ	
	Current, or, Tide will run very strong (*indicate knots per hour, if necessary*).	KOTYLE (*Measure of capacity*)ˍˍˍˍAYU	
	KGB	KRAN (*Coin*)ˍˍˍ ˍˍˍˍˍˍˍˍˍˍˍˍˍˍˍˍATV	
	How many knots an hour are you going?ˍˍˍˍˍˍˍˍˍˍˍˍˍˍˍˍˍˍˍˍˍˍˍRKJ	KREUTZER (*Coin*)ˍ ˍˍˍˍˍˍˍˍˍˍˍˍˍATX	
	How many knots an hour can you (*or vessel indicated*) go?ˍˍˍˍˍˍˍˍˍˍˍˍRTP	KRONA (*Coin*)ˍˍˍˍˍˍˍˍˍˍˍˍˍˍˍˍˍˍATY	
	I shall go about — knots an hour till daylight.ˍˍˍˍˍ ˍˍˍˍˍˍˍˍˍˍˍˍˍˍKNI	KRONE (*Coin*)ˍˍˍˍˍˍˍˍˍˍˍˍˍˍˍˍˍˍATY	
	Keep vessel going about — knots an hourˍˍˍ ˍˍˍˍˍˍˍˍˍˍˍˍˍˍˍˍˍˍˍˍPZR	KRONER (*Coin*)ˍˍˍˍˍˍˍˍˍˍˍˍˍˍˍˍZNK	
	Speed will be — knotsˍˍˍˍˍˍˍˍˍˍˍWMY		
			QCN	KRUPP GUN
QBR	KNOW TO KNOW.	KRUSCHKA (*Measure of capacity*)ˍˍAZD	
	Admiral, *or*, Senior officer wishes to knowˍˍˍˍˍˍˍˍˍˍˍˍˍˍˍˍˍˍˍˍˍˍDKZ	KVINT (*Measure of weight*)ˍˍˍˍˍˍˍˍˍBAS	
	Cargo damaged, but extent not known, HZT	KWAMME (*Measure of weight*)ˍˍˍˍˍˍBAT	
	Date not known Without dateˍˍKLW	KYBOS (*Measure of capacity*)ˍˍˍˍˍˍAZG	
QBS	—Do not know		
	Do not know the anchorage (*or*, harbor), EKU		
	Do not know the channel (*or*, passage), ILH		
	Do you, *or*, Did you get a good look at the land to know exactly whereabouts we are?ˍˍˍˍˍˍ ˍˍˍˍˍˍˍˍˍˍˍˍˍˍˍˍˍˍˍˍPY		
QBT	—Do you know?		
QBU	—Do you know anything of strange sail?		
	Do you know of any anchorage (*or*, harbor)?ˍˍˍˍˍˍˍˍ ˍˍˍˍˍˍˍˍˍˍˍˍˍˍˍˍˍˍˍRKV		
	Do you know the anchorage (*or*, harbor)?ˍˍˍˍˍˍˍˍˍˍˍˍˍˍˍˍˍˍˍˍˍˍˍˍˍEKW		
	Do you know the channel?ˍˍˍˍˍˍˍˍˍILJ		
	Do you know the coast?ˍˍˍˍˍˍˍˍˍˍDHE		
QBV	—Do you know the leading marks?		
QBW	—Do you know the pilotage?		
QBX	—Do you know the reason?		
QBY	—Do you know the tides?		
QBZ	—Do you know where the leak is? Where is the leak?	QCO	L (*Letter*) (*For new method of spelling, see page 13*)
QCA	—I do not know		
	I do not know the anchorage ˍˍˍˍˍEKY	QCP	LABEL-LED-LING
	I do not know the coast ˍˍˍˍˍˍˍˍˍDHB		
QCB	—I know nothing of the locality (*or*, place)	QCR	LABOR-ED-ING
	I know of no safe harbor ˍˍˍˍˍˍˍˍˍVDA	QCS	—Can labor be obtained?
QCD	—I know the channel (*or*, passage)		Dock laborerˍˍˍˍˍˍˍˍˍˍˍˍˍ ˍˍˍˍˍˍˍˍLIZ
	I know the coast ˍˍˍˍˍˍˍˍˍˍˍˍˍˍˍˍˍDGY	QCT	—Laborer
QCE	—Is it known?		
QCF	—Know the reason		LAC (*Money*) ˍˍˍˍˍˍˍˍˍˍˍˍˍˍˍˍˍˍˍˍATZ
QCG	—Known Knew		
	Knowledge (*see below*) ˍˍˍˍˍˍˍˍˍˍˍ QCM	QCU	LACE-D-ING
QCH	—Let him (her, or them) know		
QCI	—Let me know	QCV	LACERATE D-ING-ION.
QCJ	—Well known		
QCK	—Will let you know	QCW	LADDER
QCL	—Will you let me know?		Accommodation ladderˍˍˍ ˍˍˍˍˍˍˍDCM
			Rope, or, Jacob's ladder ˍˍˍˍˍˍˍˍˍPUJ
QCM	KNOWLEDGE		Scaling ladder ˍ ˍˍˍˍˍˍˍˍˍˍˍˍˍˍˍˍVIP
	I have (*or vessel indicated* has) knowledge of the anchorage ˍˍˍˍˍˍˍˍˍˍEKZ	QCX	LADEN WITH (*See also* LOAD)
			Deeply laden ˍˍˍˍˍˍˍˍˍˍˍˍˍˍˍˍˍˍˍKQW
	KOILON (*Measure of capacity*)ˍˍˍˍˍAYZ		
	KOKKOS. (*Measure of weight*)ˍˍˍˍˍBAG	QCY	LADING
	KONGO STATE COLORSˍˍˍˍˍˍˍˍˍˍˍJOU		Are all your bills of lading complete and signed?ˍˍˍˍˍˍˍˍˍˍˍˍˍˍˍˍˍˍGKQ
	KOP, *or*, KAN (*Measure of capacity*)ˍAZF		Bill of lading ˍˍˍˍˍˍˍˍˍ ˍˍˍˍˍ ˍˍˍˍGKP
			Bill of lading not signed ˍˍˍˍˍˍˍˍˍGKR
	KOPECK (*Coin*)ˍˍˍˍˍˍˍˍˍˍˍˍˍˍˍˍˍˍˍATW		False bill of lading ˍˍˍˍˍˍˍˍˍˍˍˍˍˍGKS

	LADING—LAND		

	LADING—*Continued*		LAND—*Continued*
	Have you bill of lading for? ....GKT		High land.........................OUS
	I have no bill of lading for ....GKU		How did the land bear when last seen?
	I will not give up the bill of lading.GKV		EH
	I will not sign the bill of lading—unless,	QDS	—How does the land bear?
	GKW	QDT	—I am standing in for the land
QCZ	LADY-SHIP	QDU	—I had a glimpse of the land about—
		QDV	—I have just lost sight of the land (or,
QDA	LAGOON		light)
			I have not seen the land...........VNQ
QDB	LAID　(*See* LAY)...................QIT		I saw the land bearing — (*add time, if*
	LAKE		*necessary*) ....VNS
QDC	LAME-D-ING-NESS.	QDW	—I see the land　Land in sight. (*Place*
			indicated, if necessary)
QDE	LAMP		I shall keep hold of the land (or, lights)
	Anchor lamp (lantern, or, light) ..BHD		or, I shall keep the land (or, lights)
	Bow lamp (lantern, or, light) .....HCG		bearing — ....QA
QDF	—Lamp cotton	QDX	—I shall not endeavor to make the land
QDG	—Lamp glass	QDY	—I shall not go into less than — fathoms
QDH	—Lamp oil		water whether I make the land or
QDI	—Lamp trimmer		not
QDJ	—Lamp wick	QDZ	—I shall stand in for the land as long as
	Mast-head lamp(lantern, or, light).RMG		I can see
	Not under command lamp (lantern, or,	QEA	—I shall stand in till I make the land
	light) ....JEC	QEB	—I shall stand off and on (*the land*)
	Port bow lamp (or, light) ....HCS	QEC	—I shall stand off the land from — (*hour*
	Side lamp (lantern, or, light) .....WBM		*indicated*) to — (*hour indicated*), or,
	Signal lamp (lantern, or, light) ...WCU		I shall stand off until —
	Speed lamp (lantern, or, light) ...WMX		I sighted land bearing — (*time, if neces-*
	Starboard bow lamp (or, light)....HCU		*sary*) ....VNS
	Stay lamp (lantern, or, light) .....WSD	QED	—I think the land is —
	Steaming lamp (lantern, or, light).WTD	QEF	—I think the land must have been about—
	Stern lamp (lantern, or, light)....WUX	QEG	—I wish to land passengers
		QEH	—Is there anything to prevent landing?
QDK	LANCET		Just lost sight of land.............QDV
			Keep a good look-out for land (or,
QDL	LAND　To LAND.		lights), or, Look out for land (or,
QDM	—Approaching land		lights) (*bearing to follow, if neces-*
	Are you going to land?　Are you going		*sary*) ....PYH
	on shore? ....FCX		Keep under the land　Keep, or, Stand
QDN	—Are you going to land all mails at — ?		closer in ....LT
	Armed men not allowed to land ...EUV	QEI	—Land (*the land*)
	Artillery has been landed..........EZV		Land breeze
	At the landing place..............FIY		Land in sight ....HGB
QDO	—At the nearest landing place		Land in sight　I see the land....QDW
	Bad landing....................FRE		Land indicated bears — ....EK
QDP	—Behind the land	QEJ	—Landed
	Boats should endeavor to land where	QEK	—Landfall
	flag is waved or light shown ......JZ	QEL	—Landing
	Can boat land?　Can a landing be ef-		Landing is impossible ....KD
	fected? ....EW	QEM	—Landing is very difficult
QDR	—Can I land my invalid passengers?	QEN	—Landing place
	Can I land my sick? ....UZ		Landing safe, but very disagreeable.LCE
	Can you give me a leading mark (or,	QEO	—Landing stage
	direction) for making the land here-		Lights, or, Fires will be kept at the best
	abouts (*or at place indicated*)? ...PW		place for landing or coming on shore,
	Do not attempt to land in your own		KE
	boats ....EY		Low land ....RBP
	Do you, or, Did you get a good look at	QEP	—Must be nearing the land　Near the
	the land to know exactly whereabouts		land
	we are? ....PY		Not easy landing on the beach　Scarcely
	Do you see the land?............. VNK		prudent to land ....PZX
	Enemy's troops land-ed-ing ....LZN		Not safe to risk making the land ...VDE
	Foreland....................... NJV		Off the land ....SOR
	Good landing ....ODY		Passenger must be landed..........TDR
	Goods are landed ....OES		Saw land ....VNX
	Had we not better run in and make the		Send suitable boats for landing pas-
	land? ....PZ		sengers ....FI
	Have you seen the land?..........VNO	QER	—Shall we keep hold of the land?
	Headland .... OQR		Stand in—for the land............WQG
			Stand off and on the land.........WQH
		QES	—Table land

LAND—LAST

QET	LAND—*Continued* —Vessel indicated, *or.* Vessel in company is standing in for the land	QFR	LARGE—*Continued* —Large size
QEU	—What do you make of the land? What land do you see? What land do you take it to be?	QFS QFT	—Larger —Largest. Making large purchase ....UCA
	When did you see the land?.........RX Where is the best landing?......GID	QFU	—Not very large Too large (big)............ .....GJN
QEV	—Will you make the land to-night? You are to land all mails at —.. ...RGC	QFV QFW	—Too large a number (*or*, quantity)—of —Which is the largest?
QEW	—You will not be permitted to land	QFX	LASCAR
	LANDFALL ....QEK	QFY	LASH-ED-ING Lash ed-ing alongside.............EBG
QEX	LANDLOCKED	LAST	(*Measure of capacity*)......AZE
QEY	LANDLORD	QFZ	LAST　THE LAST
QEZ QFA	LANDMARK　(*See also* MARK) —Are there any landmarks (*or*, lights)? What bearing shall I keep the light (*or*, landmark) on?.. ....QF		Annul the last hoist, I will repeat it .VE Annul the last signal, the last signal is annulled ....EMJ At last ....FIG At the last extremity ....MQB
QFB	LANDSMAN		Guns firing last night in —(*quarter indicated*) ....OKA
QFC	LANE		How did the land bear when last seen? EH
QFD	LANGUAGE	QGA	—Is this the last? Last account................DFH
QFE	LANIARD　LANYARD		Last autumn................... FNX
QFG	LANTERN Anchor lantern (lamp, *or*, light)...EHD		Last death on the —.... ....KOX Last evening　The last evening .. MIJ
	Bow lantern (lamp *or*, light) ......HCG		Last examination—of (*or*, at) ....MKD
	Did boat take blue light (lantern, *or*, any means of making a signal)?..GSP	QGB	—Last Friday
	Masthead lantern (lamp, *or*, light) .RMG	QGC	—Last mail
	Not under command lantern (lamp, *or*, light) .... JEC	QGD QGE	—Last Monday —Last month
	Port bow lantern (*or*, light) ......HCS	QGF	—Last night
	Side lantern (lamp, *or*, light)....WBM	QGH	—Last quarter (*of the moon*)
	Signal lantern (lamp *or*, light)....WCU	QGI	—Last quarter
	Speed lantern (lamp *or*, light) ....WMX	QGJ	—Last Saturday
	Starboard bow lantern (*or*, light) ..HCU	QGK	—Last season
	Stay lantern (lamp, *or*, light).....WSD	QGL	—Last signal Last signal is annulled ....EMJ
	Steaming lantern (lamp, *or*, light) .WTD	QGM	—Last spring
	Stern lantern (lamp, *or*, light) ....WUX	QGN	—Last summer
	LANYARD　LANIARD ....QFE	QGO	—Last Sunday
		QGP	—Last Thursday
QFH	LAPSE-D-ING　To LAPSE	QGR	—Last time
		QGS	—Last Tuesday
QFI	LARBOARD　(PORT)　SHIP'S HEAD TO GO TO STARBOARD　(*See* PORT)	QGT	—Last voyage
		QGU	—Last Wednesday
QFJ	LARD	QGV	—Last week.
		QGW	—Last winter
QFK	LARGE-LY-NESS	QGX	—Last year
	A large (great) number (*or*, quantity)—of — ....OGZ		Repeat the last signal made.......WOR The last (*the last one*) ....QFZ
QFL QFM	—A large portion—of —A larger number (*or*, quantity)—of —(*or*, than —)		The last hoist is not understood, repeat it.................. ....URK
	A very large number (*or*, quantity)—of — ....SLE		To, *or,* At the last extremity ......MQB What are the last quotations? .....UFZ
QFN QFO	—Can not take anything larger than — —How large—is (*or*, are)—?		What are the latest dates by the last mail? .... KMH
	Large arrivals of —............EYN		What is the date of your last letter (*or*, paper)? ..KMJ
	Large flare　Holme's light........NDY		When, *or,* What did you last hear of (*or*, from)...... .... ORU
	Large hawser... ....OPW	QGY	—When last seen
QFP	—Large ship (*estimated tonnage to follow*)		When will the last (*or*, end) be?...LYP Where did you take your last departure from?..... ....KTY

	LAST—LAY

QGZ	LAST-ED-ING TO LAST (*See also* CONTINUE)
	During the gale While the gale lasts, LPS
	Is illness likely to last?............PCW
QHA	—It will not last
QHB	LATE
QHC	—Before it is too late
	Compasses not adjusted lately......JHL
	I have not had observations for variation lately ....JHV
QHD	—Late in
	Lately (*recently*)......... ....ULH
QHE	—Later-est
QHF	—Latest news
QHG	—Latest posting time—is
QHI	—Too late
	What are the latest dates by the last mail? ....KMH
	What is the date of the latest news from — ? ....SE
	What is the date of your latest newspaper (or, letter)?........ ....KMJ
	What were the latest — dates at — when you last heard? ....KMN
QHJ	—Which, or, What is the latest?
QHK	LATEEN
QHL	LATH
QHM	LATHE
	LATITUDE............Code Flag over AB
	NOTE.—For *degrees of latitude, see page* 56
	Degree of latitude ....KRY
	Derelict reported passed on —, latitude —, longitude —, dangerous to navigation ....KVR
	Difference of latitude.............KZF
	Have passed ice in latitude — and longitude —, on the —............PQ
QHN	—How is your latitude obtained?
	I have been injured by the ice in latitude —, longitude — ....PBN
	I saw ice in latitude — (*date to follow, if necessary*) ....PBT
QHO	—In the latitude of —
QHP	—In what latitude?
QHR	—In what latitude did you see —?
QHS	—Is the latitude correct?
QHT	—Latitude by dead reckoning—is
QHU	—Latitude by observation—is
QHV	—Minutes of latitude
QHW	—My latitude is —
QHX	—North latitude
	Rendezvous in latitude — and longitude — ....JC
	Report the — (*vessel indicated*) all well in latitude — and longitude —, on the — (*date indicated*) ....UN
	Saw derelict in latitude —, longitude —, on the —...............JH
QHY	—Second of latitude
QHZ	—South latitude
QIA	—What is the latitude of —

QIB	LATITUDE—*Continued*
	—What is your latitude brought up to the present moment?
	What latitude and longitude did you (*or, they*) have ice in? ....PS
QIC	—What latitude did you go into?
QID	—What latitude do you intend going into?
QIE	—What was your latitude at noon?
QIF	—When were your last observations for latitude?
QIG	LATRINE (*or*, URINAL)
QIH	LATTER-LY
	At the latter end of, *or*, The latter end — of.........................FIZ
QIJ	LAUDANUM
QIK	LAUNCH-ED ING TO LAUNCH
	Countermining launch ....JYR
	Electric launch...... ....LUW
	Steam launch...................GLV
	Steam launch's engine. ....MBY
QIL	LAW-FUL-LY
	Am I compelled by law to have a pilot? JIS
	Break ing Broke the quarantine laws, UEB
	Bye-law ....P Y
	Court of law...................KAD
	Lynch law ....RDO
QIN	—Martial law
QIO	LAWLESS-LY-NESS
QIP	LAWN TENNIS
QIR	LAWSUIT
QIS	LAWYER
QIT	LAY-ING LAID TO LAY
	An embargo has been laid on at —..LWB
	Cable-laid, Cable-laid rope.........HQB
	Hawser-laid rope.................OPV
	Help me to lay out an anchor ..... EIK
	I am laying a telegraph cable Keep out of my way ....XLQ
	If you will lay out an anchor for me I can get off..................EIV
QIU	—Laid up to have a new boiler put in
	Lay-day ....KNL
QIV	—Lay fires
QIW	—Lay in
QIX	—Lay out
	Lay out an anchor ... .EIZ
	Mines laid ....RUP
	The (*nation specified*) have laid an embargo on all vessels (*of nation specified*)........... .. ..LWK
	There are no mines laid..........RUS
	When is the embargo to be laid on?..LWM
	Will you assist me (*or vessel indicated*) in laying out an anchor?........EKL
QIY	LAY ING LAID-TO TO LAY-TO
QIZ	—I can not lay-to
	I shall lay-to during the night.....LPX
QJA	—I shall lay-to off — (*or, at —*)

	LAY—LEAVE		
	Lay—*Continued*		**Leak**—*Continued*
QJB	—I shall lay to until —		Do you know where the leak is? Where
QJC	—Lay-to		is the leak?........................QBZ
QJD	—Lay-to until — •		Fire, or, Leak, want immediate assist-
			ance ....NH
	Lay-day..... ....KNL		Furnace leaky . ...NUD
QJE	Lazaretto	QJW	—Have you materials for stopping the
			leak?
	Lazy-ily-iness (*Idle*)................PCN		Have you sprung a leak? Are you, or,
			Is — (*vessel indicated*) leaking?....QJ
	Le, Cash or, Sapeque (*Coin*).......ATC	QJX	I have sprunk a leak..............,....NQ
QJF	**Lead** (*Sounding Lead*)	QJY	—Leak at the scuttles
QJG	—Arm the lead		—Leak increases
	Deep-sea lead...... ....KQS		Leak is gaining rapidly......... .....NU
	Hand lead ....OLY	QJZ	Leak is stopped.....................QK
QJH	—I shall keep the lead going regularly		—Leaked, Leaking
QJI	—Keep the lead going Keep sounding		Spring-ing, Sprang, Sprung, a leak. WOS
QJK	—Lead line		Steam pipe leaky ....WSO
QJL	—Leadsman		Vessel indicated has sprung a leak ..QL
	Patent lead .... TED		Water tanks leaky...................YWE
			Where is the leak?............ .....QBZ
QJM	**Lead-en** (*Metal*)	QKA	Lean-ness
	Red lead............................UMG		
	Sheet lead................. ....VWD		Leang, or, Tael (*Coin*) .,........ ..AVC
	White lead ....ZDI	QKB	Leap-ed-ing
QJN	**Lead-ing-er Led**		Leap year ....ZKW
	Can you give me a leading mark (or,		
	directions) for making the land here-	QKC	Learn-t ing
	abouts (or *at place indicated*)?....PW	QKD	—Could not learn name
QJO	—Can you make out the leading marks?	QKE	—Learnt by signal from —
	Convoy to keep astern of leader (or,		Might have learnt it by signal.....WCO
	escort) ....IW		
	Convoy to keep on port side of leader	QKF	Lease-d-ing Leasehold
	(or, escort)...........................IX		
	Convoy to keep on starboard side of	QKG	**Least**
	leader (or, escort)...... ....IY		At least .... FIH
	Disregard leader's movements Disre-	QKH	—The least number (or, quantity)—of
	gard motions....................LFX		What is the least depth of water we
	Do you know the leading marks? ..QBV		shall have (or, ought to have)?....VR
	Fair lead Fair leader............MRD		Without the least confusion...... ..JOS
QJP	—Leading mark		
QJR	—Leading stoker	QKI	**Leather**
QJS	—What, or, Which are the leading marks?		Pump leather .... UBK
	What are the leading marks for crossing		
	the bar? ....FWC	QKJ	**Leave-ing To leave**
	What wind leads through the passage	QKL	—Before leaving
	(or, entrance)?....................ZFD		Calls the attention of vessels leaving the
QJT	—Will you lead?		anchorage, etc......... ....DQ
	Will you lead into (or, point out) a good	QKM	—Convoy will leave at — (*time to follow*)
	berth? ....GHI	QKN	—Convoy will leave on — (*day to follow*)
			Crew have left the ship (*number to be*
QJU	**League** (*Alliance*)		*shown*).............................. KCP
			Crew will not leave the vessel ......KCX
	League (*3 miles*) ....AWH		Date of leaving . ...KLY
		QKO	—Did you leave?
QJV	**Leak-y**	QKP	—Did you leave many ships at —?
	Are you, or, Is—(*vessel indicated*) leak-	QKR	—Do not leave anything unsettled
	ing? Have you sprung a leak?....QJ		Had the mail arrived when you left?
	Boiler leaking seriously..............RF		EYG
	Boiler leaking slightly .... GWB		Has the mail left — (for —, *if necessary*)?
	Boiler leaky-ing ....GWA		QO
	Boiler leaky, must be blown off....GWC	QKS	—I am leaving
	Boiler tubes leaking .... RG	QKT	—I have left on shore
	Can you stop the leak?..... .. ..QG		I leave it to your arrangement.... EXC
	Can not stop the leak........ ....QH	QKU	—I left — in port, or, — was in port when
	Condenser tubes leaking ....JMW		I left (*Ship's name or distinguishing*
	Do you consider leak dangerous?.....QI		*signals to follow*)

	LEAVE—LEFT

	LEAVE—*Continued*
QKV	—I left port (*or place indicated*) — days ago
QKW	—I will leave a message at —
	Intend ed leaving Purpose-d leaving, ...POX
	It is expected the vessels (*or vessels indicated*) will be allowed to leave...DZJ
	Leave, or, Put a buoy on............HMP
QKX	—Leave a good margin
QKY	—Leave any papers or letters for me (or *for party indicated, or in question*) at —
QKZ	—Leave behind
	Leave buoy — at ----------------HMQ
QLA	—Leave immediately—to (or, for)
	Leave-ing, Left — for —............QLE
QLB	—Leave it for me at the railway station
	Leave it to your arrangement.......EXC
QLC	—Leave on board
QLD	—Leave the
	Leave the buoy (or, beacon) to port...LZ
	Leave the buoy (or, beacon) to starboard ----------------------MA
	Leave the matter in the hands of Lloyd's agent ....QWE
	Leave your address at —............DJN
QLE	—Leave — for —
QLF	—Left
QLG	—Left port
QLH	—Let him (or, them) leave
	Mail has left.. -------------------RFH
QLI	—Must leave
	None left (*remaining*)... ....SJB
QLJ	—Not left
	Omnibus, or, Car leaves at —.......HYB
QLK	—On leaving
	Overland mail leaves on the —......SXM
	Purpose leaving Intend leaving...POX
	Report me to owner on leave.....USB
	Ships from — did not get pratique at — when I left ----------------------TSA
	The mails leave on the — — .. RFW
	Train leaves for — at -----------XZB
	Vessel just weighed (or, leaving), show your distinguishing signals ....DZ
QLM	—Vessels of nations specified were all leaving
	Vessels wishing to be reported on leaving show your numbers (or, distinguishing signals) ---- ----------EB
QLN	—Was all well when you left?
	Was the port you left healthy?.....ORB
	Was well when I left---- ----- ---QMO
QLO	—We have left men behind (*number to be shown*)
QLP	—We left port (*or place indicated*) on the —
	What day do you leave? ----------KOA
QLR	—What men of-war did you leave at —?
QLS	—What vessels did you leave in port (*or place indicated*)?
	What was date of your (*or as indicated*) leaving? ----------------------KML
QLT	—When does (do or, did) he (she, it, or, they) leave?
	When do the trains leave — for? ...XZS
QLU	—When do (or, did) you leave?
	When does omnibus (or, car) leave for —? ---- ---------------------HYC
	When does the mail leave?.........QT

	LEAVE—*Continued*
	When ought the next mail to leave?.RGB
QLV	—Will you leave?
	Will you leave a message for me?....IL
	You have left a person behind .....GDZ
	— was well when I left...........QMO
QLW	**LEAVE** (See also **PERMISSION**)
	Absent without leave.............DAW
	Give leave — to — ....NZX
QLX	—Leave of absence
	Men on leave.....................RPF
QLY	—On leave
QLZ	**LE CLANCHÉ** BATTERY
	LED (see **LEAD**) ....QJN
QMA	**LEE**
	Keep more to leeward............PYX
	Keep on the lee beam........ ....GAQ
QMB	—Keep to leeward
	Keep to leeward of vessel chasing.. PZN
	Lee beam ....GAS
	Lee bow On lee bow ....HCN
QMC	—Lee helm
QMD	—Lee lurch
QME	—Lee quarter
QMF	—Lee scuppers
QMG	—Lee shore
QMH	—Lee side
QMI	—Lee tide
QMJ	—Leeward — of —.
QMK	—Leeway
	Look out on lee bow . ...HCJ
	Look out to leeward ....QZV
	On lee bow.......................HCN
	Shoal water, or, Danger to leeward.KLH
QML	—To leeward
	We are to leeward of harbor (or, anchorage) -------------------------ELH
	We are too far to leeward ....MTE
QMN	**LEECH**
	LEEWARD **LEEWARD OF** ....QMJ
	LEEWAY ------------------------QMK
	LEFT (See also **LEAVE**) ....QLF
	Crew have left the ship............KCP
	Had the mail arrived when you left? ....EYG
	Has the mail left—(*for —, if necessary*)? ....QO
	I have left on shore................QKT
	I left — in port, or, —was in port when I left (*ship's name or distinguishing signal to follow*) ----------------QKU
	Left port ....QLG
	Lighter has left ....QTD
	Mail has left ....RFH
	No anchors left.... ------------EJF
	Only one anchor left.............EJG
	Ships from — did not get pratique at— when I left ----------------------TSA
	Was in harbor (or, port) when I left, ...QKU
	Was well when I left..............QMO
	We have left men behind (*number to be shown*) ------------------ ---------GLO

LEFT—LET

	LEFT—*Continued* We left port (*or place indicated*) on the — ....QLP You have left a person behind ......GDZ	QNI	**LENT** (*Fast before Easter*)
QMO	— — was well when I left		LEO, *or* LEU (*Coin*) .. ..AUB
QMP	**LEFT** (*On the left*) Left arm.......................RVK On the left bank of the river (i e , *the left bank in descending the stream*), FVE		LEPTON (*Coin*).....................AUO
		QNJ	**LESSEN**-ED-ING
QMR	**LEG** (*Part of the body*)	QNK	**LESS** Do not come into less than — feet of water ....FV I shall not go into less than — fathoms of water whether I make the land or not ....QDY
QMS	**LEGAL**-LY-ITY Legal authorities..... ....FNK Not legal-ly Illegal-ly-ity ....PCY	QNL QNM QNO QNP	—Less than —Much, *or* Many less —Not less —Nothing less There is less current inshore (*or in direction indicated*)....................KQD You will have less than — feet of water, KVJ You will not have less than — feet of water ....VU
QMT	**LEGATION** Secretary of Legation ....VMN		
QMU	**LEGIBLE**-Y Illegible-y Not legible ....PCZ		
QMV	**LEGION OF HONOR**		
QMW	**LEGISLATE**-ION-URE-IVE Assembly, Congress, *or*, Parliament (*or other legislative body*) adjourns (*or adjourned*) on the — ....FEI Assembly, Congress, *or*, Parliament (*or other legislative body*) is dissolved.FEJ Assembly, Congress, *or*, Parliament (*or other legislative body*) met (*or meets*) on the —..................JPB	QNR	**LET-TING** **TO LET** (*To lend on hire*)
		QNS	**LET**-TING (*See also* ALLOW) Can you sell (*or, let me have*) any —? VOB Do not let him (*or, them*) go ......OBP Let fall.......................MRV Let go.......................OCS Let go another anchor...............KR Let go the life buoy ....HMR Let go your anchor... ....EJA Let her be ....BGV
QMX	**LEISURE**-LY	QNT	—Let him (his, her-s, it-s) Let him (her, it, *or person-s or thing-s indicated*)—have.................BWA Let him be...................BXE Let him bring on shore.............FCY Let him (*or, them*) go.............OCT Let him (her, *or,* them) know......QCH Let him (*or,* them) leave.........QLH
QMY	**LEMON** Lemon, *or*, Lime juice .... QUE		
QMZ	**LEND**-ING **TO LEND** (*See also* SPARE *and* SUPPLY) Can you lend (*or,* supply me with)?.XET Can not lend (*spare*) ....WMC I can lend you (*or, furnish you with*) —, NUH I can not lend you (*or, furnish you with*) — ....NUI	QNU	—Let me (my, mine) Let me know ... .QCI Let no boat come alongside........HBH
QNA	—I will lend what is required Lend me a boat ....GTQ	QNV QNW	—Let nothing prevent —Let run Let it run. Let the men................. .....RPE
QNB QNC	—Lent —Will you lend?	QNX	—Let them (their-s) Let them be...................BGW Let them bring ashore.............FCY Let them bring their tools with them, HJD
QND	**LENGTH** Forward following telegraphic message by signal letters instead of writing it at length................ ... ...XB	QNY	—Let us (our s) Let us be ....BGX Let us keep together for mutual support and protection ....IN
QNE	—Length of chain Measure of length (*see page* 51).....AVK Send following message through the post to owners (*or, to* Mr —) at — by signal letters instead of writing it at length ... .WZ What is the length of dock at —? ..LJP	QNZ	—Let your-s Let your boats keep to windward until they are picked up.............FD
		QOA	—Let your boats pull to — (*bearing indicated*) Let your fires down ....NAM
QNF	—What length—is —? What length are your cables (*or cable indicated*)? ....HBD	QOB	—Let your screw (*or,* wheels) revolve without disconnecting Stand by to let go.............WQE Will let you know..............QCK Will you let me know ....QCL
QNG	**LENGTHEN**-ED-ING		
QNH	**LENS**		

	LETTER—LIE

QOC	LETTER (*See also* DESPATCH *and* MAIL)
	Address my letters to me at — _____DJI
QOD	—An opportunity for sending letters to Europe (*or*, to —)
	Any letters (*or*, papers) for me (*or person indicated*)?_____EPA
	Any letters (*or*, papers) from —?____JI
	Any letters (*or*, despatches) on board? KWX
	Bring all the letters and papers ____HIL
	Call at the post office for letters __ HSY
	Call for my letters _____HSZ
	Can I take (*or*, forward) any letters for you?_____ _____NMV
QOE	—Did you hear if there were any letters for me at —?
	Forward following telegraphic message by signal letters instead of writing it at length_____ ___XB
	Forward letters—to (*or*, for) _____NMX
	Had any letters (*or*, despatches) arrived—at?_____ ___ _____ _____JM
	Have (*or vessel indicated* has) despatches (*or*, letters) for—_ ___ _____JN
	Have (*or vessel indicated* has) despatches (*or*, letters) from—_____ ___JO
	Have letters (mails, *or* despatches) for you (*or vessel or person indicated*)_ JP
	Have my letters been forwarded?__NOA
	Have you letters (mails, *or*, despatches) for me (*or vessel, or person, or place indicated*)? _____JQ
QOF	—Have you received my letter?
QOG	—I have a letter (*or* parcel) for you (*or*, for —)
QOH	—I have received your letter
	I will forward your letter on _____NOE
	Keep my letters (*or*, papers) for me till I return_____PYK
	Leave any letters(*or*, papers) for me (*or person indicated*) at —_____QKY
	Letter bag Mail bag_____FSJ
QOI	—Letter of credit
QOJ	—Letters and parcels
QOK	—Letters had (*or*, have) been received from —
QOL	—Letters must be sent—to (*or*, by)
QOM	—Letters of squadron (*or*, fleet)
QON	—Letters received from —
QOP	—Letters, *or*, Papers up to the— (*Indicate where from if necessary*)
	Mail bag Letter bag_____FSJ
	Men's letters _____ RPG
	My letters are all forwarded _____NOF
	Opportunity for letters to Europe (*or*, to —) _____ _____QOD
	Refer to my letters (*or*, despatches) JU
	Refer to your letters (*or*, despatches) JV
	Send following message through the post to owners (*or*, to Mr —) at — by signal letters instead of writing at length _____WZ
	Send for my letters_____ _____VPS
	Send for letters (*or*, despatches) ____JW
	Send my message through the telegraph by signal letters____ _____XB
	Send the letter bag on shore (*or to vessel indicated*)_____FSN
	Send your letters _____QR
	Signal letter___ _____ _____WCV
	Take letters (*or*, despatches) _____KXC

	LETTER—*Continued*
	Telegraph to my owners to send my letters to me at —_____VQP
	Tell — not to forward any more letters for me _____QS
	Were my letters properly forwarded? NOJ
	What is the date of your last letter (*or*, paper)? _____KMJ
	Will you forward my letters (*or*, despatches)—to_____ _____ _____QU
QOR	—Will you post some letters for me?
QOS	Will you take letters(*or*, despatches)?__JY
	—Your letters went by the —
	LEU, *or*, LEO (*Coin*)_____ ____AUB
	LEV, *or*, LEW (*Coin*)_____AUB
QOT	LEVANTER
QOU	LEVÉE
QOV	LEVEL-LED-LING LER-NESS
QOW	LEVER
	Air-pump levers out of order__ __ DWC
QOX	—Levers damaged (*or*, out of order)
	Pump lever links broken ____ ____UBL
	Pump levers out of order____ _____UBM
	Side lever engine_____ _____MAW
QOY	LEVY-IED-ING
	LEW (*Coin*)_____AUB
	LI (*Measure of length*)_____ ____AWI
QOZ	LIABLE-ILITY
QPA	—Not liable
QPB	LIAR LIE-D YING
QPC	LIBEL-LED-LING-LOUS
QPD	LIBERAL-LY-ITY
	Liberal contribution _____ ____ ___JTB
QPE	LIBERATE D-ING-ION
QPF	LIBERIAN LIBERIAN COLORS
QPG	LIBERTY
QPH	—Am I, *or*, Is —(*person indicated*) at liberty—to?
QPI	—Is *or*, Are at liberty—to
QPJ	—Liberty man
	LIBRA (*Measure of weight*)_____BAU
QPK	LIBRARY-IAN
QPL	LICENSE
QPM	—Are you regularly licensed, *or*, Have you a permit?
QPN	—Licensed-ing To license
QPO	—Not licensed Unlicensed
	LIE-D-YING LIAR_____QPB

LIE—LIGHT.

QPR	LIE, LAID, LYING
	How high must I lie to fetch?_____MWE
QPS	—Lying in (or, at) —
QPT	—Lying in the roadstead
	You will not lie afloat all the tide__DQG
	LIEU. IN LIEU—OF (Instead of) ___PNA
QPU	LIEUTENANT-CY
	Deputy, or, Lieutenant governor___KVN
	Lieutenant colonel-cy___ _____JBH
QPV	—Lieutenant general
	Lieutenant in the Army_____EWL
QPW	⊤—Lieutenant in the Navy
QPX	—Lieutenant in the Naval Reserve
QPY	—Senior lieutenant
	Sub-lieutenant _____XBG
QPZ	—Sub-lieutenant in the Naval Reserve
QRA	LIFE
QRB	—I have no life belts
	Life belt _____QRH
QRC	—Lifeboat
QRD	—Life buoy
QRE	—Life line
	Life-saving (Rocket) apparatus ____ERH
	Life-saving appliances _ _____ESL
	Life-saving (Rocket) station_____WRP
QRF	—No lives lost
QRG	—Several lives lost
QRH	LIFE BELT
	I have no life belts _____ _____QRB
	LIFEBOAT_____QRC
	Boat, or, Lifeboat can not come_____ER
	Do you want a lifeboat?_____EZ
	Have no lifeboat No lifeboat here_FA
	Have you any lifeboats? _____GTD
QRI	—Lifeboat Institution
	Lifeboat is going to you_____FE
QRJ	—Lifeboat sent for
QRK	—Lifeboat station
	Lifeboat unable to come _____ER
	Send lifeboat to save crew_____FJ
	Steam lifeboat _____GUY
	We have sent for lifeboat _____FL
	LIFE BUOY _____ _____QBD
	Let go the life buoy_____HMR
QRL	LIFT-ED-ING
	Can you lift your screw? _____VKE
QRM	—Lift rope
QRN	—Lift screw
QRO	LIGHT A LIGHT
QRP	—Alternating light
	Anchor light (lantern, or, lamp)___EHD
	Arc light _____EUI
	Blue light _____GON
	Bow light (lantern, or, lamp)_____HOG
	Carry a light on your bowsprit end___QY
	Catoptric light_____IFA
	Daylight At daylight _____KNB
	Dead light _____KOG
	Dioptric light _____ _____LAZ
	Electric light _ _____LUX
	Fixed light . _____ _____NCS
	Fixed and flashing light _____NCR
	Flashing light_____NEB
	Floating light _____ _____NFB

	LIGHT—Continued
	Gas, or, Light buoy _____ _____NXA
	Gaslight _____ ____NXB
	Green light _____OHT
	Group flashing light_____ _____NEC
	Group occulting light_____SOB
	Holme's light (large flare)_____NDY
	Holme's light (small flare)_____NDZ
	Incandescent light__ _____ PGA
	Intermittent light _____PQX
	Masthead light (lantern, or, lamp) RMG
	Moonlight _____ _____RYX
	Not under command light (lantern, or, lamp) ____ ___ _____JEC
	Occulting light _____SOA
	Port-bow light _____HCS
	Position light _____TFE
	Red light ___ _____UMH
	Revolving light_____UMH
	Scintillating light_____VJM
	Search light, or, Electric projectors LVI
	Side light (lantern, or, lamp) _ ____WBM
	Signal light (lantern, or, lamp) ___WCU
	Skylight _____WFM
	Speed light (lantern, or, lamp)____WMX
	Starboard bow light_____HCU
	Starlight __ _ _____WQR
	Stay light (lantern, or, lamp) _____WSD
	Steaming light (lantern, or, lamp)_WTD
	Stern light (lantern, or, lamp)_____WUX
	Sunlight _____XDU
	Towing light_____XYZ
	Very's pistol light_____ _____YPV
	White light _____ ____ _____ZDJ
	All the lights are out along the coast— of —_____ _____ ____ _____PU
	Allow no lights to be seen Extinguish all lights __ _____ _____DZB
	Are there any alterations in the lights (or, marks)?_____ECB
	Are there any lights (or, landmarks)? QFA
	At — (time indicated) light bore —__PV
	Beach your vessel where flag is waved (or, light is shown) _ _____FZT
	Boats should endeavor to land where flag is waved (or, light is shown)__JZ
	Burn blue light or, Flash powder__GOQ
	Carry a light at your bowsprit end__QY
	Did boats take a blue light (lantern, or, any means of making a signal)?_ OSP
	Do not bring the light to the — (bearing indicated) of — (bearing indicated) PX
QRS	—Do not light your fires ⏹
	Do not show a light on any account_QV
QRT	—Do not show a light on my account
	Electric light installation __ _____LUY
	Extinguish all lights Allow no lights to be seen _____DZB
	Fire and lights out _____ _____MZQ
	Fire, or, Light will be kept at the best place for landing _____KE
	Flare-d-ing—up Flare light _____NDX
	Flash powder, or, Burn a blue light_GOQ
	Have you any blue lights? _____GOR
QRU	—Have you any signal lights?
QRV	—Have you seen the light?
	Hoist a light_____ _____ _____OWC
	How far can light be seen?__ _____MSZ
QRW	—I do not think there is any light hereabouts
QRX	—I do not wish to show a light

LIGHT—LIGHTNING

	LIGHT—*Continued.*
	I have just lost sight of the land (*or*, light) .... QDV
	I shall keep hold of the land (*or*, light), or, I shall keep land (*or*, light), bearing — .... QA
QRY	—I will carry a light I will show a light
	I will show a light to-night when I alter course .... QW
QRZ	—I will show a light to-night when I make sail
	I will show a light to-night when I tack, ... QX
	Indicate bearing of light (lighthouse, *or*, light-ship) .... QB
	Keep a good lookout for lights (*or*, land), or, Look out for lights (*or*, land) .PYH
	Keep a light at your bowsprit end.... QY
	Keep a light burning .... KC
	Keep ahead and carry a light........ DVK
	Keep lights (*or*, fires) on the beach (*or*, shore) all night .... FZW
	Keep the light between — (*bearing indicated*) and — (*bearing indicated*)..QC
	Keep the search light on, *or*, Keep electric light on .... LVH
	Light-er-est .... QST
QSA	—Light-ing-ed Lit To light
QSB	—Light dues
QSC	—Lighting fires
QSD	—Light indicated has been replaced
QSE	—Light indicated is not now in its old position
QSF	—Light indicated will commence on the — (*date indicated*)
QSG	—Light, *or*, Floating light is discontinued (*or*, gone)
	Light is not to be depended upon.... QD
QSH	—Light on — is not working satisfactorily
QSI	—Light signal
QSJ	—Light your fires
	Lighted Lit To light.......... QSA
	Lighting To light.......... QSA
	Lights, *or*, Fires will be kept at the best place for landing or coming on shore, ... KE
	Off the light...................... SOU
	Shall we have daylight enough?.... KNT
QSK	—Shall you show a light?
	Temporary light shown at — (*place indicated*) . ... XNS
QSL	—There is a light on —, *or*, The lights on—
QSM	—They are only vessel's lights
	To light Lighting-ed.......... QSA
	Turn the electric light off .... LVJ
	Turn the electric light on .... LVK
	Use your search light .... YMT
QSN	—Vessel indicated is showing a light
QSO	—We did not see the light
	What bearing shall I keep the light (*or*, landmark) on?... .... QF
QSP	—What sort of a light is it?
	When I alter course to-night I will show a light...................... QW
	When you alter course (*or*, tack) to-night show a light .... JZS
QSR	—Will you carry a light?
	You must keep a light burning...... KC
	Your lights are out (*or*, want trimming)
	—Made by — — — *flashes or by blasts of a steam whistle*

QST	**LIGHT-ER EST** (*Not dark*)
QSU	**LIGHT** (*Not heavy*)
	Calm and light winds .... HTM
	Light breeze .... HGD
	Light guns...................... OKE
	Light in ballast.................. FTR
QSV	—Light sail Upper sail
	Light unsettled weather .... YZB
QSW	—Lighter-est
	Take in light sails............... XJN
	Wind light .... ZFL
QSX	**LIGHTEN ED-ING**
QSY	—Must lighten the ship
	What draught of water (*in feet*) could you lighten to? .... OB
QSZ	**LIGHTER**
	Coal lighter .... IWP
QTA	—How many lighters do you want?
QTB	—I want — (*number indicated*) lighters
QTC	—Lighter alongside
	Lighter coming off.............. QZ
QTD	—Lighter has left
	Lighter is adrift................ RA
	Powder hoy (*or*, lighter) .... OZU
	There are no lighters available...... RB
	Want a lighter (*or*, lighters) immediately (*if more than one, number to follow*) ... . YM
QTE	—Want lighters to take my cargo
	What quantity of coal remains in the collier (coal depôt, *or*, lighter)? ..IXO
	Will you send off lighters as fast as possible?... .... RC
QTF	**LIGHT-HOUSE**
	Indicate bearing of light (light-house, *or*, light-ship) .... QB
	Light on—is not working satisfactorily, ... QSH
	Light-house, *or*, Trinity Board..... GPN
	Light-house, *or*, Light-ship at — wants immediate assistance.............. CW
	What is the name of light-house (*or*, light-ship) in sight?.............. RW
QTG	**LIGHT-SHIP**
	Indicate bearing of light (light-house, *or*, light-ship) .... QB
	Is the light-ship on her station? ...WRO
	Light-ship at — is out of position, *or*, Light-ship is not at anchor on her station .. .. QE
	Light-ship, *or*, Light-house at — wants immediate assistance.......... CW
	Light-ship has been withdrawn on account of ice ... . PCB
QTH	—Steer directly for the light-ship
	What is the name of light-ship (*or*, light-house) in sight?............ RW
QTI	**LIGHTNING**
QTJ	—Lightning conductor
QTK	—Struck by lightning
QTL	—There has been, *or*, We have had a storm of thunder and lightning

LIKE—LIST

QTM	**LIKE LY** Am aground, likely to break up, require immediate assistance ------------CA Channel, *or*, Fairway is likely to be mined ------------------------ILD Is embargo likely to be taken off?--LWI Is illness likely to last?-- --------PCW
QTN	—Is it likely? Is it probable? Is it likely you (*or vessel indicated*) will be got off? --------- ----------OFJ
QTO	—Is likely to recover Is — (*vessel indicated*) likely to assist (*or,* supply) me? - -------------FGK
QTP	—Likeness (*resemblance*)
QTR	—Likely to change Likely to fail --------------------MQZ
QTS	—Likewise. Looks like----------------------ESA
QTU	—Looks likely
QTV	—Most likely (*or,* probable).
QTW	—Not likely Unlikely Pilot boat most likely — (*on bearing indicated*), *or,* off ----- -------- TK There is a likelihood of being becalmed, GCK Very likely -------------------- YPO Vessel indicated not likely to supply (*or,* assist) you-----------------FGR
QTX	—What is it like?
QTY	—What is she like? What is the sea like? ------------UX When do you think I am likely to fall in with —(*vessel indicated*)-- ----MSA Where do you think I am likely to fall in with —? --------- -----------MSB You are not likely to receive your orders for some days - ---- ------------SUX
QTZ	**LIKE-D-ING TO LIKE** Do, *or*, Does not like the appearance of, ERW
QUA	—I do not like I do not like the look of the weather--ZE I do not like to run in I prefer keeping the sea------------------- --- --VCA I do not like to trust too much to my—, YCG Natives do not like ships watering without payment -----------------YWQ
QUB	—Should like Should like to have another comparison, GV Would you like to see our newspapers? SL
QUC	**LIMB.**
QUD	**LIME**
QUE	—Lime, *or,* Lemon juice.
QUF	**LIMIT-ED-ING** Inside of limit------------------PMJ
QUG	—Outside of limit
QUH	**LINE-D-ING TO LINE**
QUI	**LINE A LINE** Along the line-------------------EAL Clew line Clew garnet -----------IUE Endeavor to send a line by boat (cask, kite, *or,* raft, etc) --------- -----KA

	LINE—*Continued* Fishing line ---------------------NBV Hauling line ------------ --------OPB Have a towline ready ------------XYF
QUJ	—Have you any means of throwing a line?
QUK	—Have you sufficient towline?
QUL	—In line Inner line ----------------------PLN Is the line fast?------- ----------KB Lead line ------------------------QJK
QUM	—Keep the two objects named in one line Life line ------ --------------------QRE Line Equator----- -----------MGK Line-of battle ship Battle ship---FYX
QUN	—Line of communication
QUO	—Line of defense
QUP	—Line of mines
QUR	—Line of torpedoes
QUS	—Log line Look out for rocket line (*or,* line) -KF Look out for the towline ----------QZU Mortar for throwing line ----------SAP Rocket to throw a line L S A. rocket, UZG
QUT	—Send a line
QUV	—Send for the towline Sending, Sent for the towline------VPX The Line Equator---- ----------MGK Towline -------------------------XZA Water line---------------------- YVU
QUW	—We will endeavor to send a line When did you cross the Lane? -----KEL Where did you cross the Line? ----KB Where will you cross the Line?----KEN
QUX	**LINEN** Bed linen ------------- - --------GCT Can I get linen (*or,* clothes) washed? IVQ Clean linen (clothes, *or,* shirts)-----ISU Dirty linen ---------------------VYO
QUY	**LINIMENT**
QUZ	**LINK-ED-ING** Pump lever links broken ----------UBL
QVA	**LINSEED**
QVB	—Linseed oil
QVC	**LINT**
QVD	**LIQUEFY-IED ING-ACTION**
QVE	**LIQUID** Floating, *or,* Liquid compass ------JHP
QVF	**LIQUOR** Malt liquor--------------------RJC
	LIRA LIRE (*Coin*) -------------AUD Lira egiziana (*Coin*)----- - - ---ATO Lira turca (*Coin*) ----------------AUB
QVG	**LIST—OF** —A list of —
QVH	Army List----------------- ------ --EWD Code List of Ships---------- -----IZD
QVI	—Is your list of ships correct? Lloyd's List---------------------QWU
QVJ	—Mercantile Navy List

LIST—LOCOMOTIVE

QVK	LIST—*Continued.*
QVL	—Passenger list
QVM	—Remittance list
	—United States, Royal, or State Navy List or Register
	Send list of defects_____KRF
	"Shipping Gazette and Lloyd's List," QWU
	Sick list_____WBF
	What is the date of your Code List?_IZH
	Your name is not on my list, spell it alphabetically (or, by alphabetical table, *page 15*)_____EBQ
QVN	LIST-ED-ING TO LIST (*To incline*)
QVO	—List to port
QVP	—List to starboard
QVR	LITHARGE
	LITRE (*Measure of capacity*)_____AZF
QVS	LITTLE-NESS
	A little farther_____MSU
QVT	—A little more
QVU	—As little as
	If you will wait a little_____YSU
	Little better_____GIR
QVW	—Little loss
	Of little importance Unimportant_FEY
	Port a little Ship's head to go to starboard_____TPG
	Sheer a little off_____VUO
	Sheer a little off when you come to the point_____VUP
	Starboard a little Ship's head to go to port_____WQU
QVX	—Too little Too small
QVY	—Very little
QVZ	LIVE D ING (*See also* LIFE)
	Live cargo_____HZG
	Live stock_____WVL
QWA	—Lively
	No boat could live_____GTW
QWB	—No means of living
QWC	LIVER
QWD	LLOYD'S
	Austro Hungarian Lloyd's_____FMJ
QWE	—Committee of Lloyd's
QWF	—Condemned by Lloyd's agent
QWG	—Consult Lloyd's agent
QWH	—Do not interfere with (or, take the matter out of the hands of) Lloyd's agent
	French Lloyd's Veritas_____NQP
	German Lloyd's_____NYD
QWI	—Have you been visited by Lloyd's agent?
QWJ	—Have you seen Lloyd's agent?
QWK	—How do you (or *vessel indicated*) stand at Lloyd's?
QWL	—I am insured at Lloyd's
QWM	—I am not classed at Lloyd's
	I require assistance from Lloyd's agent, CV
QWN	—I am not insured at Lloyd's
QWO	—I stand (or *vessel indicated* stands) — (*class indicated*) at Lloyd's.
QWP	—Inform Lloyd's agent

	LLOYD'S—*Continued*
	Intend to abandon the ship to Lloyd's agent_____CXP
QWR	—Leave the matter in the hands of Lloyd's agent
QWS	—Lloyd's agent—is
QWT	—Lloyd's agent advises you to —.
	Lloyd's Committee_____QWE
QWU	—Lloyd's List and Shipping Gazette
QWV	—Lloyd's Register of British and Foreign Shipping
QWX	—Lloyd's Registry
QWY	—Lloyd's Salvage Agreement.
QWZ	—Lloyd's Seaman's Almanac.
QXA	—Lloyd's secretary.
QXB	—Lloyd's signal station—at —
QXC	—Lloyd's surveyor
QXD	—"Lloyd's Weekly Shipping Index"
QXE	—Lloyd's Yacht Register
QXF	—Not on Lloyd's Registry book
QXG	—Please send off Lloyd's surveyor
QXH	—Removed from Lloyd's Register
	Report me by telegraph to Lloyd's_UD
	Report me to Lloyd's_____UG
	Report me (or, my communication) to Lloyd's agent at —_____UH
	"Shipping Gazette and Lloyd's List," QWU
QXI	—Surveyed by Lloyd's agent
QXJ	—Who is Lloyd's agent?
QXK	—Will Lloyd's agent come on board?
QXL	LOAD-ED-ING TO LOAD
	Breech loader-ing_____HFU
QXM	—Can load —
QXN	—Can you load any —?
	Deeply laden_____KQW
	I am loaded with combustibles_____HP
	Laden — with —_____QCX
QXO	—Loaded with
QXP	—Loading — for —
	Must load more ballast_____FTS
	Muzzle loader-ing_____SDE
QXR	LOAF
QXS	LOAN
QXT	LOBSTER.
QXU	LOCAL-LY ITY
	Have you any local attraction?_____JHR
	I know nothing of the locality (or, place)_____QCB
	Local authorities_____FNL
	Local Board_____GPI
QXV	—Local defence
	Local Government Board_____GPJ
	Local Medical Board_____GPK
QXW	—Local time
QXY	LOCK-ED ING
	Forelock_____NJW
	Lock and key_____QAI
QXZ	—Locksmith
QYA	LOCKER
	Chain locker_____IJN
QYB	LOCOMOTIVE
QYC	—Locomotion

	LOCOMOTIVE—LOOK		

	LOCOMOTIVE—*Continued*		LONG BOAT _____GTR
	Locomotive boiler _____GXD		
	Locomotive engine__ _____MAS		LONGITUDE_____Code Flag over DI
QYD	—Locomotive torpedo		NOTE—*For degrees of longitude, see*
QYE	LODGE-D-MENT		*page 56*
QYF	LODGING		Degree of longitude _____ _____KRZ
QYG	LOG-GED-GING TO LOG		Derelict reported passed on—*, latitude—, longitude—, dangerous to
	Engine-room log_____MBI		navigation_____KVR
	Log board _____ _____GQP		Difference of longitude_____KZG
	Log book _____ ___ ____GYZ	QYZ	—East longitude
QYH	—Log glass.		Have passed ice in latitude—, longitude—, on the—* _____PQ
	Log line _____QUS		I have been injured by ice in latitude—, and longitude— _____PBN
	Meteorological log book _____GZA		Ice between 30° and 35° of longitude,*
	Official log book ____ _____GZD		PF
	Patent log _____TEF		Ice between 35° and 40° of longitude,*
	Ship's log_____ _____VYC		PG
	Steam log book _____GZH		Ice between 40° and 45° of longitude,*
QYI	LOGARITHM LOGARITHM TABLE		PH
			Ice between 45° and 50° of longitude,*
QYJ	LOGWOOD		PI
QYK	LONG		Ice between 50° and 55° of longitude,*
	Any longer____ _____EPB		PJ
	As long as _____ ____ _____FAV		Ice between 55° and 60° of longitude,*
	Do not stand too long on your present		PK
	tack _____ _____JZI		Ice between 60° and 65° of longitude,*
QYL	—How long a time?		PL
	How long after?_____DRP		Ice between 65° and 70° of longitude,*
	How long—ago (*or, since*)? _____DTK		PM
QYM	—How long do you require to clean your		Ice between 70° and 75° of longitude,*
	engines?		PN
	How long have you (*or vessel indicated*) been in such distress?_____LHI		Ice between 75° and 80° of longitude,*
	How long have you (*or vessel indicated*) been on fire?__ _____NAJ	QZA	—In what longitude?
		QZB	—Is the longitude correct?
QYN	—How long have you been on the voyage?	QZC	—Longitude by dead reckoning
	How long have you had this fog?___NGK	QZD	—Longitude by observation
QYO	—How long is (*or* are)?	QZE	—Minute of longitude
	How long may we stand on our present	QZF	—My longitude by chronometer is —.
	course?_____JZK	QZG	—My longitude by — is —
	How long ought I to be getting an answer from —? _____EMY		Rendezvous in latitude — and longitude —_____JC
QYP	—How long shall you stop (remain)?		Report the — (*vessel indicated*) all well,
	How long will delay be?_____KSH		in latitude—, and longitude — (*indicated*), on the —(*date indicated*)__UN
	How long will you be getting steam up_____ _____VY		Saw derelict in latitude — and longitude — on the — * _____JH
	How long will you (*or vessel indicated*)	QZH	—Second of longitude
	be repairing damages?_____BM	QZI	—West longitude
QYR	—How much longer?	QZJ	—What is the longitude of — ?
	I fear I can not hold on much longer, OWT		What is your first meridian?_____NBL
	I shall stand in for the land as long as I can see _____ ___ QDZ	QZK	—What is your longitude brought up to the present moment?
	I will not delay you long __ _____KSJ	QZL	—What is your longitude by chronometer?
	Long after_____ _____DRQ		What latitude and longitude did you (*or*, they) have ice in?__ _____PS
	Long ago (*or*, since)_____DTL		
	Long boat_____ _____GTR	QZM	—What longitude do (*or*, did) you go into?
	Long range _____UHT		
	Long time—ago Long ago _____DTL		
QYS	—Long way	QZN	LONGITUDINAL-LY
QYT	—Longer		
QYU	—Longest		LOOD (*Measure of weight*) _____BAI
QYV	—Much longer.		
	Not long—ago_____DTN	QZO	LOOK-ED-ING TO LOOK
QYW	—Not much longer		Bad lookout _____FRG
	Rocks stretch a long way out_____GE		Be on your guard Look out _____OIV
QYX	—Too long		

* Date to be indicated if necessary

LOOK—LOSS

	LOOK—*Continued*	RAL	LOSE-ING TO LOSE
	Do you, *or*, Did you get a good look at the land to know exactly whereabouts we are? ----------------------PY		All hands lost (*or*, perished)-------DXQ
			All lost. Total loss-------------DXO
	Freights looking better------------NQF		All nets lost--------------------- --- DXP
	Good lookout ---------------------ODZ		Boat is lost------------------------ --- AQ
	Has the appearance of Looks like..ESA		Both booms with nets lost---------GZO
	Heavy weather coming, look sharp. FZ	RAM	Cargo lost ----------------------------HZY
	I do not like the look of the weather..ZE		—Do not lose sight of—it
	In anchoring, look out for moorings..EIW		Have lost — head of cattle ---------IFE
	In anchoring, look out for telegraph cable---- -----------------------WV		Have lost all boats, can you take off people?-------------------------NI
	Keep a good lookout aloft ---- --- EAG		Have lost deck cargo--------- ----IAO
	Keep a good lookout, as it is reported enemy's war vessels are going about in disguise as merchantmen-------OJ	RAN	Have lost screw, *or*, Screw disabled..MS
			—Have you lost?
			Have you lost all your boats?-------BL
	Keep a good lookout for land (*or*, light), *or*, Look out for land (*or*, light) (*bearing to follow, if necessary*)-------PYH		Heavy loss --- ---------------- -----OSX
			I have just lost sight of the land (*or*, light)---------------------------QDV
	Keep a sharp lookout --------------PYI		I have lost all my boats-------------BO
QZP	—Look at		I have lost an anchor -------------EIN
QZR	—Look in (*or*, into)	RAO	—If you lose the wind
QZS	—Look out—to (*or*, for).		It is dangerous to lose the wind-----GC
	Look out Pay attention ----------DS		Just lost sight of land (*or*, light) ..QDV
	Look out at daybreak ------------KNM		Key of propeller lost -------------QAH
	Look out for a boat bearing — -----GTS		Little loss -----------------------QVW
	Look out for a line (*or*, rocket line)..KF	RAP	—Lose no time
QZT	—Look out for a squall		Lose no time in getting to the anchorage ---------------------------ELC
	Look out for my pilot Pick up my pilot -----------------------------TJH		Lose no time in shoring up----------CF
QZU	—Look out for the tow-line	RAQ	—Lose-ing, Lost sight of — (*vessel or object indicated*)
	Look out on lee bow---------------HCJ	RAS	—Lose-ing, Lost time
	Look out on weather bow ----------HCK	RAT	—Loss.
QZV	—Look out to leeward		Loss of bulwarks-----------------HLT
QZW	—Look out to windward.	RAU	—Lost
QZX	—Lookout vessel (*or*, scout)		Lost all propeller blades-----------GLY
QZY	—Lookout will be kept for any raft (*or*, spars)		Lost an anchor -------------------EJB
			Lost his (*or*, the) certificate -------LIB
	Lookout will be kept on the beach all night ---------------------------KG		Lost number of blades (*indicated*) of screw propeller ------------------ GLZ
RAB	—Look to the slings	RAV	—Lost on the
	Looks like ------------------------ESA	RAW	—Lost overboard
	Looks likely ---- -----------------QTU	RAX	—Lost screw propeller
RAC	—Run in for pilot (*or*, pilot vessel) and look out for their signals	RAY	—Lost sight of wreck
		RAZ	—Lost the (*or* her) mast
	Ship's bottom must be looked at.....HBG		Mail steamer reported lost---------RFM
RAD	—Some one to look after stock.		My chronometer loses daily---------GT
	Vessel looks like a man-of-war------RPN		No lives lost ------- --------------QRF
	Your permit will be looked at-------MKL	RBA	—Not much loss
		RBC	One net lost
RAE	LOOK-OUT MAN (LOOKOUT)		One net with boom lost-------------GZU
			Rudder lost .. ---------------------VBJ
RAF	LOOM-ED ING		Several lives lost ------------------QRG
			Severe loss ------------------------VSY
RAG	LOOPHOLE-D		Time will be lost ------------------XTU
			Total loss All lost --------- ----DXO
RAH	LOOSE-D-ING-LY-NESS	RBD	—Two (*or number indicated*) nets lost
	Boiler loose. (*or*, adrift)----------GWD	RBE	—Vessel indicated has lost her rudder
	Broke loose ----------------------HJU	RBF	—Vessel seen (*or*, indicated) lost her masts
	Crank webs loose------------------KBR		We have lost altogether — -------EDO
	Eccentric pulleys loose -----------LSV	RBG	—What have you lost?
	Engine seating loose --------------MCO	RBH	—What loss?
	Holding-down bolts loose----------GYA		When did you lose your masts? ....RMJ
RAI	—Loose from	RBI	—Where did you lose the trade (*or*, monsoon)?
RAJ	—Loose sails		With loss of bowsprit -------------HDA
	Propeller worked loose------- ...TZC		With loss of foremast -------------NKY
		RBJ	—Without loss
RAK	LORD-SHIP		
	House of Lords Senate----------OZR		
	Lord Mayor—of — -----------------RNJ		LOSS (*See also* LOSE) -------------RAT

	LOST—LYNCH		

	LOST (*See also* LOSE) ------------RAU	RCL	LOYAL-TY
RBK	LOT	RCM	LUBBER-LY
	We have had a lot of fog ----------NGP	RCN	LUBRICANT
RBL	LOTTERY	RCO	LUBRICATE OR
RBM	LOUD-LY-ER-NESS		Lubricating oil Oil for machinery-SQC
	LOUIS, *or*, NAPOLEON (*Com*) ------AUF	RCP	LUCIFER
RBN	LOVE-D-ING	RCQ	LUCK-Y-ILY
RBO	LOW (*See also* LOWER)	RCS	LUFF-ED-ING
	At low water you will have — feet-KUZ		Close luff -----------------------IUY
	Dry at low water----------------LOP	RCT	—Luff all you can
	How many feet at low water? -----KVF	RCU	—Luff tackle
	I can not make out the lowest flag--NDF	RCV	LUG
RBP	—Low land		
RBQ	—Low pressure		
	Low-pressure cylinder -------------KIJ	RCW	LUGGAGE
	Low-pressure cylinder cover--------KAL	RCX	LUGGER
	Low-pressure cylinder cracked (*or*, out	RCY	LUGSAIL
	of order) ---------------------KAW		
RBS	—Low-pressure piston		
	Low-pressure slide jacket cover broken,	RCZ	LULL-ED-ING
	KAM		If the wind abates (*or*, lulls)-------ZEQ
RBT	—Low water	RDA	—The first lull, *or*, When it lulls.
RBU	—Low-water mark		
	Lower-ed ing To lower -----------RBY	RDB	LUMBER-ED-ING
	Lower ------------------------RCA		
RBV	—Lowest	RDC	LUMINOUS-ARY
	Markets are very bad (very low)---FRS	RDE	—Luminous buoy
	Prices are low-------------------TVR		
	Proceed at low water ------------TWZ	RDF	LUMP-ED ING
	The depth here (*or at place indicated*)		
	is — feet at low water-----------KVD	RDG	LUNAR
RBW	—The lowest price is	RDH	—Lunar rainbow.
RBX	—Too low		
	Was the tide low when you grounded?	RDJ	LUNATIC-CY
	CJ		Lunatic asylum -----------------FHZ
	What is the depth at low water?---KVF		
	What is the lowest price you will take?	RDJ	LUNCH EON TIFFIN
	TVU	RDK	—Will you lunch with me?
	You will be aground at low water --KX		
RBY	LOWER ED-ING To LOWER	RDL	LUNG
RBZ	—Lower away		Inflammation of the lungs---------PJR
RCA	LOWER.	RDM	LURCH-ED-ING
	I have sprung (*or*, damaged) lower top-		Lee lurch------------------------QMD
	sail yard------------------------KJI		Weather lurch-------------------YZP
	Lower boom---------------------GZS		
	Lower cap----------------------HVL		LYING—IN (*or*, AT) ------------------QPS
RCB	—Lower deck		Lying in the roadstead------------QPT
	Lower fore topsail yard-----------NKI		
	Lower main topsail yard ----------RGF	RDN	LYMPH VACCINE LYMPH
RCD	—Lower mast		
	Lower mizzen topsail yard---------RWO	RDO	LYNCH LAW
RCE	—Lower port		
RCF	—Lower rigging		
	Lower spanker topsail yard- -----WLU		
RCG	—Lower studding sails		
RCH	—Lower topsail		
RCI	—Lower yard		
	Premiums are lower --------------TSQ		
RCJ	—Sprung, *or*, Seriously damaged a lower		
	mast (*or vessel indicated has*)		
RCK	—Sprung *or*, Seriously damaged lower		
	yard (*or spar indicated*)		
	Top your lower yard -------------XVO		

	M—MAIL		

RDP	M (*Letter*) (*For new method of spelling, see page* 18)	·	MADE—*Continued*
			Is a prisoner, or, Made prisoner....TWH
	A M , or ANTE MERIDIAN BEFORE NOON IN THE FORENOON--------------EDU	REF	Made a breach..................HFJ
			Made an agreement with—.......DUH
			—Made an offer—of
	P M , or POST MERIDIAN AFTERNOON IN THE AFTERNOON..............DEW		Made arrangements—to (or, for)—.EXI
			Made mistake Mistake has been made,
	MAATJE (*Measure of capacity*)....AYU	REG	RIK
			Made no impression ....PFK
	MACE, or, TEIN (*Coin*).. ....AUG	REH	—Made overture
RDQ	MACHINE.		Manufactured, Made, or, Built by.HLE
	Electric machine..............LUZ		May be made efficient.........LTW
	Machine gun..................OKF		—Mistake has been made
RDS	—Patent sounding machine		Perhaps it might be made to answer.ENI
	Weighing machine............ZBA		Repeat ship's name, your flags were not
			made out...................DU
RDT	MACHINERY (*See also* ENGINES)		Repeat the last signal made.......WCR
	Accident to machinery (or, engines) (*See page* 283)	REI	Ship has not made her signal.....SLH
	Damage to machinery (*part to be indicated*) ....KJA		The signal made...............WDL
			Will an allowance be made?......DZP
	Examine (examination of) machinery, MKA	REJ	—Will not be made
			— made an unsuccessful attack on —,
	Have passed steamer with machinery disabled..................MT	REK	FKH
		REL	MADEIRA WINE
	Have you damaged your machinery? MBO		MAGAZINE
	I am going to stop, as machinery requires adjusting...............RK		After magazine ....DQZ
RDU	—I am in want of oil for machinery		—Fore magazine
	Machinery broken down..........MCH	REM	Magazine rifle................UXC
	Machinery broken down, I am disabled, BJ	REN	Powder magazine ....TRM
RDV	—Machinery broken down, irreparable at sea	REO	Small arm, or, Percussion magazine, EVF
			MAGISTRATE-CY
	Machinery broken down, under temporary repair ....MCK	REP	MAGNESIUM
	Machinery disabled ....MCN		—Magnesium wire
	Machinery out of order..........RJ		
	Machinery requires adjusting......DEF		MAGNET-IC-AL
	Machinery slightly deranged......MCQ	REQ	Electro magnet ....LVG
	Machinery supposed to have broken down ....MCA		Magnetic bearing (*i e , compass bearing corrected for deviation*).........JHX
	Machinery wants repair..........MBL	RET	—Magnetic needle
RDW	—Proceed after repairing machinery		MAGNIFICENT-CE
RDX	—Refrigerating machinery out of order	REU	MAHOGANY
	Small adjustment required to machinery ....DKG	REV	MAHOMET-AN
RDY	—Stopping only for small adjustment to machinery	REW	MAIL
		REX	—Am a mail boat
	The damage to machinery is not serious, and is such as can be repaired by the vessel's own engineers ....EL		Any news of the — mail?. ....EPJ
			Are you going to land all mails at —? QDN
	Machinist, or, Engine-room artificer, EZR		Arrange-d-ing mail ....EWT
RDZ	MACKEREL	REY	—Barely time to save the mail
			By the mail ....HPL
REA	MAD-LY-NESS	REZ	—Did not get a mail
			Expected next mail............MNI
REC	MADDER		Had the mail arrived when you left?.EVG
			Has the mail arrived—(from —, *if necessary*)? ....QN
RED	MADE (*See also* MAKE)		Has the mail left(or, sailed)—(for —, *if necessary*)? ....QO
	Every exertion has been made.....MIV		Have letters (mails, or, despatches) for you (*or vessel or person indicated*).JF
	Has arrangement been made?......EXA		
	Have, or Has made a complaint...JKB		Have the mails been delivered? ....KSU
	Have, or, Has made no calculation.HSO		Have they advertised for any vessel to carry the mail?. ....DOF
	I made the discovery............LDS		
	Inquiry has been made ....PLV		

MAIL—MAKE

	MAIL—Continued		MAIN—Continued
RFA	—Have you a mail?	RGM	—Main pump
	Have you letters (mails, or, despatches)	RGN	—Main rigging
	for me (or vessel, person, or place	RGO	—Main royal
	indicated)? _____ _____ ____JQ	RGP	—Main royal yard
RFB	—I have mails for —		Main runners_____VCB
	I will take mails for you _____ QP	RGQ	—Mainsail Main course
	Last mail _____ _____ ____QGC	RGS	—Mainsail split
	Latest posting time for mail is —_ QHG		Main shaft broken_____MR
RFC	—Mail arrived—on the —(date indicated)	RGT	—Main shaft broken in stern tube.
RFD	Mail bag _____FSJ	RGU	—Mainsheet
	—Mail closes at —.	RGV	—Mainstay.
	Mail day___ _____ _____KNO	RGW	—Main staysail
RFE	—Mail due (or, is due) on —	RGX	—Main tack
RFG	—Mails have (or, had) arrived from —	RGY	—Maintop
	(dates up to —)	RGZ	—Main topgallant mast
RFH	—Mail has left	RHA	—Main topgallant sail
RFI	—Mail has (or, had) not arrived from —	RHB	—Main topgallant yard
	(place indicated)	RHC	—Main topmast
RFJ	—Mail is now at —	RHD	—Main topmast staysail
RFK	—Mail steamer	RHE	—Main topsail
RFL	—Mail steamer arrives-d—at —	RHF	—Main topsail yard
RFM	—Mail steamer reported lost	RHG	—Main topsail yard sprung
RFN	—Mail steamer sails-ed	RHI	—Main yard
RFO	—Mail tender	RHJ	—Main yard sprung
RFP	—Mails go by sailing vessel.		Tail end of main shaft broken _____LYO
RFQ	—Mails go by steamer	RHK	—Upper main topsail yard
RFS	—Missed the mail		What are the dimensions of main yard?
RFT	—Next mail		LAO
	Overland mail leaves on the — ... SXM	RHL	MAINTAIN-ED-ING-ENANCE
RFU	—Passed the mail — days ago		
RFV	—Shall I save the mail (or, post)?	RHM	MAIZE
RFW	—The mails leave on the —		
RFX	—The next mail goes in — (vessel indi-	RHN	MAJESTY
	cated)	RHO	—His, or, Her Majesty
	What are the latest dates by the last		His, or, Her Majesty's dominions __LKH
	mail? _____KMH		His, or, Her Majesty's Government_OGB
	What is the number of mail bags you	RHP	—His, or, Her Majesty's service
	have on board? _____ _____FSP	RHQ	—His, or, Her Majesty's ship
RFY	—What vessel takes the next mail?	RHS	—On His (or, Her) Majesty's service
RFZ	—When does the — mail arrive?	RHT	—Their Majesties
	When does the mail leave?_____QT		
RGA	—When ought the next mail to arrive?	RHU	MAJOR (Army) _____EWM
RGB	—When ought the next mail to leave?		Sergeant-major _____VRP
	Will you wait for the mail? _____VTD		Town Major _____XZF
RGC	—You are to land all mails at —		
		RHU	MAJORITY—OF
RGD	MAIN	RHV	—Government majority of —
	Back main topsail _____XWC		Majority—of_____RHU
	Can I get a main yard at —?_____NYG	RHW	—Opposition majority of —
	Fill main topsail _____MYB		
	Gaff mainsail _____NVF	RHX	MAKE-ING TO MAKE (See also MADE)
	Have broken main shaft_____MR		As soon as the flood tide makes__ __FBP
RGE	—I have carried away main yard	RHY	—Be sure and make no mistake
	I have sprung (or, damaged) main yard,		Boiler maker _____GWE
	KJL	RHZ	—Can make
	Keep the main topsail shivering____PZJ	RIA	—Can make out
RGF	—Lower main topsail yard		Can make out flags (or colors)_____NCV
	Main bearing brasses broken-(or, out of		Can you make a raft?_____UGO
	order)_____ ___GBO		Can you make out her flag (or, signal)?
	Main boiler _____GXE		NCW
RGH	—Main boom ·		Can you make out the leading marks?
	Main cable _____HRL		QJO
	Main chain_____LJS	RIB	—Can not make
	Main check valves out of order.. IOE	RIC	—Can not make out
	Main clew garnet_____ILH		Can not make out her (or, your)colors
RGI	—Main deck		LGX
RGJ	—Main drain		Can not make out the flags(or, signals),
	Main hold _____OXD		NCX
RGK	—Mainmast		Can not make out your flags, come
RGL	—Mainmast sprung		nearer (or, make Distant Signals).VM

MAKE—MAN

	MAKE—*Continued.*
	Clearly make out ------------------ITK
	Convoy to disperse and make for desti-
	nation ----------------------------IS
	Do not attempt to make the anchorage
	(or, harbor) ---- -- ---------------FW
RIT	Do not make any alteration—until--ECG
RIU	Had we not better run in (or, make)for
	the land? ---- - ----------------PZ
	I can not make out the flags(or, signals),
	NCX
	I can not make out the flags, hoist the
RIX	signal in a better position-------OWL
	I can not make out the bottom flag-NDF
	I can not make out the second flag-NDG
	I can not make out the third flag--NDH
	I can not make out the top flag------NDI
	I shall make for the anchorage (or, —)
	with all despatch----------------- JT
RIY	I shall not endeavor to make the land,
	QDX
RIZ	I shall not go into less than—fathoms of
	water whether I make the land or not
	QDY
	I shall stand in till I make the land-QBA
	I will show a light to night when I
RID	make sail-------------------------QRZ
	—In making sail
	Make a demand------------------- KTB
	Make, or, Take a duplicate (or, copy),
	JWD
	Make a signal when you want a boat,
	GTU
	Make all snug---------------------DYN
	Make allowances—for (or, to)------DZK
	Make an acknowledgment ---------DGS
	Make an agreement with — --------DUH
	Make an alteration ---------------ECP
RIE	Make an appointment for—--------ESZ
	—Make an example
	Make arrangements to —(or, for—)-EXI
RIF	Make Distant Signals-------------LGH
	—Make fast—to —
RIG	Mast fast to a buoy-------------- HMS
RIH	—Make fast to the pier
	—Make for — Run for —(*name of port*
RIJ	*to follow*)
	—Make-ing, Made good—the (or, your)
	Make haste---------------------ONS
	Make haste or he will be gone -----OCU
RIK	Make inquiry ---------------------MED
RIL	—Make-ing, Made mistakes
RIM	—Make more sail Set your —.
RIN	—Make my compliments—to
	—Make no noise Keep silence
	Make out -----------------------SVX
RIO	Make out an inventory------------PSJ
	—Make provision for
	Make, or, Set sail---------------RIT
RIP	Make sail, or, Go ahead, and drop a boat
	on board. ----------------------DVN
	—Make short tacks
	Make signal—to (or, for) ---------WCM
	Make sure—of --------------------XFO
	Make sure of her staying----------WSB
	Make your number (*Distinguishing*
RIQ	*signal*) -------------------------DV
	—Maker
RIS	Makes good progress--------------TYD
	—Makes no progress
	Making large purchases -----------UCA

	MAKE—*Continued*
	Making (*number to follow*) revolutions
	per minute ---------------------UWJ
	Not able to make such short tacks--XIL
	Not safe to risk making the land. VDE
	Sail maker------------------ -----VEG
	—Set sail Make sail
	—Shall I make?
	Shall I make another attempt?------FKQ
	—They (*boats*) are making for the shore
	Try to make out name of vessel in —
	(*direction indicated*) ----------SDP
	—What do you make?
	What do you make of stranger?---WZF
	What do you make of the land?----QBU
	What harbor shall you make (or, run)
	for?------------------------------OND
	When the ebb makes---------------LSP
	When the flood makes-------------NFE
	When will the flood make?---------XJ
	—Will you make?
	Will you make an agreement°-----DUJ
	—Will you make an offer?
	Will you make the land to-night?---QEV
	Will you repeat the signals being made
	to me? ---------------------------VN
	You may make an attempt---------FKV
	You may make some abatement---OXU
	You must not (or, can not) make any
	excuse ---------------------------MLI
	MAKER ---------- ------------ RIQ
	MALARIOUS MALARIOUS FEVER ---MWQ
RJA	**MALAY**
RJB	**MALE MALE PASSENGER**
RJC	**MALT LIQUOR**
RJD	**MALTA MALTESE**
RJE	**MAN** (*See also* MEN)
	A B Able seaman-------------CXE
	Alderman ------ ---------------DWS
	Am a man-of-war ----------------RPM
	Are there any men-of-war about?---SA
	Armed men not allowed to land ---EUV
	Artilleryman---------------------EZU
	Boatman -----------------------GVL
	Boiler burst, — men killed --------AT
	Boiler burst, — men killed, — others
	wounded-------------------------AU
	Boiler burst, several men killed----AY
	Boiler burst, several men wounded--AZ
	Cattleman-----------------------IFD
	Chairman -----------------------TUQ
	Clergyman, Parson, or, Minister--ITX
	Clothing for men Slops ---------IVS
	Coastguard man------------------IYP
	Did you fall in with any men-of-war?
	MRN
	Fireman ------------------------NAB
	Fisherman ----------------------NBS
	Foreman ------------------------NJX
	Have men for --------------------RPD
	Have you men enough? ---------- MDK
	Have you seen (or, spoken to) any men-
	of-war? If so, report their names (or,
	nation)----------------------------VNM

	MAN—MANY	

	MAN—*Continued*	RJT	MANILA ROPE
	He, *or*, She is full of men ....NSX	RJU	MANŒUVRE-D-ING
	Helmsman ... .OTR	RJV	—Naval manœuvres
	How many men? ....RKI		
	Landsman ....QFB		MAN-OF-WAR. (*See* MEN-OF-WAR)...RPL
	Leadsman......................QJL		
	Liberty man ....QPJ	RJW	MANOMETER STEAM-PRESSURE GAUGE
	Lookout man (Lookout) ....RAE		
RJF	—Man-ned ning To man	RJX	MANROPE
	Man aloft......................EAH		
	Man and arm boats ....GTV	RJY	MANSION
	Man-of-war ....RPL		
	Man overboard....................BR	RJZ	MANSLAUGHTER.
	Manhole door ....LKR		
	Manrope......................RJX	P.KA	MANUFACTORY MANUFACTURE-ED-ING
	Men ....RPC		Manufactured, Made, *or*, Built by —,
	Men on leave..................RPF		HLE
	Men's clothing Slops ....IVS		Manufactured cottons Cotton goods,
	Men's letters .. ..RPG		JXY
	Merchant seaman ....RPY		
	Midshipman....................RTJ	RKB	MANURE
	Newly-raised man......... ......RPH		
	Nobleman ....SIT	RKC	MANUSCRIPT
	Officers and men..............RPI		
	Ordinary seaman..............SVD	RKD	MANY
	Postman ....TQO		A good many slight cases ....IDH
	Rifleman ....UXF		As many—as ....FAW
	Seaman........................VLG		As many as convenient............FAX
	Short manned Crew weak......KDJ		As many (*or*, much) as possible ...FAY
	Signalman ....WDE		Did you leave many ships at — ?...QKP
	Small-arm men ....EVC		He has not many hands on board..OMD
	Tradesman .... XZM	RKE	—How many?
	Vessel looks like a man-of-war......RPN		How many blades have you to your
	Want a boat, man overboard ....BT		screw? ....GLX
	Want men (*number and particulars, if*		How many boats? ....GTJ
	necessary)..................RPJ		How many chronometers have you?.IQF
	Waterman ....YWH		How many days?..............KNG
	We have left men behind (*number to be*		How many deaths? ....KOM
	shown)......................QLO		How many dozen can you spare ...LMU
	What men-of-war did you leave at —?		How many fathoms? ....MTV
	QLR		How many feet? ....MVE
	Workman ....ZIA		How many feet at low water? .....KVF
	Your men ....RPK		How many funnels has she?........NTS
	— men of crew have been ill with yellow	RKF	How many guns?
	fever (*number to follow*) .....MWY	RKG	—How many guns do you mount?
		RKH	—How many guns does she carry?
RJG	MANAGE-D-ING-ABLE-MENT		How many have you seriously ill?..KLD
	How do (*or*, did) you manage (*or*, con-		How many hours' coal have you got
	trive)—to —?..................JTF		(*state at what speed*)?..........IXC
			How many lighters do you want?..QTA
RJH	MANAGER	RKI	—How many men?
		RKJ	—How many miles an hour are you going?
RJI	MANDARIN		How many miles an hour can you (*or*
			vessel indicated) go? ....RTP
RJK	MANDRELL		How many miles from — ?..........RTQ
		RKL	—How many (*or*, much) more?
RJL	MANGER		How many passengers? ....TDO
			How many ships?.... ....VXK
RJM	MANGROVE		How many sick?................WBC
			How many times?..... ....XTM
	MANHOLE DOOR ....LKR		How many tons—of? ....XUT
			How many tons do you take?......XUV
RJN	MANIFEST		How many torpedo boats passed in
RJO	—Manifest of cargo		sight?.... ....GTK
			How many were killed? ....QAM
RJP	MANIPULATE-D-ING-TION		How many wounded?............ZIN
			In how many minutes? ....RVD
RJQ	MANNER		It is dangerous to allow too many people
	In a hostile manner ....OZD		on board at once..................GB
RJS	— In what manner?		Many expected.... ....MNO

MANY—MAST

	MANY—*Continued*
RKM	—Many minutes
	Many sick .... WBE
RKN	—Many wounded
	Much, *or*, Many less .... QNM
RKO	—Much, *or*, Many more.
	Not a great many .... OHM
RKP	—Not so many—as —
	Passed a good many vessels........ TCO
RKQ	—So many (*or*, much)
RKS	—Too many (*or*, much)
RKT	MAP-PED-ING
RKU	MARBLE
RKV	MARCH
	About March the — .... CZQ
RKW	—Beginning of March
RKX	—End of March
RKY	—Next March
RKZ	MARE
RLA	MARGIN-AL
	Leave a good margin... .... QKX
RLB	MARINE
	Marine Board .... GPL
	Marine glue .... OBE
RLC	—Marine soap
RLD	—Mercantile Marine
RLE	MARINER
RLF	—Shipwrecked Mariners' Society
RLG	MARITIME
	Commissaire d'Inscription maritime,
	JEY
	Inscription maritime .... PLY
	Inscrit maritime .... PLZ
	Préfet maritime .... TSG
	MARK (*Coin*) ... . AUH
RLH	MARK-ED-ING
	Are there any alterations in the lights
	(*or*, marks) .... ECB
	Are there any lights (*or*, landmarks)?
	QFA
	Buoy, *or*, Mark is not in its proper
	position LC
	Buoy, *or*, Mark does not appear to be in
	its proper position .... HMG
RLI	—Can see the mark (*or*, beacon)
	Can you give me a leading mark (*or*,
	direction) for making the land here
	abouts (*or at place indicated*)?..... PW
	Can you make out the leading marks?
	QJO
	Can you see the mark (*or*, beacon)?. GAF
RLJ	—Can not see the mark
	Do you know the leading mark? ... QRV
	High-water mark............... YVF
	How does the mark (beacon, *or*, buoy)
	bear? .... GAH
	How must I bring the beacon (buoy, *or*,
	mark) to bear? .... GAI
	I do not think marks (*or*, buoys) can
	be in their proper positions ..... HMK

	MARK—*Continued*
	Is there a buoy (*or*, mark) on shoal (*or*,
	on —)? .... HMO
	Is there any alteration of marks?... HCM
	Landmark .... QEZ
	Leading mark .... QJP
	Low-water mark .... RBU
RLK	—Marked in feet
RLM	—Marks are all gone There are no marks
	(*or*, buoys)
	They have a buoy (*or*, mark) on — HMW
RLN	—Trade-mark
	What, *or*, Which are the leading marks?
	QJS
	What are the leading marks for crossing
	the bar? .... FWC
	What are the marks? .... ZCB
	What bearing shall I keep the light (*or*,
	landmark) on?.................. QF
	What brand (*or*, trade-mark)? .... HDY
RLO	MARKET
RLP	—How are the markets?
	Is there any alteration in the market?
	ECN
RLQ	—Market place
	Markets are very bad (very low) ... FBS
RLS	MARLINE SPIKE
RLT	MARQUIS MARCHIONESS
RLU	MARRY-IED-ING-IAGE
	Certificate of marriage .... IBS
	Registrar of births, deaths, and mar-
	riages .... GLI
	MARSH (*Morass*) .... RZS
RLV	MARSHAL
	Provost Marshal .... UAP
RLW	MARTIAL
	Court martial .... JZY
	Martial law.................... QIN
RLX	MARTINGALE
RLY	MASK-ED-ING
	Masked battery .... FYO
RLZ	MASON-IC-RY
RMA	MASS (*Divine Service*)
RMB	MASSACRE-D-ING
RMC	MAST
	After cutting away her masts ..... KHD
	After, *or*, Jigger masts .... DRC
	Can I get a spar for — mast — at —? NYH
	Can you rig a jury mast? .... VW
	Cut away the mizzen mast .... RWN
	Cut away your masts .... KJ
	Damaged, *or*, Sprung mast, can not
	carry sail .... BH
	Do not cut away your masts ..... KHM
	Fifth mast..................... MXO
	First mast .... NAY
	Five-masted ship............... MXP
	Foremast .... NJY

MAST—ME	

MAST—Continued

Fore topgallant mast	NJG
Fore topmast	NJK
Fore topmast sprung	NJL
Four-masted barque	FXQ
Four masted schooner	NPC
Four-masted ship	NPD
Fourth mast	NPE
Fourth-mast chains	IJR
Half-mast your ensign	MEL
Have, or, Has been dismasted	LEW
Hoist masthead flags	OWG
I have (or vessel indicated has) sprung her —(mast indicated)	RCJ
Iron mast	PSZ
Jigger mast After mast	DEC

RMD —Jury mast, or, Jury masted

Lost the (or, her) masts	RAZ
Lower mast	RCD
Mainmast	RGK
Mainmast sprung	RGL
Main topgallant mast	RGZ
Main topmast	RHC

RME —Mast ed-ing   To mast
RMF —Masthead   At masthead
RMG —Masthead lantern (or, light, or, lamp)
RMH —Mast is secured
RMI —Mast rope

Mizzenmast	RWQ
Mizzen topgallant mast	RWV
Mizzen topmast	RWZ
Second mast	VMD
Sprung, or, Seriously damaged a lower mast	RCJ
Sprung, or, Damaged a mast, can not carry sail	BH
Sprung my foremast and must bear up,	NKW
Sprung my foremast, but can fish it at sea	NKX
Steel mast	WTM
Step of the — mast (tabernacle)	WUN
Third mast	XQN
Three-masted schooner	VJK
Topgallant mast	XVQ
Topmast	XVW
Under jury masts	RMD
Vessel seen (or, indicated) lost her masts,	RBF
Visible from the masthead	YQZ

RMJ —When did you lose your masts?

Will you lend me a fish for my mast?	NCI
With loss of foremast	NKY

RMK **MASTER**

Beach master	FZQ
Captain, or, Master (Mercantile Marine),	HWQ
Harbor master, Captain of the port,	HWS
Harbor master's office	OMY
Postmaster, or, Post-office official	TQK
Quartermaster	UEY
Schoolmaster	VJH
Shipping master (or, officer)	VYJ
Station master	WRQ

RML **MAT-TED-TING**

Chafing mat	IJL
Coir, or, Grass matting	IZT
Collision mat	JAT

RMN MATCH ED-ING

RMO MATCH (for lighting)

RMP **MATE**

Boatswain's mate	GVN
First mate (or, officer)	NBC

RMQ —Mate is on shore

Second mate (or, officer)	VME

RMS —Send the mate—off

Third mate (or, officer)	XQO

RMT **MATERIAL LY**

Are you, or, Is —(vessel indicated) materially damaged (or, injured)?	AO
Have you materials for stopping the leak?	QJW

RMU —Have you the materials?

Is vessel specified, or, Are you materially injured?	AO

RMV —Materials—for —

RMW MATHEMATIC-AL-LY-IAN

RMX **MATTER**

As a matter of course	FAC
Do not interfere with (or, take the matter out of the hands of) Lloyd's agent,	QWH
I decline to have anything further to say (or, do) in the matter	EQO
Is anything the matter? Has any accident happened?	BK
Leave the matter in the hands of Lloyd's agent	QWB
Money matters	RYD

RMY —What is the matter?

MATTING   (See also MAT) | RML

RMZ MATTRESS

RNA MAUL

RNB MAUVE

RNC MAXIM GUN

RND MAXIMUM

MAY   (See CAN) | CDZ

RNE MAY   (Month)

About May the —	CZR

RNF —Beginning of May
RNG —End of May
RNH —Next May

RNI MAYOR ALTY
RNJ —Lord Mayor—of —

M C MEMBER OF CONGRESS | JPA

ME | CED

After me (my, mine)	CBA
Before me (my, mine)	CEB
By me (my, mine)	BJS
For me (my, mine)	BPE
From me (my, mine)	BPQ
Give me (my, mine)	NZY
In me (my, mine)	CDK

ME—MEN

	ME—Continued	ROA	**MEASURE-ED-ING-MENT. TO MEASURE**
	Let me (my, mine) ....QNU		How many tons measurement of goods
	Me by ....BKD		can you take? ....OEV
	Me to ....CEF		Tons measurement ....XUW
	Me with ....CEG		Tons measurement of goods ......OEY
	Of me (my, mine) ....CGO		
	On me (my, mine) ....CGZ	ROB	**MEAT**
	To me (my, mine) ....CNW		Australian meat ....FMG
	With me (my, mine)............CWH		Fresh meat..................NQW
RNK	**MEAL**		Frozen meat ....NQA
	During meal hours ....LPO		Frozen meat cargo ....HZB
			Preserved meat ....TUK
	INDIAN MEAL..................PHY		Salt meat ....VFU
RNL	**MEAN-T TO MEAN** (See also MEANING	ROC	
	and INTEND)	ROD	**MECHANIC-AL-LY-ISM**
	Do you mean (intend) to —? ......POQ		—Mechanical mine.
	I mean to carry sail all night .....VEA		Stoker mechanic............WVZ
	I mean to slip my cable ....HRG	ROE	**MEDAL**
RNM	—What do you mean?	ROF	**MEDIATE-D-ING-ION-OR.**
RNO	**MEAN THE MEAN**	ROG	**MEDICAL-LY**
	In the meantime (meanwhile) .....RNX		Local Medical Board ....GPK
RNP	—Mean time		Medical assistance wanted, want a sur-
	My chronometer is fast on Greenwich		geon ....FGL
	(or, first meridian) mean time....GR		Medical authorities ....FNM
	My chronometer is slow on Greenwich		Medical Board ....GPM
	(or, first meridian) mean time . GS	ROH	—Medical comforts
	My first meridian (or, Greenwich) mean	ROI	—Medical comforts running short
	time is XL	ROJ	—Medical officer
	Will you show your Greenwich (or, first	ROK	—Medical report
	meridian) mean time? ....XN	ROL	—Medical stores
RNQ	**MEANING**	ROM	—Medical survey
RNS	—Right meaning	RON	—Medical treatment
RNT	—What is the meaning — of —?		Want immediate medical assistance.YL
	Wrong meaning..............ZKB	ROP	**MEDICINE-AL-LY**
RNU	**MEANS**		Medicine chest ....IOM
	By all means................DYK	ROQ	—Medicines have run short
	By no means ....HPE	ROS	**MEET-ING** (See also MET)
	By no means plain..........TLV		Did you meet (or, fall in) with? ...MRL
	By this means ....HPR	ROU	—Meet me at the railway station.
RNV	—By what means?		Parliament, Assembly, or, Congress met
	Can not be got off (or, afloat) by any		(or, meets) on the —...........JPB
	means now available...............CD	ROV	—Shall you meet?
	Did boat take a blue light (lantern, or	ROW	—Should you meet (or, fall in) with
	any means of making a signal)?...GSP		When you meet —........... ZLS
	Every means ....MIX		Will you (or person indicated) meet me
	Have you any means of throwing a line?		on shore to-day?..............KOD
	QUJ	ROX	**MELANITE**
	I have no means of laying out an anchor,	ROY	**MELT-ED-ING**
	EIO		Molten ....RXU
	I, or, They wish to abandon but have not		
	the means ....AJ	ROZ	**MEMBER**
RNW	—No means—of		Member of Parliament (or, Deputy),
	No means of living . ...QWB		TBO
	No means of sending ....VPL		Member of Congress............JPA
RNX	**MEANWHILE**	RPA	**MEMORANDUM MEMO**
RNY	**MEASLES.**		Admiral's, or, Senior officer's memo.DLC
RNZ	**MEASURE** (See TABLE OF BRITISH AND		Admiralty memo..............DLU
	FOREIGN MEASURES, pages 51 to 54)	RPB	**MEMORY-ABLE**
	Cubic, or, Solid measure (page 52).AYB	RPC	**MEN** (See also MAN)
	Measure of capacity (page 52) .....AYJ		Armed men not allowed to land ...EUV
	Measure of length (page 51)......AVK	RPD	—Have men for
	Measure of weight (page 53).......AZX		
	Square, or, Surface measure (page 52),		
	AXH		

MEN—MESSAGE

	MEN—*Continued*
	Have you men enough? ....MDK
	How many men? .. ..RKI
RPE	—Let the men —
RPF	—Men on leave
RPG	—Men's letters
RPH	—Newly-raised men
RPI	—Officers and men.
RPJ	—Want men (*number and particulars, if necessary*)
RPK	—Your men
RPL	**MEN-OF-WAR MAN-OF-WAR**
RPM	—Am a man-of-war
	Are there any men-of-war about? .. SA
	Are there many men-of-war now cruising ....KEX
	Did you fall in with any men-of-war? MRN
	Have you seen (*or*, spoken) any men-of-war? If so, report their names and nationalities ....VNM
	I saw war vessels disguised off — ..LEN
	Man-of-war ensign ....MEP
	Passed a war vessel disguised, funnel (*or*, funnels)—. *Here give the color or design of the line of steamers that it may be an imitation of* ....LEO
	Passed, *or*, Saw some men-of-war ..TCR
RPN	—Vessel looks like a man-of-war
	What men-of-war did you leave at —, QLR
RPO	**MEND-ED ING**
RPQ	**MENTION-ED-ING**
RPS	**MERCANTILE**
	Captain, *or*, Master (*Mercantile Marine*) ....HWQ
RPT	—Is there any mercantile news?
	Mercantile Marine ... .RLD
	Mercantile Navy List of country of vessel signaling ....QVJ
RPU	**MERCHANDISE**
RPV	**MERCHANT**
RPW	—Am a merchant vessel .
	Armed merchant ship ....EUW
	Have you the Merchant Shipping Act or Navigation Laws of the country of vessel signaling?...............-----DIG
	Keep a good look-out, as it is reported enemy's war vessels are going about disguised as merchantmen ....OJ
RPX	—Merchant Navy
RPY	—Merchant seaman
RPZ	—Merchant ship (*or*, vessel)
	Merchant Shipping Act or Navigation Laws of the country of vessel signaling......... ....DIK
	Merchant steamer........ ....RPZ
	Merchant Vessels of the United States, List of * ....QVJ
	Published annually, contains names, owners, signal flags, etc , of all United States registered vessels
RQA	**MERCURY** (*The planet*)

	MERCURY QUICKSILVER------------UFH
	Fulminate of mercury------------NTB
RQB	**MERE-LY**
RQC	**MERIDIAN. MERIDIAN OF**
	Ante-meridian ....---EDU
	Did you get a meridian altitude? ..EDH
	Ex-meridian altitude ....EDK
	Meridian altitude ....EDL
	My chronometer is fast of Greenwich (*or*, first meridian) mean time .....QR
	My chronometer is slow of Greenwich (*or*, first meridian) mean time ....GS
	My first meridian is Cadiz, west of Greenwich 6° 12′ 24″=0h 24m 49 6s , HSI
	My first meridian is Greenwich.....NBK
	My first meridian is Paris, east of Greenwich 2° 20′ 15′=0h 9m 21s ......TBL
	My Greenwich (*or*, first meridian) mean time is —† XL
RQD	—On meridian of —
	Post-meridian.... ..-----------DRW
RQE	—Vessel indicated reckons from the meridian of —
	What is your first meridian?.......NBL
	Will you show your Greenwich (*or*, first meridian) mean time?†............XN
RQF	**MERIT ED-ING ORIOUS.**
RQG	**MESS-ED-ING**
RQH	—Mess room
RQI	—Mess traps
RQJ	**MESSAGE**
	Acknowledge, *or*, Answer my signal (*or*, message) ....-----------DGO
	Can telegraph message be forwarded from —? ....-----WR
	Can you forward my message by telegraph ? ....HU
	Can not understand the message ..YHU
	Cipher message ....IQT
	Do you understand the message? ..YHV
	Following communication (message, *or*, signal) is secret and in cipher....IQV
	Forward following message by telegraph —to — ....-----NMW
	Forward following telegraphic message by signal letters, instead of writing it at length............... ----------XB
	Forward message without loss of time ....NMY
	Forward reply to my message by telegraph to —......... ....WT
RQK	—Have, *or*, Has message
	Have you any message (telegram, orders, *or*, communication) for me? ..IB
RQL	—Have you any message for your owner?
RQM	—Have you received my message?
	I will leave a message at — .. ...QKW
RQN	- I will repeat the message (*or*, what you say)
RQO	—I will telegraph message immediately
RQP	—Message is not understood
	Repeat the message ....URJ
	Secret message ....VMI

†See note to **GREENWICH**, page 258.

	MESSAGE—*Continued*	RSK	METEOROLOGICAL—*Continued*
	Send a message .. ---------- -----VPM		—Meteorological report for to-morrow gives CALM
	Send following message by post to own ers (*or*, to Mr —) at —------ - WY	RSL	—Meteorological report for to morrow gives MODERATE winds from — (*direction indicated*)
	Send following message by telegraph to owners (*or*, to Mr —) at —------XA	RSM	—Meteorological report for to-morrow gives STRONG winds from — (*direction indicated*)
	Send following message through the post to owner (*or*, to Mr —) at — by signal letters instead of writing it at length ------------------------WZ	RSN	—Meteorological report for to-morrow gives winds VARIABLE from — (*direction indicated*)
	Send my message through the tele graph by signal letters-----------XB	RSO	—Meteorological station
	Send reply to my message to signal station at —--------------------VQD		What is the meteorological forecast for to day?-------- -----------------ZK
	Shall I open your telegram and signal its contents?----:------ --------XC		What is the meteorological forecast for to-morrow?---- ----------------ZL
RQS	—Shall I send message by telegraph?		
	Telegraph following message to—(*ship or person named*) at —----------XD	RSP	METHOD-ICAL-LY
	Telegraph to my family at — the fol lowing message -----------------XMF		A bad method-------------------FQU
	Telegraphic message --------------XLN		A good method (*or*, plan)----------ODS
	Telephonic message -----•---------XMO	RSQ	METHYLATED SPIRITS
	Will you leave a message for me?----IL		
	Will you repeat the message (*or*, sig nal)?------------------------WDQ		METER. (*Measure of length*)-------AWJ
	You have abridged the message so much that it is unintelligible ----DAR	RST RSU	Cubic meter------------- -----------ZNP —Metric-al —Metrical System
RQT	MESSENGER		Square meter (*square measure*)----AXK
RQU	—By messenger	RSV	METROPOLIS-ITAN
RQV	MESSRS	RSW	MEXICAN　MEXICAN COLORS
	Messrs — and Co-----------------JGK	RSX	MICROPHONE
RQW	MET—WITH (*See also* MEET)	RSY	MICROSCOPE-IC
RQX	Has Congress (*or*, Parliament) met?-JOZ —Have you met?	RSZ	MID　MIDSHIPS
	Have you met any men-of-war? If so, report their names (*or*, nation)--VNM		About mid-channel --------------CZS Keep in mid-channel-------------LU
RQY	—Met her off (*or*, in) —	RTA	—Mid-channel
RQZ	—Met with a vessel from — (*or*, of —)	RTB	—Midway
RSA	Met with an accident ----------- ---BS —Met with — (*vessel indicated*)	RTC	MIDDAY
	Congress, Parliament, *or*, Assembly met (*or*, meets) on the —------------JPB	RTD RTE	MIDDLE　MIDST —In the middle of the —
RSB	METAL-LIC	RTF	—Middle watch
RSC	—Muntz's metal (*Number of sheets to follow*)	RTG	—The middle of
RSD	METEOR-IC	RTH	MIDNIGHT After midnight----------- ------DRA
RSE	METEOROLOGICAL		At midnight -------------------FIJ
	Meteorological log book -----------GZA	RTI	—Before midnight
	Meteorological Office reports depression approaching from the —----------ETP	RTJ	MIDSHIPMAN
	Meteorological Office reports gale ap proaching from the — ----------ETQ		MIDSHIPS-----------------------RSZ
RSF	—Meteorological report	RTK	Helm amidships -----------------RFJ —Midship section
RSG	—Meteorological report for to-day gives CALM	RTL	MIDSUMMER
RSH	—Meteorological report for to-day gives MODERATE winds from — (*direction indicated*)	RTM	MIDWAY
RSI	—Meteorological report for to-day gives STRONG winds from — (*direction indi cated*)		MIGHT (*See* COULD)---------------CHI
			MIJLE (*Measure of length*)--------AWG
RSJ	—Meteorological report for to-day gives winds VARIABLE from — (*direction indicated*)		MIL (*Danish measure of length*)---AWK

		MIL—MINUTE		

	MIL (*Swedish measure of length*)__ AWL		RUL	MINE D-ING TO MINE
	MIL (*Turkish measure of length*)__AWG			Buoyant mine ------------------------HMZ
RTN	MILD-LY-NESS			Channel, or, Fairway is likely to be mined---------------------------- ILD
RTO	MILDEW			Channel, or, Fairway is mined Beware of torpedoes-----------------------XP
	MILE (*English Statute*)----------AWN			Channel, or, Fairway is not mined__XQ
	English geographical (or, nautical) mile --------------------------------AWO			Coal mine -------------------- -----IWQ Countermine d ing ----------------JYQ
	English statute mile----------------AWN			Countermining launch ------ ------JYR
	Geographical mile-------------- AWO			Do not approach the coast because it is mined ---- -------------------ETN
	German mile---------------- ----AWM			Electric mine---------------- --LVA
	Square mile __ ----------- ------AXW			Electro contact mine --------------LVF
	About a mile (or *number indicated*) _CZI			Floating mine---------------------NFC
	After running — (*miles indicated*) _DRH			Gold mine------------------------ODM
	Current runs -- miles an hour----XFY			Ground mine--------------------OIG
	Current will run very strong (*indicate miles per hour if necessary*)-------KGB			Illuminate-d ing mine field --------PDB Is the channel mined?------------ _ILP
	How many miles an hour are you going? --------- ---------------REJ			Is there danger of mines (or, torpedo boats)?--------------------------XS
RTP	—How many miles an hour can you (or *vessel indicated*) go?			Line of mines------ ---------QUP Mechanical mine------------------ROD
RTQ	—How many miles from — ?		RUM	—Mine (*A mine*)
RTS	—I shall go — miles			Mine, or, Torpedo buoy ----------XWN
	I shall go about — miles an hour till daylight --------------------------KNI		RUN RUO	—Mine case —Mine field
	Keep vessel going about — miles an hour--- ------------------------PZR		RUP RUQ	—Mine laid —Mining district—of
	Miles from (or, distant)------------LGI			Submarine mine -----------------XBJ The entrance of the gulf (or, harbor) is mined ------ ---------------------MFU
RTU	MILITARY		RUS	—There are no mines laid
RTV	—Have military stores on board			
RTW	—Is there any military news?		RUV	MINER.
RTX	—Military invalid			
RTY	—Military officers		RUW	MINERAL
RTZ	—Military prison-er		RUX	—Mineral oil.
RUA	—Military secretary			
RUB	—Military stores			
RUC	—Military top			
RUD	—Military transport			Minim (*Measure of capacity*) -----AZH
RUE	—Naval and military news.		RUY	MINIMUM
RUF	MILITIA		RUZ	MINISTER-ED-ING-TRY
				Change of ministry----------------IKQ
RUG	MILK			Minister, Parson, or, Clergyman ---ITX
	Fresh milk -------------------------NQX			Minister, or, Ambassador----------EDV
	Preserved milk------- -----------TUL			Our Minister (or, Ambassador)—at —, EDW
RUH	MILL-ER		RVA	MINOR ITY
	MILLIER. (*Measure of weight*)------BAV			Government in a minority of — ---OFV
	MILLIGRAMME (*Measure of weight*)__BAW		RVB	MINUTE
	MILLILITER (*Measure of capacity*)__AZG			NOTE —*For minutes of Arc and Time, see page* 58
	MILLIMETER (*Measure of length*)---AWP			
	Cubic Millimeter (*Cubic Measure*)__ZNO		RVC	—Few minutes In a few minutes
	Square millimeter(*Square measure*)_AXY		RVD	—In how many minutes?
RUI	MILLION-TH			Making — (*number to follow*) revolutions per minute-------------UWJ
	MILREIS (*Coin*) __ ---------------AUI			Many minutes -----------------RKM Minute gun---------------------OKG
RUJ	MIND-ED-ING-FUL			Minute guns firing in -- (*direction indicated*) ------------------------OKH
	Change of mind--------------------IKP			Minute guns firing last night in—(*quarter indicated*) ------ ---------OKI
RUK	—Mind you get paid			Minutes of latitude ---------------QHV
	Never mind-----------------------SHE			Minutes of longitude ---------------QZE
	MINE (*See* MY) --------- -----------CED		RVE	—Minutes of time

MIRAGE—MONASTERY.

RVF	MIRAGE		RWL	MIX-ED-ING
RVG	MISCALCULATE-D-ING-ION		RWM	MIZZEN TOP
RVH	MISCARRIAGE		RWN	—Cut away the mizenmast
				Keep the mizen topsail shivering __PZK
RVI	MISCARRY-IED TO MISCARRY.		RWO	—Lower mizen topsail yard
				Mizzen chains_____LJT
RVJ	MISCELLANEOUS-LY		RWP	—Mizzen course
	Miscellaneous cargo_____HZI		RWQ	—Mizzenmast
			RWS	—Mizzen rigging
			RWT	—Mizzen royal
RVK	MISCONDUCT-ED-ING		RWU	—Mizzen royal yard.
			RWV	—Mizzen topgallant mast
RVL	MISDEMEANOR		RWX	Mizzen topgallant sail
			RWY	—Mizzen topgallant yard
RVM	MISFORTUNE		RWZ	—Mizzen topmast
			RXA	—Mizzen topmast staysail
RVN	MISLEAD-ING-LED		RXB	—Mizzen topsail
			RXC	—Mizzen topsail yard
RVO	MISMANAGE-D-ING-MENT			
				MNA *(Measure of weight)*_____BAX
RVP	MISPLACE-D-ING-MENT.			
			RXD	MOAT
RVQ	MISPRINT-ED			
			RXE	MOB
RVS	MISREPRESENT ED-ING-ATION.			
			RXF	MOBILIZE-D-ING TO MOBILIZE.
RVT	MISS *(Title)*			
			RXG	MOBILIZATION
RVU	MISS ED-ING TO MISS			
	Any missing _____EPC		RXH	MODEL-LED-LING
RVW	—Have, or, Has been missing			
RVX	—Is, or, Are missing		RXI	MODERATE-D-ING-LY-ION
RVY	—Missed fire			Are the charges moderate?_____IMN
RVZ	—Missed stays			Charges, or, Expenses are moderate,
▲	Missed the buoy _____HMT			IMP
	Missed the mail _____RFS			I shall weigh as soon as the weather
	Missed the mark, or, Passed the port			moderates (*or*, permits)_____ZO
	(harbor, *or*, entrance)_____MFV		RXJ	—If weather moderates
RWA	—None missing		RXK	—Moderate breeze
	Several missing _____VSQ		RXL	—Moderate gale
	We have missed the entrance (*or*, passed			Moderate weather _____ZF
	harbor, *or*, port) _____MFV			Unsafe to run in until weather moder-
				ates _____YKO
RWB	MISSION-ARY		RXM	—Until weather moderates
	Important mission _____PEV			Wait until weather moderates_____GK
			RXN	—When more moderate
RWC	MIST-Y		RXO	—With moderate-ion
RWD	MISTAKE-EN-ING MISTOOK		RXP	MODIFY-IED-ING-ICATION
	Any mistake?_____EPD			
	Be sure and make no mistake_____RHY			MOHAMMEDAN _____SCT
	Find out the mistake_____MYP			
	Make-ing, Made mistake_____RIK		RXQ	MOIST-URE
RWE	—Mistake *(A mistake)*			
	Mistake has been made_____REH		RXS	MOLASSES.
RWF	—No mistake			
RWG	—There is some mistake		RXT	MOLE
				Alongside the mole_____EAU
RWH	MISTER MR. —			Inside the mole _____PML
	Mister —, junior _____PXB			
			RXU	MOLTEN
RWI	MISTRESS MRS.			
			RXV	MOMENT ARY-ARILY OUS
	MISTRUST-ED-ING *(Distrust)* _____LHP			I expect every moment _____MJL
				I expect your orders every moment_MNL
RWJ	MISUNDERSTAND-ING			I shall weigh the moment the breeze
RWK	—Misunderstood			springs up _____HGA
	You have misunderstood the signal_WDR			
	MITHKAL *(Coin)* _____AUJ		RXW	MONASTERY-IC

MONDAY—MORNING

RXY	MONDAY			MOOR—*Continued*
RXZ	—Before Monday.	RZJ		—Moor head and stern
	Last Monday_____QGD			Mooring chain _____ _____LJU
RYA	—Monday morning.	RZK		—Mooring swivel
RYB	—Monday night	RZL		—Moorings
RYC	—Next Monday			Safely moored _____VDG
				Shall you moor?_____RZF
	MONEY_____ASU	RZM		—Take mooring swivel off
	NOTE.—*See* TABLE OF UNITED STATES			There are moorings on—(*bearing indi-*
	AND FOREIGN MONEYS, *pages* 47–50			*cated*) . _____EL
	Advance-d-ing To advance (*money*),	RZN		—When moored
	DNK	RZO		—You must moor, it is not safe otherwise
	Bank note, *or*, Paper money _____ASV	RZP		MOOR-ISH MOROCCO MOORISH COLORS
RYD	—Money matters			
	Paper money, *or*, Bank notes _____ASV	RZQ		MORAL-ITY
RYE	—Public money			
	Want money _____YTW	RZS		MORASS
RYF	MONITOR	RZT		MORE
				A little more _____QVT
RYG	MONKEY			Affairs appear more settled (look more
				peaceful) _____DOZ
RYH	MONOPOLY-IZE-D-ING-IST			Affairs look more unsettled _____DPB
				Any more _____EPF
RYI	MONSOON			Anything further_____EQJ
	Against the monsoon _____ _____DSF			Can not pay any more _____TEN
RYJ	—During the monsoon (*or*, trades)			Driving, I can veer no more cable, no
RYK	—Had the monsoon (*or*, trade winds) set			more anchors to let go_____NK
	in?			Have you room for more cargo?____IAR
RYL	—Next monsoon			How many (*or*, much) more?_____RKL
RYM	—Where did you fall in with the trade			I have no more anchors to let go____EIP
	(*or*, monsoon)?			Keep more away _____PYW
	Where did you lose the trade (*or*, mon-			Keep more to leeward_____PYX
	soon? _____ _____RBI			Keep more to port_____PYZ
				Keep more to starboard_____PZA
RYN	MONSTER-OUS-LY			Keep more to windward_____PZB
				Keep more toward the shore _____PZC
RYO	MONTENEGRO-IN MONTENEGRIN COLORS			Make more sail — Set your — _____RIL
				More hands are required—at — ____OMB
RYP	MONTEVIDEO-AN (*See* URUGUAY.)			More rain than wind _____ZEV
				More scope _____VJP
RYQ	MONTH THIS MONTH	RZU		—More than
RYS	—In the month of —			Much, *or*, Many more _____RKO
	Last month _____QGE			Must take in more ballast_____FTS
RYU	—Monthly	RZV		—No more Nothing more
	Month's advance _____DNG			Send more _____VPZ
	Months ago _____ _____DTM	RZW		—Some more
	Next month_____SHX			Steer more to port (*To be kept flying*
	On the — day of the month_____KNQ			*until course is sufficiently altered*).MJ
RYV	—This month			Steer more to starboard (*To be kept fly-*
				ing until course is sufficiently al-
RYW	MOON			*tered*) _____MK
	First quarter of the moon_____NBF			Veer more cable _____KU
	Full moon_____ NSP	RZX		—Want more support
	Last quarter of the moon_____QGH			—When more moderate _____RXN
RYX	—Moonlight	RZY		MORNING IN THE MORNING.
RYZ	—New moon			Every morning _____MIY
RZA	—No moon			Friday morning _____ _____NBJ
RZB	—We shall have moonlight			Monday morning _____RYA
RZC	—When moon goes down			Morning after_____DRS
RZD	—When moon rises	SAB		—Morning before
		SAC		—Morning watch
RZE	MOOR-ED-ING-WITH TO MOOR			Saturday morning _____VHW
RZF	—Do you moor? *or*, Shall you moor?			Sunday morning _____XEA
	Driven, *or*, Broke from her moorings,	SAD		—This morning
	HJT			Thursday morning _____XRU
RZG	—I have picked up moorings.	SAE		—To-morrow morning
RZH	—I shall moor			Tuesday morning _____YDK
RZI	—I shall not moor			Wednesday morning _____ZAI
	In anchoring look out for moorings_EIW	SAF		—Yesterday morning

MORPHIA—MUCH

SAG	MORPHIA

SAH	MORROW TO-MORROW ....XUO
	—Before to-morrow
	  The day after to-morrow . ...DRV
	  To-morrow afternoon.............. DRZ
	  To-morrow morning.............. SAE
	  To-morrow night ....XUP
	  To-morrow's post By to-morrow's
	    post ....TQH

SAI SAJ SAK SAL SAM	MORSE ALPHABET ....EBP —Do not understand Morse code —Do you understand Morse code? —Morse code —Use Morse code —Will use Morse code.
SAN	MORTAL-LY-ITY
SAO	MORTAR Mortar battery....................FYP
SAP SAQ	—Mortar for throwing a line —Mortar vessel
SAR	MORTGAGE-D-ING-E-E
SAT	MORTIFY-IED-ING TO MORTIFY.
SAU	MORTIFICATION
SAV	MOSQUITO

SAW	MOST LY THE MOST
	  Most certain ly ....IHL
	  Most economical speed. ....LTD
	  Most likely (or, probable).........QTV
	  Most of all .... - ....DYO

SAX	—Most of them
	  The most ....SAW
	  Winds have mostly been —.........ZFP

SAY	MOTHER-LY

SAZ	MOTION
	  Disregard motions Disregard leader's
	    movements ....LFX
	  Follow motions . ...NGT
	  Parallel-motion gear ... .TAX
	  Parallel-motion rod out of order ...TAY

SBA	—Report the motions of —
SBC SBD SBE	MOTIVE —What is the motive? —Without any motive

SBF	MOTOR
	  Electric motor ....LVB
	  Motor engine ....MAT

	MOU (Square measure) . ...AXQ
SBG	MOULD-ED-ING
SBH	MOUND

SBI	MOUNT-ED-ING TO MOUNT
	  How many guns do you mount? ...REG
	  Mounted escort ....MHC

SBJ	MOUNTAIN-OUS

	MOURN-ED-ING (Grieve)............OHX
SBK	MOUSE
SBL	MOUTH Off the mouth of the river—(or,—)_UYJ

SBM	MOVE
	  Disregard leader's movements Disre
	    gard motions..... ....LFX
	  I can not weigh until you have moved,
	    ZAV
	  I do not think we shall get any good by
	    moving ....OEG

SBN SBO SBP	—Moved —Movement —Moving
	M P MEMBER OF PARLIAMENT (or, DEPUTY) ....TBO
	MR — MISTER................. RWH
	MR —, JUNIOR ....PXB
	MRS — ....RWI

SBQ	MUCH
	  As much—as ....FAZ
	  As much as can be taken ....FBA
	  As much (or, many) as possible... FAY
	  Can not gain much................NVL

SBR	—How much?
	  How much longer? ....QYR
	  How much (or, many) more? ......RKL
	  It will not avail much ....FOJ
	  Much altered....................ECQ
	  Much better ....GIS
	  Much censure for.................IGU
	  Much cut up ....KHP
	  Much damage has been done—to (or,
	    by) —.........................KJQ
	  Much danger... ....KLF
	  Much fatigued (or, tired).......... MTX

SBT	—Much hurt
	  Much, or, Many less ....QNM
	  Much longer ....QYV
	  Much, or, Many more........... ....RKO
	  Much nearer........... .... SFI
	  Much pleased ....TMR
	  Much regret-ted..................UOA
	  Much sea Heavy sea ....UQ
	  Much surf Heavy surf ....OSZ
	  Much swell on ....UR
	  Much tired (or, fatigued)..........MTX
	  Much to blame.......... ....GMH
	  Much worse ....ZIG

SBU	—Not much
	  Not much longer . ...QYW
	  Not much loss.... ....RBA
	  Not much sea... ....US
	  Not much wanted ....YTP
	  Quite as much as...... ....UFQ
	  So much (or, many)..............RKQ
	  Take as much as .... XJK
	  Too much (or, many)..... ... RKS
	  Too much scope ....VJQ
	  Too much sea ....UV
	  Too much strain on — ... .WYI
	  Too much surf ... .XFU
	  Too much swell..................XHK

MUCH—MUST

	MUCH—*Continued.*
	Too much wind .... ZFB
SBV	—Very much
SBW	—Was much pleased
	Will there be much profit? .... TXZ
	With much pleasure . ... TMX
	MUD (*Measure of capacity*) .... AYZ
SBX	MUD DY
	Muddy bottom .... HBD
SBY	MUFFLE-D-ING
SBZ	MUG
SCA	MULATTO
SCB	MULE-TEER
SCD	MULTIPLE
SCE	MULTIPLY-IED ING-ICATION
SCF	MULTITUBULAR
	Multitubular boile .... GXF
SCG	MUMPS
SCH	MUNICIPAL-ITY
	MUNTZ'S METAL (*Number of sheets to follow*) .... RSC
SCI	MURDER-OUS-ER (*See also* ASSASSINATE)
	Accused of having murdered some one, DGI
SCJ	—Murdered
SCK	MURIATIC ACID
SCL	MUSCAT MUSCAT COLORS
SCM	MUSCLE MUSCULAR
SCN	MUSHROOM
	Mushroom anchor .... EJC
SCO	MUSIC-AL-IAN
SCP	MUSKET, SMALL ARMS, ETC. .... FVD
	—Musket shot
SCQ	—Send muskets (rifles) to party on shore
	You are within musket (*or,* rifle) range, UIA
SCR	MUSLIN
	MUSQUITO .... SAY
SCT	MUSSULMAN
	MUST .... CEJ
	He, She, It (*or* person-s *or* thing-s indicated) must .... BUY
	He, She, It (*or* person s *or* thing-s indicated) must be .... BUZ
	He, She, It (*or* person s *or* thing s indicated) must not .... BVA
	He, She, It (*or* person-s *or* thing s indicated) must not be .... BVC
	How must? .... BXP

	MUST—*Continued*
	I must .... CEK
	I must be . ... BGE
	I must not .... CEL
	I must not be .... BGF
	I must not delay (detain) .... KSI
	If I must......... .... BZJ
	If I must not................... BZK
	Letters must be sent—to —(*or,* by —), QOL
	Must anchor .... EJD
	Must be................ CEM
	Must be acknowledged .... DGT
	Must be adopted .... DMK
	Must be altered .... ECR
	Must be complete-d with — .... JKR
	Must be prevented. ... . TVJ
	Must be ready—to (*or,* for) .... UJP
	Must come..................... JDI
	Must do (*or,* be done) .... CEN
	Must examine (*or,* be examined)—for, MKE
	Must go.... .... OCV
	Must have..................... CKO
	Must have been .... CEP
	Must have had .... CEQ
	Must he (she, it, *or* person-s *or* thing-s *indicated*)?......... .... CER
	Must his (her s, it-s)? .... CES
	Must I (my, mine)?.............. CET
	Must I go?............... ..... OCW
	Must I send? .... VPK
	Must leave .... QLI
	Must not .... CEU
	Must not be .... CEV
	Must not do (*or,* be done)........ CEW
	Must obtain (*or,* be obtained)...... SNV
	Must oppose................ .... STC
	Must relinquish............... UPD
	Must shift .... VWY
	Must that (*or,* this)? .... CEX
	Must the?................... CEY
	Must there be .... CEZ
	Must these (*or,* those)? .... CFA
	Must they (their-s)? .... CFB
	Must use .... YMQ
	Must we (our-s)? .... CFD
	Must you-r-s .... CFE
	That, *or,* This must CLI
	That, *or,* This must be.......... CLJ
	That, *or,* This must not .... CLK
	That, *or,* This must not be ..... CLM
	They must .... CFG
	They must be .... BHV
	They must not .... CFH
	They must not be .... BHW
	We must .... CFI
	We must be .... BIP
	We must not .... CFJ
	We must not be .... BIQ
	What, *or,* Which must? .... CQH
	When must? .... CRO
	Where must? .... CTI
	Who must? .... CUQ
	Why must? .... CVO
	You must .... CFK
	You must be .... BJF
	You must not .... CFL
	You must not be .... BJG
	You must not (*or,* can not) make any excuse .... MLI

MUSTARD—NATION

SCU	MUSTARD
SCV	MUSTER ED-ING Muster book (or, roll). Roster ___GZB
SCW	MUSTY
SCX	MUTILATE D ING-ION
SCY	MUTINY-OUS-EER Crew have mutinied_____KCQ
SCZ	—Mutinied. Mutiny, want assistance _____VF
SDA	MUTTON Fresh mutton_____ ___NQY
SDB	MUTUAL-LY Let us keep together for mutual support (or, protection)_____ _____IN
SDC	MUZZLE-D-ING
SDE	—Muzzle loader ing.
	MY MINE _____OED After my (me, mine)_____CEA Are, or, Is my (mine)? _____BEL Before my (me, mine)_____CEB By my (me, mine) _____BJS Can, or, May my (I, mine)?_____BKG Could, or, Might my (I, mine)?___BLC Do, Does, or, Did my (mine)?_____BMH For my (me, mine)_____BPE From my (me, mine)_____ _____BPQ Give my (me, mine) _____NZY Had, Has, or, Have my (mine)?___BQS If my (I, mine)_____BYW In my (me, mine)_____CDK Let my (me, mine) _____QNU Must my (I, mine)?_____CET Need my (I, mine)?_____CEH Not my (I, mine)_____CFU Of my (me, mine)_____ _____CGO On my (me, mine)_____CGZ Ought my (I, mine)?_____CHP Shall, or, Will my (I, mine)?_____CIT Should, or, Would my (I, mine)? _CJQ That my (I, mine) _____CLD To my (me, mine)_____CNW Was, or, Were my (I, mine)?_____COS When my (I, mine) _____CRL Where my (I, mine)_____CTF With my (me, mine) _____CWH
	MYRIAGRAMME (Measure of weight)_BAY
	MYRIAMETRE (Measure of length)__AWQ
	MYSELF _____ _____CED
SDF	MYSTERY IOUS LY NESS MYSTRON (Measure of capacity) __AYR

SDG	N (Letter) (For new method of spelling, see page 13)
SDH	NAIL
SDI	NAKED-NESS
SDJ	NAME-D-ING TO NAME. Christian name _____IPZ Could not learn name _____QKD Have you seen any men-of-war? If so, report their names and nationalities, VNM
SDK	—His (or, her) name is — Keep the two objects named in one line, QUM
SDL	—My name is —
SDM	—Name (a name)
	Name on the bow _____BCL Name on the stern_____WUR
SDN	—Name unknown
SDO	—Name will be spelt No name on the bow_____HCM No name on the stern_____WUS Repeat ship's name, your flags were not made out_____DU Signal the names of places to which you wish to be reported by telegraph_WDF Spell the name _____WND
SDP	—Try to make out the name of vessel in — (direction indicated)
SDQ	—What is her name? What is the commander's name?___JEP
SDR	—What is the name—of? What is the name of admiral (or, senior officer)?_____DLO What is the name of lighthouse (light-ship, or, point) in sight?_____RW What is the name of the captain of — (ship indicated)? _____HXN What is the name of signal station (or, ship) in sight? _____VG
SDT	—What is your name? Your name is not on my list, spell it alphabetically (or, by alphabetical table, page 15) _____EBQ
SDU	NAMELESS
SDV	NAMELY
SDW	NAPHTHA
	NAPOLEON, or, LOUIS (Coin)_____AUF
SDX	NARRATE-D-ING-IVE
SDY	NARROW-LY-NESS
SDZ	—In the narrow part of the Narrow-ly escape d _____MGZ
SEA	—Not very narrow The channel (or, fairway) is narrow_ILE Very narrow-ly _____YPK
SEB	NATION-AL-ITY Have you seen (or, spoken) any men-of-war? If so, report their names and nationalities _____VNM Vessels of nations specified were all leaving _____QLM
SEC	—What nation do you (or vessel indicated) belong to?

	NATIVE—NEAT'S		

SED	NATIVE-ITY		NAVIGATE—*Continued*
	Native craft ....KBA		Person to navigate ....TGU
	Native crew ....KDC	SFC	NAVY (ROYAL, *or*, STATE)　(*See also* NA-
SEF	—Native troops		VAL.)
	Natives do not like ships watering with-		Belonging to the Navy　Naval　.. SEL
	out payment ....YWQ		Captain in the Navy...............HWF
SEG	—Vessel seen (*or*, indicated) is a native		Lieutenant in the Navy............QPW
	craft		Mercantile Navy List　. ....QVJ
SEH	NATURALIZE-D-ING-ATION		Merchant Navy ....RPX
SEI	NATURE-AL-LY		Naval ....SEL
	I should like to know the nature of the		United States, Royal, *or*, State Navy.SFC
	sickness, if any, before I send my		United States, Royal, *or*, State Navy
	boat (*or*, communicate) ....FO		List *or*, Register ....QVM
SEJ	—What is the nature—of?	SFD	NEAP TIDE
	What is the nature of the bottom? *or*	SFE	—Vessel indicated will be neaped
	What kind of bottom have (*or*, had)	SFG	NEAR ED ING　TO NEAR
	you..............................HBK		Ammunition is nearly expended ...EFR
SEK	NAUTICAL-LY		As near as.......................FRC
	Have you the Nautical Almanac for the		Can not distinguish your signal flags,
	current year (*or year indicated*)? .RQ		come nearer (*or*, make Distant Sig-
	, Nautical Almanac ....DZW		nals)............................VM
	Nautical mile, *or*, Knot...........AWO		Come nearer.........　-..........JCT
SEL	NAVAL　(*See also* NAVY.)　BELONGING		Come nearer, Stop, *or*, Heave-to, I have
	TO THE NAVY		something important to communi-
SEM	—Is there any naval news?		cate ....Code Flag over H
	Lieutenant in the Naval Reserve? .QPX		Deepest water is nearer the bank (*or*,
	Naval and military news ....RUE		reef)...........................FVB
	Naval brigade　Brigade of seamen.HIA		Deepest water is nearer the shore..KQV
SEN	—Naval court		Do not approach too near ....ETO
SEO	—Navy Department		Do not come so near ....JDE
	Naval depot ....KUR		Do not keep so near ....PYC
SEP	—Naval invalid		—How near did you pass to —?
	Naval manœuvres....... ....RJV		I will keep near you during the night,
SEQ	—Naval officer		IVF
SER	—Naval prison-er		Is adrift—near — ....DMQ
SET	—Naval Reserve　(Royal Naval Reserve)		Keep near to me...............PZD
SEU	—Naval Reserve officer　Officer of the		Keep nearer—to ....PZK
	Royal Naval Reserve	SFI	—Much nearer
SEV	—Naval station		Must be nearing the land　Near the
SEW	—Naval store		land ....QEP
	Naval Reserve ensign..... ....MEQ		Near-ly enough ....MDP
	Sub-lieutenant in the Royal Naval Re-		Near the buoy ....HMU
	serve............................QPZ		Near the engine room ....MBR
SEX	NAVIGATE-D-ING		Near the land ....QEP
	Beware of derelict, dangerous to navi-	SFJ	—Near the point
	gation........................ .....JE	SFK	—Near the rocks
	Derelict reported passed on — in lati-		Near to —　(*Close to*) IVC
	tude —, longitude —, dangerous to	SFL	—Nearer est
	navigation ....KVR	SFM	—Nearest port
	Have you a book of Navigation Tables?	SFN	—Nearly
	RN		Nearly afloat....................DQA
	Have you the Navigation Laws, *or*,		Nearly all our.........'.... ...SVQ
	Merchant Shipping Act of the coun-		Nearly clear—of ....ITO
	try of vessel signaling............DIG		Nearly enough ....MDP
SEY	—Is it necessary (i e, does the naviga-	SFO	—Not so near
	tion require me) to take a pilot?		Stand nearer the shoal (*or*, bank, *or*,
	Is the river navigation difficult?...KZW		reef) ....FWA
SEZ	—Navigable		Stop, Heave-to, *or*, Come nearer, I have
SFA	—Navigation		something important to communicate,
	Navigation Laws, *or*, Merchant Ship-		Code Flag over H
	ping Act of the country of vessel sig-		Sunk off (*or*, near) the —.........WEO
	naling...........................DIK		Too close (*or*, near) ....IVJ
SFB	—Navigation Tables		Very near-ly ....YPL
	Passed a derelict vessel apparently in	SFP	—You are too near the rocks.
	good condition, hull well out of the	SFQ	NEAT-LY-NESS-ER-EST
	water, dangerous to navigation .KVS	SFR	NEAT'S FOOT OIL.

NECESSARY—NEWS

SFT	**NECESSARY.**		SGW	**NEST.**
	Absolutely necessary `DBI`			Crow's nest `KES`
	Amputation necessary `EGS`		SGX	**NET-TED-TING**
	Complete refit necessary `JKM`			All nets lost `DXP`
	Consider it (or, them) necessary `JQH`			Boom for torpedo net `GZY`
	Despatch is necessary `JK`			Both booms with nets lost `GZO`
	Have you (or person-s indicated) the			Defence net `KRI`
	necessary certificates (or, credentials)?			Fishing net `NBW`
	`IHZ`			Hammock netting `OLU`
SFU	—I do not think it necessary			Jib net `PVO`
SFV	—If necessary, or, If you think it neces-		SGY	—Net obstruction
	sary			One boom with net lost `GZU`
	Is it necessary (i e, does the navigation			One net lost `RBC`
	require me) to take a pilot? `SEY`			Torpedo net `XWV`
SFW	—It was necessary			Two (or number indicated) nets lost,
SFX	—It will be necessary			`RBD`
SFY	—Necessary papers		SGZ	**NETT**
	Pilot not necessary `TJL`			Nett weight `ZBF`
	Regular survey necessary `UOC`			
	Repairs necessary `URE`		SHA	**NEUTER-RAL-ITY**
	Scarcely necessary `VIU`			Neutral colors (or, ensign) `MER`
SFZ	—Will it be necessary—to (or, for)?		SHB	—Neutral port
SGA	**NECESSITY**		SHC	—Neutral vessel. Vessel is a neutral
	Any necessity for? `EPG`			Port is neutral `TOS`
	In case of necessity (or, accident) `BP`		SHD	**NEVER**
SGB	—No necessity of — for —		SHE	—Never mind
	Under the necessity of `YHL`			
	Will prevent the necessity of `TVN`		SHF	**NEVERTHELESS**
SGC	**NECK** (Part of body)		SHG	**NEW-ER-EST-LY.**
				A new establishment `MHK`
SGD	**NEED-ED.**			Bend a new sail `VDS`
SGE	—In need (or, want) of			Fitting new propeller `NCO`
	Need be `BGY`			Laid up to have a new boiler put in `QIU`
	Need be under no anxiety about `EOJ`			Must have a new boiler `GXH`
	Need I (my, mine) `CEH`			Must have a new shaft `VTI`
	Need not be `BGZ`			New boiler had (or, has) been put in,
SGF	—Need not fear			`GXI`
	Need not interfere `PQL`			New book `GZC`
				New moon `RYZ`
SGH	**NEEDLE**			New system `XIO`
	Magnetic needle `REQ`			Newly-raised men `RPH`
	Twine, needles, and palms for repairing		SHI	**NEWS** (See also ACCOUNT)
	sails `YFN`			Any news—of? `EPH`
SGI	**NEGATIVE-D-ING NEGATION**			Any news of my family? `EPI`
	Answered negatively `EMS`			Any news of the—mail? `EPJ`
	No, or, Negative_D, or, Code Flag over D			Bad news—of `FRH`
SGJ	**NEGLECT-ED-ING-FUL-LY**			Good news—of `OEA`
	A great neglect of duty `LQD`			Have you any news of recent date from
SGK	—By some neglect			(or, about)—? `KMD`
SGL	—Do not neglect			Indifferent news (or, accounts)—of —,
	Great neglect `OHE`			`DFG`
SGM	—Have, or, Has been neglected.		SHJ	—Is the news official?
	Neglect of duty `LQE`			Is there any mercantile news? `RPT`
				Is there any military news? `RTW`
SGN	**NEGLIGENT-LY-CE**			Is there any naval news? `SEM`
SGO	**NEGOTIATE-D-ING-ION-BLE-OR.**		SHK	—Is there any political news?
				Latest news `QHF`
SGP	**NEGRO**			Naval and military news `RUB`
SGQ	**NEIGHBOR-HOOD**		SHL	—News expected on —
SGR	—In the neighborhood of —.			News, or, Report is confirmed `JOE`
				News, or, Report, or, Information is not
SGT	**NEITHER-OF**			correct `JXC`
SGU	—Neither of them			Newspaper `SEO`
SGV	—Neither of us			No news—of — `SD`
	Neither seen nor heard `ORM`			No news of importance `PEX`
			SHM	—Political news
				Receive d ing news—of (or, from) `ULD`

NEWS—NINE

	News—*Continued*		**Next**—*Continued*
	Some important news has arrived .EYM		Next Wednesday.................ZAH
	Telegraphic news arrived from—...JX		Next week ....ZAN
	What is date of latest news from—?.SE		Next winter ....ZGI
	What is the news from—?......... SF?		Next year.............. ....ZKX
	What will be about the latest--news		The next mail goes in—(*vessel indicated*) ....RFX
	from—?................ ....SE		What vessel takes the next mail? ..RFY
SHN	—Where, or, What source did the news come from?		When ought the next mail to arrive?
			RGA
SHO	**Newspaper**		When ought the next mail to leave?
SHP	—By a newspaper of—(*date following*)		RGB
	Can you give (*or*, lend) me a newspaper?	SHZ	**Nicaragua n Nicaraguan Colors**
	SB		
SHQ	—Have you—newspaper of—date?	SIA	**Nickel Nickel-plated**
SHR	—I am sending you some newspapers		
	Newspapers up to—(*date indicated*)	SIB	**Night**
	(*from—, if necessary*)...... ......SC		Arrived during the night ... .EXW
	What is the date of your latest newspaper (*or*, letter)?..............KMJ		As soon as it is dark......... .....FBM
SHT	—Will you send me some newspapers?		At night......................FIK
	Would you like to see our newspapers?		—Before night.
	SL		By night...... ....HPD
	New York		Do you remain all night?.........UPH
	Report me to "New York Herald" Office, London ....UI		During the night ....LPT
	Report me to "New York Herald" Office, New York ....UJ		Every night ....MIZ
	Report me to "New York Herald" Office, Paris..................UK		Friday night...................NRK
			Guns firing last night in—(*quarter indicated*) ....OKA
SHU	**Next**		I mean to carry sail all night ......VEA
	By the next ship... ....VXH		I sail to-night...................XUR
	Expected next mail.... ....MNI		I shall lie-to during the night....LPX
	Expected next week ....MNJ		I shall steer—during the night....WTZ
	Next account ... . .DFI		I will keep close to you during the night,
	Next ahead ....DVO		IVF
	Next April......... ....EUC		If the night is dark........... ... KLQ
	Next astern ....FHP		Keep a strict watch all night ......PYJ
	Next August..................FMD		Keep fires on the shore all night...FZW
	Next autumn ....FNY		Keep me in sight during the night.PYV
	Next day Day after............DRO		Last night ....QGF
	Next December ....KPN		Lookout will be kept on the beach all
	Next February.................MUS		night ... .KG
SHV	—Next flood		Midnight ... .RTH
SHW	—Next Friday		Minute guns firing last night in—(*quarter indicated*) ....OKI
	Next January ....PUW	SID	Monday night...................RYB
	Next July ....PWQ		—Night attack
	Next June ....PWY	SIE	Night clothing....................IVT
	Next mail....................RFT	SIF	—Night observation
	Next March...................RKY	SIG	—Night signal
	Next May....................RNH	SIH	—Night watch
	Next Monday ....RYO		—Nightly,
	Next monsoon ....RYL		Parted in the night from her anchors.EJK
SHX	—Next month.		Passed during the night............TCQ
	Next November ....SKX		Pay strict attention to signals during
	Next October ....SOJ		the night......................DT
	Next of kin ....QAU		Saturday night.................VHX
	Next opportunity.............SRY		Shall we get to an anchorage to night?
SHY	—Next quarter		ELF
	Next Saturday ....VHU	SIJ	Sunday night................. .....XEB
	Next season........ VIX		—This night
	Next September ....VRN		Thursday night.................XRV
	Next spring ....WOV		To-morrow night...............XUP
	Next summer ....XDO		To-night...... ....XUQ
	Next Sunday ....XDZ		To-night's post ....XUS
	Next Thursday............ .. XRT		Tuesday night..................YDL
	Next tide..................XSE		Wednesday night..............ZAJ
	Next time (*or*, opportunity) ....SRY		Will you keep close to me during the
	Next Tuesday.................YDJ		night? ....IVK
	Next voyage..................YRW	SIK	Will you make the land to-night?..QEV
			Nine-th-ly

NINETEEN—NORTH

SIL	NINETEEN-TH
SIN	NIP-PED-PING
	Freshen the nip .... NRB
SIO	—Nipper
SIP	NITRE IC ATE
	Nitrate of potassium .... TRA
	Nitrate of silver .... WEB
	Nitrate of soda .... WJA
SIQ	—Nitric acid
SIR	—Nitroglycerine
	NO, or, NEGATIVE D, or, Code Flag over D
	At no time .... FIL
	By no means .... HPE
	No accounts of — .... DFJ
	No accounts received .... DFK
	No acquisition .... DHN
	No advance .... DNH
	No advantage .... DNU
	No anchor left .... EJF
	No appearance—of .... ESC
	No arrangement—to (or, for) ...... EXJ
	No assistance can be rendered, do the best you can for yourselves CX
	No authority—to — .... FMW
	No boat available .... FG
	No boat could live .... GTW
	No boat fit for the work .... ZHV
	No boat on board .... GTX
	No bread .... HEP
	No chance (or, opportunity) ...... IKE
	No choice .... IPW
	No complaint—of .... JKC
	No danger .... KLG
	No difference .... KZM
	No difficulty .... LAC
	No doctor (or, surgeon) .... LJX
	No doctor (or, surgeon) available .. WM
	No doubt Doubtless ... . LMB
	No expectation—of — .... MNR
	No experience .... MOF
	No hope .... OYA
	No, I can not .... HUZ
	No improvement .... PFT
	No instructions (or, orders) ..... PNI
	No intelligence—of .... POF
	No interest in — .... PQJ
	No interpreter available .... PRF
	No lives lost .... QRF
	No means—of — .... RNW
	No mistake .... RWF
	No moon .... RZA
	No more Nothing more .... RZY
	No necessity—of (or, for) .... SGB
	No news—of .... SD
	No news of importance .... PEX
	No objection—to .... SMP
	No one .... SIU
	No one hurt .... PAT
	No one killed .... QAO
	No opportunity .... IKE
	No opposition .... STD
	No orders .... PNI
	No passage .... TDG
	No person .... TGR
	No proof .... TYR
	No reduction .... CXT
	No remuneration .... UQJ
	No reply .... ENH

	NO—Continued.
	No room—for .... UZQ
	No seamen to be had .... VLQ
	No surgeon (or, doctor) .... LJX
	No trace—of .... XZH
	No use .... YMR
	No variation .... JIA
	No vegetables to be obtained ..... YOI
	No water to be had .... YVH
	On no account .... DFR
	There is no obstruction .... SNL
	There is no risk .... UYD
	There is no shelter .... YWO
	There will be no profit .... TXY
	Under no anxiety—to (for, or, about), EOK
	We have no passenger .... TDU
SIT	NOBLE Y-MAN
SIU	NOBODY NO ONE NOT ANYONE
SIV	NOISE
SIW	Make no noise Keep silence ...... RIN
	—Noisy-ily
SIX	NOKTA (Measure of length) .... AWP
	NOMINAL-LY
SIX	Nominal horsepower .... OYK
SIY	NOMINATE-D-ING-ION
	NONCOMMISSIONED, or, PETTY OFFICER, THJ
SIZ	NONCONDUCTOR-ING
	NONE NOT ANY .... EPT
SJA	—I have none
	None killed No one killed .... QAO
SJB	—None left
	None missing .... RWA
SJC	—None of that size
SJD	—None received
SJE	—None to spare
SJF	—None wounded
	NOON AT NOON .... FIM
	A M Before noon In the forenoon, EDU
	P M Afternoon In the afternoon, DRW
	What was your latitude at noon? ... QIE
SJG	NOR
SJH	NORDENFELT GUN
SJI	NORMAL-LY
	NORTH (compass in degrees) .... ABC
	NORTH (compass in points) .... AQD
	—Northern-erly-ward
	Alongshore to the northward of — (distance in miles may follow) ...... EAP
SJK	—From the northward
SJL	North Atlantic .... FJL
	North cone hoisted .... JNM
	North country coal .... IXF
	North latitude .... QHX

NORTH—NOT.

NORTH—*Continued.*
North Pacific _____ _____SYW
SJM —North-east-ern-erly ing-ward-er
SJN —North-west-ern-euly-ing ward-er
On, *or,* At the north coast ot — _____IYF
On, *or,* At the NE coast of — _____IYG
On *or,* At the NW coast of —_____IYN
SJO —Pole, *or,* North star
SJP —To, *or,* For the northward
SJQ —Too far to the northward
    (*For Compass Signals see pages* 45, 46)

SJR NORWAY NORWEGIAN NORWEGIAN COLORS

SJT NOSE

NOT _____CFM
Am I not? _____ ___BED
Am, Is, *or,* Are not____ _____BEG
Am, Is, *or,* Are not able—to_____CYG
Am, Is, *or,* Are not to be _____BFD
Am, Is, *or,* Are not to have_____BQD
But not_____ _____HOM
Can, *or,* May they not?_____ ___BKM
Can not, *or,* May not __ _____BKP
Can not, or May not be ___ _____BFJ
Can not, or, May not do (*or,* be done),
                              BLX
Can not, *or,* May not do it_____BLY
Can not, *or,* May not have____ ____BQH
Can not, *or,* May not have been____BFK
Can not, *or,* May not have done (or,
   been done) _____BLZ
Can not, *or,* May not have had _____BQI
Can not, *or,* May not he (she, it, *or,* per-
  son-*s or thing-s indicated*) be? ___BTM
Can not, or May not he (she, it, *or, per-
son-s or thing-s indicated*) have been?
                          BTN
Could, *or,* Might he (she, it, *or person-s
or thing-s indicated*) not be? _____BTP
Could, *or,* Might not _____ __BLD
Could, *or,* Might not be . _____BFN
Could, *or,* Might not do (*or,* be done),
                        BMD
Could, *or,* might not have _____BQL
Could, *or,* might not have been _____BFO
Could, *or,* Might not have done (*or,* been
  done) _____ _____BMC
Could, *or,* Might not have had_____DQM
Do, does, *or,* Did not _____BMI
Do not let him (her, it, *or person indi-
cated*) have _____BTR
Had, Has, *or,* Have, he (she, it, *or person s
or thing-s indicated*) not?_____BTW
Had, Has, *or,* Have, he (she, it, *or person-s
or thing-s indicated*) not been? ___BTX
Had, Has, *or,* Have, he (she, it, *or person-s
or thing-s indicated*) not had?__ _BTY
Had, Has, *or,* Have not_____ __ __BQU
Had, Has, *or,* Have not been . _____BFS
Had, Has, *or,* Have not done (*or,* been
  done) _____BMS
Had Has *or* Have not had ____BQV
He, She, It (*or person s or thing s indi-
cated*) can not (*or,* may not) ___ BUD
He, She It (*or person s or thing-s indi-
cated*) can not (*or,* may not) be___BUE
He, She, It (*or person-s or thing-s indi-
cated*) could (*or,* might) not_____BUG

NOT—*Continued.*
He, She, It (*or person-s or thing-s indicated*) does (do, or, did) not____ _BUI
He, She, It (*or person-s or thing-s indicated*) had (has, or have) not____BUN
He, She It (*or person-s or thing-s indi-
cated*) had (has, or, have) not been _BUO
He, She, It (*or person-s or thing-s indi
cated*) had (has, or, have) not done (or,
  is, *or,* are not doing) _____BUP
He, She, It (*or person-s or thing-s indi-
cated*) had (has, or, have) not had_BUQ
He, She It (*or person s or thing-s indi-
cated*) is (*or* are) not_____ BUW
He, She, It (*or person-s or thing-s indi-
cated*) is (*or,* are) not able to ___BUX
He, She, It, etc , may   (See Can, *above*)
He, She, It, etc , might   (See Could, *above*)
He, She, It (*or person-s or thing s indi-
cated*) must not _____BVA
He, She, It (*or person-s or thing-s indi
cated*) must not be _____BVC
He, She, It (*or person s or thing-s indi-
cated*) ought not to_____ ___BVE
He, She, It (*or person-s or thing-s indi-
cated*) ought not to be. _____ _BVG
He, She, It (*or person-s or thing-s indi-
cated*) shall (*or,* will) not_____BVJ
He, She, It (*or person s or thing s indi-
cated*) shall (*or,* will) not be ___BVK
He, She, It (*or person-s or thing-s indicated*) should (*or,* would) not____BVP
He, She, It (*or person s or thing s indi-
cated*) should (*or,* would) not be_BVQ
He, She, It (*or person-s or thing-s indi-
cated*) should (*or,* would) not do (*or,*
  be done) _____BVR
He, She, It (*or person s or thing s indi-
cated*) should (or, would) not have_BVS
He, She, It (*or person-s or thing-s indi-
cated*) was (*or,* were) not_____BVU
He, She, It (*or person-s or thing-s indi-
cated*) was (*or,* were) not to be __BVX
I am not. ___ _____BEI
I can not (*or,* may not)_____ __BKR
I can not (*or,* may not) be_____BFY
I can not (*or,* may not) have_____BRJ
I could (*or,* might) not_____BLM
I could (*or,* might) not be____ ____BGA
I could (*or,* might) not have_____BRL
I do (*or,* did) not_____BMZ
I had (*or,* have) not_____BRO
I had (*or,* have) not been _____BGD
I had (*or,* have) not done _____BNC
I had (*or,* have) not had_____BRP
I must not _____ _____CEL
I must not be _____ _____BGF
I ought not to_____CHM
I shall (*or,* will) not _____CIP
I shall (*or,* will) not be _____BGI
I shall (*or,* will) not do. _____ ___BNE
I shall (*or,* will) not have_____BRS
I should (*or,* would) not_____CJN
I should (*or,* would) not be _____BGL
I should (*or,* would) not do _____BNG
I should (*or,* would) not have _ __BRU
I should (*or,* would) not have been_BGM
I was not___ _____COI
I was not able to _____ _____CYJ
If he (she, it, *or person s or thing s indi-
cated*) can not (*or,* may not) ____BYE

NOT

NOT—*Continued*

If he (she, it, *or person-s or thing-s indi-cated*) could (*or*, might) not .....BYG
If he (she, it, *or person s or thing-s indi-cated*) does (do, *or*, did) not ......BYI
If he (she, it, *or person s or thing s indi-cated*) had (has, *or*, have) not....BYK
If he (she, it *or person s or thing-s indi-cated*) is (*or*, are) not............BYN
If he (she, it, *or person s or thing-s indi-cated*) is (*or*, are) not able to......BYO
If he (she, it, *or person s or thing s indi-cated*) shall (*or*, will) not ....BYQ
If he (she, it, *or person-s or thing s indi-cated*) should (*or*, would) not ....BYS
If he (she, it, *or person-s or thing s indi-cated*) was (*or*, were) not........BYU
If I am not...............BYZ
If I am not able to —...............CYL
If I can not (*or*, may not)...........BZC
If I could (*or*, might) not ....BZE
If I do (*or*, did) not ....BZG
If I had (*or*, have) not ....BZI
If I must not..... .... BZK
If I ought not to ....BZM
If I shall (*or*, will) not ....BZO
If I should (*or*, would) not ....BZQ
If I was not ....BZU
If it can not (*or*, may not) be done...BZS
If not..............................BZV
If they are not ....CAG
If they can not (*or*, may not) ....CAI
If they could (*or*, might) not........CAL
If they did (*or*, do) not............CAN
If they had (*or*, have) not...........CAP
If they shall (*or*, will) not ....CAR
If they should (*or*, would) not.......CAT
If they were not.................CAV
If we are not...................CAY
If we can not (*or*, may not).........CBA
If we could (*or*, might) not........CBE
If we do (*or*, did) not..............CBG
If we had (*or*, have) not.......... CBI
If we shall (*or*, will) not..... .....CBK   SJU
If we should (*or*, would) not.......CBM   SJV
If we were not ....CBO
If you are not ....CBE
If you are not able to —............CYN   SJW
If you can not (*or*, may not). ....CBT
If you could (*or*, might) not........CBV
If you do (*or*, did) not............CBX
If you had (*or*, have) not ....CBZ
If you shall (*or*, will) not..........CDB
If you should (*or*, would) not......CDF
If you were not ....CDH
Is, etc (*See* Are, *above*)
It can not be ....BGO
It can not (*or*, may not) be done...BNL
It could (*or*, might) not be done...BNO
It had (*or*, has) not been done.....BNQ
It ought not to be..................BGS
It ought not to be done .... BNS
It shall (*or*, will) not be.............BGU
It shall (*or*, will) not be done.......BNU
May, etc (*See* Can, *above*)
Might, etc (*See* Could, *above*)
Must not...................CEU
Must not be...............CEV
Must not do (*or*, be done).........CEW
Need not be ....BOZ
Need not fear ....SGF

NOT—*Continued*

Need not interfere............. ....PQL
Not..................................CFM
Not a good plan ....TLY
Not a great many ....OHM
Not according to —. ....DEQ
Not according to orders ....DER
Not acquainted—with—...........DHJ
Not afloat..........................DQB
Not allowed Disallow ed.........DZL
Not always..........................EDT
Not amiss...... ... .EFL
Not amounting to......... .. ..EGK
Not any None ....EPT
Not any one . ...SIU
Not approved of ....ETW
Not armed ....YGJ
Not arrived... ....EYO
Not as soon—as ....WIN
Not at all...... ....DYP
Not at present ....TUA
Not bad ly . .. .FRT
Not be (been, being) ... .CFN
Not before ....GDJ
Not begun yet...................GDT
Not by............................CFO
Not certain ....IHM
Not classed ....ISG
Not clear.............. ....ITQ
Not come ....JDK
Not commenced ....JEV
Not correct ....JXD
Not doing.........................CFP
Not done........................CFQ
Not enough......................MDQ
Not exactly ....MJW
Not far down ....LMK
Not for ....CFR
Not from ....CFS
Not his (her-s, it-s)................CFT
Not I (my, mine)..................CFU
Not if I were in your place ....TLI
Not immediately...................PDK
—Not impossible
—Not improved
Not in ....CFV
Not in favor of. ....MUE
—Not included-ing
Not insured ....POB
Not less........................QNO
Not likely Unlikely.............QTW
Not long ago.......................DTN
Not much ... .SBU
Not much longer................QYW
Not much loss....................RBA
Not much sea ....US
Not much surf ....XFT
Not much to blame...............GMI
Not much wanted................YTP
Not my (mine)...................CFU
Not of ....CFW
Not often Seldom................SPY
Not on ....CFX
Not our-s (us)..................CFY
Not possible.....................PFB
Not quite........................UFP
Not ready.......................UJQ
Not reliable ....UOX
Not removed.....................UQE
Not room to wear ....MB
Not safe to risk making the land...VDE

NOT

NOT—Continued

Not severe ------------------------------VSX
Not signed ------------------------------WCF
Not so------------------------------------WIM
Not so many—as —----------------------RKP
Not so near------------------------------SFO
Not so often - - ------------------------SPZ
Not so well—as--------------------------ZBM
Not such a thing------------------------XPT
Not sure (certain)----------------------IHM
Not that (or, this)----------------------CFZ
Not the----------------------------------CGA
Not their-s (they) ---- -------------CGE
Not there -------------------------------CGB
Not these (or, those) ------------------CGD
Not time enough------------------------XTO
Not to ----------------------------------CGF
Not to be -------------------------------CGH
Not to be done--------------------------CGI
Not to be had---------------------------CGJ
Not to be inserted----------------------PMB
Not to exceed---------------------------MKP
Not to remain---------------------------UPL
Not too much to port-------------------TPF
Not too much to starboard-------------WQT
Not under command lights --- ----JEC
Not under control ---------------------JED
Not until -------------------------------YLF
Not very large--------------------------QFU
Not very narrow -----------------------SEA
Not wind enough------------------------MDR
Not with --------------------------------CGK
Not without permission ---------------TGF
Not worth getting------------------------OFL
Not worth getting off (or, afloat)----DQC
Not yet   By and bye ------------------HPX
Not you-r-s-----------------------------CGL
Ought not to----------------------------CHR
Ought not to be-------------------------CHS
Ought not to be able to — ---------CYP
Ought not to have----------------------CHT
Person-s, or, Thing-s, etc   (See He, above)
Shall, or, Will he (she, it, or person s or thing-s indicated) not?------BWH
Shall, or, Will he (she, it, or person-s or thing-s indicated) not be?----BWI
Shall, or, Will he (she, it, or person-s or thing-s indicated) not do (or, be done)?------BWJ
Shall, or, Will he (she it or person-s or thing-s indicated) not have? BWK
Shall, or, Will I not have?----------BRY
Shall, or, Will not -----------------CIU
Shall, or, Will not be ---------------BHD
Shall, or, Will not be able to — --CYS
Shall, or, Will not do (or, be done)-BNY
Shall, or, Will not have-------------BRZ
Shall, or, Will not have done (or, been done) -------------------------BNZ
Shall, or, Will they not?------------CIZ
Shall, or, Will they not do it?------BOC
Shall, or, Will we not?--------------CJB
Shall, or, Will you not?-------------CJE
Shall, or, Will you not be?----------BHI
Shall, or, Will you not do it?------BOE
Should or, Would he (she, it, or person s or thing s indicated) not?----BWR
Should or, Would he (she it, or person-s or thing s indicated) not be?----BWS

NOT—Continued

Should, or, Would he (she, it, or person-s or thing-s indicated) not do (or, be done)------------------------BWT
Should, or, Would he (she, it, or person-s or thing-s indicated) not have--BWU
Should, or, Would he (she, it, or person-s or thing-s indicated) not have been,------------------------BWV
Should, or, Would not----------------CJB
Should, or, Would not be ---------BHM
Should, or, Would not do (or, be done),------------------------BOH
Should, or, Would not have--------BSD
Should, or, Would not have been--BHN
Should, or Would not have done (or, been done) --------------------BOI
Should, or, Would not have had----BSE
That, or, This can not (or may not)-CKS
That, or, This can not (or, may not),------------------------CKT
That, or, This could (or, might) not be,------------------------CKW
That, or, This had (has, or have) not been ------------------------CKZ
That, or, This is not------------------CLF
That it is not-------------------------CIH
That, or, This must not --------------CLK
That, or, This must not be ----------CLM
That, or, This shall (or, will) not be-CLO
That, or, This should (or, would) not be------------------------CLQ
That, or, This was not----------------CLS
There are (or, is) not----------------CMH
There can not (or, may not)----CMK
There can not (or, may not) be----CML
There could (or, might) not be -----CMP
There had (has, or, have) not been--CMS
There shall (or, will) not ----------CMW
There shall (or, will) not be -------CMX
There should (or, would) not------CNB
There should (or, would) not be---CND
There should (or, would) not have been,------------------------CNE
There was (or, were) not------------CNG
They are not -------------------------BEV
They can not (or may not)--------BKU
They can not (or, may not) be------BHQ
They can not (or, may not) have---BSG
They could (or, might) not --------BLO
They could (or, might) not be -----BHS
They do (or, did) not----------------BOK
They had (or, have) not ---------BSJ
They had (or, have) not been------BHU
They had (or, have) not done ---BOM
They had (or, have) not had -----BSK
They must not ----------------------CFH
They must not be-------------------BHW
They ought not to--------------------CIF
They ought not to be----------------BHY
They shall (or, will) not ----------CJG
They shall (or, will) not be -------BIA
They should (or, would) not------CJZ
They should (or, would) not be --BID
They should (or, would) not have--BSM
They were not ----------------------COK
They were not to be-----------------BIF
Was, or, Were, he (she, it, or person s or thing s indicated) not--------BWY
Was I not?------------------------COU
Was, or, Were not ------------------COV

NOT—NUMBER

NOT—*Continued*	SKA NOTHING
Was, *or*, Were not done (*or*, doing)..BOP	Bring nothing .. . .HIS
We are not........... ....BEX	Let nothing prevent...............QNV
We can not (*or*, may not).........BKW	Nothing additional ....DJE
We can not (*or*, may not) be ....RJJ	Nothing belonging to you—(*or*, —).GEX
We can not (*or*, may not) have ....BSO	SKB —Nothing can (*or*, could)
We could (*or*, might) not ...:.....BLQ	Nothing decisive yet ....KPV
We could (*or*, might) not be.......BIM	Nothing else ....LVT
We do (*or*, did) not ....BOR	SKC —Nothing for you (*or*, for —)
We had (*or*, have) not.............BSE	Nothing important by last accounts.DFL
We had (*or*, have) not been........BIO	Nothing in sight..................VNW
We had (*or*, have) not done .....BOT	Nothing less ....QNP
We had (*or*, have) not had........BST	Nothing more ....RZW
We must not ....CFJ	SKD —Nothing particular
We must not be............. .....BIQ	SKE —Nothing serious
We ought not to..............CIH	SKF —Nothing shall
We shall (*or*, will) not...........CJI	Nothing to be depended upon beyond
We shall (*or*, will) not be ....BIS	your own resources ....KUM
We shall (*or*, will) not have ......BSV	SKG —Nothing to report
We should (*or*, would) not ....CKB	Say nothing......................VIK
We should (*or*, would) not be.......BIU	There is nothing ....CMT
We should (*or*, would) not have ...BSX	
We were not ....CPF	SKH NOTICE-D-ING TO NOTICE
We were not to be...............BIW	SKI —Did not notice
Were they not? .... CPI	SKJ —Did you notice?
Were we not? - ....CPK	Give notice ....OAB
Were you not? ....CPM	Notice (*warning*) has been given...YUL
Why are you not? ....CVB	SKL —Noticed nothing
Why can not (*or*, may not)? ....CVD	Short notice ....VZW
Why could (*or*, might) not?.....CVF	Stranger (*or vessel indicated*) will not
Why do (does, *or*, did) not? ....CVH	take any notice of signal... ...WDK
Why do (*or*, did) you not?.........CVJ	Take no notice . ...XJR
Why had (has, *or*, have) not?......CVL	Take notice................XJS
Why had (*or*, have) you not? .....CVN	Your zeal has been particularly noticed
Why not? ....CVP	by ....ZMD
Why shall (*or*, will) not? ....CVS	
Why should (*or*, would) not?.......CVU	SKM NOTIFY-IED-ING-ICATION
You are not ....BFZ	SKN —Have notified
You are not able to — .... CYW	SKO —Will, *or*, Shall be notified
You can not (*or*, may not) ....BKY	
You can not (*or*, may not) be .. ...BIZ	SKP NOTWITHSTANDING
You can not (*or*, may not) have....BSZ	
You could (*or*, might) not ....BLS	SKQ NOUGHT
You could (*or*, might) not be ......BJC	
You do (*or*, did) not.............BOX	SKR NOURISH-ED-ING-MENT.
You had (*or*, have) not.. ....BTD	
You had (*or*, have) not been.. ... BJE	SKT NOVEL-TY
You had (*or*, have) not done......BOZ	
You had (*or*, have) not had..... ..BTE	SKU NOVEMBER
You must not ....CFL	About November the —............CZT
You must not be ....BJG	SKV —Beginning of November
You ought not to ....CIK	SKW —End of November
You ought not to be...............BJI	SKX —Next November
You shall (*or*, will) not...........CJL	
You shall (*or*, will) not be.........BJL	NOW AT ONCE ....FIN
You shall (*or*, will) not have ......BTG	Just now........................PXL
You should (*or*, would) not... ....CKE	Mail is now at ....RPJ
You should (*or*, would) not be.....BJN	Now doing. ....LIS
You should (*or*, would) not have...BTI	
You were not ....CPO	SKY NOWHERE
You were not to be ....PJT	
	SKZ NOZZLE
SJX NOTARY	SLA NUISANCE
	SLB NUMBER-ING TO NUMBER
SJY NOTE-D-ING-ATION TO NOTE	SLC —Numbered
Advance note ....DMW	A large (great) number (*or*, quantity)—
Bank note, *or*, Paper money . .... ASV	of ....OOZ
Note (*letter*) ....QOC	A larger number (*or*, quantity)—of —
SJZ —Note of interrogation Query	than.......................QFM

NUMBER—OBEY.

SLD	NUMBER—*Continued*		SLM	NUMERAL—*Continued*
	—A small (*or*, smaller) number (*or*, quantity)—of —.			—Numeral signals (*when spoken of*) (*For use, see page* 32)
SLE	—A very large number (*or*, quantity)—of —			Numeral signal No 1 *
	An average number (*or*, quantity)—of ------------------------------FOQ			Code Flag over M
	An equal number (*or*, quantity)—of —, MGF			Numeral signal No 2 *
	Boiler burst, number of killed and wounded not yet known --------AX			Code Flag over N
SLF	—Has vessel made her number?*			Numeral signal No 3 *
	Hoist your distinguishing number (*or*, signal)*-------- ----------DV			Code Flag over O
	I, *or*, She exchanged numbers with —, MKW			*For method of using these signals, see page* 32
	Indicate the number (*or*, quantities)—of------------ ----------PIA		SLN	—Numeral table (*Pages* 32–34)
	Lost number of blades of screw (*indicated*) ---------------- --------GLZ		SLO	NUMEROUS-LY
	Make your number (*distinguishing signal*) ----------------------DV		SLP	NUN-NERY
SLG	—Number, *or*, Quantity is incorrect		SLQ	—Nun buoy
	Number of columns ------------ ..JBX		SLR	NUNCIO.
	Number of killed and wounded not yet known -----------------------QAP		SLT	NURSE-D-ING
	Numbered ---------------------SLO			Are there any sick nurses to be got?..WAZ
	Numberless---------------------SLI			Sick nurse-----------------------WBG
	Numbers-----------------------SLJ			You had better get a sick nurse...WBK
	Official number*-------- --------SPR		SLU	NUT
	Shorten in cable to the number of shackles indicated--------------HRP			Cocoanut ----------------------IYX
	Show the number (*or*, quantity)—of —, PIA		SLV	NUTMEG
	Show your distinguishing signal (*number*) ---------------------DV			
	The greater-est number (*or*, quantity)—of —------------------------OHN			
	The least number (*or*, quantity)—of —, QKH			
	Too large a number (*or*, quantity)—of —, QFV			
SLH	—Vessel has not made her number.			
	Vessels just arrived, make your numbers (*or*, distinguishing signals)*..DX			
	Vessels just weighed (*or*, leaving), show your numbers (*or*, distinguishing signals)*-------------------------DZ		SLW	O (*Letter*) (*For new method of spelling, see page* 13)
	Vessels that wish to be reported all well, show your numbers (*or*, distinguishing signals)*---------------------DYT		SLX	OAK-EN
	Vessels wishing to be reported on leaving, show your numbers (*or*, distinguishing signals)*----------------EB		SLY	OAKUM
	What is the number of mail bags you have on board?----------------FSP		SLZ	OAR.
	What is your name (*or* number)?*..SDT			Boat's oars ----------------------GRS
	What number (*or*, quantity) of — have you? ------ --------------------UDW		SMA	OATH
	See LIST OF MERCHANT VESSELS		SMB	OATMEAL
SLI	NUMBERLESS		SMC	OATS
SLK	NUMBERS		SMD	OBEDIENT-LY-CE—TO
SLJ	NUMERAL-ICAL-LY		SME	OBEY-ED-ING
	Annul the hoist shown by numeral signal---------------------- ..EMI		SMF	—I can not obey your order
	Numeral flag---- ----------------NDJ		SMG	—Obey order

OBJECT—OCEAN

SMH	OBJECT .		OBSTRUCT—*Continued*
SMI	—Is there any object?		Chain obstruction................IJO
	Keep the two objects named in one line, QUM		Channel, *or*, Fairway is obstructed .ILF
SMJ	—Object is		Floating obstruction in (*or*, at)— . NFD
SMK	—With the object (*or*, view) of		Net obstruction ....SGY
			Obstructions removed ....UQF
SML	OBJECTION-ABLE	SNK	Remove obstruction........... ....UQG
SMN	—Has it been objected to?		—Sunken obstruction in (*or*, at)—
SMO	—Have you any objection—to?		The channel is obstructed by a ship.ILT
SMP	—No objection—to	SNL	—There is no obstruction
SMQ	—Objected—to—		
	Objectionable character... ....ILY	SNM	OBTAIN-ED-ING-ABLE (*See also* GET *and*
			PROCURE)
SMR	OBLIGE-D-ING—TO (*See also* BOUND *and*	SNO	—Can be obtained—at
	COMPEL)		Can I obtain ballast at —? ....FTP
SMT	—Is obliged to undergo repairs		Can labor be obtained?............QOS
SMU	—Obligation ory		Can pratique be obtained at — ?....TRY
SMV	—Obliged to be given up	SNP	—Can you obtain (*or*, procure)?
	Obliged to cut away ....KHQ		Carpenters and caulkers may be ob-
	Obliged to discharge LCX		tained at — ....IBU
SMW	—Obliged to give it up		Fresh vegetables to be obtained.....NRB
SMX	—Obliged to scud		Fruit is to be obtained on shore .. NSE
	Obliged to stop engines.... ......WXH	SNQ	—Have, *or*, Has not obtai..ed
SMY	—Obliged to throw overboard		Have obtained soundings ....WKN
	Unless we are obliged ....YJR		Have obtained the following informa-
	Was obliged to put back........ . FQN		tion from — ....NGV
SMZ	—Will be obliged to	SNR	—Have you obtained?
SNA	—You will be obliged to		How is your latitude obtained?....QHN
SNB	—You will not be obliged to		I wish to obtain orders from my owners,
			Mr —, at — ....SW
	OBLIGED—FOR (*Grateful*) ....OGQ		Is cotton freight to be obtained?...JXW
		SNT	—It is not to be obtained
	OBOLUS (*Measure of weight*)........BAJ	SNU	—It is to be obtained
		SNV	—Must obtain (*or*, be obtained)
SNC	OBSCURE-D ING-ITY.		No fruit is to be obtained on shore .NSF
			No intelligence obtained from vessel
SND	OBSERVATORY		boarded...........................POG
			No vegetables to be obtained ....YOI
SNE	OBSERVE-D-ING TO OBSERVE (*See also*	SNW	—Obtain all the intelligence possible
	SEE)	SNX	—Shall, *or*, Will obtain
	Dead reckoning agrees with observa-		Shelter may be obtained.... ......VWM
	tion...... . . ..DTX		Water can readily be obtained ....YWS
	Did not observe (*or*, see) ....VNF		Water can not be obtained ....YVN
	Did you get an observation? ... NYR		You will have great difficulty in obtain-
	Did you get good observations? .....RM		taining water ... .LAF
	Did you observe (*or*, see)?.........VNG		
	Good observations........... ....OEC	SNY	OCCASIONAL-LY
SNF	—Have, *or*, Has been observed (*or*, seen)		
	Have you taken observations for varia-	SNZ	OCCASION-ED-ING TO OCCASION
	tion?.....................JHS		Authority sufficient for the occasion
	I have had no observation for variation		FMP
	lately ....JHV		Happened in consequence of Occa-
	Latitude by observation—is ......QHU		sioned by............... ....JPZ
	Longitude by observation—is ......QZD		
	Night observation ....SIE	SOA	OCCULTING LIGHT
	No recent observations I did not get	SOB	—Group occulting light
SNG	an observation ....RT		
SNH	—Observation	SOC	OCCUPY IED-ING-ATION
	—Observations on the —		
	Tolerable observations... ....XUM	SOD	OCCUR-RED-RING RENCE—ON (*or*, TO)
	When were your last observations for		How did it (*or*, that) occur (*or*, happen)?
	latitude?..... ....QIF		OMK
	When were your last observations for		It frequently occurs ....NQT
	time? ....RY	SOE	—When did it occur?
SNI	OBSTACLE	SOF	OCEAN-IC
			North Atlantic ....FJL
SNJ	OBSTRUCT-ED-ING ION		North Pacific.. ....SYW
	Beware of boom (*or*, obstruction) ..GJU		South Atlantic ....FJM
	Boom, *or*, Obstruction broke adrift..GZM		South Pacific....... ....SYX

OCHR—OFF

	OCHR-EL-GUERCH (*Coin*) ....AUK
	OCK (*Measure of weight*)...........BAQ
	O'CLOCK (*See* CLOCK)....... ....ZNG
SOG	OCTOBER
	About October the —.................CZU
SOH	—Beginning of October
SOI	—End of October
SOJ	—Next October
	OF...............................CGM
	By way of.. . ...HPS
	In place of ....PNA
	Not of.............................CFW
	Of course... ....JZW
	Of him (his, her s, it-s)... .. ..CGN
	Of me (my, mine)............. ....CGO
	Of no importance....... . ...PEZ
	Of our-s (us) ....CGP
	Of that (this) ....CGQ
	Of the .. ..CGR
	Of them (then-s) .... CGS
	Of these (or, those)... . ...CGT
	Of water .... YVI
	Of what (or which).................CGU
	Of whom (or, whose) ....CGV
	Of you-r-s ....CGW
	What, or, Which of the —?.........CQI
SOK	OFF
SOL	As soon as she is off (or, afloat) ...DPW
	—At —(*time indicated*) we were off —
	Beating off................... ......GCA
	Becalmed off — for —days...... ...GCI
	Blockade is not taken off ....GND
	Blockade is taken off ....GNE
	Blow-off cock (or, gear) ....GNW
	Blow-off valve ....GNZ
	Blow off your steam................GOA
	Boiler leaky, must be blown off.. ..GWC
	Bring off ....HIT
	Can be cut off ....KHF
	Can not be got off (or, afloat) by any
	means now available ....CD
	Can not save the ship, take people off.NB
	Cast off...........................IED
	Cut off............................KHJ
	Expect (or vessel indicated expects) to
	be got off (or, afloat) ....DPX
	Far-ther off (or, from).............MSV
	Fend off ....MVT
	Fire gains rapidly, take people off ..NG
	Fleet, or, Squadron when last seen
	were off —......................NEO
	Haul off...........................OPC
	Have, or, Has got off ....OPG
	Heave-to, head offshore ....LQ
	How far off? What distance?......MTA
SOM	I can get off........ ....NZB
	—I shall be off — at — (*time indicated*)
	I shall get off (or, afloat) with assist-
	ance ....DPY
	I stand off and on the land.. . ....QEB
	I shall stand off the land from —(*hour
	indicated*) to — (*hour indicated*), or, I
	shall stand off until ....QBC
	If you will lay out an anchor for me I
	can get off.. .. ..EIV

	OFF— *Continued*
	It is likely you (or, vessel indicated) will
SON	be got off?........... ....OFJ
	—Is she (*vessel indicated*) off?
	It is expected the embargo will be taken
	off..... . ...LWJ
	Keep farther off (or, from) ....MCT
	Keep off and on and await instructions,
	SX
	Lighter is coming off — to you......QZ
	May be got off ....OFK
	May be got off (or, afloat) if prompt
	assistance be given ....CG
	May be got off (or afloat) without
	assistance ....DPZ
	Met her off (or, in)—.... .. ....RQY
	Must cast off I must cast off......IEH
	Not worth getting off (or, afloat) ...DQC
	Off and on...........:...........SOW
	Offshore Off the coast ....IYD
SOP	—Off the —
	Off the buoy.....................HMV
	Off the cape ....HVR
	Off the entrance. ....MFT
SOQ	—Off the island
SOR	—Off the land
SOU	—Off the light
	Off the mouth of the river (or, —).UYJ
	Off the point ....TNM
	Off, or, From the wind......... ..NRY
	Off the wreck.... ....ZIU
	Off this port.... ... .TOR
	Pay-ing, Paid off ....TEP
	Pilot boat most likely on — (*bearing in-
	dicated*) (or, off —) ....TK
	Pilot can not get off............. .TJP
	Put off...........................UCZ
	Put off for the present.............TUD
SOV	—Put off the sale
	Send the mate—off............ ....RMS
	Several vessels becalmed off ....GCJ
	Shall I cast off?....... ....IEJ
	Sheer a little off ....VUO
	Sheer a little off when you come to the
	point ....VUP
	Sheer-ed-ing off ....VUN
	Shove d-ing off ....WAJ
	Stand off Get an offing Put to sea at
	once ....MG
	Stand, or, Haul off.................OPC
	Stand off and on the land..........WQH
	Sunk off (or, near) the — . ...WEO
	Take n-ing, Took off...............XJT
	The tide will take her off (float her),
	NFE
	The way is off my ship, you may feel
	your way past me.................HL
	Too far off. ....MTD
	Vessel (*indicated, if necessary*) has got
	off ....NZE
	What distance off is —?......... ...LGM
	When did you come off?..... ....JDN
	When do you come off?....... ....JDO
	When do you expect to be off (or, abreast
	of) —?.........................DAP
	When I have run my distance, I shall
	put her head off until —...... ...LGP
	When will the embargo be taken off?
	LWN
	You are too close in Keep farther off,
	IVL

OFF AND ON—OFFICER

SOW	OFF AND ON
	I shall stand off and on the land ...QEB
	Keep off and on and await instructions, SX
	Stand off and on the land _____WQH
SOX	OFFEND-ED ING-CE TO OFFEND
	Be careful not to give offence ____ _HYL
SOY	—Offender
SOZ	—Offensive
SPA	OFFER-ED-ING—TO TO OFFER
	Any freight offering?_____EOZ
	First offer_____NAZ
SPB	—Have you had an offer?
	Made an offer of —_____ _____REF
SPC	—Offer (an offer)
	Offer, or, Proposal not accepted ___DBW
	Offered assistance, but they declined, FGM
SPD	—What was the offer?
	Will accept your offer _____DBY
	Will you make an offer?_____RIZ
SPE	OFFICE
SPF	—Admiral's Office
	Badge of office _____FSA
	Bureau de Change Exchange office, HNC
	Call at the post office for letters____HSY
	Colonial Office_____ _____JBL
	Commander in-Chief's Office_____JEQ
	Detained by the Emigration Office_KYB
	Emigration, or, Immigration Office_LXJ
	Foreign Office, or, State Department, NKR
	Go to the Admiral's Office _____DLF
	Harbor master's office_____OMY
	Have you settled with the Emigration Office_____LXK
	Weather Bureau, or, Meteorological Office reports depression approaching from the —_____ETP
	Weather Bureau, or, Meteorological Office reports gale approaching from the —_____ETQ
	Post office, For, or, At the post office TQJ
	Post-office order_____TQL
	Report me to "New York Herald" Office, London _____UI
	Report me to "New York Herald" Office, New York___ _____UJ
	Report me to "New York Herald" Office, Paris _____UK
	Secretary's office _____VMQ
	Send to the Admiral's Office _____DLK
	Shipping office _____VYB
	Telegraph station (or, office)_____XME
	The office is closed____ _____IVH
	There are orders for you at the Admiral's Office _____DLM
	To, or, At the post office Post office, TQJ
SPG	—War office, or, War Department
	You are wanted at the — office ____YUB

SPH	OFFICER
	Admiral, or, Senior officer has commands for you ____ ____ _____DKP
	Admiral, or, Senior officer has no commands _____DKQ
	Admiral, or, Senior officer has requested, DKR
	Admiral, or, Senior officer intends__DKS
	Admiral, or, Senior officer is generally about _____ _____DKT
	Admiral, or, Senior officer is off (or, at) —, DKU
	Admiral, or, Senior officer is on board— (or, at —) _____DKV
	Admiral, or, Senior officer is on shore, DKW
	Admiral, or, Senior officer requests the pleasure of your company at dinner, DKX
	Admiral, or, Senior officer wishes__DKY
	Admiral, or, Senior officer wishes to know _____DEZ
	Admiral, or, Senior officer wishes to see you (or person indicated)_____DLA
	Admiral, or, Senior officer wishes to see you as soon as convenient_____DLB
	Admiral's, or, Senior officer's memo_DLC
	Admiral's, or, Senior officer's order__DLE
SPI	Board of Trade officer (or, inspector)_GPD
SPJ	—Boarding officer
	—Can, or, May officers?
	Chief officer_____IOS
	Commanding officer _____JDY
	Commanding officer desires _____JDZ
	Commissioned officer _____JFD
	Custom-house officer_____KGW
SPK	—Direct officer to —
	Excise officer_____ _____ ___MLC
	First officer _____ _____NBC
	Foreign officer _____NKS
	Have not seen the admiral (or, senior officer) ____ _____ ___DLG
	Have you seen the admiral (or, senior officer)?_____DLH
	Junior officer _____PXA
	Medical officer _____ROJ
	Military officer _____ _____RTY
	Naval officer _____SEQ
	Naval Reserve officer_____SEU
	Officer and men_____RPI
	Officer, or, Employee of Government, OFT
	Officer of Naval Reserve_____SEU
SPL	—Officer of the guard
SPM	—Officer of the watch
SPN	—Officer's servant
	Petty, or, Non-commissioned officer_THJ
	Quarantine officer _____UEF
	Salvage officer _____VGE
	Second officer (or, mate) _____VME
SPO	—Send an officer—to (or, for)
	Sending. Sent an officer—to (or, for)_VPO
	Senior officer_____VQZ
	Shipping officer _____VYJ
	Signal officer _____WDA
	Third officer (or, mate) _____XQO
	Under officer of ____ _____ _____YHI
	What is the name of admiral (or, senior officer)?_____ _____DLO
	Where is the admiral (or, senior officer)? SJ

OFFICER—ON

	OFFICER—*Continued* Where is the commanding officer? __JEI Where is the health officer?_____ORD	SQE	OILCAKE
		SQF	OINTMENT
SPQ	OFFICIALLY Await outside the inspection of the officials _____FPB Have you a Custom-house official on board?_____KGX Is the news official?_____SHJ Official log-book_____GZD		OKE (*Measure of weight*) _____BAQ
SPR	Official number	SQG	OLD How old? What age?_____DSI Old berth_____GFX Old canvas ____ _____HVI
SPT	Officially confirmed (*or*, certified) __LJF Official (officer) Passed the official inspection_____TCU Post-office official (*or*, postmaster) _TQK Signal the vessel's official number—to, DV Telegraph clerk (*or*, official)_____XLZ Want official confirmation _____JOH	SQH SQI SQJ	—Old rope Old system _____XID —Older —Oldest Too old _____XVE
SPU	OFFING A good offing_____ODR Cut, *or*, Slip Get an offing Weigh Wait for nothing _____KW	SQK SQL	OLIVE —Olive oil
SPV	—In the offing Put to sea at once Get an offing Stand off _____MG	SQM SQN SQO	OMIT-TING To OMIT —Do not omit —Omitted Omission
SPW	OFTEN As often as_____FBD	SQP	OMNIBUS Omnibus, *or*, Car leaves at — _____HYB When does omnibus (*or*, car) leave for —"_____HYC
SPX SPY SPZ	—How often? —Not often Seldom —Not so often Very often_____YPM		ON _____CGX All on board _____CZA Bring on board_____GPZ Bring on shore _____HIU
SQA	OHM (*Measure of electrical resistance*)		Broadside on (*or*, to) _____HJP Captain is not on board_____HWK Captain is on board_____HWL
SQB	OIL ED-ING To OIL Apparatus for casting oil on the sea to quiet it _____ERF Barrel of oil _____FXU Calm the sea by pouring oil on it__HTN Castor oil_____IEP Colza oil _____JBZ Croton oil _____KBP Gallipoli oil _____NWJ I am in want of oil for machinery__RDU Lamp oil_____QDH Linseed oil_____QVB		Captain is requested to come on board HWN Come on board _____GQA Continue on your course. _____JSM Crew not all on board_____DYA Due on _____LOW End on _____LYM Hold on _____OWR I am coming (*or*, sending) on board_CZB I have no combustibles on board____HQ I shall stand off and on (*the land*) _QEB I shall stand on until—_____WQD
SQC	—Lubricating oil Mineral Oil _____RUX Neat's-foot oil _____SFR		Is the captain on board?_____HXE Keep her bows on _____HCI Keep on and off, and await instructions, SX
SQD	—Oil-y Oil bag_____FSK Oil cake _____SQE Olive oil _____SQL Paint oil _____SZQ Paraffin oil _____TAS Petroleum oil _____THE Rangoon oil _____UIB Smooth-ed-ing sea by pouring oil on it, HTN Sperm oil_____VNI Sweet oil ____ _____XHF Vegetable oil _____YOJ Whale oil _____ZBY		No boat on board _____GTX Not on_____CFX Observations on the — _____SNH On_____CGX On account of_____DFP On and off _____SOW On board Aboard _____CYX On both sides _____HAR On condition _____JND On deck _____KQC On delivery _____KSW On demand—of _____KTD On fire. _____ _____NAP

ON—OPPORTUNE

		On—*Continued*
		On him (his, her s, it-s) _____ ___CGY
		On His (or, Her) Majesty's service_RHS
		On leave_____ ___ _ _____ QLY
		On leaving_____ ___ _ ____ _ QLK
		On lee bow Lee bow _____HCN
		On me (my, mine) ____ _____CGZ
		On our-s (us) _____CHA
		On no account _____DFR
		On port bow _____HCO
	SQV	On shore _____ _____VZM
		On starboard bow _____HCF
	SQW	On that_____CHB
		On the_____ _____CHD
		On the — day of the month ____KNQ
		On the — (*date indicated*) __ ____KMF
	SQX	On the bar _____ _____FVZ
	SQY	On the beach_____FZY
		On the beam _____GAW
	SQZ	On the coast_____ _____IYE
		On the left bank of the river _____FVE
		Note—*The right or left hand bank in descending the stream*
	SRA	On the other _____SVM
		On the port quarter _____UEQ
	SRB	On the right bank of the river _____FVG
		Note—*The right or left hand bank in descending the stream*
	SRC	On the road—to_____UYP
		On the rock _____ _____UYW
	SRD	On the same day_____KNR
		On the shoal___ _____ __VYU
		On the starboard quarter_____UER
		On the way—to _____YXF
		On the wreck_____ __ZIV
		On them (their s) _____ _____CHE
		On these (or, those)___ _____CHF
	SRE	On weather bow _____ ____ __HCQ
		On what (or, which)_____ ____CHG
	SRF	On what date did you speak the — (*vessel indicated*)? _____KMG
		On whom (or, whose) _____CHI
	SRG	On your-s _____ _____CHJ
		Run on shore _____ _____VCE
	SRH	Stand on_____ _____ _____MH
		Too much strain on—_____WYI
	SRJ	Was to sail on the —____ ___ ____VEU
SQR		—We sail on the —
	SRK	Will you (or *person indicated*) come on
	SRL	board? _____GRT
	SRM	Will your surgeon come on board?__WP
	SRN	
SQT		**Once**
	SRO	At once Now _____ _____ FIN
	SRP	I can begin at once (or, on —, *date specified*) ____ _____GDO
	SRQ	It is dangerous to allow too many people on board at once_____GB
	SRT	
SQU		**One**
	SRU	Alter course one point to port_____JZD
		Alter course one point to starboard_JZE
	SRV	At one _____ _____ FIO
		Either one Either _____LUD
		Every one Everybody_____ _____MIU

	One—*Continued*
	I have only one boat (*indicate other number, if necessary*)_____GTO
	No one Nobody _____SIU
	No one hurt_____ ___PAT
	No one killed_____QAO
	One screw disabled, can work the other,
	_____MW
	Only one anchor left_____ _____EJG
SQV	**Onion**
SQW	**Only**
	I have only — days' provisions _____TO
	I have only — days' water ___ ____ .ZA
	Only chance of safety __ _____ IKG
SQX	—Only for a short —
SQY	—Only intended as —
	—Only intended for a feint _____MVG
SQZ	—Only intend to —
	Only one anchor left___ _____ EJG
	Only to be got in small quantities__OFM
SRA	**Onward**
SRB	**Onzf** (*Measure of weight*) _____BAO
	Ooze-d-ing
SRC	**Open-ed-ing-ly**
	Keep open order_____PZH
	Open Get farther away _____ MC
SRD	—Open fire or, Commence firing
	Open sealed orders_____SUC
	Open the telegram for me and signal its contents____ _____ WX
	Shall I open your telegram and signal its contents?_____ XC
	The weather is still open (at — *to follow, if necessary*)_____YZH
SRE	**Opera**
SRF	**Operate-d-ing-ion**
	Base of operations_____FYD
SRG	—Operations have commenced
SRH	**Operator-ive**
SRJ	**Ophthalmia-ic**
SRK	**Opinion**
SRL	—In my opinion, or, My opinion is—that—
SRM	—In your opinion
	What is the general opinion? _____NXQ
SRN	—What is your opinion—of —?
SRO	**Opium-iate**
SRP	—Opium trade
SRQ	**Opodeldoc**
SRT	**Opponent**
SRU	**Opportune-ity—to** (or *for*)
	An opportunity for disposing of cargo,
	_____HZQ
SRV	—An opportunity for selling (or, disposing of) —

OPPORTUNE—ORDER

	OPPORTUNE—*Continued*
	An opportunity for sending letters to Europe (*or*, to —) ------QOD
	Every opportunity --------------MJA
SRW	—Few opportunities
	First opportunity --------------NBD
SRX	—Is there any opportunity?
SRY	—Next time (*or*, opportunity)
	No opportunity (*or*, chance) -------IKE
	Opportunity for action----- -------DIU
	Opportunity for sending letters to Europe (*or*, to —) --------------QOD
SRZ	—Opportunity to send
	Some future opportunity ---------NUS
STA	—There is an opportunity
STB	OPPOSE-D-ING-ITION
	Any opposition—to (*or*, from) -----EPK
	Great opposition—to (*or*, from) ----OHF
STC	—Must oppose
STD	—No opposition Unopposed
	Opposition majority of —--------RHW
STE	OPPOSITE
STF	OPPRESS-ED-ING-IVE-OR
STG	OPTIC-AL-IAN
STH	OPTION-AL-LY
STI	OR
	Or else------------------- ----LVU
STJ	ORANGE
STK	ORDER
	NOTE —*For machinery, etc , out of order, see under the part of machinery in question*
	According to order (*or*, direction)--DEM
	Admiral's, *or*, Senior officer's order--DLE
	After order ------------------- ----DRF
	Am, Is, *or*, Are ordered (*or*, directed) to (*or*, by) ---------------- -----LBG
STL	—Are there any orders?
STM	—Are you going to wait for orders?
	Attend to orders --------------FLH
	Bad order In bad order ----------FRI
STN	—By your owner's order (*or*, instructions)
	Call off — for orders ----------HTA
	Cargo in very good order---------HZX
	Contrary to orders.---------------JSX
STO	—Counter orders
	Disobey-ed-ing-dience of orders ----LEZ
	General orders --------------- ------NXP
	Good order In good order---------OEB
STP	—Have no orders
STQ	—Have, *or*, Has orders for
	Have orders been given—to (*or*, for)? PNE
	Have orders (*or*, telegram) for you--SM
	Have received the following communication (orders, *or*, instructions) from your agent (*or*, owner)---------IA
	Have you any orders (telegram, message, *or*, communication) for me?-------IB
	Have you received any orders—from —? SU

	ORDER—*Continued*
STR	—Have you received your orders?
	I am directed to acquaint you--- ----DHI
	I can not obey your orders---------SMF
	I expect your orders every moment-MNL
	I have instructions (*or*, orders)-----PNF
STU	—I have orders for you
	I have orders for you not to touch (*or*, call) at —--------------------SN
	I have orders to telegraph your passing. SO
STV	—I have received orders
	I have received orders for you not to proceed to —(*place next indicated*) on account of ice----- - ----------PBR
	I have received orders for you not to proceed without further instructions, SP
	I have received orders for you to await instructions from owners at — --- SQ
	I have received orders for you to discharge cargo at —----------------SR
	I have received orders for you to proceed to —---------------------TF
STW	—I have telegraphed for further orders (*or*, instructions)
	I have telegraphed for your orders --ST
STX	—I must remain till further orders
STY	—I send an order—for —
	I will telegraph for your orders if you will wait reply -------------------SV
	I wish for orders from Mr —, my owner, at —---------------- ------------SW
	NOTE —*This signal is to be followed by*
	1 *Ship's number, if not already made*
	2 *Name of owner, from* SPELLING TABLE
	3 *Owner's residence from* GEOGRAPHICAL *or* SPELLING TABLE
STZ	—If you order me
	In bad order Bad order ------ ---FRI
	In good order Good order---------OEB
SUA	—In order—to
	Keep close order -- ---------------PYQ
	Keep off and on and await instructions (*or*, orders) -------------------- SX
	Keep open order-------------------PZH
	Machinery out of order ----------- RJ
SUB	—My orders
	No instructions (*or*, orders) --------PNI
	Not according to orders Against instructions----------------------DER
	Obey orders---------------------SMG
SUC	—Open sealed orders
	Order book-----------------------GZE
SUD	—Order ed-ing by telegraph
	Order canceled-------------------HUP
SUE	—Ordered
SUF	—Ordered back
SUG	—Ordering
SUH	—Orderly
SUI	—Orders have been received
SUJ	—Orders have been sent
SUK	—Orders have not arrived yet—for —
SUL	—Out of order
	Owners desire me to order you to coal at —------------- ----------------HE
	Owners desire me to order you to discharge cargo first at —(*place next indicated*) ------------------ ---------SY

ORDER—OUGHT

	ORDER—*Continued.*		ÖRE (*Coin*) ----------------------AUL
	Pay attention to orders----------FLH		
	Post-office order -------------------TQL	SVF	ORE
	Previous order (*or*, instructions)---PNJ	SVG	ORGANIZE-D-ING-ATION
	Previous orders are all countermanded, JYP		Fleet organization------------------NEP
	Proceed in execution of previous orders, SZ	SVH	ORIENTAL
	Pumps in good order-------------UBP	SVI	ORIGIN-AL-LY
	Repeat my signals until further orders, WCP		Your original orders are canceled, I am directed to inform you to proceed
	Returns ordered -----------------UVR		to -- . -----------------------------TH
	Sailing orders ---------------------VEL		
	Sealed orders ------------ -- -------VLN	SVJ	ORLOP
SUM	Secret orders---------------------VMJ	SVK	OSCILLATE-D-ING-ION
SUN	—Send for orders		
	—Send my orders on to—(*place indicated*)	SVL	OTHER
SUO	Shall I telegraph for your orders?---TB		Any other----------------------------EPL
	—Should be ordered		Each other---- ----------------- ..LQX
	Standing orders --------- --------WQK		Get her on the other tack as fast as possible ------ ---- ---------NYU
	Stay, *or*, Stop for orders----------TE		Get her on the other tack, or you will be on shore-------- ---- ----------LF
	Telegraph orders (*or*, instructions)— to -------- ------ ----------TC		I shall not go on the other tack unless I fall off — to — ----------------MRU
SUP	Telegraph out of order (*or*, stopped), XMC	SVM	—On the other
	—The orders are coming off		Other case ----------------------IDO
	There are no orders here for you ----TD		Other cause----- ----- -----------IFS
	There are orders for you at the admiral's office ---------- -- ----------DLM	SVN	—Some other
SUQ	—Under orders — for (*or*, to)—	SVO	—The other side
SUR	—Under orders of		We shall do better on the other tack, XIN
	Until further orders--------- -----YLG	SVP	OTHERWISE
	Urgent orders (*or*, instructions) ---FNL		
	Verbal orders ---------------------YOZ		OUGHT TO------------ --------------CHK
	Wait for orders---------- ------TE		He, She, It (*or person-s or thing-s indicated*) ought to------------------BVD
SUT	—Wait off here for orders		He, She, It (*or person s or thing s indicated*) ought not to------------BVE
	Want immediate orders ----------YK		He, She, It (*or person s or thing s indicated*) ought not to be------------BVG
SUV	Weigh, and proceed as ordered -----KV		He, She, It (*or person-s or thing-s indicated*) ought to be--------------BVF
	—What are your orders?		How ought?-----------------------BXQ
	Where do you intend calling for orders? HTJ		I ought to ------------------------CHL
SUW	—Who sent the order?		I ought not to ---------------------CHM
	Written orders -------- -----------ZJU		If I ought to ---------------------BZL
SUX	—You are not likely to receive your orders for some days		If I ought not to -----------------BZM
			It ought not to be ---------------BGS
SUY	—You are ordered not to come into port		It ought not to be done -----------BNS
SUZ	—You are ordered on to —		It ought to be --------------------BGR
	You are ordered to proceed to — ----TF		It ought to be done ---------------BNR
	You must remain until further orders, NUM		Ought he (she, it, *or person-s or thing s indicated*) to?---------------CHN
	Your orders (*or*, instructions)------PNM		Ought his (her-s, it-s)? ----- -------CHO
	Your orders are at (*or*, will be at) — TG		Ought I (my, mine)?--------------CHP
	Your orders are canceled---------HUQ		Ought I to — do?---- ------- -----CHQ
	Your orders are not ready for you--UKE		Ought not to----------------- -----CHR
SVA	—Your orders have not arrived yet		Ought not to be ------------------ CHS
SVB	—Your orders will be awaiting you off (*or*, at)—		Ought not to be able to — ---------CYP
	Your original orders are canceled. I am directed to inform you to proceed to—, TH		Ought not to have--------------CHT
			Ought that (*or*, this)?-------------CHU
	Your papers are not in order-------TAP		Ought the?-----------------------CHV
	NOTE.—*For machinery out of order, look under the part of machinery in question*		Ought these (or, those)?----------CHW
			Ought they to?--------------------CHX
			Ought to be------------------------OHY
SVC	ORDINARY-ILY IN AN ORDINARY DEGREE		Ought to be able to -- -------------CYQ
SVD	—Ordinary seaman		Ought to be ashamed--------------FCV
SVE	ORDNANCE		Ought to do (*or*, be done)----------CHZ
	ORDNANCE STORES -------------- ----EZW		

OUGHT—OUT

OUGHT—*Continued*
Ought to have .... CIA
Ought we (our-s) to? .... CIB
Ought you-r-s to? .... CID
They ought to ... . CIE
They ought not to .... CIF
They ought not to be .... BHY
They ought to be .... BHX
We ought to .... CIG
We ought not to .... CIH
What, *or*, Which ought to? .... CQJ
When ought? ....CRP
When ought I to see (*or*, sight)? ..VOD
When ought the next mail to arrive? RGA
When ought the next mail to leave?.RGB
Where ought? .... CTJ
Who ought? ....CUR
Why ought? .... CVQ
You ought to .... CIJ
You ought not to .... CIK
You ought not to be .... BJI
You ought to be .... BJH

OUNCE   Oz   (*Measure of weight*)...BAZ

OUR-S-SELVES .... CIN
After our-s CIL
Are, *or*, Is our-s, *or*, Are we? . ....BEM
Before our-s (us) .... CIM
By our-s (us) .... BJY
Can *or*, May we (our-s)? .... BKN
Could, *or*, Might we (our-s)? ....BLI
Do, Did (*or*, Does) we (our-s)? ....BMP
For our-s (us) ....BPJ
From our-s (us) .... BPW
Had, has, *or*, Have we (our-s)? ....BRD
If we (our-s) .... CAW
In our s (us) .... CDL
Is, etc   (*See* Are, *above*)
May, etc   (*See* Can, *above*)
Might, etc   (*See* Could, *above*)
Must we (our-s)? .... CFD

SVQ
—Nearly all our
Not our s (us) .... CFY
Of our-s (us) .... CGP
On our-s (us) .... CHA
Ought we (our-s) to? .... CIB
Our-s-selves .... CIN
Shall, *or*, Will we (our-s)? ....CJA
Should, *or*, Would we (our-s)? ....CJW
Than we (our-s) .... CKJ
That we (our-s) .... CLW
To our-s (us) .... COB
Was, *or*, Were our-s (we) . ...CPA
When we (our-s) .... CSK
Where we (our-s) .... CUD
Why we (our-s) .... CVY
Will, etc   (*See* Shall, *above.*)
With our-s (us) .... CWM
Would, etc   (*See* Should, *above*)

SVR
OUT
All the lights are out along the coast of — .... PU
Be on your guard, look out .... OIV
Blow out .... GNU
Boiler out of order .... GWF
Bring out .... HIV
Broke out .... HJV
Can not get my boat out .... GSJ

OUT—*Continued*

SVT —Carried out
Come out .... JCV

SVU —Cut out
Do you think we can cut her (*or*, them) out? ....KHN
Do you think we can ride it out? ...UWX
Farther out .... MSX
Find out .... MYO
Fires and lights out .... MZQ
Fit-ted-ting out .... NCL
Go out (*or*, from) .... OBY
Good look-out .... ODZ
Have you got out of danger? .... KLC
Help me to lay out an anchor ...... EIK
How much cable have you out? .... HRE
I shall ride it out if I can .... UWY

SVW —In and out
Just out of port (*or*, harbor) ... ...FXM
Keep a good look out aloft .... EAG
Keep a good look-out for lights (*or*, land), *or*, Look out for land (*or*, lights) .... FYH
Keep out of the tide (*or*, current) .... XG
Lay out .... QIX
Lay out an anchor ....EIZ
Lightship at — is out of position (*or*, is not at anchor on her station) ..... QE
Look out—to (*or*, for) .... QZS
Look out at day break .... ENM
Look out for a line (*or*, rocket line) ..KF
Look out for the tow line.. ....QZU
Look-out will be kept on the beach all night .... KG
Machinery out of order .... RJ

SVX —Make out
May, *or*, Can be cut out .... KHO
Out — days .... KNS

SVY —Out of (*or*, from)
Out of dock .... IJK

SVZ —Out of harbor (*or*, port)
Out of order................... SUL
Out of port (*or*, harbor) .... SVZ

SWA —Out of quarantine
Out of range   Out of gunshot...... OKJ
Out of reach .... UJA

SWB —Out of sight
SWC —Out of soundings
Out of the course—of .... JZL

SWD —Out of the way
SWE —Outermost
Outer entrance (passage, *or*, channel)_ILS

SWF —Outer harbor
Proceed out of harbor . ... TXC
Put out ....UDA
Ran out .... UHR
Ride it out if possible..... ....KS
Rig out .... UXJ
Rocks stretch a long way out .... GE

SWG —Run out
Shift your berth farther out........ GHB
Shut out .... WAU
Stand out to sea.. ....WQI
Telegraph out of order (*or*, stopped).XMC
Vessel indicated is out of sight ..... VXR
When I am out of quarantine....... UEK
Work out .... ZHX
Your lights are out (*or*, want trimming)
Made by — — — — *flashes or, blasts of a steam whistle*

OUTBOARD—OWNER

SWH	OUTBOARD	SXI	OVERHAUL-ED-ING
SWI	OUTBREAK		The engines must be thoroughly overhauled ---------------------------MCB
	OUTERMOST ------------------------SWE	SXJ	OVERHEAD
SWJ	OUTFIT	SXK	OVERHEAT-ED-ING
SWK	OUTLET		Furnace overheated ----------------NUE
			Guns overheated-------------------OKD
SWL	OUTLINE	SXL	OVERLAND
SWM	OUTLYING	SXM	—Overland mail leaves on the —.
SWN	OUTNUMBER-ED-ING	SXN	OVERLAP-PED-PING
SWO	OUTPORT	SXO	OVERLAY-ING-ID
SWP	OUTRAGE-D-ING-OUS-LY		You will overlay my anchor -------EKO
SWQ	OUTRIGGER	SXP	OVERLOAD ED-ING
SWR	—Outrigger torpedo, or, Spar torpedo	SXQ	OVERLOOK-ED-ING
		SXR	OVERMAST-ED-ING.
SWT	OUTSIDE	SXT	OVERRULE-D-ING
	Await outside the inspection of the officials ----------------------FPB	SXU	OVERRUN-NING-RAN
	How is the wind outside? ----------ZEO	SXV	OVERSEE-D-N-ING
	Outside berth --------------------- GFY	SXW	OVERSIGHT
	Outside of limit ------------------QUG		
SWU	—Outside of the		OVERSEER (Superintendent) -------XEK
SWV	—Outside the harbour	SXY	OVERSHOOT-ING-SHOT
	Signal, or, I am going to signal state of the weather outside (or, at —, place indicated) ----- --- -----------WDE	SXZ	OVERTAKE-ING-N TOOK
	What is the weather outside (or at place indicated)?-- ------------------YZL		Do not overtake me Code Flag over T
	Wind has abated outside-----------ZFJ	SYA	OVERTHROW-N-ING-THREW
	Wind has not abated outside-------ZFK	SYB	OVERTURE
SWX	OUTWARD-LY		Made overtures ------------------REG
SWY	—Outward bound	SYC	OVERTURN-ED-ING
SWZ	OVAL	SYD	OVERWHELM-ED-ING
SXA	OVEN	SYE	OVERWORK-ED-ING
SXB	OVER	SYF	OWE D-ING
	Is — frozen over? ----------------NQB		Owing to (on account of)----------DFP
	Overalls ------- -----------------DYQ	SYG	OWN-ED-ING TO OWN
SXC	—Over the	SYH	OWN
	Put over ---------- --------------UDB		At my own disposal ---------------LFM
	What is the depth over the sill ----KVH		Do not attempt to land in your own boats ----------------------------EY
	— is frozen over------------------NQC	SYI	OWNER
SXD	OVERBOARD		By your owner's orders (or, instructions) ------------------------ STN
	Fallen overboard------------------MRQ		Consult owner - - ------------ ------JRU
	Lost overboard -------------------RAW		Have received the following communication (or, instructions) from your agent (or, owner) ----------------IA
	Man overboard --------------------BR		Have you any message for your owner?
	Obliged to throw overboard--------SMY		RQL
	Want a boat, man overboard--- ----BT		I have received orders for you to await instructions from owner at — -----SQ
	Washed overboard --- ------------YUR		
SXE	OVERCAST		
SXF	OVERCHARGE-D-ING		
SXG	OVERDUE		
SXH	OVERFLOW ED ING		

OWNER—PALM

	OWNER—*Continued*
	I wish for orders from Mr. —, my owner,
	at — .... SW
	Inform owner.. ....PKE
SYJ	—My owner is, *or*, My owners are—
SYK	—Owner desired me to inform you
SYL	—Owner's request
	Owners desire me to inform you to coal
	at —...... ....HE
	Owners desire me to inform you to dis-
	charge cargo first at — (*place indi-*
	cated). ....SY
	Report me by post to my owner (*or*, to
	Mr —) at — ....UC
	Report me by telegraph to owner (*or*,
	to Mr —) at ....UE
	Report me to owner on leaving ....USB
	Send following message by post to
	owner (*or*, to Mr —) at —.......WY
	Send following message by telegraph to
	owner (*or*, to Mr —) at —.........XA
	Send following message through the
	post to owner (*or*, to Mr —) at —, by
	signal letters instead of writing it at
	length ....WZ
	Shall I telegraph your owner? .....XLU
	Telegraph to my owner .... XMG
	Telegraph to my owner to send my let-
	ters to me at — ....VQP
	Tell my owner ship answers remark-
	ably well ....ENO
	Vessel desires to be reported by tele-
	graph to owner (Mr —) at —...USF
	What is your owner's address? ...DJS
SYM	—Who are your owners?
	Write to the owner ....ZJS
	Your owner is here ....OUH
SYN	—Your owners are
	Your port of destination is closed, your
	owners desire you to proceed to —,
	KXJ
	Ox　Bullock　Cattle............HLQ
	Oxehoved　(*Measure of capacity*) ...AZI
SYO	Oxide of iron　Oxidation
SYP	Oxygen
SYQ	Oyster
SYR	—Are any oysters to be had?
	Oyster beds ....GCU

SYT	P　(*Letter*)　(*For new method of spelling,*
	see page 13)
	P M　Post Meridian　Afternoon　In
	the afternoon ....DRW
SYU	Pacha
SYV	Pacific Islander
SYW	—North Pacific
SYX	—South Pacific
SYZ	Pacify-ied-ing
SZA	Pack-ed-ing.
	Packed tight ....XIB
SZB	—Packing ring
	Packing ring of cylinder cover broken,
	KAN
SZC	Package
SZD	Packet　(*See also* Mail)
	By the packet (*or*, mail)............HPL
SZE	Paddle
	Connect your screw propeller (*or*, pad-
	dles) ....JPH
	Paddle box.....................HDF
SZF	—Paddle float
SZG	—Paddle wheel
SZH	—Paddle-wheel steamer (*Horsepower to*
	follow, if necessary)
SZJ	Padlock
SZK	Page　(*Of a book*)
	Attention is called to page —, para-
	graph —, of the International Code.DH
	N B—*The number of the page is to be*
	shown first and then the paragraph
	(*See note as regards this signal on*
	page 87)
	Paid　(*See also* Pay)...............TEO
	Everything paid.... ....MJI
	Mind you get paid..................RUK
SZL	Pain-ed-ful
SZM	—Painless
SZN	Paint
	Black paint....................GLU
SZO	—Paint-ed-ing　To paint
SZP	—Painter
	Red paint ....UMI
	White paint .... ZDK
	Yellow paint ....ZLD
SZQ	—Paint oil
	Painter's tools (*or*, brushes)........HKP
SZR	Painter　(*A rope*)
	Pair　(*Couple of*)................JZA
SZT	Palace
	Palame　(*Measure of length*).......AVT
	Palm　(*Measure of length*)........AVT

PALM—PARREL

SZU.	PALM (*Tree*) Twine, needles, and palms for repairing sails ------------------------YFN		PARA (*Coin*) ------------------AUM
SZV	PAMPHLET	TAR	PARADE-D-ING
SZW	PAN	TAS	PARAFFIN OIL
SZX	PANE	TAU	PARAGRAPH Attention is called to page —, paragraph —, of the International Code-DH N B —*The number of the page to be shown first and then the paragraph (See note as regards this signal on page 37)*
SZY	—Pane of glass		
TAB	PANIC		
TAC	PAPACY POPE PAPAL-IST		
TAD	PAPER (*See also* DOCUMENT *and* NEWS-PAPER)	TAV	PARAGUAY-AN PARAGUAYAN COLORS
TAE	—All my papers are sent from the shore Any letters (*or*, papers) for me (*or person indicated*)?-------------EPA Any letters (*or*, papers) from —?----JI Bring all the papers and letters----HII Bring all your papers-------------LJZ Bring the papers to me -----------HIN	TAW TAX TAY	PARALLEL —Parallel-motion gear —Parallel-motion rod out of order
		TAZ	PARALYZE-D-ING-TIC-IS
TAF	—Bring your papers on board By a paper of —(*date following*)---SHP Can you lend (*or*, give) me a paper (newspaper)?------------------SB Clear-ing papers-----------------ITH Daily paper ----------------------KIR False papers ------------- -------MSF Foolscap paper -------------------NHC Has papers of following dates------KMC	TBA	PARAPET
		TBC	PARBUCKLE-D-ING
		TBD	PARCEL-LED-LING I have a letter (*or*, parcel) for you (*or*, for —) ----------------------QOG Parcels and letters--------------QOJ
TAG TAH TAI TAJ TAK	—Has she, *or*, Have you any papers? —Have you all your papers? —Have you received your papers? —I have all my papers —I have wrong papers on board Keep any letters (*or*, papers) for me till I return-------------PYK Leave any papers (*or*, letters) for me (*or for party indicated*) at —----QXY Necessary papers-----------------SFY Newspaper------------------------SHO Newspapers up to date of — (from —, *if necessary*)------------------SC Paper money Bank notes --------ASV	TBE	—Parcels Post
		TBF	PARCH-ED-ING
		TBG	PARCHMENT
		TBH	PARDON-ED-ING
		TBI	PAREGORIC
TAL TAM TAN	—Papers and documents —Papers satisfactory —Papers unsatisfactory Papers, *or*, Letters up to the —(*indicate where from if necessary*)----------QOP Shall I bring my ship's papers with me? HJF	TBJ TBK TBL	PARIS —Declaration of Paris —My first meridian is Paris, east of Greenwich 2° 20' 15"=0h 9m 21sec Report me to "New York Herald" Office at Paris----------------UK
TAO	—Suspicious papers Ship's papers (*or*, articles)---------EZK Take papers----------------------XJU The captain must bring the ship's papers with him--------------------HXJ Tracing paper --------- -----------XZI Want my papers (*or*, clearance)----ITR What is the date of your latest newspaper (*or*, letter)?-------------KMJ Will you send me some newspapers?-SHT Would you like to see our papers (newspapers)?-------------------SL Writing paper -------------------ZJT	TBM	PARK
		TBN	PARLIAMENT Act of Parliament (*or*, Congress)---DIC Both Houses of Parliament --------OZF By an Act of Parliament (*or*, Congress), DIE Has Parliament (*or*, Congress) met?-JOZ House of Commons----------------JFM House of Lords Senate --------OZR
TAP	—Your papers are not in order Your papers will be examined -----MKJ	TBO	—Member of Parliament (*or*, Deputy) Parliament, Assembly, Congress (*or other legislative body*) adjourns (*or*, adjourned) on — ---------------FEI Parliament, Assembly, Congress (*or other legislative body*) is dissolved-FEJ Parliament, Assembly, *or*, Congress met (*or*, meets)—on the — ----------JFB
			PARMAK (*Measure of length*) ------AVT
		TBP	PAROLE
TAQ	PAR AT PAR	TBQ	PARREL

PARSON—PASS

	PARSON .... ITX		PARTY—*Continued*
TBR	PART-ED-ING TO PART		Seining party ....VOJ
	Better part company ....JGL		Send muskets (rifles) to party on shore,
	Cable parted ....HQT		SCQ
	Danger of parting ....KJZ	TCF	Water-ing party................ ....YWF
	Driving, parted from anchor .....BIA		—Working party
	Have you parted from your anchor?.EIJ		
TBS	—I have parted—from (*or*, with) — .	TCG	PASS-ING TO PASS
	I shall part company ....JGM		A squadron (*give nationality*) composed
	If you do not carry more sail, we shall		of — ships passed in sight.. ....WPF
	part company ....JGN		Bar is impassable................FR
	In case we part company ....JGO		Bar passable.................. ...FVW
	Part company ....JGQ	TCH	—Can pass
TBU	—Parted from	TCI	—Can not pass
	Parted company with — on the — (*or*,		Crew will not pass ....KCY
	on the —, at —) ....IO		Did not pass survey ....XGI
	Parted from anchor, driving........EIA	TCJ	—Did you pass?
	Parted from her anchor ....EJI		Do not pass ahead of me
	Parted in the night from her anchors,		Code Flag over R.
	EJK		Do not pass too close to me ....IVE
	Permission to part company ....JGB		Do you think we ought to pass a vessel
	Should you part, endeavor to beach your		in this distressed condition?.......NE
	vessel where people are assembled on		Gone by Passed by..............HPV
	the beach (*or as pointed out by com-*	TCK	—Has not passed
	pass signal FROM *you*) ....PZU		Have passed ice in latitude —, longi-
	Vessel indicated parted from her anchor,		tude —, on the —.................PQ
	EKA		Have passed steamer with machinery
	Vessel, *or*, Ship may part company.JGU		disabled.........................MT
	We are in danger of parting .... GL		Have passed steamer with steering gear
	When did you part company—from . R		disabled ....MU
TBV	PART (*Portion*)		Have you passed any vessel (*or vessel*
	A large part (*portion*) ....QFL		*indicated*)? ....WOE
	A small part (*portion*) ....WHO		How many days since you passed —
	After part ....DRG		(*vessel's indicated*)? ....KNH
	Fore part .... NIW		How many torpedo boats passed in
	Have condemned a part of my provi-		sight?.........................GTK
	sions...........................JMO		How near did you pass to —? ....SFH
	In the narrow part of the ....SDZ		I have orders to telegraph your passing,
	Part of cargo saved ....IAW		SO
	Part of the crew (*indicate number, if*	TCL	—I passed — (*place or vessel indicated*)
	necessary) ....KDF		I shall pass within hail ....OLE
TBW	—Partly		I think I must have passed the buoy
	The greater part ....OHP		(*or*, beacon) ....HML
TBX	PARTIAL LY		I will pass ahead of you,
TBY	—Partially insured		Code Flag over X
			I will pass astern of you,
TBZ	PARTICULAR-LY-IZE		Code Flag over Z
	Every particular ....MJB		Pass ahead of me... ....DVP
	In every particular ....MJN		Pass astern of me ....FHQ
	Nothing particular ....SKD	TCM	Pass within hail ....OLG
	Very particular-ly ....YPN	TCN	Password ....TDY
	Your zeal has been particularly noticed		—Passable
	by — ....ZMD		—Passed
			Passed a derelict vessel, apparently in
	PARTLY ....TBW	TCO	good condition, hull well out of water,
TCA	PARTNER-SHIP		dangerous to navigation.........KVS
	Dissolved partnership ....LGA		—Passed a good many vessels
TCB	—Has taken a partner		Passed a war vessel disguised, funnel
	Partnership affairs .... DPE		(*or*, funnels) — (*Give the color or*
TCD	PARTY.	TCP	*design, or the line that it may be an*
	Beach party...................FZR		*imitation of*) ....LEO
	Charter party .... ING		—Passed a wreck
	Coaling party...................IWV		Passed a wreck (*date, latitude, and longi-*
	Funeral party ....NTJ		*tude to follow, if necessary*), but could
TCE	—Party interested		not render any assistance, people still
	Reconnoitering party.............ULS		on board. ....ZV
			Passed a wreck (*date, latitude, and longi-*
			tude to follow, if necessary), but was
			unable to ascertain whether any peo-
			ple were on board................ZW

PASS—PATENT

	PASS—*Continued.*	**TDM**	**PASSENGER**
	Passed a wreck (*date, latitude, and longitude to follow, if necessary*), no one on board ------ --- ------------ZX		Are your passengers all on board?..GPU
			Are your passengers all well?------DYG
	Passed by Gone by ------------HPV		Cabin passenger — for —----------HQE
	Passed disabled ship at —--------LBW		Can I land my invalid passengers? .QDR
	Passed disabled steamer at —------MX		Can you accommodate a passenger (*or, number shown*)? ----------------DCP
TCQ	—Passed during the night		Clothing for female passengers
TCR	—Passed, or Saw, some war vessels		Women's clothing --------------IVR
TCS	—Passed survey		Cuddy passenger.--------------KFM
	Passed the derelict vessel — on the —, KVT		Deck passenger ------- --------KQA
	Passed the mail — days ago ------RFU		Female passenger ---------------MVR
TCU	—Passed the official inspection		Full of passengers (*number may be indicated*) ------------------------NSQ
TCV	—Passed the — (*place or vessel indicated*)	TDN	—Have you any passenger — for — (*or, from —*)?
	Passed the port (harbor, *or,* entrance), *or,* We have missed the mark----MFV	TDO	—How many passengers?
	Passing certificate—of (*or,* for)-----IJC		I am sinking (*or,* on fire), send all available boats to save passengers and crew ----------------------------NO
	The way is off my ship, you may feel your way past me ----------------HL		
	We have missed the entrance (*or,* passed the port, *or,* harbor) ----------MFV		I can accommodate a passenger (*or number shown*)---------------------DCS
TCW	—When did you pass —?		I wish to land passenger----------QEG
	— (*number*) torpedo boats (*give nationality*) passed in sight---- -- ----XYA		Male passenger --------------------RJB
		TDP	No accommodation for passengers..DCU
TCX	**PASS** (*Ticket of admission*)		—Passenger department
	Railway pass ------- -------------UHA	TDQ	Passenger list ------------------QVK
	Requisition for passes .------------UTG		—Passenger ship
			Passengers, *or,* Crew sick with infectious complaint ----------------PCX
TCY			Passengers all on board----------GQR
TCZ	**PASSAGE**		Passengers are embarked----------LWS
TDA	—A pleasant passage	TDR	—Passengers must be landed
	—A tedious passage		Passengers not all on board-------DYE
	After a passage — of —-----------DQO	TDS	—Passengers saved
	Can you give a passage — to? -----NZS		Passengers will have to be disembarked, LEG
	Do not know the channel (*or,* passage), ILH		Saloon passenger ----------------VFQ
	Do you know the channel (*or,* passage)? ILJ		Send for passenger ----- --------VPT
	During the passage (*or,* voyage)---LPU		Send passenger--------- ---------VQB
	For passage — to (*or,* in) ----- --NHI		Send suitable boats for landing passengers — ----------------------------FI
TDB	—I have had a rough passage		Steerage passenger—for — (*or,* from—), WUD
TDC	—I (*or vessel indicated*) have had a quick passage		
	I know the passage (*or,* channel)---QCD		Vessel seriously damaged, wish to transfer passengers.----------------HM
	Inner entrance (passage *or,* channel), ILN	TDU	—We have no passenger
TDE	—Is there a passage?		What passengers? How many passengers?--------- ----------- -------TDO
TDF	—Made the passage in — days (*number to follow*)		When will passengers embark — (*or,* for—)? --------------------------LWZ
TDG	—No passage	TDV	—Where will passengers have to be disembarked?
	Outer entrance (passage, *or,* channel), ILS	TDW	—With passengers
TDH	—Passage is —		—Passengers have been ill of yellow fever (*number to follow*) ------- MWZ
	Screw passage -------------------VKG		
	Spoke several vessels during the passage, VSU	TDX	**PASSPORT**
	Take passage — to (*or,* for) -------XJV	TDY	**PASSWORD**
	There is a passage (*or,* channel)— bearing — -------------------------ILU	TDZ	—Have you the password?
	Wait for passage — to (*or,* for) ----YTA	TEA	—Password is —
TDI	—What sort of a passage have you had?		
	What wind leads through the passage (*or,* entrance)? ----------------ZFD	TEB	**PATENT PATENTEE**
TDJ	—Which passage did you take?	TEC	Patent anchor.-------------------EJH
TDK	—Will you give me a passage on shore?		—Patent fuel
TDL	—Wish you a pleasant voyage (*or,* passage)	TED	Patent fuze ---------------------NUY
		TEF	—Patent lead
			—Patent log
	PASSED (*See* PASS) ------------TCN		Patent sounding machine.--------RDS
			Thomson's (Kelvin) patent compass.JIG

PATENT—PERFORM

TEG	PATENT—*Continued* —Thomson's (Kelvin) patent sounding apparatus	TFE	PEBBLE-Y
	PATH FOOTPATH _____ NHE		PECHEUS (*Measure of length*) _____ AWJ
		·	PECK (*Measure of capacity*) _____ AZJ
TEH	PATIENT		PECUL (*Measure of weight*) _____ BCM
	Ambulance required to convey patient to hospital _____ EDY	TFG	PEN
TEI	—How is the patient?	TFH ,	PENAL-TY
TEJ	PATROL-LED-LING	TFI	—Penal servitude
TEK	PATTERN		PENDANT (*See* PENNANT) _____ TFL
TEL	PAWL	TFJ	PENDULUM
TEM	PAY-ING-MENT Bank (*name indicated*) has suspended payment _____ FUX	TFK	PENINSULA
TEN	—Can not pay any more	TFL	PENNANT, *or*, PENDANT Answering pennant (*when spoken of*), EMT
	Everything paid _____ MJI		Code pennant _____ IZF
	Forward my communication by tele- graph and pay for its transmission HY		Top-tackle pennant _____ XVN
	Have, *or*, Has paid particular atten- tion—to _____ FLE		PENNY PENCE (*Coin*) _____ AUN
	I am paying every attention _____ FLG		PENNYWEIGHT (*dwt*) (*Measure of* *weight*) __ _____ BCA
	Is freight paid? _____ NQI		
	Look out. Pay attention _____ DS		
	Mind you get paid _____ RUK		
	Natives do not like ships watering with- out payment _____ YWQ	TFM	PENSION-ED-ING-ER
TEO	—Paid	TFN	PEOPLE Can not save the ship, take people off, NB
	Pay attention to orders _____ FLH		Fire gains rapidly, take people off NG
	Pay attention to signals _____ DG	TFO	—Has, *or*, Had she any people on board?
	Pay attention to the helm _____ FLI		I am (*or vessel indicated* is) waterlogged, take people off _____ YWT
	Pay no attention to _____ FLJ		If you part, endeavor to beach where
TEP	—Pay-ing, Paid off		people are assembled (*or as pointed*
TEQ	Paymaster		*out by compass signal* FROM *you*) FZU
	Pay strict attention to signals during the night _____ DT		It is dangerous to allow too many people on board at once _____ GB
	Pay the greatest attention _____ FLK		Keep as close as you can to pick up my
TER	—Pay ticket		people ____ ____ _____ PYL
TES	—Payable at — to		People going on service to take their
TEU	—What is there to pay?		dinners with them _____ LAU
TEV	—Without payment		Put your people on an allowance DZM
	You do (*or*, did) not pay attention—to, FLM		
TEW	PAY PAID PAYING (*To Pitch*)	TFP	PEPPER
	PAY OUT (*See* VEER) _____ YOE	TFQ	PER CENT PERCENTAGE
TEX	PEA	TFR	PERCEIVE-D-ING PTIBLE Scarcely perceptible _ _____ VIW
TPY	PEACEFUL-LY Affairs look more peaceful (appear more settled) _____ _____ _____ DOZ	TFS	PERCUSSION Percussion fuze _ _____ NUZ
TEZ	PEACE	TFU	PERFECT ED-ING-LY Comprehend the communication per- fectly _____ JFW
	Is peace proclaimed? _____ _____ YT		
	No chance of peace _____ IKF		
	Peace has been proclaimed _____ YV	TFV	—Perfectly understand ing stood
TFA	PEAK ED ING TO PEAK		
TFB	PEAK (*Summit of a hill*)	TFW	PERFORM ED ING-ANCE
TFC	PEARL	TFX	—Performing quarantine
TFD	PEASANT-RY		

	PERHAPS—PHOSPHOR			

TFY	PERHAPS			PERSON—*Continued*
	Perhaps it might be made to answer, ENI			Any person on board? ---------- EPM
	Will perhaps be seized ------------HXY			Can — (*person-s or thing-s indicated*)? BTJ
TFZ	PERIL-OUS			I wish to communicate personally__JGC
TGA	PERIOD-ICAL-LY			If — (*persons or things indicated*) can (*or*, may) --------- --- ---BYD
TGB	PERISH-ED-ING-ABLE			If — (*persons or things indicated*) can not (*or*, may not) ----------------BYE
	All on board perished All hands lost, DXQ			Is — (*person indicated*) on board?__GQO
	Perishable cargo ------ ------ ------HZK	TGR		May I send a sick person to see your doctor -----------------------WL
TGC	PERJURE-D ING-Y			—No person
TGD	PERMANENT-LY-ENCE			Persons, *or*, Things indicated can__BUC
TGE	PERMIT SSION (*See also* ALLOW *and* AD-MIT)			Persons, *or*, Things indicated can not, BUD
	Ask permission_____FDE			Person indicated has not suspended business ----------------- --------HOC
	Can I get permission? ---------- ---NYK	TGS		Person indicated has suspended busi-ness --------------------- --------HOD
	Can permit Can allow ------- -- DZE	TGU		—Person indicated is wanted on shore
	Can not, *or*, Will not permit it (*or*, con sent) ------ ------ ---------------JPT			Persons, *or*, Things are not—to -- _BUW
	Have you a regular permit (*or*, license)? QPM			—Person to navigate
	I shall anchor if circumstances permit (*place or bearing to follow, if neces-sary*) -- ----------------------EIT			Person to take charge_____IME
				Personal ly-ity_____ ----TGW
				Send a responsible person on board_CZE
	I shall weigh as soon as the weather permits (*or* moderates) -----------ZO	TGV		—Send a responsible person on shore
TGF	—Not without permission			Vessel, *or*, Person indicated agrees_DUB
	Permission is requested to part com-pany ------------ ------------JGR			Vessel, *or*, Person indicated declines_KQJ
	Permission urgently requested to enter harbor ---------------------------PD			Vessel indicated was minutely exam-ined, no person to be seen, nothing visible -------------------------MKI
	Permit me to send a sick person to see your doctor ----------------------WL			You have left a person behind -----GDZ
TGH	—Permitted Permission is granted	TGW		PERSONAL-LY ITY
	Request permission—to------------USQ	TGX		PERSUADE D-ING
	Shall wear as soon as weather permits, YXS			After much persuasion--- -------DRB
	Weather permits-ted ting ----------YZI	TGY		—I am persuaded
TGI	—Will it be permitted?	TGZ		—Persuasion
	Will you permit? Will you allow?_DZQ	THA		PERU-VIAN PERUVIAN COLORS
	Wind and weather permitting -----ZFG	THB		PERVERT-ED-ING
TGJ	—With your permission			PESETA (*Coin*) -----------------AUO
TGK	—Without permission			PESO, *or*, DOLLAR (*Coin*)_____ATJ
	You will not be permitted to land_QEW			
	Your permit will be examined (*or*, looked at) -------------------------MKL	THC		PESTILENT-IAL ENCE
TGL	PERPENDICULAR I Y UPRIGHT			PETER (BLUE) -----------------------P
				Blue peter (*When spoken of*)-----GOS
	PERPETUAL-LY (*Incessant*) _ -------PGF	THD		Hoist the blue peter -- ------------KO
TGM	PERSECUTE D ING-ION			PETITION ED-ING
TGN	PERSEVERE-D-ING-ANCE	THE THF		PETROLEUM OIL
				—I am in want of petroleum
TGO	PERSIA-N PERSIAN COLORS	THG		Petroleum cargo ------------------HZL
TGP	PERSIST-ED-ING-ENCE-ENT LY			—Petroleum ship
	I shall fire into boats if they persist in coming alongside ----------------NAK	THI THJ		PETTY
				—Petty officer, *or*, Non-commissioned of-ficer
TGQ	PERSON			PFENNIG (*Coin*)- ------------------AUP
	Allow no person on board Allow no communication ----------------FM			PFUND (*Measure of weight*) -------BCD
	Any person Anybody ------------EPQ	THK		PHOSPHOR-IC-OUS ESCENCE

	PHOSPHATE—PILOT		
THL	PHOSPHATE	TIF	PIG
THM	PHOTOGRAPH-ED-ING-IC ER-Y	TIG	—Have you any pigs?
	Photographic apparatus.............ERI	TIH	PIG IRON
THN	PHYSIC-AL-LY		Pig of ballast ....FTU
THO	PHYSICIAN	TIJ	PIGEON
THP	PIANO		Carrier pigeon............ ...-.....ICH
	PIASTRE (Coin) . ...AUQ	TIK	PIKE
	PICE, or, PIE (Coin) ....AUR		Boarding pike ....GPY
			PIKE, or, DRAA (Measure of length) .AVU
THQ	PICK-ED-ING—UP	TIL	PILCHARD
	Endeavor to pick up boat...........GSV	TIM	—Pilchard fishery
	I am obliged to slip my cable, pick it up	TIN	PILE D ING
	for me.........................HRF		Pile (or, bank) of — ....FUI
	I am picking up a telegraph cable	TIO	PILL
	Keep out of my way ....XLQ	TIP	PILLAGE-D-ING
	I have picked up moorings ....RZG		Wreck is being pillaged ....ZJA
	I have picked up telegraph cable ...WU	TIQ	PILLAR.
	I shall stand about to see if I can pick	TIR	—Pillar buoy
	up anything from the wreck .....ZIT	TIS	PILOT-ED-ING TO PILOT
	Keep as close as you can to pick up my	TIU	—Pilot (A pilot)
	people ....PYL		Am I compelled by law to have a pilot?
	Keep to windward until you are picked		JIS
	up ....PZQ	TIV	—Are the pilots trustworthy?
	Let your boats keep to windward until	TIW	—Can you get (or, send) me a pilot? Is
	they are picked up. ....FD		there a pilot to be had?
	Pick up boat. ....GTY	TIX	—Can you pilot me in?
	Pick up my pilot Look out for my		Certificate of pilotage.............IHT
	pilot ....TJH		Channel dangerous without a pilot..LD
	Pieces of the wreck have been picked	TTY	—Channel pilot
	up............................TIB		Dangerous without a pilot..........TI
	Will you pick up my anchor for me?		Do not trust too much to your pilot.YCF
	EKM		Do not try (or, attempt) without a pilot,
THR	PICKAXE		FKL
THS	PICKET (or, PIQUET)		Do you know the pilotage?.........QBW
	Picket boat ....GTZ	TIZ	—Do you want a pilot?
		TJA	—Have you a pilot on board?
THU	PICKET ED-ING. TO PICKET		Have you seen a pilot vessel(or, boat)?..TJ
THV	PICRIC ACID	TJB	—I am in want of channel pilot—for —
THW	PICKLES	TJC	—I can pilot you
		TJD	—I can not pilot you
THX	PICTURE	TJE	—I have a pilot
THY	PIE, or, PUDDING		I have a pilot, steer after me......DRU
		TJF	—I have no pilot
	PIE, or, PICE (Coin) ....AUR		I want a pilot*...s, or, Code Flag over s
THZ	PIECE		*See PILOT SIGNALS, p 8.
	All to pieces ....DYC		
	Dashed to pieces ....KLU	TJG	—I will send you a pilot
	Gone to pieces ....OCF		Is it necessary, or, Does the navigation
TIA	—Not a piece to be found		require me to take a pilot?.......SEY
TIB	—Pieces of the wreck have been found		Is there a pilot to be had?.........GUB
			Make signal for a pilot.............WCN
TIC	PIER.	TJH	—Pick up my pilot Look out for my
	Alongside the pier (jetty, or, wharf),		pilot
	EAT		Pilot (A pilot).....................TIU
	Can I discharge at pier? ....LOP		Pilot boat (or, vessel) ....GUA
	Make fast to the pier ....RIG		Pilot boat (steam).................GUB
TID	—Pierhead		Pilot boat is advancing toward you,
			DNI
TIE	PIERCE-D-ING		Pilot boat most likely on — (bearing
			indicated) (or, off —).... ....TK
			Pilot boat's flag (when spoken of) ..NDK

PILOT—PLACE

	PILOT—*Continued.*
TJI	—Pilot has been sent to you
TJK	—Pilot has not come off
	Pilot jack (*when spoken of*) .. ..NDL
TJL	—Pilot not necessary
TJM	—Pilot signal flying
TJN	—Pilotage
TJO	—Pilotage is compulsory
TJP	—Pilot can not get off
	Pilots have not come off ....TJK
	River pilot ....UYK
	Run in for pilots (*or*, pilot vessels), and look out for their signals ....RAC
TJQ	—Send me (*or vessel indicated*) a pilot
	Send pilot—to..................VQC
	Steam pilot boat ....GUB
TJR	—The pilot is going to you
TJS	—The pilotage is —
	Trinity pilot.......... ....YBH
	Want a pilot*....s, *or*, Code Flag over s
	Want a pilot*..................PT
	Want you to send me a pilot........TIW
TJU	—What are the pilot fees — for — ?
	Where can I get (*or*, look for) a pilot?.TL
	Will send for a pilot................VQT
TJV	—Will you take my pilot?
TJW	—Will your pilot come on to me?
TJX	—You had better take a pilot at once
TJY	—You must take a pilot
TJZ	—You will find the pilot in — (*direction indicated*)
	*See PILOT SIGNALS, p 8
TKA	PIN-NED-NING
	Belaying pin.................... .....GED
	Crank pin..........................KBM
	Crank-pin brasses broken ....KBN
	Crank pin broken . ...KBO
TKB	PINCH-ED-ING
TKC	—Pinching pin. Pincers Pinchers
TKD	PINE
TKE	—Pitch pine
TKF	PINEAPPLE
TKG	PINION
TKH	—Pinion wheel
TKI	PINK
TKJ	PINNACE
	Steam pinnace ....GUZ
	Steam pinnace's engine...........MBZ
	PINT (*Measure of capacity*).........AZK
TKL	PINTLE
TKM	PIPE
	Air pipe-ing......................DVZ
	Blow-off pipe burst ....GNY
	Blow through pipe ....GOB
	Condenser pipe.... ......JMS
	Discharge pipe out of order........LCQ
	Eduction pipe out of order.........LTN
	Feed pipe..... ....NUY
	Feed pipe broken (*or*, out of order).MUZ
	Hawse pipe (*or*, hole) ....OPM

	PIPE—*Continued*
	Hawse pipe is broken.............OPN
	Injection pipe....................PKV
TKN	—Pipe (*for tobacco*)
	Sounding pipe (*or*, tube)..........WKU
	Steam pipe ....WSM
TKO	—Steam pipe burst
	Steam pipe damaged ....WSN
	Steam pipe leaky ....WSO
	Suction pipe.............. ....XCJ
	Tobacco pipe....................TKN
	Waste pipe ....YUV
	PIQUET (*See* PICKET)...............THS
TKP	PIRATE
	Beat off a pirate.................GBZ
TKQ	—I (*or vessel indicated*) was plundered by a pirate
TKR	—Pirate reported (*or*, seen) off —
TKS	PIRACY PIRATICAL
TKU	PISTOL (REVOLVER)
	Pistol (revolver) cartridges ICX
TKV	—Want pistol (revolver) cartridges.
TKW	PISTON.
	Crosshead (*of piston*) ....KED
	High-pressure piston.............OUW
	High-pressure piston broken.......OUX
	Intermediate pressure piston.......PQU
	Low-pressure piston .... RBS
TKX	—Piston broken
TKY	—Piston rod
	Piston rod bolts broken........ ....GYB
	Piston rod brasses broken (*or*, out of order) ....HED
	Piston rod cracked (*or*, broken) ...KAX
	Piston-rod crosshead damaged (*or*, broken) ....KEJ
TKZ	PIT
	Ash pit..........................FCS
TLA	PITCH (*Tar*)
TLB	—Pitch-ed-ing To pitch (*To smear with pitch*)
	Pitch pine ....TKE
TLC	PITCH ED-ING To PITCH (*To toss*)
TLD	PITY IED-ING PITEOUS
TLE	PIVOT
TLF	PLACE-D-ING TO PLACE
	An alteration has taken place. ....EBX
	At place specified ....FIQ
	At the landing place.............FIY
	At the nearest landing place...... QDO
	At — (*time indicated*), — (*place indicated*), bore — distant — miles...ED
	Best place.......... .... GHW
	Best watering place is — ... .GHT
	Confirmation, *or*, Ratification takes place at — (*or*, on —)............JOB
	Could the — (*vessel indicated*) have got to any place of shelter? ....OFD
	Every place, *or*, Everywhere........MJC

PLACE—PLEASE

TLG	PLACE—*Continued* Funeral takes place at — (*place*) and at — (*time and date*) ....NTK Have you cargo for more places than one (*or than place specified*) .....IAQ Have you had any communication with infected vessel (*or*, place)? ....IC Her funnels are placed — ....NTQ How are her funnels placed? ......NTR How did it take place? ....OMK In place of (*instead of*) ....PNA In the first place First of all .....NBA —In this place Indicate nearest place where I can procure coal ....HC Is there a convenient watering place? JTN	**TLZ**	PLAN (*See also* CHART)
		TMA	PLANE
		TMB	PLANET
		TMC	PLANK (*Dimensions to follow, if necessary*)
		TMD	—I want plank — feet long, — inches wide, — inches thick.
		TME	—Plank-ed-ing To plank Topsides Topside planking ......XWJ
		TMF	PLANT-ED-ING PLANTER Railway plant ....UHB
TLH	—It can not be procured at this place Landing place......................QEN Lights, *or* Fires will be kept at the best place for landing or coming on shore, KE Market place ....RLQ No alteration has taken place ......ECS	**TMG**	PLASTER-ED-ING Adhesive plaster Diachylon plaster, DJW
TLI **TLJ**	—Not if I were in your place —Place (*position*) Place indicated bears — ....EK Place indicated is blocked up by ice.GMU	**TMH**	PLATE-D-ING TO PLATE Armor plate-d-ing...............EVT Boiler plate ....GWH Boiler plates damaged........ .....GWI Carbon plate ....HYF Copper plate ....JVQ Earth plate ....LRO Iron plate-d-ing ....PTA Nickel plate-ed ....SIA Plates (*crockery*) ....TMI Steel plate-d..................WTN Template ....XNP Timplate ....XTY Zinc plate ....ZML
TLK **TLM**	—Place yourself —Proper place Repeat your signal in a better place, it can not be made out ....OWL Signal the names of places to which you wish to be reported by telegraph...WDF Take n-ing, Took place—on ....XJW The appointment has not taken place, ETB Took place in consequence of .....JQE Trial takes place on the — ....YAW Want a surgeon (*or*, doctor), send me one from the nearest place ....WO Watering place....................YWG		
TLN **TLO**	—What reliance can be placed in (*or*, on)? —When did it take place? When will funeral take place? .....NTL	**TMI**	PLATES (*Crockery*)
		TMJ	PLATFORM
TLP	—When will it take place? Where is it to take place (*or*, be done)? LIV Where is the best place to get —?...GIC Where is the best place to land?....GID Wrong place ....ZKC	**TMK**	PLATINUM
		TML	PLAY-ED-ING
		TMN	PLEASANT-LY A pleasant passage ....ICZ Wish you a pleasant voyage ....TDL
TLQ **TLR**	PLAGUE THE PLAGUE —Has the plague ceased at — (*place indicated*)? Plague flag (*when spoken of*) ..... NDM	**TMO**	PLEASE-D-ING—TO (*or*, OF) As you please ....FBX
TLS	—The plague has broken out at — (*place indicated*)	**TMP**	—Can not have the pleasure I want assistance, please remain by me, NR
TLU **TLV**	PLAIN LY-NESS —By no means plain Plain clothes ....IVU Plainly distinguishable I saw it very plainly ....LHA	**TMQ** **TMR**	—If you please —Much pleased Please send off Lloyd's surveyor ...QXG Please stand by me, I want assistance, *NR
		TMS	—Pleasure Request the pleasure—of ....USR
TLW	PLAINTIFF	**TMU**	—Shall I have the pleasure—to (*or*, of)? The admiral (*or*, senior officer) requests the pleasure of your company to dinner ....DKX The pleasure of your company .....JGS Was much pleased ....SBW
TLX	PLAN-NED NING. TO PLAN A good plan ....ODS	**TMV**	—When you please Will you give me the pleasure? ....OAP Will you give me the pleasure of your company at —? ....JGV
TLY	—Not a good plan The best plan will be...............GHX	**TMW** **TMX**	—Will you please? —With much pleasure With pleasure I will accept........DCA

PLEDGE—PORK

TMY	PLEDGE-D-ING		POLE, *or*, NORTH STAR ---------------SJO
TMZ	PLENIPOTENTIARY	TNU	POLE
			Bare poles---------------------------FWQ
TNA	PLENTY-IFUL-LY-NESS		Pole, *or*, Standard compass---------JIB
TNB	—A plentiful supply of		Running under bare poles ---------VCJ
TNC	—Are there plenty of supplies at —?		
	Plenty of coal to be got—at — .....IXG	TNV	POLICE-MAN,
	Plenty of water----------------------YWJ	TNW	—Apprehended (*arrested*) by the police
	There are plenty of fish to be caught here,		Harbor police-----------------------YVX
	NCD		Police authorities ----------------FNO
	There is plenty of room-----------UZW		Police (*or other*) authorities have ac-
			quitted -----------------------DHT
TND	PLEURISY		Police (*or other*) authorities have de-
			cided against the crew ---------KDG
	PLOMER BLOCK --------------------GMW		Police (*or other*) authorities have
TNE	PLOUGH		fined — --------------------------FNP
			Police (*or other*) authorities have inter-
TNF	PLUG GED-GING		fered-----------------------------FNB
	Hawse plug -----------------------OPQ		Police (*or other*) authorities have taken
	Shot plug -------------------------WAO		some of the crew out of the ship.KDH
TNG	PLUMB ER		Police (*or other*) authorities will not
			give any redress --------------FNQ
TNH	PLUNDER ED-ING		Police (*or other*) authorities will not
	I (*or vessel indicated*) was plundered by		interfere ----------------------FND
	a pirate -------------------------TKQ	TNX	—Police station (*or, court*)
			Want police------------------------YN
TNI	PLURAL IN THE PLURAL SENSE		Water, River, *or*, Harbor police---YVX
		TNY	POLICY
	P M IN THE AFTERNOON POST MERIDIAN,		
	DEW	TNZ	POLITIC-AL-LY-IAN
TNJ	PNEUMATIC		Is there any political news?--------SHK
			Political news----------------------SHM
TNK	PNEUMONIA		
		TOA	POLL-ED-ING
TNL	POINT (*of compass*)	TOB	POLLUTE-D-ING-ION
	Alter course one point (*or number indi-*		
	cated) to port-------------------JZD	TOC	POND
	Alter course one point (*or number indi-*		
	cated) to starboard --------------JZE		POND (*Measure of weight*) -------BAQ
	Compass signals in points and half	TOD	PONTOON
	points (*see page 48*) ------- -----JHK	TOE	PONY
	This is my best point -------------GIA		
	This is not my best point----------GIB		POOD, *or*, PUD (*Measure of weight*) BCF
	What is your best point of sailing?.VEX	TOF	POOL
	POINT, *or*, CAPE------- -----------HVQ		Whirlpool ----- -----------------XCY
	Near the point----------- ------SFJ		
TNM	—Off the point	TOG	POOP
TNO	—Round-ed-ing the point	TOH	—Pooped-ing
	Sheer a little off when you come to the		
	point --------------------------VUP	TOI	POOR-NESS POVERTY.
TNP	POINT-ED-ING OUT TO POINT OUT		POPE PAPACY PAPAL-IST---------TAC
	Can, *or*, Will you point out (*or, take me*		
	to) a good anchorage?-------------EKS	TOJ	POPULAR-LY-ITY-ACE
	Send boat to point out rock (*reef, or,*	TOK	POPULATE-D-ING-ION
	shoal) --------------------------GUO	TOL	PORE-OUS
	Will you lead into (*or, point out*) a good		
	berth?---------------------------GHI	TOM	PORK
	CARBON POINT ----------------------HYG	TON	Can you supply me with pork? (*Indi-*
			cate number of casks, if necessary)
	POINT BLANK ----------------------GML		Fresh pork-------------------------NQZ
TNQ	POISON-ED-ING-OUS		Salt pork.------------------------VFW
TNR	POLACRE		
TNS	POLAR-ITY-IZE-IZATION		

PORPOISE—POSITION

TOP	**PORPOISE**

PORT   (*Harbor*) ___ _____OMV
Are you going into port (or, harbor)? ___OBK

TOQ	—At this port

Belong-ing to this port _____GEU
Captain of the port, or, Harbor Master, ___HWS
Come out of port _____JCW
Convoy will assemble at port (*or place indicated*) _____JUO
Do you call at any port before destination? _____HTE
Do you come from any port putting you in quarantine? _____ _____TR
Entrance of the port _____MFQ
For this port _____NHR
Free port ____ _____NPU
Going into port (harbor) _____OCB
I left — in port, or, — was in port when I left (*ship's name or distinguishing signal to follow*) _____QKU
I left port (or *place indicated*) — days ago_____QKV
In, or, Into port._____ONB
Just out of port._____PXM
Left port _____QLG
Missed the mark (or, port), or, Passed the harbor (or, entrance) _____MFV
Nearest port _____SFM
Neutral port _____SHB

TOR	—Off this port

Out of port (or, harbor) _____SVZ
Port admiral _____DLI
Port authorities _____FNR

TOS	—Port is neutral

Port of call _____HTG
Ports suspected of cholera, etc._____IPT
Seaport _____VLC
The nearest port_____SFM

TOU	—This port

Was the port you left healthy?_____ORB
We left port (or, *place indicated*) on the —._____QLP
We have missed the entrance (or, passed the port, or, harbor) _____MFV

TOV	—What port?

What vessels did you leave in port (or, *place indicated*)? _____QLS
Wish to enter port or dock_____TM
You are ordered not to come into port, ___SUY
You may come into port _____JDT
Your port of destination is closed, your owners desire you to proceed to— KXJ
— was in port when I left _____QKU

TOW	**PORT PORTHOLE**

Baggage port_____FSR
Eduction port._____LTO
Lower port._____ROE
Sally port._____VFN
Steam port._____WSP
Upper port _____YMA

TOX	**PORT** (*Left hand side*) (*See also* PORT HELM*)

Breakers, Reef, Rock, or, Shoal on port bow _____HFD
Cast to port._____IEG

PORT—*Continued*
Convoy to keep on port side of leader (or, escort) _____IX
Haul your wind on port tack_____LN
Heave-to on port tack_____ _____OSE
Keep buoy (or, beacon) on your port hand _____GAK
Keep on port tack _____PZF
Keep on the port side of the channel.LV
Leave the buoy (or, beacon) to port. LZ
List to port _____QVO
On port beam . _____ _____GAU
On port bow._____HCO
On port quarter _____UEQ
Port anchor _____EJL
Port boiler _____ _____GXJ
Port boiler defective _____GXK
Port bow _____HCR
Port bow light _____HCS
Port broadside ____ _____HJQ

TOY	—Port engine
TOZ	—Port screw
TPA	—Port side
	Port tack ____ _____XIM
TPB	—Port-ed-ing To port
TPC	**PORT PORT WINE**
TPD	**PORT-FIRE**
TPE	**PORT HELM**

Alter course one point (or, — points) to port._____JZD
Hard-a-port' (*Urgent*) Head to go to starboard _____ _____MD
Keep more to port. _____PVZ
My helm is hard-a-port, head is going to starboard ___ _____W G

TPF	—Not too much to port
TPG	—Port a little Head to go to starboard

Port your helm, ship's head to go to starboard _____WI
Steer more to port   (*To be kept flying until the course is sufficiently altered*), ___MJ
You steer too much to port _____WUH

TPH	**PORTERAGE**
TPI	**PORTER. STOUT**
TPJ	**PORTER** (*Railway or other porter*)
TPK	**PORTION-ED-ING** (*See also* PART)

A large portion—of_____QFL
A small portion—of._____WHO

PORTLAND CEMENT_____IGQ

TPL	**PORTMANTEAU**
TPM	**PORTUGAL** **PORTUGUESE** **PORTUGUESE COLORS**
TPN	**POSITION**

Any alteration in the position of the army? ___ _____EBY
Any alteration in the position of the fleet (or, squadron)?_____EBZ
Buoy, or, Mark is not in its proper position _____LC

POSITION—POTASH

	POSITION—*Continued*
	Buoy, *or*, Mark does not appear to be in its proper position --------------HMG TQE
	Dangerous position-------------KLA
	I am not in a position to weigh ----ZAU
	I can not make out the flags, hoist the signal in a better position------OWL
	I do not think marks (*or*, buoys) can be in their proper positions---------HMK
TPO	—In the same position
	Indicate position (*or*, direction)—of, TQG
	LBQ
	Into position ---------------------PRQ
	Light indicated is not now in its old position ---------------------QSE
	Light-ship at — is out of position, *or*, Light-ship is not at anchor on her station----- ------QE
TPQ	—Not in the same position
TPR	—Position light
TPS	—Position, *or*, State remains unaltered
TPU	—Proper position
	What is my position by bearings?---RV
	What is my present position? -------RZ
TPV	—What was the position of—(*ship indicated*)?
	You are in a dangerous (*or*, unsafe) position Your position is dangerous, GN
TPW	POSITIVE-LY
	I am not positive (certain)---------IHG
	I am positive (certain)------- ----IHF
TPX	POSSESS-ED ING
TPY	—Possession
TPZ	—Take possession—of.
TQA	POSSIBLE-ILITY
	Abandon the vessel as fast as possible, AB
	All possible information—of (*or*, from), DXS
	As fast as possible -- ---------------FAI
	As much (*or*, many) as possible----FAY
	As quick as possible ---------------FBG
	As slow as possible----------------FBJ
	As soon as possible ---------------FBO
	Avoid, if possible----------- ----FOU
	Barely possible --------- ----------FWR
	Be as brief as possible------ -------HGU
	Bring the ship in as close as possible, HIO
	Convoy to keep as close as possible --IV
	Convoy to spread as far as possible, keeping within signal distance-----IZ
	Discharge cargo as soon as possible-IAM
	Dispose of cargo as soon as possible-IAN
	Get her head round as fast as possible, *or*, Get her on the other tack as fast as possible - -------------- ----NYU
	Get sail on your ship as fast as possible, NYZ
	Get steam up as fast as possible-----KL
	Get under weigh as fast as possible--KN
TQB	—I doubt if it is possible to—to —
	I would shift my berth as soon as possible ----------------------- ---GFT
TQC	—If possible
TQD	—Is it possible?
	Not possible (*impossible*) ---------PFB
	Obtain all the intelligence possible-SNW

	POSSIBLE—*Continued*
	—Possibly
	Quit the vessel as fast as possible - -AB
TQF	—Quite possible
	Render all the assistance possible—to-CY
	Ride it out, if possible ------------ KS
	Sell cargo as soon as possible- ---IAY
	Will you send off lighters as fast as possible? --------------- ---------RC
TQG	POSTED-ING TO POST
	By post ----------------------- -HPF
TQH	—By to-morrow's post To-morrow's post
	Call at the post office for letters----HSY
	Latest posting time—is------------QHG
	Parcels post----------------------TBE
TQI	—Post (*The mail*)
TQJ	—Post office To, *or*, At the post office
TQK	—Post-office official, *or*, Postmaster
TQL	—Post-office order
TQM	—Postage
TQN	—Postage stamp
TQO	—Postman
	Postmaster, *or*, Post-office official--TQK
	Report me by post to my owner (*or*, to Mr —) at --- ------------------UC
TQP	—Send by post
	Send the following message by post to owner (*or*, to Mr —) at - ------WY
	Send the following message through the post to owner (*or*, to Mr —) at —, by signal letters instead of writing it at length ----------------------------WZ
TQR	—Sent by post
	Shall I save the mail (*or*, post)?---RFV
	To, *or*, At the post office Post office, TQJ
	To day's post -- -----------------KNY
	To-morrow's post By to-morrow's post, TQH
	To-night's post --------------------XUS
	Will you post some letters for me?-QOR
	Yesterday's post-------------------ZLK
TQS	POST (*A post*)
	Advance post------ -------------DMX
	Body post, *or*, Propeller post ------GVR
	Rudder post----------------- - VBQ
	Samson post------------------ --- VGW
	Stern post----------------------WUY
	POSTMASTER------------------- -----TQK
	POST MERIDIAN, *or*, P M IN THE AFTER NOON ------------------------ ---- DRW
TQU	POST - MORTEM POST MORTEM EXAMINA-TION
	POSTPONE-D-ING-MENT (*Defer*) -----KRO
TQV	—Must be postponed
	Race is postponed—until ----------UGF
TQW	POT
	Teapot---------------------------XLH
TQX	POTASH SIUM
TQY	—Bromide of potassium
	Chloride of potassium------------IPG
TQZ	—Iodide of potassium
TRA	—Nitrate of potassium

POTATO—PRESENT

TRB	POTATO
	Preserved potatoes ----TUM
	POULTRY CHICKENS FOWLS----IOP
TRC	POUND ED-ING
	POUND, or, SOVEREIGN (Coin)----AUS
	POUND LB (Measure of weight)--BCE
TRD	POUR-ED-ING
	Calm the sea by pouring oil on it--HTN
	POVERTY POOR----TOI
TRE	POWDER (See also AMMUNITION)
	Blasting powder----GMP
	Burn a blue light, or, Flash a little powder----GCQ
	Cordite----JWI
	Gunpowder (weight in pounds, if necessary)----OKS
TRF	—I am in want of powder and shot
	I am taking in (or, discharging) powder, or, explosives
	Flag B, or, Code Flag over B
TRG	—I can supply you with powder
TRH	—I can supply you with powder and shot
	Powder (blasting)----GMP
TRI	—Powder and shell
TRJ	—Powder barrel
TRK	—Powder case
	Powder flag (when spoken of)----NDO
	Powder hoy (or, lighter)----OZU
TRL	—Powder is damaged
TRM	—Powder magazine
	Short of powder----VZY
	Smokeless powder----WIA
TRN	POWER
	Full power----NST
	Half power, or, Half speed----OLM
	Horsepower----OYH
	Indicated horsepower----OYI
	My horsepower (or that of vessel indicated) is----OYJ
	Nominal horsepower----OYK
	Power of attorney----FLP
TRO	—Powerful-ly
TRP	—Powerless
	Tug has not sufficient power----YDS
	What is your horsepower (or that of vessel indicated)?----OYL
TRQ	PRACTICABLE-Y (See also POSSIBLE)
TRS	—If practicable
TRU	—Is not practicable
TRV	—It is practicable
TRW	PRACTICE-D-ING
	According to the usual practice----DEO
TRX	PRATIQUE (See also QUARANTINE)
TRY	—Can pratique be obtained at—?
	Did you get pratique—at—?----NYS
	Get, or, Got pratique—at—----NYW
	Have you, or, Has — (vessel indicated) got pratique?----OFI
	I have (or vessel indicated has) pratique, TS
TRZ	PRATIQUE—Continued
	—I have (or vessel indicated has) not pratique
TSA	—Ships from — did not get pratique at — when I left
	You have pratique----TW
TSB	PRAY-ED-ING PRAYER
TSC	PRECAUTION-ARY
	Every precaution----MJD
	Every precaution has been taken----MJE
TSD	—Use every precaution
TSE	PRECIPITATE-D-ING-LY
TSF	PREFECT-URE
TSG	—Maritime Prefect
TSH	PREFER-RED-ING TO PREFER
	I do not like to run in, I prefer keeping the sea----VCA
	I prefer stopping on board----GQL
TSI	—I should prefer
TSJ	—Is preferred to
TSK	—It would be preferable
TSL	—Preference-able
TSM	PRELIMINARY
TSN	PREMIER
TSO	PREMIUM
TSP	—Have the premiums risen?
TSQ	—Premiums are lower
TSR	—Premiums have risen
TSU	PREPARE-D-ING-TO (or, FOR)
	Be prepared with affidavits----DPN
	Keep everything prepared----FYS
TSV	—Preparation
	Prepare ambulance for —----EDZ
TSW	—Prepare another hawser
	Prepare boat—to (or, for)----GUC
	Prepare for a hurricane----GD
	Prepare for action----DIV
	Prepare for coaling (to coal)----IXH
	Prepare for sailing Get ready for sea, NYX
	Prepare to be towed clear----XYS
	Prepare to disembark—at—----LEH
	Prepare to weigh----ZQ
	Preparing for sailing----NZF
TSX	PRESCRIBE-D-ING-PTION-IVE.
TSY	PRESENCE
TSZ	PRESENT-LY
	Annual present formation----EMH
	Do not stand too far on your present course----JZI
	For the present At present----NBJ
	How long may we stand on our present course?----JZK
	I do not intend to alter course at present (or, until —)----ECI
	I will answer presently----ENC
	If the wind continues as at present.ZER

	PRESENT—*Continued*
	Is my present course a correct one? Am I steering a proper course? (*The course steered, to follow, if necessary*) ___JZF
TUA	—Not at present
	Not safe to go on with the crew as at present ____ _____KDE
TUB	—Present form-ation
TUC	—Present rate
	Present state of affairs_____DPF
TUD	—Put-ting off for the present
	Stand on, continue your present course, JSM
	The course steered is the best for the present (*or until time indicated*)_IZM
	What is my present position?_____RZ
TUE	PRESENTATION
TUF	PRESERVE D-ING-ATION
TUG	—Preserve formation
TUH	—Preserve station
TUI	—Preserved fish
TUJ	—Preserved fruits
TUK	—Preserved meat
TUL	—Preserved milk
TUM	—Preserved potatoes
TUN	—Preserved soup
TUO	—Preserved vegetables
	Preserves (*Jams*)_____PUO
TUP	PRESIDE-D-ING-NCY
TUQ	PRESIDENT, *or*, CHAIRMAN
	President, Chief of State Emperor, King, etc , of — has (*or*, is) ____IOR
	President of the Republic_____USJ
	Vice President _____YQF
TUR	PRESS-ED-ING-URE
	Atmospheric pressure_____FJQ
	High pressure_____OUT
	High-pressure boiler _____GXC
	High-pressure cylinder _____KIG
	High-pressure cylinder cover. ____OUV
	High-pressure cylinder cover broken, KAI
	High-pressure cylinder cracked (*or*, out of order) _____KAU
	High-pressure piston _____OUW
	High pressure piston broken_____OUX
	High-pressure valve out of order___OUY
	Intermediate pressure_____PQS
	Intermediate-pressure cylinder_____KIH
	Intermediate-pressure cylinder cover, PQT
	Intermediate-pressure cylinder cover broken _____KAJ
	Intermediate-pressure cylinder cracked (*or*, out of order)_____KAV
	Intermediate-pressure piston_____PQU
	Low pressure _____RBQ
	Low-pressure cylinder _____KIJ
	Low-pressure cylinder cover_____KAL
	Low-pressure cylinder cracked (*or*, out of order)_____KAW
	Low-pressure piston _____RBS
	Low-pressure slide jacket cover broken, KAM
TUS	—My working pressure is
	Steam pressure _____WSQ

	PRESS—*Continued*.
	Steam-pressure gauge Manometer, RJW
	Water pressure _____YVZ
TUV	—What is your working pressure?
TUW	—What pressure have you on?
	Working steam pressure—is_____TUS
TUX	PRETEND ED-ING
TUY	—Pretence
TUZ	—Upon no pretence whatever.
TVA	—Upon what pretence?
TVB	PREVAIL-ED-ING-VALENT
	Calms prevail _____HTO
TVC	—Could not prevail—upon —
	Fever very prevalent at —_____MWH
TVD	—It is very prevalent
TVE	—The prevailing sickness is —
TVF	PREVENT-ING-ION
TVG	—Can you prevent?
TVH	—Can not, *or*, Could not prevent (*or*, be prevented)
TVI	—Get a preventer on
	Is there anything to prevent landing? QEH
	Let nothing prevent_____QNV
TVJ	—Must be prevented
	Prevent advance—of _____DNJ
TVK	—Prevented by —
TVL	—Shall *or*, Will prevent
	Weather prevents-ed-ing _____YZJ
TVM	—Will it prevent?
TVN	—Will prevent the necessity of.
TVO	PREVIOUS-LY
	Previous order (*or*, instruction)____PNJ
	Previous orders are all countermanded, JYP
	Proceed in execution of previous orders, SZ
TVP	PRICE
	An advance in price_____DMZ
	Famine prices._____MSP
	No reduction in prices No abatement, CXT
TVQ	—Price current
	Prices are fair _____MRF
	Prices are high_____OVE
TVR	—Prices are low
	Prices have gone down (fallen) ___MRW
	Prices have not fallen _____MRX
TVS	—Prices have risen
	The highest price is _____OVG
	The lowest price is_____RBW
	What is the highest price you will give? OAJ
TVU	—What is the lowest price you will take?
TVW	—What is the price? What is the value of — ?
	What price are coals at — ?_____IXN
TVX	PRIEST
TVY	—A visit from a Roman Catholic priest would be much valued
TVZ	PRIME-D-ING PRIMING OF BOILER
	Boiler primes _____GWJ
	Priming charge_____IMJ

	PRINCE—PROCURE			

TWA	PRINCE-LY
	  H R H The Prince of Wales....OVM
				PROCEED—*Continued*
				Are you (*or vessel indicated*) in a condi-
				  tion to proceed? .. . .-.----.AN

TWB	PRINCESS H R H The Princess of Wales..OVX			Collier to proceed to — ....JAN Engines broken down; obliged to pro-
TWC	PRINCIPAL-LY			ceed under sail ... .MCI Has the vessel with which you have
TWD	PRINCIPLE			been in collision proceeded on her voyage?.. ....HG
TWE	PRINT-ED-ING-ER.			I have received orders for you not to proceed to — (*place indicated*) on
TWF	PRIOR-ITY			account of ice...... ....PBR
TWG	—Prior claim			I have received orders for you not to proceed without further instructions,
	PRISMATIC COMPASS ....JIC			SP I have received orders for you to pro-
	PRISON GAOL JAIL ....PUK			ceed to — ....TF
TWH	—Is a prisoner, *or*, Are made prisoners			It is not prudent to proceed without
	Military prison-er ....RTZ		TWY	regular sounding ....WKS
	Naval prison er...................SER			—May, *or*, Can proceed
TWI	—Prisoner			Proceed after repairing machinery.RDW
	Send for prisoner.................VPU			Proceed at high water.............OVF
TWJ	PRIVATE LY		TWZ	—Proceed at low water
	Following communication is confiden-			Proceed in execution of previous orders,
	tial (*or*, private) ....HX			SZ
	I have private accounts...........DFE		TXA	—Proceed into harbor
	I have private instructions Private		TXB	—Proceed on your voyage
	instructions ....PNG		TXC	—Proceed out of harbor
	Private affairs ....DPG			Proceed to (*port indicated*) ....TA
	Private flag ....NDP		TXD	—Proceed to rendezvous
	Private information..............PKG		TXE	—Proceed to sea
	Private instructions I have private			Proceed under one engine...........MBS
	instructions ....PNG		TXF	—Proceed under steam
TWK	—Private signal		TXG	—Proceed when ready
	Private soldier..................WJH		TXH	—Proceedings
	This communication is private......JGE		TXI	—Procedure
TWL	PRIVATEER			Proceeds of the cargo ....IAX
	Fell in with a privateer ....MVK			Proceeds under her own steam .....MBT
	Have been chased by a privateer ...INM			Ready to proceed ....UJW
	I have had (*or vessel indicated* has had)			Weigh, and proceed as ordered .....KV
	a fight with a privateer ....MXT			What became of the proceeds—of —?
	Vessel seen is a privateer ....VXW			GCO
TWM	—Was taken by a privateer.			You are ordered to proceed to — ....TF
				You can proceed at any speed...WMZ
TWN	PRIVATION			Your original orders are cancelled, I am
				directed to inform you to proceed to—,
TWO	PRIVILEGE-D-ING			TH
				Your port of destination is closed, your
TWP	PRIVY-ILY PRIVY TO			owners desire you to proceed to —.KXJ
TWQ	—Privy Council			
	Privy to the whole affair..........DPH		TXJ	PROCESS
TWR	PRIZE-D-ING THE PRIZE		TXK	PROCESSION
	Prize agent-cy............ ....DSU		TXL	PROCLAIM ED-ING-MATION
				Is peace proclaimed?...............YT
				Peace has been proclaimed ....YV
TWS	PROBABLE ILITY (*See also* LIKELY) Is an armistice probable? ....EVP		TXM	PROCURE-D-ING (*See also* GET *and* OB- TAIN)
	Is it probable? Is it likely?.. .....QTN			Can fresh beef be procured?........GCZ
	Most probable (*or*, likely) ....QTV			Can I procure any —?..... ....NYI
TWU	—Probably			Can I procure any assistance in the way
	Storm center is in — (*direction indi-*			of?............................ON
	cated), probable course toward —			Can I procure charter?.............INO
	(*direction indicated*)............IGY		TXN	—Can I procure stock?
	Very probably ....YPO		TXO	—Can procure
				Can you obtain (*or*, procure)?......SNP
TWV	PROCEED-ED-ING		TXP	—Can not procure
TWX	—Am I to proceed?			Can not procure wood ....ZHO
				Charter is to be procured...........IND

PROCURE—PROTECT

	PROCURE—*Continued*
	Freight is to be procured .... NQE
	I can procure .... TXO
	I can not procure.................. TXP
	Indicate nearest place where I can procure coal .... HC
	Is cotton freight to be procured? __JXW
	It can not be procured at this place.TLH
	You can procure caulkers and carpenters at — ....IBU
TXQ	**PRODUCE-D-ING-TIVE**
	Agricultural produce............. DUN
TXR	**PRODUCT-ION**
TXS	**PROFESS-ED-ING**
TXU	**PROFESSION-AL-LY**
TXV	**PROFIT-ED-ING-ABLE-Y**
	Can not gain much profit .... NVM
TXW	—Profit (*Gain.*)
TXY	—There will be no profit
TXZ	—Will there be much profit?
TYA	**PROGRAMME**
TYB	**PROGRESS-ED-ING**
TYC	—Slow progress
TYD	—Makes good progress
	Makes no progress .... RIS
TYE	**PROHIBIT ED ING-ION-ORY**
TYF	**PROJECT-ED-ING-ION OR**
	Search light, *or*, Electric projector _LVI
TYG	**PROJECTILE**
TYH	**PROMINENT-LY-ENCE**
TYI	**PROMISE D-ING**
TYJ	—Can not promise
TYK	—Promissory.
TYL	**PROMONTORY.**
TYM	**PROMOTE-D-ING-ION**
TYN	**PROMPT-ED-ING-LY-NESS**
	May be got off (*or*, afloat) if prompt assistance be given .... CG
TYO	**PRONOUNCE-D-ING.**
TYP	**PROOF**
	Fireproof ....MZV
TYQ	—I have no proof
TYR	—No proof
	Waterproof-ed-ing.... ....YWA
TYS	—What proof is there?
TYU	**PROP-PED PING**
TYV	**PROPEL-LED-LING. TO PROPEL**
	PROPELLER (*Screw*) ....VKH
	Connect your screw propeller ..... JPH
	Fitting new propeller ....NCO
	Hawser fouled propeller...........OPU

	PROPELLER—*Continued*
	How many blades have you to your propeller? ....GLX
	Key of propeller lost ....QAH
	Lost all propeller blades...........GLY
	Lost screw propeller ....RAX
	Lost — blades (*number indicated*) of screw propeller........ ....GLZ
	One blade of propeller broken .....GMA
TYW	—Propeller broken
TYX	—Propeller damaged
TYZ	—Propeller frame broken (*or*, gone)
TZA	—Propeller jammed
	Propeller lost .... RAX
TZB	—Propeller out of order
	Propeller, *or*, Body post...........GVR
	Propeller shaft broken .... MR
TZC	—Propeller worked loose
	Screw (*Propeller*). ....VKH
	Screw propeller damaged.... ......TYX
TZD	—There is a spare propeller on board
	Three blades of propeller broken...GMB
	Two blades of propeller broken....QMC
TZE	**PROPER-LY RIETY**
	Am I steering a proper course?.....JZF
	Buoy, *or*, Mark is not in its proper position .... LO
	Buoy, *or*, Mark does not appear to be in its proper position ....HMG
	Have you, *or*, Has — (*person indicated*) the proper certificate of competency? IJA
	I do not think the marks (*or*, buoys) can be in their proper positions.......HMK
	In proper form .... NLZ
	Proper certificate .... IJD
	Proper place ....TLM
	Proper position ....TPU
TZF	—Properly dressed
	Were my letters properly forwarded?_NOJ
TZG	—With great propriety
TZH	**PROPERTY**
TZI	**PROPORTION-ATE-LY**
TZJ	**PROPOSE-D-ING-AL**
TZK	—It is proposed
	Offer, *or*, Proposal not accepted...DBW
	When do you propose sailing?.......ZT
TZL	**PROPRIETOR**
TZM	**PROROGUE-D-ING ATION**
TZN	**PROSECUTE-D-ING-ION**
TZO	**PROSECUTOR**
TZP	**PROSPECT-IVE-LY**
	Any prospect of war?..............EOQ
TZQ	—Is there any prospect—of?
TZR	**PROSPER-ED-ING TO PROSPER**
TZS	**PROTECT-ED-ING-IVE-ION**
	Let us keep together for mutual protection ....IN
	We can defend ourselves if we keep together .... PYE

| | PROTEST—PUNT | |

TZU	PROTEST ED-ING	UAY	PULL-ED-ING
TZV	—Did you protest?		Let your boats pull to —(bearing indicated) .... QOA
TZW	—Have, or, Has protested		The boats can not pull ahead........ GRK
TZX	PROTESTANT		
	A visit from a Protestant clergyman would be much valued............ITY	UAZ	PULLEY
			Eccentric pulleys loose .... LSV
TZY	PROVE-ING		
	Chase prove-d-ing to be .... INL	UBA	PUMICE STONE
UAB	—Proved		
		UBC	PUMP-ED-ING. To PUMP
UAC	PROVENDER		Air-compressing pump.... .... DVY
			Air pump .... DWA
			Air-pump gear .... DWB
UAD	PROVIDE-D-ING—WITH.		Air-pump levers out of order .... DWC
UAE	—Can you provide?		Air pump out of order........ ....DWE
UAF	—I will provide		Air-pump rod broken .... DWF
UAG	—If provided—with		All hands at the pumps .... DXM
			Bilge pump... .... GJS
UAH	PROVIDENT-IAL-LY-CE		Bilge pump (donkey) .... GJT
			Bilge pump (main)................. GJU
UAI	PROVINCE-IAL		Bilge pump out of order .... GJV
			Bilge-pump valves out of order.....GJW
UAJ	PROVISION-ED-ING		Brine pump .... HIJ
	Are you in want of provisions?......TN		Can not keep free with pumps......NPT
UAK	—Can you spare me provisions?		Circulating pump............... ....IRB
	Cook-ed provisions ....JVD		Diver's pump .... LHZ
	Fresh provisions .... NRA		Donkey pump disabled .... LKP
	Have condemned a part of my provisions, JMO		Downton pump .... LMR
	I am (or vessel indicated is) short of provisions .... VZS		Feed pump .... MVA
	I have only — days' provisions ...... TO		Feed-pump check valves out of order, IOD
	Make provision for ....RIO		Feed pump out of order..........MVB
UAL	—Provisions		Feed-pump valves out of order....MVC
	Salt provisions ....VFX	UBD	Hand pump ....OLZ
	Short of provisions Starving ......NV		—Hand pump out of order
	Want provisions immediately ...... YO		Hydraulic pump............... ....PAZ
	Want provisions ....YTU		I can keep free with pumps.........NPY
			Main pump........................BGM
UAM	PROVISIONAL-LY PROVISO	UBE	—Pump (A pump)
			Pump brakes ....HDV
UAN	PROVOKE-D-ING-CATION	UBF	—Pump choked
		UBG	—Pump crosshead
UAO	PROVOST	UBH	—Pump crosshead out of order
UAP	—Provost Marshal	UBI	—Pump gear
		UBJ	—Pump gear worn out (defective)
UAQ	PROW	UBK	—Pump leather
		UBL	—Pump lever links broken
UAR	PROXY	UBM	—Pump levers out of order
		UBN	—Pump rod
UAS	PRUDENT-LY-CE	UBO	—Pump well
	It is not prudent to proceed without regular sounding .... WKS	UBP	—Pumps in good order
	Scarcely prudent to land, not easy landing on the beach .... FZX	UBQ	—Pumps out of order
		UBR	—Pumps want repair
			Well supplied with pumps .... XEY
UAT	PUBLIC-LY		Work pump ....ZHY
	Are there any public conveyances— to —? .... JTY	UBS	PUNCH (Tool)
UAV	—Public house		PUNCHEON (Measure of capacity)...AZL
	Public money..................... RYE		
UAW	—Public service	UBT	PUNCTUAL-LY-ITY
			PUND (Measure of weight)..........BCG
UAX	PUBLISH-ER-CATION-ICITY		
		UBV	PUNISH-ED-ING-MENT
	PUD (Measure of weight)........... BCF		
		UBW	PUNT
	PUDDING, or, PIE..................THY		Copper punt ....JVR

PURCHASE—QUALITY

UBX	PURCHASE (*For lifting*) Purchase block ....GMX	UCY	PUT—*Continued* · —Put into
UBY	—Purchase falls Purchases Tackle falls Tackle, *or*, Purchase carried away..XIT	UCZ	—Put off Put-ting off for the present ....TUD Put off the sale......................SOV Put on board by — ....GQS
UBZ	PURCHASE-D-ING (*See also* BUY) Heavy purchase...... ....OSY I can purchase (*buy*) ....HOW I have purchased (*bought*)... ......HBN I intend to purchase (*buy*) ....HOX	UDA	—Put out Put out fires ....NAQ
UCA	—Making large purchases	UDB	—Put over
UCB	—No purchases made	UDC	—Put to sea
UCD	—Want to purchase Was not to be purchased (*bought*) .HBO		Put to sea at once Get an offing Stand off .----- ---------------------MG Put your people on an allowance...DZM
UCE	—Will not purchase		Was obliged to put back..........FQN
UCF	—Will you (*or person indicated*) purchase?	UDE	—What is the cause of your putting back? When I have run my distance I shall put her head off until —.........LGP
UCG	PURE-ITY	UDF	—Will put
UCH	PURGE-ATIVE	UDG	—Will you put?
UCI	PURIFY-IED-ING-ICATION	UDH	—You are to put You had better put your affairs in the hands of —...........................DPK You will be put in quarantine .....UEL
UCJ	PURPLE		
UCK	PURPORT-ED-ING What is the purport (*or*, meaning)— of —? ----------- ----------------RNT	UDI	PUTREFY-IED-ING-ACTION PUTRID
		UDJ	PUTTY
UCL	PURPOSE-D-ING—TO Answer the purpose...............EMR	UDK	PUZZLE D ING TO PUZZLE.
	Fit for the purpose—of............NCK For the purpose—of................NHK I think you might get something to an- swer the purpose .. ..ENB Purpose leaving Intend leaving...POX Unfit for the purpose ....YIQ Will it answer the purpose? ....ENR	UDL	PYRAMID-IDAL
		UDM	PYROTECHNIC-AL-IST
UCM	PURSE		
UCN	PURSER		
UCO	—Purser's steward (*or*, assistant)		
UCP	PURSUE-D ING PURSUIT TO PURSUE In pursuit (*or*, chase)—of...........INQ		
UCQ	PURSUANCE IN PURSUANCE OF		
UCR	PUSH-ED-ING		
UCS	PUT-TING TO PUT. Attempted to put to sea .... FKJ Authorities will put an end to the affair, DEJ	UDN	Q (*Letter*) (*For new method of spell ing, see page* 13)
	Damage can be put to rights in — hours, BD	UDO	QUADRANT
	Do you come from any port putting you in quarantine?.... - ....TR	UDP	QUADRUPLE Quadruple-expansion engine ......MAU
UCT	—Have, *or*, Has put (*or*, run)—into —		
UCV	—I shall put in for shelter, *or*, Put in for shelter Laid up to have a new boiler put in..QIU	UDQ	QUAIL (*Bird*) ·
	Leave, *or*, Put a buoy on — ....HMP New boiler has been put in ....GXI Put a compass in the boat ....JID	UDR	QUAKE-D-ING Earthquake ....LRP
UCW	—Put back		
UCX	—Put back damaged Put in for shelter, *or*, I shall put in for shelter.................... --------UCV	UDS	QUALIFY-IED-ING-ICATION
	Put in, in distress ....LHJ	UDT	QUALITY Of a good quality ....OEK

QUANTITY—QUICKSILVER

UDV	QUANTITY—OF	QUARTER (*Measure of weight*) ------BCH	
	A large quantity (*or*, number)—of —, OGZ	UEO	QUARTER (*Division of time, etc*)
	A larger number (*or*, quantity)—of (*or*, than) — -----------------------QFM		First quartei ------------- -------NBE
			First quarter of the moon ---------NBF
	A small (*or*, smaller) number (*or*, quan- tity) of — ---------- -----------SLD		For the quarter ------------------NHL
			In a quarter of an hour ---- ------OZM
	A very large number (*or*, quantity)— of — --- -------- ------------SLE	UEP	—In what quarter?
			Last quarter --------------------QGI
	An average numbei (*or*, quantity)— of — ------------------------- FOQ		Last quarter of the moon---------QGH
			Lee quarter ---------------------QME
	An equal number (*or*, quantity)—of —, MGF		Next quarter --------------------SHY
		UEQ	—On port quarter
	Coal can not be got in any quantity— at —----------------------- -----IWM	UER	—On starboard quarter
			Quarter boat --------------------GUD
	Indicate the quantity (*or*, number)— of ------------------- ------------PIA		Quarter-deck --------------------KQD
			Quarter gallery ------------------NWG
	Number *or*, Quantity is incorrect---SLG		Quarter speed --------------------WMV
	Only to be got in small quantities---OFM		Quarter of an hour -------- -------OZM
	Show the number (*or*, quantity)—of —, PIA	UES	—Quarterly
			Quarterly account ----------------DFM
	Supply of water is small, not exceed- ing (*quantity indicated*) a day-- YWR	UET	—This quarter.
			Weather quarter-- -------------YZQ
	The greater est number (*or*, quantity)— of --------------- ---- -- ----OHN		
	The least number (*or*, quantity)—of —, QKH	UEV	QUARTER ED-ING TO QUARTER.
			At quarters-------------------FIB
	Too large a number (*or*, quantity)— of — ------------------------QPV		Can I send my sick to hospital (*or*, sick quarters)? -----------------OYQ
UDW	—What number (*or*, quantity) have you?		Close quarters----------------IVA
UDX	—What quantity—of?		Deck quarters-----------------KQB
	What quantity of ammunition have you? -----------------------EGB		Headquarters-----------------OQI
		UEW	—Quarters
	What quantity of coal have you left?--HF	UEX	—Sick quarters
	What quantity of coal remains in the collier (coal depot, *or*, lighter)?---IXO		Winter quarters ------------ZGL
UDY	—What quantity of water?	UEY	QUARTERMASTER
	You can get any quantity of firewood, NAT	UEZ	QUAY
UDZ	QUARANTINE IN QUARANTINE	UFA	QUEEN
	Am, *or*, Is in quarantine (*number of days to be indicated, if necessary*)--TP		Chief of the State, King, Queen, Em- peror, President, etc , of — has (*or*, is) — ----- ----------------- ----IOR
UEA	—Are you in quarantine?		King's Bench, *or*, Queen's Bench --GFH
UEB	—Break-ing, Broke the quarantine laws	UFB	QUELL ED ING
	Do you come from any port putting you in quarantine? --------- ---------TR		Disturbance is quelled------------LHR
	Hoist quarantine flag---------------TV		QUERY NOTE OF INTERROGATION (?)-SJZ
	I have a clean bill of health, but am liable to quarantine--Code Flag over Q	UFC	QUESTION-ED-ING-ABLE
	Out of quarantine ----------------SWA		Authority is (*or*, will be) questioned, FMO
	Performing quarantine -------------TFX		
UEC	—Quarantine at — (*plo ? indicated*) is — days for vessels from — (*place indi- cated*)	UFD	QUICK-LY-NESS
			Are you a quick sailer? -----------VDP
			As quick as ----------------------FBE
	Quarantine flag (*when spoken of*) --NDQ		As quick as possible------- --------FBG
UED	—Quarantine is taken off		I (*or vessel indicated*) have had a quick passage --------- ------------TDO
UEF	—Quarantine officer		
UEG	—Quarantine station		Make haste--- ------------------ONS
UEH	—Vessel indicated is in quarantine.		Quick firing ammunition --------EFX
UEI	—What are you in quarantine for?		Quick-firing gun -----------------OKL
UEJ	—What quarantine shall I get?	UFE	—Quicker-est
UEK	—When I am out of quarantine		Quicksand ----------------------UFG
UEL	—You will get (*or*, be put in) quarantine		Quicksilver ----------------------UFH
			Very quickly---------------------YPQ
UEM	QUARREL-LED-LING-SOME	UFG	QUICKSAND
UEN	QUARRY		
	QUART (*Measure of capacity*) ------AZM	UFH	QUICKSILVER
	QUARTER. (*Measure of capacity*) ----AZN		

	QUIET—RAN		
UFI	QUIET-LY-NESS	UGI	RACK-ED-ING
	All quiet—at (*or*, in) — (*or*, on shore), DXT	UGJ	RADIATE-D ING-ION
	All quiet with —_____DXU	UGK	RADICAL-LY
UFJ	QUININE	UGL	RADIUS
			Destructive radius _____KXU
	QUINTAL, *or*, CENTNER (*Measure of weight*) _____ BCI		Radius bars _____ ___FWL
			Radius of action _____ _____DIW
	QUINTAL METRIC (*Measure of weight*)_BCJ	UGM	—Radius shaft
UFK	QUIT-TED-TING (*See also* LEAVE)	UGN	RAFT
	Do not quit the ship till the tide has ebbed _____AD	UGO	—Can you make a raft?
UFL	—Have, *or*, Has quitted	UGP	—Do what you can in the way of rafts
UFM	—Quit the (*or*, your) station		Endeavor to send a line by boat (cask, kite, *or*, raft) _____ _____KA
	Quit the vessel as fast as possible____AB		Fire raft (*or*, ship)_____MZW
UFN	QUITE		Lookout will be kept for any rafts (*or*, spars)_ _____QZY
UFO	—Do quite well		
UFP	—Not quite	UGQ	RAFTER
UFQ	—Quite as much as		
	Quite, *or*, Totally different _____ KZO	UGR	RAG RAGGED.
	Quite enough (*or*, sufficient) _____MDX		
	Quite full _____NSY	UGS	RAGE-D ING
	Quite intelligible_____POK	UGT	Cholera is raging at—_____IPS
	Quite possible _____TQF		—The — is raging at —
UFR	—Quite recovered		Yellow fever is raging at — _____MWX
UFS	—Quite right		
UFT	—Quite sure of —	UGV	RAIL
	Quite useless_____YMS		By rail_____HPG
UFV	—You are quite safe where you are		Came, Coming by rail _____JCH
		UGW	—Head rail
UFW	QUITTANCE	UGX	—Send by rail
UFX	QUORUM	UGY	RAILWAY
			Leave it for me at the railway station, QLB
UFY	QUOTE-D-ING-ATION		Meet me at the railway station ____ROU
UFZ	—What are the last quotations?		Railway carriage _____ICE
		UGZ	—Railway iron
		UHA	—Railway pass
		UHB	—Railway plant
			Railway porter _____TPJ
		UHC	—Railway sleeper
		UHD	—Railway station
		UHE	—To, *or*, At the railway station.
		UHF	RAIN-ED-ING TO RAIN
			More rain than wind_____ZEV
		UHG	—Rain
			Rainy weather _____ _____YZC
		UHI	RAINBOW
			Lunar rainbow_____RDB
		UHJ	RAISE-D-ING TO RAISE
			Has the blockade been raised (*or*, taken off)? _____GNL
UGA	R (*Letter*) (*For new method of spelling, see page* 13)		Newly-raised men_____RPH
UGB	RABBET-ED-ING	UHK	RAISIN
UGC	RABBIT	UHL	RAM-MED-MING
		UHM	—Ram bow
UGD	RACE-D-ING TO RACE	UHN	RAN (*See also* RUN)
	Boat race _____GRT		Ran against (*or*, foul of) _____DSG
UGE	—Race course	UHO	—Ran, *or*, Run away
UGF	—Race is postponed—until —	UHP	—Ran clear
		UHQ	—Ran into
UGH	RACE (*Tideway*)	UHR	—Ran out

RANDOM—READY

	RANDOM AT RANDOM _____ _____FIT	UIS	RAVE D ING
UHS	RANGE-D ING TO RANGE		RAVINE (*Gully*)_____OJN
	Cooking range, *or*, Copper _____ __JVE	UIT	REACH (*Tack*)
UHT	—Long range	UIV	—A short reach
	Out of range _____ _____OKJ		Forereach-ed-ing_____NKA
UHV	—Range finder.		
UHW	—Range is — yards	UIW	REACH-ED ING TO REACH
	Range of battery _____FYQ	UIX	—Can, *or*, May reach
	Range your cable _____HRN	UIY	—Can you reach?
UHX	—Rifle range		Guns will not reach _____OJX
	Short range_____ _____WAC	UIZ	—I can reach
	Take care range is clear.___ _____HYM	UJA	—Out of reach
	Vessel is in gunshot_____ _____OKN		You are within reach of guns (*or*, batteries) _____ ___ _____ ____GO
UHY	—What is the range?		
UHZ	—Within range		Your orders have not yet reached here, SVA
UIA	—You are within musket (*or*, rifle) range		
	You are within range of the guns (*or*, batteries) _____GO	UJB	REACT-ED-ING-ION
		UJC	READ ING TO READ
UIB	RANGOON OIL.	UJD	—Have you read?
		UJE	—I have read
UIC	RANK-ED-ING		
	Not of high rank _____OVC	UJF	READY
	Of high rank _____OVD	UJG	—Are you ready—to (*or*, for)?
		UJH	—Are you ready for sea?
UID	RANSOM-ED-ING		Be all ready for slipping Be ready to slip your cable _____ _____HQO
UIE	RAPID-LY-ITY	UJI	—Can be ready—to (*or*, for)
	Fire gains rapidly, take people off __NG	UJK	—Can you be ready by (*time indicated, if necessary*)?
	Leak is gaining rapidly _____NU		
	The barometer has fallen very rapidly, FXH	UJL	—Can you be ready for sea by (*or*, in) —?
UIF	RARE-LY-NESS		Fenders ready _____ ____ _____MVU
			Get ready for sea Prepare for sailing, NYX
UIG	RASH-NESS-LY.		Get steam up, report when ready __KM
UIH	RAT		Have a tow line ready _____XYP
			Have all your anchors and cables ready for use _____EIG
UIJ	RATCHET		Hoist a waft (*or*, wheft) when ready, OWE
UIK	RATE-D-ING TO RATE.	UJM	—I am (*or*, They are) ready for sea (*or*, to sail)
	Any rate _____EPN		
	Bank rate_____FUY		I have the boats ready_____GTL
	Chief rate-ing_____IOT		I shall not be ready until — _____YLD
	Dying at the rate of — a day_____LQN	UJN	—In readiness—to (*or*, for)
	My chronometers were rated —days ago at — (*indicate place, if necessary*), IQJ	UJO	—Is refitted and ready for sea
			Is, *or*, Are to be held ready—to (*or*, for), OWU
	Present rate_____TUC		Keep steam ready (*or*, up) _____EQ
UIL	—Rate (*of speed*)	UJP	—Must be ready—to (*or*, for) — by
UIM	—Rate of steaming	UJQ	—Not ready
	The rate of exchange is —_____MKY	UJR	—Not ready for sea
	Try rate of sailing _____YCU		Proceed when ready_____TXG
	What is the set and rate of current (*or*, tide)? _____ ____ _____KGJ	UJS	—Readily-ness
			Ready for action_____ _____DIX
UIN	—What rate? (*Speed*)	UJT	—Ready for sea
	When were your chronometers last rated?_____ _____ __ IQK	UJV	—Ready for use
		UJW	—Ready to proceed
	Wish to get a rate for my chronometer Will you give me a comparison? - GU		Ready to weigh _____ __ZR
		UJX	—Report when ready—to (*or*, for),
UIO	RATHER	UJY	—Report when ready for sea
		UJZ	—Ships indicated are (*or*, were) ready for sea
UIP	RATIFY-IED-ING ICATION		
	Confirmation, *or*, Ratification takes place at (*or*, at —, on —) _____JOB		Steam is not ready, will be up in —__WB
			Steam is ready_____WC
UIQ	RATIONS		Troops are ready to disembark. ___YBO
			Troops will be ready—to (*or*, for)__YBQ
UIR	RATLINE		Was getting ready for sea __ ___NZF
			Water can readily be obtained ___YWS

READY—RECKON

UKA	READY—*Continued* —When ready—to (*or*, for) When will dinner be ready? ..... LAW	ULA	RECEIVE—*Continued* —Have received Have, *or*, Has received accounts ...DFC
UKB	—When will you be ready for sea?		Have, *or*, Has received damage (*or*, injury)—to ....PLA
UKC	—When will you (*or*, it) be ready for (*or*, to) —?		Have received the following communication (*or*, instructions) from your agent (*or*, owner)... .... IA
UKD	—Will be ready to (*or*, for) You will hold yourself in readiness—to (*or*, for) ....OXB Your despatches are ready ....KXE		Have you received any damage (*or*, injury)—to? ....PLB Have you received any orders—from —? SU
UKE	—Your orders are not ready for you		Have you received my letter? ..... QOF
	REAL REIS (*Coin*) ....AUT		Have you received my message? ...RQM Have you received your orders? ... STR
UKF	REAL-LY-ITY		I can not stop to have (*or*, receive) any communication ....IF
UKG	REALIZE-D-ING-ATION		I have received information — (*place next indicated*) is clear of ice ....PBO
UKH	REAPPOINT-ED ING-MENT		I have received information — (*place next indicated*) is ice bound .....PBQ
UKI	REARMOST From the rear ....NRX		I have received orders ....STV I have received orders for you not to proceed to — (*place next indicated*) on account of ice ....PBR
UKJ	—In the rear Rear admiral....................DLJ		I have received orders for you not to proceed without further instructions, SP
UKL	—Rear column		I have received orders for you to await instructions from owner at —.....SQ
UKM	—Rear guard		I have received orders for you to discharge cargo at — .. ..SR
UKN	—Rear ship		I have received orders for you to proceed to— ....TF
UKO	—To the rear		I have received your letters.........QOH
	REAS, *or*, REIS (*Coin*) ....AUT		Letters had (*or*, have) been received from.......................QOK
UKP	REASON-ED-ING TO REASON (*See also* CAUSE) Better not ask the reason (cause) ..FDN		Letters received from —...........QON No accounts received...............DFK None received ....SJD
	Can give no reason ....NZQ Do you know the reason? .... QBX		Orders have been received..........SUI
	Have you reason to suppose?.......XFE	ULB	—Receipt
	Know the reason ....QCF	ULC	—Received
	Reason must be assigned............IFO		Received an answer ....ENJ
UKQ	—Reason-able-y The reason (cause) assigned—is ....IFT		Receiver ....ULF
	The reason (cause) assigned is sufficient, IFU	ULD	—Receive-d-ing news—of (*or*, from)
	What is the reason (cause)—of —? .IFV	ULE	—Send receipt—to (*or*, for) Telegram received—from XLV
UKR	REASSEMBLE-D-ING Convoy to disperse and reassemble at rendezvous (*or place indicated*) ....IT		Wait to receive orders ....TE What advice have you received? ...DGJ You are not likely to receive your orders for some days ....SUX
UKS	REBEL-LIOUS LION	ULF	RECEIVER
UKT	—Rebel-led-ling To rebel	ULG	—Receiver of wreck Submarine signal receivers..........ZNS
UKV	RECALL-ING Boats' recall All boats to return to the ship.......... .....Code Flag over w "General recall" signal..Code Flag over y Hoist the general recall............OWH	ULH	RECENT-LY Delayed through the recent gale ...KBF No recent observations I did not get an observation ....RT Recent Act (*of Parliament*).......DIL
UKW	Recalled	ULI	RECESS
UKX	RECAPTURE D ING	ULJ	RECKON-ED ING Dead reckoning.....................KOH
UKY	RECEIVE-ING TO RECEIVE Acknowledge-d-ing the receipt—of..DGP Arrange d-ing to receive (*or*, for reception of) ....EWU		Dead reckoning agrees with observations.......................DTX Latitude by dead reckoning—is ....QHT Longitude by dead reckoning—is ..QZO
UKZ	—Did you receive? Did you (*or vessel indicated*) receive any damage? ....KJF Expect I have to receive ....MNG Give a receipt................. ....OAC Has the receipt been acknowledged?.DGQ	ULK	—Reckoning (*a reckoning*)

	RECKON—REFUND			

	RECKON—*Continued* ·		**REDUCE**—*Continued*	
	Reckoning not to be depended upon..RU		Reduce speed, you are towing me too	
	Vessel indicated reckons from the me-		fast.. _____XYT	
	ridian of — _____RQE		Reduced charge _____ _____IMK	
	What current (*rate and direction*) do			
	you expect (*or*, reckon on)? _____KGH	UMN	**REEF**-ED ING TO REEF	
			Close reef-ed ing_____IVB	
ULM	RECOGNIZE-D ING-TION		Double reef-ed _____ _____LKY	
		UMO	—I shall close reef	
ULN	RECOIL (of a gun)		Reef earing _____ _____ ____ ..LRE	
			Reef tackle _____XIR	
	RECOLLECT-ED ING (*See* REMEMBER).UPV	UMP	—Shake the reefs out	
ULO	—Do you recollect?	UMQ	—Take in a reef (*Indicate if more than*	
	Recollection (*Remembrance*) _____UPY		*one*)	
ULP	RECOMMENCE-D-ING	UMR	**REEF** (*See also* ROCK)	
	I am going to recommence (repeat) sig-		Breakers, Reef, Rock, *or*, Shoal ahead	
	nal from — (*word indicated*) _____VH		(ahead of you)_____HFB	
	Recommence, *or*, Repeat signal from —		Breakers, Reef, Rock, *or*, Shoal on your	
	(*word indicated*) _____VI		port bow_____HFD	
			Breakers, Reef, Rock, *or*, Shoal on your	
ULQ	RECOMMEND-ED-ING-ATION (*See also* AD-		starboard bow_____HFE	
	VISE)		Coral reef _____ _____ ____ JWF	
	I strongly recommend you not to— DOP		Deepest water is nearer the bank (*or*,	
	I would advise (*or*, recommend) you—		reef) _____FVB	
	to — ___ _____DON	UMS	—Reef, Rock, *or*, Shoal extends from —	
	Recommended to — _____ ____DOL		to —	
	What would you recommend? _____DOQ		Reef, *or*, Shoal is steep to the —..VYW	
	Would recommend_____DOR		Reef, *or*, Shoal stretches a long way out,	
			GE	
ULR	RECONNOITER-D ING-AISSANCE		Reefs ahead of you_____ _____HFB	
ULS	—Reconnoitering party		Send boat to point out reef (rock, *or*	
			shoal) _____GUO	
ULT	RECONSIDER-ED-ING-ATION		Stand nearer the reef (bar, shoal, *or*	
			bank) _____ _____FWA	
ULV	RECORD-ED-ING			
ULW	RECOURSE	UMT	**REEL**	
ULX	RECOVER-ED-ING	UMV	**RE-EMBARK**-ED-ING-ATION	
	I must recover my anchor_____EIS			
	Is likely to recover _____QTO	UMW	**RE-ENGAGE**-D-ING-MENT	
	Quite recovered_____ _____UFR	UMX	**REEVE**-D-ING ROVE *	
ULY	—Recovery-able			
	Tell them to recover my anchor ...EJW	UMY	**RE-EXAMINE** ING-ATION	
	Try to recover _____ _____YCX			
		UMZ	**REFER**-RED RING—TO TO REFER	
ULZ	RECRUIT-ING		Refer to my despatch (*or*, letter)____JU	
			Refer to the Geographical Signals..NYA	
UMA	RECTIFY-IED-ING-ICATION		Refer to your despatch (*or*, letter) __ JV	
		UNA	—Reference	
UMB	REOUR-RED-ING-RENCE.			
		UNB	**REFIT**-TED TO REFIT	
UMC	RED	UNC	—Can I refit at?	
UMD	—Red buoy		Complete refit necessary_____JKM	
	Red ensign _____MES		I intend to refit at_____POY	
UME	—Red flare		In want of refitting_____YTO	
UMF	—Red hot		Is refitted and ready for sea_____UJO	
UMG	—Red lead	UND	—Refitting	
UMH	—Red light			
UMI	—Red paint	UNE	**REFRIGERATE**-ING	
			Refrigerating chamber _____ __ DWL	
UMJ	REDEEM-ED-ING-PTION		Refrigerating machinery out of order,	
			RDX	
UMK	REDRESS-ED-ING	UNF	—Refrigerator	
	Police (*or other*) authorities will not			
	give any redress_____FNQ	UNG	**REFUGE** REFUGEE	
			Harbor of refuge.. _____OMZ	
UML	REDUCE-D-ING TION			
	No reduction (*in price*) _____CXT	UNH	**REFUND**-ED-ING	
	Reduce speed Ease her_____MF			

	REFUSE—REMAND		
UNI	REFUSE-D-ING	UON	REINSTATE-D-ING-MENT
	Crew refused to go on board........KCU	UOP	—Is, or, Are reinstated
UNJ	—Refusal		
	Refused assistance Assistance was re-		REIS (Coin)...................AUT
	fused ....FEZ		
UNK	REGARD-ED-ING	UOQ	REJOIN-ED-ING
UNL	—Regardless-ly		All ships of the convoy are to rejoin
			company...........Code Flag over Y
UNM	REGATTA	UOR	—Am, Is, or, Are to rejoin
UNO	REGENT CY	UOS	RELAPSE-D-ING.
UNP	REGIMENT-AL	UOT	RELATION
UNQ	REGION	UOV	RELEASE-D-ING
UNR	REGISTER (A list)	UOW	RELIANCE. (See also DEPEND)
	Italian Register (of ships)..........PUA		What reliance can be placed in (or,
	Lloyd's Register of British and Foreign		on) — ?........ ....TLN
	Shipping...... ....QWV	UOX	—Not reliable
	Lloyd's Registry..................QWX	UOY	—Reliable
	Lloyd's Yacht Register............QXE		
	United States, Royal, or, State Navy	UOZ	RELIEVE D-ING
	List, or, Register.......... ....QVM		A great relief................OHG
	Not on Lloyd's Register Not on Lloyd's		Relieve-d-ing, Relief crew ....KDI
	Registry book.......... ...... QXF		Relieve vessel in distress ....LHK
UNS	—Register-ed-ing-ation To register		Relieving tackles ....XIS
UNT	—Registered tonnage	UPA	—Relief
	Removed from Lloyd's register ...QXH		
UNV	—What is your (or vessel indicated) regis-	UPB	RELIGION-IOUS
	tered tonnage?	UPC	RELINQUISH-ED-ING-MENT
UNW	REGISTRAR	UPD	—Must relinquish
UNX	—Registrar-General of Shipping and Sea-	UPE	RELOAD
	men		
	Registrar of births, deaths, and mar-		RELY-1ED-ING (See DEPEND)........KUB
	riages ....GLI	UPF	REMAIN-ED-ING (See also STOP and WAIT)
UNY	REGRET-TED-TING		Am I (or is person indicated) to wait
UNZ	—Do not regret		or, remain)?.. ....YEN
UOA	—Much regret-ted		Do not quit the ship until the tide has
			ebbed ....AD
UOB	REGULAR LY ITY	UPG	—Do you remain?
	Are you regularly licensed, or, Have	UPH	—Do you remain all night?
	you a permit?... ....QPM		How long shall you remain? ....QYP
	Have sounded regularly . .. .WKO		I must remain till further orders.. STX
	Have you sounded regularly? ......WKQ	UPI	—I shall not remain
	I shall keep the lead going regularly I	UPJ	—I shall remain
	shall sound regularly.............QJH		I want assistance, please remain by me,
	Irregular-ly-ity- ....PTD		NR
	Is the communication with — regular?		I will remain by you (or vessel indi-
	JGD		cated) ... .AI
	It is not prudent to proceed without	UPK	—I will remain by you if you wish it
	regular soundings...............WKS		None remaining (left).............SJB
	Regular soundings..............WKT	UPL	—Not to remain
UOC	—Regular survey necessary		Please remain by me , I want assistance,
UOD	—Regular survey unnecessary		NR
	We have sounded regularly........WYK		Position remains unaltered.........TPS
	Winter has regularly set in at — ..ZGK		Remain by the ship ....KH
UOE	REGULATE-ING-ION	UPM	—Remain, or. Stay where you are
UOF	—According to regulations		Unsafe to remain until — ....YKN
UOG	—Contrary to regulations		What quantity of coal remains in col
UOH	—Regulated		her (coal depôt, or, lighter)?......IXO
			Will remain by you.... ....UPK
UOI	REIGN-ED ING		You must remain till further orders,
UOJ	REINFORCE-D-ING MENT		NUM
UOK	—In want of reinforcements		
UOL	—Reinforcements are expected	UPN	REMAINDER
UOM	—Reinforcements constantly arriving	UPO	REMAND ED

REMARK—REPEAT

UPQ	REMARK-ED ING ABLE Y Tell my owner ship answers remarkably well .... ENO	UQX	REPAIR-ED ING-ABLE Boiler can not be repaired .... RE
			Boiler has been repaired .... GVY
UPR	REMEDY-IED-ING		Boiler has not been repaired .... GVZ
UPS	—Can it be remedied?		Boiler must be repaired .... GWF
UPT	—Can not be remedied		Boiler requires repairs .... GWL
		UQY	—Cable is repaired
UPV	REMEMBER ED-ING Do you remember (or, recollect)? ... ULO	UQZ	—Can be repaired Can damage be repaired at sea? ..... BA
UPW	—Does, or, Do not remember	URA	—Can it be repaired?
UPX	—It must be remembered		Can the damage be repaired by yourselves? .... BC
UPY	—Remembrance	URB	—Can not be repaired
			Canvas to repair sails .... HVE
UPZ	REMIND-ED ING-ER		Damage can be repaired at sea ...... BE
			Damage, or, Defects can not be repaired without assistance .... BG
UQA	REMIT TED-TING TO REMIT		Damage is repaired .... KIX
UQB	—Remittance		Defects can not be repaired at sea (or at place indicated) .... KIX
	Remittance list .. .. QVL		place indicated) .... BF
	Very good remittance. .... OEM		Defects may be set right in — hours.BD
			Engines broken down, under temporary repair .... MCK
UQC	REMONSTRATE D-ING-ANCE—WITH		Engines disabled, can repair in — hours .... MCL
UQD	REMOVE D ING AL		Engines under temporary repair...MBK
UQE	—Not removed		Execute repairs. .... MLK
UQF	—Obstruction removed		How long will you (or vessel indicated) be repairing damages? .... BM
UQG	—Remove obstruction Removed from Lloyd's register .... QXH		I am repairing a telegraph cable Keep out of my way .... XLQ
UQH	—To be removed		
		URC	—I am under repair
UQI	REMUNERATE D-ING ION-IVE	URD	I shall repair damage in — .... KJO
UQJ	—No remuneration		—In want of repairs
			Is obliged to undergo repairs .... SMT
UQK	RENDER ED ING		Machinery wants repair .... MBL
	Could not render assistance .... CQ		Must discharge cargo to repair damage, KJR
	No assistance can be rendered, do the best you can for yourselves CX		Proceed after repairing machinery .RDW
	Passed a wreck (date latitude, and longitude to follow), but could not render any assistance, people still on board, ZV	URE	Pumps want repair .... UBR —Repairs necessary
	Render all the assistance possible — to, CY		Repair on board.................... GQT Repairs of engines can not be done here, MBU
UQL	—Rendered unserviceable		Repairs of engines may be done in — hours .. .. MBV
UQM	RENDEZVOUS		Stopping only for small adjustment to machinery .... RDY
	Convoy to disperse, and reassemble at rendezvous (or at place indicated) .IT		The damage to machinery is not serious, and is such as can be repaired by the vessel's own engineers ... . RL
	Convoy will rendezvous at — on the —, JUP		Twine, needles, and palms for repairing sails ... . YFN
UQN	—Convoy will rendezvous in — (latitude and longitude indicated)		Under temporary repair........... YHK
	Proceed to rendezvous..... .... TXD		Where will you go to repair damage? BV
	Rendezvous at — (port indicated or on bearing indicated) from — (port indicated) distant — miles ....JB	URF	REPAY AID-ING
	Rendezvous in latitude and longitude (indicated) .... JC	URG	—Repayment
UQO	—To rendezvous at	URH	REPEAL-ED-ING
UQP	—To, or, At the rendezvous		
UQR	—When shall we be at the rendezvous?	URI	REPEAT-ED-ING ETITION
UQS	—Where is the (or, Where shall we) rendezvous?		Annul the last hoist, I will repeat it.VE
UQT	—Will rendezvous at (or, in)		Discontinue repeating signals ..... LDG
			I am going to repeat signal from — (word indicated) .... VH
UQV	RENEW-ED-ING-AL Can you renew the action? .... DIR		I will repeat the message (or, what you say) .... RQN
UQW	RENT-ED-ING-AL (To hire)		I will repeat the signal............ WCK

	REPEAT—REPORT.	

	REPEAT—*Continued*	REPORT—*Continued*
	Repeat my signals until further orders, ...WCP	I can not take you in tow, but will report you at —, and send immediate assistance.................................XYQ
URT	Repeat ship's name, your flags were not made out........................DU	—I have reported you to the vessel I spoke, she is bound to — (*place indicated*)
URV	Repeat your signal, *or*, Repeat your signal from (*hoist indicated*).........VI	—I will report you at —
URW	Repeat the last hoist, it is not understood..............................URK	—It is reported
URJ	Repeat the last signal made........WCR	Mail reported lost..................RFM
	—Repeat the message	Medical report....................ROK
	Repeat the semaphore signal......WCQ	Meteorological or weather report...RSF
	Repeat your distinguishing signal, the flags were not made out..........DU	News, *or*, Report, *or*, Information is confirmed .... - ..JOE
URK	Repeating rifle...................UXD	No weather report .. - - ....ZG
	—The last hoist is not understood, repeat it	Nothing to report..................SKG
	Will you repeat the signal (*or*, message)?.........................WDQ	Pirate reported (*or*, seen) off —....TKR
URX	Will you repeat the signals being made to me?........................VN	—Report-ed-ing　To report
		Report by signal ....WCS
URY		Report by telegraph..............XLT
URL	REPLACE-D-ING	Report confirmed ....JOE
URZ	Light indicated has been replaced .QSD	—Report expenditure—of
URM	REPLENISH-ED-ING	Report is not correct..............JXC
		—Report me all well
USA	REPLY-IED ING—TO　TO REPLY (*See also* ANSWER)	Report me by post to my owner (*or*, to Mr —) at —.....................UC
	Can you await reply (*or*, answer)?.EMV	—Report me by telegraph to —
	Forward reply to my message by telegraph to — ....WT	Report me by telegraph to Lloyd's..UD
URO	—I will send a reply	Report me by telegraph to owner (*or*, to Mr —) at —..................UE
	I will telegraph for your orders if you will await reply ....SV	Report me by telegraph to "Shipping Gazette"...................... .UF
URP	—In reply (*or*, answer)—to	Report me to — ....UM
	No reply　Not answered.........ENH	Report me to Lloyd's............ . UG
	Replied, *or*, Answered by telegraph.EMQ	Report me (*or*, my communication) to Lloyd's agent at — ....UH
	Reply (*An answer*).... ....EMO	Report me to Maritime Association, New York........................ZNA
	Reply by telegraph............. ...WQ	Report me to Maritime Exchange, Philadelphia..... ....ZNB
	Send a reply (*or*, answer)........ENK	Report me to "New York Herald" office, London............................UI
	Send reply to my message to signal station at —....................VQD	Report me to "New York Herald" office, New York......................UJ
	Telegraph reply　Answer by telegraph, WQ	Report me to "New York Herald" office, Paris ....UK
USB	Wait for a reply (*or*, answer). .....TX	—Report me to owner on leaving
	What reply (*or*, answer)?.........ENP	Report me to Pacific Shipowners' Association, San Francisco..........ZNC
	What was the answer (*or*, reply)?...TY	Report me to the Borsenhalle at Hamburg ....UL
USC	Will await reply (*or*, answer)......TZ	—Report my arrival (*or*, arrival of —)
	Will you await reply? ....UA	Report requires confirmation .....JOF
URQ	REPORT (*See also* ACCOUNT)	Report the — (*vessel indicated*) all well in (*latitude and longitude indicated*) on the — (*date indicated*) ..... .UN
	Alarming report (*or*, account).....DEV	Report the motions—of...........SBA
	Confidential report(*or*, communication), JFX	Report when finished (*or*, complete)— with ....JKS
	Conflicting reports ....JOM	Report when ready—to (*or*, for)...UJX
	Contradict the report ....JST	Report when ready for sea...... ..UJY
	Daily report ....KIS	—Report when ship denoted is sighted
USD	Derelict reported off — (*or*, in the vicinity of —) ....JF	—Reporter
USE	Derelict reported passed on —, latitude —, longitude —, dangerous to navigation...... ....KVR	Send report of....................VQE
	Do you wish to be reported?........UB	Signal the name of place to which you wish to be reported by telegraph.WDF
	Do you wish your arrival reported?.EYC	—Vessel desires to be reported by telegraph to owner, Mr —, at —
USF	Get steam up, report when ready...KM	Vessels that wish to be reported all well, show your numbers (*or*, distinguishing signals) ....DYT
URS	—Have you anything to report?	
	Have you seen any men-of-war?　If so, report their names (*or*, nationalities), VNM	

REPORT—RESIDENCE

	REPORT—*Continued*	USZ	REQUIRE—*Continued*
	Vessels that wish to be reported, show your distinguishing signals _____EA		—Do you require?
	Vessels wishing to be reported on leaving, show your numbers (*or*, distinguishing signals) _____EB		Do you require any assistance (*or* help)—from? _____ _____OS
			Do you require further assistance?___CT
	What is the meteorological report for to-day?_____ ZK		Do you require immediate assistance?_CU
	What is the meteorological report for to-morrow?___ .. _ ....ZL		How long do you require to clean your engine?_____QYM
USG	REPRESENT-ED-ING ATION		I am going to stop, machinery requires adjusting____ . _____RK
USH	—Representative		I can not get at goods required____OZW
	REPRIMAND-ED-ING (*Censure*) _____IGT		I require assistance from Lloyd's agent, CV
USI	REPUBLIC-AN-ISM		I will lend what is required_____QNA
USJ	—President of the Republic		Is it necessary (*or*, Does the navigation require me) to take a pilot?_____SEY
USK	REPUDIATE-D-ING ION		Machinery requires adjusting_____DKF
			More hands are required—at —____OME
USL	REPULSE D-ING		My engines require cleaning_____MBQ
			Report requires confirmation _____JOF
USM	REQUEST ED-ING—THAT TO REQUEST (*See also* ASK)	UTA	—Require an interpreter
	Admiral, *or*, Senior officer has requested, DXR		Require assistance (*or*, help)—of (*or*, from) _____CZ
	Admiral, *or*, Senior officer requests the pleasure of your company to dinner, DKX		Require cleaning_____ _____ISX
			Require diver to clear screw_____LID
	Agent's request _____DSN	UTB	Require diver to examine bottom___LIE
	As requested_____ _____FBH		—Require explanation
	Captain is requested to come on board, HWN	UTC	Require hospital treatment_____OYU
			—Require steam—for
	Captain's request_____HWT		Require water boat (*or*, tank)___ _ GUJ
	Comply with request (*or*, signal)___JLD		Should you require any assistance (*or*, help) _____FGO
USN	—Have, *or*, Has been requested		Small adjustment required to machinery_____DKG
	Owner's request_____SYL		Vessel indicated does not require further assistance ____ _ _____FGQ
	Permission urgently requested to enter harbor _____PD		What assistance do you require? What do you want?_____FGT
USO	—Request (*a request*)	UTD	—Will you require?
USP	—Request can be complied with		Will you require shoring up?___ __VZP
	Request captain to come here (*or to place indicated*) _____HXF	UTE	—You are required
USQ	—Request permission—to	UTF	REQUISITE-ITION
	Request the captain of — (*vessel indicated*) __ _____HXG		As requisite_____FBI
			Caution is requisite Take care, *or*, Be careful—of _ _____FT
USR	—Request the pleasure of	UTG	—Requisition for passes
UST	—Request you will.		
USV	—Request you will not	UTH	RESCUE-D-ING
USW	—Shall, *or*, Will request		
USX	—Sorry I am unable to comply with your request	UTI	RESEMBLE-D-ING-ANCE
	The Admiral (*or*, senior officer) requests the pleasure of your company to dinner__ _ _____DKX	UTJ	RESERVE-D-ING-ATION
			Lieutenant in the Naval Reserve __QPX
			Officer of Naval Reserve___ _____SEU
	Vessel, *or*, Person indicated is not able to comply _____JLG	UTK	—Reserve squadron
			Naval Reserve_____SET
			Naval Reserve ensign.____ _____MBQ
USY	REQUIRE-D-ING-MENT		Sub-lieutenant in the Naval Reserve, QPZ
	Am on shore, likely to break up, require immediate assistance_____ .. ____CA	UTL	RESERVOIR
	Ambulance required to convey patient to hospital ____._____RDY		Air reservoir . _____DWG
	Article indicated can be supplied, but it will require fitting_____ _____RZF	UTM	RESHIP-PED-PING
			Cargo has been reshipped . _____HZW
	Boiler requires cleaning___ _____GWK		Cargo must be reshipped_____IAB
	Boiler requires repair_____ _____GWL		Is reshipping her cargo _____IAT
	Coal requires much draft._____IWR	UTN	RESIDE-D-ING
	Do not require further assistance Do not require (*or*, want) assistance__CR	UTO	RESIDENCE

RESIDENT—RIDER

	RESIDENT (*Inhabitant*) ----------- PKO	UVT	RETURNS
UTP	RESIGN-ING-ATION		Daily returns ------------------- KMY
UTQ	—Resigned	UVW	—Returns ordered
		UVX	—Returns ready
UTR	RESIN-OUS ROSIN		Send returns — of — ----------- VQF
UTS	RESIST-ED ING ANCE	UVY	REVENGE-D-ING-FUL
UTV	—Resistance coil	UVZ	REVENUE
UTW	RESOLVE-D-ING-UTION		Is a revenue cutter ----------- KHV
UTX	RESOURCE		Revenue boat Custom-house boat KGU
	Nothing to be depended upon beyond		Revenue vessel (*or*, cruiser) ------- KFC
	your own resources ----------- KUM	UWA	REVERE-D-ING-ENCE
		UWB	—The Reverend
UTY	RESPECT-ED-ING-ABLE-Y	UWC	REVERSE-D-ING-AL
	In every respect ------------------ MJO	UWD	REVIEW-ED-ING
	My best respects ----------------- GHY	UWE	REVISE-D-ING-ION
UTZ	—Respectful-ly	UWF	REVIVE-D-ING-AL
UVA	RESPONSIBLE-ILITY	UWG	REVOKE-D-ING TO REVOKE REVOCA
	Send a responsible person on board CZE		TION
	Send a responsible person on shore TGV	UWH	REVOLT-ED-ING
	REST (*Remainder*) ----------- UPN	UWI	REVOLUTION
UVB	REST-ED-ING TO REST		Can work up to — revolutions ----- ZHU
	RESTAURANT CAFÉ ----------- HSJ	UWJ	—Making (*number to follow*) revolutions
UVC	RESTORE-D ING-ATION-ATIVE		per minute
	Communication by telegraph is re-	UWK	REVOLVE-D-ING
	stored, *or*, Communication is restored,		Let your wheels (*or*, screw) revolve
	JFT		without disconnecting ----------- QOB
UVD	RESULT-ED-ING		Revolver ------------------- TKU
	What happened (*or*, resulted)? ----OMN		Revolver ammunition ----------- ICX
UVE	RETAIL ED-ING	UWL	—Revolving light
UVF	RETAIN-ED-ING		Want pistol (revolver) cartridges -- TKV
UVG	RETAKE-ING-TOOK	UWM	REWARD-ED-ING
UVH	—Must be retaken	UWN	—Shall, *or*, Will be rewarded
UVI	RETIRE-D-ING	UWO	RHEUMATIC-ISM
UVJ	RETRACT-ED-ING-ION		RHINE WINE ----------------- ZGC
UVK	RETREAT-ED-ING	UWP	RHUBARB
	Army is retreating ---- ----------- EWC	UWQ	—Tincture of rhubarb
	Enemy is retreating ----------- LZE		RI (*Measure of length*) ----------- AWR
UVL	RETURN-ED ING TO RETURN	UWR	RIB-BED
	Boats' recall All boats to return to	UWS	RICE
	the ship ---- -----Code Flag over w	UWT	RICH-LY
	Captain will not return until — HWU	UWV	RIDE-ING RODE TO RIDE (*See also* TO
	Certificate returned ----'-------- IHV		WEATHER)
UVM	—Do not return		Can you ride out the gale? - ------NVW
UVN	—Has not returned	UWX	—Do you think we can (*or*, may) ride it
UVO	—I shall return in		out?
	Immediately on my return (*or*, on the	UWY	—I shall ride it out if I can
	return of —, *person indicated*) --- PDJ		Ride it out, if possible ------------- KS
	Keep any letters (*or*, papers) for me till		Shall you ride out the gale? ------- NWD
	I return ---------------------- PYK		Tide rode -------- -- ----------- XSN
UVP	—Return as soon as	UWZ	RIDER (*An additional clause*)
	Return cargo --------------------- HZM		
UVQ	—Return salute		
UVR	—Return to the (*or*, your) station		
UVS	—When do you (*or*, does he, *or*, did you,		
	or, did he) return?		

RIDGE—ROB

UXA	RIDGE-D	UXP	RINGLEADER
	RIDICULOUS-LY-NESS (*Absurd*) -----DBK	UXQ	RIOT There has been a serious riot at —-VRW
UXB	RIFLE		
UXC	—Magazine rifle	UXR	RIPE LY-NESS
UXD	—Repeating rifle		
	Rifle ammunition ... .RFY	UXS	RISE-ING RISEN ROSE
	Rifle, *or*, Small arm magazine .....RVF		Barometer is rising ... ...FXJ
	Rifle range ... .UHX		Have premiums risen? ....TSP
UXE	—Rifled gun		Premiums have risen ....TSR
UXF	—Rifleman		Prices have risen ....TVS
	Rounds of rifle ball cartridge ....ICY	UXT	—Rise and fall is about — feet
	Send muskets (rifles) to party on shore, SQC		Spring tides rise — feet...........WOY Tide has risen — feet ....XSJ
	Service-rifle cartridge ICV		Tide is rising .... .XSL
	Service rifle............. .....LZX		Tide rises — feet...............XSM
	You are within musket (*or*, rifle) range, UIA		What is the rise and fall of the tide?..XH When moon rises.. ... .RZD
UXG	RIG-GED GING TO RIG		
	A square-rigged vessel... ....WPM	UXV	RISK-ED-ING TO RISK
	Can you rig a jury mast?..... ......VW		Beach your vessel at all risks.........KI
	Fore rigging..... .... NIX	UXW	—Do not risk
	Full rigged ship.. ....NSU		Do not risk an anchorage unless you
UXH	—How is she rigged?		have very good ground tackle......FX
	Jury rigged.... .... PXH		Do not risk without a pilot ....FKL
	Lower rigging...................RCF	UXY	—Do not run any risk
	Main rigging...................RGN		Great risk - ....OHI
	Mizzen rigging ....RWS		Great risk in sending a boat.........GSZ
UXI	—Rig in	UXZ	—I will run no risk
UXJ	—Rig out	UYA	—If it can (*or*, may) be done without risk
UXK	—Rigging (*the ropes*)	UYB	—Is there any risk?
	Running rigging...................VCI		Not safe to risk making the land...VDE
	Set up rigging....................VSF	UYC	—Risk y
	Square-rigged ship.... ....WPM	UYD	—There is no risk
	Standing rigging ....WQL	UYE	—Without risk
	Topgallant rigging.......... ....XVR		You run the risk of seizure (*or*, capture), HXZ
	Topmast rigging ....XVZ		
	Wire rigging................. ....ZGP	UYF	RIVER
UXL	RIGHT-ED-ING TO RIGHT	UYG	—Can you go up the river?
	She is righted , she is all safe ......DYR	UYH	—Do you go into the river? Down the river ....LME
UXM	RIGHT-LY (*Correct*)		Falls of the — (*river*) . ...MRS
	All right.... ....DXV	UYI	—In the river
	Buoys have just been examined, they		Is the river navigation difficult?...KZW
	are all right.... . ...HMJ	UYJ	—Off the mouth of the river — (*or*, —)
	Damage can be put right in — hours BD		On the left bank of the river........FVE On the right bank of the river.....FVG
	If I am rightly informed ....PKD		
	Is she all right? ....-..............DYM		NOTE —*The left or right hand bank in*
	Keep right astern...................FHN		*descending the stream*
	Quite right ....UFS		
	Right meaning.....................RNS	UYK	—River pilot
	With some assistance I shall be able to		River police Water police........YVX
	set things right.......... ....DF	UYL	—Up the river
UXN	RIGHT (*On the right*)	UYM	RIVET-ED-ING
	On the right bank of the river (*The*	UYN	—Started a rivet
	right-hand bank in descending the		
	stream) ....FVG	UYO	ROAD
	Right arm ....EVL		Along the road...................EAM
		UYP	—On the road to —
	RIN (*coin*) .... AUV		Rule of the road (*at sea*)...... . VBW
		UYQ	—Which is the road to —?
UXO	RING-ING RUNG		
	Packing ring ....SZB	UYR	ROADSTEAD (*See also* ANCHORAGE)
	Packing ring of cylinder cover broken, KAN		Lying in the roadstead ....QPT Vessel in the roadstead ....YPZ
	Ring bolt ....GYC		
	Thrust-block rings defective........GNB	UYS	ROB-BED BING BER BERY

ROCK—ROPE

UYT	ROCK Y (*See also* REEF)
	A rocky beach ------------------FZM
	Apparently shoal water to the —----ERO
UYV	—Are there any rocks?
	Breakers, Reef, Rock, *or*, Shoal ahead
	of you ------------------------HFB
	Breakers, Reef, Rock, *or*, Shoal on your
	port bow ---------------------HFD
	Breakers, Reef, Rock, *or*, Shoal on your
	starboard bow ----------------HFE
	Drove on the rocks ---- ---------- LOE
	Foul ground Rocky bottom-------HAZ
	Keep the rocks on — -------------PZL
	Near the rocks -------------------SFK
UYW	—On the rocks
UYX	—Rock under water
	Rocks, Reefs, *or*, Shoals extend from —
	to — ------- ----------- --------UMS
	Rocks stretch a long way out--------GE
	Rocky bottom Foul ground------HAZ
UYZ	—Rocky coast
	Send boat to point out reef (rock, *or*,
	shoal) ------- ----- ----- ------GUO
	Very rocky ------- --------- ----YPR
UZA	—Vessel is on the rocks
	You are too near the rocks---------SFP
UZB	ROCKET
UZC	—Can you spare some signal rockets?
	Explosive rocket. War rocket ----MOZ
UZD	—Gun cotton rocket
UZE	—Have you any rockets?
	Look out for rocket line (*or*, line)---KF
	Rocket (*Life-saving*) apparatus----ERH
	Rocket (*Life saving*) apparatus coming,
	ERJ
	Rocket brigade ----- -------- ----HIB
UZF	—Rocket staff
	Rocket (*Life-saving*) station ----WRP
UZG	—Rocket to throw a line L S. A rocket
UZH	—Rocket tube
	Signal rocket -------------------WDB
	Sky rocket------ -- ---------------WFN
UZI	ROD
	Air pump rod broken ------- ----DWF
	Connecting rod-------------------JPK
	Connecting-rod bolts broken ------GXV
	Connecting-rod brasses broken (*or*, out
	of order)----- ----- -----------HEB
	Cylinder side rod ------------ ----KIF
	Eccentric rod bent ------- ------LSW
	Eccentric rod broken ---- ---------LSX
	Parallel-motion rod out of order----TAY
	Piston rod ----- ---------------TKY
	Piston-rod bolts broken ------- ---GYB
	Piston-rod brasses broken ----- ---HED
	Piston rod cracked (*or*, broken) ---KAX
	Piston-rod crosshead damaged (*or*,
	broken)----- -----------------KEJ
	Pump rod----- --------- -- ----UBN
	Slide-valve rod --------- --------WGN
	Slide-valve rod broken------------WGO
	Sounding rod ------ ------------WKV
	RODE (*See* RIDE) -------------UWV
	ROEDE (*Measure of length*) -------AVS
UZJ	ROLL-ED-ING
	Muster roll (*or*, book) Roster----GZB
UZK	—Rolling batten

UZL	ROLLER
	Rollers setting in----------------- UT
	ROMAN CATHOLIC ------ ----------IEZ
	A visit from a Roman Catholic priest
	would be much valued ---------TVY
	ROMAN CEMENT -----------------IGR
UZM	ROOF-ED-ING
UZN	ROOM (*Space*)
UZO	—Barely room to wear.
	Enough, *or*, Sufficient room ----- -MDC
UZP	—Have I room to wear?
	Have room for no more cargo ------IAP
	Have you room enough—for (*or*, to) —?
	MDL
	Have you room for more cargo? -- IAR
	I have not room to heave-to ----- --OSJ
	I will heave-to directly I have room OSN
UZQ	—No room—for
UZR	—Not room to swing
	Not room to wear ----------------MB
UZS	—Not sufficient room
	Room enough---------------------MDC
UZT	—Room to swing
	Sea room--------------- ---------VLD
	Shall I, in my present berth, have room
	for weighing, if the wind shifts?-GFZ
	Take room enough----------------MDY
UZV	—There is not room to —
UZW	—There is plenty of room
	You have barely room ------ ----FWT
UZX	ROOM (*A room*)
	Boiler room ---------------------GWM
	Bread room-------------- --------HEM
	Engine room --------------------MBE
	Engine room full of water-------- MBG
	Engine room has — feet of water in it,
	MBH
	Engine-room telegraph ------------MBJ
	Gun room------- ----------------OJT
	In the engine room----------------MBP
	Mess room-----------------------RQH
	Near the engine room----------------MBR
	Sail room ----------------------VEH
	State room------- ----- ----------HQJ
	Store room ---------------------WXS
	Under the engine room -----------MCE
	Ward room--------- ------------YUH
UZY	ROPE
	Boat rope-----------------------GRU
	Bolt rope -----------------------GXU
	Breast rope------ --------------HFQ
	Buoy rope (*of anchor*) ------------EHR
	Cable laid rope------------------HQR
	Can you spare some rope? (*Size to*
	follow) ---------------------WMB
	Coil of rope---------------------IZO
	Coir, *or*, Grass rope--------------IZU
	Drag rope ----------------------LMY
	European rope--------------------MHX
	Hawser Large rope------------OPR
	Hawser-laid rope-----------------OPV
	Is the cable (*or*, rope) belayed (fast)?
	HRK
	Lift rope-- --------------- ----QRM
	Man ropes------- ----------------RJX

	ROPE—RUN		
	ROPE—*Continued*	VAR	ROW　(*A line or series*)
	Manila rope ....RJT	VAS	ROW-ED-ING-ER
	Mast rope ....RMI	VAT	—To row guard
	Old rope ....SQH		
	Rope, or, Jacob's ladder ....PUJ	VAU	ROWLOCK
	Send a rope on board..................GQV		
	Slip rope ....WGY	VAW	ROYAL-LY-TY-IST
VAB	—Tiller rope		Fore royal............................NIY
	Top rope ....XVM		Fore-royal yard......................NIZ
VAC	—Veer a rope astern		Main royal........ .. ..RGO
	Wheel ropes........ ....ZCJ		Main-royal yard.....................RGP
VAD	—Wire rope.		Mizzen royal ....RWT
	Yard rope..............................ZKQ		Mizzen-royal yard....................RWU
		VAX	—Royal Exchange
	ROSE　RISE　RISEN ....UXS		Royal, or, State Navy List ...-....QVM
			Royal Navy, or, State Navy.........SFC
	ROSIN　RESIN-OUS ....UTR	VAY	—Royal sail
		VAZ	—Royal salute
	ROSTER ....GZB	VBA	—Royal sheet
			Royal standard　Royal flag ......WQN
VAE	ROT-TED-TING	VBC	—Royal yacht
	Dry rot ....LOQ		Royal Yacht Squadron...............WPG
VAF	—Rotten	VBD	—Royal yard
VAG	ROTATE-D-ING-ION　ROTARY.	VBE	RUB-BED-BING
	Rotary Engine ....MAV		Rubbed off the copper..............JVT
	ROTTOLO　(*Measure of weight*)......BCK	VBF	RUBBER.
			India rubber.　Gutta-percha ... ..OKU
	ROUBLE (PAPER)　(*Money*) .... AUW	VBG	RUBBISH.
	ROUBLE (SILVER)　(*Coin*) ....AUX	VBH	RUDDER
VAH	ROUGH LY-NESS		Balanced rudder....................FTA
	I have had a rough passage ....TDB	VBI	—Damaged rudder
VAI	—Very rough		Damaged rudder, can not steer.......BI
			Jury rudder ....PXI
VAJ	ROUMANIA-N　ROUMANIAN COLORS	VBJ	—Loss of, or, Lost rudder
			Rudder chain..... ....LJV
VAK	ROUND-ED-ING	VBK	—Rudder chock.
	Brace round ....HDN	VBL	—Rudder coat
	Brace round your head yards.......HDO		Rudder disabled, will you assist me
	Come round ....JCX		into — (*port indicated*)?..........DA
	Get her head round as fast as possible,	VBM	—Rudder head
	NYU	VBN	—Rudder iron
	Get her head round, or, Get her on the	VBO	—Rudder post
	other tack, or you will be on shore.LF		Vessel indicated has lost her rudder.RBE
	Going round ....OCD		
	Round about ....DAI	VBP	RUDE-LY-NESS
	Round-ed-ing the point..............TNO		
	Rounding the Cape....................HVS	VBQ	RUG
	Wear short round....................YXT		
		VBR	RUGGED
VAL	ROUND　(*Shape*)		
VAM	—Round shot	VBS	RUIN-ED-ING-OUS
	ROUNDS OF AMMUNITION—FOR —....EFZ	VBT	RULE-D-ING　TO RULE
	Rounds of ammunition for heavy guns,	VBU	—Rule (*regulation*)
	EGA	VBW	—Rule of the road (*at sea*)
	Rounds of rifle ball cartridge........ICY		
		VBX	RUM
VAN	ROUT-ED-ING		
			RUMOR (*report*)....................URQ
VAO	ROUTE		
	En route........... ....LYA	VBY	RUN-NING　*See also* RAN
			A running fight.....................MXS
VAP	ROUTINE		After running — (*miles indicated*) .DRH
			Blockade runner ......: ....GNF
	ROVE　REEVE-D-ING..................UMX		Chronometer has run down.........GQ
VAQ	Row　(*Address*)		Current runs — miles an hour......KFY

RUN—SAFE

	RUN—*Continued*
	Current, *or*, Tide will run very strong (*indicate miles per hour, if necessary*), KGB
	Do not run any risk...........UXY
	Fore runner................NJA
	Had we not better run in and make the land?...............PZ
	Have, *or*, Has run (*or*, put) into ——..UCT
VBZ	—I can not run, I must see it out.
VCA	—I do not like to run in, I prefer keeping the sea
	I have (*or vessel indicated* has) run the blockade................GNM
	I have (*or vessel indicated* has) sunk (*or*, run down) a vessel.........HK
	I will run no risk............UXZ
	Let run Let it run...........QNW
VCB	—Main runner
	Medical comforts running short....ROI
	Medicines have run short........ROQ
	My chronometer has run down......GQ
	Ran.....................UHN
	Ran against (*or*, foul of)........DSG
	Ran, *or*, Run away............UHO
	Ran clear.................UHP
	Ran into..................UHQ
	Ran out..................UHR
	Run for —— Make for ——.......RIH
VCD	—Run in (*or*, into)
	Run in for pilots (*or*, pilot vessels), and look out for their signals.........RAC
	Run on board by.............GQU
VCE	—Run on shore
	Run on the beach Beach the vessel, FZS
	Run out..................SWG
VCF	—Run out a warp
VCG	—Runners and tackles
VCH	—Running before the wind
VCI	—Running rigging
VCJ	—Running under bare poles
	Shall we run for —— (*place indicated*)? WUA
	Tide, *or*, Current will run very strong (*miles per hour may follow*).....EGB
VCK	—Underrun-ning-ran
	Unsafe to run in until weather moderates...............YKO
	Want warp run out..........YQ
	What does, *or*, What will the tide run? XSP
	What harbor shall you run for?....OND
	When I have run my distance I shall put her head off until ——......LGP
VCL	—You run away from me
	RUNG RING-ING............UXO
VCM	RUNG (*A Rung*)
	RUPEE (*Coin*).............AUY
VCN	RUPTURE-D ING
VCO	RUSH-ED-ING
VCP	RUSSIA-N RUSSIAN COLORS
	Bank of Russia.............FUR
VCQ	RUST-ED-ING-Y

VCR	RYE
VCS	S (*Letter*) (*For new method of spelling, see page* 13)
	SAADING, *or*, YOUNG (*Measure of length*), AWS
VCT	SABBATH SUNDAY
VCU	SACK-ED-ING.
VCW	SACRIFICE-D-ING
VCX	SAD LY-NESS
VCY	SADDLE-D-ING SADDLERY
VCZ	SAFE-ER-EST. (*See also* SAVE)
	All safe (saved)............DXW
	Are your friends safe?.........NRO
•	Arrived safely..............EXY
	Bar, *or*, Entrance not safe except just at slack water (*or*, at ——, *time indicated*), FVT
	Boat is safe...............GRP
	Can we anchor with safety?.....EHS
	Fears are entertained for the safety of ——, MUI
	For the safety of...........NHM
	Harbor, *or*, Anchorage is safe with all winds...............OZ
VDA	—I know of no safe harbor
VDB	—Is it safe?
VDC	—Is, *or*, Are not safe
	Is the anchorage (*or*, my anchorage) safe with all winds (*specify which wind, if necessary*)?...............PB
	It is not safe to go so fast.......LS
	It is not very safe where you are anchored................EIX
	Landing safe, but very disagreeable..LCE
	Not safe to go on with the crew at present................KDE
VDE	—Not safe to risk making the land
	Only chance of safety.........IKG
	Safe anchorage.............ELD
VDF	—Safely.
VDG	—Safely moored
VDH	—Safety
VDI	—Safety valve
VDJ	—Safety valves out of order
	She is righted, she is safe.......DYR

SAFE—SAIL

SAFE—Continued

Shift your berth Your berth is not safe .... KT

VDK —Shorten sail, it is not safe to go so fast
The berth you are now in is not safe, ... GHE

VDL —Will it be safe?
Will the soundings be a safe guide?..WKZ
You are quite safe where you are...UFV
You must moor, it is not safe otherwise .... RZO

VDM SAG-GED-GING

SAGENE, or, SASCHENE (*Measure of length*) .... AWT

VDN SAGO

SAID (*See* SAY) .... VLJ

VDO SAIL-ED-ING TO SAIL
Am about to sail...P, or, Code Flag over P

VDP —Are you a quick sailer?
VDQ —Before sailing
VDR —Before sail
VDS —Bend a new sail
VDT —Bend sails
Boat's sails .... GSB
Book of Sailing Directions (*indicate place, if necessary*) .... GYX

VDU —Can not carry more sail
Canvas to repair sails .... HVE
Damaged, *or*, Sprung mast, can not carry sail .... BH

VDW —Do not furl sails
Do you know anything of strange sail? QBU
Easy sail Keep under easy sail...LRW
Engines broken down, obliged to proceed under sail .... MOI
Fill main topsail .... MYB
Fore and aft sail .... NIU
Foresail .... NIV
Fore staysail .... NJD
Fore-topgallant sail .... NJH
Fore-topmast staysail .... NJM
Fore topsail .... NJO
Furl sails (*or sails indicated*) ...... NTX
Gaff foresail .... NKG
Gaff mainsail .... NYF
Gaff topsail .... NVG
Get sail on your ship as fast as possible, NYZ
Great circle sailing .... OHA
Has the mail sailed—for —? .... QO
Have you a book of Sailing Directions? RO
Headsail .... OQJ
Hoist windsails .... OWK

VDX —How does — (*vessel indicated*) sail?
VDY —How much sail?
I am about to sail,
Flag P, *or*, Code Flag over P

VDZ —I am going to sail (*date may follow*)
I am *or*, They are ready for sea (*or*, to sail) .... UJM

VEA —I mean to carry sail all night
I sail to-night .... XUR
I shall keep under easy sail .... LRY
I shall not furl sails .... NTY

SAIL—Continued

VEB —I shall shorten sail
I will show a light to-night when I make sail .... QRZ
If you do not carry more sail we shall part company .... JGN
In making sail .... RID
Intend to sail. .... POZ
Jigger royal yard and sail .... PVS
Jigger topgallant yard and sail ....PVT
Jigger topsail yard and sail..........PVU
Jigger yard and sail .... PVW
Keep main topsail shivering ......PZJ
Keep mizzen topsail shivering ......PZK
Light sail. Upper sail .... QSV
Loose sails .... RAJ
Lower studding sail............................ RCG
Lower topsail............................ RCH
Lugsail .... RCY
Mail has sailed .... RFH
Mail steamer sails-ed .... RFN
Mails go in sailing vessels .... RFP
Mainsail Main course............................ RGQ
Mainsail split .... RGS
Main staysail .... RGW
Main topgallant sail .... RHA
Main topmast staysail .... RHD
Main topsail .... RHE
Make more sail — Set your — ...RIL
Make, *or*, Set sail .... RIT
Make sail, *or*, Go ahead and drop a boat on board .... DVN
Mizzen topgallant sail .... RWX
Mizzen topmast staysail............. RXA
Mizzen topsail .... RXB

VEC —My sails are all blown away (*or*, split)
VED —No use carrying so much sail
Prepare for sailing Get ready for sea, NYX
Royal sail .... VAY

VEF —Sail (*A sail*)
Sail cloth............................ IVO

VEG —Sailmaker
VEH —Sail room
VEI —Sailed for —
VEJ —Sailed from —
Sailing collier .... JAP

VEK —Sailing directions (*indicate place, if necessary*)
VEL —Sailing orders
Sailing vessel .... VXE

VEM —Sails from
VEN —sails shivering
VEO —Set studding sails
VEP —Set topgallant sails
VEQ —Set — sails
Shall I furl sails? .... NTZ

VER —Shall, *or*, Will not sail
VES —Shall, *or*, Will sail
VET —Shorten sail
Shorten sail, it is not safe to go so fast, VDK
Small sail .... QSV
Split all my sails............................ WOB
Spritsail............................ WOZ
Sprung, *or*, Damaged a mast, can not carry sail .... BH
Square sail .... WPN
Staysail .... WSE
Stop the sailing of —............................ WXI
Storm sail............................ WXV

SAIL—SAND

	SAIL—*Continued*	VFN	SALLY PORT
	Storm staysail ___WSF		
	Storm trysail ___WXY	VFO	SALMON
	Strange sail ___WYT		
	Strange sail (*or vessel indicated*) is	VFP	SALOON
	bound to — ___HBT	VFQ	—Saloon passenger
	Strange sail (*or vessel indicated*) is		
	from — ___WYU	VFR	SALT SALINE
	Strange sail (*or vessel indicated*) is		Boiler tubes salted ___GWQ
	(*or, was*) in distress___LHM		Can you supply me with salt beef
	Studding sail ___XAY		(*number of casks indicated*)? ___GDA
	Studding sail boom ___GZX		Condenser tubes choked (*or, salted*),
	Take in light sails ___XJN		JMV
	Take in sail ___VET		Epsom salts ___MGD
	Take in studding sails ___XJO	VFS	—Salt beef
	Take in topgallant sails ___XJP		Salt butter ___HOT
	There is a strange sail on — (*bearing*	VFT	—Salt fish
	indicated) ___EM	VFU	—Salt meat
	This is my best point of sailing __GIA	VFW	—Salt pork
	This is not my best point of sailing.GIB	VFX	—Salt provisions
	Topgallant sail ___XVS	VFY	—Salted-ing To salt.
	Topgallant studding sail ___XVT		
	Topmast staysail ___WSG	VFZ	SALTPETRE
	Topmast studding sail ___XWA		
	Topsail ___XWB	VGA	SALUTE-D-ING
	Trysail ___YDA		Return salute ___UVQ
	Try rate of sailing ___YCU		Royal salute ___VAZ
	Twine, needles, and palms for repairing	VGB	—Shall I salute?
	sails ___YFN		
	Unbend sails ___YGN	VGC	SALVAGE
	Under all sail ___YHC		Lloyd's Salvage Agreement ___QWY
	Under easy sail ___LRZ	VGD	—Salvage demanded
	Upper sail ___QSI	VGE	—Salvage officer
	Upper topsail ___XWH	VGF	—Salvage steamer
	Was to sail immediately ___PDN	VGH	—Salvage voucher
VEU	—Was to sail on the —		
VEW	—We, *or*, Vessel indicated sailed on the —	VGI	SAME
	We sail on the — ___SQR	VGJ	—In the same
VEX	—What is your best point of sailing?		In the same position ___TPO
	When did you sail? ___SG	VGK	—In the same state as before
	When do you propose sailing? ___ZT	VGL	—Is not the same
VEY	—When does he (she, *or indicated vessel*)	VGM	—It is, *or*, They are the same
	sail?		Not in the same position ___TPQ
VEZ	—Why do you not sail?		On the same day ___KNR
	Windsail ___ZFV	VGN	—The same as —
VFA	—With all sail set		The same as usual ___YMZ
			The same color ___JBP
	SAILMAKER ___VEG		The same course ___JZN
VFB	—Am in want of sailmakers		The same description ___KWG
	Can I get any sailmakers? ___NYJ		The same kind—as ___QAY
		VGO	—The same thing
VFC	SAILOR	VGP	—The same time At the same time
VFD	—Sailors' Home	VGQ	—The same way
		VGR	—Will do the same
VFE	SAINT ST SAN-TA		
		VGS	SAMPAN
VFG	SAKE		
	For the sake of ___NHO	VGT	SAMPLE D-ING
		VGU	—Samples can be seen
VFH	SALICINE.		Send sample—of ___VQG
VFI	SAL-AMMONIAC	VGW	SAMSON POST
VFJ	SALARY-IED		
		VGX	SANATIVE SANITARY
VFK	SALE (*See also* SELL)		
	Advertise cargo and day of sale ___DOB	VGY	SANCTION-ED-ING
	Cargo not salable ___IAE	VGZ	—Can sanction
VFL	—Have you anything for sale?	VHA	—Is, *or*, Are sanctioned by
	Put off the sale ___SOV		
	Sale by auction ___FLT	VHB	SAND-Y
VFM	—Salable		A sandy beach ___FZN

SAND—SCHOOL

SAND—*Continued*
　Bar has, *or*, Sands have shifted....FVU
　Coarse bottom　Coarse sand and stones,
　　　　　　　　　　　　　　　　　　HAX
　On a sand bank ....FVD
　Quicksand ....UFG
　Sand bag..........................FSL
　Sand bank.........................FVH
VHC　—Sand dredger
VHD　—Sand spit
　Sands, *or*, Bar shifted.............FVU
　Sandy bottom......................HBF
　Send for sand.....................VPW
VHE　—Shifting sand

VHF　SANE　SANITY

　SANITARY　SANATIVE............ ....VGX
VHG　—Sanitary inspector

VHI　SANK　SINK ing　(*See* SINK)

　SAPEQUE, CASH, *or*, LE　(*Coin*)... ..ATC
VHJ　SARDINE

VHK　SARDINIA-N

　SASCHENE, *or*, SAGENE　(*Measure of
　　length*) ....AWT
VHL　SAT　SIT-TING

VHM　SATISFY IED-ING
VHN　—A very satisfactory —
　　Credentials are not satisfactory... KBZ
　　Credentials are satisfactory....!...KBY
VHO　—I am not satisfied
　　Light on — is not working satisfactorily,
　　　　　　　　　　　　　　　　　　QSH
　　Papers satisfactory................TAM
　　Papers unsatisfactory ....TAN
VHP　—Satisfaction
VHQ　—Satisfactory ily
　　The appearance is not satisfactory..ESF
　　The appearance is satisfactory .....ESG

VHR　SATURATE D-ING-ION

VHS　SATURDAY
VHT　—Before Saturday
　　Last Saturday...... ....QGJ
VHU　—Next Saturday
VHW　—Saturday morning
VHX　—Saturday night

VHY　SATURN

VHZ　SAUCE　TABLE SAUCE

VIA　SAUSAGE

VIB　SAVAGE

VIC　SAVE-ING
　All on board saved ....DXR
　All safe (saved) ....DXW
　Barely time to save the mail....REY
　Can not save the ship, take people off,
　　　　　　　　　　　　　　　　　NB
　Cargo expected to be saved......HZU

SAVE—*Continued*
　Cargo saved ....IAF
　Crew saved .. ..KCV
　I am sinking (*or*, on fire), send all avail-
　　able boats to save passengers and
　　crew ....NO
VID　—It will be a great saving
VIE　—It will be no saving
　Life-saving (*rocket*) apparatus.....ERH
　Life saving appliances............ ESL
　Part of cargo saved ....IAW
　Passengers saved ....TDS
VIF　—Saved
　Send life-boat to save crew...........FJ
　Shall I save the mail (*or*, post)?....RFV

VIG　SAVINGS BANK

　SAW　(*See also* SEE)..... ....VOA

VIH　SAW-ED ING　SAWN
　Cross-cut saw ....KEB

VIJ　SAY ING　SAID
　I decline to have anything further to do
　　(*or*, say) in the matter .... EQO
　I will repeat the message (*or*, what you
　　say) ....RQN
　It is impossible to say ....PFE
VIK　—Say nothing
VIM　—What did he say?
　Who told (*or*, said so)?............XMZ

VIN　SCALD-ED ING

VIO　SCALE-D-ING
VIP　—Scaling ladder

VIQ　SCANT-Y-INESS
　Supply scanty ....XEW

VIR　SCANTLING

VIS　SCARCE-ITY-NESS
　Fish very scarce ....NBR
　Scarcely (*barely*) ....FWP
　Scarcely any ....EPV
VIU　—Scarcely necessary
VIW　—Scarcely perceptible
　　Scarcely prudent to land　Not easy
　　landing on the beach .... FZX
　　Scarcely visible .... VIW
VIX　—There is a great scarcity
VIY　—There is (*or*, are) scarcely —
VIZ　—Was very scarce

VJA　SCARF, *or*, SCARPH-ED ING　TO SCARPH

VJB　SCARLATINA

VJC　SCARLET
　Scarlet fever ....MWR

VJD　SCATTTER-ED-ING

VJE　SCHEDULE D ING
VJF　—Forward a schedule

　SCHEPEL　(*Measure of capacity*)... AYT

VJG　SCHOOL

SCHOOLMASTER—SEA

VJH	SCHOOLMASTER
VJI	SCHOONER
	Four-masted schooner_____NPC
VJK	—Three-masted schooner.
VJL	SCIENCE-TIFIC-ALLY
VJM	SCINTILLATING LIGHT
VJN	SCISSORS
VJO	SCOPE
	Enough scope_____MDE
VJP	—More scope
	Not enough cable　Give more scope,
	HRM
VJQ	—Too much scope
VJR	SCORBUTIC-ALLY
	Anti-scorbutic _____ EOC
VJS	SCORCH-ED-ING
VJT	SCORN-ED-ING-FUL LY
VJU	SCOTCH-MAN
VJW	SCOUT-ED-ING
	Scout, or, Lookout vessel_____QZX
VJX	SCRAP
VJY	—Scrap iron
VJZ	SCRAPE-D-ING.
VKA	SCRAPER.
VKB	SCRATCH ED-ING
VKC	SCREEN-ED-ING　To screen
VKD	—Screen　(A screen)
	SCREW　(PROPELLER) _____VKH
	Auxiliary screw_____FOB
	Boss of screw ___ ____ _____HAO
VKE	—Can you lift your screw?
	Connect your screw (or propeller)_JPH
	Disconnect_____ _____LDA
	Does disconnecting affect your steering?
	LDC
	Have lost screw, or, Screw disabled_MS
VKF	—Have you a spare screw?
	Hawser fouled screw_____OPU
	How many blades have you to your
	screw?_____GLX
	Let your screw (or, wheels) revolve
	without disconnecting_____ QOB
	Lift your screw_____ _____QRN
	Lost all propeller blades of screw__GLY
	Lost number of blades (indicated) of
	screw propeller _____ _____GLZ
	Lost screw propeller _____RAX
	One blade of propeller broken _____GMA
	One screw disabled, can work the
	other_____MW
	Port screw _____ ___TOZ
	Require diver to clear screw _____LID
	Screw frame (or, banjo) _____FUG
VKG	—Screw passage
VKH	—Screw (propeller)

	SCREW—Continued
	Screw propeller damaged _____TYX
VKI	—Screw shaft
VKJ	—Screw steamer (horsepower to follow, if
	necessary)
	Screw well　Screw aperture _____ERA
	Starboard screw_____WQY
	Three blades of propeller broken___GMB
	Twin screw _____ ___YFK
	Twin screw steamer_____YFL
	Two blades of propeller broken_____GMC
VKL	—What weight is your screw?
VKM	SCREW　(For fastening)
	Clamping screw___ _____ISA
VKN	—Screw-ed-ing　To screw
	Screw bolt____ _____ _____GYD
VKO	SCRUB BED-ING
	Scrubbing brush _____ ___HKQ
VKP	SCRUPLE-ULOUS LY
VKQ	SCRUTINY ISE-D-ING
VKR	SCUD-DED DING
	Obliged to scud _____SMX
VKS	SCUPPER
	Lee scuppers_____QMF
VKT	SCURRILOUS LY
VKU	SCURVY
VKW	SCUTTLE D-ING　TO SCUTTLE
	Leak at the scuttle _____QJX
VKX	SEA
	Apparatus for casting oil on the sea to
	quiet it _____ _____ERF
	Are you ready for sea?_____UJH
	At sea_____ _____ ___FIU
	Attempted to put to sea_____FKJ
	Calm sea _____ _____UO
	Calm the sea by pouring oil on it __HTN
	Can be fished at sea _ _____NCG
	Can damage be repaired at sea? _____BA
	Can you be ready for sea by (or, in) —?
	UJL
	Cargo so badly stowed, I am not sea-
	worthy_____IAG
	Considerable sea_____UP
	Damage can be repaired at sea _____BE
	Damage can not be repaired at sea (or
	at place indicated) _____BF
	Deep-sea lead._____KQS
	Defects can not be repaired at sea (or
	at place indicated)_____ BF
	Enemy is at sea_____ _____LZB
	Get ready for sea　Prepare for sailing,
	NYX
	Gone to sea _____ ___OCG
	Have you tried the temperature of the
	sea? _____ _____XNF
	Head sea_____OQK
	Heavy sea　Much sea ___ _____UQ
	I am, or, They are ready for sea (or, to
	sail)_____UJM
	I do not like to run in, I prefer keeping
	the sea_____ _____VCA

	SEA—SECT

	SEA—*Continued*		SEARCH—*Continued*
	Is refitted and ready for sea .... UJO		'Keep the search light—on, *or*, Keep
	Machinery broken down, irreparable at		electric light—on .... LVH
	sea .... RDV		Search light, *or*, Electric projector .. LVI
	Much swell on .... UR		Use your search light .... YMT
	Not much sea .... US	VLW	SEASON
	Not ready for sea .... UJR		A healthy season .... OQX
	Proceed to sea .... '........... TXE		Fishing season .. .. NBX
	Put to sea .... UDC		Good season for fish .... NBZ
	Put to sea at once, get an offing .... MG		Last season .... QGK
	Ready for sea .... UJT	VLX	—Next season
	Report when ready for sea .... UJY	VLY	—Not this season
	Rule of the road (*at sea*) .... VBW		
	Sea anchor Floating anchor ...... EIC	VLZ	SEAT-ED ING
	Sea boots .... HAC		Boiler working on seating .... GWR
	Sea breeze .... HGC		Engine seating loose .... MCO
VKY	—Sea cock		Engines working on seating .... MCP
VKZ	—Sea going gone down		
	Sea inlet choked . ... IPQ	VMA	SEAWEED
VLA	—Sea inlet valve (*or*, cock) broken (*or*,		
	damaged)		SEAWORTHY VLI
	Sea mile (*Knot*) .... AWO		Cargo so badly stowed, I am not sea-
VLB	—Sea of —		worthy .... IAG
VLC	—Sea port		
VLD	—Sea room.	VMB	SECOND-ED-ING TO SECOND
VLE	—Sea service		
VLF	—Sea sick-ness.	VMC	SECOND (*The second*)
VLG	—Seaman		I can not make out the second flag. NDG
VLH	—Seaward		Second class .... ISH
VLI	—Seaworthy		Second flag in hoist indistinct ..... NDG
	Ships indicated are (*or*, were) ready for	VMD	—Second mast
	sea .... UJZ	VME	—Second mate (*or*, officer)
VLJ	—Shipped a sea		Second week in — .... ZAO
	Smooth the sea by pouring oil on it. HTN	VMF	Secondary
	Stand out to sea .... WQI		
	Struck by a heavy sea .... XAC	VMG	SECOND (*A second of time*)
	The temperature of the sea is — ... XNJ		*For seconds of arc and time, see page*
	Too much sea .... UV		*58*
	Very heavy sea .... UW		Second of latitude .... QHY
	Was getting ready for sea .... NZF		Second of longitude .... QZH
	'What is the sea like? .... UX		
	When the sea has gone down ...... LMP	VMH	SECRET
	When will you be ready for sea? ... UKB	VMI	—Secret message
		VMJ	—Secret order
VLK	SEAL (*Animal*)	VMK	Secretly Secrecy
			The following communication (signal,
VLM	SEAL-ED-ING TO SEAL.		*or*, message) is private (*or*, secret),
	Open sealed orders .... SUC		HX
VLN	—Sealed orders		The following signal is in cipher and
			secret * .... IQV
VLO	SEAM-ED ING		To be kept secret .... PZY
	SEAMAN .... VLG		**The cipher will be continued until ended*
	Able seaman A B .... CXE		*by the person signaling making the*
	Brigade of seamen Naval Brigade HIA		*hoist IQS = cipher is ended*
	Distressed seaman .... LHG		
VLP	—In want of seamen	VML	SECRETARY-SHIP
	Lloyd's Seaman's Almanac .... QWZ		Admiral's secretary .... DKO
	Merchant seaman .... RPY		Colonial Secretary .... JBM
VLQ	—No seamen to be had		Military secretary .... RUA
	Ordinary seaman .... SVD	VMN	—Secretary of Legation
	Registrar General of Shipping and Sea-		Secretary of Lloyd's .... QXA
	men .... UNX		Secretary of the Navy's, *or*, Admiralty
	Seamen's and Firemen's Union .... NAR		flag .... DLS
		VMO	—Secretary of State—for
	SEAPORT VLC		Secretary of the Admiralty .... DLX
VLR	SEARCH-ED-ING TO SEARCH	VMP	—Secretary to (*or*, of)
VLS	—Have been searched	VMQ	—Secretary's office
VLT	—Have you been searched?		
VLU	—Have you searched?	VMR	SECT

	SECTION—SEED		

VMS	Section-al		See—*Continued*
	Midship section ------------------RTK		Have you seen the light? .. ----- QRV
			Hoist your flags where best seen ---NDE
VMT	Secure-ly-ity		How did the land bear when last seen?
VMU	—Are you secured?		EH
VMW	—Have, or, has been secured		How far can light be seen?---------MSZ
VMX	—Have you secured?		I can not distinguish (or, see) ---- LGZ
VMY	—I have secured.		I can not run, I must see it out----VBZ
VMZ	—Is it sufficiently secure?	VNP	—I have not seen
	Mast is secured--------- --------RMH	VNQ	—I have not seen the land
VNA	—Secure d-ing To secure	VNR	—I have seen
VNB	—Shall, or, Will secure (or,,be secured)		I saw it very plainly Plainly distin-
			guishable --------------------LHA
VNC	See-ing To see	VNS	—I saw the land bearing — (and time, if
	Admiral, or, Senior officer wishes to see		necessary)
	you (or person indicated) ------ . DLA		I saw war vessel disguised off —. .LEN
	Admiral, or, Senior officer wishes to see	VNT	—I saw — (vessel indicated) on —
	you (or, —) as soon as convenient,		I see the land Land in sight (place in-
	DLB		dicated, if necessary) -------- ---QDW
VND	—Can upon, or, See------ --- --------HTB		I shall stand in for the land as long as I
	—Can, or, Could see		can see. ---------------- -------QDZ
VNE	Can see the mark (or beacon)------RLI		I wish to see you before you go on shore,
	—Can, or, Could you see?		OCN
	Can you see the mark (or, beacon)?.GAF	VNU	—I wish to see you on shore
	Can not see the mark------------ RLJ		In what latitude did you see? ------QHR
	Clearly seen ---------- ----- -----ITL		Neither seen nor heard--- -------ORM
	Desire to see (or, speak)—to------KWN	VNW	—Nothing in sight
VNF	—Did not see (or, observe)		Passed, or, Saw some war vessels ..TCR
VNG	—Did you see (or, observe)?		Pirate reported (or, seen) off —----TKR
	Did you see a brig?----------------HGW		Samples can be seen --------------VGU
VNH	—Did you see a ship on the (or, off the)?		Saw, Seen --------------------VOA
	Did you see a vessel in distress?------ND		Saw boat (indicate place and time if
VNI	—Did you see anything of — (ship indi-		necessary) --------- --- ----GUE
	cated)?		Saw derelict in latitude —, longitude —,
	Did you see her capsize?-----------HVX		JH
VNJ	—Did you see the signal?		Saw enemy's ship (or, cruiser) off —(or
	Do you intend to see?- -------------POR		in latitude and longitude indicated),
VNK	—Do you see the land? (direction indi-		KFD
	cated if necessary)		Saw enemy's ship (or, cruiser) to the —,
	Extinguish all lights Allow no lights		KFE
	to be seen ----------------------DZB	VNX	—Saw land
	Flags seen, but signals not understood,		Saw the enemy's fleet off —-------- OK
	WCX		Saw the enemy's fleet steering —-- LZS
	Foresee-n-ing ---------------- ---- NKB		Saw the fleet (or, squadron) off —, steer-
	Glad to see—it (or, that) ----------OAS		ing —. ---------------------- NET
	Glad to see you (or, person named)-OAT		Saw torpedo boat (or, boats) (number,
	Have (or, has) been seen (or, observed),		if necessary) at — (or, near —) -- XT
	SNF	VNY	—See a ship — (indicate direction)
	Have not seen the captain --------HWZ	VNZ	—See smoke in — (direction indicated)
	Have seen the captain------ ----HXA	VOA	—Seen Saw
VNL	—Have you seen?		Seen nothing of your boat---------GUF
	Have you seen a (or, the) collier?--JAO		Seldom seen--------- --------VON
	Have you seen a derelict vessel?-----JG		Should you see (or, hear)----------ORN
	Have you seen a pilot boat (or, vessel)?	VOB	—To see you
	TJ		Vessel seen (or, indicated) is —----VXU
VNM	—Have you seen (or, spoken) any men of-		Vessel seen (or, indicated)lost her masts,
	war? If so, report their names and		RBF
	nationalities		Vessel seen is a privateer---------VXW
	Have you seen any torpedo boats?...XR		We did not see the light .. -------QSO
	Have you seen any vessels since your		What land did you see? What land do
	departure? - --------------------KTU		you take it to be?----------------QEU
	Have you seen (or, heard) of a vessel	VOC	—When did you see?
	wrecked (or, in distress)?----- ----ZU		When did you see the land?-------- RX
	Have you seen (or, heard) anything of		When first seen-----------------NBM
	my boat? ----------------------FB		When last seen------------------QGY
	Have you seen disabled steamer? ...MV	VOD	—When ought I to see (or, sight) —?
	Have you seen Lloyd's agent? -----QWJ	VOE	—Wish to see you
	Have you seen the captain?--------HXB		
	Have you, or, Has —(vessel indicated)	VOF	Seed
	seen the convoy? ----------------JUQ		Cotton seed --------------------JXT
VNO	—Have you seen the land?		

SEEK—SEND

VOG SEEK-ING SOUGHT

VOH SEEM ED ING
It seems-ed (or, appears-ed) ....ESB

SEER (*Measure of capacity*) ....AZQ

VOI SEINE (*Net*)

VOJ SEINING PARTY

VOK SEIZE-D-ING URE
VOL —Has been seized
Vessel and cargo have been seized..HXU
Will be certainly seized (or, captured), HXW
Will perhaps be seized ....HXY
You run the risk of seizure (or, capture), HXZ

SELDOM NOT OFTEN ....SPY
VOM —Seldom happens
VON —Seldom seen

VOP SELECT-ED-ING-ION

SELF-ACTING........ ....DIM

VOQ SELL-ING TO SELL (*See also* SOLD and SALE)
An opportunity for selling (or, disposing of) — ....SRV
VOR —Can you sell (or, let me have) any —?
Is the cargo sold? ....IAU
Must be sold............ ....WJE
Sell cargo as soon as possible........IAY
VOS —Sold
Sold by auction ....FLU
The cargo is not yet sold . ...IBA
VOT —Vessel will be sold
Where would my cargo sell? ......IBM
Would my cargo sell here (or *at place indicated*)? ....IBP
Wreck will be sold ....ZJB

VOU SEIVAGE

VOW SEMAPHORE SEMAPHORE STATION
VOX —I am going to semaphore to you
Repeat the semaphore signal.... .WCQ
You may work the semaphore ......VO

VOY SEMIDIAMETER

SEN (*Com*)..........AUZ

SENATE (or, *House of Lords*) ....OZR

VOZ SENATOR-IAL

VPA SEND
Account sent is correct ....DET
Account sent is not correct ....DEU
An opportunity for sending letters to Europe (or, to —)..........QOD
Anything to send..........EQN
Better not send..........GIN
Bring-to, I will send a boat..........HJX
Can I send my sick to hospital (or, sick quarters)? ....OYQ
VPB —Can send
Can send boat..........GSH

SEND—*Continued*
Can you get (or, send) me a pilot? Is there a pilot to be had?..........TIW
VPC —Can you send?
Can you send a boat? ....GSI
Can not send a boat ....EX
Demand has been sent..... .....KTA
VPD —Do not send
Do not send boats ashore after dark..GST
Endeavor to send a line by boat (cask, kite, or, raft) ....KA
Great risk in sending a boat..........GSZ
VPE —Have, or, Has sent
VPF —Have you sent?
Heave-to, I will send a boat...... GTE
I am aground, send what immediate assistance you can ....CE
I am sending you some newspapers..SHR
I am sinking (or, on fire) send all available boats to save passengers and crew ....NO
VPG —I have not sent
VPH —I have sent
I send an order for —..........STY
I will send a boat..........GTM
I will send a reply ....URO
I will send you a pilot ....
VPI —Is, or, Are to be sent
Letters must be sent—to (or, by)..QOL
Lifeboat sent for ....QRJ
VPJ —May I send?
May I send a sick person to see your surgeon? ....WL
VPK —Must I send?
VPL —No means of sending
Opportunity to send ....SRZ
Orders have been sent..........SUJ
Pilot has been sent to you ....TJI
Send a boat (or, boats) (*number, if necessary*)..........FH
Send a boat within hail ....GUI
Send a duplicate ....LPG
Send a line..........QUT
VPM —Send a message
Send a responsible person on board..CZE
Send a responsible person on shore.TGV
Send a rope on board ....GQV
VPN —Send a steamer
Send a — dozen cases—of — ....IDE
Send a tug (*indicate if more than one*) to me (or *to vessel in direction pointed out*) ....YDF
Sent a water boat ....GUJ
Sending a water boat..........YVJ
Send account..........DFO
Send all available spars for — to —..FOK
Send an (or, your) account ....DFN
Sending, Sent an anchor and cable .EJN
Send an anchor and cable off immediately (*indicate if more than one*) ......EJM
Send an answer (or, reply) ....ENK
Send an armed boat... ....GUK
Send an escort—to (or, for) ....MHD
Send an excuse ....MLR
Send an express with the following intelligence ....POH
Send an officer—to (or, for)........SPO
VPO —Sending, Sent an officer — to (or, for)
Send ashore—to (or, for)— ....FCZ
VPQ —Send back
Send bag—for — ....FSM

SEND—SENIOR

SEND—*Continued*

Send boat—at	GUL
Send boat—to (*or*, for)	GUN
Send boats fit (*or*, suitable) for landing passengers	FI
Send boat for telegram	GUM
Send boat to assist	FGN
Send boat to point out reef (rock, *or*, shoal)	GUO
Send boat to take off the crew	GUP
Send boat to tow	GUQ
Send boat with hawser	GUR
Send by post	TQP
Send by rail	UGX
Send carpenters on board—(*ship indicated*)	IBY
Send cot	IXO
Send descriptions	KWF
Send dinner—for	LAV
Send flats to discharge cargo	IAZ
Send following message by post to owners (*or*, to Mr —) at —	WY
Send following message by signal letters through the telegraph, instead of at length	XB
Send following message by telegraph to owners (*or*, to Mr —) at —	XA
Send following message through the post to owners (*or*, to Mr —) at —, by signal letters, instead of writing it at length	WZ
Send following telegram to owners (*or*, to Mr —) at —	XA

VPR	—Send for
	Send for despatches (*or*, letters) — JW
VPS	—Send for my letters
	Send for orders — SUM
VPT	—Send for passenger
VPU	—Send prisoner
VPW	—Send for sand
	Send for the towline — QUV
VPX	—Sending for the towline
VPY	—Send for water
	Send immediate assistance — DB
	Send immediately—to (*or*, for) — PDL
	Send instructions — PNK
	Send lifeboat to save crew — FJ
	Send list of defects — KRF
	Send me a boat from the shore (*a shore boat*) — GUH
	Send me (*or vessel indicated*) a pilot, TJQ
	Send me a police officer Want police YN
VPZ	—Send more
	Send muskets (rifles) to party on shore, SCQ
VQA	—Send my keys (*or keys indicated*)
	Send my message through the telegraph by signal letters — XB
	Send my orders on to—(*place indicated*), SUN
	Send on board—to (*or*, for) — GQW
	Send on shore—to (*or*, for) — FCZ
VQB	—Send passenger
VQC	—Send pilot—to — (*vessel indicated*)
	Send receipt—to (*or*, for) — ULE
VQD	—Send reply to my message to signal station at —
VQE	—Send report—of
VQF	—Send returns—of
VQG	—Send samples—of

SEND—*Continued*

	Send ship's articles — RZJ
	Send steamer — VPN
VQH	—Send telegram
VQI	—Send tents on shore
VQJ	—Send the —
	Send the duplicate — LPH
	Send the following articles — EZH
	Send the letter bag on shore (*or to vessel indicated*) — FSN
	Send the mate—off — RMS
VQK	—Send the steward
VQL	—Send to — (*ship or place indicated*) for — •
	Send to the Admiral's office — DLK
	Send to the hospital — OYV
	Send tug (*indicate if more than one*) to me (*or to vessel in direction pointed out*) — YDP
	Sending, Sent tug — YDQ
	Send your address to — — DJO
	Send your demands—to (*or*, for) — KTE
	Send your despatches — KXB
	Send your letters — QR
VQM	—Sender
VQN	—Sending. To send
	Sent — VRD
	Sent by post — TQR
	Shall I send message by telegraph? RQS
VQO	—Shall I send you a tug?
	Surgeon, *or*, Doctor shall be sent for, LJW
VQP	—Telegraph to my owners to send my letters to meet at —
	Want you to send me a pilot — TIW
	We have sent for lifeboat — FL
	We will endeavor to send a line — QUW
	Who sent the orders? — SUW
VQR	—Will be sent
VQS	—Will send
	Will send boat — GVI
VQT	—Will send for a pilot
	Will send you a pilot — TJG
VQU	—Will you send?
VQW	—Will you send for him (*or*, it, *or*, her)?
	Will you send me some newspapers? SHT
	Will you send off lighters as fast as possible? — RC
VQX	—You are to send—to (*or*, for)
	Your letters were sent by the — QOS

VQY SENIOR-ITY

	According to seniority — DEN
	Admiral *or*, Senior officer has commands for you — DKP
	Admiral, *or*, Senior officer has no commands for you — DKQ
	Admiral, *or*, Senior officer has requested, DKR
	Admiral, *or*, Senior officer intends DKS
	Admiral, *or*, Senior officer is generally about — DKT
	Admiral, *or*, Senior officer is off (*or* at) DKU
	Admiral, *or*, Senior officer is on board (*or*, at —) — DKV
	Admiral, *or*, Senior officer is on shore, DKW
	Admiral, *or*, Senior officer wishes DKY

	SENIOR—SET		

SENIOR—*Continued*
Admiral, *or*, Senior officer wishes to
know ------------------------------DKZ
Admiral, *or*, Senior officer wishes to see
you (*or person indicated*) --------DLA
Admiral, *or*, Senior officer wishes to see
you as soon as convenient to yourself,
DLB
Admiral's, *or*, Senior officer's memo. DLC
Admiral's, *or*, Senior officer's order. DLE
Have you not seen the admiral (*or*, senior
officer)? ----------- ---- ----------DLG
Have you seen the admiral (*or*, senior
officer)? ---------------------------DLH
Mr —, senior --------- ---- -----RWH
Senior commander -- --------------JEO
Senior lieutenant ----------- ------QPY

VQZ —Senior officer
The admiral (*or*, senior officer) requests
the pleasure of your company to din-
ner --- ----------- -------------DKX
What is the name of the admiral (*or*
senior officer)? --------- ------DLO
Where is the admiral (*or*, senior officer)?
SJ

VRA SENNIT

VRB SENSE-IBLE-Y.

VRC SENSELESS-LY

VRD SENT   (*See* SEND)

VRE SENTENCE-D-ING

VRF SENTRY   SENTINAL.

VRG SEPARATE-D-ING-ION
Better separate -------------------GIO
If we separate we may have a chance of
escape ---------------------- ----MGY

VRH —Separator
VRI —Was, *or*, Were separated
VRJ —When did you (*or*, they) separate?

VRK SEPTEMBER
About September the— -----------CZV
VRL —Beginning of September
VRM —End of September
VRN —Next September

SER   (*Measure of capacity*)----'-----AZF

SER   (*Measure of weight*)-----------BAQ

SERGE   BLUE SERGE --- ------------GOP

VRO SERGEANT
VRP —Sergeant-major

VRQ SERIES

VRS SERIOUS-LY-NESS
Badly *or*, Seriously wounded----- FRL
Boiler burst, no one seriously hurt AW
Boiler leaking seriously. - --------RF
Damage serious (*particulars to follow,
if necessary*) -------------------KIZ
How many have you seriously (*or*,
dangerously) ill? ------------,----KLD
I am very seriously damaged--------BN

SERIOUS—*Continued.*
I heard that some of your family were
seriously ill ----------- ----------MSI
VRT —Is it serious?
Nothing serious -----------------SKE
Serious attack --- -----------------FKA
Serious disaffection in —--- -------LCB
VRU —Seriously ill
Stem seriously damaged -- ------WUK
VRW —There has been a serious riot at —
Vessel seriously damaged, wish to trans-
fer passengers ---------------------HM

SERGEANT ------------ ------VRO
Sergeant-major -------------------VRP

VRX SERVANT
Officer's servant --------- --------SPN

VRY SERVE-D-ING

VRZ SERVITUDE
Penal servitude--------------------TFI

VSA SERVIAN   SERVIAN COLORS

VSB SERVICE
Active service ---- ------- - - - DIZ
Army Service Corps ----- ----------EWF
Certificate of service--- ----------IHU
Civil Service ---------------------IRQ
Divine service --------------------LIH
Had only — boilers serviceable ----GXA
Have only one (*or number shown*) boiler
serviceable --------------------GXB
His, *or*, Her Majesty's Service-----RHP
It will be of great service (*or*, use) . OHL
On His (*or*, Her) Majesty's Service. RHS
People going on service to take their
dinners with them --------------LAU
Public service ---- ------------ UAW
Sea service--------- -----------VLR

VSC —Serviceable   (*Useful*)

VSD SESSION

VSE SET-TING   TO SET
Current sets offshore -------------KFZ
Current sets on shore-------------KGA
Defects may be set right in — hours. BD
Ebb sets ------------ ------------LSM
Engines can be set right in a short time,
MBC
Flood sets ----- ----------------NFM
Had the monsoon (*or*, trade wind) set
in? --- -------------------------RYK
Has winter set in at — ?-----------ZGF
How will the ebb set? - ----------LSO
How will the flood set? ---- -------NFQ
Hurricane not set in at — (*date to fol-
low, if necessary*) ------------- PAM
Make more sail   Set your — ------RIL
Make, *or*, Set sail ------------RIT
Rollers setting in ----- -------- ----UT
Set sail   Make sail ------------RIT
Set studding sails -----------------VEO
Set the courses ----- ------------JZU
Set topgallant sails-------------VEP
Set — sails--------------------VEQ

VSF —Set up rigging
Tide, *or*, Current sets to the — -----KGE

SET—SHALL

	SET—*Continued*	SHACKLE—*Continued*	
	What is the set and rate of current (*or*, tide)?..........................KGJ	Unshackle-d-ing ....YKV	
VSG	—Will set in	Unshackle cables........ ....HBZ	
	Winter has regularly set in at - ..ZGK	Veer more cable, *or*, Veer to number of shackles specified...................KU	
	With all sail set.......................VFA		
	With some assistance I shall be able to set things to rights ... .DF	VTG	SHADE-D-ING Y
			The temperature in the shade is — XNH
VSH	SET　(*A set of*)		What is the temperature in the shade? XNM
VSI	—Set of flags		
		VTH	SHAFT
VSJ	SETTLE-D-ING-MENT　TO SETTLE　(*See also* ARRANGE)		Crank shaft ....KBP
	Affairs appear more settled (look more peaceful) ....DOZ		Crank shaft broken (*or*, defective)..KBQ
			Engine shaft..........................MBA
	Amicably settled (*or* arranged) ....EFG		Flaw in shaft.........................NRH
VSK	—Have, *or*, Has been settled		Intermediate shaft................PQV
	Have you settled with the Emigration Office... ....LXK		Main shaft broken.... ....MR
			Main shaft broken in stern tube ...RQT
	Is all settled (*or*, arranged)?........EXD	VTI	—Must have a new shaft
	Is the appointment settled? .. ..ESY		Propeller shaft broken................MR
	Settle everything..... ....MJP		Radius shaft ....UGM
			Screw shaft . ... VKI
VSL	SETTLEMENT　(*A colony*)		Shaft bearing ....GBP
		VTJ	—Shaft broken
VSM	SETTLER		Shaft coupling bolts broken ....GYF
		VTK	—Shaft trunk
VSN	SEVEN TH-LY		Spare shaft.................... ....WMF
VSO	—Seventeen-th		Starting shaft ....WRH
			Tail end of main shaft broken .....LYO
VSP	SEVERAL-LY		Tail shaft . ...XIZ
	Boiler burst, several men killed ....AY		Thrust shaft..... . ...XRM
	Boiler burst, several men wounded..AZ		Thrust shaft broken..................XRN
	Several engagements.... ....MAH		Tunnel shaft .... YDW
	Several lives lost ....QRG		Tunnel shaft broken . ...YDX
VSQ	—Several missing		
	Several ships. ....VXN	VTL	SHAKE N-ING　SHOOK
	Several (*or number indicated*) sick on board ....GQX		Shake the reefs out........ ....UMP
	Several vessels becalmed off — ....GCJ		SHAKU　(*Measure of length*)........AWU
VSR	—Several vessels in sight		
VST	—Several years		SHALL (*or*, WILL)........... ....CIR
VSU	—Spoke several vessels during the passage		He, She, It (*or person-s or thing-s indicated*) shall (*or*, will).............BVH
			He, She, It (*or person-s or thing-s indicated*) shall (*or*, will) be...........BVI
VSW	SEVERE-LY-ITY		He, She, It (*or person-s or thing-s indicated*) shall (*or*, will) not ....BVJ
	Crops have suffered severely......KDW		He, She, It (*or person s or thing-s indicated*) shall (*or*, will) not be ......BVK
	Heavy, *or*, Severe gale ....NVZ		How shall (*or*, will) he (she, it, *or person s or thing s indicated*)? ......BXB
VSX	—Not severe		
	Severe engagement..................MAI		How shall (*or*, will) I?.............BXS
	Severe fracture ....NPK		How shall (*or*, will) we?..........BXT
VSY	—Severe loss		How shall (*or*, will) you?..........BXU
	Severe squall not far off, look sharp.FZ		I shall (*or*, will) ....CIO
	Severely bruised ....HKJ		I shall (*or*, will) be ....BGH
			I shall (*or*, will) do ....BND
VSZ	SEW-ED-ING		I shall (*or*, will) have...........BRQ
VTA	—Has she sewed?		I shall (*or*, will) if I can ....BKS
			I shall (*or*, will) not.............CIP
VTB	SEWER-AGE		I shall (*or*, will) not be.........BGI
			I shall (*or*, will) not do ......BNE
VTC	SEX		I shall (*or*, will) not have ...BRS
			If he (*it, or person s or thing-s indicated*) shall (*or*, will)............BYP
VTD	SEXTANT		If he (*she, it, or person-s or thing-s indicated*) shall (*or*, will) not ....BYQ
VTE	SHACKLE		
	Shackle bolt ....GYE		If I shall (*or*, will)............BZN
VTF	—Shackling		If I shall (*or*, will) not ....BZO
	Shorten in cable to the number of shackles indicated................HRP		If they shall (*or*, will) ....CAQ

SHALL—SHANK

SHALL—*Continued*
If they shall (*or*, will) not _____CAR
If we shall (*or*, will) _____CBJ
If we shall (*or*, will) not_____CBK
If you shall (*or*, will)_____CDA
If you shall (*or*, will) not _ _____CDB
It shall (*or*, will)_____CIQ
It shall (*or*, will) be _____BGT
It shall (*or*, will) be done _ _____BNT
It shall (*or*, will) not be _____BGU
It shall (*or*, will) not be done_____BNU
Nothing shall _____ . _____SKF
Shall _____ _____CIR
Shall, *or*, Will be_____BHA
Shall, *or*, Will do (*or*, be done)____BNW
Shall, *or*, Will have ___ _____BRW
Shall, *or*, Will have done (*or*, been
    done) _____BNX
Shall, *or*, Will he (she, it, *or person s
    or thing-s indicated*)?_____BWD
Shall, *or*, Will he (she, it, *or person s
    or thing indicated*) be?_____BWE
Shall, *or*, Will he (she, it, *or person-s
    or thing s indicated*) do (*or*, be done)?
                                       BWF
Shall, *or*, Will he (she, it, *or person-s
    or thing-s indicated*) have?___ __BWG
Shall, *or*, Will he (she, it, *or person-s
    or thing-s indicated*) not? _____BWH
Shall, *or*, Will he (she, it, *or person-s
    or thing s indicated*) not be?____BWI
Shall, *or*, Will he (she, it, *or person s
    or thing-s indicated*) not do (*or*, be
    done)?_____BWJ
Shall, *or*, Will he (she, it, *or person-s
    or thing-s indicated*) not have?__BWK
Shall, *or*, Will his (her-s, it s)?_____CIS
Shall, *or*, Will I (my, mine)?_____CIT
Shall, *or*, Will I be?_____ _____BHC
Shall, *or*, Will I have_____BRX
Shall I make?_____RIU
Shall, *or*, Will I not have?_____BEY
Shall, *or*, Will it._____BWD
Shall, *or*, Will not _____CIU
Shall, *or*, Will not be?_____BHD
Shall, *or*, Will not do (*or*, be done)__BNY
Shall *or*, Will not have_____ _____BBZ
Shall, *or*, Will not have done (*or*, been
    done) _____BNZ
Shall, *or*, Will that (*or*, this)?_____CIV
Shall, *or*, Will the?_____CIW
Shall, *or*, Will there be?_____BHE
Shall, *or*, Will these (*or*, those)? ___CIX
Shall, *or*, Will they (their-s)? _____CIY
Shall, *or*, Will they be?_____ ___BHF
Shall, *or*, Will they do it?_____BOA
Shall, *or*, Will they not?_____ _____CIZ
Shall, *or*, Will they not do it?_____BOC
Shall, *or*, Will we (our-s)? __ _____CJA
Shall, *or*, Will we be? _____BHG
Shall, *or*, Will we not?_____CJB
Shall, *or*, Will weigh_____ZAX
Shall, *or*, Will you-r s?_____CJD
Shall, *or*, Will you be?_____ _____BHI
Shall, *or*, Will you do it?_____BOD
Shall, *or*, Will you moor?_____RZF
Shall, *or*, Will you not?_____CJE
Shall, *or*, Will you not be?__ _ ____BHJ
Shall, *or*, Will you not do it?_____BOE
That, *or*, This shall (*or*, will) be____CLN
That, *or*, This shall (*or*, will) not be_CLO

SHALL—*Continued*
There shall (*or*, will)_____CMU
There shall (*or*, will) be _____CMV
There shall (*or*, will) not _____CMW
There shall (*or*, will) not be _____CMX
They shall (*or*, will) _____ _____CJF
They shall (*or*, will) be _ _ _____BHZ
They shall (*or*, will) not _____CJG
They shall (*or*, will) not be_____BIA
We shall (*or*, will) _____CJH
We shall (*or*, will) be _____ _____BIR
We shall (*or*, will) have ____ _____BSU
We shall (*or*, will) not _____CJI
We shall (*or*, will) not be_ _____BIS
We shall (*or*, will) not have_____BSV
What, *or*, Which shall (*or*, will)—I?_CQK
What *or*, Which shall (*or*, will) I (*or*,
    we) do?_____ _____ ____CQL
What, *or*, Which shall (*or*, will) I (*or*,
    we) have _____CQM
What *or*, Which shall (*or*, will) you
    (*or*, they) do?_____ _____CQN
What *or*, Which shall (*or*, will) you
    (*or*, they) have? _____CQO
When shall (*or*, will)?_____ _____CRQ
When shall (*or*, will) be_____CRS
When shall (*or*, will) he (she, it *or*, per-
    son-s or thing-s indicated)? _____CRT
When shall (*or*, will) I (*or*, we)?____CRU
When shall (*or*, will) I (they, we, or,
    you) do?_____CRV
When shall (*or*, will) I have?_____CRW
When shall (*or*, will) it be done?____CRX
When shall (*or*, will) they?_____CRY
When shall (*or*, will) they (or, you)
    have? _ _____CRZ
When shall (*or*, will) we have?____CSA
When shall (*or*, will) you?____ _ ___CSB
When shall (*or*, will) you be?_____CSD
Where shall (*or*, will)? _____CTK
Where shall (*or*, will) be?_____CTL
Where shall (*or*, will) he (she, it, *or*
    person-s or thing-s indicated)? _ _CTM
Where shall (*or*, will) I (*or*, we)___CTN
Where shall (*or*, will) I have?_____CTO
Where shall (*or*, will) they?_____CTP
Where shall (*or*, will) they (or, you)
    have?_____CTQ
Where shall (*or*, will) we have?____CTR
Where shall (*or*, will) you?_____CTS
Where shall (*or*, will) you be?_____CTU
Who shall (*or*, will)?_____ZEF
Why shall (*or*, will)?_____CVR
Why shall (*or*, will) not?_____OVS
You shall (*or*, will) _____CJK
You shall (*or*, will) be _____BJK
You shall (*or*, will) have _____BTF
You shall (*or*, will) not _ _____CJL
You shall (*or*, will) not be_____BJL
You shall (*or*, will) not have _____BTG

VTM	**SHALLOW ING-EST** (*See also* **SHOAL**)
VTN	—Is the water shallow?
VTO	—Shallow water—is
VTP	—Shallowest water is —
VTQ	—The water is shallow
VTR	**SHAME-FUL-LY**
	SHANK **SHANK OF ANCHOR** _____EHG

SHAPE—SHIP

VTS	SHAPE-ING		
	Irregular shape----------------PTE		SHELL (*Bomb*)----------------GYE
VTU	—Not under command shape		I can supply you with powder and shell----------------------TRH
	N B —*The "not under control or command" signal is two black balls, or shapes vertical*		Powder and shell----------------TRI
		VWF	—Shell and shot
VTW	—What shape—is?	VWG	SHELL (*A covering*)
VTX	SHARE-D-ING TO SHARE	VWH	—Shellfish
VTY	—Share (*A share*)	VWI	SHELTER-ED-ING
VTZ	—Shareholder		Could she (*vessel indicated*) have got to any place of shelter?----------OFD
VUA	—Shares in the —	VWJ	—Good shelter from
VUB	SHARK		I shall put in for shelter----------UCV
VUC	SHARP-LY-NESS	VWK	—I shall try for shelter in — (*port indicated*)
	Heavy weather coming, look sharp--FZ	VWL	—Is there any shelter?
	Keep a sharp lookout----------------PYI		Put in for shelter, or, I shall put in for shelter------------------------UCV
	Severe squall not far off, look sharp--FZ	VWM	—Shelter may be obtained
VUD	SHATTER-ED-ING TO SHATTER	VWN	—There is good shelter at (*or, in*).
		VWO	—There is no shelter
	SHE (*See* HE, SHE, IT ETC)--------BWL	VWP	SHELVE-D-ING SHELF
	She is (*or, has*) not yet----------ZLO		
	What is she like?----------------QTY		SHENG (*Measure of capacity*)------AZM
VUE	SHEARS		
VUF	SHEATH-ED-ING	VWQ	SHERIFF
VUG	—Sheathing for bottom (*number of sheets to follow*)	VWR	SHERRY
			SHEW (*See* SHOW)----------------WAL
VUH	SHEAVE	VWS	SHIELD-ED-ING.
	Eccentric sheave out of order------LSY	VWT	SHIFT-ED-ING TO SHIFT.
VUI	—Sheave hole		Ballast, *or,* Cargo has shifted------FTN
VUJ	SHED		Bar has, *or,* Sands have shifted----FVU
VUK	SHEEP	VWU	—Did not shift
VUL	SHEER (*Curve of the hull*)	VWX	—Have, *or,* Has shifted
VUM	Sheer strake		I shall shift my berth----------------GFS
VUN	SHEER-ED-ING OFF TO SHEER OFF		I would shift my berth as soon as possible----------------------GFT
VUO	—Sheer a little off		If, *or,* When the wind shifts------ZFE
VUP	—Sheer a little off when you come to the point		Is it worth while shifting berth?---GFV
			May I shift (*or, change*)?----------IKT
VUQ	SHEERS (*For lifting*)	VWY	—Must shift
VUR	—Sheer hulk		Sands have, *or,* Bar has shifted----FVU
			Shall I, in my present berth, have room for weighing if the wind shifts?--GFZ
VUS	SHEET (*Rope*)		Shift a course----------------JZV
	Fore sheet----------------------NJB		Shift a topsail----------------XWE
	Head sheet----------------------OQL	VWZ	—Shift of wind.
VUT	—Jib sheet		Shift your berth—further to the —-GEC
	Main sheet----------------------RGU		Shift your berth further in--------GHA
	Royal sheet----------------------VBA		Shift your berth further out------GHB
VUW	—Sheet home		Shift your berth Your berth is not safe------------------------------KT
VUX	—Staysail sheet		
VUY	—Tacks and sheets	VXA	Shifting sand----------------------VHE
VUZ	—Topgallant sheets		—Will shift
VWA	—Topmast staysail sheet		SHILLING (*Coin*)----------------AVB
VWB	—Topsail sheet		Coals — shillings a ton at —--------GZ
VWC	SHEET (*A sheet or piece of*)	VXB	SHINE ING SHONE
VWD	—Sheet lead	VXC	SHINGLE LY
	SHEET ANCHOR----------------EJQ	VXD	SHIP-PING (*See also* VESSEL)
	SHEET CABLE------------------HRO		Armed merchant ship----------BUW
VWE	SHELF PIECE		Armored ship----------------EVY

SHIP

	Ship—*Continued*		Ship—*Continued*
	Aviso, or, Despatch vessel --------KWY		Torpedo catcher (or, chaser, or, destroyer)----------------------XWR
	Barque ------------------------------FXO		
	Barquentine .. ------------------FXP		Torpedo cruiser------- ----------XWS
	Battle ship------------------------FYX		Torpedo gunboat_ -------------- XWU
	Brig------- --- -------------------HGV	VXF	—Training ship
	Brigantine ------------------------HID		Transport --- ----------------------YAH
	Clipper ---------------------------IUO		Troopship --- --------- ----------YBM
	Coast-defense ship----- .. --------IXV		Tug .. _. ---- -- ------ ----------YDR
	Coasting-trade ship Coasting vessel ----------------------------IYB		Turretship ------------------------YES
			Twin-screw steamer ------------YFL
	Collier ------ - ------------------ IWZ	VXG	—Whaler Whaleship
	Corvette----------------------------JXI		Yacht---------------------------------ZKI
	Cruiser _ . ------------- ---------KEY		A vessel (or *ship indicated*) aground near (or, in) — ------------ ---- CH
	Dhow ------------------------------KYO		
	Enemy's ship (or, cruiser) . -------LZK		Abandon the ship as fast as possible.AB
	Felucca -------- -- ----------------MVO		Ask stranger (or *ship indicated*) if he will communicate ----- ----------HT
	Fire ship (or, raft) ---------------MZW		
	First ship--------------- ---------- NBG		*Asks* name of ship (or, signal station) in sight ----------------------------VG
	Fishery steamer --------------------WTG		
	Five-masted ship---- .. ----------MXP		Athwart ship --- ---------- ----FJH
	Flagship ---------------------------NCZ		Board steamer (or, ship)----------GPV
	Foreign-going ship -----------------NKQ		Bring the ship in as close as possible, HIO
	Four-masted bark-------------------FXQ		
	Four-masted schooner ------------NPC	VXH	—By the next ship
	Four-masted ship-------------------NPD		Cannot save the ship, take people off, NB
	Frigate -------- -------------------NRT		
	Full-rigged ship----------------- ---NSU		Cargo must be shipped-------------IAC
	Government-hired vessel ----------OFU		Chamber of Shipping --------------IKB
	Government ship ---- -----------OFW		Clean ship --------------- --------IST
	Guard ship ------------------------OIZ		Could she (ship *indicated*) have got to any place of shelter?-- - ------OFD
	Guarda costa------------------ -- OJB		
	Gunboat -------- -- -------------GTB		Crew have left the ship---- -------KCP
	Gun vessel ---------------- -----OJV		Crew will not leave the ship-------KCX
	Hired cruiser------------- ----KFB		Deadworks of a ship ----------KOL
	Hired steamer---------------------WTH		Defend ship ---- ------ ----------KRM
	His, or, Her Majesty's ship--------REQ		Did you leave many ships at — ?---QKP
	Hospital ship----------------------OYS		Did you (or *ship indicated*) receive any damage?---------------------KJF
	Ironclad Armor-clad ship--------EVS		
	Ketch---- -------------------------QAE		Did you see a ship in distress? -------ND
	Large ship (*Estimated tonnage to follow*) -------------------------QFP		Did you see a ship on the (or, off the)? VNH
	Light-ship Light-vessel----------QTG		Did you see anything of —(ship *indicated*)?-------------------- ----VNI
	Mail steamer ----------------------RFK		
	Man-of-war ------ .-----------------RPL		Do not abandon the ship ------------AC
	Merchant ship (or, steamer)-------RPZ		Do not abandon the ship until the tide has ebbed ---------------------AD
	Mortar vessel --------------------SAQ		
	Passenger ship ------------ ----------TDQ		Do not have anything to do with stranger (or *ship indicated*) --------WYR
	Petroleum ship --------------------THG		
	Pirate ----------------------------TKP		Do you think we ought to pass a ship in this distressed condition?---------NE
	Privateer-------------------- ----TWL		
	Rear ship --------------- ----------UKN		Does embargo extend to ships under flag?------------------------------LWC
VXE	—Sailing ship		
	Salvage steamer ------------------VGF	VXI	—Dress ship
	Schooner --------------------------VJI		Each ship--- ---- - --------------LQY
	Scout, or, Lookout vessel----------QZX		Enemy's ship (or, cruiser)----------LZK
	Screw steamer --------------------VKJ		Flagship ---------------------------NCZ
	Ship of war -- --------------------RPL		Foreign-going ship--- --- ----------NKQ
	Slaver ------------------------ ----WGB		From the guardship ----------- ---OIX
	Sloop --------------------------- ----WHB		Funnel of ship is — ----------------NTP
	Small ship Small craft (*Estimated tonnage to follow*)- ------- ----KBC		Goods are shipped ----------------OET
			Has she (or *ship indicated*) been long abandoned?------- ---- -- ..CXN
	Square-rigged ship ---------------WPM		
	Steamship Steamer ----- ------ WTE	VXJ	—Have, or, Has shipped
	Steam trawler --------------------WTI		Have you had any communication with infected ship (or, place)?----------IC
	Steel-built ship ------------------WTL		
	Storeship----- --------------------WXT		Have you met any ship of war?----MRN
	Tank steamer--------------------XKL		Have you passed (or, spoken) any ships (or *ship indicated*)---------------WOE
	Telegraph ship -------------------XMD		
	Three masted schooner ----------VJK		Have you seen any ship (or, ships) since your departure?----------------KTU
	Timber ship ----------- ---- XTG		

	SHIP	

	SHIP—*Continued*	
	Have you seen (*or*, heard) anything of a ship wrecked (*or*, in distress)?...ZU	
	Have you the Navigation Laws, *or*, Merchant Shipping Act of the country of vessel signaling?.. .... DIG	
	Her, *or*, His Majesty's ship ....RHQ	
	How are you, *or*, How is — (*person or ship indicated*)? .... ---BXI	
	How does — (*ship indicated*) sail?.VDX	
	How is she rigged?.......... ....UXH	
	How is (*or*, was) stranger (*or ship indicated*) steering? ....WTY	
	How many days since you passed — (*ship indicated*)? ....----------KNH	
VXK	—How many ships?	
	I belong to the convoy of — (*ships indicated*)..... ... .JUS	
	I do not intend to abandon the ship.-AF	
	I have sunk (*or*, run down) a ship....HK	
	I intend to ship —............... POV	
	I must abandon the ship............AG	
	I shall abandon my ship unless you will keep by me ....AH	
	I will not abandon you (*or ship indicated*) or, I will remain by you....AI	
	I (*or*, They) wish to abandon, but have not the means...................AJ	
	Indicate the name of ship (*or*, signal station) in sight..... ....VG	
	Intend to abandon ship to Lloyd's agent ....CXP	
	Is not ship seen (*bearing* —) in distress?........NT	
	Is she (*or ship indicated*) abandoned? AK	
	Is the — light-ship on her station?.WRO	
	Is your list of ships correct?........QVI	
	Is your ship entered at the Customhouse?.. ....KGZ	
	Is — (*ship indicated*) likely to supply me?...................FGK	
	Keep a good look out, as it is reported enemy's ships are going about disguised as merchantmen............OJ	
	Keep by the ship ....PYN	
	Merchant Shipping Act............DIK	
	Must lighten the ship ....QSY	
	Natives do not like ships watering without payment ....YWQ	
VXL	Neutral ship　　Ship is a neutral....SRC	
VXM	—Not yet shipped	
	—Of how many ships was the squadron seen composed?	
	Passed a good many ships... ....TCO	
	Passed a war ship disguised, funnel (*or*, funnel-s) — (*Here give the color or design of the line it may be an imitation of*) ....LEO	
	Passed disabled ship at —........LEW	
	Passed, *or*, Saw some war ships ...TCR	
	Police (*or other*) authorities have taken some of the crew out of the ship KDH	
	Remain by the ship - .... KH	
	Repeat ship's name, your flags were not made out...............DU	
	Report when ship denoted is sighted ....USD	
	Reship-ped-ping...................UTM	
	Saw enemy's ship (*or*, cruiser) off — (*or* in —, *latitude or longitude indicated*), KFD	

	SHIP—*Continued*	
	Saw enemy's ship (*or*, cruiser) to the —, KFE	
	See a ship — (*Indicate where*)...VNY	
	Send a steamer.................VPN	
VXN	Send ship's articles ....EZJ	
	—Several ships	
	Several ships in sight................VSR	
	Shall I bring my ship's papers with me?HJF	
	Ship and cargo have been seized...HXU	
	Ship chandler ....IKM	
	Ship disabled, will you tow me? (*into port indicated if necessary*)........MY	
	Ship has no signals ....WDO	
	Ship has not made her number.....SLH	
	Ship having the guard.............OJA	
VXO	—Ship in sight is	
	Ship in sight is an enemy ....OL	
VXP	—Ship indicated — is —	
	Ship indicated appears in distress...NX	
	Ship indicated (*bearing, if necessary*) appears to be...... ....ESH	
	Ship indicated bears...... . ...GBL	
	Ship indicated can lend you carpenter's tools....................IBZ	
	Ship, *or* Person indicated declines.KQJ	
	Ship indicated does not require further assistance ....FGQ	
	Ship indicated has been wrecked ...ZIX	
	Ship (*indicated, if necessary*) has got off, NZE	
VXQ	—Ship indicated has struck on a shoal	
	Ship indicated is (*or*, I am) in want of —, YTN	
	Ship indicated is in quarantine ....UEH	
	Ship indicated is not abandoned ....CXR	
	Ship indicated is on fire.............OT	
VXR	—Ship indicated is out of sight	
VXS	—Ship indicated is sinking	
VXT	—Ship indicated is (*or*, was) standing (*or*, steering)	
	Ship, *or*, Person indicated is unable to comply........................JLG	
	Ship indicated to examine strange ship and report ....MKF	
	Ship indicated wants boat ....FK	
	Ship indicated (*or in direction indicated*) wants immediate assistance. NW	
	Ship is a foreigner ----- ....NKV	
	Ship is insured ....YPX	
	Ship is on the rocks ....UZA	
	Ship (*or ship indicated*) is very crank, KBT	
	Ship, *or*, Vessel may part company.JGU	
VXU	—Ship seen (*or*, indicated) is —	
	Ship seen (*or*, indicated) is a native craft ---------------------SEG	
VXW	—Ship seen is a privateer	
	Ship will be sold ....VOT	
VXY	—Ship yard　　Ship-building yard	
VXZ	—Shipment	
VYA	—Ship-ped-ing　　To ship	
	Shipped a sea....................VLJ	
	Shipped on board - ....GQY	
VYB	—Shipping office	
	Ship's bottom foul................HAW	
	Ship's bottom must be looked at...HBG	
	Ship's head by compass—is — .....JIE	
	Ship's husband..................PAV	

SHIP—SHOE

VYC	SHIP—*Continued* —Ship's log Ship's papers (*or, articles*)..........BZK

VYD	—Ship's side
	Ships from — did not get pratique
	  when I left ... .TSA
	Ships in the roadstead.............YPZ
	Ships indicated are (*or*, were) ready for
	  sea..... ....UJZ
	Ships just arrived show your ensigns,
	                 DY
	Ships just arrived make your numbers
	  (*or*, distinguishing signals)......DX
	Ships just weighed (*or* leaving) make
	  your numbers (*or*, distinguishing sig-
	  nals) ....DZ
	Ships that wish to be reported make
	  your distinguishing signals .... EA
	Ships wishing to be reported all well
	  make your numbers (*or*, distinguish-
	  ing signals) ....DYT
	Ships wishing to be reported on leaving
	  make your numbers (*or*, distinguish-
	  ing signals) ....EB
	Spoke several ships during the passage,
	                VSU
	Spoke the — (*ship indicated*)..... WOG
	Steer directly for the light-ship... QTH
	Strange ship hove in sight..........OZS
	Tell my owner ship answers remarkably
	  well ....ENO
	The captain must bring the ship's papers
	  with him ....HXJ
	The channel is obstructed by a ship,
	                ILT
	The light-ship — is not at anchor on her
	  station, *or*, Light-ship at — is out of
	  position ....QE

VYE	—The ship at (*or*, in)
	The ship's bags are sent ashore.....ISO
	The way is off my ship, you may feel
	  your way past me.............. HI.
	Waterlogged ship ....YWU
	What is the name of light-ship (*or*
	  lighthouse) in sight?.............RW
	What is your ship's (*or vessel indicated*)
	  registered tonnage?.............UNV

VYF	—What ship? What ship is that (*bearing indicated, if necessary*)?....................EC What ship takes the next mail?...RFY
VYG	—What ships did you speak? When was your ship last coppered? JVU Who is captain of ship signaled?..HXN
VYH	—Will your ship stay? You will be stopped by the blockading ships ....GNR

VYI	SHIPPER
	SHIPPING FEDERATION ....MUY
	"SHIPPING GAZETTE AND LLOYD'S LIST,"
	                QWU
	Lloyd's Register of British and Foreign
	  Shipping ....QWY
	"Lloyd's Weekly Shipping Index,"
	                QXD
	Report me by telegraph to "Shipping
	  Gazette"......................UF

VYJ	SHIPPING MASTER (*or*, OFFICER)
	Commissioner of Navigation, Treasury
	  Department, *or*, Registrar-General of
	  Shipping and Seamen ....UNX
	SHIPPING OFFICE................ ....VYB

VYK	SHIPWRECK (*See also* WRECK) Have shipwrecked crew on board, will you let me transfer them to you?.KDA Shipwrecked Mariners' Society ....RLF
VYL	—Shipwrecked
VYM	SHIPWRIGHT Skilled shipwright................WFE Want shipwright..................YTZ
	SHIP YARD (*See also* YARD).........VXY
VYN	SHIRT Clean shirts (*or*, linen)..........ISU
VYO	—Dirty shirts (clothes, *or*, linen)
VYP	SHIVER ED ING Keep main topsail shivering ......PZJ Keep mizzen topsail shivering.....PZK Sails shivering ....VEN
VYQ	SHOAL (*of fish*)

VYR	SHOAL (*See also* REEF)
	Apparently shoal water to the — ERO
	Breakers, Reef, Rock, *or*, Shoal ahead
	  of—you..................... HFB
	Breakers, Reef, Rock, *or*, Shoal on your
	  port bow.....................HFD
	Breakers, Reef, Rock, *or*, Shoal on your
	  starboard bow ... ...HFE

VYS	—Have you shoaled your water? I am in danger (*or*, shoal water), direct me how to steer ....NL
VYT	—I am in shoal water Is there a buoy (*or*, mark) on shoal (*or*, on —)?..................... ...HMO
VYU	—On the shoal Reef, *or*, Shoal extends from — to —.UMS

VYW	—Reef, *or*, Shoal is steep to the —
	Reef, *or*, Shoal stretches a long way
	  out ....GE
	Send boat to point out shoal.......GUO
	Shoal water, *or*, Danger in — (*direction
	  indicated*) ....QF
	Shoal water, *or*, Danger to leeward,
	                KLH
	Shoal water, *or*, Danger to windward,
	                KLI
	Stand nearer the shoal (*or*, bank, *or*,
	  reef) ....FWA

VYX	—The shoal is buoyed

VYZ	—To shoal Shoal ed-ing
	Vessel indicated has struck on a shoal,
	                VXQ
	Water is shoaling (*Depth in feet at
	  last cast may be shown*).... YWK
	Water shoaling, I must tack ....YWM

VZA	—You will get into shoal water, *or*, You will shoal your water
VZB	SHOE-ING SHOD Boot, *or*, Shoe. ....HAB
VZC	—Shoemaker

SHONE—SHORT

	SHONE SHINE-ING ---------------- VXB
	SHOO (*Measure of capacity*) -------- AZR
VZD	SHOOT-ING SHOT
	Ash shoot . ------- -------------- FCT
VZE	SHOP PING
VZF	SHORE (*The shore*)
	All my papers are sent from shore__ TAE
	All quiet—at (*or*, in —, *or*, on shore), DXT
	Along shore-. -------------------- EAN
	Along shore to the eastward of —__EAO
	Along shore to the northward of —_EAP
	Along shore to the southward of — EAQ
	Along shore to the westward of—__EAR
	Am on shore, likely to break up, require immediate assistance-------- CA
	Are you going on shore?----- -- _ FOX
VZG	—Been on shore
	Boat is on shore Boats are on shore, GRO
	Bring on shore ----- -------------- HIU
	Calls attention of shore signal station on the bearing (FROM *the person signaling*) pointed out by compass signal ------- -------------- DO
	Captain is on shore-------------- HWM
	Come on shore ---------------- JCU
	Current sets offshore -------------- KFZ
	Current sets on shore-------------- KGA
	Deepest water is nearest the shore_KQV
	Disturbance on shore (*or*, at —)--- LHS
VZH	—Driven on shore near — Drove on shore off —
	Endeavor to get a line ashore by a boat (cask, kite, *or*, raft, etc)--------- KA
	For the shore ------ ------- _ ---NHP
	Fruit is to be obtained on shore----NSE
	Get her head round or you will be on shore ------------------------ LF
	Go on shore-------------------- OBX
	Have no communication with the shore (*or vessel indicated*) -------------- HZ
VZI	—Have you been on shore?
VZJ	—Have you done with the shore?
	Have you had any communication with the shore?----- ---- ---------- JGA
	Heave-to, head offshore_ -------- LQ
	I am a stranger here; will you let me go ashore with you?------------- OCH
	I am going on shore ------------ OCK
	I am not going on shore----------- OCM
VZK	—I have been on shore
	I have left on shore----------- ---- QKT
VZL	—I have not been on shore
	I wish to see you before you go on shore ------ --------------------- OCN
	I wish to see you on shore--------- VNU
	In shore ----------- ---- -------- PMG
	Is the captain coming on shore?---HXD
	Keep more toward the shore-- ----- PZO
	Keep on port shore (*or*, side) of channel----------------------- LV
	N B—*Port side of the ship signaled to*
	Keep on starboard shore (*or*, side) of channel ------------------- LW
	N B—*Starboard side of the ship signaled to*

	SHORE—*Continued*
	Lee shore---- ------------------QMG
	Let him bring on shore (ashore)---FCY
	Lights, *or*, Fires will be kept at the best place for coming on shore----KE
	Mate is on shore ------- --------RMQ
	No fruit is to be obtained on shore_NSF
	Offshore--- ----------------------IYD
VZM	—On shore
	Person indicated is wanted on shore_TGS
	Run on shore--------- --------------VCE
	Scarcely prudent to land Not easy landing on the beach ---- --------FZX
	Send a responsible person on shore_TGV
	Send me a boat from the shore (*a shore boat*) --------------------------GUH
	Send on shore—to (*or*, for)---------FCZ
	Send rifles to party on shore--------SCQ
	Send tents on shore----- ----------VQI
	Send the letter bag on shore (*or to ship indicated*)----------------------FSN
	Shore boat --- ----- ------------GUS
	Some squabble (*or*, fight) on shore with crew ---- --------------------MXU
	Suitable boats for landing must be sent off from the shore----------------FI
	The admiral (*or*, senior officer) is on shore.--------------------------DKW
	They are(*boats*)making for the shore_RIV
	Weather shore --------------------YZR
	When did you come off?----------JDN
	When do you go on shore?---------ODC
	Will you be on shore to day? (*Specify hour, if necessary*) ------ -------KOC
	Will you (*or person indicated*) come on shore to-day? (*Specify hour, if necessary*) ------------------------JDB
	Will you give me a passage (*on shore*)? TDK
	Will you go (*or*, come) on shore?---JDS
VZN	—Will you go on shore in my boat?
	Will you (*or person indicated*) meet me on shore to day? (*Specify time*)__KOD
VZO	SHORE-D ING UP To SHORE UP
VZP	Lose no time in shoring up----------CF
	—Will you require shoring up?
VZQ	SHORT-NESS
	A short crop of —----------------KDU
	A short reach --------------------UIV
VZR	—Are you short of —?
	Have been on short allowance for some time---- ---- --------------- DZI
	Heave short ------------ --- ----HRD
	I am getting short of water and must endeavor to get same ----------YVG
VZS	—I am (*or vessel indicated* is) short of provisions
	I shall shorten sail ----------------VEB
VZT	—In a short time
	Make short tacks ------------------RIP
	Medical comforts running short----ROI
	Medicines have run short----------ROQ
	Not able to make such short tacks__XIL
	Not short enough ---------------MDT
	Only for a short—----------------SQX
	Short allowance------- -- --------DZN
VZU	—Short circuit
	Short manned, crew weak (*Number may be indicated*) --------------KDJ

SHORT—SHOULD

	SHORT—*Continued*
VZW	—Short notice
VZX	—Short of
	Short of complement ....JKE
VZY	—Short of powder
	Short of provisions Starving......NV
WAB	—Short of water
WAC	—Short range
	Short time ....XTP
	Short weight Under weight......YIA
WAD	—Shorten-ed-ing To shorten
	Shorten in both hawsers_ ....OPX
	Shorten in cable—to the number of
	shackles indicated.............HRP
	Shorten in port hawser (*See page* 35)
	Shorten in starboard hawser (*See*
	page 35)
	NOTE —*For signals between ships towing*
	and being towed see page 35 .
	Shorten sail...............VET
	Shorten sail It is not safe to go so fast,
	VDK
WAE	—Shorter est
	Shortly after...............DRT
	Too short ....XVF
	Wear short round............YXT
WAF	**SHOT**
	By a shot ....HPA
	Case shot..................IDA
	Gunshot ....OJU
	I am in want of powder and shot ..TRF
	I can supply you with powder and shot,
	TRH
	Musketshot.................SCP
	Round shot.................VAM
	Shot and shell ....VWF
WAG	—Shot plug
WAH	—Shotted
	SHOULD, or, WOULD ....CJO
	He, She, It (*or person-s or thing-s indi-*
	cated) should (or, would).........BVL
	He, She It (*or person-s or thing-s indi-*
	cated) should (or, would) be ... BVM
	He, She, It (*or person-s or thing s indi-*
	cated) should (or, would) do (or, be
	done) ....BVN
	He, She, It (*or person-s or thing-s indi-*
	cated) should (or, would) have ..BVO
	He, She, It (*or person-s or thing-s indi-*
	cated) should (or, would) not....BVP
	He, She, It (*or person-s or thing s indi-*
	cated) should (or, would) not be.BVQ
	He, She, It (*or person-s or thing-s indi-*
	cated) should (or, would) not do (or,
	be done) ....BVR
	He, She, It (*or person-s or thing s indi-*
	cated)should(or, would)not have.BVS
	How should (or, would)?... ...BXV
	I should (or, would) ....CJM
	I should (or, would) be..........BGJ
	I should (or, would) do.........BNF
	I should (or, would) have ....BRT
	I should (or, would) have been ....BGK
	I should (or, would) not.........CJN
	I should (or, would) not be ....BGL
	I should (or, would) not do........BNG
	I should (or, would) not have .....BRU
	I should(or, would)not have been..BGM

	SHOULD—*Continued*
	If he(she, it, *or person-s or thing-s indi-*
	cated) should (or, would) ....BYR
	If he(she, it, *or person-s or thing-s indi-*
	cated) should (or, would) not ....BYS
	If I should (or, would) ....BZP
	If I should (or, would) not ....BZQ
	If I were you I should not do it ....LIQ
	If they should (or, would)... .....CAS
	If they should (or, would) not CA1
	If we should (or, would) ....CBL
	If we should (or, would) not ......CBM
	If you should (or, would) ....CDE
	If you should (or, would) not.......CDF
	It, etc (*See* He, She, It)
	It is important that it should be so.PEW
	Should, or, Would.............CJO
	Should, or, Would be.........BHK
	Should, or, Would do (or, be done).BOF
	Should, or, Would have ....BSA
	Should, or, Would have been......BHL
	Should, or, Would have done (or, been
	done) ....BOG
	Should, or, Would have had ......BSC
	Should. or, Would he (she, it, *or per-*
	son-s or thing-s indicated)? ......BWM
	Should, or, Would he (she, it, *or per-*
	son-s or thing s indicated)? be?....BWN
	Should, or, Would he (she, it, *or per-*
	son-s or thing-s indicated) do (or, be
	done)?...................BWO
	Should, or, Would he (she, it, *or per-*
	son-s or thing-s indicated) have?.BWP
	Should, or, Would he (she, it, *or per-*
	son-s or thing s indicated) have been?
	BWQ
	Should, or, Would he (she, it, *or per-*
	son-s or thing s indicated) not?..BWR
	Should, or, Would he (she, it, *or per-*
	son-s or thing-s indicated) not be?.BWS
	Should, or, Would he (she, it, *or per-*
	son-s or thing-s indicated) not do (or,
	be done)?.................BWT
	Should, or, Would he (she, it, *or per-*
	son-s or thing-s indicated) not have?
	BWU
	Should, or, Would he (she, it, *or per-*
	son-s or thing s indicated) not have
	been?....................BWV
	Should, or, Would his (her-s, it-s)?.CJP
	Should, or, Would I (my, mine)?...CJQ
	Should, or, Would not...........CJR
	Should, or, Would not be ... ...BHM
	Should, or, Would not do (or, be done),
	BOH
	Should, or, Would not have ....BSD
	Should, or, Would not have been ..BHN
	Should, or, Would not have done (or,
	been done) ....BOI
	Should, or, Would not have had ...BSE
	Should, or, Would that (or, this)? .CJS
	Should, or, Would the? ....CJT
	Should, or, Would there be? .... BHO
	Should, or, Would these (or, those)?.CJU
	Should, or, Would they (their-s) ..CJV
	Should, or, Would we (our-s)? ... CJW
	Should, or, Would you-r-s? ....CJX
	That, or, This should (or, would) be,
	CLP
	That, or, This should (or, would) not
	be ....CLQ

SHOULD—SICK

SHOULD—*Continued*
There should (*or*, would) _____CMY
There should (*or*, would) be_____CMZ
There should (*or*, would) have been_CNA
There should (*or*, would) not_____CNB
There should (*or*, would) not be___CND
There should (*or*, would) not have been,
    CNE
They should (*or*, would) _____CJY
They should (*or*, would) be _____BIC
They should (*or*, would) have _____BSL
They should (*or*, would) not_____CJZ
They should (*or*, would) not be ____BID
They should (*or*, would) not have _BSM
We should (*or*, would)_____CKA
We should (*or*, would) be _____BIT
We should (*or*, would) have _____BSW
We should (*or*, would) not _____CKB
We should (*or*, would) not be_____BIU
We should (*or*, would) not have ___BSX
What, *or*, Which should (*or*, would)?_CQP
What, *or*, Which should (*or*, would) I
  (*or*, we) do?_____CQR
What, *or*, Which should (*or*, would)
  you (*or*, they) do?_____CQS
When should (*or*, would)?_____CSF
Where should (*or*, would) the —?_CTV
Where should (*or*, would) the —?_CTW
Where should (*or*, would) they (their-s)?
    CTX
Who should (*or*, would)?_____CUV
Why should (*or*, would)?_____CVT
Why should (*or*, would) not?____ __CVU
You should (*or*, would) _____CKD
You should (*or*, would) be ____ ____BJM
You should (*or*, would) have _____BTH
You should (*or*, would) not _____CKE
You should (*or*, would) not be ____BJN
You should (*or*, would) not have__BTI

WAI | SHOULDER

WAJ | SHOVE-D-ING—OFF

WAK | SHOVEL

WAL | SHOW-ING   TO SHOW
Annul the hoist shown by Numeral
  Signal_____EMI
Beach the vessel where flag is waved
  (*or*, light is shown) _____FZT
Boats should endeavor to land where
  flag is waved (*or*, light is shown) _JZ
Do not show a light on any account_QV
Do not show a light on my account_QRT
Do not show your ensign ___ _____MEI
He has shown his colors _ _____JBT
I do not wish to show a light _____QRX
I will show a light   I will carry a light,
    QRY
I will show a light to-night when I alter
  course_____QW
I will show a light to-night when I
  make sail _____QRZ
I will show a light to-night when I tack,
    QX
Shall you show a light?_____QSK
She (*or* vessel indicated) has not shown
  her ensign _____ _____MET
She (*or* vessel indicated) has shown her
  ensign_____MEU

SHOW—*Continued*
Show the number (*or*, quantity)—of —,
    PIA
Show your ensign   Hoist your colors,
    DW
Show your Code Signal _____IZG
Show your number (*or*, distinguishing
  signal)_____DV
Show your soundings_____VS

WAM | —Shown, Showed
Temporary light shown at — (*place in-
  dicated*) _____XNS
Vessel has not shown her number_SLH
Vessel in sight has shown — colors__JBT
Vessel indicated is showing a light_QSN
Vessels just arrived, show your en-
  signs_____DY
Vessels just arrived, show your numbers
  (*or*, distinguishing signals)_____DX
Vessels just weighed (*or*, leaving), show
  your numbers (*or*, distinguishing
  signals) _____DZ
Vessels that wish to be reported all well,
  show your numbers (*or*, distinguishing signals) _____ _____DYT
Vessels wishing to be reported on leaving, show your numbers (*or*, distinguishing signals) _____EB
Vessels wishing to be reported show
  your distinguishing signals _____EA
What colors has he (*or*, she) shown
  (hoisted)?_____JBU
When I alter course to-night I will show
  a light_____QW
When I tack to-night I will show a
  light_____QX
When you alter course(*or*, tack) to-night
  show a light_____JZS
Will you show me your Greenwich (*or*,
  first meridian) mean time?* __ __XN

* *See note to Greenwich mean time
(page 258)*

WAN | SHOWER-Y

WAO | SHRINK-ING   SHRANK   SHRUNK

WAP | SHROUD
Futtock shroud _____NUP

WAQ | SHUNT ED-ING

WAR | SHUT-TING   TO SHUT
WAS | —Shut in,
WAT | —Shut-off valve
WAU | —Shut out

WAV | SIAM   SIAMESE   SIAMESE COLORS

WAX | SICILY-IAN

WAY | SICK   (*See also* ILL)
Any sick on board ?_____EPO
Any sickness at —? _____UY
WAZ | —Are there any sick nurses to be got?
Can I land my sick ?_____UZ
Can I send my sick to hospital (*or*, sick
  quarters)? _____ _____OYQ
Captain is sick _____HWO
Crew sick _____KCW
Disease, *or*, Sickness is contagious __VB

SICK—SIGNAL

	SICK—*Continued*			SIGHT—*Continued*
	Disease, *or*, Sickness is not contagious, ... VC			*Asks* name of signal station (*or*, ship) in sight .... VG
WBA	For the sick .... NHQ			Bills at sight .... GKH
WBA	—Have you had any sickness on board?	WBU		—Can you keep sight—of?
WBC	—How many sick?	WBV		—Can not keep sight—of
	I should wish to know the nature of the sickness, if any, before I send my boat (*or*, communicate) .... FO			Do not lose sight of it .... RAM
				Enemy is in sight .... OE
				Foresight .... NKC
	Is the disease (*or*, sickness) contagious? VA	WBX		—Have you sighted?
WBD	—Is the sickness on the increase?			How many torpedo boats passed in sight? .... GTK
WBE	—Many sick			I have just lost sight of the land (*or*, light) .... QDV
	May I send a sick person to see your doctor? .... WL			I have not sighted land .... VNQ
	Passengers, *or*, Crew sick with infectious complaint .... PCX	WBY		—I shall keep sight of —
	Sea-sick-ness .... VLF			I sighted land about — I sighted land bearing — (*time, if necessary*) ... VNS
	Several (*or number indicated*) sick on board .... GQX	WBZ		—In sight In sight of —
WBF	Sick berth .... GHD	WCA		—Is land in sight?
WBG	—Sick list			Just lost sight of the land (*or*, light) . QDV
	—Sick nurse			Keep me in sight during the night (*Bearing to follow, if necessary*) . PVV
	Sick quarters .... UEX	WCB		—Keep within sight—of
WBH	—Sickly			Land in sight I see the land ..... QDW
WBI	—Sickness			Lose-ing, Lost sight of — (*vessel or object indicated*) .... RAQ
	Sickness is contagious . ... VB			
	Sickness is not contagious VC			Lost sight of wreck .... RAY
	The prevailing sickness is — ... TVE			Nothing in sight .... VNW
WBJ	—Very sickly			Out of sight—of .... SWB
	What is the sickness? .... VD			Report when ship denoted is sighted . USD
	When did sickness break out? .. HEZ			Several vessels in sight .... VSR
WBK	—You had better get a sick nurse	WCD		—Shall you sight —?
				Ship in sight is — .... VXO
WBL	SIDE			Ship in sight is an enemy .. .. OL
	Both sides .... HAQ			Ship indicated is out of sight...... VXR
	By the side of . ... HPM			Sight your anchor Make sure your anchor is clear .... EJR
	Convoy to keep on port side of leader (*or*, escort) IX			Strange vessel hove in sight........ OZS
	Convoy to keep on starboard side of leader (*or*, escort) .... IY			Vessel in sight—is .... VXO
	Cylinder side rod .... KIF			Vessel in sight is an enemy .... OI
	Keep on the port side of the channel. LV			What is the name of lighthouse (*or*, light-ship, *or*, point) in sight?.... RW
	Keep on the starboard side of channel . LW			What is the name of ship (*or*, signal station) in sight? .... VG
	Keep on the — (*compass signal*) side of the channel .... ILQ			When did you sight land? .... RX
	Lee side .... QMH			When ought I to see (*or*, sight)? ... VOD
	On both sides .... HAR			— (*number*) torpedo boats (*give nationality*) passed in sight... ... XYA
	Port side .... TPA			
	Ship's side .... VYD	WCE		SIGN-ING TO SIGN
	Side lever engine.............. MAW			Are your bills of lading complete (signed)? .... GKQ
WBM	—Side light (lantern, *or*, lamp)			Bills of lading not signed........... GKR
	Starboard side .... WQZ			I will not sign the bills of lading unless —, GKW
	The other side............... SVO	WCF		—Not signed
WBN	—This side			Signed articles .... EZL
	Weather side .... YZS	WCG		—Signed
WBO	—Which side?	WCH		SIGNAL
	Wrong side .... ZKD			Acknowledge, *or*, Answer my signal (*or*, message) .... DGO
WBP	SIDEWAYS			Alphabetical signal No 1, Code Flag over E*
WBQ	SIDING			Alphabetical signal No 2, Code Flag over F*
WBR	SIEGE			Alphabetical signal No 3, Code Flag over G*
WBS	SIEMENS-MARTIN STEEL			
WBT	SIGHT-ED-ING TO SIGHT (*See also* OB-SERVE)			*For manner of using these signals see page 18*
	A squadron (*give nationality*) composed of — ships passed in sight. ..... WPF			
	Anything in sight? .... EQK			

SIGNAL

SIGNAL—*Continued*

Annul the hoist shown by numeral signal .... EMI
Annul the last hoist, I will repeat it .VE
Annul the last signal　The last signal is annulled .... ---EMJ
Annul the whole signal............VF
Asks name of ship (*or*, signal station) in sight...........................VG
Attend to signals ....DG
Calls attention of shore signal station on the bearing pointed out by compass signal (FROM *the person signaling*) .DO
Calls attention of vessel whose distinguishing signal will immediately be shown ... .DR
Can you furnish (*or*, lend me) a Code of Signals?............. .. ...IZE
Can you make out her flag (*or*, signal)? ....NCW
Can you spare some rocket signals? .UZO
Can not distinguish your signal flags, come nearer (*or*, make distant signals), ...VM
Can not make out the flags (*or*, signals), ...NCX
Communicate by distant signals ...LGC
Compass signal in degrees...........JHI
Compass signals in points or half points, ...JHK
Comply with request (*or*, signal)...JLD
Convoy to spread as far as possible, keeping within signal distance...........IZ

WCI —Could you understand their signals?
Could you understand what they were signaling about?
Did boat (*or*, boats) take blue light (lantern, *or*, any means of making a signal)? ....GSP
Did you see the signal?..... .....VNJ
Discontinue repeating signals.......LDG
Distant signals　(*When spoken of*) .LGE
For Distant Signals see page 539
Distress Signal　(*When spoken of*) ....WCY

　　(*See page* 7)

Do, *or*, Did you understand their signals?......................WCI
Enquires name of ship (*or*, signal station) in sight.................. ...VG
Fire a signal gun ....MZP
Flag signal-ing ....NDA
Flags seen, but signal not understood, ...WOX

WCJ —Fog signal
Following signal is private (*or*, confidential)........................HX
Following signal (communication, *or*, message) is secret and in cipher ..IQV

　　(*The cipher will continue until the person signaling hoists* IQS)

Forward answer by telegraph to signal station at —..................EMW
Forward following telegraphic message by signal letters instead of writing it at length................ ....XB
Forward reply to my message by telegraph to —......................WT
Gun fog signal ....OJS
Has vessel signaled her number? ...SLF

SIGNAL—*Continued*

Have you any signal lights?........QRU
Helm signal.... ....OTQ
Hoist the signal in a better position, I can not make out the flags..... ..OWL
Hoist your distinguishing signal.....DV
I am going to communicate by distant signals.... ....LGF
I am going to repeat signal from — (*hoist indicated*) ....VH
I am going to signal the contents of an important telegram to be communicated to you ....IE
I can not make out the flags, hoist the signal in a better position.......OWL
I can not make out the bottom flag .NDF
I can not make out the second flag .NDG
I can not make out the signal .....NCX
I can not make out the third flag. .NDH
I can not make out the top flag.....NDI
I will attend to your signal.... ...FLB
I will repeat the message (*or*, what you say) ....RQN

WCK —I will repeat the signal
I wish to signal, come within easy signal distance ....VJ
Indicate name of signal station (*or*, ship) in sight..................VG
International Code of Signals.......PRB

WCL —Is there a signal station at —?
Keep within signal distance (*bearing, if necessary*) ....PZU
Last signal ....QGL
Last signal is annulled......... ...EMJ
Learnt by signal from — ....QKE
Light signal ....QSI
Lloyd's signal station—at —........QXB
Make a signal when you want a boat, ...GTU
Make distant signals ....LGH

WCM —Make distant signals—to (*or*, for)
WCN —Make signal for a pilot
WCO —Might have learnt it by signal
Night signals ... .SIF
Numeral signal No 1,
　　　　　Code Flag over M*
Numeral signal No 2,
　　　　　Code Flag over N*
Numeral signal No 3,
　　　　　Code Flag over O*

* *For method of making numeral signals see page* 32

Numeral signal (*when spoken of*) ..SLM
Open telegram for me and signal its contents .. ..WX
Pay attention to signals.............DG
Pay strict attention to signals during the night.........................DT
Pilot signal flying ....TJM
Private signal ... .TWK
Refer to the Geographical Signals .NYA
Repeat your signal —, *or*, Repeat your signal from —(*hoist indicated*)....VI

WCP —Repeat my signals until further orders
WCQ —Repeat the semaphore signal
WCR —Repeat the last signal made
Repeat your distinguishing signal, the flags were not made out...........DU
Repeat your signal in a better place .OWL

SIGNAL—SINK

	SIGNAL—*Continued*
WCS	—Report-ed-ing by signal.
	Run in for pilot (*or*, pilot vessel) and look out for their signals ....RAC
	Send following message through the post to owner (*or*, to Mr —) at — by signal letters.....................WZ
	Send my message through the telegraph by signal letters....... .. :........XB
	Send reply to my message to signal station at— ....VQD
	Shall I open your telegram and signal its contents?....................... XC
	Shall signal with siren (*or*, steam whistle) during fog (*or*, darkness) . WE
	Ship has not made her signal.......SLH
	Show your Code Signal. ....IZG
	Show your distinguishing signal .. DV
	Signal book ....GZF
WCT	—Signal gun
	Signal is answered............... ...VX
	Signal is not answered .... ENM
WCU	—Signal lantern (light, *or*, lamp)
WCV	—Signal letter
WCX	—Signal not understood, though the flags are distinguished
WCY	—Signal of distress
WCZ	—Signal of distress in — (*direction indicated*)
WDA	—Signal officer
WDB	—Signal rocket.
WDC	—Signal staff, or, Signal station
WDE	—Signal, or, I am going to signal state of the weather outside (*or at place indicated*)
WDF	—Signal the names of places to which you wish to be reported by telegraph—to
	Signal your vessel's official number..DV
WDG	—Signaled-ing To signal
WDH	—Signalman
	Socket signal ....WIX
WDI	—Sound signal
	Steer course indicated as per compass signal ....JIF
WDJ	—Stranger does not understand your signal.
WDK	—Stranger (*or vessel indicated*) will not take any notice of signal.
	Submarine signal.................ZNQ
	Submarine signal bell ....ZNR
	Submarine signal receivers.........ZNS
	Submarine signal station ....ZNT
	The following signal is in cipher and secret ....IQV
	The following signal (*or*, communication) is private....................HX
WDL	—The signal made
WDM	—The signals were not intended for you
WDN	—Use flashing (*light*) signals
WDO	—Vessel has no signals
	Vessel has not made her number (*or*, distinguishing signal).........SLH
WDP	—Vessel indicated is making signals of distress (*bearing, if necessary*).
	Vessels just arrived make your numbers (*or*, distinguishing signals)....DX
	Vessels just weighed (*or*, leaving) make your numbers (*or*, distinguishing signals) ....DZ

	SIGNAL—*Continued*
	Vessels wishing to be reported all well make your numbers (*or*, distinguishing signals) ....DYT
	Vessels wishing to be reported on leaving make your numbers (*or*, distinguishing signals) ....EB
	Vessels wishing to be reported make your distinguishing signals ....EA
	Vocabulary signal ....YRI
	What is the date of your signal book? VL
	Will signal with steam whistle (*or*, siren) during fog (*or*, darkness)..... ...WE
	Will you forward (*or*, communicate) the following signal for me?........IK
WDQ	—Will you repeat the signal (*or*, message)?
	Will you repeat the signals being made to me?........ ....VN
WDR	—You have misunderstood the signal.
	Your signal flags can not be distinguished .. ..LHB
	Your signal is not distinct, interpretation doubtful......................LGT
WDS	SIGNATURE
WDT	SIGNIFY-IED-ING-ICATION
WDU	SILENCE SILENT SILENTLY
	Make no noise. Keep silence ......RIN
WDV	SILICA-TE-D
WDX	SILK
WDY	SILL
	What is the depth of water over the sill? KVH
WDZ	SILTING, SILTED UP
WEA	SILVER.
WEB	—Nitrate of silver
	Quicksilver ....UFH
WEC	SIMILAR-LY-ITY
WED	SIMPLE-Y-ICITY SIMPLIFY IED-ING-ICATION.
WEF	SIMULTANEOUS-LY.
WEG	SINCE
	A fortnight since ....NMK
	Have you seen any vessel since your departure?........ .. ..KTU
	How long since (*or*, ago)? ....DTK
	Long time since (*or*, ago)..........DTL
	Not long since (*or*, ago) ....DTN
	Some time since (*or*, ago) ....DTP
WEH	SINCERE-LY-ITY
WEI	SINEW
WEJ	SING ING SANG SUNG
WEK	SINGLE-D-ING
	Form single column in line abreast.NLW
	Form single column in line ahead ..NLX
WEL	—Singly
	SINK-ING SANK To SINK.........VHI
	Boat sunk.....................GRQ
WEM	—I am sinking.
	I am sinking (*or*, on fire), send all available boats to save passengers and crew, NO

	SINK—SLIME.		
	SINK—*Continued*	WFM	**SKY**—*Continued*
	I have sunk (*or*, run down) a vessel..RK	WFN	—Skylight
WEN	—Sunk-en		—Skyrocket
	Sunken obstruction in (*or*, at) —...SNK		
WEO	—Sunk off (*or*, near) the —	WFO	**SLACK**
	Vessel indicated is sinking ....VXS		Bar, *or*, Entrance not safe except just
WEP	**SIR**		at slack water (*or*, at —, *time indi-*
			cated)......................FVT
	SIREN **STEAM SIREN** ....WSR	WFP	—Slacked-ing
	Blow siren (*or*, steam whistle) at inter-		Haul in the slack..............OPA
	vals ....WD	WFQ	—Slack away
	Shall signal with siren (*or*, steam whis-		Slack water..................YVK
	tle) during fog (*or*, darkness) ....WE	WFR	—Very slack
	Use your siren.. ....YMU	WFS	—When the tide slacks
WEQ	**SIROCCO**	WFT	**SLAIN** **SLAY**-ING **SLEW.**
WER	**SISTER**	WFU	**SLANDER**-ED-ING-OUS
	SIT-TING **SAT** ....VHL	WFV	**SLANT**-ED-ING
WES	**SITUATE**-D-ING **SITUATION**		A good slant ....ODT
	SIX TH-I Y ....WEU	WFX	**SLATE**
WET	**SIXTEENTH**	WFY	—Slate colored
WEU	**SIXTH** **SIX** **SIXTHLY**	WFZ	**SLAVE** D-ING-RY
WEV	**SIZE**-D-ING **TO SIZE**	WGA	—Slave trade
WEX	**SIZE**	WGB	—Slaver
	About the size...................CZY		**SLAY**-ING **SLAIN** **SLEW**...........WFT
	Different size ....KZI	WGC	**SLEDGE**-D-ING, *or*, **SLEIGH** ED-ING
	Indicate the size—of ....PIB	WGD	**SLEEP**-ING-Y **SLEPT**
	Large size ....QFR		Railway sleeper...............UHC
	None of that size...... ....SJC	WGE	**SLEEPLESS**-NESS
WEY	—Size unknown	WGF	**SLEET**-Y
	Small size ....WHP	WGH	**SLIDE** ING **SLID**
WEZ	—What size do you want?		D slide ....KIP
	SKETCH-ED-ING (*Draw*) ....LNG		Gun slide ....OJV
WFA	**SKELETON.**		Low-pressure slide jacket cover broken,
WFB	**SKID**		KAM
WFC	**SKIFF**	WGI	—Slide valve
WFD	**SKILL**-ED-FUL LY	WGJ	—Slide valve broken
	Skilled artificer..................EZS	WGK	—Slide-valve gear
WFE	—Skilled shipwright	WGL	—Slide-valve gear damaged (*or*, out of
WFG	**SKIN**-NED		order)
		WGM	—Slide valve out of order
	SKIN (*Hide*) ....OUQ	WGN	—Slide-valve rod
WFH	**SKIP**-PED-PING	WGO	—Slide-valve rod broken
WFI	**SKIPPER**	WGP	—Sliding faces
WFJ	**SKIRMISH**-ED-ING-ER	WGQ	**SLIGHT**-ED ING **TO SLIGHT**
	SKOINIS (*Measure of length*).......AWQ		A good many slight cases........ IDH
			Boiler burst, very slight damage..GVW
WFK	**SKULK**-ED-ING **SKULKER**		Boiler leaking slightly..........GWB
WFL	**SKY**		Cargo slightly damaged..........IAD
	Clear sky ....ITD		Machinery slightly damaged ..... MCQ
	Cloudy sky..................IWB	WGR	—Slight-ly
			Slight attack.................FKC
			Slight breeze ....HGD
			Slight case ....IDF
			Slight damage ....KJB
			Slight engagement ....MAJ
		WGS	—Slightly wounded
		WGT	**SLIME** Y

	SLING—SO	

Code	SLING—SO (left column)	Code	SLING—SO (right column)
WGU	SLING-ING　SLUNG		SMALL—*Continued.*
			Small craft ....KBC
WGV	SLING　(*A sling*)		Small hawser ....OPY
	Boat's sling ....GSC		Smallpox　Case of smallpox......IDN
	Fore yard gone in the slings........NJR		Small ship　Small craft............KBC
	Look to the slings...................RAB	WHP	—Small size
		WHQ	—Small span
WGX	SLIP-PING	WHR	—Smaller-est.
	Be all ready for slipping　Be ready to		Steward's small stores ....WVE
	slip your cable .... HQO		Stream, *or*, Small anchor............EJV
	Cut, *or*, Slip　Get an offing　Weigh		Supply of water is small, not exceeding
	KW		— (*quantity indicated*) a day...YWB
	Cut, *or*, Slip your cables............HRA		Too little, *or*, Too small............QVX
	I am obliged to slip my cable, pick it		
	up for me................. . ...HRF		
	I have slipped my anchor, but it is	WHS	—SMART-LY-NESS-ER-EST
	buoyed ....EIG	WHT	—Very smart-ly
	I intend to slip cable ....HRG		
		WHU	SMELL-ING　SMELT
WGY	—Slip rope		
WGZ	—Slip way	WHV	SMELT ED-ING
	Slip your cable and buoy the end..HRQ		
WHA	—Slipped	WHX	SMITH-Y
	Weigh, Cut, *or* Slip　Wait for nothing		Blacksmith ....GLV
	Get an offing ....KW		Chimney smith (*or*, *sweep*) ....IPB
			Coppersmith ....JVS
WHB	SLOOP		Goldsmith ....ODN
			Locksmith ....QXZ
WHC	SLOPE-ED-ING		Tinsmith ....XTZ
	Slope awnings ....FPY		Whitesmith..........................ZDN
	SLOPS　MEN'S CLOTHING..............IVS	WHY	SMOKE D ING-LY
			See smoke in — (*direction indicated*),
WHD	SLOW-LY-NESS		VNZ
	As slow as possible ....FBJ	WHZ	—Smoke joint
	Keep going ahead slowly...........DVM	WIA	—Smokeless powder
	My chronometer is slow of Greenwich		
	(*or*, first meridian) mean time....GS	WIB	SMOOTH-LY-NESS-ER-EST
	Slow progress ....TYC		—Smooth the sea with oil ....HTN
WHE	—Slow speed		
WHF	—Slower-est	WIC	SMUGGLE-D-ING
	Very slow-ly ....YPS	WID	—Smuggler
	Your chronometer must be slow ...IQM		
		WIE	SNAKE-D
WHG	SLUE-D-ING		
		WIF	SNAKE　(*Reptile*)
WHI	SLUICE-D-ING		
WHJ	—Sluice valve	WIG	SNATCH-ED-ING
WHK	—Sluice valves out of order		Snatch block ....GMY
	SLUNG　(*See* SLING)..............WGU	WIH	SNIFTING VALVE
WHL	SLUSH	WIJ	SNOW-ED-ING-Y
			Traffic interrupted by snow....... . XZP
WHM	SMACK-ED-ING		
	Fishing smack (*or*, *craft*)...........KAZ	WIK	SNUG LY-NESS
			All snug.............. ....DXY
WHN	SMALL-NESS		Make all snug.......................DYN
	A small establishment .... MHL		
WHO	—A small part (portion)—of —	WIL	SO
	A small (*or*, *smaller*) quantity (*or*, num-		I am (*or person indicated* is) trying to
	ber)—of — ....SLD		do so.... . ...YOQ
	Only to be got in small quantities. OFM		I think so ....XQG
	Small adjustment required to machin-		If so............... ....BZW
	ery ....DKG		Is it so?........ ... .CDU
	Small-arm magazine ....EVF	WIM	—Not so
	Small-arm men ....EVC		Not so often...... ....SPZ
	Small arms　(*Muskets, rifles, etc*).EVD	WIN	—Not so soon—as
	Small er boat ....GUT		Not so well as......... ....ZBM
	Small bower.......................EJS	WIO	—So as to
	Small case.... ....IDF		So much (*or*, *many*)................REQ

SOAK—SORT

WIP	SOAK ED ING		SOME—*Continued*
WIQ	SOAP		Some future opportunity .... NUS
	Marine soap .... RLC		Some more .... BZW
			Some other................. .... SVN
WIR	SOBER-LY-NESS SOBRIETY		Some swell .... XHJ
WIS	SOCIAL-LY ABLE-Y ILITY		Sometime ago (*or*, since) .... DTO
		WJP	—Somebody Someone
WIT	SOCIALIST		Someone else...................... LVW
			Someone to look after stock .... RAD
WIU	SOCIETY	WJQ	—Something
	Co operative society............... JVM		Something amiss................... EFM
	Shipwrecked Mariners' Society .... RLF	WJR	—Something else
		WJS	—Sometime
	SOCKS (*Stockings*) .... WVS	WJT	—Somewhat
		WJU	—Somewhere
WIV	SOCKET		Stop, Heave-to or Come nearer, I have
WIX	—Socket signal		something important to communicate,
			Code Flag over H
WIY	SODA IUM		There is some mistake .... RWG
WIZ	—Bicarbonate of soda		
WJA	—Nitrate of soda	WJV	SON
WJB	SODA WATER	WJX	SONG
WJC	SOFT-LY-NESS	WJY	SOON
	Soft bottom .... HBI		As soon as .... FBK
	Soft bread Fresh bread........... BEO		As soon as alongside .... EAV
			As soon as convenient .... FBL
WJD	SOIL-ED-ING		As soon as it is dark............. FBM
			As soon as it is dusk............. FBN
	SOL, or, DOLLAR (*Coin*)............. ATJ		As soon as possible .... FBO
			As soon as she is off (*or*, afloat)...DPW
	SOLD. (*See also* SELL *and* SALE) ..... VOS		As soon as the flood tide makes ....FBP
	Cargo expected to be sold........... HZV		As soon as there is a breeze........ FBQ
	Cargo sold well -- .... IAH		Discharge cargo as soon as possible.IAM
	Everything is to be sold. .... MJH	WJZ	Dispose of cargo as soon as possible.IAN
	Is the cargo sold?.................. IAU		—How soon?
WJE	—Must be sold		I shall weigh as soon as the weather
	Sold by auction -- .... FLU		permits .... ZO
WJF	—Sold her cargo		Not so soon—as . .. .WIN
	The cargo is not yet sold........... IBA		Return as soon as ....UVP
	Wreck will be sold .... ZJB		Sell cargo as soon as possible....... IAY
		WKA	—Sooner
WJG	SOLDIER		The sooner the better .... GIT
WJH	—Private soldier	WKB	—Too soon
		WKC	—Will soon
WJI	SOLITARY-ITUDE		
		WKD	SORE-LY-NESS
	SOLICITOR ATTORNEY .... FLO		Eyesore .... MQH
			Footsore .... NHF
WJK	SOLID-LY-ITY	WKE	SORROW
	Solid, *or*, Cubic measure (*See page* 52),		
	AYB	WKF	SORRY
	The ice is so solid, I can not break		Sorry am (is, or, are) engaged ....MAK
	through, send help................ PCF		Sorry I am unable to comply with your
WJL	SOLUTION SOLUBLE		request ....USX
			Sorry to hear—it (*or*, that) .... ORP
WJM	SOLVE-D-ING	WKG	—Sorry to inform you
		WKH	—Very sorry
WJN	SOLVENT-CY		
		WKI	SORT-ED ING To SORT (*See also* KIND)
WJO	SOME		All sorts .... DXZ
	By some neglect .... SGK	WKJ	—Sort (*Kind*)
	I heard that some of your family were		What sort (*or*, kind)—of? .... QAZ
	seriously ill...................... MSI		What sort of a light is it?.......... QSP
	I think you ought to get something to		What sort of a voyage (*or*, passage)
	answer the purpose................ ENB		have you had? .... TDI
	Some demand springing up......... KTF		What sort of an entrance is it (*or*, is
			there to —)? .... MFW

	SORTIE—SPARE		

WKL	SORTIE	WLN	SOVEREIGN-TY　(*Ruler*)
	SOUGHT　SEEKING ------------------VOG		SOVEREIGN, *or*, POUND (*coin*) ---------AUS
WKM	SOUND-ED-ING　TO SOUND	WLO	SOW-ED ING　SOWN
WKN	—Have obtained soundings		
WKO	—Have sounded regularly	WLP	SPACE-JOUS-LY　(*See also* ROOM)
WKP	—Have you had soundings?		Air space (*or*, cell) ---------------DWH
WKQ	—Have you sounded regularly?		
WKR	—I have sounded — (*place indicated*)　I	WLQ	SPADE
	know the soundings		
	I intend sounding — (*place indicated*)	WLR	SPAIN　SPANIARD　SPANISH COLORS
	before I try it ----------------POT		Bank of Spain--------------------FUS
	I shall heave-to to sound(*time indicated,*		
	if necessary) ----------- -------- OSL	WLS	SPAN-NED-NING.
	I shall sound regularly　I shall keep		
	the lead going------------------QJH	WLT	SPANNER
	Irregular soundings --------------PTF		
WKS	—It is not prudent to proceed without		SPANKER, *or*, DRIVER ----- -------LNX
	regular soundings		Brail up the spanker --------- -----HDT
	Keep the lead going　Keep sounding,	WLU	—Lower spanker topsail yard
	QJI		Spanker, *or*, Driver boom ----------GZW
	Out of soundings----------------SWC		Spanker boom sprung-- ------------GZW
	Patent sounding machine---------RDS		Upper spanker topsail yard -------XWG
WKT	—Regular soundings		
	Show your soundings ----- ---------VS	WLV	SPAR
	Sounding apparatus----------- ----- ERK		Can I get a spar for — mast at —?.NYH
WKU	—Sounding pipe (*or*, tube)		Can I get a spar for a fore yard at —?
WKV	—Sounding rod		NIS
WKX	—Soundings　Depth of water　(*Depth of*	WLX	—Can you supply a spar?
	water to be shown in feet)		Have you any spars that would do for —?
	Thomson's (Kelvin) sounding apparatus,		VX
	TEG	WLY	—I have no spars suitable—for —
WKY	—We have sounded regularly		Lookout will be kept for any rafts (*or*,
	What are your soundings? ----------VS		spars) --------- -------- ---------QZY
WKZ	—Will the soundings be a safe guide?		Send all available spars for — to —FOK
WLA	—You may go by your soundings if you		Small spar ----- ----- ----- -----WHQ
	use care	WLZ	—Spar buoy
			Spar, *or*, Outrigger torpedo - -- ---SWR
WLB	SOUND ED ING　TO SOUND　(*To make a*		Spare spar - ------------------WMG
	noise)		Spars are all adrift----- ------- --DMR
WLC	—Sound (*a noise*)		Spars for nets, *or*, Torpedo booms--GZY
	Sound signal---------------------WDI		Sprung *or*, Seriously damaged lower
WLD	—Sounding buoy		yard (*or spar indicated*)----------RCK
			Sprung spar (*indicated*)----- ---- WOR
WLE	SOUP		
	Preserved soup----- ------------TUN	WMA	SPARE-D ING
			Can spare (*or*, dispense with) ------LFD
WLF	SOURCE		Can spare ammunition------------EFS
	Where *or*, What source did the news		Can you spare (*or*, dispense with)?-LFE
	come from? --- ----------------SHN		Can you spare (*or*, supply me with) — ?
			XET
	SOUTH　(*Compass in degrees*)---------AIO		Can you spare a hawser?------ -----OPS
			Can you spare ammunition? -------EFT
	SOUTH　(*Compass in points*)-- ------ARL		Can you spare any biscuit? --------GLM
WLG	—Southern erly-ing-ward		Can you spare me a chart of —? - .IMC
	Alongshore to the southward of — (*dis-*		Can you spare me a chronometer...IQD
	tance in miles may follow) -------EAQ		Can you spare me a hand or two?..OMC
WLH	—From the southward		Can you spare me coal?------- ---- GY
	On *or*, At the South coast of — ...IYK		Can you spare me provisions? -- --UAK
	South Atlantic --------------------FJM		Can you spare me some rocket signals?
	South cone hoisted ----------- -----JNO		UZC
	South latitude- --------------------QHZ	WMB	—Can you spare some rope (*size to follow*)?
	South Pacific---- ----- --- ---------SYX	WMC	—Can not spare
WLI	—South east-ern-erly-ing-ward-er	WMD	—Have no spare —
WLJ	—South west-ern-erly-ing ward-er		Have you a spare screw? ----------\KF
WLK	—To, *or*, For the southward		How many dozen can you spare? .LMC
WLM	—Too far to the southward	WME	—I can spare you
			None to spare ----- ----- -- ----SJE
	NOTE —*See* COMPASS SIGNALS, *pages* 45		Spare anchor ------------------EJT
	and 46	WMF	—Spare shaft

	SPARE—SPOKE			

WMG	SPARE—*Continued* —Spare spar There is a spare propeller on board_TZD		SPELL—*Continued* Spelling Signal No 1_Code Flag over E* Spelling Signal No 2_Code Flag over F* Spelling Signal No 3_Code Flag over G*	
WMH	SPARK		** For method of using these Signals, see page 13*	
WMI	SPEAK ING-ER SPOKE-N (*See also* SPOKE) Desire to see (*or* speak)—to_____KWN		Spelling (Alphabetical) Table (*See page* 15)_____ _____ _____EBO	
WMJ	—Did you speak—to—? Have applied (*or*, spoken) to _ _____ESQ Have you applied (*or*, spoken) to? ESR	WNF	—Use Spelling (Alphabetical) Table Your name is not on my list, spell it alphabetically (*or*, by alphabetical table)_____EBQ	
WMK	—I wish to speak to you On what date did you speak the — (*ves sel indicated*)? _____ _____KMG Shall apply (*or*, speak) to ____ _____EST What ships did you speak? _____VYG Will you apply (*or*, speak) to? __ESU Wish to speak—to (*or*, with) _____ZGW	WNG	SPEND-ING SPENT	
		WNH WNI	SPERMACETI, *or*, SPERM —Sperm oil	
WML	SPEAR-ED-ING	WNJ WNK	SPHERICAL —Spherical buoy.	
WMN WMO WMP	SPECIAL-LY IST —Special survey. —Special train.	WNL	SPICE	
WMQ	SPECIE	WNM	SPIKE Handspike_____OMG Marlingspike_____RLS	
WMR	SPECIFY-IED-ING At place specified _____FIQ At the time specified _____FJA Vessels of nations specified were all leaving _____ _____QLM	WNO	SPILL-ED-ING SPILT	
		WNP	SPIN-NING SPUN Spun yarn _____WPB	
WMS	SPECULATE-D-ING ION	WNQ	SPINNAKER	
WMT	SPEECH	WNR	SPINDLE	
WMU	SPEED-Y-ILY Different speeds_____KZJ Full speed_____ ____ _____NSV Go ahead full speed_____ LI Go astern full speed __ _____LM Go full speed___ _____OBS Go half speed ___ _____OBT Half speed, *or*, Half power _____OLM I am going full speed __ __ _____OCI I am going half speed_____OCJ I am on full speed trial, Code Flag over A I have — hours' coal, full speed (*or at speed indicated*) _____ _____IXD Increase speed_____PHC Is steam up? Are you at full speed?_VZ Most economical speed_____LTD	WNS	SPIRITS Methylated spirits_____RSQ	
		WNT WNU	—Spirits of wine —Spirituous	
		WNV	SPIT Sand spit _____VHD	
			SPLENDID (*Magnificent*)_____RET	
		WNX	SPLICE-D-ING	
WMV	—Quarter speed Reduce speed Ease her _____MF Reduce speed, you are towing me too fast_____ ___ ____XYT Slow speed _____WHE	WNY	SPLINT	
		WNZ	SPLINTER-ED-ING	
		WOA	SPLIT-TING Mainsail split _____RGS My sails are all blown away (*or*, split), VEC	
WMX WMY	—Speed light (lantern, *or* lamp) —Speed will be — knots What is your consumption of coal per hour, full speed? _____ _____IXM	WOB	—Split all my sails	
		WOC WOD	SPOIL ING TO SPOIL —Spoiled	
WMZ	—You can proceed at any speed		SPOKE-N SPEAK ING-RR (*See also* SPEAK), WMI	
WNA	SPELL-ING SPELT TO SPELL Name will be spelt _____ _____SDO		Have you seen (*or*, spoken) any men-of- war? If so, report their names and nationalities _____VNM	
WNB WNC WND WNE	—Spell it —Spell the address —Spell the name —Spell the surname	WOE	—Have you spoken any vessels (*or vessel indicated*)?	
		WOF	—Have you spoken to — (*or*, the —)?	

SPOKE—SQUALL

	SPOKE—*Continued* I have reported you to the vessel I spoke, she is bound to — (*place indicated*), URT Spoke a brig. .... HGX Spoke several vessels during the passage, VSU
WOG	—Spoke the — (*vessel indicated*).
WOH	SPONGE
WOI WOJ	SPONTANEOUS-LY —Spontaneous combustion.
WOK	SPOON
WOL	SPOT-TED-TING-TY
	SPOUT Water spout .... YWB
WOM	SPRAIN-ED-ING
WON	SPRAY
WOP	SPREAD-ING Convoy to spread as far as possible, keeping within signal distance ....IZ Spread awning ....FPZ
WOQ	SPRING-ING SPRANG SPRUNG Bowsprit sprung ------------------HCZ Damaged, *or,* Sprung mast, can not carry sail ....----BH Fore topmast sprung -------------NJL Fore yard sprung ---------- ----- NJS I have sprung (*or,* damaged) ------KJG I have (*or vessel indicated* has) sprung — mast (*indicated*)--------------RCJ I have sprung (*or,* damaged) fore yard, KJH I have sprung (*or,* damaged) lower topsail yard------------------------KJI I have sprung (*or,* damaged) main yard, KJL I have sprung (*or,* damaged) topsail yard---------------------------KJM I have sprung (*or,* damaged) upper topsail yard------------------------KJN I shall weigh the moment the breeze springs up--- --------------------HGA Jib-boom sprung--------- --------GZR Main mast sprung-----------------RGI Main topsail yard sprung----------RHG Main yard sprung-----------------RHJ Some demand springing up--------KTF Spanker boom sprung-------------GZW Spring on the cable--------------HRS Sprung, *or,* Seriously damaged a lower mast ------------- -- ----------RCJ Sprung a lower yard --------------RCK Sprung fore mast, but can fish it at sea, NKX Sprung, *or,* Seriously damaged lower yard (*or spar indicated*)----------RCK Sprung, *or,* Damaged mast, can not carry sail ------------ --------- BH Sprung my fore mast, and must bear up, NKW
WOR	—Sprung spar (*indicated*)
WOS	SPRING-ING SPRANG SPRUNG A LEAK. Are you leaky? Have you sprung a leak? QJ I have sprung a leak----------------NQ Vessel indicated has sprung a leak ..QL
WOT WOU	SPRING (*Season*) —In the spring Last spring ----------------------QGM
WOV WOX WOY	—Next spring —Spring tide —Spring tides rise — feet
WOZ	SPRIT SPRITSAIL Spritsail yard (*or,* gaff) -----------ZKO
WPA	SPROCKET SPRUNG (*See* SPRING) --------------WOQ SPUN SPIN-NING--------------------WNP
WPB	—Spun-yarn
WPC	SPY-IED-ING Spyglass Telescope--------------XMQ
WPD	SQUABBLE Some squabble (*or,* fight) on shore with crew ----------------------------MXU
WPE WPF	SQUADRON —A squadron (*give nationality*) composed of — ships, passed in sight Any alteration in the position of the squadron (*or,* fleet) --------------EBZ Blockading squadron --------------GNI Channel squadron (*or,* fleet) --------ILB Detached squadron----- ----- ---KXW Did the squadron (*or,* fleet)? ------NEK Enemy's fleet (*or,* squadron)-------LZG Fleet, *or,* Squadron when last seen were off — --------------------------NEO For the fleet (*or,* squadron) --------NEQ In the fleet (*or,* squadron) --------- NES Inshore fleet (*or,* squadron)-------PMH Letters of squadron (*or,* fleet). ---QOM Of how many ships was the squadron seen composed? -----------------VXM Of, *or,* From the squadron (*or,* fleet)-NEK Reserve squadron -----------------UTK
WPG	—Royal Yacht Squadron (*Club*) Saw the enemy's fleet (*or,* squadron) off — ---------------------------OK Saw the enemy's fleet (*or,* squadron) steering — -----------------------LZS Saw the fleet (*or,* squadron) off — steering — -------------------------- NET Squadron, *or,* Fleet is (*or,* was)—at-NEM To the squadron (*or,* fleet) --------NEV What is the state of health of the fleet (*or,* squadron)? --------------- ORC
WPH	—When is the squadron (*or,* fleet)?
WPI	—Where is our fleet (*or,* squadron)? Where is the — fleet (*or,* squadron)?-SK
WPJ	—Where was the fleet (*or,* squadron) on the —?
WPK	SQUALL Y In a heavy squall------------- ------OTE Look out for a squall ----------------QZT Severe squall not far off, look sharp-FZ Squally weather Blowing hard....ZH

SQUARE—STARBOARD

WPL	SQUARE (*Shape*)		STAND—*Continued*
WPM	—A square-rigged vessel		I shall stand in for the land as long as I
	Not square enough ....MDU		can see............................QDZ
	Square centimeter (*Square measure*),		I shall stand in till I make the land..QEA
	AXR	WQC	—I shall stand in until —
	Square decameter (*Square measure*),	WQD	I shall stand off and on (*the land*)..QEB
	AXJ		—I shall stand on until —
	Square decimeter (*Square measure*),		I shall stand off from —(*hour indicated*)
	AXS		to — (*hour indicated*), or, I shall stand
	Square foot (*Square measure*) ...AXT		off until?....................QEC
	Square hectometer (*Square measure*),		I stand (*or vessel indicated* stands) —
	AXP		(*class indicated*) at Lloyd's......QWO
	Square inch (*Square measure*) ..AXU		Keep, or, Stand close in, or, Keep under
	Square kilometer (*Square measure*),		the land ....LT
	AXV		Please stand by me, I want assistance,
	Square, or, Surface measure........AXH		NR
	Square meter (*Square measure*)...AXK	WQE	—Stand by to let go
	Square mile (*Square measure*) ..AXW	WQF	—Stand closer in
	Square millimeter (*Square measure*),	WQG	—Stand in for the land
	AXY		Stand nearer the shore (or, bank, or,
WPN	—Square sail		reef) ....FWA
	Square yard (*Square measure*) ...AXZ		Stand, or, Haul off ....OFC
		WQH	—Stand off and on the land
WPO	SQUARE (*Place of address*)		Stand off Get an offing Put to sea at
			once .... ---- .... NG
WPQ	STAB-BED-BING		Stand on........ ....MH
	Accident happened, cut, stab .....DCG		Stand on Continue your present course,
			JSM
WPR	STABLE-ILITY	WQI	—Stand out to sea
		WQJ	—Stand towards—the
WPS	STACK-ED-ING	WQK	—Standing order
		WQL	—Standing rigging
	STADION (*Measure of length*) .....AWG		Vessel indicated, or, Vessel in company
			is standing in for the land........QET
WPT	STAFF		Vessel indicated is (or, was) standing
	Flagstaff.........,---------------- ..DB		(or, steering)— ..... ----------VXT
	Jackstaff .. ..PUG		Vessel indicated is standing into danger,
	Rocket staff... ....UZF		KLJ
	Signal staff (or, station) ....WDO	WQM	—Will stand by you
			You are standing into danger ....JD
WPU	STAGE	WQN	STANDARD ROYAL STANDARD ROYAL
WPV	—Floating stage		FLAG
	Landing stage...................QEO	WQO	STANDARD, or SUPPORT
	STAID (*See* STAY) ....WRX		STANDARD, or, POLE COMPASS ....JIB
WPX	STAIN-ED-ING		
		WQP	STAR-RY
WPY	STAMP-ED-ING		Pole, or, North star....... ....SJO
	Postage stamp ....TQN	WQR	—Starlight
WPZ	STANCHION	WQS	STARBOARD (For STARBOARD HELM, *see*
			below)
WQA	STAND-ING STOOD		Breakers, Reef, Rock, or, Shoal on star-
	Do not stand too far on your present		board bow ....HFE
	course ....JZI		Cast to starboard ....IEF
WQB	—Do not stand too far to the —		Convoy to keep on starboard side of
	Do not stand too long on your present		leader (or, escort)... ----- ..... IY
	course ....JZI		Hard-a-starboard! (*Urgent*) Head to
	Do not stand too long on your present		go to port ....ME
	tack ... .JZI		N B —*To be kept flying until the course*
	How do you (or *vessel indicated*) stand		*is sufficiently altered*
	at Lloyd's?..... ....QWK		Haul your wind on starboard tack ..LO
	How is (or, was) stranger (or *vessel in-*		Heave-to on starboard tack .. ..OSF
	dicated) standing (or, steering)?.WTY		Keep buoy (or, beacon) on your star-
	How long may we stand on our present		board hand ....GAL
	course?. . ...JZK		Keep more to starboard...........PZA
	I am standing in for the land ......QDT		Keep on starboard side of channel ..LW
	I shall stand about to see if I can pick		Keep on starboard tack ....PZG
	up anything from the wreck......ZIT		

STARBOARD—STATION

	STARBOARD—*Continued*	WRI	STARVE-D-ING ATION.
	Leave buoy (*or*, beacon) to starboard, MA		Short of provisions Starving......NV
WQT	List to starboard ....QVP	WRJ	STATE (*See also* CONDITION)
	—Not too much to starboard		In the same state as before ....VGK
	On starboard beam..............GAV		Present state of affairs. ....DPF
	On starboard bow ....HCP		Signal, *or*, I am going to signal the
	On starboard quarter............UER		state of the weather outside (*or at*
	Pay out starboard hawser (*See* TOW-		place *indicated*).... ....WDE
	ING SIGNALS, *page* 35)		State of affairs—at..................DPI
	Shorten in starboard hawser (*See* TOW-		State remains unaltered.............TPS
	ING SIGNALS, *page* 35)		What is the state of health of the fleet
	Starboard Helm to starboard Head		at —? ....ORC
	to port....................WJ		What is the state of health of the troops
WQU	—Starboard a little. Head to go to port		at —?.......... ....YBR
	Starboard anchor..................EJU		What is the state of the market?...RLF
	Starboard boiler ....GXL		What state does (*or*, do) the battery
	Starboard boiler defective.........GXM		(*or*, batteries) appear to be in°. _FYV
	Starboard bow .... HCT	WRK	STATE-D-ING, STATEMENT TO STATE
	Starboard bow light (lantern, *or*, lamp), HCU		Muster statement ....GZB
	Starboard broadside ....HJR	WRL	STATE (*Government*)
WQV	—Starboard engine		Chief of the State (President, Em-
WQX	—Starboard gun		peror, King, Queen, etc) in — has
WQY	—Starboard screw		(*or*, is) ....IOR
WQZ	—Starboard side		Royal Navy, *or*, State Navy.........SFC
WRA	—Starboard tack		Secretary of State.................VMO
	Steer more to starboard ....MK	WRM	—Statesman-ship
	N B —*To be kept flying until the course is sufficiently altered*		State Department, *or*, Foreign Office, NKR
	Veer starboard hawser (*See* TOWING SIGNALS, *page* 35)		STATE ROOM (*Cabin*)............HQJ
WRB	STARBOARD HELM	WRN	STATION ED-ING ARY
	Alter course one point (*or number indicated*) to starboard............JZE		*Asks* name of ship (*or*, signal station) in sight...........................VG
	Hard-a-starboard! (*Urgent*) Head to go to port ....ME		At, *or*, In station...............FIP
	N B —*To be kept flying until the course is sufficiently altered*		Calls attention of shore signal station, DO
	Keep more to starboard.. ....PZA		Coaling station....................IWX
	My helm is hard-a-starboard, ship's head going to port ....WH		Coastguard station .... IYQ
	Not too much to starboard ....WQT		Enquires name of ship (*or*, signal station)..........................VG
	Starboard. Helm to starboard Head to go to port ....WJ		Forward answer by telegraph to signal station at —...............EMW
	Starboard a little Head to go to port, WQU		Forward reply to my message by telegraph to.......................WT
	Starboard helm..................WRB		In, *or*, At station ....FIP
	Steady your helm..................MI		Indicate name of ship (*or*, signal station) in sight........ .. ..VG
	Steer more to starboard ....MK	WRO	—Is the — light-ship on her station?
	N B —*To be kept flying until the course is sufficiently altered*		Is there a signal station at —?....WCL
	You are steering too much to starboard, WUI		Keep station........ ....PZI
WRC	START-ED-ING		Leave it for me at the railway station, QLB
	Breakfast before you start.........HFJ		Lifeboat station.................QRK
WRD	—Have started	WRP	—Life saving (*Rocket*) station
	I shall not weigh until I get a favorable start ... .ZAW		Light-ship at — is out of position, *or*, Light-ship is not at anchor on her station ....QE
	Prepare for sailing. Get ready for sea ....NYX		Lloyd's signal station—at —.......QXB
WRE	—Start at —		Meet me at the railway station ....ROU
	Start-ed-ing engine ....MBW		Meteorological station ....RSO
	Started a rivet ....UYN		Naval station.....................SEV
WRF	—Starting gear		Police station ....TNX
WRG	—Starting gear out of order		Preserve station...... ....TUH
WRH	—Starting shaft		Quarantine station ....UEG
			Quit the (*or*, your) station ....UFM
			Railway station........ ....UED
			Return to the (*or*, your) station...UVR

STATION—STEAMER

	STATION—*Continued*		STEAM—*Continued*	
	Semaphore station ---------------- vow		Keep steam up (*or* ready) ---------- kq	
	Send reply to my message to signal sta		How long will you be getting steam up?	
	tion at — ------ -- ------------ vqd			yy
	Signal station (*or*, staff)_ -------- wdc		Is steam up? Are you at full speed_vz	
WRQ	—Station master		Proceed under steam_ -------- ---txf	
WRS	—Stationed at —		Proceeds under her own steam ----- mbt	
	Submarine signal station---------znt		Rate of steaming ----- ----- -------- uim	
	Telegraph station (*or*, office)----- xme		Require steam—for ----- ------------ utc	
	The light-ship—is not at anchor on her		Shall I blow off my steam?--------goj	
	station (*or*, is out of position).. _ qe		Shall I get up steam? -------------- wa	
	To, *or*, At the railway station------uhe		Shall I send you a steam tug?------vqo	
	Torpedo station ------------ ----- xwy		Shall signal with siren(*or*,steam whistle)	
	What is the name of the signal station		during fog (*or*, darkness)---- ----we	
	(*or*,ship) in sight? ----- ---------- vg	WSL	—Steam. (*Vapor of boiling water*)	
WRT	STATIONERY		Steam capstan------------------ ---hwb	
WRU	STATUTE-ORY		Steam chest-------------------------ion	
	Statute adult_----------------------dmt		Steam coal ----- ----------- --------ixj	
WRV	STAVE-ING		Steam collier ------------ ---- ---jaq	
	Boat is stove ----------- ----- ----- ar		Steam cone-------------------------jnp	
	Stove ---------------------- ----- wya		Steam crane ----------------- --- kbg	
WRX	STAY ED-ING TO STAY (*See also* REMAIN		Steam cutter's engine -----------mbx	
	and WAIT)		Steam gauge (*See* MANOMETER)_rjw	
	Stay on board ----------------------gqz		Steam is not ready, will be up in —__wb	
	Stay, *or*, Remain where you are----ufm		Steam is ready-_ -- --------------wc	
WRY	—Will you stay, *or*, wait?		Steam is up_-- ---------------------wc	
			Steam launch's engine ------------mby	
WRZ	STAY ED-ING TO STAY (*In tacking*)		Steam log book _ ----------------gzh	
WSA	—In stays		Steam pinnace's engine-----------mbz	
WSB	—Make sure of her staying	WSM	—Steam pipe	
	Missed stays ---------------------- r\z		Steam pipe burst---- -------------tko	
	Will your vessel stay?--------------vyh	WSN	—Steam pipe damaged	
		WSO	—Steam pipe leaky.	
WSC	STAY (*Support, or Rope*)	WSP	—Steam port	
	Backstay----------- -------fqo	WSQ	—Steam pressure	
	Diagonal, *or*, Triatic stay------- --kyq		Steam pressure gauge Manometer,	
	Forestay----- ---- ---------------njc			rjw
	Jackstay ---------------- -----pui	WSR	—Steam siren	
	Mainstay -------------------------rgv	WST	—Steam steering gear	
WSD	—Stay light (lantern, *or*, lamp)		Steam steering gear disabled--------mz	
		WSU	—Steam steering wheel	
WSE	STAYSAIL.	WSV	—Steam surface	
	Fore staysail ----------------------njd	WSX	—Steam surveyor	
	Fore topmast staysail - - - --------njm	WSY	—Steam trial	
	Main staysail------------- -- - ----rgw		Steam tug ----------------- -----ydr	
	Main topmast staysail-- ----------rhd	WSZ	—Steam whistle	
	Mizzen topmast staysail__ --------rxa	WTA	—Steam will(*or*, can)be up in—minutes	
	Staysail sheet---- -------------- - vux	WTB	—Steam winch	
WSF	—Storm staysail.	WTC	—Steam windlass	
WSG	—Topmast staysail	WTD	—Steaming light (lantern, *or*, lamp)	
	Topmast staysail sheet----------- vwa		Under steam --------------- ----yhj	
			With steam up -------- --------ylw	
WSH	STEADY-ILY-INESS		Working steam pressure—is --------tus	
	Barometer is steady----------------fxk			
	Steady your helm-------------------mi	WTE	STEAMER STEAMSHIP	
	Under steady helm------------------otu		Board (*or, visit*) steamer (*or*, vessel)_gpv	
	Wind appears to be steady --------zfh	WTF	—By steamer	
			Came, Coming, by steamer----- -----jci	
WSI	STEAL-ING. STOLE-N.	WTG	—Fishery steamer	
			Government steamboat-----------ofx	
WSJ	STEAM-ED-ING TO STEAM		Have passed steamer with machinery	
	Blow off your steam_ -------------goa		disabled --------------------------mt	
	Blow siren (*or*, steam whistle) at inter-		Have passed steamer with steering gear	
	vals ---------------------------wd		disabled ----- --------------------mu	
WSK	—Can you steam ahead—of?----------dve		Have you seen disabled steamer? ___mv	
	—Can not get sufficient steam	WTH	—Hired steamer	
	Get steam up as fast as possible----kl		Mail steamer---------------------rfk	
	Get steam up, report when ready---km		Mail steamer arrives-d — at -----rfl	
			Mail steamer reported lost --------rfm	
			Mail steamer sails-ed-------------rfn	
			Mails go by steamer---------------rfq	

STEAMER—STERN

	STEAMER—_Continued_		**STEER—**_Continued_
	Merchant steamer (or, ship) _____RPZ		Is my present course a correct one?
	Paddle-wheel steamer (_Horse-power to_		Am I steering a proper course? (_The_
	follow, if necessary)___ _____SZH		_course steered to follow, if necessary_),
	Passed disabled steamer at — _____MX		JZF
	Pilot boat (_steam_) _____ GUB		Saw the enemy's fleet off — steering —,
	Salvage steamer _____VOF		NET
	Screw steamer (_Horse-power to follow,_		Saw the enemy's fleet steering —___LZS
	if necessary) _____ VKJ	WUA	—Shall we steer (or, run) for — (_place_
	Send a steamer _____VPN		_indicated_)?
	Steamboat (_large_), Launch _____GUV		Steam steering gear___ _____WST
	Steamboat (_small_), Cutter, or, Gig_GUW	WUB	Steam steering gear disabled _____MZ
	Steamboat, Steamer is aground ___DUW		Steam steering wheel _____WSU
	Steam collier_____JAQ		—Steer after me
	Steam cutter, or, gig _____ _____ GUW		Steer after me, I have a pilot _____DRU
	Steam guard-boat_____GUX		Steer course indicated as per compass
	Steam launch_____GUV		signal _____JIF
	Steam lifeboat_____GUY		Steer directly for the buoy (or, beacon),
	Steam pilot boat _____ _____GUB		GAM
	Steam pinnace _____GUZ		Steer directly for the light-ship____QTH
WTI	—Steam trawler		Steer more to port. Head to go to port*,
	Steam vessel (ship) Steamer ___ WTE		MJ
	Tank steamer_ __ _____XKL		Steer more to starboard Head to go
	Twin-screw steamer _____ _____YFL		to starboard* _____MK
	Use steamboat—for _____ _____GVD		
	You can not have a steamboat_____GVJ		* _To be kept flying until the course is_
			sufficiently altered.
WTJ	**STEEL**	WUC	—Steerage
WTK	—Bessemer steel	WUD	—Steerage passengers—for —
	Cast steel _____IEL	WUE	—Steered-ing To steer
	Siemens-Martin steel_____WBS		Steerageway Headway _____OQT
WTL	—Steel built vessel	WUF	—Steering gear
WTM	—Steel mast		Steering gear disabled_____MZ
WTN	—Steel plate	WUG	—Steering to the —
WTO	—Steel-wire hawser		The course steered is the best for the
WTP	—Steelyard		present (_or until time indicated_)__JZM
	Tincture of steel _____XUB		Vessel is (or, was) standing (or, steer-
WTQ	—Whitworth steel		ing) — _____VXT
			What course are you (or, they) steering?
WTR	**STEEP-NESS**		JZO
	Reef, or, Shoal is steep to the —__VYW		What course shall you (or, they) steer?
			JZP
WTS	**STEEPLE**		What course were you (or, they) steer-
			ing?_____JZQ
WTU	**STEER** (_See also_ COURSE _and_ HELM)		You are steering right into the center
	Am I steering a proper course?_____JZF		of the hurricane (or, cyclone) ___PAN
	At — o clock I shall steer — _____IUQ	WUH	—You are steering too much to port
	Damaged rudder, can not steer —___BI	WUI	—You are steering too much to starboard
WTV	—Direct me how to steer		
	Does disconnecting affect your steering?	WUJ	**STEM** (BOW)
	LDC		Carried away stem_____ICP
	Enemy's cruisers have been seen to the	WUK	—Stem seriously damaged
	—, steering to the —_____OF		
	Enemy's fleet have been seen to the —,	WUL	**STEM-MED-MING** TO STEM
	steering to the _____ _____OG		Can not stem the tide (or, current) .KFX
	Enemy's torpedo boats have been seen		
	to the —, steering to the _____ .OH	WUM	**STEP-PED-PING**
	Hand steering gear _____OMA	WUN	—Step of the — mast (tabernacle)
	Have passed steamer with steering gear		
	disabled _____ _____MU		**STERE** (_Cubic measure_) _____AYH
	Headway, Steerage way _____OQT		
WTX	—How are you steering?	WUO	**STERLING**
WTY	—How is (or, was) stranger (or, vessel		£ (pound) sterling _____ _____AUS
	indicated) steering?		
	How was convoy steering? _____JUR	WUP	**STERN LY-NESS**
	I am in danger (or, shoal water), direct		
	me how to steer _____ _____NL	WUQ	**STERN** THE STERN STERNMOST
	I am in difficulties, direct me how to		By the stern, — inches by the stern.HPN
	steer_____ _____KZU		Come under the stern_____JCO
	I have a pilot, steer after me_____DRU		I have stern way ____Code Flag over K
WTZ	—I shall steer — during the night		Main shaft broken in stern tube ___RGT

STERN—STORE.

	STERN—*Continued*	WXB	STONE-Y
	Moor head and stern .... RZJ		Coarse bottom Coarse sand and stones,
WUR	—Name on the stern		HAX
WUS	—No name on the stern		Have you sand and holystones to spare?
WUT	—Stern frame		OXI
WUV	—Stern gland defective		Pumice stone .... URA
WUX	—Stern lantern (light, *or*, lamp)		
WUY	—Sternpost		STOOD (*See* STAND) .... WQA
WUZ	—Stern tube		
WVA	—Stern tube damaged	WXC	STOP-PING-PAGE (*See also* REMAIN *and*
	Stern way Going astern .... FHR		WAIT)
			Can you stop the leak? .... QG
		WXD	—Can you stop till?
WVB	STEVEDORE		Cannot stop the leak .... QH
			Communication by telegraph is stopped,
WVC	STEWARD		WS
	Purser's steward (*or*, assistant) .... UCO	WXE	—Full stop
	Send the steward .... VQK		Have you materials for stopping the
WVD	—Stewardess		leak? .... QJW
WVE	—Steward's small stores		How long shall you stop? .... QYP
			I am going to stop, machinery requires
WVF	STICK ING-Y-INESS STUCK.		adjusting .... RK
		WXF	—I can not stop
WVG	STIFF-LY-NESS-EN-ED		I can not stop to have (*or*, receive) any
	A stiff breeze .... HFX		communication .... IF
	Is, *or*, Are not stiff under canvas .. HVF		I prefer stopping on board .... GQL
	Is, *or*, Are very stiff under canvas . HVG	WXG	—I will stop
	Stiff clay .... ISO		Leak is stopped .... QK
			My engines are stopped,
WVH	STILL		Code Flag over U
	Still under examination .... MKG	WXH	—Obliged to stop engines
			Stop engines to adjust towing cables,
WVI	STINK-ING. STANK STUNK		DKH
			Stop her .... MP
WVJ	STIPULATE-D-ING-ION.		Stop her instantly (*Urgent*) ...... MN
			Stop, Heave-to, *or*, Come nearer, I have
WVK	STOCK-ED (*See also* SUPPLY)		something important to communicate,
	Can I procure stock? .... TXN		Code Flag over H
WVL	—Live stock.	WXI	—Stop the sailing of —
	Someone to look after stock .... RAD	WXJ	—Stop until (*or*, for).
	Stock (*supply*) — of — .... XEU	WXK	—Stop valve.
WVM	—Stock all gone	WXL	—Stopped
WVN	—Stock of goods	WXM	—Stopper.
WVO	Taps, Stocks, and Dies .... KZB		Stopping only for small adjustment to
	—Well stocked with		machinery .... RDY
			You will be stopped by the blockading
	STOCK ANCHOR STOCK .... EHI		ships .... GNR
WVP	STOCK (*Shares*)		
	Stock-jobber (*or*, broker) .... PVY		STOPA (*Measure of length*) .... AWV
WVQ	STOCKADE-D-ING	WXN	STORE-D-ING To STORE
			Artillery, *or*, Ordnance stores ..... EZW
WVR	STOCKHOLM TAR		Boatswain's stores .... GVO
			Carpenter's stores ... . IBV
WVS	STOCKINGS (*Socks.*)		Commissariat stores .... JFB
			Condemned stores .... JMN
WVT	STOKE-D-ING		Engineer's stores..... .... MCU
WVU	—Stoker		Government stores............... OFY
	Leading stoker .... QJR		Gunner's stores .. .. OKE
	Stokehole .... OXE		Have military stores on board .... RTV
	Stokehole compartment .... JHA	WXO	—Have stores for you
WVX	—Stokehole full of water		Medical stores..... .. .. ROL
WVY	—Stokehole ventilator carried away		Military stores .... RUB
WVZ	—Stoker mechanic		Naval stores .... SBW
			Steward's small stores .... WVE
	STOLEN STEAL-ING .... WSI	WXP	—Store. (*A store*)
		WXQ	—Storehouse
WXA	STOMACH	WXR	—Storekeeper
		WXS	—Storeroom
	STONE (*Measure of weight*) .... BCL	WXT	—Storeship

	STORM—STRICT				

WXU	STORM-Y (*See also* WEATHER *and* GALE)			STRANGE—*Continued*	
	Storm center in — (*direction indicated*), probable course toward — (*direction indicated*)............IGY			Strange sail (*or vessel indicated*) is bound to — ....HBT	
WXV	—Storm sail		WYU	—Strange sail (*or vessel indicated*) is from —	
	Storm staysail ... .WSF			Strange sail (*or vessel indicated*) is in distress.. ....LHM	
WXY	—Storm trysail				
WXZ	—Storm warning		WYV	—Strange vessel following convoy	
	Stormy, Boisterous weather from the —,			Strange vessel hove in sight.........OZS	
	.........ZI		WYX	—Stranger belongs to —	
	There has been, or, We have had a storm of thunder and lightning ....QTL			Stranger does not understand your signal... ....WDJ	
			WYZ	—Stranger has	
	STOUT PORTER. ....TPI		WZA	—Stranger is	
WYA	STOVE			Stranger is chasing ....INT	
	Boat is stove ....AR		WZB	—Stranger is gaining	
			WZC	—Stranger is not gaining	
WYB	STOVE (*Fireplace*)			Stranger (*or vessel indicated*) is suspicious..... ....GH	
WYC	STOW-ING-AGE			Stranger (*or vessel indicated*) will not take any notice of signal.........WDK	
	Badly stowed ....FRK		WZD	—Stranger (*or vessel bearing* —) wishes to communicate	
	Cargo so badly stowed I am not seaworthy . ...IAG			There is a strange sail on —(*bearing indicated*) ....EM	
	Stow anchor, *or*, Anchor is stowed.EHJ		WZE	—Very strange-ly	
WYD	—Stowed .		WZF	What do you make of stranger?	
WYE	STOWAWAY				
WYG	STRAIN-ING		WZG	STRAP-PED-PING	
	Great strain (*or*, stress) on — ......OHJ			Eccentric strap broken...............LSZ	
WYH	—Strained				
WYI	—Too much strain on —		WZH	STRAW	
WYJ	STRAINER		WZI	STREAK-ED-Y	
WYK	STRAIGHT		WZK	STREAM	
WYL	—Straightforward			Against the current (*or*, tide)...DSE	
				Athwart the tide (*or*, stream)......FJI	
	STRAIT, *or*, CHANNEL OF...............ILG		WZL	—Gulf Stream	
	In the strait (*or*, channel) of ....ILM		WZM	—In the stream	
WYM	—Through the strait of			Stream anchor, *or*, Small anchor (*Weight in cwts , if necessary*).. EJV	
WYN	STRAIT-EN-ED-ING			Stream cable Stream chain ......HRT	
	STRAKE ....ZNJ			STREEP (*Measure of length*).......AWP	
	Sheer strake ....VUM		WZN	STREET	
WYO	STRAND-ING			STREMMA (*Square measure*).........AXJ	
WYP	—Stranded				
	Towing cable is stranded (*or*, damaged), BRX		WZO	STRENGTH-EN-ED-ING	
WYQ	STRANGE-R-LY			Feather strengthening ....MUO	
	Ask stranger (*or vessel indicated*) if he will communicate ....HT			What is the strength of the tide?...XSP	
	Be very careful (*or*, guarded) in your intercourse with strangers ....FS		WZP	STRESS	
WYR	—Do not have anything to do with stranger (*or vessel indicated*)			Driven by stress of weather.........YZA	
				Great stress (*or*, strain) on..........OHJ	
	Do you know anything of strange sail? QBU			Stress of weather ....YZE	
	Hail the strange vessel, and ask if —, FDO		WZQ	STRETCH-ED-ING	
	How is (*or*, was) stranger (*or vessel indicated*) steering?...............WTY			Reef Rock, *or*, Shoal stretches a long way out. ....GE	
	I am a stranger here, will you let me go ashore with you?...............OCH		WZR	STRETCHER	
WYS	—It is very strange		WZS	STRICT-ER-EST-LY	
	Ship indicated to examine strange vessel and report.. ....MKF		WZT	Be very strict	
				Keep a strict watch all night......PYJ	
WYT	—Strange sail			Pay strict attention to signals during the night...............DT	

STRICTURE—SUDDEN

WZU	STRICTURE	XBD	STUN-NED-NING.
WZV	STRIKE-ING. TO STRIKE		STUNK STINK (*See* STINK)_____WVI
WZX	—Are you striking?		
	Dolphin striker _____LKD	XBE	STUPEFY-IED-ING
WZY	—Striking heavily		
XAB	—Struck — on —		STUPID-LY-ITY (*Foolish*) _____NBH
XAC	—Struck by a heavy sea		
	Struck by lightning _____QTK		SUBDUE-ED-ING (*Conquer*)_____JPM
	Vessel has struck on the bar_____FWB		
	Vessel indicated has struck on a shoal,		With assistance fire can be subdued,
	VXQ		FGV
XAD	STRIKE-ING STRUCK TO STRIKE (*to cease*	XBF	SUBJECT
	work for higher wages)		
XAE	—Is there a strike at —?	XBG	SUB-LIEUTENANT-CY
XAF	—Striker		Sub lieutenant in the Royal Naval Re-
	Struck for more wages_____YSH		serve _____ _____QPZ
XAG	—The strike at —	XBH	SUBMARINE
XAH	—The strike is over	XBI	—Submarine boat
XAI	—There is a strike of — at —		Submarine cable _____HRU
		XBJ	—Submarine mine
	STRIKE-ING STRUCK TO STRIKE (*to*		Submarine signal____ _ _____ ____ZNQ
	lower) _____RBY		Submarine signal bell_____ZNR
XAJ	—Strike your colors		Submarine signal receivers_____ZNS
			Submarine signal station_____ .ZNT
XAK	STRINGER		
		XBK	SUBMERGE-D-ING
XAL	STRIP-PED-PING TO STRIP.	XBL	—Submerged torpedo buoy
XAM	STRIPE D	XBM	SUBMIT-TED-TING-ISSION-ISSIVE-LY
XAN	—Striped buoy	XBN	SUBORDINATE-D-ING
XAO	STROKE	XBO	SUBPŒNA
	Sunstroke_____XED		
		XBP	SUBSCRIBE-D-ING-PTION
XAP	STRONG-LY	XBQ	SUBSEQUENT-LY
	Current, *or*, Tide will run very strong,		
	KGB	XBR	SUBSIDY.
XAQ	—Not strong enough	XBS	SUBSIST-ED-ING-ENCE
	Strong breeze _____HGE		
XAR	—Strong fort	XBT	SUBSTANCE-TIAL-LY
XAS	—Strong gale		
XAT	—Stronger est	XBU	SUBSTITUTE-D ING-ION.
XAU	—Strongly fortified	XBV	SUBTRACT-ED-ING-ION
	Sufficiently strong _____XCV		
	Wind strong _____ZFM	XBW	SUBURB-AN.
		XBY	SUCCEED-ED-ING
	STRUCK (*See* STRIKE)		Assault did not succeed _____FDX
XAV	STRUCTURE		I (*or vessel indicated*) made the attempt
			but did not succeed _____FKN
XAW	STRUGGLE-D-ING	XBZ	—Will not succeed
		XCA	—Will succeed
	STUCK (*See* STICK)_____ _____WVF		
		XCB	SUCCESS SUCCESSFUL-LY
XAY	STUDDING SAIL	XCD	—What success have you (*or vessel indi-*
	Lower studding sail_____RCG		*cated*) had?
	Set studding sails _____VEO		With complete success _____JKX
	Studding sail boom_ _____GZX		
	Take in studding sail_____ KJO	XCE	SUCCESSIVE-LY
	Topgallant studding sail ___ _____XVT		
	Topmast studding sail_____XWA	XCF	SUCCESSION-OR
XAZ	STUDENT	XCG	SUCH
XBA	STUDY-IED-ING	XCH	SUCK-ED-ING SUCTION
		XCI	—Suction hose
XBC	STUFF-ED-ING	XCJ	—Suction-hose pipe
	Bread stuffs _____HEN		
	Stuffing box _____ ____ _____HDG	XCK	SUDDEN-LY-NESS
	Stuffing box broken___ _____ HDI		
	Stuffing-box glands out of order____HDJ		

	SUE—SUPERSEDE		

XCL	SUE-D-ING.	XDJ	SULTAN-A
XCM	SUET		SULTCHEK *(Measure of capacity)*....AZF
XCN	SUFFER-ING TO SUFFER	XDK	SULTRY-INESS
	Crops have suffered severely....._KDW		
XCO	—Have you suffered?	XDL	SUM-MED-MING
XCP	—Suffered		
XCQ	—Sufferer	XDM	SUMMARY-ILY
XCR	SUFFICIENT-LY-CY *(See also* ENOUGH *)*	XDN	SUMMER
	Authority sufficient for the occasion._FMP		Last summer............................QGN
	Barely sufficient .... FWS	XDO	—Next summer
	Can not get sufficient steam ......WSK		Summer clothing ....IVW
	Certificate is sufficient.............IHX	XDP	SUMMON-ED-ING
	Have you sufficient towline? ....QUK		
	Have you sufficient (*or*, enough)?._MDJ	XDQ	SUMMONS-ED ING
	I have not sufficient (*or*, enough)._MDQ		Sumner's line, *or*, Double altitudes_EDI
	I have sufficient (*or*, enough).... ._MDN		SUN *(Measure of length)* ....AWX
	Is answer suitable (*or*, sufficient)?_END		
	Is it enough(suitable,*or*, sufficient)?_MDO	XDR	SUN-NY.
	Is it sufficiently secure?............VMZ		After sunrise...................... ...DRI
XCS	—Is, *or*, Are not sufficient-ly.		After sunset ....DRJ
XCT	—Is, *or*, Are sufficient ly		At sunrise ....FIV
	Is there sufficient depth of water?._KVC		At sunset..........................FIW
	Not sufficient distance.............LGJ	XDS	—Before sunrise
	Not sufficient room ....UZS	XDT	—Before sunset
	Not sufficiently cautious............IGB		By the sun........................HPO
	Not sufficiently comprehensible It is	XDU	—Sunlight, sunshine
	impossible to comprehend ......JLP		Sunrise At sunrise.............FIV
	Quite sufficient (*or*, enough)......MDX		Sun's altitude ....EDM
	Sufficient caution.. ....IGC		Sunset At sunset....... ....FIW
	Sufficient distance ....LGK		Sunstroke.......................XED
	Sufficient, *or*, Enough room......MDC	XDV	—The heat of the sun
XCU	—Sufficient water	XDW	—The temperature in the sun is —
XCV	—Sufficiently strong	XDY	—What is the temperature in the sun?
	The cause, (*or*, reason) assigned is suffi-		
	cient ....IFU		SUNDAY SABBATH.................VCT
	Tug has not sufficient power.......YDS		Last Sunday ....QGO
	Will be sufficient (*or*, enough) ....MEB	XDZ	—Next Sunday
XCW	SUFFOCATE-D-ING-ION	XEA	—Sunday morning
		XEB	—Sunday night
XCY	SUGAR	XEC	—Sunday week
XCZ	—Can you supply me with sugar? (*Pounds*		
	weight may be indicated)		SUNK-EN *(See* SINK*)* ....WEN
	Hogsheads of sugar ....OVZ		
	Sugar crop..... ....KDY		SUNRISE *(See* SUN*)*.................FIV
	SUGGEST-ED-ING-ION *(Propose)*......TZJ		SUNSET *(See* SUN*)* ....FIW
XDA	SUICIDE-AL	XED	SUNSTROKE.
XDB	SUIT-ED-ING-ABLE	XEF	SUP-PED PING *(See also* SUPPER *)*
	I have no spars suitable for — ....WLY	XEG	SUPERCARGO
	Is answer suitable (*or*, sufficient)?_END		
	Is it enough (suitable, *or*, sufficient)?_MDO	XEH	SUPERHEAT-ED-ING
	It is not suitable It will not answer_ENF	XEI	—Superheater
	Suitable boat for landing must be sent		
	from the shore.. .... FI	XEJ	SUPERINTEND ED ING-ENCE
	When will the tide suit?............XST	XEK	SUPERINTENDENT OVERSEER
	Will the tide suit?.................XSU	XEL	SUPERIOR-ITY
XDC	SUIT *(A suit of)*	XEM	—Superior force
	Coaling suit .... _IWY		
		XEN	SUPERNUMERARY
XDE	SULPHUR-OUS-IC	XEO	SUPERSEDE-D-ING
XDF	—Sulphate of copper		
XDG	—Sulphate of iron		
XDH	—Sulphate of zinc		
XDI	—Sulphuric acid.		

SUPERSTRUCTURE—SURPRISE

XEP	SUPERSTRUCTURE		
	SUPERVISE-D ING-ION (*Superintend*)_XEI		SUPPOSE—*Continued*
			Vessel indicated has not been heard of since — (*date indicated*), she is supposed to be — _____ _____ORS
XEQ	SUPPER		
	After supper _____ _ _____DRK		Vessel indicated (*or in direction indicated*) is supposed to have gone down, NOY
XER	—Before supper		
		XFI	SUPPRESS ED-ING-ION
		XFJ	SURE (*See also* CERTAIN)
XES	SUPPLY-IED-ING TO SUPPLY ' (*See also* FURNISH)	XFK	—Am, Is, *or*, Are sure
	A plentiful supply—of_____TNB	XFL	—Are you sure—of —?
	Are there plenty of supplies at —?__TNC		Be sure and make no mistake_____RHY
	Article indicated can be supplied, but it will require fitting_ _ ___ _ EZF	XFM	—I am sure I can
		XFN	—I am sure I could
	Boats are supplied with — (*articles indicated*)_ __ _____ ___ QSD	XFO	—Make sure—of
			Make sure of her staying_____WSB
XET	—Can you supply me—with (*or*, spare)—?		Not sure (certain)_____IHM
	Can you supply a spar? _____WLX		Quite sure of—_____UFT
	Can you supply me with a fish for — (*spar indicated*)?_____ __ ___NOH	XFP	—Surely
	Can you supply me with anyone to take charge (*or*, act) as engineer? _____RH	XFQ	SURETY
	Can you supply me with casks of —? (*Number, if necessary*)_____IDX	XFR	SURF
		XFS	—How is the surf?
	Can you supply me with pork? (*Number of casks if necessary*) _ _____TON		Is there much surf on the beach?___FZV
		XFT	—Not much surf
	Can you supply me with salt beef? (*Number of casks indicated*)___ _ _____GDA		Surf is heavy Much surf_____OSZ
		XFU	—Too much surf
	Can you supply me with sugar? (*Pounds weight may be indicated*)_____XCZ	XFV	SURFACE
			Air surface _____ __ _____ ____DWI
	I can supply (*or*, spare) you ——__WME		Area of surface_____EUM
	I can supply you with duplicate of engine (*part mentioned*) _____LPE		Steam surface _____WSV
		XFW	—Surface condenser
	I can supply you with powder ___ _ TRG	XFY	—Surface condenser out of order
	I can supply you with powder and shot, TRH		Water surface _____ _____YWC
	I will supply you with ammunition for — _____EFV		SURFACE, *or*, SQUARE MEASURE _____AXH
	Is — (*vessel indicated*) likely to supply me?_____FGK	XFZ	SURGEON
			Accident, want a surgeon (*or*, doctor), AM
XEU	—Supply — of —		Have you a surgeon (*or*, doctor)?___WK
	Supply of water is small, not exceeding — (*quantity indicated*) a day_YWR		I am in want of a surgeon (*or*, medical assistance) _____ _____FGL
XEV	—Supply abundant		
XEW	—Supply scanty		May I send sick person to see your surgeon? _____ _____ __ WL
	Vessel indicated not likely to supply (*or*, assist) you_____FGR		No surgeon (*or*, doctor) _ _ _ ___LJX
			No surgeon (*or*, doctor) available.___WM
	Well supplied with Well stocked with, WVO		Surgeon, *or*, Doctor is away_____LJV
XEY	—Well supplied with pumps		Surgeon, *or*, Doctor shall be sent for, LJW
XEZ	SUPPORT-ED-ING		Surgeon, *or*, Doctor wants assistance, FGP
	Let us keep together (*or*, in company) for mutual support and protection_IN		Surgeon, *or*, Doctor will come immediately _____ ____ __ ___ _____WN
	Standard, *or*, Support__ _____WQO		
XFA	—Support my attack	XGA	—Surgical Surgery
	Want more support _____RZX		Want a surgeon (*or* medical assistance), FGL
XFB	—Will you support us?		
			Want a surgeon (*or*, doctor), send me one from the nearest place _____WO
XFC	SUPPOSE-D-ING-ITION		
XFD	—Do, *or*, Did you suppose?		Will your surgeon (*or*, doctor) come on board? _____WP
XFE	—Have you reason to suppose?		
XFG	—It is supposed	XGB	SURNAME
	Machinery supposed to have broken down_____ _____MCA		Spell the surname _____ _____ ___WNE
XFH	—Supposed to be	XGC	SURPLUS
	Supposed with engines broken down, MCA	XGD	SURPRISE-D-ING-LY

SURRENDER—SYSTEM

XGE	SURRENDER-ED-ING.	XHA	SWEAR-ING SWORE-N
XGF	SURROUND-ED-ING	XHB	SWEDE-N · SWEDISH COLORS
XGH	SURVEY-ED-ING TO SURVEY.	XHC	SWEEP-ING SWEPT.
	Attend survey ....FKY		Chimney sweep (or, smith) ....IPB
	Board of survey...................GPB		Sweeping tubes (flues, or, funnels).NFZ
XGI	—Did not pass survey.		
	Engineer surveyor ....MCT	XHD	SWEEPS
XGJ	—Has been surveyed		
XGK	—Has not been surveyed	XHE	SWEET-LY-NESS
	Lloyd's surveyor ....QXC	XHF	—Sweet oil
	Medical survey ....ROM		
	Passed survey ....TCS	XHG	SWELL-ED-ING SWOLLEN TO SWELL.
	Please send off Lloyd's surveyor ...QXG	XHI	SWELL (Movement of the sea)
	Regular survey necessary ....UOC		A heavy swell.....................OTA
	Regular survey unnecessary.......UOD		Ground swell ....OIH
	Special survey....................WMO		Much swell on ....UR
	Steam surveyor...................WSX	XHJ	—Some swell
XGL	—Survey (A survey)	XHK	—Too much swell
	Survey chain cable................HRV		
	Surveyed by Lloyd's agent..........QXI		SWIFT-LY-NESS (Speedy) ....WMU
XGM	—Surveyor.		
XGN	SURVIVE-D-ING	XHL	SWIFTER (Shrouds)
XGO	—Survivor		
		XHM	SWIM-MING SWAM
XGP	SUSPECT-ING (See also SUSPICIOUS)	XHN	—Does he, or, Does it still swim?
	Ports suspected of cholera, etc.......IPT	XHO	—Swam to the —
XGQ	—Suspected Suspicious		
		XHP	SWING-ING SWUNG TO SWING
XGR	SUSPEND-ED-ING		Must be swung to adjust compasses,
	Bank indicated has suspended payment,		JHY
	FUX		Not room to swing ....UZR
	Certificate suspended ....IBW		Room to swing ....UZT
	Person indicated has not suspended busi-		Was swung for adjustment of compasses
	ness ....HOC		at — ....JIM
	Person indicated has suspended busi-		
	ness ....HOD	XHQ	SWISS
	Suspended business ....HOE		
	Your agent (or agent specified) has sus-	XHR	SWIVEL
	pended business..................DTG		Mooring swivel ....RZK
			Take mooring swivel off...........RZM
XGS	SUSPICION		
			SWOON ED ING (Faint) ....ZNI
	SUSPICIOUS (Suspected) ....XGQ		
	Stranger (or vessel indicated) is sus-	XHS	SWORD
	picious........ ....GH		
	Suspicious craft........ .. ..KBD	XHT	SYLLABLE-IC
	Suspicious papers ....TAO		
XGT	—Very suspicious	XHU	SYMBOL ICAL
XGU	SUSTAIN-ED-ING	XHV	SYMPTOM-ATIC.
	Did you (or vessel indicated) receive		
	any damage? .... KJF	XHW	SYNDICATE
	What damage have you sustained?.KJS		
		XHY	SYPHILIS-ITIC
XGV	SWALLOW-ED-ING		
		XHZ	SYPHON
	SWAM (See SWIM) ... .XHM		
	Swam to the — ....XHO	XIA	SYRINGE-D-ING
XGW	SWAMP-ED-ING	XIB	SYSTEM-ATIC-AL-LY
	Boat swamped....................AS		Metrical system ....RSU
	SWAMP (Morass) ....RZS		
XGY	SWAY-ED-ING		
	Sway-ed-ing across ....DHX		
XGZ	—Sway beam		

SYSTEM—TAKE

XIC	SYSTEM—*Continued*
XID	—New system
	—Old system

XIE	T (*Letter*) (*For new method of spelling, see page 18*)
	TABERNACLE (*Step of the mast*) ___WUN
XIF	TABLE-ULAR
	Alphabetical table (*see page 15*) ____EBO
	Azimuth table _____ _____FQG
	Deviation table _____JHO
	Have you a book of navigation tables? RN
	Logarithm—table _____ _____QYI
	Navigation tables_____ _____SFB
	Numeral table (*see page 33*)_____SLN
	Spelling table (*see page 15*) _____EBO
	Table-land _____QES
	Table sauces ___ _____VHZ
	Tide table_____XSO
XIG	—Time table
	Turn table _____ _____YEL
	Use alphabetical (spelling) table __WNF
XIH	TACIT-LY
XIJ	TACK-ED-ING To TACK.
	Do not stand too long on your present tack _____JZI
	Fetch on the other tack_____MWC
	Fetch on this tack_____ _____MWD
	Fore tack _____NJE
	Get her on the other tack as fast as possible. _____NYU
	Get her on the other tack or you will be on shore_____LF
	Haul your wind on port tack_____LN
	Haul your wind on starboard tack___LO
	Heave-to on port tack_____OSE
	Heave-to on starboard tack _____OSF
	I shall not go on the other tack unless I fall off—to — _____MRU
XIK	—I shall tack—at — o'clock
	I will show a light to-night when I tack, QX
	Keep on port tack _____PZF
	Keep on starboard tack_____PZG
	Main tack . _____ _____RGX
	Make short tacks __ _____RIP
XIL	—Not able to make such short tacks
XIM	—Port tack
	Starboard tack_____WRA

XIN	TACK—*Continued*
XIO	Tack instantly ___ _____MO
	Tacks and sheets_____VUY
	Water shoaling, I must tack_____YWM
	—We shall do better on the other tack
	—When do you tack?
	When I tack to-night I will show a light, QX
	When you alter course (*or*, tack) to-night show a light _____JZS
XIP	TACKLE
	Do not risk an anchorage unless you have very good ground tackle_____FX
XIQ	—Ground tackle
	Ground tackle not good enough ____OIJ
	Ground tackle very good_____OIK
	Luff tackle_____ _____RCU
XIR	—Reef tackle
XIS	—Relieving tackle
	Runners and tackles_____VCG
XIT	—Tackle, *or*, Purchase carried away
	Tackle falls Purchase falls_____UBY
	Top tackle pendant_____XVN
	Winding tackle_____ZFR
XIU	TACT FUL
XIV	TACTIC-AL
	TAEL, *or*, LEANG (*Coin*) _____AVC
XIW	TAFFRAIL
XIY	TAIL
	Tail end of main shaft broken _____LYO
XIZ	—Tail shaft
XJA	TAILOR
XJB	TAINT-ED-ING
XJC	TAKE-ING
	An alteration has taken place _____EBX
	As much as can be taken ___ _____FBA
	Blockade is not taken off ___ _____GND
	Blockade is taken off _____GNE
	Can I take (*or*, forward) any letters for you? _____NMV
XJD	—Can, *or*, Will take
	Can, *or*, Will you point out (*or*, take me to) a good anchorage?_____EKS
	Can you supply me with anyone to take charge (*or*, act) as engineer?_____RH
XJE	—Can, *or*, Will you take—any —?
	Can you take me in tow? _____XU
	Can not save the ship, take people off.NB
	Can not take anything larger than___QFN
	Can not take you in tow_____XV
	Caution is requisite, take care _____FT
	Did boat take —? _____ _____GSO
	Did boat take blue light (lantern, *or*, any means of making a signal)? ___GSP
	Did boat take water?_____GSQ
	Do not interfere with (*or*, take the matter out of the hands of) Lloyd's agent._____ _____ ____QWH
	Embargo is not taken off_____LWD
	Embargo is taken off_____LWE
	Every precaution has been taken___MJE
	Fire gains rapidly, take people off__NG

TAKE—TAP

TAKE—Continued

Funeral takes place at — (place) and at — (time and date) _____ NTK

XJF —Has been taken—into —

Has embargo been taken off at —?_LWG

Has taken a partner _____ TCB

Has the blockade been raised (or, taken off)? _____ GNL

Have lost all boats, can you take off crew? _____ NI

Have you taken observations for variation? _____ JHS

Heave to, or take the consequences_OSG

How did it take place? _____ OMK

How many tons do you take? _____ XUV

How many tons measurement of goods can you take? _____ OEV

I am taking in (or, discharging) powder, or, explosives,
    Flag B, or, Code Flag over B

I am (or vessel indicated is) waterlogged, take people off _____ YWT

I can not take you in tow, but will report you at —, and send immediate assistance _____ XYQ

I have taken — (quantity indicated) of fish _____ NCB

I took advantage of _____ DNS

XJG —I will take her in tow

I will take mails for you _____ QP

XJH —I will take you in tow

I wish to be taken in tow _____ XW

Is embargo likely to be taken off?_LWI

Is it necessary (or, Does the navigation require me) to take a pilot? _____ SEY

It is expected embargo will be taken off _____ LWJ

Must take in more ballast _____ FTS

No alteration has taken place _____ ECS

People going on shore to take their dinners with them _____ LAU

Person to take charge _____ IME

Police (or other) authorities have taken some of the crew out of the ship_KDH

Quarantine is taken off _____ UED

Send boat to take off the crew _____ GUP

Shall I take you in tow? _____ XY

XJI —Shall, or, Will take

Take, or, Make a copy (or, duplicate) _____ JWD

Take advantage—of _____ DNV

XJK —Take as much as

Take care, or, Be careful—of, caution is requisite _____ FT

Take care range is clear _____ HYM

Take charge—of _____ IMF

Take command—of _____ JEG

Take despatches _____ KXC

XJL —Take her in tow

XJM —Take in

Take in a reef (indicate if more than one) _____ UMQ

XJN —Take in light sails

Take in sail _____ VET

XJO —Take in studding sails

XJP —Take in topgallant sails

XJQ —Take in tow

Take letter (or, despatches) _____ KXC

Take mooring swivel off _____ RZM

XJR —Take no notice

XJS —Take notice

TAKE—Continued.

XJT —Take-n, Took off

XJU —Take papers

XJV —Take passage—to (or, for)

XJW —Take-n, Took place—on

Take possession—of _____ TPZ

Take room enough _____ MDY

XJY —Take telegram—to (or, for)

Take your choice _____ IPX

XJZ —Taken

Taken aback _____ CXG

The tide will take her off _____ NFE

XKA —Took

Took place in consequence of _____ JQE

Trial takes place on the — _____ YAW

Utmost care must be taken _____ HYQ

Want lighters to take my cargo_ QTE

Was taken by a privateer _____ TWM

What has taken? _____ ZOD

What is the lowest price you will take? TVU

What vessel takes the next mail? _RFY

XKB —What will you take?

When did it take place? _____ TLO

When did the collision take place?_JBA

When is the embargo to be taken off? LWN

When will funeral take place? _____ NTL

When will it take place? _____ TLP

Where did the collision take place?_JBC

Where did you take your last departure from? _____ KTY

Where is it to take place (or, be done)? LIV

Which channel shall I take? _____ ILW

Which passage did you take? _____ TDJ

Will take time—to complete _____ JKW

XKC —Will take you (or ship indicated) in tow

Will you take a bill on — ? _____ GKN

Will you take letters (or, despatches)_JY

XKD —Will you take me (or vessel indicated) in tow?

Will you take my bill? _____ GKO

Will you take my pilot? _____ TJV

You had better take a pilot at once_TJX

You must take a pilot _____ TJY

TALE

XKE —Telltale

XKF TALK-ED-ING

XKG TALL

XKH TALLOW

XKI TALLY-IED-ING

TAN (Measure of weight) _____ BCM

XKJ TANK

Cable tank _____ HQU

Filter tank _____ MYE

Require water tank (or, boat) _____ GUI

XKL —Tank steamer

Water ballast tank _____ FTW

Water tank _____ YWD

Water tank is empty _____ LXZ

Water tank leaky _____ YWE

XKM TAP-PED PING

Taps, Stocks, and Dies _____ KZB

TAPE—TELEGRAPH

XKN	TAPE
XKO	TAPIOCA
XKP	TAR-RED-RING-Y TO TAR
	Coal tar ....IWS
	Stockholm tar ....WVR
XKQ	—Tar barrel
	Tar brush ....HKR
XKR	TARE
XKS	TARGET
XKT	TARIFF
XKU	TARPAULIN
XKV	TARTAR-IC TARTARIC ACID
XKW	—Cream of tartar
XKY	—Tartar emetic
XKZ	TASK
XLA	TASTE-D-ING-FUL-LY
XLB	TATTOO
XLC	TAUNT-ED-ING
XLD	TAUT
	Haul taut Tauten hawser........OPD
	Heave taut ....OSB
XLE	TAVERN
XLF	TAX-RD-ING-ATION
	Income tax............... ....PCS
	TCHEKI (Measure of weight)........BAV
	TCHO, or, CHO (Measure of length)..AVR
XLG	TEA
XLH	—Teapot
XLI	TEACH-ING TAUGHT
XLJ	TEAK
XLK	TEAR-ING TORE-N
XLM	TEDIOUS
	A tedious passage ....TDA
XLN	TELEGRAM (See also under TELEGRAPH below)
XLO	TELEGRAPH-ED-ING TO TELEGRAPH
	Admiralty, or, Navy Department telegraph-ed-ing ....DLV
	Answer by telegraph Telegraph reply, WQ
	Answered, or, Replied by telegraph..EMQ
	By telegraph ....HPI
	Cablegram.................HQV
XLP	—Can I telegraph — to?
	Can telegraph message be forwarded from —?.................WR
	Can you forward my communication by telegraph? ....HU

TELEGRAPH—Continued.
Communicate by telegraph ....JFP
Communication by telegraph is ...JFS
Communication by telegraph is restored, JFT
Communication by telegraph is stopped, WS
Does telegraphic communication go all the way?.................JFY
Engine-room telegraph ....MBJ
Foreign telegram..................NKT
Forward answer by telegraph to signal station at —.................EMW
Forward following message by telegraph — to — ....NMW
Foward following telegraphic message by signal letters, instead of writing it at length.................XB
Forward my communication by telegraph and pay for its transmission.HY
Forward reply to my message by telegraph to — ....WT
Have orders (or, telegram) for you...SM
Have you any message(telegram,orders, or, communications) for me? ......IB
I am going to signal the contents of an important telegram, to be communicated to you ....IE

XLQ —I am laying (repairing, or, picking up) a telegraph cable Keep out of my way
I have orders to telegraph your passing, SO
I have picked up telegraph cable ...WU
I have telegraphed for further orders (or, instructions) ....STW
I have telegraphed for your orders...ST

XLR —I will telegraph
I will telegraph for your orders if you will await reply ....SV
I will telegraph message immediately, RQO
In anchoring, look out for telegraph cable ... .WV
Is there a telegraph cable near me?.HRJ

XLS —No telegraphic communication
Open telegram for me and signal its contents ....WX
Order-ed-ing by telegraph (telegram), SUD
Replied by telegraph.................EMQ

XLT —Report by telegraph
Report me by telegraph to — ....USA
Report me by telegraph to Lloyd's...UD
Report me by telegraph to my owner (or, to Mr —) at — ....USE
Report me by telegraph to owner (or, to —) at.................UE
Report me by telegraph to "Shipping Gazette" ....UF
Report me to owners on leaving....USB
Send boat for telegram..... .....GUM
Send following message by telegraph to owner (or, to —) at —...........XA
Send following message through the telegraph by signal letters instead of writing it at length ....XB
Send following telegram to owners (or, to Mr —) at — ....XA
Send my message through the telegraph by signal letters ....XB

TELEGRAPH—TENSILE

	TELEGRAPH—*Continued*	XMS	**TELL-ING TOLD**
	Send telegram ....VQH		Can, *or*, Will you tell(*or*, make known)?
	Shall I open telegram for you and sig-		QCL
	nal its contents XC	XMT	—Can not tell you
	Shall I send message by telegraph?.RQS		Desires me to tell you ....KWO
	Shall I telegraph for your orders ....TB	XMU	—Did you tell?
XLU	—Shall I telegraph owner?		Foretell-ing-told ....NKE
	Signal the name of place where you wish	XMV	—He (*or* person *indicated*) told
	to be reported by telegraph.......WDF		Tell my owner ship answers remarkably
	Take telegram — to (*or*, for) ......XJY		well ....ENO
		XMW	—Tell —(*person indicated*)—to
	NOTE—*In signaling a telegram to be*	XMY	—Tell —(*person indicated*) not to —
	sent, signal—		—Tell —(*person indicated*)not to forward
	1 *Ship's name*		any more letters for me...........QS
	2 *Name of person to whom telegram*	·	Telltale.............................XKE
	is to be sent.		Tell them to recover my anchor ...EJW
	3 *Text of telegram*	XMZ	—Who told (*or*, said so)?
		XNA	—You were told
	Telegram arrived from— ....JX		
XLV	—Telegram received from —	XNB	**TEMPER**
	Telegraph address ....DJP		
XLW	—Telegraph Admiralty, *or*, Navy Depart-	XNC	**TEMPERATE-LY-ANCE**
	ment to —.	XND	—Temperate zone
XLY	—Telegraph buoy		
	Telegraph cable ....HRW	XNE	**TEMPERATURE.**
XLZ	—Telegraph clerk (*or*, official)	XNF	—Have you tried the temperature of the
XMA	—Telegraph code		sea?
	Telegraph following message to— (*ship*	XNG	—The temperature by wet bulb is —, and
	or person named) at — ....XD		by dry bulb is —
	Telegraph instruction (order) to —..TC	XNH	—The temperature in the shade is —
XMB	—Telegraph instruments		The temperature in the sun is —..XDW
XMC	—Telegraph out of order (*or*, stopped)	XNI	—The temperature is at —
	Telegraph reply Answer by telegraph,	XNJ	—The temperature of the sea is —
	WQ	XNK	—What is the temperature? What is the
XMD	—Telegraph ship		thermometer?
XME	—Telegraph station (*or*, office)	XNL	—What is the temperature by wet and
	Telegraph to my agents ....DSV		dry bulb?
XMF	—Telegraph to my family at — the fol-	XNM	—What is the temperature in the shade?
	lowing message		What is the temperature in the sun?
XMG	—Telegraph to my owner		XDY
	Telegraph to my owners to send my let-	XNO	**TEMPEST UOUS**
	ters to me at —....................VQP		
	Telegraph wire.... ... .ZGO	XNP	**TEMPLATE**
	Telegraphic communication LVC		
	Telegraphic despatch arrived from —,	XNQ	**TEMPLE**
	JX		
	Telegraphic message ....XLN	XNR	**TEMPORARY-ILY**
	Telegraphic news—arrived from —..JX		Engines broken down, under temporary
	There is a telegraph cable on — (*bearing*		repair .... .MCK
	indicated) ....EN		Engines under temporary repair....MBX
XMH	—To whom shall we telegraph that all is	XNS	—Temporary light shown at — (*place*
	well on board your vessel?		*indicated*)
	Vessel desires to be reported by tele-		Under temporary repair ....YHK
	graph to owner (Mr —) at —....USF		
	Will you telegraph to—(*ship or person*		
	named) the intelligence I am about to		
	communicate? ....IM	XNT	**TEMPORIZE-D ING**
XMI	—You can telegraph as far as —	XNU	**TFN-TH-LY**
XMJ	—You can telegraph from		
		XNV	**TENANT**
		XNW	**TEND-ED-ING**
XMK	**TELEPHONE-D-ING TELEPHONIC**		
XML	—Can I telephone to —?	XNY	**TENDER-ED-ING TO TENDER**
	Diver's telephone ....LIA		Mail tender ....RFO
XMN	—Telephonic communication		
XMO	—Telephonic message		**TENNIS LAWN TENNIS** ....QIP
XMP	—You can telephone to —		
		XNZ	**TENON**
XMQ	**TELESCOPE-IC SPYGLASS**		
XMR	—Water telescope	XOA	**TENSILE**

	TENSION—THE			
XOB	TENSION			THAT—*Continued*
XOC	TENT			Shall, *or*, Will that (*or*, this)?_____CIV
	Send tent on shore_____VQI			Should, *or*, Would that (*or*, this)?__CJS
XOD	TENURE			That_____CKP
XOE	TERMS			That, *or*, This can (*or*, may) _____CKQ
XOF	TERMINATE-D-ING-ION			That, *or*, This can (*or*, may) be____CKR
	Termination of articles_____EZM			That, *or*, This can not (*or*, may not)_CKS
XOG	TERMINUS			That, *or*, This can not (*or*, may not) be,
				CKT
	TERRACE, *or*, BOULEVARD_____HBP			That *or*, This could (*or*, might)___CKU
XOH	TERRIBLE-Y			That, *or*, This could (or might) be_CKV
XOI	TERRITORY-IAL			That, *or*, This could (*or*, might) not be,
XOJ	TEST-ED-ING			CKW
	Test battery _____FYR			That, *or*, This had (has, *or*, have)_CKX
XOK	—Test coil			That, *or*, This had (has, *or*, have) been,
XOL	TESTAMENT, *or*, WILL			CKY
XOM	TESTIFY-IED-ING-MONY			That, *or*, This had (has, *or*, have) not
XON	TESTIMONIAL			been _____ _____ _____CKZ
XOP	TEXTURE			That he (she, it, *or person-s or thing-s*
				indicated) _____CLA
	THALER (*Coin*)_____AVD			That his (her-s, it-s) _____CLB
	THAN _____,_____CKF			That I (my, mine)_____CLD
	Less than _____QNL			That, *or*, This is_____ CLE
	More than _____ _____RZU			That, *or*, This is not_____ _____CLF
	Than the _____CKG			That it is _____ CLG
	Than these (*or*, those) _____CKH			That it is not _____ _____CLH
	Than they (their-s) _____CKI			That, *or*, This must_____ ____ CLI
	Than we (our s)_____CKJ			That, *or*, This must be_____CLJ
	Than which (*or*, what)_____CKL			That, *or*, This must not _____CLK
	Than you-r-s_____CKM			That, *or*, This must not be _____CLM
XOQ	THANK			That, *or*, This shall (*or*, will)be____CLN
XOR	—Thank you			That, *or*, This shall (*or*, will) not be,
XOS	—Thanked			CLO
XOT	—Thanking			That, *or*, This should (*or*, would) be_CLP
XOU	THANKFUL-LY-NESS			That, *or*, This should (*or*,would) not be,
				CLQ
	THAT _____CKP			That the ___ _____ __ _____CLT
	After that (*or*, this)_____CKN			That these (*or*, those) _____CLU
	Before that (*or*, this)_____CKO			That they (their-s)_____ ___ _____CLV
	By that (*or*, this)_____BJT			That, *or*, This was _____CLR
	Can, *or*, May that (*or*, this)?_____BKI			That, *or*, This was not_____ _____CLS
	Could, *or*, Might that (*or*, this)?___BLE			That we (our-s) _____CLW
	Do, Does, *or*, Did that (*or*, this)?___BMJ			That which (*or*, what) _____CLX
	For that (*or*, this)_____BPF			That with_____CLY
	From that (*or*, this)_____BPR			That you-r-s_____CLZ
	Had, Has, *or*, Have that (*or*, this)_BQW			This _____ ____ __ CMA
	If that (*or*, this) ___ _____BZX			To that (*or*, this)_____COA
	In that (*or*, this)_____ _____CDM			Was, *or*, Were that (*or*, this)_____COZ
	Is that (*or*, this)? _____ _ ___ ___CDV			With that (*or*, this)_____ _____CWJ
	May, etc (*See* Can, *above*)	XOV	THATCH-ED-ING	
	Might, etc (*See* Could, *above*)	XOW	THAW-ED-ING	
	Must that (*or*, this)? _____CEX			
	Not that (*or*, this)_____CFZ			THE _____CME
	Of that (*or*, this)_____ ___ _CGQ			After the _____CMB
	On that (*or*, this) _____CHB			And the _____AMD
	Ought that (*or*, this)?_____CHU			Are, *or*, Is the_____BEN
				Be the _____ ____ ____BFE
				By the _____BJU
				Can *or*, May the?_____HUE
				Could, *or*, Might the?_____ _____BLF
				Do, Does *or*, Did the?_____BMK
				Each of the (*or*, them)_____LQU
				Either of the (*or*, them)_____ ____LUE
				For the _____BPG
				From the_____BPS
				Had, Has, *or*, Have the?_____BQX
				If the _____BZY
				In the_____ ___CDN
				Is, etc (*See* Are, *above*)

THE—THESE

THE—*Continued*
May, etc (*See* Can, *above*.)
Might, etc (*See* Could, *above*)
Must the?...CEY
Not the...CGA
Of the...CGB
On the...CHD
Ought the?...CHV
Shall, *or*, Will the?...CIW
Should, *or*, Would the?...CJT
Than the...CKG
That the...CLT
The...CME
The most...SAW
The same as...VGN
The same description...KWG
The same kind—as...QAY
To the...CNX
To the rear...UKO
Was, *or*, Were the?...COW
What, *or*, Which of the?...CQI
When the...CSF
Where should (*or*, would) the?...CTW
Where the...CTY
Will the? (*See* Shall, *above*)...CIW
With the...CWI
Would the? (*See* Should, *above*)...CJT

XOY THEATRE-ICAL.

XOZ THEFT THIEVING

THEIR (*See* THEY)...CNL

THEM (*See* THEY)...CNL

XPA THEN

XPB THENCE

XPC THEODOLITE

THERE...CMF
Are, *or*, Is there—any?...BEP
By there...BJW
Can, *or*, May there be?...BFI
Could, *or*, Might there be?...BFP
Do, Does, *or*, Did there?...BML
From there...BPU
Had, Has, *or*, Have there?...BQY
If there...CAB
Is, *or*, Are there — any —?...BEP
Must there be?...CEZ
Not there...CGB
Shall, *or*, Will there be?...BHE
Should, *or*, Would there be?...BHO
There...CMF
There are (*or*, is)...CMG
There are (*or*, is) not...CMH
There can (*or*, may)...CMI
There can (*or*, may) be!...CMJ
There can not (*or*, may not)...CMK
There can not (*or*, may not) be...CML
There could (*or*, might)...CMN
There could (*or*, might) be...CMO
There could (*or*, might) not be...CMP
There had (has, *or*, have)...CMQ
There had (has, *or*, have) been...CMR
There had (has, *or*, have) not been...CMS
There is nothing...CMT
There shall (*or*, will)...CMU

THERE—*Continued*
There shall (*or*, will) be...CMV
There shall (*or*, will) not...CMW
There shall (*or*, will) not be...CMX
There should (*or*, would)...CMY
There should (*or*, would) be...CMZ
There should (*or*, would) have been...CNA
There should (*or*, would) not...CNB
There should (*or*, would) not be...CND
There should (*or*, would) not have been,
CNE
There was (*or*, were)...CNF
There was (*or*, were) not...CNG
Was, *or*, Were there?...COX

XPD THEREABOUTS

XPE THEREBY

XPF THEREFORE

XPG THERMOMETER (*For Thermometer Table according to Fahrenheit, Centigrade, and Réaumur graduations, see page 60*)

XPH —Have you a thermometer?

XPI —I have no thermometer
The thermometer by wet bulb shows —, and by dry bulb shows —...XNG
The thermometer in the shade is —...XNH
The thermometer in the sun is—...XDW
The thermometer is at —...XNI
What is the height of the thermometer?
What is the temperature?...XNK
What is the height of the thermometer by wet and dry bulbs?...XNL
What is the height of the thermometer in the shade?...XNM
What is the height of the thermometer in the sun?...XDY

THESE THOSE...CNI
After these (*or*, those)...CNH
Are these (*or*, those)?...BEQ
By these (*or*, those)...BJX
Can, *or*, May these (*or*, those)?...BKJ
Could, *or*, Might these (*or*, those)?...BLG
Do, Does, *or*, Did these (*or*, those)?...BMN
For these (*or*, those)...BPI
From these (*or*, those)...BPV
Had, Has, *or*, Have these, (*or*, those)?
BQZ
If these (*or*, those)...CAD
In these (*or*, those)...CDP
May, etc (*See* Can, *above*)
Might, etc (*See* Could, *above*)
Must these (*or*, those)?...CFA
Not these (*or*, those)...CGD
Of these (*or*, those)...CGT
On these (*or*, those)...CHF
Ought these (*or*, those)?...CHW
Shall, *or*, Will these (*or*, those)?...CIX
Should, *or*, Would these (*or*, those)?
CJU
Than these (*or*, those)...CKH
That these (*or*, those)...CLU
These...CNI
Those...CNI
To these (*or*, those)...CNZ
Were these (*or*, those)?...CPG
When these (*or*, those)?...CSG

THESE—THICK

THESE—*Continued*
Will, etc   (*See* Shall, *above*)
With these (*or*, those) ----- -----CWL
Would, etc   (*See* Should, *above*)

THEY   THEM   THEIR-S------------CNL
After they (them, their s) ---------CNJ
And they--------- - ---------- -ELM
Are, *or*, Is their s, *or*, Are they? --BEO
Before them (their-s, they) ---------CNK
By them (their-s) ----- -----------BVJ
Can, *or*, May they (their-s)?-------BKL
Can, *or*, May they not? ----------BKM
Could, *or*, Might they (their-s)? ---BLH
Do, Does, *or*, Did they (their-s)?---BMO
For them (their-s) ------ ---------BPH
From them (their-s)-------------BPT
Had, Has, *or*, Have they (them, their-s)?
----------------------------BEA
Had, *or*, Have they been?--------BFT
Had, *or*, Have they done?--------BMU
Had, *or*, Have they had?--------BRC
How are they?----------------BXH
How will they?-----------------BXY
If they (their-s)-----------------CAE
If they are ---------------------CAF
If they are not -----------------CAG
If they can (*or*, may) ----- --------CAH
If they can not (*or*, may not)-----CAI
If they could (*or*, might) --- ------CAJ
If they could (*or*, might) not ------CAL
If they do (*or*, did) ------------CAM
If they do (*or*, did) not ----------CAN
If they had (*or*, have) ----------CAO
If they had (*or*, have) not --------CAP
If they shall (*or*, will)-----------CAQ
If they shall (*or*, will) not--------CAR
If they should (*or*, would)--- ------CAS
If they should (*or*, would) not-----CAT
If they were --------------------CAU
If they were not -----------------CAV
In them (their-s)----------------CDO
Is, etc   (*See* Are, *above*)
Let them be------- -----------BGW
Let them have-----------------BRV
Most of them--------------------SAX
Must they (their-s)?---- ----------CFB
Not they (their-s) --------- -----CGE
Of them (their-s)----- ---- ------CGS
On them (their-s) ---------------CHE
Ought they to? --------- -----CHX
Shall, *or*, Will they (their-s)?-------CIY
Shall, *or*, Will they be? --------BHF
Shall, *or*, Will they do it? ---------BOA
Shall, *or*, Will they not?---------CIZ
Shall, *or*, Will they not do it? -----BOC
Should, *or*, Would they (their-s)---CJV
Than they (their-s) ------------CKI
That they (their-s)---------------CLV
They, Them, Their-s ------------CNL
They are-----------------------BEU
They are for ------------- ------CNM
They are from------------------BET
They are not—to ----- ---------BEV
They can (*or*, may)----- --------BKT
They can (*or*, may) be-----------BHP
They can (*or*, may) have---------BSF
They can not (*or*, may not)--------BEU
They can not (*or*, may not) be -----BHQ
They can not (*or*, may not) have---BSG
They could (*or*, might)------------BLN

XPJ

THEY—*Continued.*
They could (*or*, might) be---- ----BHR
They could (*or*, might) not--------BLO
They could (*or*, might) not be -----BHS
They do (*or*, did)-- --- ----------BOJ
They do (*or*, did) not------------BOK
They had (*or*, have)-------------BSH
They had (*or*, have) been----- ----BHT
They had (*or*, have) done---------BOL
They had (*or*, have) had - -- -----BSI
They had (*or*, have) not ---------BSJ
They had (*or*, have) not been --- -BHU
They had (*or*, have) not done -----BOM
They had (*or*, have) not had------BSK
They must ------------------ ----CFG
They must be---- -------------- - -BHV
They must not -----------------CFH
They must not be ------------ --BHW
They ought to -----------------CIE
They ought not to ----- ---------CIF
They ought not to be--------- ----BHY
They ought to be-----------------BHX
They shall (*or*, will)------------CJF
They shall (*or*, will) be-----------BHZ
They shall (*or*, will) not----------CJG
They shall (*or*, will) not be--------BIA
They should (*or*, would)----------CJY
They should (*or*, would) be --- ---BIC
They should (*or*, would) have------BSL
They should (*or*, would) not ------CJZ
They should (*or*, would) not be-----BID
They should (*or*, would) not have---BSM
They were---- ------------------COJ
They were not---- ----------------COK
They were not to be ----- -------BIF
They were to be------------------BIE
To them (their-s) ---------------CNY
Was, *or*, Were they (their-s)? -----COY
Were they?---------------------CPH
Were they not?-------------------CPI
What, *or*, Which do (*or*, did) they (we,
or, you)?------- -------------- CQE
What, *or*, Which shall (*or*, will) they
(*or*, you) do?---- -----------CQN
What, *or*, Which shall (*or*, will) they
(*or* you) have -----------------CQO
What, *or*, Which should (*or*, would)
they (*or*, you) do? ------- -- -CQS
When do (*or*, did) they (*or*, you)? -CRF
When shall (*or*, will) I (they, we, *or*,
you) do? ---------- ------------CRV
When shall (*or*, will) they?-------CRY
When shall (*or*, will) they (*or*, you)
have ------- -------- ------CRZ
When they (their-s) -------------CSH
When they are --------------------CSI
Where are they?------ -----------CST
Where do (*or*, did) they (*or*, you)?--CSZ
Where shall (*or*, will) they?-------CTP
Where shall (*or*, will) they (*or*, you)
have? -------------------------CTQ
Where should (*or*, would) they (their-s)?
----------------------------CTX
Where they (their s)---- ---------CTZ
Where they are -----------------CUA
Why they (their-s) --------------CVW
With them (their-s) ----- -------CWK

THICK
Delayed by thick fog Detained by thick
fog ------ -------------------KSE

THICK—THROAT

	THICK—*Continued.*	XQL	THINK—*Continued*
	During a thick fog ---------------LPN		—What do you think of doing?
	Fog too thick------ ----------------NGJ		What do you think of the weather?..YZK
	I have had thick fog on the bank ..FVO		What do you think of vessel chasing?
	Shall we have thick weather?------YZD		INX
	Thick fog----------------------------NGM		Which do you think the best? ------ GIE
	Thick fog coming on---------- ---- GJ	XQM	THIRD
	Thick weather ---- - ------- ------ZJ		I can not make out the third flag..NDH
XPK	—Thickness		Third class------- ------ ----------ISJ
XPL	THICKEN ED-ING	XQN	—Third mast
		XQO	—Third mate (*or,* officer)
XPM	THIEF		Third week in — -----------------ZAP
	THIEVING THEFT ---------------- XOZ	XQP	THIRST-ED-ING-Y
XPN	THIGH		Dying of thirst ----------------------NF
XPO	THIMBLE	XQR	THIRTEENTH
XPQ	THIN-LY-NESS	XQS	THIRTIETH
XPR	—Too thin	XQT	THIRTY FIRST
XPS	THING.		THIS (*See also* THAT) ------------- CMA
	Anything ------------------------EQD		This afternoon ---------------------DRY
	Everything --------------------,-- MJG		This day---- ------------------- -- KNV
	Everything is to be sold ------- -- MJH		This evening ------- -- ----------MIK
XPT	—Not such a thing		This month ---------------------- - RYV
	Nothing ---------------------------SKA		This morning --------- -- ---- ----SAD
	Something---- ------------------WJQ		This night ----- - --- ----------SIJ
	The same thing --------------------VGO		This quarter -----------------------UET
XPU	THINK-ING TO THINK		This side----------------- -------- ----WBN
XPV	—Do not think		This tide ------- --- ------- - ---- XSF
XPW	—Do you think?		This time ---- ---- - ------------XTR
	Do you think I could get through the		This week --------- --------------ZAQ
	ice?----------------- ----------PBH		This year ----------------- --------ZKY
XPY	—Do you think it will be —?	XQU	THITHER
	Do you think the gale is over? ------NVX		
	Do you think we can ride it out?..UWX		THOMSON'S (KELVIN) PATENT COMPASS..JIG
XPZ	—Do you think we can get away from —?		THOMSON'S PATENT SOUNDING APPARATUS,
	Do you think we ought to pass a vessel		TEG
	in this distressed condition?-------NE	XQW	THOROUGH-LY
	Do you think you can cut her (*or,* them)		The engines must be thoroughly over
	out? ---- ----------- ---------KHN		hauled ---------- - ----------MCB
XQA	—I do not think—it —		Thorough caulking necessary ----- IFM
	I do think it is necessary-----------SFU	XQY	THOROUGHFARE
	I do think it will be-- ----------LIO		
XQB	—I do not think it will do		THOSE (*See* THESE) ------ ----------CNI
	I do not think marks (*or,* buoys) can be		
	in their proper positions .. ---HMK		THOUGH ALTHOUGH ----- -------- EDF
	I do not think there is any light here-		
	abouts --------------- -----------QRW	XQZ	THOUGHT-FUL-LY-NESS
XQC	—I do not think they will	XRA	THOUSAND-TH
	I do not think we shall get any good by		
	moving ---------------------------OEG	XRB	THREAD
XQD	—I think—it —	XRC	THREATEN-ED-ING TO THREATEN
XQE	—I think I can		Appearances are threatening, be on
XQF	—I think I could		your guard. --------- ---- ------FN
	I think it will blow----------------GOE		Is, *or,* Are threatened by the enemy.LZQ
XQG	—I think so		
	I think the land is-----------------QED	XRD	THREE THRICE
	I think the land must have been about —,		Three blades of propeller broken ..GMB
	QRF		Three-masted schooner -----------VJK
XQH	—I think the weather will be —		
	I think you might get something to		
	answer the purpose--------------ENB	XRE	THROAT
XQI	—If you think—it —		
	If you think it necessary-----------SFV		
XQJ	—Thought		
XQK	—What do you think—of ?		

THROTTLE—TIFFIN

XRF	THROTTLE-D ING
XRG	—Throttle valve
XRH	THROUGH
	Can one break through the ice?....HEY
	Do you think I could get through the ice?........PBH
	The ice is so solid I can not break through, send help.........FCF
	Through, or, Into the channel ILV
	Through the day ....KNW
	Through the fog ....NGO
	Through the strait ....WYM
	Through the water ....YVL
	Through the winter ....ZGJ
XRI	THROUGHOUT
XRJ	THROW-N ING THREW
	Enemy is throwing up batteries....FYS
	Have you any means of throwing a line?.............QUJ
	Mortar for throwing a line.........SAP
	Obliged to throw overboard.......SMY
	Rocket to throw a line L S A rocket, UZG
	Was thrown on her (or, my) beam ends, GAX
XRK	THRUM-MED
XRL	THRUST
	Thrust bearing...............GBQ
	Thrust block broken...........GMZ
	Thrust block damaged..........GNA
	Thrust block rings defective. .....GNB
	Thrust collars damaged ....JAD
XRM	—Thrust shaft
XRN	—Thrust shaft broken
XRO	THUMB
XRP	THUNDER ED-ING
	There has been, or, We have had a storm of thunder and lightning..QTL
XRQ	THURSDAY
XRS	—Before Thursday
	Last Thursday ....QGP
XRT	—Next Thursday
XRU	—Thursday morning
XRV	—Thursday night
XRW	THUS
XRY	THWART-ED-ING TO THWART
XRZ	THWART (Seat of a boat)
	TICAL (Coin) ....AVE
XSA	TICKET
	Pay ticket .... TER
	Ticket of admission ....DMF
XSB	TIDE
	Against the tide (current, or, stream), DSE
	As soon as the flood tide makes ....FBP
	Athwart the current (tide, or, stream), FJI

	TIDE—*Continued*
	Bar impassable for boats on the ebb tide......................BQ
XSC	—By to-morrow's tide
	Can not stem the tide (or, current).KFX
	Do not abandon the vessel until the tide has ebbed...............AD
	Do you know the tides?...........QBY
	Drop down with the tide.........LOC
	Ebb tide ....LSN
	Eddy tide......................LFG
	Flood tide ....NFO
	Half tide.......................OLN
XSD	—High tide.
	How is the tide? What tide have we now?..........................XF
	Keep out of the tide (or, current)....XG
	Lee tide.......................QMI
	Neap tide ....SFD
XSE	—Next tide
	Race Tide race.................UGH
	Spring tide ....WOX
	Spring tides rise — feet............WOY
	The tide will float her off..........NFE
XSF	—This tide
XSG	—Tidal
	Tidal dock .... LJM
XSH	—Tide flows at —
XSI	—Tide has fallen — feet
XSJ	—Tide has risen — feet
XSK	—Tide is falling
XSL	—Tide is rising
XSM	—Tide rises — feet
XSN	—Tide rode
	Tide, or, Current sets to the — ....KGE
XSO	—Tide tables
	Tide, or, Current will run very strong (*Miles per hour may follow*).....KGB
	Try the current (or, tide)..........KGF
	Was the tide high when you grounded? CI
	Was the tide low when you grounded? CJ
	Weather tide..................YZT
	What is the rise and fall of tide?....XH
	What is the set and rate of tide (or, current)? ....KGJ
XSP	—What is the strength of the tide?
	When is the highest tide?.........OVH
XSQ	—When the tide
	When the tide slacks.............WFS
XSR	—When the tide turns
	When will the tide ebb? XI
	When will the tide flow?...........XJ
XST	—When will the tide suit?
	What will the tide turn?...........XK
XSU	—Will the tide suit?
XSV	—With the tide
	You will not be afloat all the tide..DQG
XSW	TIDINGS
	Have you any tidings of the crew?.KDB
XSY	TIE-D-ING TO TIE
XSZ	TIER
	Ground tier... ....OIL
	TIERCE (*Measure of capacity*)......AZS
	TIFFIN LUNCH...................RDJ

TIGHT—TO

XTA	**TIGHT**-LY-NESS-ER
	Air-tight ... .DWJ
XTB	—Packed-tight
XTC	—Water-tight
	Water-tight compartment..........JHB
	Water-tight door....................LKS
XTD	**TIGHTEN**-ED-ING
	TILL (*See also* UNTIL)............. YLC
	Can you stop till —?....... ......WXD
	Will hold together till —..........OWY
XTE	**TILLER**
	Tiller rope ....VAB
XTF	**TIMBER**
	Baulk of timber...................FZG
	Timber cargo ....HZN
XTG	—Timber ship
XTH	**TIME**
	(*For Hours and Minutes see page 58*)
	About the time...................DAB
	Another time ....EML
	Apparent time...................ERM
	Asks precise time of high water and minimum depth at that time . .. XE
	At all times....................DYJ
XTI	—At any time
	At no time ....FIL
	At the same time The same time.VGP
	At the (*or* that) time specified ....FJA
	At what time (*or*, hour)?...........FJC
	At what time do you expect?..... .MND
	Barely time to save the mail........REY
	Best time ....GHY
	By what time?...................HPT
	Can I depend upon your time? ......RS
	Can you depend upon your time? ..KUE
XTJ	—Can you fix any time for —?
XTK	—Can not fix any time—for —
	During the time in harbor.........LPV
	Enough time... ....MDZ
	Every time ....MJF
	First time ....NBH
	Forward message immediately without loss of time... ....NMY
XTL	—From time to time
	Have been on short allowance for some time...................... ....DZI
	How long a time?.................QYL
XTM	—How many times?
	I can not depend upon my time....KUH
	If time will admit ....DMC
	In a short time...................VZT
XTN	—In time
	Is there time ball at —? ....FTH
	Last time....................QGR
	Latest posting time—is ....QHG
	Local time ....QXW
	Long time—ago Long ago ....DTL
	Lose no time......................RAP
	Lose no time in getting to the anchorage ....ELC
	Lose no time in shoring up...........CF
	Lose-ing, Lost time . ...RAS
	May I depend upon your time? ....RS
	Mean time... ....RNP
	Minute of time ....RVE

	TIME—*Continued*
	My chronometer is fast of Greenwich (*or*, first meridian) mean time ....GR
	My chronometer is slow of Greenwich (*or*, first meridian) mean time ....GS
	My first meridian (*or*, Greenwich) mean time is — XL
	Next time (*or*, opportunity) ....SRY
XTO	—Not time enough .
XTP	—Short time
	Sometime ....WJS
	Sometime ago (*or*, since) ....DTO
	The best time.................GHY
	The same time At the same time .VGP
XTQ	—There will be time—to (*or*, for)
XTR	—This time
	Time appointed ....ETC
	Time ball ....FTI
	Time ball drops at — ....FTJ
	Time enough ....MDZ
	Time fuze ....NVA
	Time-table ....XIG
XTS	—Time will be gained—by
XTU	—Time will be lost
XTV	—Timely
	What are the time arrangements for —? EXL
	What is the precise time of high water, and minimum depth at that time?.XE
	What time is dinner? .... LAW
	When does time ball drop?..........FTK
	When were your last observations for time? ....RY
	When will be the best time for crossing the bar?... ....XM
	Will take — time to complete .. .JKW
	Will you show your Greenwich (*or*, first meridian) mean time? ....XN
	TIMELY ....XTV
XTW	**TIN**
	Tinned, *or*, Canned.................HUC
XTY	—Tinplate
XTZ	—Tinsmith
	TINCTURE
XUA	—Tincture of henbane
	Tincture of rhubarb ....UWQ
XUB	—Tincture of steel
XUC	**TIP**-PED-PING
	Coal tip.. ....IWT
XUD	**TIRE**-D-ING
	Much tired (*or*, fatigued) ....MTX
	Too tired (*or*, fatigued) ....MTY
XUE	**TITLE**
	TO ....ONO
	Am, Is, *or*, Are to...................BBH
	Am I to? ....BEF
	As to................... ......FBS
XUF	—Does not belong to
	Heave to ....OSC
	Heaving, Hove-to.... ....OSD
	Lay to ....QJC
	Me to ....CEF
	Not to ....CGF
	Not to be ....CGH

TO—TOO

	To—*Continued*	**XUO**	**To-morrow**
	Not to be done......................CGI		Before to-morrow . ...SAH
	Not to be had ....CGJ		By to-morrow's post To-morrow's post,
	To ....CNO		TQH
	To be ... .CNP		By to-morrow's tide ....XSC
	To be at ....CNQ		The day after to-morrow . . ..DRV
	To be done.... ....CNR		To-morrow afternoon............... DRZ
	To be had ....CNS		To-morrow evening.... ....MIL
	To-day ....KNX		To-morrow morning................SAE
	To-day's post ....KNY	**XUP**	—To-morrow night
	To do ....CNT		To-morrow-'s post By to-morrow's post,
	To have...........................CNU		TQH
	To him (his, her-s, it s)... .. ..CNV		What is the meteorological forecast for
	To leeward ....QML		to-morrow? ....ZL
	To me (my, mine) ....CNW		
	To-night's ... .XUQ	**XUQ**	**To-night**
	To-night's post .... XUS	**XUR**	—I sail to-night
	To rendezvous at — ....UQO	**XUS**	—To-night's post
	To see you ....VOB		
	To the .. ..CNX		**Ton** *(Measure of cubic capacity)* ...AYI
	To the buoy ....HMX		
	To, or, For the eastward...........LSE		**Ton** *(Measure of weight)* ....BAV
	To the island.....................PTR		Coals — shillings a ton at — ....GZ
	To, or, For the northward ....SJF	**XUT**	—How many tons—of?
	To, or, At the rendezvous ... ..UQP	**XUV**	—How many tons do you take?
	To, or, For the southward .... ..WLK		How many tons dead weight of cargo
	To, or, For the westward ....ZBS		must you have?................. IAS
	To, or, By the wind ....HPQ		How many tons measurement of goods
	To them (their-s) ....CNY		can you take?..... ... .OEV
	To these (or, those) .... CNZ	**XUW**	—Ton measurement
	To this (or, that).. .... COA		Ton of coal.......... ....IXK
	To us (our-s) .... COB		Ton measurement of goods ....OEY
	To which (or, what) .... COD	**XUY**	—Ton of water
	To whom (or, whose).............COE		Want — tons dead weight of cargo.. IBE
	To windward ....ZFY		
	To you-r-s. .. ..COF		**Tonde** *(Measure of capacity)* ......AZT
	Went—to ....ZBQ		
	Wrote to — on —.................ZJV	**XUZ**	**Tongue**
	You are ordered to —.......... .. SUZ		
XUG	**Tobacco-nist**	**XVA**	**Tonic**
	Tobacco pipe....................TKN	**XVB**	**Tonnage**
			Registered tonnage—is ....UNT
XUH	**To-day**... ....KNX		Tonnage burden ....HNB
	—To-day's		Tonnage gross..... .. .. .OID
	To-day's post . ...KNY		What is your (or vessel's indicated) reg-
	What is the meteorological report for		istered tonnage?................UNV
	to-day? ....ZK		
			Tonne *(Measure of weight)* ....BAV
XUI	**Toe**		
XUJ	**Together**	**XVC**	**Too**
	Let us keep together for mutual protec-		Too big ....GJN
	tion ....IN		Too close (or, near) ....IVJ
	We can defend ourselves if we keep		Too dear...... ... .KOV
	together ... .PYE		Too deep...........................KRB
	Will hold together till — ....OWY		Too far off ....MTD
XUK	**Toggle-d-ing**		Too far to the eastward ....LSF
			Too far to the northward .. ..SJQ
	Told *(See Tell)* ....•.....XMS		Too far to the southward ... .WLM
	Who told (or, said so)? ....XMZ	**XVD**	Too far to the westward ....ZBT
			—Too few
XUL	**Tolerable-y**		Too large ....GJN
XUM	—Tolerable observations		Too large a number (or, quantity) of,
	Tolerably healthy Tolerably well.OQZ		QFV
XUN	**Toll**		Too late... . ... QHI
	Toman *(Coin)*.................AVF		Too little Too small.............QVX
			Too long... .. ..QYX
	Tomme *(Measure of length)* ....AWY		Too low..... .. ..RBX
			Too much (or, many)..............RKS
			Too much disabled ....LBX
			Too much scope ....VJQ

TOO—TORPEDO

	Too—*Continued*
	Too much sea .... UV
	Too much surf .... XFU
	Too much swell.................... XHK
	Too much wind .... ZFB
XVE	Too near....... .... IVJ
XVF	—Too old
	—Too short
	Too small (*or*, little) .... QVX
	Too soon .. .. WKB
	Too thin .... XPR
	Too tired (*or*, fatigued) .... MTY
XVG	Took (*See* Take) .... XKA
	Tool
	Carpenter's tools .... IBW
	Caulker's tools........... .... IFK
	Cooper's tools ....JVK
	Engineer's tools.................. MCV
	I have not carpenter's tools Want carpenter's tools..................... IBX
XVH	—I have not tools I have no tools
	Let them bring their tools with them, HJD
	Painter's tools (*or*, brushes)........ HKP
	Vessel named can lend you carpenter's tools........................... IBZ
XVI	—With tools
XVJ	**Tooth**
XVK	**Top-ped-ing To top**
	At top of .... FJB
	Foretop .... NJF
	I can not make out the top flag..... NDI
	Maintop .... BGY
	Military top .... RUC
	Mizzentop .... RWM
XVL	—Top (*The summit of*)
	Top up boat...................... GVB
XVM	—Top rope
XVN	—Top-tackle pennant
XVO	—Top your lower yard
XVP	**Topgallant**
	Fore topgallant mast .... NJG
	Fore topgallant sail .... NJH
	Fore topgallant yard .... NJI
	Jigger topgallant yard and sail . PVT
	Main topgallant mast .... RGZ
	Main topgallant sail .... RHA
	Main topgallant yard.............. RHB
	Mizzen topgallant mast........... RWV
	Mizzen topgallant sail ... . RWX
	Mizzen topgallant yard .... RWY
	Set topgallant sails................ VEP
	Take in topgallant sails............ XJP
XVQ	—Topgallant mast
XVR	—Topgallant rigging
XVS	—Topgallant sail
	Topgallant sheet .... VUZ
XVT	—Topgallant studding sail
XVU	—Topgallant yard
XVW	**Topmast**
	Fore topmast ....NJK
	Fore topmast sprung ....NJL
	Fore topmast staysail .... NJM
	Main topmast....... ....RHC
	Main topmast staysail.............RHD

	Topmast—*Continued*
	Mizzen topmast ....RWZ
	Mizzen topmast staysail........ ...RXA
	Topmast backstay ....FQP
	Topmast cap...... .... HVM
XVY	—Topmast crosstrees
XVZ	—Topmast rigging
	Topmast staysail .... WSG
XWA	—Topmast staysail sheet............. VWA
	—Topmast studding sail
XWB	**Topsail**
XWC	—Back the main topsail
	Double topsails.. .... LKZ
XWD	—Double topsail yard
	Fill the main topsail .... MYB
	Fore topsail....... .... NJO
	Fore topsail yard ....NJP
	Gaff topsail. .... NVG
	I have carried away topsail yard...ZKN
	I have sprung (*or*, damaged) lower topsail yard.... ....KJI
	I have sprung (*or*, damaged) topsail yard ....KJM
	I have sprung (*or*, damaged) upper topsail yard . ...KJN
	Jigger topsail yard and sail........PVU
	Keep main topsail shivering ......PZK
	Keep mizzen topsail shivering .....PZK
	Lower fore topsail yard ....NKI
	Lower main topsail yard ....RGF
	Lower mizzen topsail yard.........RWO
	Lower spanker topsail yard ....WLU
	Lower topsail . ...RCH
	Main topsail ....RHE
	Main topsail yard ....RHF
	Main topsail yard sprung............REG
	Mizzen topsail ....RXB
	Mizzen topsail yard ....RXC
XWE	—Shift a topsail
	Topsail sheet .... VWB
XWF	—Topsail yard
	Upper fore topsail yard ....NKJ
	Upper main topsail yard ....RHK
XWG	—Upper spanker topsail yard
XWH	—Upper topsail
XWI	—Upper topsail yard
XWJ	**Topside Topside planking**
	Tore Torn .... XLK
XWK	**Tornado**
XWL	**Torpedo**
	Beware of torpedo boats.............XO
	Beware of torpedoes, channel is mined, XP
	Boom for torpedo net... ....GZY
XWM	—Directible torpedo
	Enemy's torpedo boat .. ..LZM
	Enemy's torpedo boats have been seen to the —, steering to the —........OH
	First-class torpedo boat.............GSX
	Have you seen any torpedo boat?....XE
	How many torpedo boats passed in sight? GTK
	Is there danger of mines (*or*, torpedo boats)? ....XS
	Line of torpedoes ....QUR
	Locomotive torpedo.................QYD

TORPEDO—TRADE

XWN	TORPEDO—*Continued* —Mine, *or*, Torpedo buoy. Outrigger torpedo, *or*, Spar torpedo.SWR Saw torpedo boat (*number, if necessary*) at — (*or*, near —)................XT Submerged torpedo buoy..........XBL		Tow—*Continued.* Engines broken down, towed in by —, MCJ Fasten your chain to towing cable..HRC
XWO XWP	—Torpedo attack —Torpedo boat	XYP XYQ	—Have a towline ready —I can not take you in tow, but will re- port you at — and send immediate assistance
	Torpedo boom.....................GZY Torpedo (*or*, Mine) buoy..........XWN Torpedo carriage.................ICF		I will take her in tow .... XJG I will take you in tow ....XJH I wish to be taken in tow ....XW
XWQ XWR	—Torpedo case —Torpedo catcher (*or*, chaser, *or*, de- stroyer)	XYR XYS XYT	—In tow —Prepare to be towed clear. —Reduce speed, you are towing me too fast
XWS XWT XWU XWV XWY XWZ	—Torpedo cruiser —Torpedo gun —Torpedo gunboat (*or*, vessel) —Torpedo net —Torpedo station —Torpedo tube		Send boat to tow...................GUQ Shall I take you in tow?............XY Ship disabled, will you tow me into — (*port indicated*)?................MY Stop engines to adjust towing cables, DKH
XYA	Whitehead torpedo........ ....ZDM — — (*number*) torpedo boats (*give nation- ality*) passed in sight		Take her in tow ....XJL Take in tow........................XJQ Tow astern........... ... .FHS
XYB	TORRENT	XYU	—Towed by, *or*, In tow of —. Towing bollard ....GXR Towing cable is damaged (*or*, stranded), HRX
XYC	TORRID. TORRID ZONE		
XYD	TORTURE-D-ING	XYV XYW XYZ	—Towing cable is fast —Towing hawser. —Towing light
XYE	TOSS-ED-ING To TOSS		Tugboat ....YDR Will take you (*or ship indicated*) in tow, XKC
XYF	TOTAL-LY Total loss All lost...............DXO Totally, *or*, Quite different-ly......KZO What is the total amount?..........EGO		Will you take me (*or vessel indicated*) in tow? ....XKD
	TOU (*Measure of capacity*)..........AZU	XZA	TOWLINE Have a towline ready..............XYP
XYG	TOUCH-ED-ING To TOUCH Do you (*or vessel indicated*) call any- where (*or*, at —)?... ....HW		Have you sufficient towline?........QUK Look out for the towline ....QZU Send for the towline ....QUV
	Does, *or*, Do not touch at (*or*, call any- where) ....HTD		Sending for the towline............VPX Shorten in both hawsers...........OPX Towing cable is fast...............XYV
	Have you touched anywhere (*or*, at—)? EQT I have (*or vessel indicated* has) not		Towing hawser. ....XYW
	touched the ground .... DUT I have orders for you not to touch at —, SN	XZB	TOWARD Stand toward the —..............WQJ
	I shall not touch anywhere (*or*, at).EQU I shall touch at —.................HTF	XZC	TOWER. Conning tower ....JPF
	TOUNG, *or*, SAADING (*Measure of length*), AWS	XZD XZE XZF	TOWN —Town hall —Town major.
XYH	TOURNIQUET		
	TOW, *or*, COTTON WASTE..............JXV	XZG XZH XZI	TRACE D-ING —No trace—of —Tracing paper
XYI	TOW-ED-ING (*For signals between ships towing and being towed, see page* 85) —Are towing cables, *or*, Is cable fast?	XZJ	TRACK-ED-ING Track chart—of...................IMY
XYJ	Be on your guard against long tows.OIW —Can tow		TRACTION ENGINE ....MAX
XYK	Can you take me in tow? ....XU —Can you tow?	XZK	TRADE-D-ING To TRADE Board of Trade....................GPC
XYL XYM	—Can you tow me (*or vessel indicated*) off? Can not take you in tow...........XV		Board of Trade officer (*or*, inspector), GPD
XYN XYO	—Cast off towing cable (*or*, cables) —Do you want to be towed?		Coasting trade....................IYA

	TRADE—TRINITY			
	TRADE—*Continued* Coasting vessel Coasting trade ship, IYB	YAI	**TRAP** Mess traps........... ....RQI
XZL	—Foreign trade Free trade-r......... ....NPV Home trade................ ....OXK Opium trade ....SRP Slave trade............. ......WGA Trades union. ....YJD	YAJ	**TRAVEL-ED-ING-ER** **TRAVELER** (*Horse*) ....OYG
XZM	—Tradesman	YAK	**TRAWL-ED-ING ER.** Steam trawler.....................WTI
XZN	**TRADER**	YAL	**TREACHERY-OUS-LY**
		YAM	**TREASON-ABLE**
	TRADE-MARK ....RLN What trade-mark (*or*, brand)?.....HDY	YAN	**TREASURE-ER.**
		YAO	**TREASURY**
	TRADE WIND...................ZFC During the monsoon (*or*, trades)...RYJ Had the monsoon (*or*, trade winds) set in? ....RYK Where did you fall in with the trade (*or*, monsoon)?.......................RYM Where did you lose the trade (*or*, mon- soon)?.......................RBI	YAP	**TREAT-ED-ING-MENT** Hospital treatment..................OYT Medical treatment................RON Require hospital treatment.........OYU
		YAQ	**TREATY** **TREBLE** (*Triple*) ....YBI
XZO XZP	**TRAFFIC-ED-ING** —Traffic interrupted by the snow.	YAR	**TREE** Cross-trees.... ....KRH Trestle-trees ....YAU
XZQ	**TRAIN** (*A railway train*) Armored train ....EVZ Special train ....WMP	YAS	**TREMBLE-D-ING TREMOR.**
XZR XZS	—Train leaves for — at — —When do the trains leave for —?	YAT	**TREND-ING**
		YAU	**TRESTLE. TRESTLE-TREES**
XZT	**TRAIN-ED-ING TO TRAIN** Training ship......................VXF	YAV	**TRIAL** I am on full-speed trial. Code Flag over A Steam trial.....................WSY
XZU	**TRAMWAY**	YAW	—Trial takes place on the —
XZV	**TRANSACT-ED-ING-ION** Business transaction....... ....HNW	YAX	**TRIANGLE-ULAR-LY** Triatic, or, Diagonal stay..........KYQ
XZW	**TRANSATLANTIC**	YAZ	**TRIBE-AL**
XZY	**TRANSFER-RED-RING-ENCE-ABLE** Have shipwrecked crew on board (*num- ber to follow*), will you let me trans- fer them to you? ....KDA Vessel seriously damaged, wish to trans- fer passengers......... ....HM	YBA	**TRICE-D-ING**—**UP.**
		YBC	**TRIFLE-ING** Damage trifling ....KJB
YAB	**TRANSHIP-PED-PING-MENT**	YBD	**TRIM** Coal trimmer ....IWU Lamp trimmer. ....QDI
YAC	**TRANSIT**	YBE	—Trimmed Trimmed by the head, — inches by the head....... ....HPK Trimmed by the stern, — inches by the stern ....HPN
YAD	**TRANSLATE-D-ING-ION-OR**		
YAE	**TRANSMIT-TED-TING-SSION** Forward my communication by tele- graph and pay for its transmission.HY	YBF YBG	—Trimmer —Trimming Your lights are out (*or*, want trimming).
YAF	**TRANSOM**		
YAG	**TRANSPORT** Military transport.................RUD Transport agent...................DSX Government transport (*or*, ship) ..OFW		*Made by* — —— — *flashes, or by blasts of a steam whistle* **TRINITY HOUSE,** *or,* **LIGHTHOUSE BOARD,** GFN
YAH	**TRANSPORT-ED-ING-ATION**	YBH	—Trinity pilot

TRIP—TUESDAY

	TRIP		
	Tripped her anchor	EJX	
YBI	TRIPLE TREBLE		
	Triple expansion engine	MAY	
YBJ	TRIPOLI TRIPOLI COLORS		
YBK	TROLLEY		
YBL	TROOP		
	Enemy's troops embark-ed-ing	LWQ	
	Enemy's troops land-ed-ing	LZN	
	Native troops	SEF	
	To embark troops	LWT	
	Troop boat	GVC	
YBM	—Troopship		
	Troops are all embarked	LWU	
	Troops are embarking	LWV	
YBN	—Troops are in good health.		
YBO	—Troops are ready to disembark		
YBP	—Troops on board		
YBQ	—Troops will be ready—to (or, for)		
YBR	—What is the state of health of the troops at —?		
	When do troops embark?	LWX	
YBS	TROPIC-AL		
YBT	TROUBLE-D-ING-SOME		
YBU	TROUGH.		
YBV	TROUSERS		
YBW	TRUCE.		
	Flag of truce	NCY	
YBX	TRUCK		
YBZ	TRUE-LY		
	Are your bearings true or magnetic?	GBD	
	True bearings	JIH	
YCA	TRUMPET		
YCB	TRUNK		
	Shaft trunk	VTK	
YCD	TRUST-ED-ING TO TRUST		
YCE	—Trusty Trustworthy		
	Are the pilots trustworthy?	TIV	
	Do not trust to the weather, it has not done yet	FY	
YCF	—Do not trust too much to your pilot		
	Harbor trust	ONA	
YCG	—I do not like to trust too much to my —		
	May be trusted (or, depended upon)	KUJ	
YCH	—Not to be trusted		
	Trustworthy Trusty	YCE	
YCI	TRUSTEESHIP		
YCJ	TRY-ING TO TRY (See also ATTEMPT and ENDEAVOR)		
YCK	—Tried		
	Channel has altered, do not try it	ECD	
YCL	—Do not try		
	Do not try to wear	YXP	
	Do not try without a pilot	FKL	
	Experiment has been tried	MOH	

	TRY—Continued	
	Experiment will be tried	MOI
YCM	—Fully tried	
YCN	—Has it been tried?	
YCO	—Have, or, Has been tried	
	Have, or, Has tried	FKM
YCP	—Have you tried?	
	Have you tried the temperature of the sea?	XNF
YCQ	—I am (or person indicated is) trying to do so	
	I shall try for shelter in — (port indicated)	VWK
	I shall try the anchorage	ELA
YCR	—In trying to	
	Shall I try (or, attempt)—to?	FKR
	Shall, or, Will try	FKT
YCS	Tried	YCK
YCT	—Try again	
YCU	—Try if you can	
	—Try rate of sailing	
	Try the current (or, tide)	KGF
	Try to club haul	IWG
	Try to enforce it	MAC
YCV	—Try to effect	
	Try to enter	MFD
	Try to escape	MHA
YCW	—Try to find out	
	Try to make out the name of vessel in — (direction indicated)	SDP
YCX	—Try to recover	
YCZ	—Try your best	
	Useless to try	FKS
	Will try (or, attempt)—to	FKT
	Will you try to —?	FKU
YDA	TRYSAIL	
	Storm trysail	WXY
	TSCHETWERIK. (Measure of capacity)	AZV
	TSIN, or, MACE (Coin)	AUG
	TSUN. (Measure of length)	AWZ
YDB	TUB	
YDC	TUBE-ULAR	
	Boiler tube	GWN
	Boiler tube burst	GWP
	Boiler-tube ferrule	GWO
	Boiler tubes leaking-y	RG
	Boiler tubes salted	GWQ
	Condenser tube	JMT
	Condenser tubes choked (or, salted),	JMV
	Condenser tubes leaking	JMW
	Condenser tubes out of order	JMX
YDE	—Condensing boiler tubes burst	
YDF	—Friction tube	
	Main shaft broken in stern tube	RGT
	Rocket tube	UZH
	Stern tube	WUZ
	Stern tube damaged	WVA
	Sweeping tubes (flues, or, funnels)	NFZ
	Torpedo tube	XWZ
YDG	—Tubes out of order	
	Tubular boiler	GXN
YDH	TUESDAY	
YDI	—Before Tuesday	

TUESDAY—ULLAGE

	TUESDAY—*Continued*	YES	TURRET TURRET SHIP
	Last Tuesday _____QGS	YET	—Turret gun
YDJ	—Next Tuesday		
YDK	—Tuesday morning	YEU	TURTLE
YDL	—Tuesday night		
		YEV	TUTOR-ITION
YDM	TUG-GED-GING TO TUG (*For signals between ships towing and being towed, see page 86*)	YEW	TWELFTH
YDN	—Do you want a tug?	YEX	TWENTIETH
	Government tug _____OFZ	YEZ	—Twenty-first.
YDO	—Is tug (*indicated*) engaged?	YFA	—Twenty-second.
YDP	—Send a tug (*indicate if more than one*) to me (*or to vessel in direction pointed out*)	YFB	—Twenty-third
		YFC	—Twenty fourth
		YFD	—Twenty-fifth
YDQ	—Sending, Sent tug	YFE	—Twenty-sixth
	Shall I send you a tug? _____VQO	YFG	—Twenty-seventh
	There are no tugs available _____XZ	YFH	—Twenty-eighth
YDR	—Tug Tugboat	YFI	—Twenty-ninth
YDS	—Tug has not sufficient power		
	Tug is disabled _____LBY		TWICE TWO _____YFP
	Tug is going to you _____YA		
	Want a tug (*if more indicate number*), YP	YFJ	TWILIGHT
	Want another tug (*indicate if more than one*) _____YTR	YFK	TWIN SCREW
		YFL	—Twin-screw steamer.
YDT	TUMBLER (*Glass*)	YFM	TWINE-D-ING
		YFN	—Twine, needles, and palms for repairing sails
YDU	TUNIS-IAN TUNISIAN COLORS		
		YFO	TWIST-ED-ING
YDV	TUNNEL		Cotton twist _____JXU
YDW	—Tunnel shaft		
YDX	—Tunnel shaft broken	YFP	TWO TWICE
			Between the two _____GIW
YDZ	TUNNY		Two blades of propeller broken _____GMC
YEA	TURBAN	YFQ	TYPE
		YFR	—Typewriter
YEB	TURBINE		
			TYPHOID FEVER _____MWS
YEC	TURBOT		
			TYPHOON HURRICANE _____PAI
YED	TURK-ISH TURKISH COLORS		
	Bank of Turkey _____FUV	XFS	TYPHUS FEVER.
YEF	TURN		
YEG	—In turn		
YEH	—It is your turn.		
YEI	TURN-ED-ING TO TURN		
YEJ	—Can turn		
YEK	—Can not turn		
	Turn electric lights—on _____LVK		
	Turn off electric light _____LVJ		
	Turnscrew _____YEQ		
YEL	—Turntable		
YEM	—Turning circle		
YEN	—Turning gear		
	When the tide turns _____XSR		
	When will the tide turn? _____XK		
YEO	—Will turn.		
YEP	TURNIP	YFT	U (*Letter*) (*For new method of spelling see page 18*)
YEQ	TURNSCREW		
	TURNTABLE _____YEL	YFU	ULCER-ATE-D-ING-ION
YER	TURPENTINE	YFV	ULLAGE

ULTIMATUM—UNDERRATE

YFW	ULTIMATUM	YGV	UNCONSCIOUS-LY
YFX	UMBRELLA	YGW	UNCOVER-ED-ING
YFZ	UMPIRE-D-ING	YGX	UNCULTIVATED
YGA	UNABLE—TO	YGZ	UNDECEIVE-D-ING
	Am, Is, *or*, Are unable—to ....CYG	YHA	UNDEFENDED.
	If—(*person-s or thing s indicated*) is (*or*, are) unable to — (can not) ..BYO	YHB	UNDER
	Inability .... PFW		Buried under ....HNO
	Lifeboat unable to come .... ER		Cargo under weight...............IAJ
	Passed a wreck (*date, latitude, and longitude to follow, if necessary*) but was unable to ascertain whether any of the people were remaining ......ZW		Come under stern...................JCO
			Do not anchor Keep under weigh.EHW
			Engines under temporary repair....MBK
	Shall, *or*, Will be unable to .....CYS		Is, *or*, Are not stiff under canvas..HVF
	Sorry I am unable to comply with your request ....USX		Is, *or*, Are very stiff under canvas.HVG
YGB	—Unable for want of		Keep under the land Keep, *or*, Stand closer in ....LT
	Unable to arrange ....EXK		Keep under weigh Do not anchor.EHW
	Unable to ask ....FDR		Need be under no anxiety about .. EOJ
	Unable to comply—with, can not..JLC		Not under control (*Two black balls or shapes vertical*) .... JED
	Unable to weigh anchor ....EJY		
	Vessel, *or*, Person is not able (unable) to comply ....JLG		Not under control lights (lamps, *or*, lanterns)...................JEC
YGC	UNACQUAINTED—WITH		Rock under water ....UYX
YGD	UNADULTERATED	YHC	—Under all sail
YGE	UNAIDED		Under arrest ... .EXR
YGF	UNALTERABLE Y UNALTERED		Under blockade...................GNP
	Position remains unaltered ....TPS		Under bond.......................GYQ
YGH	UNANIMOUS-LY		Under care (*or*, charge)—of.......HYO
YGI	UNANSWERED-ABLE-Y		Under command—of ....JEH
	Signal is not answered ....ENM	YHD	—Under consideration
YGJ	UNARMED		Under cover—of ....KAO
	UNAUTHORIZED, *or*, WITHOUT AUTHORITY, FMZ	YHE	—Undercurrent
			Under easy sail ....LRZ
YGK	UNAVOIDABLE-Y	YHF	—Under experiment
YGL	UNAWARE		Under fire—of...............NAS
YGM	UNBEND-ING-BENT	YHG	—Under ground
	Unbend cables...................HRY		Under jury masts.. ....RMD
YGN	—Unbend sails		Under no anxiety—to (for, *or*, about), EOK
YGO	UNBIT-TED		Under no circumstances...........IRJ
	UNCERTAIN-TY NOT CERTAIN ......IHM	YHI	—Under-officer of
			Under orders of ....SUR
YGP	UNCIVIL-LY-IZED		Under orders—for (*or*, to)—.......SUQ
YGQ	UNCLAIMED		Under steady helm...............OTU
YGR	UNCLE	YHJ	—Under steam
YGS	UNCOMFORTABLE-Y	YHK	—Under temporary repair
	UNCOMMON-LY (*Unusual*) ....YLN		Under the battery...............FYU
YGT	UNCONCERN-ED LY		Under the circumstances ....IRK
			Under the engine room...........MCE
		YHL	—Under the necessity of
YGU	UNCONDITIONAL-LY	YHM	—Under water (*Depth to be indicated, if necessary*)
			Under way ....YXG
			Under weigh Aweigh...............FPR
			Under weight ....YIA
			Under your care Under your charge, HYP
			UNDERCURRENT ....YHE
		YHN	UNDERGO-ING NE
			Is obliged to undergo repairs.......SMT
		YHO	UNDERMINE-D-ING
		YHP	UNDERNEATH
		YHQ	UNDERRATE-D-ING

	UNDERRUN—UNKNOWN		
	UNDERRUN-NING-RAN ....VCK	YIN	UNFAVORABLE LY
YHR	UNDERSELL-ING-SOLD	YIO	UNFINISHED
YHS	UNDERSTAND-ING-STOOD	YIP	UNFIT-TED-TING NESS
YHT	—Can understand	YIQ	—Unfit for the purpose
YHU	—Can not understand the message		
	Could you understand their signals?		UNFORESEEN COULD NOT FORESEE...NLA
	Could you understand what they were		
	signaling about? ....WCI	YIR	UNFORTIFIED
	Do not understand Morse code ......SAI	YIS	UNFORTUNATE-LY
	Do, or, Did you (or person indicated)	YIT	UNFREQUENTED
	understand?......................JLO		
	Do you understand Morse code?....SAJ	YIU	UNFULFILLED
YHV	—Do you understand the message?		
	Do, or, Did you understand their signals?	YIV	UNGUARDED-LY
	WCI	YIW	UNHANDY
	Does, or, Did he (or person indicated)		
	understand?......................JLO		UNHEALTHY-ILY-INESS (Unwholesome),
	Flags seen, but signal not understood,		YLQ
	WCX		UNHESITATINGLY.. .... OUM
	Last hoist not understood, repeat it. URK		
	Message is not understood..........RQP	YIX	UNHOOK-ED-ING
	Perfectly understand-ing-stood ....TFV		
	Signal not understood though the flags	YIZ	UNHURT
	are distinguished ....WCX		
	Stranger does not understand your sig-	YJA	UNIFORM (Official dress)
	nal ....WDJ		
	The signal is not understood, repeat it		UNIMPORTANT-CE OF LITTLE IMPORT-
	(from hoist indicated) VI		ANCE............................PEY
	Understand the communication per-		
	fectly ....JFW	YJB	UNINHABITED-ABLE
	You have misunderstood the signal.WDR		Uninhabited island (or, coast) ......PTS
YHW	UNDERTAKE-ING-N-TOOK		
YHX	—Will you (or person indicated) under-		UNINJURED ....YIZ
	take it?		
			UNINTELLIGIBLE ....POJ
YHZ	UNDERVALUE-D ING		You have abridged the message so much
YIA	UNDER WEIGHT SHORT WEIGHT		that it is unintelligible ....DAR
YIB	UNDERWRITE-ING-TEN UNDERWROTE		UNINTENTIONAL-LY (Accidental)....DCK
YIC	UNDERWRITER	YJC	UNION
	Board of underwriters.............GPE		Dockers' Union ....LJA
	Consult underwriters ....JRV		Seamen's and Firemen's Union ....NAR
	For the benefit of the underwriters.GFL	YJD	—Trades union
			Union des Yachts.......... ....ZKJ
YID	UNDO-ING-NE UNDID	YJE	UNION JACK
	UNDOCK ED-ING.........................LJN		Ensign union downwards...........MBJ
	UNDOUBTED-LY.........................LMB	YJF	UNIT
YIE	UNDRESS-ED-ING.	YJG	UNITE-D-ING
		YJH	—United force
YIF	UNEASY-ILY-NESS	YJI	—United Kingdom
			United States of America...........EFC
YIG	UNEMPLOYED	YJK	UNIVERSE-AL-LY
YIH	UNEQUAL-LED-LY	YJL	UNIVERSITY
YIJ	UNEVEN-LY-NESS	YJM	UNJUST-LY-IFIABLE-Y
			Injustice ....PLE
YIK	UNEXPECTED	YJN	UNKNOWN
YIL	UNFAIR-LY		Cause unknown...................IFR
			Name unknown ....SDN
YIM	UNFATHOMED-ABLE.		Size unknown ....WRY

	UNLAWFUL—UNTRUSTWORTHY		
YJO	UNLAWFUL-LY-NESS		UNSAFE—*Continued*
			You are in a dangerous (*or*, unsafe) position Your position is dangerous, GN
YJP	UNLESS		
	I shall abandon my vessel unless you will keep by us _____AH	YKP	UNSATISFACTORY ILY
	I will not sign the bill of lading unless — _____GKW		Papers unsatisfactory _____TAN
	Unless a change takes place _____IKX	YKQ	UNSATISFIED-YING
YJQ	—Unless I hear		
	Unless it fails _____MRA	YKR	UNSEAWORTHY INESS
YJR	—Unless we are obliged		Cargo so badly stowed I am not seaworthy _____IAG
	Unless your communication is very important I must be excused _____LJ		Condemned as unseaworthy _____JML
	UNLICENSED NOT LICENSED _____QPO	YKS	UNSEEN
	UNLIKELY NOT LIKELY _____QTW	YKT	UNSERVICEABLE USELESS
YJS	UNLIMITED		Rendered unserviceable _____UQL
YJT	UNLOAD-ED TO UNLOAD	YKU	UNSETTLE D
YJU	—Unloading		Affairs are very unsettled _____DPA
YJV	UNLOCK-ED-ING		Affairs look more unsettled _____DPB
			Do not leave anything unsettled __QKR
YJW	UNLOOKED—FOR		Light, unsettled weather _____YZB
	UNLUCKY-ILY (*Unfortunate*) _____YIS	YKV	UNSHACKLE-D-ING
YJX	UNMANAGEABLE		Unshackle cables _____HRZ
	I am unmanageable _____NP	YKW	UNSHIP-PED-PING
YJZ	UNMOOR-ED ING	YKX	UNSKILLED UNSKILLFUL-LY
YKA	UNNECESSARY-ILY	YKZ	UNSOUND-NESS
	Do not expose yourself (*or*, people) unnecessarily _____MPF		UNSTABLE INSTABILITY _____PMW
	Regular survey unnecessary _____UOD	XLA	UNSTEADY-ILY-INESS
YKB	UNOCCUPIED	YLB	UNSUCCESSFUL-LY
	UNOPPOSED NO OPPOSITION _____STD		— made an unsuccessful attack on_FKH
YKC	UNPACK-ED-ING	YLC	UNTIL
YKD	UNPAID		Bar can not be crossed until — ____FVR
			Do not alter — until _____EOF
YKE	UNPRECEDENTED		Do not make any alteration—until __ECG
YKF	UNPREPARED		Heave to until — _____LR
		YLD	—I shall not be ready until —
	UNPROTECTED (*Undefended*) _____YHA	YLE	—I shall not wear—until —
YKG	UNPROVIDED—WITH		I shall stand off the land from — (*hour indicated*) to — (*hour indicated*), or, I shall stand off until — _____QEC
YKH	UNREEVE-ING-ROVE.	YLF	—Not until
			Race is postponed—until —_____UGF
	UNRELIABLE _____UOX		Stop until (*or*, for) _____WXJ
YKI	UNRESERVED LY.		Unsafe to remain until —_____YKN
YKJ	UNRIG-GED-GING		Unsafe to run in until weather moderates _____YKO
YKL	UNRIPE-NESS	YLG	—Until further orders
		YLH	—Until I
YKM	UNSAFE-NESS	YLI	—Until she
	Shift your berth, it is unsafe _____KT		Until weather moderates _____RXM
	Unsafe anchorage _____PE	YLJ	—Until you
YKN	—Unsafe to remain until —		Wait until weather moderates ___ GK
YKO	—Unsafe to run in until weather moderates		When I have run my distance I shall put her head off until —_____LGP
		YLK	UNTO
		YLM	UNTRUE-TRUTH
			UNTRUSTWORTHY (*Not to be trusted*) _YCH

UNUSUAL—USUAL

YLN	UNUSUAL		YME	UPRIGHT LY-NESS
YLO	—Very unusual Is very unusual			Perpendicular Upright .... TGL
YLP	UNWELL		YMF	UPSET-TING
YLQ	UNWHOLESOME-NESS			UPSIDE DOWN .... LMO
YLR	UNWILLING-LY-NESS			Your flags seem to be incorrectly hoisted
YLS	—They are unwilling—to			(or, hoisted upside down) . ..... OWM
YLT	UP		YMG	UPTAKE
	Bear up—for .... GBR			UPWARDS .... YLV
	Bear up instantly .. .. LB		YMH	URGE-D-ING TO URGE
	Beating up .... GCB		YMI	URGENT-LY-NCY
	Bore up—for .... GBS			Permission urgently requested to enter
	Bound to be given up .... HBW			harbor .... PD
	Bound to give it up .... HBX			Urgent orders (or, instructions) ... PNL
	Brace up .... HDP			URINAL (Latrine) .... QIG
	Brail up .... HDS		YMJ	URUGUAY-AN URUGUAYAN COLORS.
	Breaking up .... HEX			US (See also OUR-S and WE). ..... COG
	Bring up . ... HIW		YMK	USAGE
	Clearing up .... ITJ		YML	USE-ING
	Come up . .. . JCP			Boat can not be used .... GBJ
	Get steam up, report when ready .. KM			Have all your anchors and cables ready
	Haul up .. .. OPF			for use .... HIG
	Heave up .. .. OSH			I am using variation of — degrees... JHT
	Helm hard up .... OTN		YMN	—In use
	Hoist up .... OWJ		YMO	—Is not used
	I shall bear up .... GBT			It will be of great use (or, service) .. OHL
	I shall not bear up .... GBU		YMP	—May, or, Can use (or, be used)
	Is steam up? Are you at full speed?		YMQ	—Must use
	VZ		YMR	—No use
	Keep steam up (or, ready) .... KQ			No use appealing ... . ERS
	May, or Can I bear up. .... GBV			No use carrying so much sail...... VED
	Much cut up... .... MZX			Of any use .... EPU
	Pick up boat . ... GTY		YMS	—Quite useless
	Shall I get up steam? .... WA			Ready for use ... . UJV
	Shall we bear up? .... GBW			To be used (or, employed) .... LXT
	Steam is not up, will be ready in —,			Use Alphabetical (Spelling) Table.. WNF
	WB			Use every precaution .... TSD
	Steam is up .... WC			Use flashing signals ... . WDN
	Up and down—the................. LMN			Use Morse Code .... SAL
YLU	—Up channel			Use steamboat—for .... GVD
	Up the river .... UYL			Use utmost endeavors (exertions, or,
YLV	—Upwards			efforts)—to —.................... LTY
	Vessel indicated was compelled to bear			Use your own discretion............ LDX
	up................ .... GBX		YMT	—Use your search light
	When do you intend to bring up?.. EKF		YMU	—Use your siren
	Where do you intend to bring up?.. EKH		YMV	—Used — for —
	Will you require shoring up?...... VZP			Useful Serviceable .... VSC
YLW	—With steam up			Useless Unserviceable.. .. ... YKT
YLX	UPON			Useless to attempt (or, try)—to —.. FKS
	Call upon . .. . HTB		YMW	—Usefully
	Decided upon .... KPR			What coins are used? .... IZQ
	Fire upon .... MZX			Will it be of any use? .... EPX
	Upon condition .... JND			Will use Morse code. .... SAM
	Upon no pretense whatever......... TUZ			You may go by soundings if you use
	Upon what pretense? .... TVA			care ....,.... WLA
YLZ	UPPER			USHR-EL GHIRSH (Coin) .... AUK
	Upper deck .... KQE		YMX	USUAL LY
	Upper fore topsail yard............ NKJ			According to the usual practice. . DEO
	Upper main topsail yard .... RHK			As usual .... FBT
YMA	—Upper port		YMZ	—The same as usual Usual way
	Upper sail Light sail .... QSV			
	Upper spanker topsail yard ... ... XWG			
	Upper topsail .... XWH			
YMB	—Upper works			
YMC	—Upper yard			
YMD	—Uppermost			

UTMOST—VEDRO

YNA	UTMOST Do the utmost you can to ensure holding on, everything will be wanted, MEX
YNB	—Do your utmost Of the utmost importance (or, consequence) --------------------JQC
YNC	—To the utmost Use utmost endeavors (exertions, or, efforts)—to — ------------- --LTY Utmost care must be taken --------HYQ Utmost endeavors----------------LYV

YND	V (Letter) (For new method of spelling, see page 18)
YNE	VACANT-CY—FOR
YNF	—Have no vacancy
YNG	—Have you vacancy—for?
YNH	VACCINATE-D-ING-ION Vaccine lymph -----------------RDN
YNI	VACUUM
YNJ	—Can not get a vacuum.
YNK	—Good vacuum
YNL	—Indifferent vacuum.
YNM	—Vacuum gauge
YNO	VAIN-LY
YNP	VALID-ITY
YNQ	VALLEY
YNR	VALUE-D-ABLE A visit from a Protestant clergyman would be much valued-----------ITY A visit from a Roman Catholic priest would be much valued ---------TVY Of no value .Worthless -----------ZIK Valuable cargo-------------------IBC What is the price? What is the value of? ----------------------- TVW
YNS	VALUATION
YNT	VALVE Air valve------------------------DWK Bilge-pump valves out of order-----GJW Blow-off valve ------------------GMZ Clack valve---------------------IRU

	VALVE—Continued Discharge valve out of order-------LCR Escape valve --------------------MGW Expansion valve out of order---- -MNA Feed-check valve out of order - ----IOD Feed valve out of order-----------MVC High-pressure valve out of order---OUY Kingston valve-------------------QBD Main-check valve out of order -----IOB Safety valve---------------------VDI Safety valve out of order -- -------VDJ Sea inlet valve (or, cock) broken (or, damaged)-----------------VLA Shut-off valve ------------------WAT Slide valve----------------------WGI Slide valve broken --------------WGJ Slide-valve gear ----------------WGK Slide-valve gear damaged (or, out of order) -----------------------WGL Slide valve out of order----------WGM Slide-valve rod -----------------WGN Slide-valve rod broken-----------WGO Sluice valve -- -----------------WHJ Sluice valve out of order---------WHK Snifting valve ------ ------------WIH Stop valve----------------------WXK Throttle valve-------------------XRG
YNU	VANE Distinguishing vane-------------LGY
YNV	VANG
YNW	VAPOR-IZE-D-ING
	VARA (Measure of length)---------AXB
	VARIABLE (See VARY)-------------YOC
YNX	VARIATION Have you taken observations for variation? ------------------------JES I am using variation of — degrees..JHT I have had no observations for variation lately - ---- ------ ------------JHV Is there any variation?-----------JHW No variation----------- ---------JIA Variation by azimuth -----------FQJ Variation of the compass---------JIK What is the variation? ------ ------JIN What variation do you allow? ------JIP
YNZ	VARIOUS-LY VARIETY
YOA	VARNISH-ED-ING
YOB	VARY-IED-ING Meteorological report for to-day gives winds VARIABLE from — (direction indicated) ----------------------RSJ Meteorological report for to-morrow gives winds VARIABLE from — (direction indicated) --------- ---------RSN
YOC	—Variable
YOD	—Variable wind
	VAT (Measure of capacity)--------AYZ
	VEAL --------- ----- --------------ZND
	VEDRO (Measure of capacity)------AZW

VEER—VESSEL

YOE	**VEER-ED-ING**		
	About to veer cable	HQM	
YOF	—Can not veer any more		
	Can not veer any more cable	HQY	
	Driving, I can veer no more cable, no more anchors to let go	NK	
YOG	—I shall veer to		
	Veer a breaker astern	FXZ	
	Veer a rope astern	VAC	
	Veer cable to — (*number of shackles indicated*)	HSA	
	Veer boat astern	GVE	
	Veer both hawsers	OPZ	
	Veered-ing my cable (*Indicate which, and number of shackles, if necessary*)	HSB	
	Veer more cable	KU	
YOH	**VEGETABLE**		
	Fresh beef and vegetables	GDC	
	Fresh vegetables to be obtained	NRB	
YOI	—No vegetables to be obtained		
	Preserved potatoes	TUM	
	Preserved vegetables	TUO	
YOJ	—Vegetable oil		
YOK	**VEGETATION**		
	VELOCITY (*Speed*)	WMU	
	VENEREAL DISEASE	LEB	
	VENEZOLANO, *or* **DOLLAR** (*Coin*)	ATJ	
YOL	**VENEZUELA-N VENEZUELAN COLORS**		
YOM	**VENOMOUS VENOM**		
YON	**VENTILATE-D-ING-ION-OR**		
	Stokehold ventilator carried away	WVY	
YOP	**VENTURE-D-ING**		
YOQ	—Did you venture?		
YOR	—Do not venture		
YOS	—I have ventured to		
YOT	—I shall venture		
YOU	—I will not venture		
YOV	—Will you venture?		
YOW	**VENUS** (*Planet*)		
YOX	**VERBAL-LY**		
YOZ	—Verbal order		
YPA	**VERDICT**		
YPB	**VERIFY-IED ING-ICATION**		
	VERITAS FRENCH LLOYD'S	NQP	
YPC	**VERMILION**		
	VERSHOK (*Measure of length*)	AXC	
	VERST (*Measure of length*)	AXD	
YPD	**VERTICAL-LY**		
YPE	**VERY**		
	A very satisfactory—	VFN	
	Affairs are very unsettled	DPA	

	VERY—*Continued*		
	Be very strict	WZT	
	Boiler burst, very slight damage	GVW	
	He, She, *or*, It is very bad	FRN	
	Is, *or*, Are very stiff under canvas	HVG	
	Is, *or*, Are very zealous	ZMC	
	It is very strange	WYS	
	Not very narrow	SEA	
	Very acceptable	DBX	
	Very anxious—about	DAJ	
	Very bad-ly	FBU	
	Very bad draught	FRY	
	Very bad weather	FRW	
	Very busy-ily	HOG	
YPF	Very difficult (*or*, great difficulty)	KZT	
	—Very excusable		
YPG	Very fair draught	LNE	
	—Very few		
	Very few arrivals	EYT	
YPH	—Very fine		
YPI	—Very fortunate		
YPJ	—Very glad		
	Very good	OEL	
	Very good chart	IMZ	
	Very good draught	LNF	
	Very happy-ily	OMT	
	Very healthy	ORA	
	Very heavy sea	UW	
	Very likely	QAX	
	Very little	YPO	
	Very much	QVY	
YPK	—Very narrow-ly	SBV	
YPL	—Very near-ly		
YPM	—Very often		
YPN	—Very particular-ly		
YPO	—Very probable-y (*or*, likely)		
YPQ	—Very quickly		
YPR	—Very rocky		
	Very rough	VAI	
	Very sickly	WBJ	
	Very slack	WFR	
YPS	—Very slow-ly		
	Very smartly	WHT	
	Very sorry	WKH	
	Very strange-ly	WZE	
	Very suspicious-ly	XGT	
	Very unusual-ly	YLO	
YPT	—Very weatherly		
	Very well	ZBN	
	Very well done	LIT	
YPU	—Very wrong-ly		
	Work-ed-ing very badly	FRY	
	You are in a very fair berth	GHJ	
YPV	**VERY'S PISTOL LIGHT**		
YPW	**VESSEL** (*See also* **SHIP**)		
	A square-rigged vessel	WPM	
	A vessel (*or vessel indicated*) aground near (*or*, on) —	CH	
	Abandon the vessel as fast as possible,	AB	
	Am a merchant vessel	RPW	
	Are you (*or*, Is —, *vessel indicated*) aweigh?	ZM	
	Are you (*or vessel indicated*) in a condition to proceed?	AN	
	Are you (*or*, Is —, *vessel indicated*) in want of coal?	GW	
	Armed merchant vessel	EUW	

VESSEL

VESSEL—*Continued*

Aviso, or, Despatch vessel .... KWY
Beach the vessel   Run on the beach,
           FZS
Beach the vessel where flag is waved
 (or, light is shown) .... FZT
Beach your vessel at all risks .. .. KI
Belayed to vessel .... GEC
Board (or visit) vessel (or, steamer) . GPV
Bomb vessel .... GYI
Calls attention of vessel whose distin-
 guishing signal will immediately be
 shown .... DB
Can, or, Would — (*vessel indicated*)
 weather the gale? .... NVU
Can not save the ship, take people off,
           NB
Coast-defence ship............... IXV
Coasting trade vessel   Coasting vessel,
           ILB
Could any vessel have got away after
 you? .... OFC
Could she (*vessel indicated*) have got to
 any place of shelter? .... OFD
Crew (*number to be shown*) have left the
 vessel .... KCP
Crew will not leave the vessel ..... KCX
Despatch vessel   Aviso ... . KWY
Did you leave many vessels (*or vessel
 indicated*) at — ? .... QKP
Did you (*or vessel indicated*) receive any
 damage? .... KJF
Did you see a vessel in distress? ..... ND
Did you see anything of—(*vessel indi-
 cated*)? .... VNI
Do not abandon the vessel .... AC
Do not abandon the vessel until the tide
 has ebbed .... AD
Do not have anything to do with stranger
 (*or vessel indicated*) .... WYR
Do you (*or vessel indicated*) call any-
 where (or, at —)? .... HW
Do you think we ought to pass a vessel
 in this distressed condition? .... .. NE
Does embargo extend to vessels under —
 flag? .... LWC
Each vessel.................. LQY
Enemy's vessel (*or, cruiser*) .... LZK
Fire ship (or, raft)..... .... MZW
Fishing boat (vessel, *or, craft*)...... KAZ
Foreign-going ship.......... .... NKQ
Four-masted ship .. .. NPD
Full-rigged ship.................. NSU
Funnel of — (*vessel indicated*) is — . NTP
Government hired-vessel .... OFU
Government vessel.............. OFW
Hail the strange vessel and ask if —,
           FDO
Has she (*or vessel indicated*) been long
 abandoned? .... CXN
Has the vessel with which you have been
 in collision proceeded on her voyage?
           HG
Has vessel signaled her number ? ...SLF
Have any vessels been captured?....HXT
Have (*or vessel indicated* has) not pra
 tique .... TRZ
Have they advertised for any vessel to
 carry the mails?............. DOF
Have you, or, Has — (*vessel indicated*)
 got pratique? .... OFI

VESSEL—*Continued*

Have you passed any vessels (*or vessel
 indicated*)? .. .. WOE
Have you seen a pilot vessel (*or, boat*)?
           TJ
Have you seen (*or, heard of*) a vessel
 wrecked (or, in distress)?......... ZU
Have you seen any vessel since your
 departure? .... KTU
Have you spoken any vessels (*or vessel
 indicated*)? .. .. WOE
Hospital ship .... OYS
How does — (*vessel indicated*) sail?
           VDX
How many days since you passed —
 (*vessel indicated*)? ....KNH
How many vessels? .... VXK
I do not intend to abandon the vessel,
           AF
I have been chased by a disguised war
 vessel .... INP
I have reported you to the vessel I spoke,
 she is bound to — (*place indicated*),
           URT
I have sunk (or, run down a vessel) . HK
I must abandon the vessel ....AG
I saw war vessel disguised off — ...LEN
I shall abandon my vessel unless you
 will keep by us.................. AH
I shall assist you (or, vessel in distress),
           FGI
I will not abandon you (*or vessel indi
 cated*)   I will remain by you......AI
If you part, endeavor to beach your ves-
 sel where people are assembled (or as
 pointed out by compass signal FROM
 you) .... FZU
Intend to abandon ship to Lloyd's agent,
           CXP
Is it confirmed that — (*vessel indicated*)
 has foundered? .... JOC
Is not vessel seen (*bearing* —) in dis-
 tress? .... NT
Is she (*or vessel indicated*) abandoned?
           AK
Is the vessel (object, sail, etc) worth
 examination? .... MKC
Is (*vessel indicated*) likely to supply me?
           FGK
Is — (*vessel indicated*) on fire?.....NAL
It is expected the vessel (*or vessel indi-
 cated*) will be allowed to leave ....DZJ
Keep a good lookout, as it is reported
 enemy's ships are going about dis-
 guised as merchantmen........ ....OJ
Keep by the vessel .... PYN
Keep the vessel under command LX
Keep to leeward of vessel chasing.. PZN
Keep to windward of vessel chasing.PZO
Large ship   (*Estimated tonnage to fol-
 low*) .... QFP
Light-ship   Light-vessel........... QTG
Lose-ing, Lost sight of — (*vessel or ob-
 ject indicated*) .... RAQ
Merchant steamer (vessel, *or, ship*).. RPZ
Met with a vessel from —, or, Met with
 a vessel of — .... RQZ
Met with — (*vessel indicated*) ...... RSA
Mortar vessel.................. SAQ
Natives do not like vessels watering
 without payment........ .. ..YWQ

VESSEL

VESSEL—*Continued*

Neutral vessel　Vessel is a neutral,
　　　　　　　　　　　　　　SHC
No intelligence obtained from vessel
　boarded.. ....POG
Passed a good many vessels........TCO
Passed a war vessel disguised, funnel
　(*or*, funnels) — (*Here give the color
　or design, or line that it may be an
　imitation of*) - - ....LEO
Passed, *or*, Saw some war vessels . TCR
Passed the derelict vessel — on the —,
　　　　　　　　　　　　　　KVT
Passed the (*vessel or place indicated*),
　　　　　　　　　　　　　　TCV
Passenger vessel. ....TDQ
Petroleum vessel ....THG
Quarantine at — (*place indicated*) is —
　days for vessels from — (*place indi-
　cated*) ....UEC
Quit the vessel as soon as possible...AB
Relieve vessel in distress ....LHK
Remain by the vessel...............KH
Report — (*vessel indicated*) all well, in
　— (*latitude and longitude indicated*)
　on the — (*date indicated*).........UN
Report when vessel denoted is sighted,
　　　　　　　　　　　　　　USD
Revenue vessel (*or*, cruiser) ....KFC
Run on the beach　Beach the vessel,
　　　　　　　　　　　　　　FZS
Sailing vessel ....VXE
Scout, *or*, Lookout vessel........ ....QZX
See a vessel — (*indicate where*)....VNY
Send me (*or vessel indicated*) a pilot,
　　　　　　　　　　　　　　TJQ
Several vessels becalmed off —......GCJ
Several vessels in sight ....VSR
Signal your vessel's official number
　to —...........................DV
Small ship　Small craft (*Estimated
　tonnage to follow*)...............KBC
Spoke several vessels during the pas-
　sage ....VSU
Spoke the — (*vessel indicated*) ....WOG
Square-rigged vessel. ....WPM
Steam vessel......... ....WTB
Steel-built vessel... ....WTL
Strange vessel following convoy...WYV
Strange vessel hove in sight.......OZS
Stranger (*or vessel indicated*) is suspi-
　cious ....GH
Stranger (*or vessel indicated*) will not
　take any notice of signal........WDK
Stranger (*vessel bearing* —) wishes to
　communicate....................WZD
Telegraph following message to — (*ves-
　sel or person named*) at —.........XD
Telegraph vessel...................XMD
There is a strange vessel bearing —..EM
Timber vessel...... ....XTG
To whom shall I telegraph that all is
　well on board your vessel? .....XMH
Torpedo gunboat (*or*, vessel) ......XWU
Troop vessel ....YBM
Try to make out the name of vessel in —
　(*direction indicated*) .... - ....SDP
Vessel and cargo have been seized.HXU
Vessel bearing — appears to be — .ESH
Vessel chasing is a friend...........INU
Vessel chasing is an enemy ....INV

VESSEL—*Continued*

Vessel desires to be reported by tele-
　graph to owner (Mr —) at — ...USF
Vessel disabled　Will you tow me into
　— (*port indicated*)?...............MY
Vessel (*indicated if necessary*) has got
　off .... - ....NZE
Vessel has no signals .... WDO
Vessel has not made her number...SLH
Vessel has struck on the bar ......FWB
Vessel in distress (*or indicated*) wants
　immediate assistance ....NW
Vessel in sight is —.......... ....VXO
Vessel in sight is an enemy ....OL
Vessel, *or*, Person indicated agrees.DUB
Vessel indicated appears in distress.NX
Vessel indicated (*bearing, if necessary*)
　appears to be — ....ESH
Vessel, *or*, Vessel indicated ashore near
　(*or*, on) ....CH
Vessel indicated bears — ....GBL
Vessel, *or*, Person indicated declines.KQJ
Vessel indicated does not require fur-
　ther assistance ....FGQ
Vessel indicated—is — ....VXP
Vessel indicated has — funnels ....NTU
Vessel indicated has altered course.ECU
Vessel indicated has been detained —
　days windbound at—(*place indicated*),
　　　　　　　　　　　　　　KYE
Vessel indicated has been in collision,
　　　　　　　　　　　　　　JAZ
Vessel indicated has been wrecked.ZIX
Vessel indicated has got off.. . ..NZE
Vessel indicated has hove-to.......OSQ
Vessel indicated has lost her rudder.RBE
Vessel indicated has not been heard of,
　　　　　　　　　　　　　　ORQ
Vessel indicated has not been heard of
　since — (*date indicated*), she is sup-
　posed to be — ....ORS
Vessel indicated has (*or*, I have) not
　pratique ....TRZ
Vessel indicated has pratique ....TS
Vessel indicated has sprung a leak ..QL
Vessel indicated has struck on a shoal,
　　　　　　　　　　　　　　VXQ
Vessel indicated is —...............VXP
Vessel indicated is at anchor (*or*, an
　choring) ....EJZ
Vessel indicated is aweigh ....FPS
Vessel indicated (*bearing, if necessary*)
　is capsized... ....HVY
Vessel indicated is in quarantine ..UEH
Vessel indicated is, *or*, I am in want of,
　　　　　　　　　　　　　　YTN
Vessel indicated is not abandoned. CXR
Vessel indicated is on fire..........OT
Vessel indicated is out of sight....VXR
Vessel indicated is showing a light.QSN
Vessel indicated is sinking.........VXS
Vessel indicated is (*or*, was) standing
　(*or*, steering) — ....VXT
Vessel indicated, *or*, Vessel in company
　is standing in for the land ......QET
Vessel indicated is standing into danger,
　　　　　　　　　　　　　　KLJ
Vessel indicated is supposed to have
　foundered. .... -- NOY
Vessel, *or*, Person indicated is unable
　to comply. ....JLG

	VESSEL—VINGERHOED

VESSEL—*Continued*
Vessel indicated not likely to supply (*or*, assist) you ----- ---------- ----FGR
Vessel indicated parted from her anchors --------------- ---------EKA
Vessel indicated reckons from the meridian of ----- - -- --------- -RQE
Vessel indicated to examine strange ship, and report---------------MKF
Vessel indicated wants assistance --FGS
Vessel indicated wants boats -------FK
Vessel indicated wants immediate assistance ------------------------NW
Vessel indicated was compelled to bear up ----------------------------GBX
Vessel indicated was minutely examined, no person to be seen, nothing visible ---------- ------- - --- MKI
Vessel indicated was most likely burnt, HNG
Vessel indicated will be neaped ----SFB
Vessel is a foreigner --------------NKV
Vessel is in gunshot---- - --------OKN'

YPX —Vessel is insured
Vessel is on the rocks ------------ UZA
Vessel is very crank --------------KBT
Vessel looks like a man-of-war-----RPN
Vessel on— (*bearing indicated*) is making signals of distress ---- -------WDP
Vessel seen (*or*, indicated) is-------VXU
Vessel seen (*or*, indicated) is a native craft --------------------------- -- SEG
Vessel seen is a privateer ---------VXW
Vessel seen (*or*, indicated) lost her masts -- -------------------------RBF
Vessel seriously damaged, wish to transfer passengers -- -----------HM
Vessel will be sold ----------------VOT
Vessels from --- did not get pratique at --- when I left -----------------TSA

YPZ —Vessels in the roadstead
Vessels indicated are (*or*, were) ready for sea --------------------------UJJ
Vessels, *or*, Ships indicated may part company--------------------- -----JGU
Vessels just arrived show your ensigns, DY
Vessels just arrived make your numbers (*or*, distinguishing signals)-------DX
Vessels just weighed (*or*, leaving) make your numbers (*or*, distinguishing signals)------ -- --------------------DZ
Vessels of — (*nations specified*) were all leaving ------ ------------------QLM
Vessels that wish to be reported all well make your numbers (*or*, distinguishing signals) - ---- ----------DYT
Vessels wishing to be reported on leaving make your numbers (*or*, distinguishing signals) ---------------EB
Vessels wishing to be reported show your distinguishing signals - -----EA
Vessels with combustibles not allowed in -------------------------- ----- HS
Waterlogged vessel --------------YWU
We hold our own with vessel chasing, INW
Were there any vessels, *or* Was — (*vessel indicated*) expected at — '----MNU
Whaler   Whaling vessel - - - - VXG
What do you think of vessel chasing?-INX

VESSEL—*Continued*
What is your vessel's registered tonnage?- ---- ---------------------UNV
What vessel? -------- -------- .VYF
What ship did you speak?--------VYG
What ship is that (*indicate bearing, if necessary*)?------ ----------EC
What vessels have you boarded?-_GRD
What vessel takes the next mail?-_RFY
What vessels did you leave in — (*port or place indicated*)? --------- -- ----QLS
Where do you (*or*, does —. *vessel indicated*) belong?--- ---------------GEY
Where is the vessel that collided with you?--------------------- -------JBD
Where shall I find — (*vessel indicated*)? MYU
Will you telegraph to — (*vessel or person named*) the intelligence I am about to communicate? -----------DM
Will your vessel stay? ------------VYH
With what vessel have you been in collision? ------ ----- ------ --------HN

YQA	**VETERINARY**
YQB	**VETO**
	VEX-ED-ING-ATIOUS-LY (*Annoy*) --- EMC
YQC	**VIADUCT**
YQD	**VIBRATE D-ING-ION**
YQE	**VICE-IOUS-LY**
YQF	**VICE-PRESIDENT** Vice-admiral----------------DLN Vice-consul -- -----------------JRQ
YQG	**VICEROY-REGAL**
YQH	**VICINITY** Derelict reported off — (*or*, in the vicinity of —*) ----------------------JF
YQI	**VICTIM**
YQJ	**VICTOR-IOUS-LY** **VICTORY** Gained a victory -- ---------------NVO
YQK YQL YQM	**VICTUAL-LED LING** (*See also* PROVISION) —Am, Is, *or*, Are victualled for —Victualling yard
YQN	**VIEW-ED-ING** With the view (*or*, object)--------SMK
YQO	**VIGILANT-LY-CE** Cruisers are very vigilant --- -- . KEZ The greatest vigilance (care, *or*, caution), HYN
YQP	**VILLA**
YQR	**VILLAGE**
YQS	**VINE** **VINEYARD**
YQT	**VINEGAR**
	VINGERHOED (*Measure of capacity*) -AYB

	VINTAGE—WAIT		

YQU	VINTAGE		VOYAGE—*Continued*
YQV	VIOLATE-D-ING-ION		Last voyage ---------------------QGT
		YRW	—Next voyage
YQW	VIOLENT-LY VIOLENCE		Open, *or,* Proceed on your voyage..TXB
			Pleasant voyage (*or,* passage) ------TCZ
YQX	VISIBLE-Y		What sort of voyage (*or,* passage) have
	Scarcely visible ------------------VIW		you had? --------------------------TDI
	Vessel indicated was minutely exam-		Wish you a pleasant voyage (*or,* pas-
	ined, no person to be seen, nothing		sage) ---- ----------------------TDL
	visible ------------------- ----MKI	YRX	VULCANIZE-D-ING
YQZ	—Visible from the masthead		
YRA	VISION		
YRB	—Color vision		
YRC	VISIT-ED-ING-OR (*See also* BOARD)		
	A visit from a Protestant clergyman		
	would be much valued ------------ITY		
	A visit from a Roman Catholic priest		
	would be much valued . ------TVY		
	Board (*or,* visit) steamer (*or,* vessel),		
	GPV		
	Have you been visited by Lloyd's agent?		
	QWI		
	Have you boarded (*or,* visited)? ....GQI		
YRD	—Will visit (*or,* board)		
YRE	VITRIOL	YRZ	W. (*Letter*) (*For new method of spell*
			ing, see page 18)
YRF	VIZIER	YSA	WAD-DED-DING
YRG	VOCABULARY.	YSB	WADE-D-ING
YRH	—In the General Vocabulary.		
YRI	—Vocabulary signals	YSC	WAFT, *or,* WHEFT.
		YSD	—A waft (*or,* wheft) above
YRJ	VOICE.	YSE	—A waft (*or,* wheft) at the
		YSF	—A waft (*or,* wheft) under
YRK	VOLCANO-IC		Hoist a waft (*or,* wheft)----------OWD
YRL	VOLLEY		Hoist a waft (*or* wheft) when ready.OWE
YRM	VOLT (*Measure of electrical pressure.*)	YSG	WAGES
		YSH	—Struck for more wages.
YRN	VOLTAIC	YSI	—Wages very high
		YSJ	—What wages?
YRO	VOLUME-INOUS.	YSK	WAGON.
YRP	VOLUNTARY-ILY	YSL	WAIST
YRQ	VOLUNTEER-ED-ING	YSM	WAIT (*See also* AWAIT, REMAIN, *and* STOP)
YRS	VOMIT-ED-ING	YSN	—Am I, *or,* Is — (*person indicated*) to
			wait (*or* remain)?
YRT	VOTE-D-ING-R		Are you going to wait for orders?..STM
		YSO	—Can wait
YRU	VOUCH-ED-ING		Can you wait for reply (*or,* answer)?.EMV
		YSP	—Do not wait
YRV	VOUCHER		I am waiting for your answer------EMZ
	Salvage voucher---- -------------VGH	YSQ	—I can not wait
			I wait for advice------------------DOI
	VOYAGE--------------------------TCY	YSR	—I will wait until
	A pleasant passage ---------------TCZ	YST	—I will wait whilst you write
	During the voyage (*or,* passage) . .-LPU	YSU	—If you will wait a little
	First voyage.--------------------NBI	YSV	—Shall, *or,* Will wait
	Has the vessel with which you have		Wait for arrival—of. ----------FPC
	been in collision proceeded on her voy-	YSW	—Wait for (*or,* to)
	age? - -------------------------HG		Wait for answer (*or,* reply)--------TX
	How long have you been on the voyage?		Waiting for cargo ---------------IBD
	QYN		Wait for despatches.--------------KXD

WAIT—WANT

	WAIT—*Continued*
YSX	—Wait for high water
YSZ	—Wait for me
	Wait for orders................................TE
	Wait for the attack ....FKE
	Wait off here for orders ....SUT
YTA	—Wait for passage to — (*or*, for).
	Wait until weather moderates ....GK
YTB	—Waited-ing To wait
	Weigh, Cut, *or*, Ship, wait for nothing,
	KW
YTC	—What are you waiting for?
	Will await answer (*or*, reply)........TZ
	Will you wait (*or*, stay)?..........WRY
YTD	—Will you wait for the mail?
	WAKE ING WOKE.....................ZHL
YTE	WAKE WAKE—OF
YTF	—In my wake
YTG	—In the wake—of
YTH	—In your wake
	WALES
	Wales, H R H The Prince of .....OVM
	Wales, H R H The Princess of ..OVK
YTI	WALK-ED-I ?G
YTJ	WALL-ED-ING.
YTK	WALNUT.
YTL	**WANT-ING**
	Accident, want a surgeon (*or*, doctor),
	AM
	Aground want immediate assistance,
	NA
	Am drifting, want assistance........CM
	Am in want of sailmakers ....VFB
YTM	—Are you in want—of —?
	Are you, *or*, Is — (*vessel indicated*) in
	want of coal........................GW
	Are you in want of provisions?......TN
	Are you in want of water?..........YZ
	Copy will be asked for (*or*, wanted).JVZ
	Describe exactly what you want (*size*,
	weight, *etc*) ....KWC
	Distressed for want of coal..........LHD
	Distressed for want of food.........LHE
	Distressed for want of water ......LHF
	Do not require (*or*, want) assistance.OR
	Do the utmost you can to insure hold-
	ing on, everything will be wanted,
	MEX
	Do you want a lifeboat?................EZ
	Do you want a pilot?.................TIZ
	Do you want a tug? ....YDN
	Do you want (require) assistance (*or*,
	help)—from? ....CS
	Do you want coal?.....................GW
	Do you want to be towed?..........XYO
	Dying for want of water...............NF
	Engine wants repair ....MBL
	Fire, *or*, Leak, want immediate assist-
	ance ....NH
	How many lighters do you want?..QTA
	I am attacked, want assistance Help,
	I am attacked ....NJ
YTN	—I am in want of —, *or*, Vessel indicated
	is in want of —

	WANT—*Continued*
	I am in want of a doctor (*or*, surgeon),
	FGL
	I am in want of channel pilot—for —,
	TJB
	I am in want of oil for machinery..RDU
	I am in want of petroleum ....TBF
	I am in want of powder and shot ..TRF
	I want a pilot....s, *or*, Code Flag over s
	See PILOT SIGNALS, *page* 8
	I want assistance, please remain by me,
	NR
	I want plank — feet long, — inches
	wide, — inches thick ....TMD
	I want — (*number indicated*) lighters,
	QTB
	In distress, want assistance.........NS
	In distress, want immediate assistance,
	NC
	See DISTRESS SIGNALS, *page* 7
	In need (*or*, want) of..............SGE
	In want of ammunition ....KFW
	In want of an anchor. ....YC
	In want of artillery ....EZX
	In want of biscuit........ ....GLN
	In want of cartridges ....ICW
YTO	In want of coal (*State amount*)....HB
	—In want of refitting
	In want of reinforcement ....UOK
	In want of repair . ...URD
	In want of seamen ......VLP
	Light-house, *or*, Light-ship at — wants
	immediate assistance.............. CW
	Machinery wants repair ....MBL
	Make a signal when you want a boat,
	GTU
	No want of artillery................EZY
YTP	—Not much wanted
	Person indicated is wanted on shore.TGS
	Pump wants repairs ....UBS
	Ship indicated wants boat .... FK
YTQ	—Should you want
	Surgeon, *or*, Doctor wants assistance,
	FGP
	Unable for want of ....YGB
	Vessel in distress (*or indicated*) wants
	immediate assistance .... NW
	Vessel indicated is (*or*, I am) in want
	of —...................................YTN
	Vessel indicated wants assistance ..FGS
	Vessel indicated wants boat .... FK
	Want a boat ....GVF
	Want a boat immediately (*if more than*
	one, number to follow) ....YG
	Want a boat, man overboard ....BT
	Want a cable (*size to follow*) ....YB
	Want a cable and anchor...........YD
	Want a chain cable.................YB
	Want a hemp cable..... ....HSC
	Want a pilot....s, *or*, Code Flag over s
	Want a pilot ....PT
	Want a police officer ....VN
	Want a surgeon Medical assistance
	wanted ....FGL
	Want a surgeon, send one from the
	nearest place ....WO
	Want a tug (*if more than one, indicate*
	number) ....YP

WANT—WAS

WANT—*Continued*
Want a warp run out ....YQ
Want an anchor ....YC
Want an anchor and cable ....YD

YTR —Want another tug (*indicate if more than one*)
Want assistance ....YE
Want assistance, mutiny ....YF
Want bread ....GLN
Want carpenter ....IBT
Want carpenter's tools ....IBX
Want chart of harbor (channel, etc) .YH

YTS —Want clothes
Want coals immediately ....YI
Want confirmation ....JOG

YTU —Want food
Want food immediately ....YO
Want food   Starving ....NV

YTV —Want gear—for —
Want hands (*specify number and particulars, if necessary*) ....YJ
Want immediate instructions ....YK
Want immediate medical assistance ..YL
Want immediately ....PDM
Want lighter immediately (*if more than one, number to follow*) ....YM
Want lighter to take my cargo ......QTE
Want men ....RFJ

YTW —Want money
Want more support ....RZX
Want my papers ....ITR
Want No — canvas (*length indicated*), HVJ

YTX —Want of judgment
Want of funds : ....NTG
Want official confirmation ....JOH
Want pilot ....PT
Want pilot ....s, or Code Flag over s

See PILOT SIGNALS, *page 8*

Want pistol (revolver) cartridges ..TKV
Want police ....YN
Want provisions ....YTU
Want provisions immediately ....YO

YTZ —Want shipwrights
Want to come into harbor ....ONO
Want — tons dead weight of cargo ..IBE
Want to purchase ....UCD
Want tug (*if more than one, number to follow*) ... .YP
Want water boat (or, tank) ....GUJ
Want water, dying of thirst ....NF
Want water immediately ....YR
Want you to send me a pilot ....TIW

YUA —Wanted
What assistance do you require? What do you want? ....FGT
What cargo do you want? ....IBF
What size do you want? ....WEZ

YUB —You are wanted at the — office
Your lights are out (or, want trimming) (*Made by — — — flashes, or by blasts of a steam whistle*)

YUC WAR (*See also* HOSTILE)
Am a man-of-war ....RPM
Any chance of war? ....EOQ
Are there any men-of-war about? ....SA
Are there any men-of-war now cruising? ....KEX

WAR—*Continued*
Civil war ....IRS
Council of war ....JYC
Did you fall in with any men-of-war? ....MRN
During the war..... ....LPW
Has war commenced? or, Is war declared? ....YU
Have you seen (or, spoken) any men-of-war? If so, report their names (or, nation) .... NM
Hostilities commenced—between — .OZC
I have been chased by a disguised war vessel ....INP
I saw disguised war vessels off — ..LEN
Ironclad men-of-war ....EVS
Is war declared? or, Has war commenced? ....YU
Keep a good lookout, as it is reported enemy's war vessels are going about in disguise of merchantmen . ...OJ
Man-of-war ....RPL
Man-of-war ensign ....MEP
Passed a war vessel disguised, funnel (or, funnels) — (*Here give the colors or the design, or the line that it may be an imitation of*) ....LEO
Passed, or, Saw some war vessels ..TCR
Ship of war ....RPL
Vessel looks like a man-of-war ......RPN
War between — and — ....YW

YUD —War expected
War has commenced (or, is declared) ..YX
War office ....SPG
War rocket   Explosive rocket .....MOZ

YUE —Warlike
What men-of-war did you leave at —? QLR

YUF WARD

YUG WARDER

YUH WARDROOM

YUI WAREHOUSE

YUJ WARM-ED-ING-LY-TH
Warm, or, Winter clothing—for ...IVX

YUK WARN-ED-ING
Storm warnings ....WXZ
—Warning (*notice*) has been given — of —

YUL

YUM WARP-ED-ING   To WARP
Guess warp ....OJG
Run out a warp ....VCF
Want a warp run out ....YQ

YUN —WARP (*A warp*)
Warp out to fair way ....MRH

YUO WARRANT-ED

WAS, or, WERE ....COM
He, She, It (*or person-s or thing-s indicated*) was (or, were) ....BVT
He, She, It (*or person-s or thing-s indicated*) was (or, were) not ......BVU
He, She, It (*or person s or thing-s indicated*) was (or, were) not to be ...BVX
He, She, It (*or person-s or thing-s indicated*) was (or, were) to be ....BVW

WAS—WATER

WAS—*Continued*			
How was (*or*, were)?	BXW	**YUP**	**WASH-ED-ING-ER.**
I was	COH	**YUQ**	—Black wash
I was not	COI		Can I get clothes (*or*, linen) washed?..IVQ
If he (she, it, *or person-s or thing-s in*		**YUR**	—Washed overboard
dicated) was (*or*, were)	BYT		Whitewash	ZDO
If he (she, it, *or person-s or thing-s in*		**YUS**	—Whitewash-ed-ing To whitewash
dicated) was (*or*, were) not	BYU		
If I was	BZT	**YUT**	**WASTE D ING-FUL-LY-NESS**
If I was not	BZU		Cotton waste (*or*, tow)	JXV
If they were	CAU	**YUV**	—Waste pipe
If they were not	CAV		
If we were	CBN		**WATCH** (*Timepiece*)	ZNH
If we were not	CBO		
It you were	CDG	**YUW**	**WATCH**
If you were not	CDH		Afternoon watch	DRX
That, *or* This was	CLR		Anchor watch	EHK
That, *or* This was not	CLS		Dog watch	LKB
There was (*or*, were)	CNF		First watch	NBJ
There was (*or*, were) not	CNG		Forenoon watch	NKZ
They were	COJ		Keep a strict watch all night	PYJ
They were not	COK		Middle watch	RTF
They were not to be	BIF		Morning watch	SAC
They were to be	BIE		Night watch	SIG
Was, *or*, Were	COM		Officer of the watch	SPM
Was, *or*, Were done (*or*, doing)	BON	**YUX**	—Watch-ed-ing To watch
Was, *or*, Were for	COP	**YUZ**	—Watch and watch
Was, *or*, Were from	COQ	**YVA**	—Watch below
Was, *or*, Were, he (she, it, *or person-s*			Watch the barometer	FXM
or thing s indicated)?	BWX		
Was, *or*, Were, he (she, it, *or person-s*		**YVB**	**WATER**
or thing-s indicated) not?	BWY		Are you in want of water?	YZ
Was, *or*, Were his (her-s, it-s)?	COR		*Asks precise time of high water and*
Was, *or*, Were I (my, mine)	COS		*minimum depth at that time*	XE
Was I?	COT		
Was I not?	COU		NOTE —*Reply will be —*
Was, *or*, Were not	COV		1 *By an hour signal* (*See page* 58)
Was, *or*, Were not done (*or*, doing)	BOP		2 *By a numeral signal denoting feet*
Was, *or*, Were the?	COW		
Was, *or*, Were their-s (they)?	COY		Bar, *or*, Entrance not safe except just
Was, *or*, Were there?	COX		at slack water (*or*, at — *time in-*
Was, *or*, Were this (*or*, that)?	COZ		*dicated*)	FVT
Was, *or*, Were to be	BIG		Between wind and water	GIY
Was, *or*, Were we (our-s)?	CPA		Bilge water	GJX
Was, *or*, Were with	CPB		Bread and water	HBL
Was, *or*, Were you-r-s?	CPD		Breakwater	HFM
We were	CPE		Cask of water	IDY
We were not	CPF		Complete with water	JKO
We were not to be	BIW		Deepest water is in the — channel	ILB
We were to be	BIV		Deepest water is nearer the reef (*or*,
Were there?	COX		bank)	FVB
Were there any	EPW		Deepest water is nearer the shore	VXG
Were these (*or*, those)?	CPG		Did boat take water?	GSQ
Were they?	CPH		Distressed for want of water	LHF
Were they not?	CPI	**YVC**	—Draught of water
Were we?	CPJ		Dying from want of water	NF
Were we not?	CPK		Engine room full of water	MBG
Were you?	CPL		Engine room has — feet of water in it,
Were you boarded?	GBC			MBH
Were you not?	CPM		Fresh water	NBC
What, *or*, Which was (*or*, were)?	CQT		Full of water	NSB
When was (*or*, were)?	CSJ		Good water Water is good	YVS
Where was (*or*, were)?	CUB	**YVD**	—Good water can be got at —
Who was (*or*, were)	CUW	**YVE**	—High water
Why was (*or*, were)?	CVX	**YVF**	—High-water mark
Why were you?	CVZ		Hold on till high water	KP
You were	CPN	**YVG**	—I am getting short of water, and must
You were not	CPO		endeavor to get some
You were not to be	BJP		I have only — days' water	ZA
You were to be	BJO		Is the water shallow?	VTN
			Low water	RBT
			Low-water mark	RBU
		YVH	—No water to be had

WATER

	WATER—*Continued.*
YVI	—Of water
	Plenty of water ....YWJ
	Require, *or*, Send water tank (*or*, boat),
	GUJ
	Send for water ....VPY
YVJ	—Sending a water boat
	Shallow water is ....VTO
	Shallowest water is —...........VTP
	Short of water...................WAB
YVK	—Slack water
	Soda water ....WJB
	Stokehole full of water ....WVX
	Sufficient water Water enough...XCU
	Supply of water is small, not exceeding
	— (*quantity indicated*) a day...YWR
	The water is shallow ....VTQ
YVL	—Through the water
	Tons of water........ .... XLY
	Under water (*Depth to be indicated,*
	if necessary) ....YHM
	Wait for high water...............YSX
	Want water, dying of thirst........ NF
	Want water immediately.. ....YR
	Water-ed-ing To water.........YWP
	Water ballast ....FTV
	Water-ballast tank.................FTW
YVM	—Water barrel
	Water bottle.....................HAU
	Water can readily be obtained......YWS
YVN	—Water can not be obtained
YVO	—Water course
	Water enough ....XCU
YVP	—Waterfall
YVQ	—Water gauge
YVR	—Water gauge out of order
YVS	—Water is good
YVT	—Water is not good
YVU	—Water line
YVW	—Waterlogged
	Watermark ....YWV
YVX	—Water police River, *or*, Harbor police
YVZ	—Water pressure
YWA	—Waterproof-ed-ing
YWB	—Waterspout.
YWC	—Water surface
YWD	—Water tank.
	Water tank is empty ....LXZ
YWE	—Water tank leaky
	Water telescope ....XMR
	Water-tight........ ....XTC
	Water-tight compartment ....JHB
	Water-tight door...................LKS
YWF	—Watering party
YWG	—Watering place
YWH	—Waterman
YWI	—Waterway
	What is the precise time of high water
	and minimum depth at that time?..XE
	What quantity of water?... ......UDY
	You will have great difficulty in obtain
	ing water ....LAF
	DEPTH OF WATER (SOUNDINGS)
	NOTE —*Depth of water is always to be*
	shown in feet
	Apparently shoal water to the —... ERO
	At high water you will have — feet,
	KUY

	WATER—*Continued*
	At low water you will have — feet,
	KUZ
	At — (*time specified*) there will be —
	feet of water over the bar ....FVQ
	Best anchorage (*or*, berth) is in — fath-
	oms...... ....OV
	Deep water ....KQT
	Deepest water is in the — channel...ILR
	Deepest water is near the bank (*or*, reef),
	FVB
	Deepest water is nearer the shore...KQV
	Did boat take water? ....GSQ
	Do not come into less than — feet of
	water ....FV
	Draught of water ....YVO
	Drawing too much water ....NZ
	Dry at low water... ... .LOP
	Entrance, *or*, Bar is not safe except at
	slack water (*or*, at —, *time specified*),
	FVT
	Got bottom—with — feet ....HBA
	Have you shoaled your water?.....VYS
	High-water mark ....YVF
	Hold on till high water ....KF
	How many feet at low water?.....KVF
	I am in danger (*or*, shoal water), direct
	me how to steer ....NL
	I am in shoal water ....VYI
	I am in — feet of water............VYP
	I shall not go into less than — fathoms
	of water, whether I make the land or
	not ....QDY
	Indicate the depth of What depth?
	KVB
	Is there sufficient depth of water?...KVO
	Low-water mark ....RBU
	My draught of water is — feet aft and
	— feet forward..... ... .OA
	My draught of water is — feet — inches
	aft ....LNC
	My draught of water is — feet — inches
	forward ....LND
	No bottom with — feet.............HBB
	Not water enough ....MDV
YWJ	—Plenty of water
	Proceed at high water...... .. ....OVF
	Proceed at low water . ...TWZ
	Rise and fall of water is about — feet,
	UXT
	Rock under water ....UYX
	Shallow water—is...... ....VTO
	Shallowest water is —...... ....VTP
	Shoal water, *or*, Danger in — (*direction*
	indicated) ....GF
	Shoal water, *or*, Danger to leeward..KLH
	Shoal water, *or*, Danger to wind-
	ward ....KLI
	The deepest water ....KRA
	The depth of water here (*or at place*
	indicated) is — feet at low water.KVD
	Under water (*Depth to be indicated,*
	if necessary).... ....YHM
	Wait for high water .. ..YSX
	Water is shallow.......................VTQ
YWK	—Water is shoaling (*Depth of water at*
	last cast may be shown)
YWL	—Water is too deep
	Water is too deep to anchor .. ...EKB
YWM	—Water shoaling I must tack.
YWN	—Water's edge

WATER—WE

	WATER—*Continued*
	What are your soundings? What water are you in? (*Answer in feet*) ---- VS
	What depth? Indicate the depth of_KVB
	What draught of water (*in feet*) could you lighten to? ----------- --OB
	What is depth of water alongside the—? KVI
	What is the depth at high water?__KVE
	What is the depth at low water? __KVF
	What is the depth of water in feet at — ? KVG
	What is the depth of water on the bar (*in feet*)? ------------------VQ
	What is the depth of water over the sill? ------------------- ---KVH
	What is the least water we shall have (*or*, ought to have)? ------VR
	What is the rise and fall of the tide?_XH
	What is your draught of water? ----OC
	What water do we (*or*, shall I) have at high water? ------------KVE
	What water do we (*or*, shall I) have at low water? ------------KVF
YWO	—What water shall I come into?
	When will it be high water? -------OVI
	You will be aground at low water . KX
	You will get into shoal water, or, you will shoal your water ----------VZA
	You will have less than — feet of water, KVJ
	You will have water enough over the bar (*Depth in feet may be shown*)_VT
	You will not have less than — feet of water -- ------- -------VU
	You will not have water enough over the bar (*or*, into the harbor) (*Depth in feet may be shown*)-- ------FWE
	— feet of water in the hold--------QM
YWP	WATER-RD-ING TO WATER
	WATERING PLACE ------------ ----YWG
	Beach indicated is good for watering, FZO
	Beach indicated is not good for watering,----------- ----------FZP
	Best watering place—is---- ----GHT
	Is there a convenient watering place? JTN
YWQ	—Natives do not like ships watering without payment
YWR	—Supply of water is small, not exceeding —(*quantity indicated*) a day
YWS	—Water can readily be obtained
	Water can not be obtained -------YVN
	You will have great difficulty in obtaining water ---------------------LAF
	WATERLOGGED ------- -- ----- YVW
YWT	—I am (*or vessel indicated is*) waterlogged, take people off
YWU	—Waterlogged vessel
YWV	WATERMARK
	WATERSPOUT ----------------YWB
	WATERWAY------------------ -- ------YWI

YWX	WAVE-D-ING TO WAVE
	Beach the vessel where flag is waved (*or*, light is shown)--- ----- ----FZT
	Boats should endeavor to land where flag is waved (*or*, light is shown)------JZ
YWZ	WAVE (*Sea*)
YXA	WAX-EN
YXB	WAY
	Best way ----------------------- GHZ
	By way of ----------------------- HPS
	Can I procure any assistance in the way of —? ------------------------CN
	Cross way ----------------- ------KEI
	Does the telegraph communication go all the way? ------------------JFY
	Each way -- ----------------- LQZ
	Fairway, or, Channel------------ IKZ
	Gangway ------ --------- -----NWR
	Gather ed-ing way----------------NXH
	Get under way as fast as you can----KN
YXC	—Give way
	Headway Steerageway ----------OQT
	I am laying (repairing, or, picking up) a telegraph cable Keep out of my way ----------- -------------XLQ
	I have headway ------Code Flag over J
	I have sternway------Code Flag over K
	In a bad way------------------FRO
YXD	—In the way
	Leeway -----------------------QMK
	Long way----------------------QYS
YXE	—Make way
	Midway -- --------------------RTM
YXF	—On the way—to
	Out of the way --- -- ------- --SWD
	Rocks Reefs, or, Shoals stretch a long way out ------------------------GE
	Slip way - ------- ------------ WGZ
	Sternway Going astern----------FHR
	The best way ---- -----------------GHZ
	The same way --------------------VGQ
	The way is off my ship, you may feel your way past me ----- -------HL
YXG	—Under way
	Usual way The same as usual----YMZ
	Waterway ----------------------YWI
YXH	—What way?
YXI	—Which way?
YXJ	—Wrong way
	WE OUR-S (*See also* OUR) --------CPQ
	After we (our-s, us) ----- --------CIL
	Are, or, Is our-s, or, Are we?------BEM
	As we (our-s)----------------------FBU
	Can, or, May we (our-s)?--------- BKN
	Could, or, Might we (our-s)?-----BLI
	Do, Does, or, Did we (our-s)? -----BMP
	Had, Has, or, Have we (our's)?----BRD
	Had, or, Have we been ?----------BFU
	Had, or, Have we done ? --------BMV
	Had, or, Have we had ?----------BRE
	How shall (*or*, will) we ?--------BXT
	If we (our-s) ... -------------- ---- CAW
	If we are ----------------------- CAX
	If we are not----------------------CAY
	If we can (or, may) ---------------CAZ
	If we can not (or, may not)--------CBA
	If we could (or, might) -----------CBD

WE—WEATHER

WE—Continued

If we could (or, might) not	CBE
If we do (or, did)	CBF
If we do (or, did) not	CBG
If we had (or, have) ...	CBH
If we had (or, have) not	CBI
If we shall (or, will)	CBJ
If we shall (or will) not	CBK
If we should (or, would)	CBL
If we should (or, would) not	CBM
If we were	CBN
If we were not	CBO
May, etc *(See Can, above)*	
Might, etc *(See, Could, above)*	
Must we (our-s)?	CFD
Ought we (our s)—to?.........	CIB
Shall or, Will we (our s)?..........	CJA
Shall, or, Will we be?	BHG
Shall, or, Will we not?	CJB
Should, or, Would we (our s)?	CJW
Than we (our-s)	CKJ
That we (our-s)	CLW
Was, or, Were our-s (we)?	CFA
We	CFQ
We are........	BEW
We are not	BEX
We can (or, may)	BKV
We can (or, may) be	BIH
We can (or, may) have	BSN
We can not (or, may not)	BKW
We can not (or, may not) be......	BIJ
We can not (or, may not) have	BSO
We could (or, might).........	BLP
We could (or, might) be........	BIK
We could (or, might) not	BLQ
We could (or, might) not be......	BIM
We do (or, did)	BUQ
We do (or, did) not	BOR
We had (or, have)...........	BSP
We had (or, have) been........	BIN
We had (or, have) done	BOS
We had (or, have) had ...	BSQ
We had (or, have) not.........	BSR
We had (or, have) not been........	BIO
We had (or, have) not done	BO1
We had (or, have) not had	BST
We may, etc *(See Can, above)*	
We might, etc *(See Could, above)*	
We must	CFI
We must be	BIP
We must not	CFJ
We must not be	BIQ
We ought—to	CIG
We ought not—to	CIH
We shall (or, will)	CJH
We shall (or, will) be	BIR
We shall (or, will) have.......	BSU
We shall (or will) not.......	CJI
We shall (or, will) not be..........	BIS
We shall (or, will) not have........	BSV
We should (or, would).........	CKA
We should (or, would) be.........	BIT
We should (or, would) have	BSW
We should (or, would) not.......	CKB
We should (or, would) not be.. ..	BIU
We should (or, would) not have	BsX
We were—to..........	CPE
We were not—to	CPF
We were not to be	BIW
We were to be..........	BIV
We who.......	CPR

WE—Continued

Were we—to?........	CPJ
Were we not—to?	CPK
What, or, Which do (or, did) they (we, or, you)?	CQE
What, or, Which shall (or, will) I (or,) we do?	CQL
What, or, Which shall (or, will) I (or, we) have......	CQM
What, or, Which should (or, would) I (or, we) do?.......	CQB
When are we?.......	CQX
When do (or, did) I (or, we)?......	CRE
When shall (or, will) I (or, we)? ..	CRU
When shall (or, will) I (they, we, or, you) do?.......	CRV
When shall (or, will) we have? ...	CSA
When we (our-s)	CSK
When we are	CSL
Where am I (or, are we)?......	CSQ
Where do (or, did) I (or, we)?......	CSY
Where I am (or, we are)?........	CTG
Where shall (or, will) I (or, we)?..	CTN
Where shall (or, will) we have?....	CTR
Where we (our s)	CUD
Why we (our-s)........	CVY
Will, etc *(See Shall, above)*	
Would, etc *(See Should, above)*	

YXK	**WEAK-LY-NESS**	
	Weak crew　Short manned.......	KDJ
YXL	**WEALTH-Y**	
YXM	**WEAPON**	
YXN	**WEAR ING　To wear**	
YXO	—Are to wear	
	Barely room to wear	UZO
YXP	—Do not try to wear	
	Have I room to wear?........	UZP
	I shall not wear—until —......	YLE
YXQ	—I shall wear at — (or, when —)	
YXR	—In wearing	
	Not room to wear	MB
YXS	—Shall wear as soon as weather permits	
	Wear instantly........	MP
YXT	—Wear short round	
YXU	—When do you wear?	
	Wore	ZHS
YXV	**WEAR ING　WORE　WORN** (of clothing)	
YXW	**WEATHER**	
	A regular break-up　An end to the fine weather	HEV
	Bad weather........	ZB
	Bad weather is expected .... . ---	FP
YXZ	—Changeable weather	
	Cold weather	IZX
	Do not trust to the weather, it has not done yet......	FY
YZA	—Driven by stress of weather	
	Every appearance of foul weather..	ERZ
	Fine weather	MYZ
	Foggy weather	ZC
	Heavy weather coming, look sharp .	FZ
	Hot weather	OZF
	I do not like the look of the weather.	ZE
	I shall weigh as soon as the weather permits	ZO

WEATHER—WEIGHT.

	WEATHER—*Continued*	ZAE	WEDGE D ING.
	I think the weather will be —_____XQH		
YZB	If weather moderates____ _____RXJ	ZAF	WEDNESDAY
	—Light unsettled weather	ZAG	—Before Wednesday
	Moderate weather - _ _____ZF		Last Wednesday _____QGU
	No weather report (forecast) _____ZG	ZAH	—Next Wednesday
	On account of the fog (*or* weather)_DFQ	ZAI	Wednesday morning
YZC	—Rainy weather	ZAJ	—Wednesday night
YZD	—Shall we have thick weather?		
	Shall wear as soon as weather permits,	ZAK	WEED-Y
	YXS		Weedy bottom _____HBJ
	Signal (*or*, I am going to signal) state		
	, of the weather outside (*or at place in-*	ZAL	WEEK
	dicated) _____WDE		Day of the week _____KNA
	Squally weather Blowing hard____ZH		Expected next week_____MNJ
	Stormy, *or*, Boisterous weather from	ZAM	—First week in —
	the —__ _____ZI		Fourth week in —._____NPF
YZE	—Stress of weather		Last week _____ _____QGV
	Thick weather _____ZJ		"Lloyd's Weekly Shipping Index"_QXD
	Unsafe to run in until weather moder-	ZAN	—Next week
	ates _____ _____YKO	ZAO	—Second week in —
	Until weather moderates_____RXM		Sunday week_____XEC
	Very bad weather _____ _____FRW	ZAP	—Third week in —
	Very weatherly _____YPT	ZAQ	—This week
	Wait until weather moderates _____GK	ZAR	—Weekly
YZF	—Weather bound—at —		
YZG	—Weather has been —	ZAS	WEEP ING WEPT
YZH	—Weather is still open — at —		
YZI	—Weather permit-ted-ting	ZAT	WEIGH-ED-ING TO WEIGH (*See also*
YZJ	—Weather prevents-ed-ing		AWEIGH)
YZK	—What do you think of the weather?		Cut, *or*, Ship Get an offing Weigh,
	What is the weather forecast for to-day?		KW
	ZK		Do not intend to weigh _____PON
	What is the weather forecast for to-	ZAU	—I am not in a position to weigh
	morrow? _____ - _____ZL	ZAV	—I can not weigh until you have moved
	What is the weather forecast —?___NKM		I find great difficulty in weighing_KZV
YZL	—What is the weather outside (*or*, at —)?	ZAW	—I shall not weigh until I get a favorable
YZM	—What weather have you had?		start
	Wind and weather permitting _____ZFG		I shall weigh as soon as the weather
			permits (*or*, moderates)_____ZO
YZN	WEATHER (*toward the wind*)		I shall weigh immediately (*or at time*
	Keep on weather beam _ _____GAR		*indicated*) _____ZP
	Look out on weather bow _____HOK		I shall weigh the moment the breeze
	On weather bow _____HCQ		springs up _____HGA
	Weather beam_____GAY		Prepare to weigh _____ZQ
	Weather gauge ___ _____NVI		Ready to weigh_____ __ _____ZR
YZO	—Weather helm		Shall I in my present berth have room
YZP	—Weather lurch		for weighing if the wind shifts___GFZ
YZQ	—Weather quarter	ZAX	—Shall, *or*, Will weigh
YZR	—Weather shore		Unable to weigh anchor _____EJY
YZS	—Weather side		Vessels just weighed (*or*, leaving) show
YZT	—Weather tide		your distinguishing signals _____DZ
			Weigh, and proceed as ordered_____KV
	WEATHER SIGNALS (*See* METEOROLOGI-		Weigh, Cut, *or*, Ship Get an offing
	CAL)		Wait for nothing ___ _____KW
			Weigh immediately (*or at time indi-*
YZU	WEATHER-ED-ING TO WEATHER (*See*		*cated*) _____ZS
	also RIDE)		
	Can — (*vessel indicated*) weather the	ZAY	WEIGH-ED-ING TO WEIGH (*To ascertain*
	gale? _____ _____NVU		*the weight of*)
YZV	—Can weather	ZBA	—Weighing machine
YZW	—Can not weather		
YZX	—Could not weather	ZBC	WEIGHT—OF
ZAB	WEAVE-ING-R WOVE-N		(*For Weights and Measures, see Tables,*
			page 51)
ZAC	WEB-BED-BING		
	Crank webs loose _____KBR		Cargo under weight_____IAJ
		ZBD	—Counterweight ·
ZAD	WED-DED-DING		Dead weight. _____KOI
			Dead weight cargo ____ _____KOJ
		ZBE	—Heavy weight

	WEIGHT—WHAT		
	WEIGHT—*Continued*	ZBT	**WEST**—*Continued.*
	How many tons dead weight of cargo must you have? _____IAS		—Too far to the westward
	Indicate weight—of _____ ____PIC		West India-n cargo_____HZO
	Measure of weight (*See page* 53)_AZX		West longitude _____ ____ ___QZI
ZBF	—Net weight	ZBU	—Western-erly-ing-ward
ZBG	—The weight is		(*See* COMPASS SIGNALS, *pages* 45, 46)
	Under weight Short weight_____YIA		
	Want — tons dead weight of cargo _IBE	ZBV	**WET-TED-TING**
	What weight (*how heavy*) is —?____OTB		Wet-ted clothes_____IVY
	What weight is your screw?_ __ __VKL		Wet dock _____LJO
ZBH	**WELCOME-D-ING**	ZBW	**WHALE**
			Whale boat Whaler_____GVH
ZBI	**WELD-ED-ING**	ZBX	—Whale fishery
		ZBY	—Whale oil
ZBJ	**WELL** (*of water*)		Whaler. Whale ship _____ _____VXG
	Pump well_____UBO		
	Screw well Screw aperture _____KRA	ZCA	**WHARF-AGE**
			Alongside the wharf (jetty, or, pier)_EAT
ZBK	**WELL**		Gun wharf _____ ___OJW
	All well—at_____ _____ _____DYE		
	Are you all well?_____DYF		**WHAT** _____ _____CPU
	Are your passengers all well?_____DYG		After what (*or*, which)? _____CPS
	As well as_____FBV		At what hour?_____FJC
	Captain is well _____ _____HWP		Before what (*or*, which)? _____CPT
	Cargo sold well _____IAH		By what (*or*, which)?_____BJZ
	Cattle all well _____ ____IFC		By what means? _____ ___RNV
	Do quite well _____ _____ _____UFO		By what time? _____ _____HPT
	Doing well ___ _____ ___ ____ _LIN		Do we feel any current? What is the
	Fit-ted-ting very well _____ _____NCM		current?_____KGI
	Is my family well? _____ _____MSK		For what (*or*, which)? __ _____BPK
	It will do very well _____LIR		From what (*or*, which)?_____BPX
ZBL	—Not doing well		How is the tide? What tide have we
ZBM	—Not so well—as		now?_____NXF
	Report me all well _____ ____URZ		In what condition—is— ? _____JMZ
	Report — (*vessel indicated*) all well in —		In what manner? ____ _____RJS
	(*latitude and longitude indicated*) on		In what quarter? _____UEP
	the — (*date indicated*) _____UN		Of what (*or*, which)?_____CGU
	To whom shall I telegraph that all is		On what (*or*, which)?_____ _____CHG
	well on board your vessel? _____XMH		Somewhat _____ __ _____WJT
	Tolerably healthy (*or*, well) _____OQZ		Then what (*or*, which) _____CKL
ZBN	—Very well		That what (*or*, which) _____CLX
	Very well done _____LTT		To what (*or*, which)_____ _____COD
	Vessels that wish to be reported all well		What age?_____DSI
	show your distinguishing signals_DYT		What alteration?_____HOW
	Was all well when you left? _____QLN		What, *or*, Which am I ?_ _____CPV
	Was well when I left _ _____QMO		What amount—of?____ _____EGM
	Well done_____LIU		What answer (*or*, reply)?_____ENP
	Well known_____QCJ		What, *or*, Which are (*or*, is)?_____CPW
ZBO	Will do his (*or*, its) duty very well _LQF		What, *or*, Which are (*or*, is), he (she, it,
	—Work-ed-ing well		or person-s or thing-s *indicated*)?_CPX
	Your family are all well____ _____MSL		What, *or*, Which are (*or*, is) you-r-s?_CPY
	— was well when I left_____QMO	ZCB	—What are the marks?
			What are you about?_ _____DAK
ZBP	**WELSH-MAN**		What are you waiting for? _____YTC
	Welsh coal _____IXL		What are your orders?____ _____SUV
			What arrangements—to (*or*, for)? _EXM
ZBQ	**WENT—TO**		What assistance do you require? What
	Was on board, *or*, Went on board__GBB		do you want?_____FOT
	Your letters went by the —_____QOS		What bearing shall I keep the light (*or*,
			landmark) on?_____ _____QF
	WERE (*See* WAS) _____COM		What brand?_____HDY
			What, *or*, Which can (*or*, may)? _CPZ
	WEST (*Compass in degrees*)_____AMH		What color____ _____JBQ
			What, *or*, Which could (*or*, might)?_CQA
	WEST (*Compass in points*) _____ASD		What condition?_____JNA
	Alongshore to the westward of —_EAR		What current (*rate and direction*) do you
ZBR	—From the westward		expect (*or*, reckon on)?_____KOH
	On, *or*, At the west coast of — ____IYM		What day? ___ _____KNZ
ZBS	—To, *or*, For the westward		What depth? Indicate the depth__KVB

WHAT—WHEN

WHAT—*Continued*

What did he say?---------- ----VIM
What difference?------------------KZR
What distance? How far?----------MTA
What distance off is —?----------LGM
What, or, Which do (does, or, did)..CQB
What, or, Which do (or, did) they (we or, you)?---- -- --------- ----CQE
What do you make?----- -----------RIX
What, or, Which does (do, or, did) he (she, it, or person-s or thing-s indicated)?---------------- -- --CQD
What else?-------- ---- --------LVX
What force?---------------------NHZ
What, or, Which had (has, or, have)?..CQF
—What has taken?
What, or, Which is? (*See* A1e, *above*), CPW
What is amiss?----- --------------EFN
What, or, Which is (or, are) he (she, it, or person-s or thing-s indicated)?--CPX
What is her name?-----------------SDQ
What, or, Which is it?------------CQG
What is the amount of?------------EGN
What is the cause of your putting back? UDE
What is the current? Do you, or, Do we feel any current?----------- .KGI
What is the depth of water on the bar (in feet)?--- ------------------VQ
What is the matter?----- ---------RMY
What is the motive?------- - - --SBD
What is the name—of?-------------SDR
What is the name of the ship (or, signal station) in sight?----.----------VG
What is the nature of?------- --- SEJ
What is the meaning of?---- --- RNT
What is the range?----------- -- .UHI
What is the reason—of?----------- I. V
What is the sea like?-- -----------UX
What is the sickness?------------VD
What is the state (condition) of?-- JNB
What is the temperature? What is the height of the thermometer?-----XNK
What is the variation?----- ------JIN
What is there to pay?------------TEU
What is your complement?------.JKF
What is your name (or, number)?--SDT
What is your opinion—of?--------SRN
What kind (or, sort) of---------QAZ
What length—is — ?------- --- QNF
What loss?-----------------------RBH
What, or, Which may? (*See* Can, above)---------------------CPZ
What, or, Which might? (*See* Could, above)-------------------------CQA
What, or, Which must?----- - COH
What number (or, quantity—of — have you?----------- ------ ------UDW
What, or, Which of the —?------CQI
What, or, Which ought to?-------CQJ
What port?----------------------TOV
What proof is there?-- -----------TYS
What quantity—of?----------- ----UDX
What quantity of water?-- --- UDY
What rate (speed)?----------------UIN
What, or, Which shall(or, will)—I?.CQK
What, or, Which shall (or, will) I (or, we) do?----------------------CQL
What, or, Which shall (or, will) I (or, we) have?-----------------------CQM

ZCD

WHAT—*Continued*

What, or, Which shall (or, will) they (or, you) do?----- ---------------CQN
What, or, Which shall (or, will) they (or, you) have?-----------------CQO
What shape—is?--------------------VTW
What ship?---------- ------------VYF
What ship is that? (*Indicate bearing, if necessary*)----------------------EC
What, or, Which should (or, would)?..CQP
What, or, Which should (or, would) I (or, we) do?---------------- -- --CQR
What, or, Which should (or, would) they (or, you) do?----------------CQS
What size do you want?----------WEZ
What sort (or, kind) of —? ... ----QAZ
What success have you (or vessel indicated) had?----- -------------------XOD
What variation do you allow?------JIP
What vessel have you boarded?----GRD
What vessel takes the next mail?--RFY
What wages?-------- -------------YSJ
What, or, Which was (or, were)?--CQT
What way?-------------------------YXH
What width (or, breadth)?------ .HET
What, or, Which will—It? (*See* Shall, above)------- ------ - --------CQK
What will you take?-------------XKB
—What winds have you (or vessel indicated) had?
What, or, which would? (*See* Should, above)-------------------------CQP
—Whatever-soever
Which------------------------CUG
With what (or, which)?-----------CWN

ZCE

ZCF

WHEAT
Wheat cargo-- --------------------HZP

ZCG

WHEEL-ED-ING
Cogwheel--------------------------IZM
Handwheel.------------------------OMA
Let your screw (or, wheels) revolve without disconnecting------------QOB
Paddle wheel------------------------SZG
Paddle-wheel steamer (*Horsepower to follow, if necessary*)-------------SZH
Pinion wheel----------------------TKH
Steam steering wheel-------------WSU
—Wheel (*The wheel*)
—Wheel rope

ZCH

ZCI
ZCJ

WHEFT (*See* WAFT)-------- ------YSC

WHEN------------------------------CQU
When am I to?---------------------CQV
When anchored-------------------EKD
When are (or, is)? .. -----------CQW
When are we?--------- ----------CQX
When are you?------- ------------CQY
When can (or, may)?------------CQZ
When could (or, might)?--------CRA
When did he (she, it, or, they) leave? QLT,
When did it happen?-------------OMQ
When did the accident happen?-----BU
When did you arrive?-------------EYV
When did you ground?-------------CK
When, or, What did you hear last—of (or, from)?----------------------ORU

WHEN—WHERE

WHEN—*Continued*
When did you leave? _____ QLU
When did you pass? _____ TCW
When did you part company—from — ?
                                    IR
When did you sail? _____ SG
When did you see — ? _____ VOC
When did you (or, they) separate? . VRJ
When did you write? _____ ZJP
When do (does, or, did)? _____ CRB
When do (or, did) I (or, we)? ___ CRE
When do (or, did) they (or, you) CRF
When do (or, did) you expect? ... MNW
When do you go—to? _____ ODB
When do you intend to bring up? .. EKF
When do you leave? _____ QLU
When do you sail? . _____ _____ ZT
When do you tack? _____ XIO
When do you think I am likely to fall in
  with — (*vessel indicated*)? ......MSA
When do you wear? . _____ YXU
When does (do, or, did) he (she, it, *or
  person-s or thing-s indicated*)? ...CRD
When does he (she, it, *or vessel indi-
  cated*) sail? ....-.-................ VEY
When does it begin (or, commence)?
                                    GDU
When does the blockade commence?
                                    GNQ
When done—with _____ CRG
ZCK —When empty-ied
When finished _____ MZI
When finished coaling _____ IXP
When first seen . _____ NBM
When going—to _____ ODE
When had (has, or, have)? _____ CRH
When had (or, have) you? _____ CRI
When he (she, it, *or person-s or thing-s
  indicated*) is (or, are) _____ CBJ
When his (her-s, it-s) _____ .. CRK
When I (my, mine) _____ CRL
When I can (or, may) _____ CRM
When I have run my distance I shall put
  her head off until —_____ LGP
When is? (*See* Are, *above*) . ____ CQW
When is (or, are) he (she, it, *or person-s
  or thing-s indicated*)? .... _____ CRN
When it lulls.. .. . _____ RDA
When last seen ... _____ QGY
When may? _____ _____ _____ CQZ
When may the bar (or, entrance) be
  attempted? _____ FWD
When might? _____ CRA
When moored? ... .. _____ RZN
When more moderate _____ RXN
When most convenient .. _____ JTP
When must? _____ . _____ __ CRO
When ought? _____ CRP
When ought I to see (or, sight)? ... VOD
When ought the next mail to leave? RGB
When ready—to (or, for) _____ UKA
When shall (or, will)? _____ CRQ
When shall (or, will) be? _____ CRS
When shall (or, will) he (she it, *or per-
  son-s or thing-s indicated*)? ... ... CRT
When shall (or, will) I (or, we)? .. CRU
When shall (or, will) I (they, or, we) do?
                                    CRV
When shall (or, will) I have? ____ CRW
When shall (or, will) it be done? .. CRX
When shall (or, will) they? _____ CRY

WHEN—*Continued*
When shall (or, will) they (or, you)
  have? _____ CRZ
When shall we be at the rendezvous?
                                    UQR
When shall (or, will) we have? ___ CSA
When shall (or, will) you? _____ .. CSB
When shall (or, will) you be? _____ CSD
When should (or, would)? _____ CSE
When the _____ _____ ..CSF
When the breeze goes down _____ HGI
When the captain _____ HXL
When the sea has gone down _____ LMP
When the tide _____ XSQ
When the tide slacks _____ WFS
When the tide turns _____ XSR
When these (or, those) _____ CSG
When they (their-s) .. _____ CSH
When they are _____ CNI
When was (or, were)? ... ____ . . CSJ
When was the battle fought? _____ FZB
When was the last death? _____ KOY
When we (our-s) _____ CSK
When we are _____ CSL
When were your last observations for
  latitude? _____ QIF
When will? (*See* Shall, *above*) ____ CRQ
When will be the best time for crossing
  the bar? _____ XM
When will it be high water? _____ OVI
When will it end? . _____ _____ LYP
When will it take place? _____ TLP
When will the end (or, last) be? ... LYP
When will the tide ebb? _____ XI
When will the tide flow? . _____ XJ
When will the tide suit _____ XST
When will the tide turn? _____ XK
When will you be afloat? _____ DQE
When will you (or, it) be finished? MZI
When will you be ready—for? ____ UKC
When will you be ready for sea? ... UKB
When will you board? _____ GRE
When will you complete—with? .. JKU
When will you have the cargo? _____ IBK
When wind changes _____ ZFE
When would? (*See* Should, *above*) . CSE
When you r-s ____ ____ __ __ __ CSM
When you are _____ CSN
When you are complete—with _____ JKV
When you float _____ NFG
When you meet _____ ZLS
When you please _____ TMV

ZCL WHENCE
From whence (or, where) _____ BPY

ZCM WHENEVER

WHERE _____ _____ CSO
Anywhere _____ _____ EQR
Anywhere else _____ EQS
Beach the vessel where flag is waved (or,
  light is shown) _____ FZT
Elsewhere _____ LVS
Everywhere _____ MJC
From whence (or, where) _____ BPY
Nowhere .... _____ SKY
Somewhere ____ . _____ .. __ WJU
Where am I (or, are we)? _____ CSQ
Where am I? What is my present posi-
  tion? _____ _____ _____ _____ RZ

WHERE—WHITWORTH

WHERE—*Continued.*

Where are (*or* is)? ------------------CSR
Where are they? -------------------CST
Where are you bound? ---------------SH
Where are you from? ------ ----- . SI
Where are you going? -------------CSU   ZCN
Where can (*or*, may)? -------------CSV
Where could (*or*, might)? ---------CSW   ZCO
Where did you cross the equator?--KEM
Where do (does, *or*, did)? ----- ----CSX   ZCP
Where do (*or*, did) I (*or*, we)?-- --CSY
Where do (*or*, did) they (*or*, you)?-OSZ   ZCQ
Where do you (*or vessel indicated*) be-
  long? ----------------- ------GEY   ZCR
Where do you intend to bring up?--EKH
Where for? ----------------------- CTA
Where from? From where?-------BPY
Where had (has, *or*, have)? ---------CTB
Where he (she, it, *or person s or thing s*
  indicated) is (*or*, are) -----------CTD
Where his (her-s, it-s)-- ---- ------CTE
Where I (my, mine)----------------CTF
Where I am (*or*, we are) ------------CTG
Where is? (*See* Are, *above*)--------CSR
Where is (*or*, are) he (she it, *or per-
son s or thing-s indicated*)? ------CTH
Where is he (she, *or*, it) gone (*or*, go-
  ing)?----- ------ ------------------ODF   ZCS
Where is it to take place (*or*, be done)?   ZCT
   LIV
Where is the admiral (*or*, senior officer)?   ZCU
   SJ   ZCV
Where is the captain? -------------HXM
Where is the Commander-in Chief?--IOW
Where is the commanding officer? --JEI
Where is the, *or*, Where shall we ren-
  dezvous? ------------ ---------UQS   ZCW
Where is the — army? ------------EWP
Where may? (*See* Can, *above*) ------CSV   ZCX
Where might? (*See* Could, *above*)--CSW   ZCY
Where must? ---- - ----------------CTI   ZDA
Where ought? ------------------- -CTJ
Where shall (*or*, will) ------------CTK   ZDB
Where shall (*or*, will) be? ---------CTL
Where shall (*or*, will) he (she, it, *or*,
  person-s or thing-s indicated)? ---CTM
Where shall (*or*, will) I (*or*, we)?--CTN
Where shall I find — (*vessel indicated*)?
   MYU   ZDC
Where shall (*or*, will) I have? -----CTO   ZDE
Where shall (*or*, will) they? -------CTP
Where shall (*or*, will) they (*or*, you)
  have? ------------------ ----------CTQ   ZDF
Where shall (*or*, will) we have?----CTR   ZDG
Where shall (*or*, will) you? --- -- --CTS   ZDH
Where shall (*or*, will) you be?-----CTU   ZDI
Where should (*or*, would)? ---------CTV   ZDJ
Where should (*or*, would) the?----CTW   ZDK
Where should (*or*, would) they (their-s)?   ZDL
   CTX
Where the ----- --------- ----CTY
Where they (their-s) ------------CTZ   ZDM
Where they are ---- ------------CUA
Where was (*or*, were)? -----------CUB   ZDN
Where was the battle fought?------FZO
Where we (our-s) ----- ----------- CUD   ZDO
Where we are ----- ---------------CTG
Where will you cross the equator?--KEN
Where you-r-s-------- ----------CUE   ZDP
Where you are - ------------------CUF
Whereabouts ----------------------CSP

WHEREABOUTS------- ------------ --- CSP
  Do you, *or*, Did you get a good look at
  the land to know exactly whereabouts
  we are? ----------------------- PY

WHEREAS

WHEREIN   WHEREOF

WHEREVER

WHERRY

WHETHER .

WHICH (*See* WHAT) ----------------CUG
  Before which-----------------------CPT
  Which do you think the best? ------GIE
  Which is (*or*, are) the best? Which, *or*,
    What is the best?--- --- --------GIF
  Which is the easiest? --------------LSA
  Which is the road to — ? ----------UYQ
  Which, *or*, What of the — ?----,-- CQI
  Which side? ----------------------WBO
  Which way?----------------------YXI

WHILE-ST
  Between while ------------------ GIX
  —It is not worth while
    While he -------------- - -------OQD
  —While I
  —While you (*or*, your)
    Whilst in harbor -----------------ONE

WHIP

WHIRL-ED-ING

WHIRLPOOL

WHIRLWIND

WHISTLE
  Blow steam whistle (*or*, siren) at inter-
    vals --------------------------WD
  Shall signal with steam whistle (*or*,
    siren) during fog (*or*, darkness)--WE
  Steam whistle --- ------- ------WSZ
  —Whistle-ing   To whistle
  —Whistling buoy

WHITE
  —White buoy
  —White flare
  —White lead
  —White light
  —White paint.
  —Whiten-ed-ing
    Whitesmith-----------------------ZDN

WHITEHEAD TORPEDO

WHITESMITH

WHITEWASH
  Whitewashed-ing   To whitewash-YUS

WHITING (*Cleaning substance*)

WHITWORTH STEEL --------------------WTQ

WHO—WILL

	WHO-M ----------------------------CUH
	By whom (or, whose)---- --------BKA
	For whom (or, whose) -----------BPL
	From whom (or, whose) - - -------BPZ
	Of whom (or, whose) ------------CGV
	On whom (or, whose) -----------CHI
	To whom (or, whose) ----------- COE
	To whom shall I telegraph that all is
	well on board your vessel? ------XMH
	We who ------ ---- -------------CPR
	Who, Whom----------------------CUH
	Who are (or, is)?----------------CUI
	Who are your owners? ---------- SYM
	Who can (or, may)?--------------CUJ
	Who could (or, might)?----------CUM
	Who do (does, or, did)? - --------CUO
	Who else? ----------------------- LVY
	Who had (has, or, have)?--------CUP
	Who is?------------------------ - ----CUI
	Who is dead? --- ----------- -----KON
	Whom is it for? Whom for? ------NHS
	Who is Lloyd's agent?.. --- ------QXJ
	Who is the captain of—(ship indi-
	cated)? or, Who is your captain?_HXN
	Who may? ----------------------CUJ
	Who might?---------------------CUM
	Who must?------------ ---------CUQ
	Who ought? --------------------CUR
	Who shall (or, will)?-------------ZEF
	Who should (or, would)?---------CUV
	Who told (or, said so)?-----------XMZ
	Who was (or, were)?------------CUW
	Who will (or, shall)?--------------ZEF
	Who would?---------------------CUV
	Whom does it belong to? --------GEZ
	Whom does she (or vessel indicated)
	belong to? --------------------GFA
	Whose?---------------- ---------ZDU
	With whom (or, whose) ---------CWO
ZDQ	WHOLE-LY
	Annul the whole signal-----------VF
	Privy to the whole affair---- ------DPH
ZDR	—Whole gale
ZDS	WHOLESALE.
ZDT	WHOLESOME-LY-NESS
	Not wholesome. Unwholesome-
	ness -------- -----------------YLQ
	WHOOPING (HOOPING) COUGH--------OXT
ZDU	WHOSE (See also WHO)
	WHY? ----------------------------CUX
	Why am I to?-------------------CUY
	Why are (or, is)?-----------------CUZ
	Why are you----------------------CVA
	Why are you not?-----------------CVB
	Why can (or, may)? -------------HUF
	Why cannot (or, may not)?--------CVD
	Why could (or, might)?-----------CVE
	Why could (or, might) not?-------CVF
	Why do (does, or, did)?----------CVG
	Why do (does, or, did) not?-------CVH
	Why do (or, did) you?------------CVI
	Why do (or, did)you not?---------CVJ
	Why do you not sail?-------------VEZ
	Why do you put back?------------UDE
	Why had (has or, have)?---------CVK

	WHY—Continued
	Why had (has, or have).not? ------CVL
	Why had (or, have) you?----------CVM
	Why had (or, have) you not?------CVN
	Why is?---- ----------------------CUZ
	Why might?---------------------CVE
	Why must?----------------------CVO
	Why not?-----------------------CVP
	Why ought? -------------------- ---CVQ
	Why shall (or, will)?--------- ----CVR
	Why shall (or, will) not?---------CVS
	Why should (or, would)?----------CVT
	Why should (or, would) not?------CVU
	Why they (their-s)---------------CVW
	Why was (or, were)? ------- -----CVX
	Why we (our-s)------ ------------CVY
	Why were you? ------------------CVZ
	Why will (or, shall)?. . ----------CVR
	Why would? --------------------CVT
	Why you-r-s--- --- --------------CWA
ZDV	WICK
	Lamp wick----------------------QDJ
ZDW	WIDE WIDTH
	How broad—is— ?----------------EJN
ZDX	—It is not wide
ZDY	—It is wide.
	What width (or, breadth)?-- ------HET
	Wide berth ---------------------GHF
ZEA	—Widely
ZEB	WIDEN-ED-ING
ZEC	WIDOW-ER
ZED	WIFE
	WIGTJE (Measure of weight)-------BAN
	WILL (See also SHALL)------------CWB
	How will they? ------------------BXY
	I will give you ------------------OAH
	I will if I can ------------------BKS
	I will wait until-----------------YSR
	It will do ----------------------BNV
	Not willing Does not agree ------DTZ
	Request you will ----------------UST
	Request you will not -------------USV
	Shall, or, Will wait -------------YSV
	There will be time—to (or, for)----XTQ
	Weather will be ----------------- YZW
	What, or, Which shall (or,will)—I?-CQK
	When will it be done?------------CRX
	When will you (or, it) be ready for (or,
	to)—?-------------------------UKC
ZEF	—Who will, or, shall
	Will, or, Shall arrive—at ---------EYW
	Will assist ---------------------FGU
	Will be ready—to (or, for) --------UKD
	Will be sent ------ --------------VQR
	Will be sufficient (enough)--------MEB
	Will be unable—to ---------- -----CYS
ZEG	—Will close (or, be closed)
	Will it be? - --------------------BIX
	Will it be safe? -----------------VDL
	Will it do?----------------------ROU
	Will it have? -------------------CWD
	Will it prevent? -----------------TVM
	Will not----------------------- ------CIU
	Will not do? ----------- ---------BOV

WILL—WIND

WILL—*Continued*
Will not succeed .... XBZ
Will put .... UDF
Will remain by you .... UPK
Will rendezvous at (*or, in*) .... LQT
Will set in .... VbG
Will shift .... VXA
Will soon .... WKC
Will stand by you .... WQM
Will succeed .... XCA
Will they act? .... DIO
Will visit .... YRD
Will you act for me (*or,*—)? .... DIP
Will you assist me (*or vessel indicated*)? DE
Will you be on shore today? (*Specify hour, if necessary*) .... KOC
Will you breakfast with me? .... HFL
Will you (*or person indicated*) buy? .. UCF
Will you come? .... JDQ
Will you (*or person indicated*) come on board? .... GRF
Will you do it? .... CWE
Will you get? .... NZG
Will you give me (*or, my*)—? .... OAN
Will you go? .... ODG
Will you leave? .... QLV
Will you make? .... RIY
Will you make an offer? .... RIZ
Will you permit? .... DZQ
Will you please? .... TMW
Will you put? .... UDG
Will you send? .... VQU
Will you send for him (*or, it, or, her*)? VQW
Will you stay (*or, wait*)? .... WRY
Will you support us? .... XFB
Will you (*or person indicated*) undertake it? .... YHX
Will you venture? .... YOV
Will you wait for the mail? .... YTD
Will you write? .... ZJQ
Yes, I will .... ZLH
Yes, it will .... ZLI

WILL, *or*, TESTAMENT .... XOL

ZEH WILLING-LY NESS
Unwilling-ly ness .... YLR

ZEI WIN-NING WON
— has won a great battle at —, with a loss of killed and wounded reported at — .... FZD

ZEJ WINCH
Steam winch .... WTB

ZEK WIND-ING WOUND To WIND
Winding tackle .... ZFR

ZEL WIND (*See also* FORCE OF WIND)
Anemometer Wind gauge .... ELO
Baffling wind .... FSC

ZEM —Before the wind
Between wind and water .... GIY
By the wind .... HPQ
Calm and light winds .... HTM
Delayed by contrary winds .... KSD
Detained by contrary winds .... KSD
Fair wind .... MRE
Force of wind .... NIA

WIND—*Continued.*
Foul wind .... NOM
Gale of wind .... NVB
Harbour, *or*, Anchorage is good enough with winds from— .... OX
Harbour, *or*, Anchorage is safe with all winds .... OZ
Haul wind (*or, to the*) wind .... OPG
Haul your wind on port tack .... LN
Haul your wind on starboard tack .. LO
Head to wind .... OQM
Head wind .... OQN
ZEN —How have you had the wind?
ZEO —How is the wind outside?
ZEP —How shall we have the wind?
ZEQ —If the wind abates (*or*, lulls)
ZER —If the wind continues as at present
ZES —If the wind shifts
ZET —If the wind should freshen
If you lose the wind .... RAO
In a gale of wind .... NWB
Is the anchorage (*or*, my anchorage) safe with all winds? (*Specify which, if necessary*) .... PB
It is dangerous to lose the wind .... GC
ZEU —Keep before the wind
Keep head to wind .... PYT
Meteorological or weather report for to-day gives MODERATE winds from— (*direction indicated*) .... RSH
Meteorological or weather report for to-day gives STRONG winds from— (*direction indicated*) .... RSI
Meteorological or weather report for to-day gives winds VARIABLE from— (*direction indicated*) .... RSJ
Meteorological or weather report for to-morrow gives MODERATE winds from — (*direction indicated*) .... RSL
Meteorological or weather report for to-morrow gives STRONG winds from — (*direction indicated*) .... RSM
Meteorological or weather report for to-morrow gives winds VARIABLE from —(*direction indicated*) .... RSN
ZEV —More rain than wind
Not wind enough .... MDR
Nothing but gales of wind .... NWC
Off, *or*, From the wind .... NRY
Running before the wind .... VCH
Shift of wind .... VWZ
ZEW —Should the wind
ZEX —The wind is —
ZEY —The wind is going to change
ZFA —The wind will come from the —
To, *or*, By the wind .... HPQ
ZFB —Too much wind
ZFC —Trade wind
Variable wind .... YOD
ZFD —What wind leads through the passage (*or*, entrance)?
What winds have you (*or vessel indicated*) had? .... ZCE
When it lulls The first lull .... RDA
When the wind (*or*, breeze) .... HGJ
ZFE —When wind changes
Where did you fall in with the trades (*or*, monsoon)? .... RYM
Where did you lose the trades (*or* monsoon)? .... RBI
Whirlwind .... ZDA

	WIND—WISH		

	WIND—*Continued*	ZGE	WINTER
	Wind a gale............ ....NVS	ZGF	—Had, *or*, Has winter set in at — ?
ZFG	—Wind and weather permitting		Has the ice broken up ? Has the winter
ZFH	—Wind appears to be steady		broken up—at — ?...... ....PBJ
ZFI	—Wind fresh	ZGH	—In the winter
	Wind gauge Anemometer ....ELO		Last winter......................QGW
ZFJ	—Wind has abated outside	ZGI	—Next winter
ZFK	—Wind has not abated outside	ZGJ	—Through the winter
	Wind is going to change........... ZEY		Warm, *or*, Winter clothing for......IVX
ZFL	—Wind light	ZGK	—Winter has regularly set in at —
ZFM	—Wind strong	ZGL	—Winter quarters
ZFN	—Wind will freshen		
ZFO	—Winds from the — With wind from	ZGM	WINTER-ED-ING TO WINTER
	the —		
ZFP	—Winds have mostly been	ZGN	WIRE
	Windsail....................ZFV		Insulate-d-ing wire (*or*, cable)......PNT
			Magnesium wire...................RFO
			Steel-wire hawser.................WTO
	WIND GAUGE ANEMOMETER ....ELO	ZGO	—Telegraph wire
			Wire cable (*or*, hawser) ....HbE
ZFQ	WINDBOUND—AT	ZGP	—Wire rigging
	I have been, *or*, Vessel indicated has		Wire rope....................VAD
	been detained—days windbound at—		
	(*place indicated*)................KYE	ZGQ	WISE-LY WISDOM
		ZGR	WISH-ING—TO (*or*, FOR).
ZFR	WINDING TACKLE (*See also* TACKLE)		Admiral, *or*, Senior officer wishes..DKY
			Admiral, *or*, Senior officer wishes to
ZFS	WINDLASS		know ....DKZ
	Bring it to the windlass ....HIM		Admiral, *or*, Senior officer wishes to see
	Damaged the windlass...... ......KJE		you (*or person indicated*) ....DLA
	Steam windlass ....WTC		Admiral, *or*, Senior officer wishes to see
			you as soon as convenient.........DLB
ZFT	WINDMILL.	ZGS	—Do you wish ?
			Do you wish to be reported?..... .UB
ZFU	WINDOW		Do you wish your arrival reported ?.EYC
	Cabin window ....HQG	ZGT	—Does not wish— it (*or*, that)
			I do not wish to show a light.......QRX
ZFV	WINDSAIL		I should wish to know the nature of the
	Hoist windsails..... ....OWK		sickness, if any, before I send my boat
			(*or*, communicate) ....FC
ZFW	WINDWARD		I wish for orders from Mr —, my owner,
ZFX	—Beat (work) to windward		at — ....SW
	Far enough to windward...........MDG		I wish to be taken in tow ... .XW
	Keep more to windward ....PZB		I wish to buy ....HOY
	Keep to windward..................LY		I wish to communicate, close......IUX
	Keep to windward of vessel chasing,		I wish to communicate personally...JOC
	PZO		I wish to land passengers...... ...QEG
	Keep to windward until you are picked		I wish to obtain orders from my owner,
	up ....PZQ		Mr —, at —....................SW
	Let your boats keep to windward until		I wish to see you before you go on shore,
	they are picked up..............FD		OCN
	Look out to windward ....QZW		I wish to see you on shore..........VNU
	Shoal water, *or*, Danger to windward,		I wish to signal, come within easy sig-
	KLI		nal distance .... --- .---VJ
ZFY	—To windward		I wish to speak to you ....WMK
	We are to windward of the harbor (*or*,	ZGU	—If you wish—it (*or*, that)
	anchorage) ....ELI	ZGV	—It is my wish—that.
	Work-ed-ing (beat) to windward...ZFX		Owners wished me to inform you..SYK
			Signal the name of place where you wish
ZGA	WINE		to be reported by telegraph......WDF
	Australian wine ....FMH		Stranger, *or*, Vessel bearing — wishes
ZGB	—Burgundy wine		to communicate ....WZD
	Champagne ....IKC		They, *or*, I wish to abandon, but have
	Claret ....ISD		not the means ....AJ
	Madeira wine.....................RDJ		Vessel seriously damaged, wish to trans-
	Port wine ....TPC		fer passengers............ ....HM
ZGC	—Rhine wine		Vessels that wish to be reported all well,
	Sherry ....VWR		show your distinguishing signals .DYT
	Spirits of wine................. WNT		Vessels that wish to be reported, show
ZGD	—Wine glass		your distinguishing signals.........EA

WISH—WORK

WISH—*Continued*
Vessels that wish to be reported on leaving, show your distinguishing signals .... EB
Wish to communicate—by (*or*, with) . JGF
Wish to enter port or dock .... TM
Wish to get a rate for my chronometer, GU
Wish to see you .... VOE

ZGW —Wish to speak—to (*or*, with)
Wish you a pleasant voyage (*or*, passage), TDL

ZGX —Wish you would
ZGY —Wished

WITH .... CWF
Are, *or*, Is with .... BER
Me with .... CEG
Not with .... CGK
That with .... CLY
Was, *or*, Were with .... CPB
When done—with .... CRG
With .... CWF
With all sails set .... VFA
With arms .... EVG
With complete success .... JKX
With him (his, her s, it-s) ..... CWG
With me (my, mine) .... CWH
With moderate-ion .... RXO
With much pleasure .... TMX
With passengers .... TDW
With pleasure I will accept ... DCA
With steam up .... YMC
With that (*or*, this) .... CWJ
With the .... CWI
With the captain .... HXO
With the tide .... XSV
With the view (*or*, object) ..... SMK
With them (their-s) .... CWK
With these (*or*, those) .... CWL
With tools XVI
With us (our-s) .... CWM
With what (*or*, which) .... CWN
With whom (*or*, whose) .... CWO
With you-r-s .... CWP
With your permission .... TGJ

ZHA WITHDRAW-N-ING-DREW
ZHB —Have, *or*, Has been withdrawn
- Light-ship has been withdrawn on account of ice .... PBC
ZHC —Shall, *or*, Will be withdrawn

ZHD WITHHOLD-ING-HELD

ZHE WITHIN
Come within hail .... JDA
I shall pass within hail .... OLE
Keep within hail .... OLF
Keep within sight—of .... WCB
Pass within hail .... OLG
Send a boat within hail .... GUI
Within range .... UHZ
You are within range of the batteries (*or*, of the guns) .... GO

ZHF WITHOUT
Date not known Without date .. KLW
If it can (*or*, may) be done without risk, UYA
Not without permission .... TGF

WITHOUT—*Continued*
Without a doubt .... LMB
Without any motive .... SBE
Without arms .... YQJ
Without assistance .... FGX
Without authority, *or*, Unauthorized, FMZ
Without damage .... KJT
ZHG —Without fail
ZHI —Without favor
Without hesitation .... OUM
Without loss .... RBJ
Without payment .... TEV
Without permission .... TGK
Without risk .... UYE

ZHJ WITHSTAND-ING-STOOD

ZHK WITNESS-ED-ING
Eyewitness .... MQI

ZHL WOKE WAKEING (*See also* AWAKE)

ZHM WOMAN
Clothing for female passengers
Women's clothing .... IVE
Females on board .... MVQ
Have you any women on board? .. GQH
I have no women on board ..... GQK

WON (*See* WIN) .... ZEI

ZHN WOOD-ED
ZHO —Can not procure wood
Dyewood .... LQK
Firewood .... NAC
You can get any quantity of firewood, NAT

ZHP WOOL-LY

ZHQ WOOLLEN

ZHR WORD-ED-ING
Byword .... HPU
Pass word .... TDY

ZHS WORE (*See also* WEAR)

ZHT WORK-ED-ING TO WORK
Bell buoy damaged, not working .. GEN
Board of Works .... GPF
Boiler working on seating ..... GWR
ZHU —Can work up to—revolutions
Crew disaffected, will not work .. KCN
Crew discontented, will not work . KOO
Day's work .... END
Dead works (*of a ship*) .... KOL
Earth work .... LRQ
Engine does not work well ..... MBD
Engine works very well .... MBN
Engines working badly .... FRM
Engines working on seating ... MCP
Fireworks .... NAD
Have you anyone that can work at a forge? .... NLE
Light on—is not working satisfactorily, QSH
My working pressure is .... TUS
ZHV —No boat fit for the work
One screw disabled, can work the other, MW

WORK—WRITE

	WORK—*Continued*
	Propeller worked loose ....TZC
	Upper works....................YMB
	What is your working pressure?...TUV
	Work badly.....................FRX
	Work expansively ....MNB
ZHW	—Work-ed ing in
ZHX	—Work out
ZHY	—Work pump
	Work, *or*, Beat to windward.......ZFX
	Worked-ing very badly...........FRY
	Worked-ing well ....ZBO
	Working her own engines.........MCF
	Working one engine....... .....MCG
	Working party ....TCF
	Working steam pressure—is ....TUS
	You may work the semaphore ......VO
ZIA	WORKMAN-SHIP
ZIB	WORLD
ZIC	WORM
	Worm eaten ....LSJ
	WORN....... ....YXV
	Pump gear worn out(*or*, defective)..UBJ
ZID	WORSE. WORST
ZIE	—Could, *or*, Might be worse
ZIF	—It will be worse
ZIG	—Much worse
ZIH	WORTH
	Is it worth while shifting berth?...GFV
	It is not worth while.......... ....ZCT
ZIJ	—It is worth
	Not worth getting.................OFL
	Not worth getting off (*or*, afloat) ..DQO
	Seaworthy VLI
ZIK	WORTHLESS-NESS
	WOULD (*See* SHOULD)...........CJO
	I would shift my berth as soon as possible ....GFT
	Wish you would ....XGX
	Would recommend Would advise..DOB
ZIL	WOUND-ED-ING
	Anyone hurt (*or*, wounded)?......EQB
	Badly, *or*, Seriously wounded......FRL
	Boiler burst, — men killed, — others wounded.....................AU
	Boiler burst, number of killed and wounded not yet known...........AX
	Boiler burst, several men wounded..AZ
	Dangerously wounded (*or*, hurt), accident ....AL
	Gunshot wound Accident happened, DCI
ZIM	—How are the wounded?
ZIN	—How many wounded?
	Is anyone wounded (*or*, hurt)?... .EQB
ZIO	—Killed and wounded
	Many wounded ....RKN
	None wounded...................SJF
	Number of killed and wounded not yet known...................QAP
	Slightly wounded................WGS
ZIP	—The wounded are

	WOUND—*Continued*
	— has won a great battle at — with a loss of killed and wounded reported at — ....FZD
	—'s losses in killed and wounded reported at —... . ...QAR
	WOVE (*See* WEAVE)...........ZAB
ZIQ	WRAP-PED-PING-ER.
ZIR	WRECK
ZIS	—A complete wreck
	Have you seen (*or*, heard of) a vessel wrecked (*or*, in distress)...........ZU
	I am (*or* vessel indicated is) a complete wreck.......................JKP
ZIT	—I shall stand about to see if I can pick up anything from wreck
	Is, *or*, Was there anyone on board? .EQC
	Lost sight of wreck................RAY
ZIU	—Off the wreck
ZIV	—On the wreck.
	Passed a wreck...... ... .TOP
	Passed a wreck (*date, latitude and longitude to follow, if necessary*), but could not render any assistance, people still on board................ZV
	Passed a wreck (*date, latitude and longitude to follow, if necessary*), but was unable to ascertain whether any people were on board .. ..ZW
	Passed a wreck (*date, latitude and longitude to follow, if necessary*), no one on board ....ZX
	Pieces of the wreck have been found (*or*, picked up) ....TIB
	Receiver of wreck................ULG
	Shipwrecked.....................VYL
ZIW	—The wreck of —
ZIX	—Vessel indicated has been wrecked
	What has become of wreck?........GCQ
	Will you go to the assistance of wreck? (*Bearing to follow, if necessary*)..ZY
ZIY	—Wreck buoy
ZJA	—Wreck is being pillaged
ZJB	—Wreck will be sold
ZJC	—Wreck-ed ing To wreck
ZJD	—Wrecked on the coast
ZJE	WRECKER
ZJF	WRIST
ZJG	WRIT
ZJH	WRITE-ING-TEN WROTE To WRITE
	Do you intend to write from —? ...POS
	Forward following telegraphic message by signal letters instead of writing it at length....................XB
ZJI	—Has he (*or*, she) *or*, Have they written?
ZJK	—Have, *or*, Has not written
ZJL	—Have, *or*, Has written
ZJM	—I have written
	I intend writing ....POW
	I will wait whilst you write...... . YST
ZJN	—In writing
	Send following message by post to owners (*or*, to M₁ —) at —, by signal letters instead of writing it at length.wz

WRITE—YARD	

	WRITE—*Continued*
	Send following telegraphic message by
	signal letters instead of writing it at
	length ---------------------------- XB
ZJO	—Shall I write?
ZJP	—When did you write?
ZJQ	—Will you write?
ZJR	—Write immediately
ZJS	—Write to the owner
	Writing (*Manuscript*) ----------- RKC
ZJT	—Writing paper
ZJU	—Written orders
ZJV	—Wrote to — on —
ZJW	WRITER
	Typewriter --------------------- YFR
ZJX	WRONG-ED-ING TO WRONG
ZJY	WRONG-LY
	Anything wrong with the engines?--RD
	I have wrong papers on board --- -TAK
ZKA	—I was wrong
	Very wrong-ly ------------------- YPU
ZKB	—Wrong meaning
ZKC	—Wrong place
ZKD	—Wrong side
	Wrong way ---------------------- XXJ
ZKE	WROUGHT
ZKF	—Wrought iron.
ZKG	X (*Letter*) (*For new method of spelling, see page 13*)

ZKH	Y (*Letter*) (*For new method of spelling, see page 13*)
ZKI	YACHT-ED-ING
	Lloyd's Yacht Register ------------QXE
	Royal yacht. -------------------- VBC
	Royal Yacht Squadron -----------WPG
ZKJ	—Union des Yachts
	Yacht club ---------------------- IWE
ZKL	YAM
	YARD (*Measure of length*) ---- ----AXE
	Cubic yard-----------------------AYE
	Square yard------ - -- -----------AXZ
ZKM	YARD (*Ship's*)
	Brace round your head yards -- --HDO
	Can I get a mainyard at —? --- .NYG
	Can I get a spar for a fore yard at—?.NIS
	Crossjack yard ------- --------- KEF
	Dimensions of — yard are — feet in
	length, — inches in the slings. -LAK
	Double topsail yards--------------XWD
	Fore royal yard----------------------NIZ
	Fore topgallant yard ----------- .NJI
	Fore topsail yard -----------------NJP
	Fore yard ----------------------NJQ
	Fore yard gone in the sling --------NJR
	Fore yard sprung------------- ---- NJS
	Haulyard, or, Hallaird -------- ----OLQ
	Head yard ----------------- --------OQP
	I have carried away fore yard ----- NKH
	I have carried away main yard -- .RGE
ZKN	—I have carried away topsail yard.
	I have sprung (*or*, damaged) fore yard,
	KJH
	I have sprung (*or*, damaged) lower top
	sail yard----------------------- KJI
	I have sprung (*or*, damaged) main yard,
	KJL
	I have sprung (*or*, damaged) topsail
	yard-- . -------------------------KJM
	I have sprung (*or*, damaged) upper top-
	sail yard ------------------------KJN
	Iron yard ---------------------- ------PTB
	Jigger royal yard and sail----------PVS
	Jigger topgallant yard and sail------PVT
	Jigger topsail yard and sail--------PVU
	Jigger yard and sail --------------PVW
	Lower fore topsail yard ------- ----NKI
	Lower main topsail yard ---------RGF
	Lower mizzen topsail yard------ ..RWO
	Lower spanker topsail yard -------WLU
	Lower yard ---------------------- RCI
	Main royal yard-------------------RGP
	Main topgallant yard-------- ----RHB
	Main topsail yard----------------RHF
	Main topsail yard sprung--- ------RHG
	Main yard----------------------RHI
	Main yard sprung ------- --------RHJ
	Mizzen royal yard --------------- RWU
	Mizzen topgallant yard-----------RWY
	Mizzen topsail yard---------------RXC
	Royal yard -------- ------------VBD
ZKO	—Spritsail yard (or, gaff)
	Sprung, or, Seriously damaged lower
	yard (*or spar indicated*) ---------RCK
	Steel yard------------------------WTP
	Top your lower yard -------------XVO
	Topgallant yard ---- -------------XVU

YARD—YOU		

YARD—*Continued*
Topsail yard ....XWF
Upper fore topsail yard ....NKJ
Upper main topsail yard ....RHK
Upper spanker topsail yard.... ..XWG
Upper topsail yard... ....XWI
Upper yard...........................-YMC
What are the dimensions of main yard?
                  LAO
What are the dimensions of your fore
    yard? ....LAP
ZKP —Yard arm
ZKQ —Yard rope

YARD　SHIP YARD　SHIPBUILDING YARD,
                       VXY
Dock yard....... ... .LJT
Victualling yard.............. ....YQM

ZKR YARN
      Spun yarn............................WPB

ZKS YAW-ED-ING

ZKT YAWL

ZKU YEAR
ZKV —Half year-ly.
      Last year ....QGX
ZKW —Leap year
ZKX —Next year
      Several years.......................VST
ZKY —This year
ZLA —Yearly

ZLB YEAST

ZLC YELLOW
      Has yellow fever diminished at —? MWI
      Has yellow fever disappeared at —? MWJ
      I have had cases of yellow fever on
        board ....MWK
      I have had no cases of yellow fever on
        board ....MWL
      Is there yellow fever at —?........MWO
      Yellow fever ....MWT
      Yellow fever has diminished at — MWU
      Yellow fever has disappeared at — MWV
      Yellow fever is raging at — ......MWX
ZLD —Yellow paint
      — men of the crew have been ill of yel-
        low fever (*number to follow*) ..MWY
      — passengers have been ill of yellow
        fever (*number to follow*) ....MWZ

      YEN　(*Coin*) ....AVG

ZLE YEOMAN-RY

      YES.............c, *or*, Code Flag over c
ZLF —Yes, I am
      Yes, I can ....HUG
ZLG —Yes, I have
ZLH —Yes, I will
ZLI —Yes, it will

ZLJ YESTERDAY
      The day before yesterday ....KNU
      Yesterday afternoon................DSA
      Yesterday evening..................MIN
      Yesterday morning ....SAF
ZLK —Yesterday's post

ZLM YET
      By and bye　Not yet..............HPX
      Do not begin yet ....GDQ
ZLN —It is not yet
      Not begun yet ....GDT
      Not yet　By and bye ....HPX
      Nothing decisive yet ....KPV
ZLO —She is (*or*, has) not yet
      The cargo is not yet sold...........IBA

ZLP YIELD-ED-ING

      YIN　(*Measure of length*) ....AXF

ZLQ YOKE-D-ING.

ZLR YONDER

      YOU-R-S ....CWT
      After you-r-s ....CWR
      Are, *or*, Is you-r-s ....BES
      As you-r-s ....FBW
      Before you-r-s ....CWS
      By you-r-s ....BKC
      Can, *or*, May you-r-s? ....DKO
      Could, *or*, Might you-r-s?...........BLJ
      Do, Does, *or*, Did you-r-s? ....BFM
      For you-r-s ....BPM
      From you-r-s.........................BQA
      Had, Has, *or*, Have you-r-s? ....BRF
      Had, *or*, Have you been? ....BFV
      Had, *or*, Have you done?...........BMW
      Had, *or*, Have you had? ....BRG
      How are you?........................BXI
      How do (*or*, did) you? ....BXM
      How shall (*or*, will) you?..........BXU
      If you-r-s ....CBP
      If you are ....CBQ
      If you are not ....CBR
      If you can (*or*, may) ....CBS
      If you can not (*or*, may not).........CBT
      If you could (*or*, might) ....CBU
      If you could (*or*, might) not.........CBV
      If you do (*or*, did) ....CBW
      If you do (*or*, did) not.............CBX
      If you had (*or*, have) ....CBY
      If you had (*or*, have) not...........CBZ
      If you shall (*or*, will).............CDA
      If you shall (*or*, will) not ....CDB
      If you should (*or*, would) ....CDE
      If you should (*or*, would) not ......CDF
      If you were ....CDG
      If you were not ....CDH
      In you-r-s ....CDR
      May, etc　(*See* Can, *above*)
      Might, etc　*See* Could, *above*)
      Must you-r-s? ....CFE
      Not you-r-s..........................CGL
      Of you-r-s ....CGW
      On you-r-s...........................CHJ
      Ought you-r-s to? ....CID
      Shall, *or*, Will you-r-s?............CJD
      Shall, *or*, Will you be? ....BHI
      Shall, *or*, Will you do it? ....BOD
      Shall, *or*, Will you not? ....CJE
      Shall, *or*, Will you not be? ....BHJ
      Shall, *or*, Will you not do it?.......BOE
      Should, *or*, Would you-r-s?.........CJX
      Than you-r-s ....CKM
      That you-r-s .... - ....CLZ
      To you-r-s ....COF

YOU—ZANZIBAR

	You—*Continued*	
	Was, *or*, Were you r-s? ....CPD	
	Were you? ....OPL	
	Were you not? ....CPM	
	What, *or*, Which are (*or*, is) you-r-s?	
	CPY	
	What, *or*, Which do (*or*, did) you (they,	
	or, we)? ....CQE	
	What, *or*, Which shall (*or*, will) you (*or*,	
	they) do? . ... CQN	
	What, *or*, Which shall (*or*, will) you (*or*,	
	they) have? ....CQO	
	What, *or*, which should (*or*, would) you	
	(*or*, they) do? ....CQS	
	When are you? .... CQY	
	When do (*or*, did) you (*or*, they)?..CRF	
	When had (*or*, have) you? ....CRI	
	When shall (*or*, will) I (you, they, *or*,	
	we) do?.. ....CRV	
	When shall (*or* will) you? ....CSB	
	When shall (*or*, will) you be? ..... CSD	
	When shall (*or*, will) you (*or*, they)	
	have? . ...CRZ	
ZLS	When you-r-s .... CSM	
	When you are ....CSN	
	—When you meet	
	Where are you from? ....SI	
	Where are you going? ....SU	
	Where do (*or*, did) you (*or*, they)? ..CSZ	
	Where shall (*or*, will) you? ....CTS	
	Where shall (*or*, will) you be? .....CTU	
	Where shall (*or*, will) you (*or*, they)	
	have? ....CTQ	
	Where you-r-s....................CUE	
	Where you are ....CUF	
	While you (*or*, your) ....ZCV	
	Why are you? ....CVA	
	Why are you not? ....CVB	
	Why do (*or*, did) you? CVI	
	Why do (*or*, did) you not? ....CVJ	
	Why had (*or*, have) you? ....CVM	
	Why had (*or*, have) you not?.....CVN	
	Why were you ? ....CVZ	
	Why you-r-s?............. .. ...CWA	
	Will you do it? ....CWE	
	With you-r-s.................. ....CWP	
	You-r-s.............. . ...CWT	
	You are.............. ... .BEY	
	You are not ....BEZ	
	You are to put ....UDH	
	You can (*or*, may) . ...BKX	
	You can (*or*, may) be ....BIY	
	You can (*or*, may) have...........BSY	
	You can not (*or*, may not) ....BKY	
	You can not (*or*, may not) be......BIZ	
	You can not (*or*, may not) have ...BSZ	
	You could (*or*, might)...........BLR	
	You could (*or*, might) be ....BJA	
	You could (*or*, might) not ....BLS	
	You could (*or*, might) not be ......BJC	
	You do (*or*, did) ....BOW	
	You do (*or*, did) not...............BOX	
	You had (*or*, have) ....BTA	
	You had (*or*, have) been...........BJD	
	You had (*or*, have) done ....BOY	
	You had (*or*, have) had...........BTC	
	You had (*or*, have) not ....BTD	
	You had (*or*, have) not been ....BJE	
	You had (*or*, have) not done.......BOZ	
	You had (*or*, have) not had.......BTE	
	You may, etc (*See Can, above*)	

	You—*Continued.*	
	You might, etc (*See Could, above*)	
	You must ....CFK	
	You must be ....BJF	
	You must not ....CFL	
	You must not be..................BJG	
	You ought to .... CIJ	
	You ought not to .. ..CIK	
	You ought not to be..... ....BJI	
	You ought to be...................BJH	
	You shall (*or*, will)........... .....CJK	
	You shall (*or*, will) be ....BJK	
	You shall (*or*, will) have............BTF	
	You shall (*or*, will) not ....CJL	
	You shall (*or*, will) not be ....BJL	
	You shall (*or*, will) not have.......BTG	
	You should (*or*, would)............CKD	
	You should (*or*, would) be ....BJM	
	You should (*or*, would) have.......BTH	
	You should (*or*, would) not ... .CKE	
	You should (*or*, would) not be......BJN	
	You should (*or*, would) not have....BTI	
	You were ....CPN	
	You were not ....CPO	
	You were not to be ....BJP	
	You were to be ....BJO	
	You will, etc (*See Shall, above*)	
	You would, etc (*See Should, above*)	
	Your ability to....................CYD	
	Your absence ....DBG	
	Your boat ....GVK	
	Your departure . ...KTZ	
	Your instructions (*or*, orders).......PNM	
	Your man ... ...RPK	
	Your order (*or*, instructions).......PNM	
ZLT	**YOUNG**	
ZLU	—Younger-est	
ZLV	**YOURSELF**	
	Avail yourself of — ....FOD	
	Place yourself. ....TLK	
ZLW	**YOUTH-FUL-LY-NESS**	
ZLX	**YOUYOU** (*Small boat*)	
ZLY	**Z** (*Letter*) (*For new method of spelling, see page* 13)	
	Zac (*Measure of capacity*).... ... AYZ	
ZMA	**ZANZIBAR-ER** **ZANZIBAR COLORS**	

ZARF—APPENDIX

	ZARF (*Measure of capacity*) --------AYR		APPENDIX
ZMB	ZEAL-OUS-LY	ZNA	—Report me to Maritime Association, New York
ZMC	—Is, *or*, Are very zealous		
ZMD	—Your zeal has been particularly noticed by —	ZNB	—Report me to Maritime Exchange, Philadelphia
		ZNC	—Report me to Pacific Shipowners' Association, San Francisco
ZME	—ZENITH	ZND	—Veal
ZMF	—In the zenith	ZNE	—Hardware
		ZNF	—Chapel
ZMG	ZEPHYR	ZNG	—O'clock
	ZER, *or*, GUZ (*Measure of length*)..AWC	ZNH	—Watch (*Timepiece.*)
		ZNI	—Faint-ed-ing (*Swoon*)
ZMH	ZERO	ZNJ	—Strake
ZMI	— — degrees below zero	ZNK	—Kroner (*Coin*)
		ZNL	—Doppelzentner (*Measure of weight*)
ZMJ	ZIGZAG	ZNM	—Cubic centimeter
		ZNO	—Cubic millimeter
ZMK	ZINC	ZNP	—Cubic meter
	Chloride of zinc -----------------IPH	ZNQ	—Submarine signal.
	Sulphate of zinc -----------------XDH	ZNR	—Submarine signal bell
ZML	—Zinc plate	ZNS	—Submarine signal receivers
		ZNT	—Submarine signal station
ZMN	ZODIAC		
ZMO	—Sign of the zodiac		
	ZOLOTNICK (*Measure of weight*) ----BCN		
ZMP	ZONE		
ZMQ	—Frigid zone		
	Temperate zone----------------XND		
	Torrid zone Torrid------------XYC		
ZMR	ZOOLOGY-ICAL.		

GEOGRAPHICAL LIST.

NAMES OF PLACES ARRANGED IN THEIR ALPHABETICAL ORDER.

Notations

B Bay	Hr Harbor	Pt Point, Pointe
C Cape, Capo, Cap	I Island	R River
Ch Channel	Lt Light	Sh Shoal
F Fort	Lt V Light vessel	Str Strait
G Gulf	Mt Mount, Mont	
Hd Head	P Port, Porto, Puerte	

AALBÆK—ALBATROSS

Name	Code	Name	Code	Name	Code
Aalbæk	ADQM	Adelaide	ARUX	Aguilar, C D'	ANYX
Aalborg	ADPV	Adelaide, P	ARUY	Agulhas, C	ALWY
Aale Hd	ADHM	Adelaide R	ARIF	Agweh	ALNS
Aalesund	ADOJ	Aden	AMSE	Ahlbeck	ACZJ
Aardals	ABQX	Aden, G of	AMOL	Ahorcados I	AHNV
Aarhus	ADKC	Adige R	AITM	Ahurni Bluff	ATLE
Aas Vær	ABKE	Adin Kerke	AEHP	Ahus	ACFQ
Aarö I	ADGN	Adler Grund Lt V	ADBN	Ahwaz	AMUR
Aarösund	ADGM	Admiralty G	ARKP	Aian, P	AOSD
Aasgaardstrand	ABIT	Admiralty Hd	ATZM	Aichiu	AOLE
Abaco	AXIR	Admiralty I	AQUS	Aigues-Mortes	AHQV
Abaco I, Great	AXIN	Admiralty Inlet	ATZG	Aiguille, C de l'	AKYG
Abbé, Pont L'	AGLU	Adour R	AGTP	Aiguillon	AGPC
Abbeville, P d'	AFYN	Adrianople	AKBN	Ailly, Pt d'	AFYV
Abd-al-kuri	AMOK	Adriatic	AIQL	Ailsa Craig	AEXZ
Aberaeron	AEST	Adra	AEID	Air Pt	AEUD
Aberbenoat, L'	AGJK	Adrampatam	ANFO	Aird R	AQTB
Aberdeen	AFIP	Adramyti	AKNU	Aitodor, C	AKHP
Aberdeen Docks, Hongkong	ANYR	Adventure B	ASGL	Aix, Ile d'	AGQV
Aberdovey	AESV	Adventure, P	ATHC	Ajaccio	AIBR
Abergele	AETY	Æbelö	ADHB	Ajano, P	AKNX
Abervrac'h R	AGJH	Ægina	AJOD	Ajighiol Spit Lt V	AKGJ
Aberystwith	AESU	Æröskiöbing	ADIE	Ajuda	AHCM
Abo (Africa)	ALOV	Ætna Mt	AIKZ	Akabah G	AMBB
Abo (Baltic)	ACQG	Afghanistan	AMWN	Akaroa Hr	ATJC
Abomeh, or, Abomey	ALOB	Aflandshage	ADOC	Akashi	AFBN
Abrolhos Rocks	AVWM	Africa	AXTB	Akassa	ALOT
Absecon	AZMF	Africa, British Central	AMBS	Akba	ALMI
Abu Ail Is	AMRW	Africa, British East	AMLU	Akedo	ALOR
Abukir B	AKUB	Africa, German East	AMJV	Akerman	AKFU
Aburatani B	AOVG	Africa Rock	AHZB	Akishi	APLJ
Aburatsu Hr	AOYM	Agadir	ALCF	Akka	AKSN
Abu Thabi	AMTR	Agalega I	AMIW	Akkra	ALNH
Abydos Castle	AJXE	Agaña	AQVN	Akroteri, C	AJRV
Abyssinia	AMPS	Agastira B	AFOU	Akyab	ANEZ
Acajutla	AUQH	Agate Hr	BEOH	Alabama	AXCU
Acapulco	AUOB	Agay Road	AHUS	Alabama City	AXCV
Aceitera Sh	AHFK	Agde	AEQG	Alagoas Province	AVYZ
Acheh Hd, or, Acheen Hd	APQJ	Agdenes	ABLX	Aland Isles	AGPX
Achill Hd	AFRV	Agerness	ADJI	Ålandskär	ACJE
Aci Reale	AIKY	Agerso	ADMB	Alas Str	AQNS
Acklin I	AXKH	Agger	ADRB	Alaska	ATPI
Acom Pt	BCTI	Aghaliman	AKQM	Alassio	AHWP
Acre	AKSN	Agiabampo	AUMH	Aláya	AKQG
Acul	AXTQ	Agön	ACLS	Albai	AQFZ
Adália	AKQC	Agon, Pt d'	AGEJ	Albani R	ALMP
Adams, P	AOKH	Agropoli	AIJF	Albania	AJGM
Adams Pt	AUDE	Aguada B	ANBJ	Albany	ARFQ
Adda	ALNM	Aguadilla	AXWB	Albany F	BEYX
Addison Pt	BAUA			Albatross I	AMIV

ALBEMARLE—ANKATSAOKA

Albemarle Canal	AYWF	Almadabra	AHKF	Amour Pt.	BEQJ
Albemarle I	AUVJ	Almadabra (Morocco)	ALBC	Amoy	AOBI
Albemarle Sound	AYFL	Almadi, or, Almadies Pt.	ALGB	Ampenan B	AQNM
Albenga	AHWI	Almagrundet Lt V	ACIV	Amping	AOCK
Alberoni	AITR	Al Manáma	AMTZ	Amposta	AHLK
Albert, P (N Z)	ATBL	Almeida B	AMCN	Ampurias	AHND
Albert, P (Victoria)	ASBM	Almeria	AHIJ	Amrun I	ADSF
Alberton (Pr Ed I)	BDEV	Almina Pt	ALBF	Amsterdam	AEBD
Alberton (Queensland)	ASQZ	Almirante B	AWOF	Amsterdam I	AMJU
Alberton (Victoria)	ASBO	Almissa	AJDT	Amur B	AOPS
Albino Pt	ALLE	Almuñecar	AHGT	Amur R	AORG
Albir Pt	AHJV	Alnæs	ABOL	Anaa atoll	ARGV
Alboran I	AHIB	Alnmouth	AFKY	Anacapa I	AUJY
Albrahoa B	AVTI	Alnö	ACMB	Anadolu Hissári I	AKCN
Albreda	ALHW	Alonyi	AJZL	Anadyr R	APOE
Albue Pt	ADMS	Alpena	BEGL	Anaga Pt	ALFC
Albufera	AHKN	Alprech, C	AFYI	Anam	ANVW
Alcala B	AKZA	Alps	AIUT	Anama Hr	APLY
Alcanada Islet	AHOT	Als	ADPQ	Anama saki spit Lt V	APKT
Alcantara	AWOZ	Als I	ADGE	Anamaboe	ALNF
Alcatraz I (Africa)	ALIX	Alsentia	AFSD	Anamba Is	APXU
Alcatraz I (Cal)	AUGJ	Als Sound	ADGH	Anamur C	AKQH
Alcudia	AHOS	Altagracia	AWLM	Anápa	AEJR
Aldabra I	AMJI	Altacarry Hd	AFTL	Anaphi	AJRX
Aldan B	AGZL	Altamaha R	AYKX	Anastasia B	AJKE
Aldborough	AFOM	Altata	AUMN	Anastasia I (Fla)	AYJE
Alderney	AGCW	Altea	AHJW	Anastatia I (Black Sea)	AKDO
Alegro, P	AVRK	Alten Fiord	ABGF	Anatoli, C	AKCZ
Aleutian Is	ATRO	Alten	ABGT	Anatolia	AKNM
Alexander, F	ATRK	Althorpe I	ARTM	Anclam	ACZS
Alexander, P	ALUD	Altona	ADUW	Anclote cays	AXFE
Alexandretta	AKRO	Altwarp	ACZI	Ancon	AUYR
Alexandria (Egypt)	AKUC	Alvarado	AWSV	Ancona	AISE
Alexandria (Va)	AZCE	Al Wakra	AMTS	Ancud	AVIH
Alexandrina, Lake	ARVW	Amahlongwana R	ALYV	Andaman Is	ANMT
Alexandrovsk	AKGW	Amaboi	AQPZ	Andenæs Lt	ABHL
Alexandrovsk (Kola)	ABFQ	Amai, P	AFCJ	Andramatuba	AMGW
Alexandrovski	AORF	Amalfi	AIJB	Andrava B	AMEB
Alfaques, P	AHLE	Amaliopolis	AJVD	Andreievsky	ATQN
Alfred, P (Africa)	ALYD	Amaraçao	AWCT	Andreas, C	AKRH
Alfred, P (S Australia)	ARUL	Amargoso R	AWCP	Andros	AJQP
Algajola	AICB	Amastra	AKNC	Andros I	AXJC
Algarrobo	AVFQ	Amazon R	AWEJ	Androth I	ANDG
Algeciras	AHFQ	Ambavarane	AMDW	Ane, L', Bank	AHTQ
Alger	AKYN	Ambavatobi B	AMGN	Anegada	AXZQ
Algeria	AKWY	Amber, C	AMDR	Aneityum	AQYG
Alghero	AIEY	Amberno R	AQPX	Ange Gardien, L'	BDSU
Algier	AKYN	Amboina	AQPR	Angeles, Los (Cal)	AUIY
Algiers	AWYK	Ambondro R	AMFN	Angeles, P (Mexico)	ALOL
Algoa B	ALXY	Ambong	AFYW	Angeles Pt	AVFE
Algodonales B	AVCT	Ambre	AMDR	Angelos P (Wash)	AUCF
Alguada Reef	ANLW	Ambriz B	ALSV	Anger	AFUS
Alhucemas	AKZU	Ambrizette	AISR	Angermannaelf	ACMH
Ali-Agha, P	AKOB	Amédée I	AQYR	Angers	AGPM
Alibagh	AMZI	Ameland	ADYV	Angiang	ANVP
Ahcante	AHJQ	Amelia I	AYJZ	Anglesea	AETJ
Alicudi	AIOM	America B	AOQS	Angoche R	AMBZ
Alistro	AHZX	America, Central	AUQF	Angola	ALSQ
Ahwal	ALXK	America, North	ATOZ	Angra	ALDC
Ahwal Sh	ALYW	America, South	AUWO	Angra dos Reis	AVFK
Alkmaar	ADZO	American Shs	AXHG	Angra Pequeña	ALUQ
Al Katif	AMUC	Amet Isle	BCWX	Angsa Pulo	ANBO
Al Kuweit Hr	AMUE	Amherst (Moulmein)	ANOJ	Anguilla	AYBC
Allegranza	ALEC	Amherst Hr (N S)	BCAT	Anhatomirim Islet	AVSB
Aleppi	ANCS	Amherst (Pr Ed I)	BDFN	Anholt I	ADPH
Alligator R	ARJD	Amica Pt	AJCT	Anholt Knob Lt V	ADPI
Alligator Reef	AXHM	Amirante Is	AMJD	Anjer	APUS
Alloa	AFJU	Amok	AQNI	Anjer, New	APUV
Allor	AQOL	Amokwa	ALMG	Anjer, Old	APUT
Allsta, or, Hallsta	ACKI	Amorgo	AJRP	Ankatsaoka B	AMFH

ANKOBRA—AUSTRALIA

Ankobra R	ALMS	Archer Pt	ASYH	Arthur R	ASFC
Ann, C...	BAJF	Arctic Ocean	ABCD	Arvoredo I	AVSF
Anna B.	AONV	Ardglass Hr	AFUE	Arzeu, or, Arzew	AKYX
Anna de Chaves B	ALQG	Ardmore	AFWN	Arzic Pt	AGNO
Anna Maria B	ARGL	Ardnakinna Pt	APQB	Arzila	ALBN
Annapolis (Md)	AZDL	Ardnamurchan	APBZ	Arzobispo	AQVF
Annapolis (N S)	BCFD	Ardrossan (Clyde)	AEYK	Asab.	AMPJ
Annatám	AQYG	Ardrossan (S Australia).	ARUN	Asaba ,	ALOZ
Annatto B	AXRE	Arecibo	AXWF	Asafi.	ALBZ
Annesley B	AMPO	Arena Pt (Cal)	AUFQ	Asan Anchorage	AOLV
Annisquam.	BAJL	Arena Pt (Cent Am)..	AUWK	Ascension I	ALUX
Annobon I	ALQH	Arenas Pt	AURY	Ashanti	ALNG
Annunziata, P Torre del	AHBQ	Arendal	ABWF	Ashburton.	ATIZ
Año Nuevo Pt	AUHM	Arendsee.	ADOG	Ashley R	AYPE
Anse, L'	BEOT	Arensburg	ACUD	Ashrali Reef.	AMQI
Anse de Perros.	AGHR	Arenys de Mar	AHML	Ash-Shehr	AMSI
Anse du Pouldu	AGMO	Arequipa	AVBP	Ashtabula	BEDV
Ansiola Pt	AHPB	Ares	AGXS	Asia	AMQU
Antibes..	AHVE	Argao Pt	AQIS	Asia Minor	AKJM
Anticosti I	BDLH	Argentario Mt	AIFR	Asinara I	AIFB
Antifer, C d'	AFZE	Argentiera C	AIFD	Askevold	ABFZ
Antigonish	BCVW	Argentina	AVKX	Askold I	AOQM
Antagua	AYCN	Argos	AJNL	As Ness	ADLS
Antioch	AKRQ	Argostoli.	AJKG	As Rocas.	AWBY
Antioche	AGQS	Arguello Pt	AUIL	Aspra Spitia	AJLI
Antiparos.	AJRL	Arguin B	ALFV	Assateague.	AZGX
Anti-Paxo	AJHW	Argyle.	BCOZ	Assens	ADHQ
Antipodes Is	ATON	Arholma	ACKD	Assini R	ALMK
Antivari..	AJOX	Arica	AVCD	Asso.	AJIZ
Antofagasta	AVDE	Arichat	BCFZ	Assumption I	AMJH
Antongil B	AMEF	Arildsläge	ACDI	Astoko	AJKU
Antonio, P	AXRC	Arizona	AULQ	Astoria (N Y)	AZPO
Antsirana..	AMDV	Arkhangel	ABEO	Astoria (Or).	AUDG
Antwerp (Anvers)	AEGS	Arklow	AFVK	Astrolabe B	AQTR
Anvil Pt	AEMS	Arklow Bank, North, Lt V	AFVP	Astrolabe Road	ATEN
Anzio.	AIGJ	Arklow Bank, South, Lt V	AFVR	Asturias	AGWI
Aomori	AOTQ	Arko	ACIP	Asuncion	AVFE
Aotea.	ATMF	Arkona, C.	ADBM	Atacames	AUVB
Aotea Hr.	ATGF	Arles	AHRS	Atáfu atoll	ARES
Apaiang	AQWC	Armeghon	ANGM	Atalaia Pt	AWDQ
Apalachicola	AXET	Ar-Men Rock	AGLD	Atami	APHB
Aparri	AQEZ	Armenia	AKLZ	Atchafalaya R.	AWXS
Apenrade	ADGI	Armenisti, C	AJRC	Athens.	AJOF
Apia	AHEJ	Armi, C dell'	AIKG	Athos, Mt	AJWD
Apollo H	ARXU	Arnam R	AOSJ	Atico.	AVBK
Apollonia.	AKST	Arnaut Kioi	AKBY	Atkinson Pt	ATUY
Apple Hd	BCAZ	Arnauti C	AKQW	Atlantic City	AZMC
Appledore	AEPO	Arnel Pt	ALDI	Atlantic Ocean	ALCK
Appomattox R	AYWU	Arnemuden	AEFW	Atsuta	APFU
Apruague R	AWFQ	Arnhem B	ARIP	Attu I	ATRS
Aquin	AXVC	Arnis	ADFE	Aucanda Islet	AHGT
Arabat B	AKIM	Arno R	AHXW	Auckland	ATMS
Arabia	AMSD	Arosa	AGYW	Auckland Is	ATOJ
Arablar	AJZF	Ar-Rabbali	AMOS	Aucutta I	ANDI
Aracaju, P	AVYP	Arran	AEZY	Audierne	AGLN
Aracati	AWCJ	Arran Is	AFRL	Aue	AQZL
Arafura Sea.	AQFH	Aranmore (Aran I)	AFSP	Augusta (S Australia).	ARSL
Arago, C	AUBH	Arrecife Pt	ALEM	Augusta (Me)	BALH
Arakan R.	ANLC	Arroroi	AQHK	Augusta, P (Sicily)	AILE
Arama	AQZO	Arroyo	AXWZ	Augusta, P (B C)	ATVW
Aranci B	AIDF	Arru Is	AQPI	Augustenburg	ADGF
Arang Arang Anchorage	ANRX	Arsa Canal	AIXU	Aujouan I	AMDE
Aransas.	AWTY	Arsachena G	AJGS	Auray	AOOB
Arari	APGS	Arsila	ALBN	Aurich	ADXU
Arauco	AVHI	Arsui	AKST	Auskerry	AFEX
Arbe P	AJBN	Art I	AQZK	Aussen Jade Lt V	ADWM
Arbroath	APIV	Arta G	AJIG	Austral Is	ARHS
Arcachon	AGTI	Artaki B	AJZT	Australia	ARHW
Arcadia G	AJLV	Arthur, P	AOKM	Australia, North.	ARHX
Arcadins	AXTI	Arthur, P (Tasmania)..	ASGZ	Australia, South	ARQG

AUSTRALIA—BARQUE

Australia, West . .	...ARKG	Bagnara . . .	AIJZ	Banda.. .	...AQPN
Austria... . . .	... AIVB	Bagnkop. . . .	.ADIL	Bander Abbas	. AMVP
Ava.	.. ANMO	Bagno P . .	AIHE	Bander Marsyeh.	....AMOR
Avatcha B .	.....AFNH	Bagot Pt	.BDLN	Bandol. . .	. AHSV
Aveiro .. .	. AHBK	Bagur, C. .	.AHMS	Bandon	. ANTY
Avenza .	.AHXS	Bahama Bank . . .	AEXH	Banes, P . .	. AXMD
Avernakö. .	... . ADHV	Bahama Is	AXIL	Banff.	. AFIE
Avery.. . .	. . AWXM	Bahia . . .	.AVXS	Bangaai Archipelago	. AQMI
Avery Rock..	.... BAVI	Bahia Blanca . .	.AVMZ	Bangalong	. AIJE
Avignon.. .	. . ᴀHRJ	Bahia Blanca Lt V .	. AVMY	Bangkok	. ANUJ
Avilés..	. AGWK	Bahia de Cadiz cay . .	AXNG	Bangkok River Lt V	. ANUK
Avola	. .AILO	Bahia Honda (Colombia)	AWLQ	Bangor (Wales).	. .AETQ
Avolos I (Torre Avolos)	. . AILF	Bahia Honda (Cuba) .	.AXOD	Bangor (Ireland)	. AFTV
Avon R (England). .	. AEQG	Bahia Honda (Panama). .	AUSR	Bangor (Me)	. BAQD
Avon R (N S) .	...BCBL	Bahret el bu Grara .	.AXVG	Bang pa Kong R.	. ANUL
Avonmouth. . .	. AEQI	Bahram Hr .	AMTX	Banguey I	. .AFZD
Avranches.	.AGES	Bahydarat B . .	ABCU	Banholm . .	. ADMO
Avreas B . .	... AQWX	Baia . . .	AIHF	Banjarmasin	. AQBN
Awaji Sima . .	AFDL	Bail, P .	. . AGED	Banjuwangi . .	. APWX
Awanui R. .	. ATNY	Baila... .	. .AMWI	Banka .	. AFRZ
Awarua.. . ,.	. ATGN	Barnsdale. . .	ASCB	Banka. I. .	. AQMB
Awarua Lt V .	.ATGO	Bajoli, C . .	AHFD	Bankot P	. AMZV
Awatere R. .	. ATJW	Baker I (Me) . .	BASI	Banks, C .	. ARWS
Axim	....ALMT	Baker I (Mass) .	BAIW	Banks Is .	... AQWT
Axmar	. ACLJ	Bakit B . .	.AQCH	Banks Peninsula	. ATJE
Ayamonte	AHDO	Baklar, P . .	.AJWQ	Banks Str .	... ASDG
Ayas.. . .	.AKRJ	Baku . .	AKLV	Bannow B . .	. .AFWB
Ayer Bangies . . .	APTZ	Balábac I . .	APZH	Bansering .	. APWU
Aylen B. . . .	. AOIN	Balade . .	AQZS	Bantam .	. APUX
Ayr.... .. .	. .AEYG	Balaklava . .	AKHO	Bantayan I	. AQIL
Ayre I . .	AHPG	Balambángan I	APZO	Bantry B .	. APFW
Ayre Pt.. .	. AEWX	Balang Nipa. .	.AQMT	Banyuls. .	. AHPO
Azamushi .	. AOTN	Balasor .	ANIM	Bar Hr .	. BASO
Azimur	..ALBV	Balbriggan..	. AFUS	Baracoa, P. .	.AXLM
Azizieh	. .AJPI	Balclutha.. .	.ATHW	Baracoon Pt .	.ALOT
Azores	. .ALCM	Bald Head Ch .	.AYQO	Barahona .	. AXUV
Azov.	...AKIZ	Balearic Is. .	AHNO	Baranju Hd .	. ASLU
		Balemes Pt . .	.AGQH	Baranof I .	.ATSJ
Bâ B.	AQZX	Balete, P .	AQKP	Barataria B .	AWYE
Baagö . .	.DHO	Balga . .	ACXE	Baratti P. .	.AHYJ
Baba I.... . .	. AQFE	Bal-haf .. .	AMSG	Barbacoas B .	. AWMY
Baba, C. .	... .AKNR	Bali	AQNG	Barbados .	.AYGP
Babac I	AJCX	Baleira Pt. .	AHDE	Barbary Coast .	ALFZ
Bab el Mandeb C. . .	-AMSB	Balique I . .	. AWFG	Barbas, C .	. ALFS
Bab el Mandeb Str	AMPE	Baljik . .	AKDW	Barberyn I .	.ANER
Babi I .	AFUZ	Ballang R . .	APYI	Barbuda .	. AYCH
Babushkin G.. ..	. AOSK	Ballast Pt . .	. . AUJM	Barcelona (Spain) .	. AHMC
Babuyan Is .	. AQFG	Ballina (Ireland) .	. AFSE	Barcelona (Venezuela)	AWJO
Baccalieu I .	..BRVR	Ballina (N S W)..	ASOP	Barclay Sound. .	ATXS
Baccaro Pt. .	BCHR	Ballinskellig .	APQJ	Bardsey I .	. AETD
Bachian .. .	AQRO	Ballona P . .	AUIX	Barfleur. .	.AGCH
Bacht Nmh .	ANWX	Ballycastle .	AFTN	Barfleur, C .	. AGCI
Back R.... .	.AYXG	Ballycottin I . .	AFWP	Bari .	AIQT
Backofen Lt. .	ACVB	Ballyshannon .	.AFSI	Bari B .	AQOF
Backstairs Passage .	.ARVM	Balmain . .	ASKX	Barrdi, C .	AMBF
Bacton.	AFNS	Balsö . .	.ACLZ	Barili .	.AQIN
Badagri . . .	.ALOC	Balstad . .	ABIV	Barisal .	... ANJX
Badalona . . .	...AHMP	Balta I .	AFGU	Banito R .	. . AQBM
Badia P. .	. AJES	Baltic Sea . .	ACEU	Barques, P des	. AQQU
Bado P. .	. AIXB	Baltic, P . .	.ACTJ	Barka .	. AMTD
Badong B .	.AQNK	Baltic & North Sea Canal.	.ADUF	Barletta .	. AIRB
Baffin B .	... BEZM	Baltimore (Ireland).	.AFXH	Barmouth . . .	. AESX
Baffin Land .	.. BEYP	Baltimore (Md).	... AZDF	Barnegat .	. AZMO
Baffle Creek .	.. ASTP	Baltrum I .	. ADXQ	Barnstable .	. BAGO
Bafra C .	. AAMQ	Baluchistan .	AMWC	Barnstaple. .	. AEPR
Bagacai Pt. . .	. AQIY	Bafía Pt. .	. AHLG	Baroda .	. AMYI
Bagamoyo .	AMKX	Banana Creek .	ALSE	Baron Korfa G .	AFNV
Bagasse Pt . .	. AKQO	Banana Is . .	.. ALIY	Bar Lt V (Mersey) .	AEUL
Bagdad .. .	. .AWTG	Banaré B .	AQZJ	Baros. .	. APUG
Baghdád. .	. .AMUN	Banche, La. . .	AGPB	Barque Cove .	. . AYDK

BARQUERO—BESSIN

Barquero	AGXB	Batu Is	AFLD	Belleville	BDWS
Barra	AFDK	Batúm	AKLP	Bellinger R	ASOD
Barra Str.	BCTU	Batur	ALIP	Bellingham B	ATYF
Barracouta Hr	AORB	Bauld, C.	BERD	Bello, P (Colombia)	AWNO
Barra Grande	AVZE	Bavaria	ACVS	Bello, P (Philippines)	AQJH
Barragua Lt V	ANLY	Bay City	BEGA	Belmonte	AVWY
Barram R.	APYK	Bay of Islands	BESF	Belmore Hr	AMCJ
Barrancas	AWIJ	Bay R	AYSN	Belo Pulo I	AJNF
Barrancas F	AXEB	Bay Rock	ASWV	Belvedere Tower	AILK
Barranquilla	AWMD	Bayas	AKRM	Bembridge	AEMX
Barrels Rock Lt V	AFVW	Bayona	AGZN	Bénard Pt	AGEZ
Barren Is	AOEF	Bayonne	AGTQ	Bénat, C.	AHTW
Barrier I , Great	ATMF	Bazaruto C	ALZX	Bender Erekli	AKNF
Barrington B.	BCHO	Bant Pt.	AKRS	Benevente	AVUT
Barrow	AEWG	Beachport	ARWP	Bengal B	ANIO
Barrow Pt.	ATPK	Beachy Hd	AELP	Ben Ghazi	AKUT
Barry Docks	AERB	Beale, C.	ATXU	Bengo B	ALSZ
Barseback	ACEQ	Bear, C (France)	AHPQ	Benguela	ALTS
Barsö	ADGJ	Bear, C (Pr Ed I)	BDAL	Bengut, C	AKYH
Bartika	AWIH	Bear I	ABDM	Benicarló	AHLB
Barth	ADBU	Béarn	AHPQ	Bendorme	AHJU
Bartlett Reef Lt V	AZTF	Beaver, P	AMKJ	Benin R	ALOR
Barton, P	AQCE	Beaufort (N C)	AYRG	Beni Saf Hr	AKZH
Barunguba I	ASIJ	Beaufort (S C)	AYNU	Benkulen	APTH
Baruva (Barwa)	ANHO	Beaufort I	AOMT	Bénodet	AGLX
Bas, Ile de	AGIQ	Beaufort, P	ALXD	Benti Pt	ALJQ
Bascuñan, C	AVDY	Beaujeu Ch	BDRE	Benzert	AKWM
Baseleghe P	AIUO	Beaumaris	AETU	Beppu	AOZC
Bashi R.	ALYQ	Beauport	BDSX	Berau G	AQSB
Bashee Is	AQFK	Beaver Hi	BAXY	Berbera	AMOU
Bashika B.	AKNP	Beaver I	DCMZ	Berbice	AWHK
Básdu	AMVJ	Bebek	AKBZ	Berck	AFYM
Basilan I	AQLJ	Bec de l'Aigle	AHSR	Berdiansk	AKIR
Basin of Mines	BCDN	Bec Melen Pt	AGMV	Berdugi I.	AJPB
Basque, P	BESU	Bedeque B	BCZI	Berehaven	AFFZ
Basques Roads	AGQU	Bedford Basin	BCLS	Berenice (Tripoli)	AKUT
Basra	AMUI	Bedloe I	AZOH	Berenice P	AMFZ
Bass Hr	ANQH	Bedout I.	ARLO	Beresow	ABCF
Bass R	BAED	Bega	ASIH	Berezan I.	AKFZ
Bass Str.	ASCN	Bégleguer Pt.	AGHY	Berg	ABRI
Bassas da India	AMHB	Beglitzkaia Lt V.	AKIV	Bergen (Germany)	ADBI
Bassein R (Burma)	ANLU	Beg Meil Pt	AGLZ	Bergen (N J).	AZOE
Bassein (India)	AMZB	Beg Morg Pt	AGML	Bergen (Norway)	ABRL
Basseterre (Guadaloupe)	AYDM	Behkeimisaki	APMX	Bergen-op-Zoom	AEFK
Basse Terre (St Kitts)	AYBZ	Beian	ABLV	Bergudden Pt	ACMX
Bastia	AHZU	Beikos	AKCR	Bergues	AFXN
Bastion, C.	ANXT	Beipur	ANCO	Berguglie B	AJCK
Basto	ABXV	Beira	AMBI	Berhala Str	AFRV
Bata	ALQT	Beirut	AKSP	Berhampur	ANHS
Batabanó	AXOZ	Beit Hr	AMXI	Bering I.	AFNQ
Batan Is	AQFI	Bekkervig	ABRT	Bering Sea	AFNW
Batan P (Panay)	AQRX	Bel Sound	ABDC	Bering Str	AFOM
Batanga	ALQM	Bel-Air Pt	AMBI	Beruich Spit	AKIP
Batangas	AQOY	Belem	AHCB	Berlevaag	ABGC
Batavia	AFVE	Belfast (Ireland)	AFTR	Berlin	ACVX
Batavia R	ARIB	Belfast (Me)	BAPO	Bermeo	AGUT
Bateman B	ASIT	Belfast (Victoria)	ARXA	Bermuda	ATHQ
Bath (Netherland)	AEGR	Belgium	AEGO	Bernam R	ANRK
Bath (Me)	BAMR	Belgrano, P	AVMZ	Berne	AHVR
Bath (N C)	AYST	Beliling Roadstead	AQNH	Bernieres	AGBS
Bathurst (Canada)	BDIZ	Belillo	AWMJ	Berry Hd	BCOE
Bathurst (Cape Colony)	ALYF	Belimbing	APTB	Bersimis R	BDNY
Bathurst (Gambia)	ALHU	Belitung I	AFSI	Bertheaume	AGJZ
Bathurst, C.	ATPC	Belize	AWRC	Berville	APZN
Batonga Roadstead	ALQP	Bell Rock	AFIW	Berwick	AFKR
Battery Pt.	BCSZ	Bellavista, C	AIDS	Besós R.	AHMD
Batticaloa Roads	ANFD	Bellechasse I	BDSF	Bessaker	ABLN
Battle I	BEMF	Belle-Ile (France)	AGNK	Bessan	AHQL
Batu Bara	AFRB	Belle Isle	BEQS	Bessarabia	AKFQ
		Belle Isle Str	BEPO	Bessin, P en	AOBT

BESUKI—BOTAFOCH

Besuki	.APWO	Blaavand Pt .	ADRP	Bois Rouge Pt . .	AMHK
Betanzos . .	. AGXV	Black Deep Lt V .	.AEIR	Bojador, C. .	ALFQ
Bethune Pt . .	.AOGH	Black Head... . .	.AFTS	Bojana R	AJGP
Betsiboka R ...	. AMFX	Blackpool. . .	AEVU	Bojeador, C..	AQEX
Betts Cove. .	BEXH	Black R (Jamaica)	AXSB	Bokel cay	AWQV
Betty I .	BCLD	Black River B... .	.AMIQ	Bolinao. .	AQEL
Beundo R. .	.ALQO	Black Rock Hr (Conn) .	.AZSX	Bolivar Pt (Tex) .	AWVI
Buzéval Pt	AGBH	Black Rock (N S)	BCDH	Bolivar (Venezuela)	AWIN
Bezerta Grande.	.ALDY	Blackrock Pt (C Br I)	BCTR	Bologna	.AISR
Béziers.. .	AHQC	Black Sea . .	AKDB	Bol Shuzhmui I .	. ABET
Bhagwadandi . .	AMYP	Blacksod B .	.AFRX	Bolt Hd ..	ABNR
Bhamo. . .	ANMR	Blackstrap B .	AHGB	Boma, P	AXLK
Bhatkul Cove . .	ANBY	Blackwall .	AEIX	Bom Abrigo I	AVSR
Bhaunagar Creek .	AMYE	Blackwater Bank Lt V	APVM	Bomarsund	ACQO
Biafra, Bight of. . .	.ALPW	Blair Hr .	ANTH	Bombah G	AI UJ
Bianca I .	AIDK	Blair, P (Andaman) .	ANMW	Bombay . .	.AIIZD
Blanche Pt . ..	AJCP	Blakeney, P. .	ATUM	Bombay Hook . .	AZIT
Biarritz. . .	AGTS	Blakistone I .	AYZW	Bombetoke B . .	AMFU
Bias B . .	ANZI	Blanc, P.. .	.AGHQ	Bombori F . .	AKLH
Bic I . .	BDOT	Blanc Nez, C . . .	AFXZ	Bomma.. . .	AL8I
Bicquette I .	BDQQ	Blanche B (New Pom)	AQUH	Bommelö .	ABSR
Bidassoa. . . .	AGUF	Blanche P (S Australia) .	ARQS	Bon, C .	AKWD
Bideford	AEPN	Blanco, C (Cent Am).	AURQ	Bona. . .	AKXD
Bidston . . .	AEUN	Blanco, C (Or). .	AUEL	Bonacca I .	.-AWPI
Bielosarai Spit . .	AKIT	Blanco, C (Palma).	AHOF	Bonaire I .	AWKF
Bienhoa .	ANVO	Blanco, C (W Africa)	ALFU	Bonaventura Cove .	AWOB
Big Arrow I. . .	BCQO	Blanes .	-AHMQ	Bonavista (C Verde Is) .	ALHF
Bihit Pt .	AGIC	Blankenberghe .	-AEHK	Bonavista, C .	BEWC
Bijonga Is .	ALIU	Blankenese .	.ADUR	Bondé R .	.AQZP
Bilbao. . .	.AGUZ	Blavet. .	AGNF	Bondulan Pt. .	AQIG
Billhook I	BDEJ	Blaye .	AGSX	Bône. . . .	AKXD
Billingsgate I . .	BAGL	Blegen Lille	ABSV	Bo'ness .	ATKC
Billingsport.	AZLB	Blenheim .	ATKB	Bongao I. . .	AQLV
Billiton I .	APSI	Blève Reef .	AMEL	Bonham I .	AOEY
Biloxi. . ..	AXBZ	Blexen .	.ADVS	Boni .	AQMR
Bima. .	AQNV	Blight Boat Entrance .	ASZG	Bonifacio .	.AIBM
Bimlipatam. . .	ANBI	Blixvær. . . .	.ABJO	Bonifati, C .	AIJQ
Bingo nada .	APDY	Block I .	.AZUT	Bonin Is .	AQVP
Binue R . .	.ALPC	Blonde Is .	AOKU	Bonita Pt .	AUGD
Bio Bio R . .	AVGT	Bloody Foreland .	AFSN	Bonne B. .	.-BERZ
Björkö. .	ACRP	Bloscon Pt . .	AGIN	Bonny R .	ALPJ
Björn .	ACKZ	Bluefield .	.AWOV	Bon Portage I .	BCHI
Björnabben Rock . .	ACGX	Bluefields B .	AXRY	Booby I .	ASZU
Björneborg . .	ACPO	Blue Hill B. .	BARQ	Boom .	AEGV
Björnö.	ABJH	Blue Mountain .	AXSV	Boompjes I .	APVJ
Birch Pt .	BDHY	Blueskin . .	-ATIL	Boon I. .	BAKS
Bird Island Lt (C Br I)	BCUA	Bluff Hr .	AFGN	Borcea. . .	-AKEU
Bird I (Cape Colony).	ALYC	Bluff, The . .	ALYZ	Borda, C..	ARTW
Bird I (C Verde Is).	ALGX	Blunt I .	ATZE	Bordeaux .	-AGTO
Bird I (Mass). .	AZYU	Blyth .	AFLE	Bordentown .	AZKP
Bird Rock (Bahamas)	AXKF	Bö .	ABQK	Borgholm .	ACGW
Bird Rock (C Br I)..	.BCTF	Boar I (Burgeo) .	-BETG	Borgå .	-ACRH
Birkenhead... .	AEVK	Boars Hd .	BCFP	Borinquen, C .	AXWD
Birmingham. .	AEVT	Boca Chica .	AWMF	Borja B .	-AVJH
Biscay, B of. ...	.AGJS	Boca del Toro .	AWOH	Borkum Flat Lt V .	ADYP
Bisceglie.. .	.AIQX	Bodalla .	ASIN	Borkum I.. .	-ADYE
Bishop and Clerks .	BADU	Bodega B .	AUFW	Bormes Road .	AHUE
Bishop and Clerks (Wales)	AESM	Bodie I .	AYTI	Borneo .	APXF
Bishop Rock (Scilly Is)	AEOW	Bodö .	ABJC	Bornholm .	ACEY
Bishop Sound . .	ARBT	Bodö .	ADHE	Böröholm .	ABMJ
Bismarck Archipelago..	AQUC	Bogense .	AKGS	Borrowstounness .	AFKC
Bissao I.	ALIQ	Boggonosky .	-AJTS	Bosa .	AIEV
Bitter Lake, Great .	AKTQ	Boghazi P .	AWMB	Bösch .	ADUG
Bitter Lake, Little .	AKTR	Bogota .	AWMB	Bosnia .	AJGE
Biurö (Bjuro) Klubb	ACNF	Bogoyavlensk .	AKGP	Bosphorus .	AJZX
Bizerta . .	.AKWM	Bogskär. .	ACQK	Bossiljina .	AJDN
Björnsund .	ABLI	Bohemia .	ACVR	Boston (England) .	AFNI
Björö . --- .	ACNF	Bohul I. .	-AQIZ	Boston (Mass) .	BAHS
Bjurö .	.ACNF	Bohus B.. .	-ABZI	Boston Lt V .	BAGY
		Boigu I .	AQSN	Botafoch I .	AHNY
		Boisdale Loch . .	-AFDJ		

BOTANY—BURNETT

Botany	ASJQ	Breton, (	AGTO	Brusa	AJYW
Bothnia, G of	ACKO	Breton, P	AGPV	Brusc	AHSZ
Bouc	AHRP	Breves	AWEH	Brussels	AEGZ
Bougainville Str	AQXF	Brevik	ABWX	Bruster Ort	ACWT
Bougainville I	AQWN	Brevilacqua	AJCH	Buac	AQGT
Bougaroni, C	AKXV	Brewei	BAQG	Buaro I	AQWF
Boughton R	BDCG	Brewer Str.	APRF	Buccaneer Archipelago	ARLD
Bougie	AKYB	Bridgeport (C Br I)	BCSK	Buccari B	AIYT
Boulogne	AFYG	Bridgeport (Conn)	AZTD	Buchan Ness	AFIN
Boundary B	ATYB	Bridgetown	AYGJ	Bucharest	AAFH
Bounty Is.	ATOP	Bridgewater (England)	AEPZ	Buchupero	AVGI
Bouquet, P	ARBI	Bridgewater (N S)	BCJQ	Buckminster I	AOGT
Bourchier Group	AOKY	Bridlington	AFMI	Bucksport	BAQJ
Boursefranc	AGEL	Bridport (England)	AENB	Buctouche R	BCYK
Bouzaréah	AKYL	Bridport (Tasmania)	ASDO	Bud	ABNG
Bovalino	AIPM	Brielle	AEDH	Budapest	AIYN
Bovbierg	ADBL	Brier I	BCFS	Budd Inlet	AUBZ
Bowen	ASWO	Brighton (England)	ABLI	Buddon Ness	AFIX
Bowen, P	ASVJ	Brighton (N Y)	AZQD	Budrum	AKPF
Bowling Green, C	ASWQ	Brigus B	BEVI	Budua	AJGC
Boyanna B	AMPR	Brill Rk	AQMY	Buenaventura	AUTQ
Boyne R (Ireland)	APUQ	Brin Factory	ALID	Buena Vista	AQHT
Boyne R (Queensland)	ASTZ	Brinatauri R.	AQSW	Bueno Rio	AXRM
Boz Burnu	AJYR	Brindisi	AIQK	Bueno R	AVHX
Brabant, C	AMIJ	Brioni Is	AIWZ	Buenos Aires	AVOD
Bradford	AFMT	Brisbane	ASQP	Buffalo	BDZS
Bradley Hd.	ASKH	Brisbane Road Pile Lt Ho	ASQL	Buffalo R	ALYM
Braganza Shoal Lt V	AWDU	Bristol (England)	AEQC	Bug R	AKGL
Brahestad	ACOM	Bristol (R I)	AZWS	Bugia	AHCL
Brahmaputra R.	ANKC	British Columbia	ATSX	Bugio (Madeira)	ALDZ
Braila	AKES	British Guiana	AWHI	Buholm	ABKJ
Brake	ADVX	British Honduras	AWQL	Buitenzorg	AFVG
Brameya R	ALJG	British New Guiana	AQRV	Buk Pt	ADCI
Brandaris	ADZB	Briton Ferry	AERN	Bukas Is	AQJZ
Brandinchi, P	AIDM	Brixham	AENK	Bulama	ALIV
Brandso	ADGQ	Broach	AMYO	Bulari B	AQYS
Brandy Pots	BDPO	Broadhaven	AFSC	Bulbineh	ALJI
Brandywine Sh	AEIE	Broad Mount Hr.	ASUY	Buleleng Roadstead	AQNH
Brantevik	ACFT	Broad Sound	ASVN	Bulgaria	AKDT
Bras d'Or	BCTL	Broadstairs	AEKI	Bulhar	AMOV
Brass R	ALPI	Brockton Pt.	ATVD	Bulk Pt	ADEW
Bratholm	ABQJ	Brockville.	BDVU	Bull B	AYPQ
Brava I	ALHS	Broken B	ASLY	Bull Pt (C Verde Is)	ALGR
Brawa	AMNY	Bronnö	ABEN	Bull Pt (England).	AEPS
Brazil	AVQT	Bronnösund	ABEL	Bull Rock	APQG
Brazos R	AWUT	Brooklyn	AZPF	Buller R	ATFI
Brazos Santiago	AWTP	Broome	ARLM	Bulsar	AMYT
Brazza I	AJDV	Brothers (China).	AOBE	Bulusan	AQGE
Brazzaville	ALSJ	Brothers Lt (N Z)	ATDU	Bulwer Pilot Station	ASPY
Breaker Pt	ANZQ	Brothers, The (Red Sea)	AMQP	Bunbury	AROP
Breaksea Pt	AERC	Broughton Hr (Korea)	AOMX	Buncrana	APSV
Breaksea I	AERI	Broughton, P	ARSI	Bundaberg	ASTL
Breaksea Lt V	AERD	Broughty Ferry	AFIZ	Bundoo	ASIV
Breakwater Hr	AZHV	Brouwershaven	AEFC	Bunga B	AFTQ
Bredasdorp	ALXC	Brow Hd.	AFPV	Bungo Ch	AOYT
Bredskar	ACMK	Brownsville	AWTM	Buol	AQLO
Breede R	ALXE	Bruceport	AUCV	Burai, P	AQYZ
Bréhat I	AGHJ	Bruges	AEHL	Burano	AITZ
Bremen	ADWG	Bruit R	APYO	Burges	ASEN
Bremer B	ARPV	Brulé, C	BDRW	Burgess Islet	ATNE
Bremerhaven	ADVR	Brulos	AKTX	Burghaz B	AKDM
Bremerton	AUBQ	Brunet I	BETM	Burght	AEGU
Bremö	ACLW	Brunci	AFYN	Burias I	AQGO
Brenton Reef Lt V	AZXK	Brunsbuttel	ADUE	Burin I	BEUH
Brescou I	AHQK	Brunsbüttelkoog	ADUC	Burketown	ARIQ
Breskens	AEPZ	Brunshausen	ADUM	Burleigh	ASPI
Bressay I	AFGN	Brunswick	AYKU	Burling I	AHBP
Brest	AGKH	Brunswick Dock	AEVH	Burlington	AZKS
Brest, Rade de	AGKM	Brunswick R	ASOY	Burma	ANKY
Bretagne	AGET	Bruny I	ASFW	Burnett Hr	ASGT

BURNETT—CARLISLE

Burnett R	ASTJ	Calafat	AKFD	Canaveiras	AVXB
Burnham	AEPX	Cala Grande I	AIFQ	Canaveral, C	AXIH
Burnt I (G of Aden)	AMOS	Calagua	AQFO	Cancale	AGEV
Burnt I (Me)	BANP	Calais	AFXY	Candas	AGWE
Burnt Is (India)	ANBH	Calamianes Group	AQCN	Candia	AJFD
Burntcoat Hd	BCEP	Calandorang B	AQBC	Candia Town	AJPM
Burntisland	AFIP	Calarasi	AKBX	Candolu Islet	AQJT
Burra	AIZV	Calaspan	AQCZ	Cani Is	AKWJ
Burrard Inlet	AFVB	Calbayoc	AQJM	Cannes	AHUW
Burriana	AHKS	Calbuco Ch	AVIJ	Canning (N S)	BCER
Burrow Hd	AEXQ	Calcasieu R	AWXD	Canning F (Singapore)	ANSU
Burrum R	ASTH	Calcutta	ANJI	Canning, P (India)	ANJT
Burry P	ABRU	Caldera, P.	AVDQ	Canonnier Pt	AMIP
Buru I	AQRE	Caldwell	ALKI	Canso	BCOJ
Burwell, P	BEAI	Caldy I	AERY	Canso, Gut of	BCOY
Bushire	AMVB	Caledonia Hr	AWNG	Cantick Hd	AFEW
Bustos C	AVUB	Caledonian Canal	AFBR	Cantin, C	AUBX
Buskar I.	ACBQ	Calella	AHMN	Canton	ANYI
Busselton	AROW	Caleta Buena	AVCN	Canton R	ANYJ
Bustard Hd	ASTR	Calf of Man	AEWY	Cantyre, Mull ot	AEZP
Buso, P	AIUY	Calicut	ANCM	Canzirri	AIKR
Busto, C.	AGWO	California	AUEX	Capalonga	AQFP
Busuanga	AQCR	California G	AULD	Cape Breton I	BCPN
Busum	ADTL	California, Lower	AUKJ	Cape Colony	ALVM
Butang Group	ANQD	Calimere, Pt	ANFP	Cape Town	ALWI
Bute Docks	AEQX	Callao	AUYT	Capel rosso Pt	AHZK
Buton	AQML	Calle, La	AKWZ	Cape Verde Is	ALGO
Butt of Lewis	AFCV	Calliope R	ASUG	Capis	AQHY
Butuan	AQKH	Calonge	ACSQ	Capocesto	AJDH
Buyuk B	AKCI	Calonna R	AXOV	Capones Pt	AQEB
Buyuk Chekmejeh	AJXS	Calpe	AHJY	Caporosso Pt	AHZG
Buyukdéré	AKCG	Calshot Lt V	AELY	Cappuccini, Mt	AISG
Buzzards B	AZXW	Calumbo	ALTI	Capraia	AHZL
By Lt V	AOSN	Calumet	BEJO	Caprara, C	AIFR
Bynoe Hr	ARJT	Calvi	AIBY	Caprara I	AIRP
Byron, C	ASOT	Camaguan	AWIP	Caprera I	AICW
		Camamú	AVXM	Capri	AIHX
Caballeria, C	AHPL	Camana	AVBJ	Capricorn, C	ASUK
Caballo I	AQDO	Camarat, C	AHUJ	Capucins Pt	AGKT
Caballo Pt	AGVI	Camaret	AGKU	Carabobo Province	AWKB
Caballos Pt	AWPU	Camargue	AHRC	Caracas (Ecuador)	AUVD
Cabañal	AHKR	Camariñas	AOYF	Caracas (Venezuela)	AWJY
Cabañas P	AXOB	Cambay G	AMYG	Caracol	AXTU
Cabedello	AWBC	Cambodia	ANVD	Caracoh	AWMH
Cabello, P.	AWKH	Cambridge G.	ARKI	Caraga B	AQAL
Cabeza de San Juan	AXWL	Cambrils	AHLQ	Cara-Irman	AKEI
Cabonico	AXLW	Camden (Me)	BAPF	Caramifial	AGYU
Cabot I	BEWM	Camden (N J)	AZKV	Caraquette I	BDIK
Cabot Str	BCUV	Camden Haven	ASNI	Caravellas	AVWK
Cabra I	AQDG	Camiguin	AQFH	Carbon, C	AKYD
Cabral B	AVWS	Caminha	AGZU	Carbonara, C	AIDT
Cabrera	AHOZ	Cammin	ACYR	Carbonear I	BEVO
Caccia, C	AIFC	Campana	AVOL	Carboneras	AHIQ
Cacheo R	ALIK	Campanella Pt	AIHW	Carcar	AQIV
Cacilhas Pt	AHCN	Campbell, C	ATIU	Cárdenas	AXNQ
Cadaqués	AHNG	Campbell I	AFOL	Cardiff	AEQW
Cades Pt	AVCF	Campbell, P (Andaman)	ANMX	Cardigan (Wales)	AESQ
Cadiz	AHEP	Campbell, P (Victoria)	ARXQ	Cardigan B Lt V	AESR
Caen	AGBN	Campbelltown (Canada)	BDJR	Cardigan B (Pr Ed I)	BDAR
Caermarthen	AERX	Campbelltown (N Z)	AFOL	Cardwell	ASXI
Cæsarea	AKSR	Campbelton	AEZN	Carena Pt	AIHY
Cagliari	AIDY	Campeche	AWRY	Carentan	AGBX
Caicos Is	AXKN	Campen	ADYN	Cargados Carajos	AMIV
Cairns	ASXY	Campo alle Serre	AHYQ	Cargodo I	AOMW
Cairo	AKFD	Campo R	ALQR	Cariaco	AWJG
Cala, C.	AHOJ	Campobello I	BAXJ	Caribou	BCWU
Calabar	ALPN	Campos	AVUJ	Carimata Str	APXJ
Calabazas Isles	AQHZ	Canada	BAWL	Carleton	BDKA
Calabria	AIPK	Cananea	AVST	Carlingford	AFUJ
Calaburras Pt	AHGL	Canary Is	ALEB	Carlisle	AEWT

CARLISLE—CHEMULPO

Carlisle B (Barbados)	AYGH	Catalina Hi.	BEVX	Chamatla R	AUMT
Carlisle B (Jamaica).	AXSH	Catanduanes	AQFX	Chamé B	AUSV
Carljohansværn	ABXY	Catania.	AILB	Chamela.	AUNE
Carlo Forte	AIEQ	Catanzaro.	AIPR	Champborne Pt	AMHN
Carloway.	AFCU	Catastrophe, C	ARSB	Champerico	AUFM
Carlsö	ACHG	Catbalogan	AQJN	Champion B	ARMW
Carmanah	ATXQ	Catherine Hill B	ASMD	Champlain	BDTS
Carmel	AUHW	Cathlamet	AUDI	Champoton	AWSD
Carmel, C	AKSP	Catlin R	ATHO	Chanak	AJXB
Carmen I	AWSF	Catoche, C	AWRM	Chafiaral	AVDM
Carnarvon (Wales)	AETF	Cattaro	AJFZ	Chancay.	AUYP
Carnarvon B Lt V	AETG	Cattewater	AENX	Chandbali.	ANIK
Carnarvon (W Australia)	ARMK	Cattolica P	AISM	Chandeleur Is.	AWZC
Carnero	AHFO	Catuáma	AVZW	Chandernagore.	ANJP
Carnsore Pt	AFVU	Caucasus	AKJN	Chandler B	BAVC
Carolina Is.	AQUX	Caudebec	AFZR	Chang I	ANUQ
Carolinensiel	ADXM	Cauto R	AXQF	Channel Is	AGCU
Carouge R	BDTM	Cavalaire	AHUI	Channel Rk Lt V	ASYQ
Carpentaria G	ARHJ	Cavalière, C	AKQL	Chanonry Pt	AFHY
Carraca	AHEV	Cavalleria, C	AHPL	Chapel I	AOBF
Carranza, C	AVGF	Cavallo I	AIBH	Chapeo Virado Lt	AWDX
Carriacou	AYFW	Cavallo Bianco Pt .	AIPD	Chaptico	AZBD
Carrickfergus	AFTU	Cavallo Pass	AWUE	Chapu.	AOEN
Carril	AGYX	Cavalo, C	AIBV	Chaquaınegon Pt	BENE
Carrizal Bajo B	AVDU	Cavetta Canal	AIUK	Charak	AMVF
Cartagena (Colombia)	AWMR	Cavite.	AQDS	Chardak Pt	AKJE
Cartagena (Spain)	AHIZ	Cavoli I.	AIDU	Charente, La, R	AGQZ
Cartaja	AHDS	Cavvanba	ABOX	Charles, C.	AZGO
Carteret, C de	AGEB	Caxias (Observatory)	AHCJ	Charles C Lt V	AZGF
Carthage, C	AKWG	Caxine, C	AKYO	Charleston	AYOS
Cartwright Hr	BEXV	Cayharien, P.	AXNC	Charleston Lt V	AYOQ
Carupano	AWJC	Cayemites	AXVO	Charlestown (Mass)	BAHV
Carvoeiro, C (Portugal, S)	AHDC	Cayenne	AWPU	Charlestown (W I)	AYCD
Carvoeiro, C (Portugal, W)	AHBR	Cayes	AXYQ	Charlotte, P	AEZT
Carysfort Reef	AXHP	Cayeux	AFYR	Charlotte Amalia	AXZD
Casablanca	AVFI	Caymanera.	AXQT	Charlotte Hr	AXPW
Cascaes B	AHBX	Cay Sal Bank	AXJG	Charlotte Town	BDAC
Casco B	BALZ	Cayutlan Lagoon.	AUNM	Charters Corner.	BCAE
Cascumpeque Hr	BDES	Ceará.	AWCN	Chasseloup B.	AQZD
Casilda, P	AXPO	Cedar Keys	AXFB	Chassiron	AGRJ
Casma B	AIYF	Cefalù	AINT	Château, Le	AGRI
Caspian Sea	AKLS	Celebes	AQLY	Château B	BEQF
Casquets	CCX	Celestun.	AWRU	Château d'If I	AHSG
Cassanha	ALTN	Central America.	ALQF	Châteaudin Road.	AXVI
Cassie Pt	BCYH	Centre I	ATGD	Chateaulin	AGXQ
Cassis	AHSQ	Cépéron F	AWGB	Chatham (Canada)	BDEG
Castellamare (Italy)	AIHS	Cephalonia	AIJX	Chatham (England)	AEJY
Castellamare (Sicily)	AINH	Ceram I	AQPS	Chatham (Mass)	BABG
Castelli, Pt du	AGOX	Cernavoda	AKEW	Chatham I (Galapagos, Is)	AIVL
Castello Mt	AHZM	Cerigo I	AJMX	Chatham Is (New Zealand)	ATOR
Castellon de la Plana	AHKT	Cerigotto	AJNC	Chatte, C	BDNJ
Castel nuovo (Spalato)	AJDQ	Cerro Azul	AIZE	Chauchu fu.	ANZV
Castelnuovo (Cattaro, G)	AJFV	Cerros I	AIKV	Chauda, C	AKHX
Castel Sardo	AIFO	Cervera P	AIWN	Chaul, P	AMZR
Castiglione	AHYO	Cervia P	AISQ	Chausey Is	AGEN
Castillo B	AVQR	Cesenatico P.	AISF	Chauveau Pt.	AGQJ
Castine	BAQS	Cette.	AHQP	Chay	AGRZ
Castle B.	AFDL	Ceuta	ALBE	Chebek.	AKUW
Castle Haven	AZFH	Ceylon.	ANEG	Chabucto Hd	BCLM
Castle I	AXKJ	Chabarova	ABDY	Chedabucto B	BCOS
Castlemaine	AFQO	Chabrol Hr	AQVB	Chedúha Str	ANLM
Castletown	AEXC	Chacahua B	AIOH	Chefoo	AOIW
Castries B.	AORL	Chagos Archipelago	AMJL	Chefuncte	AXBD
Castries, P	AIII	Chagres	AWNY	Chejin	AOMP
Castropol	AGWU	Chagu Chien Dogu	AONK	Chelindreh P	AKQJ
Castro Urdiales	AGVC	Chahbár B	AMVX	Cheltau	AOLG
Castua.	AIYK	Chaki Chaki.	AMLK	Chelyuskin, C	ABCE
Cat I (Bahamas)	AXJW	Chala	ABVF	Chemoulin Pt	AGPI
Cat I (Miss)	AXBT	Chaleur B.	BDIQ	Chemulpo	AOLR
Cataingan	AQHF	Chalmers, P	ATIK		

CHENAMPO—COMOX

Chenampo	AOLG	
Chentabun R . ..	ANUP	
Chepo R . ..	.AUTE	
Chepstow	.AEQR	
Cherbourg ..	.AGCM	
Cherchell	.AKYR	
Cheribon. .	APVM	
Cherryfield . .	.BAUE	
Cherrystone Inlet .	AZGL	
Cherso..	AIYV	
Chesapeake and Albemarle Canal ...	AYWP	
Chesapeake B . ..	AYUV	
Chesme	AKOT	
Chestakof, P .	.AONR	
Chester (England) .	AEUI	
Chester (N S) .	..BCKL	
Chester (Pa) . ..	AZJU	
Chester R .	AZES	
Chesterfield Inlet	BEZI	
Chetican Pt .	BCUY	
Chetko R .	AUEV	
Chévre C ..	AGKY	
Chiappa Pt	.AIBE	
Chiaruccia Tr	AIFY	
Chiavari . ..	.AHXF	
Chiba... .	APIH	
Chicago	BEJI	
Chichagoff I .	ATSJ	
Chichakoff C .	AOYE	
Chichester. .	AELO	
Chicken Rock. .	AEWZ	
Chierjima (Beechy Group)	AQVT	
Chico Bank Lt V .	AVNQ	
Chico P.. .	AUVP	
Chico R .	AVLI	
Chicoutimi. ...	BDFF	
Chidley C.... .	BEYK	
Chignecto B .	BAZC	
Chilachap Inlet	APWZ	
Chilaw .	ANRJ	
Chilca .	AUZC	
Chile .	AVBZ	
Chiletu B . .	APXD	
Chilka Lake .	.ANHW	
Chiloango R .	ALRU	
Chiloe I . ..	.AVIB	
Chiluán I .	AMBE	
Chimbote, P .	AUXY	
Chumo F ..	REYN	
China .. .	ANXJ	
China Bakir	ANMB	
Chincha Is .	.AUZI	
Chinchu .	AOBJ	
Chincoteague Inlet .	AZGU	
Chinde R .	AMBR	
Chin Do .	AOMN	
Chinhai .	AOEL	
Chinkiang fu .	AOFZ	
Chioggia P . .	AITN	
Chipiona . .	AHEL	
Chiriqui Lagoon	.AWOD	
Chita .	APFV	
Chitando Inlet . ..	APXC	
Chittagong .	ANKH	
Choiseul I .	.AQWM	
Choiseul P .	AMEG	
Choptank R .	AZFE	
Chorillos B .	.AUYZ	
Chorrera. ..	AUSX	
Chowan R .	- AYTX	
Christchurch (England)	.AEMP	
Christchurch (N Z) .	ATJI	
Christiana Creek .	.AZJI	
Christianburg	.AWHS	
Christiania. .	ABYI	
Christiansand .	.ABVT	
Christianeo .... f	.ACFI	
Christianstad .	ACFS	
Christiansted ...	AXZW	
Christmas I (China)	AOHD	
Christmas I (Indian Ocean)	APUN	
Christmas I (Pacific) ...	AREW	
Chuen I ..	ANUO	
Chungking .	.AOHY	
Chukiang . .	ANYJ	
Chupat R .	AVMF	
Church Pt .	.BCGH	
Churchill F .	BEZF	
Churubash. .	AKIG	
Chusan Archipelago. .	AOED	
Cibil Dock .	AVQB	
Ciboux I. .	.BCUA	
Cienfuegos .	.AXPM	
Cies I	AGZM	
Ciotat. .	AHSU	
Circassia. .	AKJO	
Circeo (Circello) Mt .	AIGK	
Circle City .	ATQU	
Cispata, P .	AWMZ	
Cittanuova .	AIWL	
Citta Vecchia ...	AJEF	
City Pt .	AYWX	
City, The . .	.AEHX	
Ciudad de David .	AUSF	
Ciudadela .	.AHPK	
Civita Vecchia .. .	AIFV	
Clare .. .	.AFRC	
Clare I .	AFRP	
Claremont I Lt V ..	ASYT	
Clarence B .	ALUZ	
Clarence, P .	.ATPW	
Clarence R .	.ASOH	
Clarence Str. ...	.AMVN	
Clark Pt. .	.AZYF	
Clayton . .	BDWA	
Clear, C .	.AFFR	
Cleveland, C .	ASWR	
Cleveland (Erie) .	BEDS	
Cleveland (Queensland)	ASQT	
Clew B .	AFET	
Cliffy I .	ASBE	
Clifton (Canada) .. .	BDIT	
Clifton (N Z) .	ATER	
Clinton .	ARUP	
Clinton, P .	BECZ	
Cloch Pt .	.AEYM	
Clonakilty .	AFXC	
Clonmel I. .	ASRK	
Cloudy B .	ATKF	
Clyde Canal .	AFJY	
Clyde R (N S W) .	ASIV	
Clyde (Scotland) .	AEYF	
Coacoacho B .	.BEFU	
Coal I .	AVTX	
Coanza R .	.ALTG	
Coast Castle, C. .	.ALNE	
Coatzacoalcos R .	AWSP	
Cobden . .	ATFQ	
Cobequid B....	.BCDT	
Cobija B	.AVCU	
Coburg . .	BDXN	
Cocague .	BCYJ	
Cocanada... .	.ANIID	
Cochin . .	ANCR	
Cochin China .	ANVJ	
Cochmos B .	. AXPI	
Cockatoo I .	. ASKZ	
Cockburn Hr .	. AXKF	
Cockburn Sound .	AROE	
Cockle Lt V .	. AFNZ	
Cockspur I	.AYMW	
Cocos Is .	APTX	
Cod, C .	. BAPW	
Codera, C .	AWJS	
Codling Bank Lt V .	.AFVJ	
Codolar Pt .	AHNR	
Codrington. .	. .AYCJ	
Coffins I	.BCJI	
Cognena G .	.AIDB	
Cohansey. .	AZLK	
Coilnapatam .	ANDX	
Coilsnipatam	ANDW	
Colastine .	AVOP	
Colberg. . .	ACYN	
Colborne, P .	BDZV	
Colchester .	AFPI	
Cold Spring Hd . .	BCXM	
Coleraine .	AFTI	
Colerun .	ANGB	
Colima .	AUNO	
Coll .	AFBX	
Collingwood (Huron).	BEHP	
Collingwood (N Z)	ATEV	
Collinson, C .	ANYZ	
Collioure .	.AHPT	
Collo .	AKXS	
Colombia .	AWLO	
Colombo. . .	ANEM	
Colon .	AWNQ	
Colon, P .	AHOW	
Colonia. .	AVPG	
Colonna, C. . .	AJOS	
Colonne, C. . .	AIPT	
Colonsay .	AFBL	
Colorado G. .	AURW	
Colorado R (Argentina)	AVMI	
Colorado R (Mexico)	AULO	
Colorado R (Tex)	AWUK	
Columbaia I .	.AIND	
Columbia . .	AZCK	
Columbia, R (Or)	.AUCZ	
Columbia (S C)	.AYQF	
Columbia Falls .	BAUQ	
Columbretes Is	AHKW	
Colville R .	BDCJ	
Comacchio. .	AISY	
Combahee R. .	AYOD	
Combermere B .	ANLE	
Comboyuro Pt .	ASPW	
Combrit .	.AGLV	
Comino I .	AIOY	
Comino, C .	.AIDP	
Comisa. .	AJEM	
Commandatubu .	AVXC	
Commercial Docks .	.AEJM	
Comoro I (Comores)	AMCY	
Comorin, C	ANDB	
Comox	ATVY	

COMPARE—DABHOL

Compare, C	.AIXR	Corny Pt	ARTI	Creux Hr	AGDM
Comptroller B	ARGK	Coro Province	AWKL	Crib Pt	ARZW
Conanicut I	AZVL	Coromandel Hr	.ATMH	Crimea	AKHE
Concarneau	AGMD	Coromandel Coast	ANFR	Crinan Canal	.AFBI
Conceicao R	AMBQ	Corona Hd	AVIF	Croatia	AIYP
Concepcion (Chile)	AVGP	Coronados, Los	AUKL	Croc Hr	BEXD
Concepcion B (Mexico)	AULI	Coronel	AVHC	Crocus B	AYBD
Concepcion (Uruguay)	AVPQ	Coronie	AWGZ	Croisette, C	AHSN
Conception Pt	AUIN	Corosal	AWRE	Croisic, Le	AGOY
Conchali B	AVEN	Corpus Christi	AWTS	Croix	.AGMU
Concord	BAKJ	Corral	AVHS	Croix de Vie	AGPY
Concordia, La (Cent Am)	AUQN	Corregidor I	AQDN	Croix Rocks	AGHI
Concordia (Uruguay)	.AVPU	Correnti I	AILT	Cromarty	AFHU
Conducia B	AMCH	Corrientes	AVOX	Cromer	AFNQ
Cone, The	ANOV	Corrientes, C (Argentine)	AVNE	Crooked Island Passage	AXKD
Conejera I	AHOB	Corrientes, C (Cuba)	AXON	Crookhaven	AFPT
Conero, Mt	AISD	Corrientes, C (Mexico)	AUMZ	Cros P	ARUB
Coney I	AZPX	Corrientes, C (Africa)	ALZT	Crosby Lt V	AEUO
Conference Group	AOLZ	Corrobedo, C	AGYT	Cross I	BCJZ
Congaree R	.AYQF	Corrotoman R	AYZK	Cross Ledge Sh	AZIN
Conil	AHFG	Corse	AHZP	Cross R	ALPT
Conimicut Pt	AZVR	Corse, C	AHZQ	Cross Rip Lt V	BAES
Coningbeg Lt V	.AFVX	Corsen Pt	AGJU	Cross Sand Lt V	AFOB
Connahs Quay	AEUF	Corsewall Pt	AEXU	Crouchers I	BCKX
Connecticut	AZSO	Corsica	AHZP	Crow Hr	BCOP
Conquet P	AGJV	Corsini, P	.AI8V	Crowdy Hd	AENG
Constantine	AKXP	Corso, C	AHZQ	Crozet I	AMJQ
Constantinople	AKBJ	Cortellazzo P	AIUJ	Cruz, C	AXQL
Constantza	AKEH	Cortez, P	AWPY	Cruz del Padre Cay	AXNK
Constitucion	AVGB	Corton Lt V	AFOD	Cruz Pt	ALDV
Constitucion Hr	.AVCZ	Coruña	AGXY	Cuarenta Dias P	AVBF
Contarina	AITJ	Corvo	ALCN	Cuba	AXLC
Contas	AVXK	Cosmoledo Is	AMJG	Cuddalore	ANGD
Conte, P	AIEZ	Cossack	ARLX	Cudillero	AGWN
Contis	AGTM	Costa Rica	AWOL	Curassier Bank Lt V	.AVNO
Conway	.AETR	Cotinguiba R	AVYN	Culebra, P (Costa Rica)	AURO
Cook, B	AQXY	Cotrone	.AIFV	Culebra (W I)	.AXYQ
Cook Is	ARBO	Cottage City	BADO	Culebrita	AXYS
Cook Str	.ATDQ	Coubre Pt	AGRQ	Culiacan R	AUML
Cooktown	ASYN	Coudres I	BDQJ	Culion I	AQCP
Cooper R	.AYPH	Country Hr	BCNU	Cullera	ATRL
Cooea R	AXDJ	Couronne, C	AHRS	Cumaná	AWJI
Cope	AHIV	Courseulles	AGBP	Cumarebo	AWKJ
Copenhagen	ADOM	Covelong	ANGI	Cumberland	AEWH
Copiapo	.AVDS	Covesea	AFID	Cumberland B	AVPO
Coppename R	AWGV	Cowan Cowan Pt	ASQB	Cumberland Basin	BCAN
Copper Hr	BEOK	Cowcolli	ANIZ	Cumberland Sound	AYIW
Copper I	APNR	Cowes	AEMC	Cumbraes	AEYR
Coquet P	AFEZ	Cowitchin	ATWV	Cumplida Pt	ALFN
Coquille R	AUEJ	Cox Bazár	ANKR	Cuneno R	ALUG
Coquimbo	AVEF	Coxen Hole	AWPO	Cunhahu R	AWBK
Corabia	.AKFC	Coy Inlet	AVLE	Curaçao I	.AWKT
Corbay Pt	BELZ	Cozumel I	AWRI	Curanipe	AVGK
Corbière, La	AGDR	Cozzo Spadaro	AILR	Curaumilla Pt	AVFG
Corcubion	AGYL	Craiova	AKFJ	Curritia Pt	.AIPJ
Cordouan	AGRU	Cranberry I	.BCOM	Curtis I	ASCU
Core Sound	AYRP	Crane I	BDRN	Curtis, P	ASTX
Corentyn R	AWHC	Craney I	AYYQ	Curzola	AJEO
Corfu	.AJBO	Cranmar	AVKR	Cuttack	ANIC
Coringa B	ANHB	Crance I	AJMT	Cuttyhunk	BACF
Corinth	AJLD	Cranz	ACWS	Cuvier I	ATMD
Corinto, P	AGRC	Crapaud	.BCZU	Cuxhaven	ADTY
Corisco B	ALQW	Créach-ar-Maout	AGHN	Cnyuni R	.AWBY
Cork	AFWR	Créac'hmeur Pt	AGKB	Cygnet, P	ASGH
Cork Lt V	.AFPD	Creac'h Pt	AGJR	Cypres Mort	AWXP
Corme	.AGYC	Creighton Hd	BCQF	Cyprus	.AKQT
Corner Inlet	.ASBG	Crescent City	AUFC	Cyrene	AKUR
Cornfield Pt Lt V	.AZTO	Creston I	AUME		
Cornwall (Canada)	BDUW	Crete	AJFD	Dabai	AMTQ
Cornwall (England)	AEOD	Creus C	AHNI	Dabhol	.AMZX

DACCA—DORNOCH

Dacca . .	. ANKG	Dead Sea .	. AKSZ	Difnein I	. AMPR
Dædalus Sh	. AMQC	Deal . .	. . AEKN	Digby	BCFG
Dagerort C	. ACTW	Deal I .	. ASCZ	Digdequash R	. BAWX
Dagö . . .	ACTS	Décollé Pt	AGFO	Dikili . . .	AKNV
Dáhanu .	AMYW	Dédéagatch	. AJWM	Dilam	AMUX
Dahomey	. ALNZ	Dee Lt V	AEUC	Dil Burnu	AJYO
Daila P .	A1WH	Dee R	AEUB	Dilhi	. AQOS
Diantree R .	ASYF	Deer I (La) .	AWYN	Dillon B	AQXU
Dairi .	AOVY	Deer I (Me) .	BARH	Dinagat I	AQJV
Dakar .	ALGK	Deer I (Mass)	BAHD	Dinan	. AGFK
Dalarö ..	ACJF	Delagoa B	. ALZK	Dinard	AGFL
Dale . .	. ABQC	Delaware Breakwater	AZBY	Dinding R	ANRC
Dalhousie C .	. ATPE	Delaware City	AZJC	Dingle B	AFQN
Dalhousie, P (Ontario)	BDXZ	Delaware (State)	. AZHM	Dingwall	AFHW
Dalhousie (St Lawrence)	BDJL	Delaware (Va)	. AYXP	Dinlleyn Porth	ABTH
Dalmatia .	AJCL	De Lemmer	AEBU	Diomede Is	ATPU
Dalni Pt	APNI	Delfshaven	AECZ	Dirk Hartog I	ARMN
Dalrymple Hr (Jolo) .	AQLN	Delfzyl	ADYK	Dirkoomsduin	. AEBY
Dalrymple Pt	ASWM	Delgada Pt (Azores) .	. ALDG	Dirschau	. ACXO
Dalrymple P (Tasmania)	ASDQ	Delgada Pt (Argentina)	. AVMJ	Discovery I	ATXC
Daly, P	ARJH	Delgada Pt (Canary Is)	. ALED	Discovery, P	ATZK
Daman, P	AMYV	Delgada Pt (Magellan)	AVJP	Disko B	BEZS
Damaraland .	ALUH	Delgado, C	AMJW	Dismal Swamp Canal	AYUG
Damariscotta .	. BAND	Deli River Lt V	. AFQV	Diu .	AMXS
Damascus . .	AKSG	Della Bocca I	AIDH	Dives . .	AGBJ
Dame Hd .	ADCV	Dellys ,	AKYE	Divi Pt .	. ANOT
Dame Marie, C	AXVM	Delos	AJRF	Dix Cove . .	. ALMY
Dame Pt	AYJN	Demata B	AJIL	Dixon Entrance	ATUD
Dames, Pt des	AGFU	Demarara .	AWHN	Djerba	AKVF
Damietta	AKTU	Demerara Lt V	AWHO	Djibouti	AMOZ
Damkut	AMSF	Denia	AHKE	Djidjelli	AKXW
Damma I .	AQFB	Denison, P	ASWK	Djursten	ACKU
Dammám	. AMUB	Denmark	ADGS	Dnieper R	AKFY
Damman	ACGL	Dennis I . . .	AMJZ	Dniester B	AKFR
Dampier Archipelago .	ARMC	Dent Haven	APZT	Dobbo	. AQPL
Dampier Str	AQRU	Dent I . .	ASWI	Doboy Sound	AYLD
Dana Pt	AJNZ	D'Entrecasteaux (h	ASFQ	Dobrogea .	AKPO
Dandé Pt	. ALSW	Departure B	ATWF	Doce R	AVWG
Danger Pt (Africa)	ALWX	Deptford	AEJC	Dodd I	AOBK
Danger Pt (N S W)	. ASPE	Derby (B C) . .	ATVR	Dodman Pt .	. AEOI
Danube . .	AKFG	Derby (W Australia)	. ARLJ	Doel	. AEGJ
Danzig	ACXP	Dernah .	AKUO	Dog I	ATGR
Dapitan B	AQKD	Derwent R	ASGN	Doha . .	AMTV
Dardanelles .	AJWH	Desecheo I	AXVS	Dohan Aslan Bank Lt V	AJXI
Dar el Beida	ALBT	Désirade .	AYDU	Dolga Spit	. AKJG
Dar es Salaam .	. AMKW	Desire, P . .	AVLM	Dolin I	AJBO
Dargaville . '	ATBJ	Destruction I	AUCN	Dolma	ABKS
Darien, G of	AWNC	Détour Passage	BECB	Domburg	AEFN
Darien Hr .	AUTI	Detroit	BECH	Domesness . .	. ACUS
Darling Hr	ASKT	Deuthero Cove	AJWG	Dominica	AYBF
Dare Pt	ADBX	Devi R . .	ANHZ	Don R .	. AKJC
Dartmouth (England)	AENL	Devil I	BCMH	Donaghadee	AFTY
Dartmouth (N S)	BCLV	Devonport	. AENY	Doncella Pt	AHGF
Dartuch, C .	AHFJ	Dhamra R	ANIH	Dondo (Africa)	ALTK
Darvel B	AFZU	Diamant, Le .	. AYFC	Dondo B (Celebes)	AQNF
Darwin, P .	ARJK	Diamond Hr .	ANJE	Dondo B (Flores)	AQOJ
Dassen I	ALWC	Diamond I	ANLV	Dondra Hd	ANEW
Dassow	ADCP	Diamond Pt	APQS	Donegal	AFSJ
Daudai	AQSF	Diamond Rock	. AYFC	Dongara	ARNJ
Daunt Rock Lt V	AFWX	Diamond Shoal Lt V	. AYTB	Donnæso	. ABKG
Dauphin, F (Haiti) .	AXTW	Diana cay	AXNM	Donsol	AQGL
Dauphin, F (Madagascar)	AMFB	Dianna	ALIG	Doob Pt	. AKJS
Davao . .	AQKS	Dibba	. AMTI	Dorchester (England)	ASED
Davey, P .	ASFM	Dice Hd .	B4QP	Dorchester (N S)	BCAH
Daviken . .	ABPK	Diego Garcia	. AMJN	Dordogne R	AGTB
Davis Inlet . .	BEYF	Diego Ramirez	AVKG	Dordrecht . .	AEDX
Davis Str	BEZK	Diego Suarez B .	AMDS	Dorei, P .	AQUB
Dawson . .	. ATQS	Dielette, P de	AGDY	Doris G	. AKPN
Dayyir . .	. AMVC	Dieppe	AFYU	Dornbusch Pt	ADBF
Dazol B	AQEK	Dievenow	. . ACYQ	Dornoch	AFHQ

DORO—ELLEN

Doro Ch	AJQU	Dungeness (S America)	AVJQ
Douarenez	AGLB	Dunkerque	AFXL
Double I	ANOQ	Dunkirk	BRFD
Double Island Pt	ASRK	Dunmore	AFWJ
Douélan	AGMN	Dunnet Hd	AFEB
Douglas Inlet (Korea)	AOMY	Duperré, P	ARBC
Douglas (Isle of Man)	AEXF	Durazzo	AJGT
Douglas P (Queensland)	ASYB	Durban	ALZB
Douglas, P (S Australia)	ABQY	Durian Sabatang	ANRH
Dourga Str	AQSI	Durian Str	APRH
Douro R	AHBG	Durnford B	ALZE
Dove Pt Beacon	AOHI	Durnford, P	AMNT
Dover (England)	ADKV	Duroch, C	AONJ
Dover Hr Lt V	AEKW	D'Urville I	ATEH
Dover (N S)	BCLA	Dury Voe	AFGQ
Dovercourt	AFOY	Dutch Guiana	AWGN
Downs, The	AEEO	Dutch I	AZVI
Dowsing, Inner, Lt V	AFND	Duxbury Pier	BAGU
Dowsing, Outer, Lt V	AFNC	Dvale Ground Lt V	ADQF
Dragons Mouth	AYHM	Dwarka Pt	AMXL
Dragonera I	AHOK	Dwina, N , Lt V	ABEM
Dragør, or, Drogden Lt V	ADOG	Dwina R (Riga)	ACUL
Drammen	ABYD	Dwina R (White Sea)	ABEN
Drammond, C	AHUR	Dyck Lt V	AFXT
Dranske	ADBS	Dyer B	BATP
Drayton Hr	ATYD	Dyre Fiord	BFAM
Drepano, C (Candia)	AJPF	Dzunan Is	AFIR
Drepano P (Greece)	AJIT		
Dresden	ACWD	Eads, P	AWYT
Drewin	ALME	Eagle Hr	BEOD
Dröbak	ABYL	Eagle Hawk B	ASGW
Drogden Ch	ADOE	Eagle I Beacon (China)	AOHC
Drogheda	AFUR	Eagle I (Ireland)	AFRZ
Dubh Artach	AFBU	Eagle I (Me)	BARE
Dublin	AFVB	Eagle Nest Pt	ARXZ
Duddon R	AEWI	East Arm (Canada)	BCWL
Dudgeon Lt V	AFNG	East C (Asia)	APOS
Due Odde	ACFM	East Ch (Kattegat)	ACBL
Duén Cove	AMDC	East Chop Pt	BADI
Duhnen	ADTW	East Goodwin Lt V	AEKR
Dui	AORN	East Pascagoula	AXCI
Duino	AIVL	East Pt (Pr Ed. I)	BDCP
Duiven I	AFWT	East Pt (Queensland)	ASCW
Dukati, P	AJGY	East Pt (Saturna I)	ATWR
Dukato, C	AJIR	East R (N Y)	AZPI
Duke Town	ALPS	Eastbourne	AELH
Dulce, G of	AUSF	Easter I	ARHU
Dulce, R	AWQF	Eastern Ch Lt V	ANIS
Dülcigno	AJGO	Eastern Grove Lt	ANMK
Duluth	BEMX	Eastern Pt (Mass)	BAJC
Dumaguete	AQIR	East India Docks	AEJH
Dumankilas B	AQLE	East Ironbound I	BCKO
Dumaran I	AQCL	East London	ALYK
Dumbarton	AEZB	East Main F	BEYT
Dumfries	AEXJ	Eastport	BAWC
Dumpling Rock	AZYC	Ebeltofte	ADKJ
Dunamund F	ACVG	Ebro R	AHLJ
Dunamund I	ACUO	Ecija, Nueva	AQFC
Dunbar	AFKN	Eckernforde	ADEY
Duncan City	BEGO	Ecréhos, Les	AGDI
Duncannon	AFWH	Ecuador	AUTW
Duncansby Hd	AFEG	Edam (Holland)	ADZW
Dundalk	AFUO	Edam I (Java)	AFVC
Dundee	AFIC	Eddy Pt	BCPE
Dundrum B	AFUI	Eddystone	AENT
Dunedin	AILJ	Eddystone Pt	ASHO
Dungan R	AONQ	Eden	ASHY
Dungarvan	AFWK	Edenton	AVTU
Dungeness (England)	ARLB	Edgar Town	BADL
Dungeness (Queensland)	ASXE	Edie Rajut R	APQT
Edinburgh	AFJW		
Edinburgh Ch	AEIF		
Edisto R	AYOM		
Edithburgh	ARUP		
Edö	ABMR		
Edwardsburg	BDVI		
Eeragh	AFRI		
Efate, or, Sandwich I	AQXN		
Egedesminde	BEZU		
Egense	ADPS		
Egg Hr	AVLW		
Egg I (Bahamas)	AXJQ		
Egg I (N S)	BCMN		
Egg I (Ontario)	BDXH		
Egg I (St Lawrence)	BDMX		
Egg Rock (Mass)	BAIH		
Egg Rock (Me)	BATH		
Eggegrundet	ACLI		
Eggemoggin Reach	BARB		
Egmond aan Zee	AECP		
Egmont, C (N Z)	ATCV		
Egmont, C (Pr Ed I)	BCZF		
Egmont cay	AXPH		
Egrilar	AKOV		
Egypt	AKTC		
Ehrenberg	AULY		
Eider Canal	ADTG		
Eider Galliot Lt V	ADTE		
Eider, Outer, Lt V	ADTC		
Eider R	ADSZ		
Eight Degree Ch	ANDP		
Eisk spit	AKJE		
Eierland	ADZG		
Eke Fiord	ABPU		
Eken Sound	ADFT		
Ekero	ABUK		
Ekersund	ABUL		
Ekerums	ACGU		
Eknås	ACQT		
El Amaid	AKUE		
Elaputi B	AQFX		
El Araish	ALBO		
El Arish	AKTF		
Elba I	AHYR		
Elbe No 2 Lt V	ADTS		
Elbe No 3 Lt V	ADTU		
Elbe No 4 Lt V	ADTV		
Elbe Pilot Galliot	ADTQ		
Elbe, Outer, No 1 Lt V	ADTP		
Elbe R	ADTO		
Elbing	ACXK		
Elbow cay	AXIP		
Elburg	AEBM		
Elena, P.	AURM		
Eleos I	AJTZ		
Elephant B	ALTW		
Elephant Pt (Burma)	ANMF		
Elephant Pt (India)	ANKT		
Eleusis	AJOL		
Eleuthera I	AXJS		
Elizabeth B	ALUT		
Elizabeth, C (Me)	BALW		
Elizabeth, C (Asia)	AORX		
Elizabeth City	AYUD		
Elizabeth, P	ALXZ		
Elizabeth R	AYVI		
Elizabethtown	AZNY		
Elk R.	AZEJ		
El Kamela	AKWH		
Ellen, P	AEZS		

ELLENBOGEN—FATHER PT

Ellenbogen... . ..	ADSK
Ellensburg	AUEP
Ellesmere Pt.	AEVC
Ellewoutsdijk	AEZY
Ellice Is	AREU
Ellinggaard Kilen ..	ABYR
Elliot, P . . .	ARVQ
Ellis, B	BDLZ
Ellsworth	.BARZ
Elmina.	.ALNC
Elobey I . .	ALQX
Elopura . .	APZQ
El Portillo.	.AXQN
El Rincon . . .	AUSJ
El Roque, P.	AWKD
Elsfleth . . .	ADVY
Elsinore	.ADOY
Elysée	AGBC
Embiez I . . .	.AHTB
Embomma. . . .	.ALSI
Emden. . . .	ADYG
Emineh C. . .	.AKDQ
Emma Hr . .	.APOJ
Empedocle, P. .. .	AIME
Empire City . .	AUEP
Empress Augusta R. .	..AQTW
Ems Jade Canal. .	ADWX
Ems R	..ADYB
Endau R . .	ANTI
Endeavor R	ASYL
Endeavor Str	ASZQ
Endelave I	ADJQ
Endermo Hr . .	APKZ
Endœrodde . .	ADJF
Enfant Perdu. .	..AWFY
Engaño, C (Haiti) .	AXIJ
Engaño, C (Philip) .	AQFD
Engelholm	ACDK
England.	.AEHR
English and Welsh grounds Lt V . .	.AEQU
English B	ATVF
English Bank Lt V .	.AVNM
English Ch	AEHS
English Hr . . .	AYCT
English R	.ALZN
English Road	ALHG
Enkhuzen	ADZT
Enos . . .	AJWN
Enragé, C . .	BAYZ
Ensenada (Mexico).	AUKN
Ensenada de Barragan	AVNS
Enskar . . .	ACPW
Entrance I . . .	ATLW
Entry I . . .	BDFX
Ephesus. .	AKPB
Epi I .	AQXL
Equitorial Ch	ANDS
Ercole P .	AIPU
Erekli. .	.AJXQ
Eriboll Loch . .	AFDV
Ericeira.. .	AHBT
Erie City	BEDY
Erie, Lake	BDZP
Erith .	AEIU
Eritra .	AKOR
Erknó .	ABOG
Eromanga .	.AQXI
Erqui .	AGFS
Erronan. .	..AQYE

Ervily, P de l' ...	AGLM
Erzerum .	AKMD
Esbjerg .	ADRY
Escala, La .	AHNC
Escocés, P. .	AWNE
Escombrera .	.AHJC
Escondido B . .	.AUOJ
Escondido P	AXUR
Esconil L'. .	.AGLT
Escudel .	AHLO
Escuminac Pt	BDGI
Esmeralda .	.AUTY
Espalamaca Pt .	ALCT
Española Pt .	AUWI
Esperance B.	.ARPX
Esperance, P	ASGD
Espevær .	ABBW
Espichel, C .	AHCP
Espiguette Pt	AHQX
Espirito Santo B	AVUZ
Espiritu Santo I	AQXD
Espoir, C d'	BDKM
Esquimalt Hr	ATXO
Essequibo R . .	.AWHU
Essington, P	ATUI
Estaca Pt	.AGXE
Estacio .	AHJI
Estanques B	AWLI
Estepona .	AHGI
Estéro .	AUST
Esteros	AUIF
Etang de Thau .	.AHQR
Etang du Nord ...	BDFW
Etang, L', Hr	.BAXG
Etaples .	AFYJ
Eten Pt	.AUXE
Etretat	AFZC
Eubœa I	AJUW
Eucla .	ARQE
Eugénie Archipelago	AOQI
Eume R	AGXR
Eupatoria .	.AKHG
Euphrates R .	AMUJ
Eureka (Cal).	AUFI
Eureka (Mexico) .	AULW
Euripo Ch.	AJOY
Europa I . .	AMGZ
Europa Pt	AHFX
Europe . . .	ABDT
Evangelistas	AVJE
Evanston	BEJF
Everard C .	ASCD
Eversand .	ADWH
Execution Rocks .	AZSI
Exeter .	ASEF
Exholm ..	ACSY
Exmouth	AEND
Expedition B	AOPI
Exuma I .	AXJY
Eyemouth	APKQ
Eyre, P	ARQI
Ezaro	AGYO
Faaborg	ADHU
Faber, Mt	ANSW
Fæno ..	ADHJ
Fæsö	ABTK
Færder	ABXO
Fagnet Pt	AFZB

Faichaburi .	ANUB
Fair I. . . .	AFGJ
Fairhaven (Maes) . .	.AZYO
Fair Haven (Ontario)	.BDYU
Fairhaven (Wash) .	ATYJ
Fairy, P .	ARXI
Fakarava atoll	AROW
Fakir Pt	ANLD
Fakkebierg	ADIO
Falase, La	AGRW
Falcon, C. .	..AKZL
Falcone Pt	AICN
Falconera P	AIUN
Falifa Hr	AREL
Falkenberg .	ACBZ
Falkland Is	AVKL
Falkner I. .	.AZTM
Fall River . .	AZXB
Falluden .	ACHY
Falmouth (England).	ABOJ
Falmouth (Jamaica)	AXRO
Falmouth (Tasmania)	ASHK
False B (Africa) . .	ALWT
False B (India) .	.ANIF
False C (New Guinea).	AQBH
False C (Va)	AYUS
False Duck I	BDWY
False Pt .	AIND
Falster	ADMZ
Falsterbo	ACEQ
Falsterboref Lt V.. .	ACES
Famagusta .	AXRF
Fame Pt .	DDLE
Fanad Pt	AFSU
Fanar Adasi Islet	AJZH
Fanaraki. . .	.AKCJ
Fanar Burnu	AJXW
Fanemotra B	AMFK
Fangaloa B	AREN
Fangar, P. .	AHLF
Fango Pt .	AHLO
Fano P (Adriatic) .	.AISJ
Fanö (Denmark). .	ADBX
Fano I (Ionian Is)	AJHK
Fanrang B ..	.ANVZ
Fao .	AMUG
Farafangana .	AMEW
Farallon I .	AVPI
Farallon Sucio.. .	AWNM
Farallones Islets. .	AUHI
Faraman. . .	AHRC
Furasina .	AIYF
Faré Hr .	ARHJ
Farewell, C	.BEZY
Farewell Spit C	.ATBX
Farin	ALIN
Farina P .	AKWI
Farm Cove .	ASKR
Farn I .	.AFKX
Fåro (Baltic) .	.ACHO
Faro, C (Italy)	AHWX
Faro, C (Sicily)	AIKM
Faro (Portugal)	.AHDF
Faroe Is .	AFHE
Farquhar Is .	AMJE
Farsund	.ABUX
Fasana	AIXD
Fastnet .	AFPS
Father Pt .	.BDOE

FATOUVILLE—FREDRIKSHALD

Fatouville	AFZM	Fishbourne I	.AOFT	Formby Lt V	.AEUR
Favignana	AIMQ	Fisher I	AZUH	Formentera I	AHNQ
Favone	AIIZY	Fishguard	AESO	Formenton, C.	AHOP
Faxö	ADNU	Fitzhugh Sound	ATUX	Formica	.,AIMY
Fayal	ALCR	Fitzmaurice R	.ARJZ	Formigas	ALDP
Fear, C	.AYQL	Fitzroy I	AOGW	Formosa (Brazil)	AWBI
Fécamp	AFYZ	Fitzroy Dock	.ASLC	Formosa (China Sea)	AOBS
Fedderwarder Sicl.	.ADWJ	Fitzroy R Lt V	ASUQ	Formosa, Mt (Singapore)	ANSI
Federal Pt	AYQU	Fiumara Grande	AIGE	Formosa, C.	ALPG
Fehmarnbelt	ADMT	Fiume	.AIYR	Fornæs	.ADKP
Fehmarnbelt Lt V	...ADEL	Fiumicino	AIGB	Fornelli Road	AIFK
Fejan	.AGJW	Five Fathom Bank Lt V	AZHP	Fornells, P	.AEPN
Felixstowe	AFPB	Fjeldö	ABTS	Fort de France	AYEN
Femern I	.ADEC	Fjeldvik	.ABKR	Fort de France B	AYEL
Femern Sound	ADEK	Flaavær	ABOS	Fort Pt (Cal)	AUGH
Femris	ABJS	Fladda Hr	.AFBJ	Fort Pt (E India)	ANQB
Fenaio Pt	AHZF	Fladen Lt V	ACBW	Forth, Firth of	.AFJH
Fenar Pt.	AJWL	Fladholm	ABTY	Forth Bridge	.AFJQ
Fénérive	AMEN	Fladö	ABHZ	Forth & Clyde Canal	AFJY
Feno, C	AIBN	Flagstad	.ABIJ	Fortune B	BETP
Fenwick I Shoal Lt V	AZHJ	Flamanville, C de	AGDZ	Foujes	AKOR
Fer, C de	.. AKXJ	Flamborough Hd	AFMG	Foula	AFGY
Fermo	AISC	Flamenco, P	.AVDO	Foulepointe	ANEP
Fernanda Pt.	ALFY	Flannan Is	AFCY	Foulweather, C	AUDS
Fernandina	.AYKC	Flat C	.APSZ	Foulwind, C	ATFM
Fernando Noronha I	AWBU	Flat I	AMIP	Fouquets Isle	.AMIO
Fernando Po	.ALPX	Flat Pt	BOSN	Four, Le (Bretagne)	AGJA
Fernand Vaz	ALRK	Flatholm	AEQB	Four, Le (Quiberon B)	AGOT
Ferraione, C.	.AHZO	Flattery, C	AUCJ	Fouras	.AGQW
Ferrajo, P	.AHYS	Flat-Top I	ASVX	Fouches Mt	AHTY
Ferrara	.AIIK	Fleetwood	ARVW	Fourchu, C	BCGN
Ferraria Pt	.ALDK	Flekkefiord	.ABUP	Fourteen-feet Bank	..AZIH
Ferrat, C	.AHVJ	Flensborg	ADFR	Fourth Pt	APLR
Ferret, C	.AGTH	Flinders	ARQU	Foveaux Str.	AIGB
Ferro (Canary Is)	ALFJ	Flinders B	ARFH	Fowey	AEOH
Ferro, C (Italy)	.AHWX	Flinders I	ASDC	Fowey Rocks Beacon	AXHS
Ferro, C (Sardinia)	AICZ	Flint	AETZ	Fowler B	.ARQK
Ferro Castel	AHIC	Flint Ch (Flintrannan)	ACEF	Fox I	BDGO
Ferrol	.AOXL	Flint I	BCRV	Foxton	ATDX
Ferrol B	AUYB	Florence	AHYB	Foyle, Lough	.AFTB
Ferryland Hd	BRUW	Flores (Azores)	ALCO	Foynes Hr	AFQY
Fez	ALCD	Flores (E Archipelago)	AQOB	Foz	.AHRM
Fianona	.AIYF	Flores I (Uruguay)	AVQF	Francavilla	.AIRY
Fidonisi	AKEN	Floriana	AIOS	France	AFXJ
Fidra	AFKM	Florida	AXDV	Frances cay.	AXMZ
Fife	AFIL	Florida, C	.AXHV	Frances I	AVUR
Fife Ness	AFJG	Florida Str	AXHY	François.	AYRX
Figari, C.	AIDC	Florida, East Coast.	.AYIL	Franklin Hr	ARSH
Figeholm	ACGP	Florö	ABPR	Franklin I	BANY
Figueira	AHBN	Flugge	ADEI	Franklin Sound.	ASDE
Fiji Is	.ARBZ	Flushing	AEFU	Franz Josef Land	.ABCW
Fiksdal	ABNX	Fly R	AQSZ	Fratelli Rocks	AKWN
Fila Mt	.AQXO	Focardo F	AHYV	Fraser R	.ATYL
Fildtvedt	ABYH	Foggia Nova	AKOD	Fraserburgh	AFIJ
Filey	AFME	Fogo Hr (Newfoundland)	BEWE	Frauenburg	ACXB
Filsand	ACUB	Fogo I	.ALHQ	Fray Bentos	AVFO
Fingal Pt	.ASOZ	Fohr	ADST	Frederichsted	.AXZU
Finisterre, C	AGYK	Folgerö	ABRV	Frederich Wilhelm Hr	AQTS
Finkenwerder	ADUZ	Folkestone	AERY	Fredericia	ADGU
Finland	ACOD	Folly Pt	.BAZO	Frederick, P	.ASRP
Finland, G of	ACQY	Fonseca, G	AUQR	Frederick Henry B	AGV
Finmarken	.ABFT	Fönskov Pt	ADIIK	Frederick Henry I	.AQSF
Finngrundet Lt V	.ACLK	Fontana, C	AKFV	Frederiksberg	ADOP
Finngrund West Lt V	.ACLN	Fontane	AIWQ	Frederiksburg	AYZN
Fino P	AIIXB	Froa I	ALDR	Frederikshavn	ADQJ
Finsch Hr	AQTO	Forcados R	.ALOJ	Frederikssund	ADLC
Fire I	AZQJ	Foreland Bluff	ALSF	Frederikstad.	ABYX
Fire I Lt V	AZMS	Foreland The	.ARPY	Fredrichstadt	ADTI
First Pt	APUR	Forest, La	AGMC	Fredriksborg	.ACJK
Fischausen.	ACWY	Forio	AIHB	Fredrikshald	ABZE

FREDRIKSHAMN—GIBRALTAR

Fredrikshamn	ACRL	Gabovacatrida	AJBP	Gasparilla I	AXFT
Fredriksværn	ABXG	Gacholle, La	AHQZ	Gaspé	BDKV
Freeman Ch	ASQH	Gadaro I	AJUO	Gaspé B Lt V	BDKU
Free Town	ALJW	Gadji	ARBP	Gaspé C	BDKT
Fréhel, C	AGPQ	Gaet, La	AGST	Gata, C (Cyprus)	AKQY
Freiburg	ADUI	Gaeta	AIGN	Gata, C de (Spain)	AHIL
Fréjus	AHUO	Gage Road	ARNT	Gatcombe Hd	ASUC
Fremantle..	ARNX	Gaidaro Islet	AJQX	Gatteville	AGCK
French Pass	ATEF	Gainsborough	AFMY	Gavazzi Pt	AJOD
French Pt	ALRZ	Galantry Hd	BBUD	Gavdo I	AJPW
French Guiana	AWFO	Galápagos Is	AUVH	Gaviotas Lt V	AWDV
Frenchman B	BASR	Galata	AKBR	Gavrion P	AIQS
French Town	AZEM	Galata, C.	AKDU	Gawler, P.	ARUV
Fresvik	ABQS	Galatz	AKER	Gay Hd	BACX
Friedrichsberg	ADPI	Galaxidi	AJLP	Gaya B	APFT
Friedrichsort	ADEU	Galdobin, C	AOQD	Gayo, P	AJHU
Friedrichsschleuse	ADXL	Galea	AGUX	Gebi I	AQRM
Frikes P	AJKO	Galera Pt (Chile)	AVHW	Gedney Ch	AZND
Frio, C (Africa)	ALUI	Galera Pt (Trinidad)	AYHK	Geelong	ARYQ
Frio, C (Brazil)	AVTY	Galera de Zamba B	AWMP	Geelvink B	AQFY
Frioul, P du	AHSJ	Galéria	AIBU	Geestemünde	ADVW
Frontera	AWSL	Galets, Pt des.	AMHX	Gefle	ACLF
Frontignan	AHQT	Galibi Pt	AWGR	Gellen	ADWT
Fruholm	ABGN	Galiola Is	AIZC	Gellibrand Pt	ARYV
Frying-pan Shoals Lt V	AYQR	Galita (Galite, La)	AKWR	Gellibrand Pt Lt V	ARYW
Fuchau fu	AODI	Gallegos, P	AVLC	Gelsa	AJEI
Fuengirola	AHGM	Galley Hd	APXD	Gemlik	AJYS
Fuenterrabia	AGUI	Gallipoli (Italy)	AIQB	Genesee R	BDYL
Fuerteventura	ALFP	Gallipoli (Turkey)	AJXH	Geneva	ADXC
Fuglenæs	ABGQ	Gallipoli, P (Asia Minor)	AKFI	Genius Bank Lt V	ADXC
Fukabori	AOXI	Gallo, C (Adriatic)	AIQN	Genoa	AHWR
Fukai	AOWZ	Gallo, C (Sicily)	AINL	Génois, F	AKXF
Fukuoka	AOWD	Galloo I	BDZG	Genoves	AHIM
Fuk ura	APDN	Galloper Lt V	AFPN	Gensan	AONL
Fuku shima	APKO	Galloway, Mull of	AEXR	Géographe B	AROU
Fukuyama (Honshu)	APBH	Galveston	AWVC	George B (Canada)	BCVN
		Galveston Lt V	AWVD	George B (Tasmania)	ASHM
Fukuyama (Hokoshu)	APKM	Galway	AFRK	George, C	BCVZ
Fulehuk	ABXS	Gambia	ALHT	George R	ABJU
Fuller Rock	AZWG	Gamla (Old) Karleby	ACOR	George I	BCLY
Fumboni B	AMDC	Gamova B	AOPQ	George Town (Africa)	ALXO
Funabashi	APIG	Gamvig	ABGD	Georgetown (Ascension)	ALUY
Funakoshi	APKB	Gana Gana	ALOM	Georgetown (Demerara)	AWHQ
Funchal	ALDT	Gandia	AHKI	Georgetown (Malay Pen)	ANQO
Fundukli	AKBU	Ganges R	ANJY	George Town (Pr Ed I)	BDAU
Fundy, B of	BAYE	Ganjam	ANHT	Georgetown (S C)	AYQC
Fungasar Hr	AREB	Gannet Rock	BAXF	Georgetown (Tasmania)	ASDW
Funk I.	BEWN	Gap Rock	ANYU	Georgia (Asia Minor)	AKLX
Funkenhagen	ACYM	Garcia de Avila	AVYH	Georgia (State)	AYKF
Funta B.	ALSX	Garde, C de	AKXG	Georgia, Str of	ATUW
Fuokbinkiang Lt	ANVS	Garden I (N S W)	ASEN	Georgian B	BEHM
Furni	AFFB	Garden I (W Australia)	ARNO	Georgina Pt	ATWP
Furon	ACGO	Garden Reach	ANJG	Geraldine	ARMS
Furugrund	ACNII	Gardiner I	AZQS	Geraldton (Queensland)	ASXQ
Fusan	AOND	Gare Loch	AEZC	Geraldton (W Australia)	ARNB
Fusan Kai	AONF	Gargano Hd	AIRG	Geresik	APWH
Fushiki	AOUB	Garonne R	AGSZ	Germandö	ACNS
Fusi yama	APIT	Garoupe	AVHC	Germein, B.	ARSN
Futsu saki	APIM	Garpen	ACGH	Germany	ACVP
Futuna	AQYB	Garraway	ALKW	Gersö	ADJH
Fuyan I	AODP	Garrison Pt	AEIK	Ghelenjik B	AKJY
Fuyen Hr	ANWF	Garrucha	ABIS	Ghenitchesk Str	AXIN
Fyen	ADGZ	Garston	ABUW	Ghent	AEGY
Fyne, Loch	AEZG	Gascogne, G of	AGIS	Ghijinsk B	AOSL
		Gascoyne Road	ARMI	Ghubbet Hashish	AMSU
Gabbard, Outer, Lt V	AFOZ	Gåsören	ACNG	Giannutri	AHZI
Gabes	AKVI	Gaspar Str	APSF	Gibara, P	AXMJ
Gabo I	ASCH	Gaspar, Lower, Lt V	ANIU	Gibbs Hill	AYHS
Gabo P	AQJW	Gaspar, Upper, Lt V	ANIV	Gibraltar	AHFU
Gaboon R	ALQZ			Gibraltar (Erie)	BEAR

GIEDSER—GRECALE

Giedser Pt Lt V	. ADNE	Gold Coast. .	. ALMW	Gran I.. . . .	...ACLV
Giedser Reefs Lt V	. ADNC	Golden Horn	AKBO	Granatello. . .	. AIHN
Gienner Fiord	. ADQK	Golden Horn Hr	AOQB	Granada	. AHGY
Giens	. . AHTR	Golden I. . . .	AOGE	Granada, Nueva	. AUTM
Gighiga .	. AOSN	Goldsboro .	AYSE	Gran Canaria. . .	. ALBR
Gaglio	. AHZD	Goletta .	AKWE	Grand B (Mauritius)	. . AMIG
Grijon . . .	. AQWD	Golfito .	.AUSL	Grand I. (St Lawrence)	.BDPU
Gilbert Is.	. AQVZ	Gombé Pt	. ALRC	Grand I (Lake Superior)	BEPF
Gilbierg Hd. . .	. . ADKW	Gomen B .	. AQZF	Grand Banc Lt V. .	AGRS
Gill B	.AVLY	Gomena, C. .	AJEW	Grand Bassa	ALKO
Gilleskaal .	. ABJV	Gomenizza P	...AJHP	Grand Bassam . .	ALMJ
Gillolo I. . . .	. AQRI	Gomera. . .	..ALPH	Grand Bourg. .	AYED
Gillolo Passage.	. AQRT	Gonaives. . . .	. AXIK	Grand Cayman . .	AXQV
Gimso .	ABIR	Gooch B .	ATJS	Grand Haven . .	BEKM
Giorosa Jonica	. AIPN	Goode I	.ASZK	Grand Jardin Islet .	ACFR
Gioja .	. AIJY	Good Hope, C of .	ALWQ	Grand Léjon . .	AGFV
Giova Pt . .	AKPG	Good Hope, C (China) .	ANZR	Grand Manan I. .	BAXS
Giovinazzo	AIQU	Goodwin, East, Lt V .	AEKR	Grand Mont Pt. . . .	AGOK
Giraglia I	.AHZS	Goodwin, North, Lt V	.AEKL	Grand Port (Mauritius) .	AMIK
Girdle Ness . . .	AFIR	Goodwin, South, Lt V	.AEKS	Grand Port (New Cal). .	AQYP
Girdler Lt V	.. .AELJ	Goodwin Sands.	. AEKM	Grand Ribeau I . . .	AHTU
Girgenti	AIMC	Goole. . .	AFMS	Grand Rustico	BDCY
Gironde R. .	. AGRP	Goolwa. . . .	ARVY	Grand Sesters . . .	ALKV
Girvan . .	.AEYB	Goose C.	BDQO	Grand Surrey Canal	AEJQ
Gisborne	ATIS	Goose I . . .	. ASDF	Grand Turk I . .	AXKT
Gisund .	. ABHT	Gopalpur . .	. ANHR	Grande B . .	. AYBJ
Giudecca Canal	AIUB	Gorda Pt. .	. AXPE	Grande Banc . . .	. AGRS
Giurgiu. . .	AKEZ	Gordon R . .	. ASFK	Grande P (C Verde Is)	ALGW
Gjerrild Lt . .	ADPO	Gore B . .	. ATJM	Grande P (Stephens B)	ALVN
Gjesvier .	.ABGM	Gorée . .	ALGJ	Grands Cardinaux	AGOU
Gjoeslingerne	ABKX	Gorey . .	AGDT	Grangemouth .	AFKB
Gladstone .	.ASUE	Gorgona I . .	AHYF	Granite I . .	BEOZ
Glandore . .	.AFXG	Gorino . .	. AJTE	Granitola, Cape .	AIMK
Glarenza .	AJLO	Gorizia or, Goritz .	. AIVK	Grant Beach Lts . .	BDHA
Glas I . . .	.AIVW	Goro, P . . .	AQYL	Granton Hr. .	AFKO
Glasgow (Ontario)	.BDYR	Goro Pt	. AJTD	Granville (New Guinea) .	AQIG
Glasgow (Scotland) .	AEYZ	Gorontalo . .	AQMF	Granville (France)	AGEO
Glasgow, P (N Guinea)	. AQTI	Goshkevitch B. .	AONZ	Grao de Valencio	AHKP
Glasgow, P (Scotland)	AEYW	Gota (Gotha) Canal	. ACIM	Grasgard . . .	ACGZ
Glavina Pt	AIYX	Gotland .	.ACHD	Grassy B . .	AVID
Glenelg	. ARVC	Goto, C .	AOWY	Gratiot, F .	BFFW
Glenelg R	ARWZ	Goto Is	AOWT	Grave Pt . .	AGNB
Glengariff Hr . . .	.AFQE	Gotska Sandon . .	ACHQ	Grave Pt (France) .	AGSF
Glopen .	ARIZ	Gottenburg	ACBR	Grave Pt (B C). . .	ATWJ
Glorieuse Isle . .	AMDN	Gough I .	ALVK	Gravelines	AFXC
Glosholm .	ACRF	Goulburn . .	ASJI	Gravesend	AEIP
Gloucester (England)	AEQN	Gouldsborough. . '.	BATM	Gravosa, P .	AJFN
Gloucester (Mass)	BAIZ	Gouletto, La . .	AKWE	Gray Hr . .	AUCP
Gloucester (N J) .	.AZKY	Goulfar B (Talut Pt)	AGNR	Great Basses	ANFB
Gluckstadt	ADUJ	Goulven . .	AGIX	Great Belt . .	ADLN
Goa .	ANBM	Gourock . .	AEYN	Great Bird Rock .	BDFZ
Goapnáth Pt	AMXZ	Government Hill	ANQM	Great Borris Hd .	ADPV
Goat I (Cal) . . .	AUGL	Governors I . .	AZPC	Great Britain . .	AEHQ
Goat I (Me) .	BAIK	Goyder R .	ARIU	Great Captain I .	AZSL
Goat I (N S W). .	ASLG	Goza Hr	AFFC	Great Duck I .	BASP
Goat I (R I). .	AZXH	Gozier I	AYDQ	Great Fish B	ALUF
Goché .	.AOCF	Gozo . .	AIOZ	Great Fish R .	ALYF
Godavari R. .	ANGZ	Graa Deep . .	. ADRV	Great Hr	AXWN
Goderich. . .	.BEHJ	Gracias a Dios, Cape, Hr	AWOZ	Great Inagua I . .	.AXKV
Godhavn. . .	.BEJQ	Graciosa (Azores) .	ALCX	Great Isaac. .	AXIW
Godley Hd. . . .	.ATJK	Graciosa (Canaries). .	ALEG	Great Kapuas R .	AFXM
Godnatt Rock . .	.ACFY	Gradaz P . .	ALXV	Great Namaqualand	ALUV
Godrovy I . . .	.AEFG	Grado P . .	AIVD	Great Orme Hd	AETW
Goedcreede . . .	AEDK	Græsholmen . .	ABON	Great Ribaud I . .	AHTU
Goeree	AEDJ	Grafton . .	ASOL	Great Round Str Lt V .	BAFK
Goes . .	ABFM	Grafton, Cape . .	ASXW	Great Sandy Str .	ASRO
Gogaligyun . .	ANOU	Graham P . . .	ATSF	Great Skarcies R	.ALJR
Gogha . . .	AMRC	Grahams Town (Africa) .	ALYQ	Great Swan P .	ASHO
Göhren . .	ADBG	Grahamstown (N Z) . .	ATML	Great Wall of China.	AOJW
Goiana R. . .	AVZY	Grain Coast	ALMR	Grecale, C . .	AIPE

GRECO—HAMPTON

Greco, C.	AKRE	Grundkallen Lt V	ACKV	Haaks Lt V	AECD
Greco, Torre del.	AJHP	Grunendeich	ADUO	Haaler Is.	ABZF
Greece	AJIE	Gryto	ABHY	Haarbjerget	ABHM
Greek Archipelago	AJPX	Guadalcanar I	AQWI	Haarlem	AECK
Green B.. ...	BEID	Guadalquiver R	AHEG	Haber Ness	ADFQ
Green, C.. ...	ASHU	Guadalupe R	AWUB	Habana	AXNW
Green Hr.	BCIJ	Guadeloupe..	AYDC	Habibas Is.	AKZF
Green I (C Br I).	BCQI	Guadiana B.	AXOJ	Habitants Hr	BCFW
Green I (China)	ANYV	Guadiana R	AHDM	Hacha La	AWLS
Green I (N S)	BCOA	Guahyba R	AVRI	Hachken R	APCE
Green I (Lunenburg)	BCKA	Guaira, La	AWJU	Hacking, P	ASJO
Green I (St Lawrence)	BDFI	Guardafui, C.	AMOG	Hadersleben	ADGO
Green Pt.	ALWP	Guajan I..	AQYL	Hadlock, P	AIZQ
Greenhithe.	AEIS	Guajara R	AWED	Haedik I	AGOL
Greenland	BEZO	Guam I	AQVL	Hafrnnge	ACIR
Greenly I	BEQG	Guama R.	AWEB	Hagi	AOVC
Greenock	AEYV	Guañape Is	AUXS	Hague, C de la	AGCQ
Greenore (Carlingford)	AFUK	Guanica, P	AXYK	Hague, The	AECO
Greenport	AZRH	Guanta B	AWJM	Haidar Pasha.	AKBE
Greenspond	BEWJ	Guantanamo	AXQR	Haiderabad	AMWU
Greenwich	AEIY	Guarapari	AVUX	Haifa.	AKSO
Greenwich (R I)	AZVU	Guardia, La	AGZR	Haifong	ANXC
Greenwich Observatory	AEJB	Guardia Pt	AIGP	Hainan.	ANXL
Gregory, C (Or).	AUEH	Guardiana I.	AJKI	Hains Pt	AOGP
Gregory, C (Magellan).	AVJN	Guardia Vecchia Hill	AICV	Haisborough Lt V	AFNR
Gregory, P	ARMU	Guatemala	AUPI	Haitan Str	AOBR
Greifswald	ACZV	Guaytulco, P.	AUOF	Haiti	AXTC
Greifswalder Oie	ACZP	Guayanilla, P	AXYI	Haiti, C	AXTS
Grenaa.	ADKO	Guayaquil.	AUVZ	Haiyang	ACIL
Grenada (Spain)	AHGY	Guaymas	AUMD	Haiyuntan I	AOKV
Grenada (W I)	AYGB	Guduat	AKLI	Hakata	AOWE
Grenadines, The.	AYFU	Gudvangen.	AKUH	Haken.	ACDX
Grenville Hr	BDEA	Guerbaville	APZS	Hakluyts Hd	ABUG
Grepen Lt V	ACKW	Guérin I	AOMB	Hakodate	APKS
Grey, P	ARNF	Guernsey	AGDB	Hales Bluff	ASMB
Grey R	ATPO	Guetaria	AGUM	Half-Moon cay	AWQN
Greymouth	ATFU	Guia	AHCG	Half-Moon Shoal Lt	AWVL
Grignano B	AIVM	Guia F	ANYH	Halfway Rock	BAMF
Griko, P.	AJFB	Guiana	AWFM	Halifax.	BCLP
Grimsby	AFMP	Guldford.	AROC	Halifax B	ASNX
Grimskar.	ACGI	Gulfinec	AGLE	Halki	AJYB
Grimstad.	ABWE	Guimarás	AQIE	Hallam.	ADYW
Grindel Pt	BAPL	Guinchos cay.	AXMV	Hallands Wadero	ACDG
Grindstone I (N B)	BAZI	Guinea G	ALMX	Hållo	ABZR
Grindstone I (P Ed I).	BDFQ	Guisanburg.	AWFS	Hall Sound.	AQTE
Grino.	ABJX	Guiuan.	AQJO	Halmstad.	ACDF
Grip.	ABMU	Guldborg Sound	ADMX	Halong B	ANXF
Gris Nez C	AFYB	Gulf Stream	AYIJ	Hals	ADFW
Grissellhamn	ACKG	Gull Lt V...	AEKQ	Halskov	ADLW
Gröhara	ACQV	Gull I	BEWY	Halten Is.	ABLM
Groix I	AGMR	Gull Rock	BCIP	Halys Pt	AKMQ
Groningen	ADYU	Gumaka	AQFM	Hamada	AOUY
Gronningen..	ABVQ	Gun cay	AXIV	Hamamatsu	APGE
Gronskar.	ACIH	Gunfleet	AFPII	Hamanaka	APLK
Gronskaren.	ABZY	Gunieh	AKLR	Hambantotti	ANEZ
Gron Sound	ADNJ	Gureina	AKUR	Hamburg.	ADUV
Gros, C.	AHLU	Guruni Hd	AJVL	Hamelin B	ARPD
Grosa Pt (Iviza)	AHNZ	Gurupa R	AWEN	Hamilton (Bermuda)	AYIB
Grosa Pt (Majorca)	AHON	Gurupi R	AWDM	Hamilton (Ontario).	BDXW
Gross Ziegenort.	ACYO	Gustaf Hr	AYBP	Hamilton (Queensland)	ASIJ
Grossa I	AJCM	Gutvik	ABEQ	Hamilton, P.	AOMU
Grossbruch.	ACXB	Gutzlaff I..	AOFB	Hamilton Inlet.	BEYA
Grosse I	BDFT	Guyon I	BCRG	Hammamet	AKVZ
Grosse Isle..	BDRT	Guysborough	BCOV	Hammerö	ABHX
Grosses Haff.	ACYV	Guzerat	AMXJ	Hammer Pt	ACFG
Grosso, Mt.	AIRZ	Gwa	ANLQ	Hammerfest.	ABGR
Grosso Pt..	AJMX	Gwádar	AMWD	Hammershus.	ACFE
Grouin Pt	AGEX	Gwatar B	AMVZ	Hamnskar.	ACBE
Grovenor P	ALYS	Gythium.	AJMS	Hampton	BAJX
Gruizza, or Grucia I	AIZR			Hampton Roads	AYVN

HAN—HOBRO

Han R.	ANZW	Haulbowline I.	AFWT	Heppens	ADWQ
Hancock, C	AUDC	Haulover Isthmus	BCQU	Heraklea	AKNF
Handa	AFFY	Hausu B	ANXQ	Hérault, R.	AHQF
Handkerchief Lt V.	BAEM	Haut Banc du Nord	AGQI	Herbert	ASVB
Haneda	APHY	Haute I	BCDA	Herbert R	ASXC
Hangchau B	AOER	Havannah Hr	AQXS	Herbertsböh	AQUG
Hangklip, C.	AIWV	Have, La, R	BCJN	Herbes, Pt aux.	AXBE
Hangö	ACQP	Havelock	ATDY	Hereford Inlet	AZLT
Hanish Is	AMRX	Havösund	ABGK	Herkla (Hergla)	AKVY
Hanko	ABYS	Havre	AFZH	Herm	AGDJ
Hankow	AOHX	Harve Bouche	BCPK	Hermit Is.	AQUW
Hano I	ACFU	Havre de Grace	AZDX	Hermopolis	AJQY
Hanobugten	ACbP	Havre du Robert	AYEV	Hermoso, Mt	AVMX
Hanoi	ANWV	Hawaii	ARBZ	Hernöeand	ACMF
Hanois	AGDE	Hawaiian Is	AREY	Heron I	BDJI
Hanover	ADVC	Hawke, C	ASNC	Heron Neck (Green I).	BARK
Hanstholm	ADQV	Hawkes B	ATLI	Herradura, P (Cent Am).	AUSB
Hanswest (Hansweerd)	AEGD	Hawkesbury, P	BCPT	Herradura, P (Chile).	AVEH
Haonbaikan I.	ANVM	Hawkesbury R	ASLW	Hervey B	ASTP
Haparanda	ACNV	Hawksbury	ATIP	Herzegovina	AJGF
Hapsal	ACTP	Haya saki Ch	AOXN	Hesselö I	ADKS
Haradsskar	ACLJ	Hayatomo Str	AOVX	Heste Hd	ADNC
Harbour Grace	BEVL	Hayle	ABPF	Hestskjær.	ABMZ
Harbour I	AXJO	Hayti	AYSK	Heugh	AFLV
Harburg	ADVB	Heard I	AMJS	Hève, C la	AFZG
Harcourt B	AQZN	Heath Pt.	BDLK	Hever R	ADSY
Hardanger Fiord	ABRZ	Héaux de Bréhat	AGHM	Hewittville	ASVP
Harderwijk	AEBL	Hebert, P	BCIS	Hexham	ASMR
Hardwick	AYMB	Hebrides	AFCQ	Heyst	AEHJ
Hardwicke B	ARTG	Heceta Hd	AUDY	Hibi Ch.	ACWM
Hardy, P	ATED	Hector	ATFG	Hiddensee I	ADBQ
Hare B	BEXN	Heda Hr	APGR	Hiechechin B.	ANZL
Hare I.	ANEB	Hedjaz Province.	AMQY	Hielm	ADKN
Harfleur	AFZJ	Heiligenhafen.	ADBM	Hierapetra	AJPT
Harg	ACKQ	Heisternest	ACXY	Hierro	ALFJ
Haringháta R	ANJU	Hekkingen	ABHB	Highlands	BAPW
Harlem	AZOR	Hela Pt	ACXW	Higuera, C	AGUH
Harlem Ship Canal	AZPS	Helder	ADZK	Higuerita Bar	AHDQ
Harlingen	ADZC	Helensburg	ARZF	Hiji	AOZD
Haro, C	AUMB	Helensville	ATBH	Hijo	AQKR
Haro Str	ATWZ	Helgeraaen	ABXD	Hiku sima	AOVQ
Harrington (England)	AEWM	Helgoland	ADVJ	Hillsborough.	BAZR
Harrington Inlet	ASNE	Helle.	ABWR	Hillsborough B	BCZX
Harrington (Me)	BAUH	Hellehavn Pt	ADNL	Hillsborough I (Baily Is).	AQVS
Harstad	ABHU	Helleholm.	ADMB	Hillswick	AFGV
Hartland Pt.	AEPK	Helles, C	AJWT	Hilo	ARFB
Hartlepool	AFLU	Hellevik	AROC	Hilton Head I	AYNR
Harvey Corner	BAZF	Hellevoetsluis	AEDO	Hime sima	AOZG
Harwich.	AFOS	Hell Gate	AZFR	Hinchinbrook	ASWZ
Hasenore Hd.	ADKL	Helliar Holm.	AFBP	Hindo	ABQF
Hasle	ACFD	Hellig Vær	ABFL	Hingham	BAHM
Hassel	ABHR	Hellsö	ABRF	Hinlopen Str.	ABUB
Haste Aftera Sh.	ABES	Hellman	ACQD	Hino misaki	APEK
Hastellic Pt.	AGNQ	Hellville	AMGR	Hiogo	APBS
Hastings (Sussex).	AELC	Helsingborg	ACDV	Hirado	AOWR
Hastings.	ASGB	Helsingfors.	ACQX	Hirado no Seto	AOWQ
Hastings Hr	ANPC	Helsingkallan	ACOX	Hirosima	AOZU
Hastings, P	RCPQ	Helsingör.	ADOY	Hirota B	APJS
Hastings R	ASNO	Helwick Lt V	AERS	Hirsova, or, Hirasova.	AKEV
Hasvig.	ABGU	Hempstead.	AZSC	Hirtshals.	ADQR
Haszard Pt.	BDAF	Hen Pt	AOGZ	Hirtsholm.	ADQK
Hatanga R	APQG	Hen & Chickens Reef Lt V	AZXT	Hisaka sima	AOWX
Hatholm.	ABVE	Hendrick Hd	BAMX	Hitokatpu B	APMC
Hatien.	ANUY	Henlopen, C	AZHB	Hittero	ABUQ
Hatinh	ANWP	Hennings Vær.	ABIQ	Hiva Oa	AROE
Hatteras, C	AYTC	Henry, C	AYVB	Hjelm	ADKN
Hatteras Inlet.	AYSZ	Henzada	ANMI	Hobart	ASGP
Hattes, Les	AWGL	Heongsan, P	AOCB	Hoboken	AZOQ
Hatzfeldt Hr	AQTU	Hepburn	AFLP	Hoborg	ACHF
Haugesund.	ABSZ			Hobro.	ADPN

HOBSON—INDIO

Place	Signal	Place	Signal	Place	Signal
Hobson B	ARYT	Horsens	ADJP	Hustad	ABNF
Hodeida	AMBV	Horst	ACYP	Husum	ADSX
Hoedekenskerke	AEGB	Horta	ALCS	Hutt, P	ATOV
Hoek.	AEBF	Horten	ABXW	Hven I	ACDY
Hog I Lt V.	AZWP	Horton	BCBU	Hvidingsö	ABTR
Hoganas	ACDR	Horup Haff	ADGB	Hvidsteen	ABYM
Hogan Group	ASCW	Hougue, La	AGCB	Hwangchau	AOHW
Hogland	ACSU	Hourtin	AGTF	Hwanghai	AOIJ
Hogsholm	ABOY	Houston	AWVF	Hyannis	BADX
Hogsten, or, Hogstenen	ABOK	Houtman Rocks.	ARMY	Hydra	AJNU
Hohe Weg Flat	ADWI	Hov Huk	ADJR	Hyeres.	AHTV
Hoien	ADQO	Hov Sand	ADIR	Hylleholt	ADNT
Hoievarde	ABTN	Howe, C	ASCJ	Hythe	ASFY
Hoihau B	ANXP	Howe Ch	ASQF		
Hokianga R	ATBD	Howrah	ANJL	Iana R	APOZ
Hokitika	ATFW	Howth Bailey	AFUX	Iba	AQED
Hokushu	APKJ	Hoy	AFDS	Ibn Hâni Pt	AKRT
Holar	BFAQ	Hoya B	AQFW	Ibo	AMCR
Holbek	ADLG	Hoyer	ADSL	Ibrahim, P	AMQP
Holdfast B	ARVE	Hoylake	AEUK	Ibraila	AKES
Hollænder Deep	ADOT	Huacho	AUYL	Ica	AUZM
Hollam's Bird I	ALUM	Huafo I	AVIM	Icacos Pt	AYHI
Holland (Mich)	BCKJ	Huaheine I	ARHG	Icacos Road	AWQX
Hollen	ABVM	Huaina Pisagua B	AVCH	Iceland	BFAC
Hollesley B.	AFOR	Huanchaco	AUXM	Ichabo	ALUP
Holmedal	ABSN	Huanillos	AVCB	Ichang fu	AOIF
Holmengraa	ABRD	Huarmey	AUYH	Ida	ALFB
Holmestrand	ABYO	Hnasco, P.	AVDW	Idb.	ACID
Holmogadd	AOMT	Hubbard Cove	BCKU	Ieraka P	AJNG
Holmön	ACMZ	Hudiksvall	ACUT	Iero P	AJUB
Holmudden	ACHF	Hudson B	BEYL	Ierös, C	AKMP
Hol Ness	ADPS	Hudson City	AZON	Iggesund	ACIU
Holsteinborg	BEZW	Hudson R.	AZOW	Iglesias	AIER
Holtenau	ADET	Hué	ANWM	Ikata, P	APCI
Holy I	AFKU	Huelva	AHDU	Iki sima	AOWG
Holyhead	AETL	Hueneme Pt	AUIT	Ikopa R	AMFY
Homborgsund	ABWC	Huertas, C	AHJR	Ilay	AVBN
Hommelviken	ABMH	Huevos I.	AVEP	Ilette Pt	AHVB
Honawar	ANDX	Hufvudskar	ACJD	Ilfracombe (England)	AEPT
Honda	AWMF	Hôgh Pt	ANJF	Ilfracombe (Tasmania)	ASDY
Hon Dau I	ANAB	Hôgh R.	ANIR	Ilha Grande	AVTG
Hondeklip B	ALVL	Huilingsan Hr	ANXZ	Ilin I	AQCU
Honduras	AWFE	Hukau	AOHM	Illinois.	BEJL
Honfleur	AGBB	Hull (England)	AFMR	Ilocos	AQER
Honga R	AZFK	Hull (Mass)	BAHJ	Iloilo.	AQID
Honghai B	ANZJ	Humacao	AXWR	Imabari	AFEB
Hongkong	ANYM	Humansdorp	ALXU	Imadzumi gawa	AFJU
Hon Kobe, P.	ANWC	Humber R	AFMN	Imari	AOWP
Honmoku Lt V	APHS	Humboldt	AUFG	Imbetiba	AVUD
Honolulu	ARFX	Humt Adjim.	AKVH	Imbituba	AVRS
Honshu	AOTB	Hunchun	AOPG	Imbros	AJUP
Hood Canal	ATZU	Hungary	AIYL	Imperatore Pt	AIHC
Hood, P	BCVK	Hunghwa Ch	AOBN	Imperial Hr	AORB
Hoogly	ANIR	Hunstanton	AFNM	Imperial R.	AVHP
Hook of Holland	AECV	Hunter Cove	AUER	Imuruk Lake	ATPY
Hook of Schouwen	AEFD	Hunter I	ASBZ	Inaccessible I	ALVH
Hook Pt	AFWG	Hunter Pt Dock (Cal)	AUGK	Inada	AKDF
Hooper Str	AZFN	Hunter R	ASMN	Inch Keith	AFJM
Hoorn	ADZV	Hunting I	AYOG	In chon	AOLN
Hopedale	BEYG	Huntington	AZRT	In Dail Loch	AIZW
Hope I	ANHC	Huntly	ATBW	Independencia B	AIZS
Hope Pt	ATPO	Huon R	ASOF	Inderamayu Pt	APVK
Hope, P.	BDXQ	Hurlock B	APTL	India.	AMWL
Hora	AJXN	Huron I	BEOW	Indian I	BAOZ
Hormigas, Las	AHJG	Huron Lake	BEFQ	Indian R.	AXIE
Horn, C	AVJY	Huron, P	BEFR	Indian Tickle	BEXY
Horn I	AXCR	Hurst Castle	AEMG	Indiana	BEJU
Hornbæk	ADPE	Hurup	ADFR	Indianola	AWUN
Horn Reefs Lt V	ADRS	Husevegg	ABUH	Indigirka R	APOX
Horsburgh Lt.	ANSZ	Hussein Pt	AKPE	Indio Pt Lt V	AVNO

INDRAGIRIE—KAILUA

Indragirie R . . .	APRT
Indus R . . .	AMWR
Ineboli, C . .	AKMX
Infreschi, P . .	AIJL
Ingonish . . .	BCUG
Ingornachoix B .	BERW
Inhamboio . .	AMBK
Inisheer . . .	APRH
Inishgort . .	AFRQ
Inishowen Hd. . .	AFTD
Inishtrahull . .	AFSZ
Injeh, C . .	AKMW
Injir B . .	AKCP
Inkerman . . .	AKHL
Inland Sea . .	AOZF
Innambán R . .	ALZW
Inner Road . .	AHTM
Inokushi, P . .	AOYP
Inskip Pt . .	ASRQ
Intermediate Lt V .	ANIT
Inuboye saki . .	APIU
Inverary . .	AEZH
Invercargill .	ATGJ
Invergordon . .	AFHV
Inverness . .	AFHZ
Investigator Str .	ARTO
Iomfruland . .	ABWS
Iona . .	AFBT
Ionian Is .	AJHG
Ipsera Islet . . .	AKPO
Ipswich (England) .	AFOV
Ipswich (Mass) .	BAJO
Ipswich (Queensland) .	ASQR
Iquique . .	AVCP
Ireland . .	AFPO
Ireland I . .	AYIF
Iris R . .	AJMU
Iro o saki . .	AFGV
Iroquois Pt . .	BELW
Irrawaddy R . .	ANLX
Irvine . .	AEYJ
Isaac Hr . .	BCNX
Isabela, P . .	AQLK
Isaccea . .	AKEQ
Isaki . .	AOZJ
Ischia . .	AIGZ
Ise Fiord . .	ADLB
Ise Sea . .	APFI
Ishikari R . .	APMQ
Ishima . .	APDC
Ishi no maki . .	AFJE
Isidoro B . .	AJLI
Isigny . .	AGBW
Iskanderún . . .	AKRO
Iskanderún G . .	AKRN
Islands, B of .	ATNM
Islay (Peru) .	AVBN
Islay (Scotland) .	AEZR
Islay Sound . .	AFBE
Isleta Pt . .	ALET
Ismailia . .	AKTM
Ismid . .	AJYL
Isola . .	AIVY
Isonzo R . .	AIVG
Ispe Pt . .	ACGT
Israelite B . .	ARQC
Isse Hd . .	ADJU
Istapa . . .	AUQB
Istria . .	AIVU
Istria, C d' . .	AIVX

Isumi Str . . .	APCB
Itabapuana . . .	AVUN
Italy . .	AHVW
Itamaraca I . .	AVZU
Itapacaroya B . .	AVSH
Itaparica I . .	AVXW
Itapemerim . .	AVUP
Itapérina B . .	AMEY
Itapuan Pt . .	AVYF
Ithaca I . .	AJKM
Ivi, C . .	AKYU
Iviza . .	AHNS
Ivory Coast . .	ALMQ
Iwaga saki . .	APIK
Iwanai B . .	APMV
Iwo sima . .	AOXH
Iyo nada . .	AOFQ
Izabel . .	AWQJ
Jablanaz, P . .	AJBR
Jackson . .	AWZU
Jackson, F . .	AYNC
Jackson, P . .	ASJW
Jacksonville . .	AYJK
Jacmel . .	AXUZ
Jacobstadt . .	ACOU
Jade R . .	ADWK
Jæderens Pt . .	ABUE
Jáfarabad . .	AMXU
Jaffa, C . .	ARWF
Jafnapatam . .	ANFK
Jaguarybe R . .	AWCL
Jaltepeque Lagoon .	AUQP
Jaluit . .	AQVY
Jamaica . .	AXQZ
Jambie B . .	AFRU
Jambu Ayer . .	APQS
James B . .	BEYR
James P . .	AZRN
James R . .	AYWL
James Town . .	ALVD
Jamestown Hr . .	AQVD
Jammer B . .	ADQS
Jammerland B . .	ADLU
Jandia Pt . .	ALEQ
Janet, C . .	AHEU
Janjira Hr . .	AMZS
Jan Mayen I . .	BFAU
Japan . .	AOSZ
Japara Islet . .	APVW
Jar Fiord . .	ABFV
Jardeheu Pt . .	AGCS
Jarman I . .	ARLT
Jarrow . .	AFLN
Jáshak, C . .	AMVT
Jasmund . .	ADBL
Jassy . .	AKFI
Java . .	APUQ
Javea . .	AHKB
Jeba R . .	ALER
Jebeil . .	AKSE
Jebel Teir . .	AMRT
Jebel Zukur . .	AMRY
Jeddah . .	AMRJ
Jeddore Rock . .	BCMX
Jekoits Hr . .	AQVD
Jelai R . .	AQBV
Jella Koffi . .	ALNF
Jerba, C el . .	AKVT
Jerba I . .	AKVF

Jeremias Anchorage .	ALBK
Jericho . . .	AKSV
Jersey . .	AGDN
Jersey City . .	AZOK
Jershoft . .	ACYJ
Jerusalem . .	AKSX
Jervis B . .	ASJE
Jervis, C . .	ARVK
Jezirat Nábiyu Tanb .	AMVL
Jezirat at Tawfla . .	AMVO
Jezireh . .	AKSJ
Jibuti . .	AMOY
Jidda . .	AMRJ
Jidjelli . .	AKXW
Jinchuen . .	AOLN
Jobabo . .	AXQB
Jobos, P . .	AXYC
Johanna I . .	AMDE
John R . .	BCXD
Johnstone R . .	ASXO
Johore R . .	ANSY
Joka sima . .	APHF
Joliette Basin . .	AHSD
Jolo . .	AQLI
Jondal . .	ABSP
Jones Pt . .	AZDC
Jonquière, C . .	AORM
Jordan R . .	BCIG
Jouan G . .	AHUZ
Jourimain I . .	BCXV
Juan de Fuca Str .	ATXK
Juan de Nova I . .	AMHC
Juan Fernandez I .	AVFM
Juana . .	APVY
Juba . .	AMNW
Jubal, Str . .	AMQJ
Jubia B . .	AGXN
Juby, C . .	ALCJ
Jucar R . .	AHXJ
Jucaro . .	AXPY
Judith Pt . .	AZUW
Jugru R . .	ANRV
Juist I . .	ADXT
Juja Pt . .	AJDZ
Júldia Hill . .	ANKL
Juneau . .	ATQK
Juniskáren . .	ACMD
Junin . .	AVCJ
Junkseylon . .	ANPM
Jupiter Inlet . .	AXIB
Jura Sound . .	AEZV
Jutland . .	ADJM
Kabelvaag . .	ABIL
Kabinda . .	ALRW
Kabes . .	AKVI
Kada B . .	APEX
Kadiak I . .	ATRW
Kadi Kioi . .	AKBD
Kadosh Pt . .	AKJZ
Kaffa . .	AKHW
Kaffir Kuyl B . .	ALXH
Kaffraria . .	ALVQ
Káge . .	ACNL
Kageno sima . .	AOXF
Kagoshima . .	AOYC
Kahului Hr . .	ARFK
Kaibobo Road . .	AQRB
Kaikoura . .	ATJQ
Kailua . . .	ARFG

KAIPARA→KEWAUNEE

Place	Signal	Place	Signal	Place	Signal
Kaipara Hr	ATBE	Kangaroo I	ABTU	Kana R.	AQSV
Kaisariyeh	AKSR	Kangio Bank	ANVR	Kauai	ARGB
Kaiserfahrt	ACZF	Kangún	AMVD	Kaukhali	ANIZ
Kaiser Hr	ADWE	Kankao R	ANUZ	Kaulung	ANYQ
Kaiser Wilhelm Canal	ADUF	Kannanur	ANCI	Kaunakakai	ARFT
Kaiser Wilhelmsland	AQTN	Kanonsaki	APHI	Kaupanger	ABQV
Kaitan	ATHY	Kansö	ACBP	Kavak Burnu	AKBF
Kakesuka minato	APGF	Kapau	AQSE	Kavala	AJWH
Kalagaan R	APZN	Kapiti I	ATDM	Kavárna	AKDX
Kalaioki	ACOQ	Kappel Pt.	ACOY	Kawaihae	ARFE
Kalama	AUDK	Kappeln	ADFC	Kawa saki	APHX
Kalamaki	AJOQ	Kapsali B	AJNB	Kawatsi	APBU
Kalamata	AJMI	Kapury	AFNG	Kawau	ATNC
Kalamuti Hr	AJWI	Kara, C	AJVX	Kawhia	ATCH
Kalantan	ANTU	Kara Sea	ABDO	Kawur	APTF
Kalazak	AJZK	Kara Str	ABDS	Kayeli B	AQRF
Kalbáden Lt V	ACRE	Kara-Agatch	AKNB	Kazakavitch	AOQE
Kalgalaksha	ABFD	Karabane I	ALIB	Kealakekua	ARFI
Kaliakra, C	AKDY	Karabuga	AJZU	Keats, P.	ARJX
Kahbia	AKWB	Kara Burnu (Black Sea)	AKDC	Kebula B.	AQOM
Kalimno	AJSQ	Kara Burnu (Levant)	AKQD	Kedah R	ANQI
Kahngapatam	ANHK	Karachi	AMWP	Ké Dulan I	AQFM
Kaliu	AQUO	Karadash Burnu	AKQR	Keeling Is	APTX
Kaliwan R	AOCT	Karaga Hr	APNS	Kefken, C	AKNH
Kalix	ACNZ	Karaghatch Hr.	AKPU	Kega Pt	ANVX
Kalk Ground Lt V	ADFF	Karaginski I.	APNU	Kei R	ALYN
Kalkgrundet Lt V	ACEK	Kara-Herman	AKEI	Keiskamma R	ALYI
Kallo I	ACPR	Karakaiki B.	AQOI	Keko Ness	ADWZ
Kalloni, P.	AJTW	Karamania	AKPS	Kelumpang B	AQBH
Kallundborg	ADLR	Karamca R	ATFC	Kelung Hr	AOBT
Kalmar	ACGJ	Karamusal	AJYN	Kem	ABFC
Kalmar Sound	ACGF	Karang Mas	APWS	Kema	AQMD
Kalo B	ADKE	Karatsu	AOWI	Kemi	ACOE
Kalolimno (Archipelago)	AJSP	Kárikal	ANFX	Kendari B	AQMJ
Kalolimno (Mammara)	AJYV	Kariya	AOWL	Kenmare	AFQH
Kalomo	AJIX	Karló	ACOL	Kennebec R.	BAML
Kalpi Anchorage	ANJC	Karlshamm	ACFV	Kennebunk	BALE
Kaltura	ANEF	Karlskrona	ACFW	Kennedy, P (Queensland)	ASZO
Kalveboderne Lt V	ADOS	Karnafuli R.	ANEI	Kennedy, P (W Aust)	AROI
Kamaishi	APJW	Kar Nicobar	ANOB	Kennedy R	ASYR
Kamao Pt.	ANVD	Karquines	AUGZ	Kenosha	BEIZ
Kamaran Hr	AMRQ	Karrebæk	ADMI	Kent Group	ASCX
Kamata saki	APDB	Kartal	AJYE	Kent I	AZEV
Kamchatka	AKDS	Karumbhar I	AMXE	Kentish Knock Lt V	AVFK
Kamchy R	ALQJ	Karun R	AMUO	Keonga B	AMJX
Kamerun R	AJZW	Karwar Hd	ANBT	Kephalo, C	AJUR
Kamir	AKIF	Kasado sima	AOZP	Kephez Pt	AJWY
Kamish	AOTW	Kasamauze R	ALHY	Keppel C	ASUM
Kamo	APMU	Kashi no saki	APER	Kerama Group	AOYK
Kamoi saki	ANOD	Kaskö	ACPJ	Keramot Saki	APLI
Kamorta	BDFX	Kasom R	ANPW	Kerassond (Kerasunda)	AKMI
Kamouraska	BDFX	Kassandra G	AJVZ	Kerdonis Pt	AGNS
Kampen (Netherland)	AEBN	Kassar	ACTV	Kerempeh, C	AKMY
Kampen (Schleswig)	ADSO	Kasserodde	ADGW	Kerguelen I	AMJR
Kampong Ton R	ANUW	Kastelorizo I	AKPX	Kerhuon	AGKO
Kamrang B	ANWB	Kastri Pt	AJHL	Kerisoc Pt	AGJD
Kanagawa	APHW	Kastro (Khios)	AJTO	Kerkenah Is	AKVO
Kanahena Pt.	ARFN	Kastro (Lemnos)	AJUI	Kermadec Is	ARBX
Kanais G	AKUF	Kastrup	ADOI	Kernic	AGIV
Kanala B	ARBD	Katakami	APBJ	Kéroman	AGNE
Kanazawa	AOUI	Katakolo	AJLP	Kerpenhir Pt	AGNZ
Kanbun	ANUF	Katang Islet	APVT	Kertch	AKII
Kandalak G	ABFG	Katau R	AQSW	Kerteminde	ADIY
Kandalaksha	ABFI	Kata ura	AOXZ	Kervasara	AJIF
Kandavu	ARCB	Katavita B	AQZC	Kerynia	AKQU
Kandeluisa	AJSL	Katland	ABUY	Kesonnuma	APJR
Kandi	ANEO	Katorei B	APTM	Ketchumville	ATQF
Kandilli	AKCL	Katsépé	AMOD	Keti	AMWT
Kaneda B	APHJ	Kattegat	ACBJ	Keum gang	AOLX
Kangan	AMVD	Kattoji saki	APKR	Kewaunee	BEIH

KEWEENAW—KRA

Place	Signal	Place	Signal	Place	Signal
Keweenaw B	BEOQ	Kingston (Jamaica)	AXSR	Koidzumi	APJQ
Key West	AXGU	Kingston (Ontario)	BDWM	Koko Reef	APWK
Keyham	AEOC	Kingston (S Australia)	ARWD	Kokotoni Hr	AMLF
Kezil Pt	AKOS	Kingstown (Ireland)	AFVG	Koksi, P	AOCH
Khaidari P	AJNP	Kingstown (St Vincent)	AYFS	Kokskär	ACSZ
Khambhalia	AMXF	Kinhon Hr (Thinai Hr)	ANWI	Kokuntau Is	AOMC
Khamír	AMVK	Kinkuwasan	APJG	Kokura	AOVZ
Khania	AJPE	Kinn	ABPS	Kolaba Pt	AMZO
Khanlijeh	AKCO	Kinnard Hd	AFIG	Kolaba Oby	AMZP
Kharag (Kharij) I.	AMUY	Kino	ACUH	Kolachel	ANCY
Khelidromi I	AJVM	Kinsale	AFWY	Kolding	ADGT
Khelong B	ANPR	Kinsale, Old Hd of	AFXB	Kollefjord	ABGE
Kherson	AKGZ	Kinservik	ABSJ	Kolmakoff F	ATRG
Khersonese, C	AKHM	Kiöge	ADNY	Kolokithia (Archipelago)	AJTQ
Khios	AJTM	Kios	AJYS	Kolokithia (Greece)	AJMP
Khor Fakán	AMTH	Kipini	AMNL	Kolorat Hr	AIZL
Khori R.	AMWY	Kirchdorf	ADCL	Kolyma, R	APOW
Khor Rabij	AMVX	Kirkcaldy	AFJO	Komandorski Is	APNM
Khorya Morya Is	AMSQ	Kirkcudbright	AEXN	Komatsushima Anch...	APDH
Khramtshenko	ATRI	Kirkeby	ADSG	Komo R	ALRE
Khristinestad	ACPL	Kirkwall	AFEQ	Konakri	ALJM
Khunbandar	AMYF	Kirpen I	AKNJ	Kondia P	AJUG
Khwa	ANLQ	Kisaradzu	APIL	Konge Deep	ADOR
Kiahtsz	ANZM	Kish Lt V	AFLZ	Kongo Free State.	ALSD
Kiama	ASJK	Kishin	AMSK	Kongo R	ALRY
Kiangyin	AOFS	Kishm.	AMVO	Kongoni	AMBP
Kuchau	AOHV	Kismayu B.	AMNV	Königsberg	ACXD
Kidderpur	ANJO	Kissa	AQOX	König Wilhelm Canal	ACWH
Kidderpur Wet Docks	ANJH	Kistna R.	ANGS	Ko-je Do	AOMW
Kidorong Pt	AFYJ	Kistrand	ABOI	Konstantinovski	AKGV
Kiel	ADEO	Kiswere Hr	AMKH	Koombanah B	AROM
Kiel and Eider Canal	ADTG	Kita kami R.	APJM	Koos R (Cocs)	AUED
Kienchufu	ANXO	Kiti, C	AKRC	Kopah Inlet	ANPI
Kieuhien	AOGS	Kitries, C	AJMK	Kopalin...	ACYB
Ku Ch	APEI	Kitta	ALNQ	Kopparstenarne Lt V	ACHS
Kikik B	ANWJ	Kiukiang	AOHR	Korableny Lt V	ACSI
Kiladia P	AJNQ	Kiungani	AMLC	Kordion B...	AJQT
Kilcradan Hd	AFQW	Kiushan Beacon	AOPP	Korea	AOLF
Kildonan Pt.	AEZM	Kiushu	AOVS	Kornilof G	AONX
Kilia Mouth.	AKEO	Kiutoan Lt V	AOFG	Koroni	AJMF
Kilia Pt	AKNL	Kiyerr Inlet	ASIB	Kororarika B	ATNS
Kilifi R	AMNH	Kizil Irmak	AKMS	Korsakovsk.	AORW
Kilimán R	AMBY	Kizu gawa	APBW	Korshagen	ADLJ
Kilindini, P.	AMNB	Kjeldsnor.	ADIM	Korshavn.	ADJB
Kiliomeh, C.	AJVB	Kjengskjær	ABLT	Korsö.	ACQI
Killala	AFSD	Klagstorp.	ACEO	Korsör	ADLY
Killary, B	AFRO	Kladesholm	ACBD	Korsor Road Lt V	ADJX
Killybegs	AFSL	Klang	ANRQ	Kos	AKPJ
Kiloarda, C	AKQF	Klazati	AJZK	Kosedo Str.	AOVN
Kilwa Kisiwani	AMEI	Kleines Haff.	ACYW	Koseir	AMQE
Kilwa Kivinje.	AMKO	Kleven	ABVH	Koshiki Is	AOXW
Kimánis B.	APYS	Klinkowstroem B	APOF	Koshiki-sima.	AOWU
Kimberley (N Australia)	ARIF	Klinte	ADJL	Kosi R	ALZI
Kimberley (W Australia)	ARLH	Klintehamn	ACHI	Koslu B	AKND
Kina Balu	AFZC	Khitmöller	ADQW	Kosounai	AORP
Kinabatangan R.	APZS	Klondike R	AFQO	Koster I...	ABZL
Kinashiri	AFLT	Klong Bagatae.	ANPL	Kota Baharu.	AQBJ
Kinburn	AKGI	Klutz Hd.	ADCN	Kota-batú.	AQKY
Kincardine	BEHG	Knivaia.	AHFH	Kota Raja	APQM
Kinchau	AOKI	Knivaniemi	ACOI	Kotawa B.	APHE
Kingani R	AMXY	Knocke.	AEGP	Kotawaringin R	AQBU
Kingchu fu	AOJY	Knubble, The	BAKY	Kotelnoi I	APQC
King George Sound	ARPJ	Knuds Hd	ADIU	Kotka I	ACRJ
King I	ASCP	Knysna Hr	ALXP	Kotonu	ALNX
King Pt	ARPN	Kobberdal	ABXH	Kottapatam	ANGP
King Road.	AEQJ	Kobber Grund Lt V	ADQG	Kotzebue Sound	ATFQ
Kingscote	ARIY	Kobé	APBR	Kourland	ACUV
Kings Lynn	AFNK	Kochi.	APCX	Kowie R	ALYE
King Sound	ARLF	Koepang	AQOP	Kowno	ACVI
Kingsport	BCEO	Koh Chang	ANUQ	Kra	ANPE

KRAGCHENBURG—LEITH

Kragchenburg . . . . . AEBT	Kwala Kangsa . . . . .ANRJ	Langeoog I . . . ADXJ
Kragerö . . . . .ABWQ	Kwala Lumpor . . . .ANRT	Langesund . . . . .ABWU
Kran Fiord . . . . . . .ABWO	Kwamera . . . . .AQYC	Langkat R. . . APQW
Krasnaia Gorka . . . ACSM	Kwangchauwan . . . . ANXV	Langkawi I . . .ANQE
Kraut Sand Lt V . . . . .ADUK	Kwikambo . . . . . ALTP	Langkiangki . . . . .AQGZ
Kraxtepellen . . . . ACWU	Kwanglotau I. . . . AOKS	Langksa R. . . .APQU
Kresta G. . . . . . . APOH	Kyauchau Hr . . . . . AOIK	Lang Ness . . AEXD
Krimea . . . . .AKHE	Kyauk Pyu Hr . . . ANLG	Langore . . . ADJV
Krimon Java Is. . . . . APVS	Kyholm . . . . . ADJY	Langotangen . . . .ABWV
Krionero Pt . . . . AJKR	Kykduin . . . . . AEBX	Langshan Crossing . . AOFM
Krishna Lt V . . . ANLZ	Kyle Akin . . . . AFCJ	Langston Hi . . . AELP
Kristiansand . . . . ACFS	Kyparissia . . . . AJLW	Langwas I . . . .APSH
Kristiansund . . . . . . .ABMW	Kyz Aul Pt . . . . .AKHY	Lannion . . . .AGHZ
Kroé . . . . APTE		Lantau . . . .ANYL
Krom B . . . . . ALXT	Laaland . . . . .ADMK	Lanvaon Height . . . AGJI
Kronborg . . . . . ADOZ	Labiau . . . . .ACWO	Lanzarote . . . . ALEH
Kronelot . . . . . ACSK	Labó . . . . .ADEN	Laon . . . . . AQJD
Kronstadt . . . . ACRV	Labrador . . . . . . BEXQ	Lapland . . . ABFK
Kru . . . . ALKS	Labúan . . . . . APYO	Lappe Grund Lt \ . .ADFC
Krusenstern B . . . . .ABCT	Labuan Amok . . . .AQNI	Lara Pt . . . . AVOB
Kuahoi R . . . . .ANWR	Labuan Tering B . . AQNP	Larangeiras . . . AVSN
Kua Kam R. . . . .ANWY	Labuk R . . . . APZL	Laraquete . . . AVHG
Kuakue B. . . . . . ARBK	Laccadive Is . . . . .ANDF	Laredo . . . . AGVE
Kuantan R . . . .ANTQ	Lacepede B . . . . .ARWB	Larnaka . . . AKRD
Kuaua B . . . .AQZY	Lachine Canal . . .BDUT	Larne . . . . . . .AFTQ
Kuchinotsu . . . .AOXJ	Lacre Pt. . . . . .AWKR	Lárut R . . . . ANQX
Kudat B . . . . . . .AFYX	Ladrone Is (China) . .ANYE	Lasserre Pt. . . . AVKF
Kuga sima . . . . .AOWX	Ladvik . . . . . ABQM	Lassan . . . . ACZR
Kuilu R . . . . ALEP	Lady Elliot I . . ASTN	Latakiyah. . . . AKRV
Kuji B. . . . . APKD	Lae o ka Laau . . . .ARFS	Latham I. . . . AMKU
Kullen . . . . . . ACDN	Læso . . . . . . ADQB	Latimer Reef. . . . AZUE
Kultuznay B. . . . .APNV	Læsö, Channel Lt V . . ADQF	Latour, P . . . BCHU
Kumai R . . . AQBS	Læsö Rende . . . . ADPZ	Lauersvælgen . . . . .ABZG
Kum Kalessi . . . .AJWV	Lagoa dos Patos . . . .AVRG	Laumubo . . . . .AOJS
Kumassi . . . . ALNI	Lagos (Africa) . . . ALOE	Launceston . . . .ASEJ
Kumpta . . . . ANBV	Lagos (Portugal) . . AHCY	Laurkollen . . . . ABYQ
Kum Pt. . . . AKPL	Lagosta . . . . AJEU	Laürvik. . . . . .ABXI
Kundapur. . . . ANBZ	Lagostini I . . . . AJFK	Lauterbach . . . .ADBH
Kundari I. . . . .AMZQ	Lågskar . . . . . ACQL	Lauwerzee . . . ADYR
Kunfida . . . . .AMRO	Laguan B . . . . AQJS	Lavaca . . . . . AWUQ
Kunsan . . . . .AOMD	Laguerre, P . . . . .AQYV	Lavaissiere B . . . . ARBE
Kuper Hr. . . . . . .AOMQ	Lagunmanoc, P . . . AQGU	Lavandou . . . . AHUF
Kure . . . . .APOS	Laguna B . . . . ASRG	Lavapié Pt. . . .AVHK
Kuri, C . . . . AKDG	Lagundi Str . . . . .APSW	Lavata B. . . . .AVDI
Kuriat Is . . . . .AKVS	Lahaina. . . . . .ARFM	Lavezzi I . . . . . .AIBJ
Kuril Is . . . . . .APLU	Laka P. . . . . AJHS	Lavrova Hr . . . APNY
Kurong . . . . . . AQOT	Lake Borgne . . . .AWZI	Lazaref, P. . . . . AONP
Kuru R . . . . . AWGF	Lake Reeve . . . .ASBU	Lazaret Basin. . . . AHSC
Kuru Obesmeb. . . . . . AKBW	Lakhpat R . . . AMWY	Lazaretto Road . . . . .AHTI
Kusaie I. . . . . .AQUY	Lakon . . . . ANTW	League I . . . AZKD
Kush. . . . . . . ABEV	Lakona B . . . . AQXC	Leander Pt. . . . ARNII
Kushiro . . . . .APLH	Lamb I . . . . BEMI	Leander Tower . . .AKBH
Kusilvak . . . . ATQJ	Lambayeque . . . AUXG	Leasowe . . . . ADUM
Kusimoto . . . . . . .APEO	Lambton Hr. . . ATKX	Leba . . . . ACYF
Kuskino . . . . . ABCI	Lamentin. . . . AYDI	Lebak, P . . . . AQKV
Kuskoquim R. . . . . .ATRE	Lamlash . . . . . .AEZK	Lebbin . . . . ACZE
Kustenjeh (Kustendji) . .AKEC	Lamock Is . . . . .AOBC	Lebesby . . . . ABGF
Kutabdia I . . . . ANKP	Lampedusa . . . . . AIFB	Lebidah . . . . AKVB
Kutali Road. AJ7O	Lampon, P . . . AQFL	Lebu P. . . . . .AVHM
Kutch G. . . . AMWZ	Lamu . . . . . AMNP	Led Sund. . . . ACQM
Kutpur I. . . . . AMXW	Lanai . . . . .ARPO	Leeds . . . . AFMX
Kuwana. . . . . . APFR	Lancaster . . . . .AEVY	Leer . . . . . ADYI
Kuweit Hr. . . . .AMUE	Landerneau . . . AGKP	Leeuwin, C . . . ARPF
Kværnholm . . . . . ABNZ	Land's End . . . AEOR	Leeward Is . . . AXYM
Kvalbein. . . . . . . AMXI	Landskrona . . . . . ACEB	Leghorn . . . AHYC
Kvilio . . . . . . .ABUW	Landsort . . . . ACIU	Légué, Le . . . AGFT
Kvindherred . . . . ABSC	Landunevés Pt . . . .AGJM	Leichau . . . ANXU
Kvitholm . . . . . . ABNE	Langat R . . . .ANRU	Leipzig . . . . . ACWB
Kvitnes . . . . .ABMY	Langeland . . . . ADIH	Leite . . . . . .AQJF
Kwacbau . . . . AOGB	Langenuen ⸍ . . . ABRY	Leith . . . . . .AFKD

LEIXÔES—LOOKOUT

Leixôes P.	AHBF	Lille Blegen	ABSV	Livisa	AXLY
Leka.	ABKP	Lille Feisten	ABUD	Lixuri.	AJKH
Lekanger.	ABUZ	Lillesand	ABVZ	Liyushan.	AOHT
Lekhanger	ABQR	Lillo	AEGQ	Lizard, The	AEOL
Lekki.	ALOU	Lima	AUYV	Lizard I.	ASYP
Lema Is	ANYT	Lima R.	AGZY	Ljugarn	ACHW
Leman & Ower Lt V	AFNV	Limasol	AKQZ	Llanelly	AERW
Lemantin Pt.	AXTG	Limeni P	AJML	Llanes.	AGVW
Leme Ch	AIWT	Limerick	AFRD	Llansá.	AHNL
Lemnos	AJUE	Limhamn.	ACEN	Llico	AHON
Lemvig	ADRK	Limon, P	AWON	Llarga Pt	AVFY
Lena R.	AFQE	Limones R	AXQJ	Llobregat	AHMB
Lenha, P da	ALSF	Limpopo R	ALZS	Lloret.	AHMR
Lensviken.	ABLY	Linano B.	AQRW	Lloyd, P	AQVL
Leonardstown B	AZBG	Linaro, C	AIFW	Loanda I	ALTE
Leones Isle	AVMB	Lincoln, P.	ARSF	Loanda, P of	AITB
Leopoldville	ALSM	Lindaas	ABRH	Loango B	ALRQ
Lepanto.	AJLE	Linde R.	AMBX	Lobito B.	ALTR
Lepreau Pt.	BAYH	Lindesnæs	ABUZ	Lobos de Afuera.	AUXF
Lepsorev.	ABOD	Lindi R.	AMKB	Lobos I (Canary Is)	ALEN
Lequeitio	AGUQ	Lindos	AJRZ	Lobos I (Uruguay).	AVQJ
Lerdalsören	ABQW	Linga.	APRL	Lobos cay	AXMT
Lerhamn	ACDQ	Lingan Hd	BCSH	Loch Fyne.	AEZG
Lerici	AHXR	Lingayen	AQSP	Loch In Dail.	AEZW
Lero (Archipelago).	AJSU	Lingi R.	ANRZ	Lockeville	AROS
Lero (Bergen)	ABRM	Linja.	AMVH	Loc-Maria	AGNM
Lervig.	ABRW	Linnhe Loch	AFBN	Lodbjerg	ADQZ
Lerwick	AFGP	Linosa	AIPF	Löddingen	ABID
Lesina.	AJED	Linschoten Is	AOYG	Lofoten Is	ABIC
Leskar.	ACNP	Lions Rump	ALWK	Löfsta	ACLB
Leton Rk.	ALHP	Lipari I	AIOK	Logan R	ASQX
Letti I	AQOZ	Lipso I.	AJSV	Loggerhead cay	AXGO
Leucate	AHPX	Liptrap C.	ARZY	Logstör	ADRG
Leukungtao	AOIT	Lisbon.	AHCE	Loheiya.	AMRF
Levanger	ABMI	Lisburne, C	ATPM	Loire, R.	AGPF
Levant.	AKPQ	Liscanor.	APRG	Loka I	AOEG
Levant Is.	AHUD	Liscia P.	AICR	Lokiang.	AOCG
Levanto.	AHXK	Liscomb Hr.	BCNI	Lolo I	AQTC
Levanzo	AIMW	Lisle	AFXM	Loma Pt	AUJK
Leven, P.	AMDZ	L'Islet.	BDIK	Lomaloma	ARDM
Levi, C.	AQCL	Lismore I	AFBQ	Lomas (Cape) Roadstead	AVBD
Levis.	BDTG	Lissa..	AJBK	Lombardy	AIUW
Levitha.	AJSR	Lister.	ABUR	Lombok	AQNL
Levuka.	ARCN	Lister Deep	ADSE	Lomma.	ACEH
Lewes.	AZIB	Lith..	AMRL	London	AEHV
Lewis	AFCS	Little Basses.	ANFC	Londonderry.	AFTE
Lewiston	BDYI	Little Belledune Pt.	BDHJ	London Docks	AEJN
Leyden	AECM	Little Belt.	ADGY	London Sh.	ACSL
Leyre R.	AGTK	Little Cayman	AXQW	Long Branch.	AZMR
Liakhov	AFQB	Little Cumberland I	AYKL	Long Hope	AFEU
Liant, C	ANUM	Little Curacao I	AWKX	Long Hr	BITS
Liauho.	AOKB	Little Denier I	BDWQ	Long I (Bahamas)	AXKB
Liautishan Promontory	AOKL	Little Egg Hr	AZML	Long Island (N Y)	AZPL
Liautung Province	AOJX	Little Fish B	ALUB	Long Island Hd	BAKG
Libau.	ACVF	Little Gull I	AZQY	Long Island Sound	AZQK
Libby I.	BAVF	Little Hope Islet	BGIV	Longone, P.	AHYT
Libeccio Pt.	AIMU	Little Karas I	APRN	Long Pt (Erie)	BEAF
Liberia	ALKH	Little Popo	ALNU	Long Pt (Mass)	BAGI
Libertad	AUQL	Little Ross	AEXM	Long Pt (Tasmania)	ASHI
Libreville.	ALRD	Little Sando	ABNC	Long Pilgrim I.	BDPR
Licata.	AILZ	Little Sea Hill	ASUD	Long Sand Lt V (Bengal)	ANIW
Licosa Pt.	AIJG	Liusne.	ACLQ	Long Sand Lt V (England)	AFPE
Lidung B	APRC	Livadhi, P	AJQE	Longships Lt.	AEOQ
Lively P.	BEZQ	Livadia B.	AJSG	Longstone.	AFKV
Liffey R	AFVE	Liverpool (England)	AEUV	Longueuil	BDUK
Lignano P.	AIUR	Liverpool (N B)	BCYQ	Lonsdale P	ARYE
Ligua B.	AVET	Liverpool (N S).	BCJE	Loo Rock.	AIDS
Lium Fiord...	ADPU	Liverpool, P.	AMGY	Loog P	AQHD
Likuri Hr	ARCP	Liverpool R	ARIW	Looké P	AMDX
Lildstrand.	ADQT	Livingstone.	AWQH	Lookout, C	AYRM

LOOP—MALEMBA

Loop Hd	AFQU	Lynn Haven	AYVE	Magdalena B (Mexico)	AUKZ
Lopakta, C	AFNG	Lynn Well Lt V	AFNE	Magdalena B.	ABDL
Lopez, C (Africa)	ALRI	Lynus Pt.	AETV	Magdalena R.	AWLY
Lopez I (Wash)	ATYZ	Lyon	AHRL	Magellan Str	AVJG
Loppen	ABQW	Lyon, G of	AHQI	Magerösund	ABGJ
Loqué	AMDX	Lyse Ground	ADKT	Magnisi	AILH
Lorain	BEDP	Lysekıl I	ACBF	Magnok B	AQGP
Lorne P	BCFA	Lyser Ort	ACUX	Magothy R	AZDR
Lourenço Marquez	ALZO	Lytham	AEVP	Mahábalıpúr Pagodas	ANGH
Lorient	AGND	Lyttelton	ATJG	Mahakan R	AQBE
Lord Howe I	ARBW	Lytton	ASQN	Mahambo Pt	AMBO
Los, Isles de	ALJH			Mahanadı R	ANIB
Los Vilos	AVEO	Maas Lt V	AELP	Mahanoro	AMES
Lota	AVHE	Maas R	AECQ	Mahé	ANCL
Louis, P (France)	AGMZ	Maasluis	AEDF	Mahé I	AMJB
Louis, P (Mauritius)	AMIB	Maatsuyker I	ASFO	Mahébourg	AMIL
Louis, P Lt V	AMIC	Maba Road	AQRJ	Mahedia	AKVR
Louis, P (W I)	AYDG	Mabou	BCVH	Mahı R.	AMYJ
Louisburg Hr.	BCRJ	Macahe	AVLP	Mahon, P	AHPL
Louisiade Archipelago	AQTM	Macao.	ANYG	Mahone	BCKP
Louisiana	AWVX	Macapa	AWEL	Mahukona	ARFD
Louttit B	ARXW	Macarco R	AWIR	Mahuwa	AMXV
Lovisa	ACRI	Macáu	AWCG	Maidens	AFTF
Low Hd.	ASDP	Macclesfield Bank	ANXI	Maikulaung	ANUD
Low I	ASYD	Macdonnell, P	ARWX	Maı Kussa	AQSM
Low Pt	AOHU	Macdonnell Sound	ARUH	Maine	BAKP
Lower California	AUKJ	Macedonia	AJVQ	Maintirano	AMPF
Lower Pt Clear	AXBK	Maceió	AVYX	Maisonneuve	BDUN
Lower Traverse Lt V	BDQS	Machias	BAVO	Martencillo	AVBL
Lowestoft	AFOI	Machiasport	BAVL	Maitland	ASMT
Lowly Pt	ARBJ	Machias Seal I	BAXM	Maitland, P.	BEAC
Loyalty Is.	ARBQ	Machichaco, C	AGUV	Majamba B	AMGB
Luanco	AGWF	Machipongo Inlet.	AZGR	Majar B	AKCV
Luarca	ACWQ	Macın	AKET	Majorca	AHOD
Lubang I	AQDE	Mackau I	AOMJ	Makalla	AMSH
Lubeck (Germany)	ADCR	Mackey	ABWE	Makambı R	AMPT
Lubeck (Me)	BAVX	Mackenzie R	ASUW	Makarska	AJDU
Lucietta I	AJDG	Mackenzie R (Alaska)	ATFG	Makassar	AQMX
Lucifer Lt V	AFVN	MacKinnon Hr	BCTX	Makassar Str.	AQMU
Lucon	AGQD	Mack Reef	AUET	Makıl	AMUX
Lucrecia Pt	AXMP	MacLean	ASOJ	Makısar	AQOX
Ludington	BEKY	MacLeay R	ASNY	Makrán Coast	AMVU
Ludlam Beach	AZLW	Macoripe Pt.	AWCP	Makrı Hr	AKPW
Ludlow, P	ATZS	Macoris	AXUN	Maksa	AMVT
Luikangtao	AOIT	Macquarie Pt	ASKD	Maktan I	AQIO
Luleå	ACNT	Macquarie Hr.	ASPG	Makung Hr	AOVC
Lundo	ABLQ	Macquarie I.	ATOX	Malabar Coast	ANCH
Lundy I	AEPM	Macquarie, Lake	ASMH	Malabar Pt	AMZE
Luneburg	ADVF	Macquarie P	ASNM	Malabata Pt	ALBG
Lunenburg	BCJW	Madagascar	AMDO	Malabrigo Road.	AIXK
Lungmun Hr.	AOIU	Maddalena I	AICT	Malacca	ANSC
Lungo, P (Dalmatia).	AJCO	Maddy, Loch	AFDB	Malacca Str	ANPZ
Lungo P (Istria)	AIYB	Madeira	ALDQ	Malago.	AHGO
Lungö	ACMG	Madeira R	AWFI	Malaita I.	AQWJ
Lupar R.	APYC	Madison, P	AUBN	Malalag P	AQEU
Lusaran Pt	AQIH	Madoc, P	AESY	Malamocco P.	AITO
Lussin Piccolo	AIZX	Madonna I.	AJHT	Malampaya Sound	AQCF
Lutke Hr	APOR	Madonna Pt	AIWC	Malangen	ABHD
Lutostrak I	AJCP	Madraga B	ALBC	Malaspina P	AVLQ
Luz, P de la	ALEU	Madras	ANGJ	Malay Peninsula	ANPH
Luzon I	AQDJ	Madryn, P	AGUD	Malcolm Pt	ARPZ
Lydo	ADHS	Madura I	AVMH	Maldive Is.	ANDK
Lyen	ABMQ	Mæsholm	AFWF	Maldon	AFPM
Lyme Regis	AENC	Maestra Pt	ADFI	Maldonado (Mexico)	AUOF
Lymington	AEMP	Mafia I	AITH	Maldonado (Uruguay)	AVQH
Lynchburg	AWVB	Magadoxa.	AMKS	Mále.	ANDM
Lyngen	ABGX	Magallanes	AMOC	Malea, C	AJND
Lyngor	ABWM	Magdalen, C	AQUI	Maleg Hr	AFCE
Lyngsbek	ADKI	Magdalen Is	BDNA	Malekula	AQXH
Lynn	BAIK		BDFH	Malemba B	ALRV

MALFATANO—MATIFU

Place	Signal	Place	Signal	Place	Signal
Malfatano, P	AIEC	Mannefjord	ABVI	Marsa Scirocco	AIOW
Malheureux C	AMIE	Manning R	ASNF	Marsa Sousah	AXUP
Malimba R.	ALQK	Manningtree	APOW	Marsa Tebruk	AKUI
Malin Hd.	AFSY	Mano R.	ALKX	Marsa Zafran	AKUW
Malindi	AMNJ	Manokin R	AZFW	Marseillan	AHQM
Malinska P.	AIZX	Manora Pt	AMWQ	Marseille	AHRW
Mallagoota Inlet	ASCF	Manta B.	AUVF	Marsha I.	AMBC
Mallawallé I	APZF	Mantanzas	AYIV	Marshall Is	AQVX
Mallicolo	AQXH	Manukau	ATBN	Marshall Pt	BAOE
Malmo	ACEJ	Manzanilla B.	AUNK	Marstal	ADIG
Maloren	ACNX	Manzanilla I	AWNS	Marsten	ABRP
Maloyama	AFMT	Manzanillo (Cuba)	AXQH	Marstrand	ACBI
Malpeque	DDEF	Manzanillo B (Haiti)	AXTY	Martaban G	ANOF
Malsoro B	AQMW	Manzo	AJCQ	Martha, C	ALTY
Malta	AIOQ	Maplin Sands	AEIB	Marthas Vineyard	BACU
Maltezana	AJSI	Maputa R.	ALZF	Martigues	AHRQ
Malwan	ANBG	Maquereau Pt.	BDKG	Martin, C	AHVN
Mamanguape	AWBG	Maracaibo	AWLK	Martin, R.	BDNG
Mamberonmo R	AQTX	Marad	AMZT	Martin Garcia I	AVPK
Mambolo	ALJS	Maranhão	AWDC	Martinique	AYEJ
Mambulao P	AQFR	Marano	AIUS	Martinscica (Martincica) P	AIZH
Mamuju B	AQMZ	Marans	AGQP	Martins Industry Lt V	AYNL
Mamutzu	AMDK	Marawang Road	APSC	Martin Vas I.	AWCE
Man, Isle of	AEWU	Marbella	AHGK	Marudu B	APZB
Manado	AQLZ	Marblehead (Lake Erie)	BEDG	Marugame	APDX
Manacle Rks	ASOK	Marblehead (Mass)	BAIQ	Mary R.	ABRW
Mana Dock	AVQD	Marchesa B.	AFZJ	Maryborough	ASRY
Managua	AURE	Marcus Hook	AZJO	Maryland	AZCQ
Mananjara R	AMET	Mare I	AUGT	Maryland Pt	AZBV
Manáos	AWFC	Marennes	AGRD	Maryport	AEWP
Manapad Pt	ANDV	Margarita I	AWJE	Marysville	AUBF
Manar	ANEF	Margate	AEKF	Masanpo	AONB
Manar, G of	ANDT	Maria, Loc.	AGNJ	Masbate I	AQHJ
Manati R.	AXWH	Maria, P.	AXRO	Máseskar	ABZT
Manawatu R	ATDI	Mariager	ADPM	Mashike	APML
Manchac, Pass.	AWER	Marianas	AQVK	Mashonaland	AMBO
Manchester	AEVS	Marias R	ASNQ	Masinglok P.	AQEH
Manchester Ship Canal	ABUX	Maria Teresa Mole.	AIYS	Masio, P	AXPQ
Manchuria	AOFC	Maria Van Diemen, C	ASZY	Masíra G	AMSV
Manda Roads	AMNQ	Marie Galante	AYEB	Maskat	AMTB
Mandal.	ABVF	Mariel, P.	AXNY	Maskhal I.	ANKQ
Mandalay	ANMP	Marien	ADEP	Maski	APML
Mandalike I	AFVX	Marifjæren	ABRC	Másknuf.	ACIZ
Mandao	AQHL	Marigot	AYBH	Masnou	AHMI
Mandarin	AYJQ	Mariguana I	AXKL	Maspalomas Pt	ALEX
Mandavi R	ANBL	Marin (Spain)	AQZJ	Massabé	ALES
Mandinga Hr	AWNK	Marin (W I)	AYBZ	Massachusetts	AZXQ
Mandinga Pt	AUWG	Marina I	AQXD	Massa Lubrense	AIHU
Mandraki	AJNX	Marinduque I	AQGR	Massangano	ALTJ
Mandri, P	AJOV	Marine City	BEFK	Massaua Hr	AMPT
Mandurah	AROK	Marine Transport Railway	AHRZ	Masuji R.	AFSL
Mandvi	AMXB	Maritime Basin	AHRZ	Masuliptam	ANGV
Manfredonia	AIRC	Maritimo.	AIMT	Mata, P	AXLI
Mangalia	AKEF	Mariupol	AKIU	Matagorda	AWUH
Mangalore	ANCF	Mariveles, P.	AQDW	Matakong I	ALJN
Mangareva Is	ARGT	Marjoribanks Hr	AOLW	Matamoros	AWTI
Mangarin	AQCV	Markeledorf Pt	ADBG	Matane	BDNP
Mangolia	AKEF	Marken I	AEBC	Matanzas	AXNU
Mangrol	AMXQ	Märket Rock	ACQB	Matanzas Caleta (Chile)	AVPT
Manhattan	AZQG	Markhab.	AKRW	Matapan, C	AJMO
Manicouagan	BDNV	Marmara I	AJZC	Matapwa R	AMNG
Manila	AQDT	Marmarice	AKPT	Mataró	AHMK
Maninzari	AMET	Marongo	AMCU	Mataura R	ATGP
Manistee	BELD	Maroni	AMCZ	Matchedash B	BEHS
Manitou R	BDMR	Maroni R	AWGJ	Maternillos Pt.	AXMP
Manitou I.	BEON	Marquesas Is	ARGC	Mathew Town.	AXKZ
Manitoulin I	BEGX	Marquette	BEFA	Mathias Pt	AZBM
Manitowoc	BEIK	Marsa el Kébir	AKZC	Mathurin B	AMIT
Manly Beach	ASLQ	Marsala	AIMO	Mati	AQKN
Manna R	ALKD	Marsa Omrakum	AKUG	Matifu, C	AKYJ

MATILDA—MISCOU

Matilda	BDVF	Megalo Kastron	AJPM	Meuangpran	ANTZ
Matinicus Rock	BAON	Megalo Nisi I	AKDJ	Mevatanana	AMFZ
Matitanana R.	AMEU	Meganisi Ch	AJIQ	Mew I	AFTW
Matochkin Shar (Strait)	ABDQ	Meghna R	ANKD	Mexico	AUKH
Matoya	APFD	Mehediya	ALBP	Mexico, G of	AWRZ
Matsugahama B	AFLX	Meraco sima Is	AOCX	Meyers Lodge	AHWG
Matsumai	APKM	Meinderts Reef	APWS	Mèze	AHQO
Matsu shima	APIY	Mejillones Cove	AVCL	Mezen	ABEG
Matsuyama	APEC	Mejillones del Sur B.	AVCX	Mgau Mwania.	AMKD
Mattapoiset	AZYR	Mekari seto	APBE	Miautau Is	AOIY
Mattinata	AIRD	Mel I.	AVSP	Michael (Pissen)	ACUW
Matupi I	AQUJ	Melabu B	APUK	Michigan (State)	BECK
Matura	ANEV	Melbourne	ARZD	Michigan City	BEJR
Maturin	AWIV	Melbourne P.	ARZH	Michigan I	BENP
Manger beach Lt	BCMB	Mele, C	AHWE	Michigan, Lake	BEHY
Manger cay	AWQT	Meleda I	AJFD	Michipicoten I	BEMA
Maui I	ARFJ	Meli Anchorage	AQXP	Middelburg	AEFV
Mauka Cove	AORS	Melilla	AKZR	Middelfart	ADHG
Maullin	AVHY	Melinka P	AVIO	Middelgrund	ADOL
Maumusson Ch	AGRM	Mellakori R	ALJO	Middelharnis	AEDR
Maurice R	AZLN	Mellum Flat	ADWL	Middle Bank Lt V	ARSI
Mauritius	AMHZ	Melncraggen	ACWG	Middle Hr	ASLO
Maurizio	AHWB	Melo, P	AVLU	Middlesbrough	AFLZ
May, C	AZLQ	Meloria Sh	AHYE	Middleton R	ALON
May I	AFJK	Melstenen	ABKO	Miho B	AOUX
Mayaguez	AXVY	Melville, C	AQBY	Miike	AOXM
Mayo	APRY	Melville, C Lt V	ASYQ	Mijelia	AJVD
Mayo I	ALHJ	Melville I	ARJB	Mikawa	APFZ
Mayor, C.	AGVP	Melville, P	ALZR	Mikindani Hr.	AMKB
Mayorga	AHFP	Memba B	AMCL	Mikomoto	APGW
Mayotta	AMDG	Memel	ACWF	Mikuni	AOUI
Maysi, C.	AXLE	Memel R	ACWL	Milan	AIUX
Mayu R	ANKV	Menadou Passage	BCRP	Milazzo	AINY
Mayumba B	ALRN	Menai Str	AETP	Mileto Pt	AIRT
Mazagon	AMPJ	Mendocino C	AUFO	Milford Haven.	AESF
Mazaroni R	AWID	Menelaus Hr	AKUL	Milk R	AXSF
Mazarron	AHIX	Mengka	AOBX	Millbridge	BATY
Mazatlan	AUMP	Menjawak	APVJ	Millier Pt	AGLC
Mazepa Pt	ALYP	Menschikoff Pt	AORC	Millwall Docks	AEJL
Mazighan	ALBW	Mentawi Is	APTI	Milna	AJDX
Mazimbwa B	AMCS	Mentone (Menton)	AHVO	Milo	AJFY
Mazzara	AIML	Menzaleh, Lake	AKTL	Milonia R	AKZN
Mazzone Pt	AIKN	Meoko Hr	AQUL	Milwaukee	BEIT
Mbau Waters	ARCL	Mercer	ATBU	Minatitlan	AWBR
Mbau B	ARCY	Mercury B	AILY	Min R	AOCY
McArthur Hd.	AFBC	Mergui	ANOX	Minch, The	AFCO
McArthur R	ARIL	Merida	AWSB	Mindanao I	AQKC
McCluer Inlet	AQSB	Merigomish	BCWF	Mindoro	AQCS
Mchinga B	AMKF	Merimbula	ASIF	Mine Hd.	AFWL
Meares, C.	AUDQ	Merka	AMNZ	Mines, Basin of	BCDN
Meat Cove	BCUM	Merlera Pt	AIXQ	Mingan Hr.	BDML
Mebondo	ALQV	Mermerjik B	AJZP	Mingan Pass	AODG
Mecattina C	BEQC	Merminji, C	AKOH	Minikoi I	ANDJ
Mekka	AMRX	Merrill Shell Bank	AXBQ	Minnesota	BEMU
Mechigme B	APON	Mersey	AEUS	Miño R	AGZV
Mecklenburg	ADBY	Mersey Bluff	ASER	Minorca	AHPC
Meda I	AHMZ	Mersin P	AKOW	Minots Ledge	BAGX
Medan	APQX	Mersina	AKQF	Minou Pt	AGKD
Medano Pt	AVNG	Mesa de Roldan	AHIP	Miquelon	BETV
Médée Rock	AGME	Mesco Pt	AHXL	Minquiers	AGDX
Medemblik	ADZS	Me Sima Group	AOYL	Minsener Old Oog	ADXF
Medford	BAHY	Messaragotsem Pt.	ACUR	Minsener Sand Lt V	ADWO
Medina	AMRH	Messemvria	AKDP	Mira	BCRS
Mediterranean	AHFZ	Messier Ch	AVIX	Miramba P.	AMGH
Medni	APNR	Messina	AIKS	Miramichi B	BDGF
Medolino G	AIXN	Messina, Str of	AIKO	Miramichi B Lt V	BDCV
Medomak R	BANG	Mestre	AIVC	Mirik, C	ALFW
Medway	BCJK	Mesurado, C	ALKJ	Mirs B	ANZH
Medway R.	AYLF	Metbie	AFJI	Misamis, P	AQKG
Medway, The.	AEJR	Metis Pt	BDNS	Miscou Hr	BDIE

MISENO—MUGERES

Place	Signal	Place	Signal	Place	Signal
Miseno, C	AIGT	Monach Is	AFDE	Morocco	AKZM
Misina nada	AOZT	Monaco	AHVL	Morondava	AMFL
Mispillion Creek	AZIK	Monastir	AKVU	Moroto Saki	APCZ
Misratah	AKUX	Moncton	BAZU	Morpeth Dock	AEVL
Missessy	AHTP	Mondaca R	AGUS	Morris, P	AYFU
Missipezza Rk	AIQJ	Mondego, C	AEBL	Morrisburg	BDUZ
Mississauga I	BEGL	Mongat Pt	AHMG	Morrison B	ANOZ
Mississippi I	AWZF	Mongunui Hr	ATNW	Morristown	BDVR
Mississippi City	AXBW	Monhegan I	BAOD	Morro (Cano) Pt	AHEB
Missolonghi	AJKY	Monie B	AZFT	Morro Ayuca	AUOT
Misumi Hr	AOXQ	Monkey R	AWQZ	Morro San Paulo	AVXO
Mitaziri	AEQO	Monmouth	AEQO	Morrosquillo G	AWMX
Mitchell R (Carpentaria)	ARIC	Monomoy Pt	BAEJ	Mors	ADRH
Mitchell R (Victoria)	ASBY	Monopoli	AIQO	Mortar I	AIZO
Mitford Hr	APZE	Monroe	BECN	Mortella Pt	AICF
Mitho	ANVE	Monroe, F	AYVH	Morter	AJDB
Mitsuga hama	APED	Monrovia	ALKN	Morup Tånge	ACBY
Mitylene	AJTV	Montagu I	ASIJ	Moruya	ASIR
Mitzura	AOZX	Mont' Alégre	AWET	Morzhovetz I	ABEI
Miwara	AOZY	Montauk Pt	AZQM	Moscenice	AIYG
Miyadsu	AOUR	Monte Christi	AXUB	Moscow	ACSB
Miyako	APKF	Monte Christo (Cent Am)	AUVR	Moskenæs	ABIY
Mizzen Hd	AFPX	Monte Christo (Italy)	AHZC	Mosquito Coast	AWOX
M'Kenzie Pt	BCQX	Monte Grosso	AHYU	Mosquito Inlet	AYIP
Mlui R	AMBZ	Montego	AXRQ	Mosoen	ABKI
Moar R	ANSE	Monteleone	AIJV	Moss	ABYO
Mohile	AXCW	Montelungo	AIQF	Mossamedes	ALUC
Mocha I (Chile)	AVHO	Montenegro	AJGH	Mossel B	ALXM
Mocha (Red Sea)	AMRZ	Montepeñ B	AMCQ	Mostaganem	AKYW
Mochima P	AWJK	Monterey	AUHS	Mostardas	AVRO
Moda Liman	AJKV	Montevideo	AVPY	Moster	ABSQ
Modeste I	AOMI	Montgomery	AXDG	Mota I	AQXB
Modyugski I	ABEX	Montpelier	AHQU	Mothoni	AJMU
Moella I	AVSZ	Montreal	BDUQ	Motrico	AGUF
Möen	ADNK	Montrose	AFIT	Motril	AHGW
Moeraki	ATIR	Monts, Pt de	BDMN	Motueka R	ATEL
Moerdyk	AEDW	Montserrat	AYCX	Mouillé Pt	ALWN
Mogador	ALCB	Montt, P	AVIL	Moule, P	AYDE
Mogdishu	AMOC	Monze, C	AMWK	Moulmein	ANOH
Mogotes Pt	AVNC	Moody, P	ATVJ	Moultrieville	AYPB
Mohaka R	ATLK	Mooen	ABNS	Mount, C	ALLF
Mohawk I	BDZY	Mooloolah R	ASRE	Mount Desert Rk	BART
Mohilla I (Mohéli)	AMDB	Möon	ACUF	Mount Hill	AYHU
Mohoro B	AMKQ	Moon Islet	AMSF	Mounts B	AEON
Mojacar	AHIR	Moonta	ARSZ	Moura B	AOTM
Mojanga	AMFV	Moorowie, P	ARTQ	Mourillon	AHTN
Moji	AOVT	Moose F	BEYZ	Mourilyan Hr	ASXM
Mokambo, P	AMCD	Moose-a-bec Reach	BAUZ	Mouro I	AGVN
Mokau R	ATCJ	Moose Peak (Mistake I)	BAUW	Mouse Lt V	AEID
Mokha	AMRZ	Morant	AXST	Mousset	AGSR
Mokihinui	ATFE	Morant Cays	AXSZ	Mouton P	BCIY
Mokpo	AOMF	Moratabas Entrance	APYB	Moutons, Ile aux	AGMH
Mola	AIQP	Moray Firth	AFHS	Moville	AFTH
Moldava	AKFN	Morayra	AHJZ	Mowila	AMRC
Molde	ABNM	Morbihan	AGOF	Moyne R	ARXM
Molde Fiord	ABNO	Morbylånga	ACGS	Moz, P do	AWEP
Moldo	ABPG	Morea	AJME	Mozambique	AMCE
Molenpolder	AEBL	Morecambe B	AEWR	Msimbati Hr	AMJZ
Molfetta	AIQW	Morehead City	AYRJ	Muara I	APYM
Molino Pt	AHMV	Morellganj	ANJW	Muar R	ANSE
Molle	ACDO	Moreno B	AVDC	Mucaras Reef	AXMB
Mollendo, P	AVBR	Moresby, P	AQTF	Mud Pt	ANJB
Moller B	ABDR	Moresby R	ASXK	Muda R	ANQJ
Moller, P	ATRM	Moreton B	ASPQ	Mudania	AJYT
Mollo	ABZV	Moreton, C	ASPU	Mudros B	AJUF
Molokai	ARFQ	Morgan City	AWXV	Muendu B	AQYW
Molucca Passage	AQRN	Moii	APKX	Mueó P	AQZB
Molyneux B	ATHU	Morlacca Ch	AJBO	Muertos I	AXYE
Mombasa	AMLY	Morlaix	AOIJ	Mugardos	AGXP
Mona	AXVQ	Mornington	ARZJ	Mugeres Hi	AWRK

MUGGIA—NEUFELD

Muggia	AIVP
Mughú	AMVG
Muhammera	AMUP
Muiden	AZBH
Muka Hd	ANQR
Mukden	AOKE
Mula, P	AXYW
Mulata	AXOF
Mulaya R	AKZP
Mulgrave, P. (Alaska)	ATSH
Mulgrave, P (N S)	BCPH
Mulgrave R	ASXU
Mulki Rocks	ANCE
Mull	AFBY
Mull Hd	AFGH
Mulo I	AJDL
Mulpy	ANCD
Mumbles	AERP
Mungopani	AMLQ
Muni	ALQY
Munich	ACVY
Munkholm	ABME
Muntok	APSB
Mura Hr	APEZ
Murano	AITX
Murchison R	ARMP
Murmagao I	ANBF
Muros	AGYS
Murray Hr	BDAO
Murray R	ARVT
Murro di Porco, C	AILN
Muscangue Pequena	AVTX
Muskegon	BEKP
Musquash Hr	BAYK
Musselburg	AFKI
Mutenia	AKFM
Mustang I	AWTV
Mutlah Lt V	ANJS
Mutsure	AOVK
Muttum Pt	ANCZ
Mvita	AMLZ
Mwambi B	AMCO
Mwana Mwana	AMLE
Myanterano R	AMFP
Myhlenfels Pt	AXZI
Mykoni	AJRB
Myriophyto	AJXM
Naaf R	ANKU
Naarden	AEBI
Nab Lt V	AELT
Nabe sima	APDT
Nadjini	AONY
Naga	AQIW
Nagahama	APDH
Nagar	ANFW
Nagara Kalessi	AJXE
Nagara Liman	AJXD
Nagasaki	AOXE
Naga sima	AOXV
Nagoya	APFS
Naguabo P	AXWP
Nahant	BAIN
Naibo Hr	APMD
Nain	BEYH
Nairn	AFIB
Naju Group	AOME
Naka Hr	APIV
Nakai	AOUF
Nakala P	AMCI
Nakashimagawa	APDG
Nakatsu	AOZH
Nakke Hd	ADPF
Nakskov	ADMR
Namaura	AOWV
Nambuckra R	ASOB
Nam Dinh	ANWT
Namkwan Hr	AODQ
Namo Hr	ANYC
Namoa I	ANZY
Namsos	ABLC
Namu Hr	ATUO
Nanaimo Hr	ATWH
Nanao	AOUE
Nandi B	ARCX
Nandi Waters	ARCO
Nangamessi B	AQNZ
Nankauri Hr	ANOC
Nankin	AOGJ
Nanoose Hr	ATWD
Naasemond R	AYWI
Nansha	AOCQ
Nantai I	AODK
Nantes	AGFL
Nanticoke R	AZFQ
Nantucket	BAEY
Nantucket Shoals Lt V	BAJE
Nanuku Passage	ARDI
Naos I	AUTC
Naos, P	ALEK
Napa	AUGX
Napanee R	BDWP
Napier	ATLG
Naples	AJHL
Napoule G	AHUV
Narainganj	ANKF
Narakal	ANCQ
Naranjo P	AXMH
Narbada R	AMYN
Narbonne	AHQB
Narendri	AMGI
Narenta Ch	AJEY
Nares Hr	AQUT
Narestö	ABWI
Nargen I	ACTF
Narragansett B	AZVF
Narraguagus R	BATV
Narrows (N Y)	AZNM
Narrows, The (Bermuda)	AYHW
Narsapur	ANGW
Naruto Passage	APDK
Narva	ACST
Narvick	ABJF
Nash I	BAUT
Nash Pt	AERF
Nasilai	ARCK
Nass B	ATUC
Nassau	AXJM
Nassau R	AYJT
Natal (Africa)	ALYX
Natal (Brazil)	AWBQ
Natal (Sumatra)	APUC
Natalie B	AFOB
Natashquan R	BDMI
Nateva B	ARDG
National Basin	AHRY
Natuna Is	APXV
Naturaliste, C	ARPB
Naufragados, Pt dos	AVRY
Nauplia	AJNX
Nauset Beach	BAFQ
Naussa P	AJRH
Naustdal	ABFY
Navallo, P	AGOC
Navarin	AJLZ
Navarin, C	APOD
Navas, P	AXLQ
Navassa I	AXTD
Navosink	AZMU
Navia	AGWR
Navibandar	AMXO
Navidad B	ALNI
Navirar Pt	AMXD
Nawanagar	AMXH
Naxia	AJRO
Naxos	AJRM
Nazareth B	ALRH
Naze, The	ABUZ
Ndoni	ALOW
Neath	AERL
Nebel	ADSQ
Neddick, C	BAKY
Neeah B	AUCH
Neebish Rapids	BELN
Needham Pt	AYGL
Needles, The	AEMH
Neerstrand	ABTG
Negapatam	ANFS
Negombo	ANEK
Negrais, C	ANLR
Negril Hr	AXRU
Negrito B	AUWX
Negritto Pt	AIZV
Negro C	AKWU
Negro I	BAPC
Negro I (N. S)	BCHX
Negro R	AWFE
Negropont	AJUW
Negros I	AQIJ
Neguac gully	BDHF
Néhoué B	AQZG
Nékété B	ARBG
Nelligan	ASIX
Nellore	ANGO
Nelson Town (N B)	BDHM
Nelson (N Z)	APEJ
Nelson, C	ARXB
Nelson, P	ARKY
Nemfña	AGYI
Nemoro Anchorage	APLQ
Nemours	AKZL
Nena, P	AVHN
Neo Kastro	AJLZ
Nepean Pt	ARZP
Neponset	BAHP
Neptune Battery	ANET
Neptune Is	ARSG
Nerang R	ASPK
Nera Pt	AIXZ
Nerbudda R	AMYN
Nerja	AHGS
Nerva Rock	ACSP
Netherlands	ADYJ
Neuendorf	ACYS
Neuenfelde	ADUS
Neuf P	BDOH
Neufahr	ACXV
Neufahrwassel	ACXQ
Neufeld	ADTM

NEUHARRLINGERSIEL—NORWAY

Neuharrlingersiel.. .... . ADXN
Neuhaus. . ... ADUB
Neuse R. . . . . . . AYRV
Neustadt . . ADCV
Neuwerk. .. ADVH
Neuzen....... . . . AEGF
Neva. . ... .ACSE
Neva Lt V . .. ACSI
Nevez, P. . . . AGIS
Nevis . . . AYCB
Nevis, Loch . .AFCG
Nevlunghavn. ... ABXZ
New Amsterdam . . . AWHM
Newark . ... AZOB
Newarp Lt V. . . . AFOE
New Bedford . . . AZYI
New Berne.. . . . . .AYSB
New Besca . . . AIBC
Newbiggin . . . . . .AFLD
New Brighton. .... AZNV
New Brunswick. . . BAWO
Newburg . . . AFIO
Newburyport . . BAJE
New Calabar R. .... . ALPK
New Caledonia .... AQYJ
New Carlisle . . BDJU
New Castle (Del) .... AZJF
Newcastle (England) . .AFLO
Newcastle (N B) .. BDHI
Newcastle (N S W).. .ASML
Newcastle Hr (Ontario) ...BDXK
Newchwang. .AOKC
Newchwang Lt V. .AOKD
New Dungeness . .AUCD
Newfoundland ... . BEQY
New Glasgow. . BCWO
New Granada. AWLO
New Guinea. . AQRV
New Guinea, British . AQRW
New Hampshire . . BAJU
New Haven (Conn) . AZTJ
Newhaven (England).. AELO
Newhaven (N Z)... ... ATHQ
Newhaven (Scotland) . .AFKH
New Hebrides . . .AQWS
New Jersey . . . . . AZKJ
New Jahore .. . . ANTE
New Langley . . ATVR
New Liverpool. .. BDTJ
New London.. . . AZTS
New Mecklenburg .... .AQUK
New Orleans.... . . AWYH
New Plymouth . . ATCR
New Point Comfort. .... AYXS
New Pomerania . . .AQUD
Newport (Bristol Ch) . . AEQS
Newport (Cal). . AUJE
Newport (R I) . . . AZXE
Newport (Wales). . . . . AESP
Newport News . . AYWR
Newport Rock Lt V . AMQO
New Providence I . . AXJK
New Ross . . . AFWC
New Rotterdam ...... AWHB
Newry. . AFUM
New Siberia Is..... ... .APQB
New Smyrna . . AYIS
New South Wales.. . ASHQ
Newtonquaddy.. ...... BCNF
Newtown ... .. .ALMO
New Westminster. . . ATVN

New Whatcom . ATYH
New York State . . . AZOS
New York City. . . . . . . AZOT
New Zealand.... .... ASZW
Nexo.... .... . ACFK
Neyba B . . .. AXUT
Neyland .. . . . AESG
Nea de Jobourg ..... . AGCT
Ngaloa Hr.. .... . ARCD
Nganking Reach. . AOHB
Ngaukiang. . . AODU
Niagara. . . BDYF
Nicaragua . . . AUQX
Nice . . . . AHVF
Nicholson, P . ... ATKL
Nickerie R. . . . . . . AWHG
Nicobar Is . . ANMY
Nicolas Ch . . AXJI
Nicosia. . . AKQV
Nicoya G. . . ..AURS
Nidden.. . . ACWR
Nidingen. . ... .ACBV
Nieuport . . . . AEHN
Nieuwe Diep . . ADZL
Nieuwe Maas.. . AEDG
Nieuwe Sluis . . AEGL
Nievre, P. . . . AMDT
Nifu . . . ALKT
Niger .... ALOP
Nightingale I. . ALVI
Nigisi . . . APHQ
Niigata Roadstead.. . AOTZ
Nikaria. . .AJTD
Nikolaev.. . AKGF
Nikolaev F . . . AKGT
Nikolaevsk . . AORJ
Nikolaistad... .. . ACPE
Nikolo, P . . . .AJPO
Nile R . . . AKTW
Nimmersatt.. .... ACWE
Nimrod Sound . . AOEC
Nine Degree Ch. . . .ANDO
Ninghai . . AOJU
Ningpo fu. .. . AOEM
Nio. . . . AJHT
Niort. . AGQR
Nipe, P. . . . AXMB
Nipsight B . . BDIW
Niseros........ . AJSM
Nisita. . . AIHK
Nisqually. . . . AUBV
Nitendi Is . . AQWO
Nitheroy. .... AVTU
Niue, P. . . . ARDV
Nivaa . . . ADOV
Nizampatam. . . ANGR
Noarlunga, P . ARVF
Nobby Hd. ... . .ASMJ
Nobska Pt . . BACO
Nodendal...... .. .ACQE
Nohechi . . AOTJ
Noire, Ile... . . AOIE
Noirmoutier I ... . .AGPT
No Kinsei.. . . ARDC
Noli . . . ARWJ
Nolloth, P . . ALVT
Noorder Hoofd. . AECS
No Point, Pt... . .AUBJ
Norddeich . . ADXS
Norden . . . ADYF
Nordenham . . . . ADVT

Norderney I. . ... ADXR
Norderney Lt V ... .ADXW
Norder Piep . . ADTK
Nord Fiord . . ABPH
Nordland . . ABHN
Nordmaling.... .. ACMP
Nordöerne. . . ABST
Nordöerne (Vigten Is) .ABKU
Nordre Röse . . ADOH
Nordstrand I .. .. ADSV
Nore Lt V .... . . AEIH
Norfolk. . . AYVW
Norfolk I. . ... ARBV
Norfolk Navy Yard . AYWC
Norman, C . ... BERK
Norman R Lt V. . ARIE
Normanton . . ARID
Norrbyskar.. . . ACMU
Norrköping.... ... . ACIN
Norrskär . . . ACOZ
Norretelje. . . . . ACIX
Norrsund. .... . ACLO
Norstromsgrund Lt V..ACNR
North America .... . ATOZ
Northampton..... . ARND
North Berwick . . AFKL
North C (C Br I).... . BCUJ
North C (N Z)..... . ATOF
North C (Norway)... ....ABGL
North Carolina.... . AYQI
North Ch . . AORI
North Dock (Liverpool) AEVJ
North-East Crossing Bn Lt V
. . . . AOHJ
North-East End Lt V.. AZHQ
North East R (Md) . AZEG
Northern Two cays. . . .AWQP
Northfleet . . AEIQ
North Foreland . . AEKH
North Friesland . . ADXP
North Goodwin Lt V... AEKL
Northhaven. . BAPI
North Hinder Lt V. . AEGH
North Holland Canal. . ADZN
North Kallan I... . ACPD
North Post Signal Station .AYBE
North Pt . . BDFE
North R (N C). . . . AYUM
North Reef Lt .... ASVH
North Rona. . . AFDQ
North Ronaldsay. . . APGJ
North Ronner. . ADQE
North Saddle I. . . . AOEZ
North Sea . . ADSB
North Sea Ship Canal. . AECI
North Shields . . .AFLJ
North Shoalhaven R. . ASJG
North Sydney... .. BCSQ
North Tree Beacon .. AOIN
North Uist .. . . APCZ
Northumberland, C . ARWU
Northumberland Str . BCWR
North Vorupör. . . ADQX
North Wall (Mersey). . AEUQ
North Watcher Lt. . APSK
North West Lt V . AEUJ
Norton B . . .. ATQB
Norwalk . . . AZSU
Norway .... . : . . ABFS
Norway Is . . . ANXD

NOSHAPP—ORSO

Noshapp Saki	AFLO	Obstruction Pt	AMPI	Omai saki	APGI
Nosi-Bé	AMGQ	Ocean City (Md)	AZHG	Omán, G of	AMSZ
No sima saki	APIQ	Oceania	AQUR	Omasaki sima	AOTE
Nosi Vauru	AMGS	Oceanside	AUJI	Ombai	AQOL
Nossa Senhora do Desterro	AVRW	Ochákov	AKGE	Ominato	AOTH
Noss Hd.	AFBN	Ocho Rios B.	AXRI	Omö.	ADMF
Nossyab, C	APMH	Ockseu I	AOBM	Omoa	AWQB
Noto Peninsula (Japan)	AOUC	Ocos R	AUPK	Omura	AOXB
Noto (Sicily)	AILP	Ocracoke Inlet	AYRS	Ona	ABNJ
Notoro, C..	AORT	Odawara.	APHC	Onagawa	APJH
Notske B	APLS	Odde	ABSM	One-and-a-Half Degree Ch	ANDR
Noumea P	AQYU	Odderö	ABVR	One-Fathom Bank Lt V	ANRP
Nourse R...	ALUG	Odemira	AHCT	Onega	ABEU
Noup Hd	AFGD	Odense	ADJG	Oneglia	AHWD
Nouvelle, La.	AHPY	Odens-holm	ACTL	Onehunga	ATBF
Novaja...	AJBT	Oder R	ACYU	Onitsha.	ALOX
Nova Redonda.	ALTO	Odessa.	AKFW	Onomichi no seto	AFBD
Nova Scotia..	BCAD	Odiel R	AHDV	Ons I	AGZH
Nova Zembla	ABDP	Odzuchi	APJZ	Ontario, Lake	BDWJ
Novgorod	AOPK	Oe Fiord	BPAO	Ontonagon.	BENV
Novi P	AJBF	Offer Wadham I	BEWF	Ooltgensplaat	AEDS
Novik B	AQQF	Ofoten Fiord.	ABJE	Oo sima	APBQ
Novo P (Africa)	ALNY	Ofunato B	APJV	Oo-Ura B.	APBM
Novo P. (India)	ANGC	Ogdensburg	BDVL	Oparo I	ARHT
Novorossısk	AKJW	Ogeechee R.	AYLV	Opobo R	ALPO
Novosilzov C	APMU	Oglak I	AKOF	Oporto	AHBJ
Nowikakat...	ATQR	Ogowé R	ALRG	Opotiki R	ATLU
Noyo	AUFM	Ohama	AOUP	Oppa B.	APJL
Nubel	ADFU	Ohata	APKI	Opua	ATNO
Nueva Carceres.	AQFU	Ohio	BECT	Opunake	ATCX
Nueva Ecija.	AQFC	Ohora R	ATOB	Or, C d'	BCDB
Nueva Granada	AUTM	Oita	AOZB	Oran.	AKZB
Nuevitas, P	AXMN	Oitavos	AHBW	Orange Free State..	ALVN
Nugget Pt.	ATHS	Okatsu	APJI	Orange R	ALUW
Nukualofa...	ARDT	Okayama	APBI	Orangestadt	AWLE
Nukuhiva	ARGJ	Okhotsk	AOSG	Orange Town	AYBV
Nulato.	ATQF	Okinawa sima	AOYJ	Oranienbaum	ACBH
Numadsu	APGN	Oki sima.	AOUS	Orcas I	ATYR
Nun Entrance	ALOS	Okkak	BEYJ	Orbetello.	AIFS
Nun R	ALCH	Okuchi B (Honshu).	APFM	Ord R	ARKL
Nuñez R	ALTY	Okuchi B (Shikoku)	APCN	Ording Pt	ADSW
Numivak I	ATRC	Okusiri	APMY	Oregon	AUCX
Nusa Hr..	AQUM	Oland	A'GQ	Oregon Inlet	AYTF
Nuyts Archipelago	ARQM	Olands Ostra	ACGY	Oregrund.	ACKS
Nwa la bo...	ANOY	Old Basin	AHSE	Oreth R	ATGH
Nya (New) Karleby.	ACOV	Old Calabar R.	ALFQ	Orford C	AUEL
Nyassa Lake	AMBV	Old Castle Hd	AESC	Orford Ness	AFON
Nyborg..	ADIV	Oldenburg	ADWS	Orford P	AUEN
Nyholm	ABJK	Old Harbour B	AXSL	Orignaux Pt	BDQA
Nykiobing (Falster)	ADNB	Old Hd of Kinsale	AFXB	Orilla	AUNS
Nykiobing (Liim Fiord)	ADRJ	Old Pt Comfort.	AYVK	Orinoco R .	AWIH
Nykiobing (Zealand).	ADLH	Old Providence Is.	AWPC	Oristano	AIET
Nyköping.	ACIT	Olden	ABPN	Orkney Is	AFEK
Nyminde Gab.	ADRO	Olehleh B.	AFQN	Orland	BAQM
Nystad (Finland).	ACPV	Olen.	ABSO	Orland B.	ABLU
Nysted (Laaland)...	ADMW	Oleneka R	APQF	Orlando, C	AINU
Nyukcha	ABEW	Oleron, Ile d'	AGRB	Orleans I	BDRZ
		Olga B	AOQV	Orlov, C'.	ABFM
Oahu	ARPV	Olifant R	ALVW	Orlovka.	ABFQ
Oak I	AYQX	Olinda	AVZS	Ormára.	AMWG
Oakland	AUGN	Olipa I	AJFL	Orne R	AGBL
Oamaru	ATIV	Olintorski C.	APNZ	Ornsay I...	AFCI
Oban	AFBM	Oliva Road	AVDF	Ornskoldsvik	ACMO
Obdorsk	ABCO	Olivenca	AVXG	Oro R	AHBF
Ob	ABCM	Oliveri	AINX	Oropesa C	AHKV
Obiat	AMOD	Olivi Islet	AIXF	Orotava P	ALFE
Obidos	AWEZ	Olongapo, P	AQDX	Orrio de Tapia I	AGWS
Obitotchna Spit	AKIQ	Oltenia.	AKFL	Orsera	AIWR
Obokh	AMPB	Oltenitza	AKEY	Orskar.	ACKY
Obrestad	ABUG	Olympia.	AUBX	Orso, C	AIJC

ORSTEN—PARAMARIBO

Place	Signal	Place	Signal	Place	Signal
Orsten	ABOU	Oyster B (N Y)	AZRW	Palmas	ALEV
Orsvaag	ABIP	Oyster B (Tasmania)	ASHE	Palmas B	AIEH
Orta Kioi.	AKBV	Oyster Beds	AYMT	Palmas, C	ALKZ
Ortegal, C.	AGXF	Oyster Hr	ATWN	Palmasola B	AXPN
Ortona	AIRX	Oyster I	ANKX	Palmeirinhas Pt.	ALTF
Oruba I	AWKZ	Oyster Rock	ANBQ	Palmer I.	AZYL
Orwell R	AFOU			Palmer Ort	ADBC
Osa	ABUC	Pabellion de Pica	AVCO	Palmerston	ARJO
Osaka	APBV	Pabou	BDKJ	Palmetto Pt.	AYCL
O Sima	AOZW	Pacasmayo	AUXI	Palmito	AUMV
Osborne	AEMB	Pacific Ocean	AQUR	Palmnicken	ACWV
Oscargrundet Lt V	ACBM	Padang	APTS	Palm Oil Rivers	ALOF
Oscarshamn	ACGM	Padaran, C	ANVY	Palmones R	AHFM
Osel	ACTY	Padilla	ATYL	Palmyra Pt	ANFI
Ose saki.	AOWY	Padrão.	ALSB	Palmyras Pt..	ANIG
Osima	AOYN	Padre, P.	AXML	Palmyre Pt	AGRV
Osima (Inland sea)	AOZS	Padre Santo Pt	AIED	Palo R	APXR
Ossabaw Sound	AYIS	Padron.	AGYZ	Palomera.	AHLW
Ossero	AIZG	Padron Pt	ALSO	Palopo.	AQMN
Ostende.	AEHM	Padstow	AEPI	Palos (Celebes).	AQNB
Ostensö	ABSI	Padua	AIUE	Palos (Spain).	AHDY
Oste Reef Lt V	ADTZ	Pagadian B	AQIB	Palos, C	AHJE
Ostergarn	ACHU	Pagania P.	AJHD	Paluan B	AQCY
Osterrenden	ACBL	Pagata	ANOK	Palumbo I	AINC
Osthasselstranden	ABUT	Paghilao B	AQGV	Pamanzi Road	AMDJ
Ostia	AIGF	Pagen Sand.	ADUL	Pamban	ANED
Ost-Pic Rock	AGHD	Pagc I	AJBS	Pambuang	AQBR
Ostrejm	ABRG	Pago Pago Hr	AREC	Pamlico	AYSQ
Ostro, Pt d'	AJFS	Pagoda Anchorage	AODH	Pampanga	AQEC
Osumi Str.	AOYF	Pahang R	ANTJ	Panama	AUSN
Oswego	BDYX	Pahoturi R	AQSV	Panarukan	APWQ
Otago Hi	ATID	Paihia	ATNQ	Panay I	AQHR
Otago Lt V	ATIG	Paimbœuf	AGPK	Panbula	ABID
Otawa	AOXD	Paimpol	AGHC	Panchur Tanjong	ANSG
Otchisi saki	APLN	Paita	AUWZ	Pandan	AQHU
Oterranai	AFMR	Paix, P	AXTO	Pan de Azucar	AVDH
Otranto	AIQG	Pajaros Islets	AVEB	Pandelemona P	AJKV
Ottawa	BEFZ	Pakan	ANTL	Panderma	AJYX
Ottawa (Canada)	BDUO	Pakchan R	ANPC	Pañela Rock Lt V.	AVPW
Otter Pt	AOHK	Pakefield	AFOJ	Panga R	ANPY
Otterö	ABKY	Paker Ort	ACTI	Pangani	AMLQ
Otway, C	ARXS	Pakhoi	ANXX	Pangasinan I	AQLP
Otway P	AVIW	Palais, Le.	AGNT	Panjang I (Borneo)	APXL
Oudon.	ANVII	Palak	AQKZ	Panjim	ANBK
Ouelle R	BDQM	Palamos	AHMU	Panmure Hd	BDCA
Ouessant, Isle d'	AGJO	Palanog, P	AQHM	Panne La	AEHP
Ouetique I	BCRA	Palao Is	AQVJ	Panomi Pt	AJVY
Ou Kiang	AODV	Palapa P.	AQIR	Pantellaria	AIFH
Oura Anchorage	AOVF	Palascia Tower.	AIQII	Panti Barat	AQNZ
Ouro R	ALFR	Palatia B	AJZD	Panzano B.	AIVH
Oussa	ABEC	Palawan I	AQCD	Paoushan Pt.	AOFK
Ousuri B	AOQL	Palazzo P	AJFE	Papas, C	AJLN
Outer Apostle I	BENM	Palembang R	APRY	Papas I	AKDO
Outer Lt V (Bombay)	AMZF	Palermo	AINF	Papa Stour I	AFHC
Out Skerries, The	AFGT	Palermo P (Turkey)	AJHB	Papayes, Butt aux	AMJJ
Oversay I	AEZX	Palestine	AKSW	Papenburg	ADXV
Owara B	APFI	Pálilug	ASZK	Paphos	AKQX
Owari	APFJ	Palima	AQMO	Papieté Hr	ARIIE
Owers Lt V	AELK	Palinuro, C	AIJH	Paposo Mines.	AVDK
Owharre	ARHJ	Palios Pt	AJZQ	Papua (New Guinea)	AQRV
Owikino	ATUQ	Palk Str	ANFI	Papua, G of	AQSX
Owls Hd	BACQ	Pallada Road	AOPL	Papudo, P	AVEX
Owyombo R	AMNI	Pallice, P La	AOQM	Pará	AWDY
Oxelö	ACIQ	Palliser, C	ATLC	Paracas	AUZO
Oxhöft Pt	ACXS	Palma, P (Adriatic).	AJFG	Paracel Is	ANXG
Oxö	ABVF	Palma (Balearic Is)	AHOE	Para Guassu	AVYD
Oyama	APGD	Palma (Canary Is).	ALFM	Paraguay	AVOZ
Oyapok R	AWFK	Palma (Sicily)	AIMB	Parahiba R	AWBE
Oyestreham	AGBK	Palmajola I	AHYW	Pará, R	AWDS
		Palmaria I	AHXN	Paramaribo	AWGT

PARANA—PICO

Parana	AVOR	Pechili G	AOJB	Percé..	BDKP
Paranagua.	AVSL	Pechiquera Pt	ALEI	Percéo Pt	AGBV
Paranahyba..	AWCR	Peconic B.	AZRL	Percy, C.	BORY
Parati	AVTE	Pedena Pt	AIXB	Percy I	ASVR
Paredon Grande cay	AXMW	Pedra Branca Lt .	ANTE	Perekop G	AKHC
Parekhia P	AJRK	Pedras, P de	AVZC	Perico I	AUSZ
Parenga renga Hr	ATOJ	Pedro Bluff	AXSD	Périgot B ..	ARGF
Parenzo P.	AIWP	Pedro Pt.	ANFG	Perim	AMPH
Parga P	AJHY	Peedee R	AYPZ	Perlas Is	AUTK
Pargo Pt	ALDU	Peel I (Beechy Group)	AQVT	Pernambuco.	AVZO
Paria G	AWIT	Peene R	ACZM	Pernau	ACUI
Paris	AFZW	Peenemunde	ACZN	Pérouse, La, Str	AORU
Paris I.	AYNO	Peerd Pt	ADBE	Perpignan	AHPU
Paros	AJRG	Pegasus, P.	ATGZ	Perroquets Lt..	BDMO
Parramatta..	ASLK	Pegolotta Pt.	AIWE	Perros-Guirrec.	AOHT
Parrsboro	BCDQ	Pegu	ANLS	Persia	AMUT
Parry	BEHV	Peiho R	AOJG	Persian G.	AMTN
Parry I	ABDK	Pekalongan	APVO	Perth (Scotland)	AFJD
Partridge I..	BAYQ	Peking	AOJF	Perth (W Australia)	ARN7
Passages, P.	AGUK	Pelago I	AJVO	Pertusato, C	AIBK
Pasakao Anchorage	AQGM	Pelagosa I	AIRQ	Peru	AUWQ
Pasewark	ACXL	Pelée I	AGCO	Pervicchio I	AJBL
Pasha I	AJTN	Pelée Pt	BEAO	Pesaro P	AISL
Pasha Liman	AJZM	Pelew Is.	AQVJ	Pescador Pt	AGVJ
Pasig R	AQDU	Pelican Spit Lt V.	AKOI	Pescadores Is.	AOCU
Pasindava B	AMGO	Peling I.	AQMH	Pescara	AIRZ
Pasir	AQBF	Pellegrino, M	AINM	Peschici	AIRK
Pasman	AJCY	Peloro, C	AIKM	Pestchanii Pt	AOQZ
Pasni	AMWE	Pelotas	AVRE	Pestchannun Pt	AOPV
Paspargo I	AJTR	Pelworm I	ADSR	Petala P.	AJKW
Paspebiac	BDKE	Pelzer Pt	ACDX	Petali G	AJOX
Passage I	BEMW	Pemaquid Pt.	BANV	Petalidi	AJMG
Pass I	BETJ	Pemba	AMLJ	Petaluma	AUGE
Pass a Loutre	AWYZ	Pembroke Dock	AESH	Petang	AOJQ
Passamaquoddy B	BAWR	Pembroke, C	AVKN	Petchora R	ABDZ
Passarge R	ACXG	Peña Blanca Cove..	AVDL	Peter Pt (N B)	BDKS
Passaro, C	AILS	Penai Pt	AKJU	Peter Pt (Ontario)	BDXE
Pastolik	ATQH	Penang	ANQL	Peterhead	AFIL
Pasuruan	APWL	Penarth	AEQZ	Peter the Great B	AOQJ
Patagonia..	AVKZ	Peñas, C	AGWJ	Petersburg	AYWO
Patapeco R	AZDU	Peñas G	AVIP	Petitcodiac R	BAZL
Patea R	ATCZ	Pencarrow Hd	ATKN	Petite Terre	AYDW
Paternoster I (Pulo Jenga)	AQNX	Penco	AVGL	Petit Manan	BATS
Paternoster I (Russia)	ACTM	Pendik	AJYF	Petit Passage	BCPM
Pater Noster (Sweden)	ACBH	Peñedo	AVVT	Petit Rocher Lt..	BDJC
Paterson Inlet	ATHE	Penerf	AGOJ	Petropavlovsk	AFNK
Patras I	AGSW	Penfeld R	AGKL	Petuni I	AJFO
Patmos	AJBW	Penfret I	AOMI	Peyriére, La	AGNI
Patos I	ATYP	Penguin I	ARWN	Pezzo	AIKD
Patras	AJLB	Peniche	ARBS	Phaeton, P	ARHF
Patrick, P (Pacific).	AQYH	Peñiscola	AHKZ	Phalerum B	AJON
Patrick, P (Scotland)	AEXT	Penjinsk G.	AOSP	Phanari P	AJIC
Patta	AMNR	Penlan Pt	AGOK	Philadelphia	AZKG
Patterson, P	A RJV	Penmarc'h	AGLQ	Philippeville	AXXM
Patteson, P	AQWY	Pen Men Pt	AGMS	Philippine Is	AQBX
Patu	AINV	Penn Grove	AZLE	Philipsburg	AYBL
Patuxent R	AZCT	Pennsylvania (State)	AZJR	Philip, P	ARYC
Pauhenehene Spit.	ATMN			Phleva I	AJOR
Pauillac.	AGSV	Penobscot	BAQV	Phoenix Is	ARET
Paulo Limań	AJYG	Penpoull	AOIK	Pi, P	AHOI
Pawan R	APXI	Pensacola	AXDY	Piana I	AKWH
Pawtucket	AZWD	Pentland Firth.	AFEH	Piankatank R	AYZB
Paxo I	AJHQ	Pentland Skerries	AFEI	Pianosa I (Adriatic)	AIRO
Paysandu	AVPS	Pentwater	BEKV	Pianosa I	AHYX
Payung I	APVB	Penwa R.	AEOM	Piave R.	AIUI
Paz, La (Argentine)	AVOT	Pepe, C.	AXOP	Piave Vecchia P	AIUG
Paz, B La (Mexico)	AULG	Pera	AKBQ	Picacho	AHDZ
Pease Creek	AXFZ	Pera, C	AHOV	Pichidanque B.	AVER
Peases I	BCGQ	Perak R.	ANRF	Pichilemu	AVPW
Pechany Lt V	AKJD	Peramo B	AJZB	Pico	ALCU

PICTON—PRIBILOFF

Picton	ATDW
Pictou	BCWI
Piedmont	AHWS
Piedra Pt	AQEM
Piedras Blancas	AUIB
Piedras cay (Cuba, N)	AXNO
Piedras cay (Cuba, S)	AXPG
Piel Hr	AEWC
Pierowall Road	AFGC
Pierre de Herpin	AGBW
Pierres Noires, Les	AGJY
Pietermaritzburg	ALZC
Pigeon I	BDWV
Pigeon Pt	AUHK
Pihmatsui	AOGR
Piju B	AQNR
Piller I	AGPR
Pillar, C.	AVJI
Pillau	ACWX
Pilot Station	BDOK
Pilots Ridge Lt V	ANIP
Pilsum	ADYM
Pinarello B	AIBD
Pine, C	BEUQ
Pine I	ASVT
Pines, Isle of (Pacific)	ARBM
Pinos, Isla de (Cuba)	AXOR
Piney Pt.	AYZT
Pingyang Inlet	AOLI
Pinos Pt.	AUHQ
Piombino Pt.	AHYK
Pioneer R	ASVZ
Piopetto Pt	AIGX
Piper Is Lt V	ASYV
Piraeus	AJOM
Piram I	AMYB
Pirangi R	AWBM
Pirano	AIVZ
Pirholm	ABRQ
Pirie, P	ARSP
Piring C	APWD
Piru B	AQRC
Pisa	AHXY
Pisang B	AQSJ
Piscataqua R	BAKM
Pisco	AUZK
Plainchau	AOOF
Pistolet B	BERH
Pitcairn I	ARGS
Pitea	ACNO
Pitre, Pt á	AYDO
Pitsunda Pt	AKLF
Pitt Passage	AQRG
Pitt R	ATVP
Pitt Water	ASGX
Pitum	ATKZ
Piura R	AUXL
Pizzo	AIJT
Pjasina R	ABCH
Placentia Hr	BEUK
Pladda	AEZL
Plana I	AHJN
Planier I	AHSM
Planumi I.	AJQN
Plata, La	AVNW
Plata, La, I	AUVF
Plata, P	AXUD
Plata, Pt.	AXLO
Plata R	AVNK
Platana	AKML
Plati	AJWE
Plaunick I	AIZM
Pleasant R	BAUK
Plettenberg B	ALXQ
Plevna	ACGG
Floumanac'h	AGHU
Plouneour	AGIY
Plouzec Pt	AGHB
Plum I	AZQV
Plum Pt	AXSP
Plymouth (England)	AENU
Plymouth (Mass)	BAGR
Plymouth (Montserrat)	AYCZ
Plymouth (N C)	AYTO
Plymouth (Tobago)	AYGX
Po R	AISZ
Pocomoke R	AZGF
Poel	ADCJ
Poge, C.	BADP
Pogon P	AJOC
Poile, La, B	BETC
Point de Galle	ANES
Pois, P des	AGKX
Pokemouche	BDHV
Pokesudi I Lt	BDIH
Pokhb I	AJBX
Pol Pt	ADGC
Pola	AIXE
Polak	AQDB
Poland	ACVT
Polenia R	AQXW
Policastro	AEJI
Poliki B	AQGC
Pollensa	AHOR
Pollock Rip Lt V	BAFN
Polonio, C.	AVQP
Pomarão	AHDN
Pomba B.	AMCO
Pomègues I	AHSL
Pomerania	ACYE
Pomona	APEO
Pomoni Hi	AMDF
Pomquet	BCVT
Pondicherri	ANGE
Ponapi I	AQVC
Ponce	AXYG
Pond I	BAMO
Pondang Islets	AQME
Ponente Pt	AJUN
Ponga R	ALJC
Ponghau Hr	AOCW
Pon-pon R.	AYOJ
Pontchartrain, Lake	AWZO
Pontevedra	AGZI
Pontianak	APXN
Pontusval	AGIZ
Ponza	AIQO
Poolbeg	AFVD
Poole	AEMQ
Pope Hr	BCMT
Porbandai	AMXN
Porer Rock	AIXK
Poririua Hr	ATDO
Porman B	AHJP
Pornic	AGPQ
Poros Pt (Black Sea)	AKDL
Poros (Greece)	AJNY
Porphyry Pt	BRML
Porpoise Hr	BALN
Porquerolles	AETZ
Porsgrund	ABWZ
Portage I	BDGU
Portage Lake Canal	BENY
Portalegre	AVWI
Portarlington	ARYO
Port au Port	BESI
Port au Prince	AXTE
Portbou	AHNM
Portendick	ALFY
Porthcawl	AERH
Porth Dinlleyn	AETH
Portici (Black Sea)	AKEJ
Portici (Naples)	AIHM
Portmao	AHCZ
Portishead	AEQF
Portiza	AKEJ
Portland (England)	AEMW
Portland (Mc)	BAMC
Portland (Oreg)	AUDM
Portland (Victoria)	ARIC
Portland Bight	AXSJ
Portland Bill	AEMX
Portland I	ATLM
Portneuf (Quebec)	BDTP
Port of Spain	AYEC
Porto Rico I	AXVU
Portree	AFCN
Portrieux	AGFX
Portsmouth (England)	AELS
Portsmouth (N H)	BAKG
Portsmouth (Va)	AYVZ
Portugal	AGZT
Portugalete	AGUY
Portzic Pt.	AOKE
Posiette B	AOPH
Possession I.	ALUT
Postillion Is	AQOG
Poti F.	AKLO
Potomac R	AYZQ
Potta Road	AQOH
Pottinger I	AOFV
Pouebo	AQZT
Pouléazec Pt	AGIU
Poulgoazec	AGLP
Povorotny Pt	AOQT
Poyang Lake	AOHN
Pozzalo	ATLV
Pozzuoli	AIHG
Prado	AVWO
Præsto	ADNS
Prague	AIYO
Prah R	ALND
Prasonisi	AJSC
Pratas I.	ANXH
Pravia R	AGWM
Prawl Pt	AENP
Praya	ALDB
Praya, P.	ALHM
Preguiza	ALBC
Premuda	AJCS
Prerow	ADBV
Prescott	BDVO
Prestenizze Pt	AIZB
Preston	AEVR
Preston Beach Lts	BDGL
Preston, C.	ARME
Prevesa	AJID
Priaman	APTV
Pribachi C	AKDH
Pribiloff Is	ATRQ

PRIETO—RAS

Prieto C	AGVY	Pulo Katak	ANRD	Quoin Is.	AMTL
Prim Pt (N S)	BCFJ	Pulo Lada	ANQF		
Prim Pt (Pr Ed I)	BDAI	Pulo Laut	AQBI	Råå	ACDW
Primaro	AISW	Pulo Leat	APSJ	Rabat	ALBS
Primeira Rocks	ANCE	Pulo Nias	AFUF	Rabaz P	AIYC
Primel Pt	AGID	Pulo Mendanao	AFSG	Race, C	BEUT
Primero P	AIVF	Pulo Pangkor	ANRB	Race Is	ATXI
Prince Edward I	BCYW	Pulo Pisang	ANSK	Race Pt	BAFZ
Prince Edward I (Ind Oc)	AMJO	Pulo Pisang Hr	APTD	Race Rock	AZTY
Prince Frederick Hr	ARKW	Pulo Sapatu	ANVT	Rachado, C	ANRY
Prince Jérôme G	AOLT	Pulo Sau	AFRQ	Racine	BEIW
Prince of Wales C	ATPS	Pulo Tikus	APTG	Racoon cay	AYPW
Prince of Wales Ch	ASZI	Pulo Tenga	AQNX	Radama, P	AMGL
Prince's Ob	AEIG	Pulo Undan	ANSH	Rade de Brest	AGKM
Prince's Dock	AEVF	Pumpkin I	BAQY	Raffles Lt	ANSE
Princes Dock (Bombay)	AMZL	Puna	AUWE	Ragai G	AQGP
Princes I (Africa)	ALQC	Pundi	ANHM	Ragged Island Hr	BGIM
Princes I (Dardanelles)	AJXZ	Pundu	APSV	Ragged Pt	AYGP
Prince's Landing Stage	AEVD	Pungue R	AMBH	Raglan	ATCD
Princess Royal Hr	ARPT	Puno	AVBX	Ragusa	AJFF
Principe Pt	AVVK	Punta Arenas	AVJM	Ragusa, Old (Vecchia),	AJFB
Prinkipo	AJYC	Puntadura	AJCG	Raiatea I	ARHK
Prior, C	AGXH	Puntales Castle	AHEX	Raine I	ASYX
Priorino, C	AGXK	Purfleet	AEIT	Rajang R	APYB
Probbernau	ACXJ	Puri	ANHX	Rajapur	ANBE
Problaka B	AJWC	Purnea B	AJUK	Rajpuri	AMZU
Probolingo	APWN	Putaatu I	AOQR	Raleigh (N Z),	ATCN
Procida	AIGV	Putlam	ANEI	Raleigh (N C),	AYSH
Proctorville	AWZL	Putzig	ACKU	Ram I Reef Lt V	AZUB
Progresso	AWRQ	Putziger	ACXT	Rambha	ANRV
Prome	ANMJ	Pnuloa	ARPZ	Ramintok B	AMGJ
Promontore C	AIXM	Puysegur Pt	AIFY	Ramkine I	AKSC
Prony B	AQYM	Pwllheli	AESZ	Ramree Hr	ANLK
Propriano	AIBQ	Pyrénées	AGTY	Ramsgate	AEKJ
Prorer B	ADBR	Pyrgos	AJIR	Rameö	ABLK
Prospect Hr	BAIJ	Pyrmont	ASKV	Rance R, La	AOFI
Prosperous B	ALVR			Randers	ADFL
Prosser B	ASHC	Quaco	BAYW	Råneå	ACNW
Proti I	AJLX	Quaker I	BCKI	Rangaunu	ATNY
Proudfoot Shoal Lt V	ASZV	Quarnero G	AIXR	Rangitikei R	ATDG
Providence	AZWJ	Quartu	AIDW	Rangitoto I	ATMX
Providence B	AFOI	Quebec	BDTA	Rangoon	ANMD
Providence Ch	AXIT	Queenborough	AEJU	Ranobé, P	AMFI
Providence & St Pierre Is	AMJK	Queen Charlotte Is	ATUE	Rapa I	ARHT
Provincetown Hr	BAGF	Queen Charlotte Sd (B C)	ATUS	Rapallo	AHXE
Province Wellesley	ANQT	Queen Charlotte Sd (N Z)	ATDS	Raphti, P	AJOW
Prudence I	AZWV	Queenscliff	ARYI	Rappahannock R	AYZE
Prussia	ACVQ	Queens Dock	AEVQ	Raritan	AZNG
Psara	AJTK	Queensferry	AFJS	Rarotonga I	ARHP
Pubnico Hr	BCIF	Queensland	ASPG	Ras Aha	AKXZ
Puchepo Pt	AVGH	Queenstown	AFWS	Ras al Bir	AMFD
Puchoco	AVGZ	Queijal Pt	AGYP	Ras al Hadd	AMSW
Puercas, Las	AIFB	Quelpart I	AOMS	Ras al Khaima	AMTN
Puercos I	AHNW	Querqueville, Pt de	AGCF	Ras al Kuh	AMVQ
Puget Sound	AUBL	Quexo C	AGVL	Ras Alula	AMOQ
Pugwash	BCXJ	Quiberon (Loc Maria)	AGNW	Ras Asir	AMOG
Pujada B	AQKM	Quidico, P	AVIIN	Raschgoun I	AKZI
Puket Hr	ANPU	Quieto P	AIWJ	Ras Engela	AKWO
Pulicat	ANGK	Quilan, C	AVID	Ras Fartak	AMSL
Pulithra P	AJNH	Quilca R	AVBL	Ras Guarib	AMQL
Pulo Angsa	ANRO	Quilimane R	AMBY	Ras Hafun	AMOE
Pulo Batam	APRJ	Quilleboeuf	AFZQ	Ras ul Hadik	ALBX
Pulo Brani	ANSP	Quilon	ANCV	Ras Kanzi	AMKV
Pulo Bras	AFQK	Quimper	AGLY	Ras Kegomacha	AMLN
Pulo Buru	AFQO	Quimperlé R	AGMQ	Ras Madraka	AMSR
Pulo Cercer de Mar	ANVU	Quindalup	AROY	Ras M'Kumbi	AMXT
Pulo Condore	ANVI	Quintero B	AVEZ	Ras Muwari	AMWK
Pulo Dapur	APSD	Quiriquina I	AVGE	Ras Nungwe	AMLI
Pulo Jarak	AFRE	Quito	AUWC	Ras Radressa	AMOJ
Pulo Katang Katang	APTR	Quoddy Hd	BAVE	Ras Rakkin	AMTW

RAS—ROSARIO

Ras Raweiya.	AMPX	Rèunion I	AMHD	Robbins I	ASEX
Ras Upemba	AMLO	Revel	ACTB	Robbins Reef Lt	AZOF
Ras Zafarana	AMQN	Revellata Pt	AIBW	Robe	ARWJ
Ras Zorug	AKUY	Revel Stone Lt V	ACTD	Robert Hr	ALUS
Rat I.	APTG	Revanæs	ADLQ	Roberts Pt	ATXY
Ratai	APSU	Revilla Gigedo Is	AUNC	Roc Pt	AGEP
Ratan	ACND	Réville	AGCF	Roca, C	AHBV
Ratbun	ANUE	Rewa	ARCJ	Roca, Partida	AWST
Rathlin I.	AFTJ	Reyes Pt	AUGB	Rocas	AWBY
Rathlin O'Birne I	AFSM	Reykjavik	BFAE	Rocchetta Ch	AITQ
Rathmullan	AFSW	Rhenea	AJRD	Roche Bernard, La	AGOP
Ratnagiri	ANBD	Rhine R	AECU	Roche Bonne Lt V.	AGQL
Ratoneau I.	AHSI	Rhio Str	APRK	Rochefort.	AGGY
Rattlesnake Sh	AYPN	Rhôda	AJZS	Rochelle, La.	AGQN
Rattray Hd	AFIK	Rhode Island	AZUQ	Roche Pt	AFWV
Raumo	ACPS	Rhodes	AJRY	Roche Ile Beacon	AMIN
Rauna	ABUJ	Rhodesia	ALZJ	Roches Bonnes	AGFB
Ravenna	AISU	Rhône R	AHRD	Roches d'Ouvres	AGHL
Ray, C	BESO	Riachuelo R	AVOF	Rochester	AEKC
Raza, I.	AVTO	Ribble R	AEVO	Rockabill.	AFUV
Raz de Sein, Bec du	AGLK	Rabnitz	ADCE	Rockall	AFDP
Razzoli I	AJCO	Rich Pt	BERQ	Rockhampton	ASVD
Ré, Ile de	AGQE	Richard	AGSM	Rockingham.	AROG
Re, P	AIYU	Richelieu I	BDUH	Rockingham B	ASXG
Real, P	AHET	Richibucto Hd	BCYN	Rock I	APGW
Real R	AVYJ	Richmond (N B)	BDJX	Rockland	BAOT
Reale Road	AIFJ	Richmond (Va)	AYXD	Rockly B	AYGB
Rebecca Sh	AXGR	Richmond Hr (Pr Ed I)	BDEM	Rockport.	BAOW
Recherche B	ASFU	Richmond R	ASON	Rocky I	ASYJ
Recife, C	ALXW	Riga	ACUM	Rod.	ABNT
Red B	BEQM	Rigas	APUL	Rodberg	ABLZ
Red Islet Bank Lt V.	BDOW	Rigolets, The	AXBH	Rodby	ADMV
Red Islet Lt	BDOX	Rigoulette, The	BEYC	Rodd Hr	ASTV
Redang Is	ANTS	Riknoku G.	AOTF	Rodkallen	ACNQ
Redes.	AGXT	Rimini P	AISN	Rodo	ABJY
Red Fish Bar Lt	AWIO	Rimmen	ADPY	Rodoni, C	AJGS
Redland	ASQV	Rimouski	BDON	Rodosto.	AJXO
Redon	AGOQ	Ringholm	ABMS	Rodriguez I	AMIS
Redonda	AYCV	Rio Colorado	AVMT	Rödsbugt	ABNK
Redoute Kalessai	AKLN	Rio de Janeiro	AVTQ	Rodskar.	ACSV
Red Sea	AMPF	Rio del Rey	ALPU	Rodtangen	ABYG
Reedy I	AZIW	Rio Grande de Cagayan	AQEY	Rodven	ABNW
Reef I	ANOU	Rio Grande do Norte.	AWBO	Rodvig	ADNW
Reersø	ADLV	Rio Grande do Sul	AVQX	Roe I	AUHC
Refunsiri I	APMJ	Rio Grande	AWTE	Roebourne	ARLZ
Roggio	AIKE	Rio Negro (Parana)	AVMN	Roebuck B.	ARLK
Régnéville	AGEK	Rib Santiago P	AVNY	Rogosnizza.	AJDI
Reine	ABIX	Riou I	AHSP	Rojo, C	AXVW
Reisen	ABHJ	Rio Vista	AUHE	Rokuron I	AOVJ
Reitz, P	AMNC	Riposto	AIKW	Röm I	ADSC
Reke Fiord	ABUO	Risano	AJFX	Romain, C	AYPT
Reko	ABLH	Risiri I	APMK	Roman Rocks.	ALWS
Relandersgrund	ACPT	Risor.	ABWN	Romanzof, C.	ATQV
Roma B.	AKNZ	Rithymno	AJPK	Romblon, P	AQIE
Rembang	APWB	Riudsu saki	APCV	Rome	AIGD
Remesvigen	ABVD	Riu Kiu	AOYI	Rompido Pt	AHDR
Remine	ASFE	Rivadeo	AGWV	Romso	ADIZ
Rendsburg	ADEQ	Rivadesella	AGVZ	Ronciglio Pt	AINF
Renesse	AEDZ	Riverhead	AZBQ	Ronehamn	ACHX
Rennes	AGOS	Riverton	ATGF	Ronne	ACEZ
Renong	ANPD	Rivière du Loup	BDFL	Ronnen.	ADLK
Renskär	ACQU	Rivoli B	ARWL	Ronze	AIXO
Repulse B	ASWG	Rixhoft	ACXZ	Roper.	ARIN
Repvaag	ABGH	Riyōishi	AFJY	Roque, Pt de la.	AFZO
Restigouche R	BDJO	Rizo	AKMB	Roquetas	AHIG
Restinga Pt	AQDL	Rizzuto, C	AIPS	Rorvig	ABKW
Retchnoi I	AOPW	Roancarrig I.	AFQC	Rosa, C	AKXC
Retimo	AJPK	Roanoke R	AYTB	Rosario	AVON
Retiro B	AWCH	Roatan I	AWPK	Rosario cay.	AXPC
Reuben Pt	ALZM	Robben I	ALWI	Rosario Str	ATYN

ROSAS—SAMBRO

Rosas	AHNE	Ruke-Ruku B	ARDB	Saigo Hr	AOUT
Roscanvel	AGKS	Rukyira B	AMKN	Saigon	ANVL
Roscoff	AGIO	Rum I	AFCD	Saiki	AOYS
Roscau	AYEH	Rumih	AKCX	Saintes	AGRB
Rose Blanche B	BESX	Rumih Hussar	AKCB	Saintes, Isle des	AYDS
Rose, C	AKXC	Rumih Kavak	AKGW	Saintes Maries	AHQY
Roselier Pt	AGFW	Runcorn	AEVB	Sakai (Isumi Nada)	AFBY
Rosetta	AKTZ	Rundô	ABOP	Sakai (Mikuni)	AOUJ
Roseway, C..	BCID	Runo I	ACUQ	Sakaide	APDW
Roeholm	ABOP	Rupert's House	BEYV	Sakata	AOTV
Rosia B	AHFW	Rusânovka	ABEF	Sakhalin	AORK
Rosatten	ACWQ	Rusaro	ACQR	Sakitsu ura	AOXS
Roskilde	ADLP	Ruso Bank	ABVG	Sakonnet Ch	AZXN
Ross Cove F	AUFS	Russ	ACWK	Sakoshi B	AFBL
Ross I	ABDL	Russell P	ATNS	Sal I	ALHE
Rossa B	AIED	Russia	ABDU	Sala	ANUV
Rossa I	AIEX	Russian Tartary	AOPD	Salamis B.	AJOH
Rossa Pt	AIRF	Ru Stoer	APDT	Salang I	ANPM
Rosello, C.	AJMP	Ruvu R	AMKY	Salaverry	AUXO
Rosslare	AFVQ	Ruytingen Lt V.	AFXQ	Salayar	AQMV
Rosso, P..	AJEV	Ryan, Loch	AEXV	Salcombe	AENQ
Rost	ABIR	Ryde	AELZ	Saldanha B	ALWB
Rostok	ADCH	Rye	ARZL	Sale	ASBW
Rostov	AKIY	Ryvarden	ABSX	Salée R	AOLF
Rota	AHEV	Ryvingen	ABVJ	Salem (Mass)	BAIT
Rothe Kliff	ADSM			Salem (N J)	AZLH
Rothersand	ADVN	Saadani	AMLP	Salengketa Pt	AQMS
Rothesay	AEYQ	Saba I	AYBR	Salerno	AIJD
Rothsay Pt.	ATSR	Sabalanga Is	AQOG	Salgar	AWMN
Rotoava Anchorage	ARGX	Sabang B	APQR	Sali	ALBS
Rotterdam.	AECW	Sabbioncello.	AJFB	Salina I	AJOL
Rotterdam Canal, New	AECR	Sabi R	AMBD	Salina Cruz	AUOV
Rotti	AQOU	Sabinal Pt.	AHIF	Salinas B (Peru)	AUYN
Rottnest I	ARNL	Sabine	AWVU	Salinas B (Cent Am)	AURK
Rotumah	ARDO	Sabine Pt	AZVX	Salinas, C.	AHOX
Rouen	AFZU	Sabinilla	AHGE	Salinas Pt	ALTU
Roumania	AKED	Sablayan	AQCW	Salisbury	BAZX
Roumelia	AJVS	Sable, C (Fla)	AXGI	Sallis (Alt Salis)	ACUJ
Round, C	BCQL	Sable, C (N S)	BCHL	Salmis	ACNU
Round I (England)	AEOZ	Sable I (N Atlantic)	BCLN	Salobreña	AHGU
Round I (Miss)	AXCO	Sable, Pt au	BEFI	Salomague, P	AQEV
Rousse I	AICE	Sables d'Olonne	AGGC	Salona G	AJLH
Rovai	ABTC	Sacatula R.	AUNQ	Saloniki	AJVT
Rovigno	AIWU	Sacketts Hr	BDZJ	Salou	AHLS
Rovondrau B	ARCT	Sackville	BCAK	Salsang	ALIM
Rovuma R	AMJY	Saco	BALF	Salt I	ALHE
Rowley Shs	ARLV	Sacramento	AUHG	Saltees	AFVZ
Roxo, C	ALIH	Sacratif, C	AHGX	Salten Fiord	ABJI
Royal Albert Docks	EJD	Sacrificios	AUON	Saltholm	ADOK
Royal Sovereign Lt V	AELD	Sada	AGXW	Salt Lake B	BDLQ
Royal Victoria Docks	AEJG	Sadashivgad	ANBR	Saltney	AEUG
Royal, P (Honduras)	AWPM	Saddleback Ledge	BARN	Saluafata Hr	ARLK
Royal, P (Jamaica)	AXSN	Sadineiro (Lobera)	AGYM	Salut Is	AWGD
Royal, P (S C)	AYNI	Sado I	AGTX	Salvador	AUQD
Royale Isle	BEMV	Smdenstrand	ADRU	Salvage Is	ALFO
Royalist, P	AQCM	Smtterriet Road	ADKZ	Salvora I	AGZD
Royan	AGSD	Safatu Hr	AREH	Salvore Pt	AIWD
Rozel	AGDU	Safi	ALBZ	Salwatti	AQRX
Rozier, C.	BDKY	Sag Hr	AZRB	Salween R	ANOL
Rua I	AGZE	Sagami ura	APGZ	Salzhorn	ADVO
Ruad I	AKRZ	Sagano seki	AOYX	Samaná B	AXUP
Ruapuke I	ATGV	Saginaw B	BEGF	Samana Pt	AJGU
Ruden I	ACZK	Sagres	AHCX	Samanco B	AUYD
Rudha Mhail	AFBD	Sagro, C	AHZT	Samar	AQJK
Rudkiobing	ADIK	Sagua la Grande, P	AXNE	Samarai I	AQTK
Rufiji R	AMKR	Saguenay R	BDOZ	Samarang	APVR
Rufisque	ALGN	Sahib P	AKOF	Samarowsk	ABCQ
Rugen I	ACZW	Saibai I	AQST	Sambar Pt	AFXH
Rugenwalde	ACYL	Said, P,	AKTO	Sambas R	AFXQ
Rugsund	ABPI	Saida	AKSI	Sambro I	BCIJ

SAMBULAUAN—SANTIAGO

Sambulauan, P	.AQLD
Saminos	AFWX
Samoa Is	.ARDX
Samos (Archipelago)	AJTF
Samos (Greece)	AJKC
Samothraki	AJUS
Samothraki I	AJHM
Sampit	AQBP
Samsa Inlet	AODL
Samshui	ANZD
Samso	ADJT
Samsun	.AKMP
San Aleixo I	.AVZK
San Andrea I	AIQC
Sanary	.AHSW
San Antioco I	AIEJ
San Antonio C (Arg)	.AVNI
San Antonio C (Cuba)	.AXOL
San Antonio C (Spain)	.AHKD
San Antonio del Petrel	AVFX
San Antonio Nuevo	AVFR
San Antonio, P (Arg)	AVML
San Antonio, P (Balearic Is)	AHNU
San Antonio, P (Egg Hr)	.AVLW
San Antonio Pt (Adriatic)	.AJCD
San Antonio Pt (Brazil)	AVXU
San Antonio Pt (Chile)	.AVFS
San Bartolomé, P	AUKX
San Benigno, C.	AHWX
San Benito	,AUPO
San Bento R.	.ALQU
San Bernardino Str.	.AQGD
San Blas Hr (Arg)	.AVMP
San Blas (Mexico)	AUMX
San Blas, C	AXEN
San Buenaventura	AUIR
San Carlos Hr	AXGF
San Carlos Pt	AHPF
San Cataldo Pt	AIQS
San Christoval I	.AQWF
San Ciprian	AGWX
San Clemente I	AUKD
Sancti Petri	.AHFE
Sand cay	AXHD
Sand I (Ala)	AXDM
Sand I (Superior)	.BENJ
Sanda	.AEZO
Sandakan Hr	APZO
Sandalo, C.	AIEP
Sandarli G	AKNY
Sandefiord.	ABXL
Sandeid	.ABFF
Sandesund.	.ABYZ
Sandgate Road	.AEKZ
Sandhammar Pt.	ACEX
Sandhamn.	.ACJI
San Diego.	AUJO
Sandnes	ABTX
Sand5	.ABOT
San Domingo (Cent Am)	AUSH
Sandoway.	ANLO
Sandridge	ARZH
Sands Pt.	.AZQP
Sandtorr	ABHP
Sandusky	.BEDJ
Sandviken	.ABYK
Sandwich B.	.BRXZ
Sandwich I	AQXN
Sandwich, P.	AQXK
Sandy C	ASTD
Sandy Hook	.AZMX
Sandy Hook Lt V.	.AZMY
Sandy I	AYCR
Sandy Pt (Queensland).	.ASPO
Sandy Pt (S America)	.AVJM
San Felice	AIGL
San Felipe B.	.AULM
San Feliu de Guixols	AHMT
San Fernando Pt	AQET
San Fernando (Spain)	.AHEZ
San Fernando (Trinidad)	.AYHG
San Francisco	AUGF
San Francisco Lt V	.AUGE
San Francisco R	AVYB
San Francisco do Norte R.	AVYR
San Gallan I	AUZQ
San Giovanni di Pelago	.AIWY
Sanguinaire I	AIBS
Sanibel I	AXGC
San Isidro, C	.AVJO
San Jeronimo	.AUPW
San Joao I.	AWDI
San Jorge.	ALCW
San José (Costa Rica)	AWOP
San José de Guatemala	AUPY
San José del Cabo B	.AULE
San José Ignacio Pt.	AVQL
San Juan Pt (Cal)	AUJG
San Juan (Mexico)	.AUPC
San Juan Bautista	AWSN
San Juan del Norte	AWOR
San Juan del Sur	AWOT
San Juan de Nicaragua R	.AWOT
San Juan I	ATYX
San Juan, P (B C)	ATXO
San Juan, P (Peru)	.AUZY
San Juan, P (P R)	AXWJ
San Juan R (Cent Am)	.AUTO
San Julian	.AHCI
San Julian, P.	.AVLK
Sankaty Hd	BAFH
San Lorenzo	AUQV
San Lorenzo, C	AUYX
San Lucar.	.AHEJ
San Lucar de Barrameda	.AHEI
San Lucas, C	AULC
San Luis (Cent Am)	AUPO
San Luis Pass	AWUZ
San Luis Pt	AUIH
San Luis D'Apra, P	AQVO
San Luis Obispo	.AUIJ
San Marco C	AIKU
San Marcos B	AWDG
San Martinho	.AHBO
San Miguel	ALDE
San Miguel B (Cent Am)	AUTO
San Miguel B .	.AQFT
San Miguel F	ALTC
San Miguel I	.AUJQ
Sanmun B.	AODY
San Nicola I.	AIRM
San Nicolas	.AVOH
San Nicolas I	.AUXF
San Nicolas, P	.AUZW
San Nicolo Del Lido P	.AITV
San Pablo B.	AUGP
San Paola I.	AIPX
San Pedro (Cal)	.AUIZ
San Pedro (Spain)	AHIN
San Pedro do Sul.	.AVQZ
San Quentin B.	AUKR
San Remo	AHVZ
San Roman, C.	.AWLG
San Roque.	AHFT
San Salvador	.AXJU
San Sebastian (Cadiz)	AHFC
San Sebastian (Canary Is)	.ALFI
San Sebastian (Spain, N C)	.AGUL
San Sebastian, C	AHMX
Sancery	AHSW
Sansego.	.AIZF
Sansego I.	..AIZP
San Simeon.	AUID
Santa	.AUXW
Santa Ana Hr	AWKV
Santa Anna I.	AWCX
Santa Barbara Ch .	AUIP
Santa Barbara B .	AUMF
Santa Barbara, P	AXUH
Santa Catalina.	AUKB
Santa Catharina I.	AVRU
Santa Clara B.	AXNI
Santa Croce, C.	.AILD
Santa Cruz (Azores)	ALCP
Santa Cruz (Cal).	AUHO
Santa Cruz (Canary Is)	ALFD
Santa Cruz (C Verde Is)	ALGT
Santa Cruz (Morocco)	ALCF
Santa Cruz (Philippines)	AQCI
Santa Cruz (W I).	AXZS
Santa Cruz B.	.AVWU
Santa Cruz del Sur	AXPZ
Santa Cruz F .	AVSD
Santa Cruz F (Rio de Janeiro),	AVTS
Santa Cruz I (Cal).	AUJW
Santa Cruz Is (Pacific)	AQWO
Santa Cruz, P (Arg)	.AVLG
Santa Elena.	AUVX
Santa Elena, P	.AVMD
Santa Fé R.	.AXOT
Santa Isabel	ALQB
Santa Luzia.	ALGZ
Santa Margherita.	AHXC
Santa Maria, (	AVQN
Santa Maria de Leuca, C	.AIQE
Santa Maria I (Azores)	ALDM
Santa Maria I (Chile)	.AVGX
Santa Marta.	.AWLU
Santa Marta Grande, C	AVRQ
Santa Maura	.AJIN
Santa Monica	AUIV
Santander	.AGVM
Santapilli	ANHJ
Santa Pola.	AHJO
Santa Quaranta	AJHC
Santarem.	AWRV
Santaren Ch.	.AXJE
Santa Rosa B.	.AXOH
Santa Rosa I	AUJS
Santa Rosa R.	AUWM
Santa Rosalia	AULK
Santa Teresa Gallura.	.AICL
Santa Venere, P.	.AIJU
San Theodoro Pt	AJKD
Santiago (Chile).	AVFK
Santiago, C	AQDK

SANTIAGO—SHA

Santiago de Compostella	AGYQ	Scaletta .	AIKU	Selve	AJCB
Santiago de Cuba	AXQF	Scalloway	AFGX	Selvi Burnu	AKCS
Santo Antonio B	ALQD	Scarbrough (England)	AFMD	Semangka B .	APSY
Santo Domingo Hr	AXUP	Scarbrough (W I)	AYGT	Semao .	AQOT
Santo, P (Madeira)	ALDX	Scario.	AIJM	Semaphore	ARUZ
Santo, P (Venezuela).	AWIX	Scarpanto..	AJSD	Semeni R	AJGU
Santoña	AGVH	Scarweather Lt V.	ABRJ	Sena	AMBT
Santorin	AJRU	Scatari I	BCRM	Sendai B	APIX
Santos	AVSX	Scattery I	AFQX	Sénégal R	ALGB
San Vicente de la Barquera	AGVT	Schank, C	ARZS	Senegambia.	ALGF
Saô Francisco	AVSJ	Scheldt.	AEFG	Sénéquet, Le.	AGEH
São José do Norte.	AVRC	Scheveningen.	AECN	Sénétose Pt	AIBO
Saolara.	AMFG	Schiedam....	ABCY	Senigallia P .	AISH
Saona I.	AXUL	Schiermonnikoog.	ADYQ	Sen Zaki uchi	AOVD
Sapangar B	AFYV	Schiewenhorst.	ACXN	Seoul.	AOLM
Sapeh B	AQNW	Schilbols Nol..	ADZI	Sepet, C....	AHTE
Sapetang..	ANQV	Schillighorn	ADWZ	Sept-Iles, Les.	AGHV
Sapelo.	AYLJ	Schleswig	ADFH	Seraglio Pt.	AJZY
Sapetiba...	AVTM	Schleswig-Holstein.	ADCU	Serantes R	AGXO
Sapienza I.	AJMB	Schokland....	AEBQ	Serdze Kamen, C	APOT
Sapri ..	AIJP	Scholpin.	ACYH	Seripe.	AVYL
Sapudi.	APWG	Schooner Cove...	BCSE	Serk..	AGDL
Saputi R	APSO	Schouwen	ARDY	Serpents I	AKEN
Saracen Hd...	AOCP	Schouwen Bank Lt V	AEFH	Serpent Mouth.	AYHO
Saraceno Mt.	AIOH	Schulau Lt V	ADUQ	Serpho	AJQC
Sarco . .	AVDJ	Schultz Ground Lt V	ADKR	Serrat, C	AKWP
Saramacca R .	AWGX	Schuylkill R .	AZJX	Servia	AJGL
Sarasota B	AXFQ	Schwarzort.	ACWF	Servola	AIVQ
Saráwak...	APXZ	Sciacca	AIMG	Serwatti Is.	AQOV
Sardao, C..	AHCU	Scilla .	AIKC	Sesajab R...	AQBD
Sardina Pt.	ALEY	Scilly Is...	AEOT	Sesecapa.	AUFQ
Sardinia.	AICJ	Scirocco Marsa	AIOW	Seskar	ACSO
Sardinierès Pt	AHUN	Scio	AJTM	Seskarö	ACOB
San Siglar.	AJWZ	Scotland	AEXI	Sesters, Grand	ALKV
Sar Kioi..	AJXK	Scotland Lt V	AZNB	Sestri Levante	AHXG
Sarnia.	BEFT	Scrabster	AFDY	Sestroretzk.	ACRU
Sarpsborg	ABZC	Scranton	AXCL	Seto Uchi . .	AOZF
Sarstoon R	AWQM	Scuso, P.	AIEM	Sestos R	ALKR
Saru.	APLB	Seaham .	AFLS	Settée R.	ALRM
Sasardi B..	AWNI	Seal I (N S)	BCGT	Setubal	AHCQ
Saseno I.	AJGW	Seal Rocks (W I)	AYBE	Seudre R	AGRN
Saska B.	AJBV	Searsport	BAPR	Sevastópol.	AKHJ
Sassafras R.	AZBP	Seattle	AUBF	Seven Is , The..	ABDI
Sassandra.	ALMF	Sea Wolf I.	BCVE	Seven I B..	BDMU
Sassari.	AIFN	Sebastian Viscaino B.	AUKT	Seven Stones Lt V	AEPB
Sassnitz..	ADBK	Sebenico	AQYN	Severn, F......	BEZA
Sastmola ...	ACPN	Sebert Cove	AQYN	Severn R (England)	AEQK
Sata no misaki.	AOYE	Sebu	AQIM	Severn R (Md)	AZDO
Saude Dock.	AVTW	Sedhin	ALJF	Seville.	AHEK
Saugatuck....	BEKG	Sedimi B	AOPU	Seychelles.	AMIY
Saugor.	ANIX	Segaar B.	AQRY	Seyne B , La	AHTK
Saunders, C.	ATIB	Segerstad	ACHB	Sfax	AKVN
Saunders, P ...	BERT	Segna	AJBI	Sgeir Maiole (Sgeir Vuile)	AFBH
Sauo B. .	AOCS	Segoro Wedi B .	AFWY	Shableh, C	AKEB
Sault Sainte Marie	BELJ	Seguin I	BAMI	Shadwan I	AMQG
Sauzon	AGNV	Segura R .	AHJM	Shakotan Hr	AFLW
Savaii I.	AREO	Seguro, P (Africa)	AJNR	Shakutien I	AOJR
Savana I.	AXZF	Seguro P (Brazil)	AVWQ	Shambles Lt V	AEMY
Savanilla.	AWML	Seidis Fiord.	BFAS	Shanghai	AOFJ
Savannah	AYMN	Sein, Ile de.	AGLF	Shanhainuan .	AOJV
Savannah-la-Mar. .	AXRW	Seine R......	AFZK	Shannon R	AFQT
Savaradrug.	AMZW	Seiro......	ADLO	Shantarski Is	AORY
Savitri R. .	AMZV	Sekondi..	ALNB	Shantung Promontory	AOIQ
Savona	AHWN	Sekru......	AQSD	Sharja ..	AMTO
Savu Savu B..	ARCW	Selángor R....	ANRL	Shark B..	ARMG
Sawai Hr.	AQPT	Selatan, C...	AQBL	Shark Pt..	ALSB
Saxbi Ness....	ACTQ	Selker Lt V	AEWN	Sharps I....	AZFB
Sarkiobing.....	ADMN	Selsea Bill .	AELN	Sharpness Docks .	AEQK
Saxony...	ACVT	Selsovik..	ABJZ	Shastri R .	ANBC
Scalambri, C...	AILW	Selva P...	AHNJ	Sha Sze.	AOIG

SHATT-AL-ARAB—SNIPE

Shatt-al-Arab....	AMUF	Sibutu I	APZX	Sivutch B . . .	AONU
Sheboygan.. .	BHN	Sibuyan I	AQHG	Siwo misaki. .	APEM
Shediac.. . .	BCYB	Sicié, C . . .	AHTD	Six-Fours	AHSY
Sheerness. .	AEJT	Sicily... . .	AIKJ	Sizepoli	AEDH
Sheet Hr...	BCMW	Sidby . . .	ACPM	Sjelanger. .	ABRJ
Sheet Rock... . .	BCMX	Sidero, C. .	AJPS	Skagen . . .	ADQN
Sheffield . .	AFMV	Sidmouth . .	ASEB	Skagens Hamn. .	ACML
Sheik ul Abu. . .	AMPN	Sienniumiao....	AOGC	Skagerrack. . .	ABXF
Shelburne. .	.BCIA	Sierra Colorado Pt	AWLC	Skagganas . . .	ACGK
Sheldrake I. . . .	BDGX	Sierra Leone.. .	ALJV	Skala.	AJSY
Shelikoff Str .. .	ATRU	Siete Pecados	AQIC	Skala Nuova G .	AKPC
Shemogue .	BCXY	Sighajik P...........	AKOZ	Skalgrund... .	ACPI
Shepherd Hr	AWOJ	Signal Mt (Mauritius)	AMID	Skalskar . . .	ACPZ
Shepstone, P.	ALYT	Sigri P. . . .	AJTX	Skano . .	ACEP
Sherbro R. . .	AIJZ	Siguapa... .	AXNS	Skaresnde	ACHN
Sherbrooke .	BCNR	Siguenza Pt . . .	AXDP	Skaw, The . .	ADQN
Sherki I .	AKVQ	Sihuatanejo. .	AUNW	Skeena R . .	ATUG
Sherm Wej	AMRD	Sikia, P .	AKOX	Skelder B. .	ACIJ
Shershel . .	AKYR	Si Kiang. . . .	ANZB	Skelleftea .	ACNK
Shetland Is . .	AFGL	Sikijor I .	AQJB	Skelligs . .	A8QK
Shidzukawa .	APJN	Silam Hr . .	APZW	Skerda I .	AJBW
Shields, North. .	.AFLJ	Silvri . . .	AJXR	Skernies .	AETN
Shields, South	AFLM	Silleiro, C .	AGZQ	Skerryvore .	AFBV
Shieldsboro. . . .	AXBN	Silloth.	AEWS	Skiatho. . .	AJVH
Shikoku . .	APCG	Silver I . .	AOPX	Skibotten ..	ABGV
Shimabara .	AOXL	Simá.. . .	ACOH	Skielskor. .	ADLZ
Shimidzu Hi .	APGM	Simoda.... .	APGX	Skien. . .	AB\C
Shimonoseki	AOVP	Simons B . .	ALWU	Skilling R . .	ABSU
Shimonoseki Str	AOVU	Simpnäsklubb .	ACKE	Skiöds Hd	ADKF
Shinagawa...	APIB	Simpson, P .	ATSZ	Skiold Ness .	ADIB
Shinás. . . .	AMTF	Simrishamn. .	ACFO	Skira .	AKVJ
Shingle Pt	AMHX	Sinai, Mt. .	....AMQX	Skokomish . .	ATZW
Shiogama	APIZ	Sind Coast .	AMWV	Skomvær .	ABJP
Shiokubi mi saki	APKU	Sines, C	AHCR	Skongsnæs. .	ABPD
Ship Hr .	BCMQ	Singapore	ANSM	Skopelo. .. .	AJVK
Ship I... . .	AXCF	Singapore New Hr	ANSV	Skoven . .	ADJC
Ship John Sh	AZIQ	Singapore, Old Strait of.	ANTD	Skraaven . .	ABIF
Shippigan. ..	.BDIN	Singatoka R .	ARCS	Skrypleff I . .	AOQH
Ship Island Sh ..	.AWXY	Singkel. . . .	APUJ	Skudesnæs . .	ABTP
Shipu Hr .	AODZ	Singo...... . .	APEV	Skutari .	AJGQ
Shipwash Lt V. .	APOQ	Singora...... . .	ANTV	Skutari (Bosphorus)	AKBI
Shiranai B .	AOTL	Singti Reach. .	AOIC	Skutari (Greece) .	AJMQ
Shirasu	AOVL	Siniscola .	AIDO	Skutskar .	ACLE
Shire R. . .	AMBW	Sinjai	AQMT	Skye, Isle of .	AFCM
Shiriya saki .	APKH	Sinkeamun Hr	AOEH	Skyro . .	AJVG
Shitama, P. .	APCM	Sinmororan .	APKZ	Slavianski B .	AOFR
Shizuoka	APGK	Sinnamar.. .	AWGH	Sletten .	ADOX
Shoals, Isle of	BAKD	Sinope (Sinub)	AKMT	Sletterhage .	.ADKH
Shoalwater... .	AUCT	Siphano . .	AJQB	Slevik. . .	ABYU
Shoalwater B ..	A8VL	Sippican	AZYX	Slema .	AIGV
Shoeburyness ..	AEIL	Sirakami, C	APND	Sli (Schlei) Fiord .	ADFJ
Shoreham .. .	AELJ	Sirakami saki	APKN	Sligo .	AFSH
Shortland Bluff	ARYG	Sirik, C. . .	APYF	Slimunde .	ADFB
Shortt I . .	ANIJ	Sir James Hall Group	AOLJ	Sliteham. .	.ACHT
Shovelful Shoal Lt V	BAEP	Siro sumas	AOVH	Slottero . .	ABRS
Shubenacadie R. .	.BCDZ	Siru unku. .	APMV	Sluis . .	.AEGN
Shummá I ...	.AMPL	Sirunoki . .	APLG	Slyne Hd . . .	AFRM
Shuya	.ABEZ	Sisal . .	AWRS	Smalls.	AESK
Sælland.. .	ADKV	Sisargas I.	AGYB	Smith I .	ATZE
Sælland Reef Lt	.ADLP	Sisamiut.	BEZW	Smiths Knoll Lt V. . .	AFNX
Siak.	APRG	Sisiran P	AQPV	Smoky C . .	ASNU
Siam	.ANCU	Sissibou R .	BCGR	Smörhavn .	.ABPM
Siargao I. . .	AQJX	Sitges.. .	AHLY	Smyge Pt	ACBT
Siassi...	.AQLT	Sitia, C.. .	.AJFR	Smyrna .	AKOM
Siberia. . . .	AOSF	Sitka .	ATSN	Smyth Ch .	AVJC
Siboga.	.APUI	Sittang R . . .	ANOG	Snares, The . .	ATHI
Sibonga .	.AQIU	Siversov .	AKGR	S Nikolo, P . .	AJIM
Sibpur .	ANJK	Sivota I . .	.AJHX	Snipan Lt V.	ACNB
Sibu Is .	ANTF	Sivriji Pt	AKNT	Snipe I . .	AODT
Sibuko B.... ..	.APZY				

SNOHOMISH—ST GEORGE

Snohomish.	AUBH	Soufrière B .	AYFM	Split, C . . .	BCDK
Snouw Lt V.	AFXR	Sound, The .	ACDM	Spotorno . . .	AHWK
Socha Bitke Pt	AKLE	Scutis	BDCM	Spotsbierg. .	ADKX
Sochinski.	AKLE	Sourop, C	ACTH	Sprogø	ADIW
Society Is	ARGZ	Sousse .	AKVW	Spuria. .	AIKQ
Socoa	AGTX	Souter Pt.	AFLQ	Spurn Lt V	AFMQ
Söderarm	ACJV	South America.	AUWO	Spurn Pt . .	AFMO
Söderhamn.	ACLR	Southampton (England)	AEIW	Squillace .	AIPO
Söderköping	ACIL	Southampton (Huron)	BEHC	Srigina I	AKXQ
Söder Skars	ACRD	Southampton Road	ANLH	Srira, La	AKVJ
Sodertelje	ACIY	South C . .	AQTJ	St Abb's Hd	AFKP
Soduspoint	BDYO	South East P .	AMIU	St Agnes .	AEOV
Sofala R	AMBG	Southend	AKIM	St Agostinho, C .	AVZM
Soggendal strand.	ABUN	Southerness .	AEXL	St Alban's Hd	AEMT
Sogndal	ABQT	South Foreland.	AEKU	St Ambrose I.	ADVX
Sogne Fiord	ABQI	South Goodwin Lt V	AEKS	St Andrew (N B).	BAWU
Sohár..	AMTE	South Haven . . .	BELC	St Andrew, C.	AMFG
Soholm	ABKD	South Head (Inner) . .	ASJY	St Andrew City (Fla)	AXEH
Sokótra I	AMOH	South Joggins. . .	BCAW	St Andrew Ch	BCTC
Solaro, Mt.	AIGW	Southold	AZRR	St Andrew Pt (Pr Ed I)	BDAX
Solent, The.	AELX	South Carolina . . .	AYNF	St Andrew R	ALMF
Soller, P .	AHOM	South Carolina I	AMBC	St Andrew Sound (Ga)	AYXO
Solomon Is	AQWB	South Joggins.	BCAW	St Andrews	AFJE
Solor I	AQOD	South Orkneys	AVKJ	St Ann, C. .	ALKB
Solo Rock.	ARCF	South Pass	AWYQ	St Anne (Alderney) . .	AGCX
Solovetzki	ABER	Southport (England)	AEYN	St Anne Hr (C Br I)	BCUD
Solvesborg	ACFR	Southport (N Australia)	ARJQ	St Anns	AXRK
Solvorn	ABQZ	Southport (Queensland)	ASPM	St Ann's Hd	AESD
Solway Firth.	AEWQ	South Pt (Amherst I)	BDFO	St Antonio	ALGP
Somali	AMON	South Pt (Anticosti)	BDLT	St Aubin	AGBO
Somahland, Italian	AMNX	South Pt (Barbados). .	AYGN	St Aubin (Jersey). . . .	AGDQ
Sombrero cay. . .	AXHJ	South Rock Lt V.	AFUB	St Augustine	AYJB
Sombrero I	AXZY	South Ronaldsay .	AFEL	St Augustine B	AMFD
Somers Cove..	AZGC	South Shetland .	AVKH	St Barbe Pt	AGTV
Somerset . . .	ASYZ	South Shields	AFLM	St Bartholomew.	AYBN
Somes Hr	BASL	South Solitary I.	ASQF	St Bees Hd .	AEWK
Somes I. . .	ATKP	South Stack.	AETI	St Benoit	AMHO
Sommars	ACSQ	South Uist	AFDG	St Blaize, C. .	ALXJ
Somme R.	AFYO	Southwest Prong Lt	AMZII	St Briac	AGFM
Somo Somo Str	ARDH	Southwest Pt .	BDIW	St Bride's B. . . .	AESL
Sonapur . .	ANHP	Southwold .	AFOL	St Brieuc	AGFU
Sonderburg	ADFX	Sövde	ABOX	St Carlos.	AHLF
Söndervig . .	ADRN	Sow and PigsLt V (N S W)	ASKF	St Cassano, P	AJCW
Songka R.. . .	ANWF	Sow and Pigs Lt V (Mass)	BACL	St Cast Pt	AGFP
Songchiu. . .	AONS	Soya saki. .	APMG	St Catherine B..	AGDV
Songhoi R	ANWR	Spadillo Pt..	AIFJ	St Catherine Pt	AEMO
Songa Songa I. .	AMKP	Spain	AGUB	St Catherine Sound	AYLM
Songvaar Lt.	ABVL	Spalato. . .	AJDR	St Christopher I	AYBX
Sonsonate .	AUGJ	Spalmadori Ch	AJEC	St Clair	BEFN
Sooke Inlet	ATXM	Sparo . .	ACHZ	St Clair Lake	BEFH
Soon	ABYN	Sparta . .	AJMW	St Clement B . . .	AYZX
Sophiko P	AJOE	Spartel, C .	ALBJ	St Croix I . .	AXZ8
Sordwana Road	ALZH	Spartivento, C (Italy).	AIXH	St Croix R.	BAWF
Sorel	BDJE	Spartivento, C (Sardinia)	AIEB	St Cruz de la Selva . .	AHNJ
Sorell, P	ASEL	Spathi, C. .	AJMY	St Denis .	AMHE
Sorelle Rocks	AKWS	Spear, C	BEUZ	St Duka F . . .	AKLD
Sörhaugo	ABTD	Spencer B.	ALUO	St Elia, C	AIDX
Sorio .	AONG	Spencer, C	BAYI	St Elmo	AIOU
Soroka .	ABEY	Spencer G	ARSD	St Esprit	BCRD
Sorrento (Italy)	AIHT	Spencer Rock	AOHG	St Eustatius	AYBT
Sorrento (Victoria).	AQGK	Sperone, C .	AIBL	St Famille	BDSO
Sorsogon	ADOB	Spex Str.	AOWQ	St Florent	AICU
Sorte Ch . .	ABJC	Spezia.	AHXP	St Francis, C (Africa)	ALXS
Sorvaag.	ABFL	Spezzia I	AJNR	St Francis, C	BEVF
Sosnovetz I.	AJKZ	Sphakia..	AJPV	St Francis, P . .	BDTY
Sosti I.	ANUT	Spitekeroog I	ADXI	St François	BDSC
Soton.	AIVT	Spinalonga, C	AJPN	St George (Grenada)	AYGD
Sottile, Pt (Austria)	AIMS	Spithead	AELE	St George (N B)	BAXD
Sottile Pt (Sicily) .	AINJ	Spit Lt V	ANME	St George Basin	ARLB
Soudan		Spitzbergen	ABCY	St George, C (Fla) .	AXEW

ST GEORGE—STENJER

St George Hr	BESL	St Marco, C	AIMJ	St Sébastien, C	AMGV
St George Hr (Bermuda)	AYIB	St Marcouf	AGBZ	St Servan	AGFE
St George I	AMCF	St Margaret B	BCKR	St Simon Sound	AYKR
St George Mouth	AKEL	St Margaret's Hope	ABJT	St Stefano	AICX
St George R	BANJ	St Marguerite Ile	AHUX	St Stephen	BAWI
St Georges Ch	AESI	St Marie	AMHF	St Suzanne	AMHJ
St Georges Hd	ASJC	St Marie, C	AMFC	St Thomas (St Lawrence)	BDRQ
St Georges, P	AGSK	St Marks	AXEZ	St Thomas (W I)	AXZB
St Georges Reef	AUEZ	St Martin (W I)	AYBF	St Thomas Bight	AWCD
St Georgio Islet	AJPZ	St Martin de la Arena	AGVQ	St Thomas Peak	AQER
St Georgio, P	AJEL	St Mary B	BCFV	St Thomé	ALQF
St Germain	AGEF	St Mary (Ga)	AYKI	St Thomé, C	AVUH
St Giacomo Pt	AHWZ	St Mary (Newfoundland)	BEUN	St Tropez	AHUK
St Gildas Pt	AGPO	St Mary, C (Africa)	ALTX	St Ubes	AETC
St Gillies sur Vie	AGPZ	St Mary, C (Madagascar)	AMPC	St Ubes	AHCQ
St Giovanni, P	AJIB	St Mary, C (Portugal)	AHDI	St Vaast	AGCD
St Helena	ALVC	St Mary I	AMEJ	St Valery	AFYQ
St Helena B	ALVZ	St Mary, P	AMEK	St Valery-en-Caux	AFYW
St Helena Sound	AYNX	St Mary R (Mich)	BELK	St Vincent (C Verde Is)	ALGV
St Helens	AEML	St Mary R (N S)	BCNO	St Vincent I (Fla)	AXEQ
St Helier	AGDF	St Marys I (Scilly)	AEOU	St Vincent (W I)	AYPQ
St Hospice Pt	AHVK	St Mary's Road	AEOY	St Vincent, C	AHCV
St Ignace	BDRH	St Mathieu Pt	AGJW	St Vincent G	ARUD
St Ives	AEFD	St Matthews I	ATQX	St Vincente, P	AQFE
St Jacques	AGIR	St Maximo	AHUM	St Vito, C (Italy)	AIPZ
St Jago	ALHK	St Michael P	ATQW	St Vito, C (Sicily)	AING
St James, C	ANVQ	St Michaels (Azores)	ALDE	St Vladimir B	AOQW
St Jean	BDSI	St Michel, Mt	AGER	Staaken	ADCZ
St Jean de Luz	AGTU	St Nazaire (Loire Inf)	AGPH	Stabben	ABFQ
St Joao da Barra	AVUL	St Nazaire (Fr Pyr Orient)	AHPW	Stade	ADUN
St John (N B)	BAYN	St Nicholas Lt V	AFOC	Stagno	AJFI
St John B	BERN	St Nicholas	ALHB	Stambul	AKBJ
St John, C	AJFN	St Nicolas	AGSE	Stamford Hr	AZSR
St John Hr (Antigua)	AYCF	St Nicolas Pt	APUW	Stampalia	AJSH
St John Hr (Staten I)	AVKD	St Nicolas, P	AHZW	Stamphani I	AJKS
St John I (China)	ANYD	St Nikolas	AJQL	Stamsund	ABIU
St John I (Red Sea)	AMQB	St Nikolo, P	AJQI	Stangate Creek	AEJX
St John I (W I)	AXZK	St Orlovsk	ATSB	Stangholm	ABHV
St John R (Africa)	ALYR	St Paul (Réunion)	AMHT	Stanislav	AKGX
St John R (Fla)	AYJH	St Paul B	BDQF	Stanley	ASBV
St Johns (Newfoundland)	AEVA	St Paul, C (Africa)	ALNO	Stanley Hr	AVKF
St Johns I	AMQB	St Paul, C (Black Sea)	AKID	Stanley Pool	ALSN
St Johns Pt	AFUG	St Paul de Loanda	ALTB	Stanley, P (Erie)	BEAL
St Jorge dos Ilheos	AVXI	St Paul Hill	ANSD	Stanley, P (Pacific Is)	AQXJ
St Joseph (Mich)	BEJX	St Paul Hr (Alaska)	ATRY	Stanley R	BDEG
St Joseph (Fla)	AXEK	St Paul I (C Br I)	BCUS	Stapleton	AZNP
St Katharine Docks	AEJP	St Paul I (Indian Ocean)	AMJT	Starhejm	ABFL
St Kilda	AFDO	St Paul Rocks	AWCB	Starkenhorst	ACIZ
St Kilda Pier (Melbourne)	ARZG	St Peter B	BCQR	Stark Pt (England)	AENM
St Kitts I	AYBX	St Peter Hr	BDCS	Stark Pt (Orkneys)	AFOB
St Laurent	BDSL	St Peter P	AGDC	Stat	ABPC
St Lawrence (Queensland)	ASVP	St Petersburg	ACRY	Staten I (N Y)	AZNJ
St Lawrence, C	BCLP	St Pierre (Newfoundland)	BETY	Staten I (S America)	AVKB
St Lawrence, G of	BDGC	St Pierre (Oleron I)	AGRH	Stathelle	ABWY
St Lawrence I	APOL	St Pierre (Quiberon)	AGNL	Statland	ABLD
St Lawrence, R	BDMF	St Pierre (Réunion)	AMHR	Stauns Hd	ADFJ
St Leonards	ASLM	St Pierre (St Lawrence)	BDSR	Stauærnsodden	ABXH
St Leu	AMHS	St Pierre (W I)	AYER	Stavanger	ABTW
St Lewis Sound	BEXT	St Pierre de Royan	AGSB	Stave	ABUS
St Louis (Africa)	ALGD	St Pietro (Adriatic)	AJDY	Stave I	BATD
St Louis (N B)	BCYT	St Pietro di Nembo	AIZQ	Stavenes	ABMV
St Louis (Réunion)	AMHS	St Pietro I (Sardinia)	AIEN	Stavoren	ABBV
St Louis B	AXVI	St Pol de Léon	AGIM	Stavros, P	AJSZ
St Louis Canal	AHRN	St Quay	AGFY	Stedsholm	ACIE
St Louis I	AYEP	St Raphael	AHUQ	Steep I	AOEV
St Lucca Hr	AXRS	St Roque, C	AWBS	Stefano Pt	AJXU
St Lucia B (Africa)	ALZF	St Rose,	AMHF	Stege	ADNQ
St Lucia (Madagascar)	AMDX	St Sampson's	AGDH	Steilacoom	AUBT
St Lucia (W I)	AYFG	St Sauveur	AGPW	Steinort, C	ACVE
St Malo	AFCG	St Sebastian B	ALXG	Stenjer	ABMF

STENKYRKEHUK—TAARS

Stenkyrkehuk .	ACHM	Stronsay	AFEM
Steno Pass	AJQO	Strontian I	BEDC
Stenskär .	ACSX	Strood. .	AEJZ
Stensündene	ABQG	Strovathi I.	AJKS
Stephansoit . .	AQUF	Strukamp Pt.	ADEJ
Stephens B	AUVN	Struys B .	ALWZ
Stephens, C	ATEH	Stryensas	AEDU
Stephens, P (Falkland Is)	AVKT	Stuart I.	ATYV
Stephens, P (N S W)	ASMX	Stubbekiobing	ADNH
Stephens Pt.	ASMV	Stylida G	AJUY
Stettin	ACYX	Suakin	AMFW
Stevns C (Klint)	ADNX	Sual	AQRO
Stewart I.	ATGX	Suances	AGVR
Staff Pt.	AGJQ	Suberbieville.	AMGC
Stig Ness	ADMG	Subig, P	AQDY
Stikine R.	ATSP	Succonnesset Shoal Lt V.	BAEV
Stinking I	BEWL	Suchau.	AOIH
Stilow Beacon	ACYD	Suda B	AJFH
Stirling.	AFJX	Sudak	AKHT
Stirrup cays	AXIZ	Sudder-ghat	ANKM
Stire Pt	ACRS	Sunderhoft	ADTB
Stockholm	ACJQ	Suderö.	AFHK
Stockton (Australia)	ASMP	Suder Piep	ADTN
Stockton (England)	AFLY	Sudzu misaki.	AOUG
Stockton (Me).	BAPU	Suez	AMQS
Stokkoen	ABLR	Suez B	AMQT
Stollergrund	ADEX	Suez Canal	AKTI
Stolmen	ABTH	Suez G	AMQK
Stolpmunde	ACYI	Sugarloaf I	ANZS
Stoncica Pt	AJEN	Sugarloaf Pt	ASMZ
Stonehaven	AFIS	Suga sima.	APFH
Stonehouse	AEVV	Sugut R	APZK
Stone Pillar.	BDQY	Sujak B.	AXJW
Stonesbotten	ABEG	Sukkúm B	AKLJ
Stonnington.	AZUK	Sule Skerries	AFDW
Stono Inlet	AYOP	Sulina Mouth.	AKEM
Stony Pt	BDZA	Sulisker.	AFDR
Stor R	ADUH	Sulitjelma Mines.	ABJM
Stora.	AKXN	Sullivan.	BASX
Stora Fiaderagg.	ACMW	Sultan Shoal Lt	ANSQ
Storjungfrun	ACLP	Sultans I	ANDM
Storkallagrund	ACPH	Suma	ABEX
Storklappen.	ACIK	Sumatra.	APQI
Storm B.	ASGR	Sumaya.	AGUO
Stornoway	AFCR	Sumba I.	AQNY
Storskar.	ACOY	Sumbawa.	AQNU
Stosch Ch	AYIT	Sumburgh Hd	AFGM
Stotsund	ABJT	Summerside.	BCZL
Strahan	ASFI	Sumter F	ATOV
Straitsmouth I.	BAJI	Sunda Str	AFSQ
Stralsund	ADBT	Sundarbans, The	ANJQ
Strandebarm	ABSE	Sunderland	AFLR
Strangford	AFUD	Sundet	ACDM
Stranraer	AEXY	Sundsvall	ACLX
Strasburg	ACVZ	Sundt.	ACDM
Stratford Shs.	AZIG	Sungi Banju Asing	APRX
Strathie Pt	AFDX	Sungi Sungsang.	APRY
Strati.	AJUC	Sungi Rokan	AFRD
Streaky B.	ARQO	Suira	ALCB
Streckelsberg	ACZT	Sunk Lt V	AFFG
Strelna.	ACSP	Sunk Rock Lt	AMZG
Strelok B	AOQN	Sunmiyani B	AMWH
Strib Pt.	ADHP	Supé	AUYJ
Stroma	AFEC	Superior City	BENC
Stromboli I	AIOJ	Superior, Lake	BELQ
Strome Ferry	AFCK	Sur (Arabia)	AMSY
Stromness	AFET	Súr (Syria)	AKSL
Stromo	AFHM	Sur Pt	AUHY
Strömstad	ABZM	Surabaya.	APWJ
Strongilo I	AJUX	Surat	AMYS

Surigao	AQKI
Surinam R Lt V	AWGP
Sur-Kenis B.	AKVM
Surry	BARW
Suruga G .	APGJ
Susa (Tunis)	AKVW
Susa, P (Japan)	AOVB
Susaki (Tokyo G)	APHO
Susaki (Shikoku)	APCT
Susan, P	AUBC
Susquehanna R	AZED
Sutherland.	AFDS
Sutherland Dock	ASLE
Sutsini	AFLC
Sutt B	APMX
Suva	ARCH
Suvarov Is	ARHN
Suvero, C	AIJR
Suvorovski	AKGD
Suwo nada	AOEK
Svaneke	ACFJ
Svangen	ABZF
Svanö	ABPV
Svartklubben	ACKH
Sveaborg	ACQZ
Svelvik	ABYE
Svendborg	ADHW
Svendseyre	BFAH
Svenöer	ABXJ
Svenska Bjorn Lt V	ACJU
Svenska Högarne.	ACJT
Sviatotroitski	AKGN
Svinbåden Lt V.	ACDS
Svinör	ABVC
Svolvær	ABIM
Svyatoi Nos	ABFP
Swakopmund Road	ALUJ
Swalfer Ort	ACUE
Swan B	ASEH
Swan I (Tasmania)	ASDM
Swan I (Victoria)	ARYK
Swan R	ARNQ
Swanage	AEMR
Swansea (England)	AERO
Swaneea (N S W)	ASHD
Swanterwitz Lt V	ACZH
Swatau.	ANZU
Sweden	ABZH
Swettenham, P	ANRS
Swilly, Lough	AFSR
Swinemunde	ACZB
Swine R	ACZD
Swin Middle Lt V	AFFL
Switzerland	AHVQ
Syauki Pt	AOBV
Syd Krogo.	ABLF
Sydney	ASKB
Sydney (C Br I)	BCSW
Sydostbrotten Lt V	ACMQ
Sylt I .	ADSE
Symi	AJSN
Synes	ABOH
Syra	AJQW
Syracuse	AILJ
Syria	AKRI
Syrsan.	ACIG
Taa hu ku B	ARGH
Taaj	AQGZ
Taars	ADMP

TAASINGE—TERRIBLES

Taasinge I	ADLJ	Taltal	AVDG	Tasman Hd	ASGJ
Tabarca I	AHJN	Talus, La	BEQV	Tasmania	ASDX
Tabarka	AKWV	Talut Pt (Lorient)	AGMY	Tatamagouche	BCXA
Tabasco	AWSJ	Taman Lake	AKJI	Tate ishi zaki	AOUK
Tablas I	AQHC	Tamandare	AVZI	Tate yama	AFIO
Tablas, Las	AWIL	Tamar, P	AVJL	Tatoosh I	AUCL
Table B	ALWG	Tamar R	ASDU	Tatsingho	AOJC
Table Bluff	AUFK	Tamatave	AMEQ	Tatsupi, C	AOTR
Table C	ASBT	Tambelan Is	APXS	Tauan	AQSR
Table I	ANMV	Tambo R	AVBT	Taukopah	ANPK
Taboria	ALJD	Tambo de Mora P	AUZG	Taunton R	AZWY
Tabu	ALMC	Tamise	AEGX	Taupiri	ATBY
Tabuan	AFWT	Tampa	AXFK	Tauranga	ATLW
Tabuia	ALJD	Tampico	AWTC	Tansk B	AOSH
Tachibana ura	APDF	Tamrida	AMOI	Tavifi	AHDJ
Tachin R	ANCG	Tamsui Hr	AOBW	Taviuni	ARDK
Tacking Pt	ASNK	Tanabe	APEL	Tavolara I	AIDL
Tacloban	AQJI	Tana Fiord	ABFY	Tavoy	ANOT
Tacna	AVCF	Tana R	AMNK	Tawi Tawi	AQLU
Taco, P	AXLS	Tanamo, P	AXLU	Tay R	AFJB
Tacoma	AUBR	Tananarivo	AMDP	Tayabas R	AQGX
Tadousac	BDPC	Tandine B	ARBU	Tayangho	AOLB
Tadri R	ANBU	Tanga B	AMLS	Taylorville	ATFS
Taganrog	AKIX	Tangalle	ANBX	Tchio	ARBH
Tagbilaran	AQJB	Tangaluma Roads	ASQD	Tearaght I	AFQP
Taghamento R	AIUP	Tanganyika Lake	AMNE	Teavarua Hr	ARBL
Tago B	APGT	Tangier	ALBI	Tecoanapa	AUOD
Tagus R	AHBZ	Tangier I	AZGI	Tecojate	AUPS
Tahiti	ARHB	Tangkiang	AOCN	Tees R	AFLW
Taiaroa Hd	ATIF	Tangoa	AQXG	Tegal	APVN
Taiho I	ANYL	Tangola tangola B	AUOR	Teheran	AMUW
Taillefer Pt	AGNF	Tangtang	AMEH	Tehmaki Str	ATMR
Taimir G	ABCF	Tanjong Bulus	ANSL	Tehuantepec	AUOZ
Tai oa, P	ARGN	Tanjong Datu	APXW	Teignmouth	AENG
Tai-o-hué	AROL	Tanjong Priok	APVP	Teignouse, La	AGNJ
Taiping fu	AOGM	Tanjore	ANFT	Tekfur Dagh	AJXO
Taipu Pt Lt V	AWDV	Tankar	ACOT	Tellicherri	ANCJ
Taitai B	AQCJ	Tankari	AMYK	Telok Anson	ANBG
Taitan I	AOBH	Tanlé B	AQZH	Telok Betung	APSR
Taitzuchi Beacon	AOGX	Tanna (India)	AMZC	Teluti B	AQFW
Taiwan fu	AOCI	Tanna (Pacific)	AQXZ	Temriuk B	AKJH
Taize Ura	APEU	Tanôn Str	AQIP	Tenacatita B	AUNG
Tajer, P	AJDB	Tanso	ABQE	Tenasserim	ANOY
Tajura	AMOZ	Tantu	AOFW	Tenby	AESB
Takamatsu	AFDS	Tani-Keli Islet	AMGU	Tendra B	AKHB
Taka Ramata	AQMY	Taormina	AIKV	Tenedos	AJUL
Taka sima	AOWJ	Tapagipe B	AVXY	Tenerife	ALEZ
Takasimacho	APHU	Tapajoe R	AWEX	Ténèz, or, Ténès	AKYT
Takau	AOCL	Tapanuli B	APUH	Tengchaufu	AOIX
Taketoyo	APFW	Tappi Saki	AOTR	Tenkitten	ACWZ
Takhkona Pt	ACTR	Tapti R	AMYR	Tennant Hi	BAOH
Takly, C	AKIB	Tapul Group	AQLS	Teodo B	AJFW
Ta ko ho	AGJF	Taranaki	ATCR	Tepa	AQPC
Taku	AOJH	Tarang	ANQB	Tequepa B	AUNY
Taku Bar Lt V	AOJK	Taranto	AIPW	Terang B	AQOE
Takush B	AKXI	Tarapur Pt	AMYX	Terceira	ALCY
Takii Shan	AOKZ	Tarascon	AHRO	Tereboli	AKMG
Talamone B	AHYP	Tarbert	AFRB	Terkolei I	AFRO
Talanta B	AJOZ	Tarbet Ness	AFHR	Termini Imerese	AINR
Talara	AUWV	Tarifa	AHFL	Terminos Lagoon	AWSH
Talauer Is	AQLX	Tarkhan, C	AKHF	Termoli	AIRU
Talautse Is	AQLW	Tarpaulin Cove	BACL	Ternate	AQRP
Talbot I	AQSN	Tarrafal B (St Antonio)	ALGS	Terneuse	AEOF
Talbot, P	AERK	Tarrafal Pt (St Jago)	ALHO	Terningen	ABMN
Talca	AVOD	Tarragona	AHLP	Terracina	AIGM
Talcahuano	AVON	Tarraville	ASBQ	Terranova (Sardinia)	AIDG
Talienwan B	AOKN	Tartary, G of	AOQX	Terranova (Sicily)	AILX
Talindak	AQCI	Tartus	AKRX	Terre-Negre	AQRY
Talkna Fiord	BFAK	Tasanho	AOJF	Terribles, The	ANLI
Tallais Bank Lt V	AGSI	Tasiko I	AQXL		

TERSCHELLING—TRANSVAAL.

Terschelling	ADYZ	Tilig, P	AQDF	Tönsberg.	ABXN
Terschelling Bank Lt V	ADYX	Tillamook Rock	AUDO	Topche B	AJYP
Terstenik I..	AIZT	Tilsit	ACWN	Topdals Fiord	ABVX
Testa, C	AICK	Tilt Cove	BEXK	Topkhana	AKBS
Teste de Buch, La	AGTL	Timaru	ATIX	Topla B	AJFT
Testigos Is.	AWIZ	Tumballiei I	AWYB	Topolobampo B.	AUMJ
Tete	AMBU	Timmendorf	ADCK	Tor	AMQW
Tete de Flandre	AEGT	Timor	AQON	Tor Pt	APTO
Tetuan	AKZY	Timor Laut	AQPF	Toragy Pt	ASIP
Tetulia R.	ANKB	Timsah, Lake	AKTO	Tor B (England)	ABNI
Teuk cham	AOBZ	Tina Mayor	AGVU	Tor B (N S)	BCOD
Teulada, C.	AIEG	Tinghai Hr	AOEI	Tordera	AHMP
Tévennec	AGLH	Tino I	AHXO	Töre	ACNY
Tewantin	ASRI	Tinos	AJQK	Torgauten..	ABYV
Texas	AWTK	Tiñoso, C..	AHIY	Toriñana, C	AGYH
Texel, The.	ADZJ	Tinto R	AHDW	Tormentine, C.	BCXS
Texel I	ADZH	Tintingue	AMBH	Toro Pt.	AWNU
Thaipeng	ANQZ	Tipara B.	ARSX	Toronto	BDXT
Thames R (Conn)	AZTV	Tipaza P.	AKYQ	Torquay	AENH
Thames R (England).	AEHU	Tiri tiri Matangi I	ATMZ	Torre	AIWM
Thames R (N Z).	ATMJ	Tistlarne.	ACBU	Torre dell' Annunziata, P.	AIHQ
Tharúa Hr.	ANPV	Tizenko B..	AONT	Torre Orlando ..	AIGQ
Thaso I.	AJUT	Tiznitz	ALCG	Torres, P	AIFM
Thayetmyo	ANML	Toamasina	AMRQ	Torres Str	ASZE
Themistocles, C.	AJOQ	Toba.	APFE	Torrevieja	ABJK
Theodosia.	AKHW	Tobacco P , R	AZBP	Torrox.	AHGQ
Therapia...	AKCF	Tobago.	AYGR	Torruella	AHMY
Thermia	AJQP	Tobako R	AQFY	Tortola	AXZM
Thessow...	ACZY	Tocantins R	AWEF	Tortoli	AIDQ
Thinai Hr ...	ANWI	Tocopilla	AVCR	Tortoralillo.	AVED
Thisted	ADRF	Todos Santos B	AUKP	Tortosa, C	AHIM
Tholen.	AEFJ	Togo	ALNT	Tortuga I (Haiti)	AXTM
Thomas I	ALQP	Tokaroíski, C	AOPY	Tortuga I (Venezuela)	AWJQ
Thomas Pt.	AZDI	Tokelau	AREP	Tortugas Is	AXGL
Thomaston	BANM	Toku shima	APDJ	Torungen.	ABWG
Thorbjörns Kjær.	ABYW	Tokuyama	AOZN	Tory Ch	ATKJ
Thornton R..	AMBL	Tokyo	APIC	Tory I	AFSQ
Thornton Haven	AOKW	Tolboukin	ACRW	Tougourski G	AOSB
Thorō.	ADHY	Toledo	BEOO	Toulinguet Pt	AGKW
Thorshavn...	AFHI	Tolero P	AJEZ	Toulinguet I	BEWV
Thousand Is , The	BDVX	Tolkemit	ACXI	Toulon	AHTF
Three Hummock I	ASFB	Tolkeehwar Pt.	AMZY	Toulouse	AHQD
Three Points, C...	AIMU	Tolle	AITF	Touquet, Pt du	AFYK
Three Points, C (Arg)	AVLO	Tolmeita.	AKUS	Tourane	ANWK
Three Rivers.	BDTV	Tolon B	AJNM	Tour de la Lande.	AGIH
Throgs Neck	A7SF	Tolstonow	ABCL	Tourneville.	AGEL
Thunder B	BEGI	Tolten	AVHQ	Tourville	AGEL
Thunder C	BEMO	Tomaga sima	APBZ	Touzla	AJYI
Thunō	ADJZ	Tomales B.	AUFY	Tova I	AVLS
Thursday I.	ASZM	Tomari ura	AOYB	Toward Pt	AEYO
Thurso	AFOZ	Tomé	AVGJ	Townsend	BANS
Thybo Rön	ADRG	Tomil B .	AQVH	Townsend, P	ATZO
Tibbett Pt	BDZM	Tomo	APBF	Townsville	ASWT
Tiber R.	AIFZ	Tompkinsville	AZNS	Toyohashi	APGB
Tiefen Hafen	ACTX	Tona Canal	APJD	Tracadie (N B)	BDHS
Tiemunkwau	AOJD	Tonala	AUPE	Tracadie (N S)	BCVQ
Tienpak Hr	ANXW	Tondi	ANFM	Tracadie (P Ed I)	BDCV
Tientsin	AOJL	Tone gawa	APIF	Trade Town	ALKP
Tierra del Fuego.	AVJU	Tonga Is.	ARDQ	Træn I	ABKC
Tiflis ..	AKLW	Tong Aing	ANOP	Trafalgar, C.	AHFI
Tigani, P...	AJTH	Tongass P	ATSV	Traiventos	AVNU
Tigil R	AOSQ	Tongatábu I	ARDS	Tralee	APQS
Tignish.	BDEY	Tongka	ANPS	Tramontana Pt	AIMP
Tignoso I.	AJHP	Tongoi, P .	AVEJ	Tranekicer	ADIP
Tigris R ...	AMUS	Tongsang Hr	AOBD	Tranholm	ABML
Tihenpien Hr	ANXW	Tougshu I....	AODM	Trani	AIQY
Tikao.	AQOI	Tongue Lt. V	AEIC	Tranö (Nyleden)	ABRU
Tiko	AFTW	Tonkin G .	ANWO	Tranö	ABHW
Tilat .	AQHT	Tonning	AOTF	Tranquebar	ANFZ
Tilbury Docks	AEIO	Tonquet	AFYK	Transvaal	ALVP

TRAPANI——URAGA

Trapani .	AINB	Tsaulianghai	AOND	Tutun Liman	AJYK
Traste	AJGB	Tsiontangkiang . .	AOES	Tuxpan	AWSZ
Trau . .	AJDP	Tsingseu I	AOBG	Tuzla (G of Ismid)	AJYI
Travancore Coast . .	ANCU	Tsu	APFN	Tuzla Bank (Kertch Str) .	AKJQ
Travemunde	ADCQ	Tsuda B .. .	APDQ	Tuzla C (Roumania) ..	AKEG
Trave R . .	ADCT	Tsugaru Str.	AOTC	Tvedestrand . . .	ABWK
Traverse C (P Ed I).	BCZO	Tsukiji .	AFID	Tweed R (England)	AFKT
Traverse, Lower Lt V	BDQS	Tsukumi. . .	AOYV	Tweed R (Queensland) .	ASPC
Traverse, Upper, Lt V	BDQV	Tsuno sima	AOLW	Twofold B	ASHW
Traverse City .	BELH	Tsuruga B	AOUN	Tybee . .	AYMQ
Trebizond .	AKMC	Tsurugi saki	APHG	Tyka..	AOCD
Tred Haven Creek	AZEY	Tsuru sima	APEG	Tylö	ACDE
Tre Forcas, C .	AKZT	Tsu sima	AONC	Tyne R . .	AFLH
Tréguier	AGHP	Tuamarina .	ATKD	Tynemouth	AFLI
Treisseny, P . . .	AGJC	Tuamotu Arch	ARGQ	Tyre? .	AKSL
Trekroner	ADOQ	Tuban .	APWC	Tyroom Road	ASRU
Trelleborg . .	ACER	Tubuai	ARHS		
Trembles Sh.	BDTN	Tucker Beach .. .	AZMI	Ua Huku I	ARGO
Tremiti Is	AIRL	Tubud Pt .	AQKE	Ua Pu I. .	ARGP
Trenton (Erie). .	BEAU	Tudor, P	AMND	Ubas Pt .	AIXW
Trenton (N J)	AZKM	Tugela R. .	ALZD	Ubati .	AQZU
Tréport Le .	AFYS	Tula	AMVU	Ubatuba .	AVTC
Tre Porti P .	AITW	Tulang Bawang R . .	APSM	Uckermunde	ACZU
Tre Pozzi P . .	AJER	Tulalip .	AUBD	Uddevalla . .	ABZU
Tres Marias Is , las . .	AUMY	Tulang R. . .	APSM	Uddyhoj Lt. .	ADPK
Tres Montes, C	AVIN	Tulcea .	AKEP	Udsire	ABTL
Tres Puntas C	AVIO	Tulcha .	AKEG	Udski B	AOSC
Trévignon Pt . .	AGMJ	Tullear .	AMFH	Uea	ARDW
Trevose Hd .	AEPJ	Tumaco, P .	AUTS	Ugi I . . .	AQWG
Triagoz .	AGHX	Tuman B .	AVFU	Ukinsk B .	AFNL
Trial B .. .	ASNW	Tumen R , or, Tumen Ula	AOPE	Ulbo	AJBZ
Trichinopoly ,.	ANFV	Tumbez .	AUWS	Uleåborg	ACOK
Trieste	AIVO	Tuma Pt	AXWV	Ulenge . .	AMIT
Trieux R .	AGHF	Tunara .	AHGC	Ulfsten .	ABOQ
Trincomali	ANFE	Tunas .	AXPU	Ulko Kalla .	ACOP
Trindelen Lt V.	ADQI	Tungao .	ANZO	Ulladulla .	ASIZ
Tringano .	ANTR	Tungchau .	AOJM	Ullanger .	ACMJ
Trinidad (W I) .	AYGZ	Tungenæs .	ABTU	Ullensvang .	ABSK
Trinidad de Cuba .	AXPS	Tunghi B .	AMCV	Ulua R	AWPS
Trinidad, G of	AVIZ	Tungkuen .	AODC	Ulvesund .	ABPE
Trinidad Hr (Cal) .	AUFE	Tunghu .	AOHE	Ulveshale .	ADNO
Trinidad Is (S Atlantic)	AWCD	Tunglotu Beacon .	AOFR	Umago .. .	AIWG
Trinité B , La	AYET	Tungsha .	AODE	Umea .	ACMS
Trinity B	AOPM	Tungsha Lt V .	AOFE	Umpanbinyoni R	ALYU
Trinity Hr.	BEVU	Tungting Lake.	AOIE	Umpqua R	AUEB
Trinkitat Hr.	AMPV	Tungyai R .	AXUR	Umur Banks Lt V	AKGT
Tripoli (Africa) . .	AKUN	Tunis .	AKVD	Una .	AVXE
Tripoli (Syria)	AKSB	Turanganui R .	ATLQ	Una R .	AVZG
Tristan da Cunha .	ALVG	Turiamo, P .	AWKF	Unalaklik F	ATQD
Tristoma. . . .	AJSE	Turin	AHWU	Understen	ACKL
Tristomos. . . .	AKPY	Turk Is.. .	AVKR	Underwood, P .	ATKH
Triumpho	AVRM	Turkey .	AJVP	Unie I	AIZD
Trivandrum	ANCW	Turnabout I .	AOBP	Unieh.	AKMO
Trocadero . .	AHEW	Turnberry Pt .	AEYD	Union B	AVMR
Troense. . . .	ADHZ	Turneffe Is .	AWQR	Union City	ATZY
Troja I. . . .	AHYN	Turner B .	BCLG	Union Iron Works Docks.	AUGM
Tromlien I .	AMIX	Turnu-Magurele . .	AKFB	Union Is .	AREP
Tromsö .	ABHC	Turnu-Severin .	AKFE	Union, P La	AUQT
Trondhjem	ABMD	Turos Inlet . .	ASIL	Union River B	BASC
Troon . .	AEYI	Turton, P .	ARTK	United States	AZCN
Troubridge Shs .	ARTS	Turyassu .	AWDK	Unkofsky B .	AONH
Trouville	AGBF	Tuscany	AHXV	Unst	AFHD
Truk Is .	AQVP	Tuskar Rk .	AFVS	Uomo-Morto Pt .	AIOC
Truro .	BCDW	Tusket R .	BCGW	Upolu I	AREG
Trussan B	APTN	Tuslinski Lt V .	AKIC	Upper Cedar Pt	AZBS
Truxillo (Honduras).	AWPG	Tuspan. . . .	AWSZ	Upper Traverse Lt V.	BDQV
Truxillo (Peru) . .	AUXQ	Tuticorin .	ANDZ	Upright, C .	AVJK
Tsangrad Mouth. .	AKFS	Tutoia .	AWCV	Upsala .	ACLG
Tryon. . .	BCZR	Tutuila I. .	ARDZ	Urado Hr	APCW
Tsauhia I	AOGI	Tutukaka Hr	ATNI	Uraga Ch .	APHK

URA—VINCENT.

Ura Kami Hr	APET	Vanua Mbalavu . . . .ARDL	Vestborg Pt . . ADJX
Urakawa . . .- . .	APLE	Varanger Fiord . . . .ABFU	Vesteraalen . . ABHQ
Ureparapara I . .	AQWV	Varzze . . . . AHWO	Vesterrenden. .. . . . ACBM
Urk . .	AEBR	Vardar Bank Lt V . . AJVU	Vesterwik . . . ACIR
Ursholm . . .	ABZN	Vardö . . . . . . ABFX	Vestmanhavn . . AFHJ
Ursvik	ACNJ	Varel . . . . ADWU	Vestra Banken Lt V ACLN
Uruguay.	AVPM	Vareler Siel. . . ADWY	Vestre Havn . . ABVW
Urup I.. . .	APMF	Varella, C . . . ANWB	Vesuvius, Mt. . . AIHO
Urzuf.	AKHS	Vares, C . . . . . AGXC	Vianna . . .AHBC
Usedom . . .	AOZQ	Varna . . . . AEDV	Viareggio. . . AHXT
Ushant . .	AGJO	Varne Lt V . . . AEKX	Vibberodden . . . . ABUM
Ushibuka Hr . . .	AOXU	Varsko. . . . . . ADAF	Viborg . . . . ACRN
Ushinish . .	AFDH	Vasili R . . . . AJMU	Vicenza . . . . . AIUF
Ushuwaia	AVJW	Vasilina, (. . . AJVC	Vicksburg . . .. AWZX
Usk . . .	AEQT	Vasto . . . . AIRV	Victor, P . . . ARVO
Uson B . .	AQHO	Vatersay I . . . . AFDN	Victoria (Australia) . . ARXC
Ustica I	AINZ	Vathi, P (Archipelago). . AJTI	Victoria (Brazil) . . AVWE
Usu, C . . .	APLN	Vathi, P (Greece) . . AJKN	Victoria (B C) . . ATXE
Usuki. . . .	AOYW	Vathy, P . . . AJRQ	Victoria (Hongkong) . . ANYP
Ut Grimden Lt V . . . ACCE		Vaticano, (. . AIJX	Victoria B . . AOKQ
Utholmen. . .	ACHK	Vatomandri . . . AMER	Victoria Dock (Bombay) . AMZK
Utilla I . .	AWPQ	Vato. . . . ACJY	Victoria Hr.. . . AFYR
Utklippan . . .	ACFZ	Vaxholm . . . ACJM	Victoria Haven . ABJF
Utlangan . . .	ACOD	Veblüngsnæs . . . . ABNP	Victoria I . . . BEMR
Utö . . .	ACQH	Vecchio, P . . . . AHYL	Victoria, Mt (Auckland) . ATMV
Utrecht.. . .	AEBJ	Vecchio, P . . . . AIBF	Victoria, Mt (Wellington). ATKR
Utsire . .	ABTL	Vedbæk . . . . ADOU	Victoria Nyanza Lake . AMNF
Uvita Hr . . .	AUSD	Vedelsborg Hd . . . ADHN	Victoria, P (England) . . AEJV
Uwajima. . .	APCR	Veere . . . AEFQ	Victoria, P (S Australia) . ARTC
		Vefring . . . ABPW	Victoria, P (Seychelles) . AMJC
		Vegesack. . . ADWB	Victoria Pier. . . BCST
Vacche I ... . AICF		Veglia. . . . . AIZU	Victoria R . . . . AREC
Vada . . .	AHYG	Veile . . . ADJN	Victorovski. . . AKGB
Vaderoarne . . . . ABZX		Veis Ness . . . .ADIF	Vicuna Hr . . . . ARDE
Vaderobod	ABZQ	Vekens B . . . APTJ	Vieille, La. . . . AGLI
Vaddo . . .	ACKB	Vela, C La . . . AWKN	Vienna . . . . AIVC
Vado	ABWM	Velasco . . . AWUV	Vieques I . . . AXYU
Vadsö.	ABFW	Velcz de la Gomera. . AKZV	Viergo I . . . AGJE
Væggen	ABIE	Velez Malaga . . ABGP	Vieste . . AIRH
Værö	ADMJ	Velo, P. . . . AWNO	Vieux F . . . AYFO
Vacröerna . . .	ABMO	Vendres, P . . . AHPS	Vifsta (Wifsta) wharf. . ACMAE
Vailoa . .	AREF	Venere, P. . . . AHXJ	Vig . . . ABUF
Vakalpudi	ANHF	Venezuela . . . AWIF	Vigan . . . AQBU
Valberg. . . .	ABIT	Vengurla . . . ANBI	Vigie Summit . . AYFK
Valdemarsvik . .	ACIH	Vengurla Rocks . . ANBH	Vigo . . . AGZF
Valdersund	ABLS	Venice . . . AITS	Vigsnæs Mines . . ABTO
Valdivia . . .	AVHU	Ventimiglia . . . AHFY	Vigten Is . . . ARET
Valença (Brazil) . .	AVXQ	Ventoso B . . . AUOX	Vikor . . . ABSG
Valença (Portugal). .	AGZX	Ventoso Mt . . . AGXJ	Viksören . . . ABQN
Valencia... . .	AHKO	Ventotene I . . . AJGS	Vilaine, La, R. . . . . AGON
Valentia	AFQL	Venus Hr . . . AEQW	Villa (Norway) . . . ABLF
Valetta . . .	AIOR	Venus Pt . . . ARHC	Villa (Finland). . . ACRM
Valle Grande. .	AJEQ	Ver, Pt de . . AGBR	Villa de Conde . . AHBE
Vallejo	AUGV	Vera Cruz . . . AWSX	Villa do Porto . . AIDN
Vallière Pt . . .	AGSJ	Verawal. . . AMXR	Villa Franca . . ALDB
Vallo	ABXQ	Verbenico P . . . AIZY	Villagarcia . . . AGZC
Valona B . .	AJGX	Verboska.. . . AJRH	Villajoyosa . . ABJS
Valparaiso. .	AVFC	Verde C . . . ALGI	Villano, C . . AGYD
Valsche, C . .	AQSH	Verde I . . . . AHFS	Villanueva . . AHLX
Vamadori Ch. . .	AFVH	Verga, C . . . ALIZ	Villa Pillar . . AVCF
Vancouver . .	ATUV	Verko Matala Lt V . . ACRQ	Villa Reale . . AHDK
Vancouver I. . .	ARIY	Vermilion . . BEDM	Villaricos . . AHIT
Van Diemen G.	AOYF	Vermilion B ... . . AWXJ	Villa Velha B . . AVWC
Van Diemen Str... .	AMLW	Vernon, Mt . . AZBY	Villaviciosa . . AGWB
Vanga.	ABQO	Vernon R... . . AYME	Ville-es-Martin. . . AOPD
Vangsnæs	AQWR	Versailles.. . . AFZY	Villefranche . . . AHVG
Vanikoro I	AGOE	Vert C . . . . ALGI	Villegagnon F . . . AVTR
Vannes	ARBO	Verte B . . . BCXP	Vin I . . . BDGR
Vao	ARCU	Veruda, P . . AIXJ	Vinaroz . . . AHLC
Vanua Levu . . .	ARCU	Ve Skerries . . . AFGZ	Vincent, C . . . . BDWG

VINCENT—WIGTON

Vincent, P	ARUJ	Waipaoa R	ATLO	Weihaiwei	AOIS
Vineyard Haven	BADF	Waipapapa Pt	ATHK	Welchpool	BAVU
Vineyard Sound Lt V	BACI	Wairau R	ATJY	Weld, P	ANQV
Vinga	ACBN	Waitapu	ATHP	Welland Canal	BDYC
Vingaskar	ACDI	Waitara R	ATCL	Wellesley Is	ARIJ
Vingorla	ANBI	Wakamatsu	AOWC	Wellington	ATKV
Virama	AXQD	Wakasa B	AOUM	Wells	AFNP
Virgin Gorda	AXZO	Wakayama	APCF	Welshpool	ASBI
Virgin Is	AXYO	Wakefield	ARUS	Wenchau B	AODX
Virginia	AYUP	Walcheren	AEFO	Wenchau fu	AODR
Virgins, C	AVJS	Walcott, P	ARLQ	Weser Lt V	ADVL
Visby	ACHL	Walde	AFXV	Weser R	ADVK
Viscaya	AGVD	Wales	AEQV	Wessel, C	ARIS
Visdal	ABNV	Walfisch B	ALUK	West C Howe	ARPI
Vistula R	ACXM	Wallace	BCXG	West Channel (Kattegat)	ACBM
Viti Levu Lt V	ARCG	Wallaroo	ARSV	West Chop Pt	BADC
Vivero	AGWZ	Wallis I	ARDW	Western P	ARZU
Vizagapatam	ANHG	Walney I	AEWD	Westerplatte	ACXR
Vizeu	AWDO	Walton	BCEI	West Hinder Lt V	AEHI
Viziadrug	ANBF	Wandelaar Lt V	AEHF	West India Docks	AEJI
Vlaardingen	AEDB	Wando R	AYPK	West Indies	AXIJ
Vladivostok	AOPZ	Wandsbek	ADUX	West Ironbound I	BCJT
Vlieland	ADZP	Wangaehu	ATDE	Westkapelle	AEFR
Vliko P	AJIP	Wanganui	ATDC	Westminster	ABHW
Vlissingen	AEFU	Wanganui Inlet	ATEZ	West Pt (Anticosti)	BDMC
Vogeeland	ADTX	Wangeroog I	ADXF	West Pt (P Ed I)	BCYZ
Vohimao B (Vohemar)	AMEC	Wani B	AQNC	West Pt York R (Va)	AYXM
Volano	AITB	Wantaokwan	AOEQ	Westport (Ireland)	AFRU
Volcano B	APKW	Warberg	ACBX	Westport (N Z)	ATPK
Volden	ABOW	Wardell	ASOR	West R	ANZB
Volga R	AKLT	Wari	ALOI	West Sand	ARYM
Volo	AJVF	Warkworth	AFLC	West Sister I	BECW
Volosca	AIYJ	Warnhro Sound	ARNV	West Volcano I	AOEW
Voloshski, or, Voloshskua	AKGO	Warnemunde	ADCF	Wetta I	AQOW
Volta R	ALNK	Warner Lt V	ABLV	Wexford	AFVO
Voltri	AHWQ	Waro	ACBT	Weymouth (England)	AEMU
Vona B	AKML	Warren	AZWM	Weymouth (N S)	BCFY
Vonitza	AJIH	Warrender, P	ARKS	Whaingaroa Hr	ATCB
Voorne Canal	AENP	Warrington	AXDS	Whale Rock	AZVC
Vordingborg	ADNP	Warrnambool	ARXO	Whalsey	AFGR
Vorona Islet	AMGS	Warsaw	ACVM	Whampoa	ANYK
Vostitza	AJLK	Warsaw Sound	AYMH	Whangarei Hr	ATNG
Vourlah	AKON	Waru B	AQPV	Whangaroa (Chatham Is)	ATOV
Vraengen	ABXP	Wasa	ACPF	Whangaroa (N Z)	ATNU
Vresen	ADIT	Wash, The	AFNB	Whangaruru Hr	ATNK
Vulcano I	AIOG	Washington (D C)	AZCH	Whidbey I	ATZC
Vungchao	ANWQ	Washington (N C)	AYSW	Whitby	AFMB
Vungkuit	ANWJ	Washington F	AZCB	White Fish Pt	BEPL
Vurko B	AJIV	Washington G	AOMK	Whitehall	BEKS
Vyl Lt V	ADRT	Washington, P	BBIQ	Whitehaven	AEWL
		Washington State	ATXW	Whitehead I (Me)	BAOK
		Washishti R	AMZX	Whitehead I (N S)	BCOG
Wada no misaki	APBQ	Wash-sheecootai B	BEFR	White I Reef Lt V	DDPM
Waddington	BDVC	Wasin	AMLX	White Sea	ABED
Wäddö	ACKB	Wassi Kussa	AQSL	Whitianga	ATMB
Wade I	AOGL	Watagheistic Sound	BEPX	Whitsand B	AEOF
Wade Pt	AYUJ	Watchet	AEPW	Whitstable	AEKD
Wadi Draa	ALCK	Watch Hill Pt	AZUN	Whyda	ALNV
Waglan Islet	ANZF	Waterford (Ireland)	AFWE	Wick (Scotland)	AFHO
Waialua	ARPW	Waterford (Queensland)	ASRC	Wickford	AZVO
Waiauoua R	ATJO	Waterview B	ASLI	Wickham, C	ASCR
Waigatch I	ABDV	Watling I	AXJU	Wicklow	AFVI
Waigiu I	AQRS	Watson B	ASKL	Wicomico R	AZBJ
Waihau Hr	ATMB	Watum	ADYO	Wide Bay Hr	ASRM
Waiheke Ch	ATMP	Waukegan	BEIA	Widnes	AEUZ
Waikato R	ATBE	Wauraltee	ARTB	Wiclingen Lt V	AEHD
Waikawa	ATHM	Wea Sisi	AQYD	Wieringen	ADZR
Waikouaiti	ATIN	Weda	AQRK	Wufsta Wharf	ACME
Wailangilala I	ARDJ	Wedge I	BCNL	Wight, Isle of	AEMJ
Wainfleet	AFNH	Weener	ADYZ	Wigton	AEXF

WIJU—ZENGG

Wiju	AOLH
Wik	ADES
Wiken	ACDU
Wild Boar Reach	AOGV
Wilhelmshaven	ADWP
Willapa R	AUCR
Willemsoord	ADZF
Willemstad	AEDT
Wilham, F	AFBP
Wilham Pitt B	AMGW
William, P.	ATHG
Williamstown	ARYX
Williamstown Docks	ARZF
Willoughby C	ARUB
Willunga	ARVG
Wilmington (Cal)	AUJC
Wilmington (Del)	AZJL
Wilmington (Ga)	AYMK
Wilmington (N C)	AYRD
Wilna	ACVJ
Wilson Promontory	ASBC
Wilson Pt	ATZI
Windau	ACUZ
Windenburger	ACWJ
Windmill Pt	AYZH
Windsor (England)	AEBZ
Windsor (Erie)	BECD
Windsor (N S)	BCEX
Windward Is	AYPE
Winga	ACBN
Wings Neck	BACD
Winnissimmet	BAIE
Winter Hr	BATG
Winterport	BAPX
Winter Quarter Sh Lt V	AZHD
Winterton	AFNU
Wisbeach	AFNL
Wiscasset	BAMU
Wisconsin	BENG
Wise Mt	AENZ
Wismar	ADCM
Withernsea	AFMK
Wittower	ADBO
Wivenhoe	AFPI
Woitzig Lt V	ACZG
Wokam I	AQFJ
Wolf I	AYLG
Wolf Rock	AEOS
Wolf Trap Shs	AYXV
Wolgast	ACZO
Wollin	ACYT
Wollongong	ASJM
Wolves Is	BAXV
Wonsan	AOLS
Wood End	BAGC
Wood I	BALQ
Wood Holl	BACR
Woody I	ASTB
Wooloomooloo B	ASKF
Woolwich	AEIW
Workington	AEWO
Worm's Hd	AERT
Wormso	ACTN
Worthing	ABLM
Would Lt V	AFNT
Wrangel, P	AFSP
Wrangell J	AFOV
Wrath, C	AFDU
Wreck B	AUVF
Wremen Tief	ADVP
Wuchang fu	AOIB
Wuching	AOBP
Wuchu fu	ANZE
Wufu	AODB
Wuhu	AOGN
Würtemberg	ACVW
Wustrow	ADCB
Wusueh	AOHS
Wusung	AOFI
Wyandotte	DEAX
Wyk	ADSU
Wyndham	AEKN
Wyre R	AEVX
Xagua, P	AXPK
Xeres	AHER
Xeros, G of	AJWP
Xingu R	AWER
Yafa	AKSU
Yalahau	AWRO
Yali B	AKQB
Yalmal Promontory	ABCS
Yalta	AKHQ
Yalu R	AOLD
Yamada (Owari B)	AFFL
Yamada	AFKE
Yamadori Ch	AFJG
Yanaon	ANGX
Yanez B	AVHL
Yangtse R	AOFD
Yankalilla	ARVI
Yap I	AQVG
Yaquina	AUDW
Yarbutenda	ALHX
Yarmouth (England)	AFNY
Yarmouth (N S)	BCGK
Yarra R	ARZB
Yate, P	ARBL
Yatsushiro	AOXR
Yawatahama Hr	APCK
Ydro Mökkelae	ABWJ
Yé	ANOR
Yebnah	AKSY
Yebosi I	AOWH
Yollaboi Sound	AIJU
Yellow I	ATWB
Yellow R	AOJC
Yellow Sea	AOIJ
Yemen Province	AMSC
Yenbo	AMRO
Yonchu	AOFU
Yengen	AQZW
Yenikale C	AKIL
Yeni-Kioi	AKCE
Yenisei	ABCJ
Yeni-Shehr, C	AJWU
Yeno ura	APGO
Yontoa B	AOKE
Yorimo saki	APLF
Yosan Saki	APKV
Yesashi	APMZ
Yotorup I	APMB
You, Ile d'	AGPX
Yingtsu	AOKC
Yinkoa	AOKG
Ylo Road	AVBU
Ymuiden	AECG
Yobuko	AOWK
Yochau fu	AOID
Yokka ichi	APFQ
Yokohama (Rikuoku G)	AOTI
Yokohama (G of Tokyo)	AFHT
Yokosuka	APHN
Yola	ALPE
Yonesiro Gawa	AOIS
Yongsangang R	AOMG
Yonodzu, P	AOYR
York (England)	AFMJ
York, C	ASZC
York Factory	BEZC
York R (Me)	BALV
York R (Va)	AYXJ
York Sound	ARKU
Yorktown	AYXL
Yoshida, P	APCO
Youghal	AFWO
Ysabel I	AQWK
Ystad	ACEV
Ytapere	AMEY
Ytterðerne	ABPO
Yucatan	AWRG
Yuensan	AONL
Yugorski Str	ABDW
Yukon R	ATQL
Yukyeri Pt	AKNQ
Yule Roads	ASQJ
Yulinkau B	ANXS
Yuma	AULS
Yumuri B	AXLG
Ynnan	ANMS
Yungching	AOIP
Yung R	AOEJ
Yura Hr	APDO
Yzeren Baak	AEDN
Zaandam	ADZX
Zafarin I	AKZQ
Zaffarano, C	AINQ
Zaglava Rock	AIYW
Zambales Province	AQEG
Zambezi R	AMBN
Zamboanga	AQLF
Zand B	APXD
Zanddijk	AEBZ
Zandvoort	AECJ
Zannone	AIGR
Zante	AJKQ
Zanzibar	AMLB
Zacle	AIVS
Zara	AJCU
Zarza R	AXPW
Zarzis	AKVE
Zaudzi Road	AMDJ
Zaverda B	AJIU
Zea	AJQG
Zealand	ADKV
Zebayir Is	AMRU
Zeberjed I	AMQB
Zebu	AQIM
Zeeland	AEGK
Zeila	AMOW
Zeitun, G	AJUY
Zengg	AJBI

ZHIZHGINSK—ZWAANTJIES

Zhizhginsk I . ABEQ	Zinnowitz. . . ACZL	Zululand . . . . . . ALVR
Zierikzee . . AEFI	Zirona. . . . AJDM	Zumaya. . . . . AGUO
Zaghinkor . ALIC	Zoutelande . AEFS	Zurva, C . . . . . AJNV
Zimnegorski . . . ABEJ	Zoutkamp . . . ADYS	Zuydcoote . . . . AFXP
Zingst . . . ADBW	Zuider Zee . . AEBP	Zwaantjies Drooghte Lt . . AFWK

INTERNATIONAL CODE OF SIGNALS.

PART III.

STORM-WARNING DISPLAY STATIONS, LIFE-SAVING STATIONS, TIME-
SIGNAL STATIONS, AND WIRELESS TELEGRAPH STATIONS OF THE
UNITED STATES; LLOYD'S SIGNAL STATIONS OF THE
WORLD; SEMAPHORE, DISTANCE, AND WIGWAG
CODES, AMERICAN, ENGLISH, AND FRENCH.

(503)

PART III.

STORM-WARNING DISPLAY STATIONS OF THE UNITED STATES.

ATLANTIC COAST

EASTPORT, ME

PORTLAND, ME (*Center*)
  * West Quoddy Head, Me
  * Cutler, Me
  * Machiasport, Me
  * Bangor, Me
  Whitehead, Me
  Marshall Point Light, Me
  Booth Bay Harbor, Me

BOSTON, MASS (*Center*)
  Portsmouth, N. H.
  * Seaveys Island Navy Yard, N H
  * Jerrys Point, N H
  * Rye Beach, N H
  * Wallis Sands, N H
  Newburyport, Mass
  Gloucester, Mass
  * Thatchers Island, Mass
  Marblehead, Mass
  Hull, Mass
  Race Point, Mass
  Provincetown, Mass
  Highland Light, Mass
  Wellfleet, Mass
  Chatham, Mass
  Monomoy, Mass
  Hyannis, Mass
  * Nantucket Shoals Light-ship, Mass
  * Cross Rip Light-ship, Mass
  East Chop, Mass
  Nobska Light, Mass
  Tarpaulin Cove, Mass
  Cuttyhunk, Mass
  New Bedford, Mass
  * New Bedford Yacht Club, Mass
  Fall River, Mass
  Newport, R I
  Point Judith, R I
  Saunderstown, R I
  Stonington, Conn
  New London, Conn

NANTUCKET, MASS

PROVIDENCE, R I (*Center*)
  Providence Yacht Club, R. I.

BLOCK ISLAND, R I (*Center*)
  Southeast Light, R I

NEW HAVEN, CONN (*Center*)
  New Haven Light, Conn
  Bridgeport, Conn

NEW YORK, N Y (*Center*)
  * Greenwich, Conn
  New Rochelle Yacht Club, N Y
  Center Island, Oyster Bay, N Y
  Sewanhaka Yacht Club, N Y
  * Montauk Point, N Y
  * Ardsley on Hudson, N Y
  * Yonkers Yacht Club, N Y
  Fort Schuyler, N Y
  Governors Island, N Y
  * Sea Gate, N Y
  * Princebay, N Y
  * Raritan Yacht Club, N Y
  * Bay Shore Yacht Club, N Y
  * Perth Amboy, N J
  Sandy Hook, N J
  Long Branch, N J

ATLANTIC CITY, N J (*Center*)
  * Little Egg Life-Saving Station, N J
  * Great Egg Life-Saving Station, N J
  * Ocean City Life-Saving Station, N J
  * Corson Inlet Life-Saving Station, N J
  * Townsend Inlet Life-Saving Station, N J
  * Hereford Inlet Life-Saving Station, N J
  CAPE MAY, N J
  * Cape May Point Life-Saving Station, N J

PHILADELPHIA, PA (*Center*)
  Port Norris, N J
  Bivalve, N J
  Reedy Island, Del
  Delaware Breakwater, Del

BALTIMORE, MD (*Center*)
  Oxford, Md
  Annapolis, Md
  The Anchorage, Baltimore, Md
  Baltimore American Building, Baltimore, Md.

NORFOLK, VA (*Center*)
  Fort Monroe, Va
  Newport News, Va

CAPE HENRY, VA
  Elizabeth City, N C
  Edenton, N C
  Columbia, N C
  MANTEO, N C
  Washington, N C
  * Diamond Shoals Light-ship, N. C.

HATTERAS, N C

* Cooperative

(505)

WILMINGTON, N C   (*Center*)
   * Newbern, N C
   * Cape Lookout Life-Saving Station, N C
    Beaufort, N C
    Morehead City, N C
    Southport, N C
   * Oak Island, Life-Saving-Station, N C

CHARLESTON, S C   (*Center*)
    North Island, S C
    South Island, S C
    Georgetown, S C
    Moultrieville, S C
   * Charleston Light-vessel, S C
    Youngs Island, S C

SAVANNAH, GA   (*Center*)
    Port Royal, S C
    Tybee Island, Ga
   * Thunderbolt, Ga
   * Montgomery, Ga
    Darien, Ga
   * St Simons Light, Ga
    Brunswick, Ga

JACKSONVILLE, FLA   (*Center*)
    Fernandina, Fla
   * Fernandina Quarantine Station, Fla
    Mayport, Fla
   * Cummers Mill (Jacksonville), Fla
    St Augustine, Fla
   * Daytona, Fla
    Miami, Fla
   * Puntarasa, Fla
   * Punta Gorda, Fla
   * Port Inglis, Fla
    Cedar Keys, Fla

JUPITER, FLA   (*Center*)
    West Palm Beach, Fla

KEY WEST, FLA   (*Center*)

SAND KEY, FLA
    Dry Tortugas, Fla

TAMPA, FLA   (*Center*)
   * St Petersburg, Fla
    Egmont Key, Fla
    Port Tampa, Fla
    Clear Water, Fla
   * Anclote, Fla

PENSACOLA, FLA   (*Center*)
    Carrabelle, Fla
    Apalachicola, Fla
   * Bagdad, Fla

MOBILE, ALA   (*Center*)
    Fort Morgan, Ala
    Scranton, Miss
   * Biloxi, Miss
    Gulfport, Miss

NEW ORLEANS, LA   (*Center*)
    Pass Christian, Miss
   * Bay St Louis, Miss
   * Salmen, La.
    Morgan City, La
    Pilottown, La
    Burwood, La.

GALVESTON, TEX   (*Center*)
   * Sabine Pass, Tex
   * Port Arthur, Tex
   * Seabrook, Tex
   * Velasco, Tex
    Matagorda, Tex
   . Port Lavaca, Tex
    Point Isabel, Tex

CORPUS CHRISTI, TEX   (*Center*)
   * Rockport, Tex
    Tarpon, Tex

GREAT LAKES

OSWEGO, N Y   (*Center*)
    Oswego Life-Saving Station, N. Y

ROCHESTER, N Y

BUFFALO, N Y   (*Center*)
    Buffalo Life-Saving Station, N. Y
    Ogdensburg, N Y
    Cape Vincent, N Y
    Sacket Harbor, N. Y
   * Stoney Point, N Y
    North Fair Haven, N Y
    Sodus Point, N Y
    Fort Ontario Life-Saving Station, N Y.
    Charlotte, N Y
    Fort Niagara, N. Y
    Tonawanda, N Y
    Dunkirk, N Y

ERIE, PA

CLEVELAND, OHIO   (*Center*)
    Soldiers' Home, Erie, Pa
    Conneaut Harbor, Ohio
    Ashtabula, Ohio
    Fairport, Ohio
    Cleveland Life-Saving Station, Ohio
    Lorain, Ohio
    Huron, Ohio

CLEVELAND, OHIO—Continued
    Marblehead, Ohio
    Kelleys Island, Ohio
    Put in Bay, Ohio.

SANDUSKY, OHIO

TOLEDO, OHIO

DETROIT, MICH   (*Center*)
   * Smiths Coal Docks, Detroit, Mich.

PORT HURON, MICH   (*Center*)
   * Kendalls Wharf, Port Huron, Mich
   * Millers Dock, Port Huron, Mich.
    Lake View Beach Life-Saving Station, Mich
    Harbor Beach, Mich
    Point aux Barques, Mich
    Bay City, Mich

ALPENA, MICH   (*Center*)
    Tawas, Mich
    East Tawas, Mich
    Oscoda, Mich
    Thunder Bay Island, Mich
    Middle Island, Mich
    Presque Isle Light, Mich
    Cheboygan, Mich

* Cooperative

GRAND HAVEN, MICH

CHICAGO, ILL   (Center)
    Mackinac Island, Mich
    Mackinaw, Mich
    Harbor Springs, Mich
    Charlevoix, Mich
    St James, Beaver Island, Mich
    Church Hill, Beaver Island, Mich
    North Manitou Island, Mich
    South Manitou Island, Mich
    Glen Haven, Mich
    Frankfort, Mich
    Manistee, Mich
    Manistee Life-Saving Station Mich
    Big Point Sable, Mich
    Ludington Life-Saving Station, Mich
    Ludington, Mich
    Muskegon, Mich
    Macatawa, Mich
    Holland, Mich
    Saugatuck, Mich
    South Haven, Mich
    St Joseph, Mich
    Michigan City, Ind
    Chicago Life-Saving Station, Ill
    South Chicago Life-Saving Station, Ill
    Red Wing, Minn
    Lake City, Minn

MILWAUKEE, WIS   (Center)
    Kenosha, Wis
    Racine, Wis
    Milwaukee Life-Saving Station, Wis
    Sheboygan, Wis

MILWAUKEE, WIS —Continued
    Manitowoc, Wis
    Kewaunee, Wis
    Sturgeon Bay, Wis

GREEN BAY, WIS

ESCANABA, MICH   (Center)
    Menominee Mich
  * Gladstone, Mich

SAULT SAINTE MARIE, MICH   (Center)
    Detour, Mich
    Whitefish Point, Mich
    Deer Park, Mich
    Grand Marais, Mich
    Munising, Mich

MARQUETTE, MICH   (Center)
    Presque Isle, Mich

HOUGHTON, MICH   (Center)
  * Pequaming, Mich
    Eagle Harbor, Mich
    Ship Canal Life-Saving Station, Mich
  * Ontonagon, Mich

DULUTH, MINN   (Center)
    Ashland, Wis
    Washburn, Wis
    Bayfield, Wis
    Superior, Wis
  * West Superior, Wis
    Two Harbors, Minn

PACIFIC COAST

TACOMA, WASH

SEATTLE, WASH

PORT CRESCENT, WASH

TATOOSH ISLAND, WASH

NORTH HEAD, WASH

PORTLAND, OREG   (Center)
  * Tacoma Tug and Barge Co Office, Wash
  * Everett, Wash
  * Bellingham, Wash
  * Bellingham, Station A, Wash
  * Blaine, Wash
    Anacortes, Wash
    Port Townsend, Wash
    Port Angeles, Wash
    EAST CLALLAM, WASH
    NEAH BAY, WASH
    Aberdeen, Wash
  * Cape Disappointment Life - Saving Station,
    Wash
  * Point Adams Life Saving Station Wash

ASTORIA, OREG
  * Tillamook, Oreg
    Marshfield, Oreg

EUREKA, CAL

SAN FRANCISCO, CAL   (Center)
  * Humboldt Bay Life-Saving Station, Cal
  * Oil Port, Cal
  * Table Bluff, Cal
    Eureka Life-Saving Station, Cal
  * Fort Bragg, Cal
  * Mendocino, Cal
    POINT REYES LIGHT CAL
    MOUNT TAMALPAIS, CAL
    Sausalito, Cal
  * Mare Island Navy-Yard, Cal
  * Goat Island, Cal
    Point Lobos, Cal
    SOUTHEAST FARALLON, CAL.
  * Port Harford, Cal
    Santa Barbara, Cal
  * Port Los Angeles, Cal
  * San Pedro, Cal

SAN LUIS OBISPO, CAL

LOS ANGELES, CAL   (Center)
  * Redondo, Cal

SAN DIEGO CAL

*Cooperative

NOTE —The attention of mariners is invited to the fact that at any regular Weather Bureau Station, by
application to the Weather Bureau officials, they may have their barometers compared This service will
always be rendered without compensation

EXPLANATION OF STORM WARNINGS.

Warnings of the approach of windstorms will be published by the display of flags by day and lanterns by night, in connection with the bulletins posted and the reports furnished to newspapers, mariners, and others interested

The warnings adopted by the United States Weather Bureau for announcing the approach of windstorms are as follows·

The STORM WARNING (a red flag, eight feet square, with black center, three feet square) indicates that a storm of marked violence is expected.

The RED PENNANT (eight feet hoist and fifteen feet fly) displayed with the flags, indicates easterly winds—that is, from the northeast to south, inclusive—and that the storm center is approaching

The WHITE PENNANT (eight feet hoist and fifteen feet fly) displayed with the flags, indicates westerly winds—that is, from north to southwest, inclusive—and that the storm center has passed

When the RED PENNANT is hoisted above the Storm Warning, winds are expected from the *northeast quadrant;* when below, from the *southeast quadrant*

When the WHITE PENNANT is hoisted above the Storm Warning, winds are expected from the *northwest quadrant,* when below, from the *southwest quadrant*

NIGHT STORM WARNINGS —By night a *red light* will indicate *easterly winds,* a *white* above a *red light* will indicate *westerly winds*

The INFORMATION WARNINGS have been discontinued at all storm-warning display stations, and such advices as were previously given in information orders will be furnished hereafter in advisory messages

The HURRICANE WARNING (two storm-warning flags, red with black centers, displayed one above the other) indicates the expected approach of a tropical hurricane or of an extremely severe and dangerous storm

No night Hurricane Warnings are displayed.

TIME SIGNALS OF THE UNITED STATES.

Signal station latitude and longitude	Place	Signal adopted	Situation of time signal	Time of signal being made		Additional details
				Greenwich mean time	Standard mean time	
o ′ ″ 42 21 30 N 71 03 30 W	Boston, Mass	Black ball (diameter 4 feet)	Roof of Ames Building 41 feet above the roof, 263 feet above water, 228 feet above ground	h m s 5 00 00	h m s 0 00 00	Ball hoisted 5 minutes before signal Ball dropped (by electricity from Washington Observatory) at noon. mean time of the 75th meridian If signal fails, ball lowered slowly 5 minutes after noon, mean time of the 75th merid [NOTE —Signal not made on Sundays and holidays]
41 29 15 N 71 19 37 W (Goat Island light)	Newport, R I	Black ball (diameter 3½ feet)	Machineshop at torpedo station, 75 feet above water, 68 feet above ground (Drop 15 feet)	5 00 00	0 00 00	, Do
40 42 38 1 N 74 00 35 7 W	New York, N Y.	Black ball (diameter 3½ feet)	Staff on tower of Western Union Telegraph Building, 286 feet above water, 277 feet above ground (Drop 25 feet)	5 00 00	0 00 00	Do
39 56 58 N . 75 08 54 W	Philadelphia Pa	Black ball (diameter 4 feet)	On a shelter house on the SE corner of the Philadelphia Bourse, height above water 189 feet above the ground 169 feet 10 inches, above the roof 42 feet (Drop 29 feet)	5 00 00	0 00 00	Do
39 17 22 N . 76 36 38 W	Baltimore Md	Black ball (diameter 47 inches)	Roof of Baltimore American Building, 235 feet above water, 212 feet above ground (Drop 27 feet)	5 00 00	0 00 00	Do
38 53 51 8 N 77 02 19 16 W	Washington D C	Black ball (diameter 2 feet 7 inches)	Department of the Navy, 197 feet above water, 150 feet above ground (Drop 21 feet)	5 00 00	0 00 00	Ball hoisted 5 minutes before signal Ball dropped at noon mean time of the 75th meridian If signal fails, ball lowered slowly 5 minutes after noon mean time of the 75th meridian [NOTE —Signal not made on Sundays and holidays]
36 59 59 N 76 18 44 W	Fort Monroe Va	Telegraphic tick	Time signal-service office, Government building Fort Monroe wharf	5 00 00	0 00 00	Telegraphic time signals received in office from 11 55 a m to noon mean time of the 75th meridian [NOTE —Signals not received on Sundays and holidays]
36 50 30 N 76 17 30 W (Approximate)	Norfolk, Va	Black ball (diameter 4 feet)	NW corner Citizens' Bank Building No 191-195 Main street, 137 feet above the land, 147 feet above the water (Drop 30 feet 1 inch)	5 00 00	0 00 00	Do
36 58 47 N 76 25 52 W (Approximate)	Newport News Va	Black ball (diameter 4 feet)	Silsby Building, corner 27th street and Washington avenue, 121 feet above the land, 151 feet above the water (Drop 31 feet)	5 00 00	0 00 00	Do
32 04 52 N . 81 05 22 W	Savannah, Ga	Black ball (diameter 3 feet)	Cotton Exchange, 110 feet above water, 98 feet above ground (Drop 16 feet)	5 00 00	0 00 00	Do

Signal station latitude and longitude	Place.	Signal adopted	Situation of time signal	Time of signal being made		Additional details
				Greenwich mean time	Standard mean time	
° ′ ″ 24 33 29 N 81 48 24 W	Key West, Fla	Black ball	Roof of Equipment Building, 91 feet above water	h m s 5 00 00	h m s 0 00 00	Ball hoisted 5 minutes before signal Ball dropped at mean noon of the 75th meridian [NOTE.—Signal not made on Sundays and holidays.]
29 56 59 N 90 03 47 W	New Orleans, La	Black ball (diameter 2 feet 8 inches)	On the roof of the American Sugar Refining Company Building, 174 feet above water, 163 feet above ground (Drop 18 feet 5 inches.)	5 00 00	11 00 00	Ball hoisted 5 minutes before signal Ball dropped at 11 a m, standard time of the 90th meridian If signal fails, ball lowered slowly at 11h 05m 00s, standard time [NOTE.—Signal not made on Sundays and holidays.]
29 18 17 N 94 47 28 W	Galveston, Tex	Black ball (diameter 4 feet)	Levy Building, SE corner 23d (Tremont) and D (Market) streets, 111 feet above the land 118 feet above the water (Drop 27 feet.)	5 00 00	11 00 00	Do
42 53 10 N 78 52 40 W	Buffalo, N Y	Black ball (diameter 3½ feet)	SE side of the Prudential Building on the SW cor of Church and Pearl streets, 188 feet above land, 227½ feet above water (Drop 18½ feet.)	5 00 00	0 00 00	Ball hoisted 5 minutes before signal Ball dropped (by electricity) at noon, standard time of the 75th meridian If signal fails, ball lowered slowly 5 minutes after noon [NOTE.—Signal not made on Sundays and holidays.]
41 30 03 N 81 41 28 W	Cleveland, Ohio	Black ball (diameter 4½ feet)	Arcade Building 239 feet above water 159 feet above ground (Drop 19 feet 1 inch.)	6 00 00	0 00 00	Ball hoisted 5 minutes before signal Ball dropped at noon, mean time of the 90th meridian If signal fails, ball lowered slowly 5 minutes after noon, mean time of the 90th meridian [NOTE.—Signal not made on Sundays and holidays.]
41 53 06 N 87 37 38 W	Chicago, Ill	Red ball (diameter 4½ feet)	Masonic Temple, 350 feet above water 334 feet above ground (Drop 28 feet.)	6 00 00	0 00 00	Do
46 30 06 N 84 20 40 W	Sault Ste Marie, Mich	Black ball (diameter 3 feet)	NW corner News Building, No 115 Ashmun street, 85 feet above the land 100 feet above the water (Drop 20 feet.)	6 00 00	0 00 00	Ball hoisted 5 minutes before signal Ball dropped (by electricity) at noon, mean time of the 90th meridian If signal fails, ball lowered slowly 5 minutes after noon, mean time of the 90th meridian [NOTE.—Signal not made on Sundays and holidays.]
46 47 00.81 N 92 06 06.54 W	Duluth, Minn	Red ball (diameter 4 feet)	SW corner of Torrey Building No 316 W Superior street 168.5 feet above land, 200.3 feet above water (Drop 31 feet 6 inches.)	6 00 00	0 00 00	Ball hoisted 5 minutes before signal Ball dropped (by electricity) at noon, mean time of the 90th meridian Should ball fail to drop, it is slowly lowered, immediately rehoisted, and dropped at 12h 05m 00s, mean time of the 90th meridian Should ball drop at wrong time it is again mastheaded and lowered at 12h 05m 00s [NOTE.—Signal not made on Sundays and holidays.]

Signal station latitude and longitude	Place	Signal adopted	Situation of time signal	Time of signal being made		Additional details
				Greenwich mean time	Standard mean time	
37 47 38 0b N 122 23 37 38 W	San Francisco, Cal	Black ball (diameter 30 inches)	New ferry building, foot of Market street, 254 feet above water, 249 feet above ground (Drop 15 feet)	h m s 8 00 00	h m s 0 00 00	Ball hoisted 5 minutes before signal Ball dropped (by electricity from Mare Island Observatory) at noon, mean time of the 120th meridian If signal fails, ball lowered slowly 5 minutes after noon, mean time of the 120th meridian [NOTE —Signal not made on Sundays and holidays]
38 05 56 50 N 122 16 19 10 W	Mare Island, Cal	Black ball (diameter 3 feet)	Building No 51, 78 feet above water, 69 feet above ground (Drop 14 feet)	8 00 00	0 00 00	Do
43 31 39 N 122 40 44 W	Portland, Oreg	Deep red ball with broad white horizontal band (diameter 4 feet)	Custom-house, 150 feet above water	8 00 00	0 00 00	

Noon-time signals are sent to every office of the Western Union Telegraph Company in the United States, and by going to any one of them corrections for chronometers may be obtained (75th meridian time)

The signals sent out by the Observatory are wholly automatic, and consist of a series of short makes, produced in an open telegraph circuit by the beats of a mean-time clock, the pendulum closing the circuit at each beat.

The signals begin at 11h. 55m. 00s and cease at 12h 00m 00s, 75th meridian, mean time

During that interval there is a make at the beginning of every second, except that in each minute the makes corresponding to the 29th second and to the 55th, 56th, 57th, 58th, and 59th seconds are omitted

Thus, the first make after the pause of five seconds always marks the beginning of a minute, and the first make after the pause of one second marks 30 seconds In order to distinguish the last minute and give time to manipulate switches to time balls, control clocks, etc , the makes cease after 11h 59m 50s and until 12h 00m. 00s, when there is a single make and the signals cease

When these signals are received at points where the time of the 90th meridian is used they will give the time from 10h 55m 00s to 11h. 00m 00s, or just one hour earlier than when representing 75th meridian time, otherwise the signals will be read in the manner above described

UNITED STATES STANDARD TIMES.

Intercolonial .	4h 00m 00s	W of Greenwich—Standard for Porto Rico
Eastern and Panama Canal Zone	5h 00m 00s	W of Greenwich—Standard from east coast United States to Long 82½° W Includes Cuba
Central. . .	6h 00m 00s	W of Greenwich—Standard from Long 82½° W to Long 97½° W
Mountain .	7h 00m 00s	W of Greenwich—Standard from Long 97½° W to Long 112½° W
Pacific . .	8h 00m 00s	W of Greenwich—Standard from Long 112½° W to west coast United States
Hawaiian . . .	10h 30m 00s	W of Greenwich—Standard for Hawaiian Islands
Tutuila	11h 30m 00s	W of Greenwich—Standard for Tutuila Island
Guam	9h 30m 00s	E of Greenwich—Standard for Guam Island
Philippine	8h 00m 00s	E of Greenwich—Standard for Philippine Islands

UNITED STATES NAVAL WIRELESS TELEGRAPH STATIONS.

	Call letter		Call letter
Annapolis, Md. (Naval Academy)	QG	Jupiter Inlet, Fla. .	RA
Arguello, Point, Cal.	TK	Key West, Fla	RD
Beaufort, N C.	QS	League I , Pa	PV
Blanco, Cape, Oreg.	TA	Loma, Point, Cal	TM
Boston, Mass (Navy-Yard)	PG	Mare Id , Cal (Navy-Yard) .	TG
Bremerton Navy-Yard .	SP	Nantucket Shoal Lt V .	PI
Cavite, P I (Navy-Yard)	UT	New Orleans	RO
Charleston Lt V	QV	Newport, R I . . .	PK
Charleston, S C	QU	New York (Navy-Yard) .	PT
Cod, Cape, Mass.	PH	Norfolk, Va	QL
Culebra, W I	SD	North Head, Wash	SX
Diamond Shoals Lt V	QP	Oahu, H I.	UC
Dry Tortugas, Fla.	RF	Panama Canal Zone	SL
Elizabeth, Cape, Me	PA	Pensacola, Fla (Navy-Yard)	RK
Farallon Ids , Cal	TH	Portsmouth, N H (Navy-Yard)	PC
Fire I , N Y	PR	St Augustine, Fla	QX
Goat I , Cal	TI	San Juan, P R	SA
Guam I	UK	Sitka	SO
Guantanamo, Cuba	SI	Table Bluff, Cal	TD
Henlopen, Cape	PX	Tatoosh I . . .	SV
Henry, Cape	QN	Washington, D C (Navy-Yard) .	QI

The following regulations governing the use of the U S Naval Coastwise Wireless Telegraph stations are hereby established

1. The facilities of the naval coastwise wireless telegraph stations (including the one on the Nantucket Shoal lightship), for communicating with ships at sea, where not in competition with private wireless telegraph stations, are placed at the service of the public generally and of maritime interests in particular, under the rules established herein, which are subject to modification from time to time, for the purpose of—

(a) Reporting vessels and intelligence received by wireless telegraphy with regard to maritime casualties, derelicts at sea, and overdue vessels

(b) Receiving wireless telegrams of a private or commercial nature from ships at sea, for further transmission by telegraph or telephone lines

(c) Transmitting wireless telegrams to ships at sea

2. For the present this service will be rendered free.   All messages will, however, be subject to the tariffs of the ship stations and land lines   Arrangements have been made with both the Western Union and Postal Telegraph companies for forwarding messages received from ships at sea   When a message is not prepaid the company delivering it will collect the charges   Shipowners should arrange with companies operating the land lines as to tariffs and the settlement therefor   Messages will not be accepted for transmission to ships whose owners have not agreed to accept unpaid messages, unless a sufficient sum is deposited to cover all charges

3 The Nantucket Shoal Lightship station will report vessels and transmit messages from them, if the signals are made by the International Code or any other known to the operators on the lightship

4. When notified by the Weather Bureau of the Department of Agriculture, naval wireless telegraph stations will give storm warnings to vessels communicating with them by wireless telegraphy   Storm warnings will soon be sent to the Nantucket Shoal lightship by wireless telegraphy, and storm signals furnished by the Weather Bureau will be displayed therefrom to warn passing vessels

5 All vessels having the use of the naval wireless telegraph service are requested to take daily meteorological observations of the weather when within communicating range, and to transmit such observations to the Weather Bureau by wireless telegraphy at least once daily, and transmit observations oftener when there is a marked change in the barometer

6 Arrangements for a time signal service by wireless telegraphy are now being made.

7 All shipowners desiring to use any special code of signals for communicating with the Nantucket Shoal Light-ship station or any of the shore stations, or make any other special arrangements, are requested to communicate with the Bureau of Equipment, Navy Department, Washington, D C

8 All chambers of commerce, maritime exchanges, newspapers, news agencies, and others desiring to have vessel reports and general marine news forwarded to them regularly are requested to communicate with the Bureau of Equipment in order that necessary arrangements for the service may be made In no case will an operator attached to a station be allowed to act as an agent for any individual or corporation, but all vessel reports and marine news not of a private nature will be supplied to all applicants so long as this service does not too greatly tax the personnel of the stations, when it will be necessary for those desiring information involving much time for its distribution to appoint agents who will be allowed access to the station bulletins

9 Naval wireless telegraph stations are equipped with apparatus of several systems, and can communicate with all the principal wireless telegraph systems now in use if tuned to the same wave length. The Department is desirous of cooperating with all shipowners wishing to avail themselves of its wireless telegraph service, and, judging from its experience with numerous systems, it is believed that there will be little or no difficulty in arranging for communication between its stations and ships equipped with apparatus of other systems if the owners of the apparatus as well as the owners of the ships are desirous of establishing such communication

10 Vessels desiring to make use of this service regularly must agree to transmit and receive all Government messages free.

The Bureau of Equipment expects to erect wireless telegraph stations at the principal points along the coast of the United States and at points in its insular possessions As fast as they are completed they will be open for public use under the regulations established herein

Notice will be given in the "Notices to Mariners" when stations are put in operation or withdrawn from the service for any reason

The Nantucket Shoal light-ship will transmit its messages to the Torpedo Station, Newport, R I All messages intended to be sent via this light-ship to ships at sea should be sent to the Torpedo Station

Arrangements have been made with the Weather Bureau for the transmission of messages between Cape Henry wireless telegraph station and Norfolk All messages intended for the Cape Henry station should be sent via the Weather Bureau, Norfolk, Va

All messages intended for Dry Tortugas should be sent via the Naval Station, Key West, Fla

The station at Goat Island, Cal., can be reached by either the Postal Telegraph or the Western Union system, and the one at Mare Island by the Western Union.

The Farallon station will communicate with Goat Island, Cal

INSTRUCTIONS TO GOVERN COMMUNICATION BY WIRELESS TELEGRAPHY BETWEEN WIRELESS TELEGRAPH STATIONS AND SHIPS

I A vessel wishing to communicate with a station, and having ascertained by "listening in" that she is not interfering with messages being exchanged within her range, should make the call letter of the station at a distance not greater than 75 miles from it

II The call should not be continuous, but should be at intervals of about three minutes, in order to give the station a chance to answer

III After the station answers, the vessel should send her name, distance from station, weather, and number of words she wishes to send, then stop until the station makes O. K , signals the number of words she wishes to send to vessel, and signals to go ahead

IV. Then the vessel begins to send her messages, stopping at the end of each 50 words and waiting until the station signals O K and go ahead, when all messages have been sent she will so indicate If the sender desires to designate the Western Union or Postal Telegraph system for further transmission of his message, he should do so immediately after the address, as, for example, "A. B. C., Washington, D C , via W U (or P T.).''

V. When a vessel has indicated that she has finished, the station will send to the vessel such messages as she may have for her, in the following order

(a) Government business, viz, telegrams from any Government Departments to their agents on board

(b) Business concerning the vessel with which communication has been established, viz, telegrams from owner to master

(c) Urgent private dispatches, limited.

(d) Press dispatches

(e) Other dispatches

VI In the case of the Nantucket Shoal light-ship, it will, immediately on receiving the vessel's call, acknowledge, and (after receiving vessel's name, distance, weather report, and number of words she wishes to send) transmit the first three to Newport, and then tell the vessel to go ahead with her messages

VII After receiving these and sending the vessel any message on file for her, the light-ship will transmit to Newport messages received from the communicating vessel in the following order

(a) Government business

(b) Urgent private dispatches, limited

(c) Press dispatches.

(d) Other dispatches

VIII A naval wireless telegraph station has the right to break in on any message being sent by a vessel at any time, and the right of way may be given at any time to a Government vessel or one in distress.

IX. When two or more vessels desire to communicate with a naval wireless telegraph station at the same time, the one whose call is first received will have the right of way, and the others will be told to wait and will be taken up in turn Vessels having been told to wait must cease calling

X In case communication is not established with any ship for which messages are on file, the naval wireless telegraph station will notify the telegraph company from which the messages were received, giving sufficient information for them to identify the telegrams and notify the sender

XI In order to obtain the best results, both sending and receiving apparatus should be tuned to wave length of 320 meters.

XII Until further notice the speed of sending should not exceed 12 words per minute

XIII In order that all messages received at naval wireless telegraph stations may be forwarded to ships for which they are intended, and in order that all ships equipped with wireless telegraph apparatus may receive storm warnings, they should always report when in signaling distance of a naval wireless telegraph station

XIV. The service being without charge at present, the Government accepts no responsibility for the reception or transmission of messages from or for passing vessels. Every effort will be made to transmit all messages without error and as expeditiously as possible. It must be remembered that errors are not uncommon in ordinary telegraph and cable messages, so that due allowance should be made

XV. In order that the service may be made as good and as useful as possible, it is requested that complaints should be promptly reported to the Bureau of Equipment as soon as possible after the cause therefor, giving date, hour, and other details, to enable the Bureau to investigate the case

XVI Information regarding the naval wireless telegraph service will be published in "Notices to Mariners.''

LIFE-SAVING DISTRICTS AND STATIONS IN THE UNITED STATES.

All manned Life-Saving Stations of the United States are equipped with International Code flags, and are prepared to send or receive signals in that code or by means of the United States Army and Navy Wigwag Code. They are in many cases, however, unprovided with means of telegraphic communication.

LIFE-SAVING SIGNALS.

The following signals, recommended by the International Marine Conference for adoption by all institutions for saving life from wrecked vessels, have been adopted by the Life-Saving Service of the United States.

1 Upon the discovery of a wreck by night, the life-saving force will burn a red pyrotechnic light or a red rocket to signify, ''You are seen, assistance will be given as soon as possible ''

2 A red flag waved on shore by day, or a red light, red rocket, or red roman candle displayed by night, will signify, ''Haul away ''

3 A white flag waved on shore by day, or a white light swung slowly back and forth, or a white rocket or white roman candle fired by night, will signify, ''Slack away ''

4 Two flags, a white and a red, waved at the same time on shore by day, or two lights, a white and a red, slowly swung at the same time, or a blue pyrotechnic light burned by night, will signify, " Do not attempt to land in your own boats, it is impossible "

5 A man on shore beckoning by day, or two torches burning near together by night, will signify, ''This is the best place to land ''

LIFE-SAVING DISTRICTS AND STATIONS IN THE UNITED STATES.

FIRST DISTRICT

COASTS OF MAINE AND NEW HAMPSHIRE

Name of station	State	Locality	Approximate position	
			Latitude, north	Longitude, west
			° ′ ″	° ′ ″
Quoddy Head	Me	Carrying Point Cove	44 48 40	66 58 50
Cross Island	Me.	Off Machiasport	44 36 45	67 16 30
Great Wass Island	Me	Off Jonesport	44 28 00	67 35 30
Cranberry Islands	Me	Little Cranberry Island, off Mount Desert	44 15 30	68 12 40
White Head	Me	On southwest end White Head Island	43 58 40	69 08 00
Burnt Island	Me .	Off mouth St Georges River	43 52 20	69 17 40
Damiscove Island	Me	On the west shore of Damiscove Harbor	43 45 20	69 37 00
Hunniwells Beach	Me	On west side mouth Kennebec River	43 45 00	69 46 55
Cape Elizabeth	Me .	Near the Lights. .	43 33 08	70 12 00
Fletchers Neck	Me .	Biddeford Pool, Fletchers Neck	43 26 30	70 20 30
Jerrys Point	N H	Southeast point Great Island, Portsmouth Harbor	43 03 30	70 42 45
Wallis Sands	N H	1¼ miles south of Odiornes Point	43 01 15	70 44 00
Rye Beach	N H	North end of Rye Beach	42 59 30	70 45 20
Hampton Beach	N H	1½ miles north of Great Boars Head	42 56 20	70 47 40

SECOND DISTRICT

COAST OF MASSACHUSETTS

Name of station	State	Locality	Approximate position	
			Latitude, north	Longitude, west
			° ′ ″	° ′ ″
Salisbury Beach ...	Mass	7 mile south of State line	42 51 40	70 49 00
Newburyport	Mass	North end of Plum Island mouth of Merrimac River	42 48 30	70 49 00
Plum Island	Mass	On Plum Island, 2½ miles from south end	42 44 00	70 47 15
Straitsmouth a	Mass	½ mile west of Straitsmouth light	42 39 30	70 36 00
Gloucester	Mass	Old House cove, westerly side of harbor, 1½ miles from town	42 35 30	70 41 10
Nahant	Mass	On the neck, close to Nahant	42 25 45	70 56 00
City Point	Mass.	Floating station in Dorchester Bay, Boston Harbor		
Point Allerton	Mass	1 mile west of Point Allerton	42 18 20	70 54 00
North Scituate	Mass	2½ miles south of Minots Ledge light	42 14 00	70 45 30
Fourth Cliff	Mass	South end of Fourth Cliff, Scituate	42 09 30	70 42 10
Brant Rock	Mass	On Green Harbor Point	42 05 30	70 38 40
Gurnet	Mass	4½ miles northeast of Plymouth	42 00 10	70 36 10
Manomet Point	Mass	6½ miles southeast of Plymouth	41 55 30	70 32 40
Wood End	Mass	½ mile east of light	42 01 15	70 11 30
Race Point	Mass	1½ miles northeast of Race Point light	42 04 45	70 13 15
Peaked Hill Bars	Mass.	2½ miles northeast of Provincetown	42 04 40	70 09 50
High Head	Mass	3½ miles northwest of Cape Cod light	42 03 55	70 06 50
Highland	Mass	½ mile northwest of Cape Cod light	42 02 55	70 04 20
Pamet River	Mass.	3½ miles south of Cape Cod light	42 00 00	70 01 15
Cahoons Hollow	Mass.	2½ miles east of Wellfleet	41 56 45	69 59 05
Nauset	Mass.	1½ miles south of Nauset lights	41 50 40	69 56 45
Orleans	Mass.	Abreast of Ponchet Island	41 45 35	69 55 55
Old Harbor	Mass.	½ mile north of Chatham Inlet	41 41 45	69 56 00
Chatham .	Mass	1½ miles south-southwest of Chatham lights	41 39 10	69 57 10
Monomoy	Mass	2½ miles north of Monomoy light	41 35 25	69 59 10
Monomoy Point	Mass. .	½ mile southwest of Monomoy light.	41 33 10	70 00 20
Coskata	Mass	2½ miles south of Nantucket (Great Point) light	41 22 00	70 01 15
Surfside	Mass	2½ miles south of the town of Nantucket	41 14 30	70 06 00
Maddequet	Mass.	6 miles west of Surfside	41 16 05	70 12 30
Muskeget	Mass	Near west end of Muskeget Island	41 20 20	70 18 50
Gay Head	Mass.	Near light.	41 21 04	70 50 08
Cuttyhunk .	Mass	Near east end Cuttyhunk Island	41 25 25	70 54 45

THIRD DISTRICT.

COASTS OF RHODE ISLAND AND FISHERS ISLAND

Brenton Point	R I	On Price's Neck	41 26 58	71 20 10
Narragansett Pier	R I	Northern part of the town	41 25 45	71 27 20
Point Judith .	R I	Near light	41 21 40	71 29 00
Quonochontaug .	R I	7½ miles east of Watch Hill light	41 19 50	71 43 10
Watch Hill ..	R I	Near light	41 18 20	71 51 30
Fishers Island	N Y	West shore of East Harbor	41 17 00	71 56 40
Sandy Point .	R I	Block Island north side near light	41 13 40	71 34 40
New Shoreham	R I	Block Island, east side, near landing	41 10 20	71 33 30
Block Island	R I	Block Island, west side, near Dickens Point .	41 09 40	71 36 40

FOURTH DISTRICT.

COAST OF LONG ISLAND

Montauk Point b	N Y	At the light	41 04 00	71 51 30
Ditch Plain	N Y	3½ miles southwest of Montauk light	41 02 10	71 54 30
Hither Plain	N Y	½ mile southwest of Fort Pond	41 01 30	71 57 50
Napeague	N Y	Abreast of Napeague Harbor	40 59 45	72 02 40

a Formerly Davis Neck   b In charge of keeper of Ditch Plain station   No crew employed

FOURTH DISTRICT—Continued

COAST OF LONG ISLAND—Continued

Name of station	State	Locality	Approximate position	
			Latitude, north	Longitude, west
			° ′ ″	° ′ ″
Amagansett	N Y	Abreast of the village	40 58 00	72 08 20
Georgica	N Y	1 mile south of village of East Hampton	40 56 40	72 11 40
Mecox	N Y	2 miles south of the village of Bridgehampton	40 54 10	72 18 00
Southampton	N Y	¾ mile south of the village	40 52 10	72 23 40
Shinnecock	N Y	2 miles east-southeast of Shinnecock light	40 50 40	72 27 50
Tiana	N Y	2 miles southwest of Shinnecock light	40 49 40	72 31 30
Quogue	N Y	½ mile south of the village	40 48 20	72 36 00
Potunk	N Y	1½ miles southwest of Potunk village	40 47 30	72 39 00
Moriches	N Y	2½ miles southwest of Speonk village	40 46 30	72 43 10
Forge River	N Y	3½ miles south of Moriches	40 44 30	72 49 00
Smiths Point	N Y	Abreast of the point	40 44 00	72 52 20
Bellport	N Y	4 miles south of the village	40 42 40	72 55 50
Blue Point	N Y	4½ miles south of Patchogue	40 40 40	73 01 20
Lone Hill	N Y	8 miles east of Fire Island light	40 39 40	73 04 20
Point of Woods	N Y	4 miles east of Fire Island light	40 38 50	73 08 10
Fire Island	N Y	½ mile west of Fire Island light	40 37 40	73 13 20
Oak Island	N Y	East end of Oak Island	40 38 10	73 17 40
Gilgo	N Y	West end of Oak Island	40 37 20	73 22 20
Jones Beach	N Y	East end of Jones Beach	40 36 40	73 26 20
Zachs Inlet	N Y	West end of Jones Beach	40 36 10	73 28 50
Short Beach	N Y	½ mile east of Jones Inlet	40 35 30	73 31 20
Point Lookout	N Y	2 miles west of New Inlet	40 35 10	73 35 40
Long Beach	N Y	Near west end of Long Beach	40 35 10	73 40 45
Far Rockaway [a]	N Y			
Rockaway	N Y	Near the village of Rockaway	40 35 30	73 47 30
Rockaway Point	N Y	West end of Rockaway Beach	40 34 10	73 51 50
Coney Island [b]	N Y	Manhattan Beach	40 34 20	73 55 30
Eatons Neck	N Y	East side entrance to Huntington Bay Long Island Sound	40 57 10	73 24 00
Rocky Point	N Y	Near Rocky Point, Long Island Sound, about 4 miles northerly from Greenport	41 08 20	72 21 10

FIFTH DISTRICT.

COAST OF NEW JERSEY

Sandy Hook	N J	On Bay side, ½ mile south of point of Hook	40 27 51	74 00 27
Spermaceti Cove	N J	2½ miles south of Sandy Hook light	40 25 40	73 59 00
Seabright	N J	About a mile south of Navesink light	40 22 50	73 58 30
Monmouth Beach	N J	About a mile south of Seabright	40 20 30	73 58 30
Long Branch	N J	Greens Pond	40 16 40	73 59 00
Deal	N J	Asbury Park	40 13 50	73 59 50
Shark River	N J	Near the mouth of Shark River	40 11 30	74 00 40
Spring Lake	N J	2½ miles south of Shark River	40 09 20	74 01 20
Squan Beach	N J	1 mile southeast of Squan village	40 07 00	74 02 00
Bayhead	N J	At the head of Barnegat Bay	40 04 00	74 02 40
Mantoloking	N J	2½ miles south of head of Barnegat Bay	40 01 40	74 03 10
Chadwick	N J	5 miles south of head of Barnegat Bay	39 59 10	74 04 00
Toms River	N J	On the beach abreast mouth Toms River	39 56 10	74 04 30
Island Beach	N J	1¼ miles south of Seaside Park	39 53 40	74 05 00
Cedar Creek	N J	5¾ miles north of Barnegat Inlet	39 51 10	74 05 10
Forked River	N J	2 miles north of Barnegat Inlet	39 48 10	74 05 40
Barnegat	N J	South side of Barnegat Inlet	39 45 30	74 06 10
Loveladies Island	N J	2½ miles south of Barnegat Inlet	39 43 50	74 07 20
Harvey Cedars	N J	5¼ miles south of Barnegat Inlet	39 41 20	74 08 30
Ship Bottom	N J	Midway of Long Beach	39 38 10	74 11 00
Long Beach	N J	1¾ miles north of Beach Haven	39 35 00	74 13 20
Bonds	N J	2¼ miles south of Beach Haven	39 32 00	74 15 20

[a] Station destroyed by sudden gale while being moved across the water to new site

[b] Not in operation

FIFTH DISTRICT—Continued

COAST OF NEW JERSEY—Continued

Name of station	State	Locality	Approximate position	
			Latitude, north	Longitude, west
			° ′ ″	° ′ ″
Little Egg	N J	Near the light north of Inlet	39 30 10	74 17 30
Little Beach	N J	South side of Little Egg Inlet	39 27 30	74 19 30
Brigantine	N J	5½ miles north of Absecon light	39 25 30	74 20 30
South Brigantine	N J	3½ miles north of Absecon light	39 24 00	74 22 30
Atlantic City	N J	At Absecon light	39 22 00	74 24 50
Absecon	N J	2¾ miles south of Absecon light	39 20 50	74 27 40
Great Egg	N J	6½ miles south of Absecon light	39 19 00	74 31 10
Ocean City	N J	South side of Egg Harbor Inlet	39 17 00	74 34 00
Pecks Beach	N J	3½ miles north of Corson Inlet	39 14 50	74 36 50
Corson Inlet	N J	Near the Inlet, north side	39 13 10	74 38 20
Sea Isle City	N J	7¼ miles north of Townsend Inlet	39 09 40	74 41 05
Townsend Inlet	N J	Near the Inlet, north side	39 07 30	74 42 45
Avalon	N J	3¾ miles southwest from Ludlam Beach light	39 05 50	74 43 10
Tathams	N J	2½ miles northeast from Hereford Inlet light	39 02 30	74 45 50
Hereford Inlet	N J	Near Hereford light	39 00 20	74 47 20
Holly Beach	N J	6 miles northeast of Cape May City	38 58 40	74 49 50
Two Mile Beach	N J	4 miles northeast of Cape May City	38 57 10	74 51 10
Cold Spring	N J	½ mile east of Cape May City	38 56 00	74 54 30
Cape May	N J	Near the light	38 55 40	74 57 30
Bay Shore a	N J	2½ miles west of Cape May City	38 56 40	74 58 10

SIXTH DISTRICT.

COAST BETWEEN DELAWARE AND CHESAPEAKE BAYS

Lewes	Del	2 miles west from Cape Henlopen light	38 46 50	75 07 10
Cape Henlopen	Del	¾ mile southerly of Cape Henlopen light	38 45 50	75 04 50
Rehoboth Beach	Del	Opposite north end of Rehoboth Bay	38 41 30	75 04 20
Indian River Inlet	Del	North of Inlet	38 37 50	75 03 40
Fenwick Island	Del	1½ miles north of light	38 28 20	75 03 00
Isle of Wight	Md	3 miles south of Fenwick light	38 24 10	75 03 30
Ocean City	Md	At village	38 20 00	75 05 00
North Beach	Md	10 miles south of Ocean City	38 11 30	75 09 20
Green Run Inlet	Md	13½ miles northeast of Assateague light	38 04 30	75 12 50
Popes Island	Va	10 miles northeast of Assateague light	38 00 20	75 15 40
Assateague Beach	Va	1½ miles south of Assateague light	37 53 40	75 21 40
Wallops Beach	Va	1½ miles south of Chincoteague Inlet	37 52 00	75 26 50
Metomkin Inlet	Va	On Metomkin Beach near the Inlet	37 40 45	75 34 50
Wachapreague	Va	South end of Cedar Island	37 35 20	75 36 40
Parramore Beach	Va	Midway of beach	37 32 20	75 37 20
Hog Island	Va	South end of Hog Island	37 22 20	75 42 45
Cobb Island	Va	South end of Cobb Island	37 17 30	75 47 00
Smith Island	Va	At Cape Charles light	37 07 00	75 53 40

SEVENTH DISTRICT

COAST BETWEEN CHESAPEAKE BAY AND THE NORTHERN BOUNDARY OF SOUTH CAROLINA

Cape Henry	Va	¾ mile southeast of Cape Henry light	36 55 10	75 54 50
Virginia Beach	Va	5½ miles south of Cape Henry light	36 51 10	75 58 40
Dam Neck Mills	Va	10 miles south of Cape Henry light	36 47 10	75 57 30
Little Island	Va	On beach abreast of North Bay	36 41 30	75 55 20
False Cape	Va	On beach abreast of Back Bay	36 36 00	75 52 50
Wash Woods	N C	On beach abreast of Knotts Island	36 32 00	75 52 10
Penneys Hill	N C	5¾ miles north of Currituck Beach light	36 27 30	75 50 40
Currituck Beach	N C	¾ mile north of Currituck Beach light	36 23 20	75 49 40

a In charge of keeper of Cape May station   No crew employed

SEVENTH DISTRICT—Continued

COAST BETWEEN CHESAPEAKE BAY AND THE NORTHERN BOUNDARY OF SOUTH CAROLINA—Continued

Name of station	State	Locality	Approximate position	
			Latitude, north	Longitude, west
			° ′ ″	° ′ ″
Poyners Hill.	N C	6½ miles south of Currituck Beach light	36 17 10	75 48 00
Caffeys Inlet	N C	10¾ miles south of Currituck Beach light	36 13 40	75 46 20
Paul Gamiels Hill	N C	5 miles north of Kitty Hawk.	36 08 00	75 43 50
Kitty Hawk	N C	On the beach abreast of north end of Kitty Hawk Bay	36 03 50	75 41 30
Kill Devil Hills	N C	4½ miles south of Kitty Hawk	36 00 10	75 39 40
Nags Head	N C	9 miles north of Oregon Inlet	35 56 00	75 36 40
Bodie Island	N C	¾ mile northeast of Bodie Island light	35 49 40	75 33 20
Oregon Inlet	N C	½ mile south of Oregon Inlet.	35 47 30	75 32 10
Pea Island	N C	2 miles north of New Inlet	35 43 10	75 29 30
New Inlet	N C	¼ mile south of New Inlet	35 40 40	75 29 00
Chicamacomico	N C	5 miles south of New Inlet	35 36 40	75 27 50
Gull Shoal	N C	11¾ miles south of New Inlet	35 29 50	75 28 40
Little Kinnakeet	N C	11½ miles north of Cape Hatteras light	35 25 00	75 29 10
Big Kinnakeet	N C	5½ miles north of Cape Hatteras light	35 20 00	75 30 20
Cape Hatteras	N C	1 mile south of Cape Hatteras light.	35 14 20	75 31 20
Creeds Hill	N C	4 miles west of Cape Hatteras light	35 14 30	75 35 15
Durants	N C	3 miles east of Hatteras Inlet	35 12 35	75 41 10
Hatteras Inlet	N C	1½ miles west of Hatteras Inlet	35 11 00	75 40 10
Ocracoke	N C	3 miles northeast of Ocracoke Inlet	35 06 55	75 59 20
Portsmouth.	N C	Northeast end of Portsmouth Island	35 04 00	76 03 05
Core Bank	N C	On Core Bank, opposite Hunting Quarters, about halfway between Ocracoke Inlet and Cape Lookout	34 51 30	76 18 30
Cape Lookout	N C	1½ miles south of Cape Lookout light	34 36 30	76 32 20
Fort Macon	N C	Beaufort Entrance, ¼ mile north of fort	34 42 00	76 40 50
Bogue Inlet	N C	Inner shore of Bogue Banks, ½ mile east of inlet	34 39 00	77 05 40
Cape Fear	N C	On Smiths Island, Cape Fear	33 50 40	77 57 30
Oak Island	N C	West side mouth Cape Fear River	33 53 20	78 01 20

EIGHTH DISTRICT.

COASTS OF SOUTH CAROLINA, GEORGIA, AND EASTERN FLORIDA

Name of station	State	Locality	Latitude, north	Longitude, west
Sullivans Island	S C	At Moultrieville, Sullivans Island, at north end of harbor jetty	32 45 30	79 51 05
Bulow a	Fla	30 miles south of Matanzas Inlet	29 26 10	81 06 25
Mosquito Lagoon a	Fla	On beach outside the lagoon.	28 51 30	80 46 20
Chester Shoal a	Fla	11 miles north of Cape Canaveral	28 36 40	80 35 50
Cape Malabar b				
Bethel Creek a.	Fla	16 miles north of Indian River Inlet	27 40 00	80 21 20
Indian River Inlet a	Fla	South side of inlet	27 29 45	80 17 50
Gilberts Bar a	Fla	At St Lucie Rocks, 2 miles north of Gilberts Bar Inlet	27 12 00	80 09 50
Jupiter Inlet c	Fla			
Orange Grove d	Fla			
Fort Lauderdale a	Fla	4 miles north of New River Inlet	26 06 45	80 06 15
Biscayne Bay a	Fla	6 miles north of Norris Cut	25 54 10	80 08 00

NINTH DISTRICT

GULF COAST OF UNITED STATES

Name of station	State	Locality	Latitude, north	Longitude, west
Santa Rosa	Fla	Santa Rosa Island, 3 miles east of Fort Pickens	30 19 00	87 14 30
Sabine Pass	Tex	West side of pass, south of light	29 42 27	93 51 10
Galveston	Tex	East end Galveston Island	29 20 10	94 46 10
San Luis.	Tex	West end Galveston Island	29 07 00	95 04 00
Velasco.	Tex	2¼ miles northeast of mouth of Brazos River	28 57 45	95 16 30
Saluria	Tex	Northeast end Matagorda Island	26 23 00	96 24 00
Aransas	Tex	Northeast end Mustang Island	27 51 00	97 08 00
Brazos	Tex	North end Brazos Island, entrance to Brazos Santiago	26 04 00	97 08 00

a House of refuge, no crew employed   c Discontinued January 21, 1899
b Discontinued March 30, 1891   d Discontinued October 1, 1896

TENTH DISTRICT.

LAKES ERIE AND ONTARIO AND A STATION AT LOUISVILLE, KY

Name of station	State	Locality	Approximate position	
			Latitude, north	Longitude, west
			° ′ ″	° ′ ″
Big Sandy	N Y	North side mouth of Big Sandy Creek, Lake Ontario .		
Salmon Creek a				
Oswego	N Y.	East side entrance of Oswego Harbor, Lake Ontario. .		
Charlotte	N Y.	East side entrance of Charlotte Harbor, Lake Ontario		
Niagara	N Y.	East side entrance of Niagara River, Lake Ontario		
Buffalo	N Y.	South side entrance of Buffalo Harbor, Lake Erie		
Erie	Pa	North side entrance of Erie Harbor, Lake Erie		
Ashtabula	Ohio	West side of Ashtabula Harbor, Lake Erie		
Fairport	Ohio	West side entrance of Fairport Harbor, Lake Erie.		
Cleveland	Ohio	West side entrance of Cleveland Harbor Lake Erie		
Marblehead	Ohio	Point Marblehead, near Quarry Docks Lake Erie		
Louisville	Ky	Falls of the Ohio River, Louisville, Ky		

ELEVENTH DISTRICT

LAKES HURON AND SUPERIOR

Lake View Beach	Mich	5 miles north of Fort Gratiot light		
Harbor Beach	Mich	Inside Harbor Beach Harbor, Lake Huron		
Pointe aux Barques .	Mich	Near light, Lake Huron		
Port Austin	Mich	About 2 miles northeast of Port Austin, and about 2 miles south east of Port Austin Reef light Lake Huron		
Tawas	Mich	Near light, Lake Huron		
Sturgeon Point	Mich	Near light, Lake Huron		
Thunder Bay Island	Mich	West side of island, Lake Huron		
Middle Island	Mich	North end of Middle Island, Lake Huron		
Hammond	Mich	Hammonds Bay Lake Huron .		
Bois Blanc	Mich	About midway east side of island, Lake Huron		
Vermilion	Mich	10 miles west of Whitefish Point, Lake Superior		
Crisps. .	Mich	18 miles west of Whitefish Point, Lake Superior		
Two Heart River	Mich	Near mouth of Two Heart River Lake Superior		
Deer Park	Mich	Near mouth of Sucker River Lake Superior		
Grand Marais	Mich	West of harbor entrance		
Marquette	Mich.	Near light, Lake Superior		
Portage	Mich	Old Portage Lake Ship Canal, ¾ mile from north end on east bank		
Duluth	Minn	On Minnesota Point, Upper Duluth		

TWELFTH DISTRICT

LAKE MICHIGAN

Beaver Island b	Mich	Near light		
Charlevoix	Mich	South side of harbor entrance		
North Manitou Island.	Mich	Near Pickards wharf		
South Manitou Island	Mich	Near light, Lake Michigan		
Sleeping Bear Point	Mich	Near Glenhaven, Michigan		
Point Betsie	Mich	Near light		
Frankfort	Mich	South side entrance of harbor		
Manistee	Mich	North side entrance of harbor		
Grande Pointe au Sable .	Mich	1 mile south of light		
Ludington	Mich	North side entrance of harbor		
Pentwater	Mich	North side entrance of harbor		
White River	Mich	North side entrance of White Lake		
Muskegon	Mich	South side entrance of harbor Port Sherman		
Grand Haven	Mich	North side entrance of harbor .		

a Destroyed by fire   b No crew employed

TWELFTH DISTRICT—Continued

LAKE MICHIGAN—Continued

Name of station	State	Locality	Approximate position	
			Latitude, north	Longitude, west
			° ′ ″	° ′ ″
Holland ..	Mich .	In the harbor, south side .		
South Haven	Mich	North side entrance of harbor.		
Saint Joseph	Mich	In the harbor, north side		
Michigan City.	Ind .	East side entrance of harbor		
South Chicago.	Ill .	North side entrance of Calumet Harbor. .		
Jackson Park.	Ill .	About 7 miles S by E of Chicago River light		
Old Chicago. ..	Ill .	In the harbor.		
Evanston	Ill	On the Northwestern University grounds.		
Kenosha.....	Wis	In the harbor, on Washington Island.		
Racine	Wis	In the harbor, adjoining light		
Milwaukee. . .	Wis	Near entrance of harbor, south side		
Sheboygan	Wis	Entrance to harbor, north side.		
Two Rivers..	Wis	North side entrance of harbor		
Kewaunee. .	Wis	North side entrance of harbor		
Sturgeon Bay Canal	Wis .	Eastern entrance of canal, north side		
Baileys Harbor	Wis	On easterly side of harbor		
Plum Island .	Wis	Near northeast point of island, 2 miles northwest of Pilot Island light		

THIRTEENTH DISTRICT

COASTS OF CALIFORNIA, OREGON, WASHINGTON AND ALASKA

Name of station	State	Locality	Latitude, north	Longitude, west
Nome . . .	Alaska..	At Nome. . .	64 30 00	165 23 00
Neah Bay a ..	Wash			
Grays Harbor	Wash	Just south of Grays Harbor light	46 53 15	124 07 15
Willapa Bay. .	Wash	Near light-house boat landing	46 43 00	124 03 00
Ilwaco Beach .	Wash	13 miles north of Cape Disappointment . .	46 27 50	124 03 25
Cape Disappointment	Wash .	Bakers Bay, ½ mile northeast of light	46 16 40	124 03 00
Point Adams .	Oregon	½ mile southeast of Fort Stevens	46 12 00	123 57 00
Yaquina Bay .	Oregon	About 1 mile south of harbor entrance	44 35 30	124 03 54
Umpqua River	Oregon	Near entrance of river, north side	43 42 00	124 10 30
Coos Bay .	Oregon.	Coos Bay, north side .	43 22 56	124 18 00
Coquille River.	Oregon	In town of Bandon	43 07 00	124 25 00
Humboldt Bay	Cal	Near old light-house tower, north side entrance Humboldt Bay	40 46 00	124 13 00
Arena Cove . .	Cal	3 miles southeast from Point Arena light	38 54 56	123 42 30
Point Reyes..	Cal	3½ miles north of light	38 02 20	122 59 30
Bolinas Bay b ..	Cal	. .		
Point Bonita.	Cal	Near Point Bonita light	37 47 50	122 31 40
Fort Point	Cal	½ mile east of light.	37 48 10	122 27 50
Golden Gate.	Cal	On beach in Golden Gate Park, San Francisco, ¾ mile south Point Lobos	37 46 10	122 30 30
Southside .	Cal	3½ miles south of Golden Gate Life-Saving Station	37 43 18	122 30 18

a Discontinued December 17 1890.   b Destroyed by fire

LLOYD'S SIGNAL STATIONS.

UNITED KINGDOM

Southend Pier
Dover (day and night station)
Sandgate
Dungeness
Beachy Head
No-Man's Fort, Spithead (day and night station)
St Catherines Point, Isle of Wight
The Needles, Isle of Wight
Portland Bill
Prawle Point (day and night station)
The Lizard (day and night station)
Penzance
Scilly Islands
Lundy Island
Barry Island
Mumbles Head and Pier
St Anns Head, Milford Haven.

Roche's Point (Queenstown)
Old Head of Kinsale (day and night station)
Fastnet (day and night station, W T Marconi)
Brow Head, County of Cork (day and night station)
Tory Island, County Donegal
Inishtrahull (day and night station, W T Marconi)
Malin Head, County Donegal (W T , Marconi)
Tor Point, County Antrim
Kildonan, mouth of the Clyde
Stornoway
Butt of Lewis
Dunnet Head, Pentland Firth
St Abbs Head
Tynemouth
Flamborough Head
Spurn Head
Aldborough

EUROPE—NORTH AND WEST COAST

Færder
Oxö
Vinga
Helsingborg
Krasnaja Gorka (G of Finland)
Fornaes
Hammershus
Hanstholm
Hirtshals
Skagen
Elsinore
Helgoland
Holtenau
Brunsbuttelkoog

Cuxhaven
Rothesand
Hoheweg
Hook of Holland
Flushing
Zeebrugge
Gris Nez
Creac'h Pt , Ushant
Cape Finnisterre
Ortavos
Peniche.
Sagres.
Tarifa

MEDITERRANEAN AND BLACK SEA

Gibraltar (Admiralty Signal Station, Windmill Hill)
Pomegues
Cape Corse
Cape Pertusato
Malta (Admiralty Signal Station, Castille)
Cape Testa
Cape d'Armi
Fort Spuria

Pantellaria I
Cape Bon
Zea Island
Dardanelles
Kom-el-Nadura (Alexandria)
Mex (Alexandria)
Port Said (W T , Marconi)

AFRICA—WEST COAST

Cape Spartel
Tenerife
Point Ferraria (St Michaels)
Point do Arnel (St Michaels)

Capellinhos Pt (Fayal)
Las Palmas
Ascension
St Helena

CAPE COLONY

Cape Point
Cape Agulhas
Cape St Francis

Cape Recife
Cape Hermes

AFRICA—EAST COAST

Bluff (P Natal)
Fort San Sebastian (Mozambique)

Flat Island . . . . .
Butte aux Sables. . ⎫
Butte aux Papayes ⎬ Mauritius
Port Louis Mountain ⎭

RED SEA

Port Tewfik (Suez) (W T Marconi)
Perim (day and night station)

Aden

INDIAN OCEAN

Jaak . . ⎫
Henjam ⎬ Persian G
Bushire ⎭
False Point
Sandheads. . . . ⎫
Saugar Island . . ⎪
Mud Point. . . ⎪
Diamond Harbor ⎬ Hughly R.
Hughly Point. ⎪
Achipur . . ⎪
Budge Budge . ⎭

Amherst . . . ⎫
Diamond Island ⎬ Burma
Elephant Point ⎭
Point de Galle (day and night station)
Sabang Bay, Pulo Weh (day and night station)
Penang
Malakka
Mt Faber ⎫
Fort Canning ⎬ Singapore
Anjer ⎭

AUSTRALIA

Rottenest Island (day and night station)
Breaksea Island (day and night station)
Cape Leeuwin (day and night station)
Cape Naturaliste (day and night station)
Point Moore (day and night station)
Cape Borda
Cape Willoughby
Cape Jervis
Cape Northumberland
Cape Nelson
Cape Otway
Point Lonsdale
Cape Schanck

Wilson's Promontory
Gabo Island
Queenscliff
Table Cape
Mersey Bluff
Low Head
Eddystone Point
Cape Sorell
Cape Wickham
Curry Harbor
Bruni
Kent Group
Goode Island

NEW ZEALAND

Cape Maria van Diemen
Farewell Spit
Nugget Point

Bluff Harbor
Norfolk Island

SOUTH AMERICA

Point Curaumilla
Point Tumbes
Cape San Antonio
Mogotes Point

Penguin Island
Cape Virgin
Fernando Noronha

WEST INDIES AND BERMUDA

Morro Castle, Habana
Monk's Hill ⎫
Goat Hill . ⎬ Antigua
Rat Island ⎭

Turks Island
Gibbs Hill, Bermuda

STATIONS MAINTAINED BY CANADIAN GOVERNMENT THAT REPORT TO LLOYD'S ON THE SAME CONDITIONS AS LLOYD'S STATIONS

Belle Isle, Labrador
Cape Race, Newfoundland
Cape Ray, Newfoundland
St Paul Island Cape Breton Island
Cape St Lawrence, Cape Breton Island
Heath Point, Anticosti
South Point, Anticosti

Southwest Point, Anticosti
West Point, Anticosti
Cape Rosier, Gaspé Coast
Fame Point, Gaspé Coast
Cape Magdalen, Gaspé Coast
Amherst Island, Magdalen Islands

SIGNALS USED AT LLOYD'S SIGNAL STATIONS.

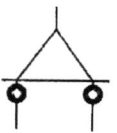

This signal indicates that the station at which it is hoisted is temporarily closed, and that no communication can be held.*

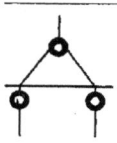

This signal indicates that telegraphic communication is interrupted, and that messages can not be forwarded by telegraph, but will be forwarded by other means as soon as possible.*

* These signals will be kept up until the signal station is again occupied, or until telegraphic communication is again possible

These stations are available to shipowners for reporting the passing of their vessels.

At these signal stations the International Code is the only code recognized, and vessels of any nation may make their names known by means of this code, and thus secure the immediate publication of the announcement in the "Shipping Gazette and Lloyd's List" and "Lloyd's Weekly Shipping Index" Cases of wreck or other accidents at sea should always be made known by shipmasters to the signal stations with which they communicate.

Night signals at special stations —The apparatus used for the signals will be a flashing lamp

A series of continuous short flashes is to call the attention of a passing vessel

A series of long-short flashes, repeated as often as may be necessary, indicates that a vessel's signals have been seen and recognized.

If a signal shown by a vessel has not been understood the lamp will be kept dark until the vessel repeats the signal

Special night signals at certain stations —Also, to denote telegraphic interruption during the night, the signal stations at the Old Head of Kinsale and Brow Head will show *3 green* lights, arranged in the shape of a triangle, apex upward, and to call attention to this signal a Roman candle will be burnt if the vessel can be recognized.

PLATE IV

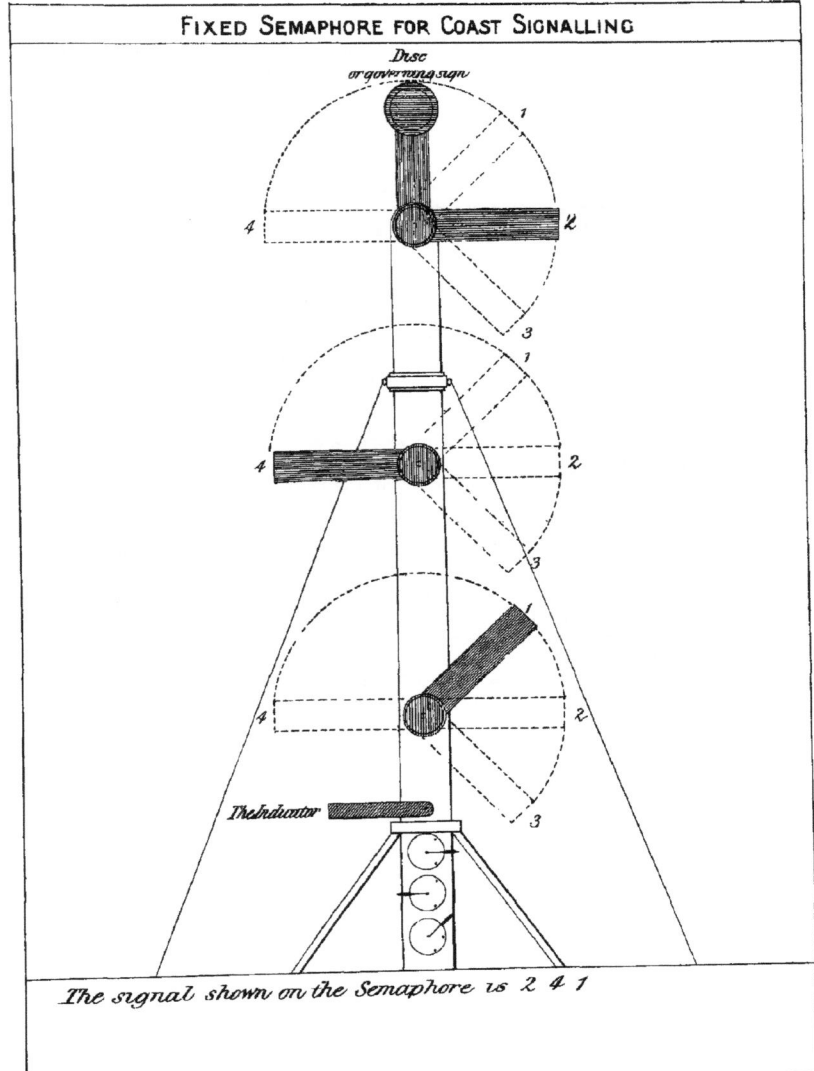

FIXED SEMAPHORE FOR COAST SIGNALLING

The signal shown on the Semaphore is 2 4 1

GENERAL ALPHABETICAL TABLE FOR MAKING THE INTERNATIONAL CODE SIGNALS BY MEANS OF DISTANT SIGNALS BY FIXED SEMAPHORE

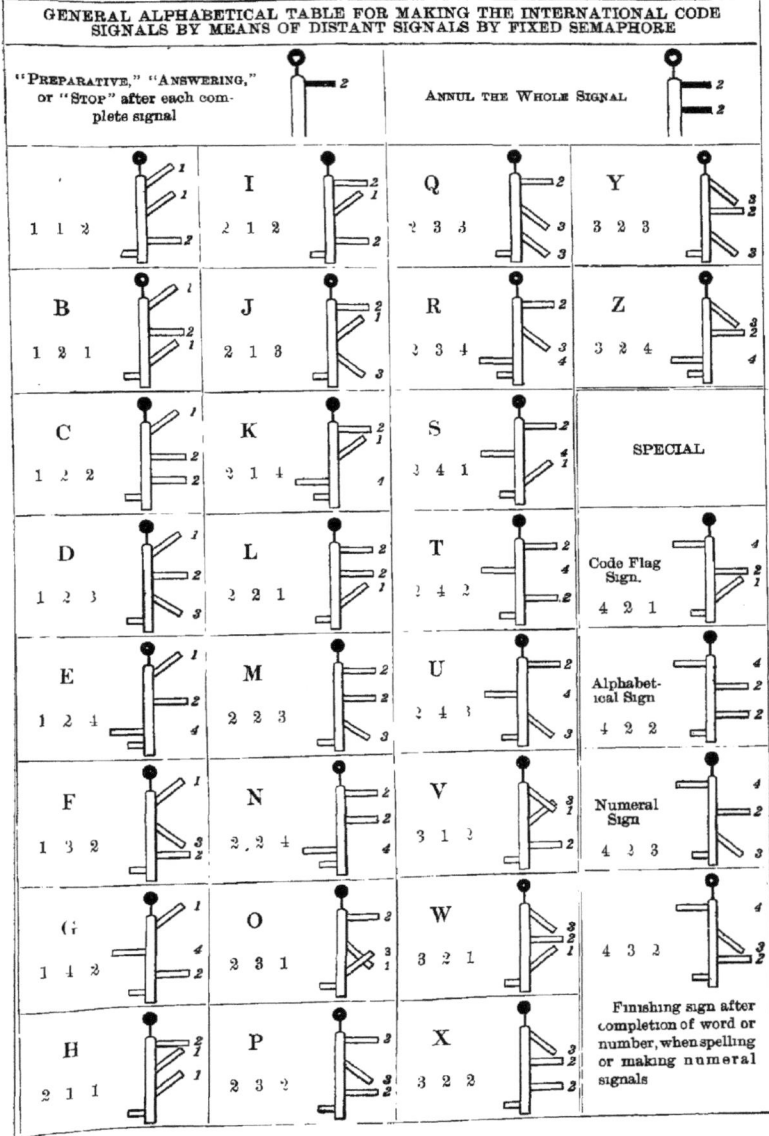

GENERAL ALPHABETICAL TABLE FOR MAKING THE INTERNATIONAL CODE SIGNALS BY MEANS OF DISTANT SIGNALS BY SHAPES

"PREPARATIVE," "ANSWERING," OR "STOP," after each complete signal		ANNUL THE WHOLE SIGNAL.	
A 1 1 2	I 2 1 2	Q 2 3 3	Y 3 2 3
B 1 2 1	J 2 1 3	R 2 3 4	Z 3 2 4
C 1 2 2	K 2 1 4	S 2 4 1	SPECIAL SIGNS
D 1 2 3	L 2 2 1	T 3 4 2	Code Flag Sign 4 2 1
E 1 2 4	M 2 2 3	U 2 4 3	Alphabetical Sign 4 2 2
F 1 3 2	N 2 2 4	V 3 1 2	Numeral Sign, 4 2 3
G 1 4 2	O 2 3 1	W 3 2 1	4 3 2
H 2 1 1	P 2 3 2	X 3 2 2	Finishing sign after completion of word or number when spelling or making numeral signals

If no cones are available, a square flag may be substituted for the cone point upward, a pennant for the cone point downward, and a wheft for the drum   (See page 529)

DISTANT SIGNALS

1 Distant Signals are required when, in consequence of distance or the state of the atmosphere, it is impossible to distinguish the *colors* of the flags of the International Code, and, therefore, to read a signal made by those flags, they also provide an alternative system of making the signals in the Code, which can be adopted when the system of flags can not be employed

2. Three different methods of making Distant Signals are explained below:

   (*a*) By Cones, Balls, and Drums

   (*b*) By Balls, Square Flags, Pennants, and Whefts

   (*c*) By the Fixed Coast Semaphore

The last method (Fixed Coast Semaphore) is not necessarily a method of making *Distant* Signals, as it can be, and is, used at close quarters and under conditions when flags could equally be employed, but it has been placed here under the heading of Distant Signals for ease of explanation

3 The *Characteristic* of Distant Signals is the *Ball*, one ball at least appearing in each hoist of the Distant Code In the case of the Semaphore the ball is replaced by a *Disk*

4 Hitherto only three Symbols have been required for Distant Signaling, but the increase made in the number of flags of the International Code renders four Symbols necessary, in order that it may be possible to provide a distinct hoist to represent each of the flags of the Code (*i e*, letters of the alphabet)

5 Distant Signals are made

  From a ship—by hoisting shapes

  From the shore—by hoisting shapes, or by the position of the arms of a Semaphore

6 The Shapes used as Symbols are.

  . (*a*) A Cone point upward,

   A Ball

   A Cone point downward, and

   A Drum (*The Drum should be at least one-third greater in height than the Ball*)

  (*b*) A square Flag may be substituted for the cone point upward

   A Ball,

   A Pennant may be substituted for the cone point downward, and

   A Pennant with the fly tied to the halyards, or a Wheft for the drum.

   (A wheft is any flag tied in the center)

As in calms, or when the wind is blowing toward or from the observer, it is impossible to distinguish with certainty between a square flag, pennant, and wheft, and as flags when hanging up and down may hide one of the balls and so prevent the signal being understood, the system of cones and drums is preferable to that of flags, pennants, and whefts.

   (*c*) In signaling by the Semaphore, the positions of the arms represent the shapes

7 To simplify the "Taking in," "Reporting," and "Reading off" of the Distant Signals, the four positions of the Semaphore Arms, and the four Symbols have been numbered 1, 2, 3, 4

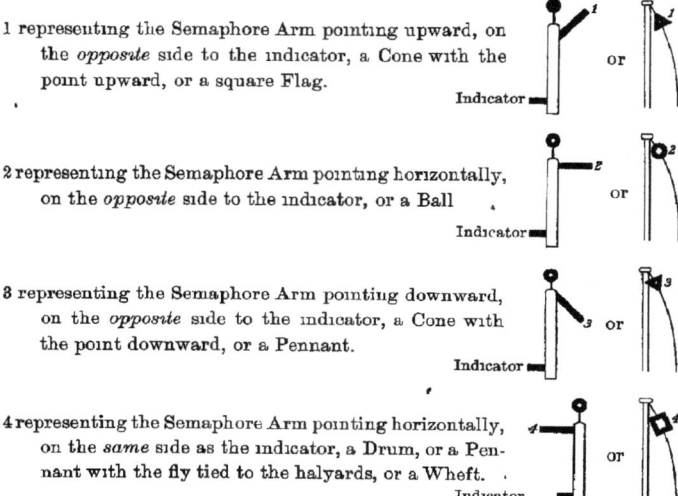

1 representing the Semaphore Arm pointing upward, on the *opposite* side to the indicator, a Cone with the point upward, or a square Flag.

             Indicator   or

2 representing the Semaphore Arm pointing horizontally, on the *opposite* side to the indicator, or a Ball

             Indicator   or

3 representing the Semaphore Arm pointing downward, on the *opposite* side to the indicator, a Cone with the point downward, or a Pennant.

             Indicator   or

4 representing the Semaphore Arm pointing horizontally, on the *same* side as the indicator, a Drum, or a Pennant with the fly tied to the halyards, or a Wheft.

             Indicator   or

8. To facilitate signaling by Semaphore or Shapes, the signals representing the letters of the alphabet have been arranged in numerical order, the figures representing the signal for the letter A being the first in numerical sequence

      Thus A is represented by 1 1 2
      Thus B is represented by 1 2 1
      Thus C is represented by 1 2 2,
          etc , etc

The signals representing the letters from A to G begin with 1,
Those from H to U begin with 2,
Those from V to Z begin with 3,
And the Special Signs (*i e ,* Code Flag, Alphabetical, Numeral, and Finishing Signs) begin with 4

        (*See plates* V and VI)

9 The Code Flag Sign 4 2 1 (*see plates* V and VI) is always to be shown before signals taken from the General Vocabulary of the International Code are commenced.

10. When signals are made by the Semaphore the disk is always to be kept up until the signal is completed, and the hoist is to be read from the top arm downward.

11. The Stop Signal 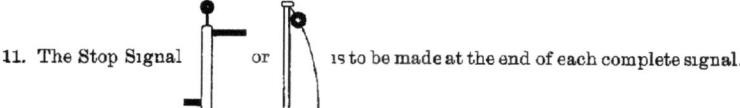 is to be made at the end of each complete signal.

12. With two Balls, two Cones, and one Drum, every signal in the International Code can be made, each hoist representing one letter of the two, three, or four letters forming the signal

EXAMPLE OF A SIGNAL FROM THE INTERNATIONAL CODE

MADE BY FIXED SEMAPHORE OR BY DISTANT SIGNALS.

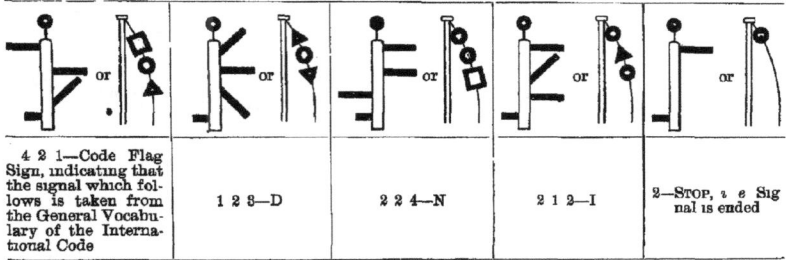

| 4 2 1—Code Flag Sign, indicating that the signal which follows is taken from the General Vocabulary of the International Code | 1 2 3—D | 2 2 4—N | 2 1 2—I | 2—Stop, i e Signal is ended |

Looking DNI out in the International Code, we find it to be "Pilot boat is advancing toward you"

ALPHABETICAL DISTANT SIGNALS.

13 When it is desired to spell a word by Distant Signals, the Alphabetical Sign 4 2 2 (*see Plates* V and VI) is to be shown first All the hoists which follow until the Finishing Sign 4 3 2 (*see Plates* V and VI) is shown are to be understood as representing the particular letters of the alphabet allotted to them in Plates V and VI, which, when combined, spell the word which it is desired to signal.

NUMERAL DISTANT SIGNALS.

14 When it is desired to signal numbers by Distant Signals, the Numeral Sign 4 2 3 (*see Plates* V and VI) is to be shown first After that sign has been shown, and until the Finishing Sign 4 3 2 (*see Plates* V and VI) is shown, the hoists representing the various letters of the alphabet (*see Plates* V and VI) are to be understood as having the numerical values which are allotted to the particular letters under the system of making Numeral Signals by flags, which is explained on page 32

Thus, after the Numeral Sign 4 2 3 has been shown, the Distant Signal hoist representing the letter A will mean the number 1, that representing B will mean 2, that representing K will mean 11, and so on as in the Numeral Table on page 32.

SPECIAL DISTANT SIGNALS.

15 As shown in the Example to Paragraph 12, above, signals from the General Vocabulary of the International Code require to be made by more than one hoist, which involves loss of time Arrangements have, however, been adopted by which thirty-seven important signals (*see pages* 527–531) can be made by one hoist only These thirty-seven signals are called "Special Distant Signals," and are represented by the numbers explained in Paragraph 7, page 524, and not by letters

16 The Special Distant Signals are distinguished from Distant Signals taken from the General Vocabulary of the International Code by the fact (1) that they are not preceded by the Code Flag Sign (*see paragraph* 9), and (2) that the Stop Signal immediately follows the single hoist representing the particular "Special Distant Signal" which is being made.

EXAMPLE OF A SPECIAL DISTANT SIGNAL

MADE BY SEMAPHORE OR DISTANT SIGNALS

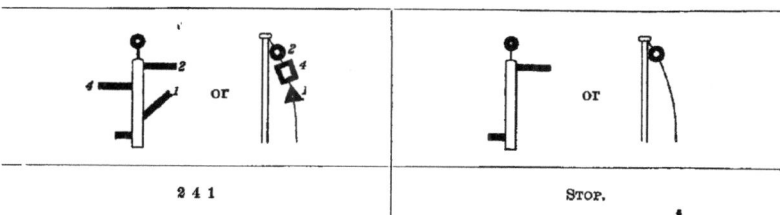

2 4 1	STOP.

Looking 2 4 1 out in the Table of Special Distant Signals (page 535), we find it to be "Can not distinguish your flags, come nearer or make Distant Signals."

SPECIAL DISTANT SIGNALS.

MADE BY A SINGLE HOIST FOLLOWED BY THE STOP SIGNAL ARRANGED NUMERICALLY FOR READING OFF A SIGNAL.

These signals may be made by the semaphore, by cones, balls, and drums, or by square flags, balls, pennants, and whefts

Signal		Meaning
	2	"Preparative," "Answering," or, "Stop," after each complete signal
	1 2	Aground; want immediate assistance a
	2 1	Fire, or, leak, want immediate assistance a
	2 2	Annul the whole signal
	2 3	You are running into danger, or, Your course is dangerous
	2 4	Want water immediately
	3 2	Short of provisions, starving.
	4 2	Annul the last hoist, I will repeat it
	1 1 2	I am on fire
	1 2 1	I am aground

a See Distress Signals (page 7)

Signal		Meaning.
	1 2 2	Yes, or, Affirmative
	1 2 3	No, or, Negative.
	1 2 4	Send lifeboat.
	1 3 2	Do not abandon the vessel.
	1 4 2	Do not abandon the vessel until the tide has ebbed.
	2 1 1	Assistance is coming.
	2 1 2	Landing is impossible.
	2 1 3	Bar, or, Entrance is dangerous.
	2 1 4	Ship disabled, will you assist me into port ?
	2 2 1	Want a pilot α
	2 2 3	Want a tug, can I obtain one ?
	2 2 4	Asks the name of ship (or, signal station) in sight, or, Show your distinguishing signal

α See Pilot Signals (page 8)

Signal		Meaning
	2 3 1	Show your ensign.
	2 3 2	Have you any despatches (message, orders, or, telegrams) for me?
	2 3 3	Stop, Bring to, or, Come nearer, I have something important to communicate
	2 3 4	Repeat signal, or hoist it in a more conspicuous position
	2 4 1	Can not distinguish your flags, come nearer, or make distant signals
	2 4 2	Weigh, Out, or, Slip, wait for nothing, get an offing.
	2 4 3	Cyclone, Hurricane, or, Typhoon expected
	3 1 2	Is war declared? or, Has war commenced?
	3 2 1	War is declared, or, War has commenced
	3 2 2	Beware of torpedoes, channel is mined
	3 2 3	Beware of torpedo boats
	3 2 4	Enemy is in sight

Signal.		Meaning
	3 3 2	Enemy is closing with you, *or*, You are closing with the enemy
	3 4 2	Keep a good lookout, as it is reported that ememy's men-of-war are going about disguised as merchantmen
	4 1 2	Proceed on your voyage

The following distant signals made with flag and ball, or pennant and ball, have the special signification indicated beneath them

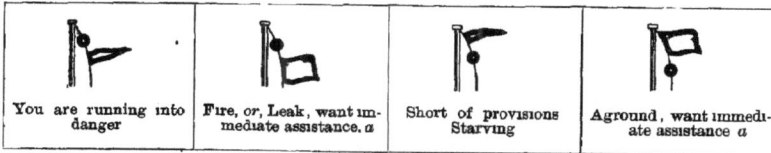

| You are running into danger | Fire, *or*, Leak, want immediate assistance. *a* | Short of provisions Starving | Aground, want immediate assistance *a* |

a See Distress Signals (page 7).

SPECIAL DISTANT SIGNALS.

MADE BY A SINGLE SIGNAL FOLLOWED BY THE STOP SIGNAL. ARRANGED ALPHABETICALLY FOR MAKING A SIGNAL.

ABANDON—IMMEDIATE

ABANDON
1 3 2 —Do not abandon the vessel
1 4 2 —Do not abandon the vessel until the tide has ebbed

1 2 2 **AFFIRMATIVE**, *or*, **YES**

AGROUND
1 2 —Aground, want immediate assistance.*
1 2 1 —I am aground

ANNUL
4 2 —Annul the last hoist, I will repeat it
2 2 —Annul the whole signal

2 **ANSWERING SIGNAL.**
1 2 3 —No, *or*, Negative
1 2 2 —Yes, *or*, Affirmative

ASSISTANCE
1 2 —Aground, want immediate assistance *
2 1 1 —Assistance is coming
2 1 —Fire, *or*, Leak, want immediate assistance *
2 1 4 —Ship disabled, will you assist me into port?
1 2 4 —Send life boat

ATTENTION, *or*, **DEMAND**
2 2 4 —Asks the name of the ship (or, signal station) in sight, *or*, Show your distinguishing signal
2 3 1 —Show your ensign

BAR
2 1 3 —Bar, *or*, Entrance is dangerous.

BOAT
1 2 4 —Send lifeboat

CAUTION
2 1 3 —Bar, *or*, Entrance is dangerous
3 2 3 —Beware of torpedo boats
3 2 2 —Beware of torpedoes, channel is mined
3 3 2 —Enemy is closing with you *or*, You are closing with the enemy
3 2 4 —Enemy is in sight
3 4 2 —Keep a good look-out, as it is reported that enemy's war vessels are going about disguised as merchantmen
2 1 2 —Landing is impossible
2 4 3 —Typhoon, Hurricane, *or*, Cyclone expected.
3 2 1 —War is declared, *or*, War has commenced
2 4 2 —Weigh, Cut, *or*, Ship, wait for nothing, get an offing
2 3 —You are running into danger

CHANNEL
3 2 2 —Beware of torpedoes, channel is mined

COMING
2 1 1 —Assistance is coming

COMMUNICATE
2 3 3 —Stop, Bring-to, *or* Come nearer, something important to communicate

DANGER.
2 1 3 —Bar, *or*, Entrance is dangerous
2 1 2 —Landing is impossible
2 3 —You are running into danger

DESPATCHES
2 3 2 —Have you any despatches (message, orders, *or*, telegrams) for me?

DISABLED
2 1 4 —Ship disabled, will you assist me into port?

DISTANT SIGNALS
2 4 1 —Can not distinguish your flags, come nearer or make distant signals

DISTRESS
1 2 —Aground, want immediate assistance *
2 1 —Fire, *or*, Leak; want immediate assistance *
1 2 1 —I am aground
1 1 2 —I am on fire
2 1 4 —Ship disabled, will you assist me into port?
3 2 —Short of provisions Starving
2 4 —Want water immediately.

ENEMY
3 2 3 —Beware of torpedo boats
3 2 2 —Beware of torpedoes, channel is mined
3 3 2 —Enemy is closing with you, *or*, You are closing with the enemy
3 2 4 —Enemy is in sight
3 4 2 —Keep a good lookout, as it is reported that enemy's vessels are going about disguised as merchantmen.

FIRE
2 1 —Fire, *or*, Leak, want immediate assistance *
1 1 2 —I am on fire

IMMEDIATE
1 2 —Aground, want immediate assistance *
2 1 —Fire, *or*, Leak, want immediate assistance *
2 4 —Want water immediately

*See Distress Signals, page 7.

LANDING—YES

	LANDING
2 1 2	—Landing is impossible
	LEAK
2 1	—Fire, or, Leak, want immediate assistance *
	LIFEBOAT
1 2 4	—Send lifeboat
	MESSAGE
2 3 2	—Have you any message (telegram, orders, or, despatches) for me?
1 2 3	NEGATIVE, or, No
	OFFING
2 4 2	—Weigh, Cut, or, Ship, wait for nothing, get an offing
	PILOT
2 2 1	—Want a pilot †
2	PREPARATIVE SIGN
	PROCEED
4 1 2	—Proceed on your voyage
	PROVISIONS AND WATER
3 2	—Short of provisions Starving
2 4	—Want water immediately
	REPEAT
4 2	—Annul the last hoist, I will repeat it
2 3 4	—Repeat signal, or hoist it in a more conspicuous position
	RUN
2 3	—You are running into danger
	SEND
1 2 4	—Send lifeboat
	SIGNALS
4 2	—Annul the last hoist, I will repeat it
2 2	—Annul the whole signal
2 2 4	—Asks name of ship (or, signal station) in sight
2 4 1	—Can not distinguish your flags, come nearer or make distant signals
1 2 3	—No, or, Negative

	SIGNALS—Continued
2 3 4	—Repeat signal, or hoist it in a more conspicuous position
2 2 4	—Show your distinguishing signal
2 3 1	—Show your ensign
2 3 3	—Stop, Bring-to, or, Come nearer, I have something important to communicate
1 2 2	—Yes, or, Affirmative
	STARVING
3 2	—Short of provisions Starving
2	STOP AFTER EACH COMPLETE SIGNAL
	TELEGRAM
2 3 2	—Have you any telegram (message, order, or, despatches) for me?
	TORPEDOES, or, MINES
3 2 3	—Beware of torpedo boats
3 2 2	—Beware of torpedoes, channel is mined
	TUG
2 2 3	—Want a tug, can I obtain one?
	WANT
1 2	—Aground, want immediate assistance *
2 1	—Fire, or, Leak, want immediate assistance *
3 2	—Short of provisions Starving
2 2 1	—Want a pilot †
2 2 3	—Want a tug, can I obtain one?
2 4	—Want water immediately
	WAR
3 1 2	—Is war declared? or, Has war commenced?
3 2 1	—War is declared, or, War has commenced
	WATER
2 4	—Want water immediately
	WEATHER
2 4 3	—Cyclone, Hurricane, or, Typhoon expected
	WEIGH
2 4 2	—Weigh Cut, or, Ship, wait for nothing, get an offing
1 2 2	YES, or, AFFIRMATIVE

* See Distress Signals, page 7   † See Pilot Signals, page 8

SEMAPHORE SIGNALS.

INSTRUCTIONS FOR THE USE OF THE BRITISH MOVABLE SEMAPHORE

THE INDICATOR.

The Indicator denotes from which side the signs are to be read, but when first shown it is to call attention, and may be considered the preparative signal   When closed it denotes the finish of the communication

HOW TO SEMAPHORE.

The person intending to Semaphore will make the International Code Signal VOX (I am going to Semaphore to you), and set his Semaphore at the alphabetical sign (*see Plate* VII) with the Indicator out, and wait until the person to whom the Semaphore signal is to be made hoists his answering pennant CLOSE UP   Then he will proceed with the communication by spelling, making a momentary pause between each sign or letter, the arms are to be dropped between each word or group, the Indicator alone remaining out

Should the answering pennant be dipped by the person taking in the signal, the last TWO words are to be repeated until the answering pennant is again hoisted CLOSE UP

When in the middle of a spelling signal numerals have to be made, the Semaphore is to be put at the numeral sign (*see Plate* VII), and the number then made   When the numeral signal is finished the alphabetical sign is to be made and the communication by spelling proceeded with

HOW TO ANSWER AND TAKE IN SEMAPHORE SIGNALS.

The answering pennant is to be hoisted CLOSE UP by the person taking in the Semaphore signal, thus denoting he is ready to read and *write down* the signal

It is to be "dipped" when a word is lost, and the person making the signal is then to repeat the TWO last words until the answering pennant is hoisted again CLOSE UP

In answering by the Semaphore the arm in position (1) represents the answering pennant at the dip   And in position (3) the answering pennant close up

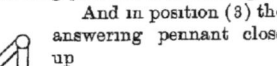

SEMAPHORING BY HAND FLAGS

(1) BRITISH METHOD

The British method of Semaphoring by flags held in the hand which is shown in plate VIII is exactly the same as the British Movable Semaphore system, which has just been explained, the positions of the apparatus which denote the letters, numbers, and special signs being, it will be seen, identical in each case, and the only difference being in the apparatus employed

(2) FRENCH METHOD

The French method of Semaphoring by flags held in the hand (*see plate* IX) is based on the same principle as the British method, but the positions in which the flags are held to denote the letters, etc, are different

THE BRITISH MOVABLE SEMAPHORE.

SEMAPHORE SIGNS.	GOVERNING SIGNS.			
Indicator......	Preparative. When closed it denotes the finish.	Alphabet-ical.	Numeral.	Annul or negative.

SIGNS.									
Alphabetical Signification.	A	B	C	D	E	F	G	H	I
Numerical Signification.	1	2	3	4	5	6	7	8	9

SIGNS.									
Alphabetical Signification.	J Also the alphabetical sign.	K	L	M	N	O	P	Q	R
Numerical Signification.	0								

SIGNS.								
Alphabetical Signification.	S	T	U	V	W	X	Y	Z

NOTE.—If a numeral signal is to be followed by words, the end of the numerical signification of the signs is shown by the alphabetical sign being made, indicating that spelling is again to commence. For instructions as to use see page 559.

BRITISH METHOD OF SEMAPHORING BY HAND FLAGS

SIGNS					
Alphabetical Signification	A	B	C	D	E	F
Numeral Signification	1	2	3	4	5	6
SIGNS						
Alphabetical Signification	G	H	I	J	K	L
Numeral Signification	7	8	9	ALSO ALPHABETICAL SIGN	0	
SIGNS.						
Alphabetical Signification.	M	N	O	P	Q	R
SIGNS						
Alphabetical Signification	S	T	U	V	W	X
SIGNS						
Signification	Y	Z		ALPHABETICAL	NUMERAL	ANNUL

For instructions as to use see page 539

FRENCH METHOD OF SEMAPHORING BY HAND FLAGS

SIGNS						
Alphabetical Signification	A	B	C	D	E	F
Numeral Signification	1	2	3	4	5	6
SIGNS						
Alphabetical Signification	G	H	I	J	K	L
Numeral Signification	7	8	9	0		
SIGNS						
Alphabetical Signification	M	N	O	P	Q	R
SIGNS						
Alphabetical Signification	S	T	U	V	X	Y
SIGNS						
Signification	Z		DO NOT UNDERSTAND	NUMBERS	ATTENTION	END OF WORD OR PHRASE

NOTE —The **W** is made by means of two consecutive **V's**. To indicate the finish of a Numerical Signal the Number sign is again made thus the number sign begins and ends Numerical Signals

FRENCH SEMAPHORE STATIONS

Semaphore stations and vessels at sea may exchange either ''semaphore telegrams'' or ''messages ''

I Semaphore Telegrams

Definition —The semaphore telegrams are those which are sent partly over the telegraph wires   Semaphore telegrams sent from vessels at sea are accepted for all countries which have signified their adherence to the Telegraphic Convention of St Petersburg and which are named in the following list

Argentine Republic	Germany	Persia
Australia	Great Britain	Portugal
Austria	Greece	Portuguese Colonies
Belgium	Holland	Roumania
Bosnia-Herzegovina	Hungary	Russia
Brazil	British India	Senegal
Bulgaria	Italy	Servia
Cape of Good Hope	Japan	Siam
Ceylon	Luxemburg	Spain
Crete	Madagascar	Sweden
Denmark	Montenegro	Switzerland
Dutch East Indies	Natal	Tasmania
Egypt	New Caledonia	Tunis
France and Algeria	New Zealand	Turkey
French Indo-China	Norway	Uruguay

Semaphore telegrams intended for ships at sea are accepted for all the French semaphore stations marked with an ''a'' in the Geographical List, Part I, of the International Code

Wording and transmission —Messages may be worded either in the language of the country for which they are intended or in the signals of the International Code   These signals are represented in writing by means of the twenty-six letters of the alphabet   The number of letters in each group should not exceed four

The groups of letters signaled from a vessel sending a semaphore telegram are translated by the signal officer into the language of the country in which the semaphore station is situated, and it is this translation which is transmitted to the addressee

Upon a request from the vessel, however, made by the signal XB or WZ (see ''Urgent and Important Signals,'' Part I), the groups of letters themselves will be transmitted without translation

The address and, if necessary, the signatures to telegrams from ships at sea are always transmitted in ordinary language by electric telegraph   The address of telegrams intended for ships at sea should specify the name or official number of the vessel and her nationality

Charges —(1) The usual telegraphic charge and any accessory charges which the circumstances may render necessary

(2) A maritime charge which has been fixed at five centimes per word with a minimum charge of fifty centimes per telegram for transmission in France   For international service this charge is one franc per telegram

All charges for semaphore telegrams intended for ships at sea must be paid by the sender   The charges for telegrams coming from vessels must be paid by the persons to whom the telegrams are addressed

II Messages

Definition —Vessels passing French semaphore stations may exchange messages (1) either by communications to be sent to their destination or to be forwarded by post without making any use of the electric telegraph wires, or (2) by communications deposited at the semaphore stations, or addressed to it by post, for transmission direct to the vessel at sea

The messages may be worded either in French or in the signals of the International Code

The charge is five centimes per word, with a minimum of fifty centimes, and is payable by the sender in the case of messages addressed to vessels at sea and by the receiver in case of messages coming from such vessels

UNITED STATES ARMY AND NAVY SIGNAL CODE

Communication may be had by this code with the United States Army, Navy,
Revenue Cutter, and Life-Saving services   (*See page 515*)

A	2 2	H	1 2 2	O	2 1	V	1 2 2 2
B	2 1 1 2	I	1	P	1 2 1 2	W	1 1 2 1
C	1 2 1	J	1 1 2 2	Q	1 2 1 1	X	2 1 2 2
D	2 2 2	K	2 1 2 1	R	2 1 1	Y	1 1 1
E	1 2	L	2 2 1	S	2 1 2	Z	2 2 2 2
F	2 2 2 1	M	1 2 2 1	T	2	tion	1 1 1 2
G	2 2 1 1	N	1 1	U	1 1 2		

NUMERALS

1	1 1 1 1	3	1 1 1 2	5	1 1 2 2	7	1 2 2 2	9	1 2 2 1
2	2 2 2 2	4	2 2 2 1	6	2 2 1 1	8	2 1 1 1	0	2 1 1 2

ABBREVIATIONS

a	after	c	can	n	not	t	the	ur	your	wi	with
b	before	h	have	r	are	u	you	w	word	y	yes

x x 3 "numerals follow" or "numerals end "
sig 3 "signature follows "

CONVENTIONAL SIGNALS

End of a word	3	Repeat after (word)	121 121 3 22 3 (word)
End of a sentence	33	Repeat last word	121 121 33
End of a message	333	Repeat last message	121 121 121 333
Error	12 12 3	Move a little to right	211 211 3
Acknowledgment, or, "I understand"	22 22 3	Move a little to left	221 221 3
Cease signaling	22 22 22 333	Signal faster	2212 3
Wait a moment	1111 3		

INSTRUCTIONS FOR USING THE SYSTEM

The whole number opposite each letter or numeral stands for that letter or numeral

TO SIGNAL WITH FLAG, TORCH, HAND LANTERN, OR BEAM OF SEARCH LIGHT.

There are but one position and three motions
The first position is with the flag or other appliance held vertically, the signalman
facing squarely towards the station with which it is desired to communicate
The first motion, or "one" or "1," the signal is waved to the right of the sender
and will embrace an arc of 90°, starting with the vertical and returning to it, and will
be made in a plane exactly at right angles to the line connecting the two stations
The second motion, or "two" or "2," is a similar motion to the left of the sender
To make the third motion, "front" or "three" or "3," the signal is waved to the
ground directly in front of the sender, and instantly returned to the first position
Numbers which occur in the body of the message must be spelled out in full
To use the torch or hand lantern, a footlight must be used as a point of reference
to the motion   The lantern is more conveniently swung out upwards, by hand, from
the footlight for "1" and "2" and raised vertically for "3"

TO SEND A MESSAGE

"To call" a station, signal its initial or call letter until acknowledged   To acknowledge a call or receipt of a message, signal "I understand."

Make a slight pause after each letter and also after "front "   If the sender discovers that he has made an error, he should make the "front" and "12 12 3," after which he proceeds with the message, beginning with the word in which the error occurred

FLASH SIGNALS WITH LANTERN, HELIOGRAPH, OR SEARCH LIGHT

Use short flash for "1," two short flashes in quick succession for "2," and a long steady flash for "3 "   The elements of a letter should be slightly longer than in sound signals

To call a station, make the initial or call letter until acknowledged.   Then turn on a steady flash until answered by a steady flash   The calling station will then proceed with the message

All other conventional signals are the same as for the flag.

SOUND SIGNALS WITH FOG WHISTLE, FOG HORN, OR BUGLE

Use one toot (about half second) for "1," two toots (in quick succession) for "2," and a blast (about two seconds' long) for "3."   The ear and not the watch is to be relied upon for the intervals

The signal of execution for all tactical or drill signals will be one long blast followed by two toots in quick succession

In the use of any other appliance, such as a bell, by which a blast can not be given, three strokes in quick succession will be given in the place of the blast to indicate "3 "

When more than two vessels are in company, each vessel, after making "I understand," should make her call letter that it may be certain which vessel has acknowledged.

MORSE SIGNAL CODE

—— indicates a LONG of about 3 seconds' duration

— indicates a SHORT of about 1 second duration

Preparative Signal to attract attention – – – – – – – – – – etc

Answering Signal, or, I understand —— – —— – – —— – —— – etc

Interval between each flash or sound ....1 second

Interval between each letter ....3 seconds

Interval between each word...........................6 seconds.

A .... – ——		N .... —— –
B .... —— – – –		O .... —— – ——
C .... —— – —— –		P .... – —— —— –
D .... —— – –		Q .... —— —— – ——
E .... –		R .... – —— –
F .... – – —— –		S .... – – –
G .... —— —— –		T .... ——
H .... – – – –		U .... – – ——
I .... – –		V .... – – – ——
J .... – – – ——		W .... – —— ——
K .... —— – ——		X .... —— – – ——
L .... – —— – –		Y .... —— – —— –
M .... —— ——		Z .... —— —— – –

(a) LIGHT AND SOUND SIGNALS, ACCORDING TO COLOMB'S FLASHING SIGNALS SYSTEM

The following urgent and important signals may be made at night or in thick weather, either by long and short flashes of light or by long and short sounds on a steam whistle, siren, fog horn, etc.

INSTRUCTIONS FOR THE USE OF FLASHING OR SOUND SIGNALS

With flashing signals the lamp must always be turned toward the person addressed

To attract attention, a series of rapid short flashes or sounds should be made and continued until the person addressed gives the sign of attention by doing the same

If, however, it is supposed that the person addressed can not reply, the signal may be made after a moderate pause, or, under certain circumstances, the communication may be made without preparatory signs

After making a few rapid short flashes or sounds as an acknowledgment, the receiver must watch or listen attentively until the communication is completed, when he must make the sign indicated below, showing that the message is understood.

If the receiver does not understand the message, he must wait until the signal is repeated.

Duration of short flashes or sounds — .... 1 second.
Duration of long flashes or sounds —— .... 3 seconds.
Interval between each flash or sound .... 1 second.
Answer, or, "I understand" — — — — — — — — etc.

SIGNALS.

You are standing into danger .... — — ——
I want assistance; remain by me .... — — — ——
Have encountered ice .... — — — ——
Your lights are out (or, want trimming) .... — —— — —
The way is off my ship; you may feel your way past me .... — —— — ——
Stop, or, heave to; I have something important to communicate ...... — —— — —
Am disabled; communicate with me .... — — —— —

When a vessel is in tow, the following signals made by flashes of light may be used between her and the tug or towing vessel:

Steer more to starboard .... —
Steer more to port .... — —
Cast off hawsers .... — — — —

(b) FLAG WAVING.

INSTRUCTIONS FOR COMMUNICATING BY FLASHING SIGNALS WITH A FLAG (GENERALLY TERMED FLAG WAVING).

The system used is the Morse Alphabet, the letters being made by groups of LONG and SHORT flashes caused by moving a flag through a long or short arc, as described below.

The signalman may work from left to right, or from right to left, as shown in figures 1 and 2, according to convenience and direction of the wind.

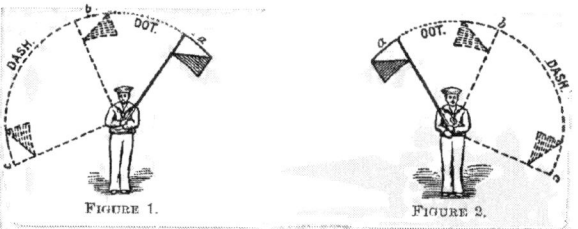

FIGURE 1.   FIGURE 2.

In the normal position (a) in the above figures, the flag should make an angle of 25° with a vertical line through the center of the body.

The pole should be kept high enough to permit seeing underneath the flag while in motion.

TO MAKE A SHORT FLASH OR DOT

The flag is waved from *a* to *b*, and without any pause back again to the normal position

TO MAKE A LONG FLASH OR DASH

The flag is waved from *a* to *c*, and after a short but distinct pause at *c* brought back to the normal position.

WHEN SIGNALING A LETTER

When signaling a letter the flashes representing it should be made in one continuous wave of the flag, taking particular care that no pause is made when at the normal position

Example —To make R - —— - wave the flag from *a* to *b* back to *a* and without a pause down to *c*, making there a short but distinct pause (*vide* instructions for long flash) back to *a*, then without a pause to *b* and back to the normal position *a*

HOW TO SIGNAL

A pause equal to the length of a long flash should be made at the normal position *a* between each letter of a word, or letter of a group of letters

When the word or group of letters is completed, the butt of the staff is to be brought to the ground, and the flag at the same moment gathered in

A slight pause should be made at the normal position before commencing a word or group of letters

In receiving a message the flag should be lowered and gathered in until required for answering.

POSITION OF THE SIGNALMAN

The signalman should stand square to the station to which he is signaling, and the pole should be kept as upright as possible while in motion, the point never being allowed to droop to the front

In order to keep the flag always exposed while moving it across the body to form the flashes, the point of the staff should be made to describe an elongated figure of 8 in the air

76564—09——35

CONVENTIONAL SIGNALS: THEIR EQUIVALENT LETTERS AND THE METHOD OF ANSWERING THEM.

Meaning	Sign	Equivalent letter and how made	How answered
Preparative.	– – – – – etc	By a succession of E's in one group	By the general answer T
Answer	—	T (singly)	
Break sign	– – – –	I I as separate letters	
Stop	– – – – – –	I I I as separate letters	
Finish of a message	– ‒ – — –	V E as one group	– — – R — – – D as separate letters
Erase sign	– – – – – etc	By a succession of E's as separate letters	By a succession of E's as separate letters
Annul	W W	W W as one group	By W W as one group
Repeat word after — (when a single word is required)	I M I W A followed by the word preceding the one required	I M I as one group W A as separate letters	By the general answer T
Repeat all after —(if more than one word is required)	I M I A A	I M I as one group A A as separate letters	By the general answer T
Repeat all (if the whole message is to be repeated)	I M I A L L	I M I as one group A L L as separate letters.	By the general answer T

THE USE OF THE SPECIAL SIGNS.

THE PREPARATIVE.

Is used to call attention, and is answered by the General Answer

THE GENERAL ANSWER.

Is made by a long flash ——, or letter T.

THE BREAK SIGN.

Is to be used between the address of the receiver and the text of the message, and, after the text, if the name of the sender is signaled

THE STOP

Is made by three separate letter I's

THE FINISH.

Is to be made by V E in one group at the completion of a message, and is to be answered, if the message is understood, by *R D* in separate letters   N B —*R D means Read*

THE ERASE

Is made by a series of E's as separate letters, and is used to erase a word or group that has been wrongly sent, and is to be answered by the Erase.

THE ANNUL

Is made by W W in one group, and is used to negative ALL the message that has gone before, and is to be answered by the Annul

METHOD OF ANSWERING.

Each word when understood is to be answered by one long flash —— (T).
If a word is not answered, the sender is to repeat it until answered by a long flash
At the end of the message, if understood, receiver will make - —— -   —— - -
(or R D), meaning Read
The Erase and Annul signs are to be answered by their own signs

METHOD OF SIGNALING NUMBERS.

All numerals, whether representing time, distance, numbers, etc , are to be spelled in full

METHOD OF ASKING FOR REPETITION.

If the receiver requires any word to be repeated, he makes

— – —— — – – —— — – —— —— — ——
　　Repeat　　　　　　　W　　　　　A

(or, repeat word after) the word — (or, if necessary, words) preceding the doubtful word.

NOTE — *When* W A *is only sent, the Repeat sign is implied*

If the receiver requires the remainder of the message, he makes

— – —— — – – —— — – — ——
　　Repeat　　　　　　A　　　　　A

(or, repeat all after) the word — preceding the doubtful ones

If the receiver wants all the message repeated, he makes

— – —— — – – —— — – — —— – – —— —— – –
　　Repeat　　　　　A　　　　　L　　　　　L

When he requires no more repetition, he makes – —— – ——— – – *(or,* R D), which means the signal is read

IF A MISTAKE IS MADE IN A WORD

The sender makes the Erase sign – – – – – etc, E's as separate letters, which is to be answered by the Erase sign

N. B —This only applies to the last word made.

IF A WHOLE MESSAGE REQUIRES TO BE NEGATIVED.

The sender makes the Annul – —— —— – —— —— *(or,* W W as one group), which is to be answered by the Annul.

O

Milton Keynes UK
Ingram Content Group UK Ltd.
UKHW051629171123
432521UK00010B/13